LECTURE NOTEBOOK for *Lehninger Principles of Biochemistry*

THIRD EDITION

DAVID L. NELSON

MICHAEL M. COX

Worth Publishers

New York

ISBN: 0-7167-4167-9

Printed in the United States of America

First Printing 2000

Contents

Introduction

This Lecture Notebook provides you with the most useful figures and tables from *Lehninger Principles of Biochemistry*, Third Edition along with plenty of room for taking notes. You will no longer need to spend valuable class time trying to draw complicated figures from the text.

ALSO AVAILABLE FOR STUDENTS

Lehninger Online Companion Site
www.worthpublishers.com/lehninger

- **3-D Structure Tutorials**, which allow you to more fully explore basic and advanced topics covered in the text—interactively and in three dimensions. The tutorials are built around the Chemscape Chime® 2.0 plug-in, which reads the atomic coordinates of a biological macromolecule and renders it as a fully interactive 3-D molecular image. New topics are being added all the time, so check back often.

- **Online support for the "Biochemistry on the Internet" problems**. The site offers support to help you solve these in-text problems by providing links to the most accessible databases and research tools on the Internet, as well as instructions for downloading the 3-D molecular viewing utilities Rasmol and Chemscape Chime®.

- **Annotated Web Links** that guide you to useful and effective sites related to topics covered throughout the textbook.

- **Database of PDB coordinates** used to create the molecular illustrations in your textbook.

- **Stereo views of molecular textbook illustrations.**

- **Further Readings**, which offer updated annotated citations to help you explore the primary literature.

Study Guide and Solutions Manual

The Absolute, Ultimate Guide to Lehninger Principles of Biochemistry, Third Edition (ISBN 1-57269-167-6) combines an innovative study guide with a reliable solutions manual in one convenient volume. Most elements of the study guide have been thoroughly class-tested and have proven effective at teaching the fundamentals of biochemistry. For each chapter of the text, the *Guide* features:

- **What to Review**, which recaps key points from previous chapters

- **Discussion Questions** used for individual review, study groups, or classroom discussions

- A **Self-Test** including "Do you know the terms?"crossword puzzles, plus "Do you know the facts?" and "Applying what you know" questions and answers

The *Guide* also contains:

- "**Biochemistry on the Internet**" problems, which teach you how to use online databases and biochemistry resources

- **Complete solutions** for all end-of-chapter problems in the text

- A poster-size **Cellular Metabolic Map**. Draw the important chemical structures and reactions within the cell!

chapter

1

The Molecular Logic of Life

CD-ROM

When you study for exams, don't forget to review The Minicourses on the UNDERSTAND! BIOCHEMISTRY CD that came with your textbook.

Minicourses that apply to this Chapter include:

Bioenergetics

Energetics of Metabolism

Monomeric subunits

Letters of English alphabet (26 different kinds)	Deoxyribo-nucleotides (4 different kinds)	Amino acids (20 different kinds)
J H T R D Z Y N Q V B F I S U A P E M O W G X K L C	C A T G	Phe Tyr Thr Gly Asn Gln Lys Asp Met Val Ala Ser Arg Ile Leu His Pro Trp Cys Glu

M	M	P	G	A	G	Gly	Arg	Asp
E	A	A	A	G	A	Val	Gly	Gly
S	S	S	T	T	C	Asp	Pro	Pro
S	S	S	C	C	T	Ser	Leu	Leu
A	A	A	A	T	A	Phe	Lys	Gly
G	G	G	T	T	C	Arg	Asn	Lys
E	E	E	A	G	G	Glu	Phe	Trp
S	S	S	C	C	T	Ala	Glu	Cys

English words | Deoxyribonucleic acid (DNA) | Protein

Ordered linear sequences

For a segment of 8 subunits, the number of different sequences possible =

26^8 or 2.1×10^{11} | 4^8 or 65,536 | 20^8 or 2.56×10^{10}

figure 1–3, page 5

Monomeric subunits in linear sequences can spell infinitely complex messages.

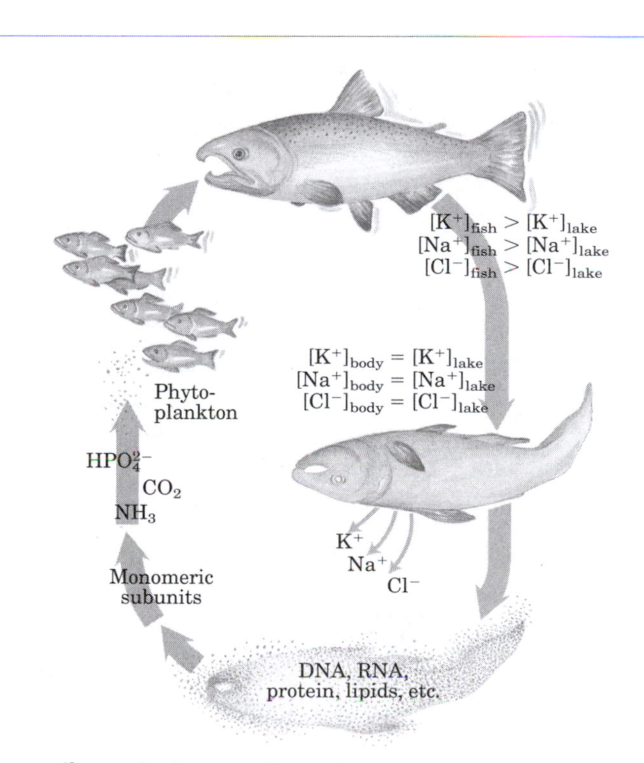

figure 1–4, page 6

Living organisms are not at equilibrium with their surroundings.

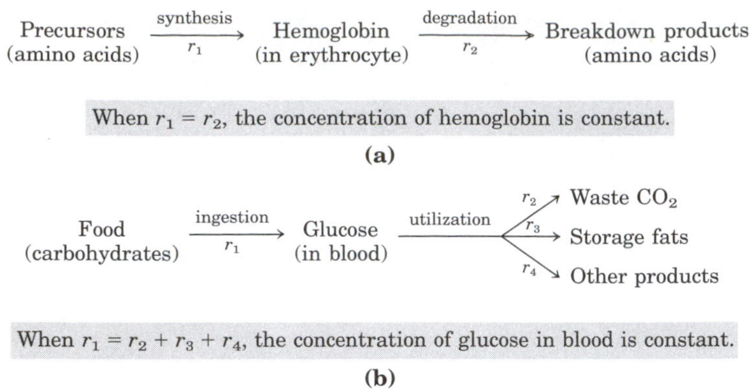

figure 1–5, page 7
The dynamic steady state.

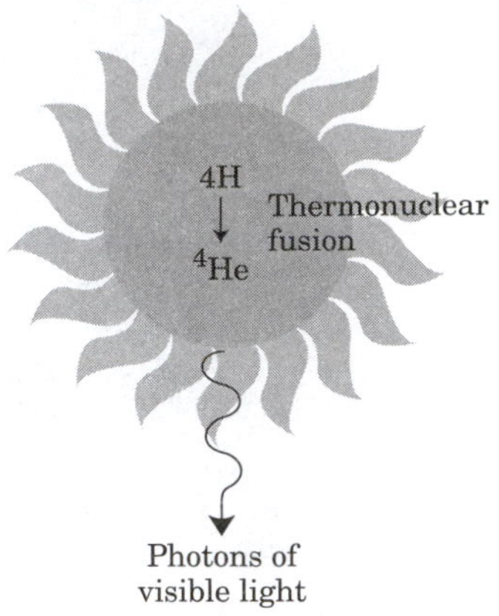

figure 1–7, page 8
Sunlight is the ultimate source of all biological energy.

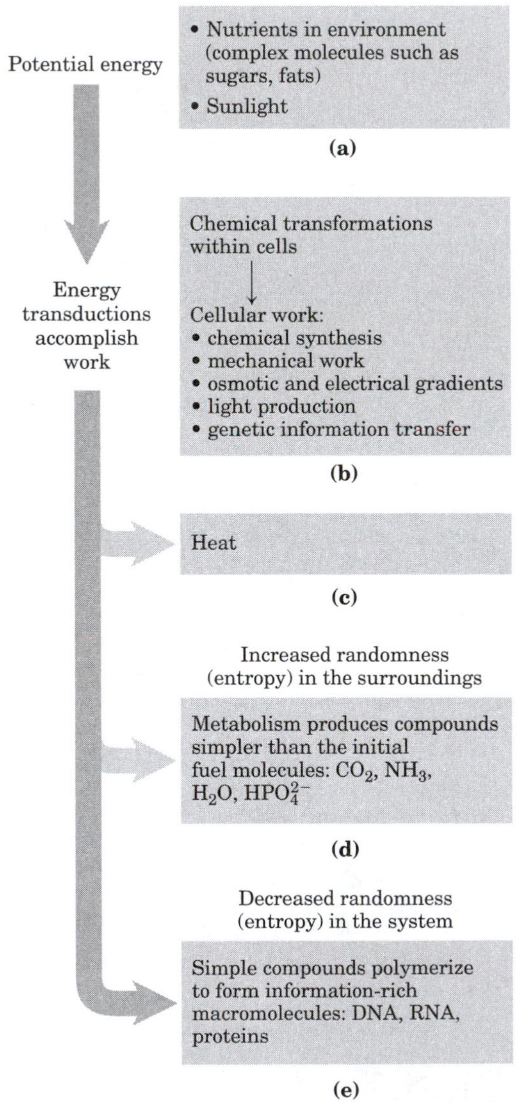

figure 1–6, page 8
During metabolic transductions, the randomness of the system plus surroundings (expressed quantitatively as entropy) increases as the potential energy of complex nutrient molecules decreases.

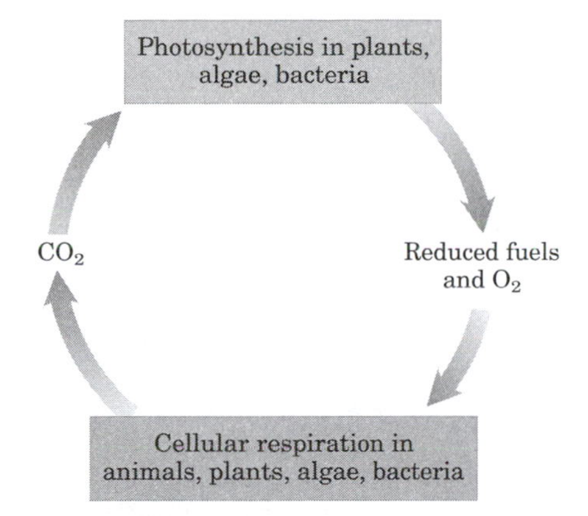

figure 1–8, page 9
Photosynthetic organisms (plants, some algae, and some bacteria) are the ultimate providers of fuels—reduced, energy-rich compounds—in the biosphere.

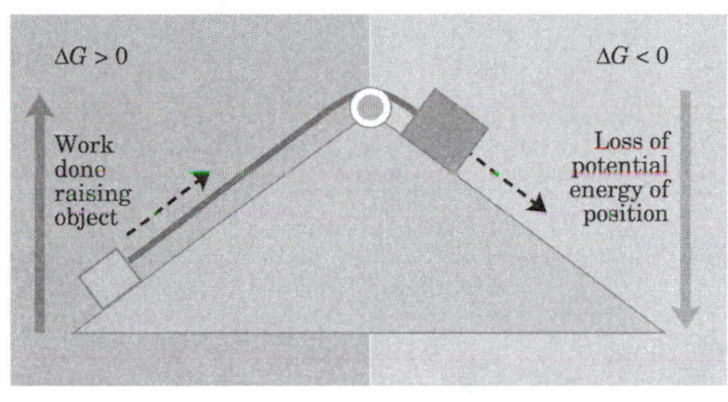

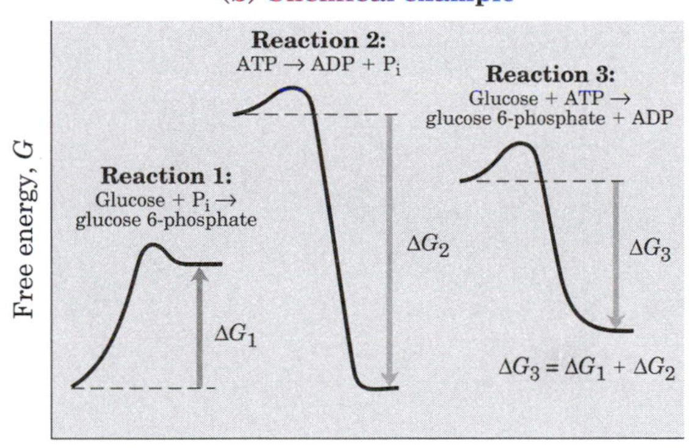

figure 1–9, page 9
Energy coupling in mechanical and chemical processes.

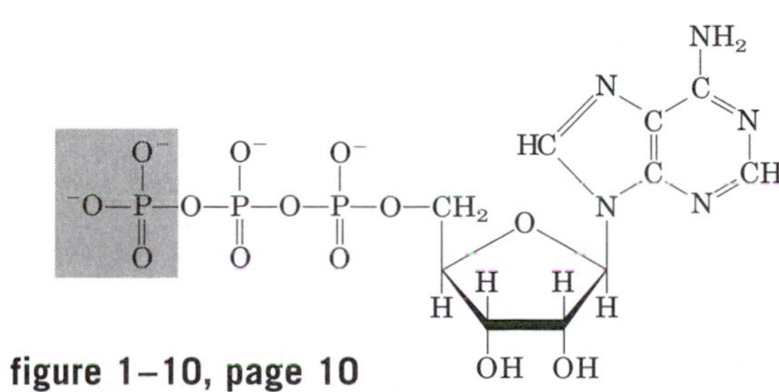

figure 1–10, page 10
Adenosine triphosphate (ATP).

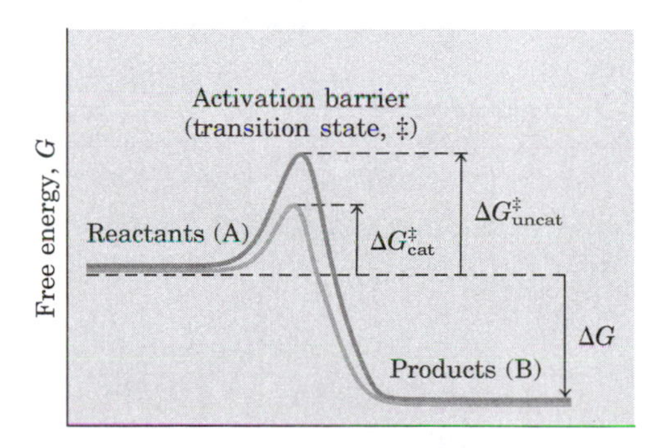

figure 1–11, page 11
Energy changes during a chemical reaction.

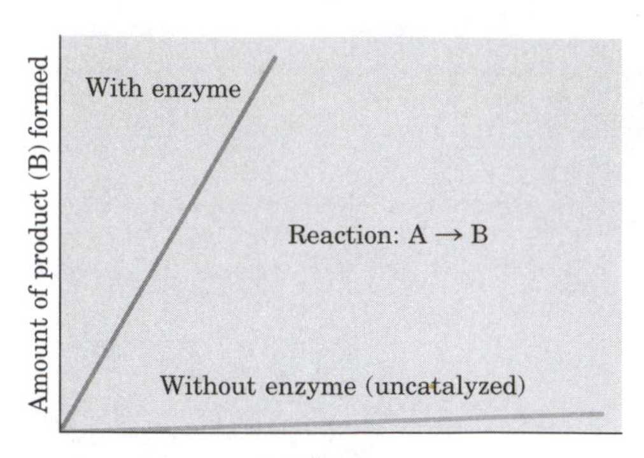

figure 1–12, page 11
An enzyme increases the rate of a specific chemical reaction.

$$A \xrightarrow{\text{enzyme 1}} B \xrightarrow{\text{enzyme 2}} C \xrightarrow{\text{enzyme 3}} D \xrightarrow{\text{enzyme 4}} E \xrightarrow{\text{enzyme 5}} F$$

figure 1–13, page 11
A linear metabolic pathway.

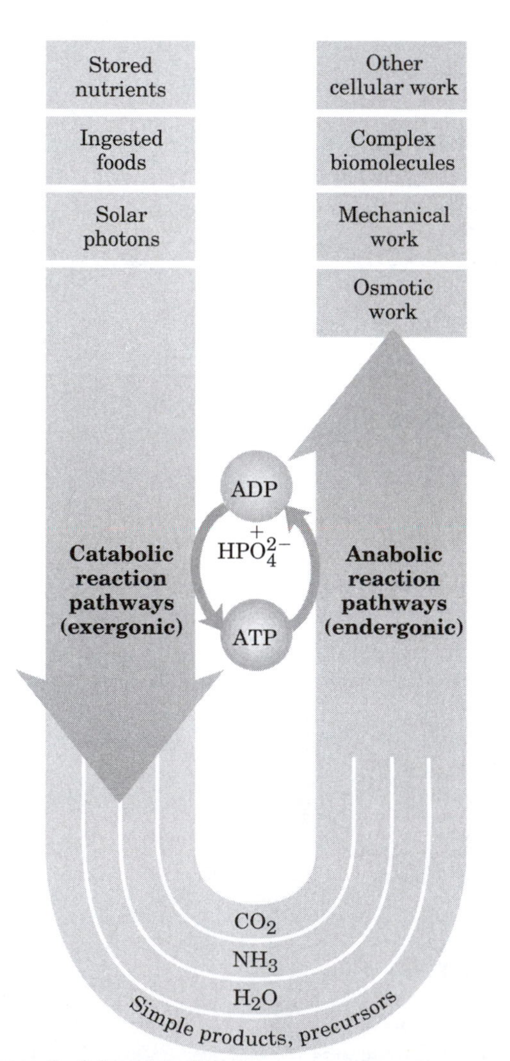

figure 1–14, page 12
ATP is the shared chemical intermediate linking energy-releasing to energy-requiring cell processes.

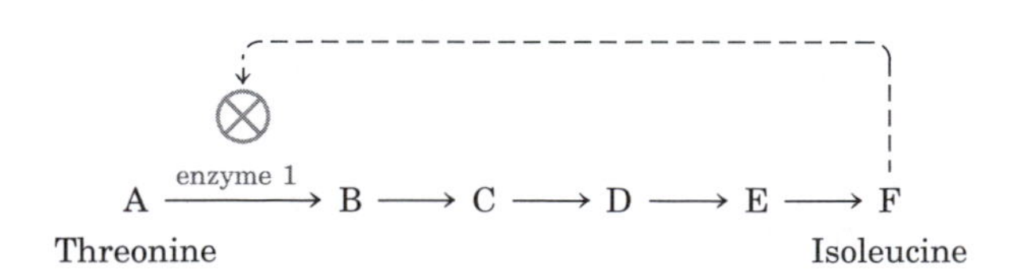

figure 1–15, page 12
Feedback inhibition.

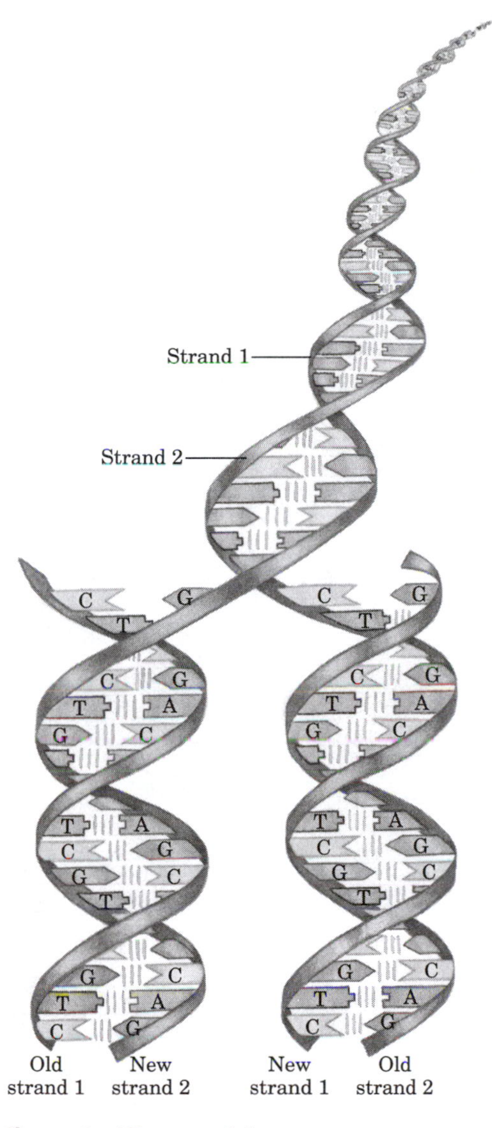

figure 1–17, page 14
The complementary structure of DNA.

table 1–1, page 16

Some Organisms Whose Genomes Have Been Completely Sequenced

Organism	Genome size (million bases)	Biological interest
Mycoplasma pneumoniae	0.8	Causes pneumonia
Treponema pallidum	1.1	Causes syphilis
Borrelia burgdorferi	1.3	Causes Lyme disease
Helicobacter pylori	1.7	Causes gastric ulcers
Methanococcus jannaschii	1.7	Grows at 85 °C!
Haemophilus influenzae	1.8	Causes bacterial influenza
Methanobacterium thermo-autotrophicum	1.8	Member of the Archaea
Archaeoglobus fulgidus	2.2	High-temperature methanogen
Synechocystis sp.	3.6	Cyanobacterium
Bacillus subtilis	4.2	Common soil bacterium
Escherichia coli	4.6	Some strains cause toxic shock syndrome
Saccharomyces cerevisiae	12.1	Unicellular eukaryote
Caenorhabditis elegans	97	Multicellular roundworm

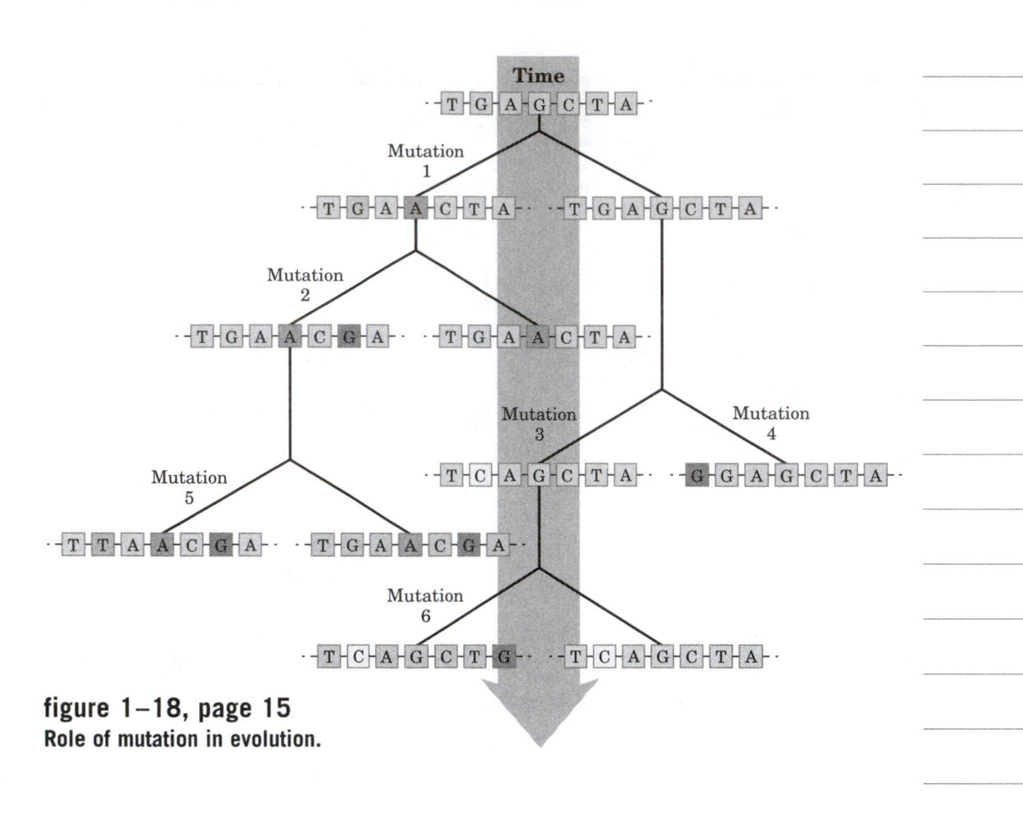

figure 1–18, page 15
Role of mutation in evolution.

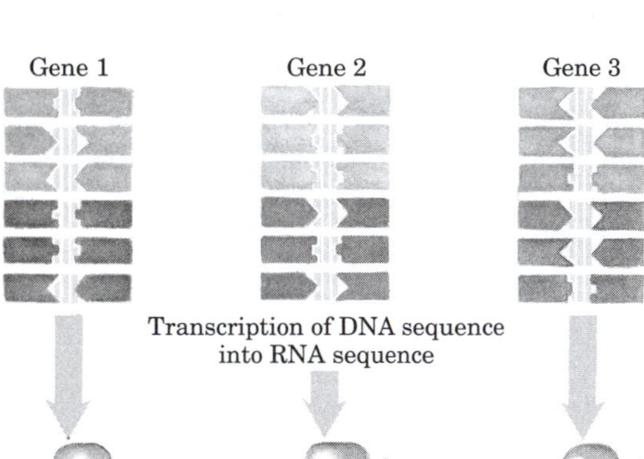

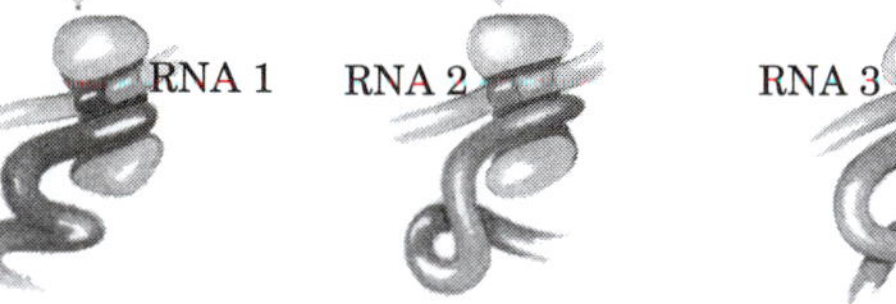

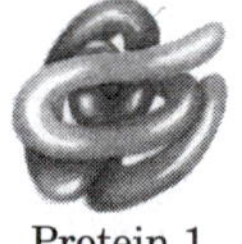
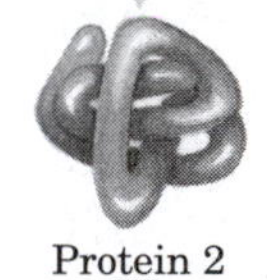
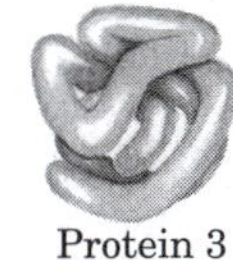

figure 1–19, page 17
Linear sequences of deoxyribonucleotides in DNA, arranged into units known as genes, are transcribed into ribonucleic acid (RNA) molecules with complementary ribonucleotide sequences.

chapter

2

Cells

When you study for exams, don't forget to review The Minicourses on the UNDERSTAND! BIOCHEMISTRY CD that came with your textbook.

Minicourses that apply to this Chapter include:

Background
- Fundamentals of Cell Architecture

Bioenergetics
- C4 Photosynthesis

Cellular Architecture and Traffic
- The Cytoskeleton
- Motor Proteins and Movement
- Protein Sorting in the ER and Golgi Apparatus

Cellular Architecture and Traffic (*continued*)
- Nuclear Transport
- Sorting Lysosomal Enzymes
- Protein Modification

The Dividing Cell
- Cancer
- Viruses

Nucleus (eukaryotes) or nucleoid (bacteria)
Contains genetic material –DNA and associated proteins. Nucleus is membrane-bounded.

Plasma membrane
Tough, flexible lipid bilayer. Selectively permeable to polar substances. Includes membrane proteins that function in transport, in signal reception, and as enzymes.

Cytoplasm
Aqueous cell contents and suspended particles and organelles.

centrifuge at 150,000 *g*

Supernatant: cytosol
Concentrated solution of enzymes, RNA, monomeric subunits, metabolites, inorganic ions.

Pellet: particles and organelles
Ribosomes, storage granules, mitochondria, chloroplasts, lysosomes, endoplasmic reticulum.

figure 2–1, page 20
The universal features of living cells.

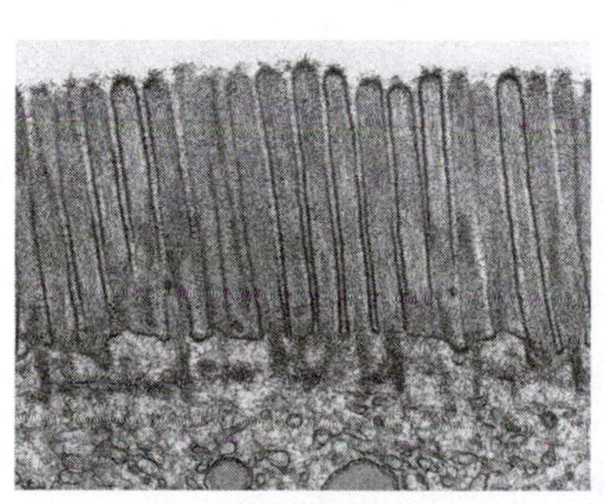

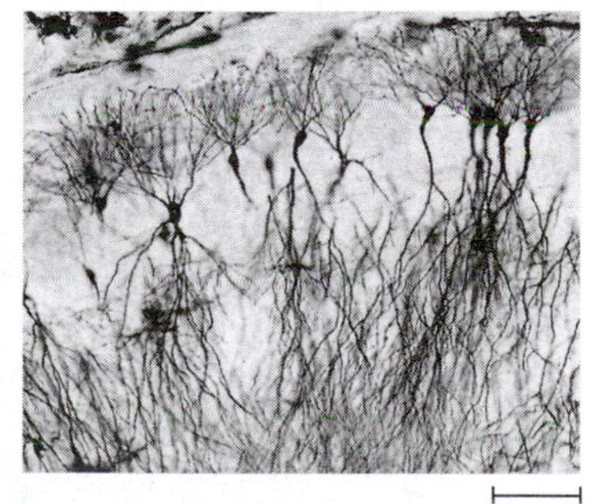

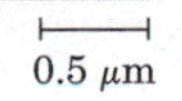

0.5 μm

50 μm

figure 2–2, page 22
Convolutions of the plasma membrane, or long, thin extensions of the cytoplasm, increase the surface-to-volume ratio of cells.

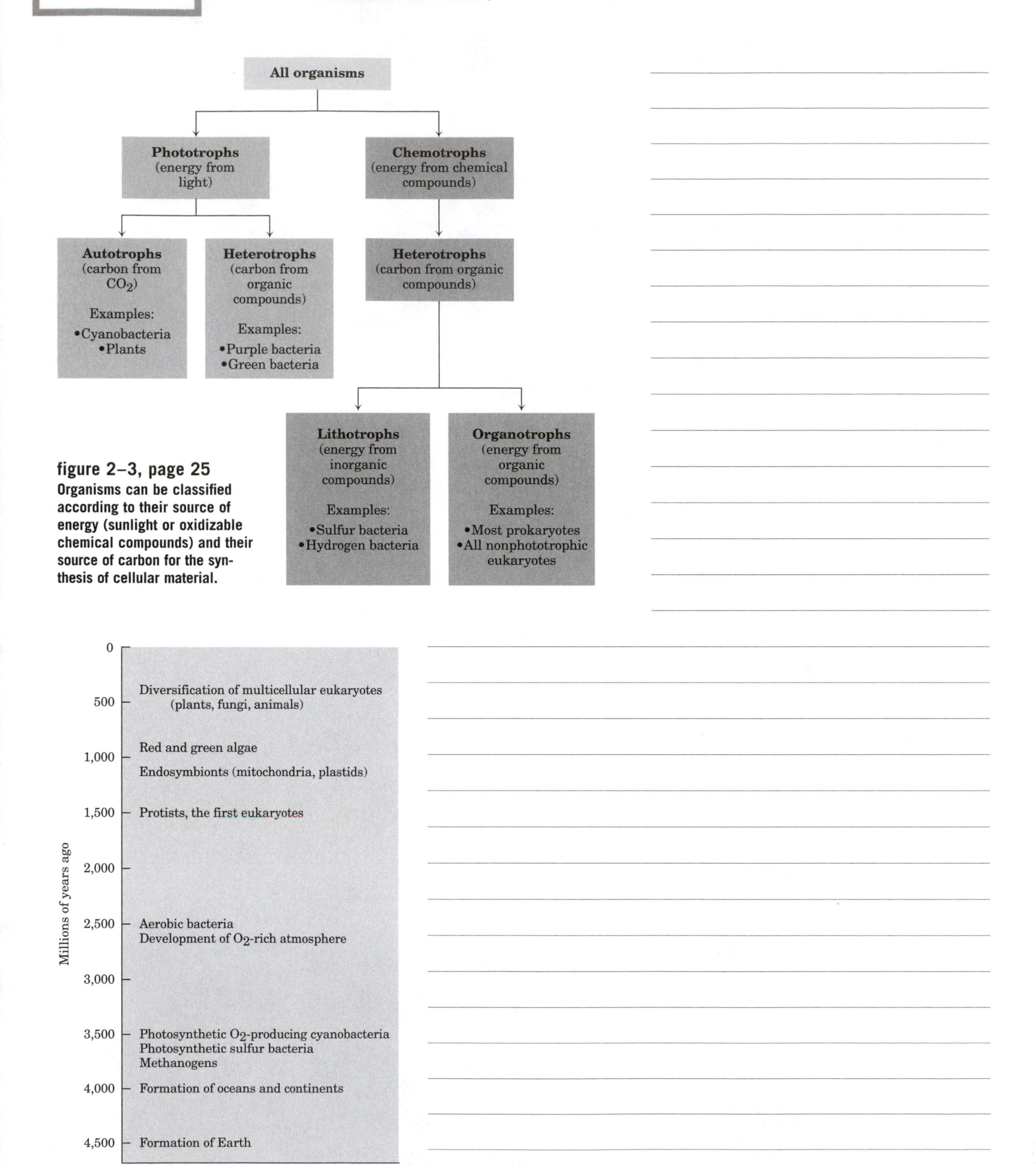

figure 2–3, page 25
Organisms can be classified according to their source of energy (sunlight or oxidizable chemical compounds) and their source of carbon for the synthesis of cellular material.

figure 2–4, page 25
Landmarks in the evolution of life on Earth.

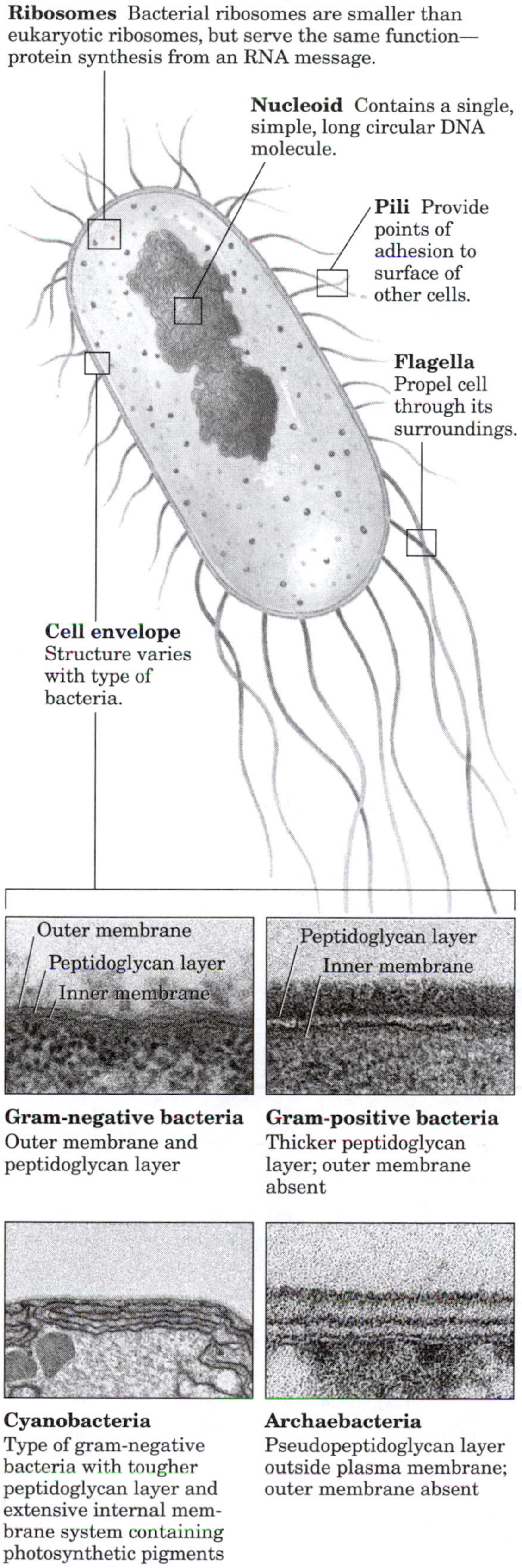

Ribosomes Bacterial ribosomes are smaller than eukaryotic ribosomes, but serve the same function—protein synthesis from an RNA message.
Nucleoid Contains a single, simple, long circular DNA molecule.
Pili Provide points of adhesion to surface of other cells.
Flagella Propel cell through its surroundings.
Cell envelope Structure varies with type of bacteria.
Outer membrane
Peptidoglycan layer
Inner membrane
Gram-negative bacteria Outer membrane and peptidoglycan layer
Peptidoglycan layer
Inner membrane
Gram-positive bacteria Thicker peptidoglycan layer; outer membrane absent
Cyanobacteria Type of gram-negative bacteria with tougher peptidoglycan layer and extensive internal membrane system containing photosynthetic pigments
Archaebacteria Pseudopeptidoglycan layer outside plasma membrane; outer membrane absent

figure 2–5, page 26
Common structural features of bacterial cells.

figure 2–6, page 28
Evolution of eukaryotes.

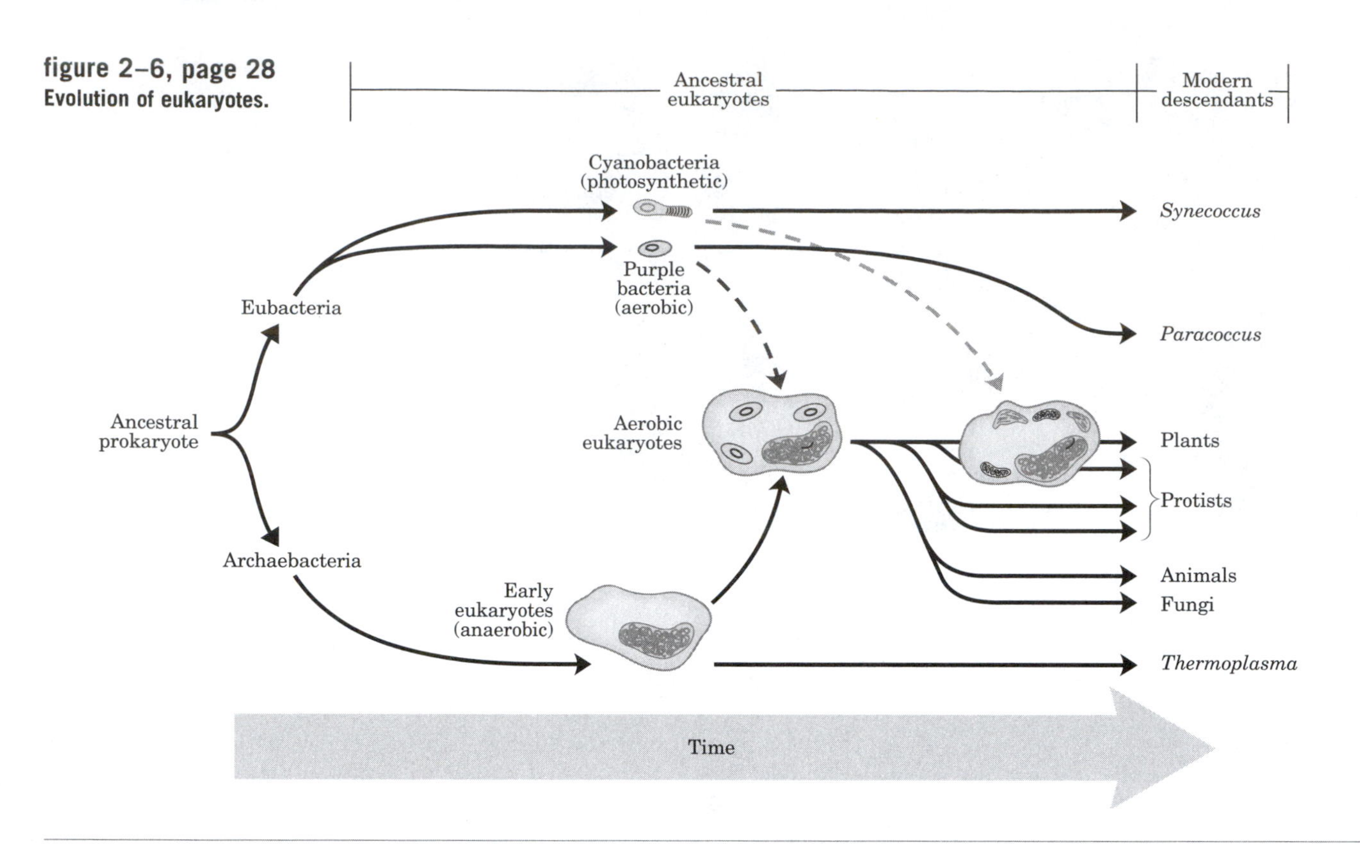

table 2–1, page 29

Comparison of Prokaryotic and Eukaryotic Cells

Characteristic	Prokaryotic cell	Eukaryotic cell
Size	Generally small (1 – 10 μm)	Generally large (5 – 100 μm)
Genome	DNA with nonhistone protein; genome in nucleoid, not surrounded by membrane	DNA complexed with histone and nonhistone proteins in chromosomes; chromosomes in nucleus with membranous envelope
Cell division	Fission or budding; no mitosis	Mitosis including mitotic spindle; centrioles in many species
Membrane-bounded organelles	Absent	Mitochondria, chloroplasts (in plants, some algae), endoplasmic reticulum, Golgi complexes, lysosomes (in animals), etc.
Nutrition	Absorption; some photosynthesis	Absorption, ingestion; photosynthesis in some species
Energy metabolism	No mitochondria; oxidative enzymes bound to plasma membrane; great variation in metabolic pattern	Oxidative enzymes packaged in mitochondria; more unified pattern of oxidative metabolism
Cytoskeleton	None	Complex, with microtubules, intermediate filaments, actin filament
Intracellular movement	None	Cytoplasmic streaming, endocytosis, phagocytosis, mitosis, vesicle transport

Source: Modified from Hickman, C.P., Roberts, L.S., & Hickman, F.M. (1990) *Biology of Animals,* 5th edn, p. 30, Mosby–Yearbook, Inc., St. Louis, MO.

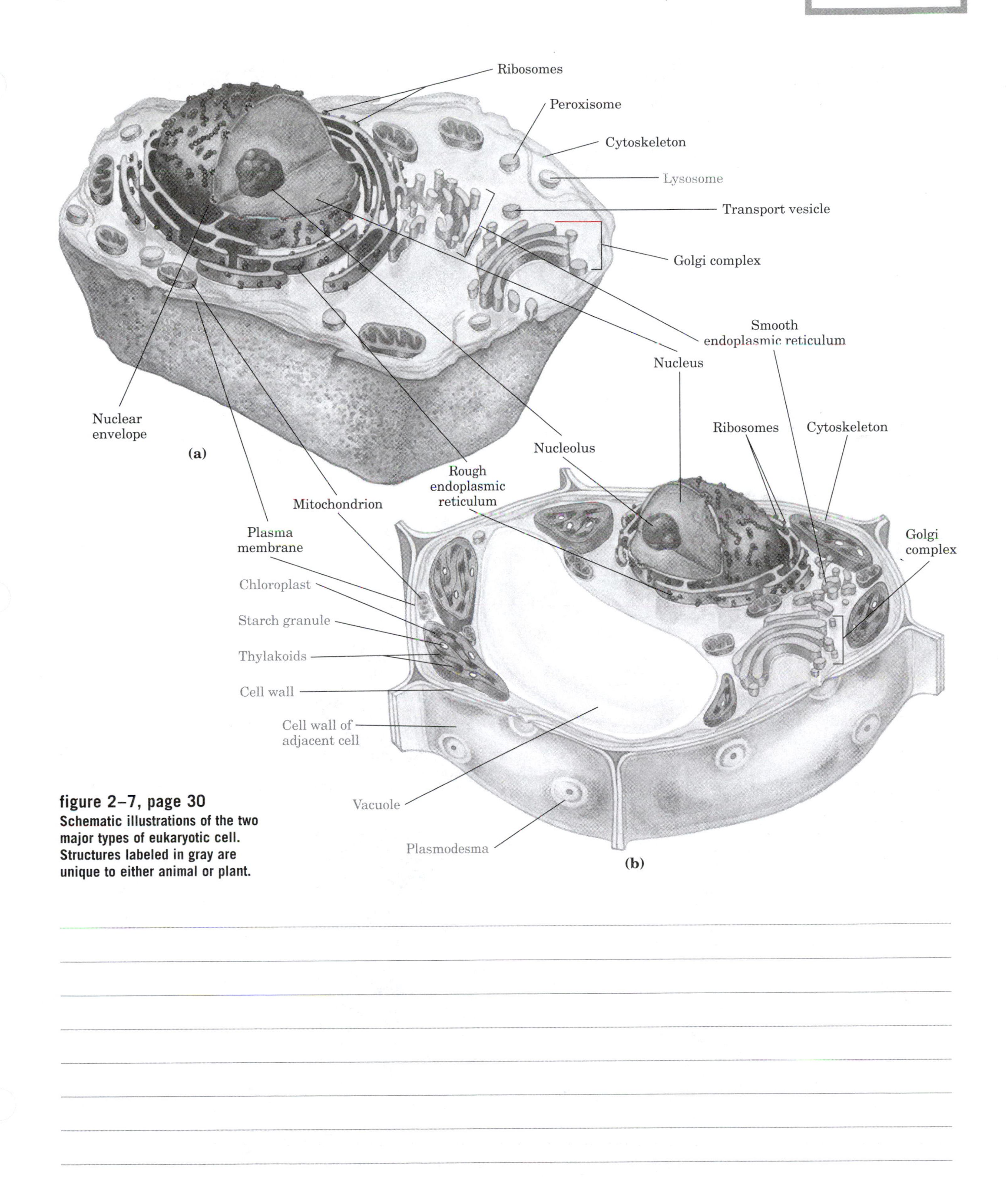

figure 2–7, page 30
Schematic illustrations of the two major types of eukaryotic cell. Structures labeled in gray are unique to either animal or plant.

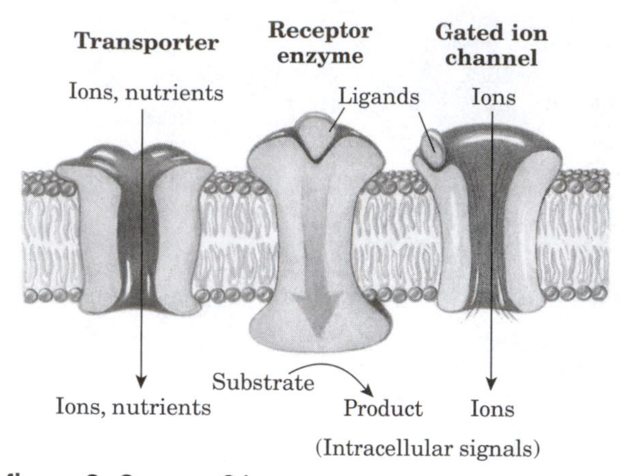

figure 2–8, page 31
Proteins in the plasma membrane serve as transporters, signal receptors, and ion channels.

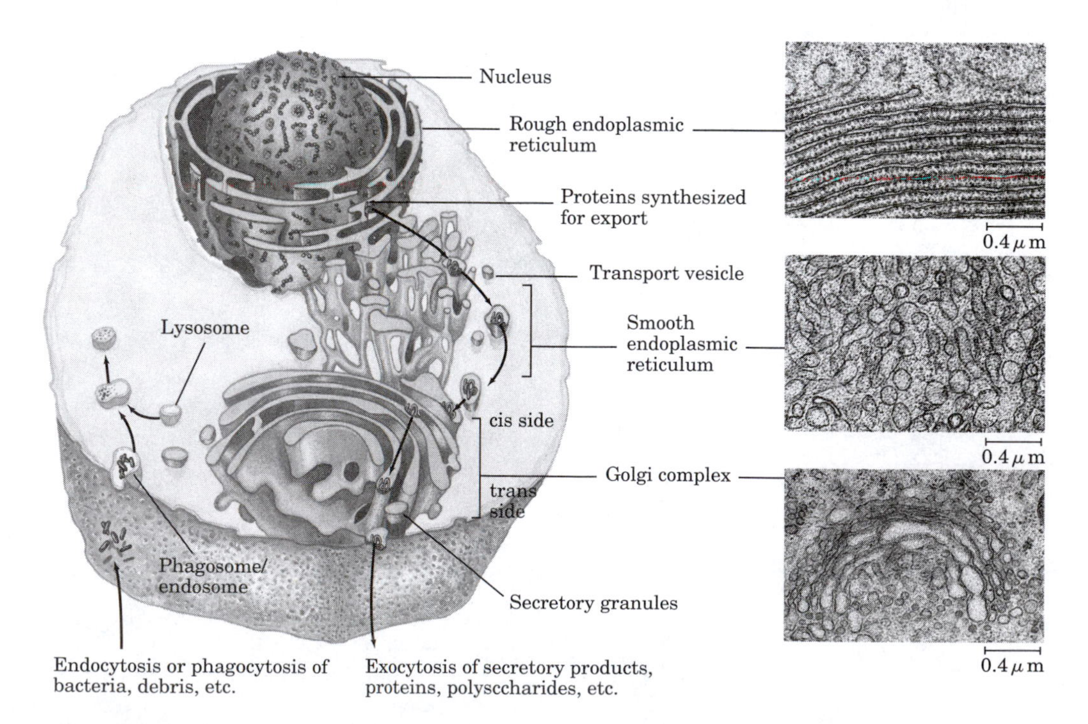

figure 2–9, page 32
The endomembrane system.

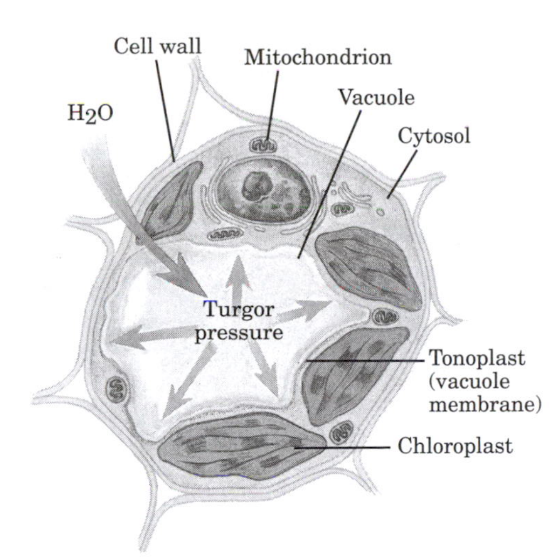

figure 2–10, page 34
The vacuole of a plant cell.

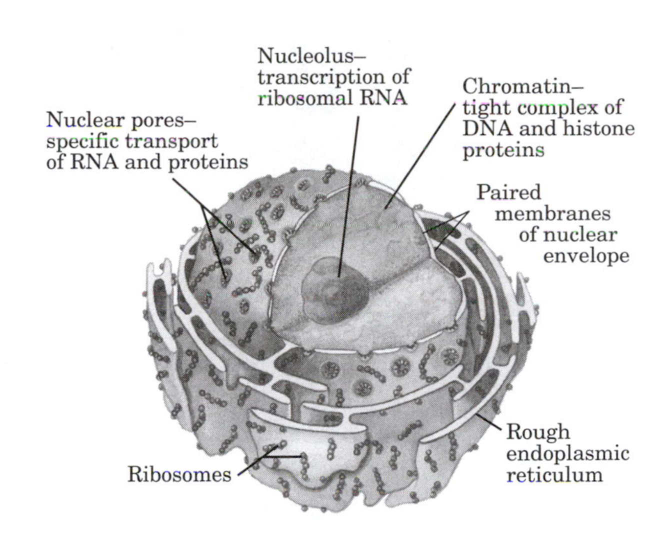

figure 2–11, page 35
The nucleus and nuclear envelope.

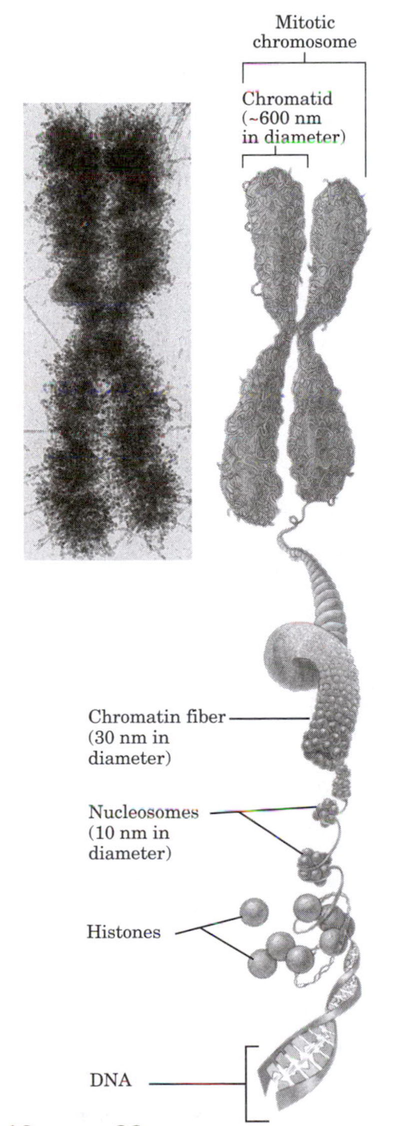

figure 2–12, page 36
Chromosomes are visible microscopically during mitosis.

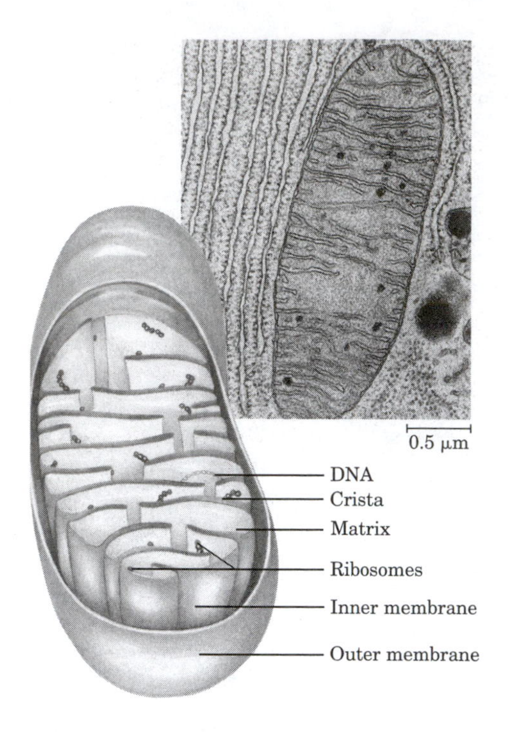

figure 2–13, page 37
Structure of a mitochondrion.

figure 2–14, page 37
Structure of a chloroplast.

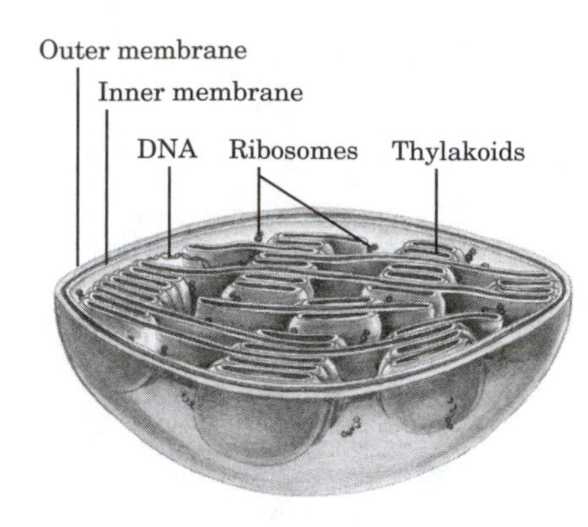

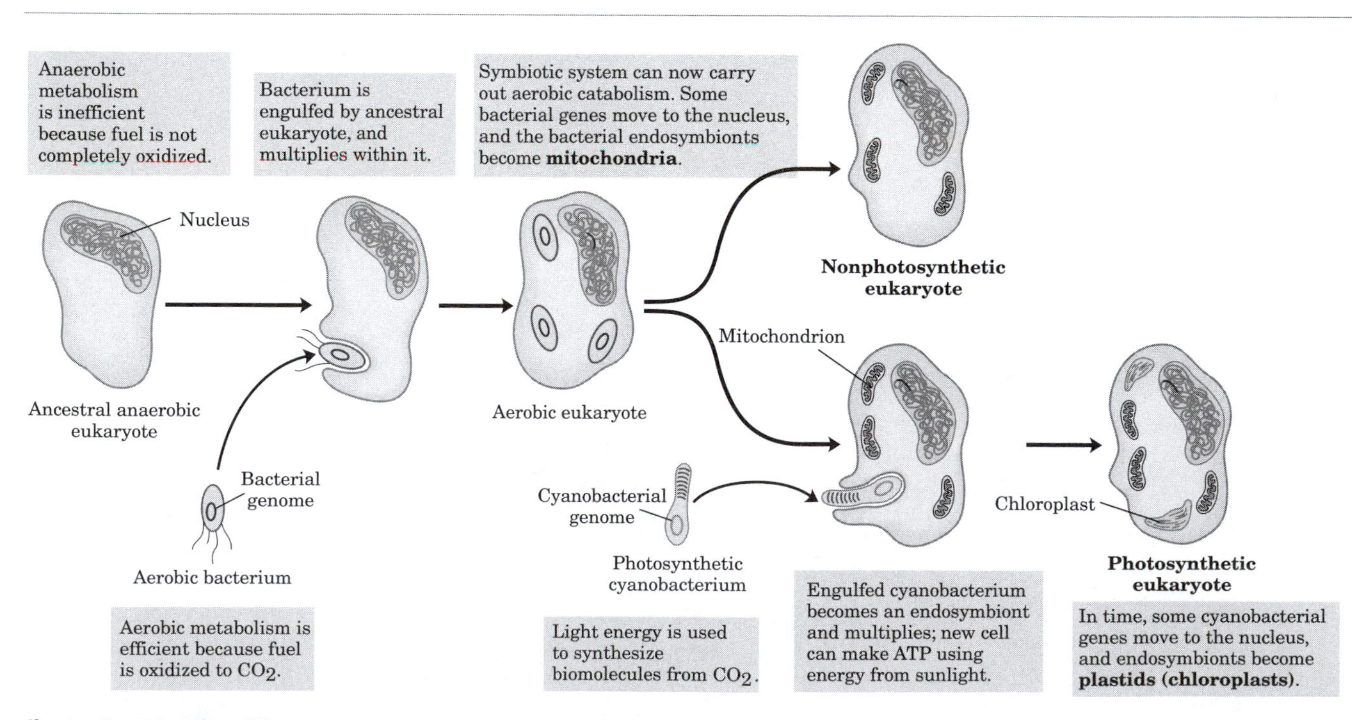

figure 2–15, page 38
A plausible theory for the evolutionary origin of mitochondria and chloroplasts.

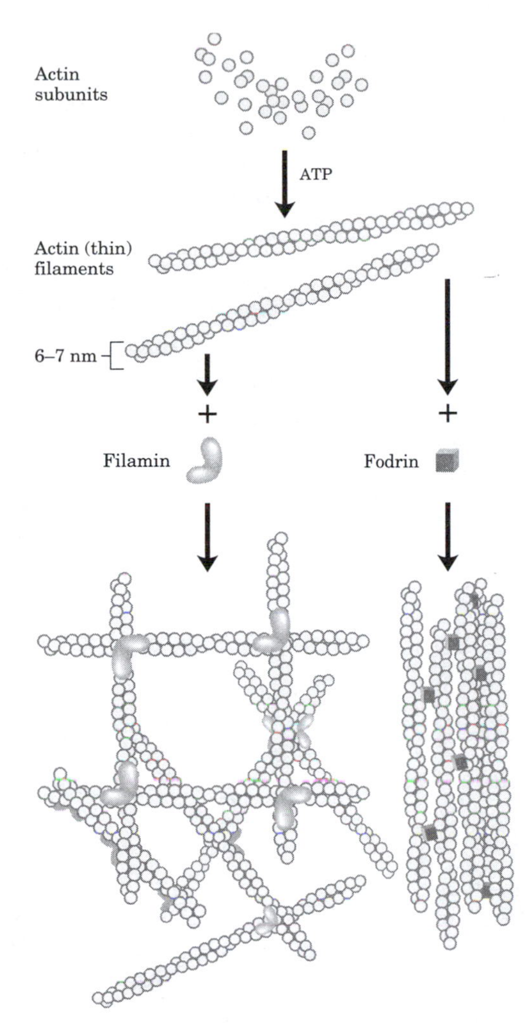

figure 2–17, page 40
Individual subunits of actin polymerize to form actin filaments.

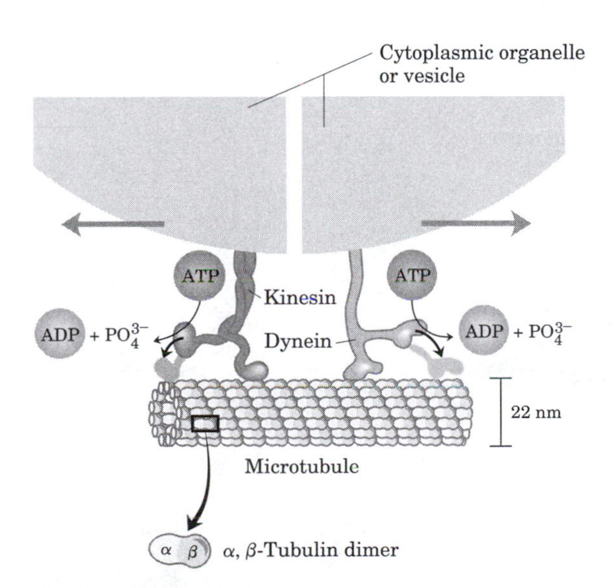

figure 2–19, page 41
Kinesin and cytoplasmic dynein.

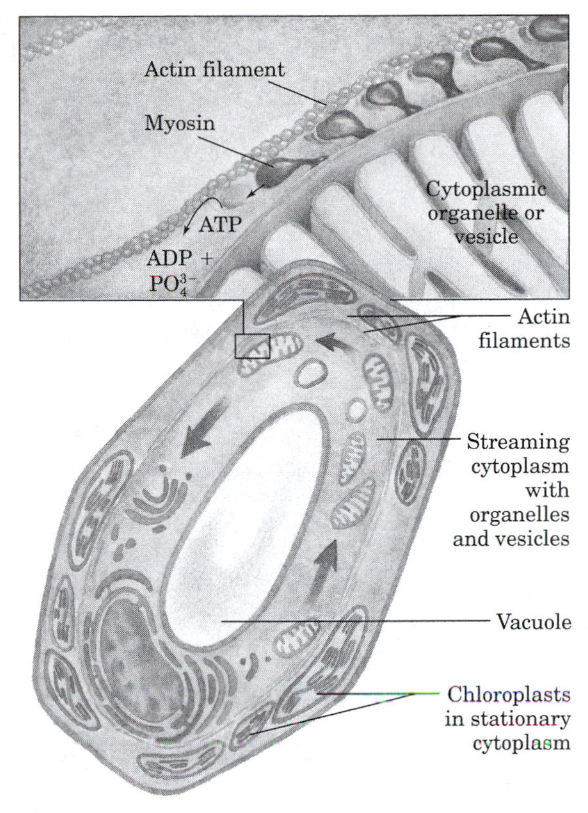

figure 2–18, page 41
Organelle transport.

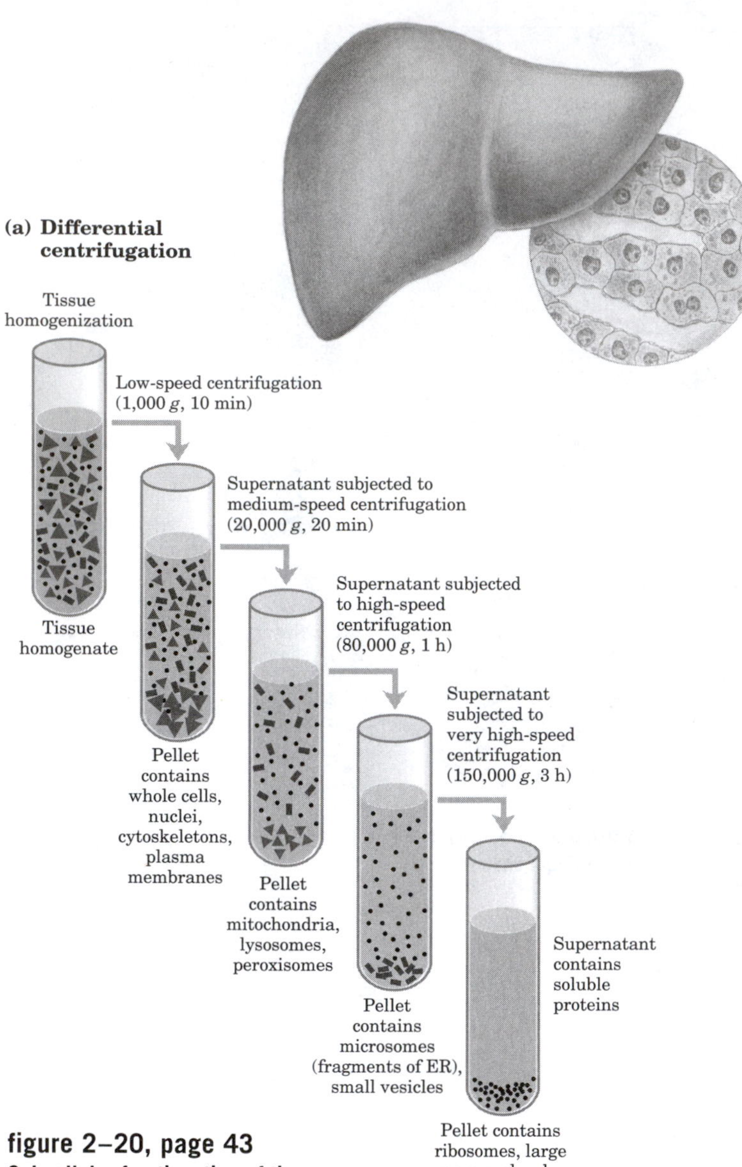

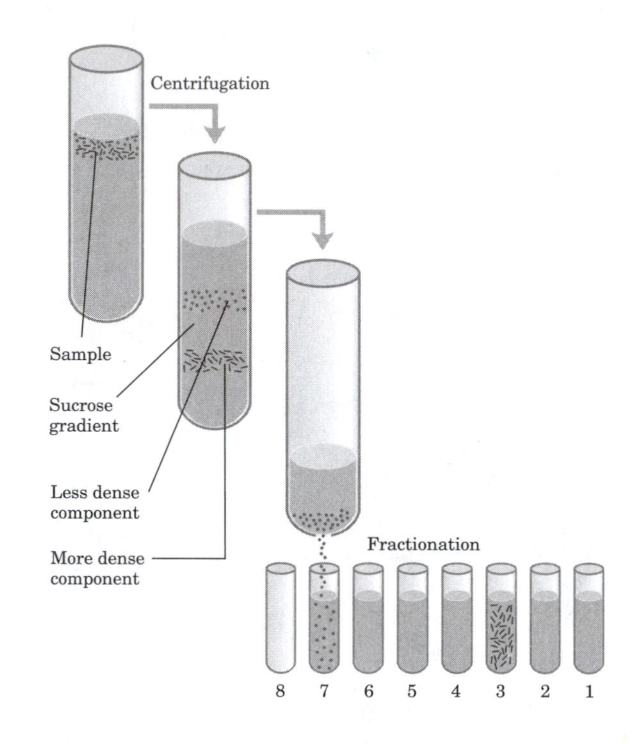

figure 2–20, page 43
Subcellular fractionation of tissue.

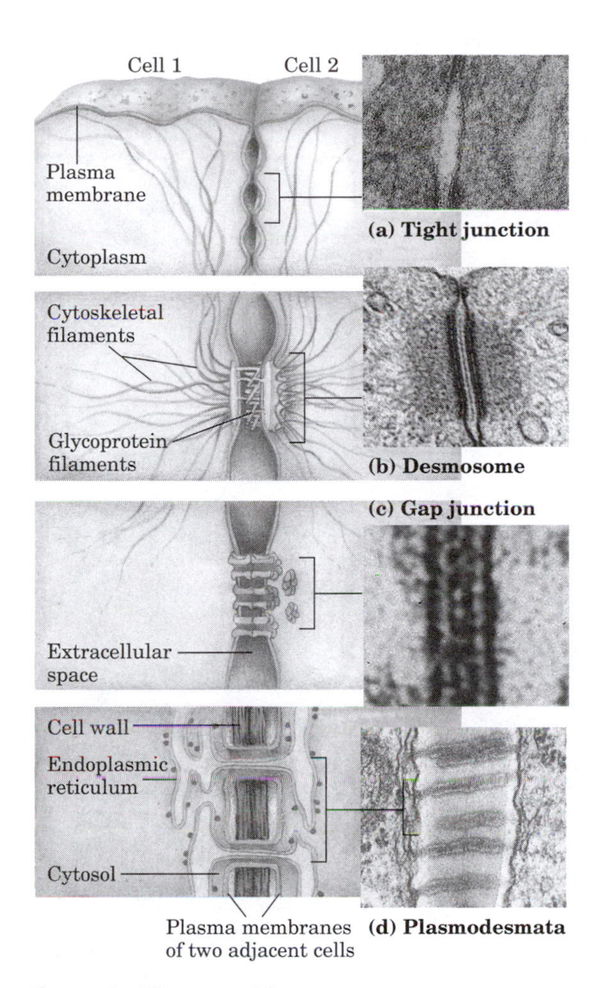

figure 2–22, page 46
Cellular connections.

PROBLEMS

5. A Strategy to Increase the Surface Area of Cells, page 51.

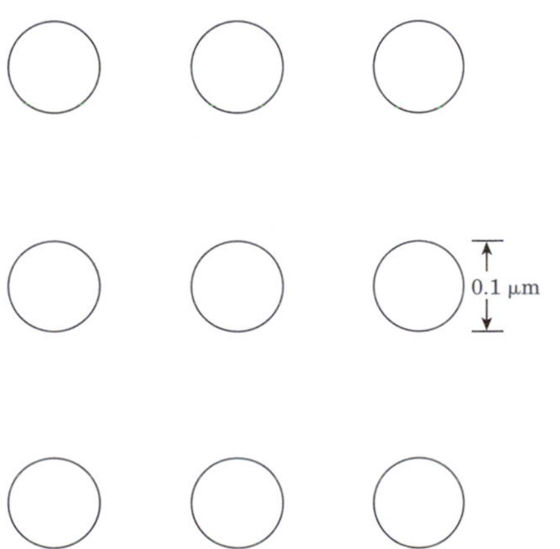

Arrangement of microvilli on the "patch"

chapter

3 Biomolecules

CD-ROM

When you study for exams, don't forget to review The Minicourses on the UNDERSTAND! BIOCHEMISTRY CD that came with your textbook.

Minicourses that apply to this Chapter include:

Background

Chemical Fundamentals

Molecules of Life

Macromolecules: Proteins and Nucleic Acids

Bulk elements
Trace elements

1 H																	2 He
3 Li	4 Be											5 B	6 C	7 N	8 O	9 F	10 Ne
11 Na	12 Mg											13 Al	14 Si	15 P	16 S	17 Cl	18 Ar
19 K	20 Ca	21 Sc	22 Ti	23 V	24 Cr	25 Mn	26 Fe	27 Co	28 Ni	29 Cu	30 Zn	31 Ga	32 Ge	33 As	34 Se	35 Br	36 Kr
37 Rb	38 Sr	39 Y	40 Zr	41 Nb	42 Mo	43 Tc	44 Ru	45 Rh	46 Pd	47 Ag	48 Cd	49 In	50 Sn	51 Sb	52 Te	53 I	54 Xe
55 Cs	56 Ba		72 Hf	73 Ta	74 W	75 Re	76 Os	77 Ir	78 Pt	79 Au	80 Hg	81 Tl	82 Pb	83 Bi	84 Po	85 At	86 Rn
87 Fr	88 Ra																

Lanthanides
Actinides

figure 3–1, page 54
Elements essential to animal life and health.

Atom	Number of unpaired electrons (in gray)	Number of electrons in complete outer shell
H·	1	2
:Ȯ·	2	8
:Ṅ·	3	8
·Ċ·	4	8
:Ṡ·	2	8
:Ṗ·	3	8

H· + H· ⟶ H:H = H—H (Dihydrogen)

:Ö· + 2H· ⟶ :Ö:H (H) = O—H (H) (Water)

:N· + 3H· ⟶ H:N:H (H) = N—H (H, H) (Ammonia)

·C· + 4H· ⟶ H:C:H (H, H) = H—C—H (H, H) (Methane)

:S· + 2H· ⟶ :S:H (H) = S—H (H) (Hydrogen sulfide)

3H· + :P· + 4·Ö: ⟶ (H:O:)(:O::P:O:H)(:O:H) = O=P—OH (OH, OH) (Phosphoric acid)

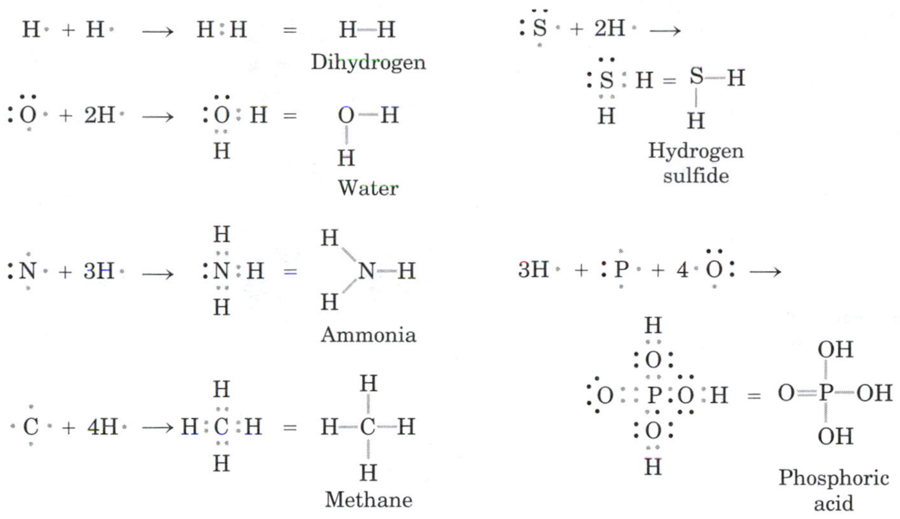

figure 3–2, page 54
Covalent bonding.

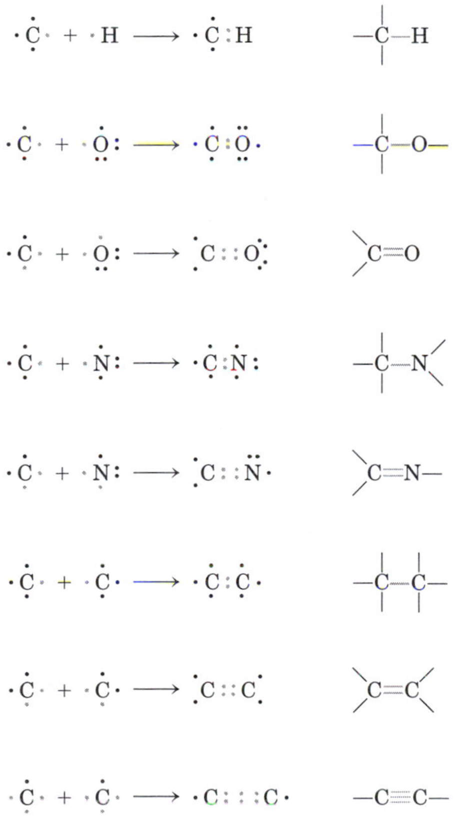

·Ċ· + ·H ⟶ ·Ċ:H —C—H

·Ċ· + ·Ö: ⟶ ·Ċ:Ö· —C—O—

·Ċ· + ·Ö: ⟶ Ċ::Ö: >C=O

·Ċ· + ·N: ⟶ ·Ċ:N: —C—N<

·Ċ· + ·N: ⟶ Ċ::N· >C=N—

·Ċ· + ·Ċ· ⟶ ·Ċ:Ċ· —C—C—

·Ċ· + ·Ċ· ⟶ Ċ::Ċ >C=C<

·Ċ· + ·Ċ· ⟶ ·C:::C· —C≡C—

figure 3–3, page 55
Versatility of carbon in forming covalent single, double, and triple bonds (in gray), particularly between carbon atoms.

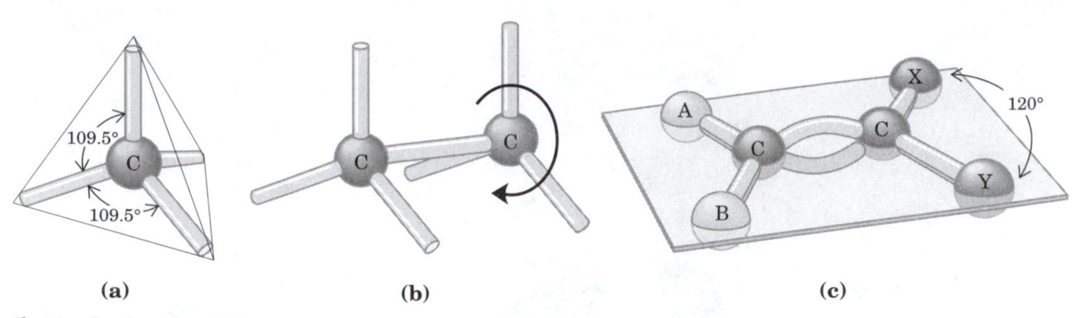

figure 3–4, page 55
Geometry of carbon bonding.

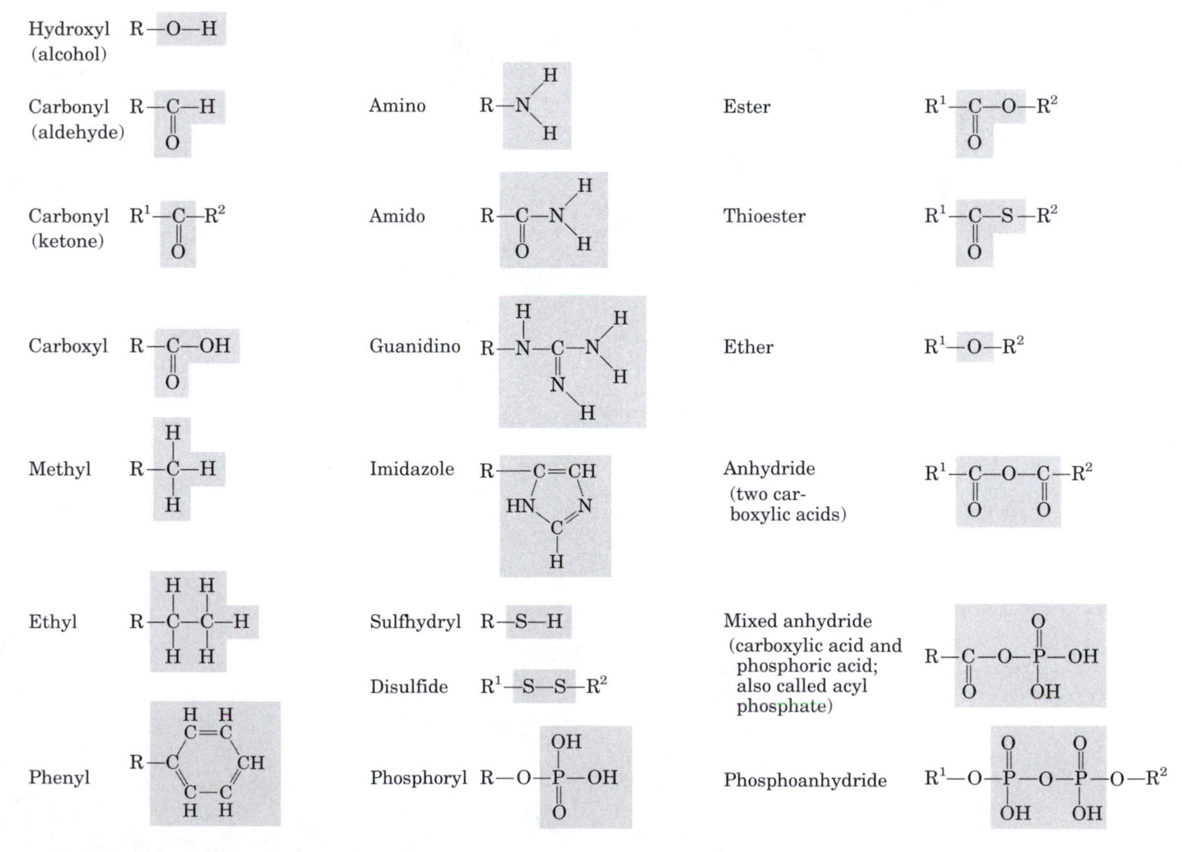

figure 3–5, page 56
Some common functional groups of biomolecules.

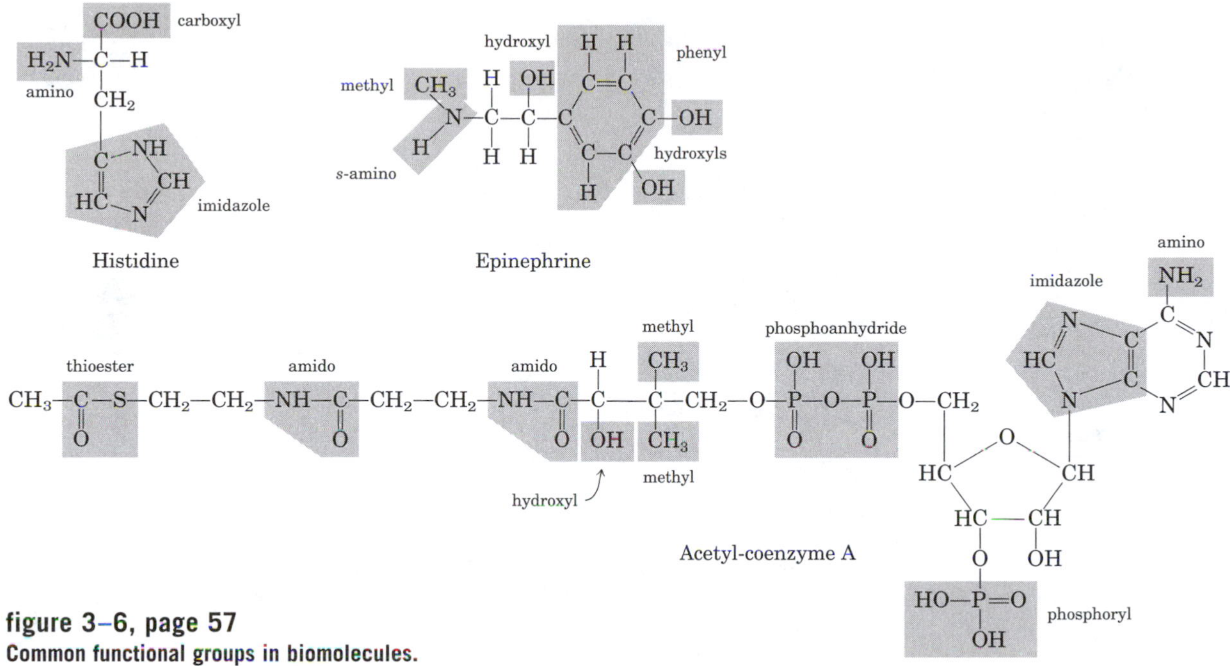

figure 3–6, page 57
Common functional groups in biomolecules.

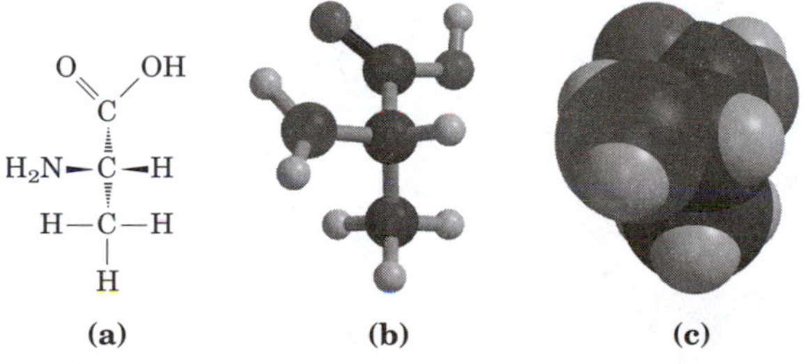

figure 3–7, page 57
Three ways to represent the structure of the amino acid alanine.

table 3–1, page 57

Van der Waals Radii and Covalent (Single-Bond) Radii of Some Elements*

Element	Van der Waals radius (nm)	Covalent radius for single bond (nm)
H	0.1	0.030
O	0.14	0.074
N	0.15	0.073
C	0.17	0.077
S	0.18	0.103
P	0.19	0.110
I	0.22	0.133

*Van der Waals radii describe the space-filling dimensions of atoms. When two atoms are joined covalently, the atomic radii at the point of bonding are less than the van der Waals radii, because the joined atoms are pulled together by the shared electron pair. The distance between nuclei in a van der Waals interaction or in a covalent bond is about equal to the sum of the van der Waals radii or the covalent radii, respectively, for the two atoms. Thus the length of a carbon–carbon single bond is about 0.077 nm + 0.077 nm = 0.154 nm.

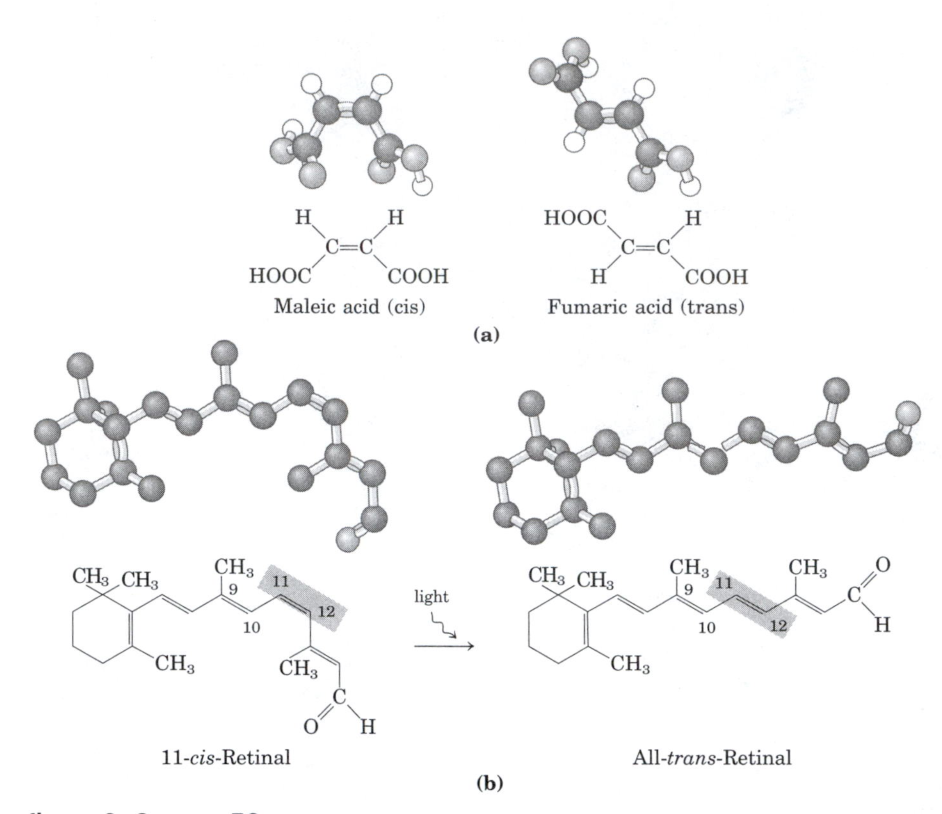

figure 3–8, page 58
Configurations of geometric isomers.

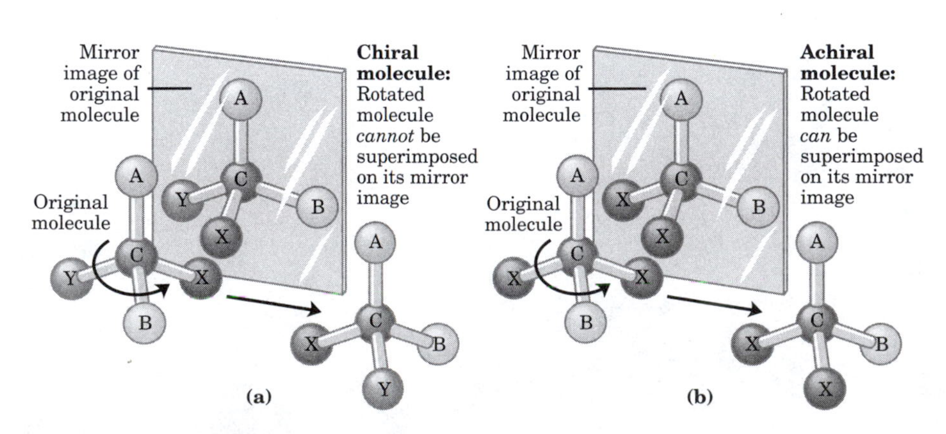

figure 3–9, page 59
Molecular asymmetry: chiral and achiral molecules.

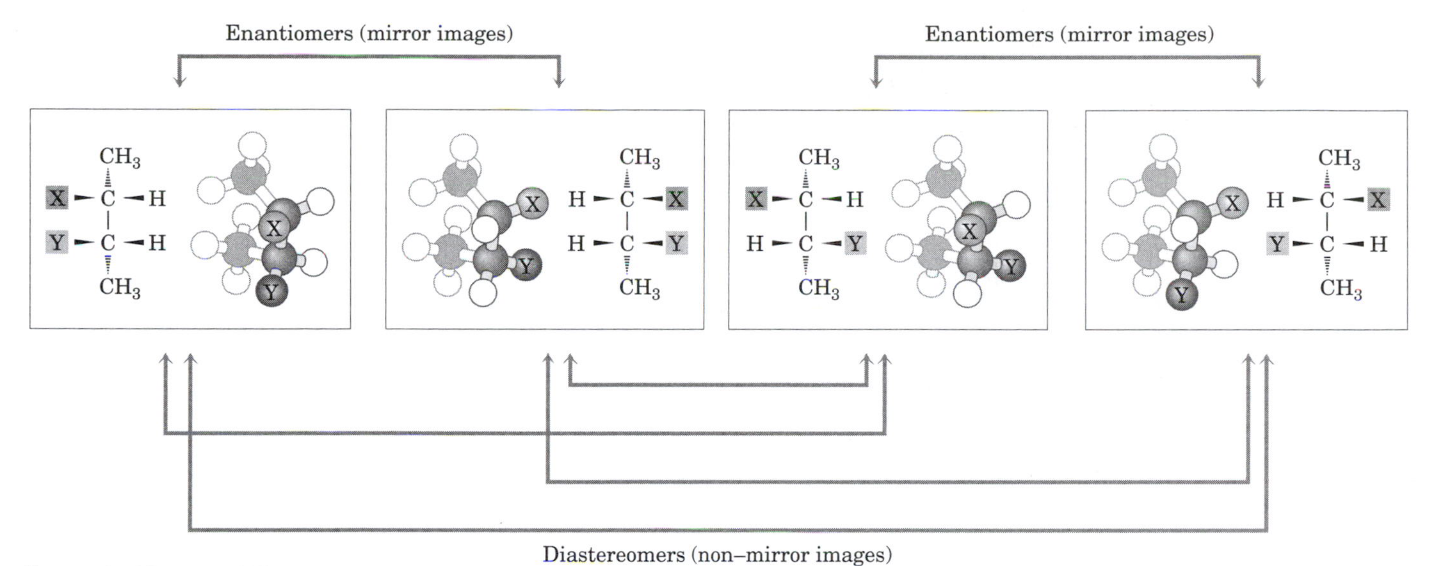

figure 3–10, page 59
Two types of stereoisomers.

$-OCH_2 > -OH > -NH_2 > -COOH > -CHO > -CH_2OH > -CH_3 > -H$

RS priority system, page 60

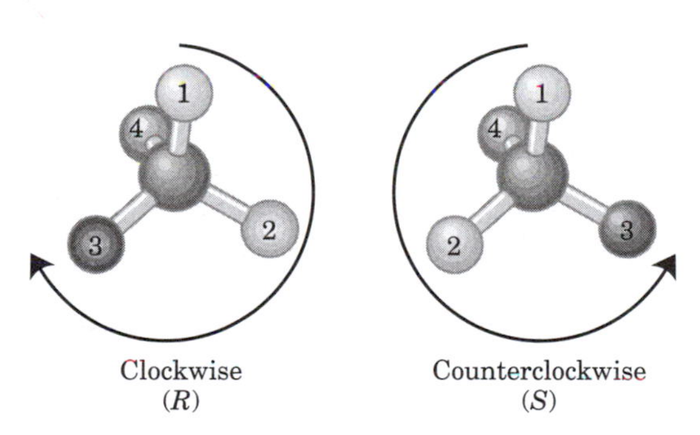

Assigning RS priority, page 60

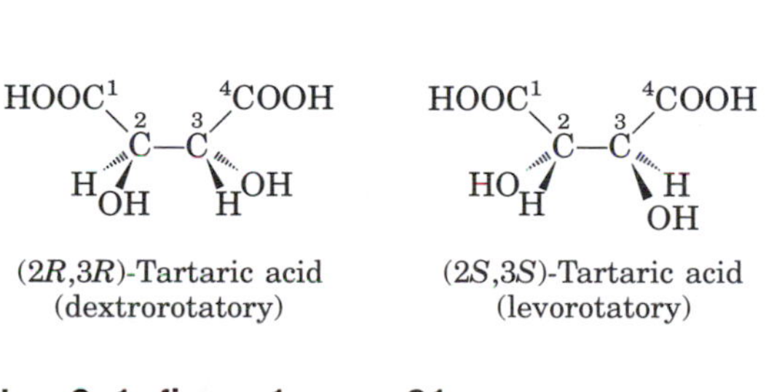

box 3–1, figure 1, page 61
Pasteur separated crystals of two stereoisomers of tartaric acid and showed that solutions of the separated forms rotated polarized light to the same extent but in opposite directions.

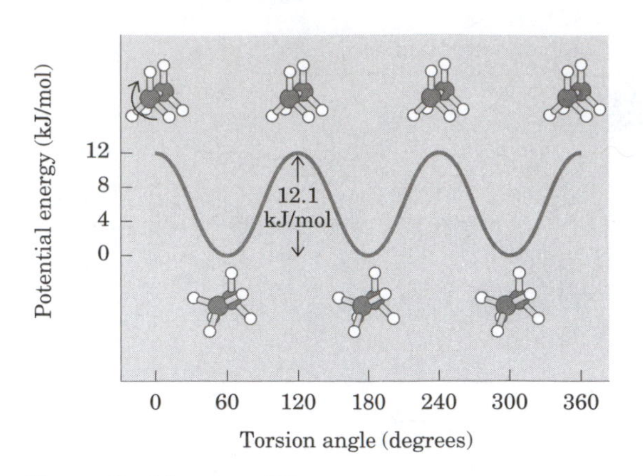

figure 3–11, page 61
Many conformations of ethane are possible because of freedom of rotation around the C—C bond.

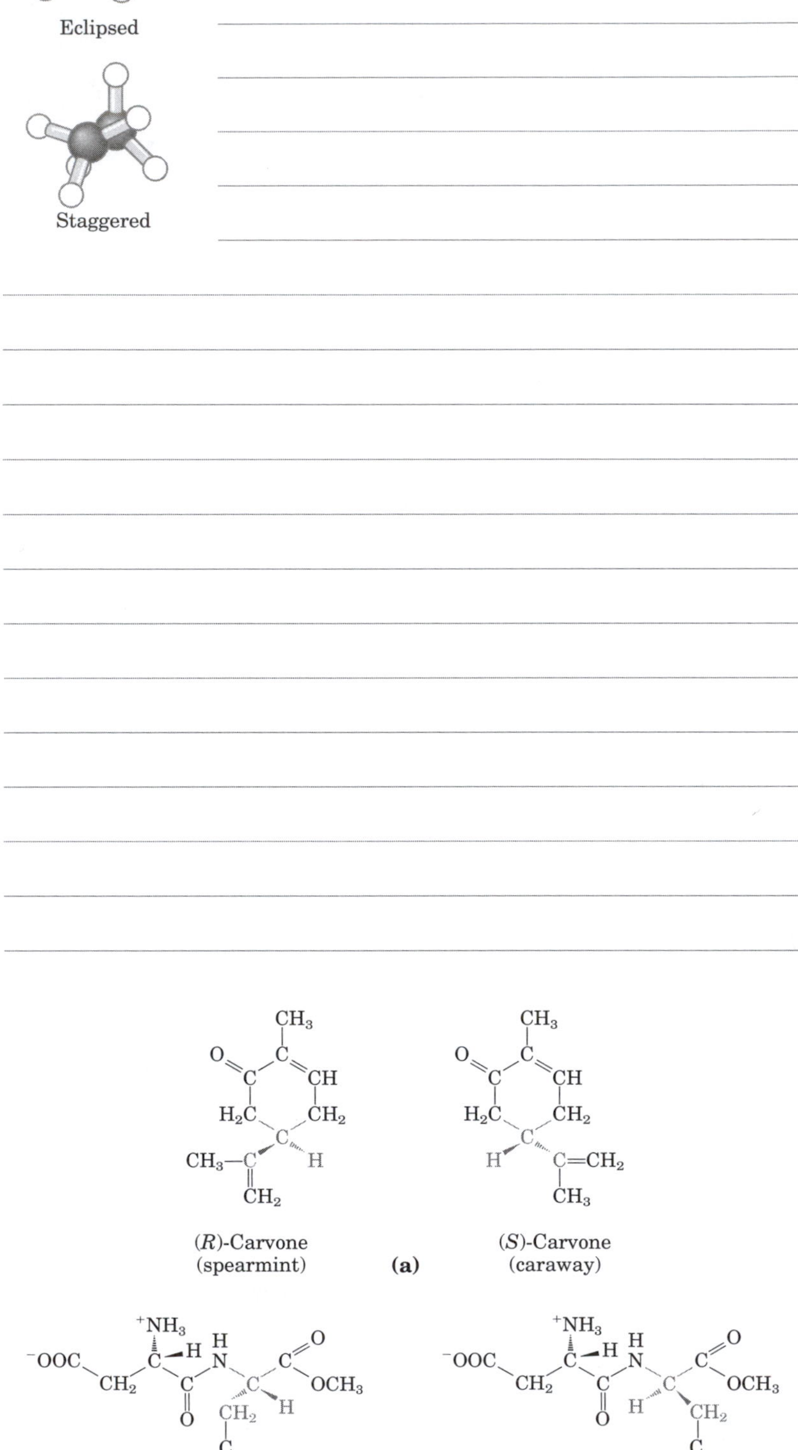

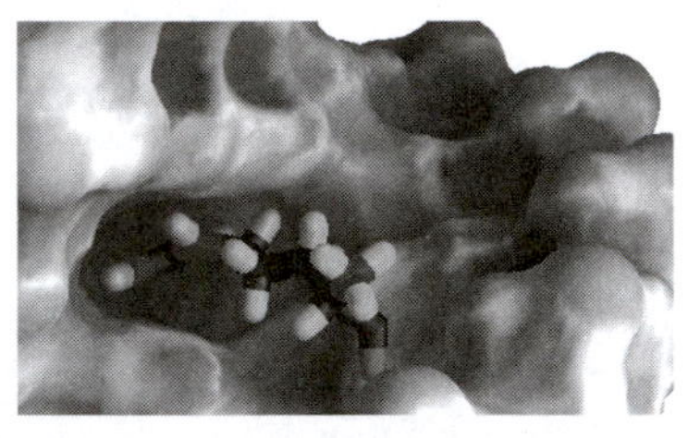

figure 3–12, page 62
Complementary fit of a macromolecule and a small molecule.

figure 3–13, page 62
Structure of the protein brazzein as determined by NMR spectroscopy.

figure 3–14, page 63
Stereoisomers distinguishable by smell and taste in humans.

L-Aspartyl-L-phenylalanine methyl ester (aspartame) (sweet)

(b)

L-Aspartyl-D-phenylalanine methyl ester (bitter)

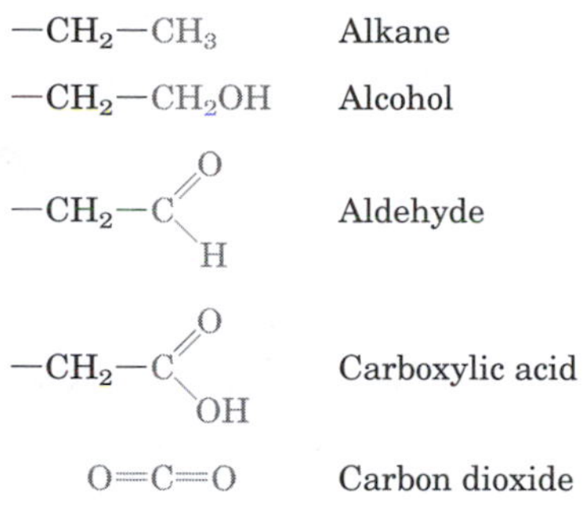

Structure	Name
$-CH_2-CH_3$	Alkane
$-CH_2-CH_2OH$	Alcohol
$-CH_2-C(=O)H$	Aldehyde
$-CH_2-C(=O)OH$	Carboxylic acid
$O=C=O$	Carbon dioxide

figure 3–15, page 65
The oxidation states of carbon in biomolecules.

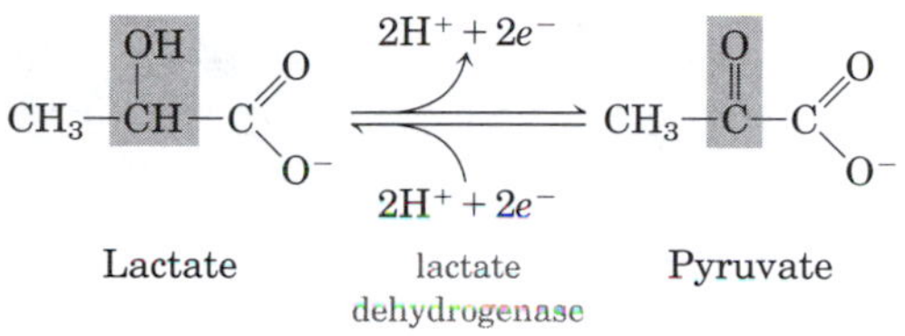

$$CH_3-CH(OH)-COO^- \rightleftharpoons CH_3-C(=O)-COO^- + 2H^+ + 2e^-$$

Lactate — lactate dehydrogenase — Pyruvate

figure 3–16, page 65
An oxidation-reduction reaction.

table 3–2, page 64

The Electronegativities of Some Elements

Element	Electronegativity*
F	4.0
O	3.5
Cl	3.0
N	3.0
Br	2.8
S	2.5
C	2.5
I	2.5
Se	2.4
P	2.1
H	2.1
Cu	1.9
Fe	1.8
Co	1.8
Ni	1.8
Mo	1.8
Zn	1.6
Mn	1.5
Mg	1.2
Ca	1.0
Li	1.0
Na	0.9
K	0.8

*The higher the number, the more electronegative (the greater the electron affinity of) the element.

table 3–3, page 64

Strengths of Bonds Common in Biomolecules

Type of bond	Bond dissociation energy* (kJ/mol)	Type of bond	Bond dissociation energy (kJ/mol)
Single bonds		**Double bonds**	
O—H	461	C=O	712
H—H	435	C=N	615
P—O	419	C=C	611
C—H	414	P=O	502
N—H	389		
C—O	352	**Triple bonds**	
C—C	348	C≡C	816
S—H	339	N≡N	930
C—N	293		
C—S	260		
N—O	222		
S—S	214		

*The greater the energy required for bond dissociation (breakage), the stronger the bond.

Homolytic cleavage $-\overset{|}{\underset{|}{C}}-\overset{|}{\underset{|}{C}}- \rightleftharpoons -\overset{|}{\underset{|}{C}}\cdot + \cdot\overset{|}{\underset{|}{C}}-$

Carbon radicals

Heterolytic cleavage $-\overset{|}{\underset{|}{C}}-\overset{|}{\underset{|}{C}}- \rightleftharpoons -\overset{|}{\underset{|}{C}}:^- + {}^+\overset{|}{\underset{|}{C}}-$

Carbanion Carbocation

figure 3–17, page 66
Two mechanisms for cleavage of a C—C bond.

$-\overset{|}{\underset{|}{C}}:W + Z: \rightleftharpoons -\overset{|}{\underset{|}{C}}:Z + W:$

Leaving group Nucleophile

figure 3–18
A nucleophilic substitution reaction.

table 3–4, page 66

Some Functional Groups Active as Nucleophiles within Cells*

Water	$H\ddot{O}H$
Hydroxide ion	$H\ddot{O}:^-$
Hydroxyl (alcohol)	$R\ddot{O}H$
Alkoxyl	$R\ddot{O}:^-$
Sulfhydryl	RSH
Sulfide	$R\ddot{S}^-$
Amino	$R\ddot{N}H_2$
Carboxylate	$R-C(=O)O^-$
Imidazole	(imidazole ring with R and NH)
Inorganic orthophosphate	$^-O-P(=O)(O^-)-OH$

*Listed in order of decreasing strength. Weaker nucleophiles make better leaving groups.

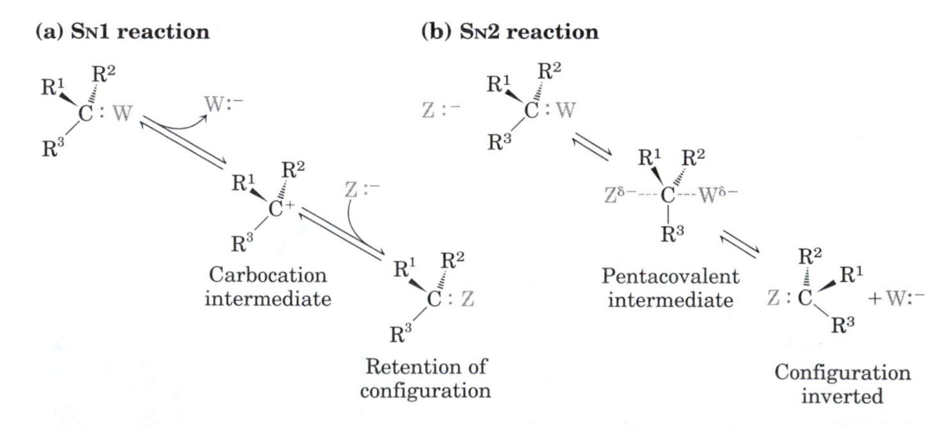

figure 3–19, page 67
Two classes of nucleophilic substitution reactions.

(a)

Glucose 6-phosphate ⇌ (phosphohexose isomerase) Fructose 6-phosphate

(b)

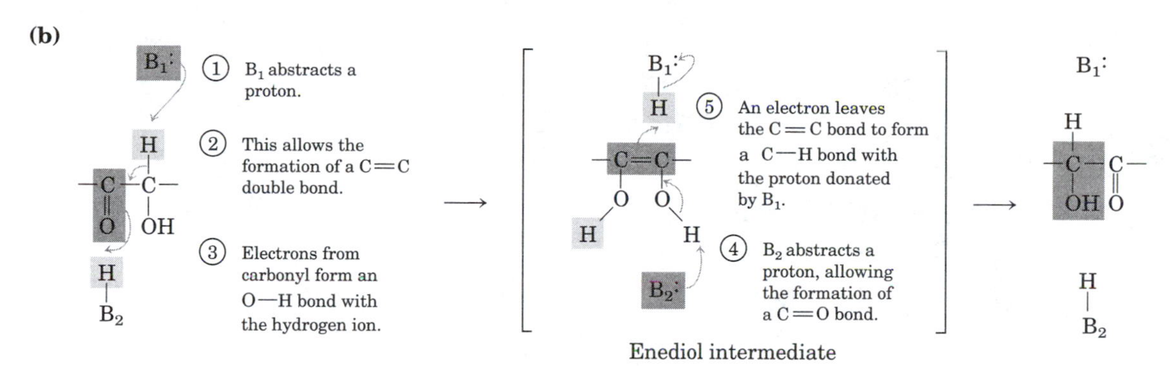

figure 3–20, page 67
An isomerization reaction.

(a)

(b)

(c) Adenine—Ribose—O—P—O—P—O—P—O⁻ (ATP) + HÖ—R (Glucose)

Adenine—Ribose—O—P—O—P—O⁻ (ADP) + ⁻O—P—O—R (Glucose 6-phosphate, a phosphate ester)

(d)

figure 3–21, page 68
Alternative ways of showing the structure of inorganic orthophosphate.

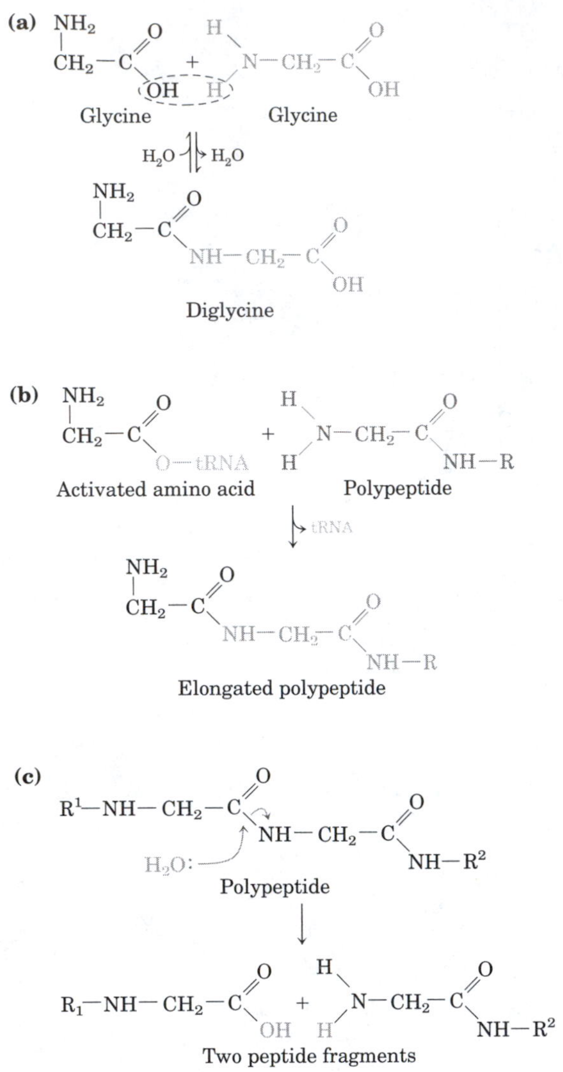

(a) Glycine + Glycine ⇌ Diglycine (H_2O)

(b) Activated amino acid + Polypeptide → Elongated polypeptide (tRNA)

(c) Polypeptide (H_2O) → Two peptide fragments

figure 3–22, page 69
Condensation and hydrolysis.

table 3–5, page 70

Molecular Components of an *E. coli* Cell

	Percentage of total weight of cell	Approximate number of different molecular species
Water	70	1
Proteins	15	3,000
Nucleic acids		
DNA	1	1
RNA	6	>3,000
Polysaccharides	3	5
Lipids	2	20
Monomeric subunits and intermediates	2	500
Inorganic ions	1	20

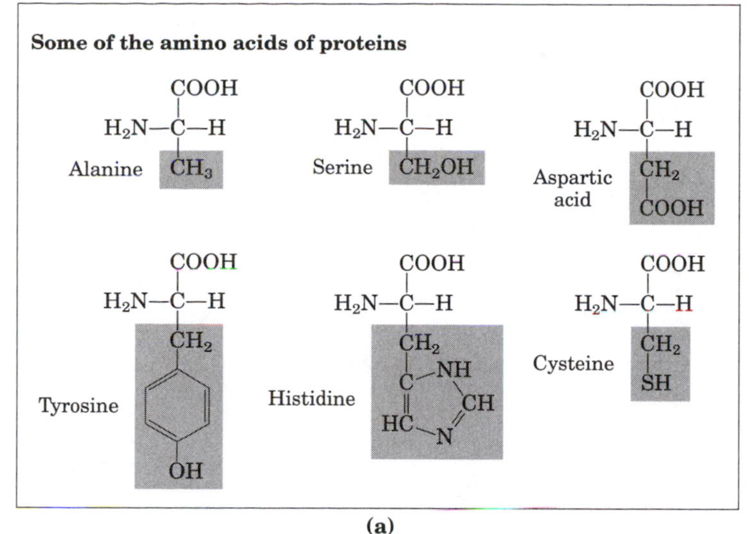

(a)

The components of nucleic acids

Uracil

Thymine

α-D-Ribose

Cytosine

2-Deoxy-α-D-ribose

Adenine

Guanine

Phosphoric acid

(b)

Some components of lipids

Choline

Glycerol

Oleic acid

Palmitic acid

(c)

The parent sugar

α-D-Glucose

(d)

figure 3–24, page 71

The organic compounds from which most cellular materials are constructed: the ABCs of biochemistry.

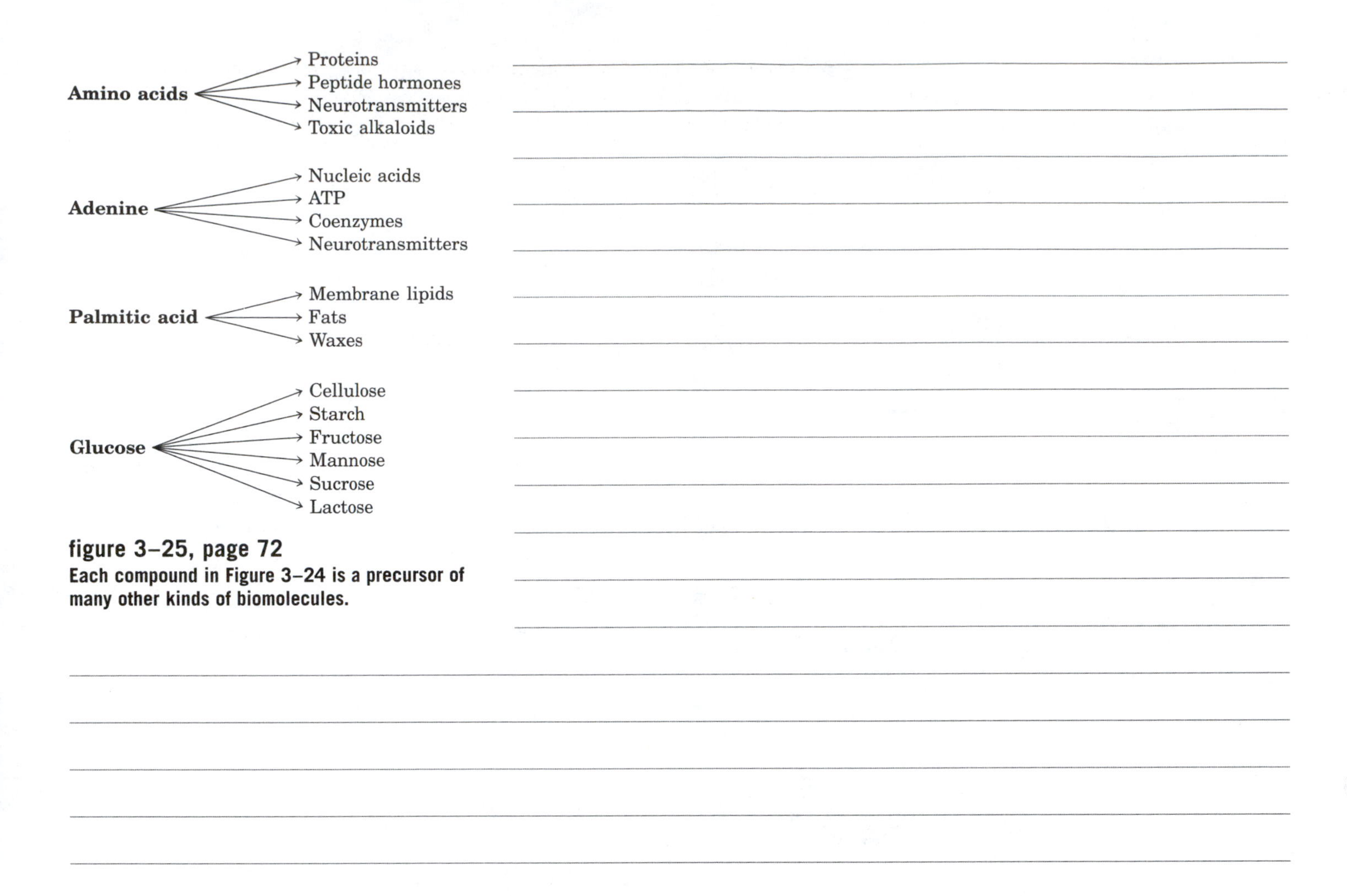

figure 3–25, page 72
Each compound in Figure 3–24 is a precursor of many other kinds of biomolecules.

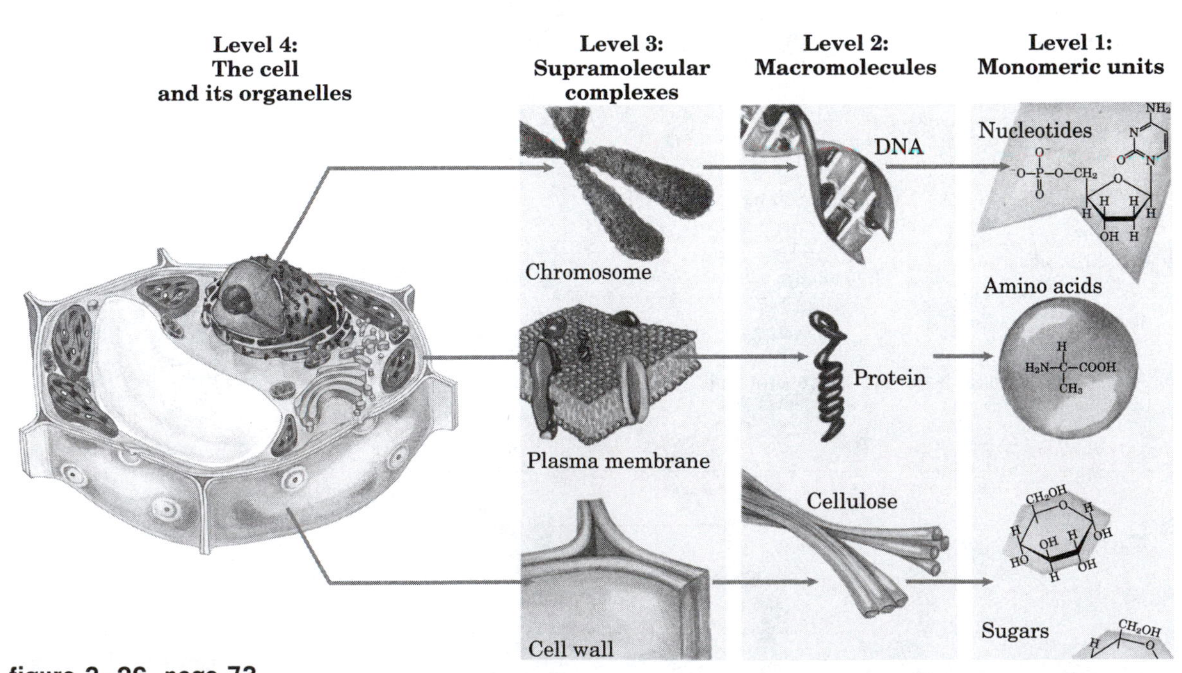

figure 3–26, page 73
Structural hierarchy in the molecular organization of cells.

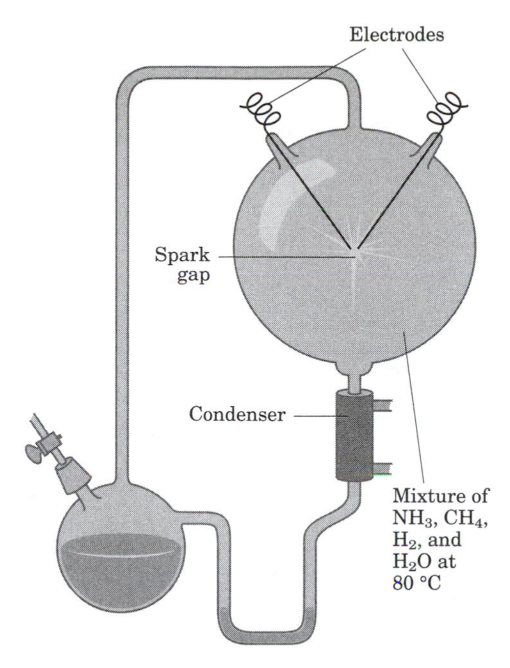

figure 3–27, page 74
Spark-discharge apparatus of the type used by Miller and Urey in experiments demonstrating abiotic formation of organic compounds under primitive atmospheric conditions.

table 3–6, page 75

Some Products Formed under Prebiotic Conditions

Carboxylic acids	Nucleic acid bases	Amino acids	Sugars
Formic acid	Adenine	Glycine	Straight and branched pentoses and hexoses
Acetic acid	Guanine	Alanine	
Propionic acid	Xanthine	α-Aminobutyric acid	
Straight and branched fatty acids ($C_4 - C_{10}$)	Hypoxanthine	Valine	
Glycolic acid	Cytosine	Leucine	
Lactic acid	Uracil	Isoleucine	
Succinic acid		Proline	
		Aspartic acid	
		Glutamic acid	
		Serine	
		Threonine	

Source: From Miller, S.L. (1987) Which organic compounds could have occurred on the prebiotic earth? *Cold Spring Harb. Symp. Quant. Biol.* **52,** 17–27.

Creation of prebiotic soup, including nucleotides, from components of Earth's primitive atmosphere

↓

Production of short RNA molecules with random sequences

↓

Selective replication of self-duplicating catalytic RNA segments

↓

Synthesis of specific peptides, catalyzed by RNA

↓

Increasing role of peptides in RNA replication; coevolution of RNA and protein

↓

Primitive translation system develops, with RNA genome and RNA-protein catalysts

↓

Genomic RNA begins to be copied into DNA

↓

DNA genome, translated on RNA-protein complex (ribosome) with protein catalysts

figure 3–28, page 76
One possible "RNA world" scenario, showing the transition from the prebiotic RNA world (shades of light gray) to the biotic DNA world (dark gray).

PROBLEMS

2. Identification of Functional Groups, page 80.

Ethanolamine
(a)

Glycerol
(b)

Phosphoenolpyruvic acid, an intermediate in glucose metabolism
(c)

Threonine, an amino acid
(d)

Pantothenic acid, a vitamin
(e)

D-Glucosamine
(f)

Problem 2

3. Drug Activity and Stereochemistry, page 80.

Isoproterenol

4. Drug Action and Shape of Molecules, page 80.

Dexedrine (Benzedrine)

5. Components of Complex Biomolecules, page 81.

GTP

Phosphatidylcholine

Methionine enkephalin

chapter

4

Water

When you study for exams, don't forget to review The Minicourses on the UNDERSTAND! BIOCHEMISTRY CD that came with your textbook.

Minicourses that apply to this Chapter include:

Background

Chemical Fundamentals I

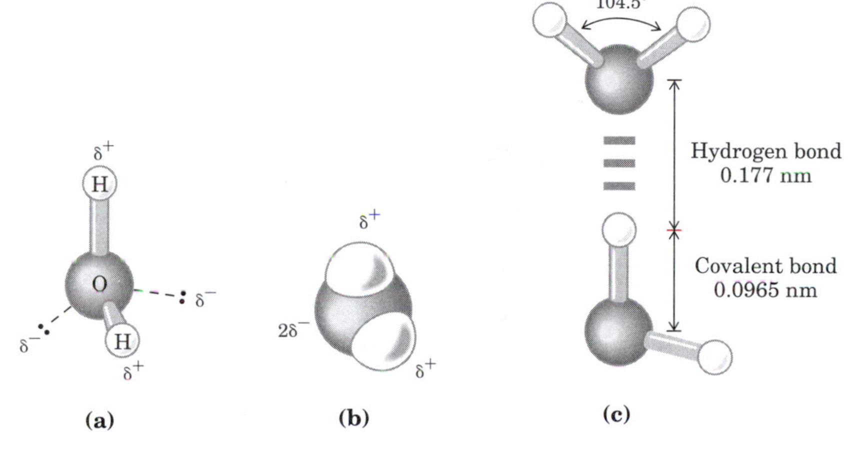

figure 4–1, page 83

Structure of the water molecule.

table 4–1, page 83

Melting Point, Boiling Point, and Heat of Vaporization of Some Common Solvents

	Melting point (°C)	Boiling point (°C)	Heat of vaporization (J/g)*
Water	0	100	2,260
Methanol (CH_3OH)	−98	65	1,100
Ethanol (CH_3CH_2OH)	−117	78	854
Propanol ($CH_3CH_2CH_2OH$)	−127	97	687
Butanol ($CH_3(CH_2)_2CH_2OH$)	−90	117	590
Acetone (CH_3COCH_3)	−95	56	523
Hexane ($CH_3(CH_2)_4CH_3$)	−98	69	423
Benzene (C_6H_6)	6	80	394
Butane ($CH_3(CH_2)_2CH_3$)	−135	−0.5	381
Chloroform ($CHCl_3$)	−63	61	247

*The heat energy required to convert 1.0 g of a liquid at its boiling point, at atmospheric pressure, into its gaseous state at the same temperature. It is a direct measure of the energy required to overcome attractive forces between molecules in the liquid phase.

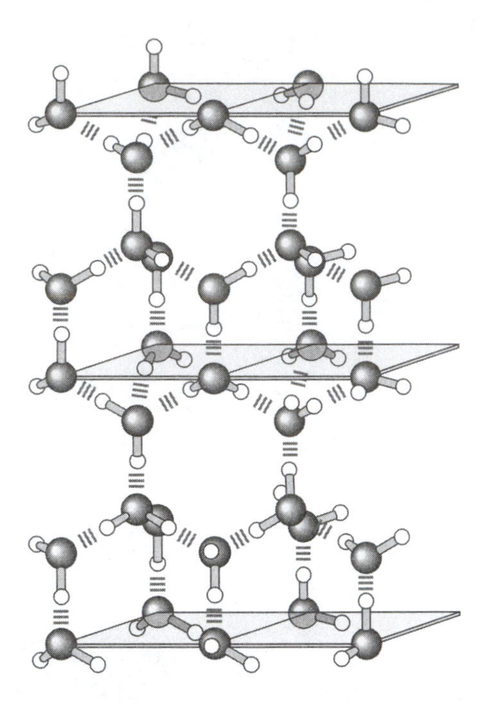

figure 4–2, page 84
Hydrogen bonding in ice.

Hydrogen acceptor

Hydrogen donor

figure 4–3, page 85
Common hydrogen bonds in biological systems.

Between the hydroxyl group of an alcohol and water

Between the carbonyl group of a ketone and water

Between peptide groups in polypeptides

Between complementary bases of DNA

Thymine

Adenine

figure 4–4, page 85
Some biologically important hydrogen bonds.

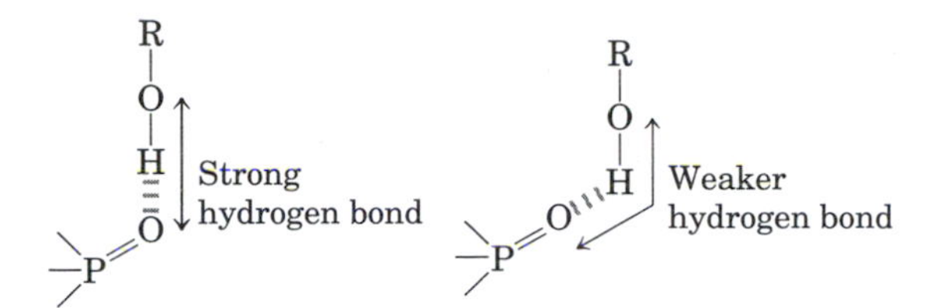

figure 4–5, page 85
Directionality of the hydrogen bond.

table 4–2, page 86

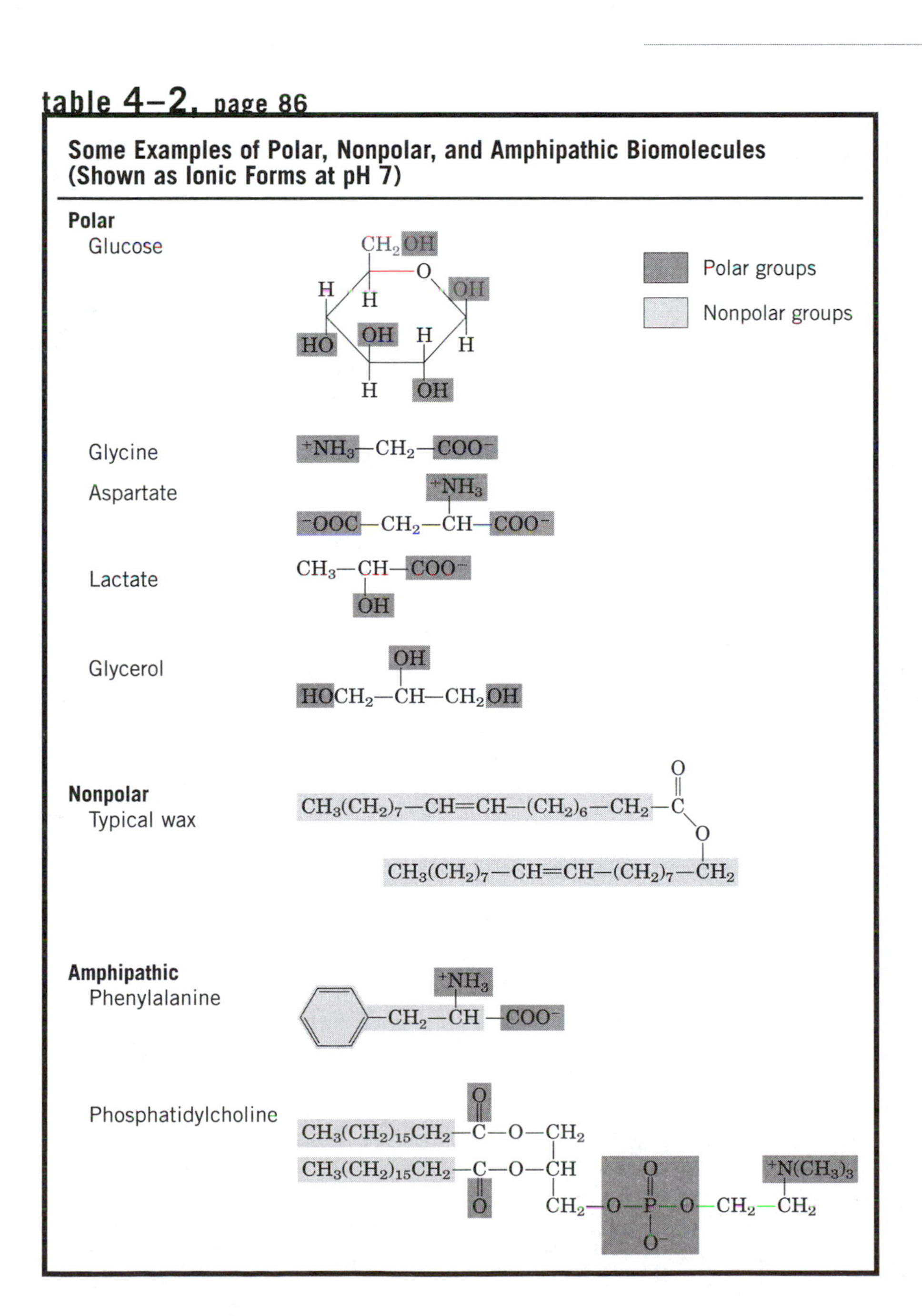

Some Examples of Polar, Nonpolar, and Amphipathic Biomolecules (Shown as Ionic Forms at pH 7)

Polar groups
Nonpolar groups

Polar

Glucose

Glycine: $^{+}NH_3—CH_2—COO^-$

Aspartate: $^{-}OOC—CH_2—CH(^{+}NH_3)—COO^-$

Lactate: $CH_3—CH(OH)—COO^-$

Glycerol: $HOCH_2—CH(OH)—CH_2OH$

Nonpolar

Typical wax: $CH_3(CH_2)_7—CH{=}CH—(CH_2)_6—CH_2—C(=O)—O—CH_2—(CH_2)_7—CH{=}CH—(CH_2)_7CH_3$

Amphipathic

Phenylalanine: $C_6H_5—CH_2—CH(^{+}NH_3)—COO^-$

Phosphatidylcholine: $CH_3(CH_2)_{15}CH_2—C(=O)—O—CH_2$; $CH_3(CH_2)_{15}CH_2—C(=O)—O—CH$; $CH_2—O—P(=O)(O^-)—O—CH_2—CH_2—{}^{+}N(CH_3)_3$

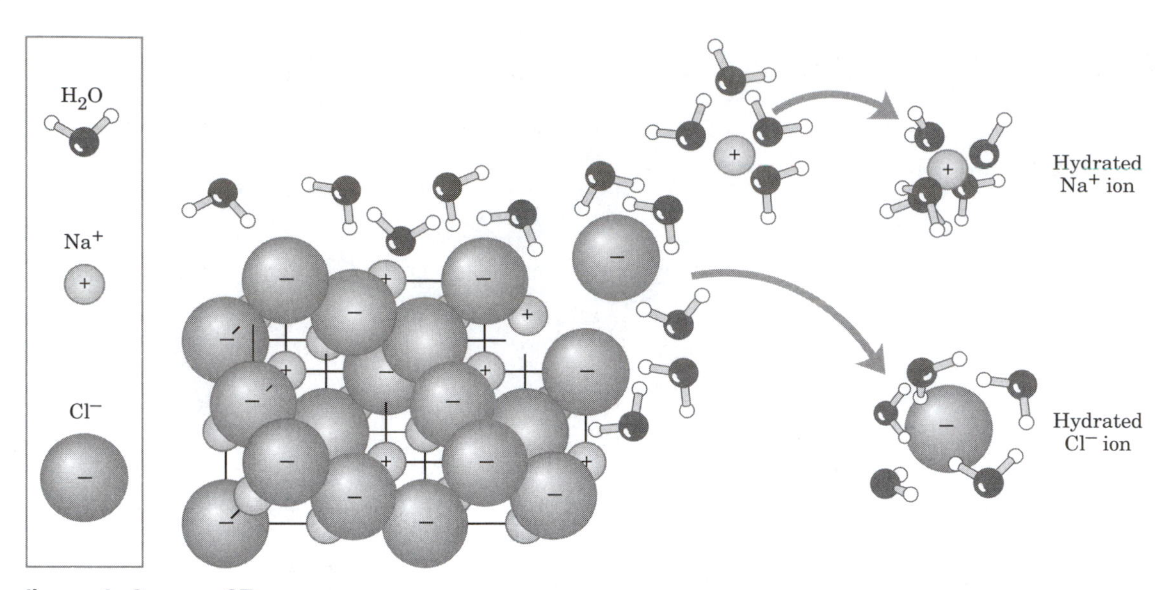

figure 4–6, page 87
Water dissolves many crystalline salts by hydrating their component ions.

$$F = \frac{Q_1 Q_2}{\epsilon r^2}$$

Strength of ionic interactions, page 87

table 4–3, page 88

Solubilities of Some Gases in Water

Gas	Structure*	Polarity	Solubility in water (g/L)[†]
Nitrogen	N≡N	Nonpolar	0.018 (40 °C)
Oxygen	O=O	Nonpolar	0.035 (50 °C)
Carbon dioxide	O=C=O (δ⁻ ← → δ⁻)	Nonpolar	0.97 (45 °C)
Ammonia	NH_3 (↓ δ⁻)	Polar	900 (10 °C)
Hydrogen sulfide	H_2S (↓ δ⁻)	Polar	1,860 (40 °C)

*The arrows represent electric dipoles; there is a partial negative charge (δ^-) at the head of the arrow, a partial positive charge (δ^+; not shown here) at the tail.
[†]Note that polar molecules dissolve far better even at low temperatures than do nonpolar molecules at relatively high temperatures.

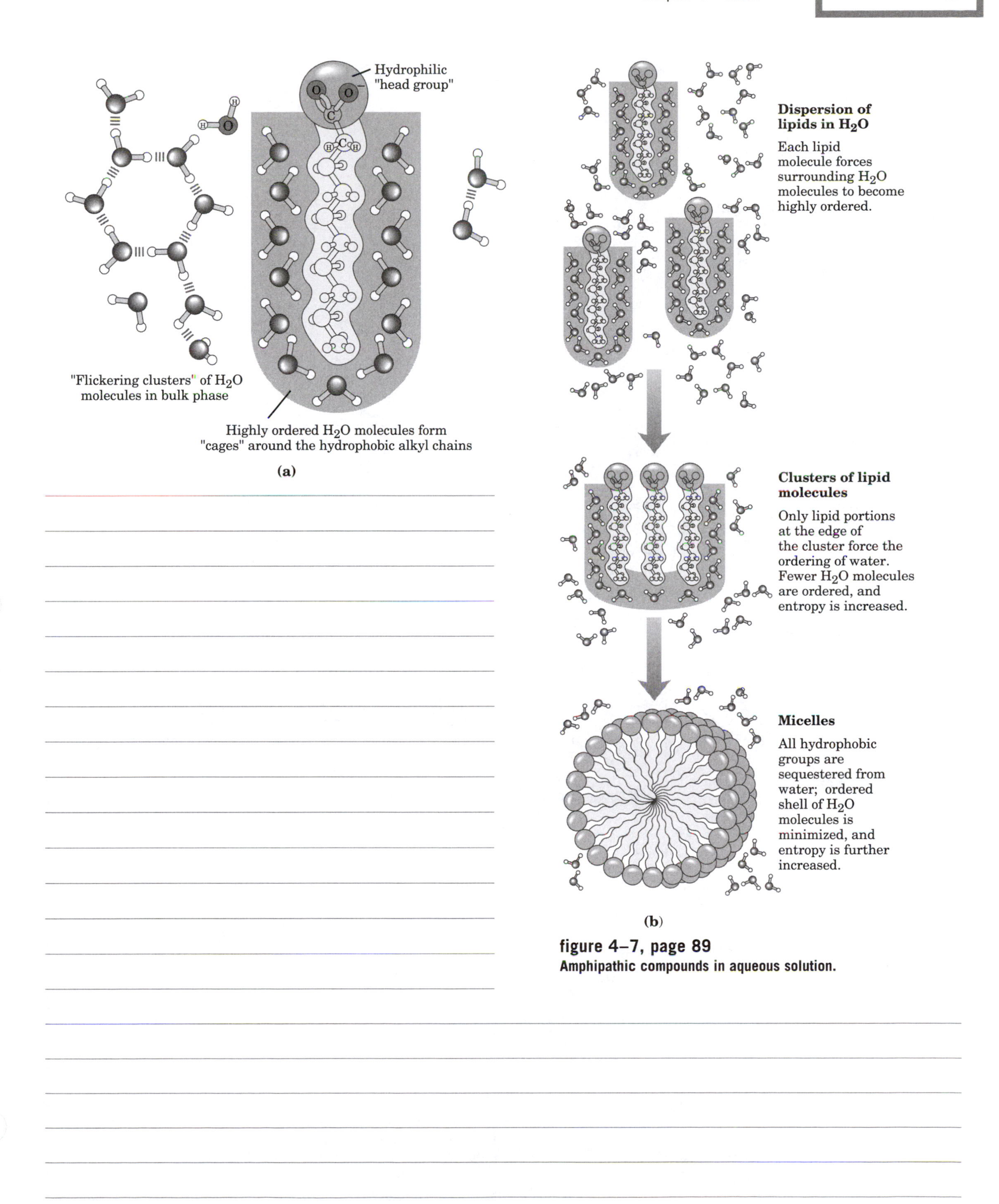

figure 4–7, page 89
Amphipathic compounds in aqueous solution.

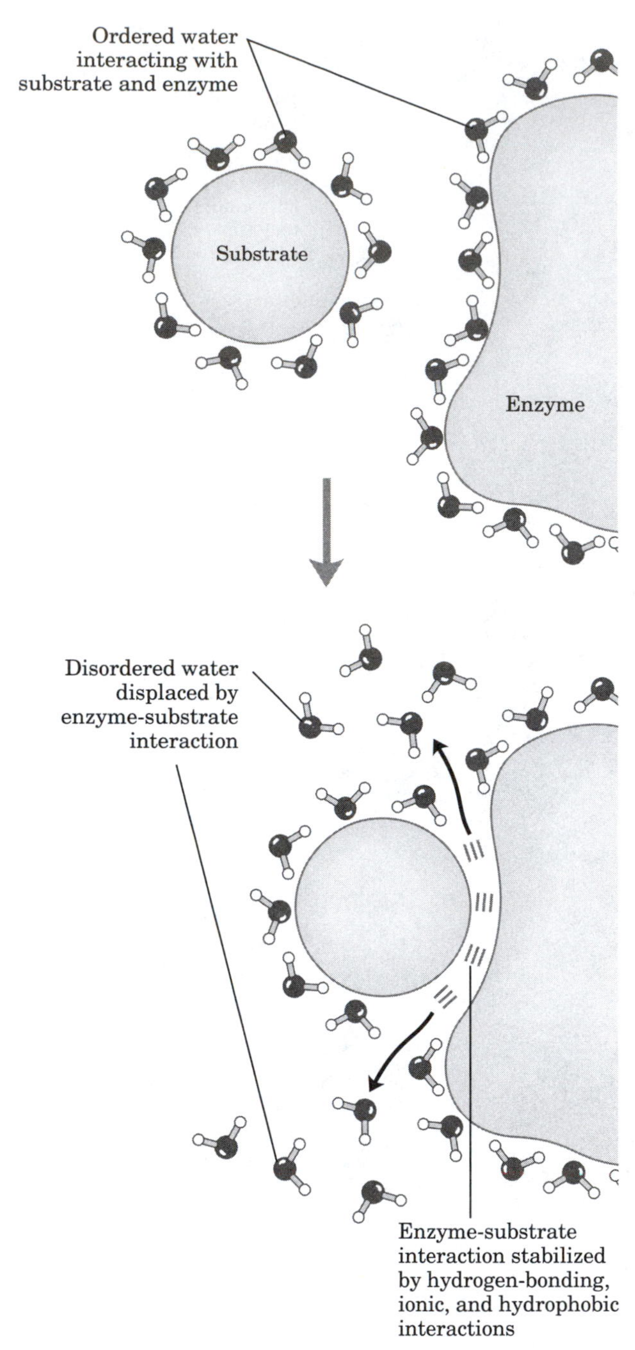

figure 4–8, page 90

Release of ordered water favors formation of an enzyme-substrate complex.

table 4–4, page 91

Four Types of Noncovalent ("Weak") Interactions among Biomolecules in Aqueous Solvent

Hydrogen bonds	
Between neutral groups	$>C{=}O \cdots H{-}O{-}$
Between peptide bonds	$>C{=}O \cdots H{-}N<$
Ionic interactions	
Attraction	$-{}^{+}NH_3 \rightarrow\leftarrow {}^{-}O{-}C({=}O){-}$
Repulsion	$-{}^{+}NH_3 \longleftrightarrow H_3N^{+}-$
Hydrophobic interactions	water; $CH_3\ CH_3$, CH, CH_2; CH_2 (benzene ring)
Van der Waals interactions	Any two atoms in close proximity

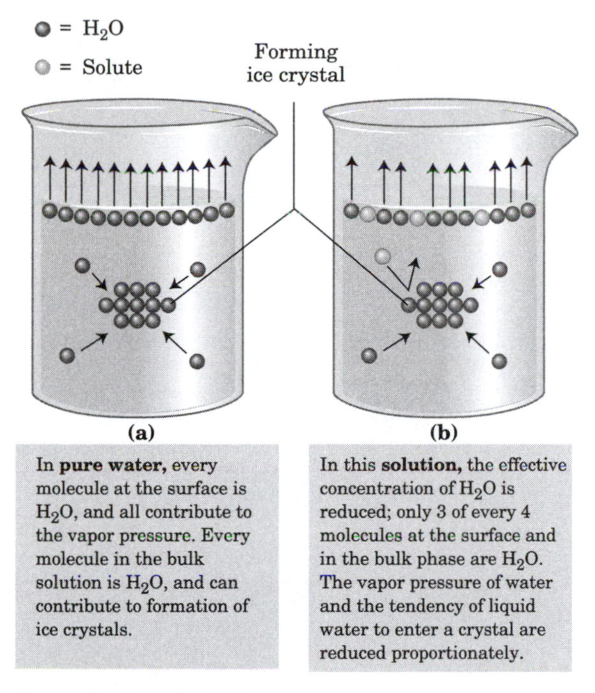

figure 4–9, page 92
Solutes alter the colligative properties of aqueous solutions.

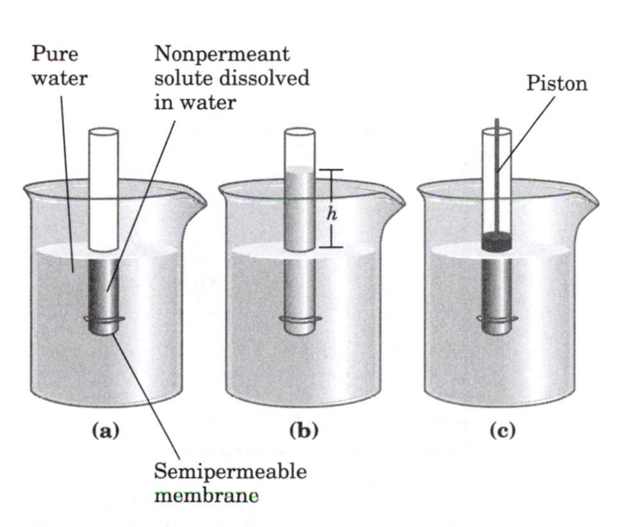

figure 4–10, page 92
Osmosis and the measurement of osmotic pressure.

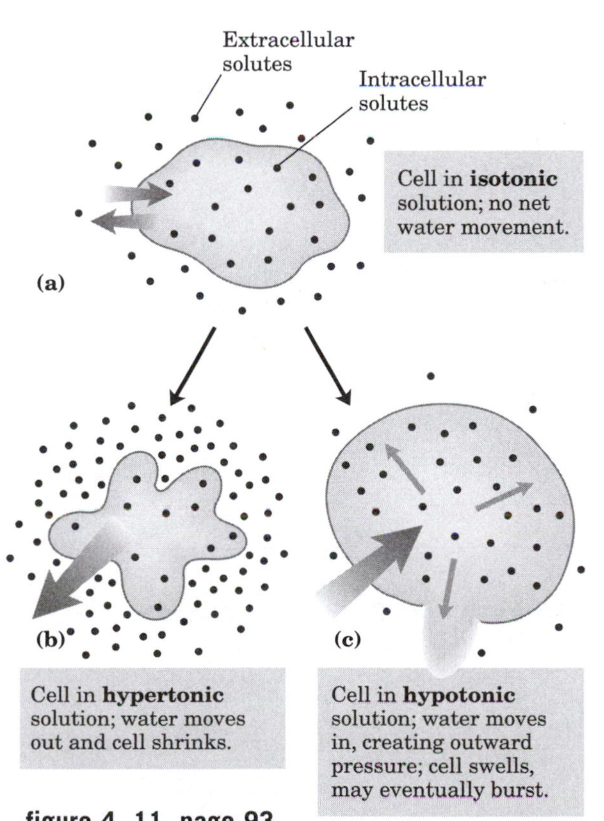

figure 4–11, page 93
The effect of extracellular osmolarity on water movement across a plasma membrane.

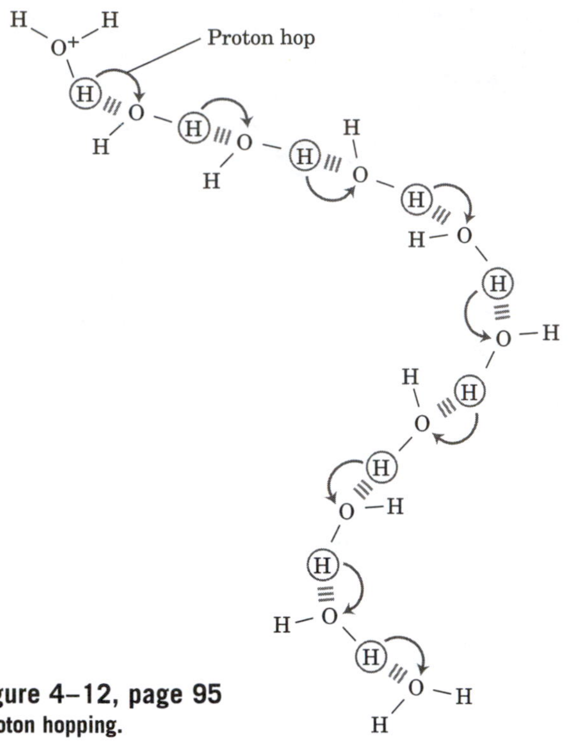

Proton hop

figure 4–12, page 95
Proton hopping.

Ionization of Water

$$H_2O \rightleftharpoons H^+ + OH^-$$ **(Equation 4–1), page 95**

Equilibrium Constant

$$A + B \rightleftharpoons C + D$$ **(Equation 4–2), page 96**

$$K_{eq} = \frac{[C][D]}{[A][B]}$$

Ionization of Water

$$K_{eq} = \frac{[H^+][OH^-]}{[H_2O]}$$ **(Equation 4–3), page 96**

$$K_{eq} = \frac{[H^+][OH^-]}{55.5\ \text{M}}$$

$$(55.5\ \text{M})(K_{eq}) = [H^+][OH^-] = K_w$$ **(Equation 4–4), page 96**

$$K_w = [H^+][OH^-] = (55.5\ \text{M})(1.8 \times 10^{-16}\ \text{M}) = 1.0 \times 10^{-14}\ \text{M}^2$$ **(page 96)**

$$K_w = [H^+][OH^-] = [H^+]^2$$ **(page 97)**

$$[H^+] = \sqrt{K_W} = \sqrt{1 \times 10^{-14}\ \text{M}^2}$$

$$[H^+] = [OH^-] = 10^{-7}\ \text{M}$$ **(page 97)**

pH

$$\text{pH} = \log \frac{1}{[H^+]} = -\log [H^+]$$ **(page 97)**

table 4–5, page 97

The pH Scale

$[H^+]$ (M)	pH	$[OH^-]$ (M)	pOH*
10^0 (1)	0	10^{-14}	14
10^{-1}	1	10^{-13}	13
10^{-2}	2	10^{-12}	12
10^{-3}	3	10^{-11}	11
10^{-4}	4	10^{-10}	10
10^{-5}	5	10^{-9}	9
10^{-6}	6	10^{-8}	8
10^{-7}	7	10^{-7}	7
10^{-8}	8	10^{-6}	6
10^{-9}	9	10^{-5}	5
10^{-10}	10	10^{-4}	4
10^{-11}	11	10^{-3}	3
10^{-12}	12	10^{-2}	2
10^{-13}	13	10^{-1}	1
10^{-14}	14	10^0 (1)	0

*The expression pOH is sometimes used to describe the basicity, or OH^- concentration, of a solution; pOH is defined by the expression pOH = $-\log [OH^-]$, which is analogous to the expression for pH. Note that in all cases, pH + pOH = 14.

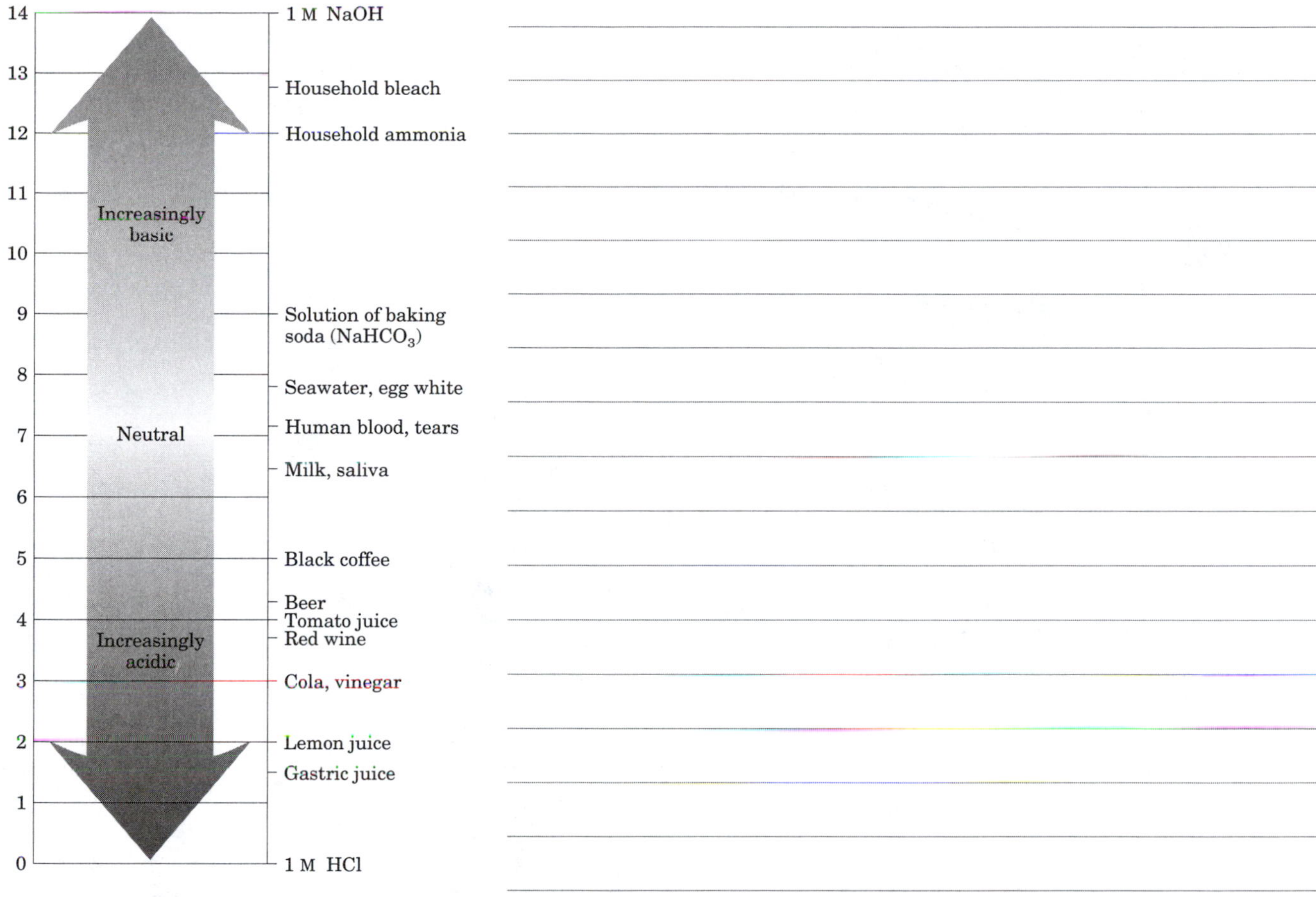

figure 4–13, page 98
The pH of some aqueous fluids.

pK_a, page 99 $pK_a = \log \frac{1}{K_a} = -\log K_a$

Monoprotic acids
Acetic acid
($K_a = 1.74 \times 10^{-5}$ M)

Ammonium
($K_a = 5.62 \times 10^{-10}$ M)

Diprotic acids
Carbonic acid
($K_a = 1.70 \times 10^{-4}$ M);
Bicarbonate
($K_a = 6.31 \times 10^{-11}$ M)

Glycine, carboxyl
($K_a = 4.57 \times 10^{-3}$ M);
Glycine, amino
($K_a = 2.51 \times 10^{-10}$ M)

Triprotic acids
Phosphoric acid
($K_a = 7.25 \times 10^{-3}$ M);
Dihydrogen phosphate
($K_a = 1.38 \times 10^{-7}$ M);
Monohydrogen phosphate
($K_a = 3.98 \times 10^{-13}$ M)

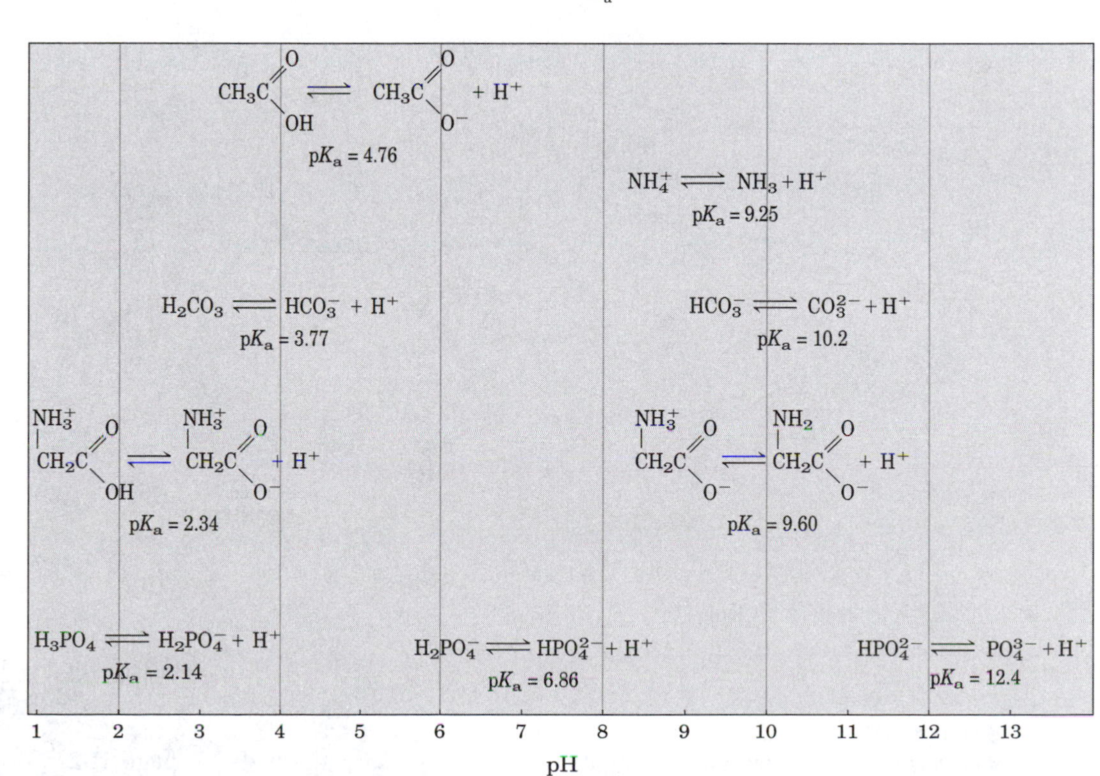

figure 4–14, page 99
Conjugate acid-base pairs consist of a proton donor and a proton acceptor.

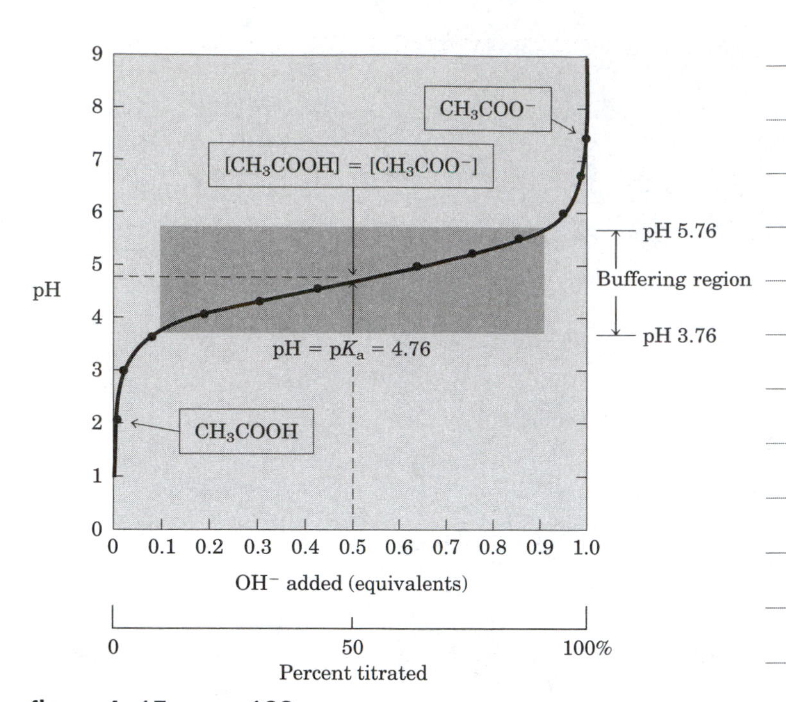

figure 4–15, page 100
The titration curve of acetic acid.

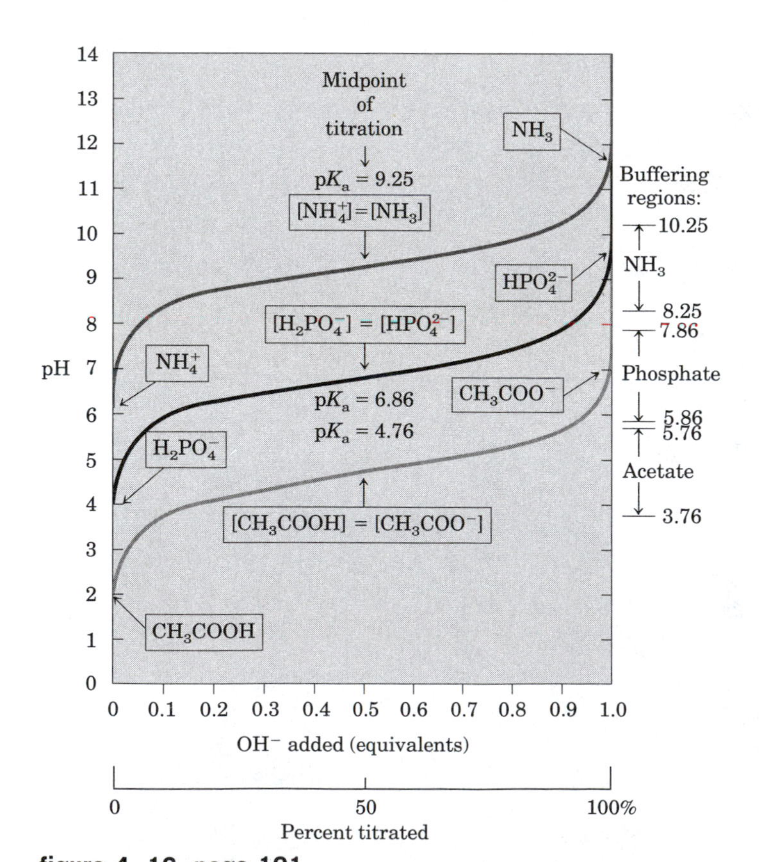

figure 4–16, page 101
Comparison of the titration curves of three weak acids, CH_3COOH, $H_2PO_4^-$, and NH_4^+.

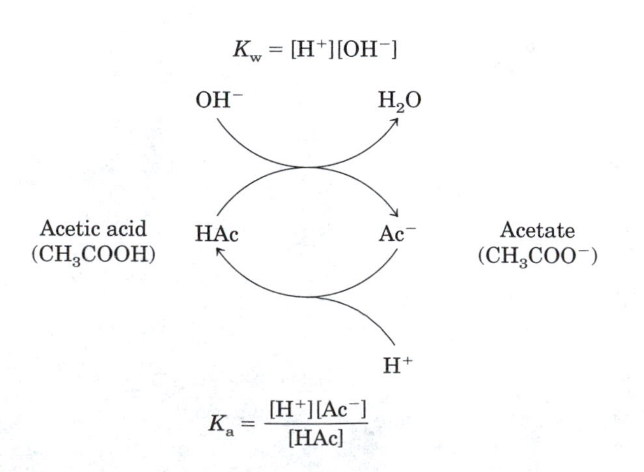

figure 4–17, page 102
The acetic acid–acetate pair as a buffer system.

$$\mathrm{pH} = \mathrm{p}K_a + \log \frac{[\text{proton acceptor}]}{[\text{proton donor}]}$$

Henderson-Hasselbalch equation, page 103

Protein—CH_2—(imidazolium, $\overset{+}{N}$—H) $\rightleftharpoons$ Protein—CH_2—(imidazole) + H^+

figure 4–18, page 104
The amino acid histidine, a component of proteins, is a weak acid.

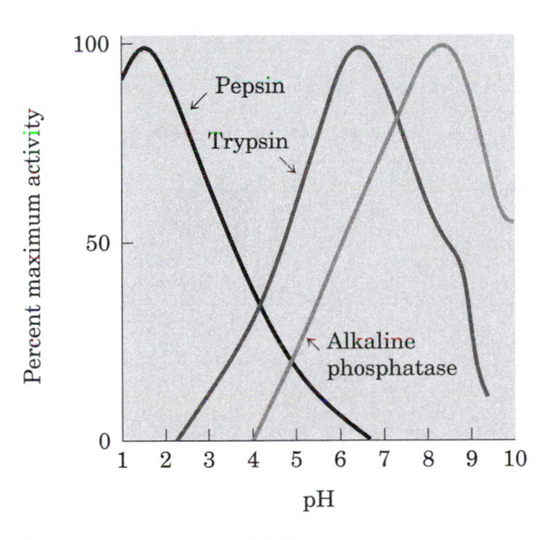

figure 4–19, page 105
The pH optima of some enzymes.

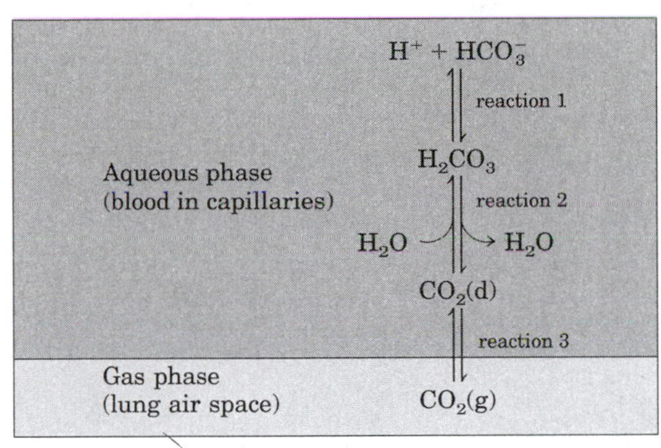

box 4–4, figure 1, page 105
The CO_2 in the air space of the lungs is in equilibrium with the bicarbonate buffer in the blood plasma passing through the lung capillaries.

$$\text{R—O—P(=O)(O}^-\text{)—O—P(=O)(O}^-\text{)—O}^- + H_2O \rightleftharpoons \text{R—O—P(=O)(O}^-\text{)—OH} + \text{HO—P(=O)(O}^-\text{)—O}^-$$

(ATP) (ADP)

Phosphoanhydride **(a)**

$$\text{R—O—P(=O)(O}^-\text{)—O}^- + H_2O \rightleftharpoons \text{R—OH} + \text{HO—P(=O)(O}^-\text{)—O}^-$$

Phosphate ester **(b)**

$$R^1\text{—C(=O)—OR}^2 + H_2O \rightleftharpoons R^1\text{—C(=O)—OH} + \text{HO—R}^2$$

Carboxylate ester **(c)**

$$\text{R—C(=O)—O—P(=O)(O}^-\text{)—O}^- + H_2O \rightleftharpoons \text{R—C(=O)—OH} + \text{HO—P(=O)(O}^-\text{)—O}^-$$

Acyl phosphate **(d)**

figure 4–20, page 106
Participation of water in biological reactions.

PROBLEMS

3. Measurement of Acetylcholine Levels by pH Changes, page 109.

$$CH_3-C(=O)-O-CH_2-CH_2-{}^+N(CH_3)_2-CH_3 \xrightarrow{H_2O}$$

Acetylcholine

$$HO-CH_2-CH_2-{}^+N(CH_3)_2-CH_3 + CH_3-C(=O)-O^- + H^+$$

Choline Acetate

4. Osmotic Balance in a Marine Frog, page 110.

$H_2N-C(=O)-NH_2$

Urea (CH_4N_2O)

5. Properties of a Buffer, page 110.

$$R-NH_3^+ \rightleftharpoons R-NH_2 + H^+$$

6. The Effect of pH on Solubility, page 110.

Benzoic acid (C—OH)
$pK_a \approx 5$

Benzoate ion (C—O⁻)

Pyridine ion
$pK_a \approx 5$

(a)

β-Naphthol
$pK_a \approx 10$

(b)

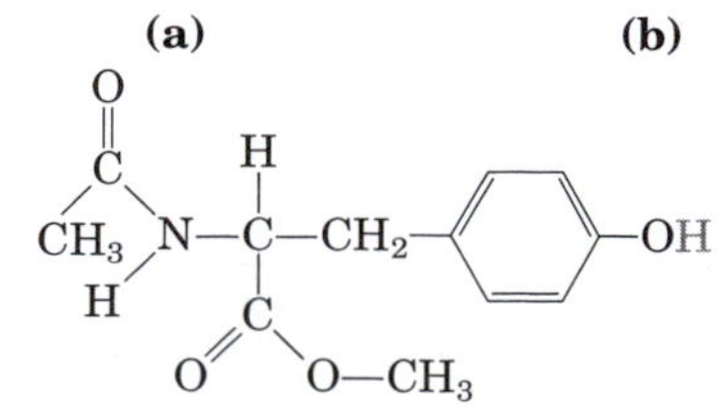

N-Acetyltyrosine methyl ester
$pK_a \approx 10$

(c)

7. Treatment of Poison Ivy Rash, page 111.

OH, OH, $(CH_2)_n-CH_3$

Catechol in Poison Ivy

$pK_a \approx 8$

8. pH and Drug Absorption, page 111.

Aspirin

chapter

5

Amino Acids, Peptides, and Proteins

CD-ROM

When you study for exams, don't forget to review The Minicourses on the UNDERSTAND! BIOCHEMISTRY CD that came with your textbook.

Minicourses that apply to this Chapter include:

Background
- Chemical Fundamentals I

Molecules of Life
- Important Biomolecules
- An Interactive Gallery of Molecular Models
- An Interactive Gallery of Protein Structures

Proteins in Action
- Catalysis and Regulation

Some Important Techniques
- Nucleic Acid Analysis
- Protein Separation and Analysis
- Protein Structure Determination

$$\begin{array}{c} COO^- \\ | \\ H_3\overset{+}{N}-C-H \\ | \\ R \end{array}$$

figure 5–2, page 116
General structure of an amino acid.

$$\begin{array}{cccccccccc} \epsilon & & \delta & & \gamma & & \beta & & \alpha & \\ 6 & & 5 & & 4 & & 3 & & 2 & 1 \\ CH_2 & - & CH_2 & - & CH_2 & - & CH_2 & - & CH & -COO^- \\ | & & & & & & & & | & \\ {}^+NH_3 & & & & & & & & {}^+NH_3 & \end{array}$$

Lysine

Carbon designations in amino acids, page 116

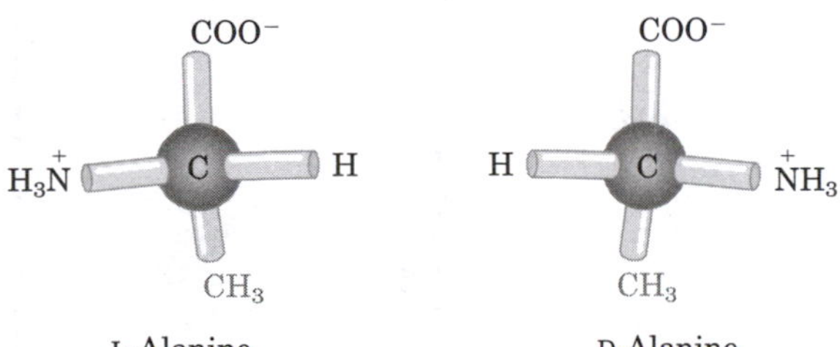

$$\begin{array}{c} COO^- \\ H_3\overset{+}{N}-C-H \\ CH_3 \end{array} \quad \begin{array}{c} COO^- \\ H-C-\overset{+}{N}H_3 \\ CH_3 \end{array}$$

L-Alanine D-Alanine

(b)

$$\begin{array}{c} COO^- \\ | \\ H_3\overset{+}{N}-C-H \\ | \\ CH_3 \end{array} \quad \begin{array}{c} COO^- \\ | \\ H-C-\overset{+}{N}H_3 \\ | \\ CH_3 \end{array}$$

L-Alanine D-Alanine

(c)

figure 5–3, page 116
Stereoisomerism in α-amino acids.

^{1}CHO
HO—^{2}C—H
$^{3}CH_2OH$
L-Glyceraldehyde

CHO
H—C—OH
CH_2OH
D-Glyceraldehyde

COO^-
$H_3\overset{+}{N}$—C—H
CH_3
L-Alanine

COO^-
H—C—$\overset{+}{N}H_3$
CH_3
D-Alanine

figure 5–4, page 117
Steric relationship of the stereoisomers of alanine to the absolute configuration of L- and D-glyceraldehyde.

table 5–1, page 118

Properties and Conventions Associated with the Standard Amino Acids

Amino acid	Abbreviated names		M_r	pK_a values			pI	Hydropathy index*	Occurrence in proteins (%)[†]
				pK_1 (—COOH)	pK_2 ($—NH_3^+$)	pK_R (R group)			
Nonpolar, aliphatic R groups									
Glycine	Gly	G	75	2.34	9.60		5.97	−0.4	7.2
Alanine	Ala	A	89	2.34	9.69		6.01	1.8	7.8
Valine	Val	V	117	2.32	9.62		5.97	4.2	6.6
Leucine	Leu	L	131	2.36	9.60		5.98	3.8	9.1
Isoleucine	Ile	I	131	2.36	9.68		6.02	4.5	5.3
Methionine	Met	M	149	2.28	9.21		5.74	1.9	2.3
Aromatic R groups									
Phenylalanine	Phe	F	165	1.83	9.13		5.48	2.8	3.9
Tyrosine	Tyr	Y	181	2.20	9.11	10.07	5.66	−1.3	3.2
Tryptophan	Trp	W	204	2.38	9.39		5.89	−0.9	1.4
Polar, uncharged R groups									
Serine	Ser	S	105	2.21	9.15		5.68	−0.8	6.8
Proline	Pro	P	115	1.99	10.96		6.48	1.6	5.2
Threonine	Thr	T	119	2.11	9.62		5.87	−0.7	5.9
Cysteine	Cys	C	121	1.96	10.28	8.18	5.07	2.5	1.9
Asparagine	Asn	N	132	2.02	8.80		5.41	−3.5	4.3
Glutamine	Gln	Q	146	2.17	9.13		5.65	−3.5	4.2
Positively charged R groups									
Lysine	Lys	K	146	2.18	8.95	10.53	9.74	−3.9	5.9
Histidine	His	H	155	1.82	9.17	6.00	7.59	−3.2	2.3
Arginine	Arg	R	174	2.17	9.04	12.48	10.76	−4.5	5.1
Negatively charged R groups									
Aspartate	Asp	D	133	1.88	9.60	3.65	2.77	−3.5	5.3
Glutamate	Glu	E	147	2.19	9.67	4.25	3.22	−3.5	6.3

*A scale combining hydrophobicity and hydrophilicity of R groups; it can be used to measure the tendency of an amino acid to seek an aqueous environment (− values) or a hydrophobic environment (+ values). See Chapter 12. From Kyte, J. & Doolittle, R.F. (1982) *J. Mol. Biol.* **157,** 105–132.

[†]Average occurrence in over 1150 proteins. From Doolittle, R.F. (1989) Redundancies in protein sequences. In *Prediction of Protein Structure and the Principles of Protein Conformation* (Fasman, G.D., ed) Plenum Press, NY, pp. 599–623.

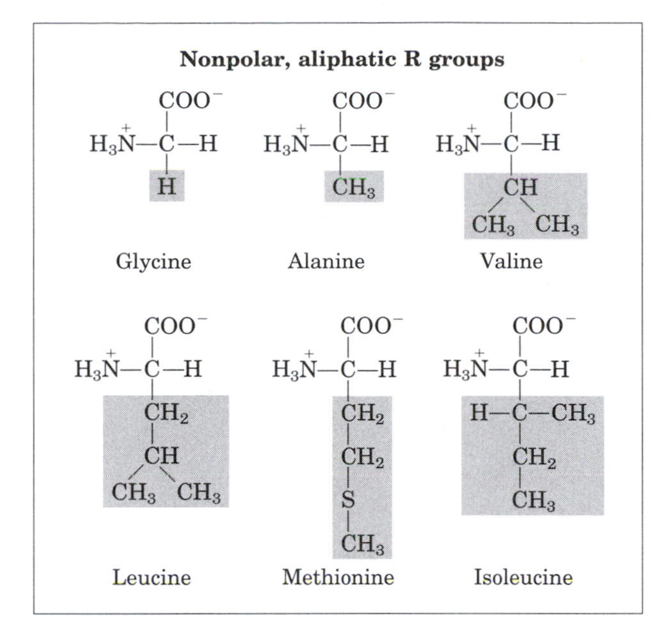

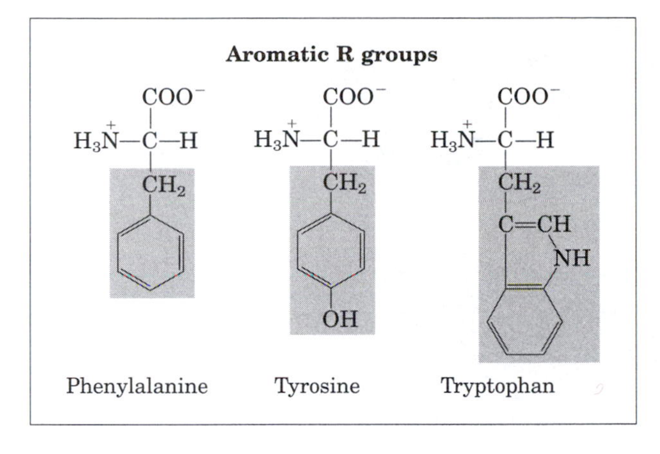

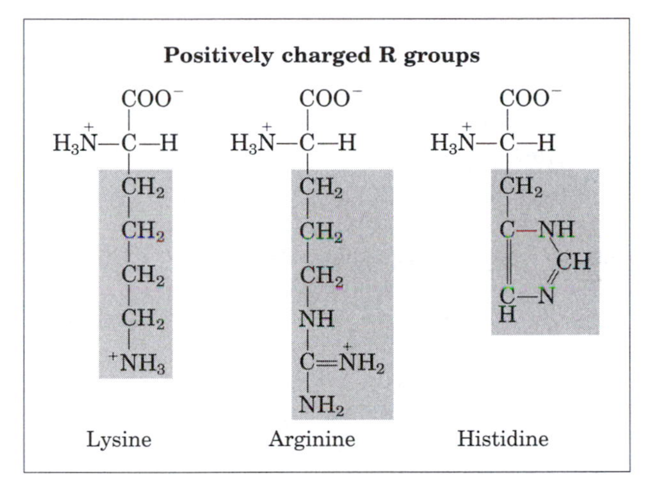

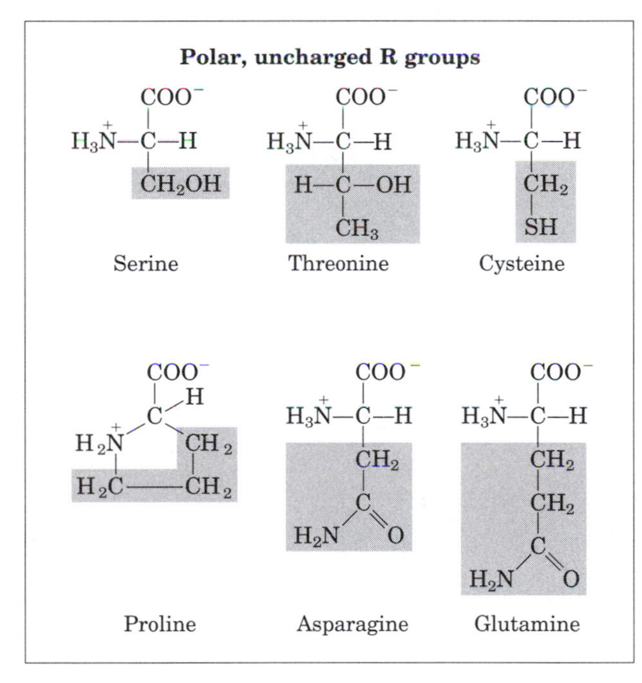

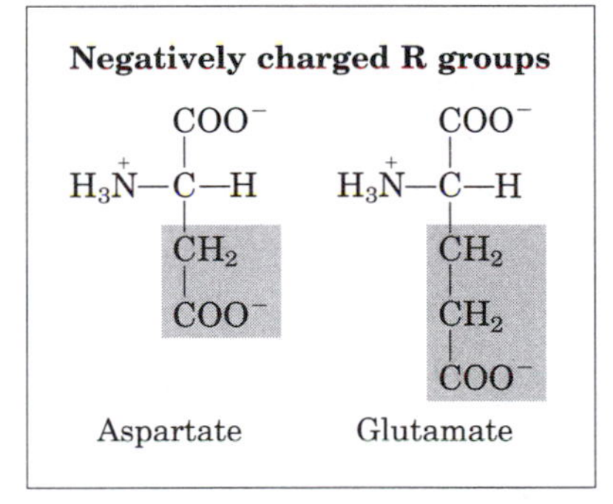

figure 5–5, page 119
The 20 standard amino acids of proteins.

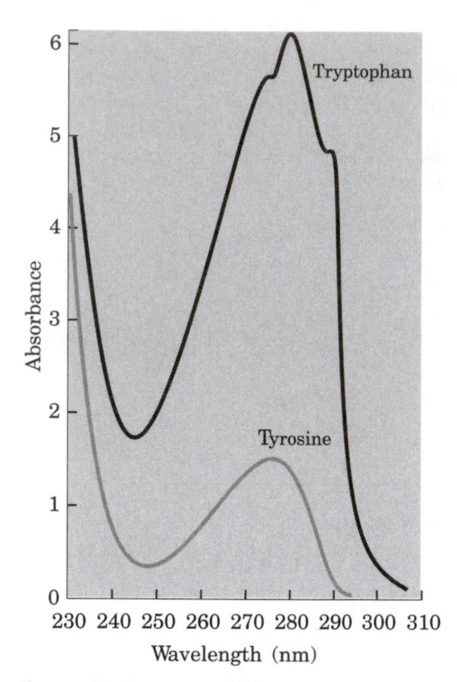

figure 5–6, page 120
Absorbance of ultraviolet light by aromatic amino acids.

Cysteine: COO^- — $H_3\overset{+}{N}$—CH — CH_2 — SH

Cysteine: SH — CH_2 — CH—$\overset{+}{N}H_3$ — COO^-

$\rightleftharpoons$ ($2H^+ + 2e^-$ / $2H^+ + 2e^-$)

Cystine: COO^- — $H_3\overset{+}{N}$—CH — CH_2 — S — S — CH_2 — CH—$\overset{+}{N}H_3$ — COO^-

figure 5–7, page 120
Reversible formation of a disulfide bond by the oxidation of two molecules of cysteine.

$\log \frac{I_0}{I} = \epsilon cl$ **Lambert–Beer law, page 121**

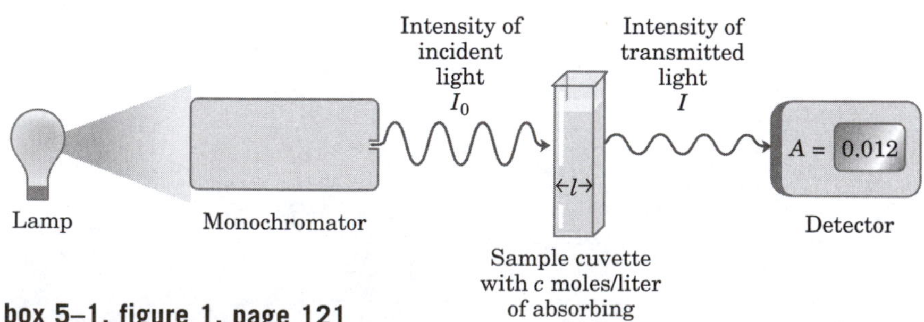

box 5–1, figure 1, page 121
The principal components of a spectrophotometer.

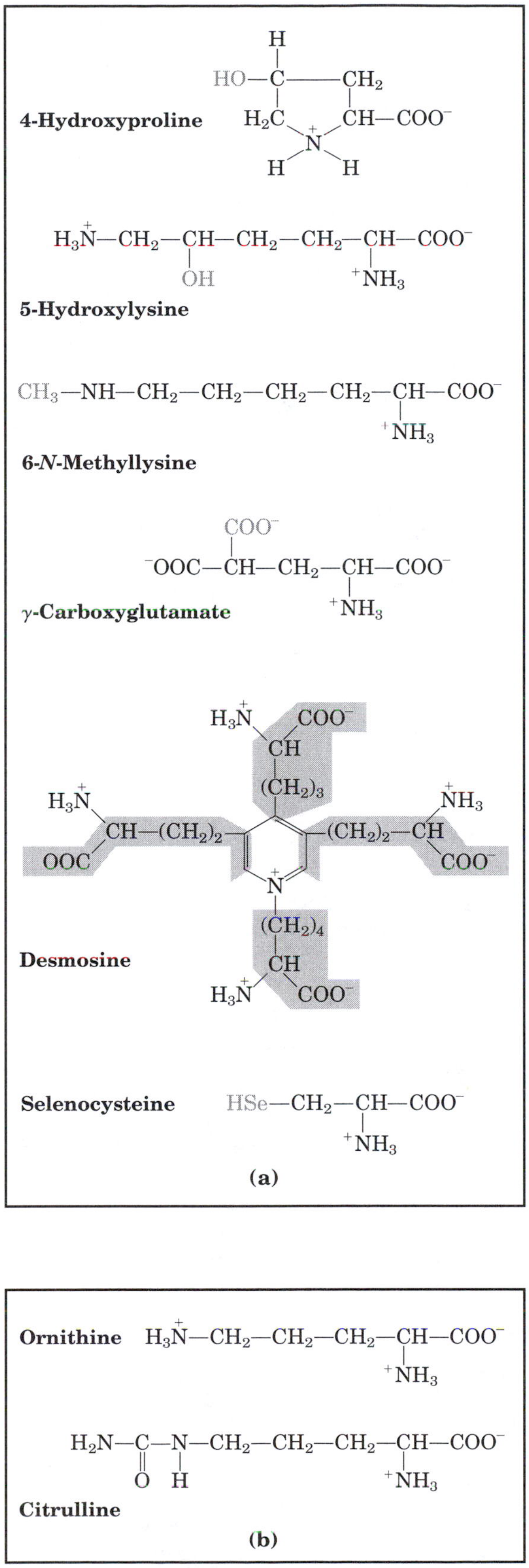

4-Hydroxyproline

5-Hydroxylysine

$H_3\overset{+}{N}-CH_2-CH(OH)-CH_2-CH_2-CH(\overset{+}{N}H_3)-COO^-$

6-*N*-Methyllysine

$CH_3-NH-CH_2-CH_2-CH_2-CH_2-CH(\overset{+}{N}H_3)-COO^-$

γ-Carboxyglutamate

$^-OOC-CH(COO^-)-CH_2-CH(\overset{+}{N}H_3)-COO^-$

Desmosine

Selenocysteine

$HSe-CH_2-CH(\overset{+}{N}H_3)-COO^-$

(a)

Ornithine

$H_3\overset{+}{N}-CH_2-CH_2-CH_2-CH(\overset{+}{N}H_3)-COO^-$

Citrulline

$H_2N-C(=O)-NH-CH_2-CH_2-CH_2-CH(\overset{+}{N}H_3)-COO^-$

(b)

figure 5–8, page 122
Nonstandard amino acids.

Nonionic form

Zwitterionic form

figure 5–9, page 123
Nonionic and zwitterionic forms of amino acids.

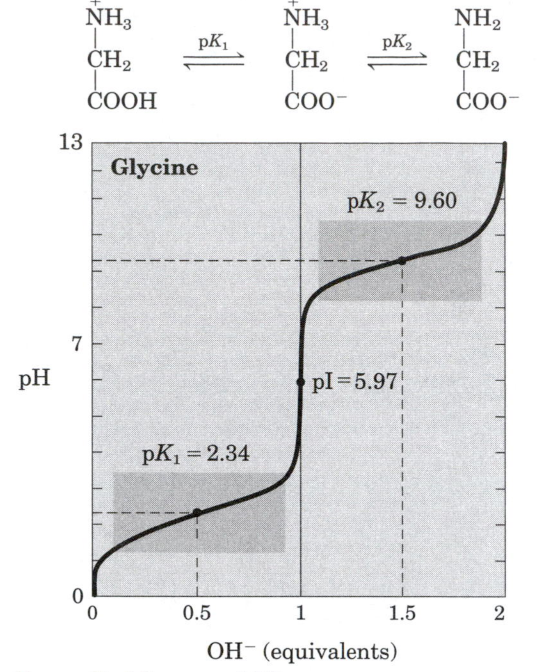

figure 5–10, page 123
Titration of an amino acid.

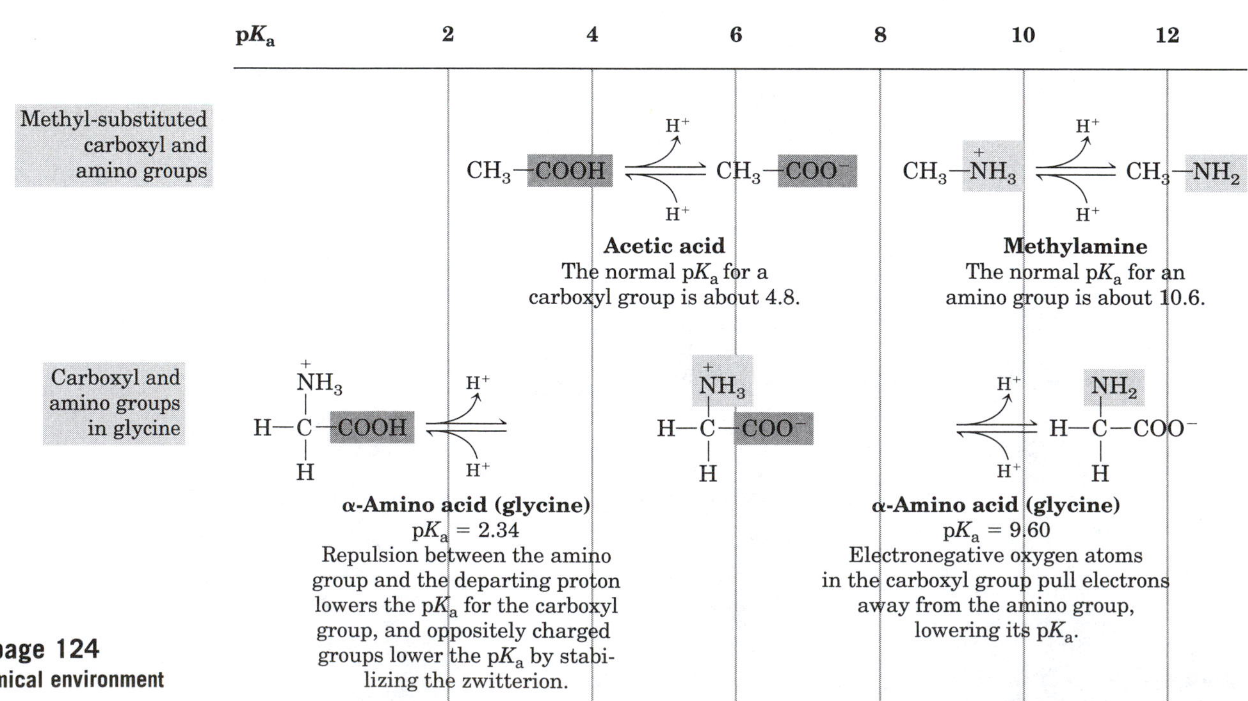

figure 5–11, page 124
Effect of the chemical environment on pK_a.

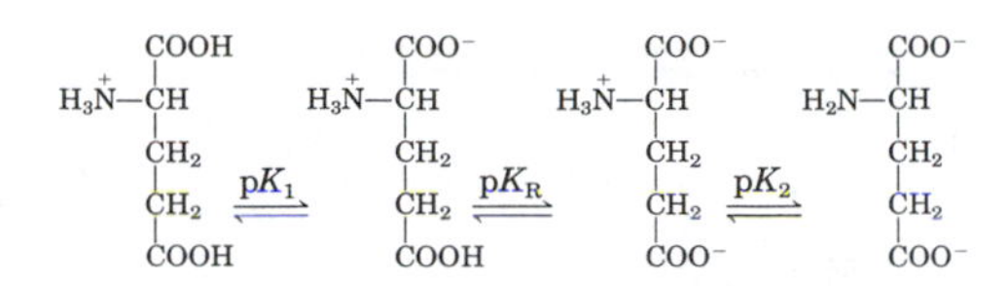

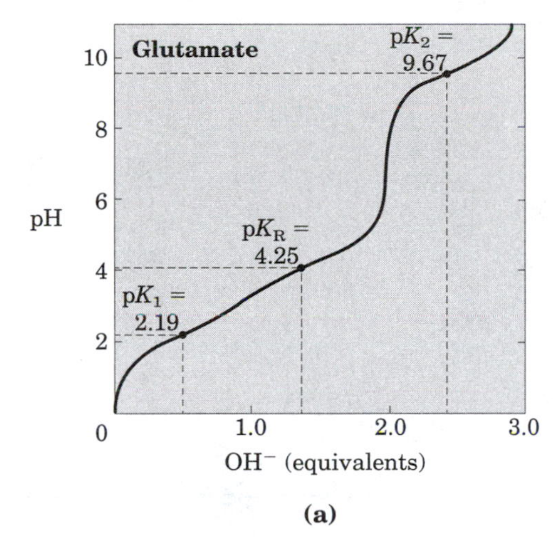

(a)

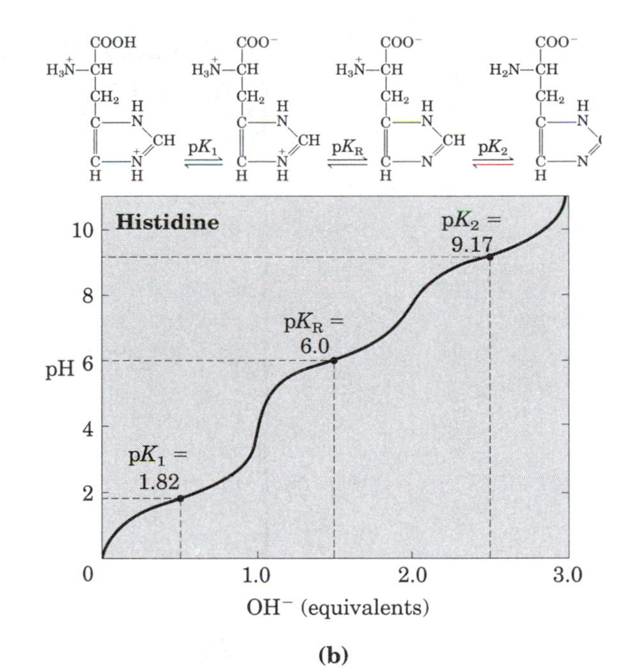

(b)

figure 5–12, page 125
Titration curves for (a) glutamate and (b) histidine.

$$pI = \frac{1}{2}(pK_1 + pK_2)$$

Isoelectric point, page 125

$$H_3\overset{+}{N}-CH(R^1)-C(=O)-OH + H-N(H)-CH(R^2)-COO^- \underset{H_2O}{\overset{H_2O}{\rightleftharpoons}} H_3\overset{+}{N}-CH(R^1)-C(=O)-N(H)-CH(R^2)-COO^-$$

figure 5–13, page 126
Formation of a peptide bond by condensation.

$H_3\overset{+}{N}$–CH(CH_2OH)–CO–NH–CH_2–CO–NH–CH(CH_2–C_6H_4–OH)–CO–NH–CH(CH_3)–CO–NH–CH(CH_2–CH(CH_3)$_2$)–COO^-

Amino-terminal end   Carboxyl-terminal end

figure 5–14, page 126
The pentapeptide serylglycyltyrosylalanylleucine, or Ser–Gly–Tyr–Ala–Leu.

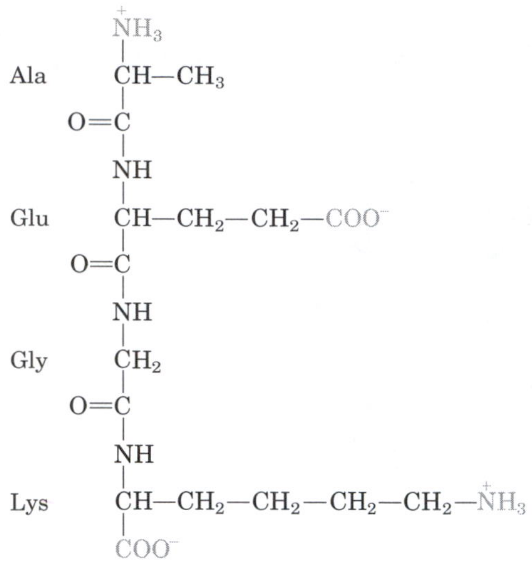

$$
\begin{array}{lll}
 & \overset{+}{N}H_3 & \\
 & | & \\
\text{Ala} & CH{-}CH_3 & \\
 & | & \\
 & O{=}C & \\
 & | & \\
 & NH & \\
 & | & \\
\text{Glu} & CH{-}CH_2{-}CH_2{-}COO^- & \\
 & | & \\
 & O{=}C & \\
 & | & \\
 & NH & \\
 & | & \\
\text{Gly} & CH_2 & \\
 & | & \\
 & O{=}C & \\
 & | & \\
 & NH & \\
 & | & \\
\text{Lys} & CH{-}CH_2{-}CH_2{-}CH_2{-}CH_2{-}\overset{+}{N}H_3 & \\
 & | & \\
 & COO^- &
\end{array}
$$

figure 5–15, page 127
Alanylglutamylglycyllysine.

table 5–2, page 128

Molecular Data on Some Proteins

	Molecular weight	Number of residues	Number of polypeptide chains
Cytochrome *c* (human)	13,000	104	1
Ribonuclease A (bovine pancreas)	13,700	124	1
Lysozyme (egg white)	13,930	129	1
Myoglobin (equine heart)	16,890	153	1
Chymotrypsin (bovine pancreas)	21,600	241	3
Chymotrypsinogen (bovine)	22,000	245	1
Hemoglobin (human)	64,500	574	4
Serum albumin (human)	68,500	609	1
Hexokinase (yeast)	102,000	972	2
RNA polymerase (*E. coli*)	450,000	4,158	5
Apolipoprotein B (human)	513,000	4,536	1
Glutamine synthetase (*E. coli*)	619,000	5,628	12
Titin (human)	2,993,000	26,926	1

table 5–3, page 128

Amino Acid Composition of Two Proteins*

	Number of residues per molecule of protein	
Amino acid	Bovine cytochrome *c*	Bovine chymotrypsinogen
Ala	6	22
Arg	2	4
Asn	5	15
Asp	3	8
Cys	2	10
Gln	3	10
Glu	9	5
Gly	14	23
His	3	2
Ile	6	10
Leu	6	19
Lys	18	14
Met	2	2
Phe	4	6
Pro	4	9
Ser	1	28
Thr	8	23
Trp	1	8
Tyr	4	4
Val	3	23
Total	104	245

*Note that standard procedures for the acid hydrolysis of proteins convert Asn and Gln to Asp and Glu, respectively. In addition, Trp is destroyed. Special procedures must be employed to determine the amounts of these amino acids.

table 5–4, page 129

Conjugated Proteins

Class	Prosthetic group(s)	Example
Lipoproteins	Lipids	β_1-Lipoprotein of blood
Glycoproteins	Carbohydrates	Immunoglobulin G
Phosphoproteins	Phosphate groups	Casein of milk
Hemoproteins	Heme (iron porphyrin)	Hemoglobin
Flavoproteins	Flavin nucleotides	Succinate dehydrogenase
Metalloproteins	Iron	Ferritin
	Zinc	Alcohol dehydrogenase
	Calcium	Calmodulin
	Molybdenum	Dinitrogenase
	Copper	Plastocyanin

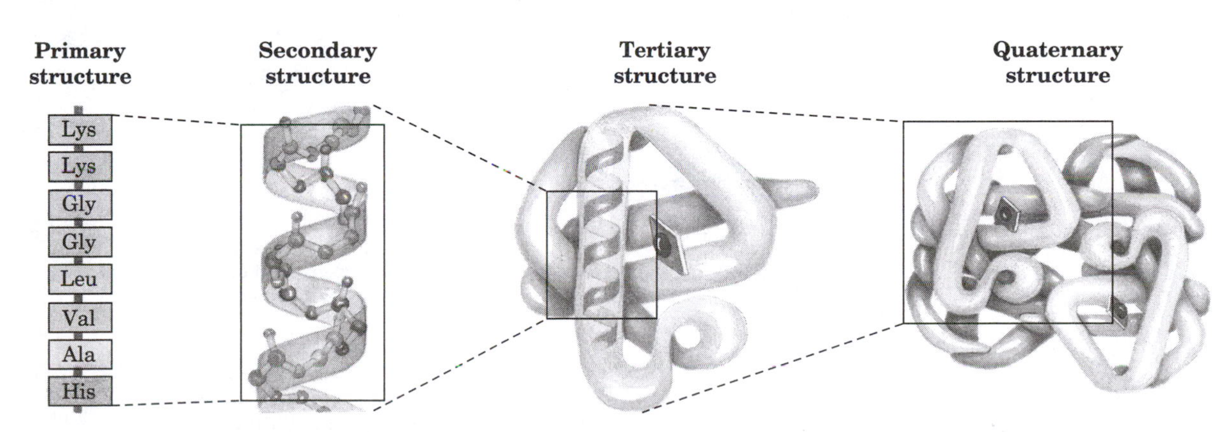

figure 5–16, page 129
Levels of structure in proteins.

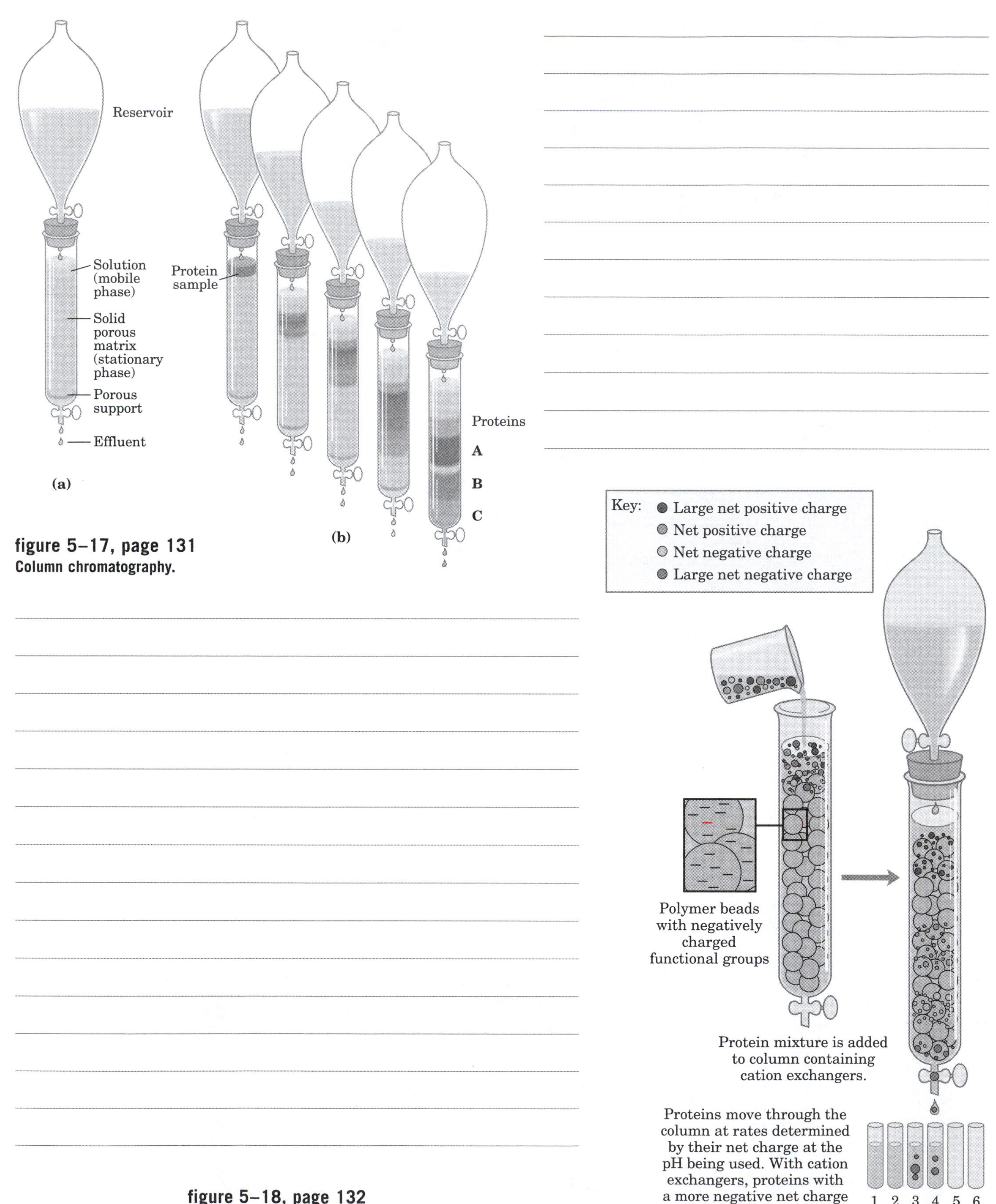

figure 5–17, page 131
Column chromatography.

figure 5–18, page 132
Three chromatographic methods used in protein purification. (a) Ion-exchange chromatography.

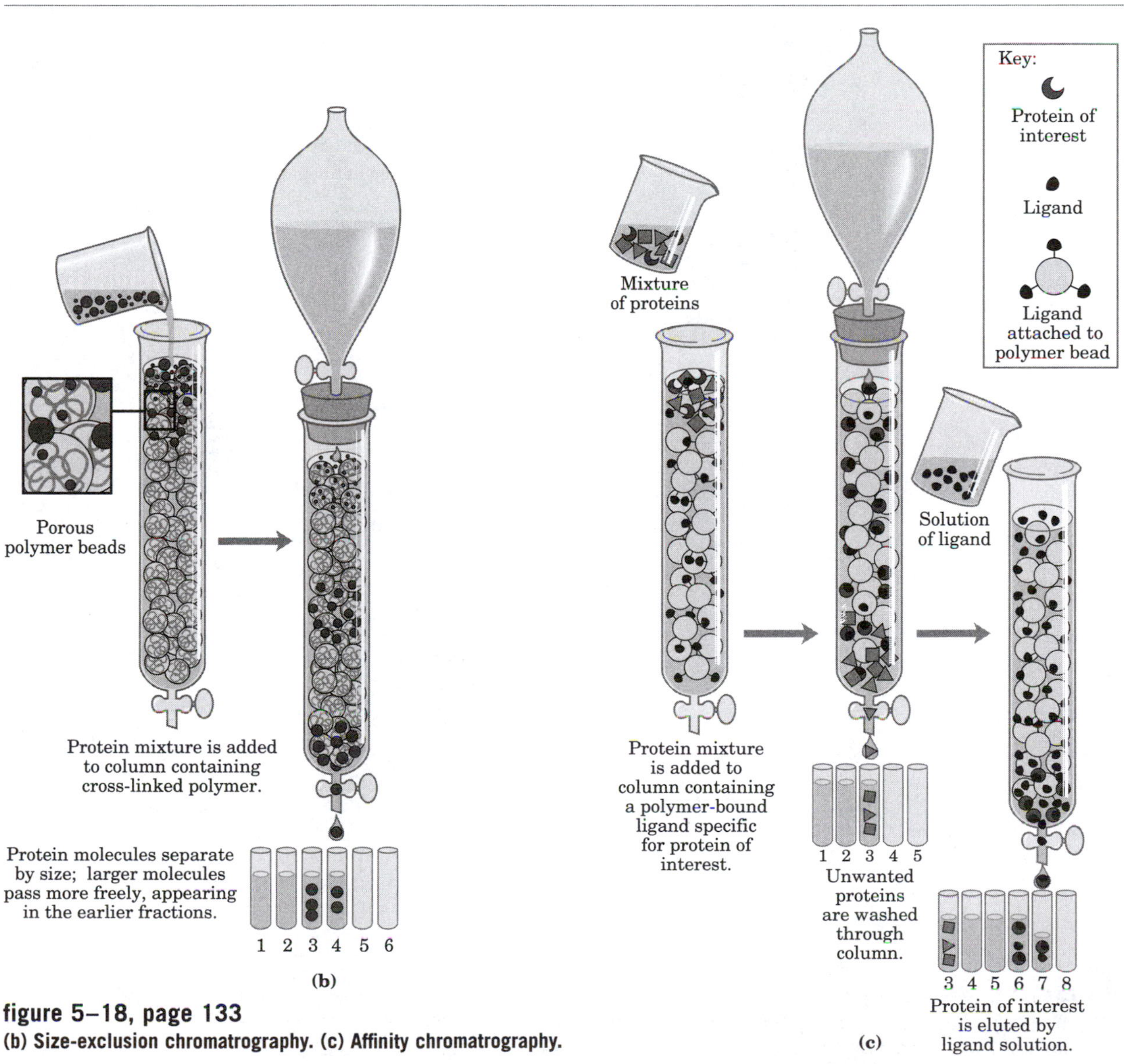

figure 5–18, page 133
(b) Size-exclusion chromatrography. (c) Affinity chromatrography.

table 5–5, page 133

A Purification Table for a Hypothetical Enzyme*

Procedure or step	Fraction volume (ml)	Total protein (mg)	Activity (units)	Specific activity (units/mg)
1. Crude cellular extract	1,400	10,000	100,000	10
2. Precipitation with ammonium sulfate	280	3,000	96,000	32
3. Ion-exchange chromatography	90	400	80,000	200
4. Size-exclusion chromatography	80	100	60,000	600
5. Affinity chromatography	6	3	45,000	15,000

*All data represent the status of the sample *after* the designated procedure has been carried out. Activity and specific activity are defined on page 137.

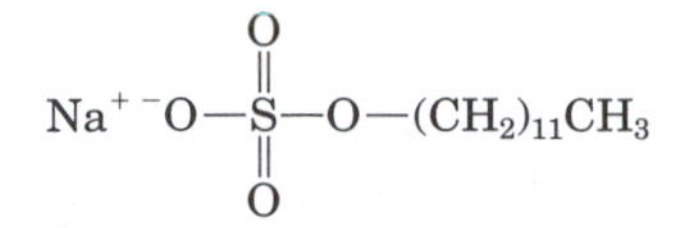

Sodium dodecyl sulfate (SDS), page 134

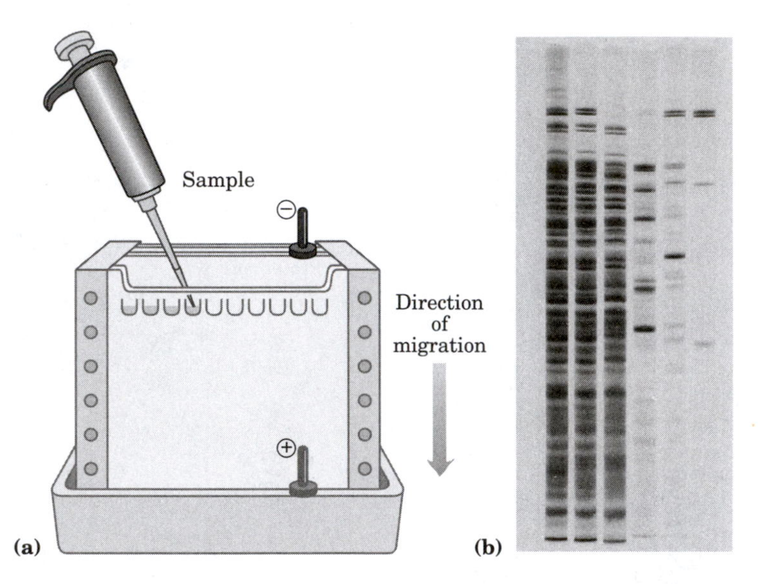

figure 5–19, page 134
Electrophoresis.

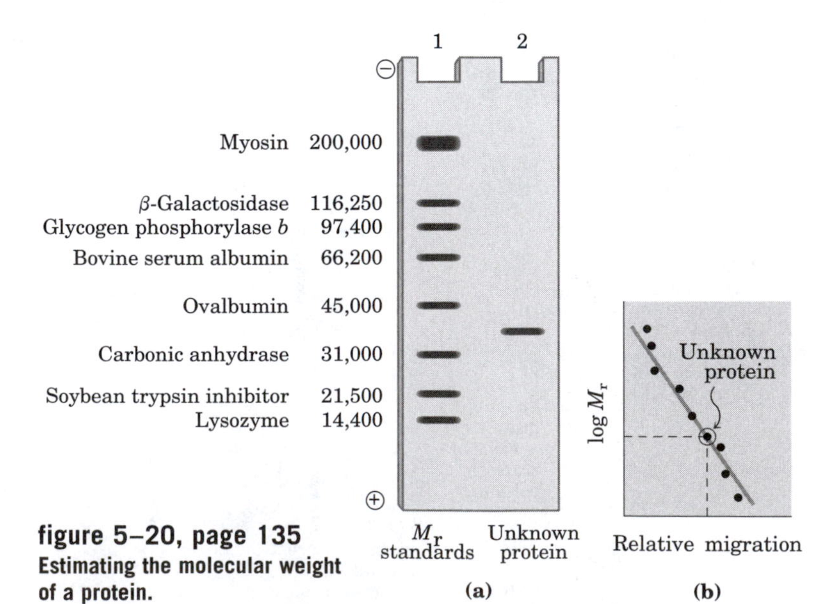

figure 5–20, page 135
Estimating the molecular weight of a protein.

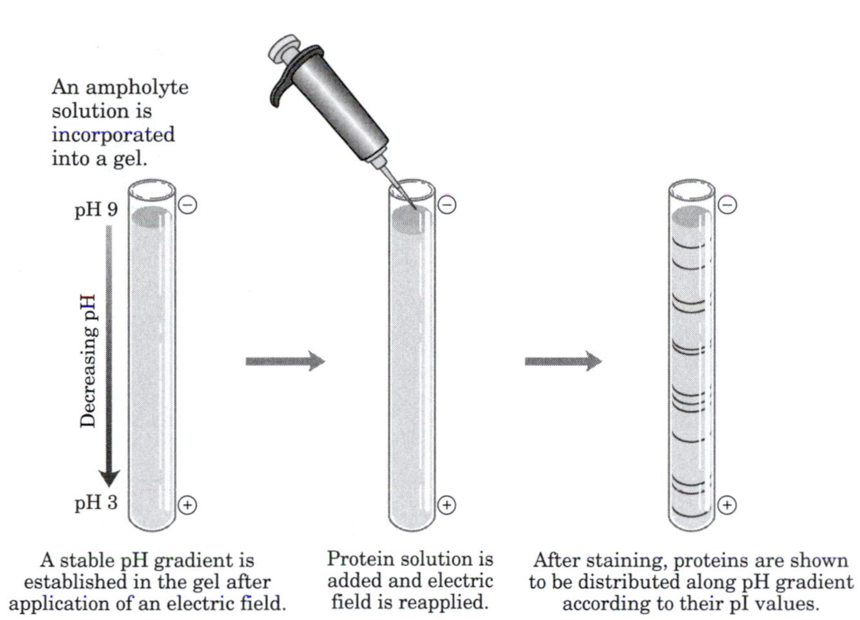

figure 5–21, page 135
Isoelectric focusing.

table 5–6, page 135

The Isoelectric Points of Some Proteins

Protein	pI
Pepsin	~1.0
Egg albumin	4.6
Serum albumin	4.9
Urease	5.0
β-Lactoglobulin	5.2
Hemoglobin	6.8
Myoglobin	7.0
Chymotrypsinogen	9.5
Cytochrome *c*	10.7
Lysozyme	11.0

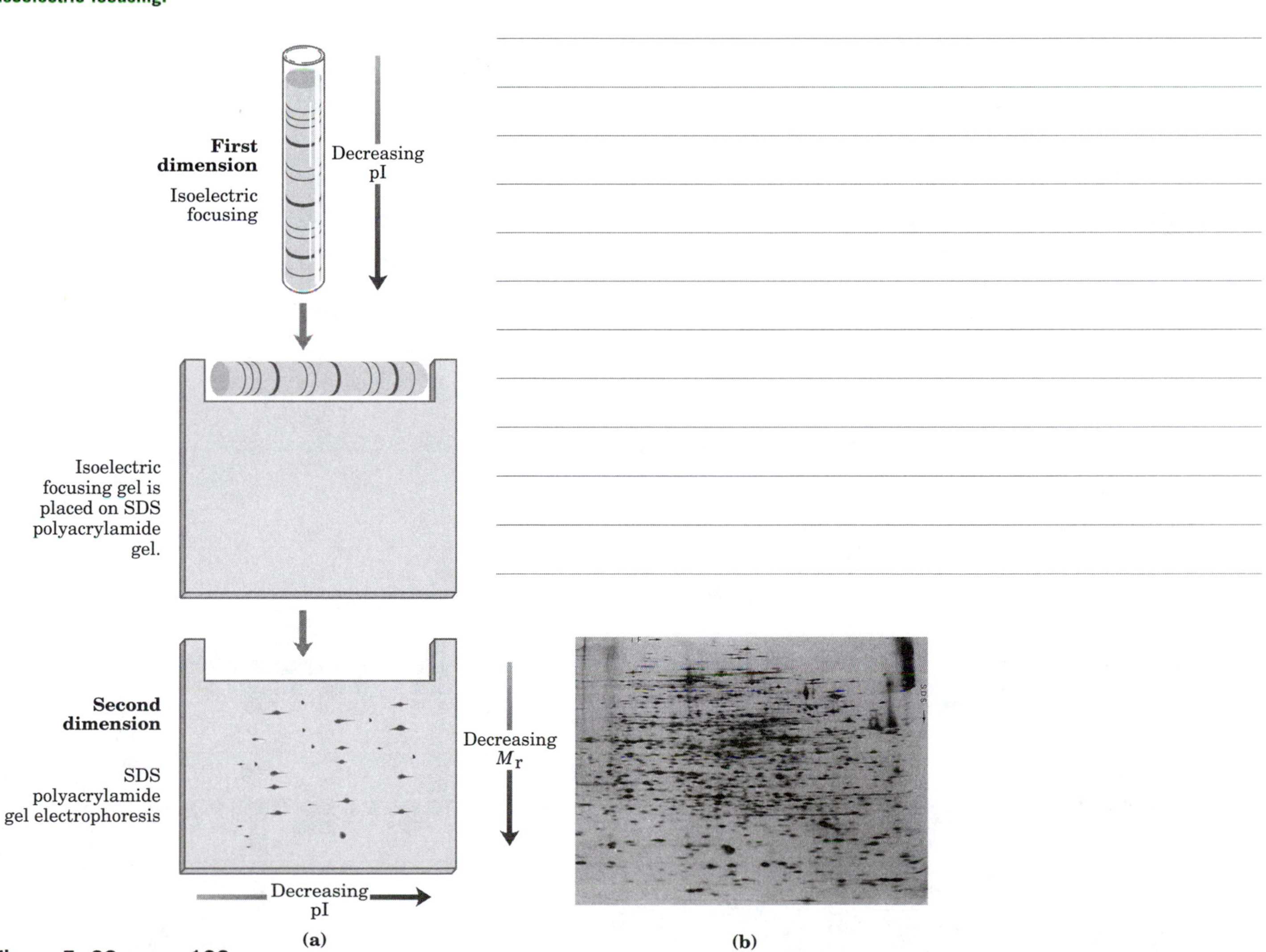

figure 5–22, page 136
Two-dimensional electrophoresis.

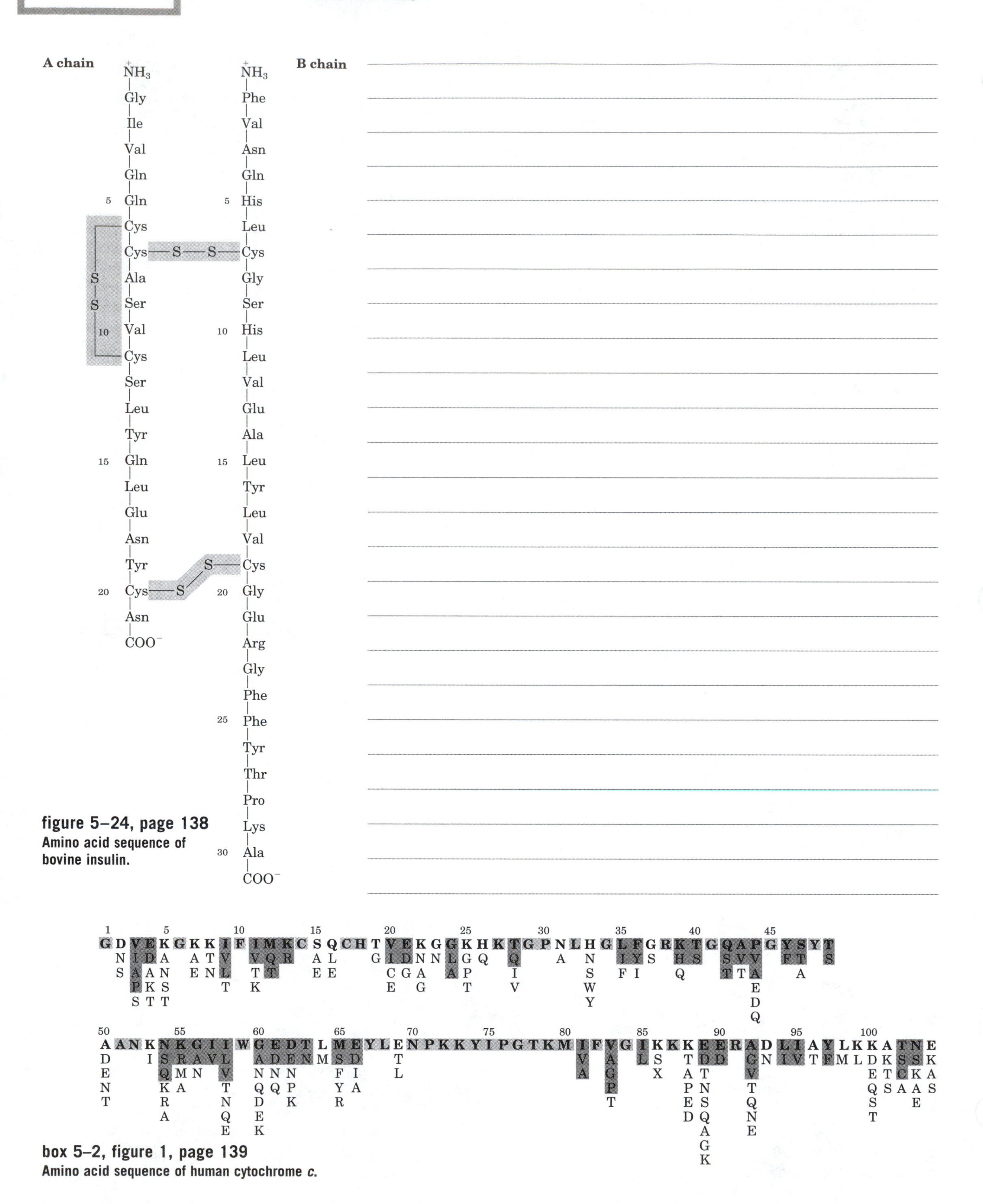

figure 5–24, page 138
Amino acid sequence of bovine insulin.

box 5–2, figure 1, page 139
Amino acid sequence of human cytochrome *c*.

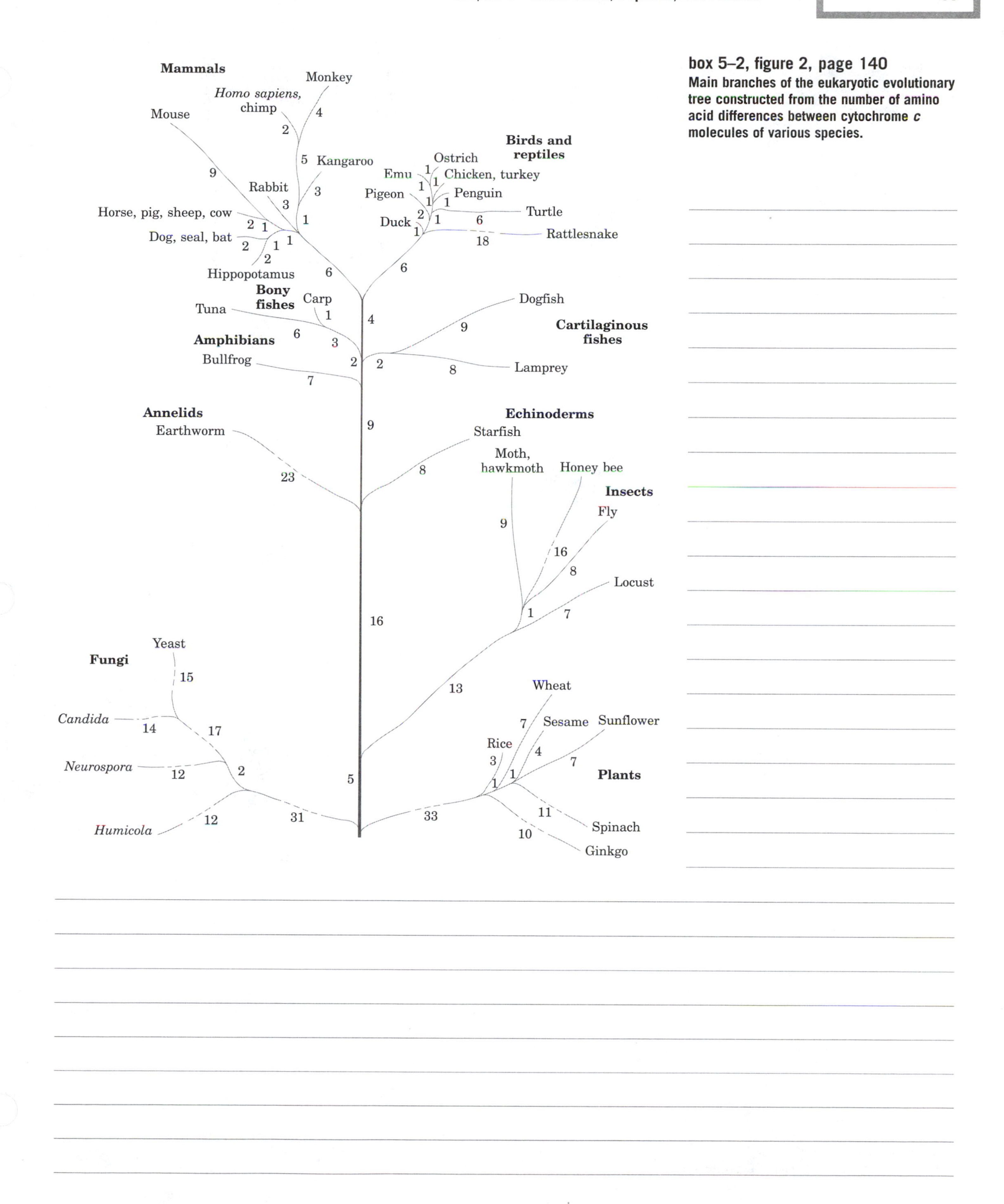

box 5–2, figure 2, page 140
Main branches of the eukaryotic evolutionary tree constructed from the number of amino acid differences between cytochrome *c* molecules of various species.

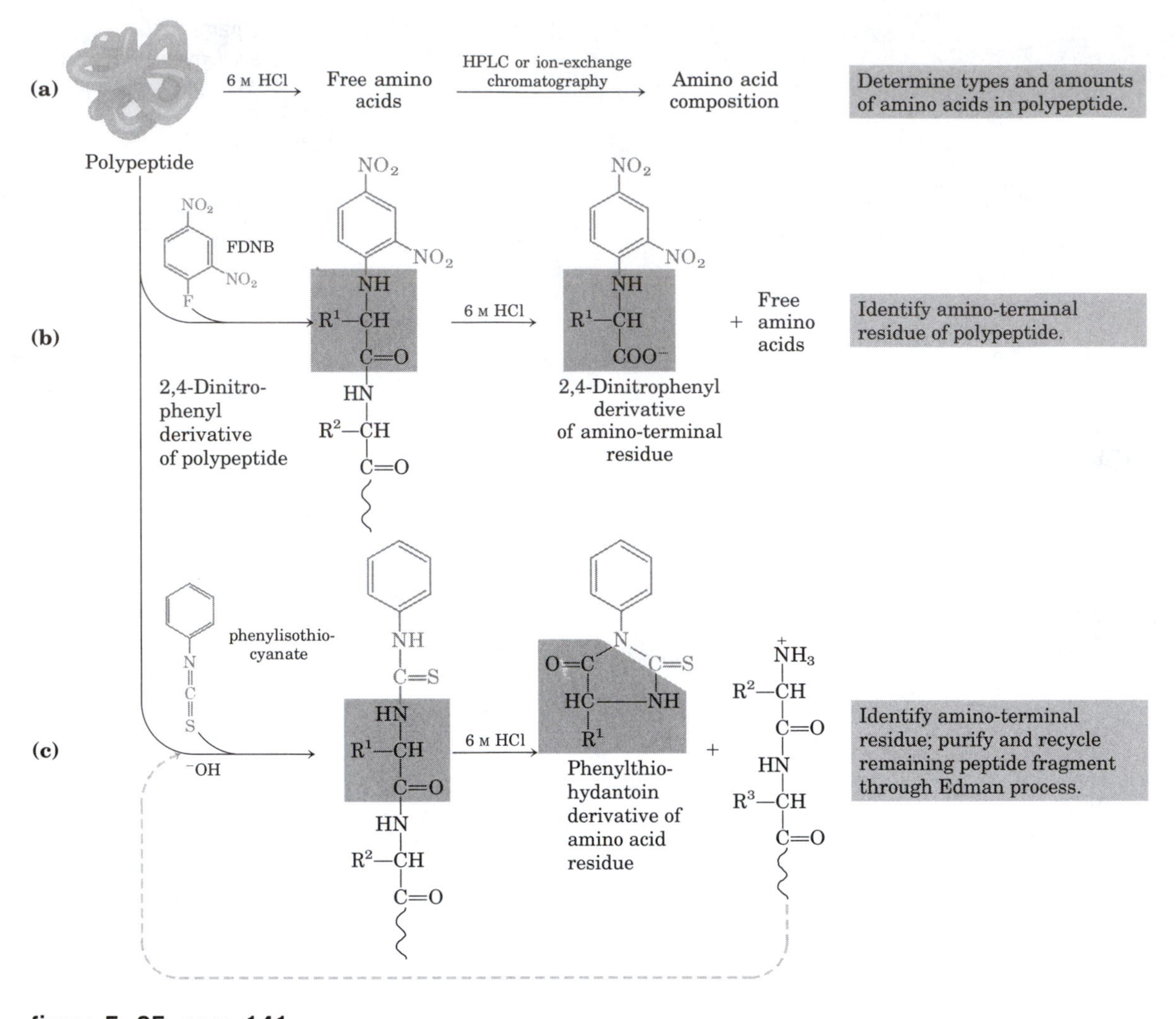

figure 5–25, page 141
Steps in sequencing a polypeptide.

Dansyl chloride

Dabsyl chloride

Sanger technique reagents, page 142

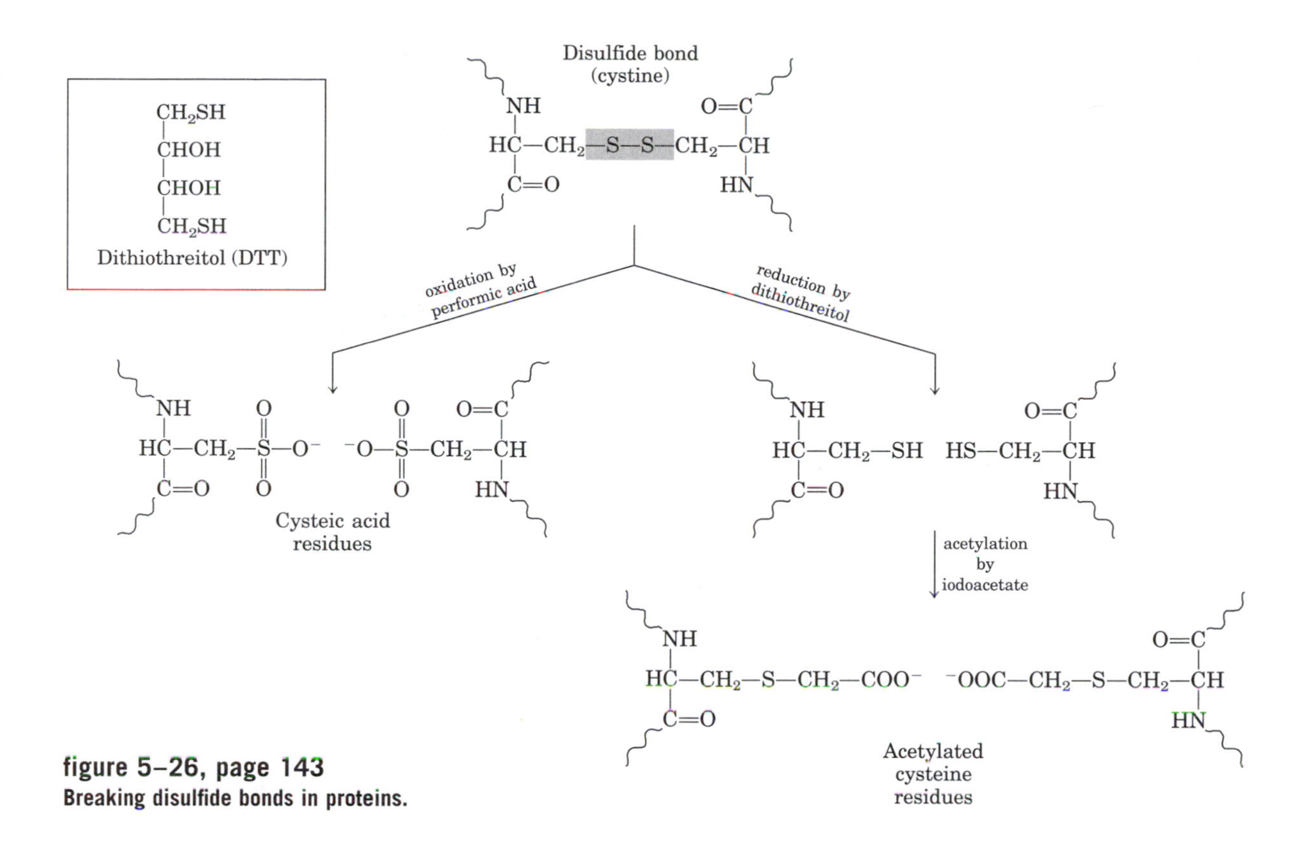

figure 5–26, page 143
Breaking disulfide bonds in proteins.

table 5–7, page 143

The Specificity of Some Common Methods for Fragmenting Polypeptide Chains

Treatment*	Cleavage points†
Trypsin	Lys, Arg (C)
Submaxillarus protease	Arg (C)
Chymotrypsin	Phe, Trp, Tyr (C)
Staphylococcus aureus V8 protease	Asp, Glu (C)
Asp-*N*-protease	Asp, Glu (N)
Pepsin	Phe, Trp, Tyr (N)
Endoproteinase Lys C	Lys (C)
Cyanogen bromide	Met (C)

*All except cyanogen bromide are proteases. All are available from commercial sources.

†Residues furnishing the primary recognition point for the protease or reagent; peptide bond cleavage occurs on either the carbonyl (C) or the amino (N) side of the indicated amino acid residues.

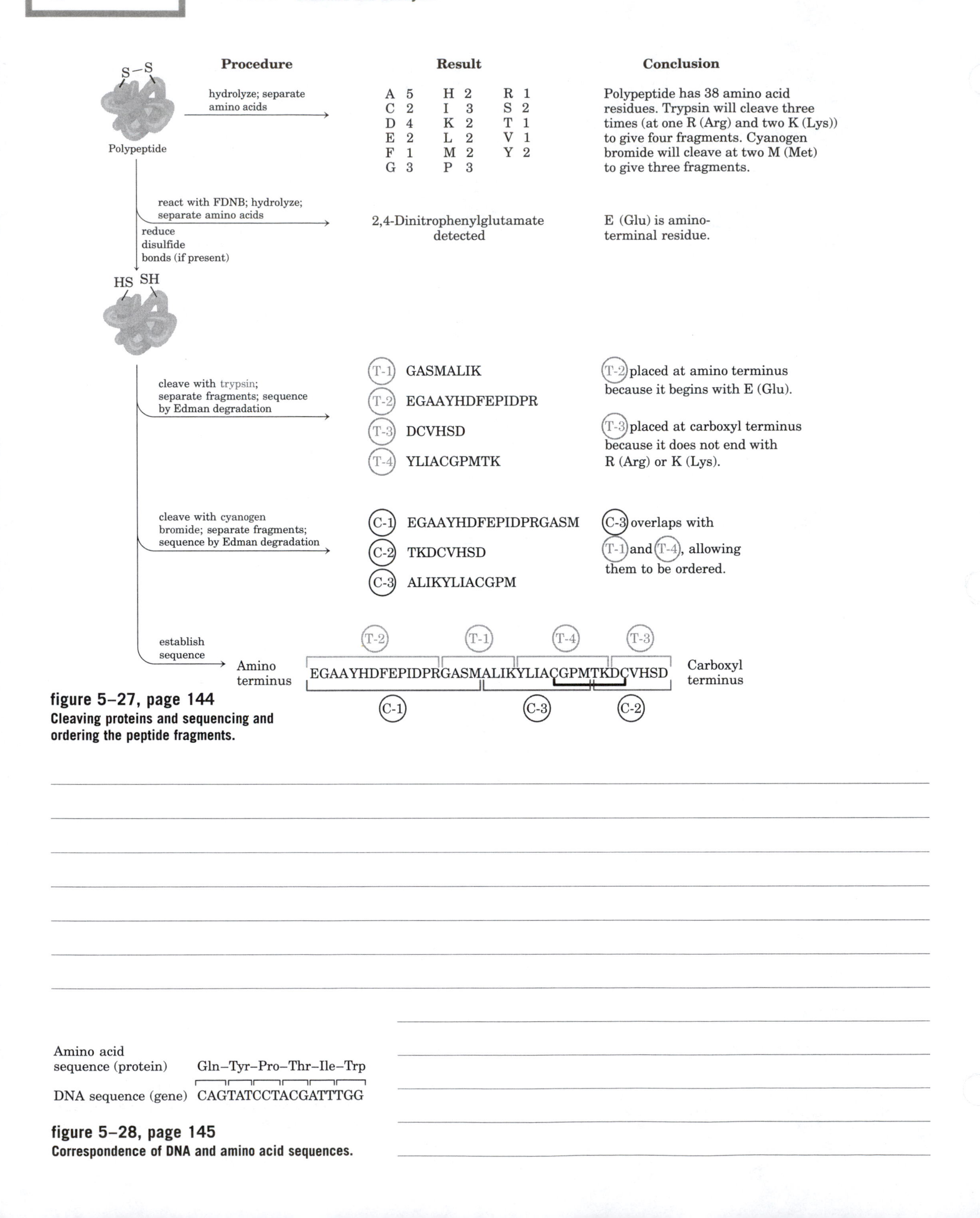

figure 5–27, page 144
Cleaving proteins and sequencing and ordering the peptide fragments.

figure 5–28, page 145
Correspondence of DNA and amino acid sequences.

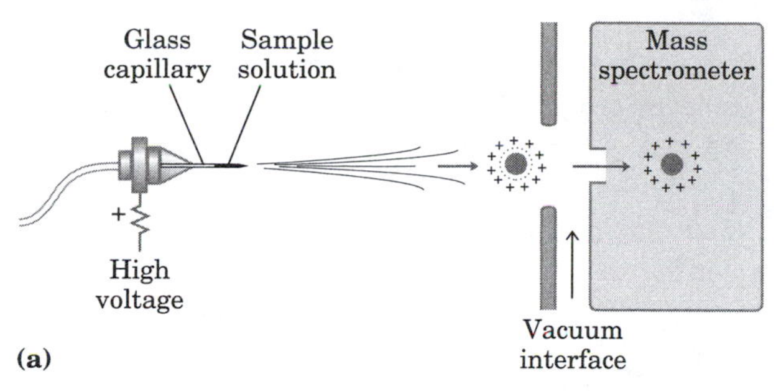

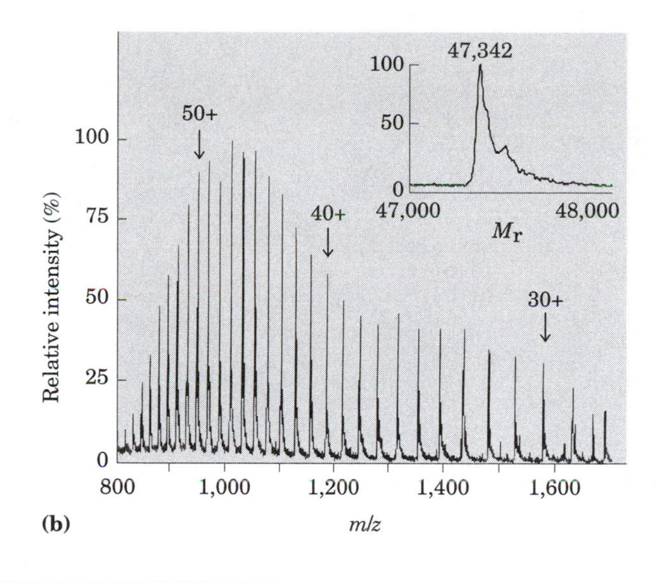

box 5–3, figure 1, page 146
Electrospray mass spectrometry of a protein.

Electrospray ionization
MS-1
Separation
Collision cell
Breakage
MS-2
Detector
b
y
(a)

box 5–3, figure 2, pages 148–149
Obtaining protein sequence information with tandem mass spectrometry.

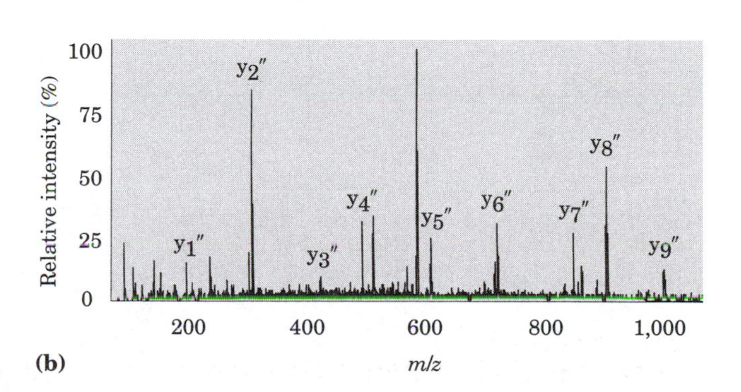

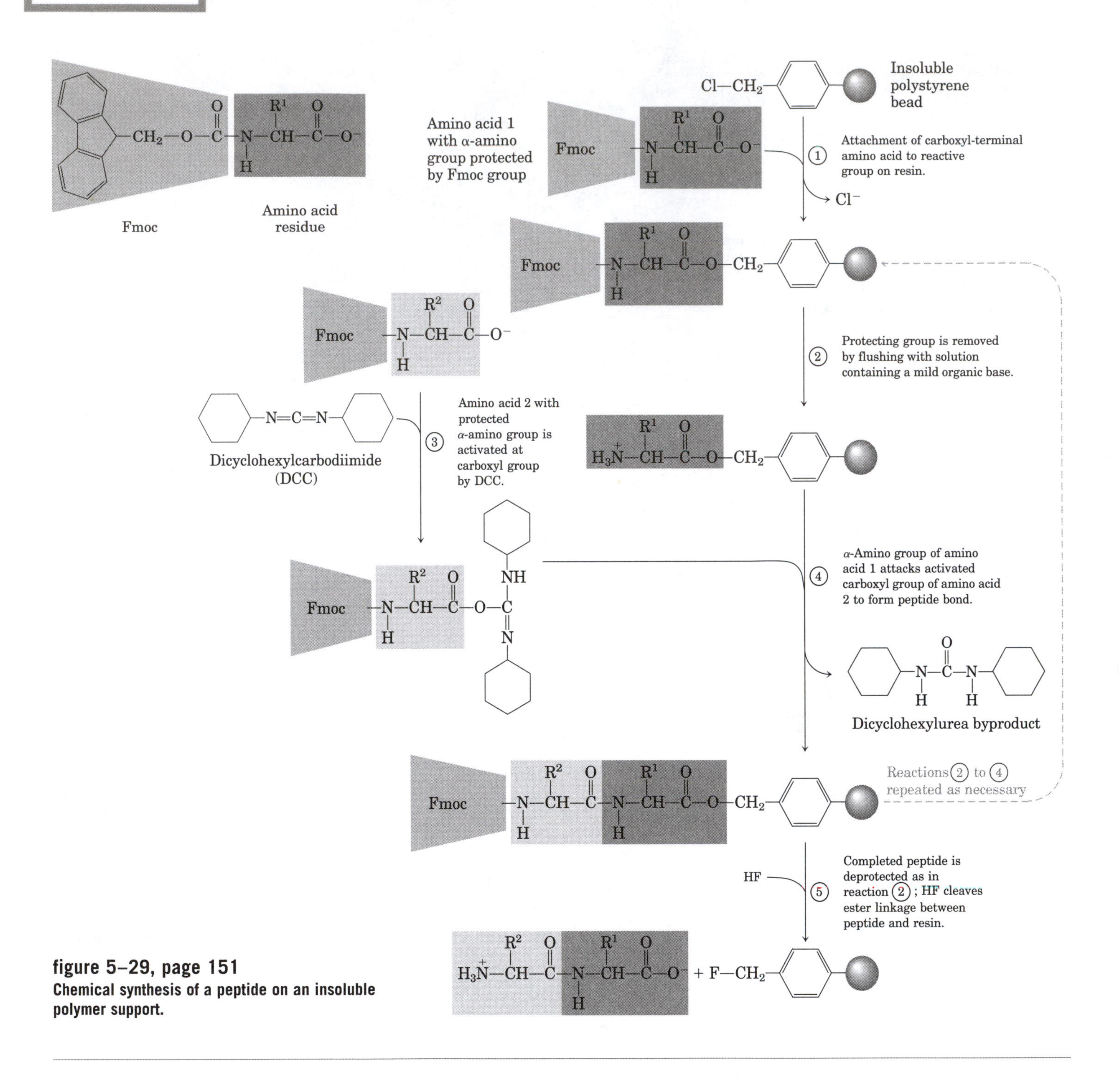

figure 5–29, page 151
Chemical synthesis of a peptide on an insoluble polymer support.

table 5–8, page 152

Effect of Stepwise Yield on Overall Yield in Peptide Synthesis

Number of residues in the final polypeptide	Overall yield of final peptide (%) when the yield of each step is: 96.0%	99.8%
11	66	98
21	44	96
31	29	94
51	13	90
100	1.7	82

PROBLEMS

1. Absolute Configuration of Citrulline, page 154.

$CH_2(CH_2)_2NH—C(=O)—NH_2$

$H—C—\overset{+}{N}H_3$

COO^-

Citrulline

2. Relationship between the Titration Curve and the Acid-Base Properties of Glycine, page 154.

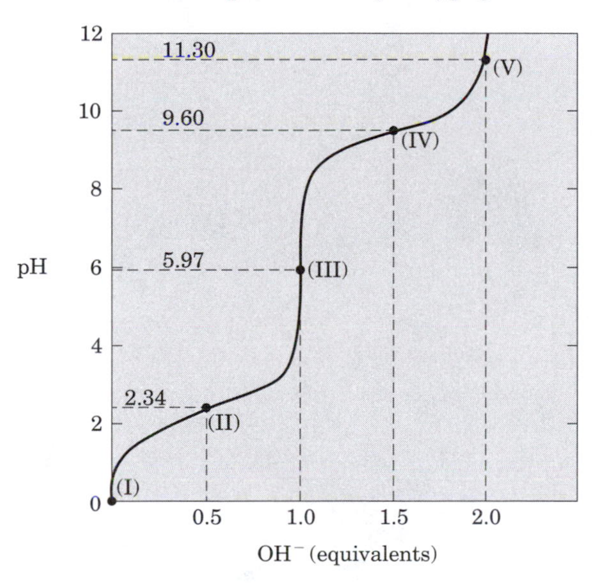

3. How Much Alanine Is Present as the Completely Uncharged Species?, page 155.

$H_3\overset{+}{N}—C(CH_3)(H)—C(=O)O^-$

Zwitterionic

$H_2N—C(CH_3)(H)—C(=O)OH$

Uncharged

6. Naming the Stereoisomers of Isoleucine, page 155.

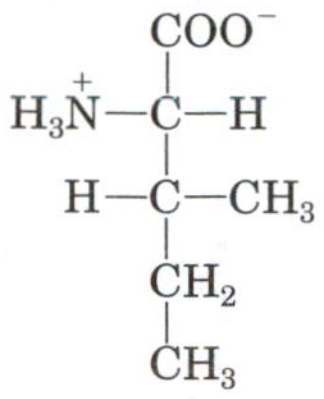

Isoleucine

$$\begin{array}{c} COO^- \\ | \\ H_3\overset{+}{N}-C-H \\ | \\ H-C-CH_3 \\ | \\ CH_2 \\ | \\ CH_3 \end{array}$$

7. Comparing the pK_a Values of Alanine and Polyalanine, page 156.

Amino acid or peptide	pK_1	pK_2
Ala	2.34	9.69
Ala–Ala	3.12	8.30
Ala–Ala–Ala	3.39	8.03
Ala–(Ala)$_n$–Ala, $n \geq 4$	3.42	7.94

14. Purification of an Enzyme, page 156.

Procedure	Total protein (mg)	Activity (units)
1. Crude extract	20,000	4,000,000
2. Precipitation (salt)	5,000	3,000,000
3. Precipitation (pH)	4,000	1,000,000
4. Ion-exchange chromatography	200	800,000
5. Affinity chromatography	50	750,000
6. Size-exclusion chromatography	45	675,000

16. Structure of a Peptide Antibiotic from *Bacillus brevis*, page 157.

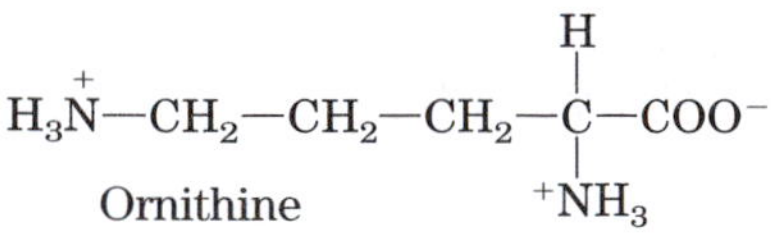

$$H_3\overset{+}{N}-CH_2-CH_2-CH_2-\underset{^{+}NH_3}{\overset{H}{C}}-COO^-$$

Ornithine

$$O_2N-C_6H_3(NO_2)-NH-CH_2-CH_2-CH_2-\underset{^{+}NH_3}{\overset{H}{C}}-COO^-$$

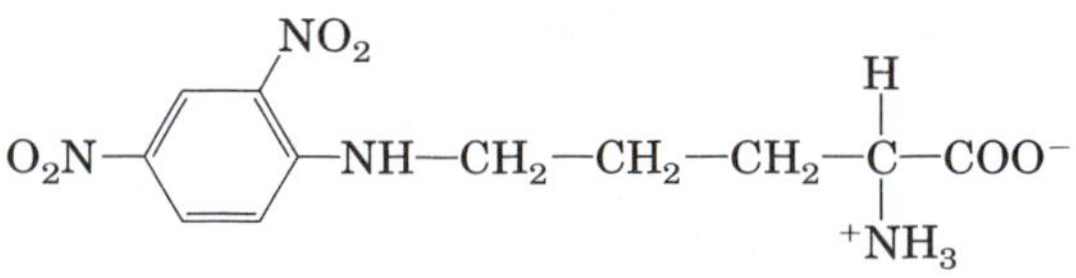

Peptide derivative after hydrolysis

chapter

6

The Three-Dimensional Structure of Proteins

When you study for exams, don't forget to review The Minicourses on the UNDERSTAND! BIOCHEMISTRY CD that came with your textbook.

Minicourses that apply to this Chapter include:

Molecules of Life
- Macromolecules: Proteins and Nucleic Acids
- An Interactive Gallery of Protein Structures

Proteins in Action
- Catalysis and Regulation

Bioenergetics
- Energetics

Cellular Architecture and Traffic
- The Extracellular Matrix

Some Important Techniques
- Protein Structure Determination

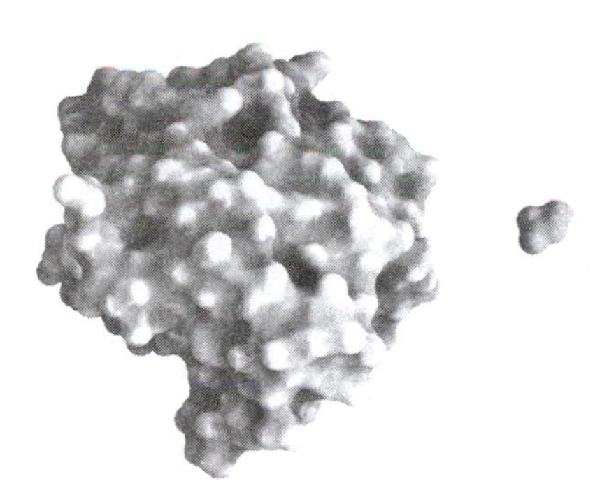

figure 6–1, page 159
Structure of the enzyme chymotrypsin, a globular protein.

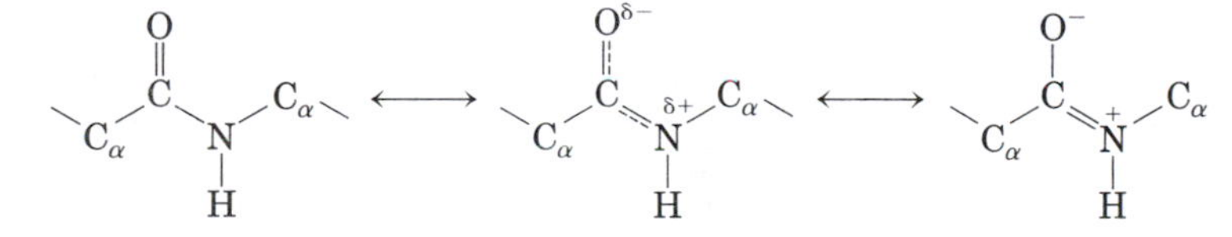

The carbonyl oxygen has a partial negative charge and the amide nitrogen a partial positive charge, setting up a small electric dipole. Virtually all peptide bonds in proteins occur in this trans configuration; an exception is noted in Figure 6–8b.

(a)

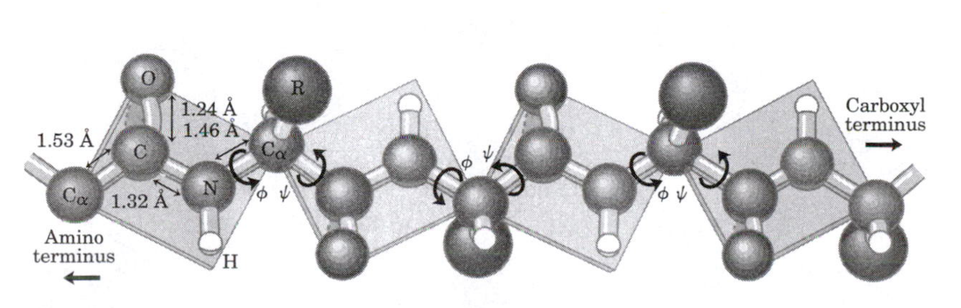

(b)

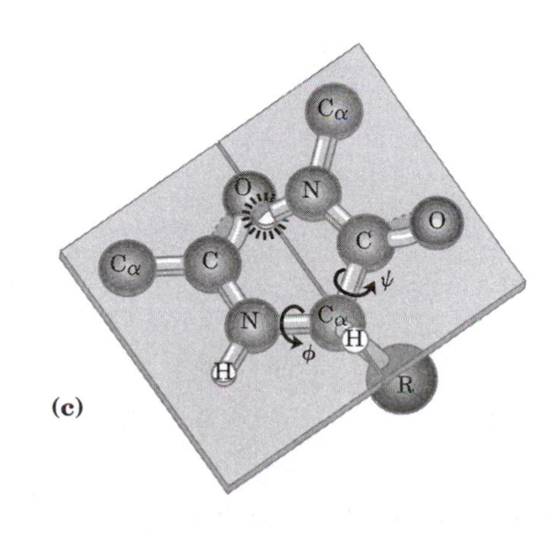

(c)

figure 6–2, page 162
The planar peptide group.

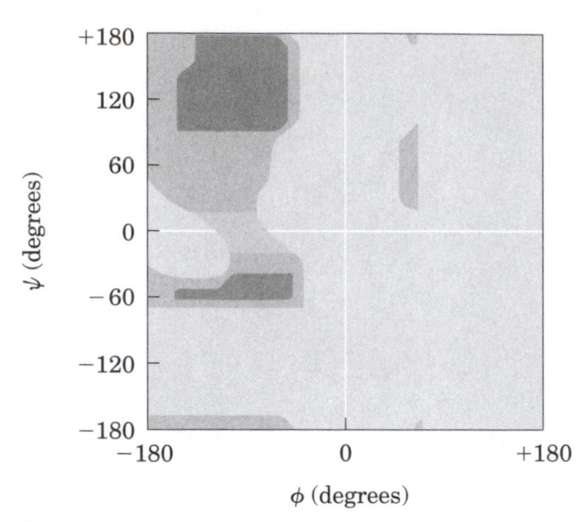

figure 6–3, page 163
Ramachandran plot for L-Ala residues.

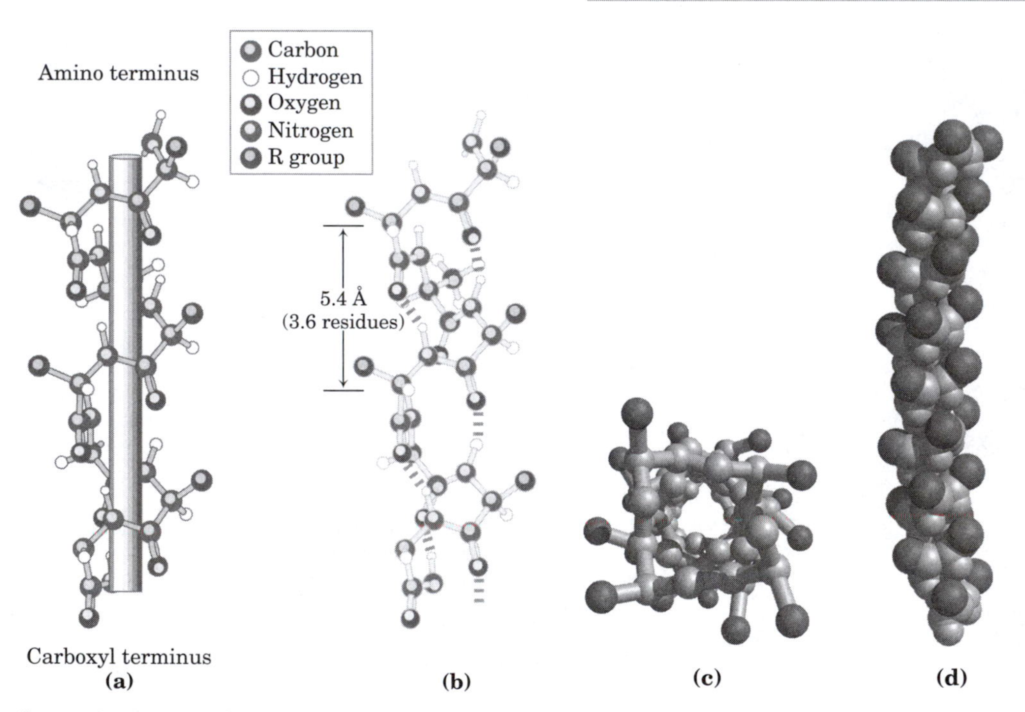

figure 6–4, page 164
Four models of the α helix, showing different aspects of its structure.

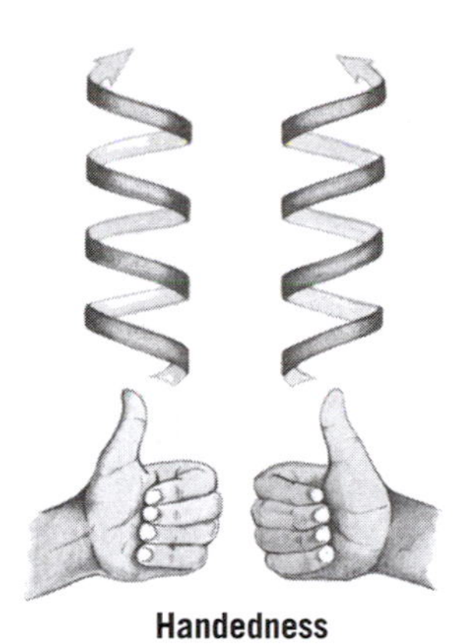

box 6–1, page 165
Knowing the right hand from the left.

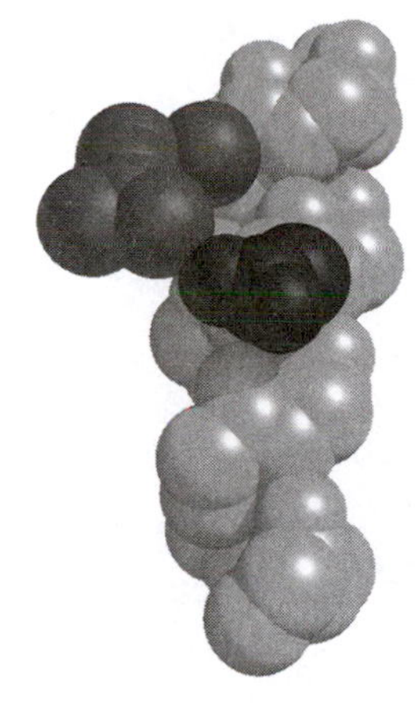

figure 6–5, page 165
Interactions between R groups of amino acids three residues apart in an α helix.

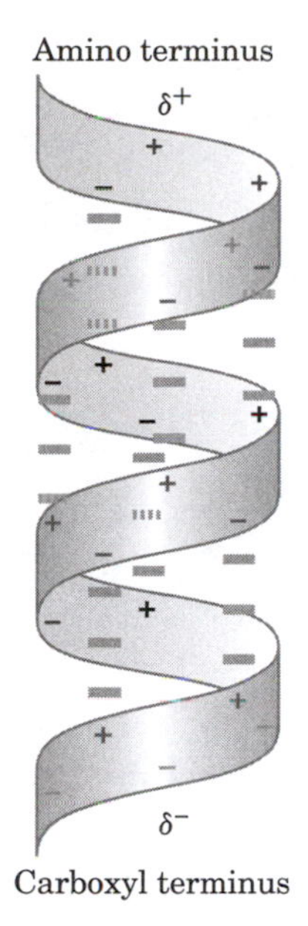

figure 6–6, page 166
The electric dipole of a peptide bond.

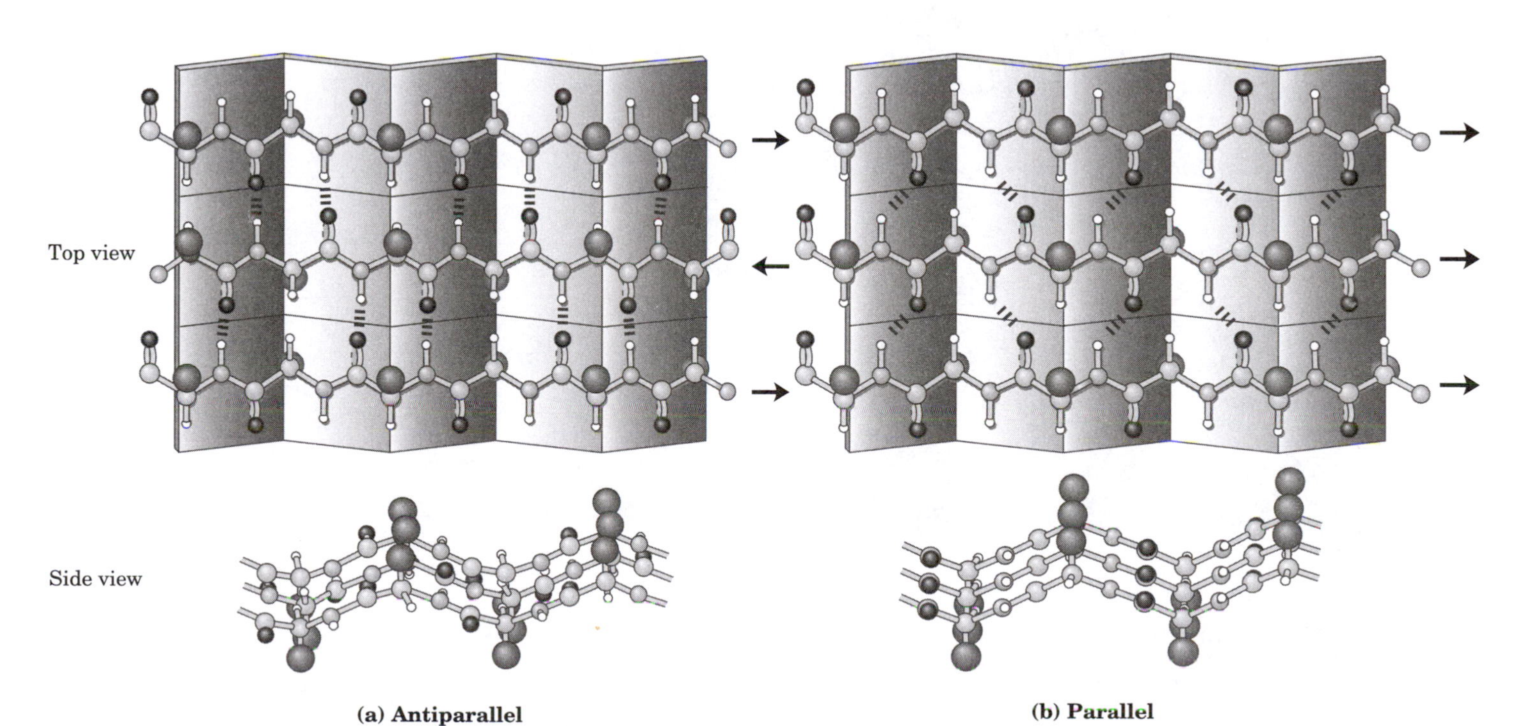

figure 6–7, page 167
The β conformation of polypeptide chains.

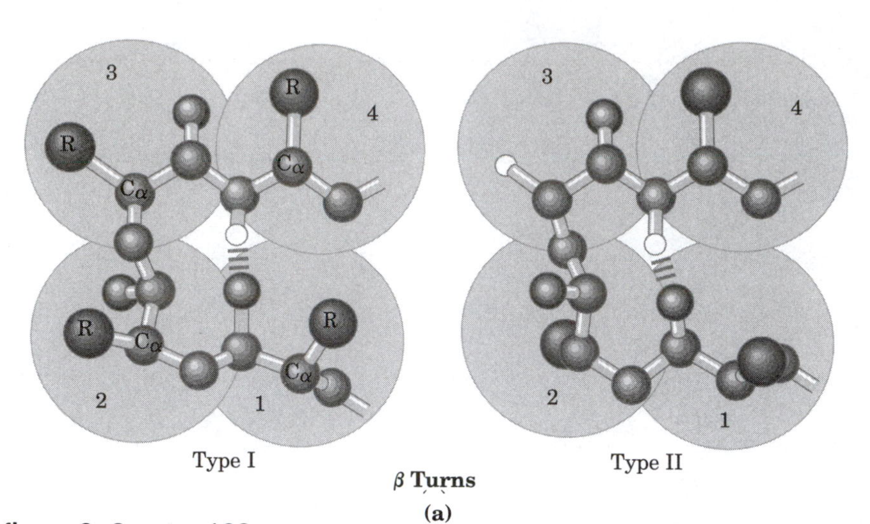

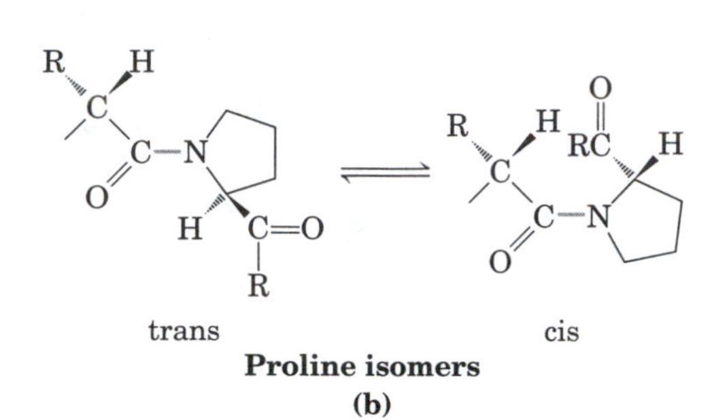

figure 6–8, page 168
Structures of β turns.

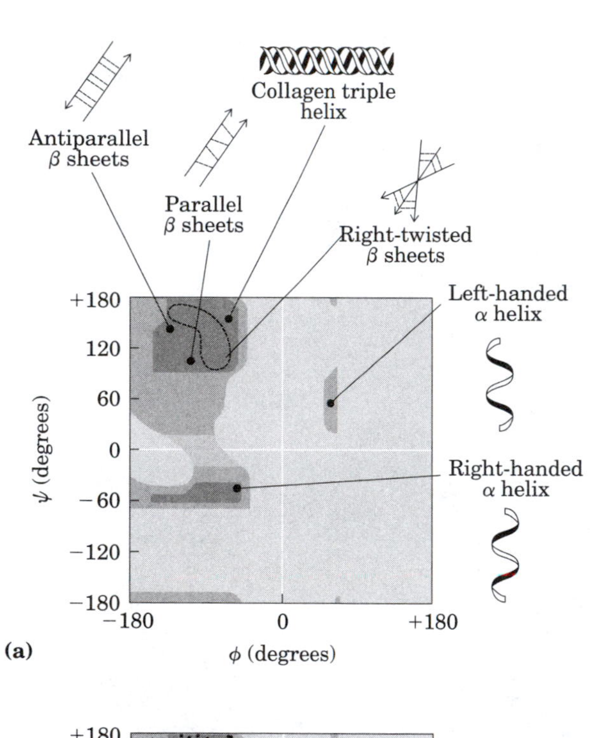

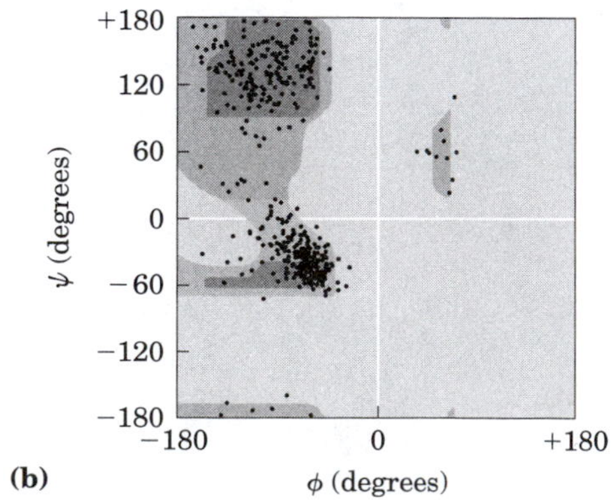

figure 6–9, page 169
Ramachandran plots for a variety of structures.

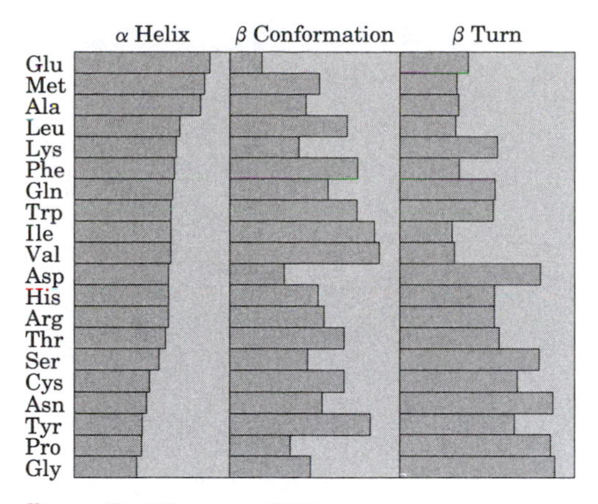

figure 6–10, page 169
Relative probabilities that a given amino acid will occur in the three common types of secondary structure.

table 6–1, page 170

Secondary Structures and Properties of Fibrous Proteins

Structure	Characteristics	Examples of occurrence
α Helix, cross-linked by disulfide bonds	Tough, insoluble protective structures of varying hardness and flexibility	α-Keratin of hair, feathers, and nails
β Conformation	Soft, flexible filaments	Silk fibroin
Collagen triple helix	High tensile strength, without stretch	Collagen of tendons, bone matrix

Keratin α helix

Two-chain coiled coil

Protofilament 20–30 Å

Protofibril 40–50 Å

(a)

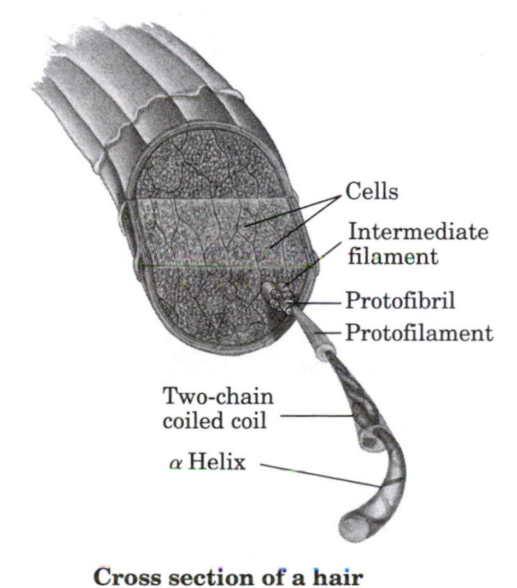

(b) Cross section of a hair

figure 6–11, page 171
Structure of hair.

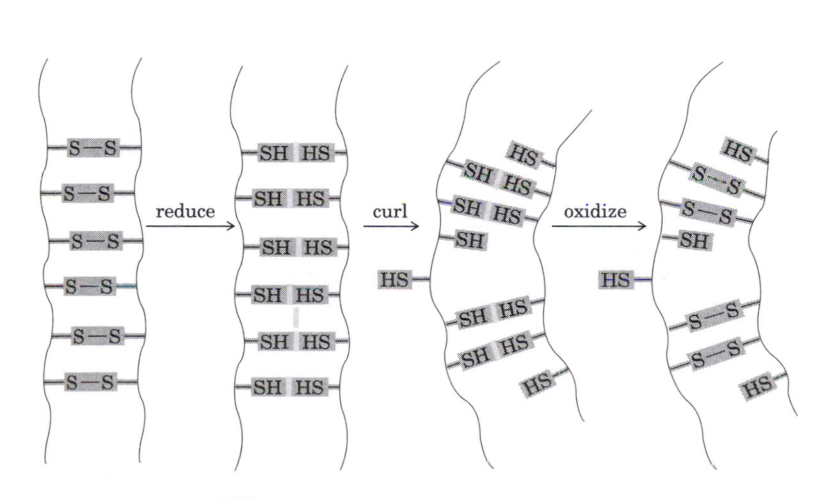

box 6–2, page 172
Permanent wave.

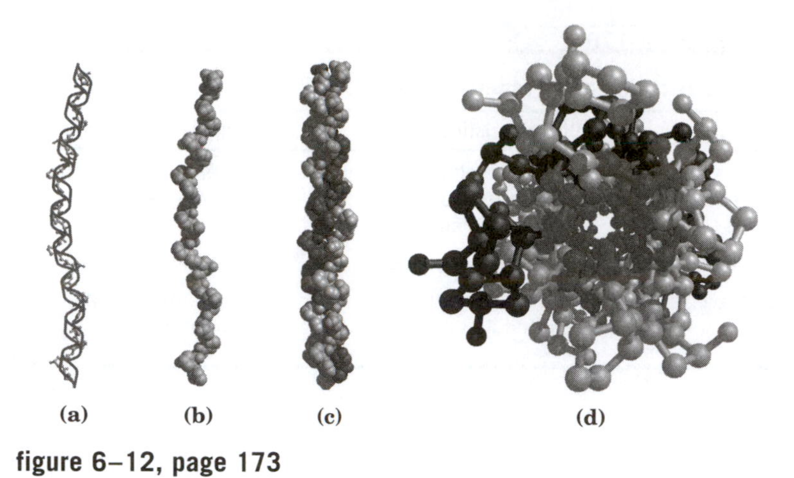

figure 6–12, page 173
Structure of collagen.

3.5 Å
5.7 Å
Ala side chain
Gly side chain
(a)

figure 6–14, page 174
Structure of silk.

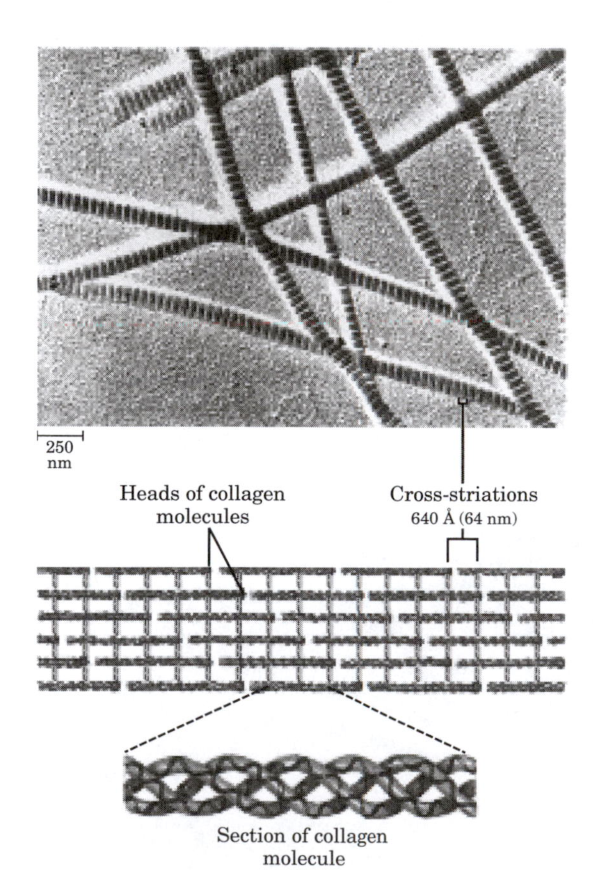

figure 6–13, page 173
Structure of collagen fibrils.

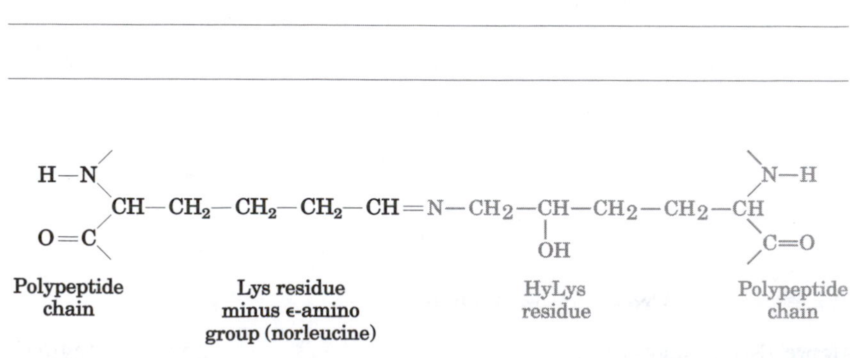

Dehydrohydroxylysinonorleucine, page 174

β Conformation
2,000 × 5 Å

α Helix
900 × 11 Å

Native globular form
130 × 30 Å

figure 6–15, page 175
Globular protein structures are compact and varied.

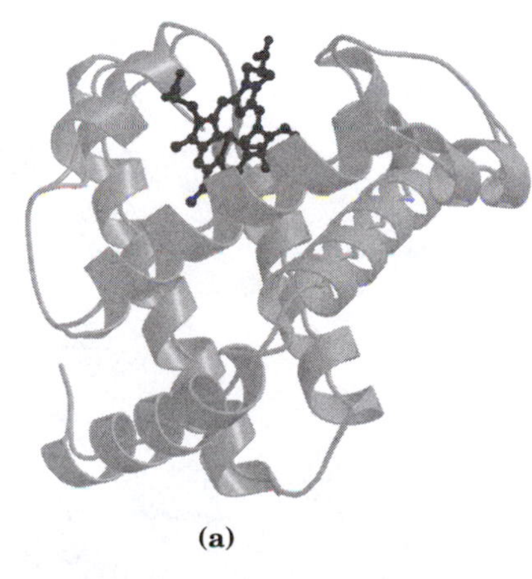

(a)

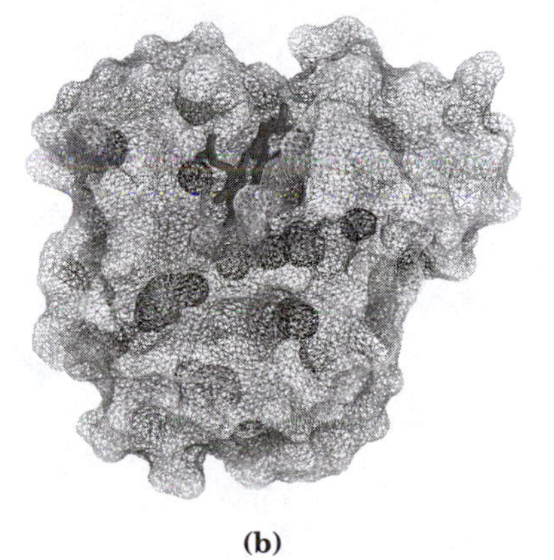

(b)

figure 6–16, page 176
Tertiary structure of sperm whale myoglobin.

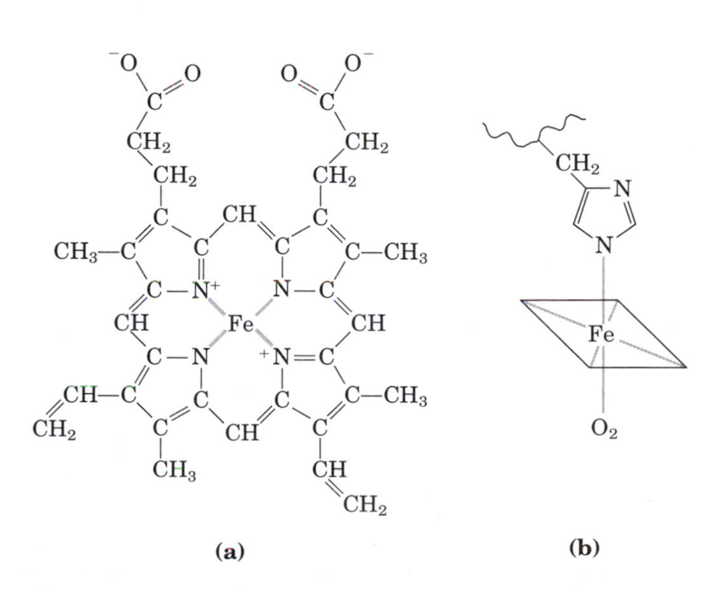

figure 6–17, page 177
The heme group.

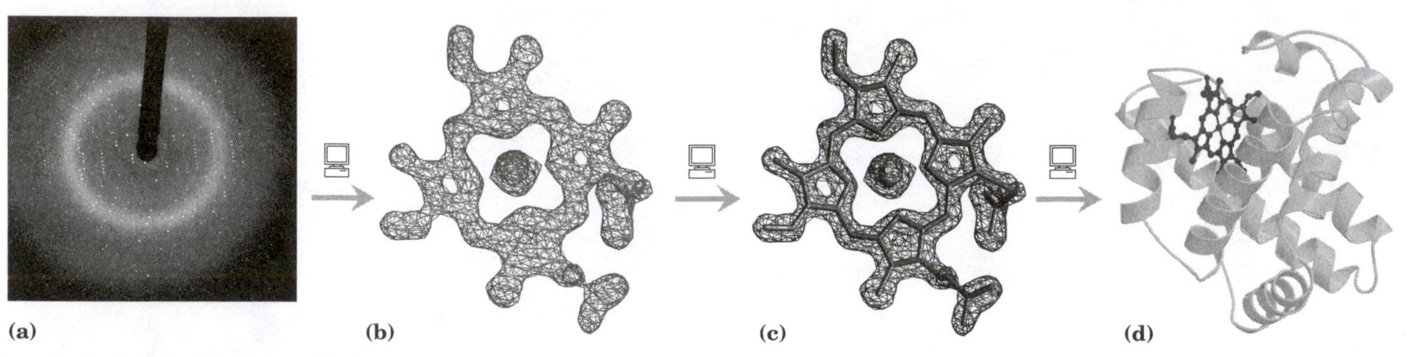

box 6–3, figure 1, pages 178-179
Steps in the determination of the structure of sperm whale myoglobin by x-ray crystallography.

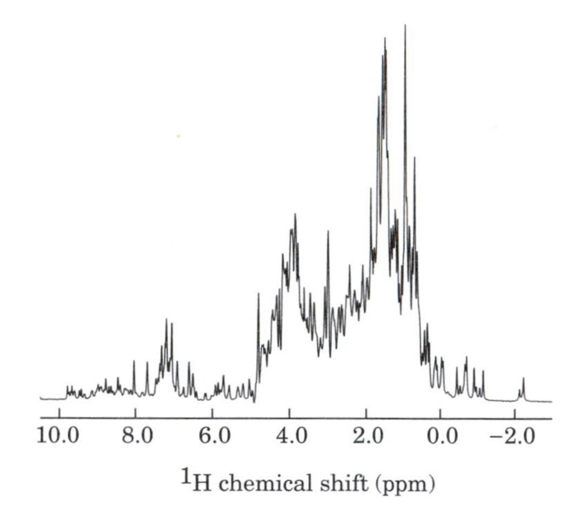

box 6–3, figure 2, page 180
A one-dimensional NMR spectrum of a globin from a marine blood worm.

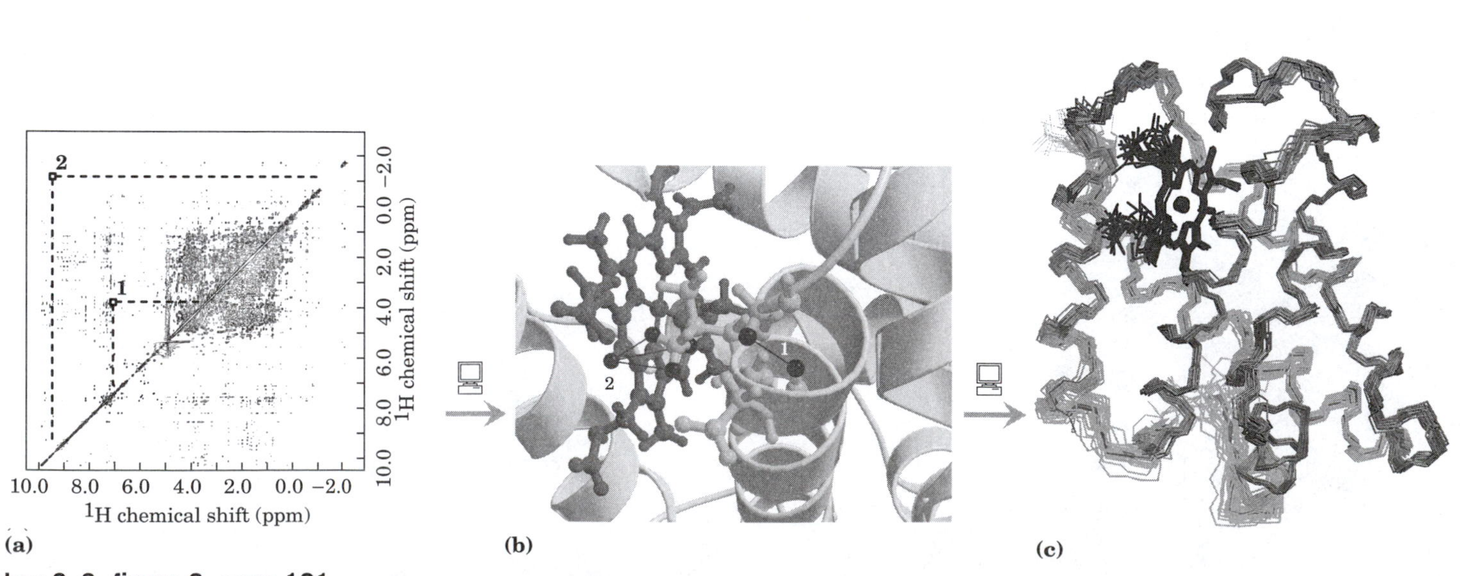

box 6–3, figure 3, page 181
The use of two-dimensional NMR to generate a three-dimensional structure of a globin, the same protein used to generate the data in Figure 2.

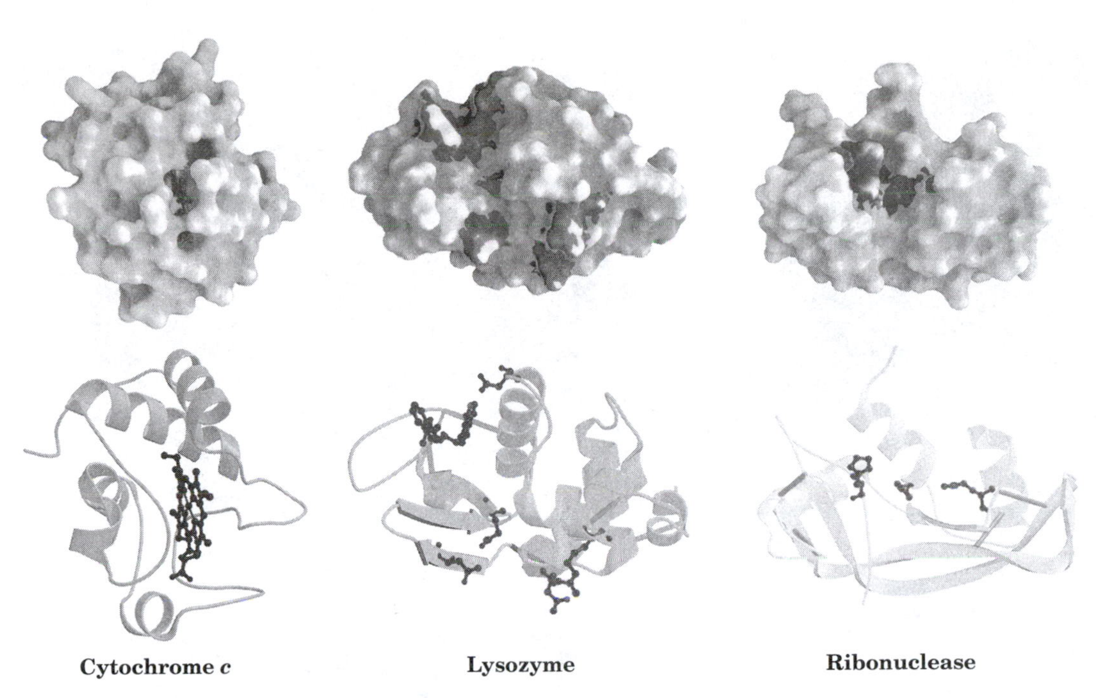

figure 6–18, page 182
Three-dimensional structures of some small proteins.

table 6–2, page 183

Approximate Amounts of α Helix and β Conformation in Some Single-Chain Proteins*

Protein (total residues)	Residues (%)	
	α Helix	β Conformation
Chymotrypsin (247)	14	45
Ribonuclease (124)	26	35
Carboxypeptidase (307)	38	17
Cytochrome *c* (104)	39	0
Lysozyme (129)	40	12
Myoglobin (153)	78	0

Source: Data from Cantor, C.R. & Schimmel, P.R. (1980) *Biophysical Chemistry,* Part I: *The Conformation of Biological Macromolecules,* p. 100, W.H. Freeman and Company, New York.

*Portions of the polypeptide chains that are not accounted for by α helix or β conformation consist of bends and irregularly coiled or extended stretches. Segments of α helix and β conformation sometimes deviate slightly from their normal dimensions and geometry.

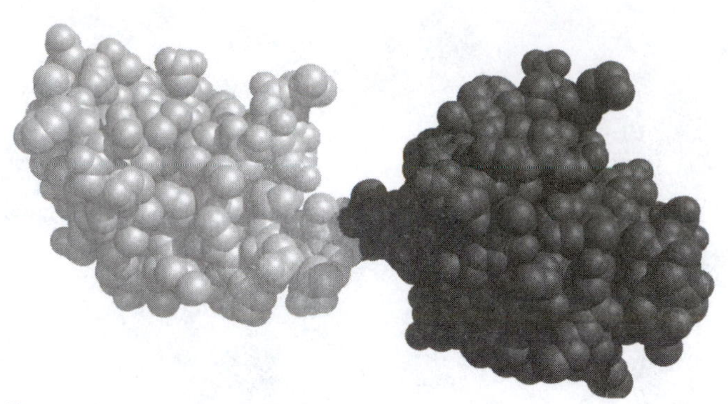

figure 6–19, page 184
Structural domains in the polypeptide troponin C.

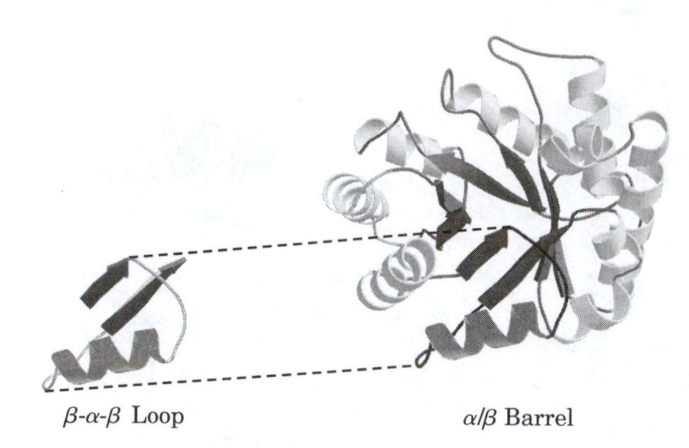

figure 6–21, page 185
Constructing large motifs from smaller ones.

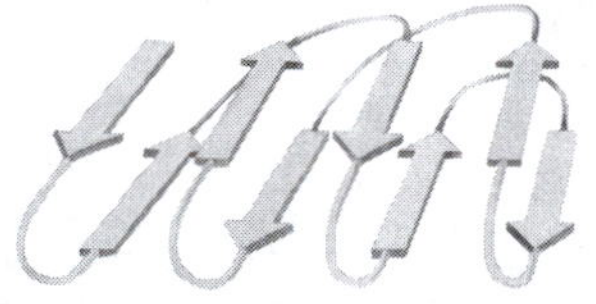

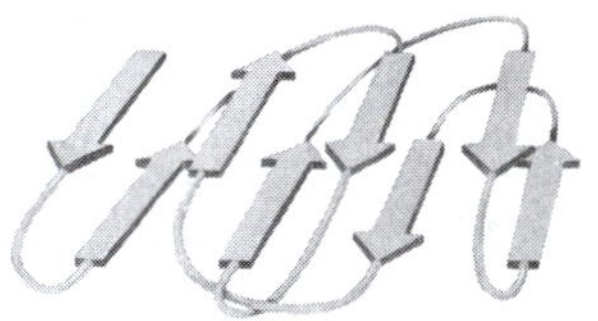

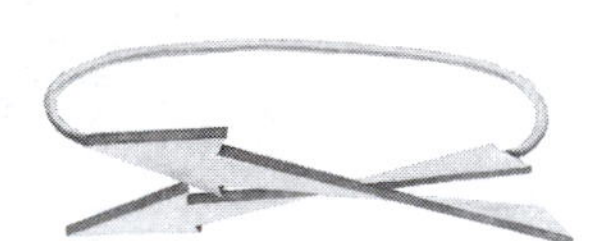

figure 6–20, page 185
Stable folding patterns in proteins.

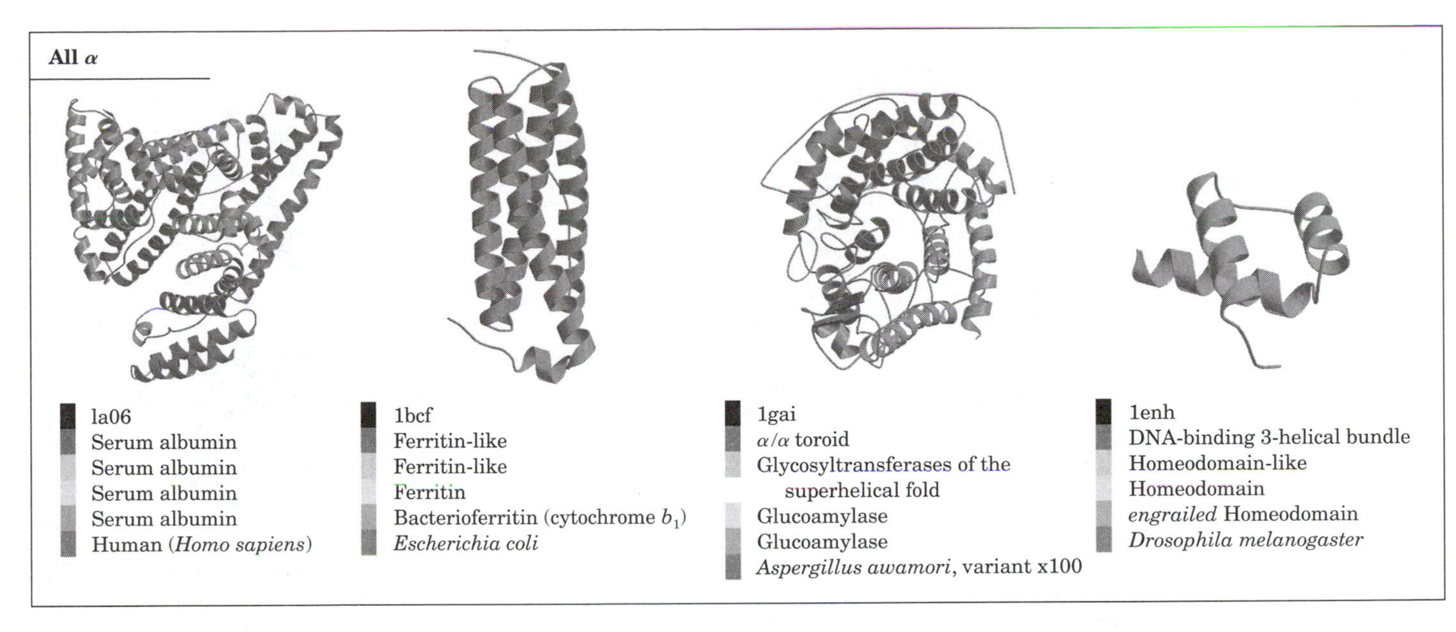

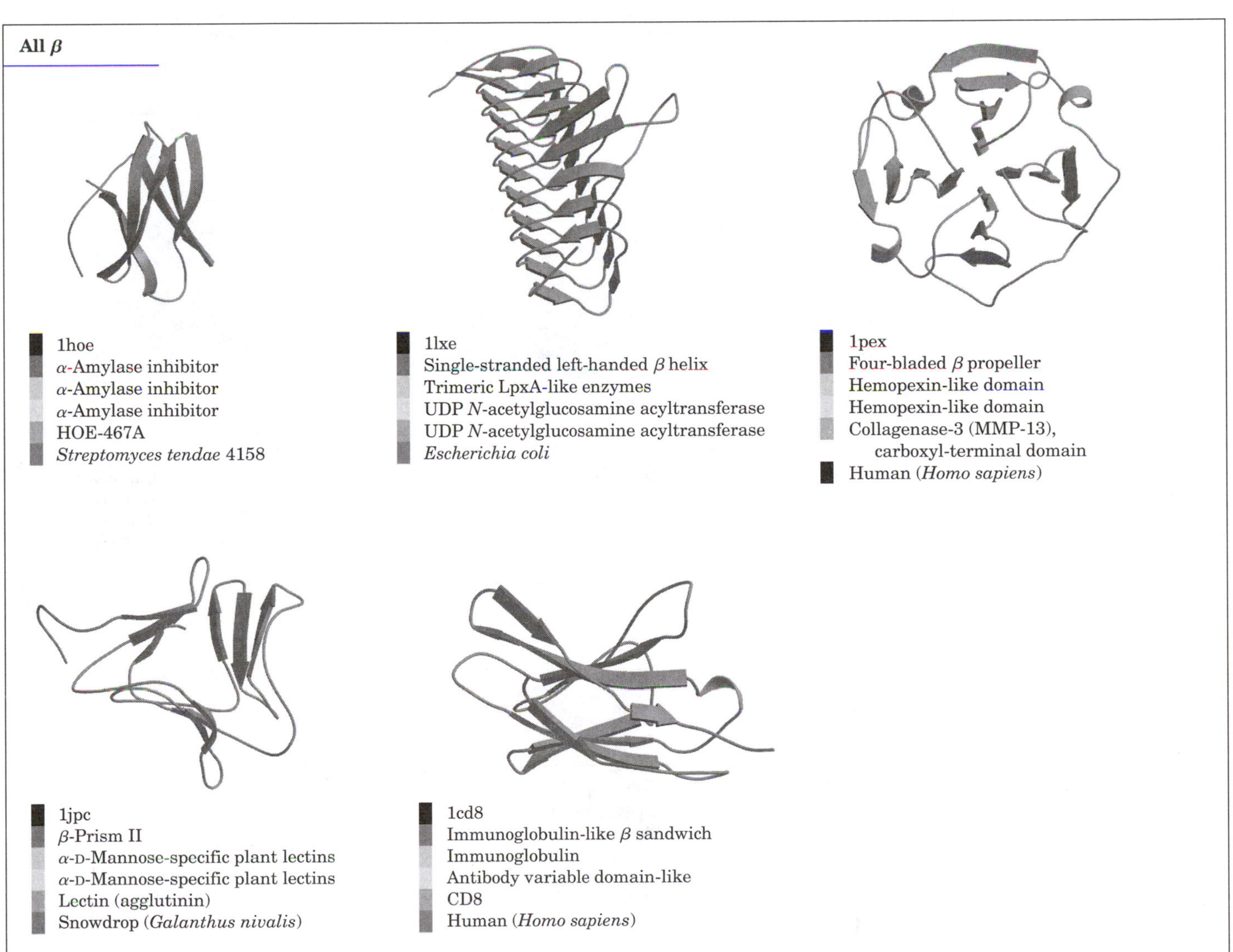

figure 6–22, pages 186–187
Organization of proteins based on motifs.

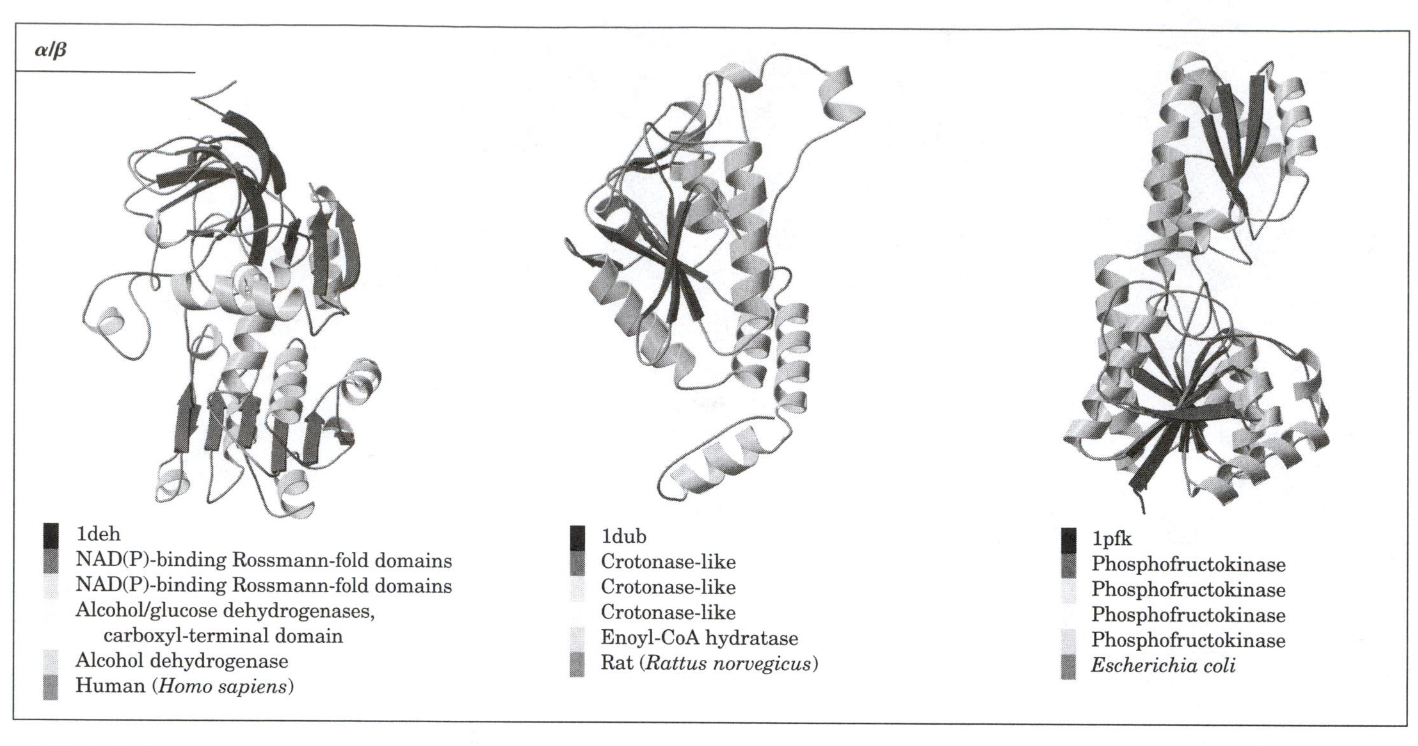

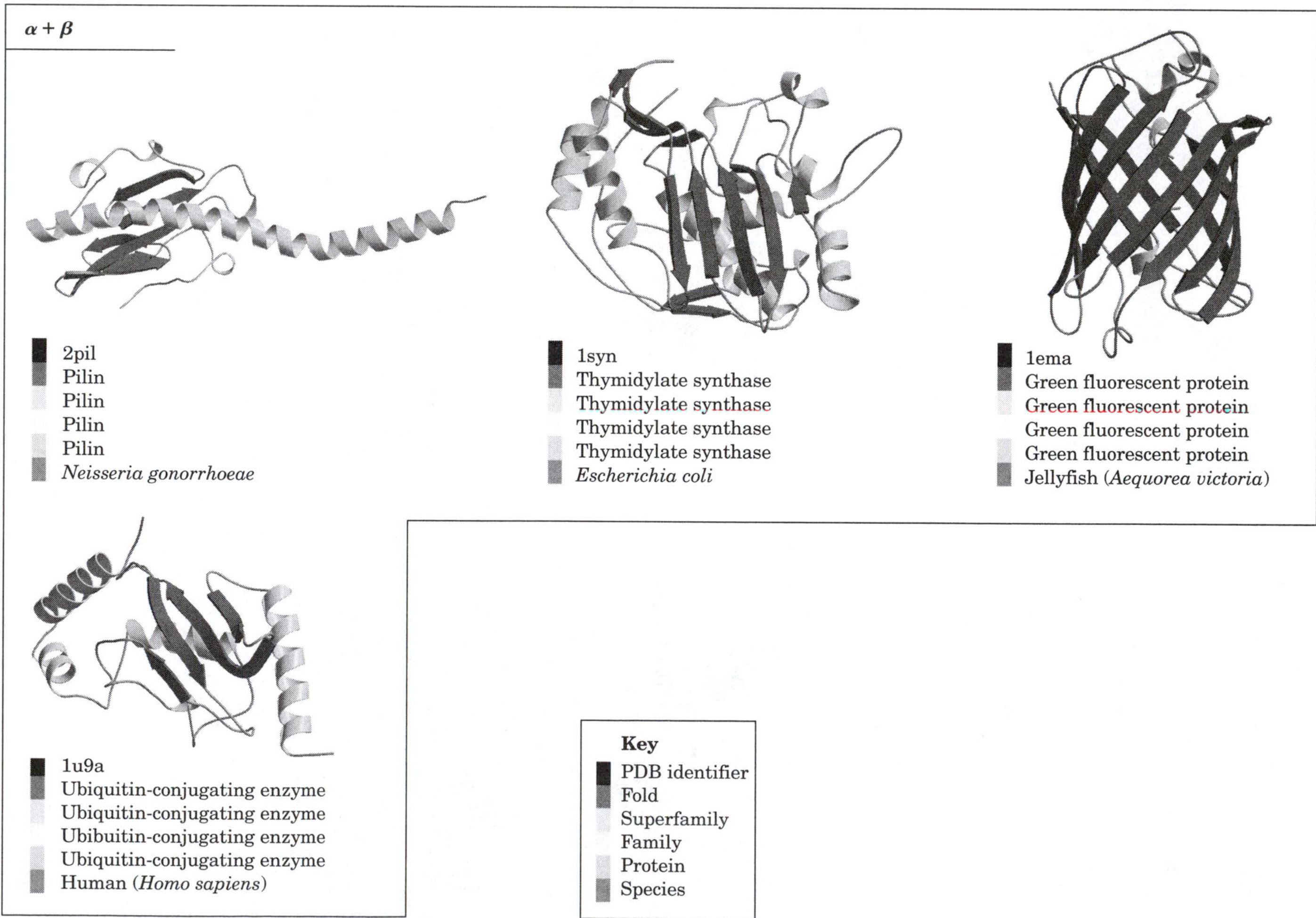

figure 6–22 (continued), pages 186–187
Organization of proteins based on motifs.

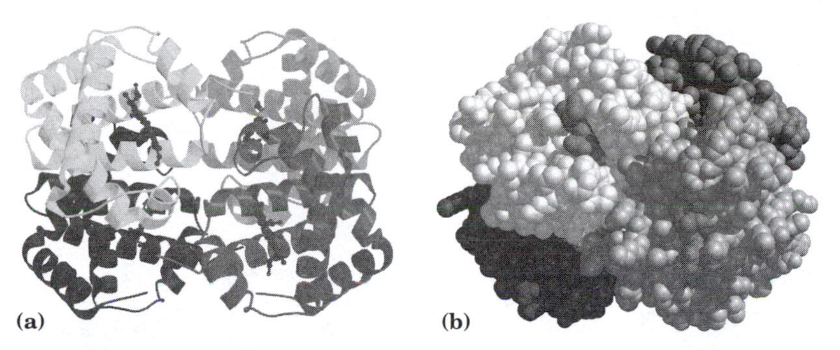

figure 6–23, page 189
The quaternary structure of deoxyhemoglobin.

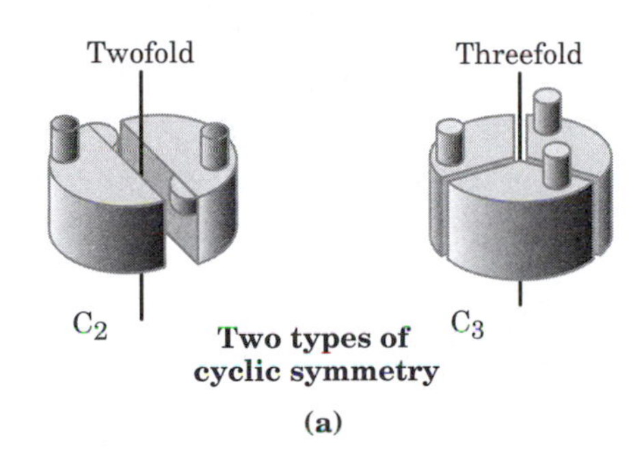

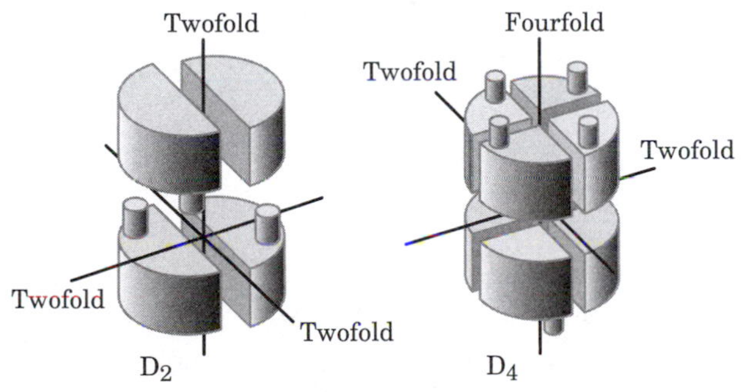

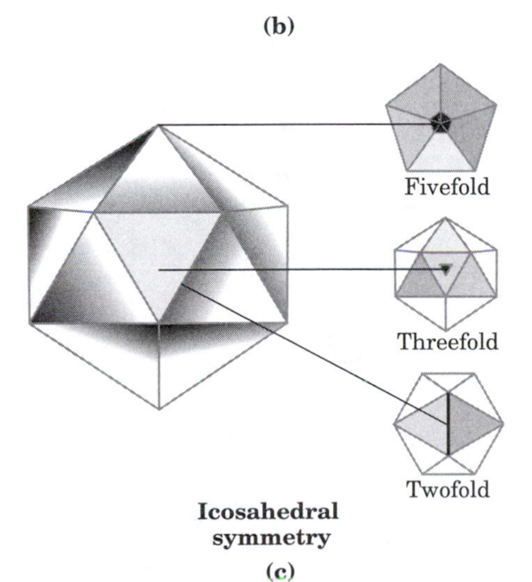

figure 6–24, page 190
Rotational symmetry in proteins.

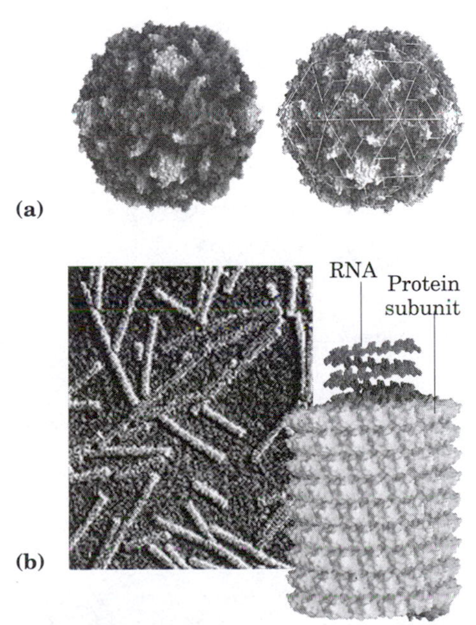

figure 6–25, page 191
Viral capsids.

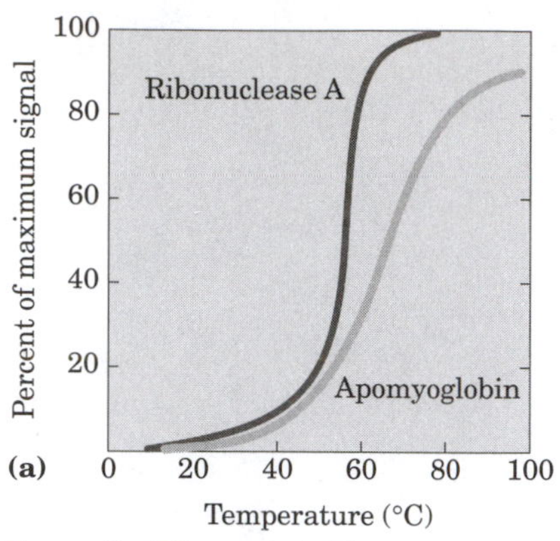

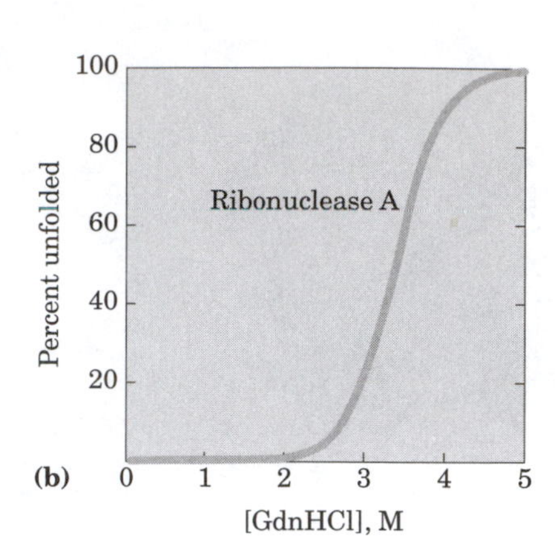

figure 6–26, page 192
Protein denaturation.

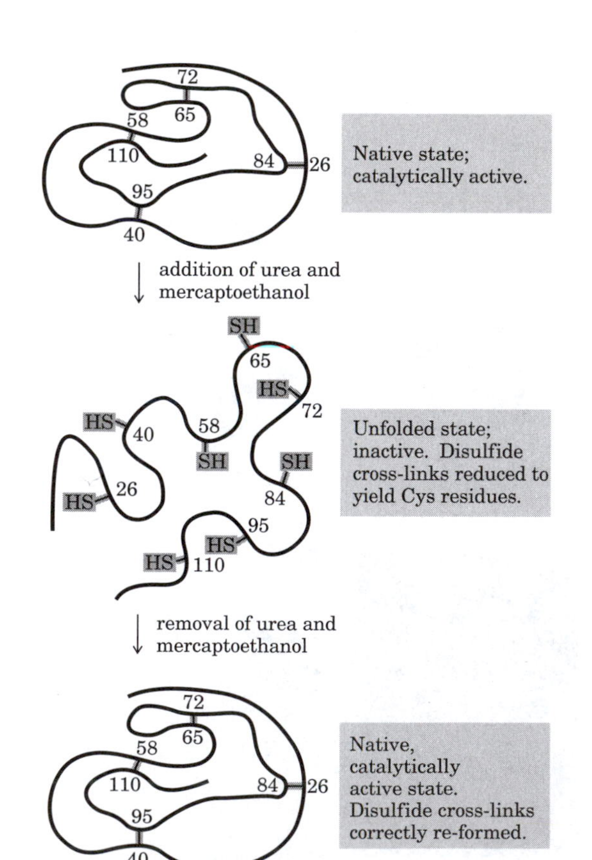

figure 6–27, page 193
Renaturation of unfolded, denatured ribonuclease.

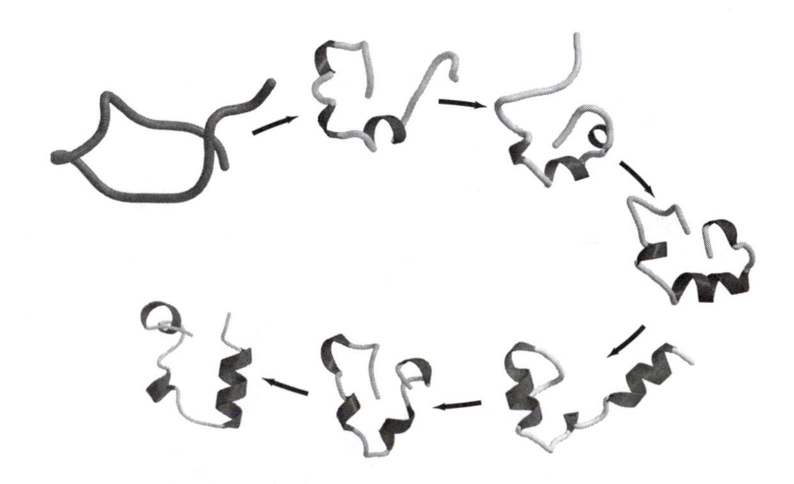

figure 6–28, page 194
A simulated folding pathway.

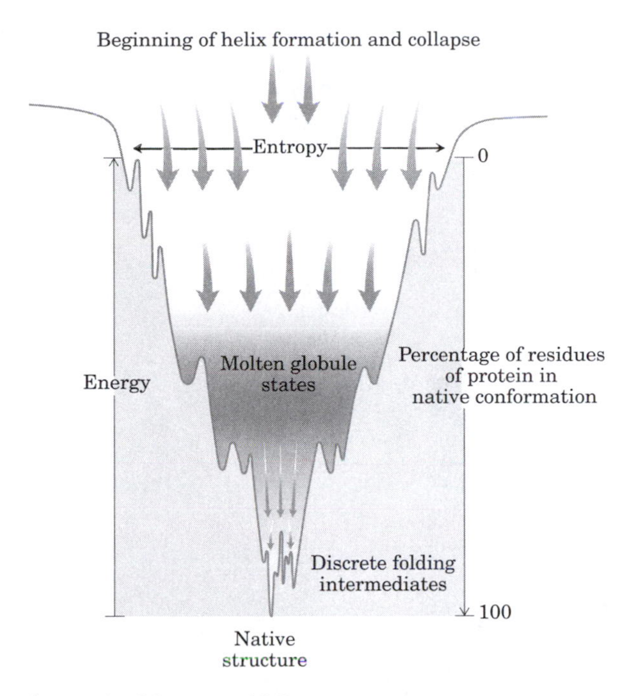

figure 6–29, page 195
The thermodynamics of protein folding depicted as a free-energy funnel.

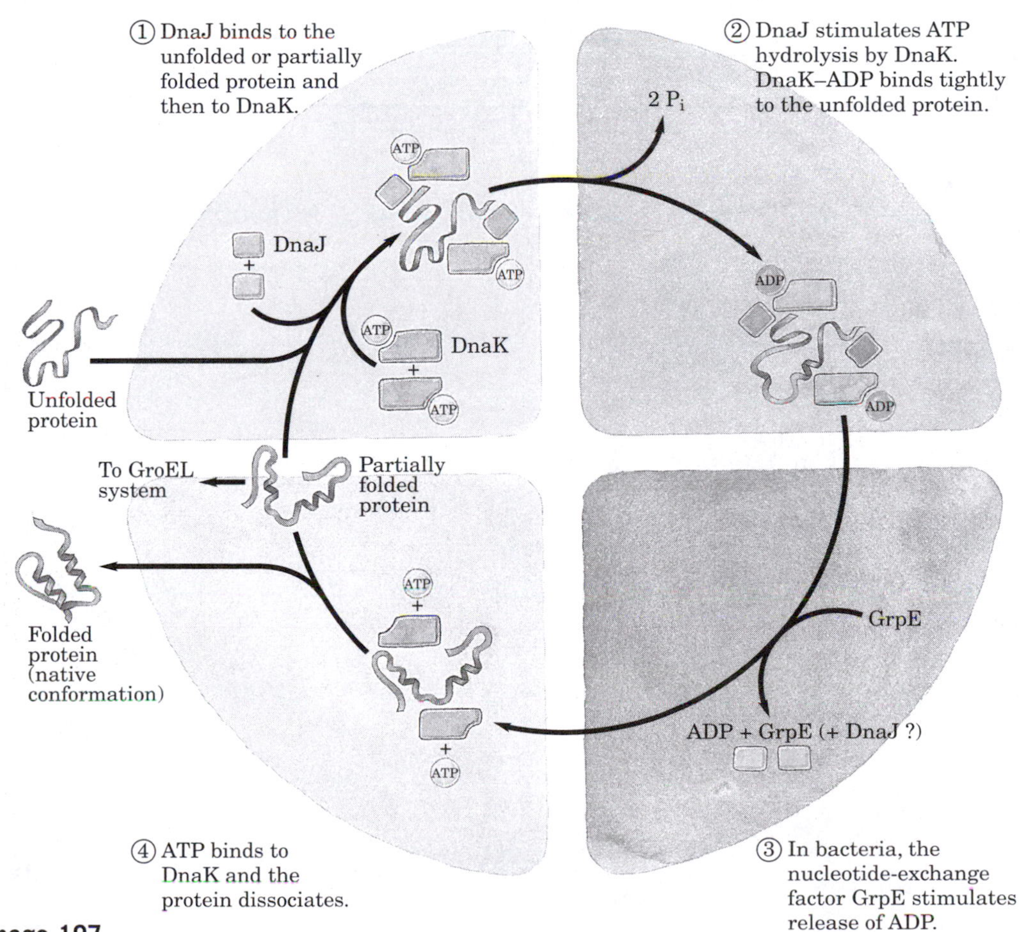

figure 6–30, page 197
Chaperones in protein folding.

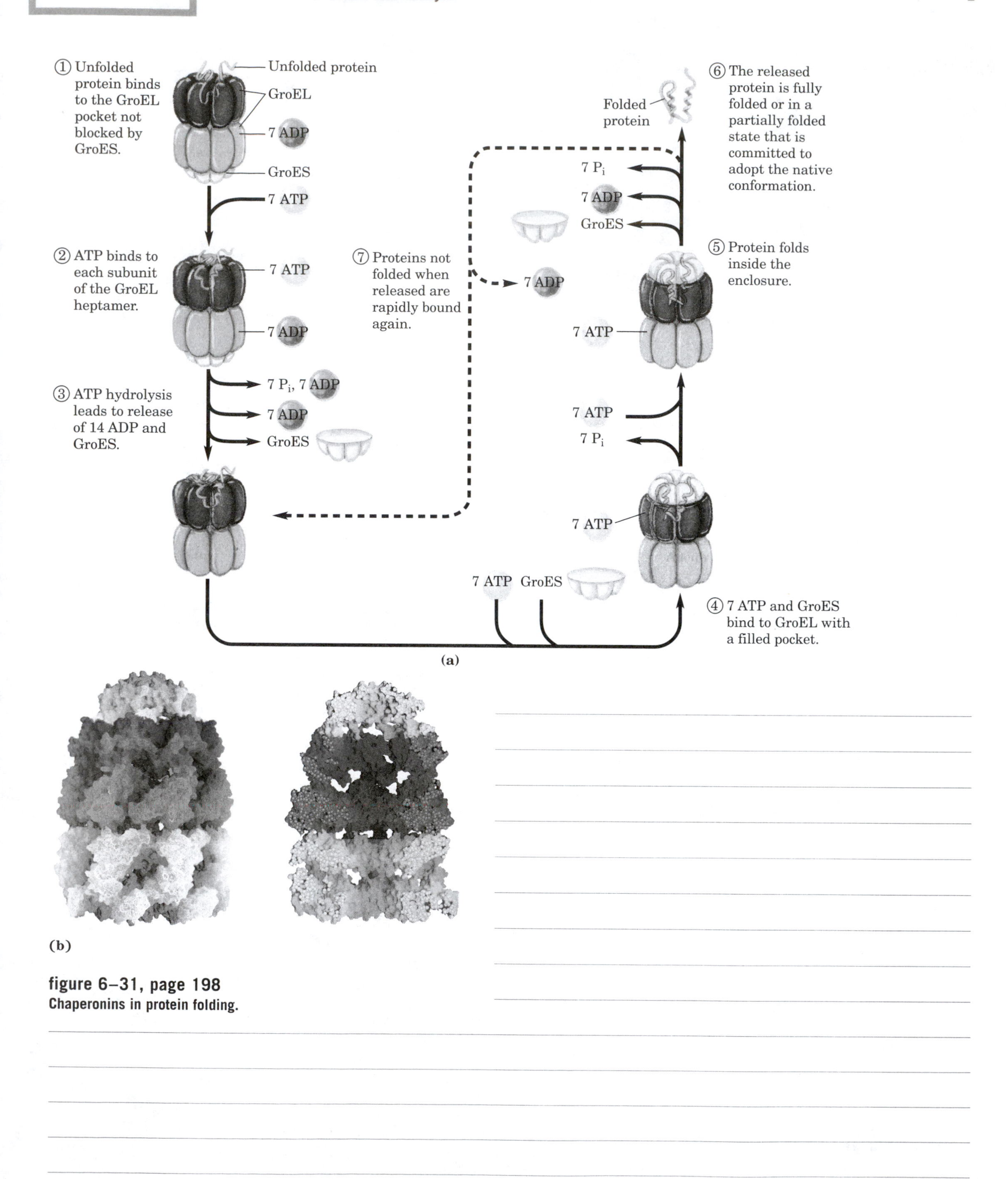

figure 6–31, page 198
Chaperonins in protein folding.

PROBLEMS

1. Properties of the Peptide Bond, page 201.

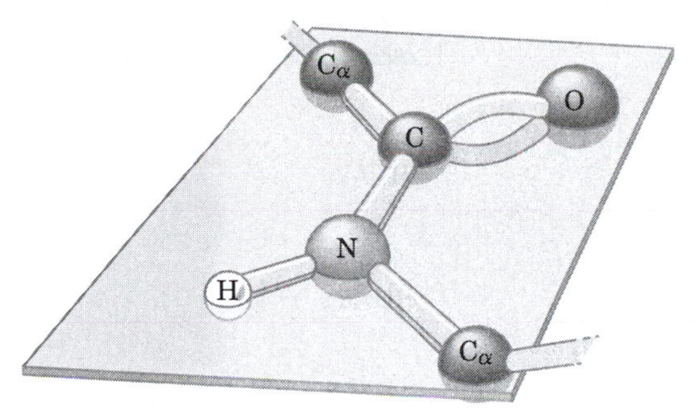

Peptide group

4. The Effect of pH on the Conformation of α-Helical Secondary Structures, page 201.

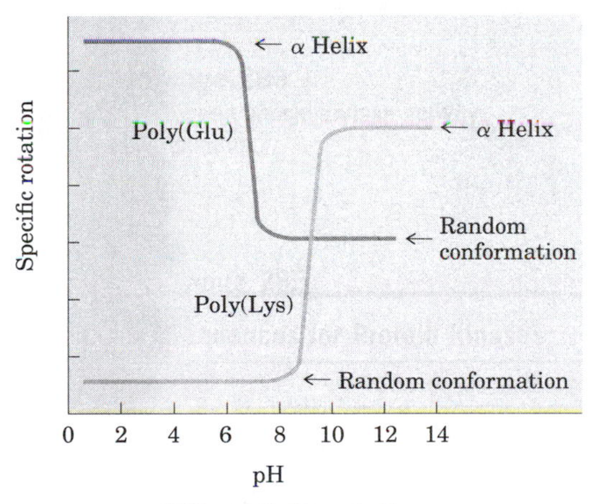

Effect of pH on helices

6. Amino Acid Sequence and Protein Structure, page 201.

1 2 3 4 5 6 7 8 9 10
Ile – Ala – His – Thr – Tyr – Gly – Pro – Phe – Glu – Ala –

11 12 13 14 15 16 17 18 19 20
Ala – Met – Cys – Lys – Trp – Glu – Ala – Gln – Pro – Asp –

21 22 23 24 25 26 27 28
Gly – Met – Glu – Cys – Ala – Phe – His – Arg

Peptide sequence

9. Number of Polypeptide Chains in a Multisubunit Protein, page 202.

NO_2 CH_3 CH_3 CH
O_2N– –NH–C–COOH
H

Peptide hydrolysate

10. Protein Modeling on the Internet, page 202.

EAELCPDRCI	HSFQNLGIQC	VKKRDLEQAI
SQRIQTNNNP	FQVPIEEQRG	DYDLNAVRLC
FQVTVRDPSG	RPLRLPPVLP	HPIFDNRAPN
TAELKICRVN	RNSGSCLGGD	EIFLLCDKVQ
KEDIEVYFTG	PGWEARGSFS	QADVHRQVAI
VFRTPPYADP	SLQAPVRVSM	QLRRPSDREL
SEPMEFQYLP	DTDDRHRIEE	KRKRTYETFK
SIMKKSPFSG	PTDPRPPPRR	IAVPSRSSAS
VPKPAPQPYP		

Unidentified protein sequence

chapter 7 Protein Function

When you study for exams, don't forget to review The Minicourses on the UNDERSTAND! BIOCHEMISTRY CD that came with your textbook.

Minicourses that apply to this Chapter include:

Molecules of Life
- An Interactive Gallery of Protein Structures

Proteins in Action
- Catalysis and Regulation
- Allosteric Enzymes

Proteins in Action (continued)
- Molecules of the Immune Systems
- The Immune System

Some Important Techniques
- Protein Separation and Analysis

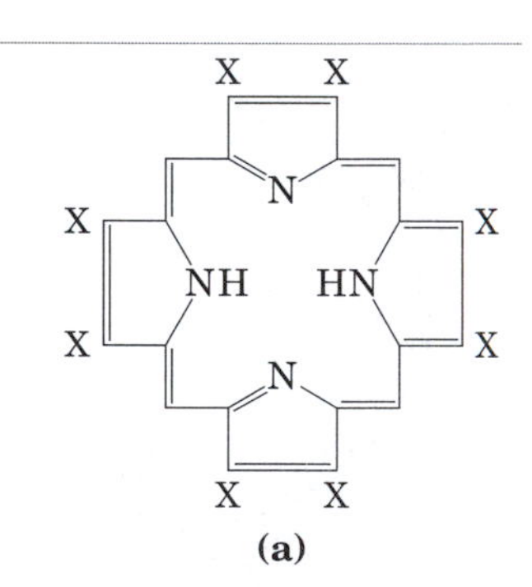

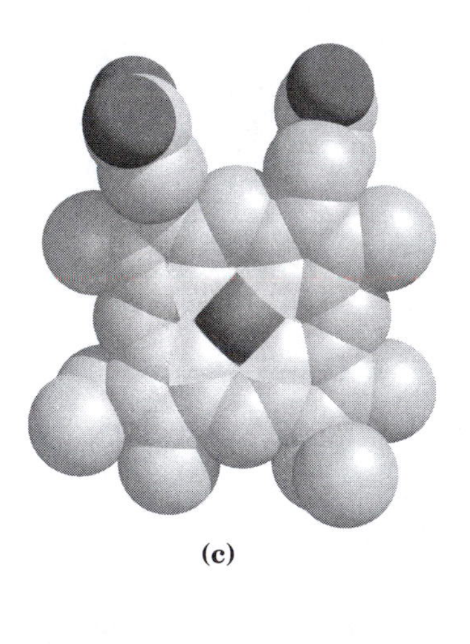

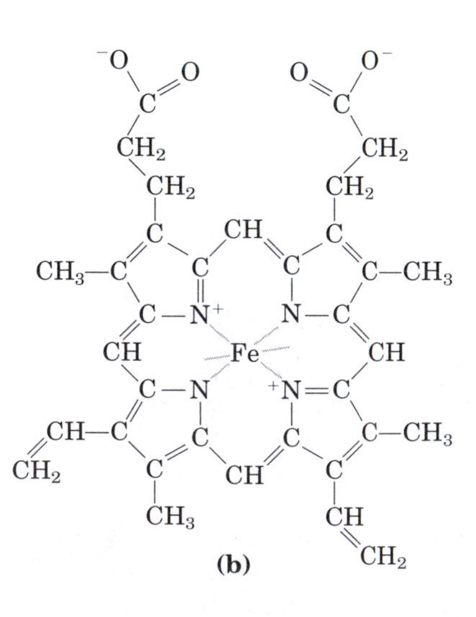

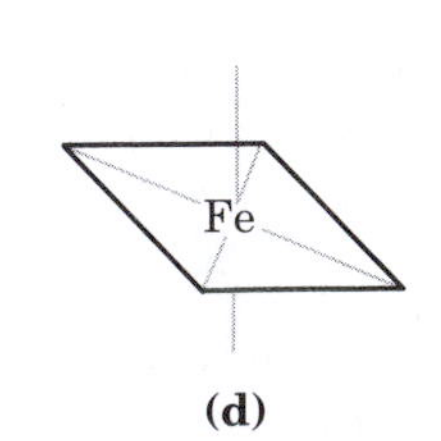

figure 7–1, page 205
Heme.

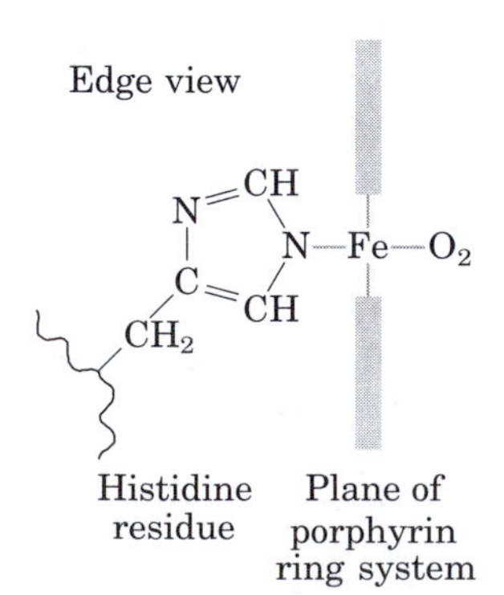

figure 7–2, page 205
The heme group viewed from the side.

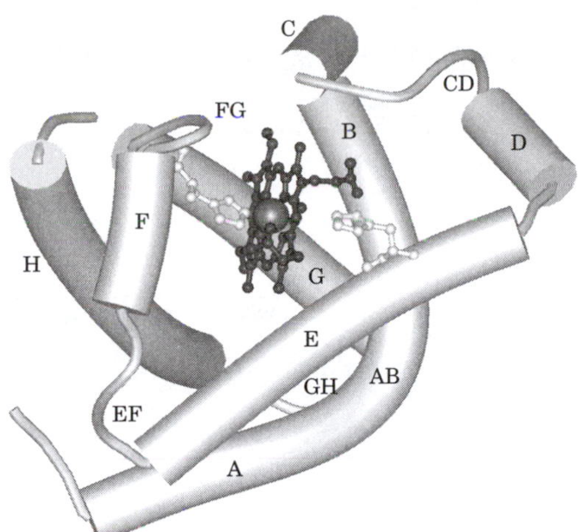

figure 7–3, page 206
The structure of myoglobin.

table 7–1, page 208

Some Protein Dissociation Constants

Protein	Ligand	K_d (M)*
Avidin (egg white)†	Biotin	1×10^{-15}
Insulin receptor (human)	Insulin	1×10^{-10}
Anti-HIV immunoglobulin (human)‡	gp41 (HIV-1 surface protein)	4×10^{-10}
Nickel-binding protein (*E. coli*)	Ni^{2+}	1×10^{-7}
Calmodulin (rat)§	Ca^{2+}	3×10^{-6}
		2×10^{-5}

*A reported dissociation constant is valid only for the particular solution conditions under which it was measured. K_d values for a protein-ligand interaction can be altered, sometimes by several orders of magnitude, by changes in solution salt concentration, pH, or other variables.

†Interaction of avidin with the enzymatic cofactor biotin is among the strongest noncovalent biochemical interactions known.

‡This immunoglobulin was isolated as part of an effort to develop a vaccine against HIV. Immunoglobulins (described later in the chapter) are highly variable, and the K_d reported here should not be considered characteristic of all immunoglobulins.

§Calmodulin has four binding sites for calcium. The values shown reflect the highest- and lowest-affinity binding sites observed in one set of measurements.

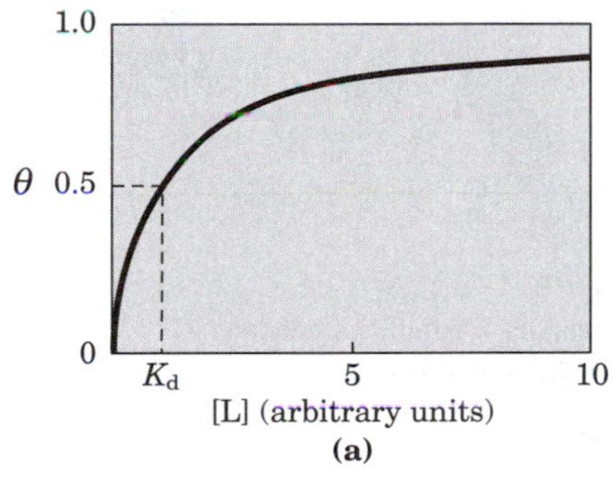

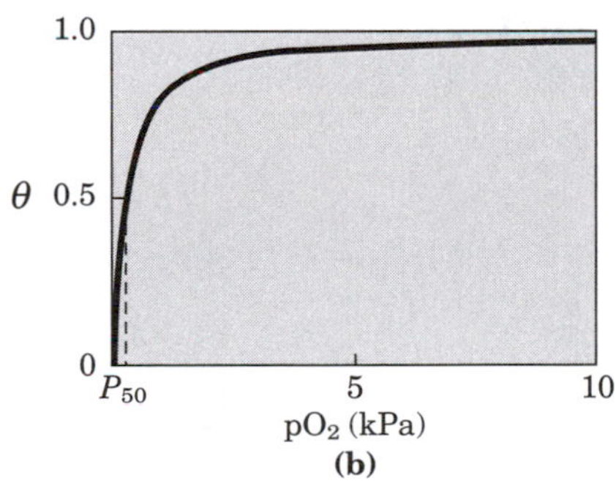

figure 7–4, page 207
Graphical representations of ligand binding.

$$K_d = \frac{[P][L]}{[PL]} \quad \text{(Equation 7–6)}$$

$$[PL] = \frac{[P][L]}{K_d} \quad \text{(Equation 7–7)}$$

$$\theta = \frac{[L]}{[L] + K_d} \quad \text{(Equation 7–8)}$$

Dissociation constant, page 208

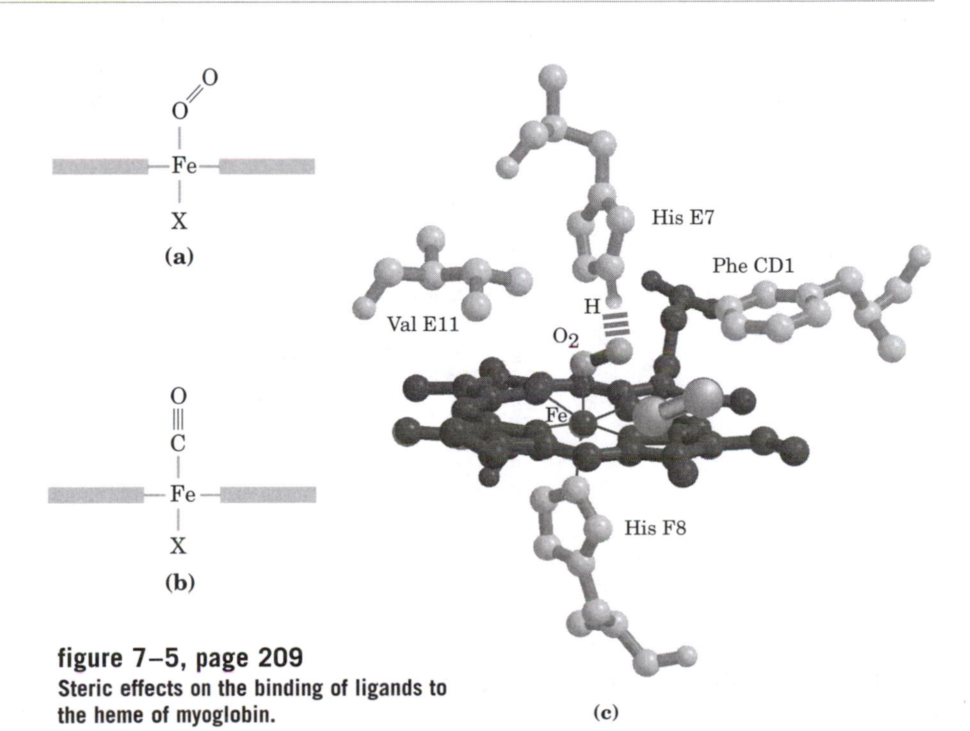

figure 7–5, page 209
Steric effects on the binding of ligands to the heme of myoglobin.

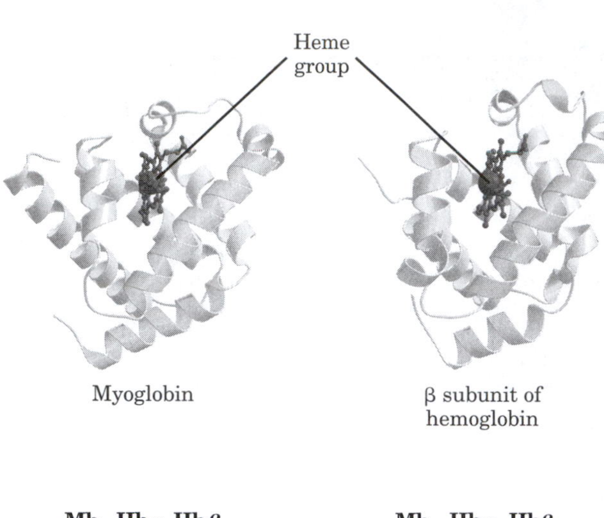

figure 7–6, page 210
A comparison of the structures of myoglobin and the β subunit of hemoglobin.

	Mb	Hbα	Hbβ
NA1	1 V	1 V	1 V
	—	—	H
	L	L	L
A1	S	S	T
	E	P	P
	G	A	E
	E	D	E
	W	K	K
	Q	T	S
	L	N	A
	V	V	V
	L	K	T
	H	A	A
	V	A	L
	W	W	W
	A	G	G
	K	K	K
	V	V	V
A16	E	G	—
	A	A	—
B1	20 D	20 H	N
	V	A	20 V
	A	G	D
	G	E	E
	H	Y	V
	G	G	G
	Q	A	G
	D	E	E
	I	A	A
	L	L	L
	I	E	G
	R	R	R
	L	M	L
	F	F	L
	K	L	V
B16	S	S	V
C1	H	F	Y
	P	P	P
	E	T	W
	T	T	T
	40 L	40 K	Q
	E	T	40 R
C7	K	Y	F
	F	F	F
	D	P	E
	R	H	S
	F	F	F
	K	—	G
	H	D	D
	L	L	L
	K	S	S
D1	T	H	T

	Mb	Hbα	Hbβ
	E	—	P
	A	—	D
	E	—	A
	M	—	V
	K	—	M
D7	A	G	G
E1	S	S	N
	E	A	P
	60 D	Q	K
	L	V	60 V
	K	K	K
	K	G	A
Distal His E7	H	H	H
	G	G	G
	V	60 K	K
	T	K	K
	V	V	V
	L	A	L
	T	D	G
	A	A	A
	L	L	F
	G	T	S
	A	N	D
	I	A	G
E19	L	V	L
	K	A	A
	K	H	H
	K	V	L
	80 G	D	D
	H	D	80 N
	H	M	L
	E	P	K
	A	N	G
	E	A	T
F1	L	80 L	F
	K	S	A
	P	A	T
	L	L	L
	A	S	S
	Q	D	E
	S	L	L
Proximal His F8	H	H	H
F9	A	A	C
	T	H	D
	K	K	K
	H	L	L
	K	R	H
	I	V	V
G1	100 P	D	D
	I	P	100 P
	K	V	E
	Y	N	N

	Mb	Hbα	Hbβ	
	L	F	F	
	E	K	R	
	F	100 L	L	
	I	L	L	
	S	S	G	
	E	H	N	
	A	C	V	
	I	L	L	
	I	L	V	
	H	V	C	
	V	T	V	
	L	L	L	
	H	A	A	
	S	A	H	
G19	R	H	H	
	H	L	F	
	120 P	P	G	
	G	A	120 K	
	D	E	E	
	F	F	F	
H1	G	T	T	
	A	P	P	
	D	120 A	P	
	A	V	V	
	Q	H	Q	
	G	A	A	
	A	S	A	
	M	L	Y	
	N	D	Q	
	K	K	K	
	A	F	V	
	L	L	V	
	E	A	A	
	L	S	G	
	F	V	V	
	R	S	A	
	140 K	T	N	
	D	V	140 A	
	I	L	L	
	A	T	A	
H21	A	S	H	
	K	K	K	HC1 (Hbα and Hbβ only)
	Y	140 Y	Y	HC2 (Hbα and Hbβ only)
	K	141 R	146 H	HC3 (Hbα and Hbβ only)
	E			
H26	L			
	G			
	Y			
	Q			
	153 G			

figure 7–7, page 211
The amino acid sequences of whale myoglobin and the α and β chains of human hemoglobin.

α2 β1 β2 α1

figure 7–8, page 212
Dominant interactions between hemoglobin subunits.

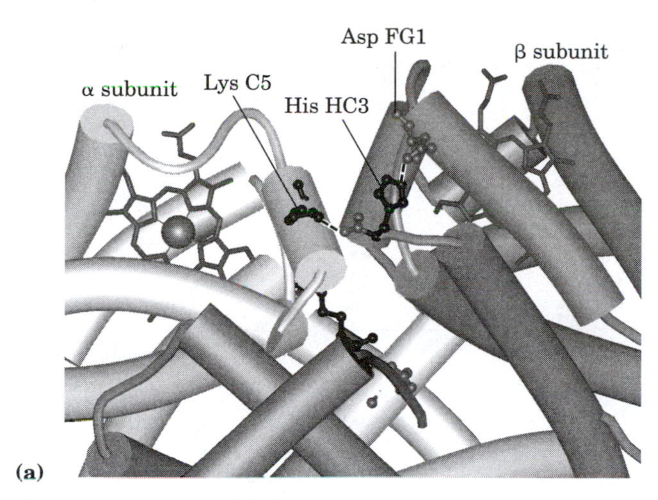

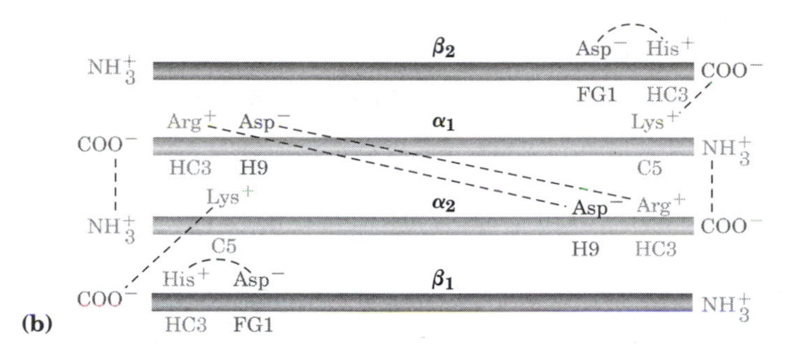

figure 7–9, page 212
Some ion pairs that stabilize the T state of deoxyhemoglobin.

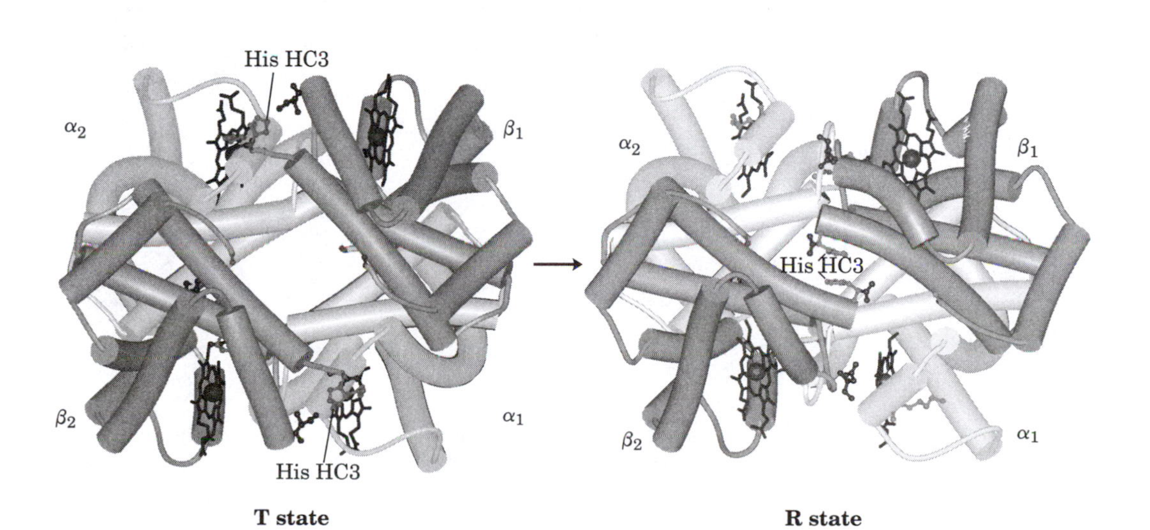

figure 7–10, page 213
The T ⟶ R transition.

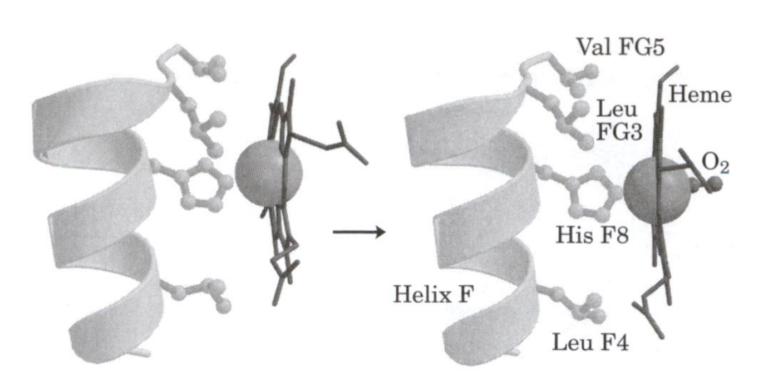

figure 7–11, page 213
Changes in conformation near heme on O_2 binding.

$$\log\left(\frac{\theta}{1-\theta}\right) = n \log [\mathrm{L}] - \log K_d \qquad \text{(Equation 7–16)}$$

Hill equation, page 215

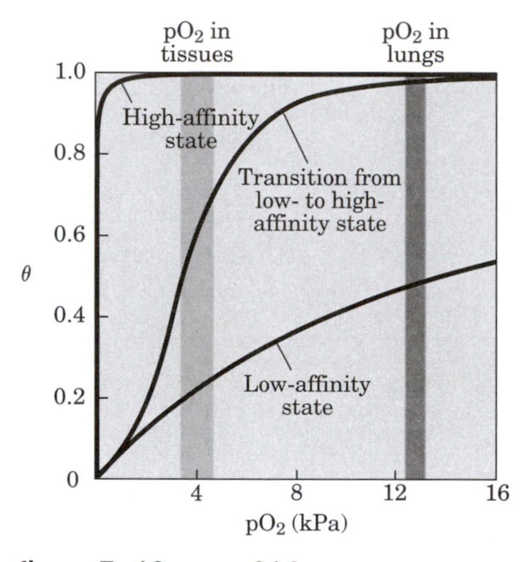

figure 7–12, page 214
A sigmoid (cooperative) binding curve.

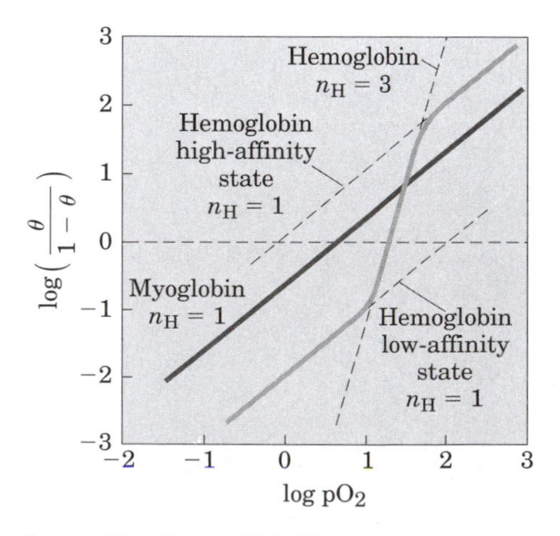

figure 7–13, page 215
Hill plots for the binding of oxygen to myoglobin and hemoglobin.

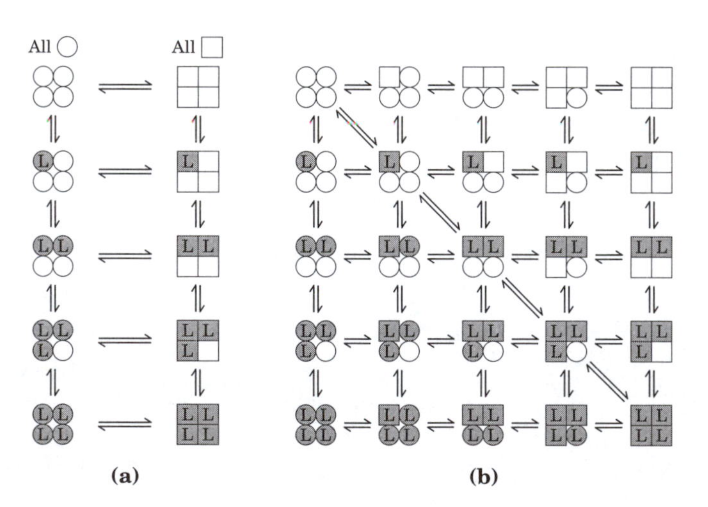

figure 7–14, page 216
Two general models for the interconversion of inactive and active forms of cooperative ligand-binding proteins.

$$CO_2 + H_2O \rightleftharpoons H^+ + HCO_3^-$$

Bicarbonate, page 216

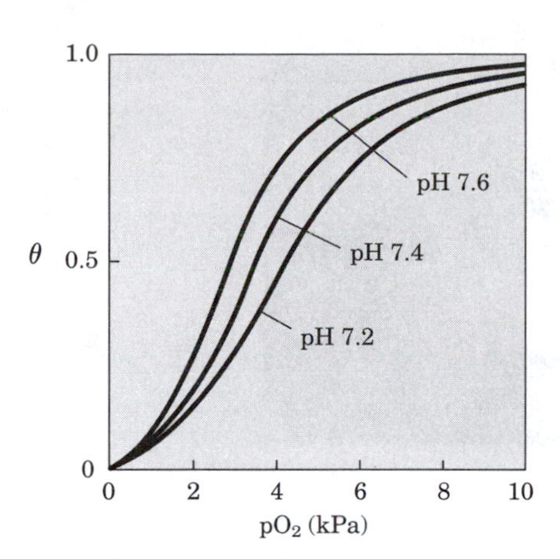

figure 7–15, page 217
Effect of pH on the binding of oxygen to hemoglobin.

Amino-terminal residue → Carbamino-terminal residue

page 217

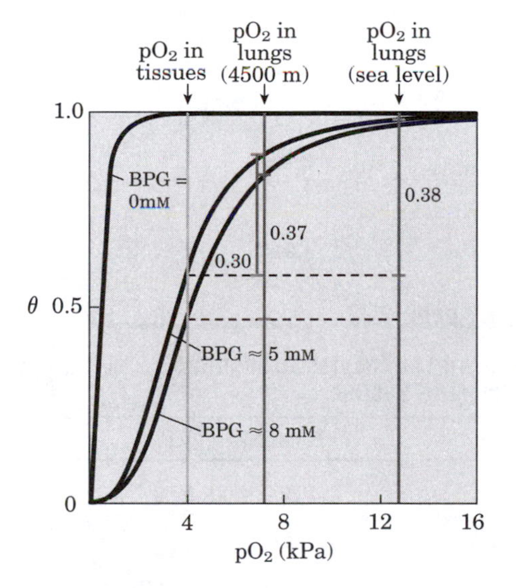

figure 7–16, page 218
Effect of BPG on the binding of oxygen to hemoglobin.

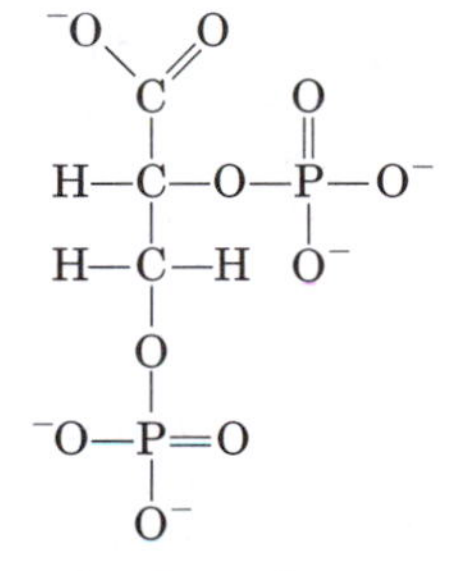

2,3-Bisphosphoglycerate, page 218

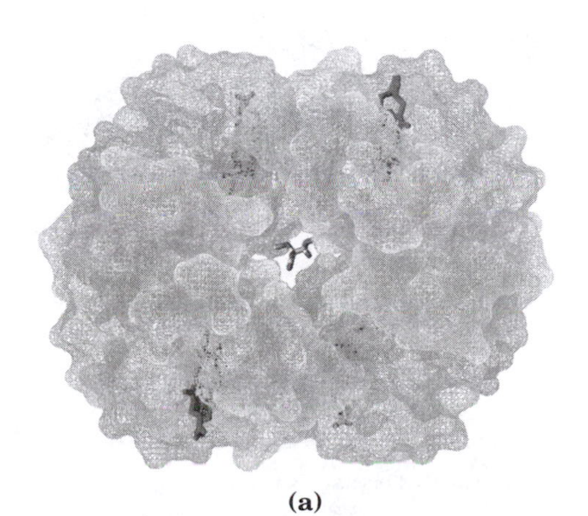

figure 7–17, page 219
Binding of BPG to deoxyhemoglobin.

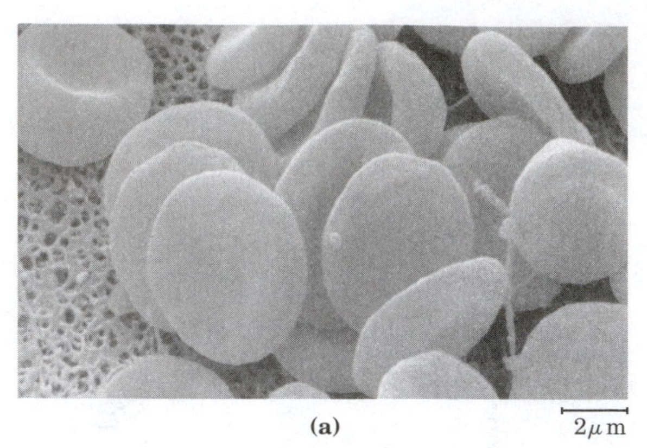

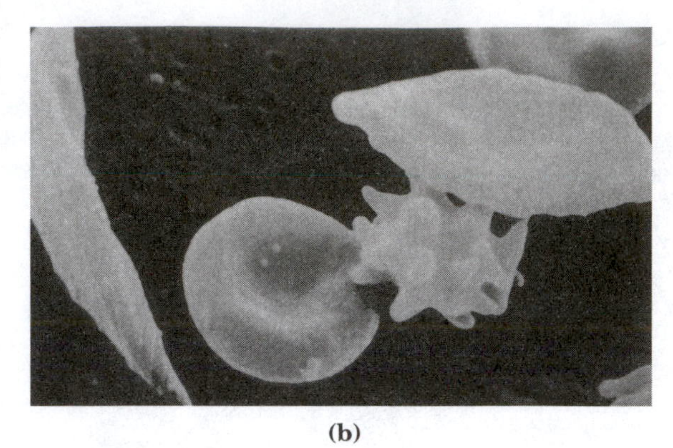

figure 7–18, page 220
A comparison of uniform, cup-shaped, normal erythrocytes (a) with the variably shaped erythrocytes seen in sickle-cell anemia (b).

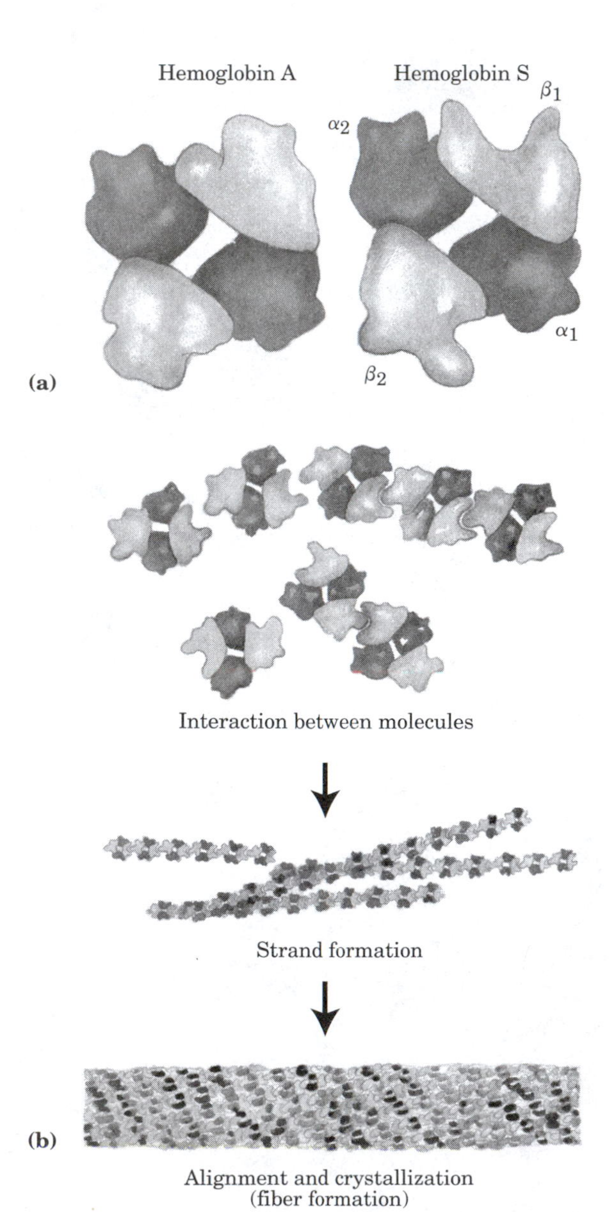

figure 7–19, page 220
Normal and sickle-cell hemoglobin.

table 7–2, page 222

Some Types of Leukocytes Associated with the Immune System

Cell type	Function
Macrophages	Ingest large particles and cells by phagocytosis
B lymphocytes (B cells)	Produce and secrete antibodies
T lymphocytes (T cells)	
Cytotoxic (killer) T cells (T_C)	Interact with infected host cells through receptors on T-cell surface
Helper T cells (T_H)	Interact with macrophages and secrete cytokines (interleukins) that stimulate T_C, T_H, and B cells to proliferate.

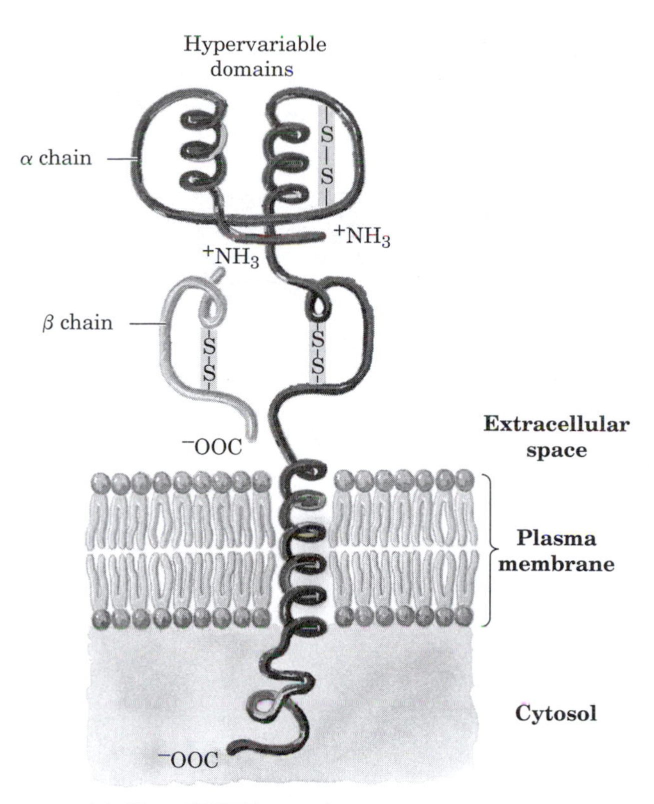

(a) Class I MHC protein

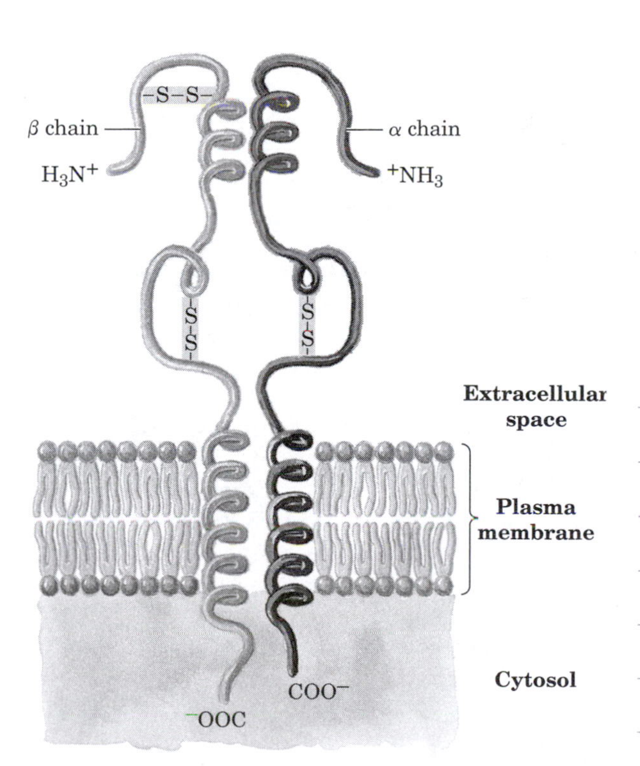

(b) Class II MHC protein

figure 7–20, page 223
MHC proteins.

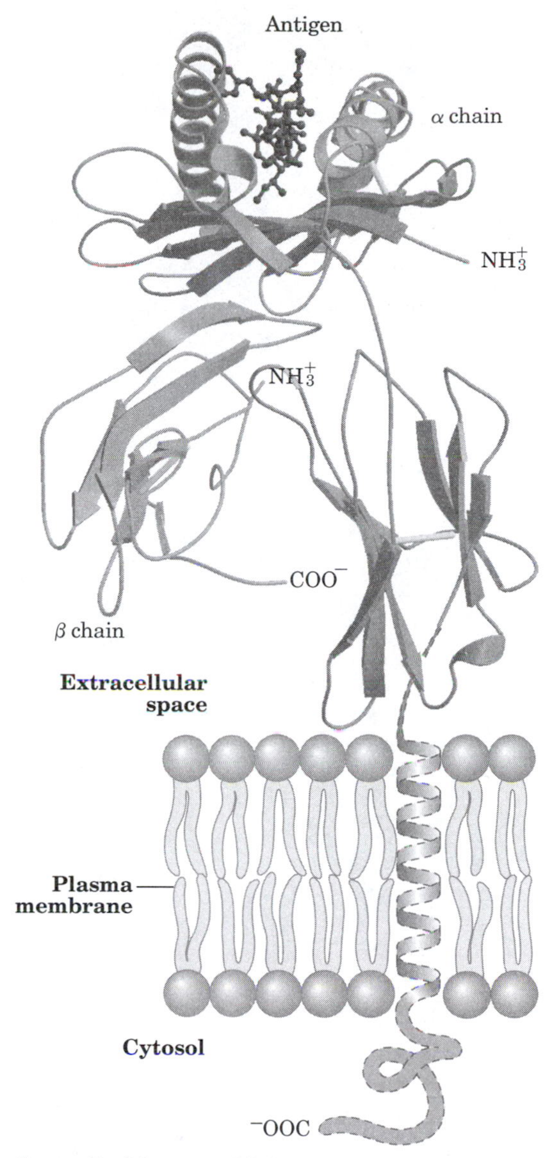

figure 7–21, page 224
Structure of a human class I MHC protein.

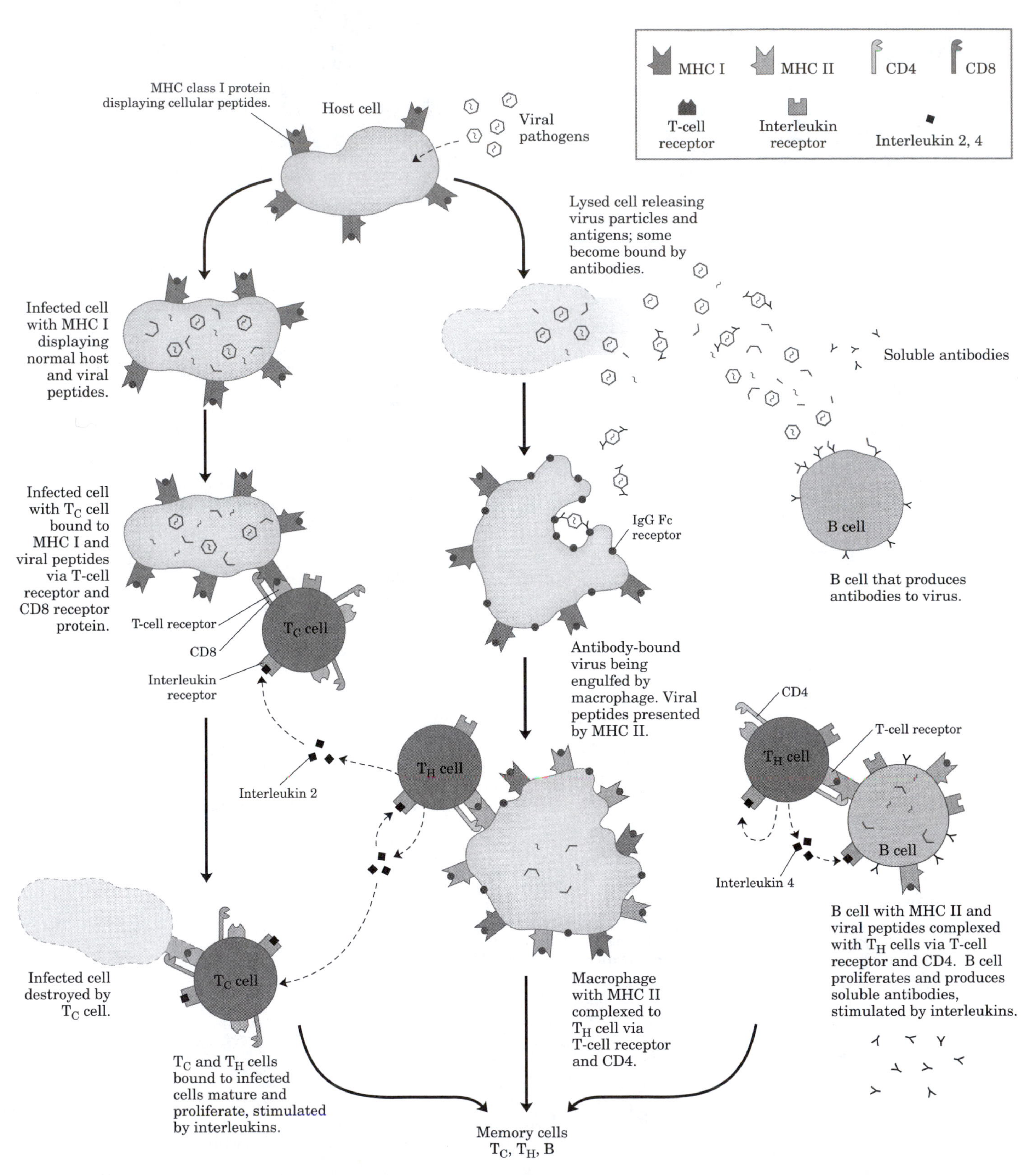

figure 7–22, page 226
Overview of the immune response to a viral infection.

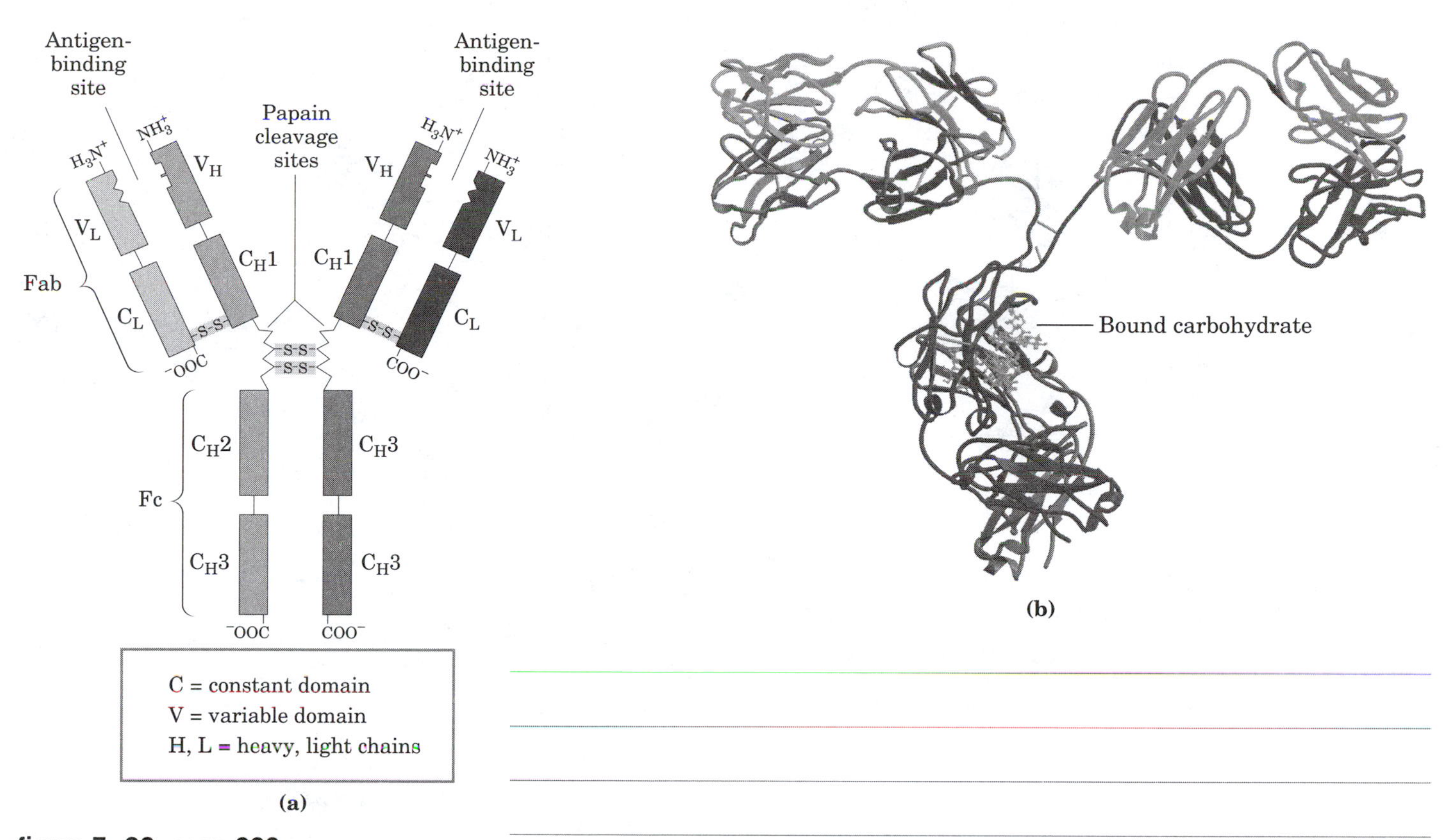

figure 7–23, page 228
The structure of immunoglobulin G.

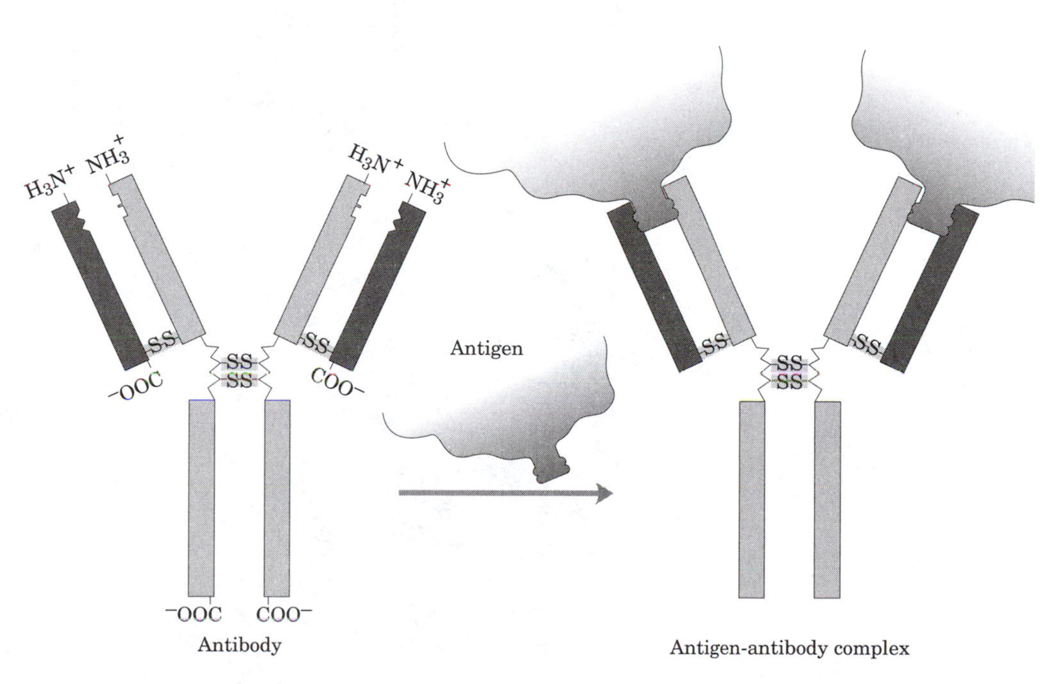

figure 7–24, page 229
Binding of IgG to an antigen.

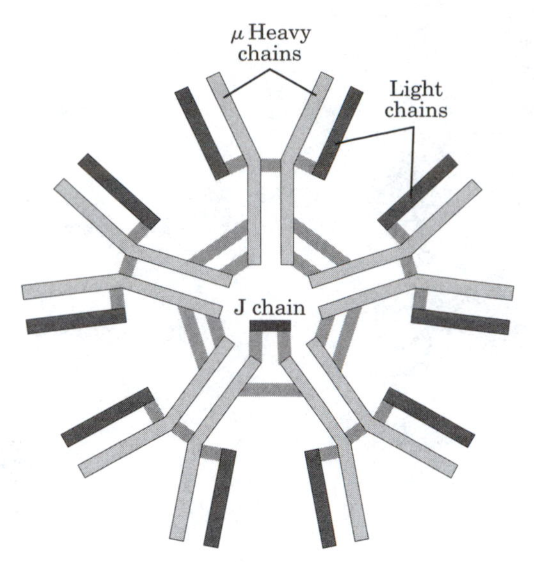

figure 7–25, page 229
IgM pentamer of immunoglobulin units.

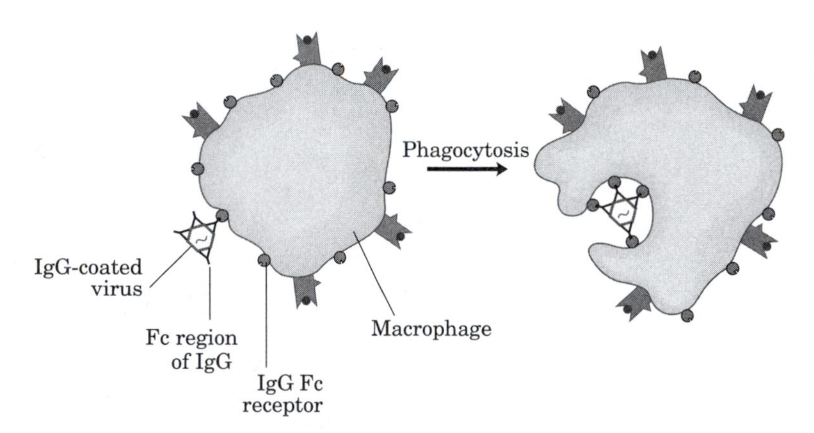

figure 7–26, page 230
Phagocytosis of an antibody-bound virus by a macrophage.

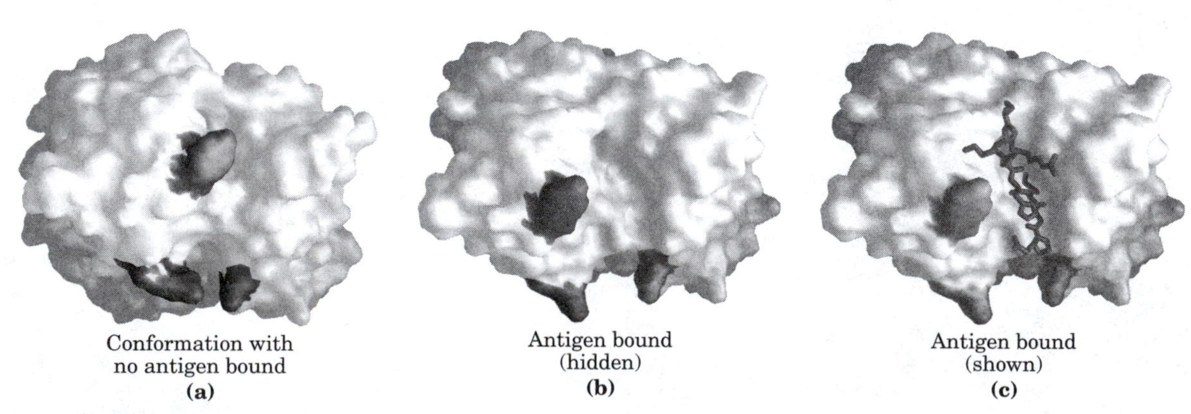

figure 7–27, page 231
Induced fit in the binding of an antigen to IgG.

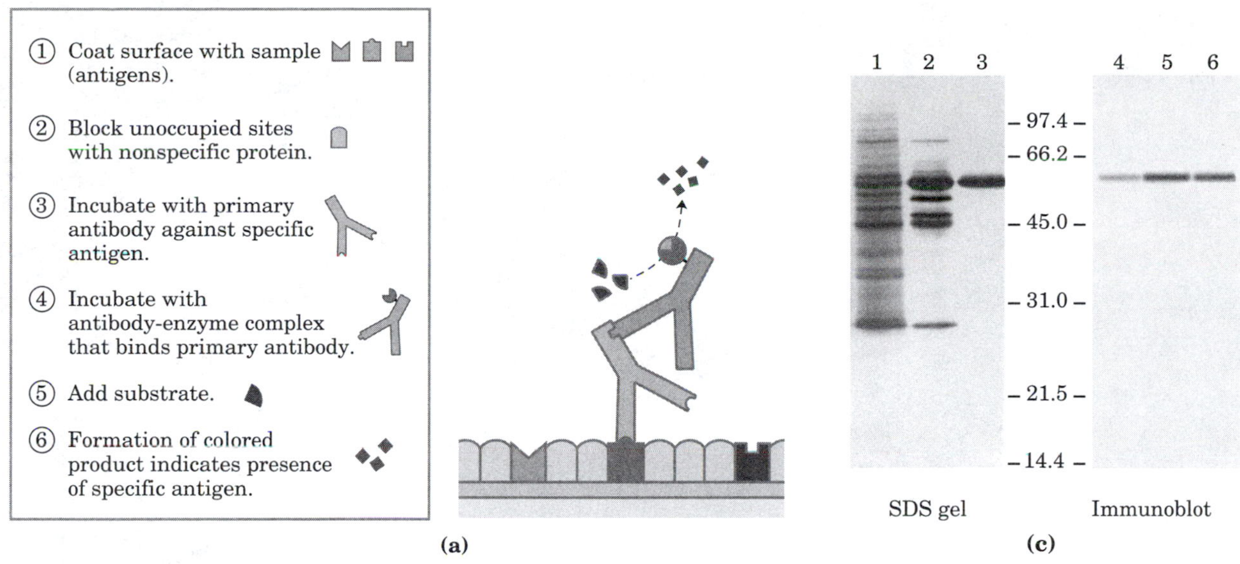

figure 7–28, page 232
Antibody techniques.

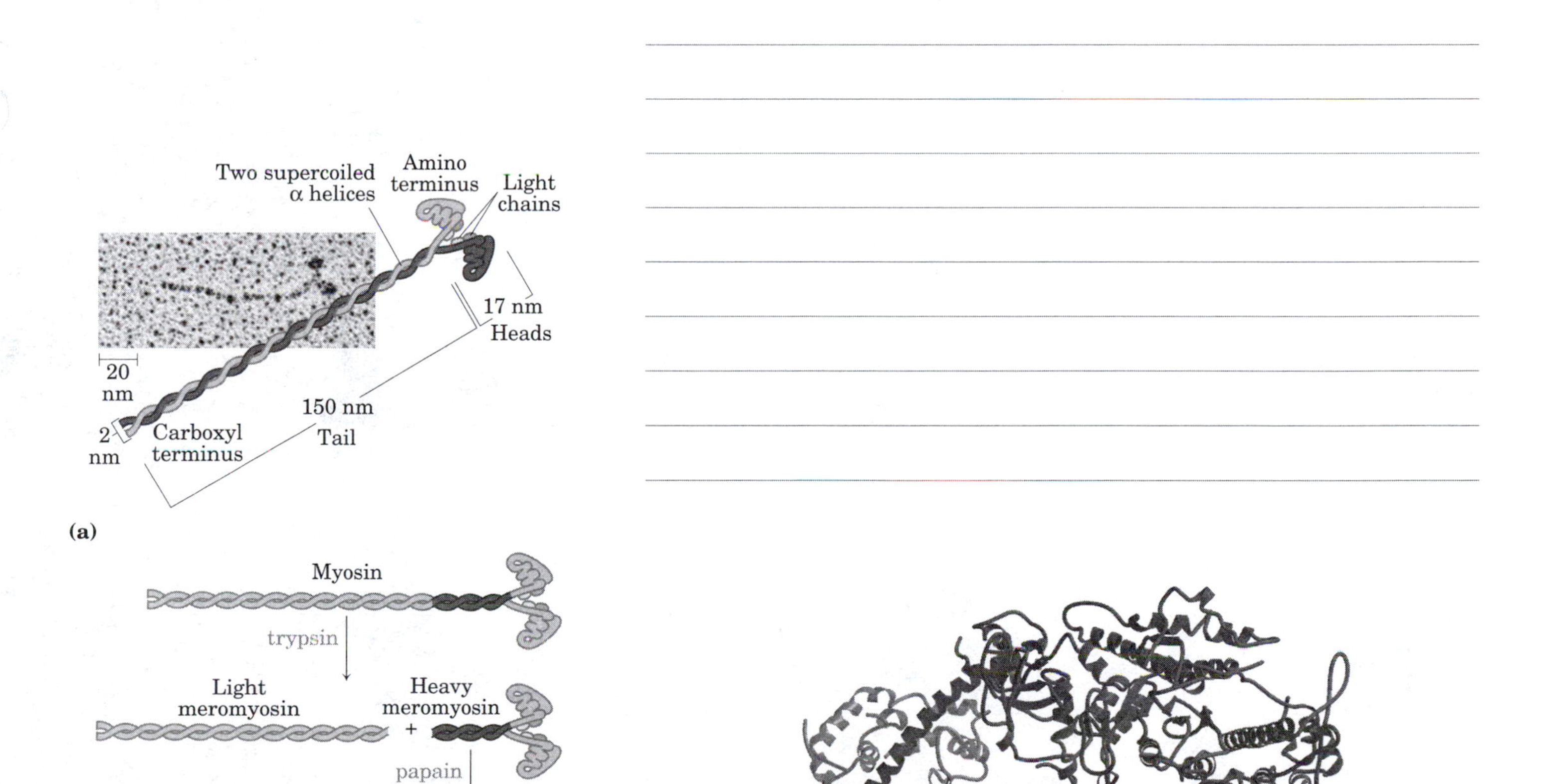

figure 7–29, page 233
Myosin.

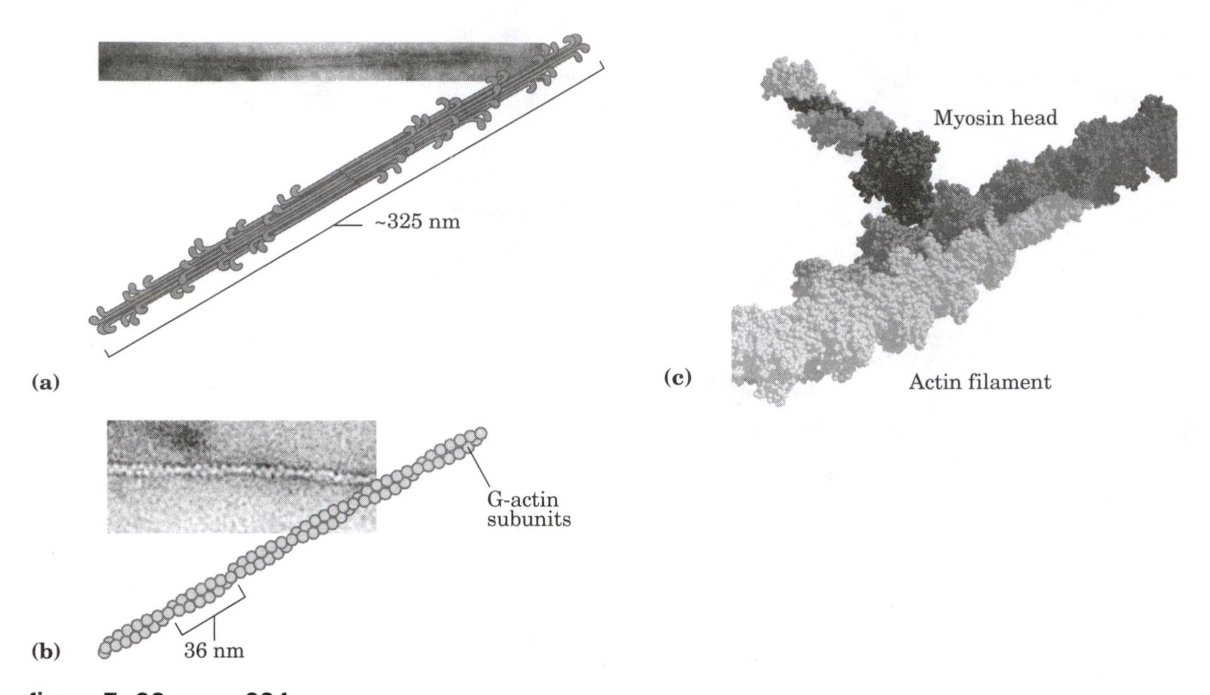

figure 7–30, page 234
The major components of muscle.

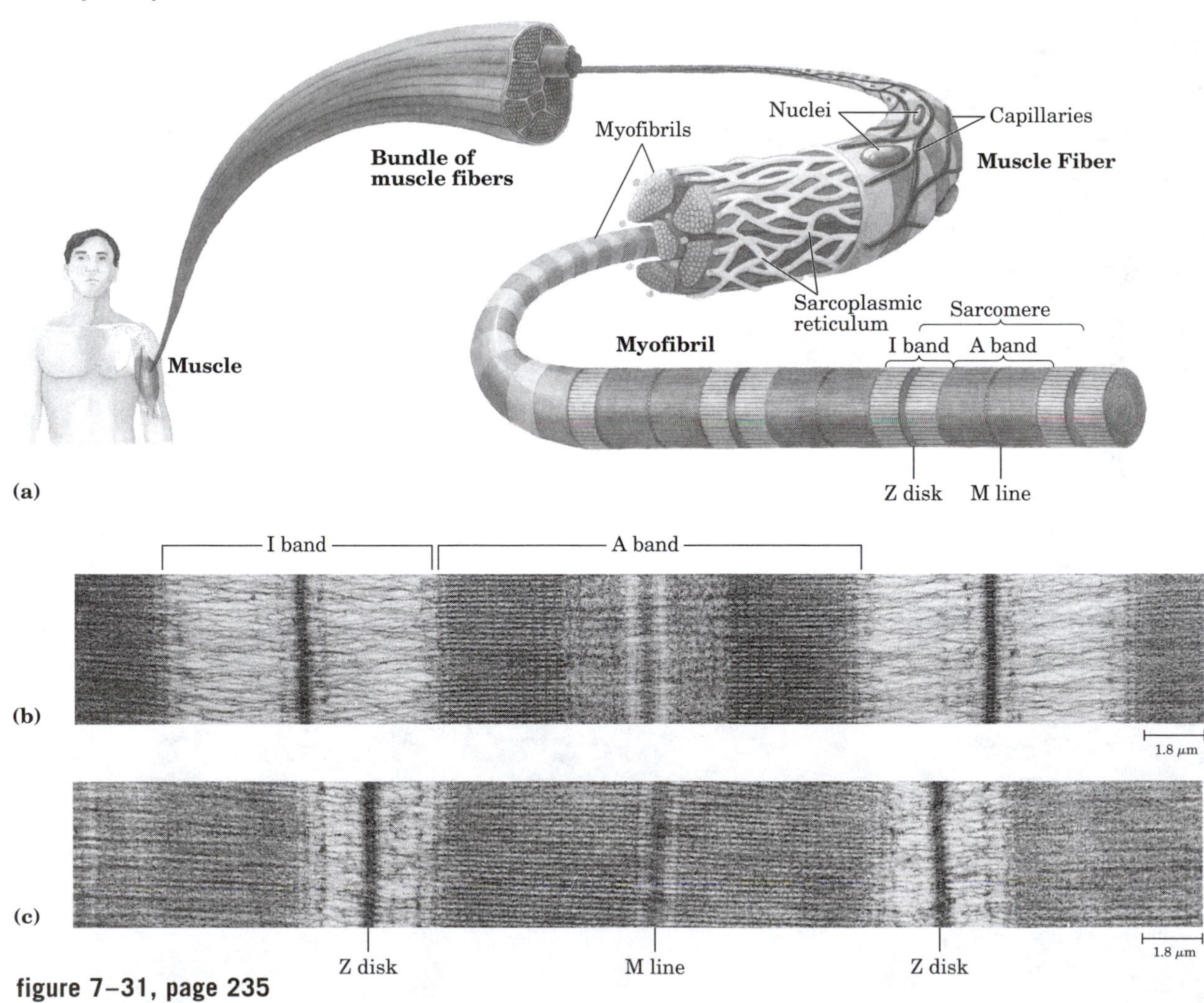

figure 7–31, page 235
Structure of skeletal muscle.

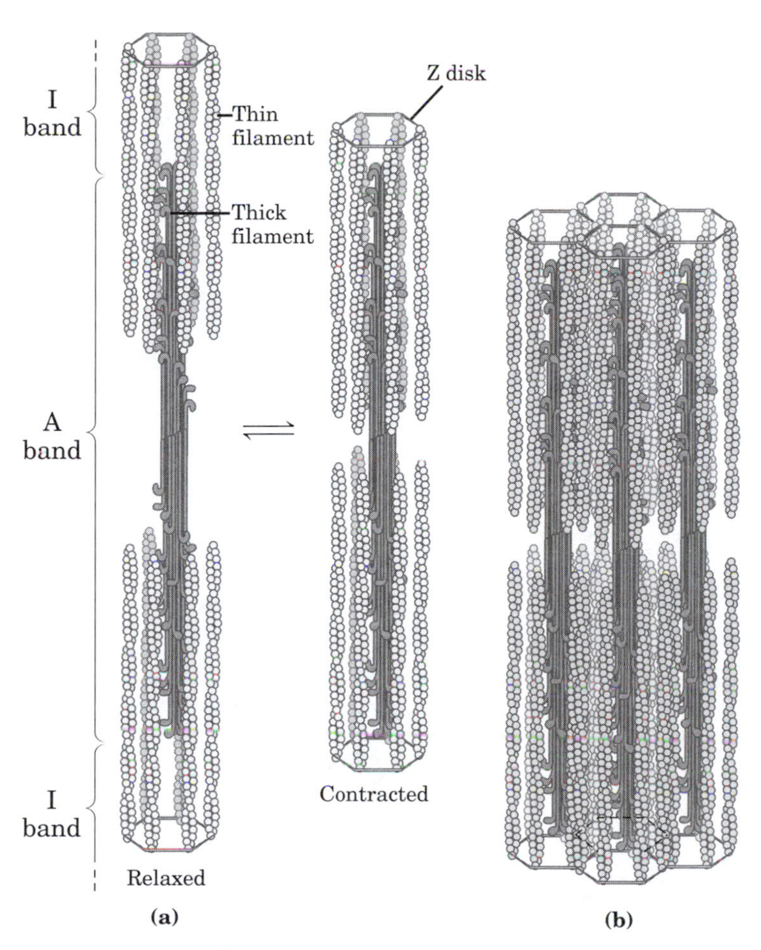

figure 7–32, page 236
Muscle contraction.

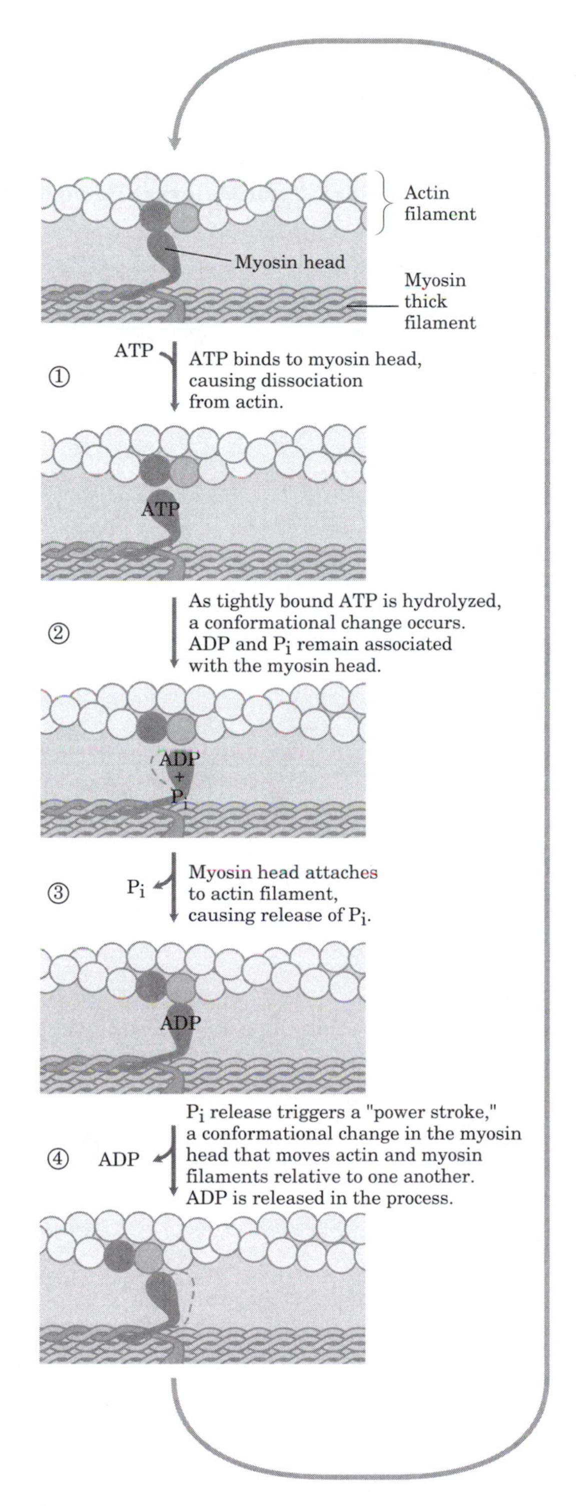

figure 7–33, page 237
Molecular mechanism of muscle contraction.

PROBLEMS

5. Comparison of Fetal and Maternal Hemoglobins, page 241.

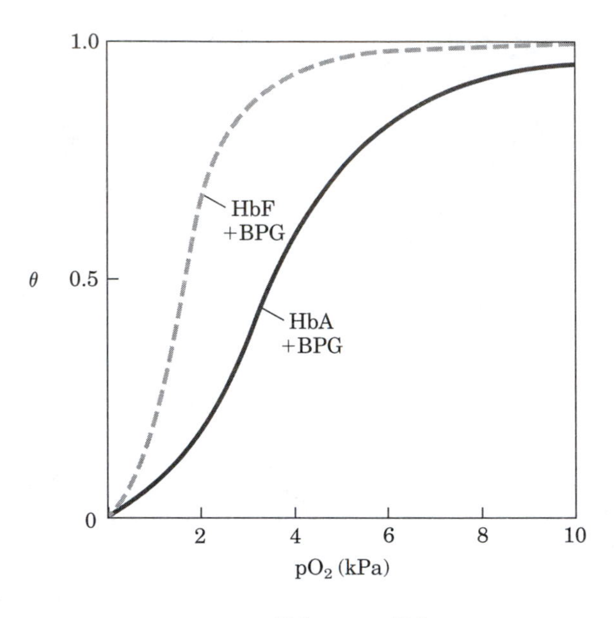

HbA versus HbF

chapter

8

Enzymes

CD-ROM

When you study for exams, don't forget to review The Minicourses on the UNDERSTAND! BIOCHEMISTRY CD that came with your textbook.

Minicourses that apply to this Chapter include:

Molecules of Life
An Interactive Gallery of Molecular Models

Proteins in Action
Catalysis and Regulation
Nonallosteric Enzyme Kinetics
Allosteric Enzymes

Biosynthesis and Catabolism
Nitrogen Metabolism

table 8–1, page 245

Some Inorganic Elements That Serve as Cofactors for Enzymes

Cu^{2+}	Cytochrome oxidase
Fe^{2+} or Fe^{3+}	Cytochrome oxidase, catalase, peroxidase
K^{+}	Pyruvate kinase
Mg^{2+}	Hexokinase, glucose 6-phosphatase, pyruvate kinase
Mn^{2+}	Arginase, ribonucleotide reductase
Mo	Dinitrogenase
Ni^{2+}	Urease
Se	Glutathione peroxidase
Zn^{2+}	Carbonic anhydrase, alcohol dehydrogenase, carboxypeptidases A and B

table 8–2, page 245

Some Coenzymes That Serve as Transient Carriers of Specific Atoms or Functional Groups*

Coenzyme	Examples of chemical groups transferred	Dietary precursor in mammals
Biocytin	CO_2	Biotin
Coenzyme A	Acyl groups	Pantothenic acid and other compounds
5′-Deoxyadenosylcobalamin (coenzyme B_{12})	H atoms and alkyl groups	Vitamin B_{12}
Flavin adenine dinucleotide	Electrons	Riboflavin (vitamin B_2)
Lipoate	Electrons and acyl groups	Not required in diet
Nicotinamide adenine dinucleotide	Hydride ion ($:H^-$)	Nicotinic acid (niacin)
Pyridoxal phosphate	Amino groups	Pyridoxine (vitamin B_6)
Tetrahydrofolate	One-carbon groups	Folate
Thiamine pyrophosphate	Aldehydes	Thiamine (vitamin B_1)

*The structure and mode of action of these coenzymes are described in Part III of this book.

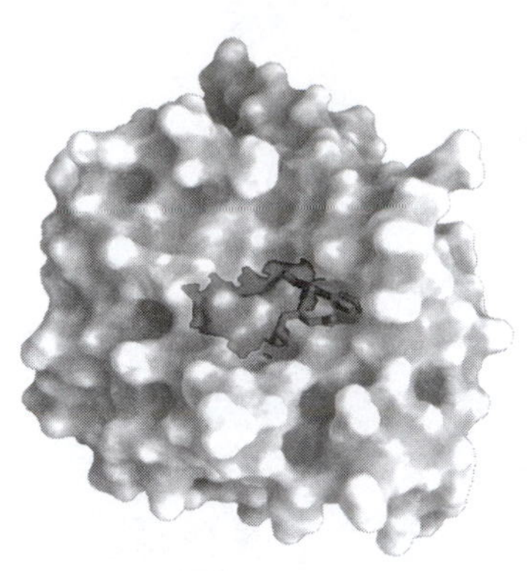

figure 8–1, page 247
Binding of a substrate to an enzyme at the active site.

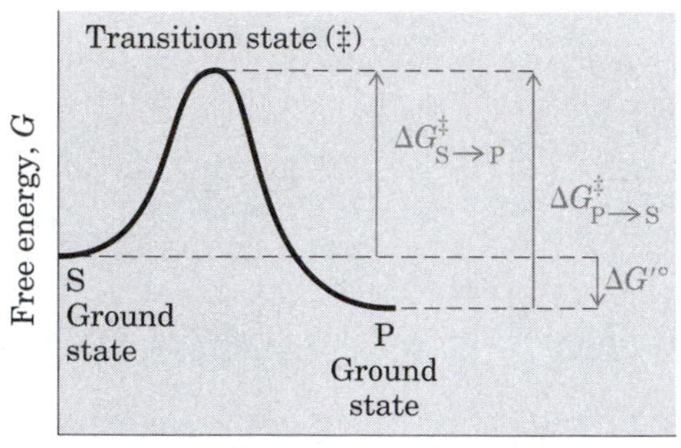

figure 8–2, page 247
Reaction coordinate diagram for a chemical reaction.

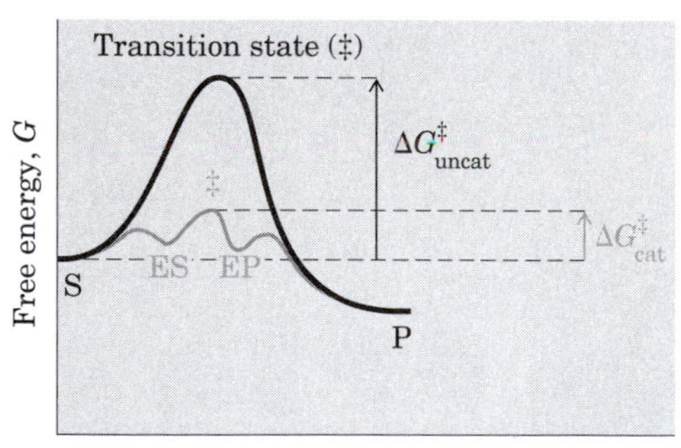

figure 8–3, page 248
Reaction coordinate diagram comparing enzyme-catalyzed and uncatalyzed reactions.

$$K'_{eq} = \frac{[P]}{[S]} \quad \text{(Equation 8–2)}$$

Equilibrium constant, page 249

$$\Delta G'^{\circ} = -RT \ln K'_{eq} \quad \text{(Equation 8–3)}$$

Relationship between $\Delta G'^{\circ}$ and K'_{eq}, page 249

table 8–3, page 246

International Classification of Enzymes*

No.	Class	Type of reaction catalyzed
1	Oxidoreductases	Transfer of electrons (hydride ions or H atoms)
2	Transferases	Group-transfer reactions
3	Hydrolases	Hydrolysis reactions (transfer of functional groups to water)
4	Lyases	Addition of groups to double bonds, or formation of double bonds by removal of groups
5	Isomerases	Transfer of groups within molecules to yield isomeric forms
6	Ligases	Formation of C—C, C—S, C—O, and C—N bonds by condensation reactions coupled to ATP cleavage

*Most enzymes catalyze the transfer of electrons, atoms, or functional groups. They are therefore classified, given code numbers, and assigned names according to the type of transfer reaction, the group donor, and the group acceptor.

table 8–4, page 249

Relationship between K'_{eq} and $\Delta G'^{\circ}$ (see Eqn 8–3)

K'_{eq}	$\Delta G'^{\circ}$ (kJ/mol)
10^{-6}	34.2
10^{-5}	28.5
10^{-4}	22.8
10^{-3}	17.1
10^{-2}	11.4
10^{-1}	5.7
1	0.0
10^{1}	−5.7
10^{2}	−11.4
10^{3}	−17.1

$$V = k[\mathrm{S}] \qquad \text{(Equation 8–4)}$$

Rate equation, page 250

$$V = k[\mathrm{S}_1][\mathrm{S}_2] \qquad \text{(Equation 8–5)}$$

Second-order rate equation, page 250

table 8–5, page 250

Some Rate Enhancements Produced by Enzymes

Enzyme	Rate enhancement
Cyclophilin	10^{5}
Carbonic anhydrase	10^{7}
Triose phosphate isomerase	10^{9}
Carboxypeptidase A	10^{11}
Phosphoglucomutase	10^{12}
Succinyl-CoA transferase	10^{13}
Urease	10^{14}
Orotidine monophosphate decarboxylase	10^{17}

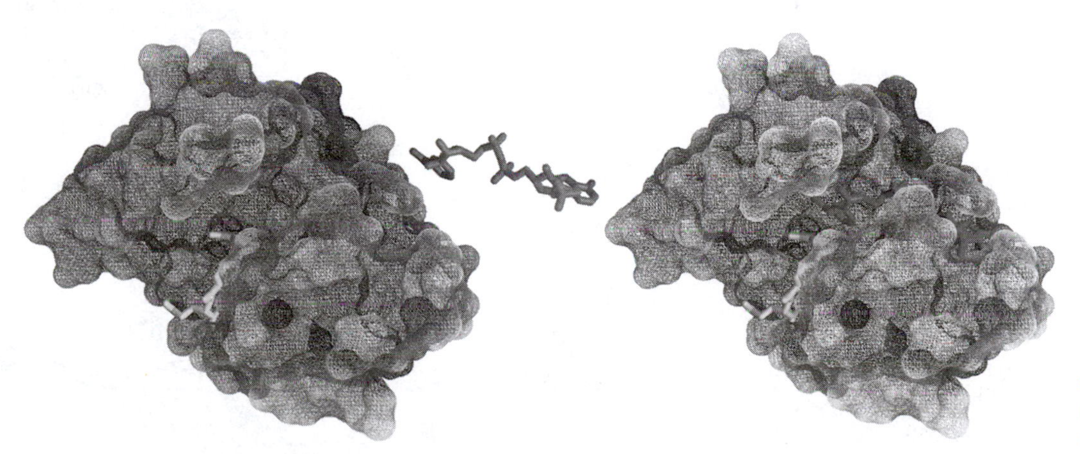

figure 8–4, page 251
Complementary shapes of a substrate and its binding site on an enzyme.

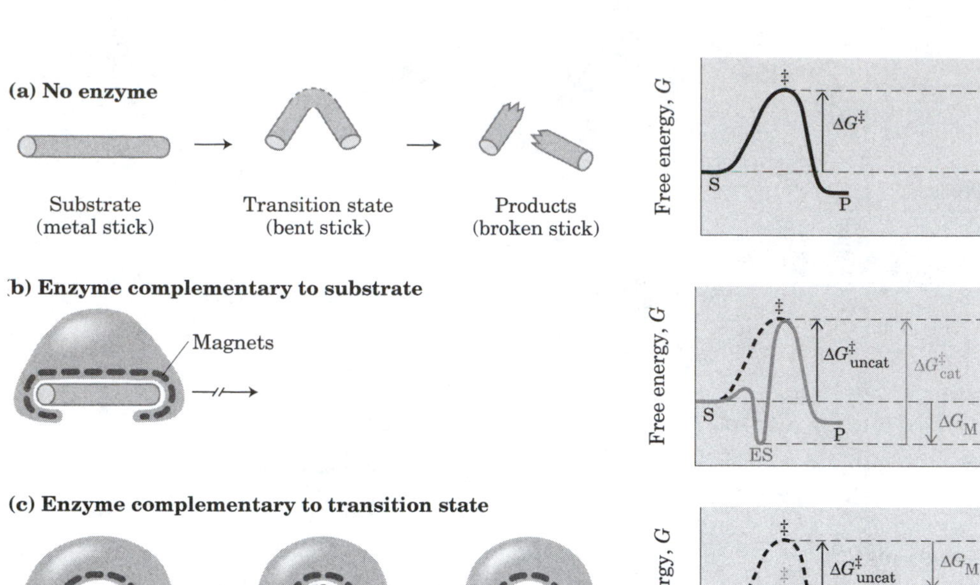

(c) Enzyme complementary to transition state

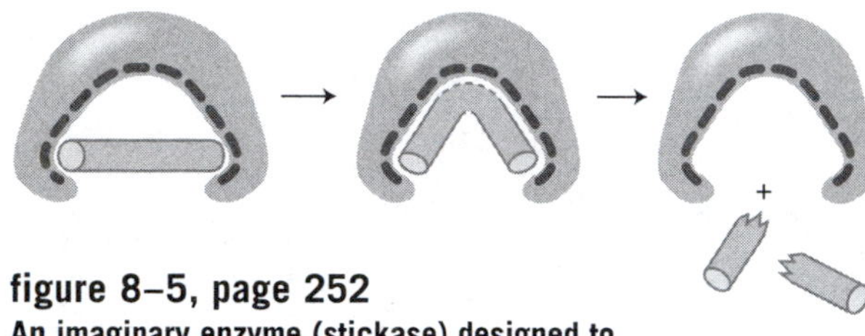

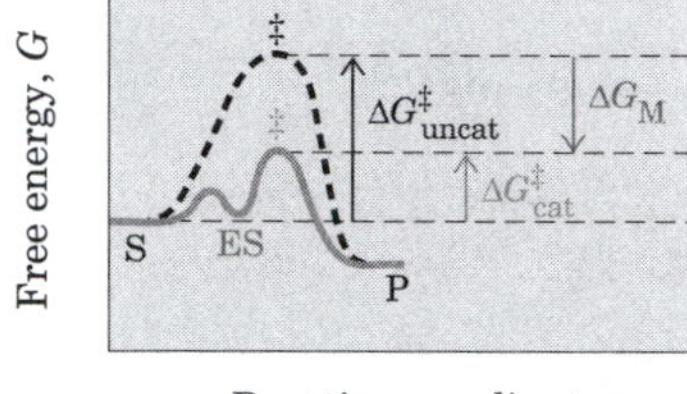

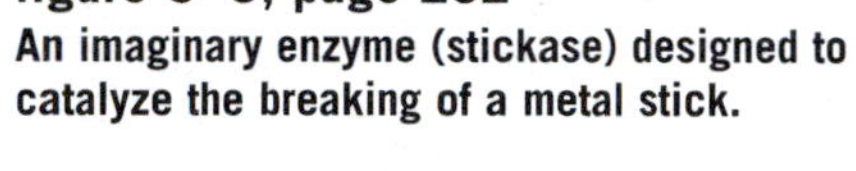

figure 8–5, page 252
An imaginary enzyme (stickase) designed to catalyze the breaking of a metal stick.

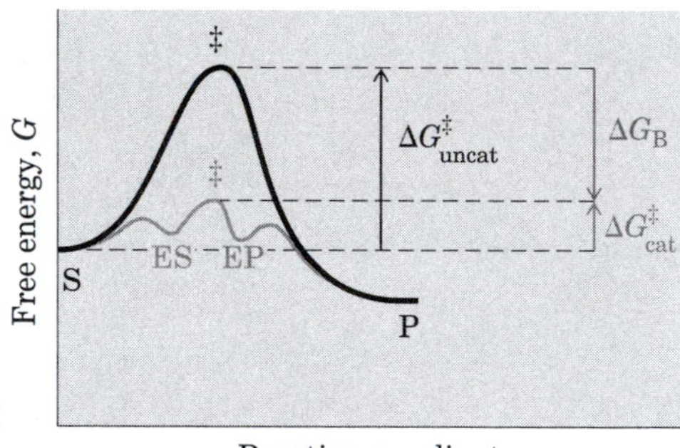

figure 8–6, page 253
Role of binding energy in catalysis.

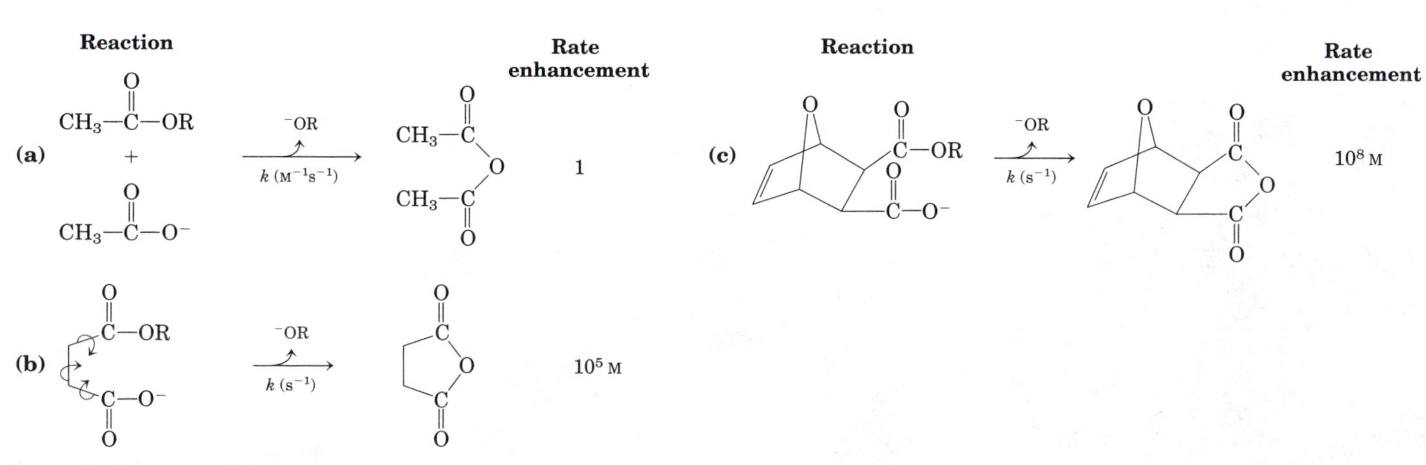

figure 8–7, page 254
Rate enhancement by entropy reduction.

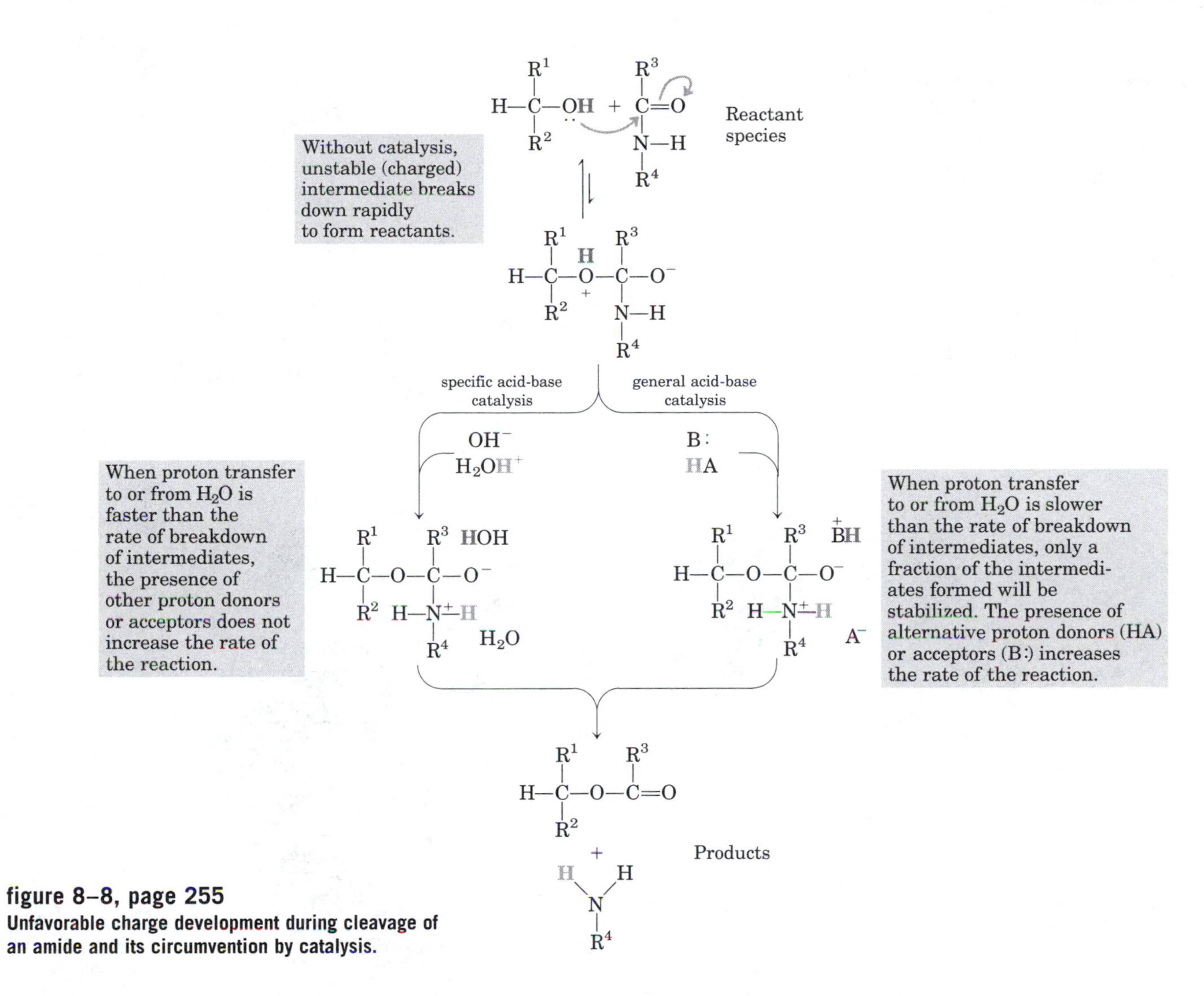

figure 8–8, page 255
Unfavorable charge development during cleavage of an amide and its circumvention by catalysis.

Amino acid residues	General acid form (proton donor)	General base form (proton acceptor)
Glu, Asp	R—COOH	$R—COO^-$
Lys, Arg	$R—\overset{+}{N}H_3$	$R—\ddot{N}H_2$
Cys	R—SH	$R—S^-$
His	imidazolium (R—C=CH, HN, $\overset{+}{N}H$, C—H)	imidazole (R—C=CH, HN, N:, C—H)
Ser	R—OH	$R—O^-$
Tyr	R—(phenyl)—OH	R—(phenyl)—O^-

figure 8–9, page 256
Amino acids in general acid-base catalysis.

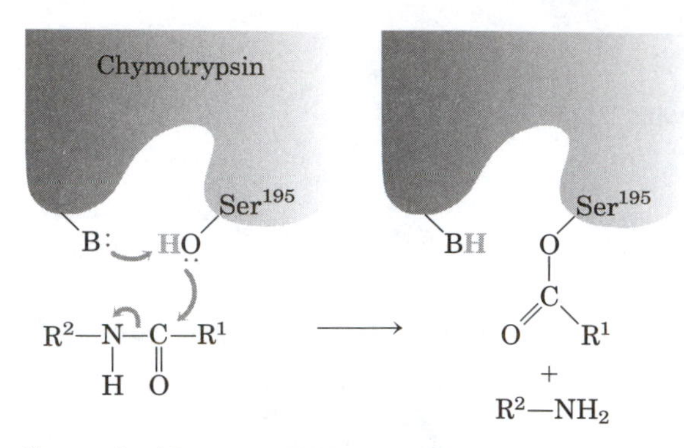

figure 8–10, page 257
Covalent and general acid-base catalysis.

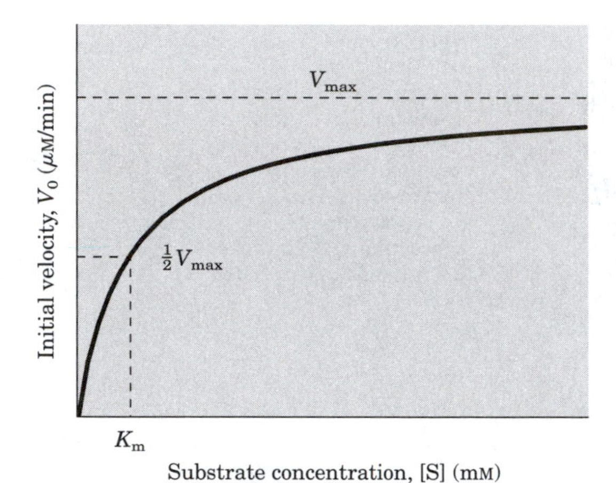

figure 8–11, page 258
Effect of substrate concentration on the initial velocity of an enzyme-catalyzed reaction.

$$V_0 = \frac{V_{max}[S]}{K_m + [S]} \qquad \text{(Equation 8–9)}$$

Michaelis-Menten equation, page 259

$$V_0 = \frac{V_{max}[S]}{K_m + [S]}$$

$$\frac{1}{V_0} = \frac{K_m + [S]}{V_{max}[S]}$$

$$\frac{1}{V_0} = \frac{K_m}{V_{max}[S]} + \frac{[S]}{V_{max}[S]}$$

$$\frac{1}{V_0} = \frac{K_m}{V_{max}[S]} + \frac{1}{V_{max}}$$

Lineweaver-Burk equation, page 261

figure 8–12, page 260
Dependence of initial velocity on substrate concentration.

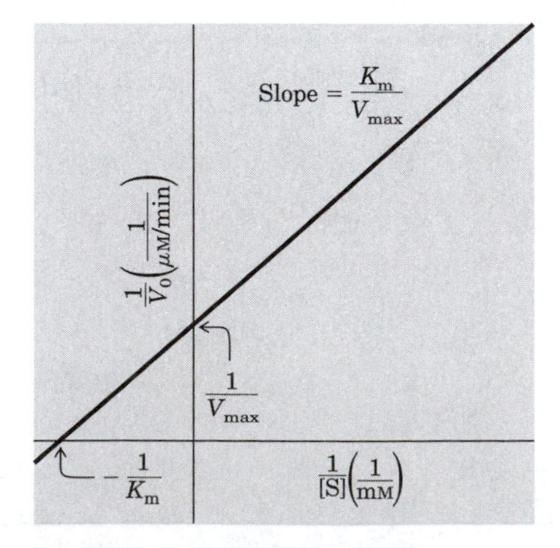

box 8–1, figure 1, page 261
A double-reciprocal or Lineweaver-Burk plot.

table 8–6, page 262

K_m for Some Enzymes and Substrates

Enzyme	Substrate	K_m (mM)
Catalase	H_2O_2	25
Hexokinase (brain)	ATP	0.4
	D-Glucose	0.05
	D-Fructose	1.5
Carbonic anhydrase	HCO_3^-	26
Chymotrypsin	Glycyltyrosinylglycine	108
	N-Benzoyltyrosinamide	2.5
β-Galactosidase	D-Lactose	4.0
Threonine dehydratase	L-Threonine	5.0

table 8–7, page 263

Turnover Numbers (k_{cat}) of Some Enzymes

Enzyme	Substrate	k_{cat} (s^{-1})
Catalase	H_2O_2	40,000,000
Carbonic anhydrase	HCO_3^-	400,000
Acetylcholinesterase	Acetylcholine	140,000
β-Lactamase	Benzylpenicillin	2,000
Fumarase	Fumarate	800
RecA protein (an ATPase)	ATP	0.4

table 8–8, page 264

Enzymes for Which k_{cat}/K_m Is Close to the Diffusion-Controlled Limit (10^8 to 10^9 $M^{-1}s^{-1}$)

Enzyme	Substrate	k_{cat} (s^{-1})	K_m (M)	k_{cat}/K_m ($M^{-1}s^{-1}$)
Acetylcholinesterase	Acetylcholine	1.4×10^4	9×10^{-5}	1.6×10^8
Carbonic anhydrase	CO_2	1×10^6	1.2×10^{-2}	8.3×10^7
	HCO_3^-	4×10^5	2.6×10^{-2}	1.5×10^7
Catalase	H_2O_2	4×10^7	1.1	4×10^7
Crotonase	Crotonyl-CoA	5.7×10^3	2×10^{-5}	2.8×10^8
Fumarase	Fumarate	8×10^2	5×10^{-6}	1.6×10^8
	Malate	9×10^2	2.5×10^{-5}	3.6×10^7
β-Lactamase	Benzylpenicillin	2.0×10^3	2×10^{-5}	1×10^8
Triose phosphate isomerase	Glyceraldehyde 3-phosphate	4.3×10^3	4.7×10^{-4}	2.4×10^8

Source: Fersht, A. (1999) *Structure and Mechanism in Protein Science*, p. 166, W.H. Freeman and Company, New York.

(a) Enzyme reaction involving a ternary complex

Random order

$$E \rightleftharpoons ES_1 \rightleftharpoons ES_1S_2, \quad E \rightleftharpoons ES_2 \rightleftharpoons ES_1S_2, \quad ES_1S_2 \longrightarrow E + P_1 + P_2$$

Ordered

$$E + S_1 \rightleftharpoons ES_1 \overset{S_2}{\rightleftharpoons} ES_1S_2 \longrightarrow E + P_1 + P_2$$

(b) Enzyme reaction in which no ternary complex is formed

$$E + S_1 \rightleftharpoons ES_1 \rightleftharpoons E'P_1 \overset{P_1}{\rightleftharpoons} E' \overset{S_2}{\rightleftharpoons} E'S_2 \longrightarrow E + P_2$$

figure 8–13, page 265
Common mechanisms for enzyme-catalyzed bisubstrate reactions.

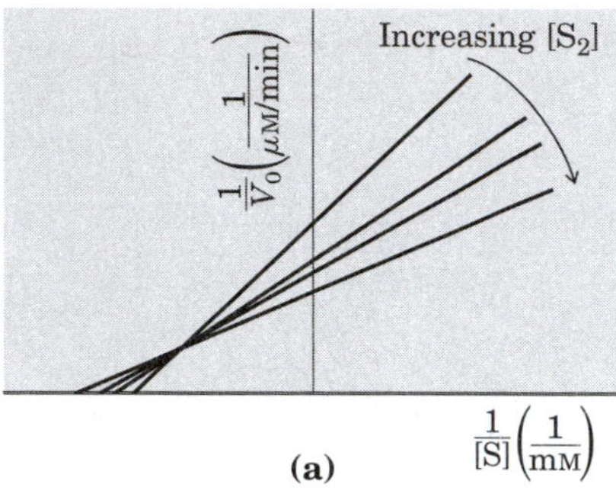

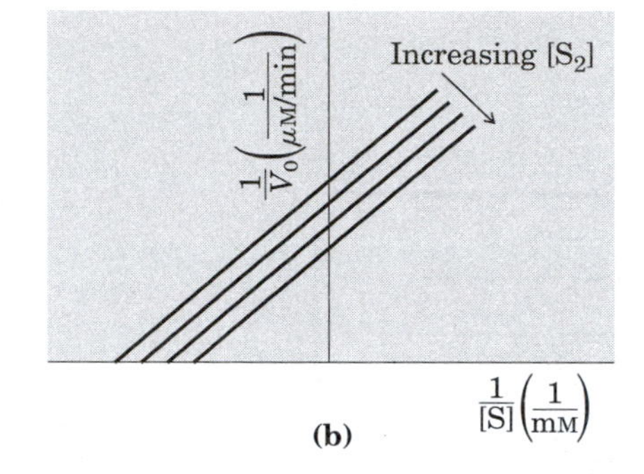

figure 8–14, page 265
Steady-state kinetic analysis of bisubstrate reactions.

$$E + S \rightleftharpoons ES \longrightarrow E + P$$
$$+ \; I \overset{K_I}{\rightleftharpoons} EI$$

S, S, I, I

(a) Competitive inhibition

$$E + S \rightleftharpoons ES \longrightarrow E + P$$
$$ES + I \overset{K_I'}{\rightleftharpoons} ESI$$

S, S, I, I, S

(b) Uncompetitive inhibition

$$E + S \rightleftharpoons ES \longrightarrow E + P$$
$$E + I \overset{K_I}{\rightleftharpoons} EI, \quad ES + I \overset{K_I'}{\rightleftharpoons} ESI$$
$$EI + S \rightleftharpoons ESI$$

S, S, I, I, S, I, S, I

(c) Mixed inhibition

figure 8–15, page 266
Three types of reversible inhibition.

$$V_0 = \frac{V_{max}[S]}{\alpha K_m + [S]} \qquad \text{(Equation 8 – 28)}$$

where $\alpha = 1 + \frac{[I]}{K_I}$

and $K_I = \frac{[E][I]}{[EI]}$

Rate equation for competitive inhibition, page 266

$$\frac{1}{V_0} = \left(\frac{\alpha K_m}{V_{max}}\right)\frac{1}{[S]} + \frac{1}{V_{max}}$$

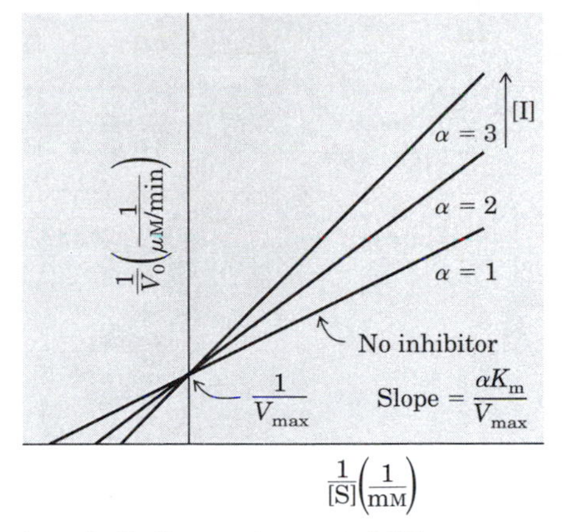

box 8–2, figure 1, page 267
Competitive inhibition.

$$\frac{1}{V_0} = \left(\frac{K_m}{V_{max}}\right)\frac{1}{[S]} + \frac{\alpha'}{V_{max}}$$

[I]
$\alpha' = 2$
$\alpha' = 1.5$
$\alpha' = 1$
$\frac{1}{V_0}\left(\frac{1}{\mu M/min}\right)$
$-\frac{1}{K_m}$
$\frac{1}{[S]}\left(\frac{1}{mM}\right)$

box 8–2, figure 2, page 267
Uncompetitive inhibition.

$$\frac{1}{V_0} = \left(\frac{\alpha K_m}{V_{max}}\right)\frac{1}{[S]} + \frac{\alpha'}{V_{max}}$$

[I]
$\frac{1}{V_0}\left(\frac{1}{\mu M/min}\right)$
No inhibitor
$\frac{1}{[S]}\left(\frac{1}{mM}\right)$

box 8–2, figure 3, page 267
Mixed inhibition.

$$V_0 = \frac{V_{max}[S]}{K_m + \alpha'[S]} \qquad \text{(Equation 8–29)}$$

where $\alpha' = 1 + \frac{[I]}{K'_I}$

and $K'_I = \frac{[ES][I]}{[ESI]}$

Rate equation for uncompetitive inhibition, page 268

$$V_0 = \frac{V_{max}[S]}{\alpha K_m + \alpha'[S]} \qquad \text{(Equation 8–30)}$$

Rate equation for mixed inhibition, page 268

$$\text{Enz}-CH_2-OH + F-\overset{\overset{O}{\|}}{P}(-O-CH(CH_3)_2)-O-CH(CH_3)_2 \xrightarrow{H^+ + F^-} \text{Enz}-CH_2-O-\overset{\overset{O}{\|}}{P}(-O-CH(CH_3)_2)-O-CH(CH_3)_2$$

(Ser[195])

DIFP

figure 8–16, page 269
Irreversible inhibition.

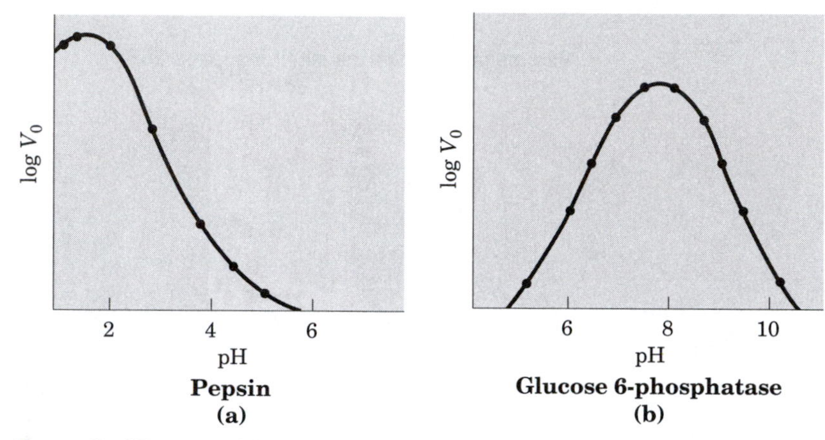

figure 8–17, page 269
The pH-activity profiles of two enzymes.

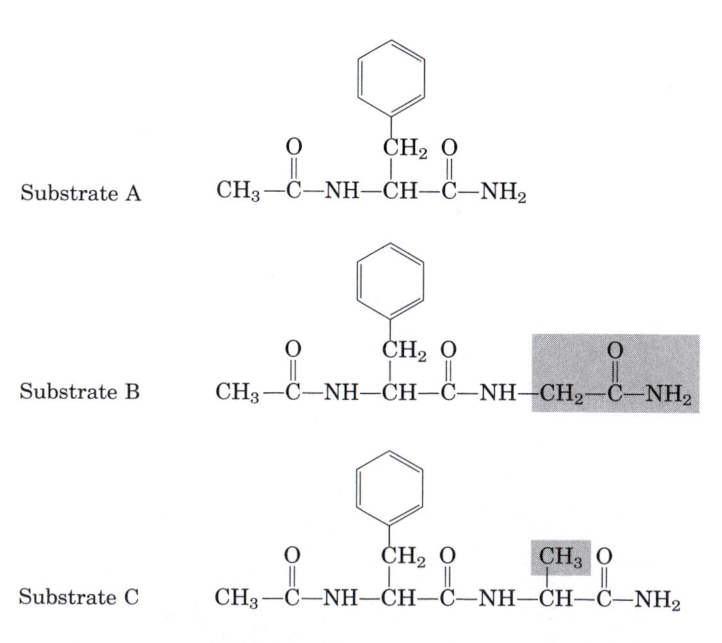

k_{cat} (s^{-1})	K_m (mM)	k_{cat}/K_m ($M^{-1}s^{-1}$)
0.06	31	2
0.14	15	10
2.8	25	114

box 8–3, figure 1, page 270
The pH-activity profiles of two enzymes.

Ester hydrolysis

R^1–C(=O)–O–R^2 + ^{-}OH → [R^1–C($^{\delta-}OH$)($O_{\delta-}$)–O–R^2]‡ →→ Products

Transition state

R^1–P($O^{\delta-}$)($O_{\delta-}$)–O–R^2

Analog (phosphonate)

Carbonate hydrolysis

H_3N^+–CH_2CH_2–O–C(=O)–O–C_6H_4–NO_2 + ^{-}OH → [H_3N^+–CH_2CH_2–O–C($^{\delta-}OH$)($O_{\delta-}$)–O–C_6H_4–NO_2]‡ →→ Products

Transition state

H_3N^+–CH_2CH_2–O–P($O^{\delta-}$)($O_{\delta-}$)–O–C_6H_4–NO_2

Analog (phosphate)

box 8–3, figure 2, page 272

The pH-activity profiles of two enzymes.

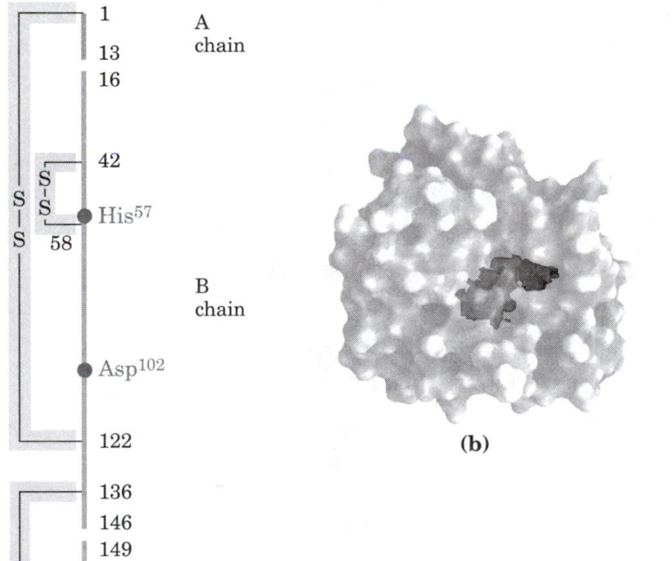

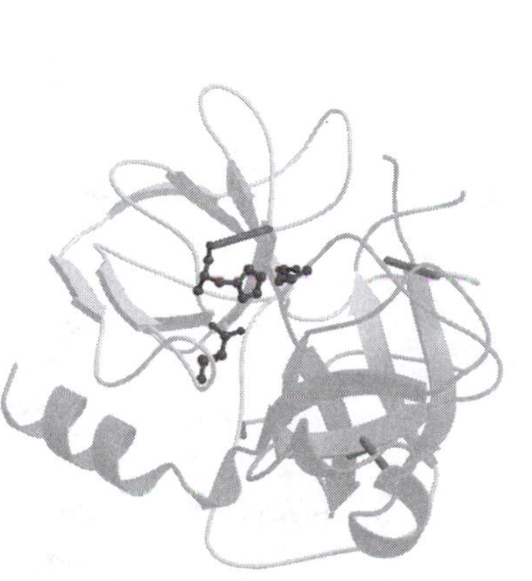

(c)

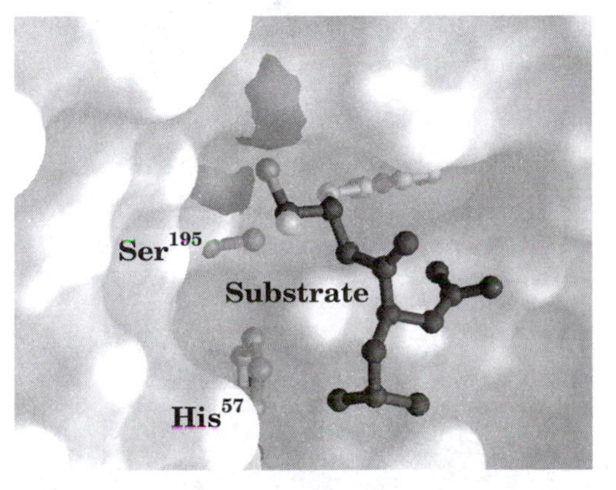

(d)

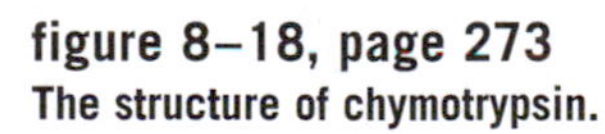

figure 8–18, page 273

The structure of chymotrypsin.

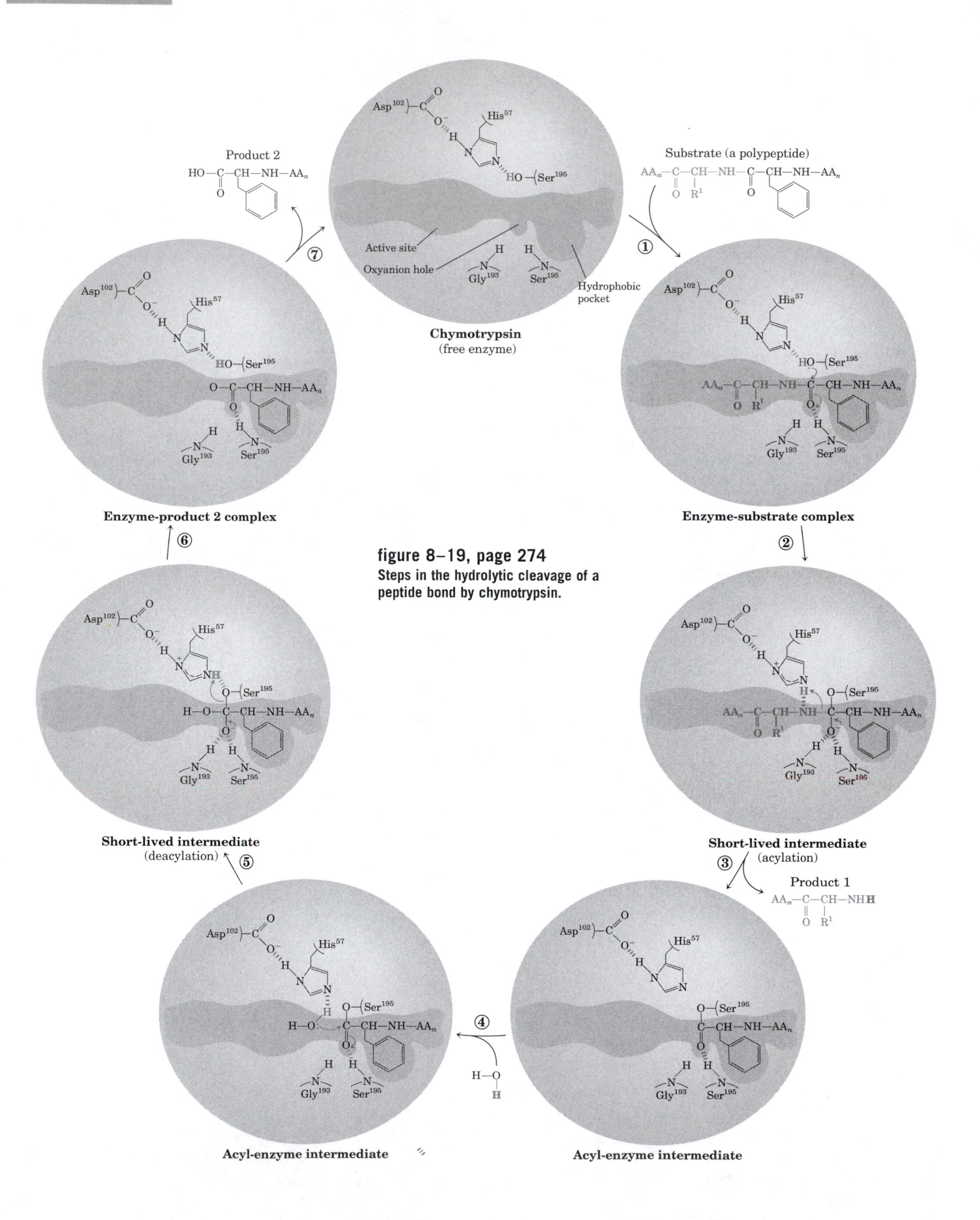

figure 8–19, page 274
Steps in the hydrolytic cleavage of a peptide bond by chymotrypsin.

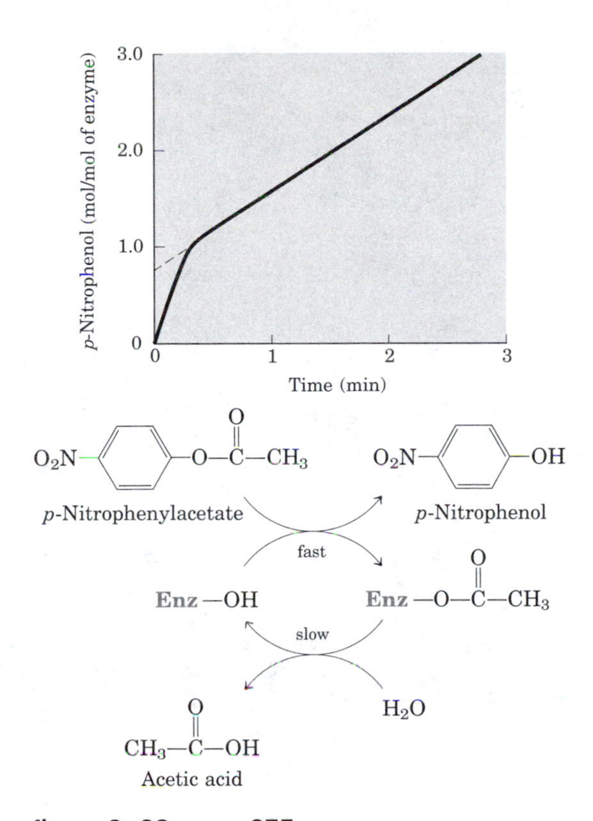

figure 8–20, page 275
Pre-steady state kinetic evidence for an acyl-enzyme intermediate.

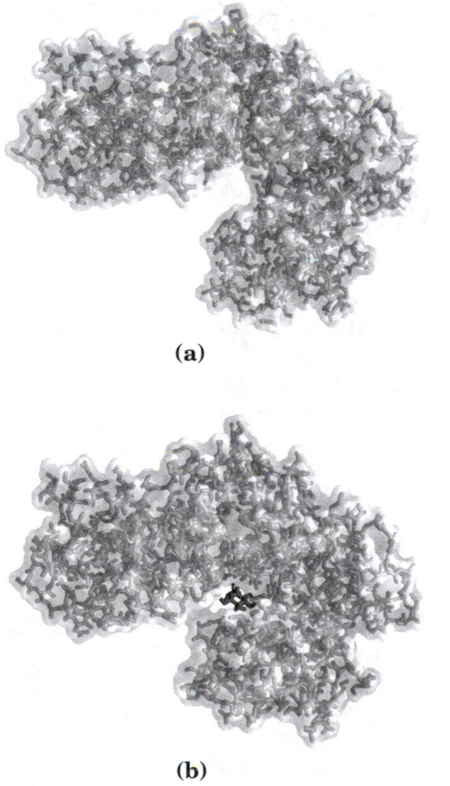

(a)

(b)

figure 8–21, page 276
Induced fit in hexokinase.

$$\text{Mg}\cdot\text{ATP} + \text{Glucose} \xrightleftharpoons{\text{hexokinase}} \text{Mg}\cdot\text{ADP} + \text{Glucose 6-phosphate}$$

Glucose (CH_2OH) — Glucose 6-phosphate ($CH_2OPO_3^{2-}$)

Hexokinase, page 275

Xylose: HC=O; H—C—OH; HO—C—H; H—C—OH; CH_2OH

Glucose: HC=O; H—C—OH; HO—C—H; H—C—OH; H—C—OH; CH_2OH

page 276

$$\text{2-Phosphoglycerate} \xrightleftharpoons{\text{enolase}} \text{Phosphoenolpyruvate} + H_2O$$

2-Phosphoglycerate: ^-O—C(=O)—CH(OPO_3^{2-})—CH_2—OH

Phosphoenolpyruvate: ^-O—C(=O)—C(OPO_3^{2-})=CH_2

Enolase, page 276

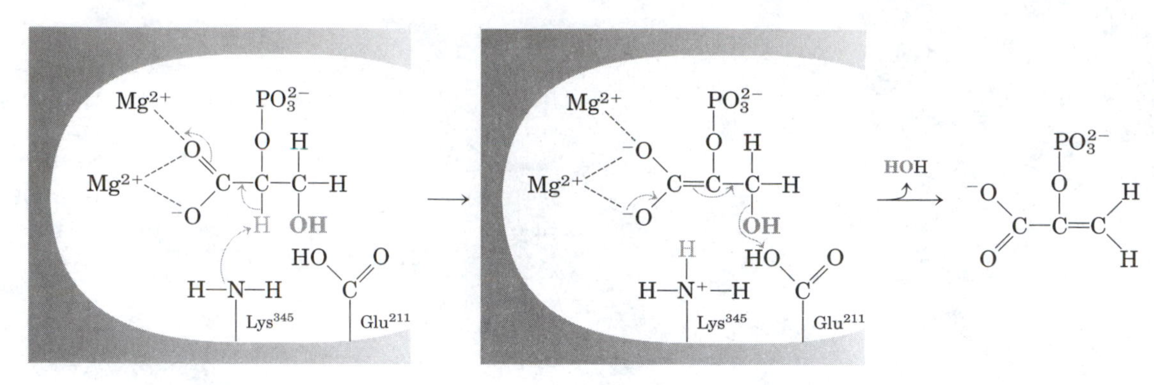

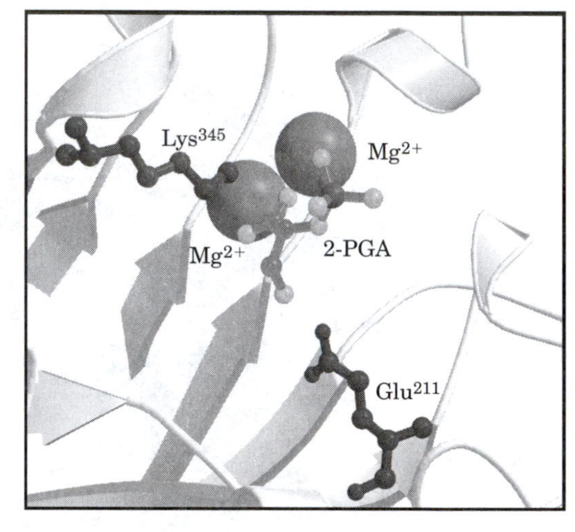

figure 8–22, page 277
The two-step reaction catalyzed by enolase.

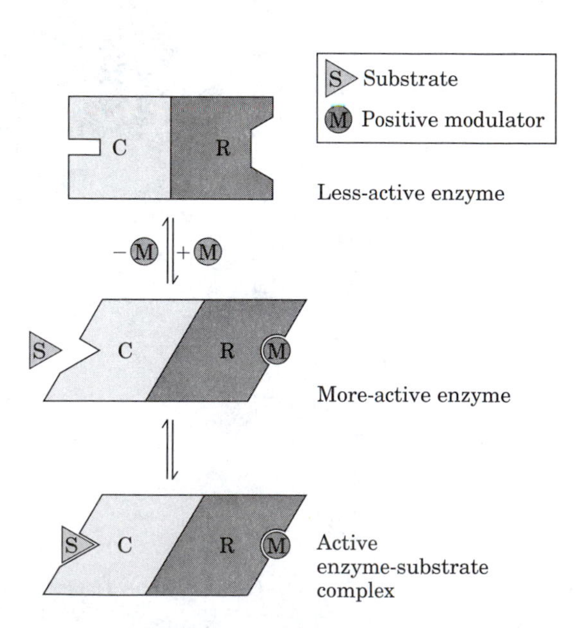

figure 8–23, page 279
Subunit interactions in an allosteric enzyme, and interactions with inhibitors and activators.

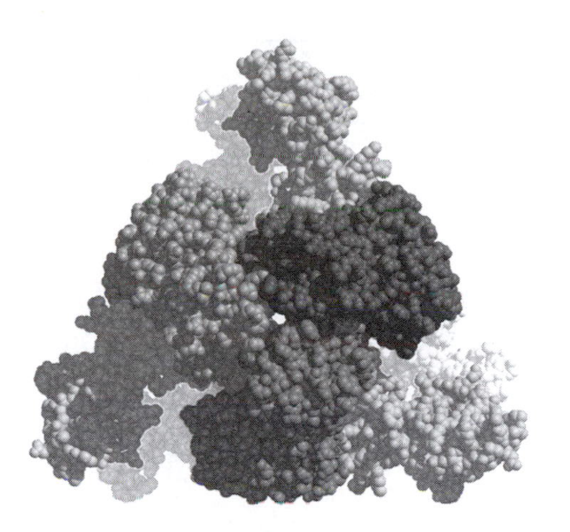

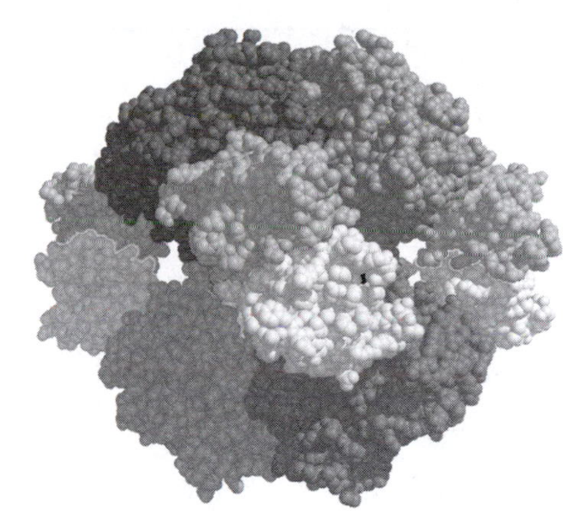

figure 8–24, page 279
Two views of the regulatory enzyme aspartate transcarbamoylase.

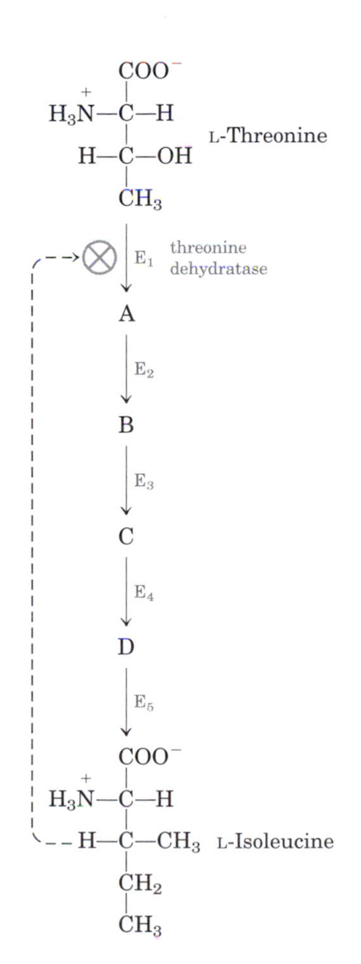

figure 8–25, page 280
Feedback inhibition.

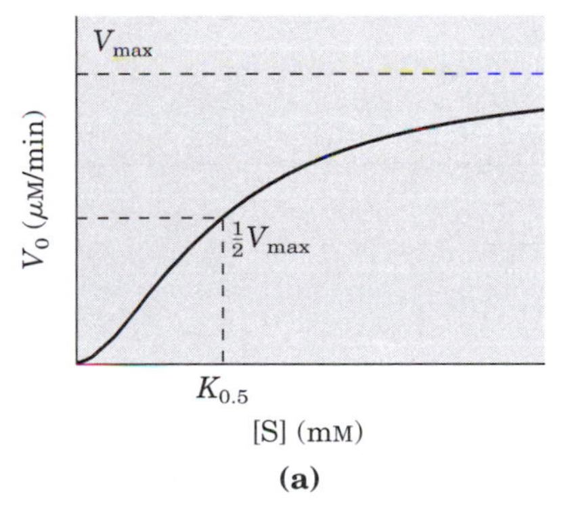

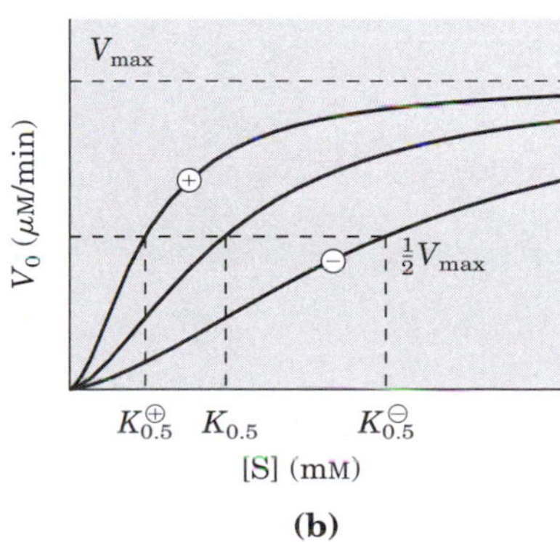

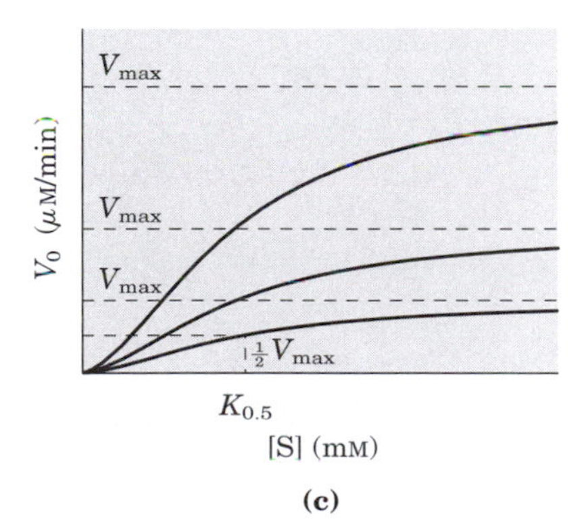

figure 8–26, page 281
Substrate-activity curves for representative allosteric enzymes.

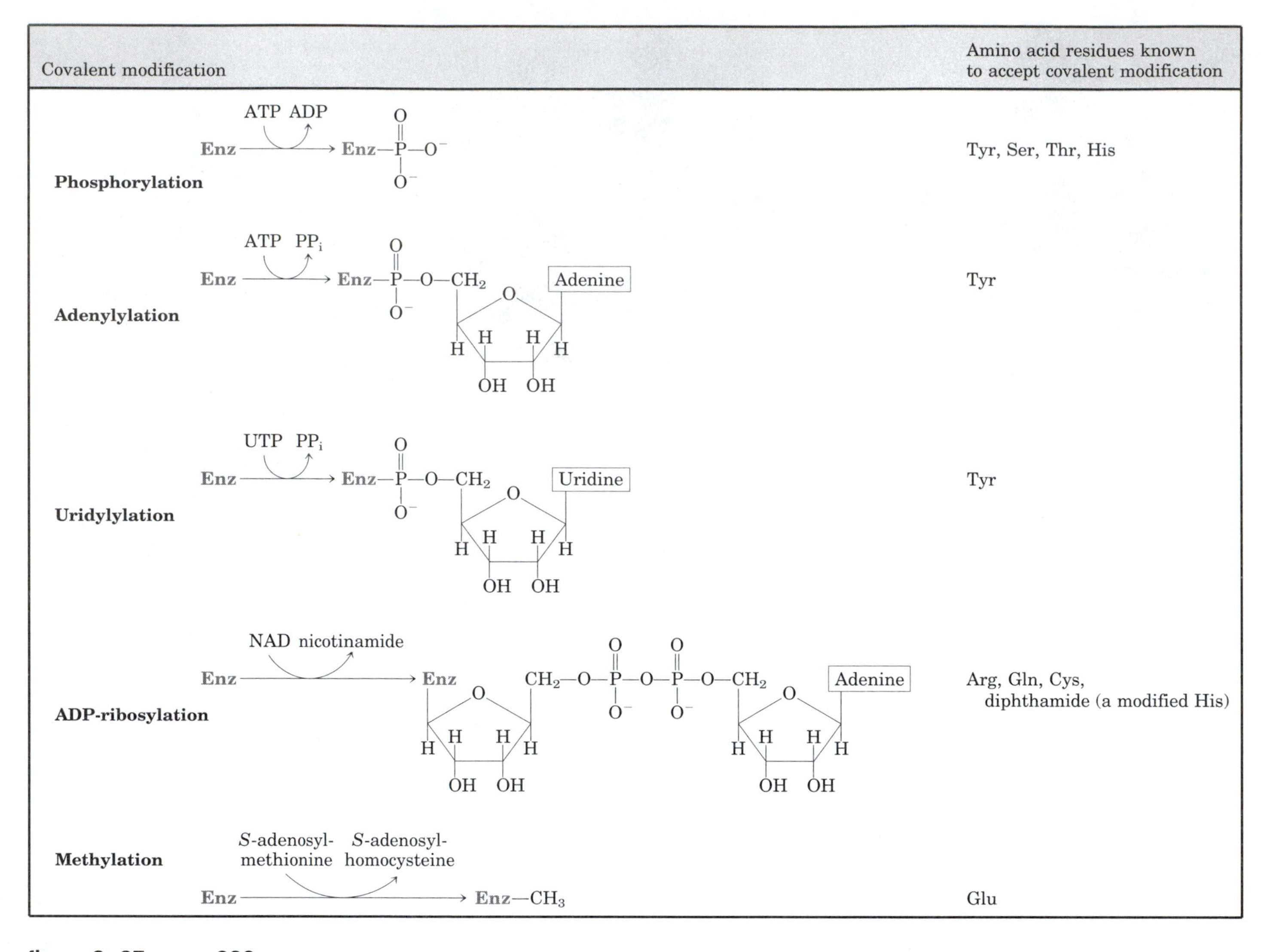

figure 8–27, page 282
Examples of enzyme modification reactions.

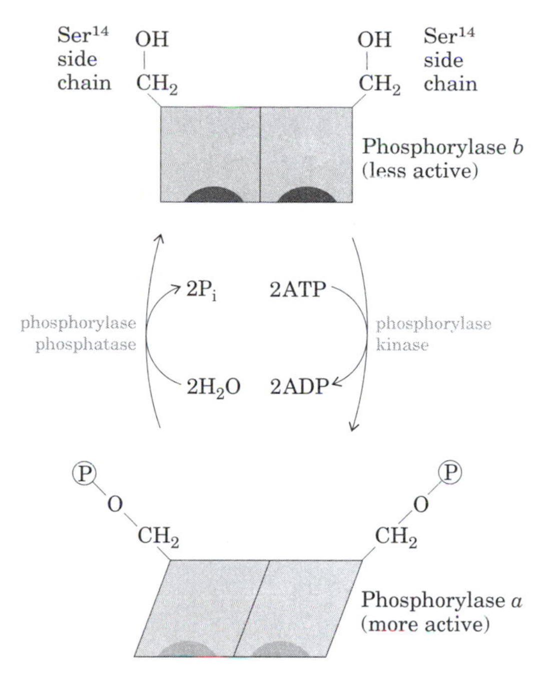

figure 8–28, page 283

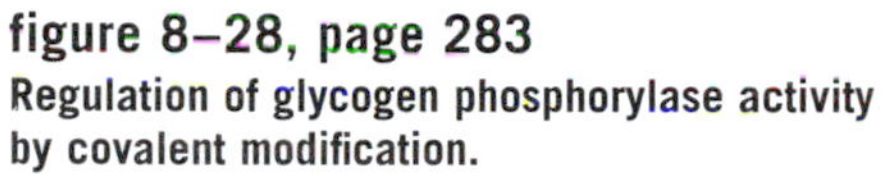

Regulation of glycogen phosphorylase activity by covalent modification.

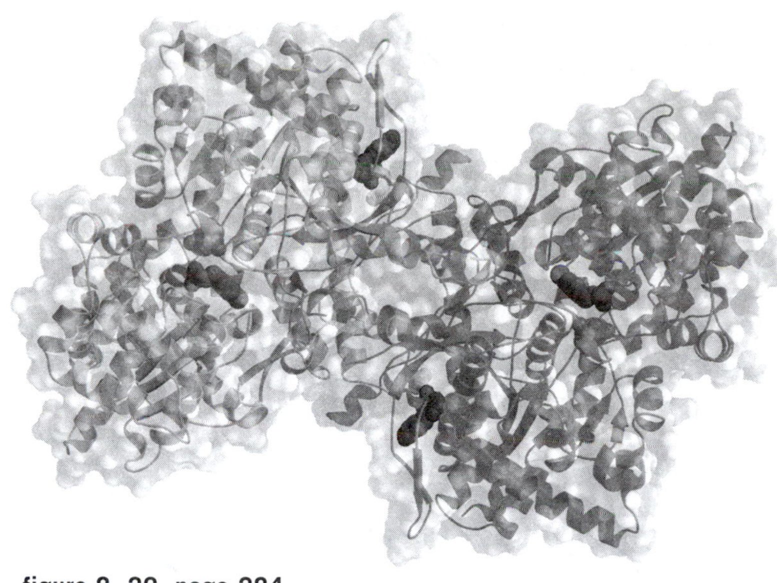

figure 8–29, page 284
Regulation of glycogen phosphorylase.

table 8–9, page 285

Consensus Sequences for Protein Kinases

Protein kinase	Consensus sequence and phosphorylated residue*
Protein kinase A	–X–R–(R/K)–X–(**S/T**)–B–
Protein kinase G	–X–R–(R/K)–X–(**S/T**)–X–
Protein kinase C	–(R/K)–(R/K)–X–(**S/T**)–B–(R/K)–(R/K)–
Protein kinase B	–X–R–X–(**S/T**)–X–K–
Ca^{2+}/calmodulin kinase I	–B–X–R–X–X–(**S/T**)–X–X–X–B–
Ca^{2+}/calmodulin kinase II	–B–X–(R/K)–X–X–(**S/T**)–X–X–
Myosin light chain kinase (smooth muscle)	–K–K–R–X–X–**S**–X–B–B–
Phosphorylase *b* kinase	*–K–R–K–Q–I–**S**–V–R–*
Extracellular signal–regulated kinase (ERK)	–P–X–(**S/T**)–P–P–
Cyclin-dependent protein kinase (cdc2)	–X–(**S/T**)–P–X–(K/R)–
Casein kinase I	–(Sp/Tp)–X–X–(X)–(**S/T**)–B
Casein kinase II	–X–(**S/T**)–X–X–(E/D/Sp/Yp)–X–
β-Adrenergic receptor kinase	–(D/E)$_n$–(**S/T**)–X–X–X–
Rhodopsin kinase	–X–X–(**S/T**)–(E)$_n$–
Insulin receptor kinase	–X–E–E–E–**Y**–M–M–M–M–*K–K–S–R–G–D–**Y**–M–T–M–Q–I–G–K–K–K–L–P–A–T–G–D–**Y**–M–N–M–S–P–V–G–D–*
Epidermal growth factor (EGF) receptor kinase	*–E–E–E–E–**Y**–F–E–L–V–*

Sources: Pinna, L.A. & Ruzzene, M.H. (1996) How do protein kinases recognize their substrates? *Biochim. Biophys. Acta* **1314,** 191–225. Kemp, B.E. & Pearson, R.B. (1990) Protein kinase recognition sequence motifs. *Trends Biochem. Sci.* **15,** 342–346. Kennelly, P.J. & Krebs, E.G. (1991) Consensus sequences as substrate specificity determinants for protein kinases and protein phosphatases. *J. Biol. Chem.* **266,** 15,555–15,558.

*Shown here are deduced consensus sequences (in roman type) and actual sequences from known substrates (italic). The Ser (S), Thr (T), or Tyr (Y) residue to undergo phosphorylation is in red; all amino acid residues are shown as their one-letter abbreviations (see Table 5–1). X represents any amino acid; B, any hydrophobic amino acid; Sp, Tp, and Yp, already phosphorylated Ser, Thr, and Tyr residues.

Kinase	Glycogen synthase sites phosphorylated	Degree of synthase inactivation
Protein kinase A	1A, 1B, 2, 4	+
Protein kinase G	1A, 1B, 2	+
Protein kinase C	1A	+
Ca^{2+}/calmodulin kinase	1B, 2	+
Phosphorylase *b* kinase	2	+
Casein kinase I	At least 9 sites	+ + + +
Casein kinase II	5	0
Glycogen synthase kinase 3	3A, 3B, 3C	+ + +
Glycogen synthase kinase 4	2	+

figure 8–30, page 286
Multiple regulatory phosphorylations.

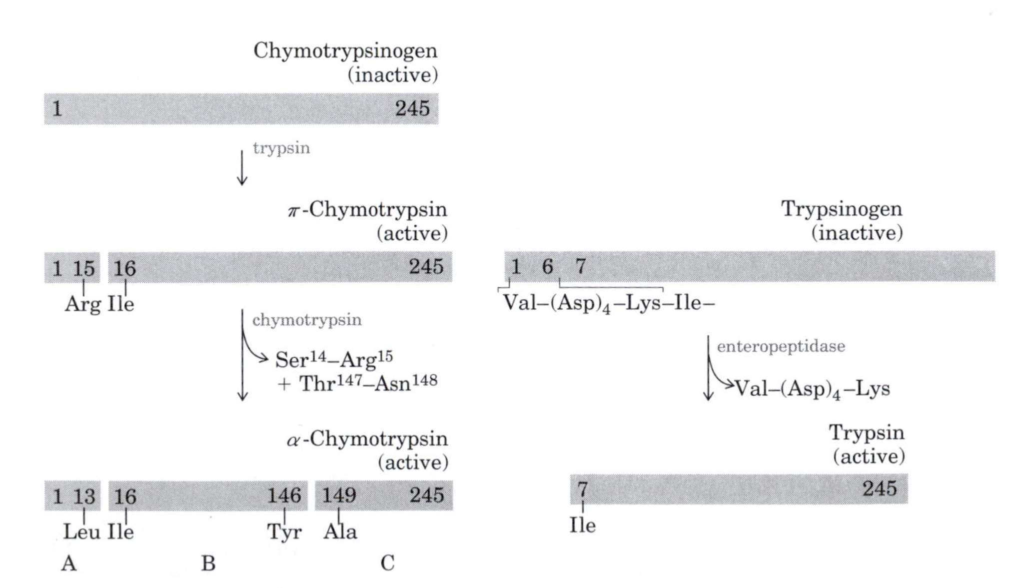

figure 8–31, page 287
Activation of zymogens by proteolytic cleavage.

PROBLEMS

6. Quantitative Assay for Lactate Dehydrogenase, page 290.

$$CH_3{-}\overset{O}{\overset{\|}{C}}{-}COO^- + NADH + H^+ \longrightarrow CH_3{-}\underset{H}{\underset{|}{\overset{OH}{\overset{|}{C}}}}{-}COO^- + NAD^+$$

Pyruvate → Lactate

8. Estimation of V_{max} and K_m by Inspection, page 291.

[S] (M)	V_0 (μM/min)
2.5×10^{-6}	28
4.0×10^{-6}	40
1×10^{-5}	70
2×10^{-5}	95
4×10^{-5}	112
1×10^{-4}	128
2×10^{-3}	139
1×10^{-2}	140

9. Properties of an Enzyme of Prostaglandin Synthesis, page 291.

[Arachidonic acid] (mM)	Rate of formation of PGG_2 (mM/min)	Rate of formation of PGG_2 with 10 mg/mL ibuprofen (mM/min)
0.5	23.5	16.67
1.0	32.2	25.25
1.5	36.9	30.49
2.5	41.8	37.04
3.5	44.0	38.91

10. Graphical Analysis of V_{max} and K_m, page 291.

Glycylglycine + H_2O $\longrightarrow$ 2 glycine

[S] (mM)	Product formed (μmol/min)
1.5	0.21
2.0	0.24
3.0	0.28
4.0	0.33
8.0	0.40
16.0	0.45

11. The Eadie-Hofstee Equation, page 291.

$$V_0 = (-K_m)\frac{V_0}{[S]} + V_{max}$$

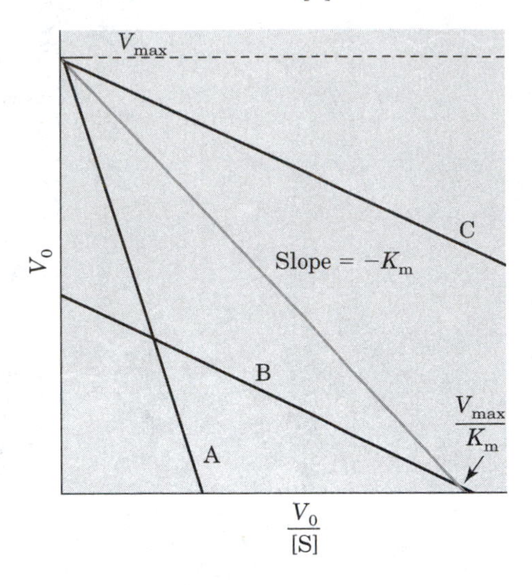

16. Inhibition of Carbonic Anhydrase by Acetazolamide, page 292.

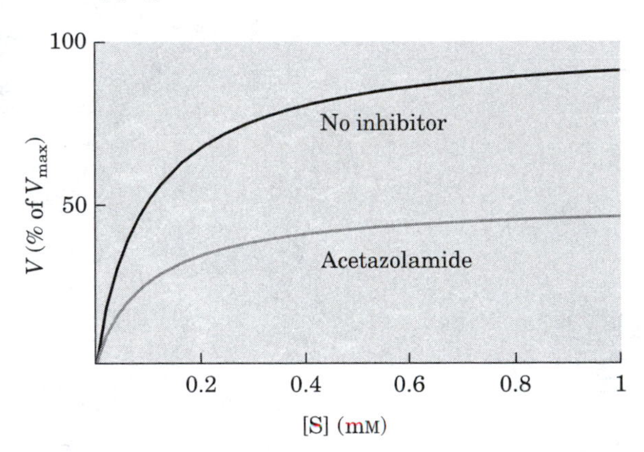

17. pH Optimum of Lysozyme, page 292.

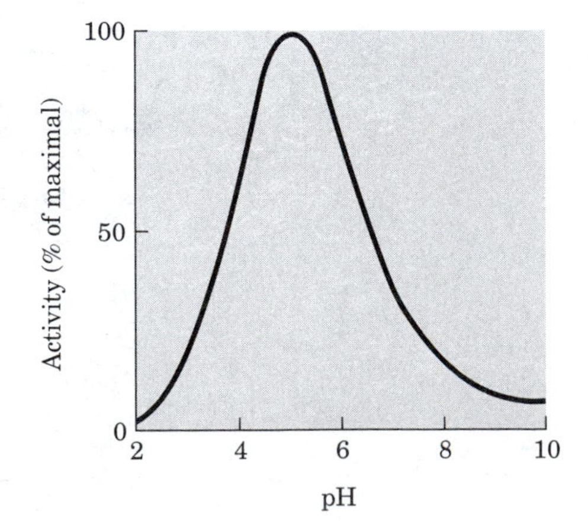

chapter

9

Carbohydrates and Glycobiology

When you study for exams, don't forget to review The Minicourses on the UNDERSTAND! BIOCHEMISTRY CD that came with your textbook.

Minicourses that apply to this Chapter include:

Background
- Chemical Fundamentals

Molecules of Life
- Important Biochemicals
- An Interactive Gallery of Molecular Models
- An Interactive Gallery of Protein Structures

Proteins in Action
- Catalysis and Regulation

Bioenergetics
- Glycogen

Cellular Architecture and Traffic
- The Extracellular Matrix

Glyceraldehyde, an aldotriose — Dihydroxyacetone, a ketotriose

(a)

D-Glucose, an aldohexose — D-Fructose, a ketohexose

(b)

D-Ribose, an aldopentose — 2-Deoxy-D-ribose, an aldopentose

(c)

figure 9–1, page 294

Representative monosaccharides.

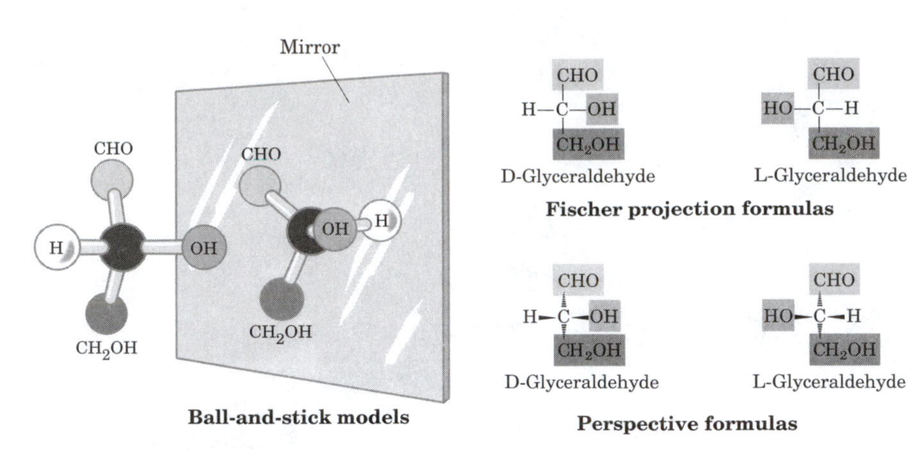

figure 9–2, page 295
Three ways to represent the two stereoisomers of glyceraldehyde.

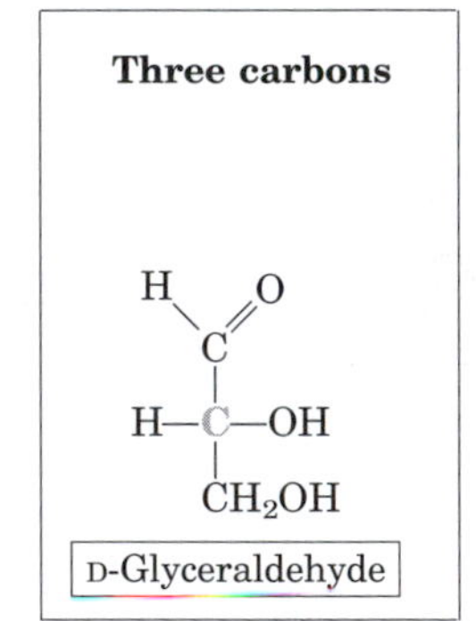

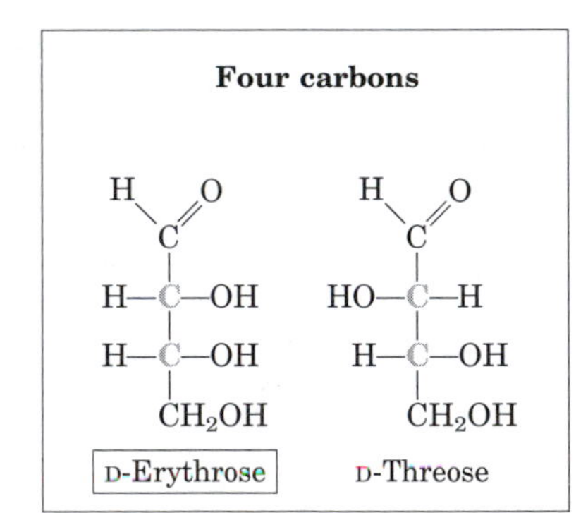

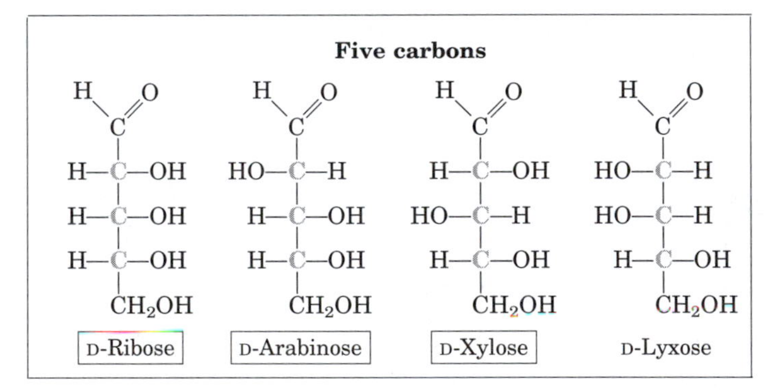

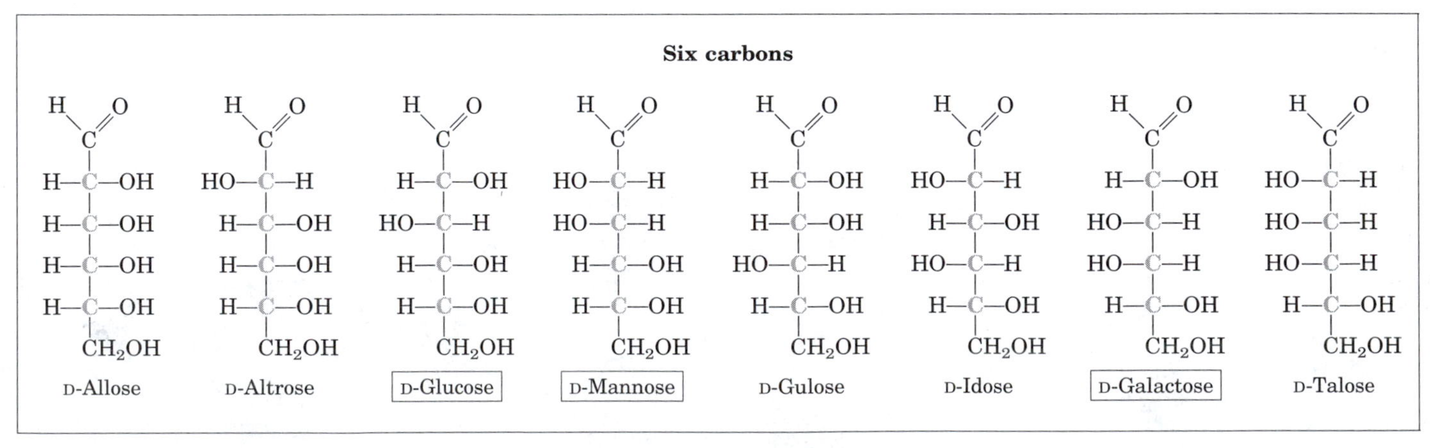

D-Aldoses
(a)

figure 9–3 (a), page 296
The series of (a) D-aldoses and (b) D-ketoses having from three to six carbon atoms, shown as projection formulas.

Three carbons

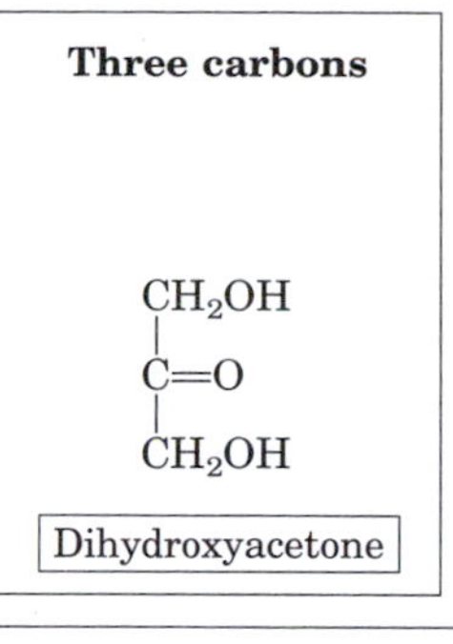

Four carbons

CH_2OH
C=O
H—C—OH
CH_2OH

D-Erythrulose

Five carbons

D-Ribulose	D-Xylulose
CH_2OH	CH_2OH
C=O	C=O
H—C—OH	HO—C—H
H—C—OH	H—C—OH
CH_2OH	CH_2OH

Six carbons

D-Psicose	D-Fructose	D-Sorbose	D-Tagatose
CH_2OH	CH_2OH	CH_2OH	CH_2OH
C=O	C=O	C=O	C=O
H—C—OH	HO—C—H	H—C—OH	HO—C—H
H—C—OH	H—C—OH	HO—C—H	HO—C—H
H—C—OH	H—C—OH	H—C—OH	H—C—OH
CH_2OH	CH_2OH	CH_2OH	CH_2OH

D-Ketoses

(b)

figure 9–3 (b), page 296
The series of (a) D-aldoses and (b) D-ketoses having from three to six carbon atoms, shown as projection formulas.

D-Mannose (epimer at C-2)	D-Glucose	D-Galactose (epimer at C-4)
¹CHO	¹CHO	¹CHO
HO—²C—H	H—²C—OH	H—²C—OH
HO—³C—H	HO—³C—H	HO—³C—H
H—⁴C—OH	H—⁴C—OH	HO—⁴C—H
H—⁵C—OH	H—⁵C—OH	H—⁵C—OH
⁶CH_2OH	⁶CH_2OH	⁶CH_2OH

figure 9–4, page 297
Epimers.

H O
C
H—C—OH
HO—C—H
HO—C—H
CH_2OH

L-Arabinose, page 297

$$R^1{-}C(=O){-}H + HO{-}R^2 \rightleftharpoons R^1{-}C(OH)(H){-}OR^2 \xrightleftharpoons[HO-R^3]{HO-R^3} R^1{-}C(OR^3)(H){-}OR^2 + H_2O$$

Aldehyde Alcohol Hemiacetal Acetal

$$R^1{-}C(=O){-}R^2 + HO{-}R^3 \rightleftharpoons R^1{-}C(OH)(R^2){-}OR^3 \xrightleftharpoons[HO-R^4]{HO-R^4} R^1{-}C(OR^4)(R^2){-}OR^3 + H_2O$$

Ketone Alcohol Hemiketal Ketal

figure 9–5, page 297
Formation of hemiacetals and hemiketals.

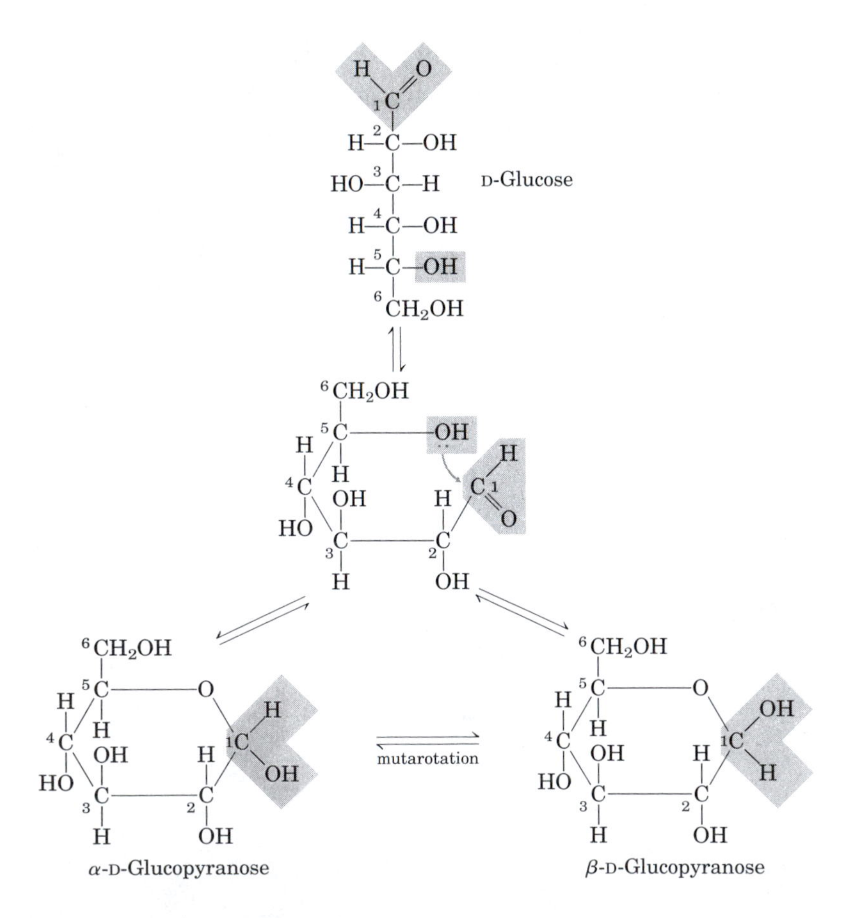

figure 9–6, page 298
Formation of the two cyclic forms of D-glucose.

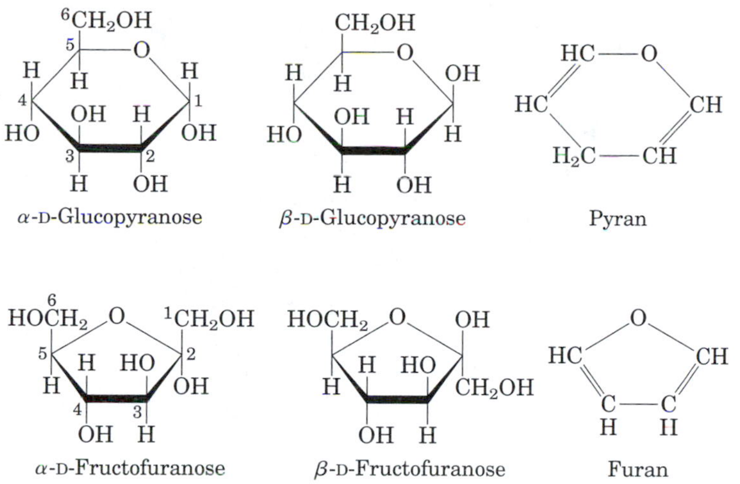

figure 9–7, page 298
Pyranoses and furanoses.

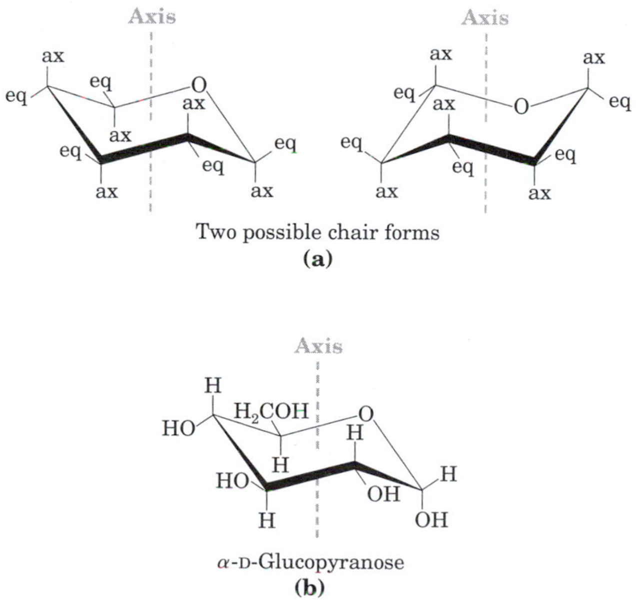

figure 9–8, page 299
Conformational formulas of pyranoses.

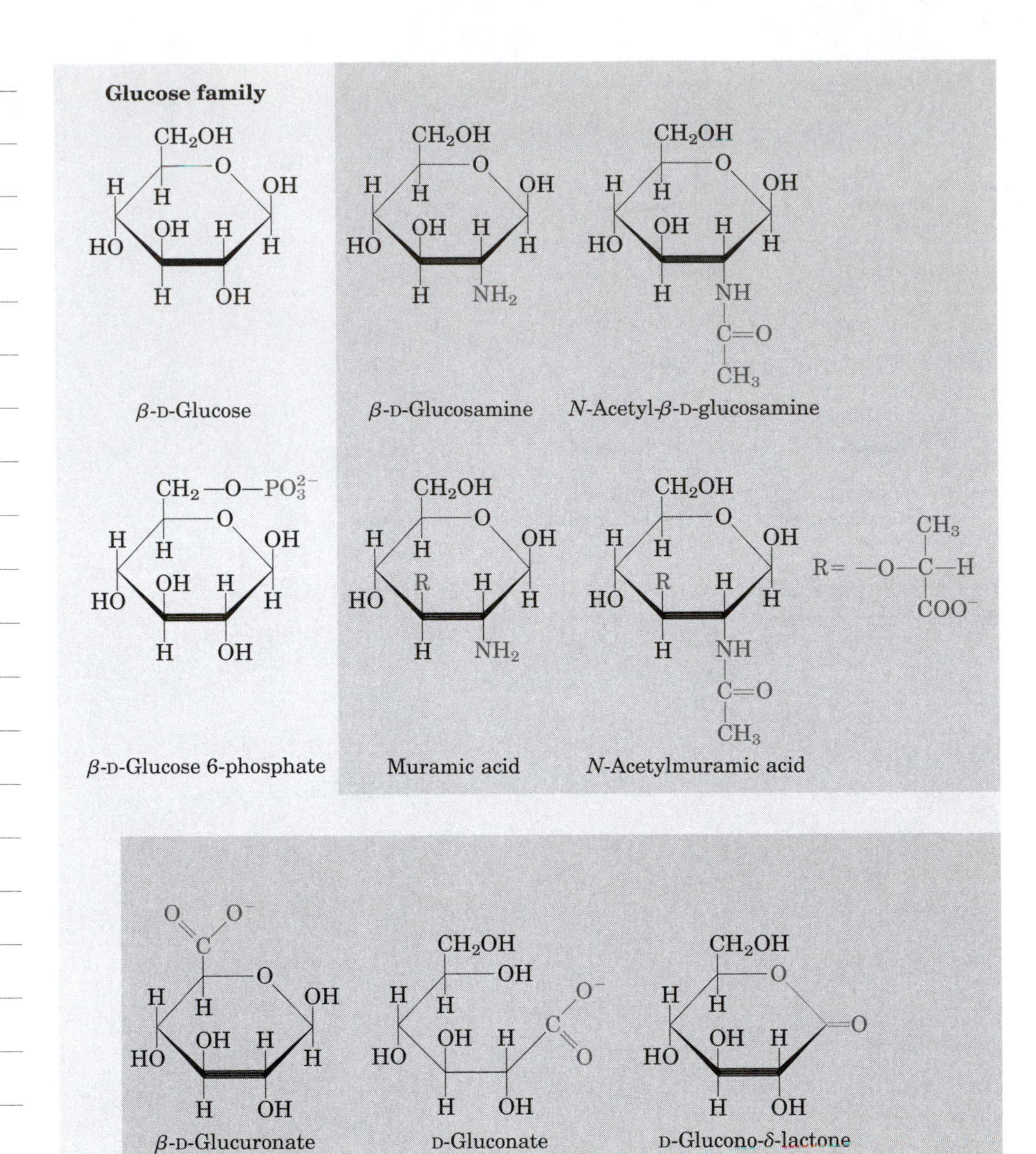

figure 9–9, page 300
Some hexose derivatives important in biology.

Amino sugars

β-D-Galactosamine

β-D-Mannosamine

Acidic sugars

N-Acetylneuraminic acid (sialic acid)

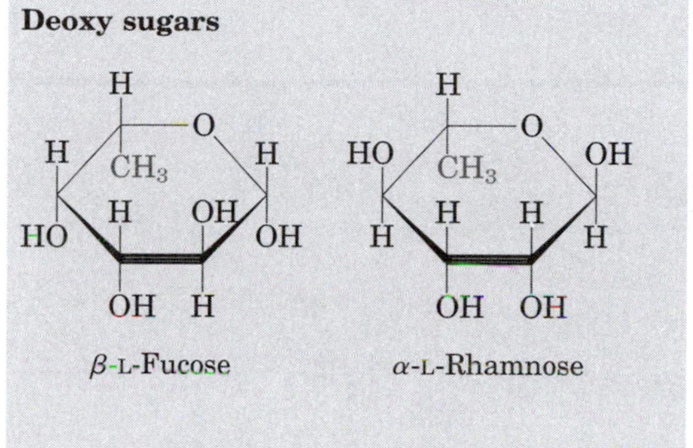

figure 9–9 (continued), page 300
Some hexose derivatives important in biology.

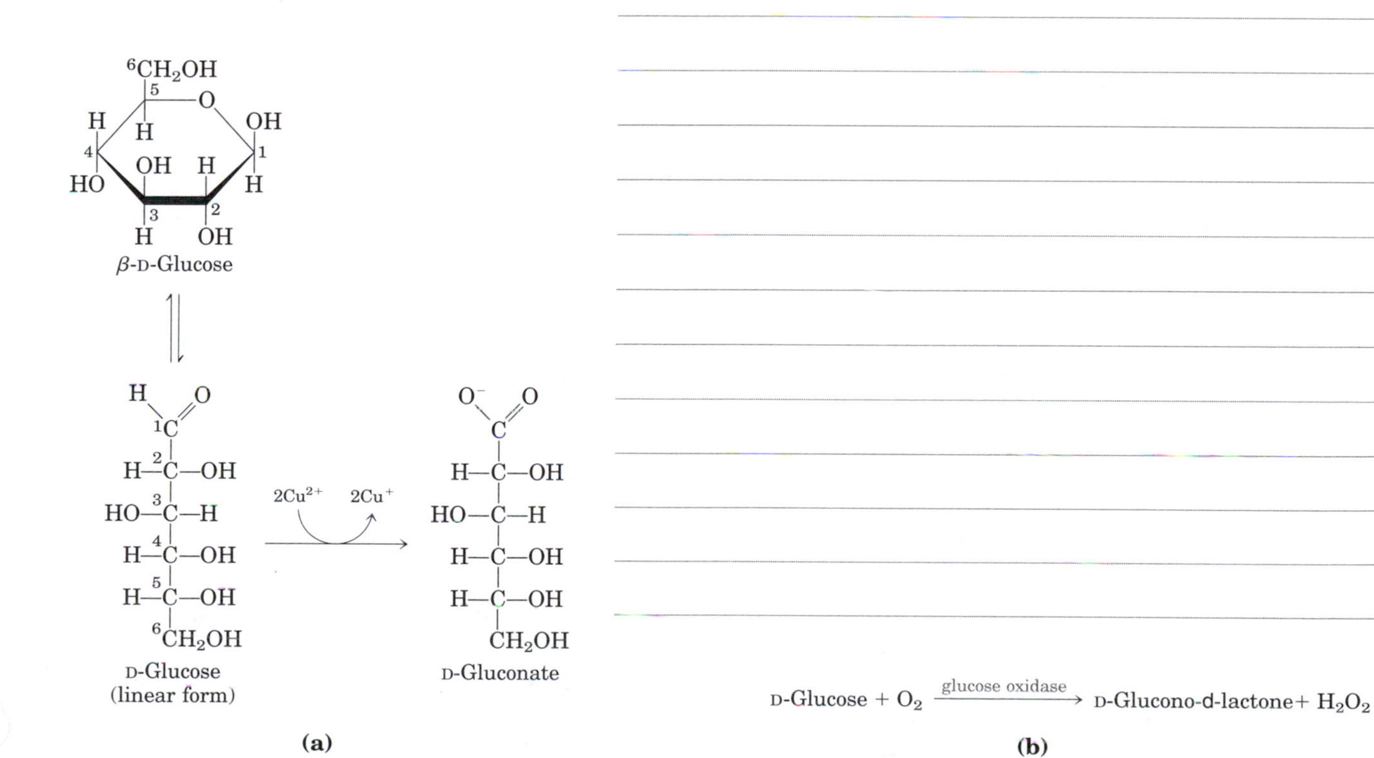

figure 9–10, page 301
Sugars as reducing agents.

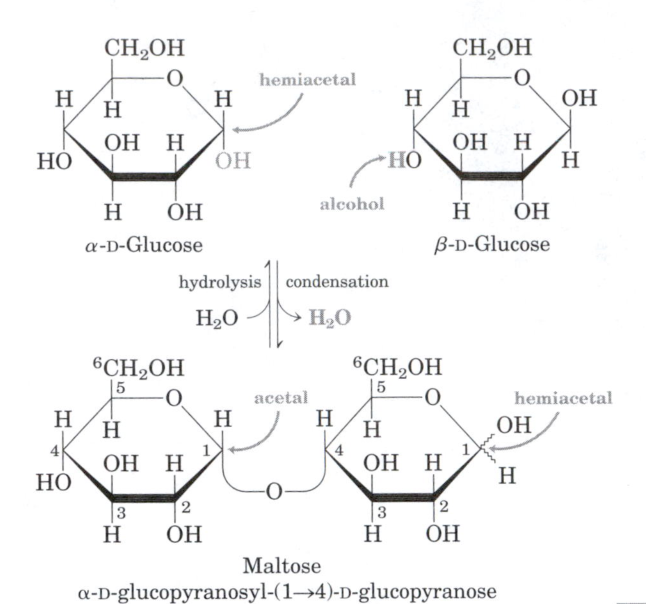

figure 9–11, page 302
Formation of maltose.

table 9–1, page 302

Abbreviations for Common Monosaccharides and Some of Their Derivatives

Abequose	Abe	Glucuronic acid	GlcA
Arabinose	Ara	Galactosamine	GalN
Fructose	Fru	Glucosamine	GlcN
Fucose	Fuc	*N*-Acetylgalactosamine	GalNAc
Galactose	Gal	*N*-Acetylglucosamine	GlcNAc
Glucose	Glc	Muramic acid	Mur
Mannose	Man	*N*-Acetylmuramic acid	Mur2Ac
Rhamnose	Rha	*N*-Acetylneuraminic acid (sialic acid)	Neu5Ac
Ribose	Rib		
Xylose	Xyl		

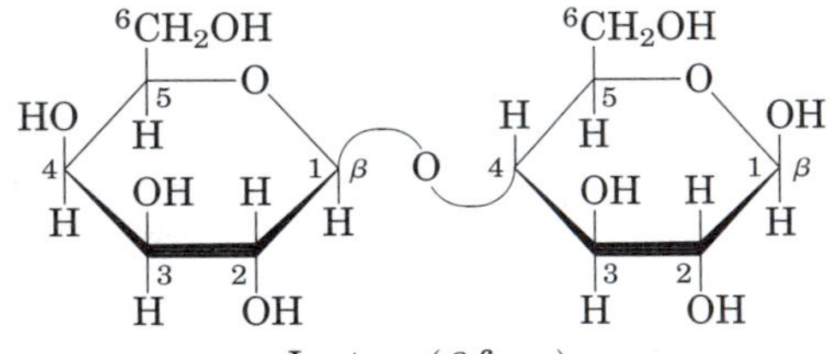

Lactose (β form)
β-D-galactopyranosyl-(1→4)-β-D-glucopyranose
Gal(β1→4)Glc

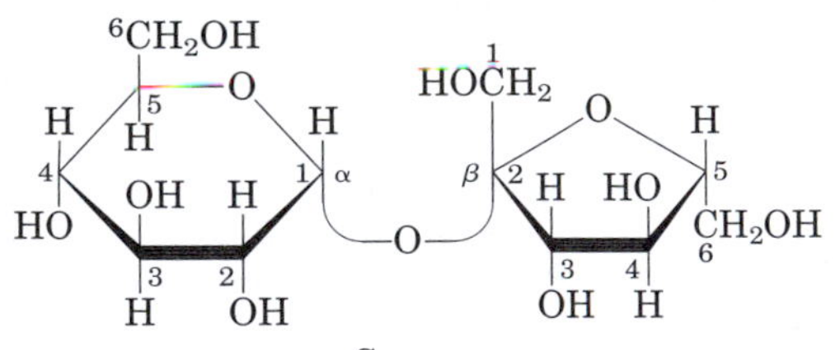

Sucrose
β-D-fructofuranosyl α-D-glucopyranoside
Fru(β2↔1α)Glc

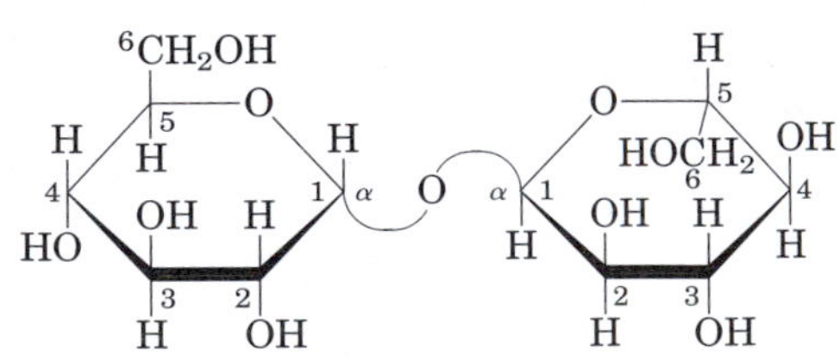

Trehalose
α-D-glucopyranosyl α-D-glucopyranoside
Glc(α1↔1α)Glc

figure 9–12, page 303
Some common disaccharides.

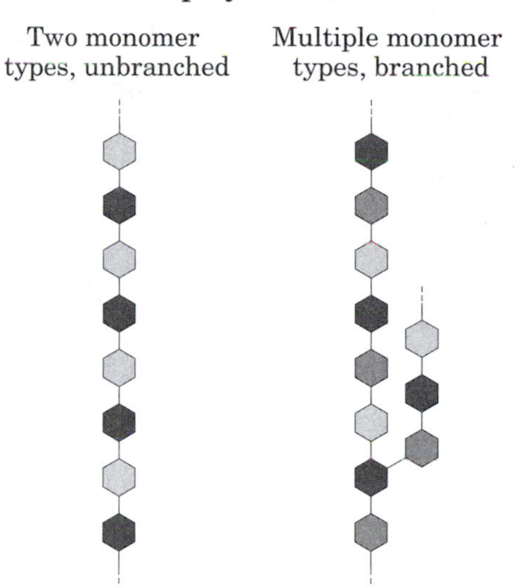

figure 9–13, page 303
Polysaccharides.

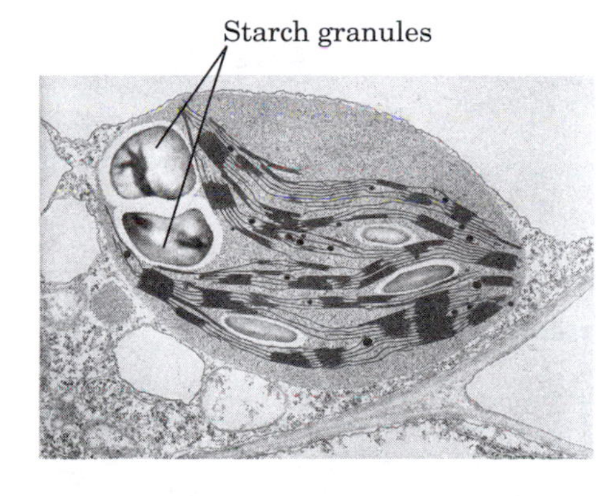

(a)

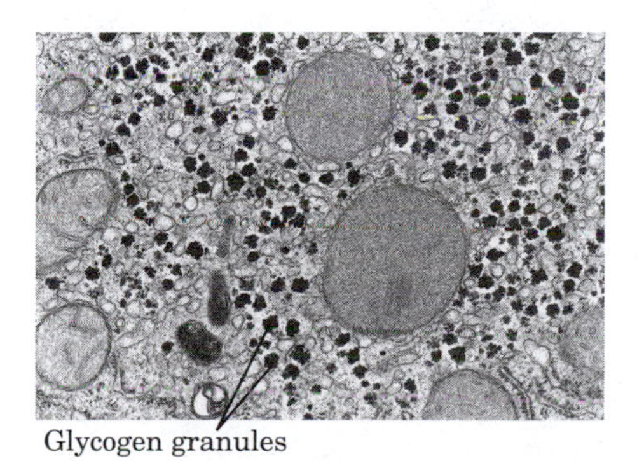

(b)

figure 9–14, page 304
Electron micrographs of starch and glycogen granules.

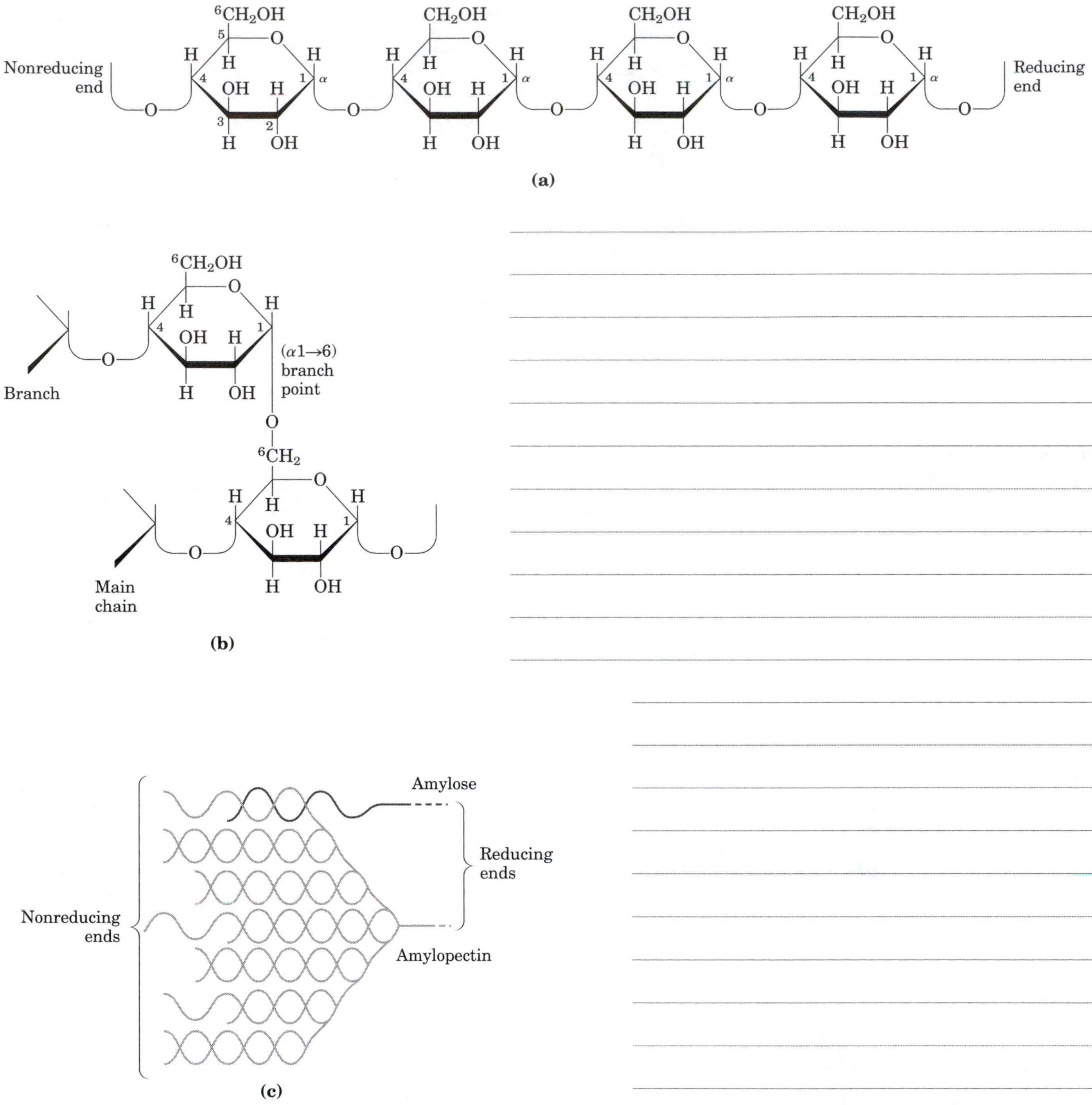

figure 9–15, page 305
Amylose and amylopectin, the polysaccharides of starch.

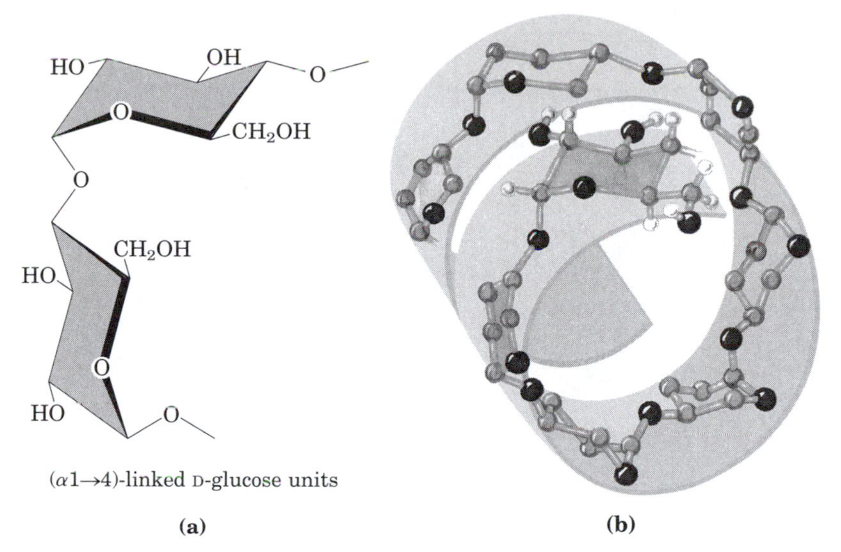

figure 9–16, page 306
The structure of starch (amylose).

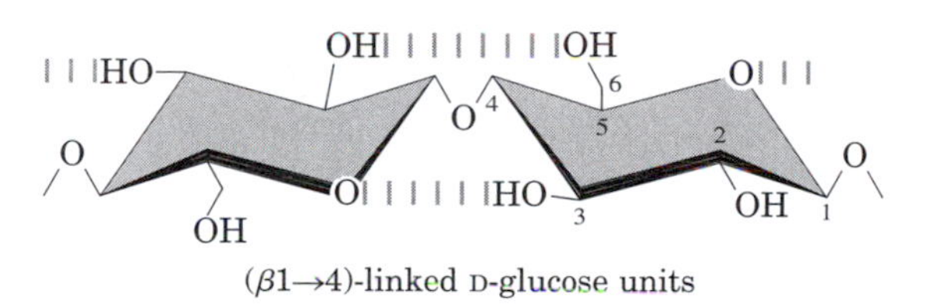

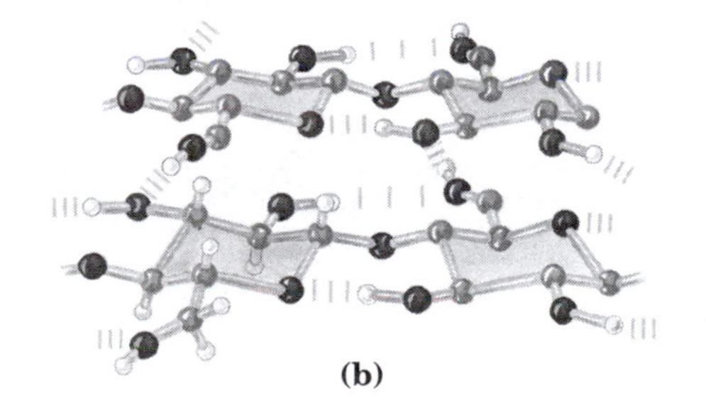

figure 9–17, page 307
The structure of cellulose.

figure 9–18, page 307
A short segment of chitin, a homopolymer of *N*-acetyl-D-glucosamine units in (β1→4) linkage.

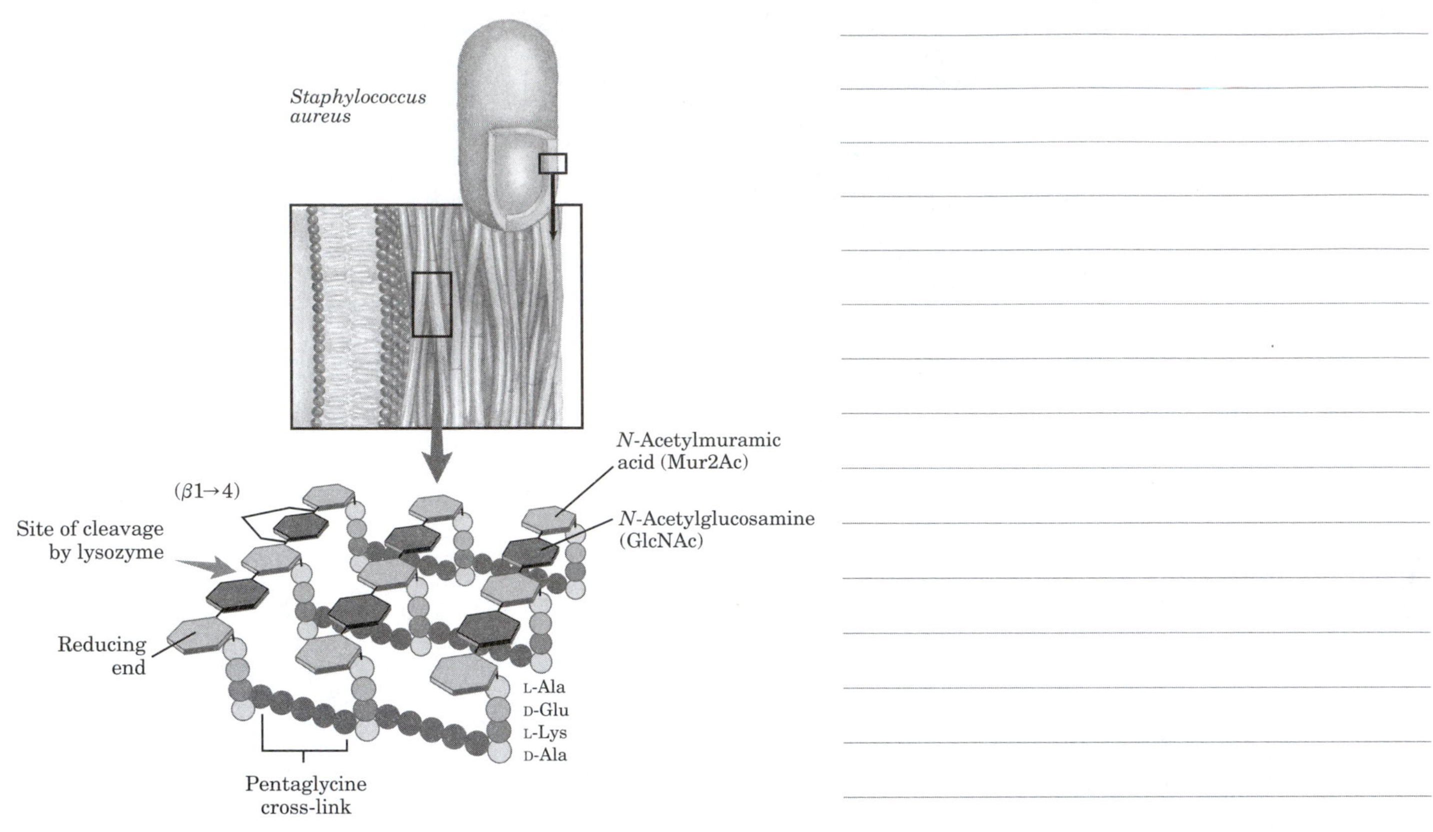

figure 9–19, page 308
Peptidoglycan.

table 9–2, page 310

Structures and Roles of Some Polysaccharides

Polymer	Type*	Repeating unit[†]	Size (number of monosaccharide units)	Roles
Starch				Energy storage: in plants
Amylose	Homo-	(α1→4)Glc, linear	50–5,000	
Amylopectin	Homo-	(α1→4)Glc, with (α1→6)Glc branches every 24 to 30 residues	Up to 10^6	
Glycogen	Homo-	(α1→4)Glc, with (α1→6)Glc branches every 8 to 12 residues	Up to 50,000	Energy storage: in bacteria and animal cells
Cellulose	Homo-	(β1→4)Glc	Up to 15,000	Structural: in plants, gives rigidity and strength to cell walls
Chitin	Homo-	(β1→4)GlcNAc	Very large	Structural: in insects, spiders, crustaceans, gives rigidity and strength to exoskeletons
Peptidoglycan	Hetero-; peptides attached	4)Mur2Ac(β1→4) GlcNAc(β1	Very large	Structural: in bacteria, gives rigidity and strength to cell envelope
Hyaluronate (a glycosamino-glycan)	Hetero-; acidic	4)GlcA(β1→3) GlcNAc(β1	Up to 100,000	Structural: in vertebrates, extracellular matrix of skin and connective tissue; viscosity and lubrication in joints

* Each polymer is classified as a homopolysaccharide (homo-) or heteropolysaccharide (hetero-).

[†]The abbreviated names for the peptidoglycan and hyaluronate repeating units indicate that the polymer contains repeats of this disaccharide unit, with the GlcNAc of one disaccharide unit linked β(1→4) to the first residue of the next disaccharide unit.

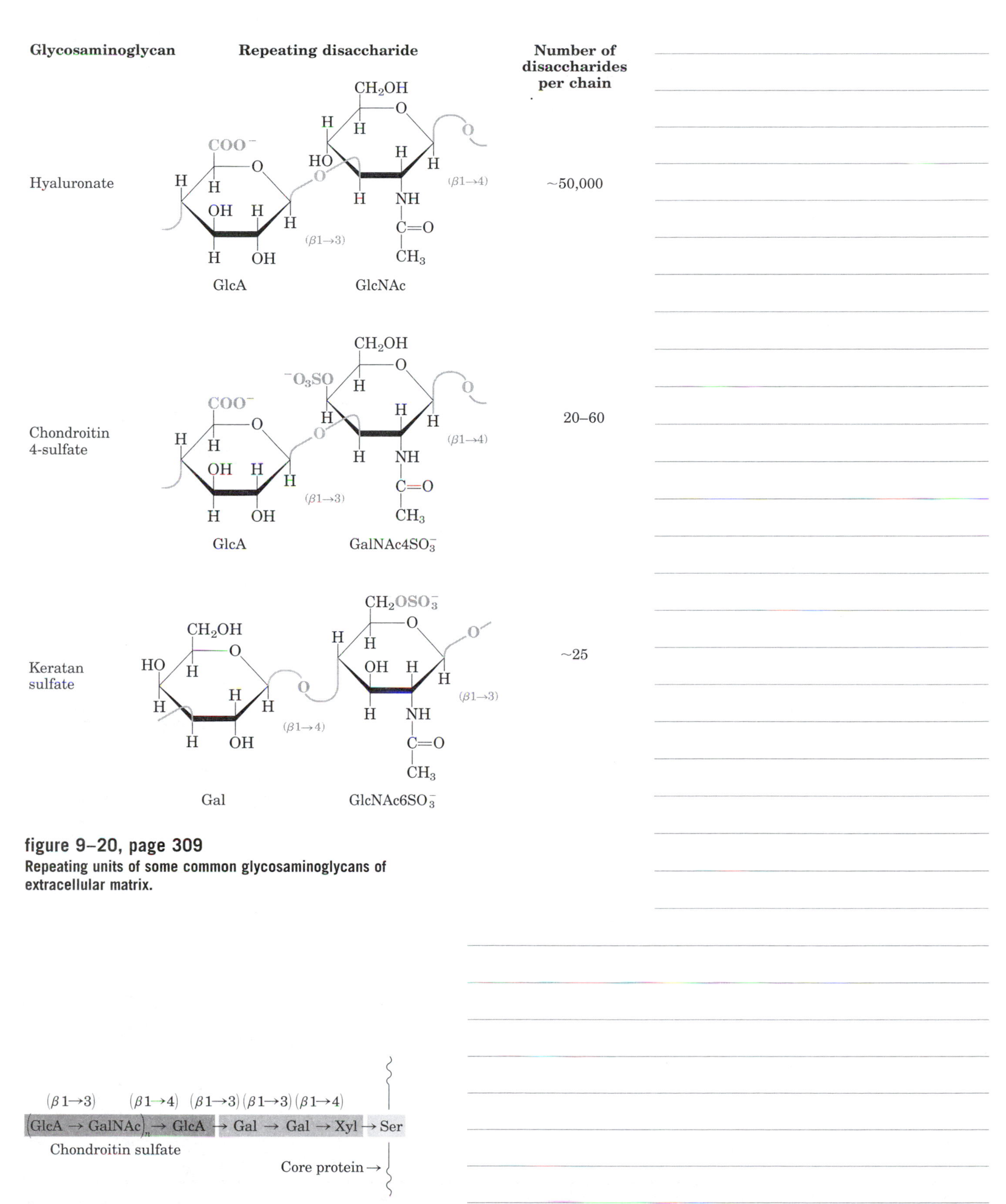

figure 9–20, page 309
Repeating units of some common glycosaminoglycans of extracellular matrix.

figure 9–21, page 311
Proteoglycan structure, showing the trisaccharide bridge.

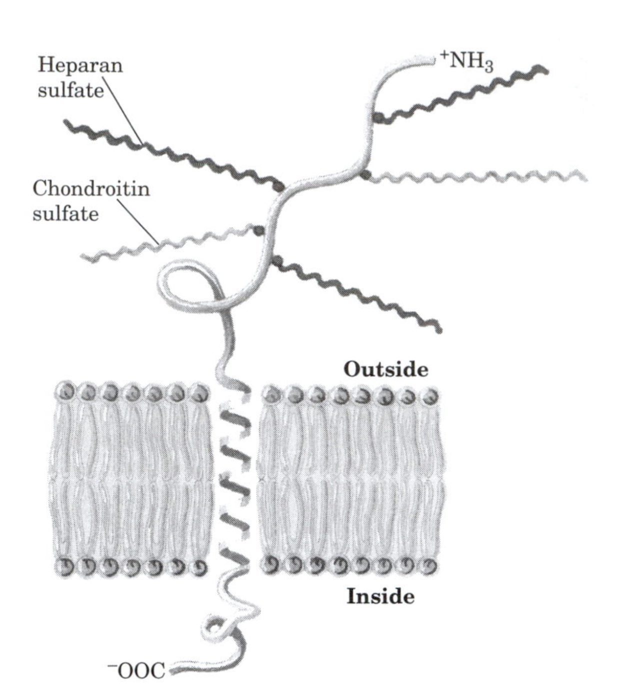

figure 9–22, page 312
Proteoglycan structure of an integral membrane protein.

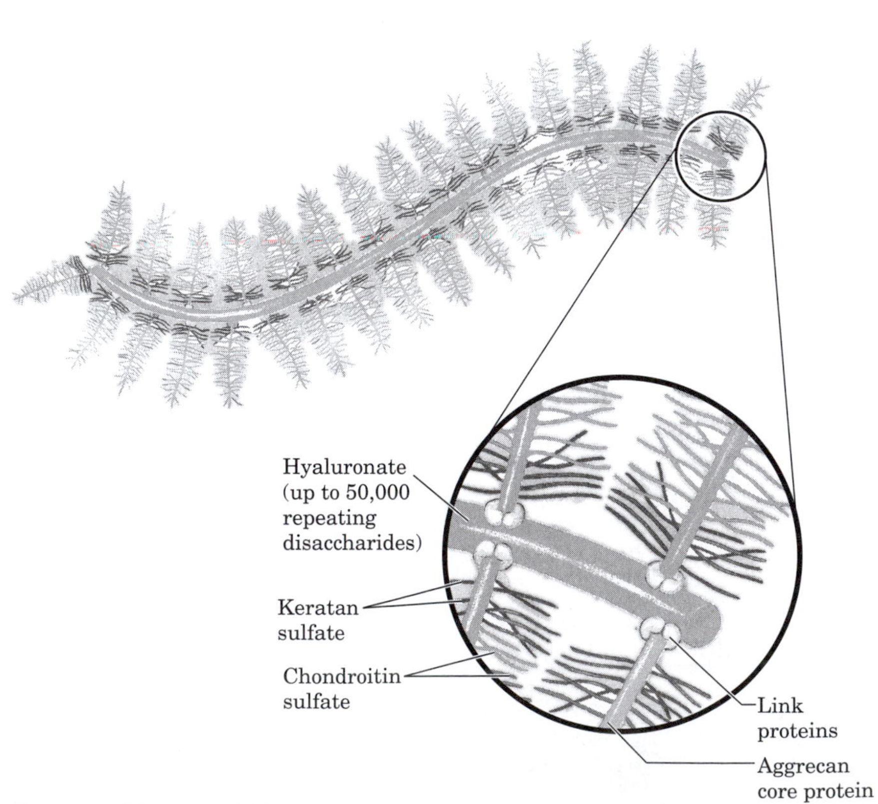

figure 9–23, page 312
A proteoglycan aggregate of the extracellular matrix.

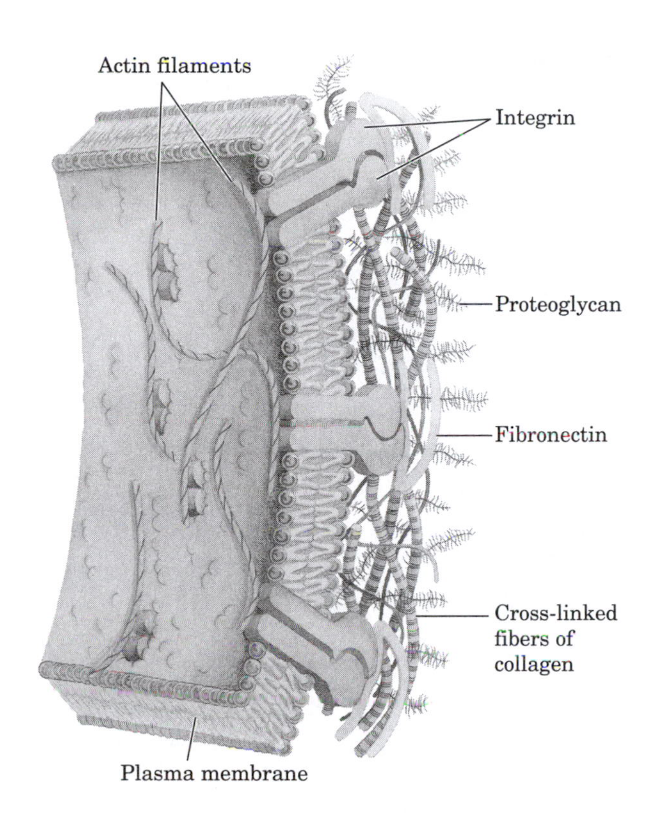

figure 9–24, page 313
Interactions between cells and extracellular matrix.

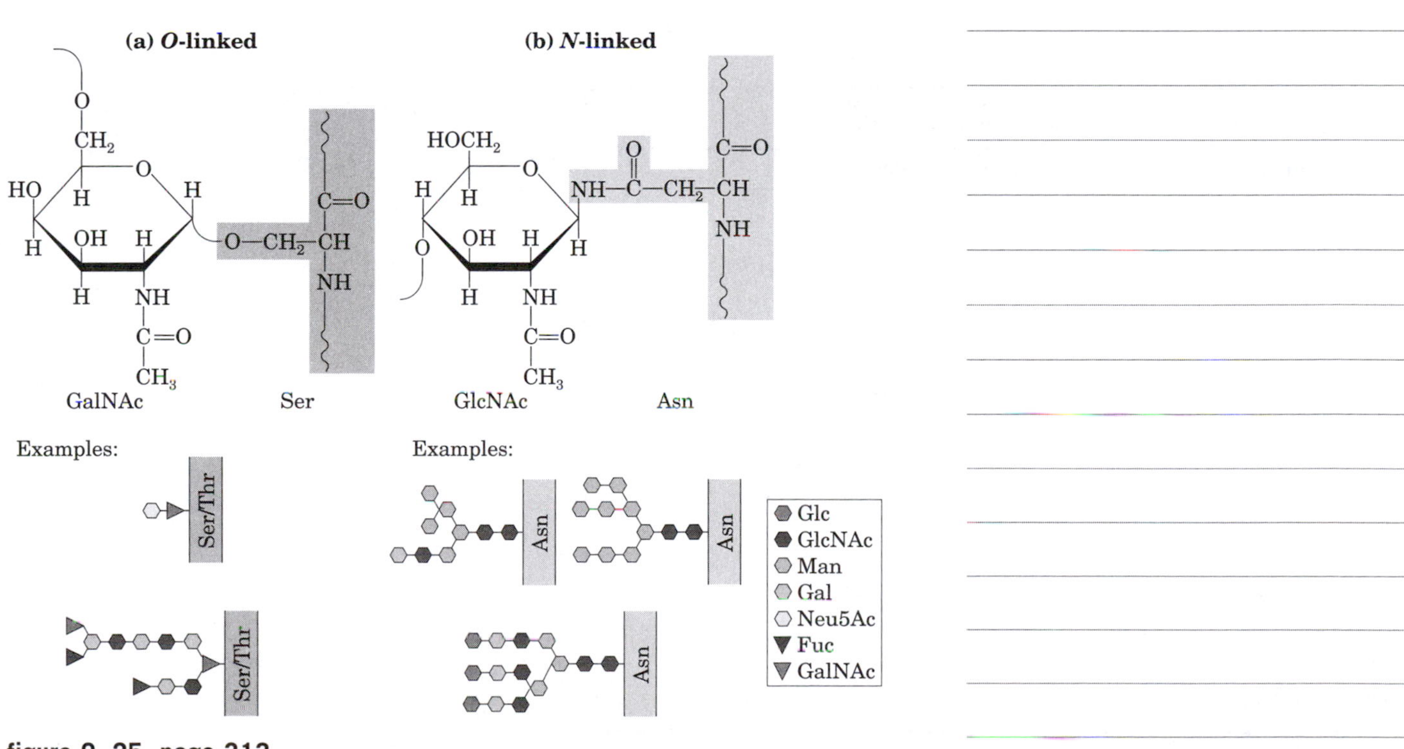

figure 9–25, page 313
Oligosaccharide linkages in glycoproteins.

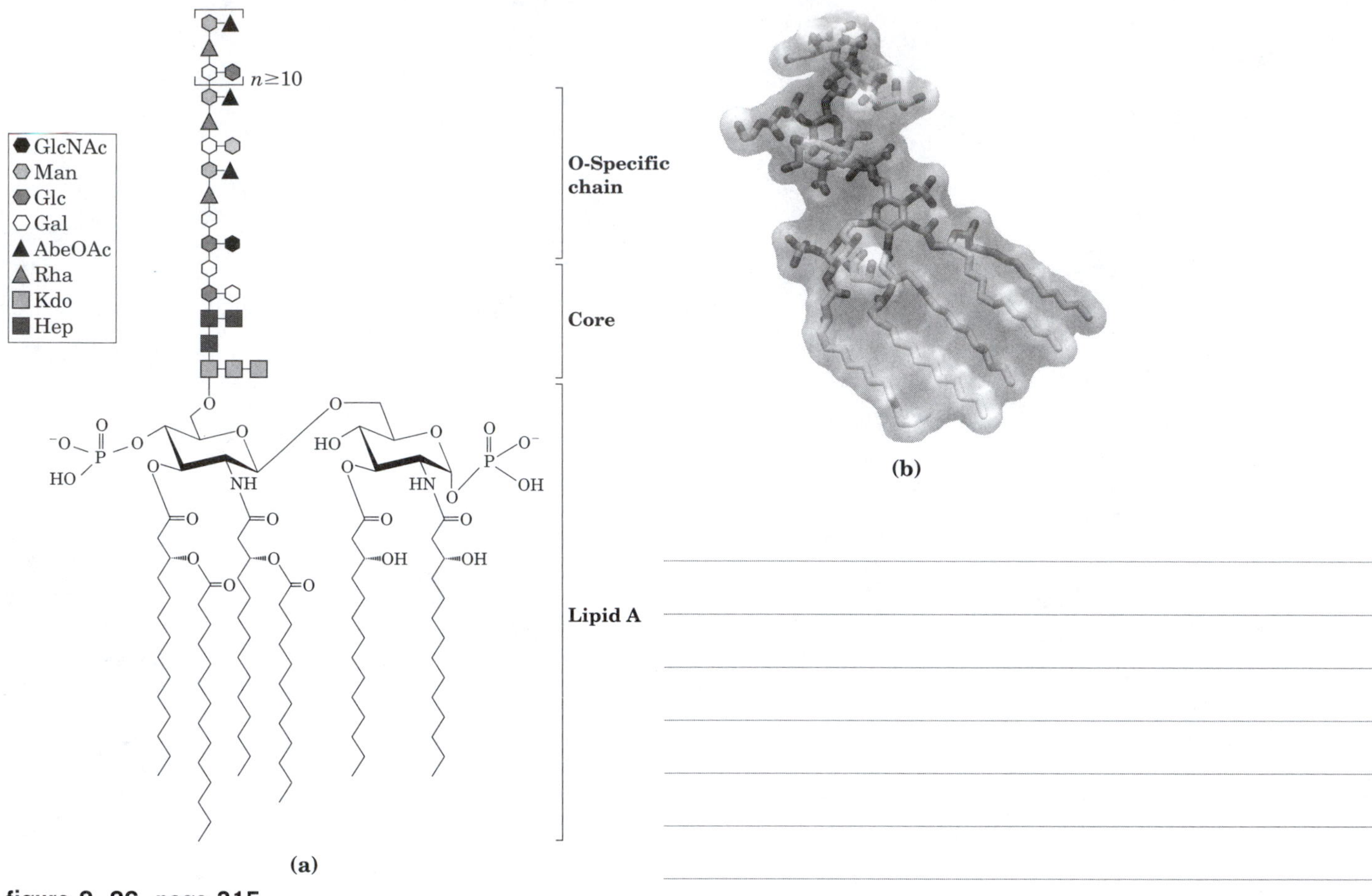

figure 9–26, page 315
Bacterial lipopolysaccharides.

table 9–3, page 316

Lectins and the Oligosaccharide Ligands That They Bind

Lectin family and lectin	Abbreviation	Ligand(s)
Plant		
Concanavalin A	ConA	Manα1–OCH$_3$
Griffonia simplicifolia lectin 4	GS4	Lewis b (Le^b) tetrasaccharide
Wheat germ agglutinin	WGA	Neu5Ac(α2→3)Gal(β1→4)Glc GlcNAc(β1→4)GlcNAc
Ricin		Gal(β1→4)Glc
Animal		
Galectin-1		Gal(β1→4)Glc
Mannose-binding protein A	MBP-A	High-mannose octasaccharide
Viral		
Influenza virus hemagglutinin	HA	Neu5Ac(α2→6)Gal(β1→4)Glc
Polyoma virus protein 1	VP1	Neu5Ac(α2→3)Gal(β1→4)Glc
Bacterial		
Enterotoxin	LT	Gal
Cholera toxin	CT	GM1 pentasaccharide

Source: Weiss, W.I. & Drickamer, K. (1996) Structural basis of lectin-carbohydrate recognition. *Annu. Rev. Biochem.* **65,** 441–473.

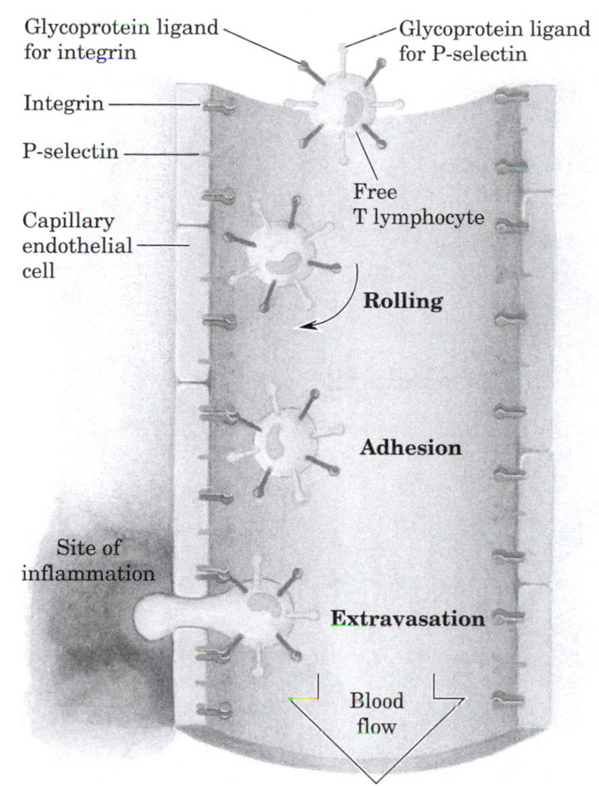

figure 9–27, page 317
Role of lectin-ligand interactions in lymphocyte movement to the site of an infection or injury.

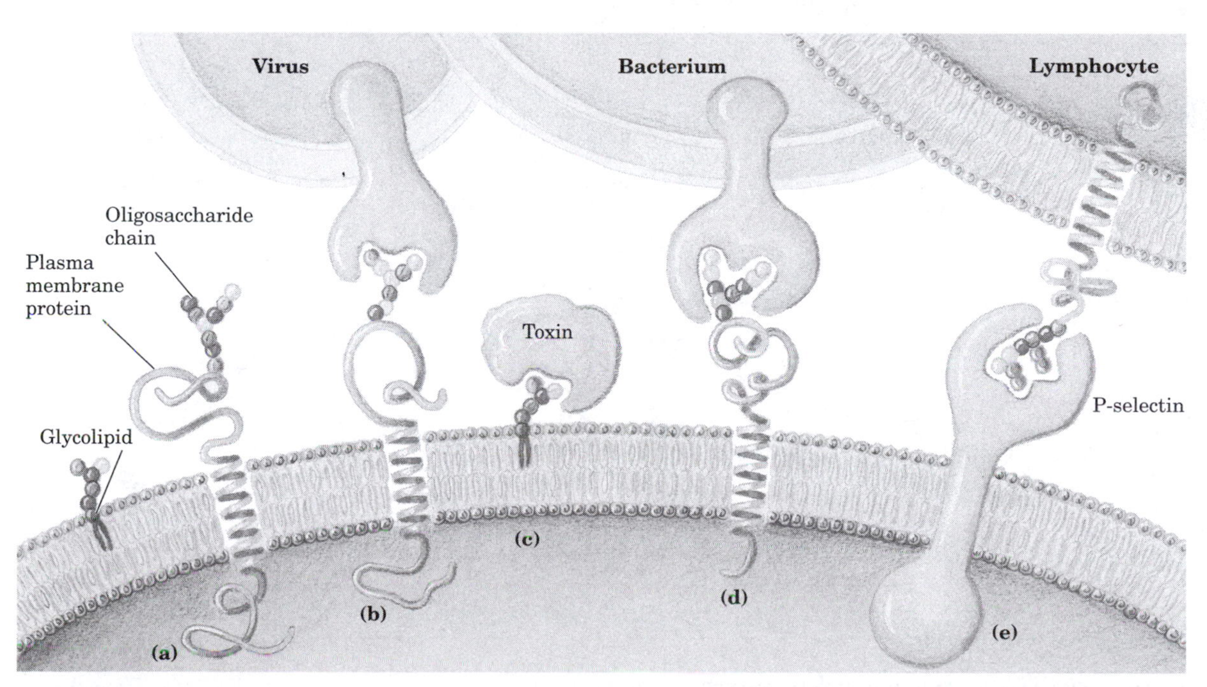

figure 9–29, page 318
Roles of oligosaccharides in recognition and adhesion at the cell surface.

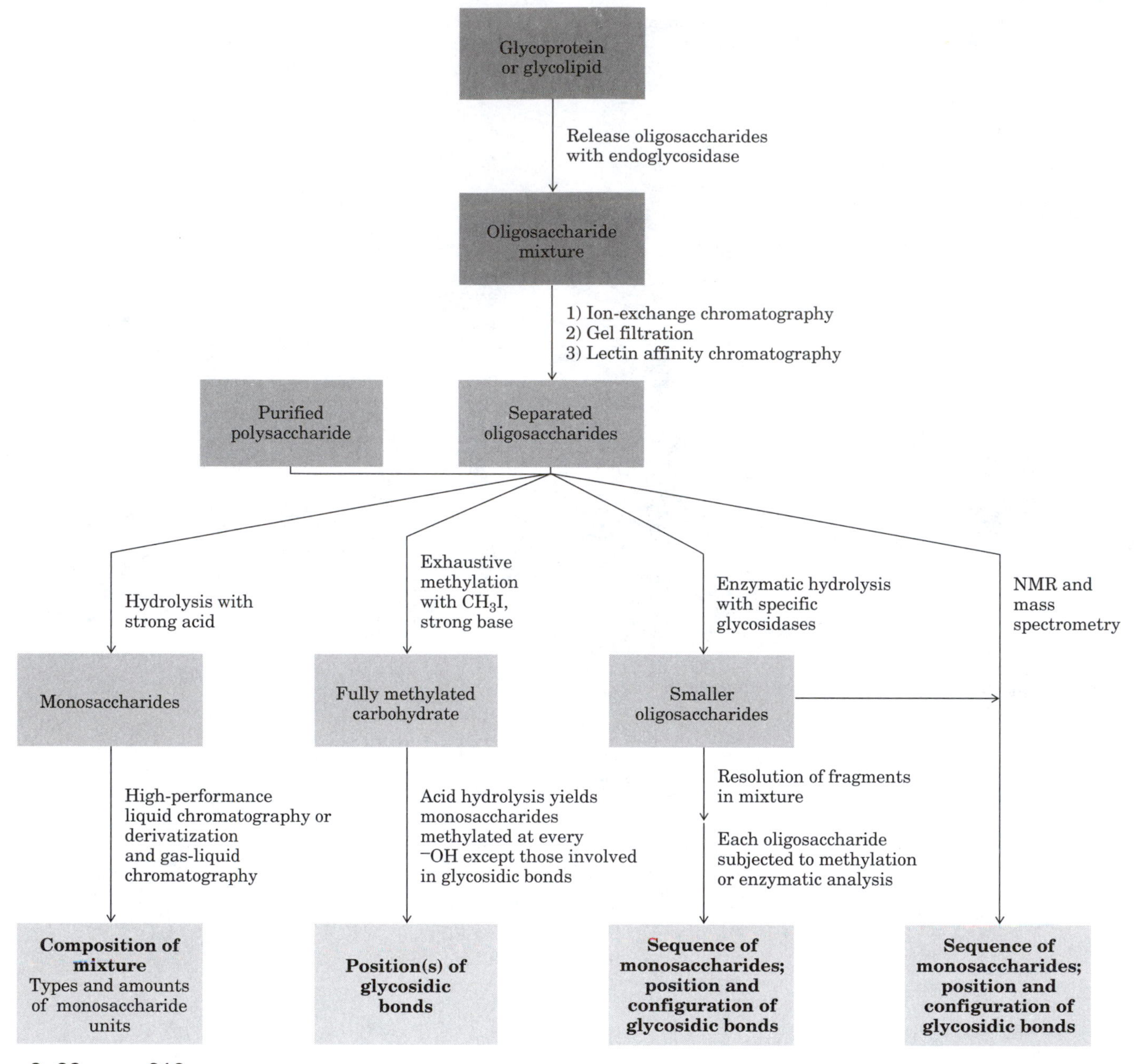

figure 9–30, page 319
Methods of carbohydrate analysis.

PROBLEMS

3. Melting Points of Monosaccharide Osazone Derivatives, page 323.

Glucose → ($C_6H_5NHNH_2$) → Osazone derivative of glucose

Monosaccharide	MP of anhydrous monosaccharide (°C)	MP of osazone derivative (°C)
Glucose	146	205
Mannose	132	205
Galactose	165–168	201
Talose	128–130	201

16. Determination of the Extent of Branching in Amylopectin, page 323.

2,3-Di-*O*-methylglucose

chapter

10 Nucleotides and Nucleic Acids

When you study for exams, don't forget to review The Minicourses on the UNDERSTAND! BIOCHEMISTRY CD that came with your textbook.

Minicourses that apply to this Chapter include:

Molecules of Life
- Important Biomolecules
- Macromolecules: Proteins and Nucleic Acids
- An Interactive Gallery of Molecular Models

Nucleic Acids and Their Expression
- Mutation
- Mutagens

Some Important Techniques
- DNA Cloning and Sequencing
- Nucleic Acid Analysis

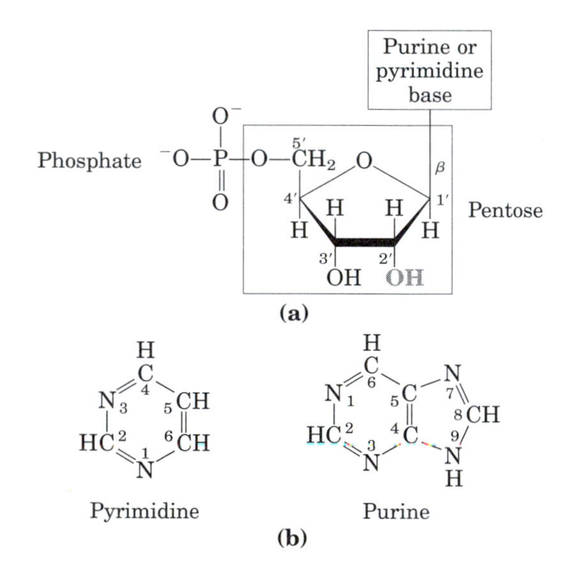

figure 10–1, page 325
Structure of nucleotides.

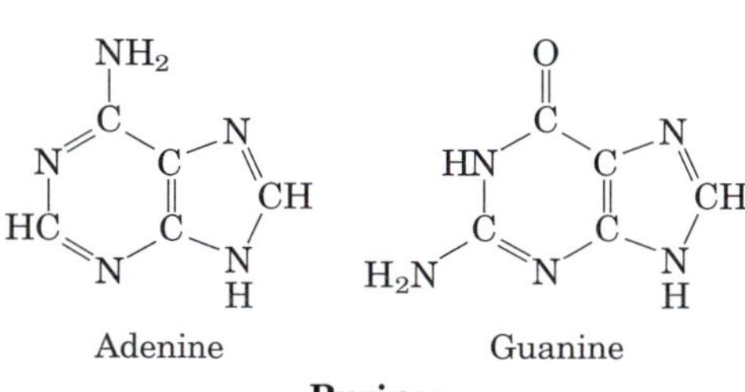

figure 10–2, page 326
Major purine and pyrimidine bases of nucleic acids.

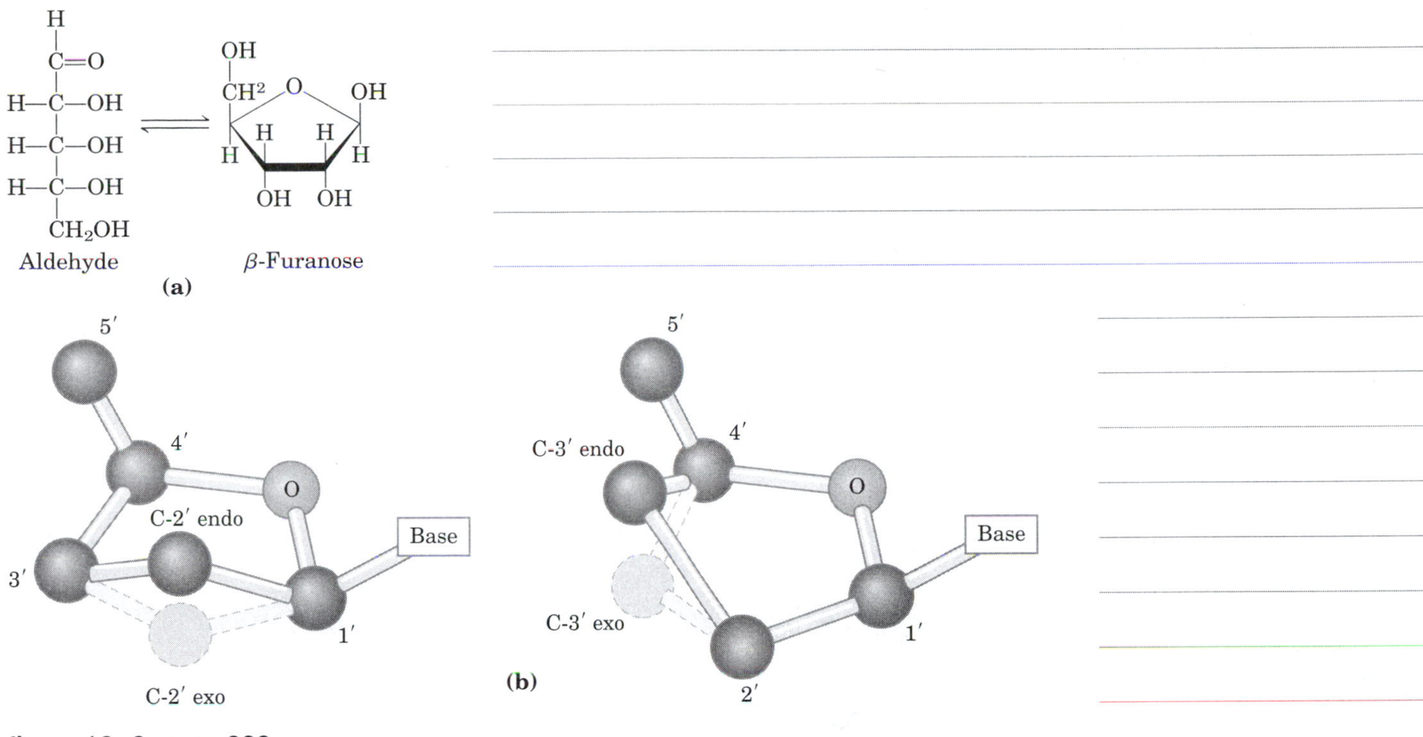

figure 10–3, page 326
Conformations of ribose.

table 10–1, page 327

Nucleotide and Nucleic Acid Nomenclature

Base	Nucleoside*	Nucleotide*	Nucleic acid
Purines			
Adenine	Adenosine	Adenylate	RNA
	Deoxyadenosine	Deoxyadenylate	DNA
Guanine	Guanosine	Guanylate	RNA
	Deoxyguanosine	Deoxyguanylate	DNA
Pyrimidines			
Cytosine	Cytidine	Cytidylate	RNA
	Deoxycytidine	Deoxycytidylate	DNA
Thymine	Thymidine or deoxythymidine	Thymidylate or deoxythymidylate	DNA
Uracil	Uridine	Uridylate	RNA

**Nucleoside* and *nucleotide* are generic terms that include both ribo- and deoxyribo- forms. Note that here ribonucleosides and ribonucleotides are designated simply as nucleosides and nucleotides (e.g., riboadenosine as adenosine), and deoxyribonucleosides and deoxyribonucleotides as deoxynucleosides and deoxynucleotides (e.g., deoxyriboadenosine as deoxyadenosine). Both forms of naming are acceptable, but the shortened names are more commonly used. Thymine is an exception; the name ribothymidine is used to describe its unusual occurrence in RNA.

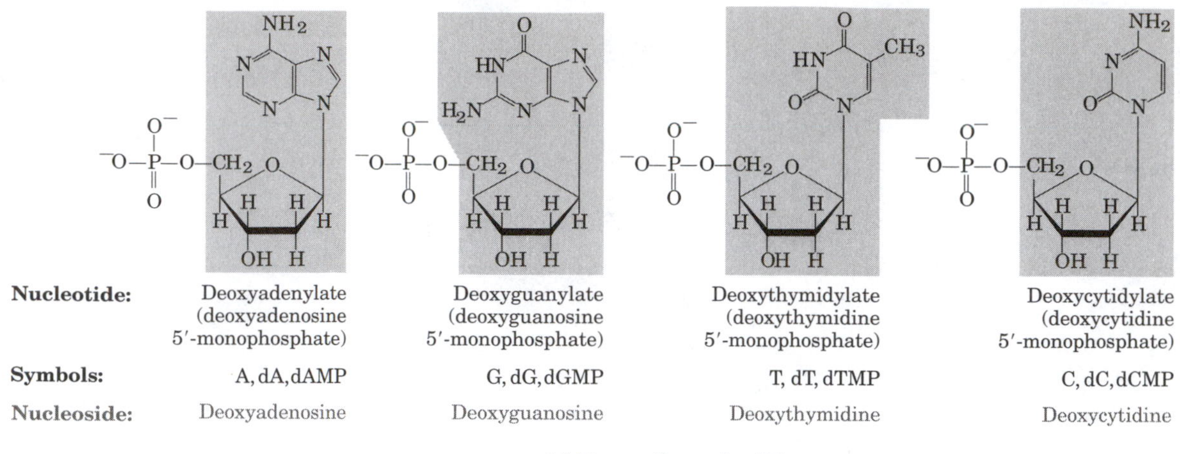

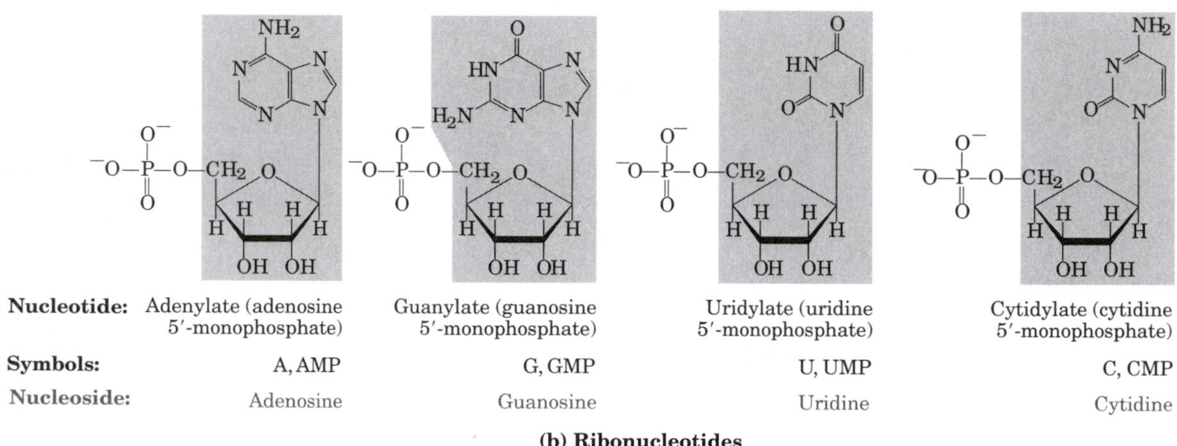

figure 10–4, page 327

Deoxyribonucleotides and ribonucleotides of nucleic acids.

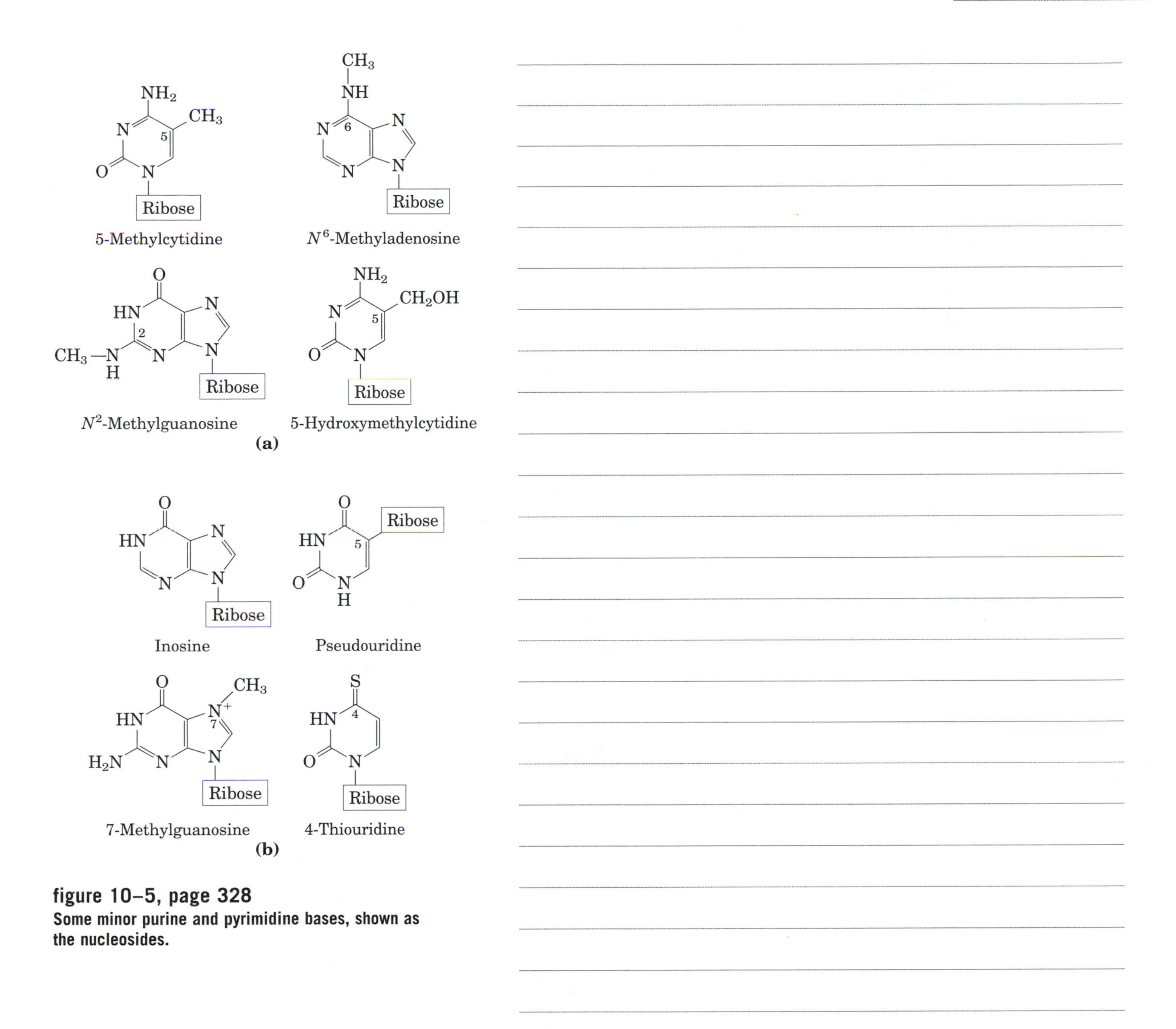

figure 10–5, page 328
Some minor purine and pyrimidine bases, shown as the nucleosides.

Adenosine 5′-monophosphate

Adenosine 2′-monophosphate

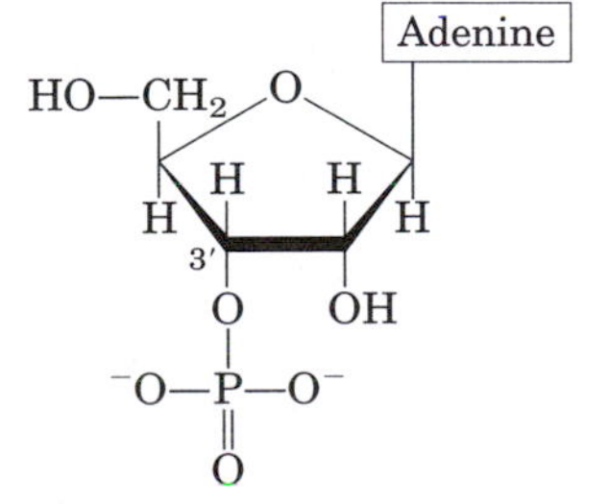

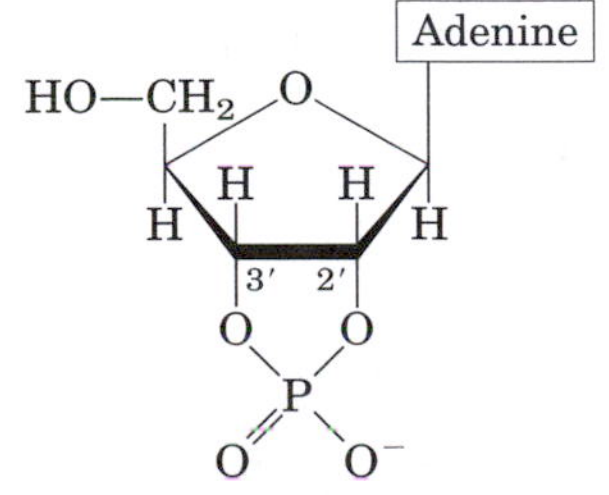

figure 10–6, page 328
Some adenosine monophosphates.

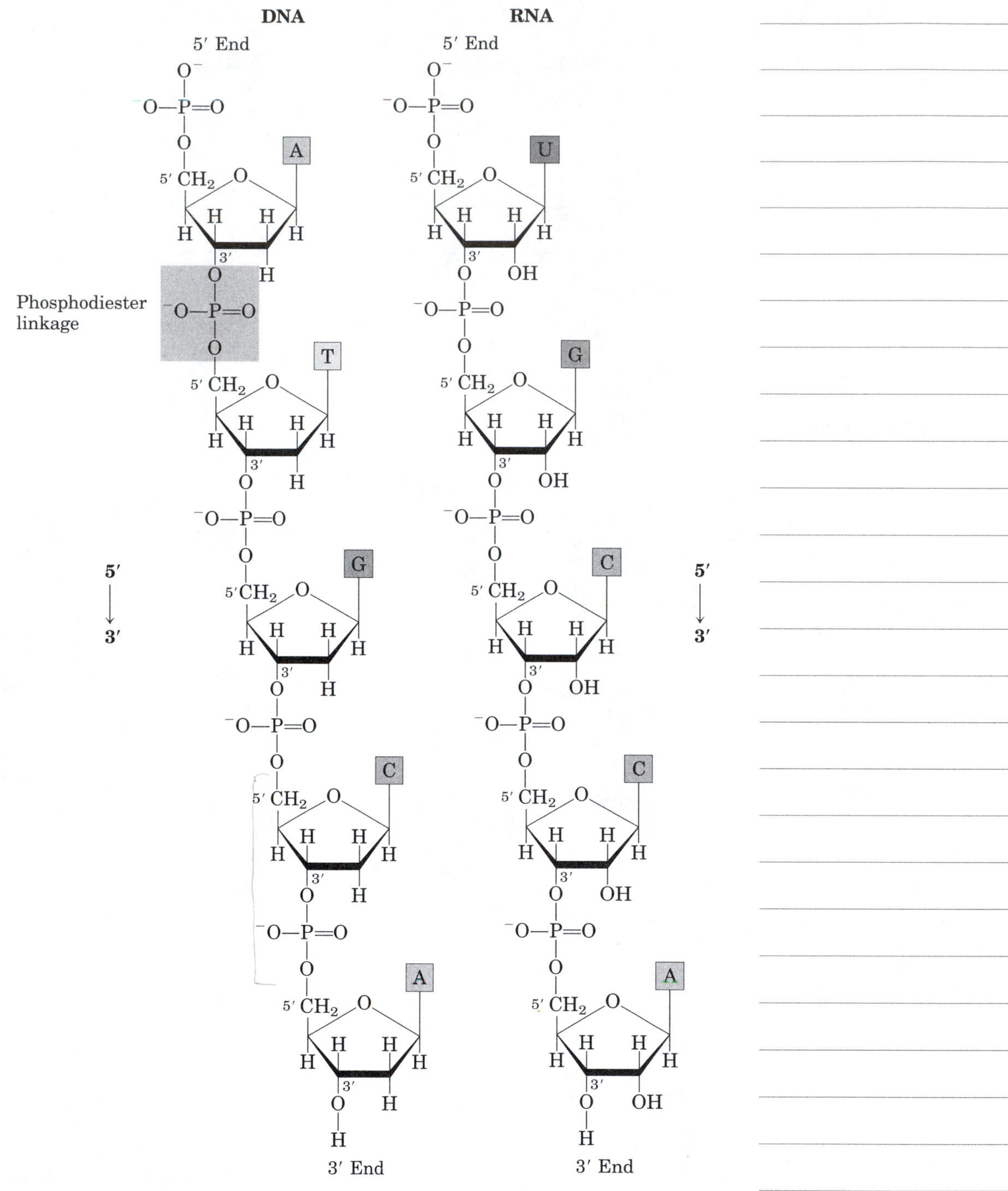

figure 10–7, page 329
Phosphodiester linkages in the covalent backbone of DNA and RNA.

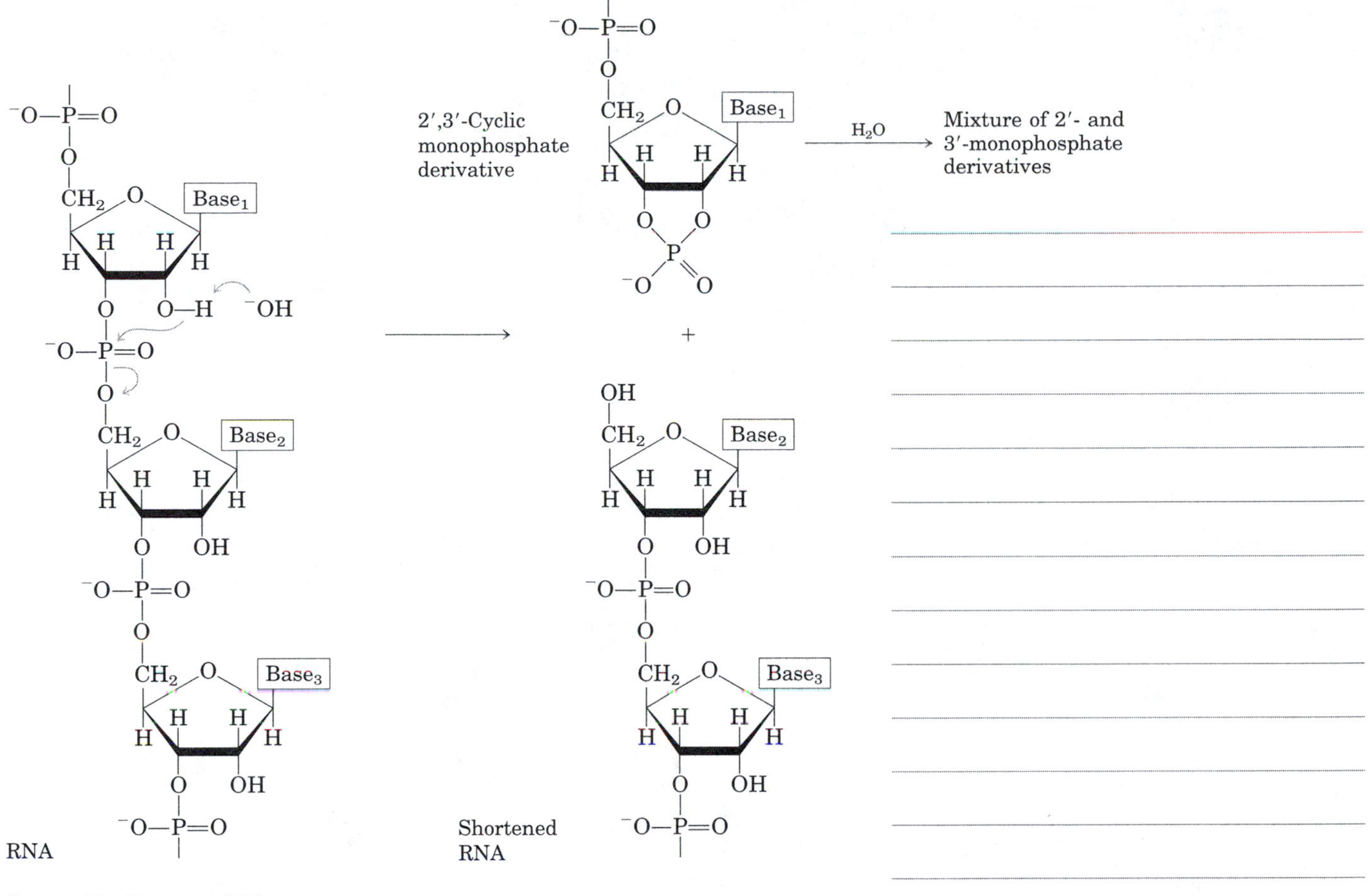

figure 10–8, page 330
Hydrolysis of RNA under alkaline conditions.

A C G T A

5′ End 3′ End

P P P P P OH

Schematic of oligonucleotide, page 330

Lactam ⇌ Lactim ⇌ Double lactim

figure 10–9, page 331
Tautomeric forms of uracil.

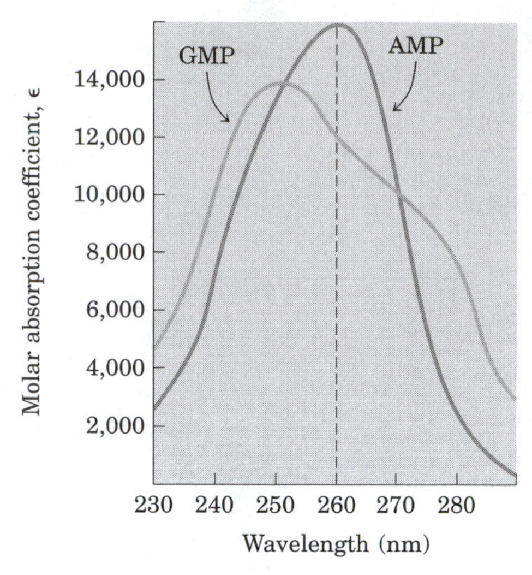

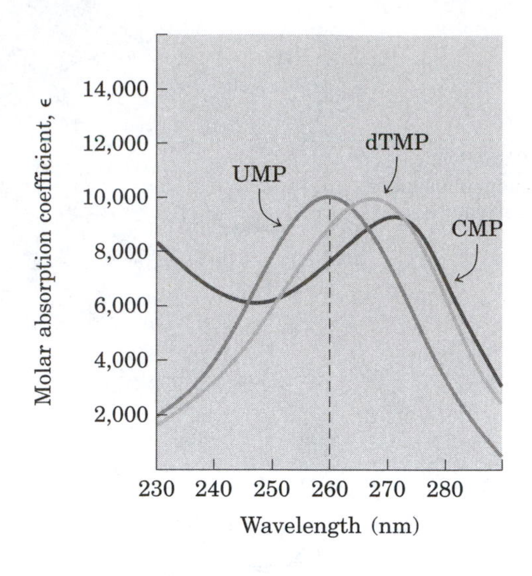

Molar absorption coefficient at 260 nm, ϵ_{260} ($M^{-1}cm^{-1}$)	
AMP	15,400
GMP	11,700
CMP	7,500
UMP	9,900
dTMP	9,200

figure 10–10, page 331
Absorption spectra of the common nucleotides.

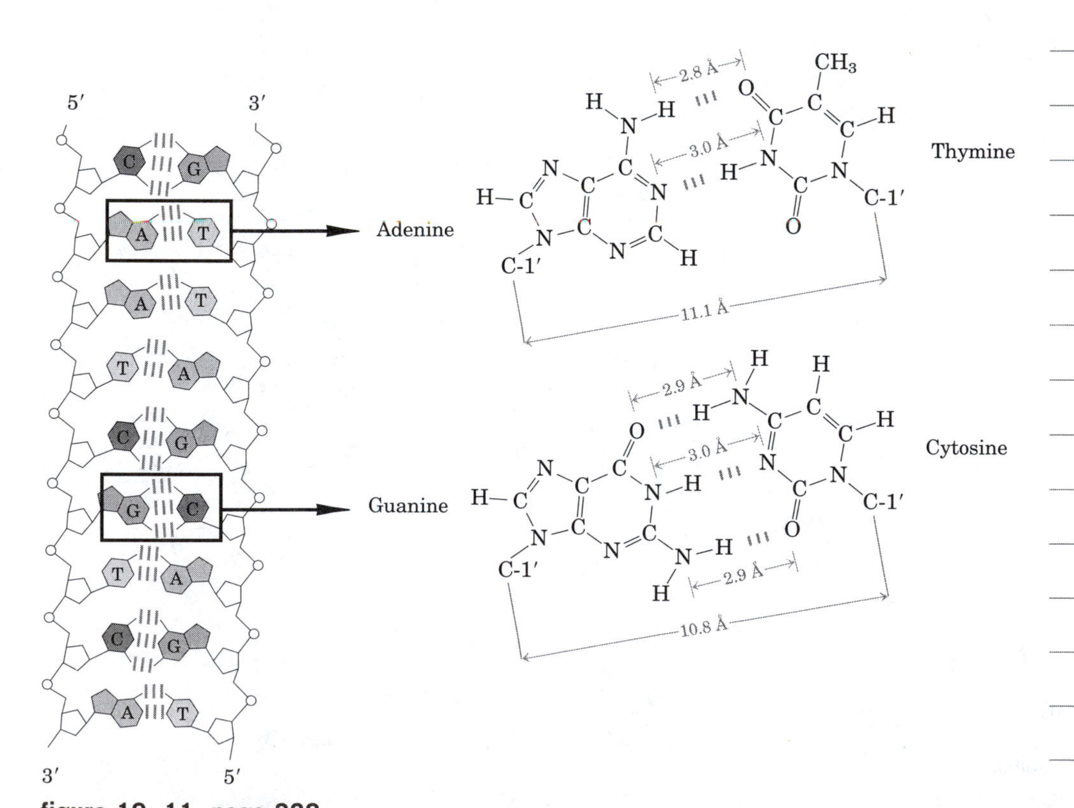

figure 10–11, page 332
Hydrogen-bonding patterns in the base pairs defined by Watson and Crick.

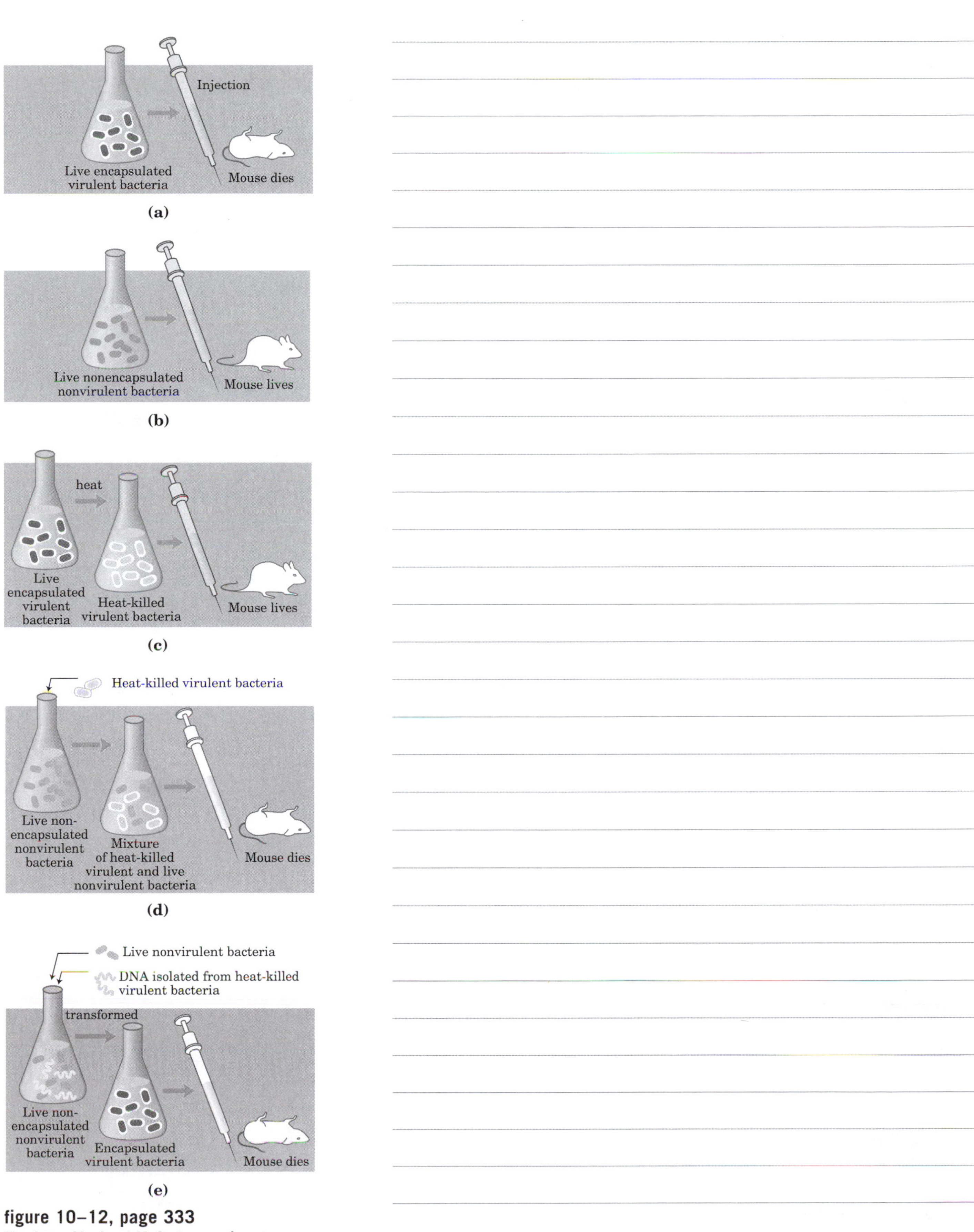

figure 10–12, page 333
The Avery-MacLeod-McCarty experiment.

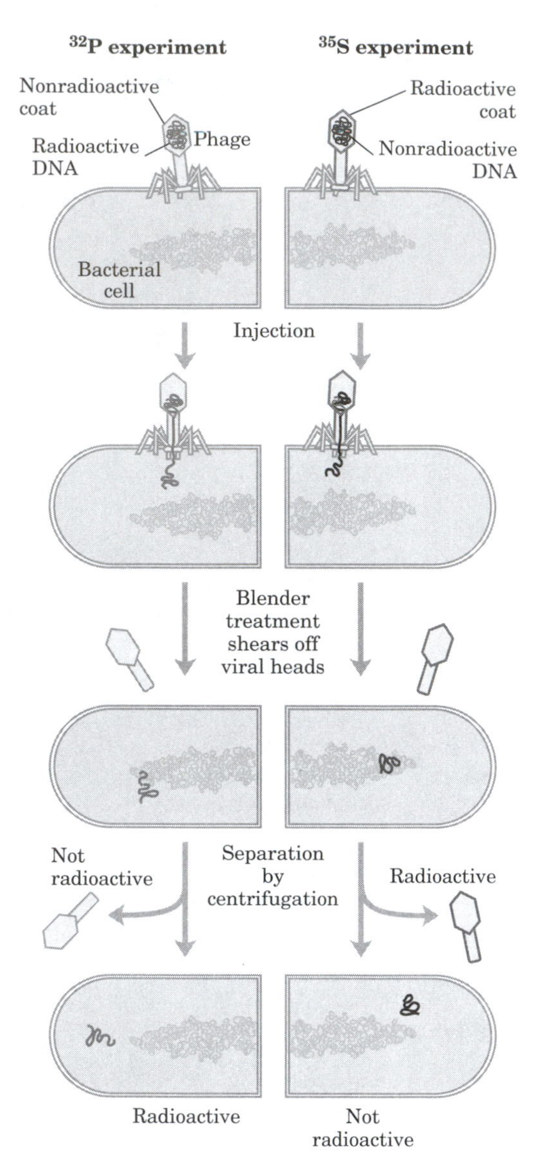

figure 10–13, page 334
The Hershey-Chase experiment.

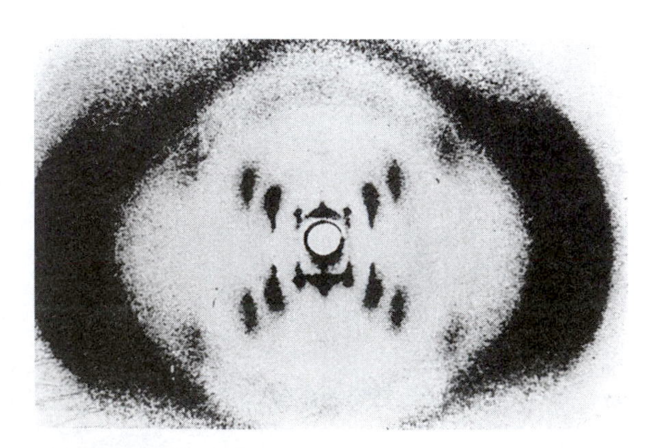

figure 10–14, page 335
X-ray diffraction pattern of DNA.

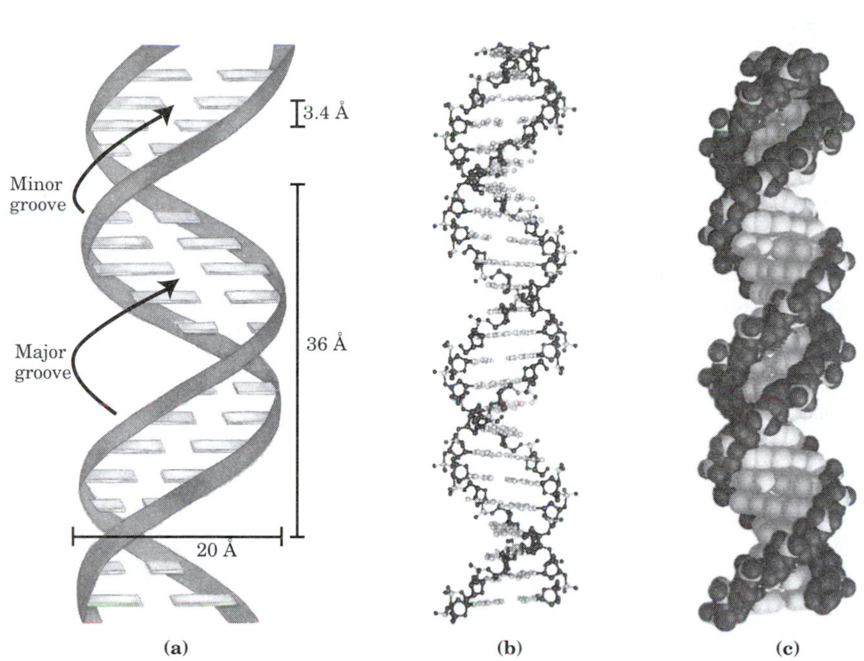

figure 10–15, page 336
Watson-Crick model for the structure of DNA.

figure 10–16, page 336
Complementarity of strands in the DNA double helix.

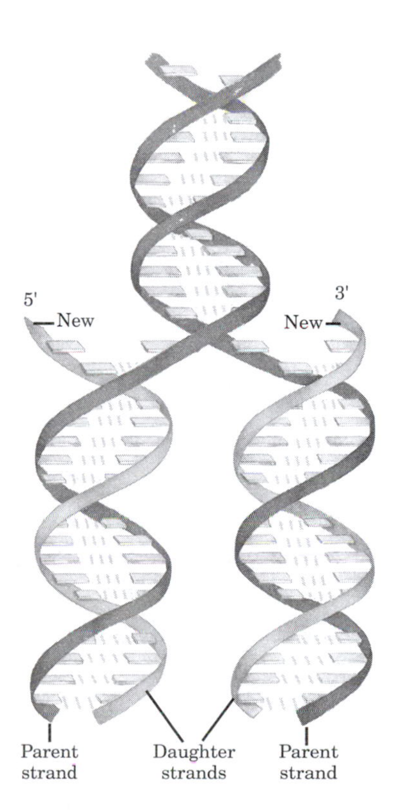

figure 10–17, page 337
Replication of DNA as suggested by Watson and Crick.

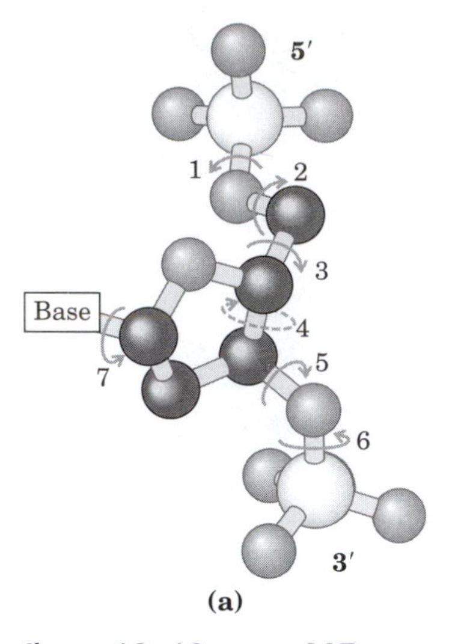

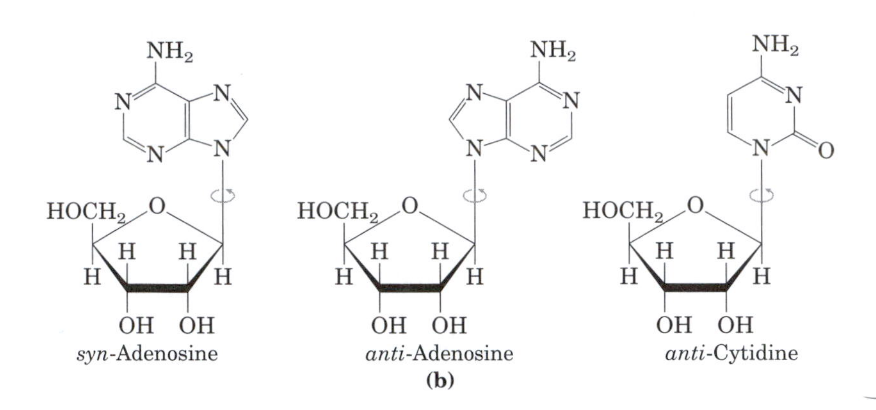

figure 10–18, page 337
Structural variation in DNA.

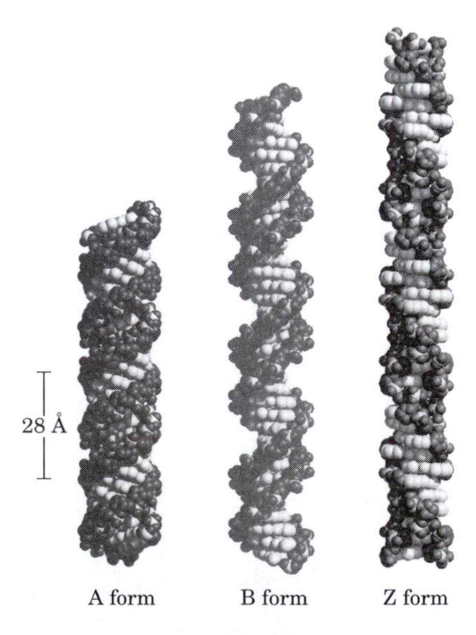

figure 10–19, page 338
Comparison of A, B, and Z forms of DNA.

	A form	B form	Z form
Helical sense	Right handed	Right handed	Left handed
Diameter	~26 Å	~20 Å	~18 Å
Base pairs per helical turn	11	10.5	12
Helix rise per base pair	2.6 Å	3.4 Å	3.7 Å
Base tilt normal to the helix axis	20°	6°	7°
Sugar pucker conformation	C-3′ endo	C-2′ endo	C-2′ endo for pyrimidines; C-3′ endo for purines
Glycosyl bond conformation	Anti	Anti	Anti for pyrmidines; syn for purines

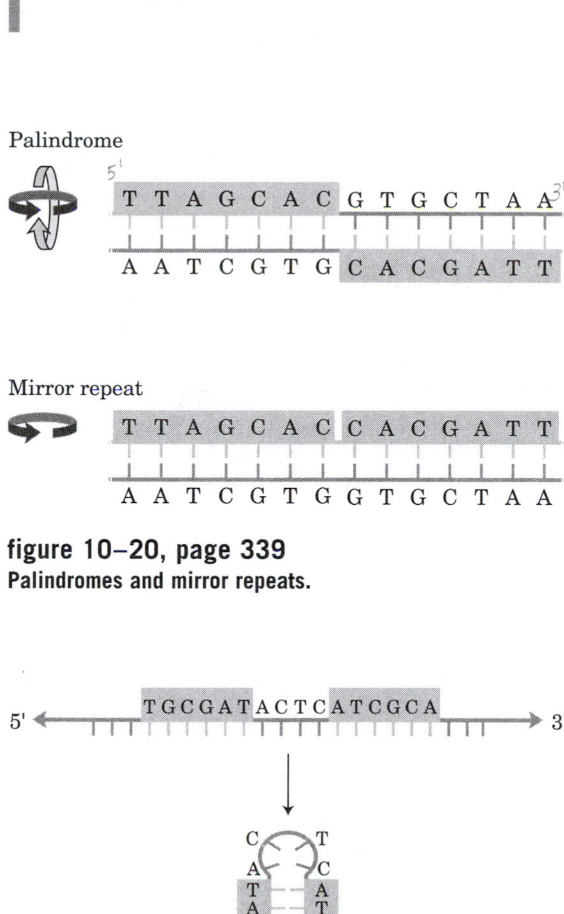

figure 10–20, page 339
Palindromes and mirror repeats.

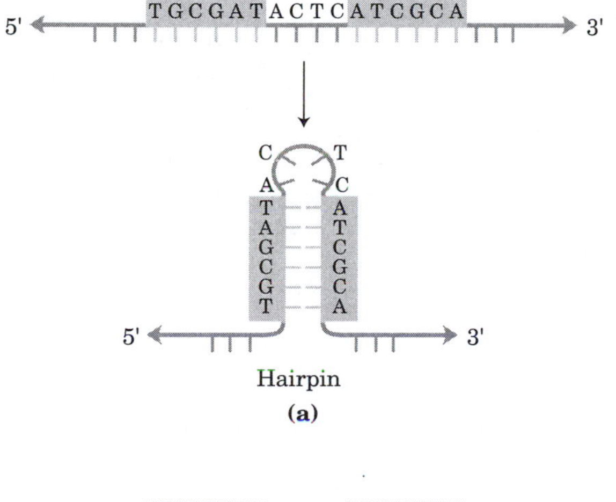

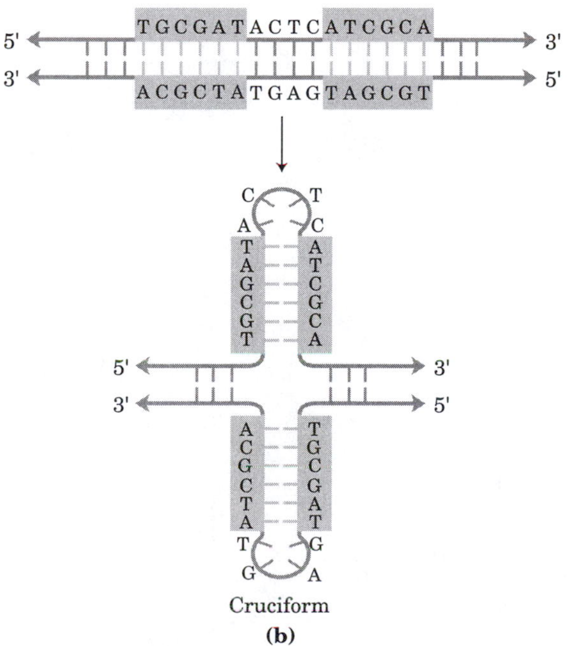

figure 10–21, page 339
Hairpins and cruciforms.

(a) T=A•T $C \equiv G \cdot C^{+}$

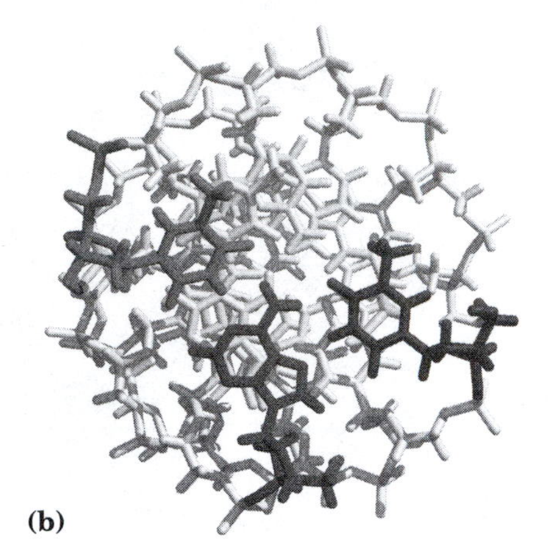

(b)

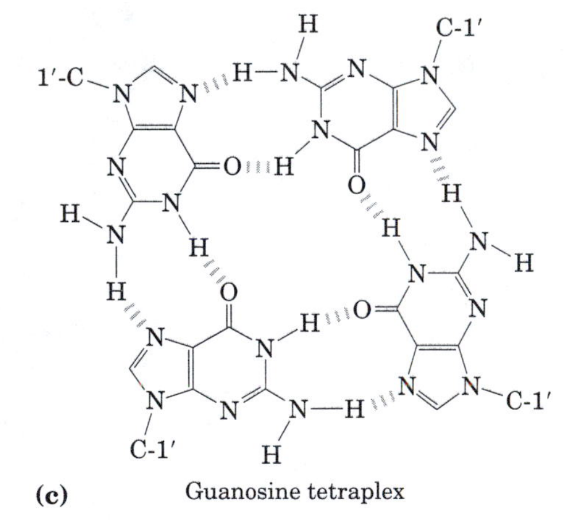

(c) Guanosine tetraplex

(d)

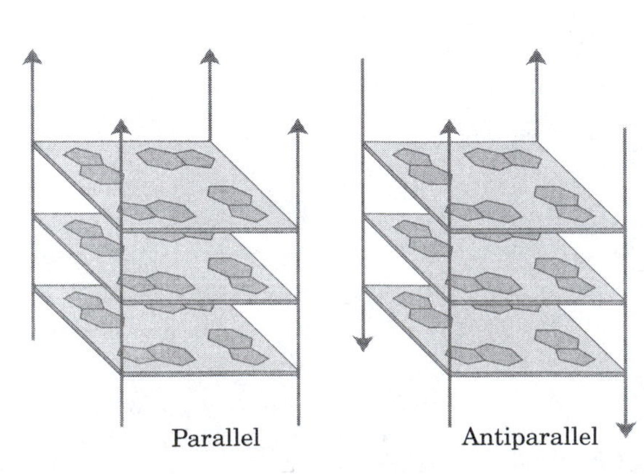

(e)

figure 10–22, page 340
DNA structures containing three or four DNA strands.

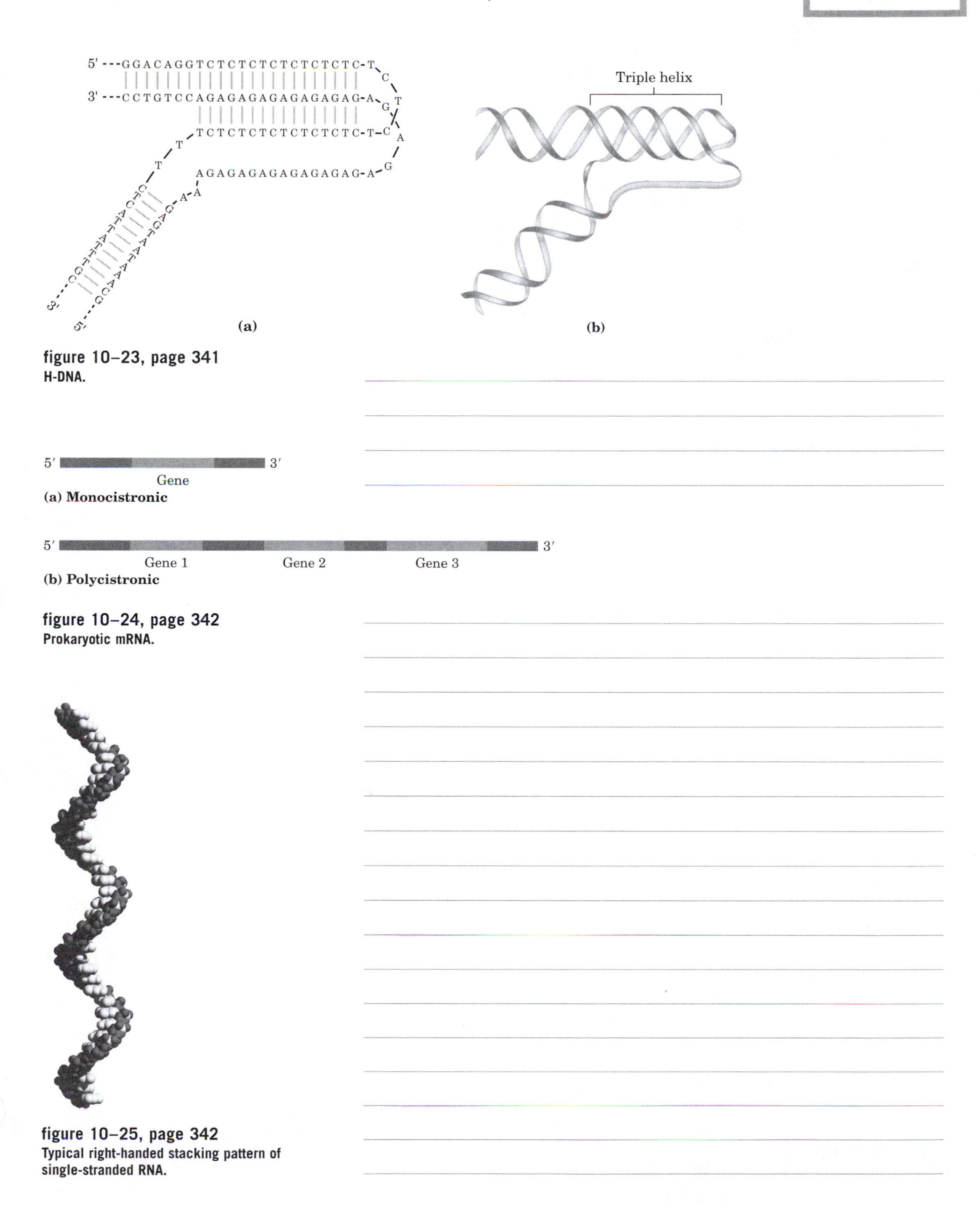

figure 10–23, page 341
H-DNA.

figure 10–24, page 342
Prokaryotic mRNA.

figure 10–25, page 342
Typical right-handed stacking pattern of single-stranded RNA.

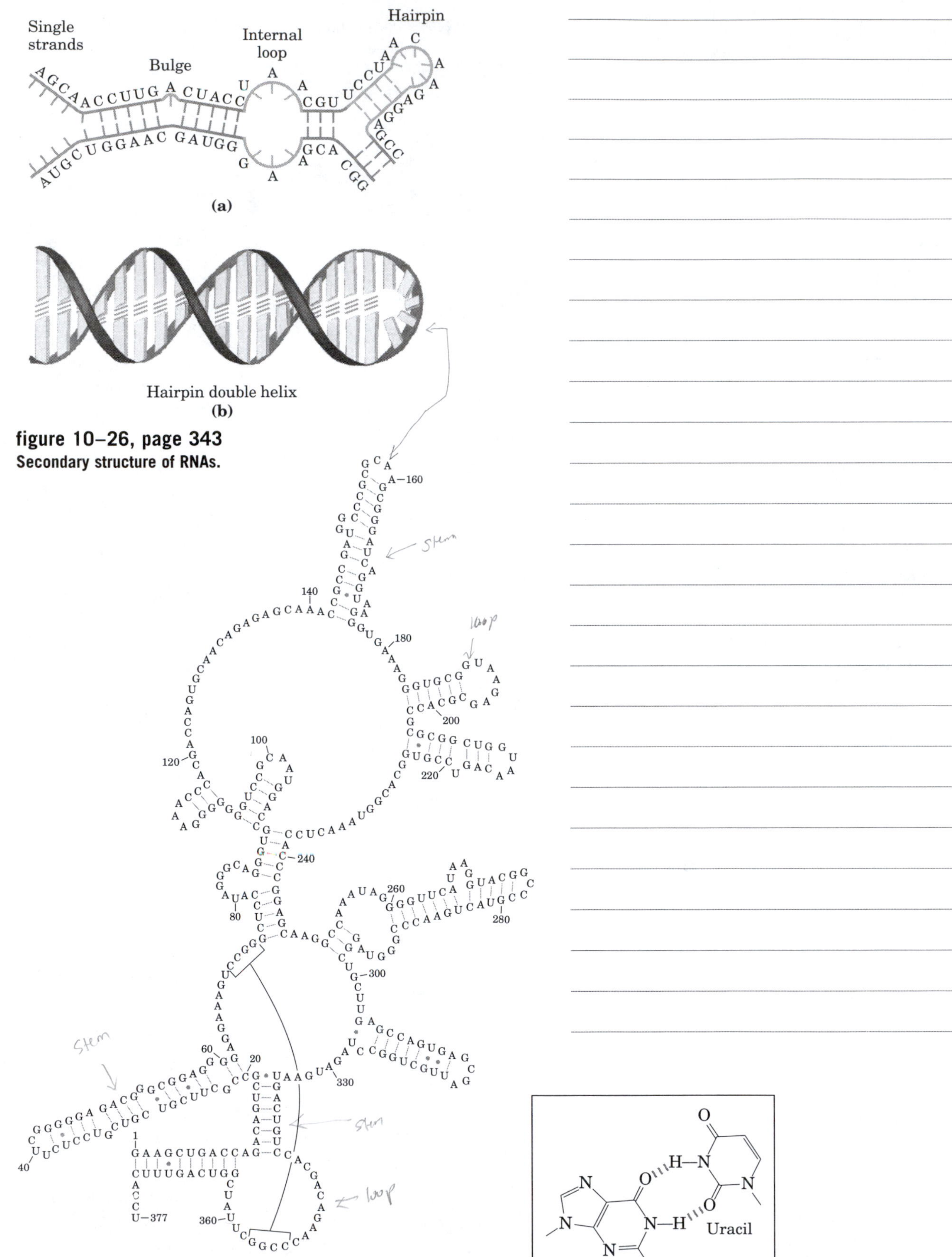

figure 10–26, page 343
Secondary structure of RNAs.

figure 10–27, page 343
Base-paired helical structures in an RNA.

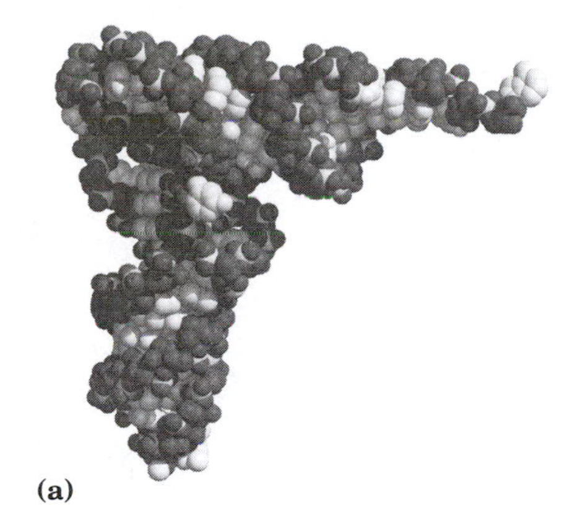

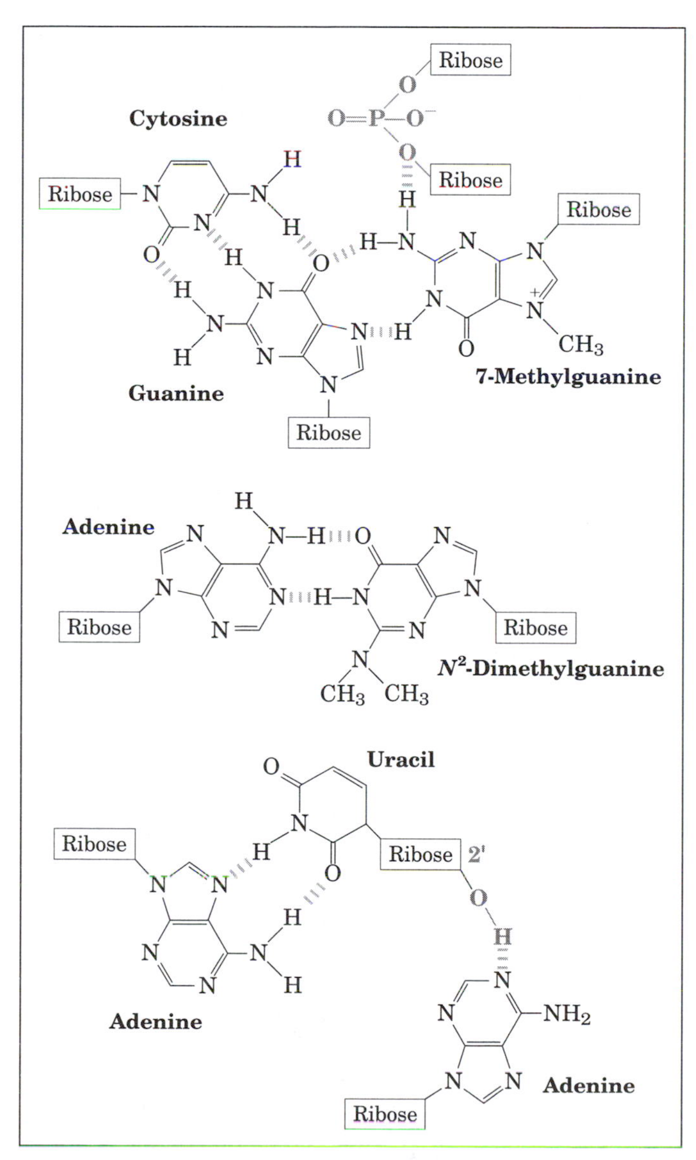

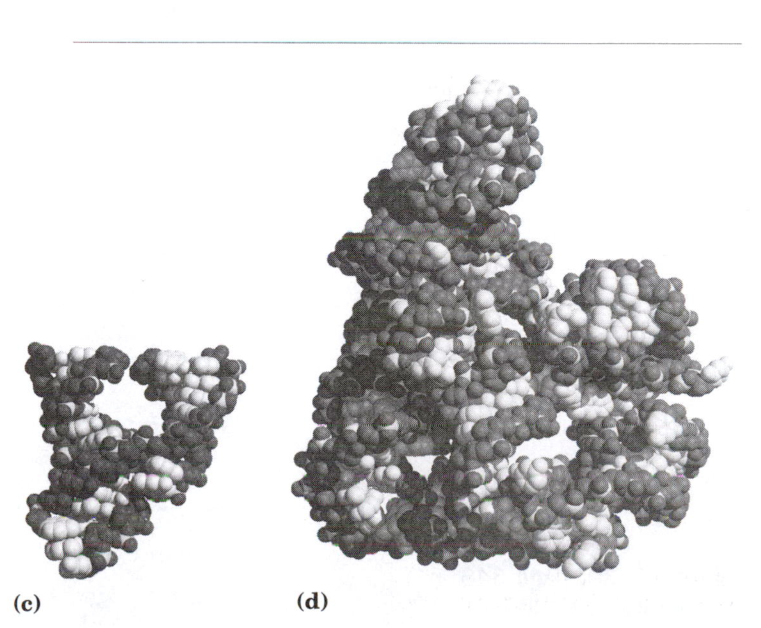

figure 10–28, page 344
Three-dimensional structure in RNA.

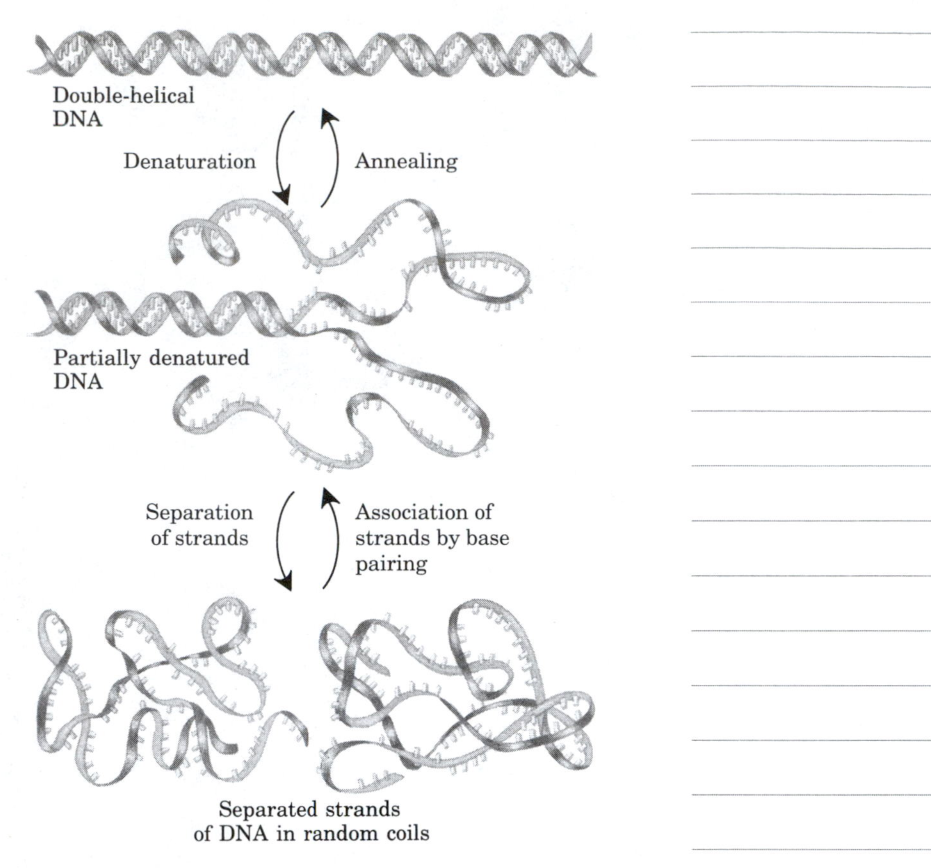

figure 10–29, page 345
Reversible denaturation and annealing (renaturation) of DNA.

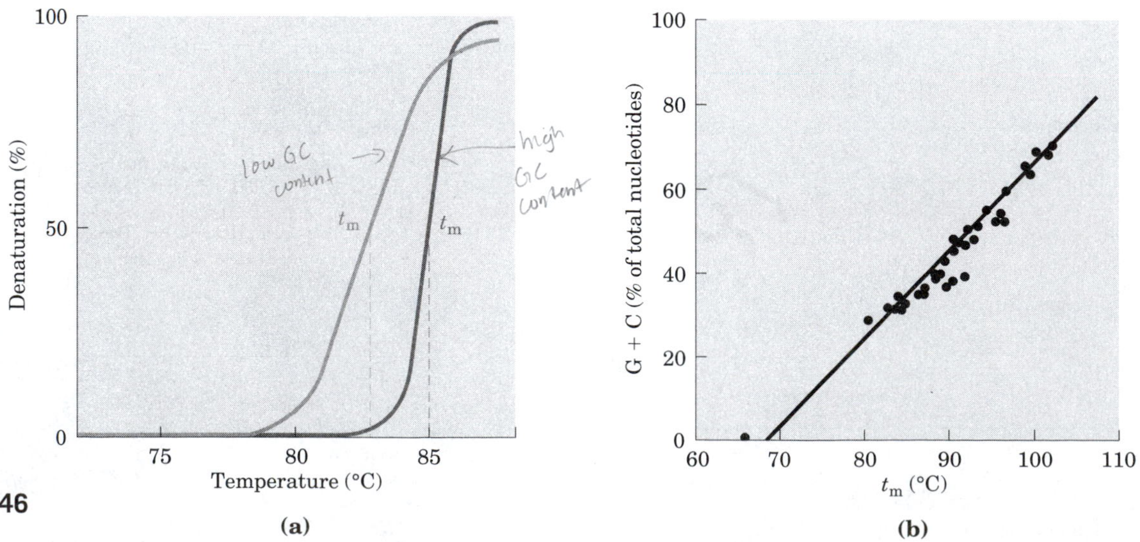

figure 10–30, page 346
Heat denaturation of DNA.

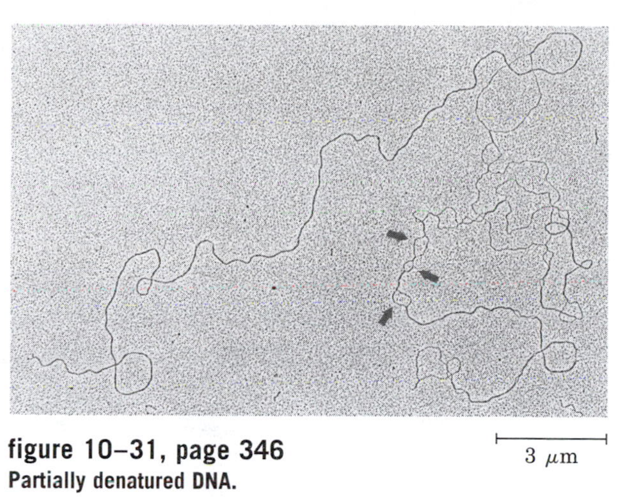

figure 10–31, page 346
Partially denatured DNA.

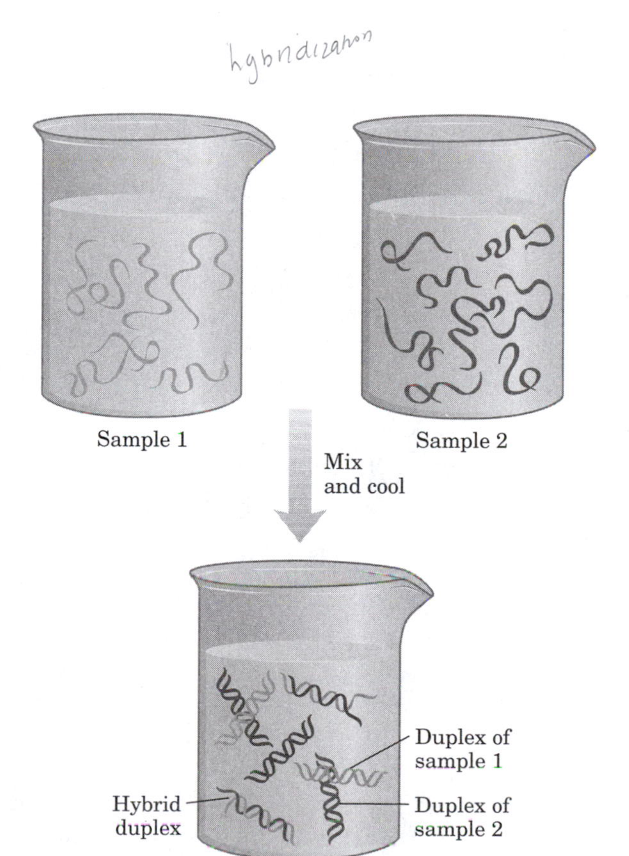

figure 10–32, page 347
DNA hybridization.

Deamination

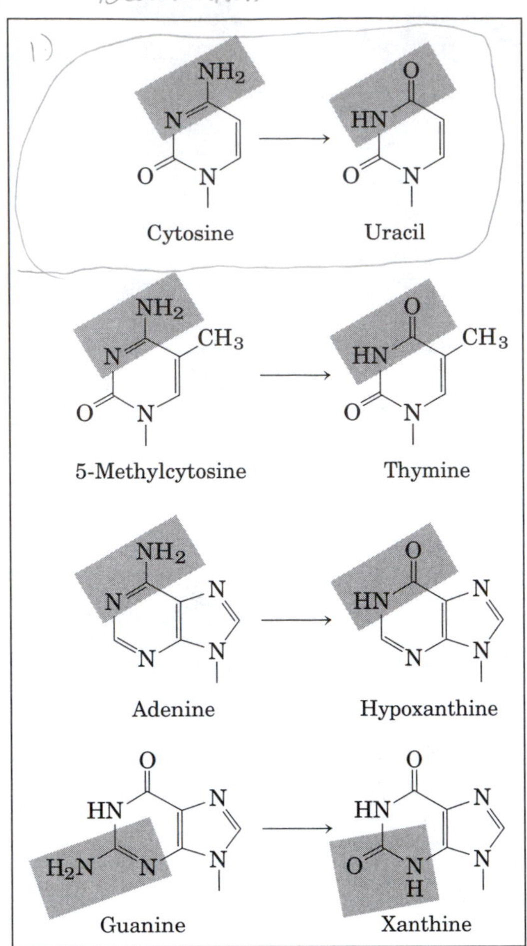

Deamination
(a)

Base Hydrolysis

O
HN
N
H_2N
N
N
O^-
—O—P—O—CH_2 O
O
H H
H H
O H
Guanosine residue
(in DNA)

H_2O
catalyzed by H^+

O
HN
N
H_2N
N
N
H
Guanine

+ O^-
—O—P—O—CH_2 O
O
OH
H H
H H
O H
Apurinic
residue

Depurination
(b)

figure 10–33, page 348
Some well-characterized nonenzymatic reactions of nucleotides.

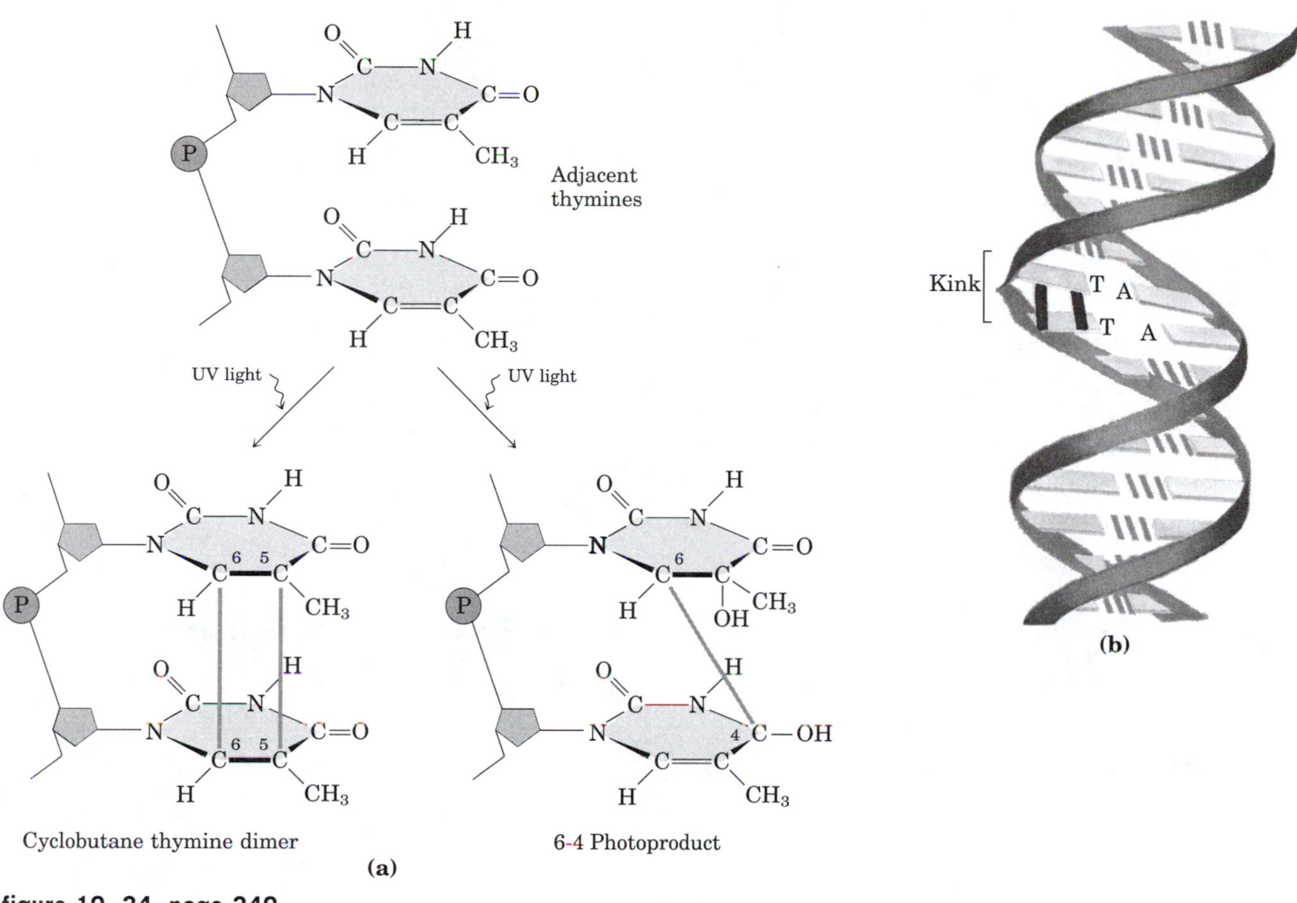

figure 10–34, page 349
Formation of pyrimidine dimers induced by UV light.

Guanine tautomers

$(CH_3)_2SO_4$

O^6-Methylguanine

Methylation of guanine, page 350

$NaNO_2$
Sodium nitrite

$NaNO_3$
Sodium nitrate

R^1, R^2 N—N=O
Nitrosamine

Nitrous acid precursors
(a)

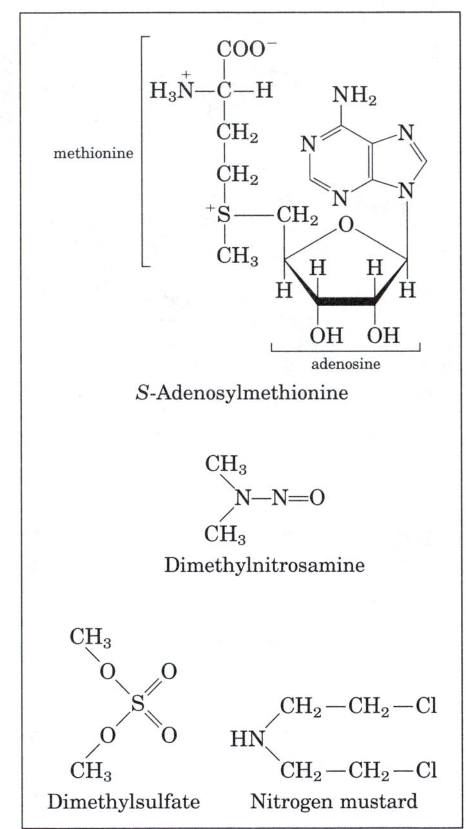

Alkylating agents
(b)

figure 10–35, page 350
Chemical agents that cause DNA damage.

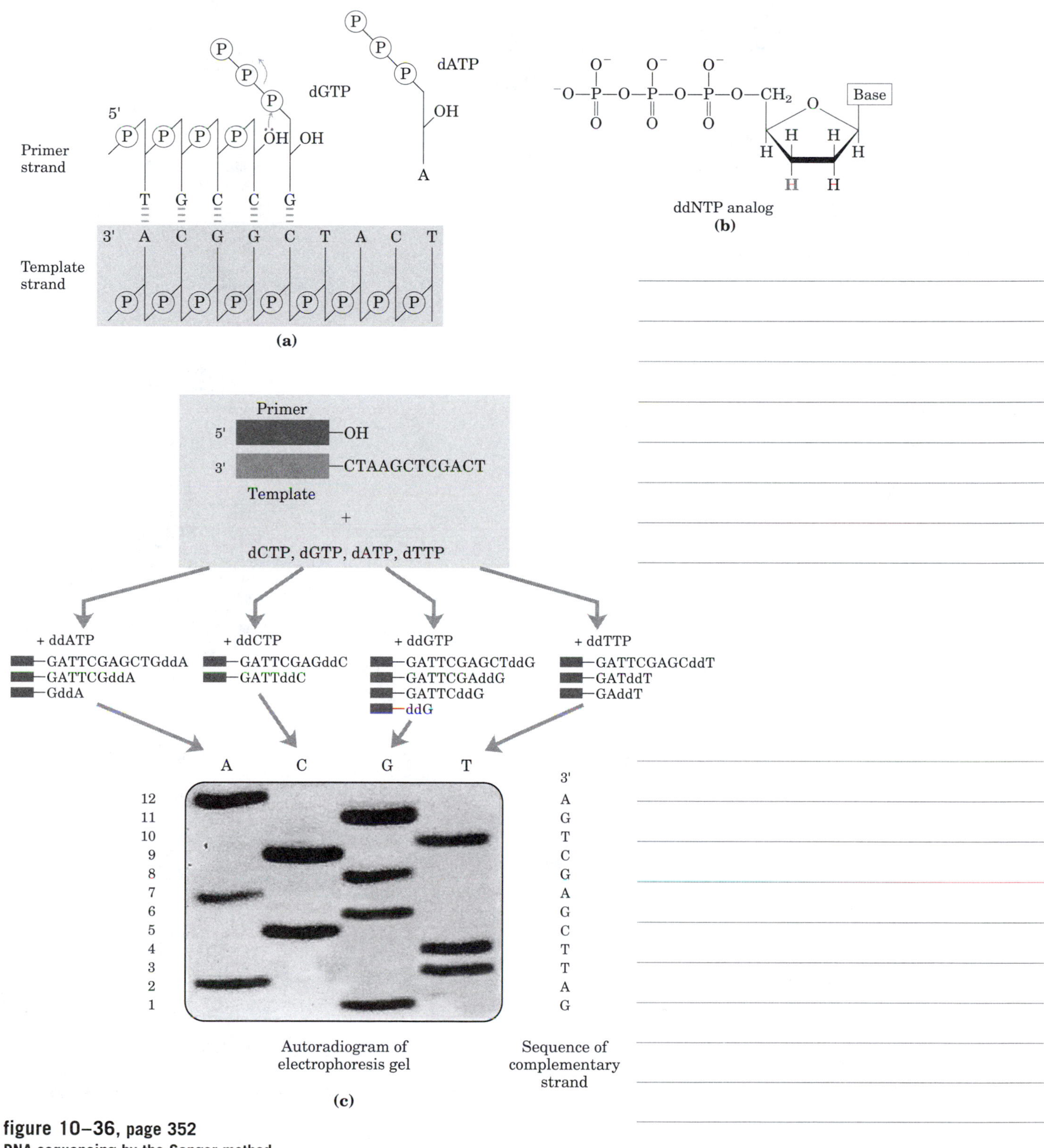

figure 10–36, page 352
DNA sequencing by the Sanger method.

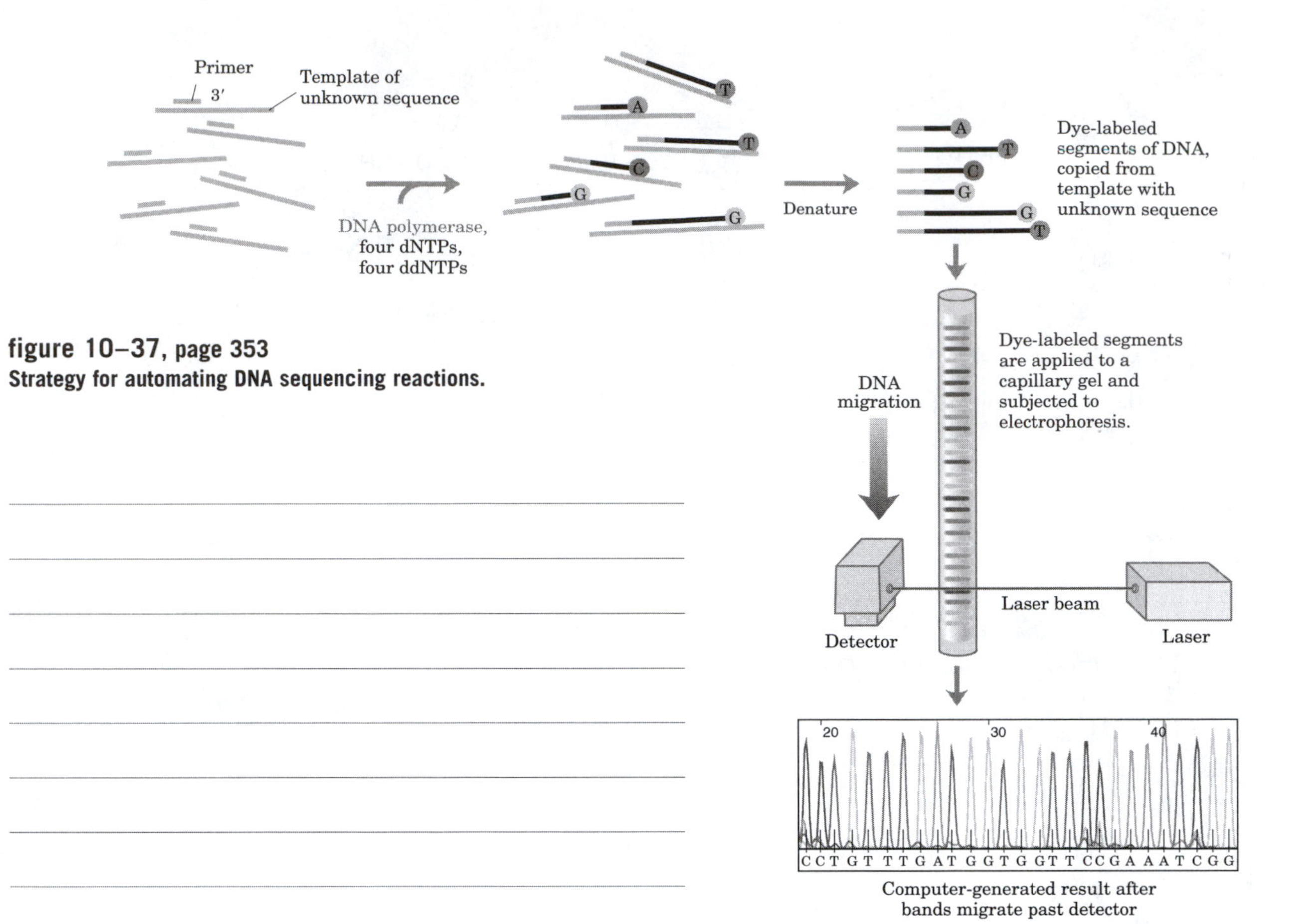

figure 10–37, page 353
Strategy for automating DNA sequencing reactions.

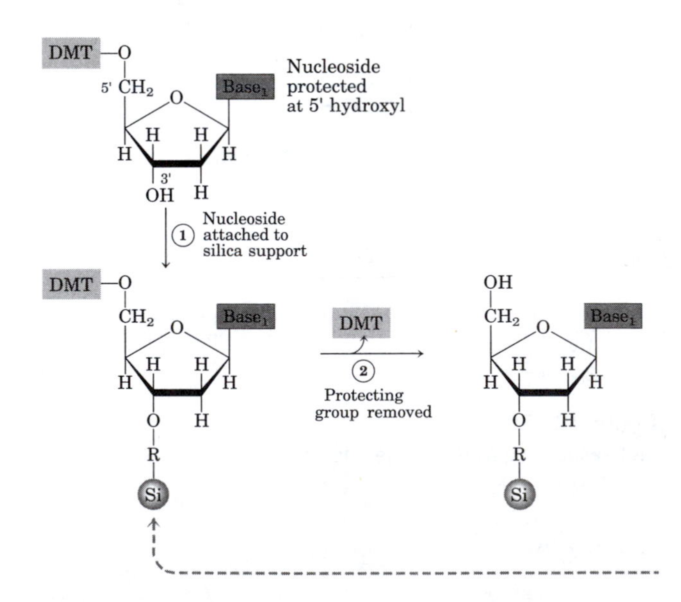

figure 10–38, page 354–355
Chemical synthesis of DNA.

γ β α
NMP
NDP
NTP

Abbreviations of ribonucleoside 5'-phosphates			
Base	Mono-	Di-	Tri-
Adenine	AMP	ADP	ATP
Guanine	GMP	GDP	GTP
Cytosine	CMP	CDP	CTP
Uracil	UMP	UDP	UTP

Abbreviations of deoxyribonucleoside 5'-phosphates			
Base	Mono-	Di-	Tri-
Adenine	dAMP	dADP	dATP
Guanine	dGMP	dGDP	dGTP
Cytosine	dCMP	dCDP	dCTP
Thymine	dTMP	dTDP	dTTP

figure 10–39, page 355
General structure of the nucleoside 59-mono-, di-, and triphosphates (NMPs, NDPs, and NTPs) and their standard abbreviations.

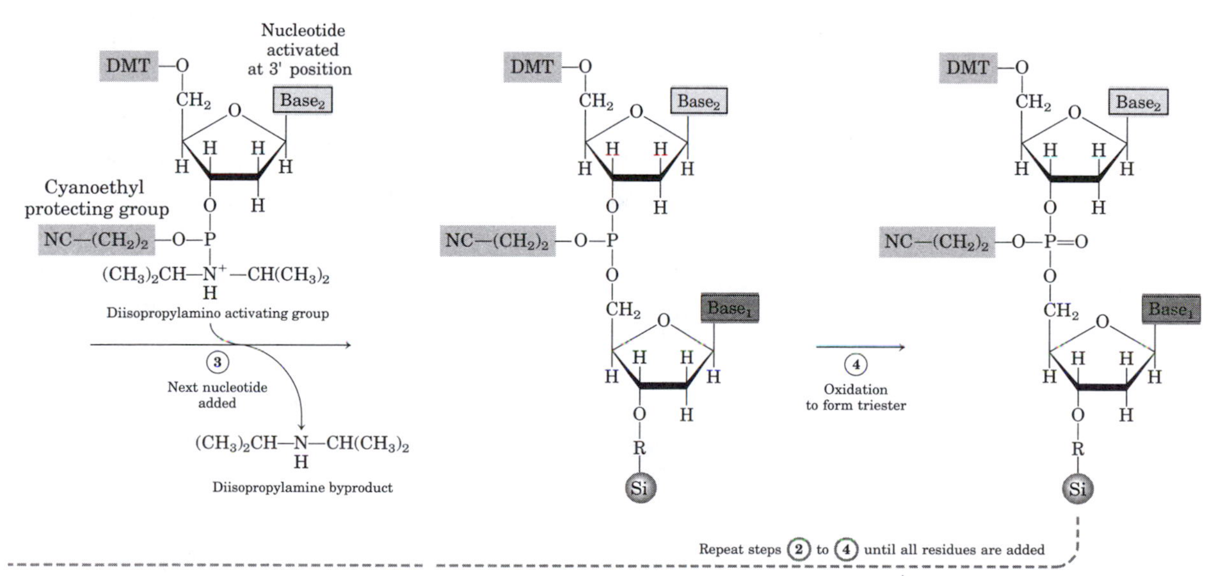

⑤ Remove protecting groups from bases
⑥ Remove cyanoethyl groups from phosphates
⑦ Cleave chain from silica support

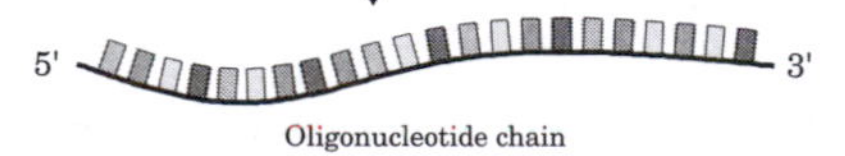

Oligonucleotide chain

Ester

Anhydride

Anhydride

Adenine

ATP

Acetic anhydride, a carboxylic acid anhydride

Methyl acetate, a carboxylic acid ester

figure 10–40, page 356
The phosphate ester and phosphoanhydride bonds of ATP.

β-Mercaptoethylamine

Pantothenic acid

3'-Phosphoadenosine diphosphate (3'-P-ADP)

Coenzyme A

Nicotinamide

Nicotinamide adenine dinucleotide (NAD+)

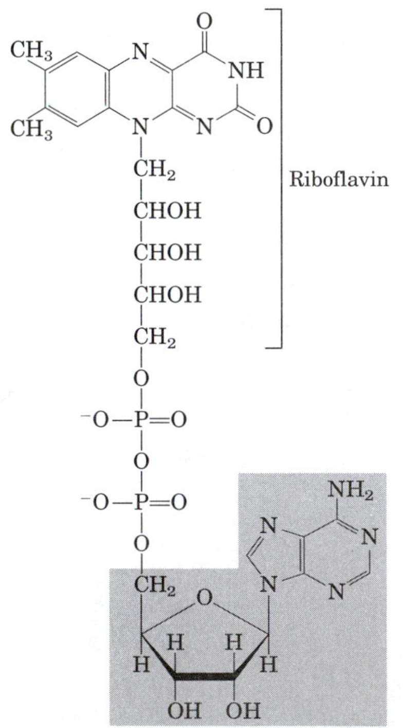

Flavin adenine dinucleotide (FAD)

figure 10–41, page 357
Some coenzymes containing adenosine.

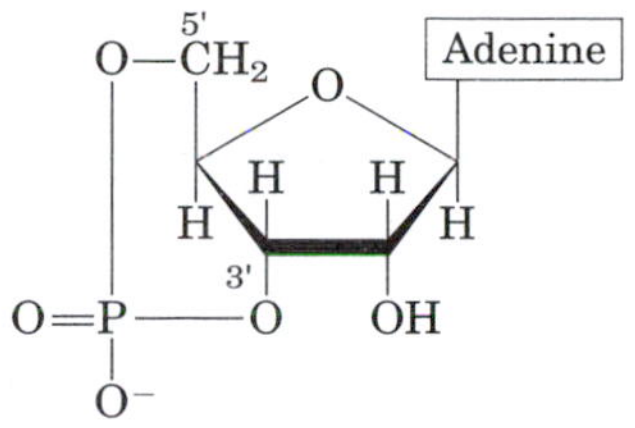

Adenosine 3',5'-cyclic monophosphate
(cyclic AMP; cAMP)

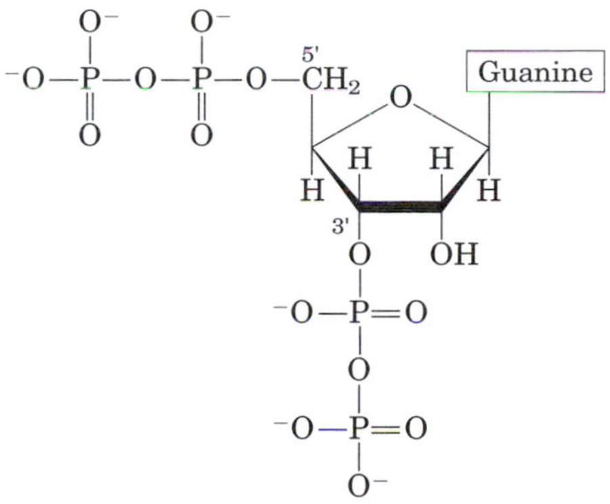

Guanosine 3'-diphosphate,5'-diphosphate
(guanosine tetraphosphate)
(ppGpp)

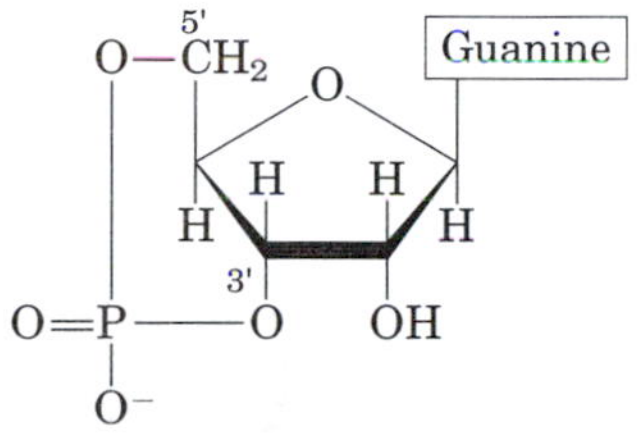

Guanosine 3',5'-cyclic monophosphate
(cyclic GMP; cGMP)

figure 10–42, page 358
Three regulatory nucleotides.

PROBLEMS

1. Determination of Protein Concentration in a Solution Containing Proteins and Nucleic Acids, page 361.

$R_{280/260}$	Proportion of nucleic acid (%)	F
1.75	0.00	1.116
1.63	0.25	1.081
1.52	0.50	1.054
1.40	0.75	1.023
1.36	1.00	0.994
1.30	1.25	0.970
1.25	1.50	0.944
1.16	2.00	0.899
1.09	2.50	0.852
1.03	3.00	0.814
0.979	3.50	0.776
0.939	4.00	0.743
0.874	5.00	0.682
0.846	5.50	0.656
0.822	6.00	0.632
0.804	6.50	0.607
0.784	7.00	0.585
0.767	7.50	0.565
0.753	8.00	0.545
0.730	9.00	0.508
0.705	10.00	0.478
0.671	12.00	0.422
0.644	14.00	0.377
0.615	17.00	0.322
0.595	20.00	0.278

10. DNA Sequencing, pages 361–362.

*5′ ——— 3′-OH
3′ ——— ATTACGCAAGGACATTAGAC---5′

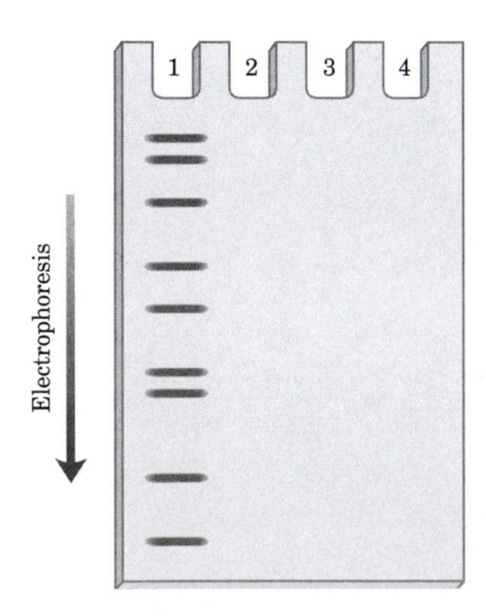

chapter

11

Lipids

CD-ROM

When you study for exams, don't forget to review The Minicourses on the UNDERSTAND! BIOCHEMISTRY CD that came with your textbook.

Minicourses that apply to this Chapter include:
Molecules of Life
Important Biochemicals
An Interactive Gallery of Molecular Models

table 11–1, page 364

Some Naturally Occurring Fatty Acids

Carbon skeleton	Structure*	Systematic name†	Common name (derivation)	Melting point (°C)	Solubility at 30 °C (mg/g solvent): Water	Solubility at 30 °C (mg/g solvent): Benzene
12:0	$CH_3(CH_2)_{10}COOH$	*n*-Dodecanoic acid	Lauric acid (Latin *laurus,* "laurel plant")	44.2	0.063	2,600
14:0	$CH_3(CH_2)_{12}COOH$	*n*-Tetradecanoic acid	Myristic acid (Latin *Myristica,* nutmeg genus)	53.9	0.024	874
16:0	$CH_3(CH_2)_{14}COOH$	*n*-Hexadecanoic acid	Palmitic acid (Latin *palma,* "palm tree")	63.1	0.0083	348
18:0	$CH_3(CH_2)_{16}COOH$	*n*-Octadecanoic acid	Stearic acid (Greek *stear,* "hard fat")	69.6	0.0034	124
20:0	$CH_3(CH_2)_{18}COOH$	*n*-Eicosanoic acid	Arachidic acid (Latin *Arachis,* legume genus)	76.5		
24:0	$CH_3(CH_2)_{22}COOH$	*n*-Tetracosanoic acid	Lignoceric acid (Latin *lignum,* "wood" + *cera,* "wax")	86.0		
16:1(Δ^9)	$CH_3(CH_2)_5CH{=}CH(CH_2)_7COOH$	*cis*-9-Hexadecenoic acid	Palmitoleic acid	−0.5		
18:1(Δ^9)	$CH_3(CH_2)_7CH{=}CH(CH_2)_7COOH$	*cis*-9-Octadecenoic acid	Oleic acid (Latin *oleum,* "oil")	13.4		
18:2($\Delta^{9,12}$)	$CH_3(CH_2)_4CH{=}CHCH_2CH{=}CH(CH_2)_7COOH$	*cis-,cis*-9,12-Octadecadienoic acid	Linoleic acid (Greek *linon,* "flax")	−5		
18:3($\Delta^{9,12,15}$)	$CH_3CH_2CH{=}CHCH_2CH{=}CHCH_2CH{=}CH(CH_2)_7COOH$	*cis-,cis-,cis*-9,12,15-Octadecatrienoic acid	α-Linolenic acid	−11		
20:4($\Delta^{5,8,11,14}$)	$CH_3(CH_2)_4CH{=}CHCH_2CH{=}CHCH_2CH{=}CHCH_2CH{=}CH(CH_2)_3COOH$	*cis-,cis-,cis-,cis*-5,8,11,14-Icosatetraenoic acid	Arachidonic acid	−49.5		

*All acids are shown in their nonionized form. At pH 7, all free fatty acids have an ionized carboxylate. Note that numbering of carbon atoms begins at the carboxyl carbon.

†The prefix *n*- indicates the "normal" unbranched structure. For instance, "dodecanoic" simply indicates 12 carbon atoms, which could be arranged in a variety of branched forms; "*n*-dodecanoic" specifies the linear, unbranched form. For unsaturated fatty acids, the configuration of each double bond is indicated; in biological fatty acids the configuration is almost always cis.

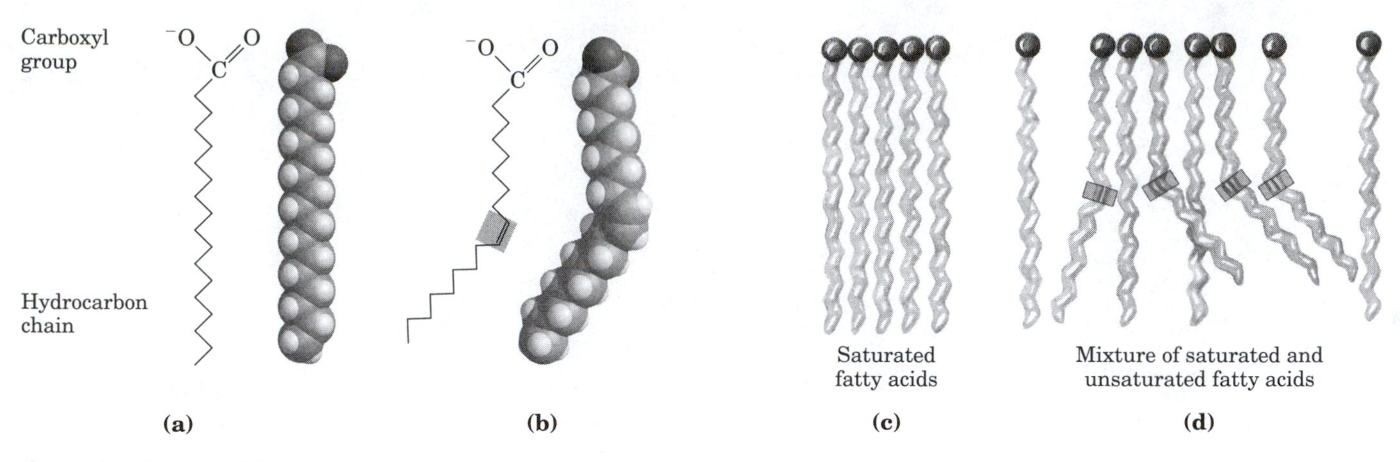

figure 11–1, page 365
The packing of fatty acids into stable aggregates.

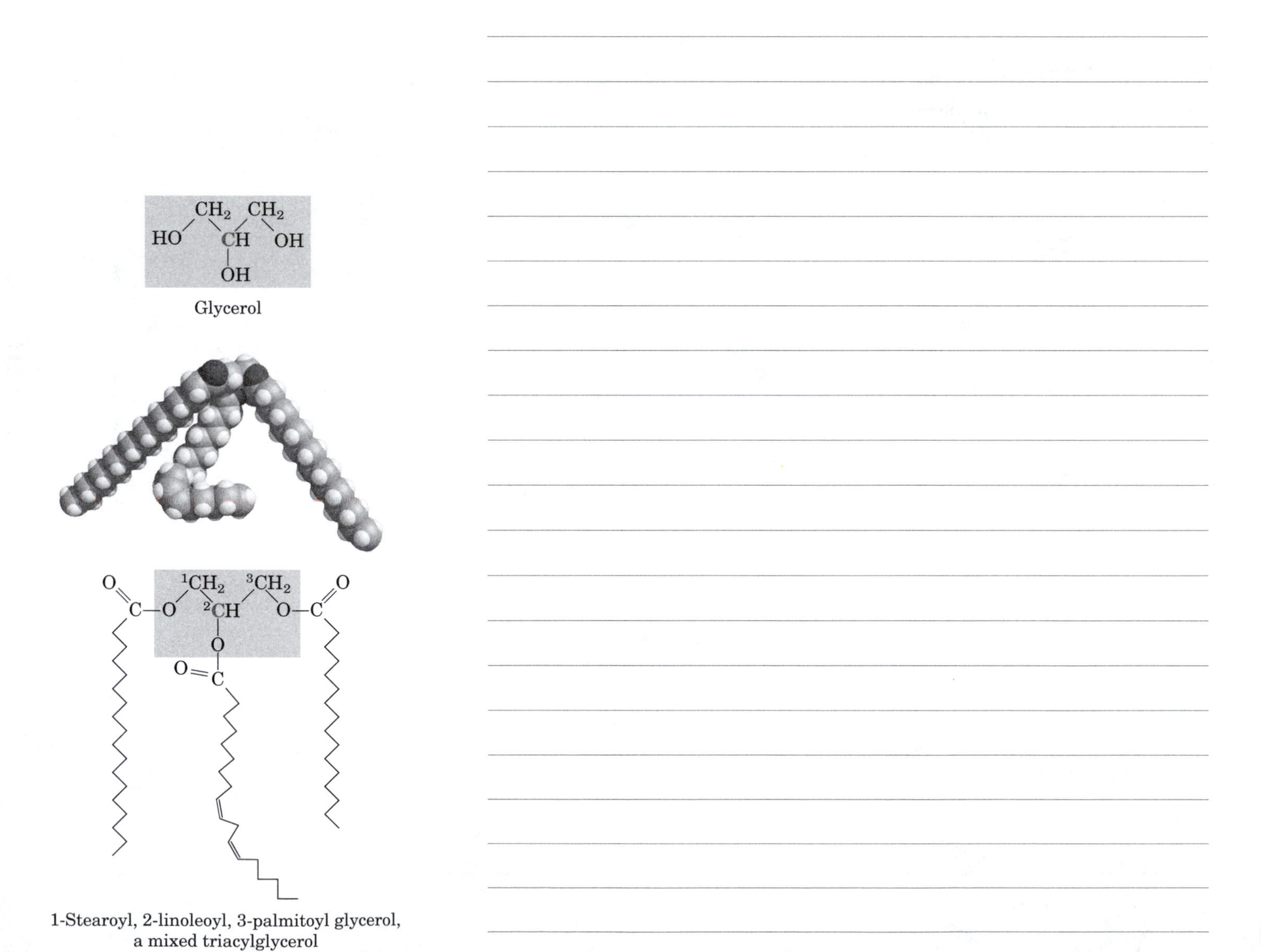

figure 11–2, page 366
Glycerol and a triacylglycerol.

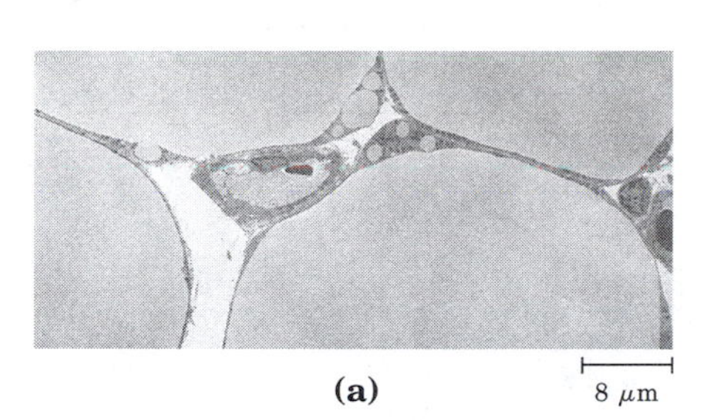

(a) 8 μm

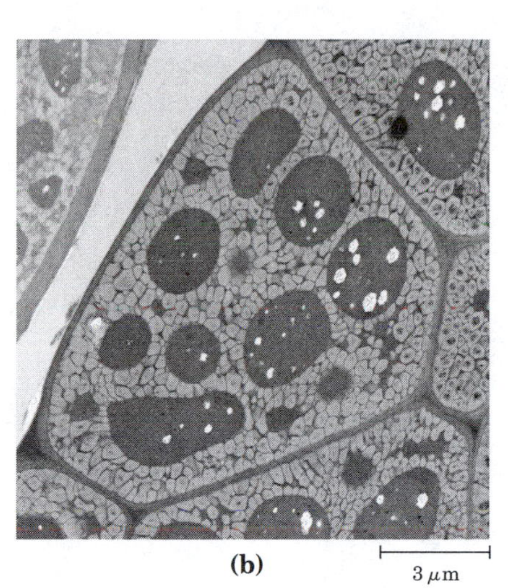

(b) 3 μm

figure 11–3, page 366
Fat stores in cells.

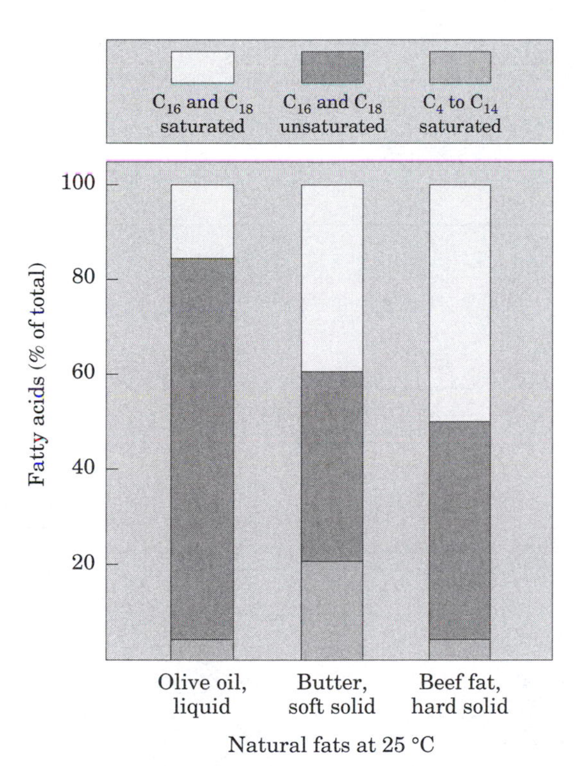

figure 11–4, page 368
Fatty acid composition of three food fats.

$$\underbrace{CH_3(CH_2)_{14}\text{—}\overset{\overset{\large O}{\|}}{C}\text{—}}_{\text{Palmitic acid}}\underbrace{O\text{—}CH_2\text{—}(CH_2)_{28}\text{—}CH_3}_{\text{1-Triacontanol}}$$

(a)

figure 11–5, page 368
Biological wax.

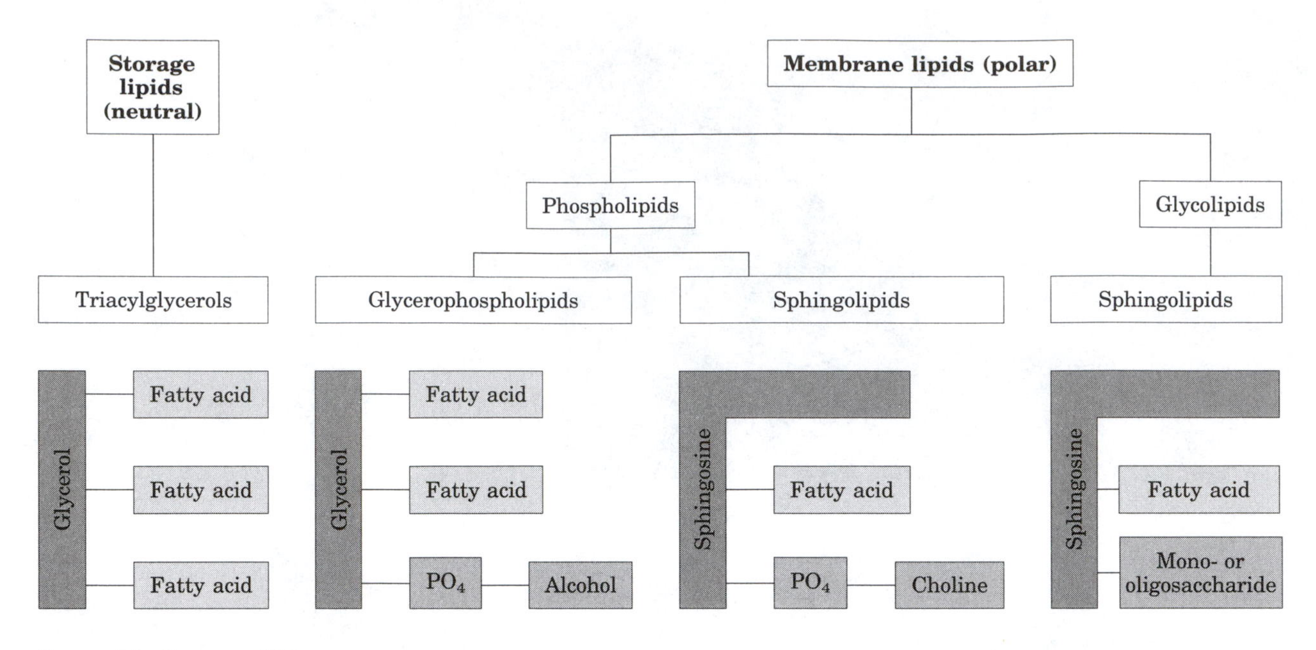

figure 11–6, page 369
The principal classes of storage and membrane lipids.

$$\begin{array}{l} \quad\;\; {}^{1}CH_2OH \\ H - {}^{2}C - OH \qquad O \\ \qquad\quad\;\; \| \\ \quad\;\; {}^{3}CH_2 - O - P - O^- \\ \qquad\qquad\quad\;\; | \\ \qquad\qquad\quad\; O^- \end{array}$$

figure 11–7, page 369
L-Glycerol 3-phosphate, the backbone of phospholipids.

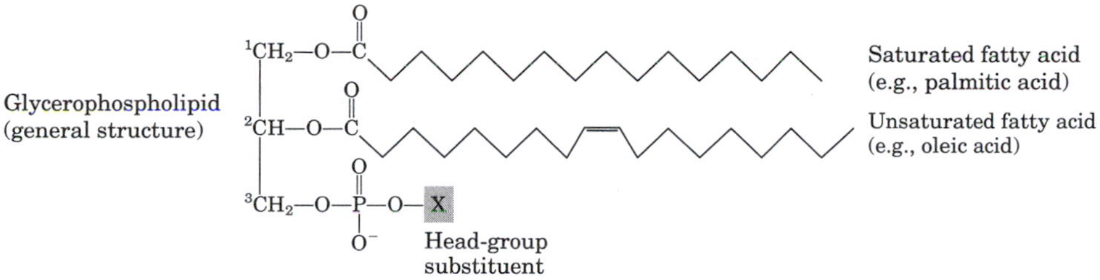

Glycerophospholipid (general structure)

$^1CH_2-O-C(=O)-$ Saturated fatty acid (e.g., palmitic acid)

$^2CH-O-C(=O)-$ Unsaturated fatty acid (e.g., oleic acid)

$^3CH_2-O-P(=O)(O^-)-O-X$ Head-group substituent

Name of glycerophospholipid	Name of X	Formula of X	Net charge (at pH 7)
Phosphatidic acid	—	$-H$	−1
Phosphatidylethanolamine	Ethanolamine	$-CH_2-CH_2-\overset{+}{N}H_3$	0
Phosphatidylcholine	Choline	$-CH_2-CH_2-\overset{+}{N}(CH_3)_3$	0
Phosphatidylserine	Serine	$-CH_2-CH(COO^-)-\overset{+}{N}H_3$	−1
Phosphatidylglycerol	Glycerol	$-CH_2-CH(OH)-CH_2-OH$	−1
Phosphatidylinositol 4,5-bisphosphate	*myo*-Inositol 4,5-bisphosphate	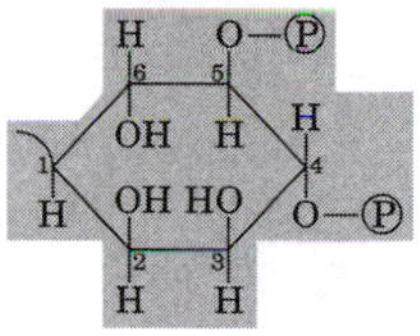 (inositol ring with O–Ⓟ at C-4 and C-5)	−4
Cardiolipin	Phosphatidyl-glycerol	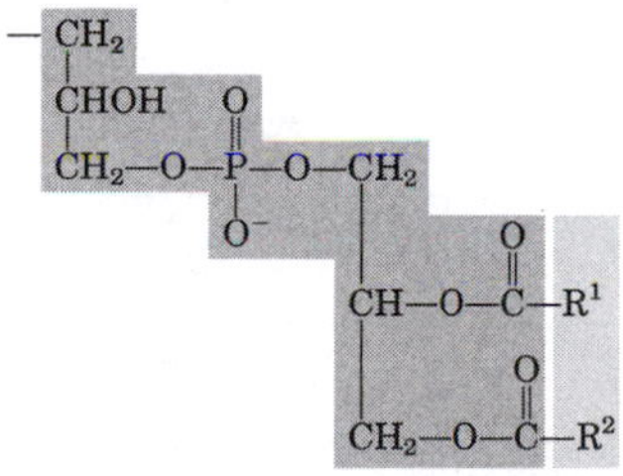 $-CH_2-CHOH-CH_2-O-P(=O)(O^-)-O-CH_2-CH(-O-C(=O)-R^1)-CH_2-O-C(=O)-R^2$	−2

figure 11–8, page 370
Glycerophospholipids.

ether-linked alkene

Plasmalogen

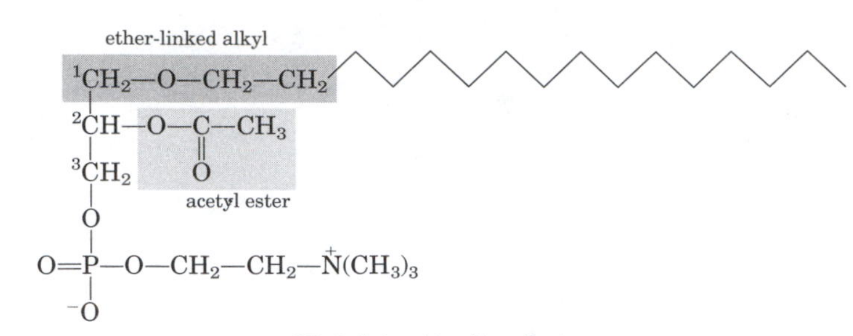

figure 11–9, page 371
Ether lipids.

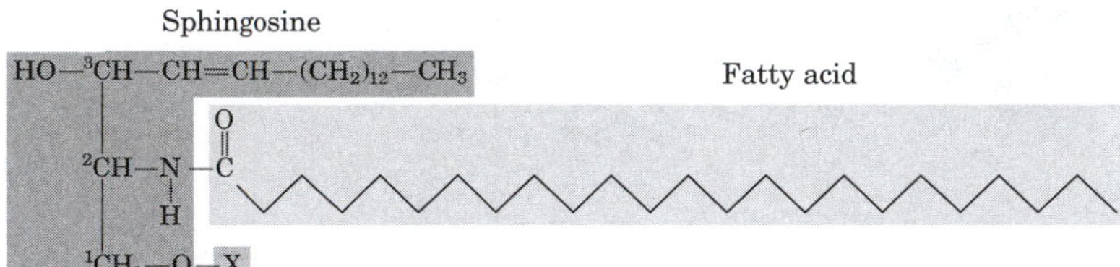

Name of sphingolipid	Name of X	Formula of X
Ceramide	—	— H
Sphingomyelin	Phosphocholine	$-P(=O)(O^-)-O-CH_2-CH_2-\overset{+}{N}(CH_3)_3$
Neutral glycolipids Glucosylcerebroside	Glucose	glucose ring
Lactosylceramide (a globoside)	Di-, tri-, or tetrasaccharide	–Glc–Gal
Ganglioside GM2	Complex oligosaccharide	–Glc–Gal(–Neu5Ac)–GalNAc

figure 11–10, page 372
Sphingolipids.

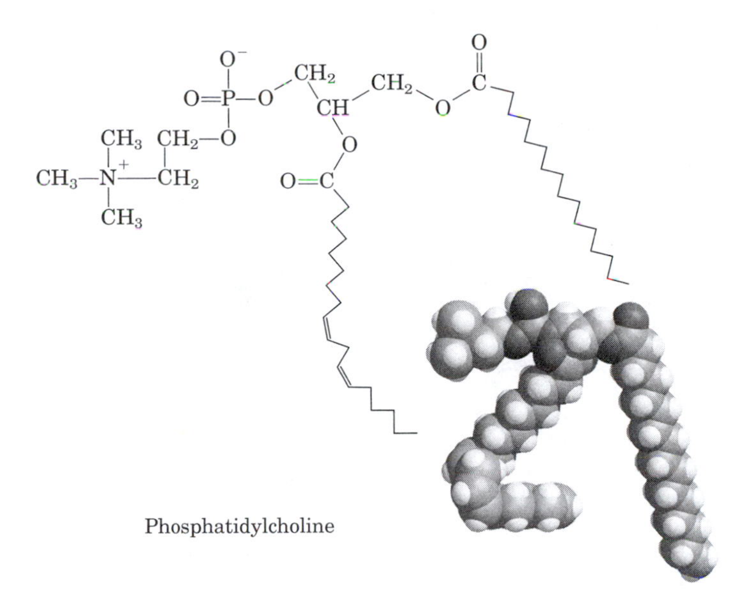

O⁻
O=P—O
CH₂
CH
OH
CH₃ CH₂—O
NH
C
H
CH₃—N⁺—CH₂
CH₃
O=C

Sphingomyelin

figure 11–11, page 373
The similarities in shape and in molecular structure of phosphatidylcholine (a glycerophospholipid) and sphingomyelin (a sphingolipid).

COO⁻
O
OH OH
H
OH
CH—CH—CH₂OH
OH H
H
H
H HN—C—CH₃
O

***N*-Acetylneuraminic acid (sialic acid) (Neu5Ac), page 373**

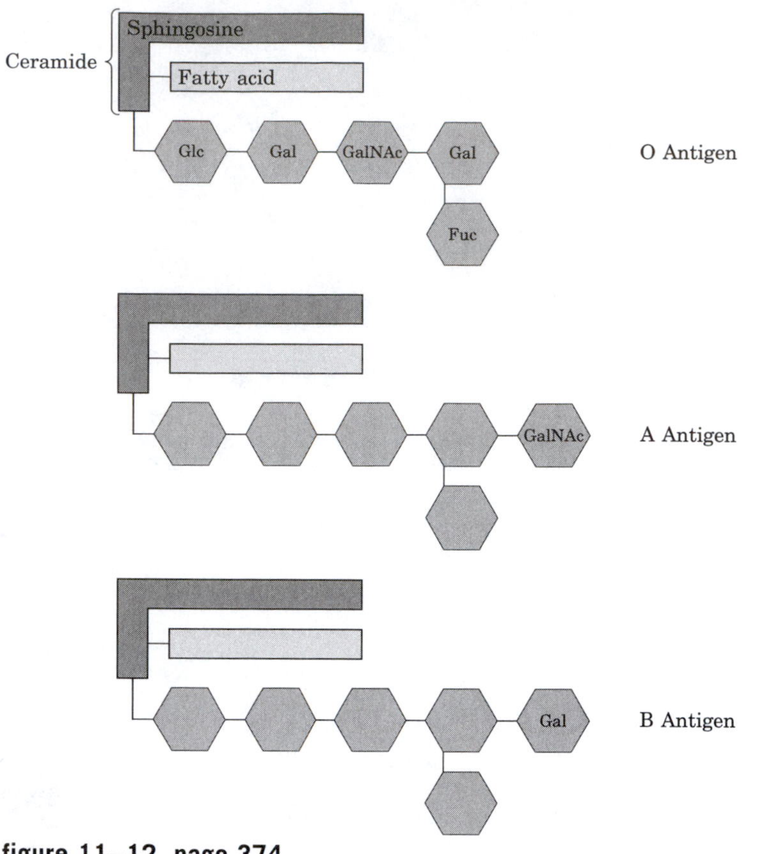

figure 11–12, page 374
Glycosphingolipids as determinants of blood groups.

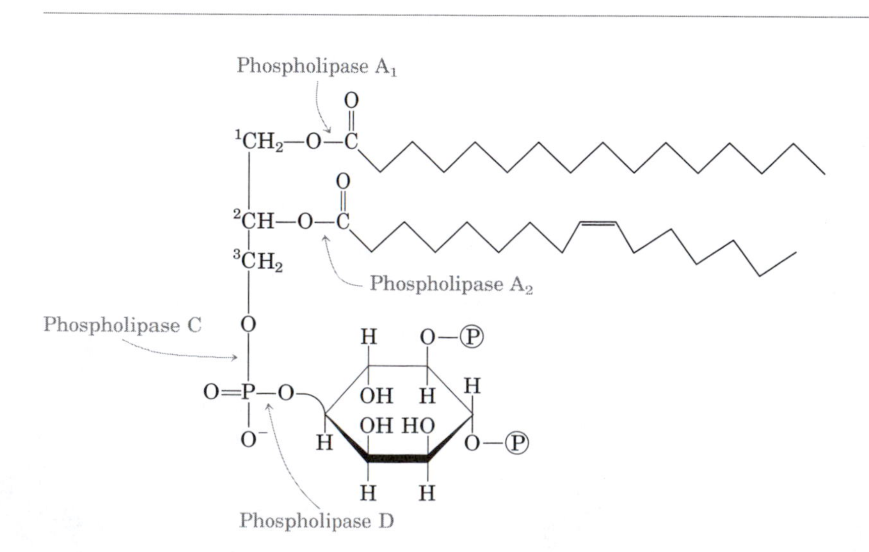

figure 11–13, page 374
The specificities of phospholipases.

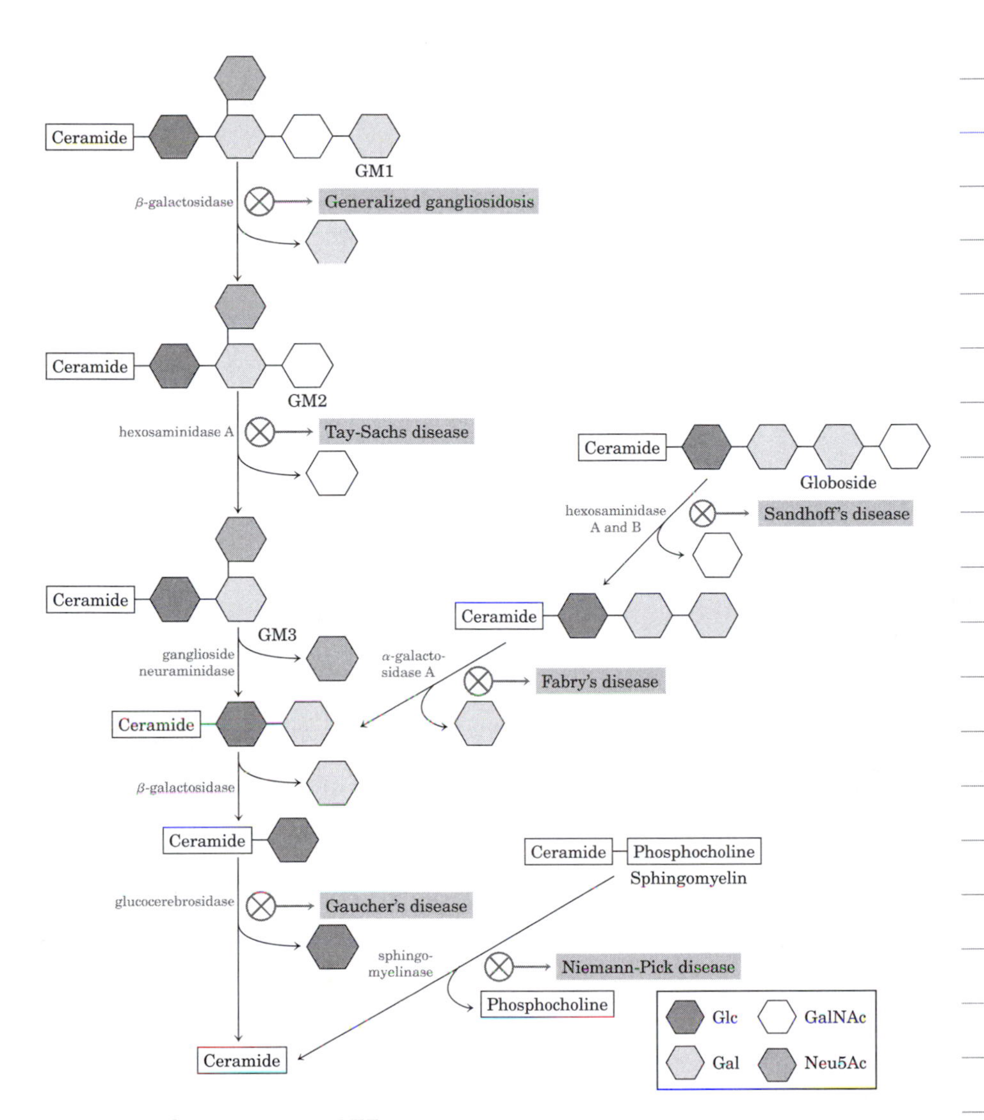

box 11-2, figure 1, page 375
Pathways for the breakdown of GM1, globoside, and sphingomyelin to ceramide.

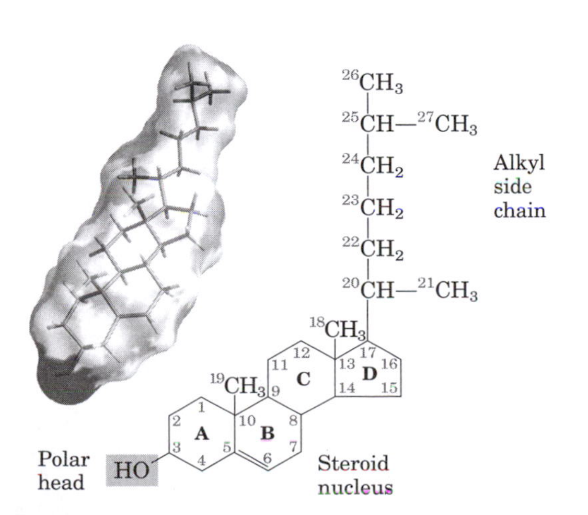

figure 11–14, page 376
Cholesterol.

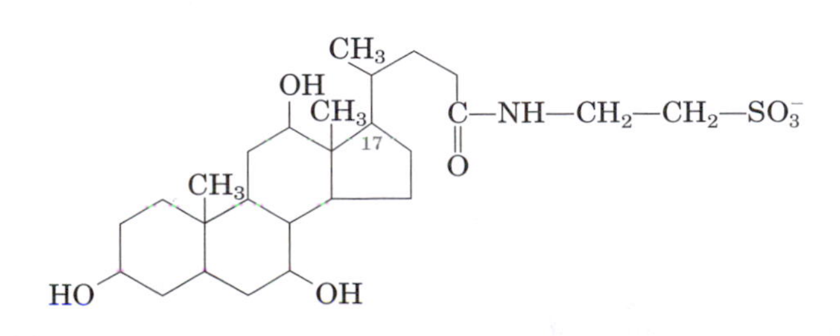

Taurocholic acid (a bile acid), page 376

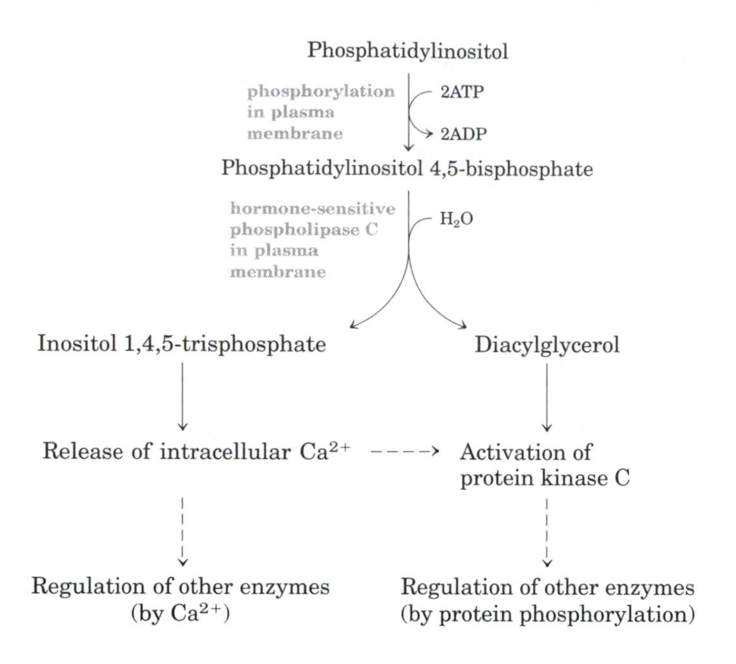

figure 11–15, page 377
Phosphatidylinositols in cellular regulation.

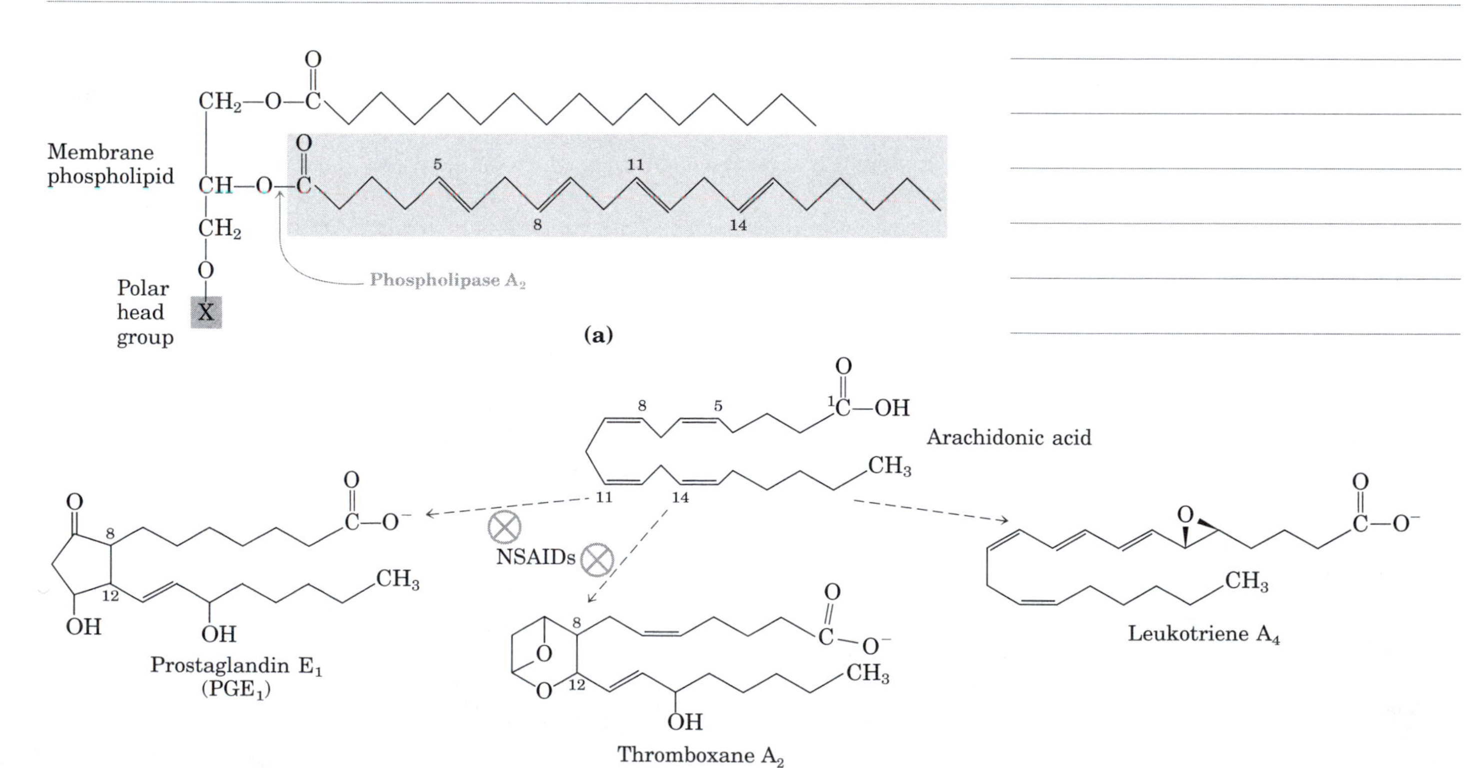

figure 11–16, page 378
Arachidonic acid and some eicosanoid derivatives.

Testosterone

Estradiol

Cortisol

Aldosterone

Prednisolone

Prednisone

figure 11–17, page 379
Steroids derived from cholesterol.

Cholecalciferol (vitamin D_3)

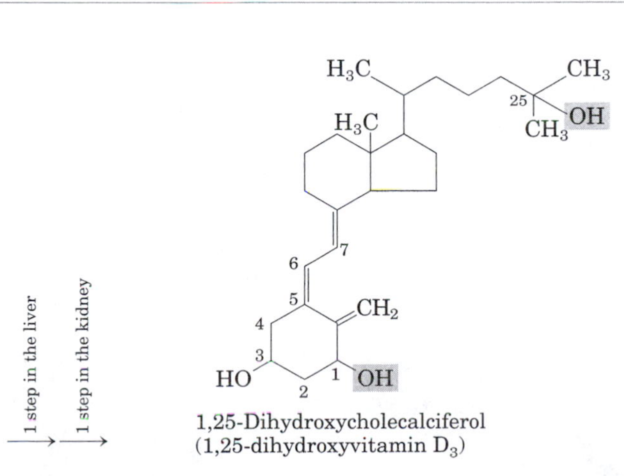

figure 11–18, page 380
Vitamin D_3 production and metabolism.

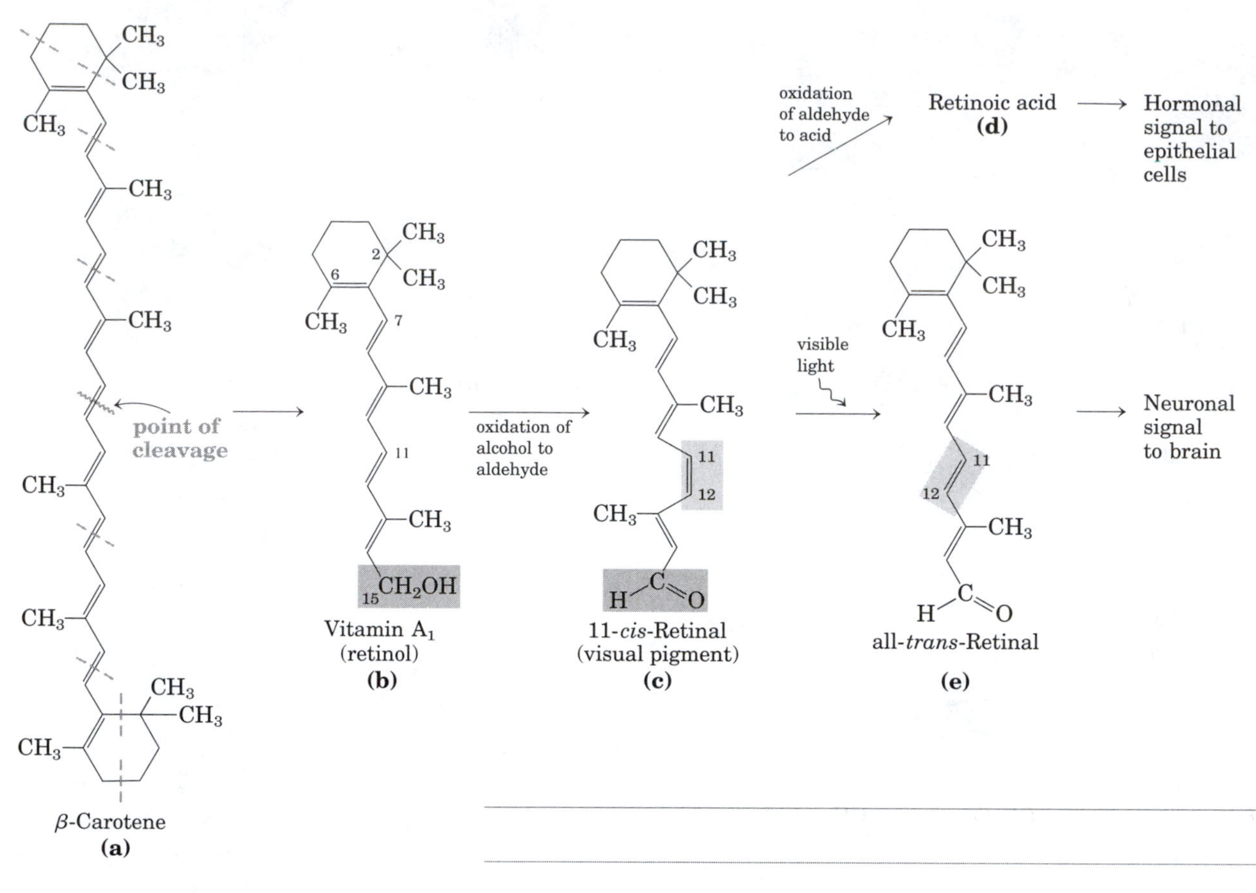

figure 11–19, page 381
Vitamin A_1, its precursor, and derivatives.

(a)
Vitamin E: an antioxidant

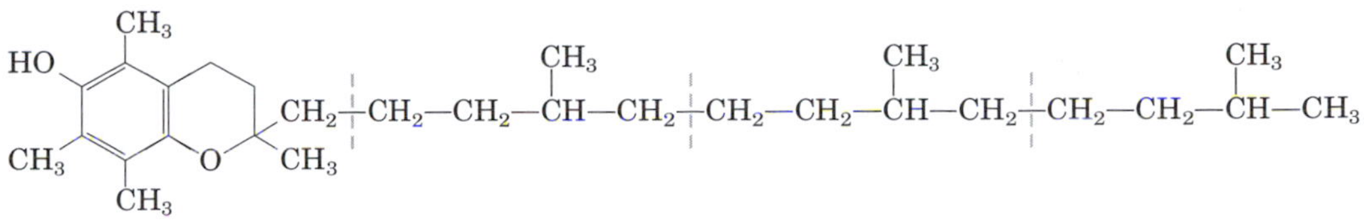

(b)
Vitamin K_1: a blood-clotting cofactor (phylloquinone)

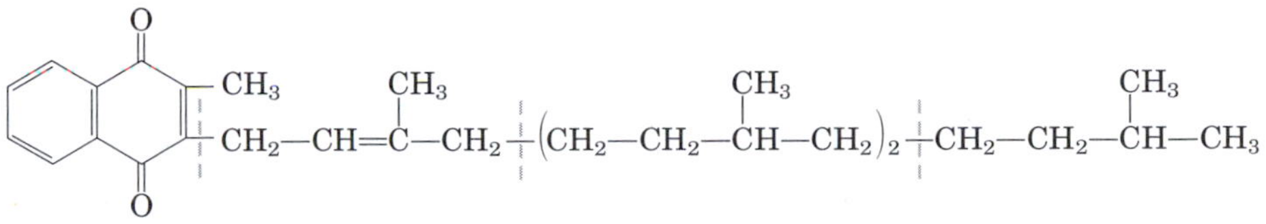

(c)
Warfarin: a blood anticoagulant

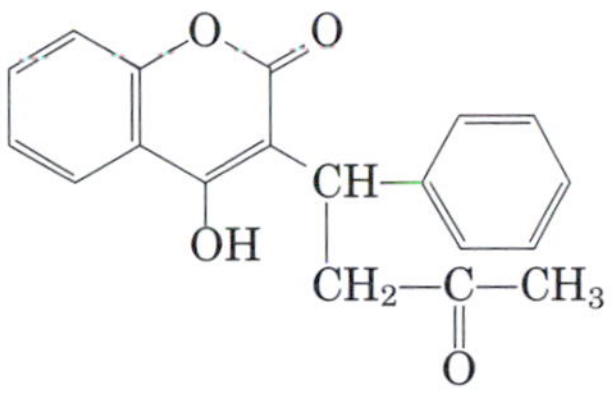

(d)
Ubiquinone: a mitochondrial electron carrier (coenzyme Q) (n = 4–8)

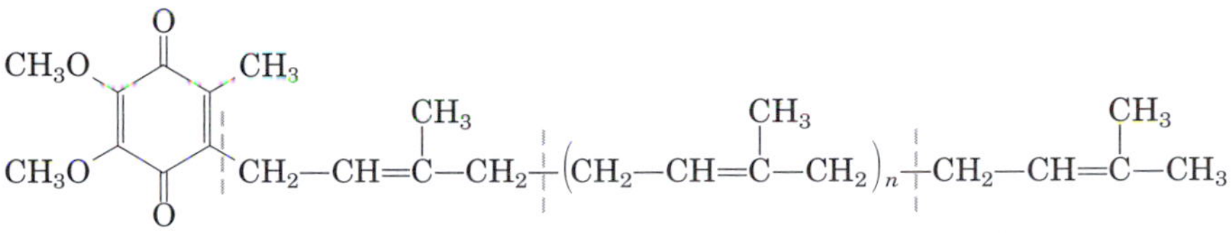

(e)
Plastoquinone: a chloroplast electron carrier (n = 4–8)

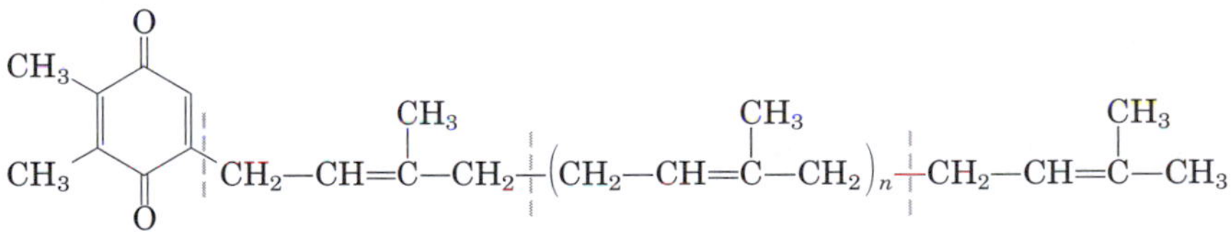

(f)
Dolichol: a sugar carrier (n = 9–22)

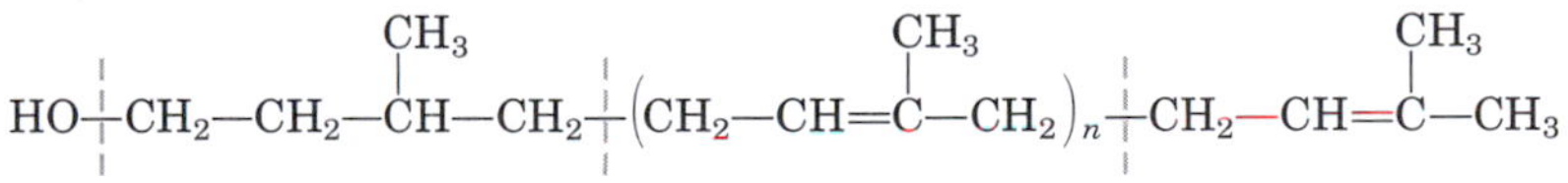

figure 11–20, page 383
Some other biologically active isoprenoid compounds or derivatives.

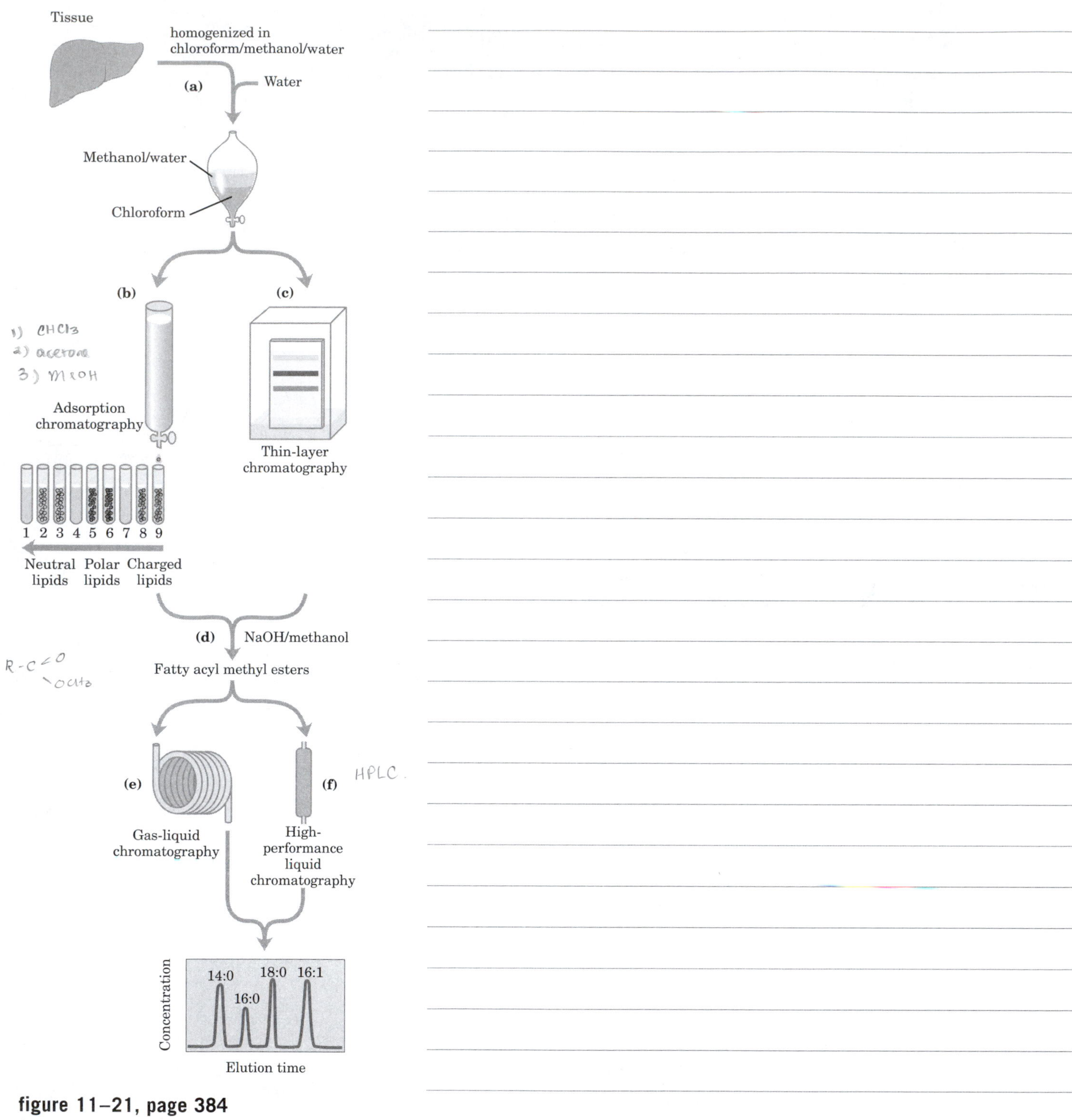

figure 11–21, page 384
Common procedures in the extraction, separation, and identification of cellular lipids.

chapter 12

Biological Membranes and Transport

When you study for exams, don't forget to review The Minicourses on the UNDERSTAND! BIOCHEMISTRY CD that came with your textbook.

Minicourses that apply to this Chapter include:

Molecules of Life
- An Interactive Gallery of Protein Structures

Bioenergetics
- Energetics

Cellular Architecture and Traffic
- The Cytoskeleton
- The Extracelllar Matrix
- Protein Modification

The Dividing Cell
- Viruses

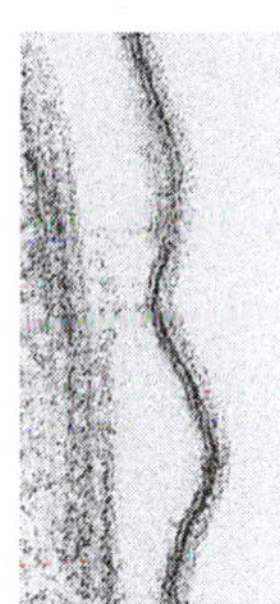
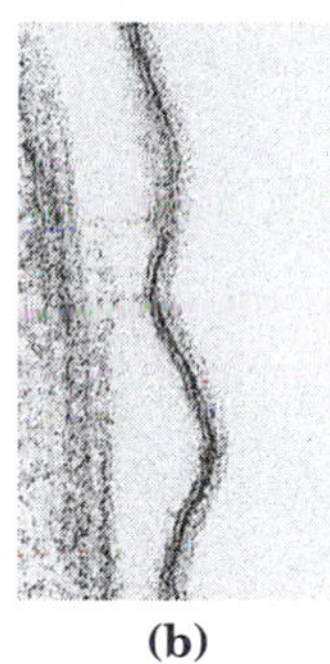
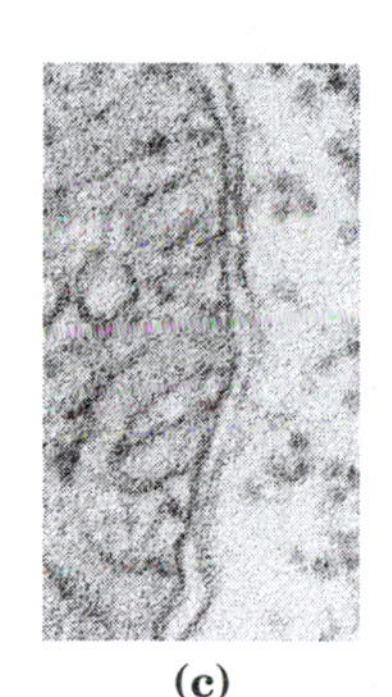
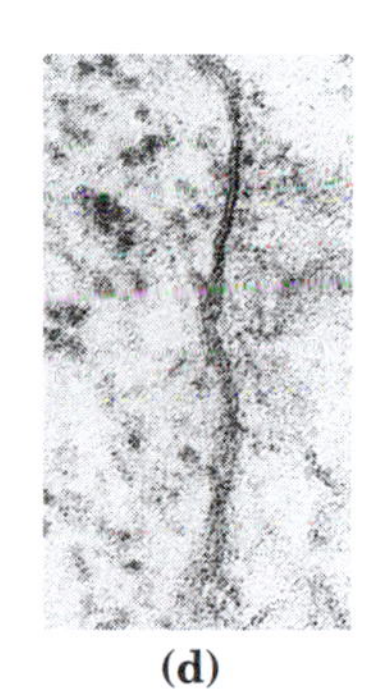
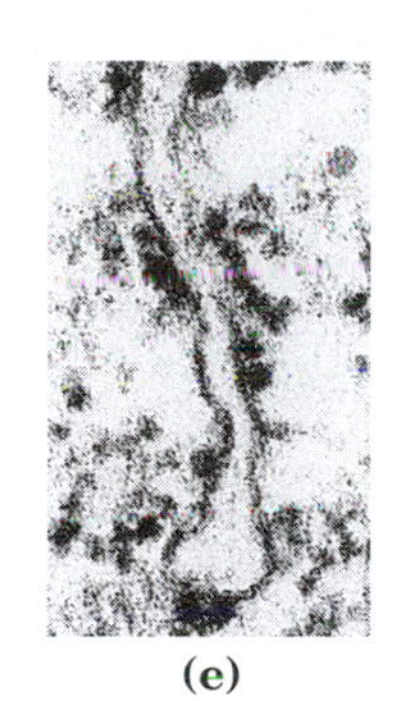
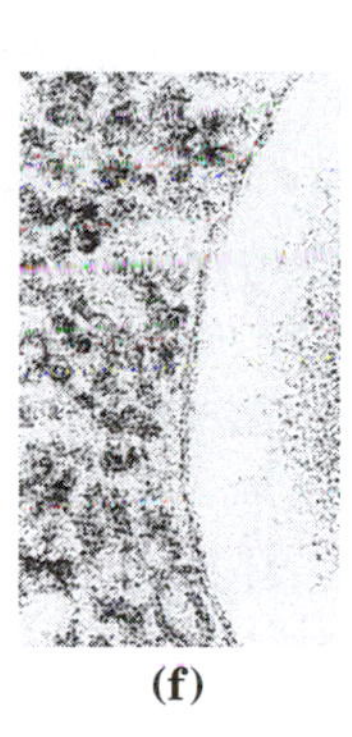

(a) (b) (c) (d) (e) (f)

figure 12–1, page 389
Biological membranes.

table 12–1, page 390

Major Components of Plasma Membranes in Various Organisms

	Components (% by weight)				
	Protein	Phospholipid	Sterol	Sterol type	Other lipids
Human myelin sheath	30	30	19	Cholesterol	Galactolipids, plasmalogens
Mouse liver	45	27	25	Cholesterol	—
Maize leaf	47	26	7	Sitosterol	Galactolipids
Yeast	52	7	4	Ergosterol	Triacylglycerols, steryl esters
Paramecium (ciliated protist)	56	40	4	Stigmasterol	—
E. coli	75	25	0	—	—

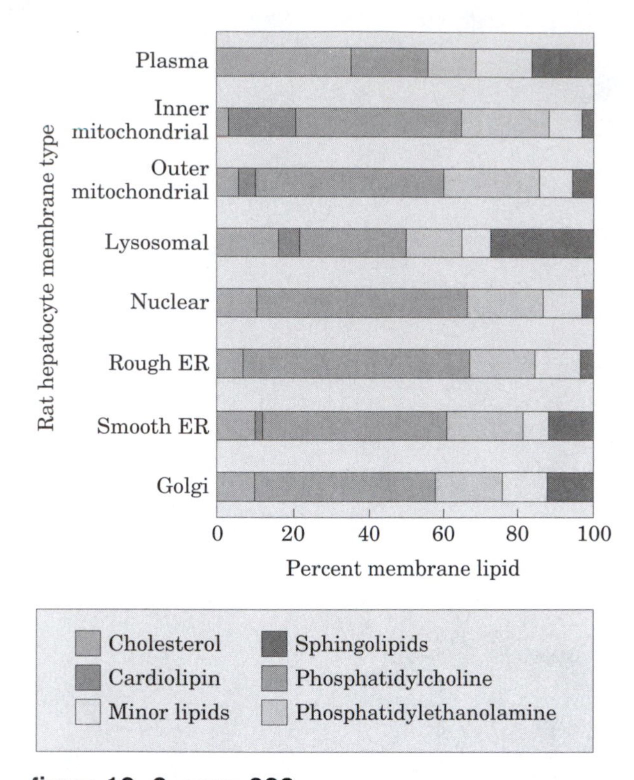

figure 12–2, page 390
Lipid composition of the plasma membrane and organelle membranes of a rat hepatocyte.

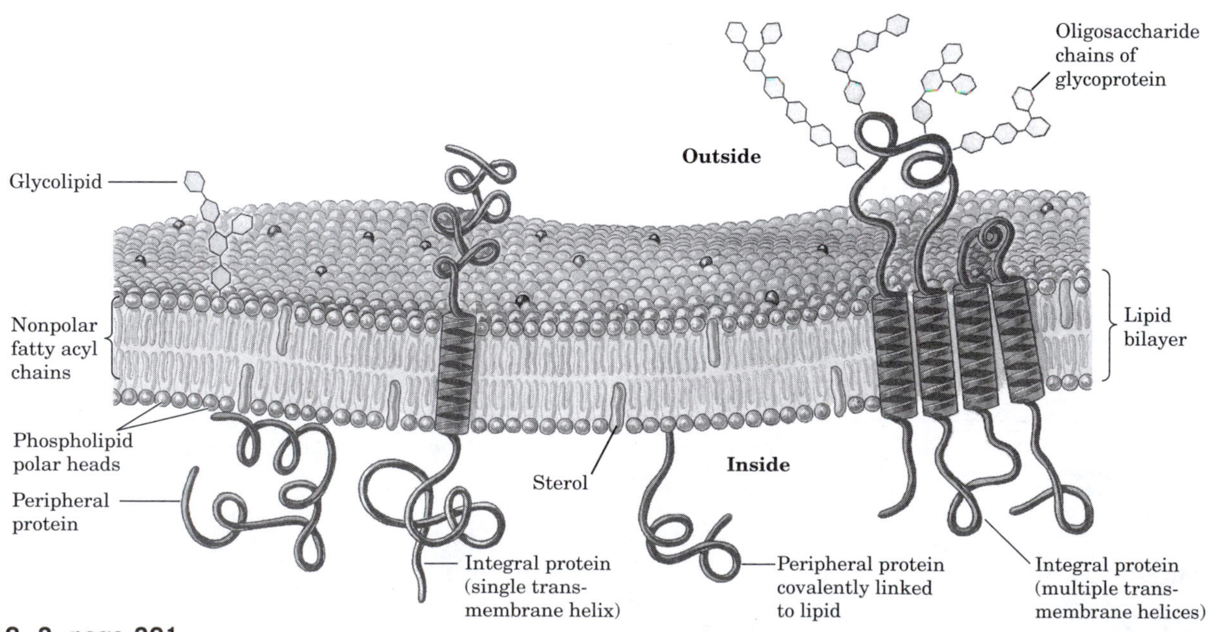

figure 12–3, page 391
Fluid mosaic model for membrane structure.

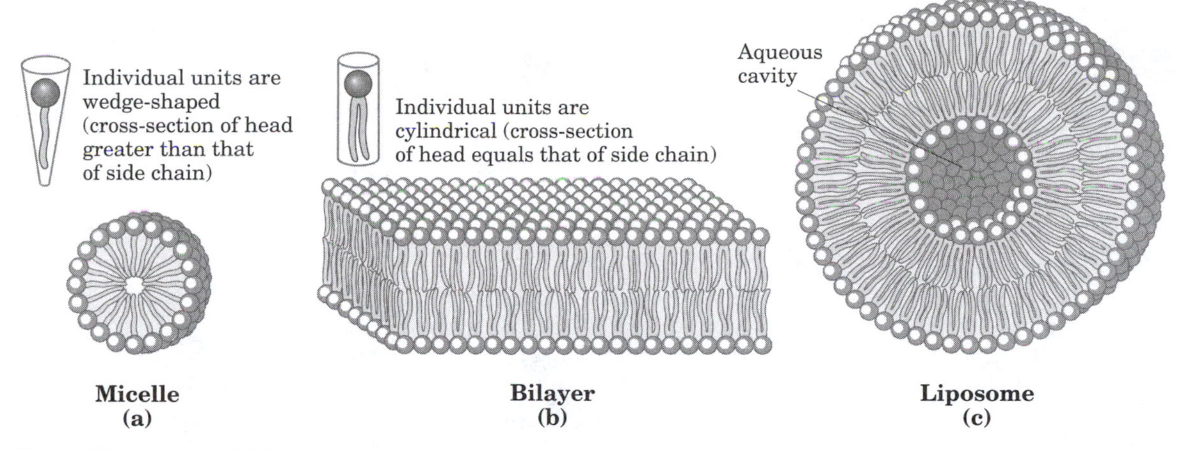

figure 12–4, page 392
Amphipathic lipid aggregates that form in water.

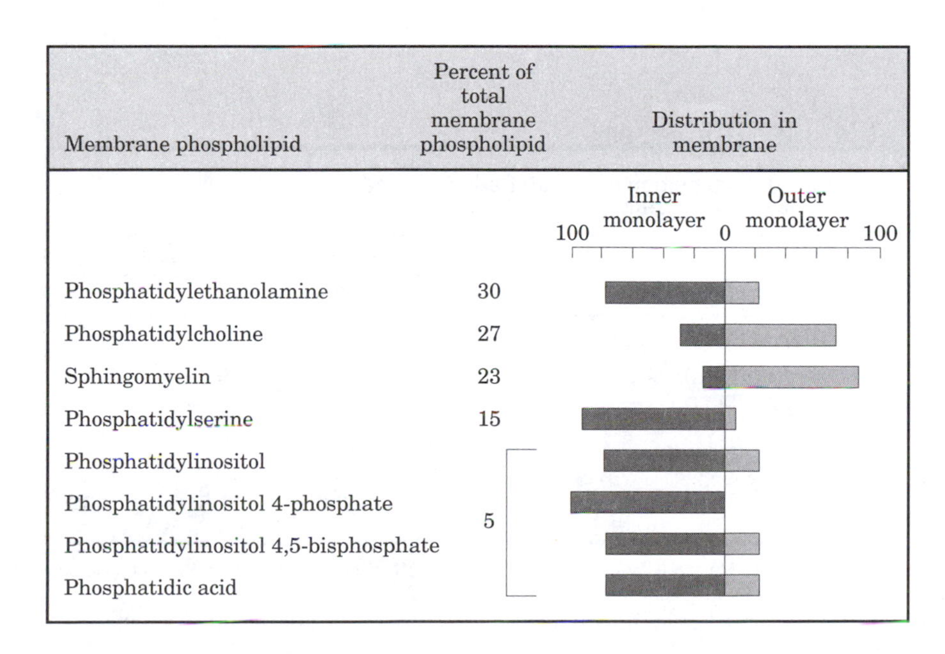

Membrane phospholipid	Percent of total membrane phospholipid	Distribution in membrane
		Inner monolayer / Outer monolayer (100 0 100)
Phosphatidylethanolamine	30	
Phosphatidylcholine	27	
Sphingomyelin	23	
Phosphatidylserine	15	
Phosphatidylinositol	5	
Phosphatidylinositol 4-phosphate		
Phosphatidylinositol 4,5-bisphosphate		
Phosphatidic acid		

figure 12–5, page 393
Asymmetric distribution of phospholipids between the inner and outer monolayers of the erythrocyte plasma membrane.

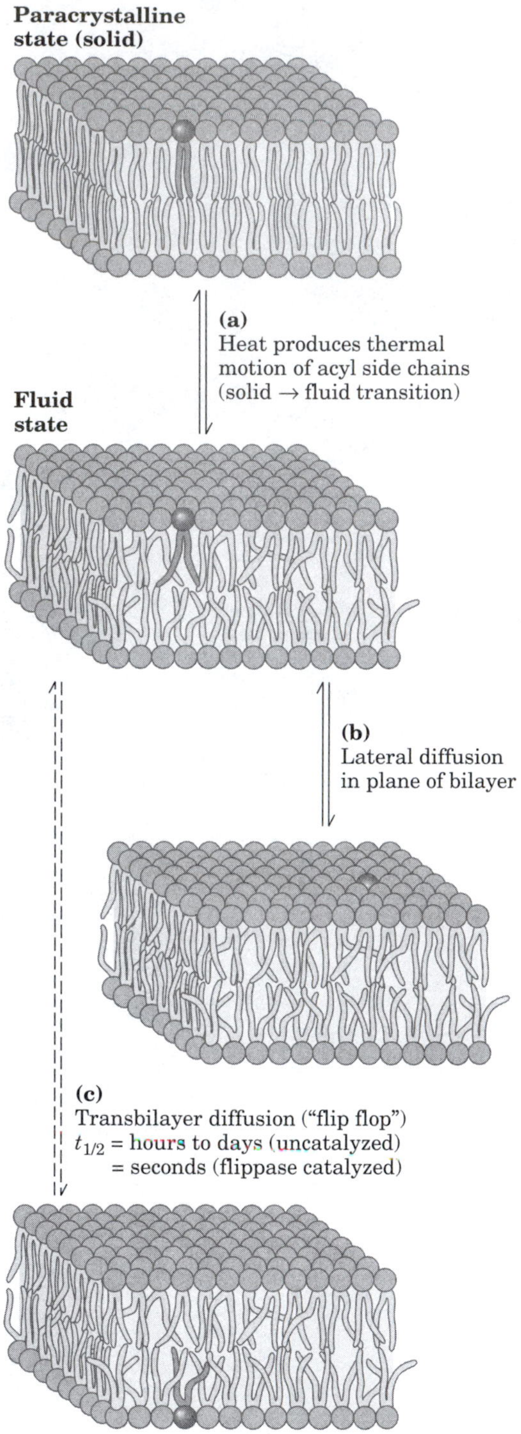

Paracrystalline state (solid)

(a)
Heat produces thermal motion of acyl side chains (solid → fluid transition)

Fluid state

(b)
Lateral diffusion in plane of bilayer

(c)
Transbilayer diffusion ("flip flop")
$t_{1/2}$ = hours to days (uncatalyzed)
= seconds (flippase catalyzed)

figure 12–6, page 394
Motion of membrane lipids.

table 12–2, page 395

Fatty Acid Composition of *E. coli* Cells Cultured at Different Temperatures

	Percentage of total fatty acids*			
	10 °C	20 °C	30 °C	40 °C
Myristic acid (14:0)	4	4	4	8
Palmitic acid (16:0)	18	25	29	48
Palmitoleic acid (16:1)	26	24	23	9
Oleic acid (18:1)	38	34	30	12
Hydroxymyristic acid	13	10	10	8
Ratio of unsaturated to saturated†	2.9	2.0	1.6	0.38

Source: Data from Marr, A.G. & Ingraham, J.L. (1962) Effect of temperature on the composition of fatty acids in *Escherichia coli*. *J. Bacteriol.* **84,** 1260.

*The exact fatty acid composition depends not only on growth temperature but on growth stage and growth medium composition.

†Calculated as the total percentage of 16:1 plus 18:1 divided by the total percentage of 14:0 plus 16:0. Hydroxymyristic acid was omitted from this calculation.

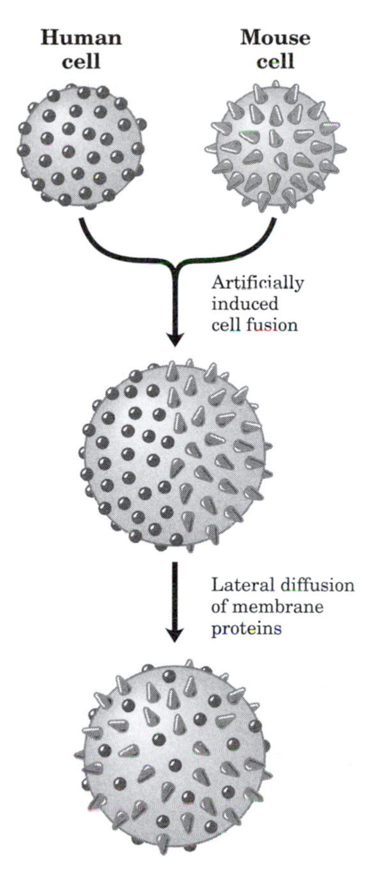

figure 12–7, page 395
Demonstration of lateral diffusion of membrane proteins.

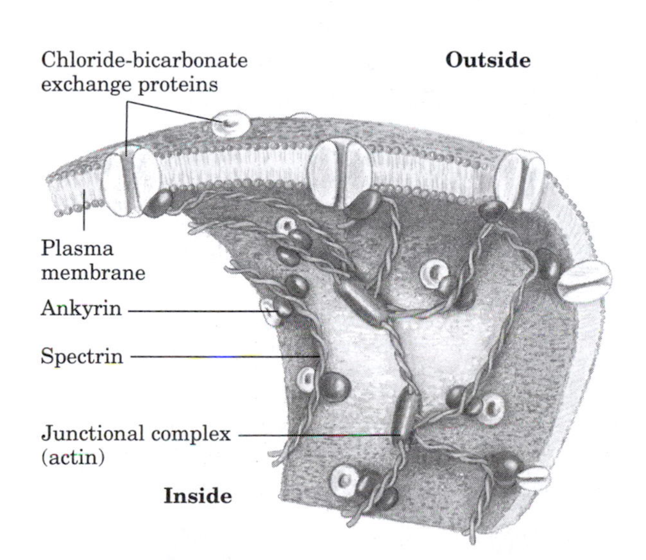

figure 12–8, page 395
Restricted motion of the erythrocyte chloride-bicarbonate exchanger.

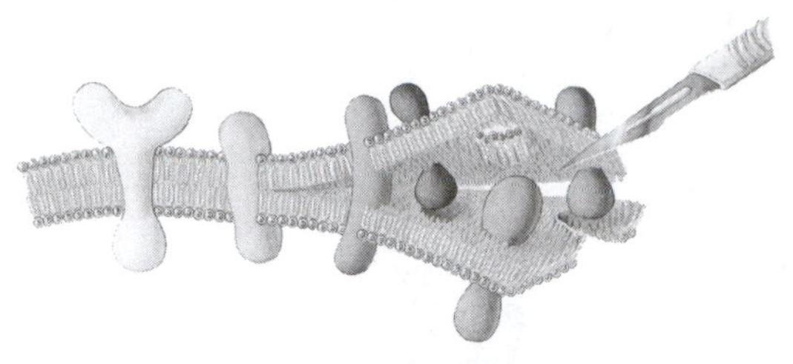

box 12–1, figure 1, page 396
Splitting of a membrane bilayer by the freeze-fracture technique.

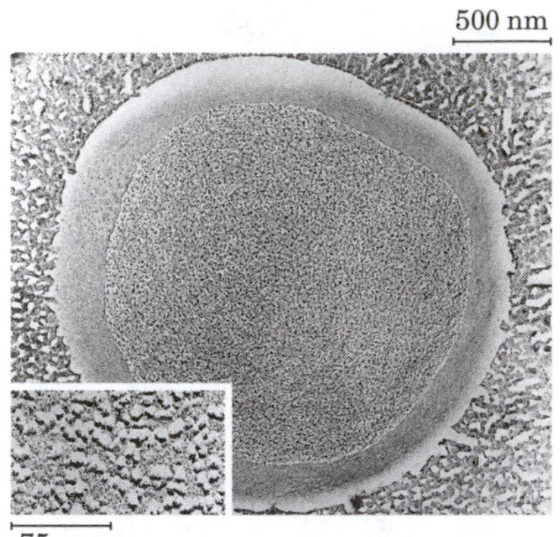

box 12–1, figure 2, page 396
An erythrocyte membrane is frozen in water, etched (by sublimation of some water), and a carbon replica is viewed in the electron microscope.

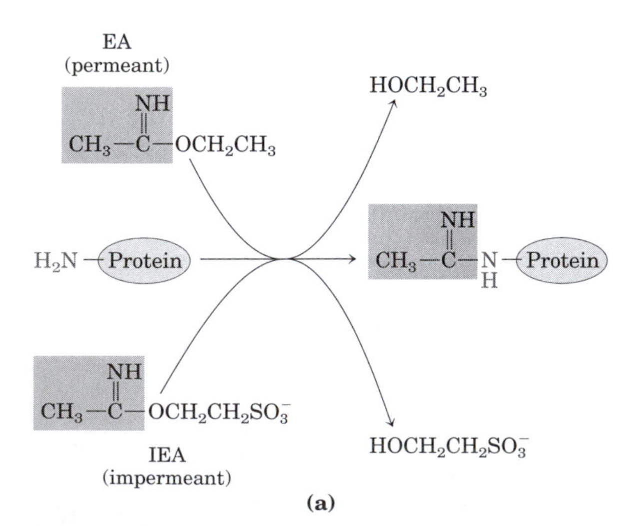

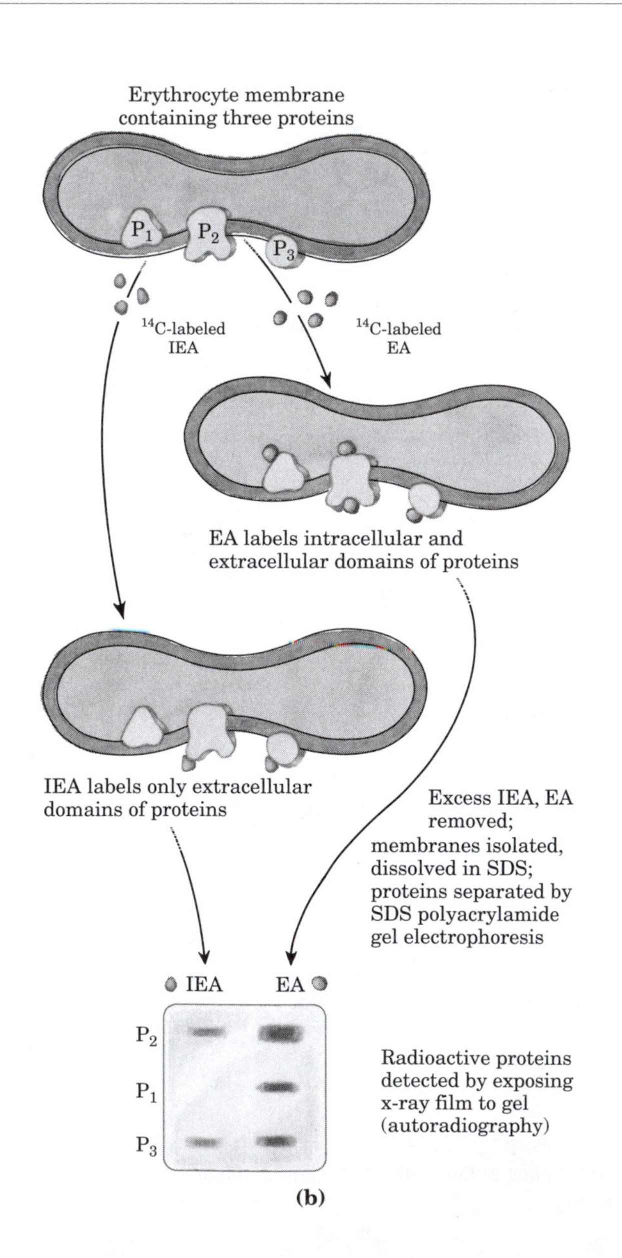

figure 12–9, page 397
Experiments to determine the transmembrane arrangement of membrane proteins.

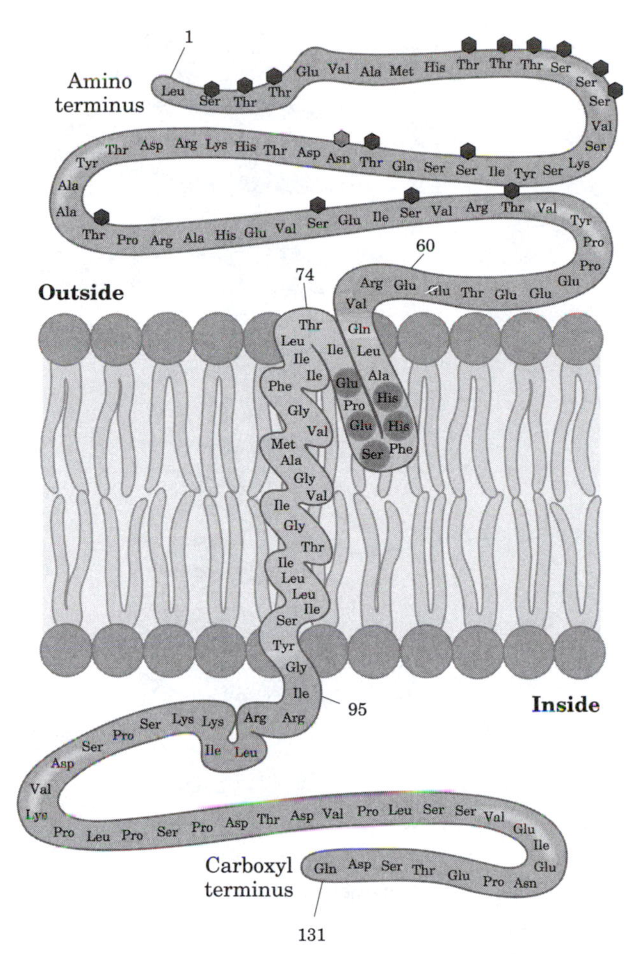

figure 12–10, page 398
Transbilayer disposition of glycophorin in the erythrocyte.

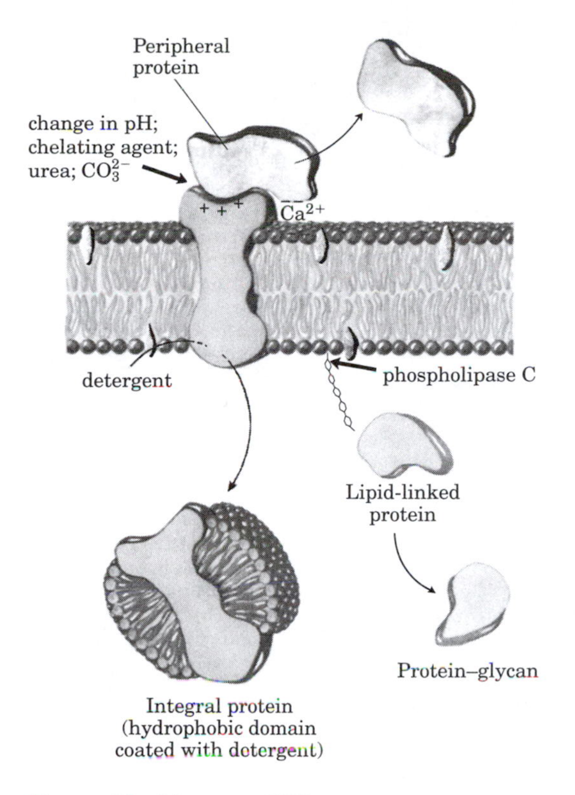

figure 12–11, page 399
Peripheral and integral proteins.

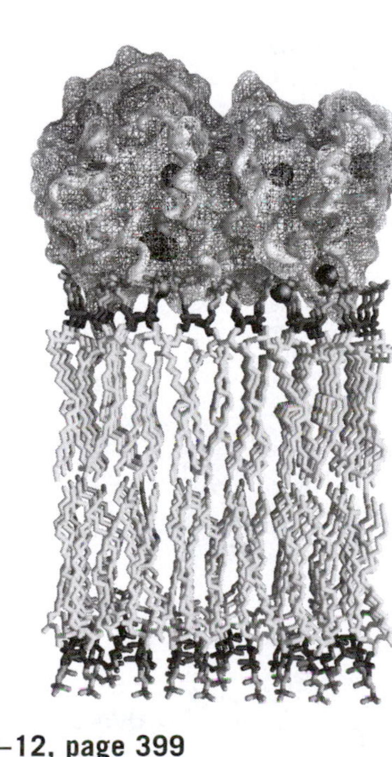

figure 12–12, page 399
Proposed structure of annexin V in association with the lipid head groups of a membrane.

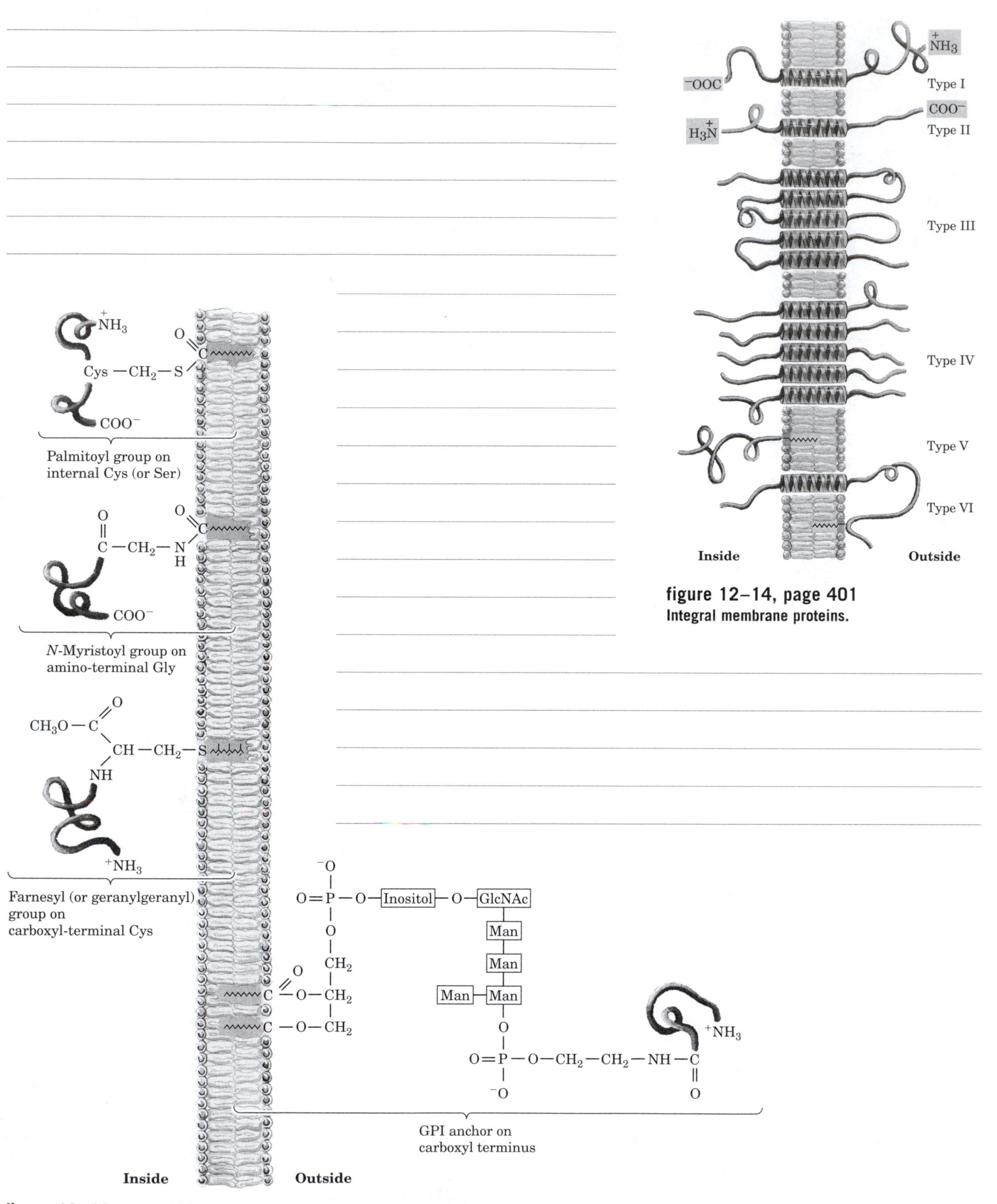

figure 12–14, page 401
Integral membrane proteins.

figure 12–13, page 400
Lipid-linked membrane proteins.

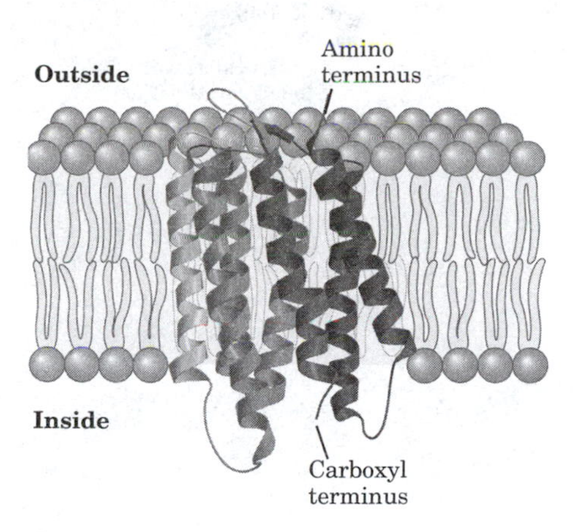

figure 12–15, page 401
Bacteriorhodopsin, a membrane-spanning protein.

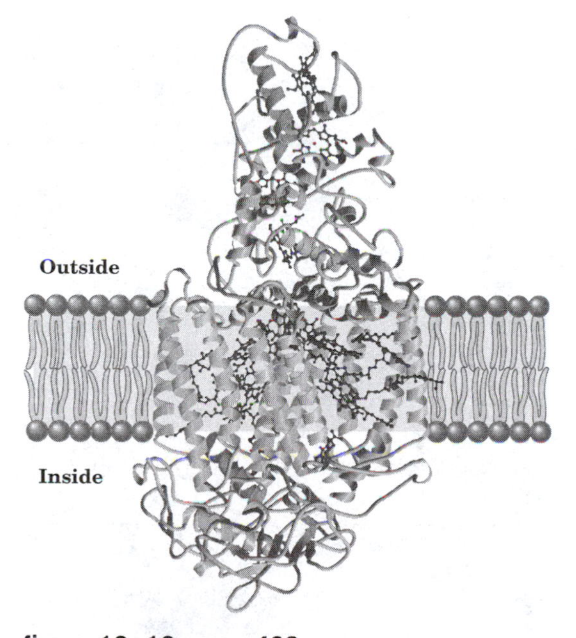

figure 12–16, page 402
Three-dimensional structure of the photosynthetic reaction center of *Rhodopseudomonas viridis,* a purple bacterium.

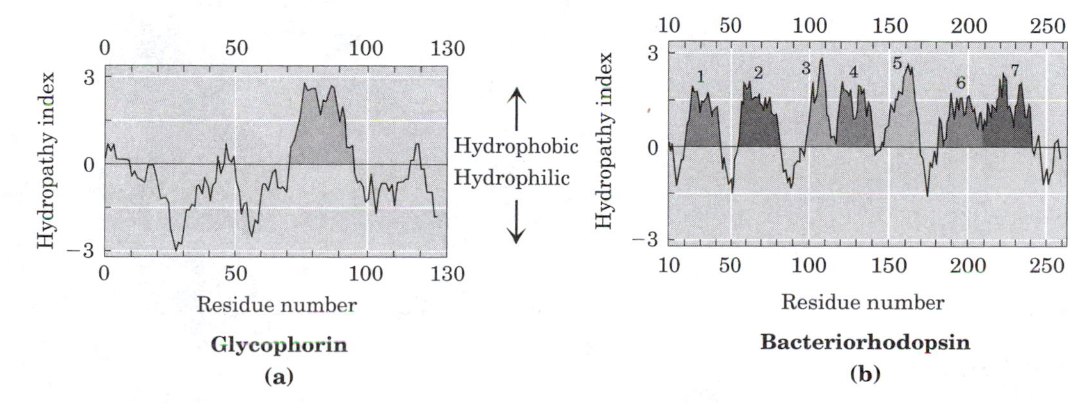

figure 12–17, page 403
Hydropathy plots.

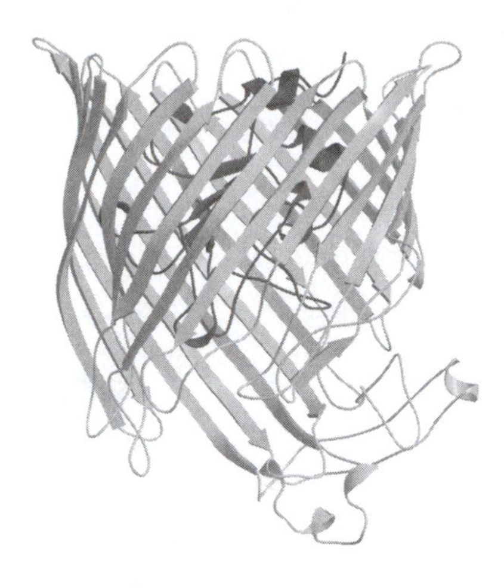

figure 12–18, page 403
Porin FhuA, an integral membrane protein with β-barrel structure.

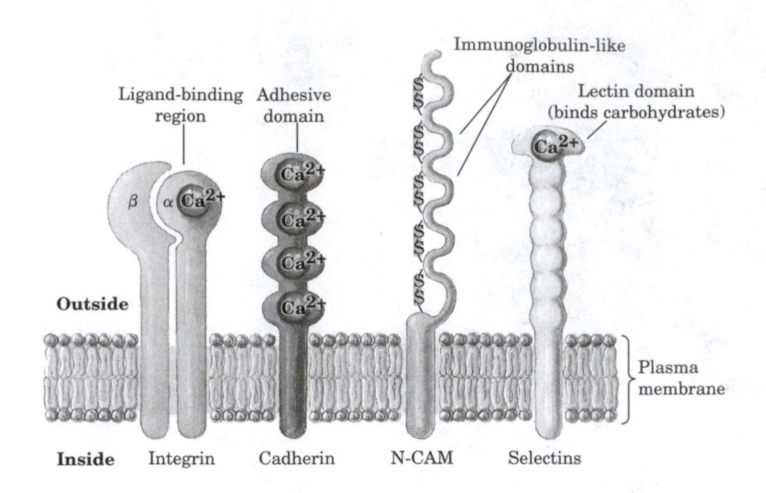

figure 12–19, page 404
Four examples of integral protein types that function in cell-cell interactions.

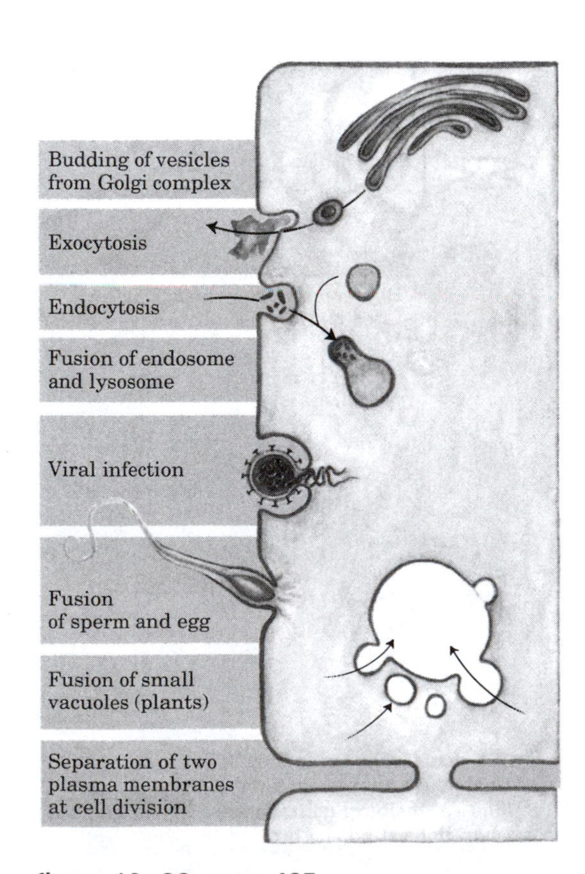

figure 12–20, page 405
Membrane fusion.

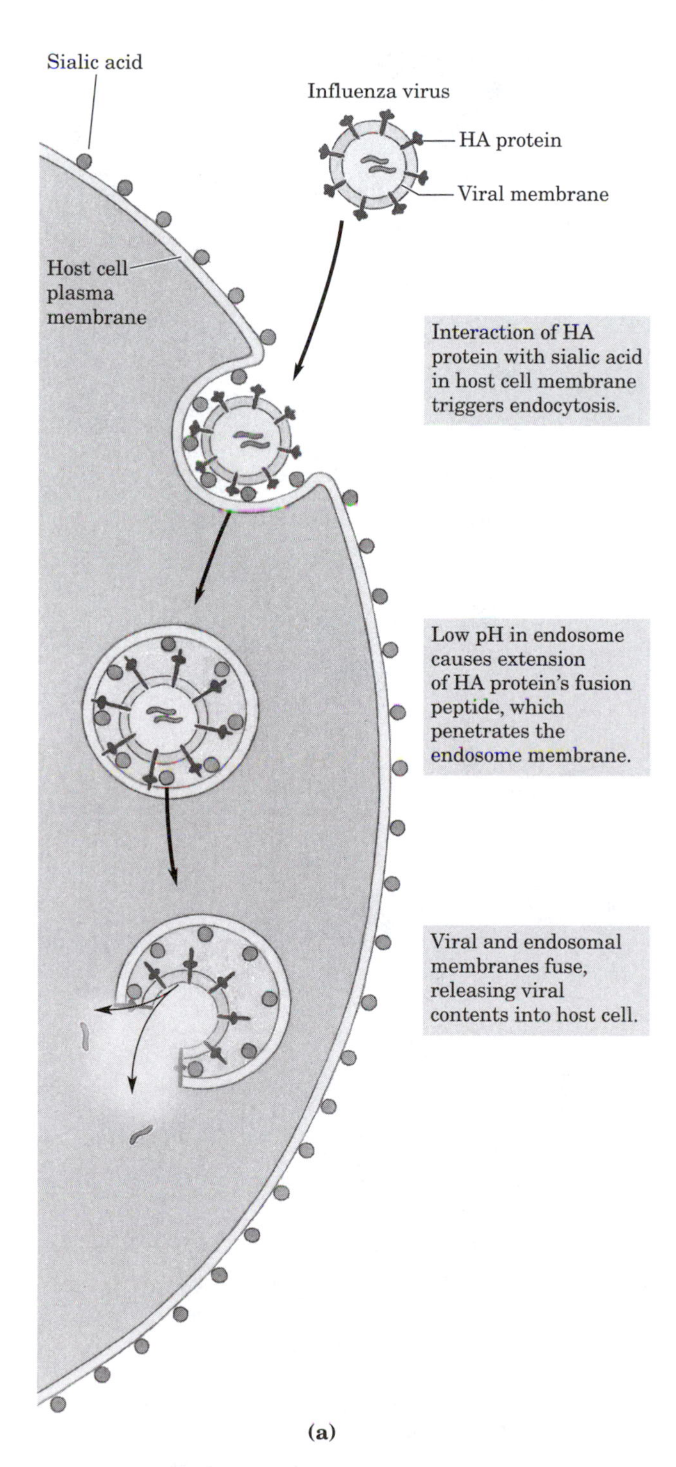

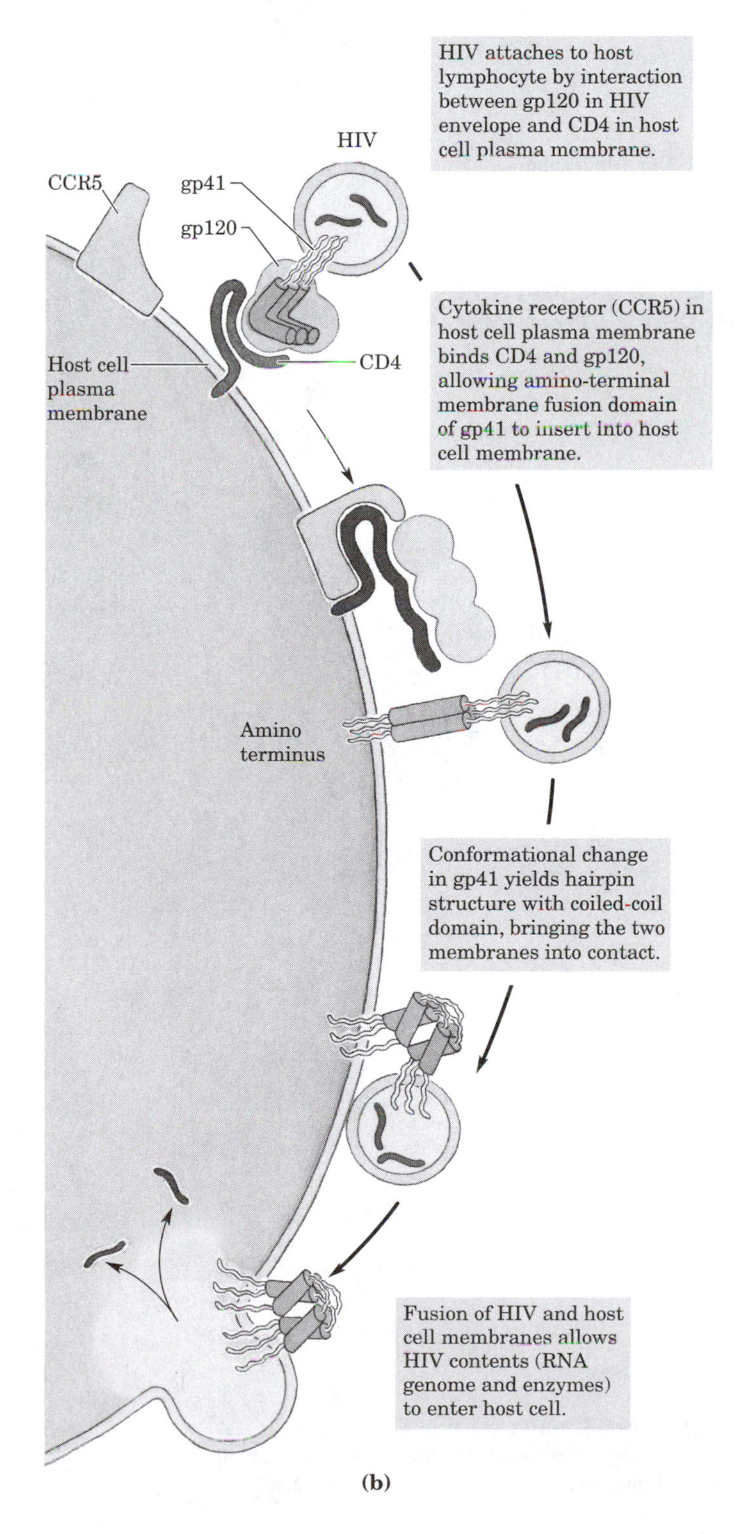

figure 12–21, page 407
Membrane fusion during viral entry into a host cell.

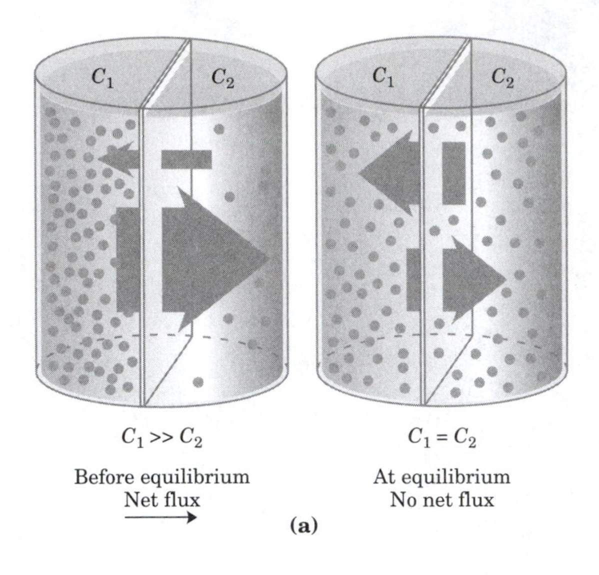

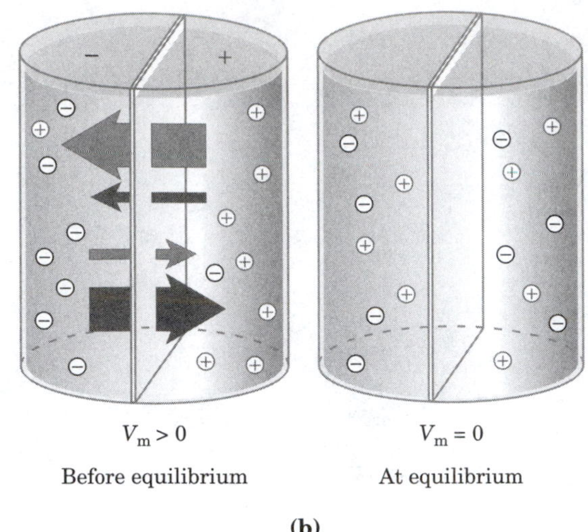

figure 12–22, page 408
Movement of solutes across a permeable membrane.

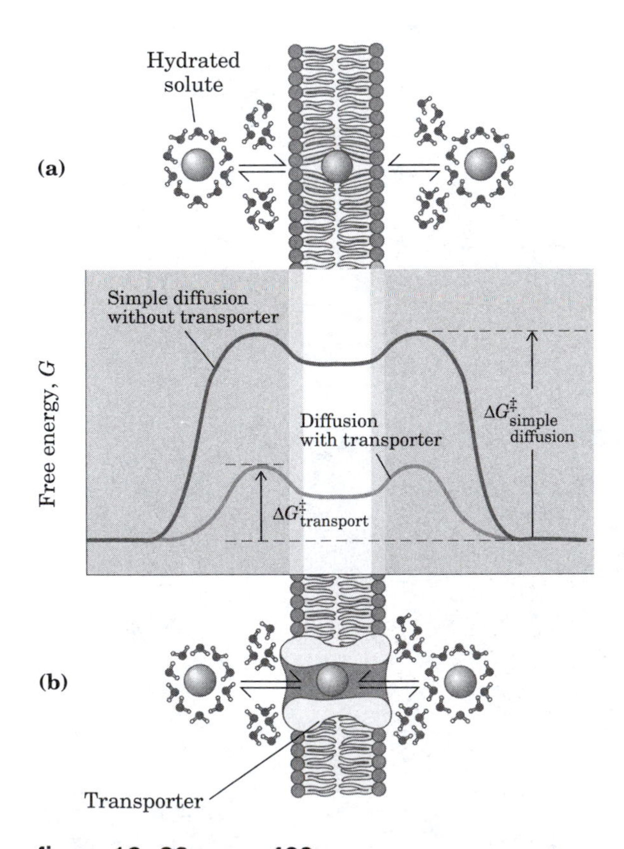

figure 12–23, page 409
Energy changes accompanying passage of a hydrophilic solute through the lipid bilayer of a biological membrane.

table 12–3, page 410

Aquaporins

Aquaporin	Roles and location
AQP-1	Fluid reabsorption in proximal renal tubule; secretion of aqueous humor in eye and cerebrospinal fluid in central nervous system; water homeostasis in lung
AQP-2	Water permeabilily in renal collecting duct (mutations produce nephrogenic diabetes insipidus)
AQP-3	Water retention in renal collecting duct
AQP-4	Reabsorption of cerebrospinal fluid in central nervous system; regulation of brain edema
AQP-5	Fluid secretion in salivary glands, lachrymal glands, and alveolar epithelium of lung
γ-TIP	Water uptake by plant vacuole, regulating turgor pressure

Source: King, L.S. & Agre, P. (1996) Pathophysiology of the aquaporin water channels. *Annu. Rev. Physiol.* **58,** 619–648.

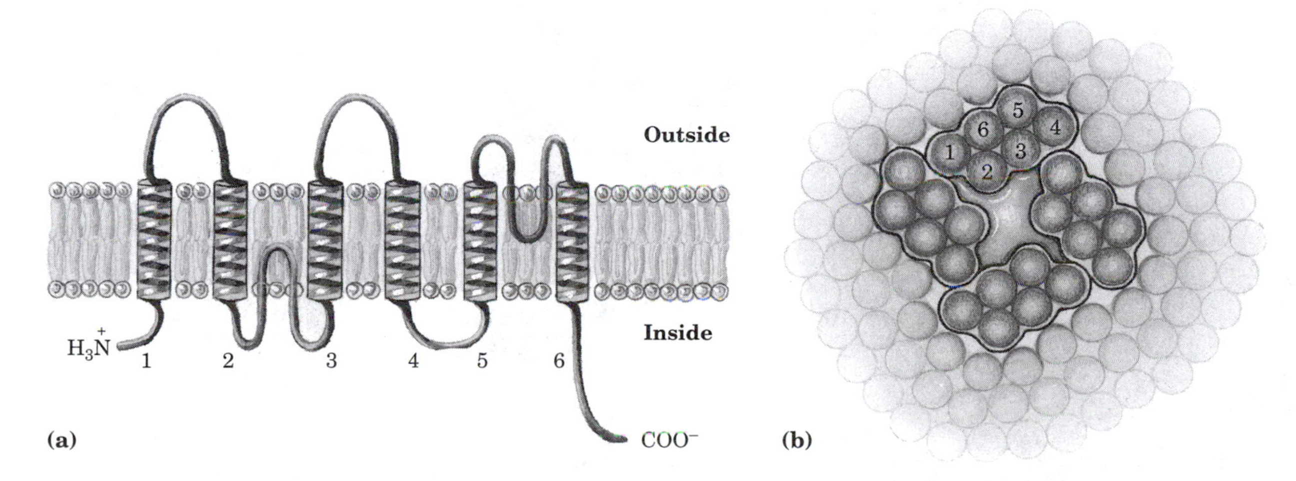

figure 12–24, page 410
Likely transmembrane topology of an aquaporin, AQP-1.

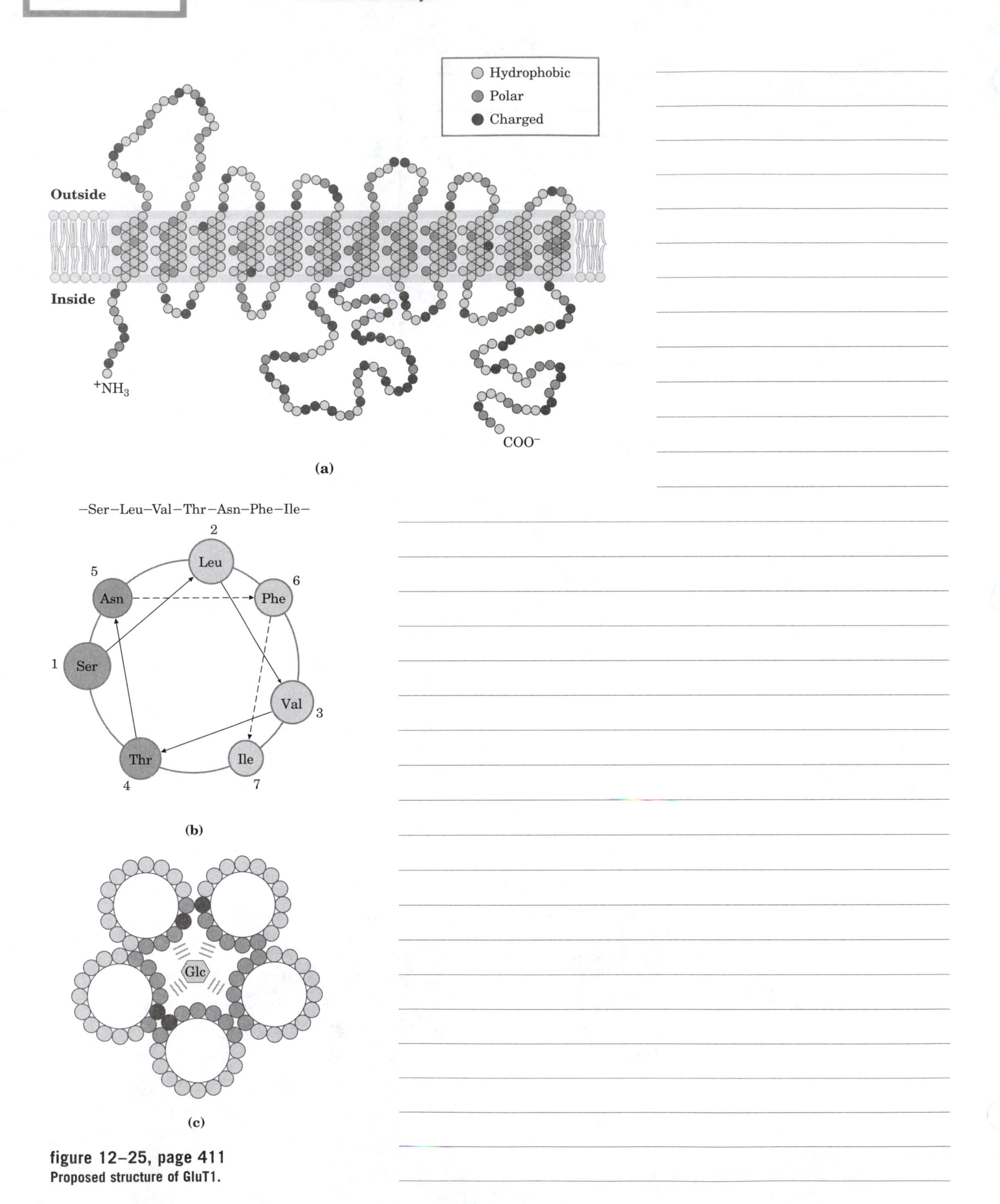

figure 12–25, page 411
Proposed structure of GluT1.

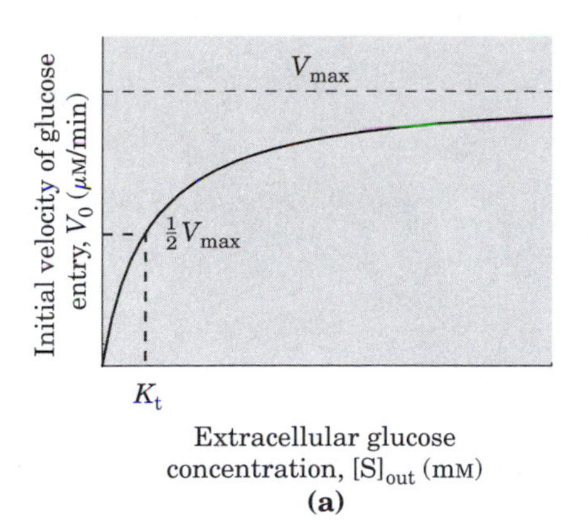

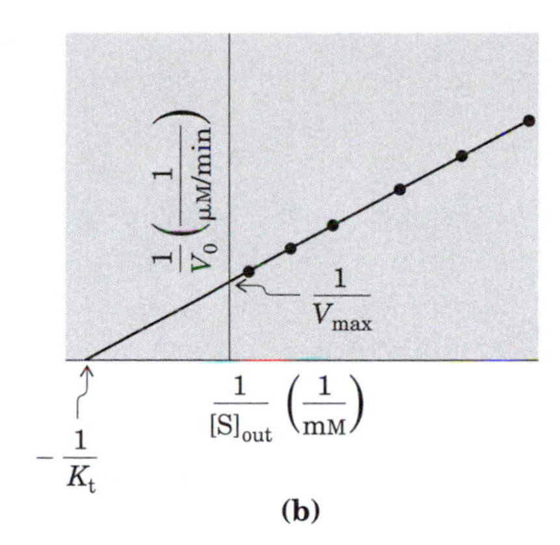

figure 12–26, page 412
Kinetics of glucose transport into erythrocytes.

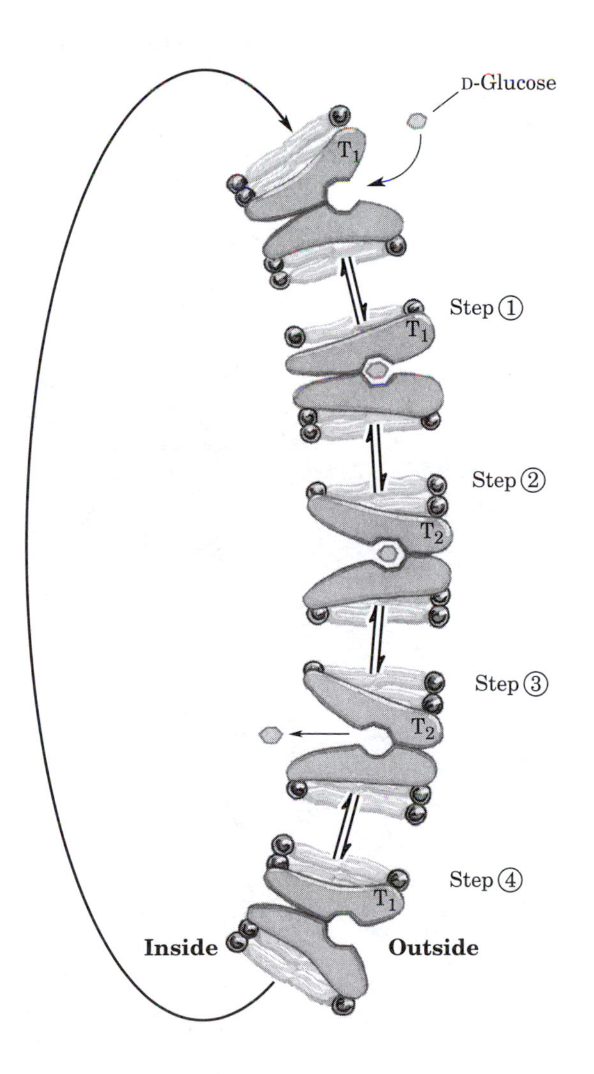

figure 12–27, page 412
Model of glucose transport into erythrocytes by GluT1.

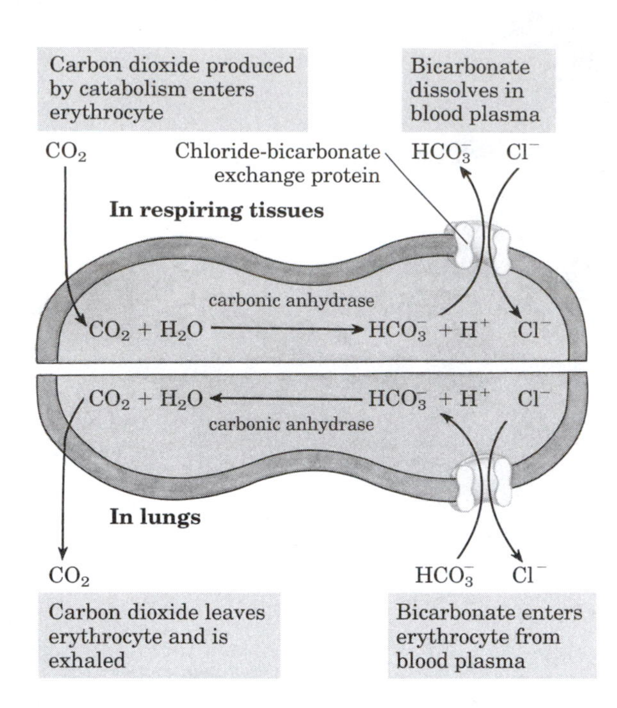

figure 12–28, page 413
Chloride-bicarbonate exchanger of the erythrocyte membrane.

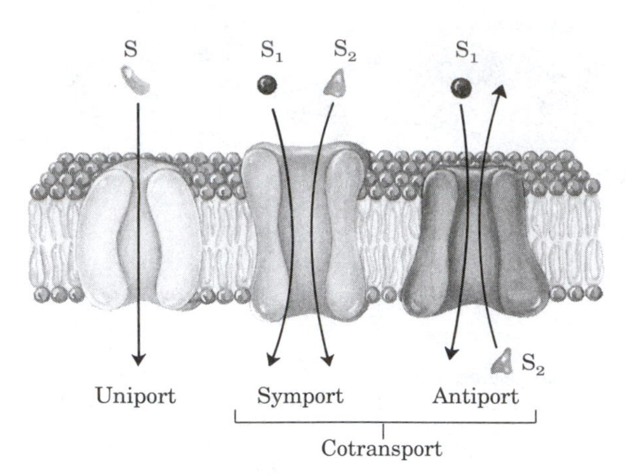

figure 12–29, page 413
Three general classes of transport systems.

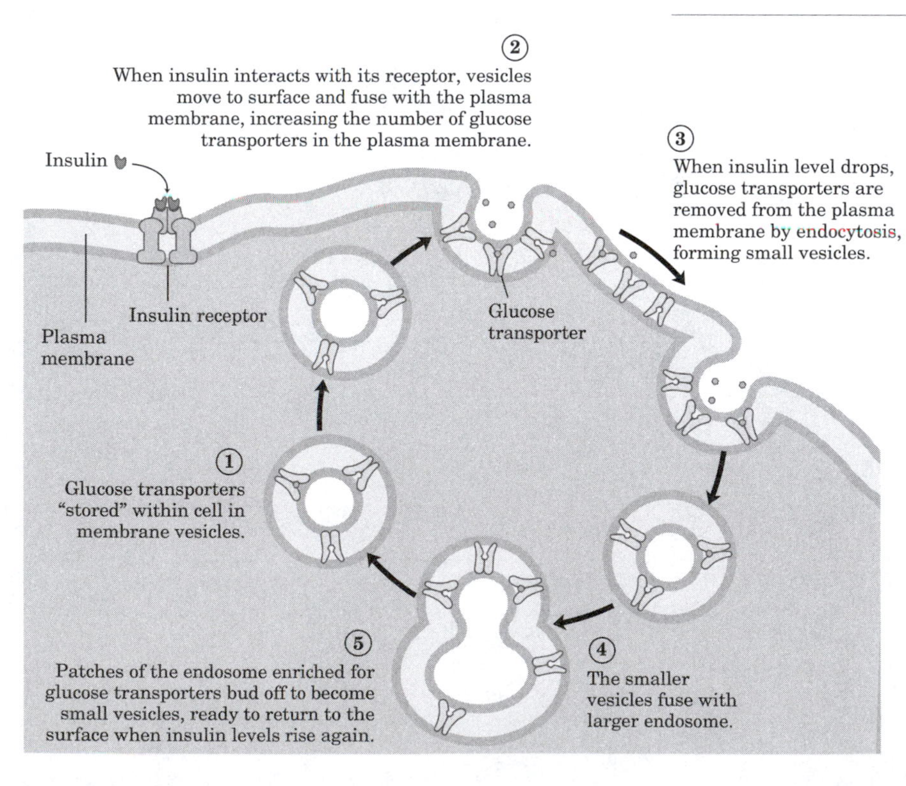

box 12–2, figure 1, page 414
Regulation by insulin of glucose transport into a myocyte.

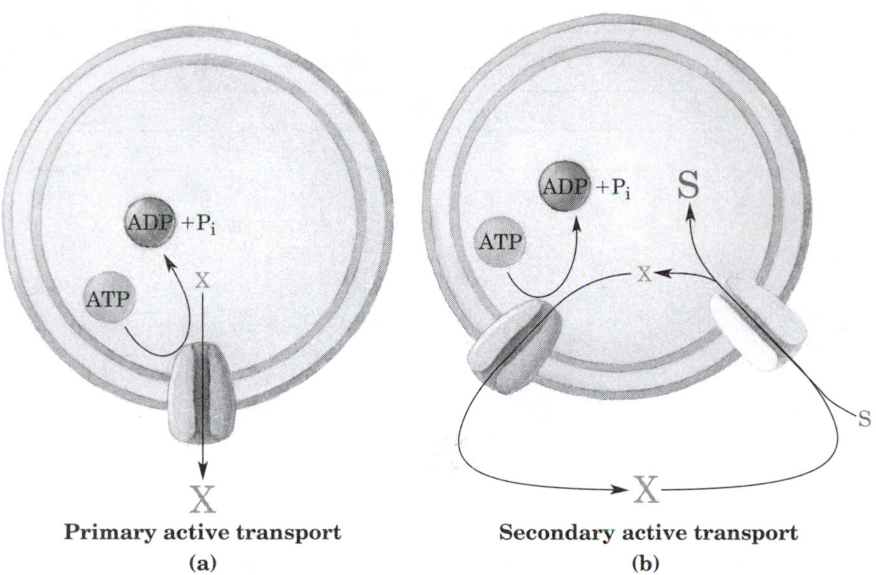

figure 12–30, page 415
Two types of active transport.

$$O{=}P(O^-)_2(OH) \qquad O{=}V(O^-)_2(OH)$$

Phosphate **Vanadate**

page 416

$$\Delta G_t = RT \ln (C_2/C_1) \quad \text{(Equation 12–2)}$$

Free-energy change for transport, page 415

$$\Delta G_t = RT \ln (C_2/C_1) + Z \mathcal{F} \Delta\psi \quad \text{(Equation 12–3)}$$

Free-energy change for electrogenic transport, page 416

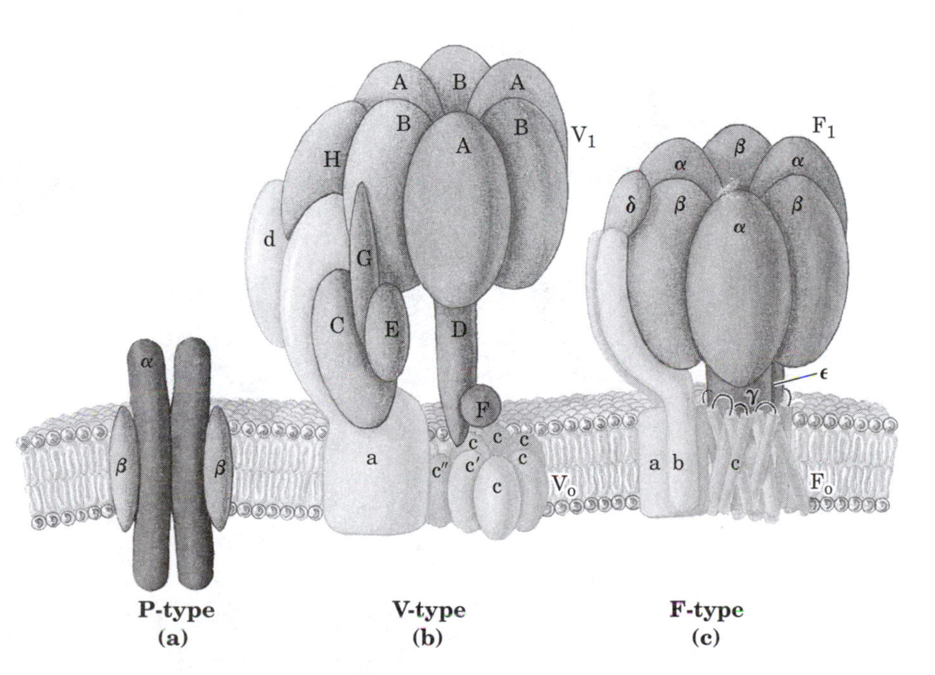

figure 12–31, page 417
Subunit structure of three types of ion-transporting ATPases.

table 12–4, page 417

Four Classes of Transport ATPases

	Organism or tissue	Type of membrane	Role of ATPase
P-type ATPases			
Na^+K^+	Animal tissues	Plasma	Maintains low [Na^+], high [K^+] inside cell; creates transmembrane electrical potential
H^+K^+	Acid-secreting (parietal) cells of mammals	Plasma	Acidifies contents of stomach
H^+	Fungi *(Neurospora)*	Plasma	Create H^+ gradient to drive secondary transport of extracellular solutes into cell
H^+	Higher plants	Plasma	
Ca^{2+}	Animal tissues	Plasma	Maintains low [Ca^{2+}] in cytosol
Ca^{2+}	Myocytes of animals	Sarcoplasmic reticulum (endoplasmic reticulum)	Sequesters intracellular Ca^{2+}, keeping cytosolic [Ca^{2+}] low
Cd^{2+}, Hg^{2+}, Cu^{2+}	Bacteria	Plasma	Pumps heavy metal ions out of cell
V-type ATPases			
H^+	Animals	Lysosomal, endosomal, secretory vesicles	Create low pH in compartment, activating proteases and other hydrolytic enzymes
H^+	Higher plants	Vacuolar	
H^+	Fungi	Vacuolar	
F-type ATPases			
H^+	Eukaryotes	Inner mitochondrial	Catalyze formation of ATP from ADP + P_i
H^+	Higher plants	Thylakoid	
H^+	Prokaryotes	Plasma	
Multidrug transporter			
	Animal tumor cells	Plasma	Removes a wide variety of hydrophobic natural products and synthetic drugs from cytosol, including vinblastine, doxorubicin, actinomycin D, mitomycin, taxol, colchicine, and puromycin

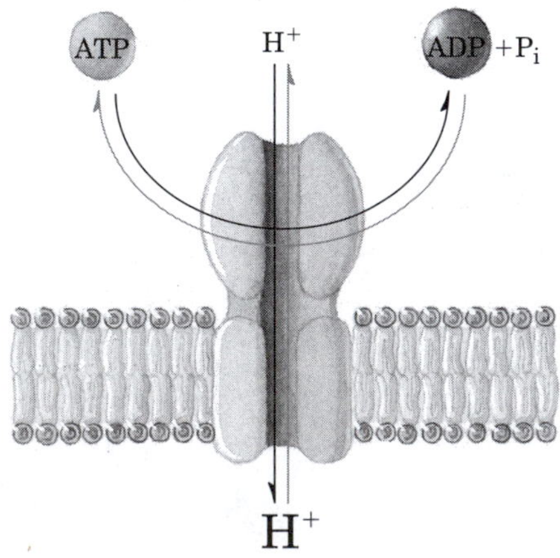

figure 12–32, page 418
Reversibility of F-type ATPases.

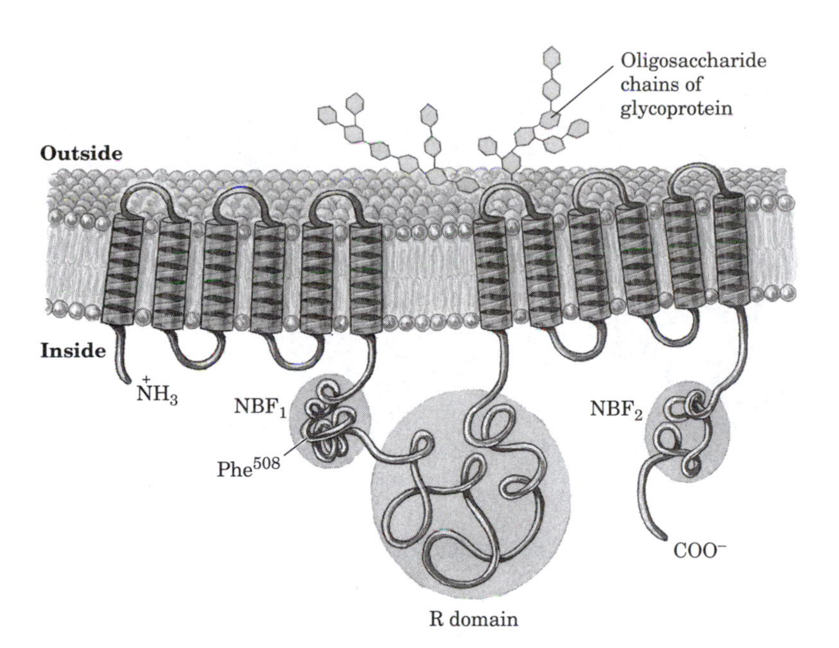

box 12–3, figure 1, page 418
Topology of the cystic fibrosis transmembrane conductance regulator, CFTR.

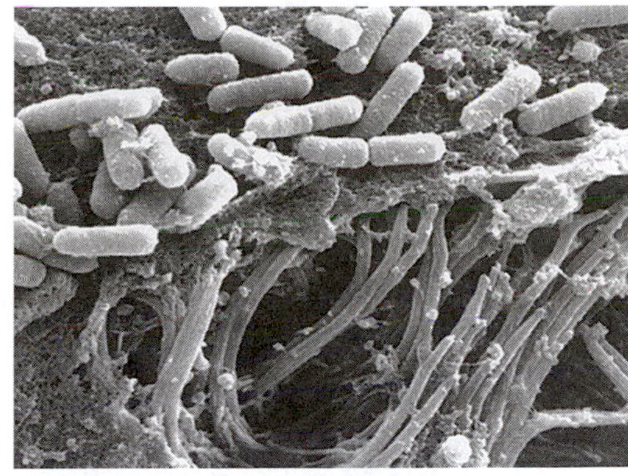

box 12–3, figure 2, page 419
Mucus lining the surface of the lungs traps bacteria.

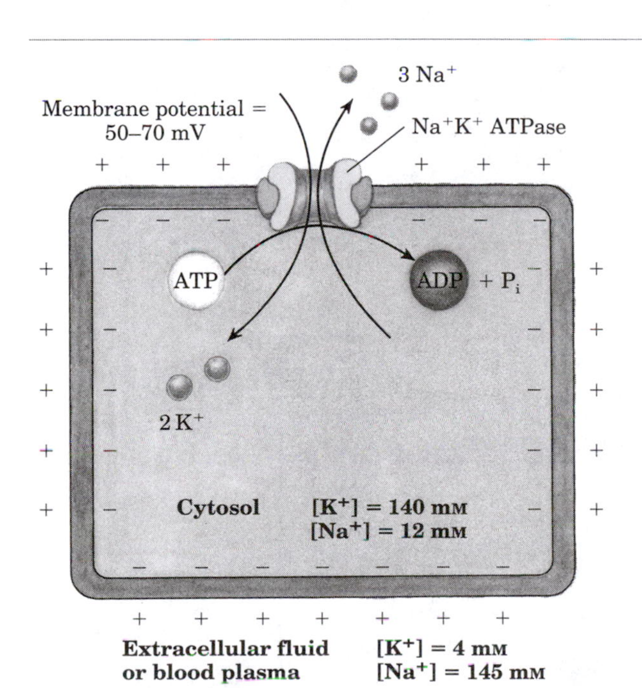

figure 12–33, page 420
Na^+K^+ ATPase.

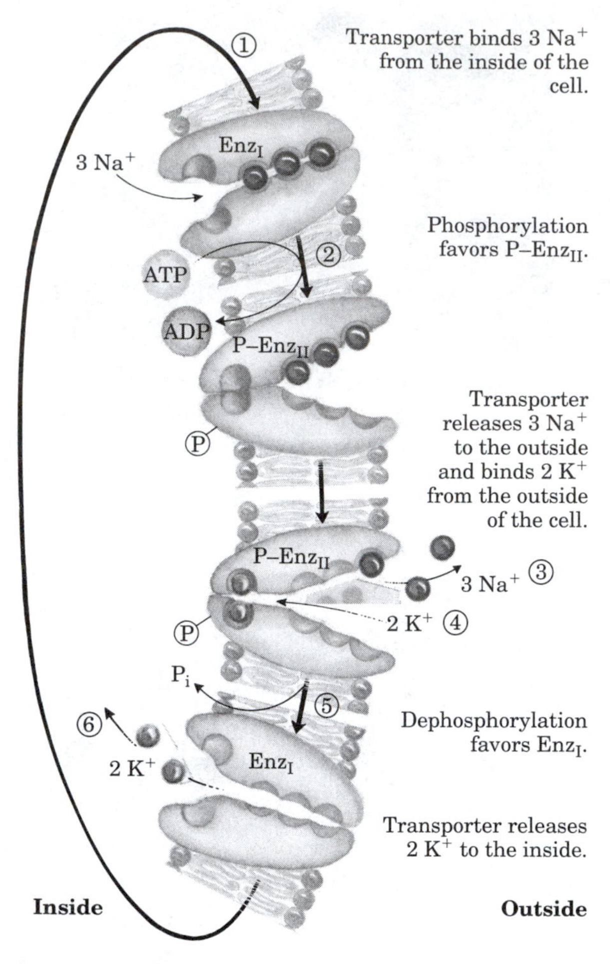

figure 12–34, page 421
Postulated mechanism of Na^+ and K^+ transport by the Na^+K^+ ATPase.

Oubain, page 421

table 12–5, page 422

Cotransport Systems Driven by Gradients of Na^+ or H^+

Organism or tissue	Transported solute (moving against its gradient)	Cotransported solute (moving down its gradient)	Type of transport
E. coli	Lactose	H^+	Symport
	Proline	H^+	Symport
	Dicarboxylic acids	H^+	Symport
Intestine, kidney of vertebrates	Glucose	Na^+	Symport
	Amino acids	Na^+	Symport
Vertebrate cells (many types)	Ca^{2+}	Na^+	Antiport
Higher plants	K^+	H^+	Antiport
Fungi (*Neurospora*)	K^+	H^+	Antiport

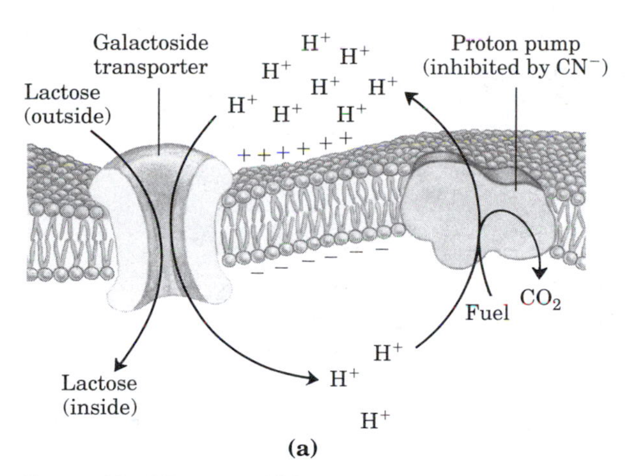

figure 12–35, page 422
Lactose uptake in *E. coli*.

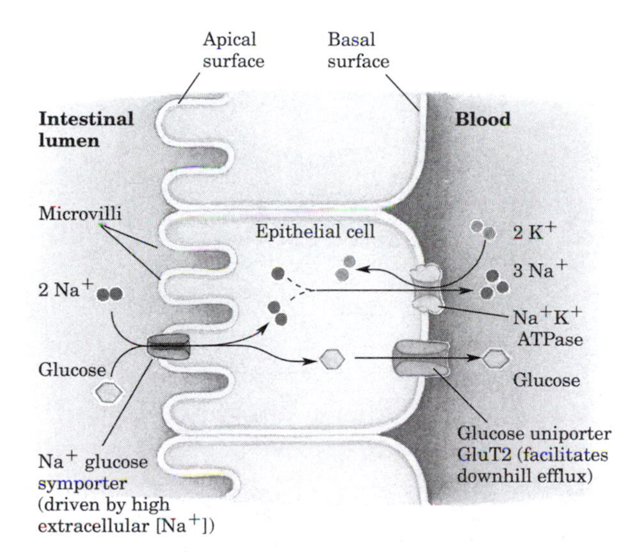

figure 12–36, page 423
Glucose transport in intestinal epithelial cells.

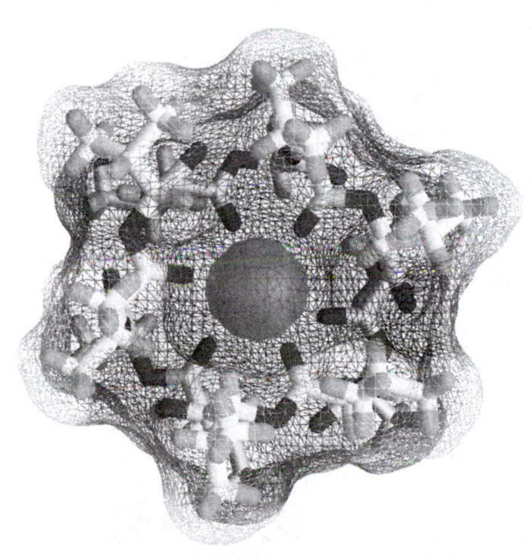

figure 12–37, page 423
Valinomycin, a peptide ionophore that binds K^+.

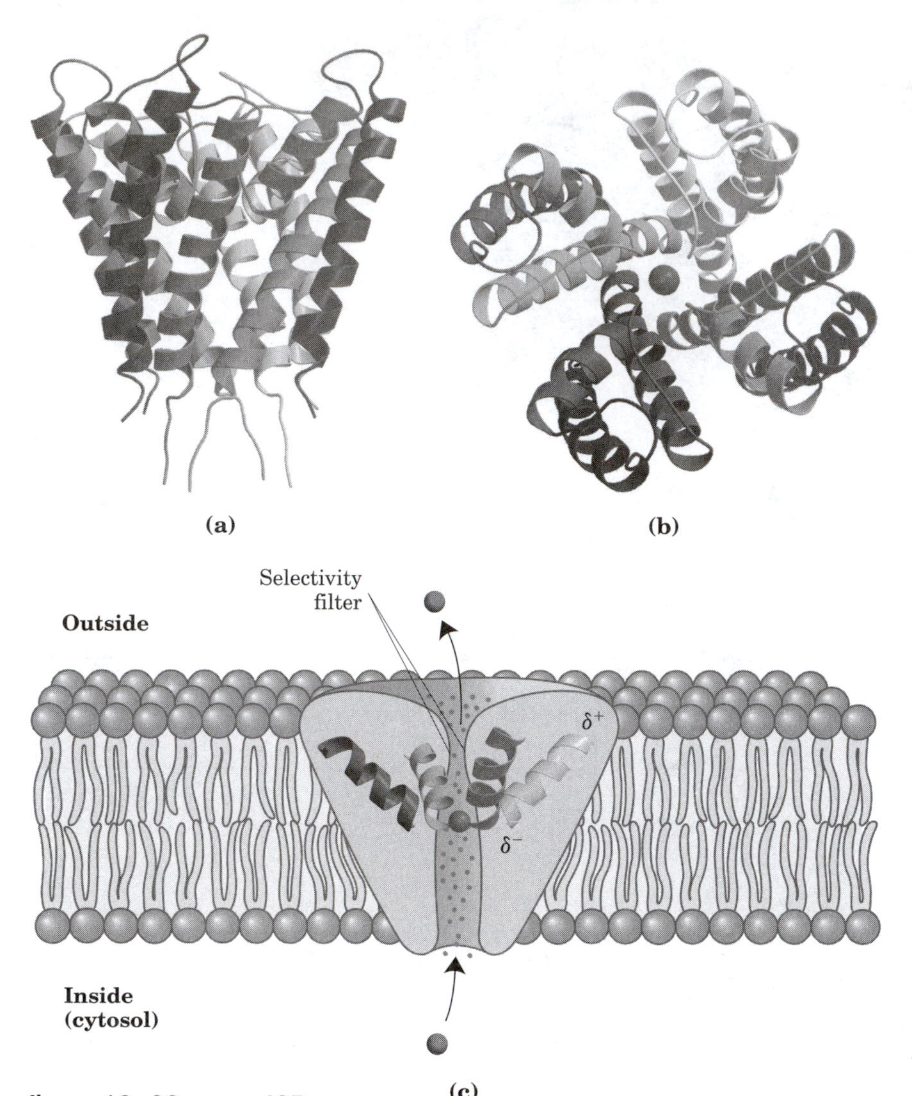

figure 12–38, page 425
Structure of the K^+ channel of *Streptomyces lividans*.

$$CH_3-\overset{\overset{O}{\|}}{C}-O-CH_2-CH_2-\overset{+}{N}(CH_3)_3$$

Acetylcholine, page 426

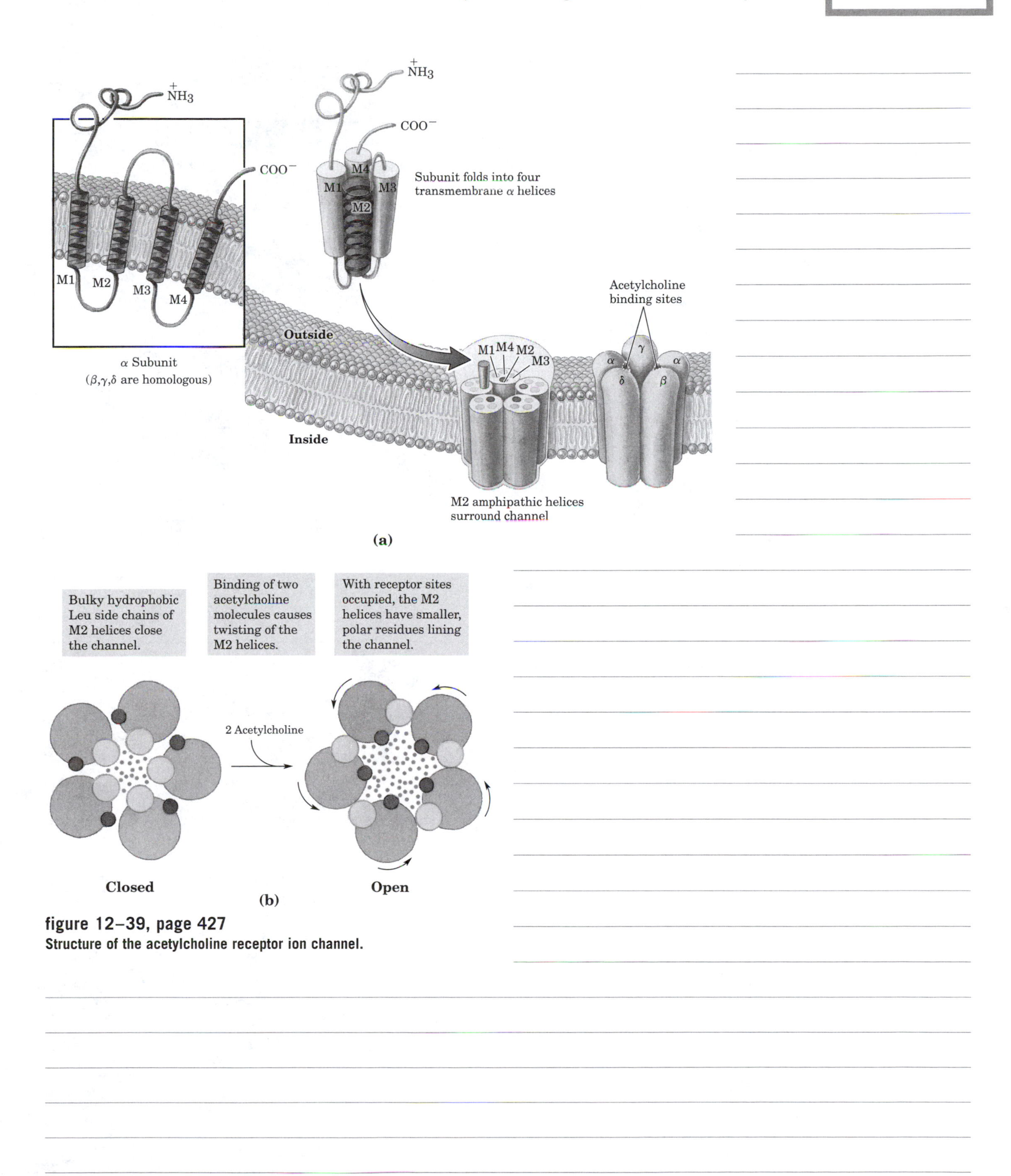

figure 12–39, page 427
Structure of the acetylcholine receptor ion channel.

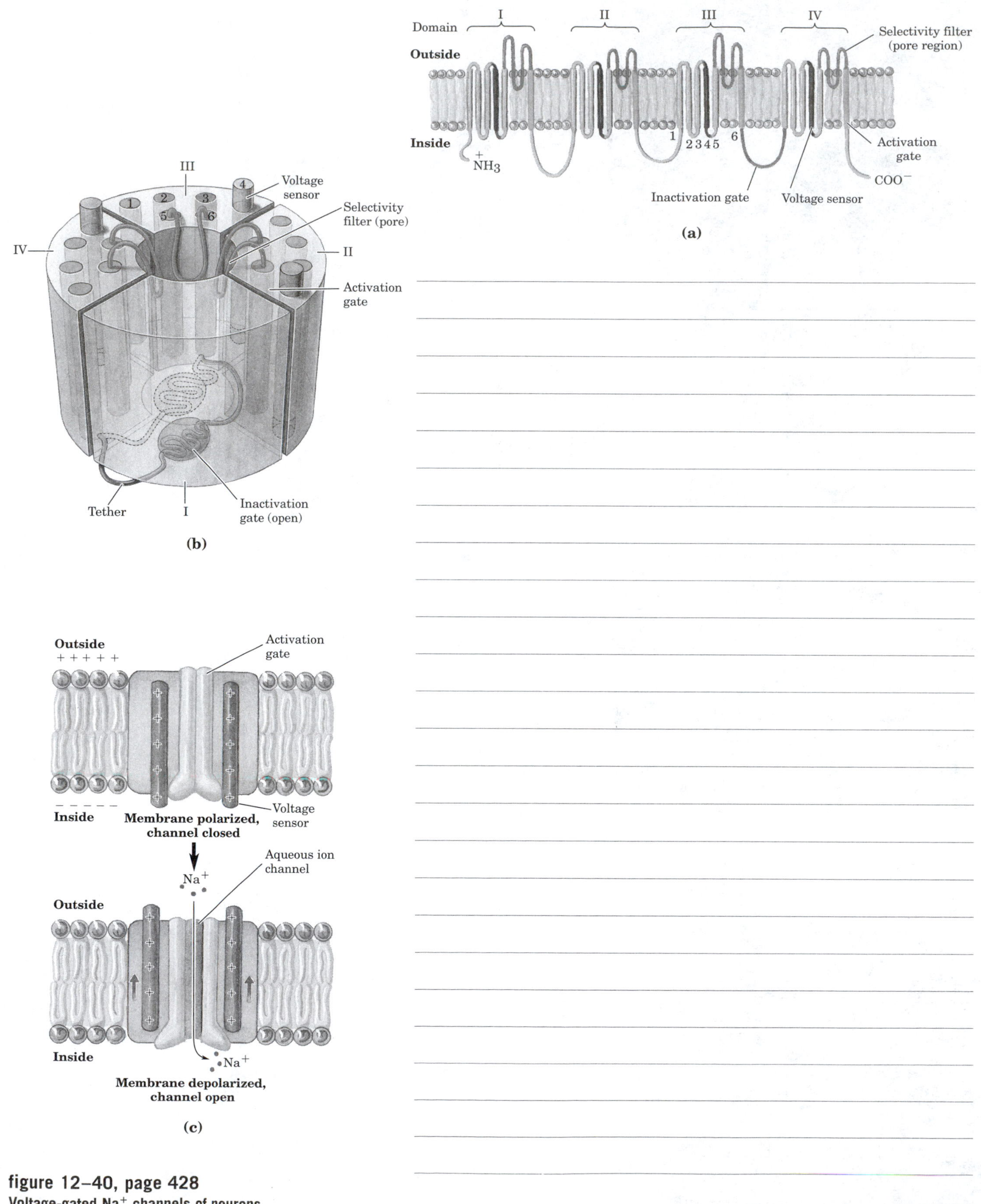

figure 12–40, page 428
Voltage-gated Na^+ channels of neurons.

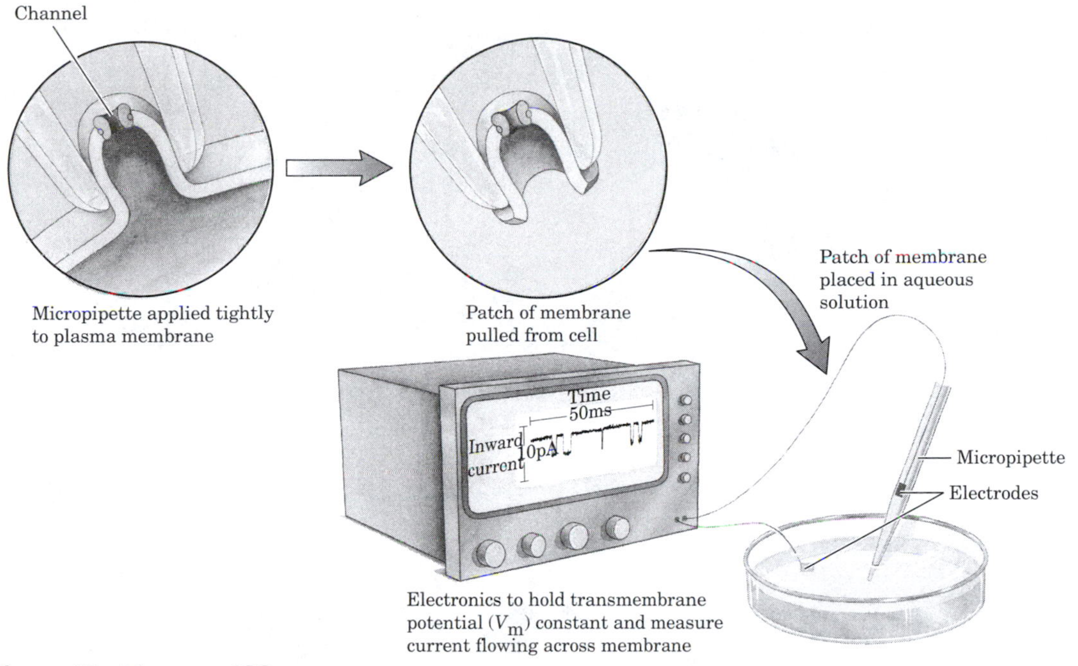

figure 12–41, page 429
Electrical measurements of ion channel function.

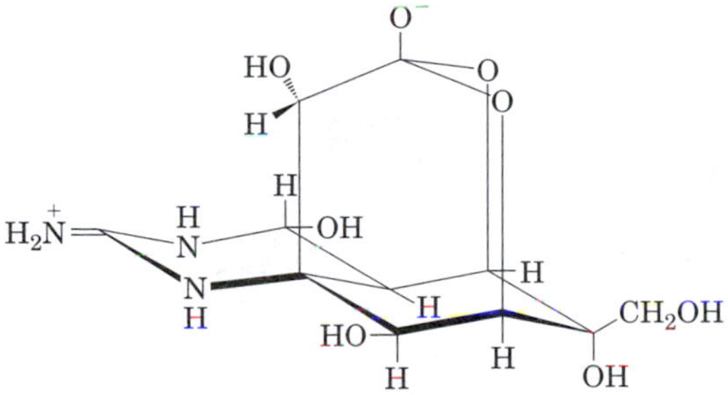

Tetrodotoxin, page 430

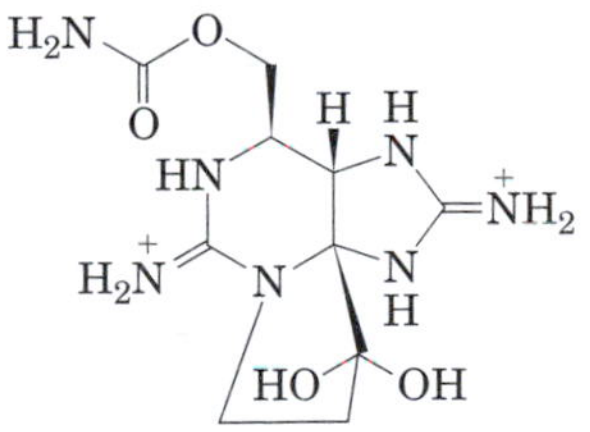

Saxitoxin, page 430

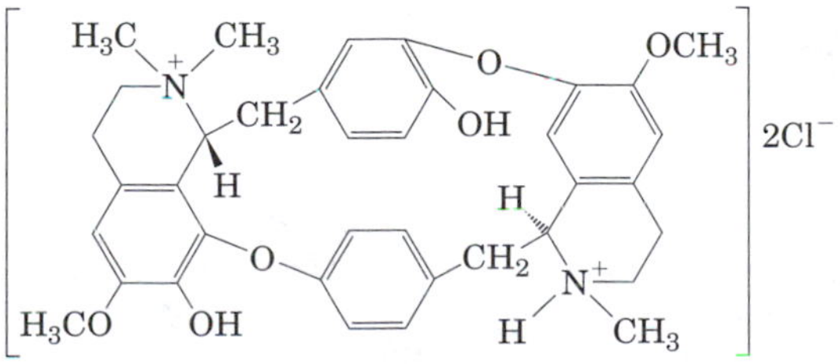

$$\left[\ \right] 2Cl^-$$

D-Tubocurarine Chloride, page 430

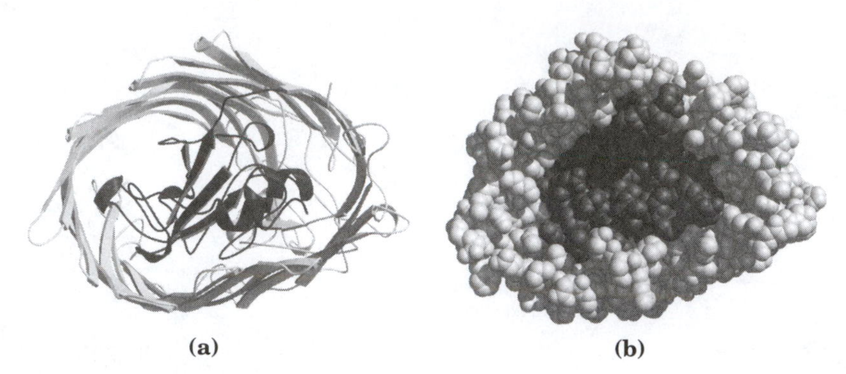

figure 12–42, page 431
Structure of FhuA, an iron transporter from *E. coli.*

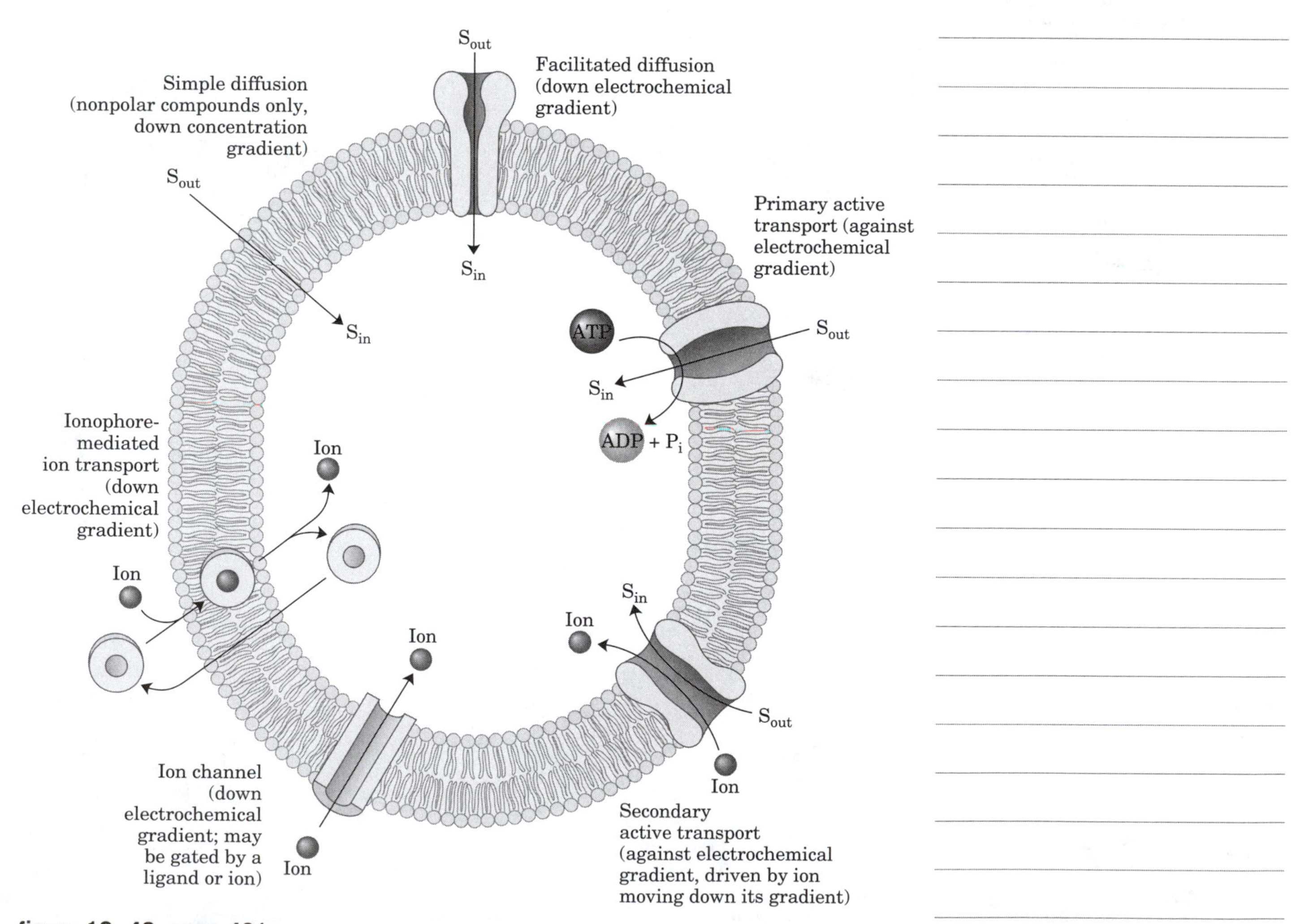

figure 12–43, page 431
Summary of transport types.

PROBLEMS

1. Determining the Cross-Sectional Area of a Lipid Molecule, page 434.

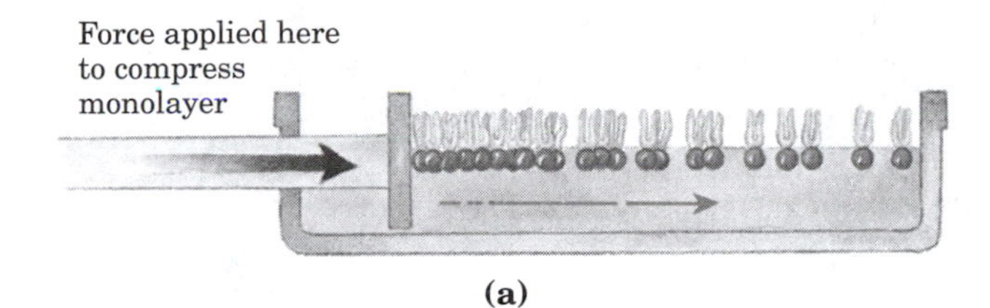

(a)

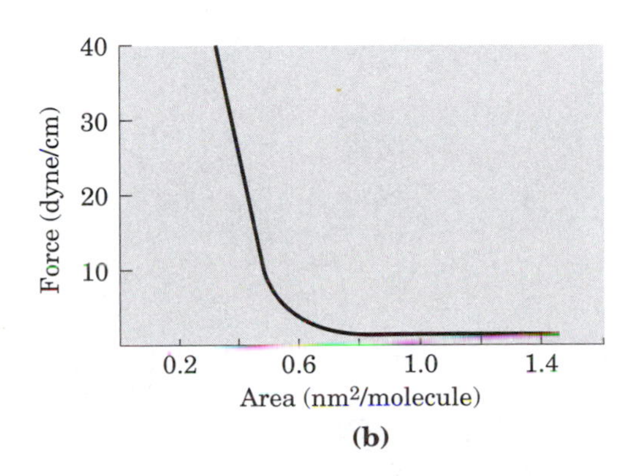

(b)

2. Evidence for Lipid Bilayer, page 434.

Animal	Volume of packed cells (mL)	Number of cells (per mm³)	Total surface area of lipid monolayer from cells (m²)	Total surface area of one cell (μm²)
Dog	40	8,000,000	62	98
Sheep	10	9,900,000	6.0	29.8
Human	1	4,740,000	0.92	99.4

Source: Data from Gorter, E. & Grendel, F. (1925) On bimolecular layers of lipoids on the chromocytes of the blood. *J. Exp. Med.* **41,** 439–443.

15. Membrane Permeability, page 436.

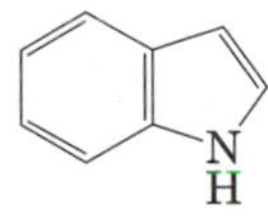

19. Intestinal Uptake of Leucine, page 436.

Substrate	Uptake in presence of Na^+		Uptake in absence of Na^+	
	V_{max}	K_t (mM)	V_{max}	K_t (mM)
L-Leucine	420	0.24	23	0.24
D-Leucine	310	4.7	5	4.7
L-Valine	225	0.31	19	0.31

chapter

13 Biosignaling

When you study for exams, don't forget to review The Minicourses on the UNDERSTAND! BIOCHEMISTRY CD that came with your textbook.

Minicourses that apply to this Chapter include:

Molecules of Life
- An Interactive Gallery of Protein Structures

Cellular Architecture and Traffic
- Cell Signaling
- Phototransduction
- Neuronal Signaling
- Receptor Kinase Signaling
- G Protein Signaling
- Second Messanger Molecules

The Dividing Cell
- Cell Division Cycle
- Proto-Oncogenes and Tumor Supressors

table 13–1, page 437

Some Signals to Which Cells Respond
Antigens
Cell surface glycoproteins/oligosaccharides
Developmental signals
Extracellular matrix components
Growth factors
Hormones
Light
Mechanical touch
Neurotransmitters
Odorants
Pheromones
Tastants

(a) Specificity
Signal molecule fits binding site on its complementary receptor; other signals do not fit.

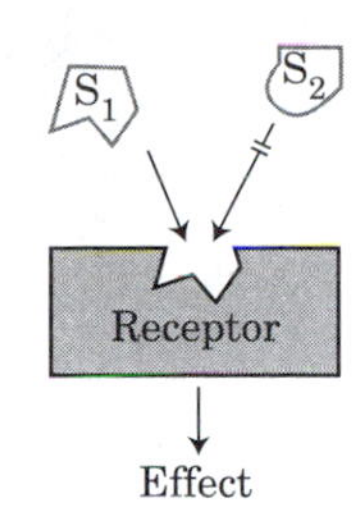

(b) Amplification
When enzymes activate enzymes, the number of affected molecules increases geometrically in an enzyme cascade.

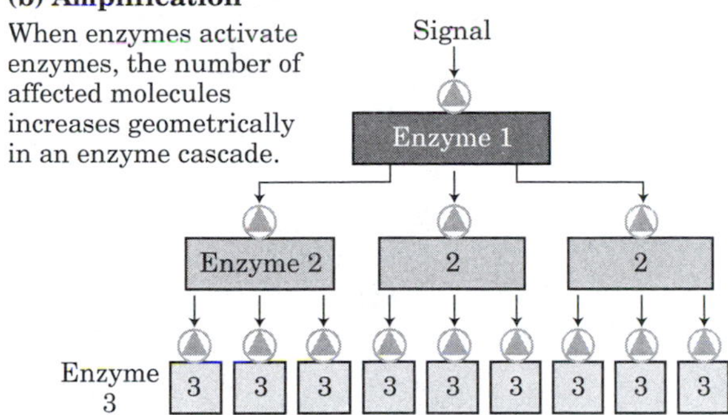

(c) Desensitization/Adaptation
Receptor activation triggers a feedback circuit that shuts off the receptor or removes it from the cell surface.

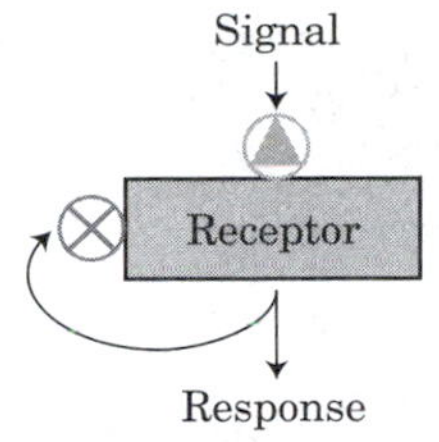

(d) Integration
When two signals have opposite effects on a metabolic characteristic such as the concentration of a second messenger X, or the membrane potential V_m, the regulatory outcome results from the integrated input from both receptors.

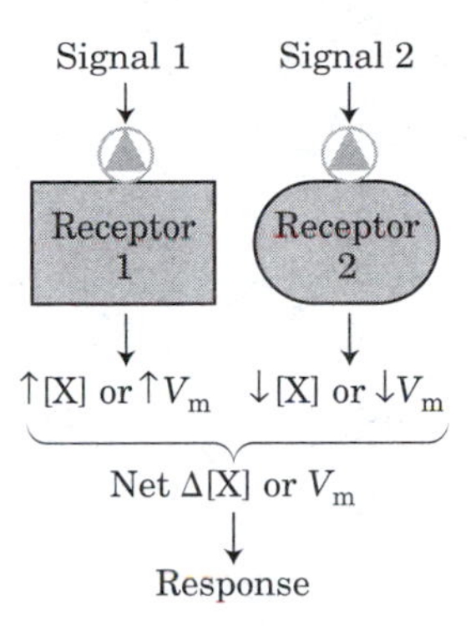

figure 13–1, page 438
Four features of signal-transducing systems.

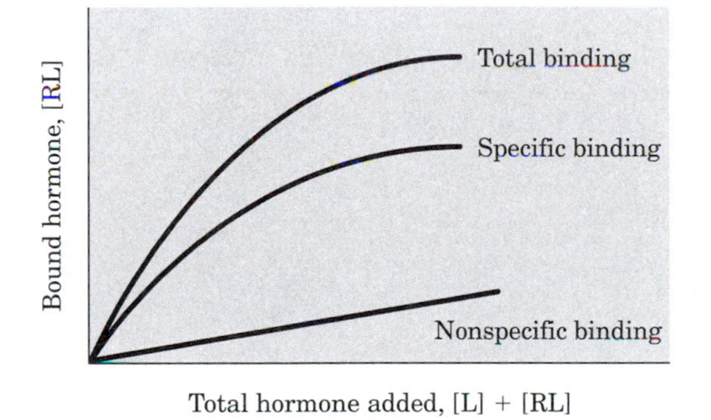

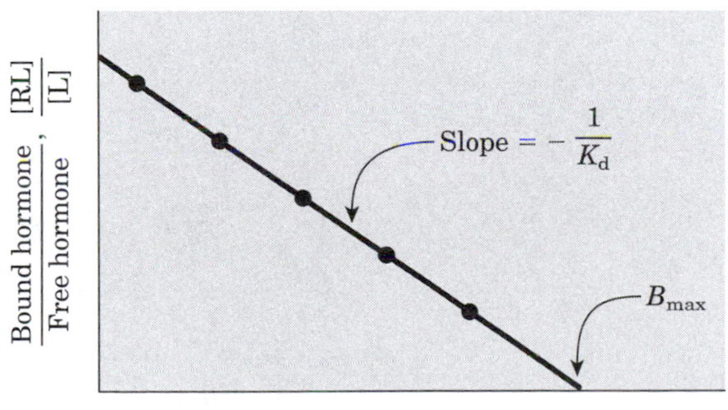

box 13–1, figure 1, page 439
Scatchard analysis of the receptor-ligand interaction.

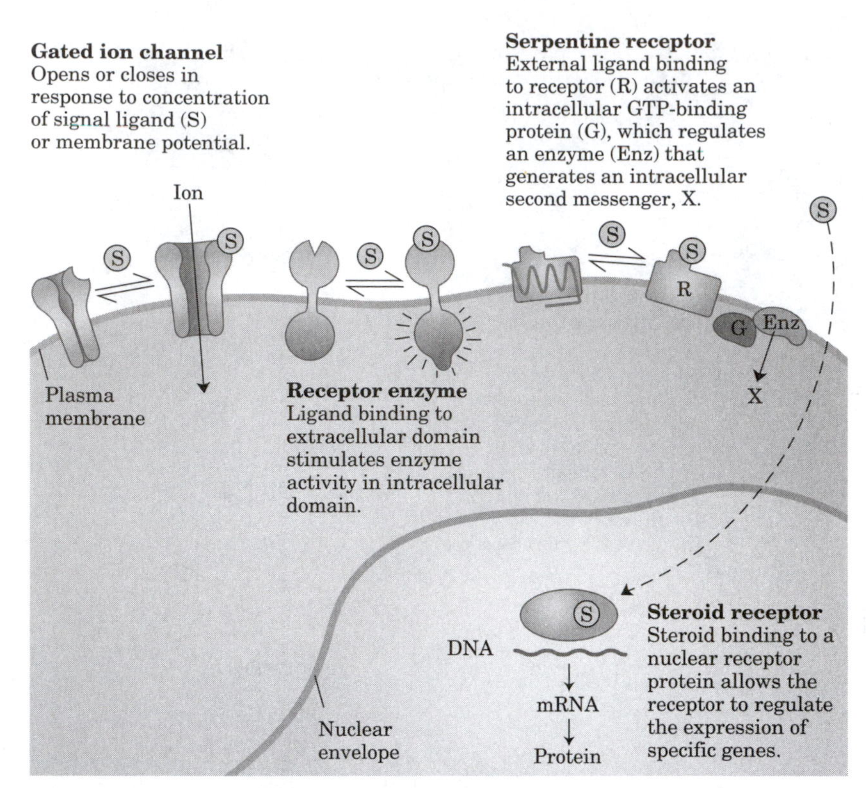

figure 13–2, page 440
Four general types of signal transducers.

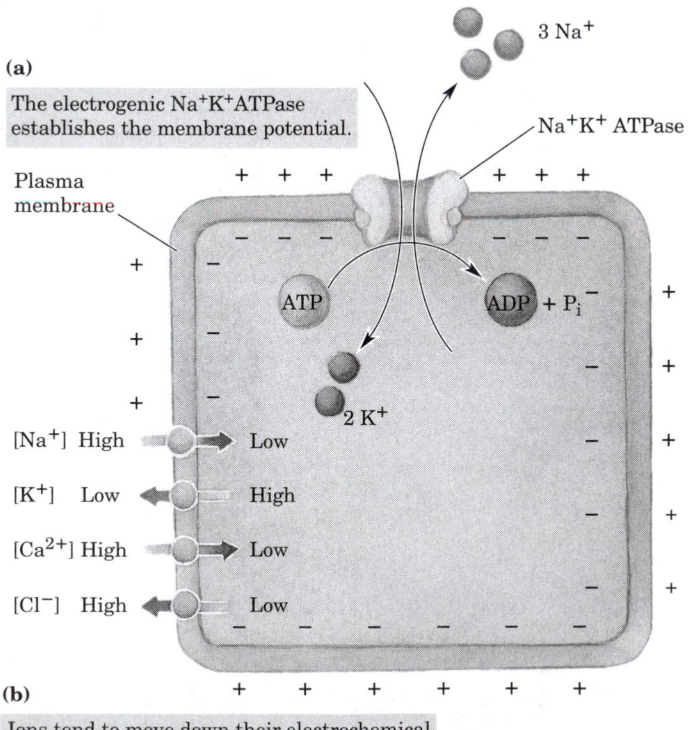

figure 13–3, page 441
Transmembrane electrical potential.

table 13–2, page 442

Ion Concentrations in Cells and Extracellular Fluids (mM)

	K^+		Na^+		Ca^{2+}		Cl^-	
	In	Out	In	Out	In	Out	In	Out
Cell type								
Squid axon	400	20	50	440	≤0.4	10	40–150	560
Frog muscle	124	2.3	10.4	109	<0.1	2.1	1.5	78

$$CH_3-\overset{+}{N}(CH_3)_2-CH_2CH_2O-\overset{O}{\overset{\|}{C}}-CH_3$$

Acetylcholine (Ach), page 443

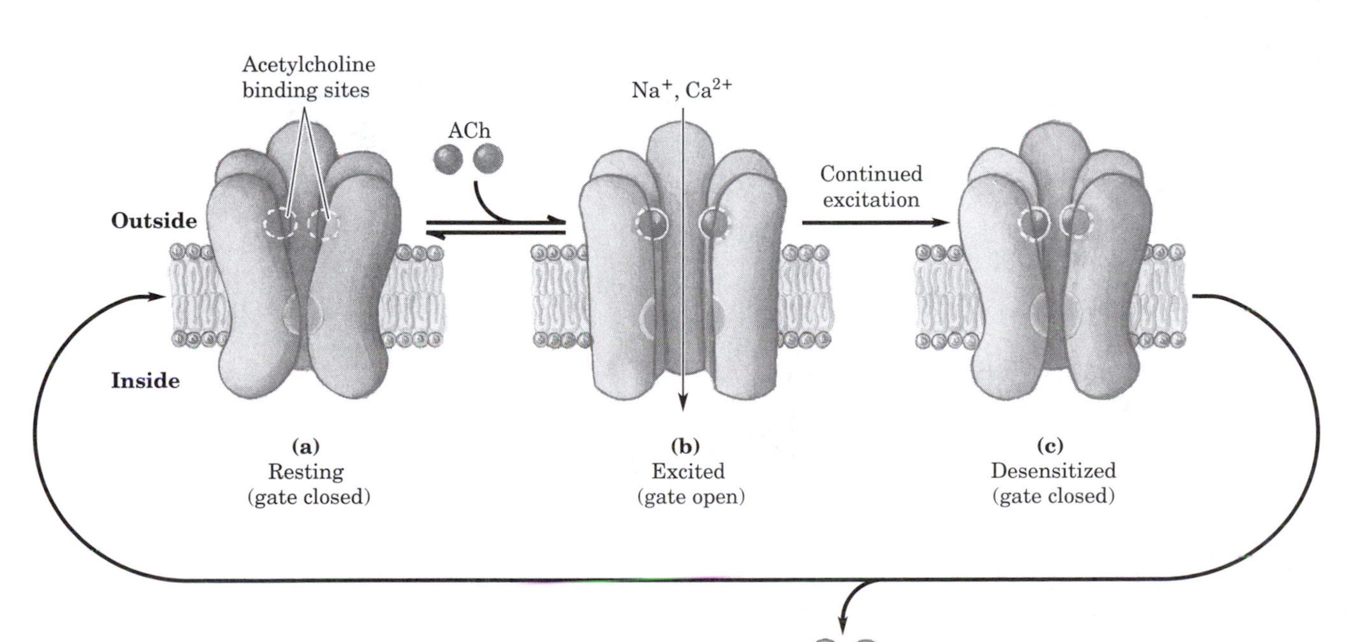

figure 13–4, page 443
Three states of the acetylcholine receptor.

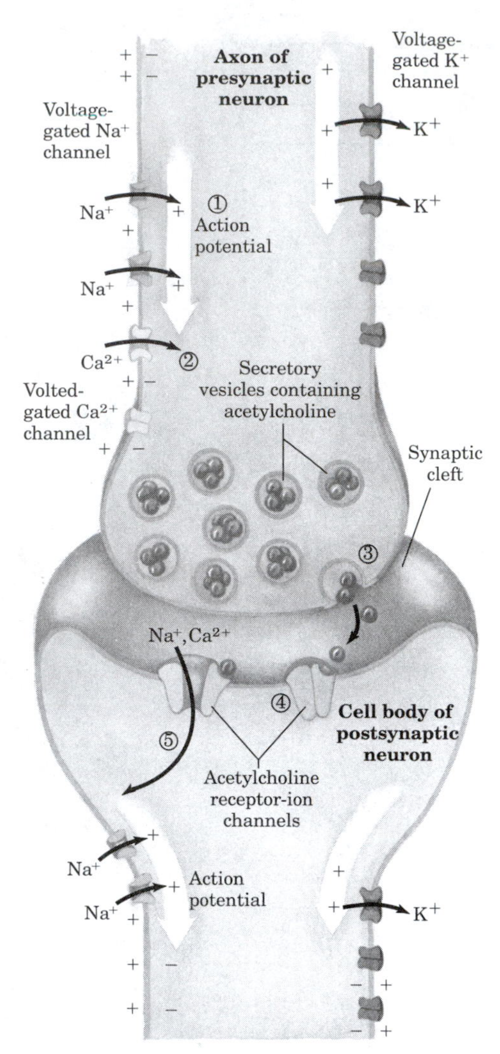

figure 13–5, page 444
Role of voltage-gated and ligand-gated ion channels in neural transmission.

$$\text{HO-indole-}CH_2CH_2\overset{+}{N}H_3$$

Serotonin (5-Hydroxytryptamine), page 445

$$H_3\overset{+}{N}-CH(COO^-)-CH_2-CH_2-COO^-$$

Glutamate, page 445

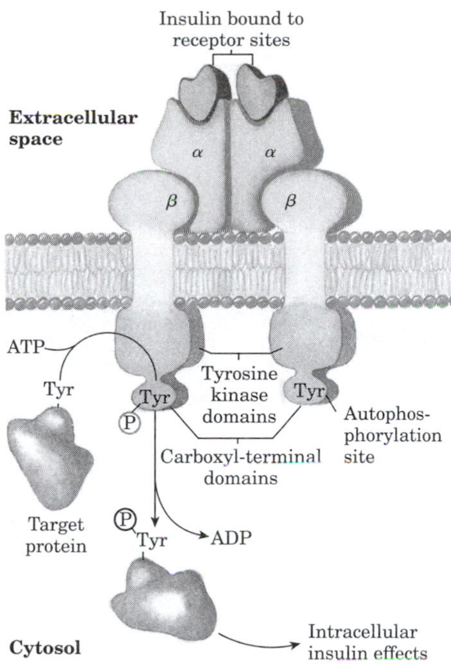

figure 13–6, page 445
Insulin receptor.

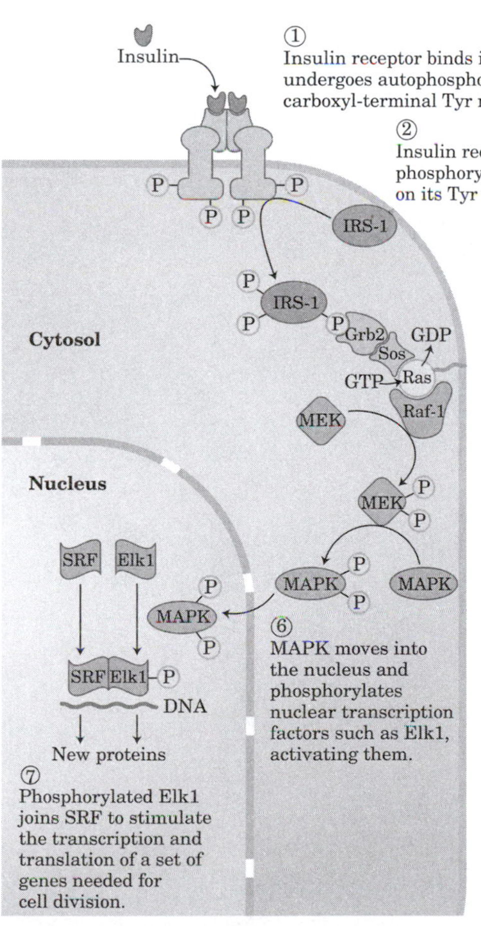

figure 13–7, page 446
Regulation of gene expression by insulin.

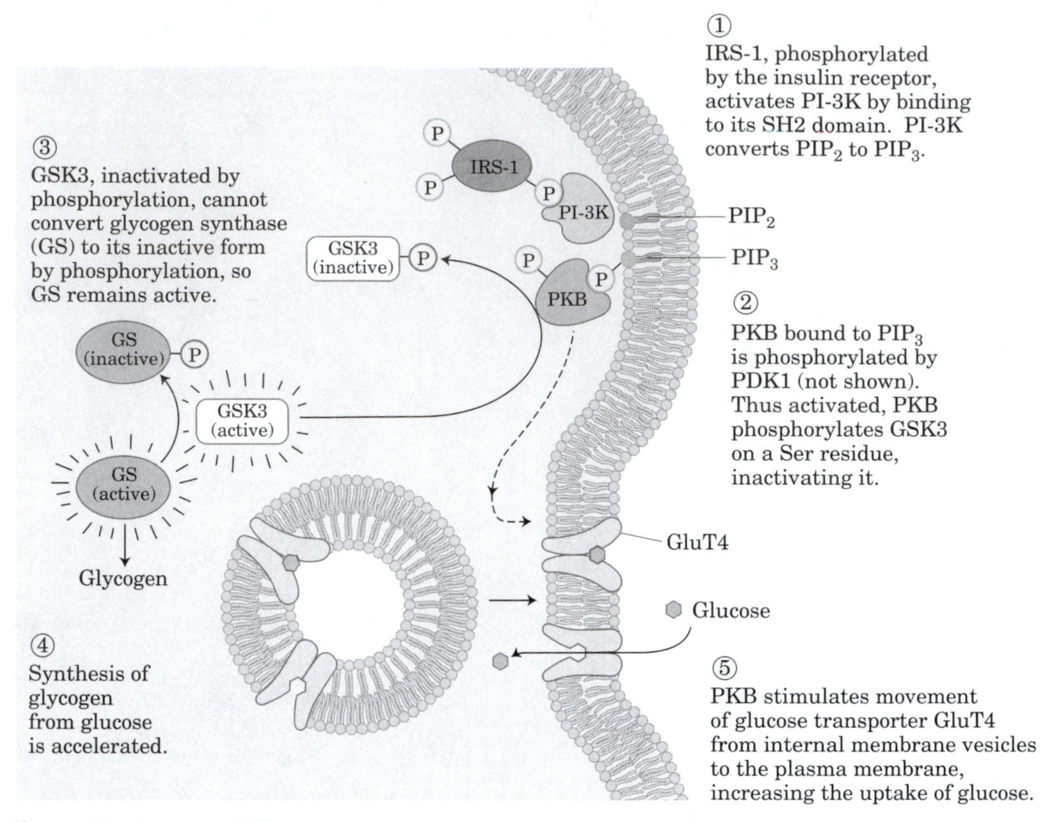

figure 13–8, page 447
Activation of glycogen synthase by insulin.

$$\text{GTP} \xrightarrow[\text{guanylyl cyclase}]{PP_i} \text{Guanosine } 3',5'\text{-cyclic monophosphate (cGMP)}$$

GTP

Guanosine 3′,5′-cyclic monophosphate (cGMP)

page 448

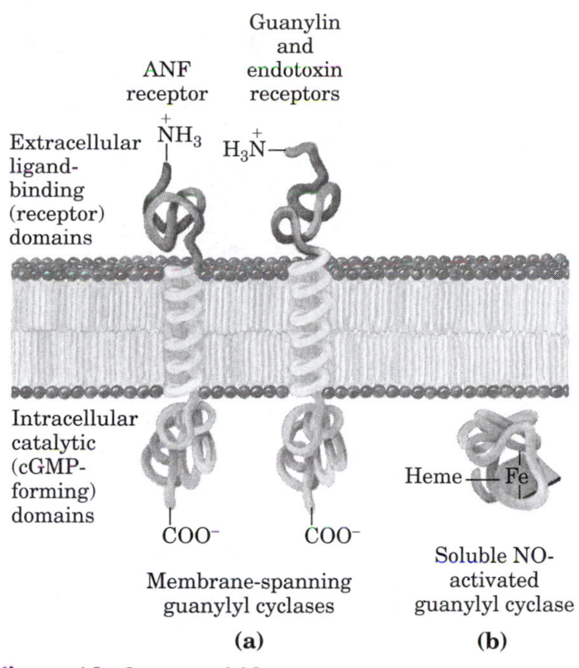

figure 13–9, page 448
Two types (isozymes) of guanylyl cyclase involved in signal transduction.

Arginine (NH_2–C(=$\overset{+}{N}H_2$)–NH–$(CH_2)_3$–CH($\overset{+}{N}H_3$)–COO^-) + NADPH + O_2 $\xrightarrow[\text{NO synthase}]{Ca^{2+}}$ Citrulline (NH_2–C(=O)–NH–$(CH_2)_3$–CH($\overset{+}{N}H_3$)–COO^-) + $NADP^+$ + NO

NO Synthase, page 449

HO, HO (ring) —CH(OH)—CH_2—$\overset{+}{N}H_2$—CH_3

figure 13–10, page 449
Epinephrine.

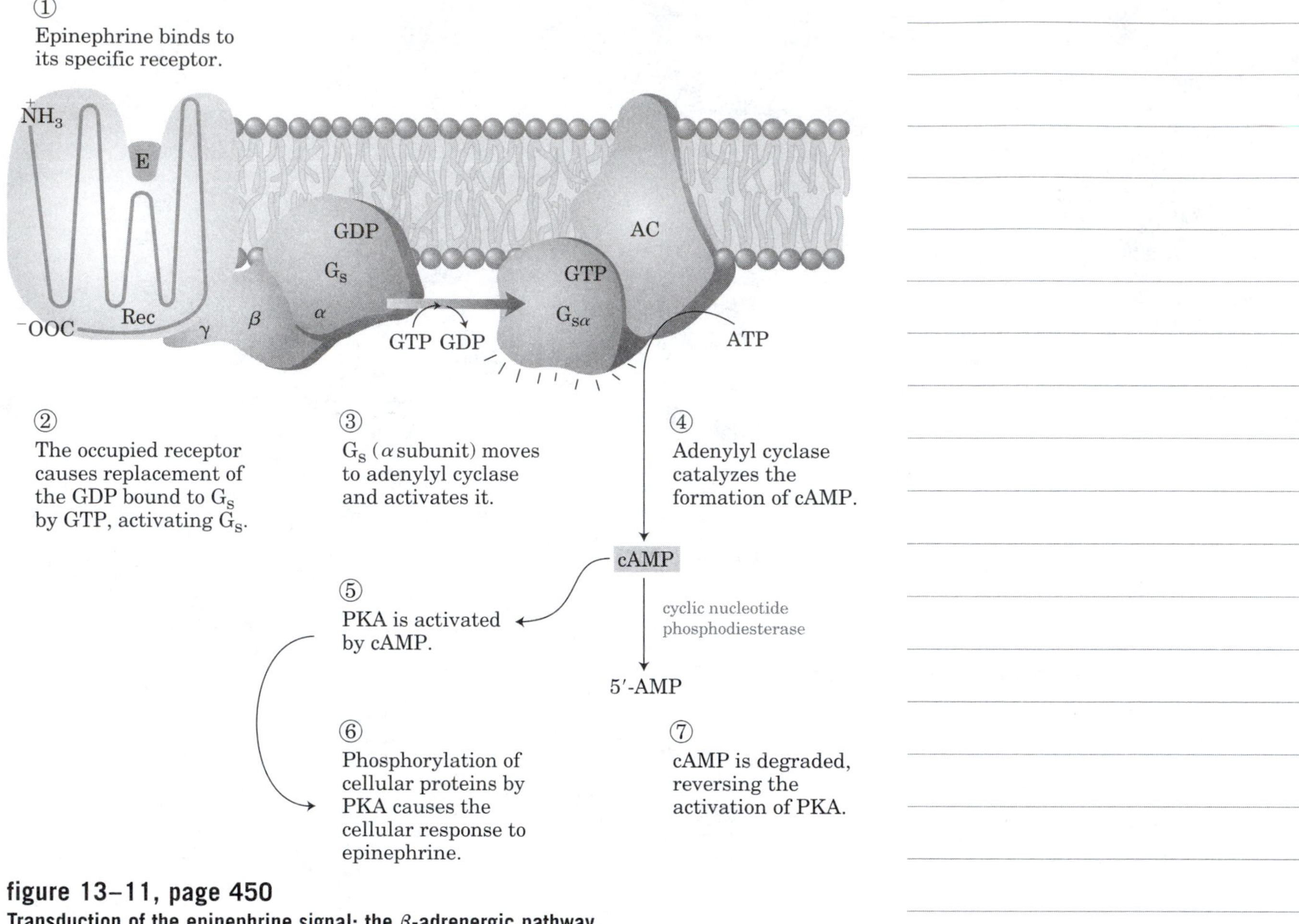

figure 13–11, page 450
Transduction of the epinephrine signal: the β-adrenergic pathway.

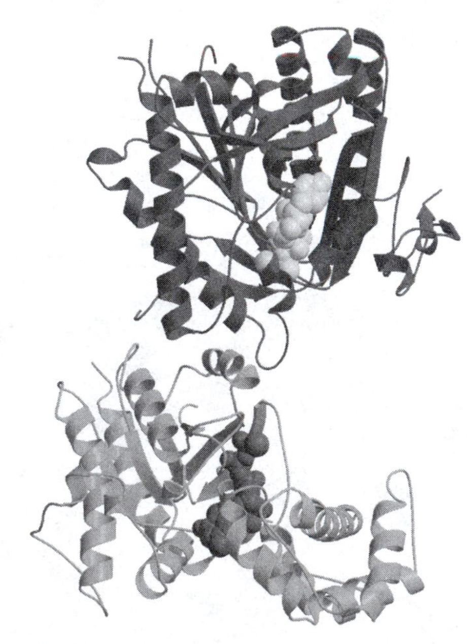

figure 13–12, page 451
Interaction of $G_{s\alpha}$ with adenylyl cyclase.

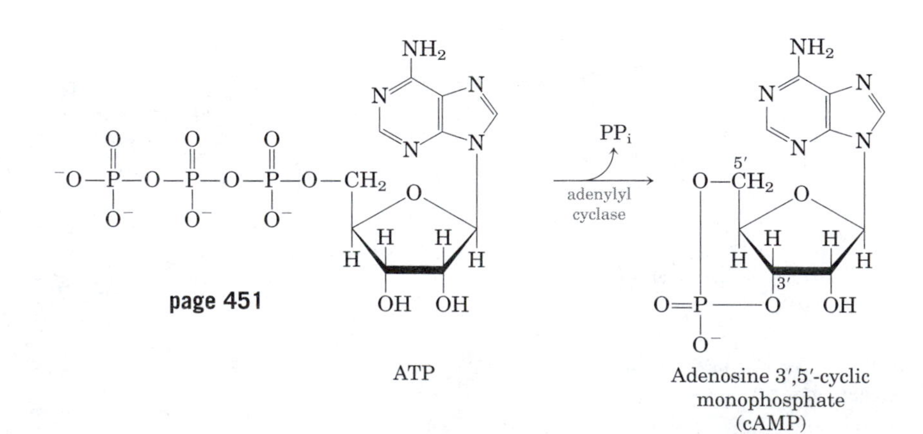

page 451

① G_s with GDP bound is turned off; it cannot activate adenylyl cyclase.

② Contact of G_s with hormone-receptor complex causes displacement of bound GDP by GTP.

③ G_s with GTP bound dissociates into α and $\beta\gamma$ subunits. $G_{s\alpha}$-GTP is turned on; it can activate adenylyl cyclase.

④ GTP bound to $G_{s\alpha}$ is hydrolyzed by the protein s intrinsic GTPase; $G_{s\alpha}$ thereby turns itself off. The inactive α subunit reassociates with the β, γ subunits.

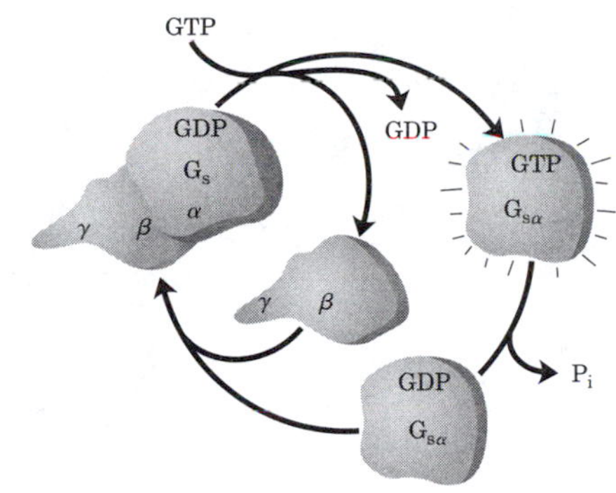

figure 13–13, page 451
Self-inactivation of G_s.

Inactive PKA

Regulatory subunits: empty cAMP sites

Catalytic subunits: substrate-binding sites blocked by autoinhibitory domains of R subunits

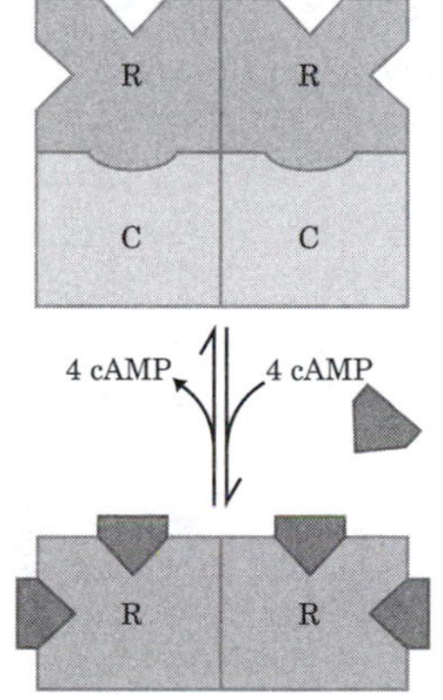

Regulatory subunits: autoinhibitory domains buried

+

Active PKA

Catalytic subunits: open substrate-binding sites

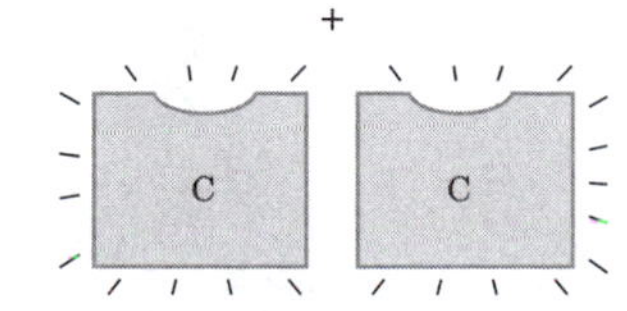

(a)

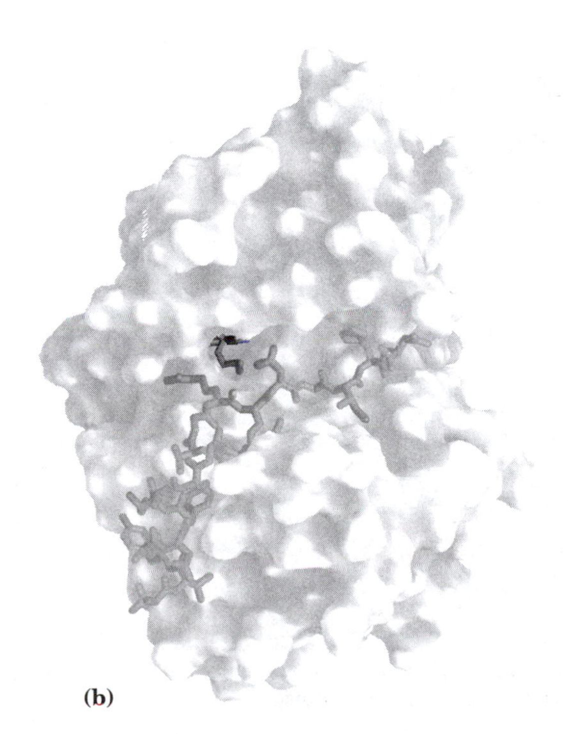

(b)

figure 13–14, page 452
Activation of cAMP-dependent protein kinase, PKA.

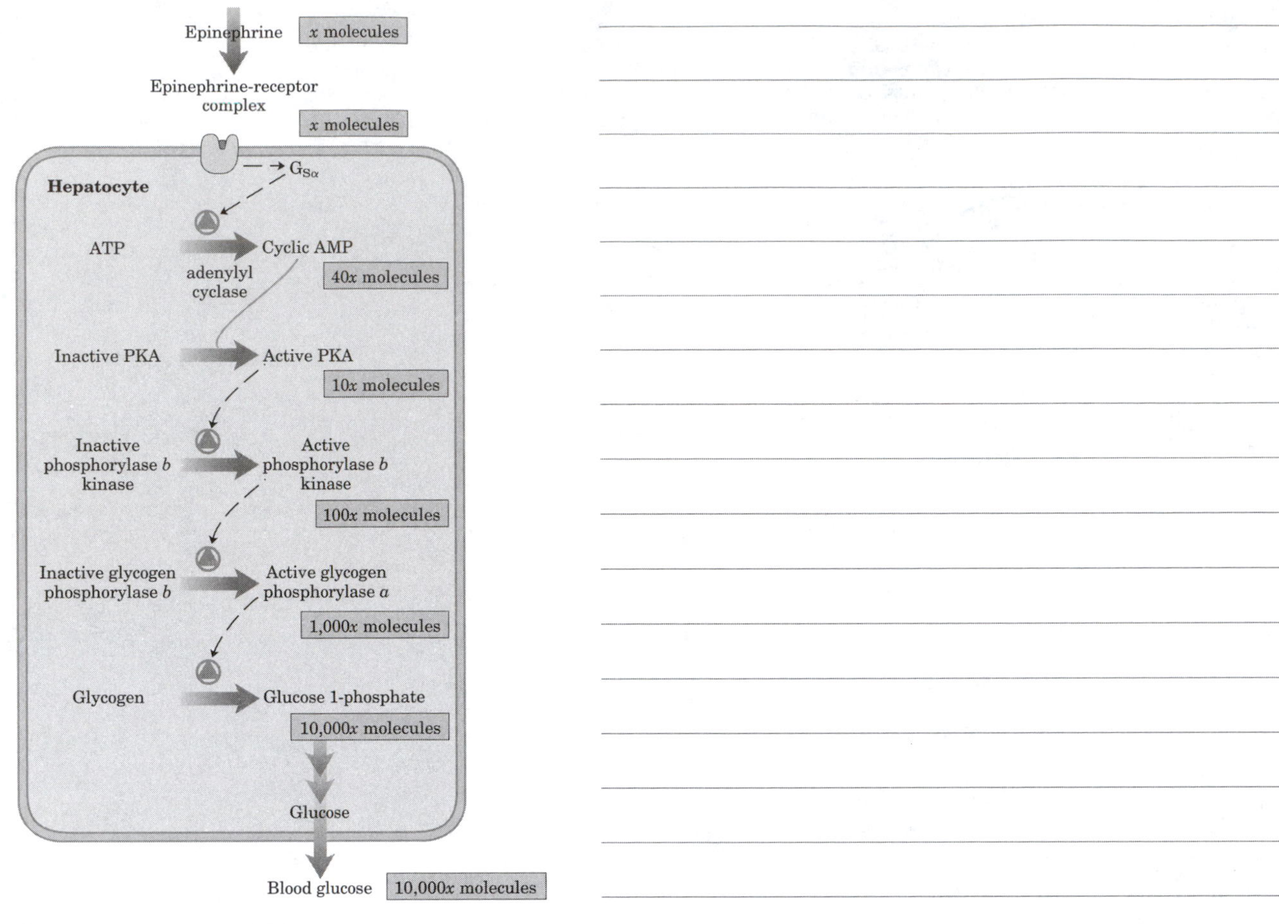

figure 13–15, page 453
Epinephrine cascade.

table 13–3, page 453

Some Enzymes Regulated by cAMP-Dependent Phosphorylation (by PKA)

Enzyme	Sequence phosphorylated*	Pathway
Glycogen synthase	RASCTSSS	Glycogen synthesis
Phosphorylase *b* kinase	α subunit: VEFRRLSI β subunit: RTKRSGSV	Glycogen breakdown
Pyruvate kinase (rat liver)	GVLRRASVAZL	Glycolysis
Pyruvate dehydrogenase complex (type L)	GYLRRASV	Pyruvate to acetyl-CoA
Hormone-sensitive lipase	PMRRSV	Triacylglycerol mobilization and fatty acid oxidation
Phosphofructokinase-2/fructose 2,6-bisphosphatase	LQRRRGSSIPQ	Glycolysis/gluconeogenesis
Tyrosine hydroxylase	FIGRRQSL	Synthesis of L-DOPA, dopamine, norepinephrine, and epinephrine
Histone H1	AKRKASGPPVS	DNA condensation
Histone H2B	KKAKASRKESYSVYVYK	DNA condensation
Cardiac phospholamban (a cardiac pump regulator)	AIRRAST	Regulation of intracellular [Ca^{2+}]
Protein phosphatase-1 inhibitor-1	IRRRRPTP	Regulation of protein dephosphorylation
CREB	ILSRRPSY	cAMP regulation of gene expression
PKA consensus sequence†	XR(R/K)X(S/T)B	

*The phosphorylated S or T residue is shown in red. All residues are given as their one-letter abbreviations (see Table 5–1).
†X is any amino acid; B is any hydrophobic amino acid.

Cyclic AMP $\xrightarrow[\text{cyclic nucleotide phosphodiesterase}]{H_2O}$ Adenosine 5′-monophosphate (AMP)

caffeine, theophylline (inhibit phosphodiesterase)

page 454

table 13–4, page 454

Some Signals That Use cAMP as Second Messenger
Corticotropin (ACTH)
Corticotropin-releasing hormone (CRH)
Dopamine [D-1, D-2]*
Epinephrine (β-adrenergic)
Follicle-stimulating hormone (FSH)
Glucagon
Histamine [H-2]*
Luteinizing hormone (LH)
Melanocyte-stimulating hormone (MSH)
Odorants (many)
Parathyroid hormone
Prostaglandins E_1, E_2 (PGE_1, PGE_2)
Serotonin [5-HT-1α, 5-HT-2]*
Somatostatin
Tastants (sweet, bitter)
Thyroid-stimulating hormone (TSH)

*Some signals have two or more receptor subtypes (shown in square brackets), which may have different transduction mechanisms. For example, serotonin is detected in some tissues by receptor subtypes 5-HT-1a and 5-HT-1b, which act through adenylyl cyclase and cAMP, and in other tissues by receptor subtype 5-HT-1c, acting through the phospholipase C–IP_3 mechanism (see Table 13–5).

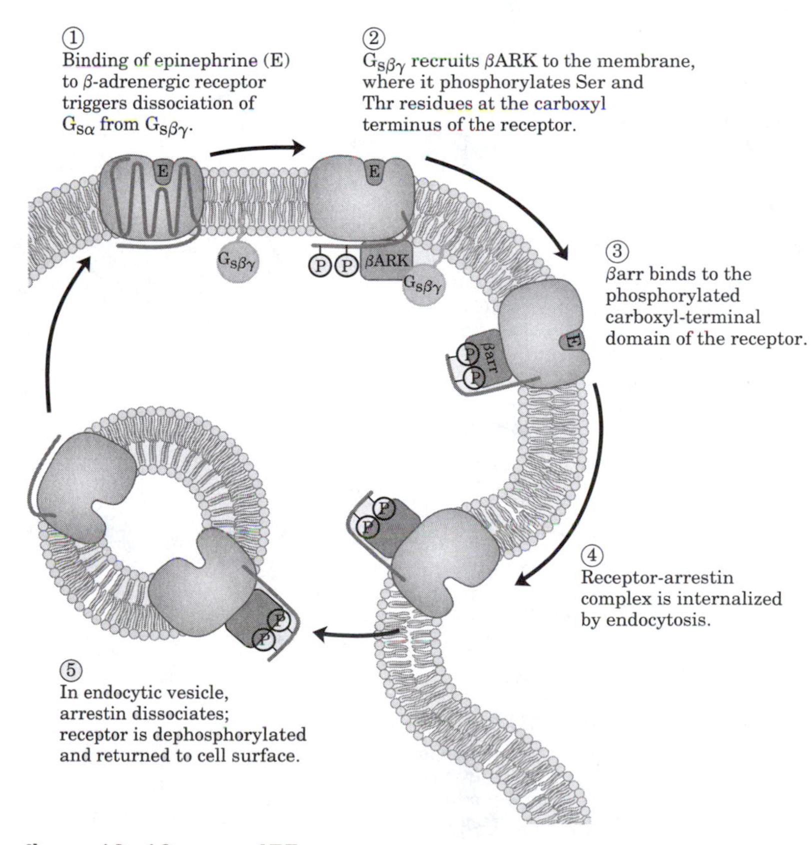

figure 13–16, page 455
Desensitization of the β-adrenergic receptor in the continued presence of epinephrine.

Inositol 1,4,5-trisphosphate, page 456

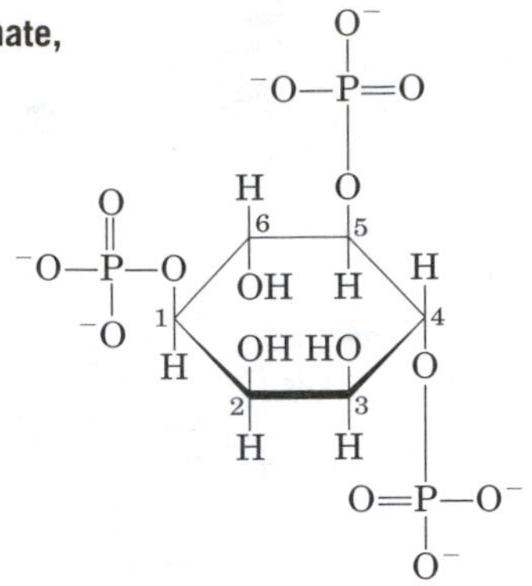

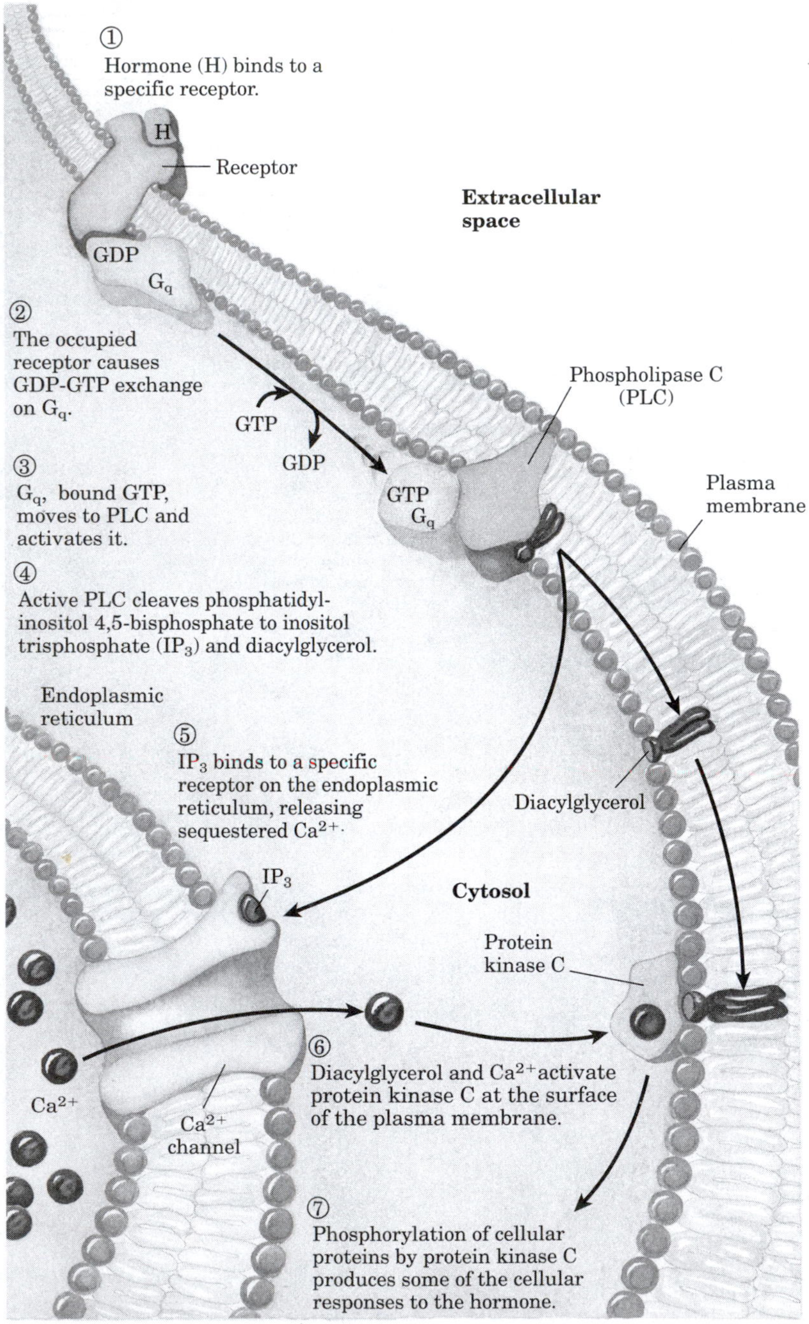

figure 13–17, page 456
Hormone-activated phospholipase C and IP_3.

table 13–5, page 456

Some Signals That Act through Phospholipase C and IP_3
Acetylcholine [muscarinic M_1]
α_1-Adrenergic agonists
Angiogenin
Angiotensin II
ATP [P_{2x} and P_{2y}]*
Auxin
Gastrin-releasing peptide
Glutamate
Gonadotropin-releasing hormone (GRH)
Histamine [H_1]*
Light (*Drosophila*)
Oxytocin
Platelet-derived growth factor (PDGF)
Serotonin [5-HT-1c]*
Thyrotropin-releasing hormone (TRH)
Vasopressin

*Receptor subtypes are in square brackets; see footnote to Table 13–4.

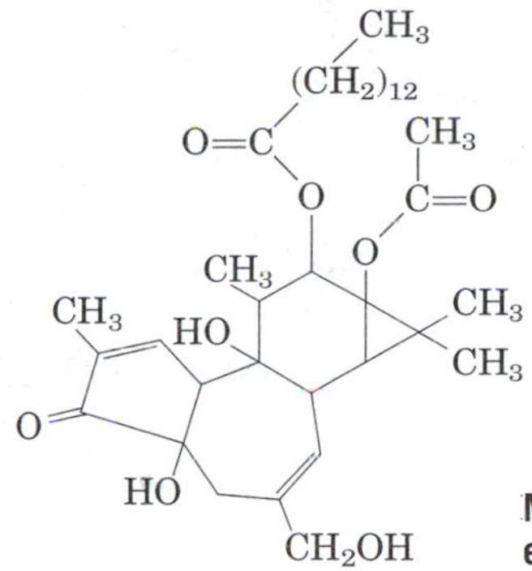

Myristoylphorbol acetate (a phorbal ester), page 457

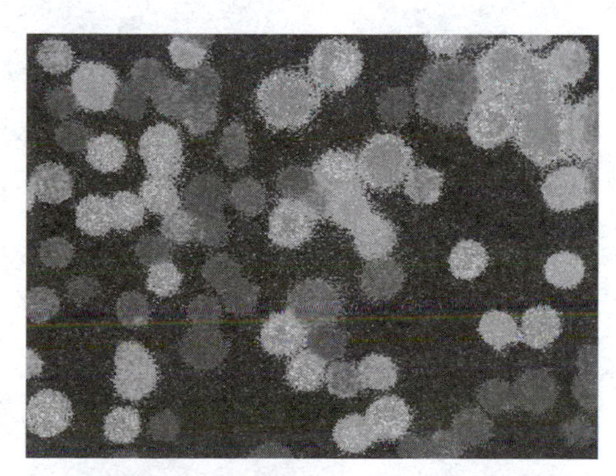

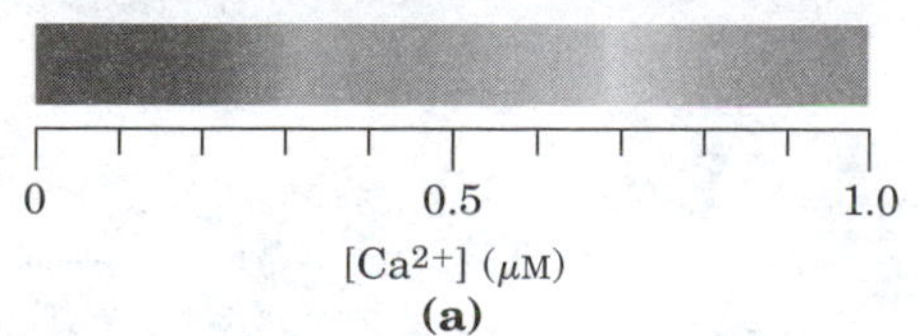

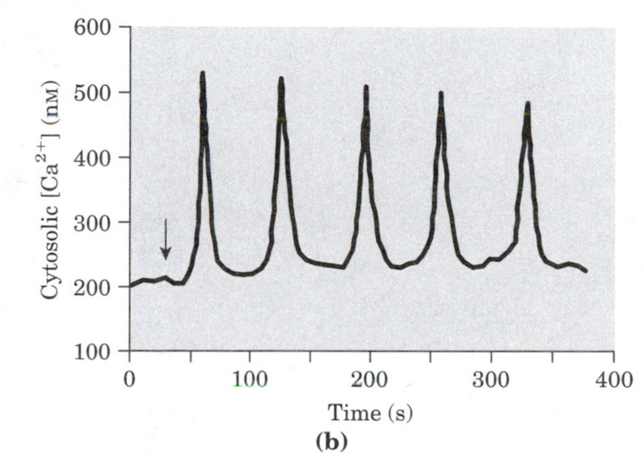

figure 13–18, page 457

Triggering of oscillations in intracellular [Ca^{2+}] by extracellular signals.

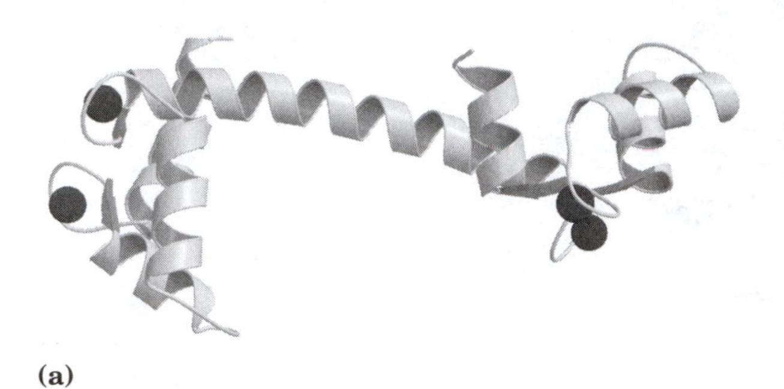

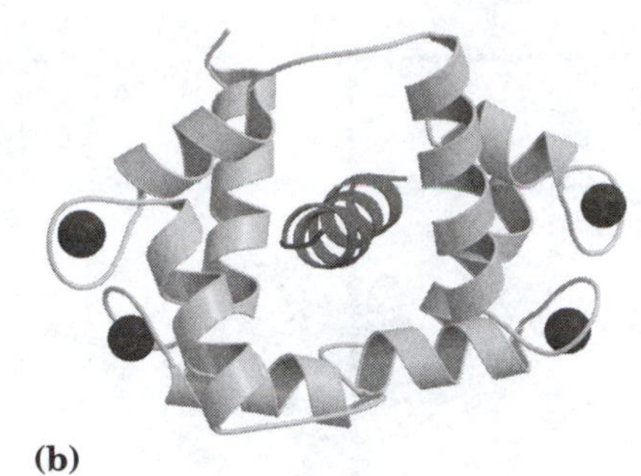

figure 13–19, page 458
Calmodulin.

table 13–6, page 458

Some Proteins Regulated by Ca^{2+} and Calmodulin
Adenylyl cyclase (brain)
Ca^{2+}/calmodulin-dependent protein kinases
Ca^{2+}-dependent Na^+ channel (*Paramecium*)
Ca^{2+} release channel of sarcoplasmic reticulum
Calcineurin (phosphoprotein phosphatase 2B)
cAMP phosphodiesterase
cAMP-gated olfactory channel
cGMP-gated Na^+, Ca^{2+} channels (rod and cone cells)
Myosin light chain kinases
NADH kinase
Nitric oxide synthase
PI-3 kinase
Plasma membrane Ca^{2+} ATPase (Ca^{2+} pump)
RNA helicase (p68)

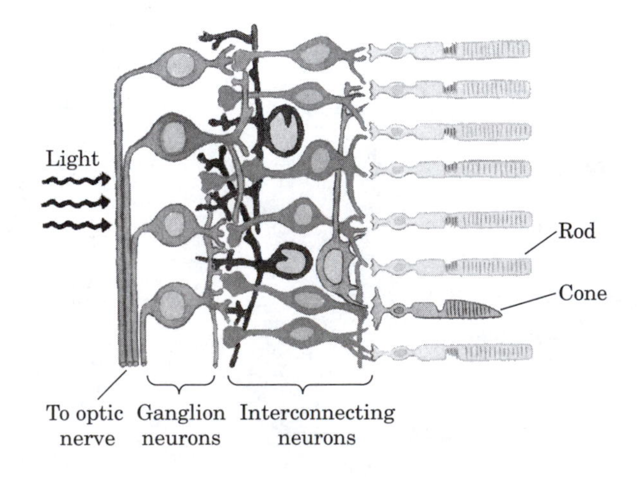

figure 13–20, page 459
Light reception in the vertebrate eye.

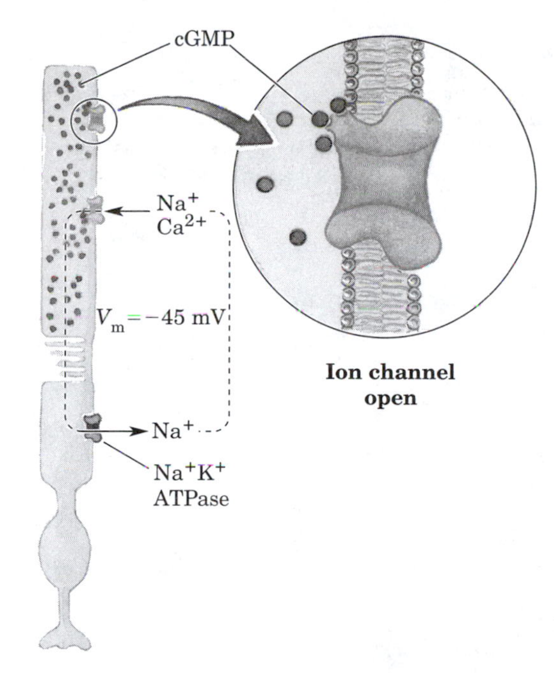

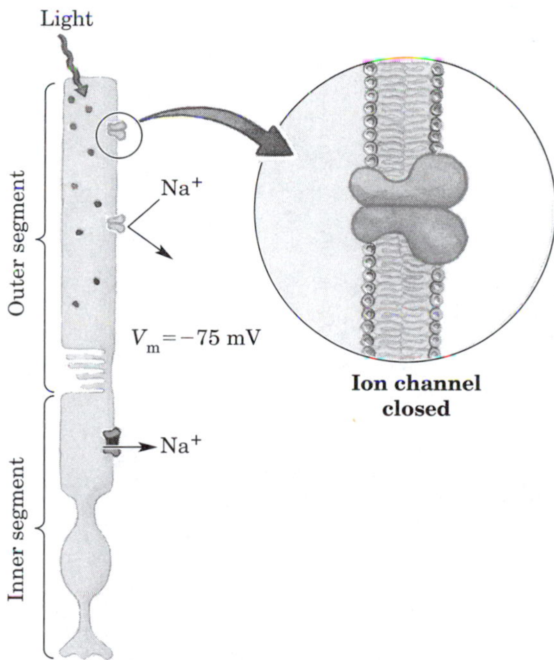

figure 13–21, page 459
Light-induced hyperpolarization of rod cells.

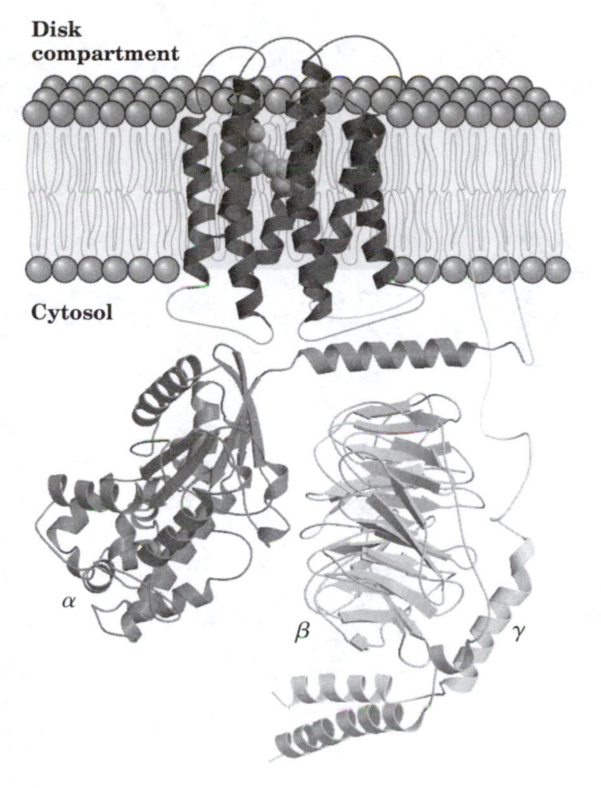

figure 13–22, page 460
Likely structure of rhodopsin complexed with the G protein transducin.

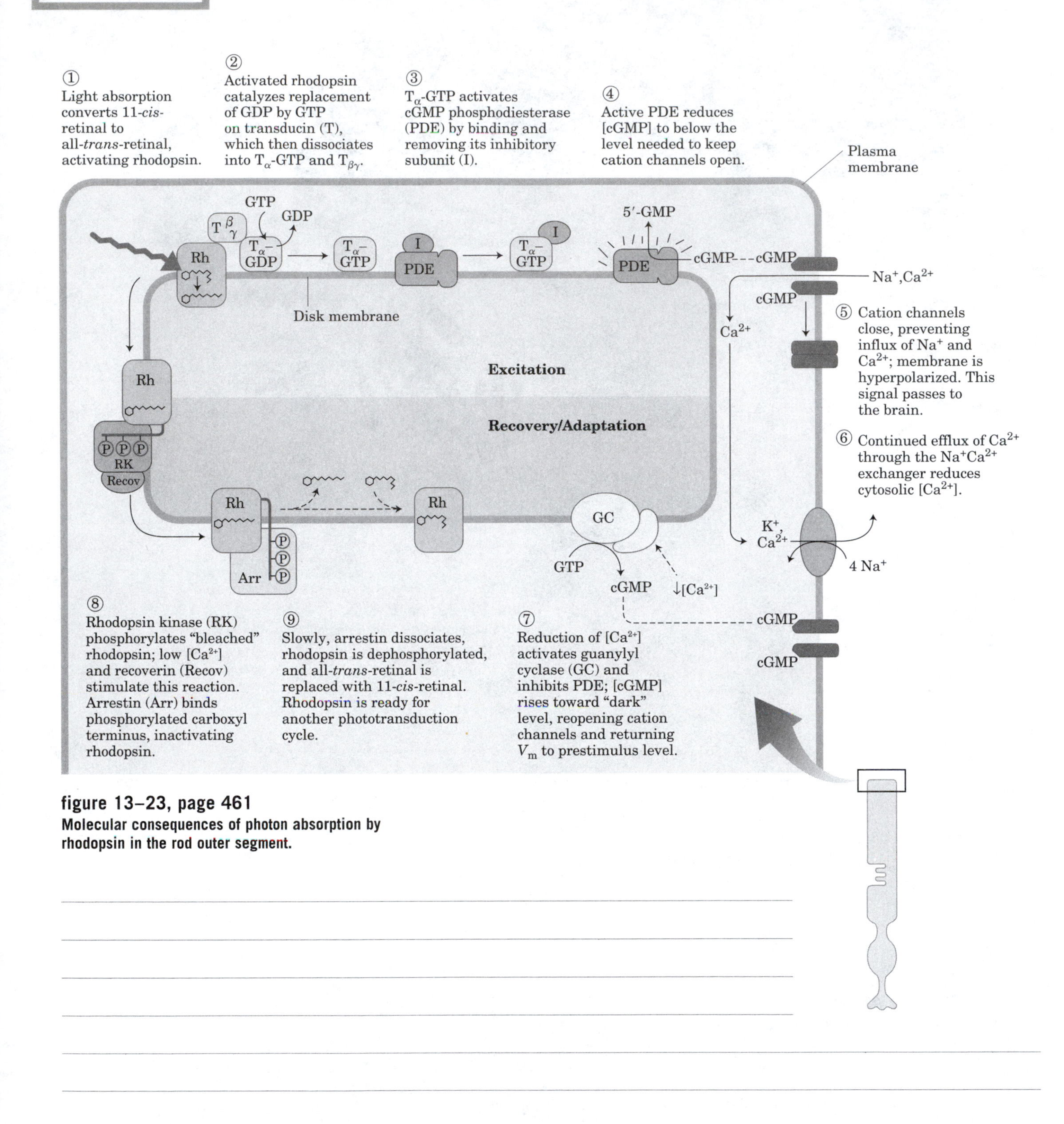

figure 13–23, page 461

Molecular consequences of photon absorption by rhodopsin in the rod outer segment.

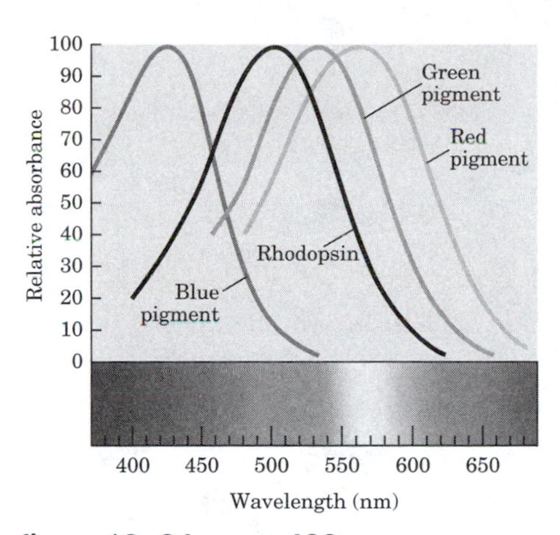

figure 13–24, page 462

Absorption spectra of purified rhodopsin and the red, green, and blue receptors of cone cells.

① Odorant (O) arrives at the mucous layer, and binds directly to an olfactory receptor (OR) or to a binding protein (BP) that carries it to the OR.

② Activated OR catalyzes GDP-GTP exchange on a G protein (G_{olf}), causing its dissociation into α and $\beta\gamma$.

Air

③ G_α-GTP activates adenylyl cyclase, which catalyzes cAMP synthesis, raising [cAMP].

④ cAMP-gated cation channels open. Ca^{2+} enters, raising internal [Ca^{2+}].

Mucous layer

O

BP

OR

AC

α GDP

G_{olf} $\beta\gamma$

α GTP

GTP

GDP

ATP cAMP

Ca^{2+}

Cl^-

Ciliary membrane

⑦ G_{olf} hydrolyzes GTP to GDP, shutting itself off. PDE hydrolyzes cAMP. Receptor kinase phosphorylates OR, inactivating it. Odorant is removed by metabolism.

⑥ Ca^{2+} reduces the affinity of the cation channel for cAMP, lowering the sensitivity of the system to odorant.

⑤ Ca^{2+}-gated chloride channels open. Efflux of Cl^- depolarizes the cell, triggering an electrical signal to the brain.

Olfactory neuron

Cilia

Dendrite

Axon

figure 13–25, page 464

Molecular events of olfaction.

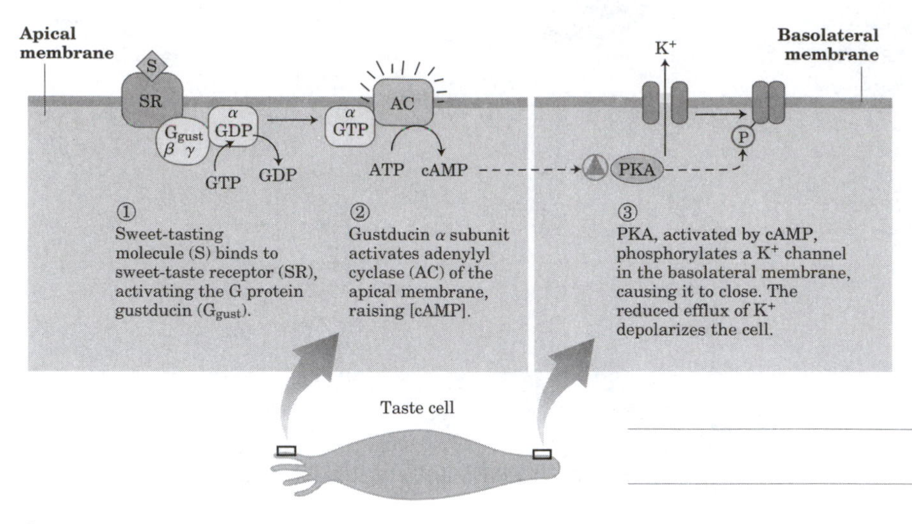

figure 13–26, page 465
Transduction mechanism for sweet tastants.

table 13–7, page 465

Some Signals Transduced by G Protein–Coupled Serpentine Receptors
Acetylcholine (muscarinic)
Adenosine
Angiotensin
ATP (extracellular)
Bradykinin
Calcitonin
Cannabinoids
Catecholamines
Cholecystokinin
Corticotropin-releasing factor (CRF)
Cyclic AMP (*Dictyostelium discoideum*)
Dopamine
Endothelin
Follicle-stimulating hormone (FSH)
γ-Aminobutyric acid (GABA)
Glucagon
Glutamate (metabotropic)
Growth hormone releasing hormone (GHRH)
Histamine
Leukotrienes
Light
Luteinizing hormone (LH)
Melatonin
Odorants
Opioids
Oxytocin
Platelet-activating factor
Prostaglandins
Secretin
Serotonin
Somatostatin
Tastants
Thyrotropin
Thyrotropin-releasing hormone (TRH)
Vasoactive intestinal peptide
Vasopressin
Yeast mating factors

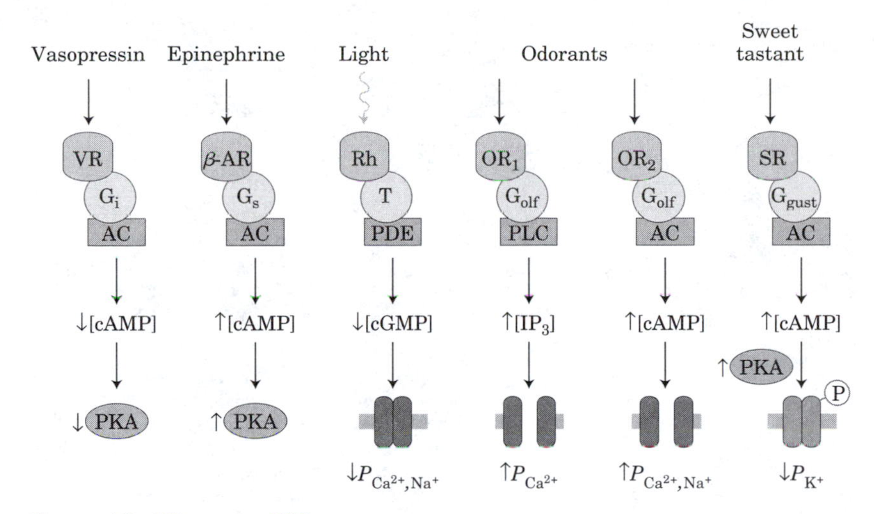

figure 13–27, page 466
Common features of signaling systems that detect hormones, light, smells, and tastes.

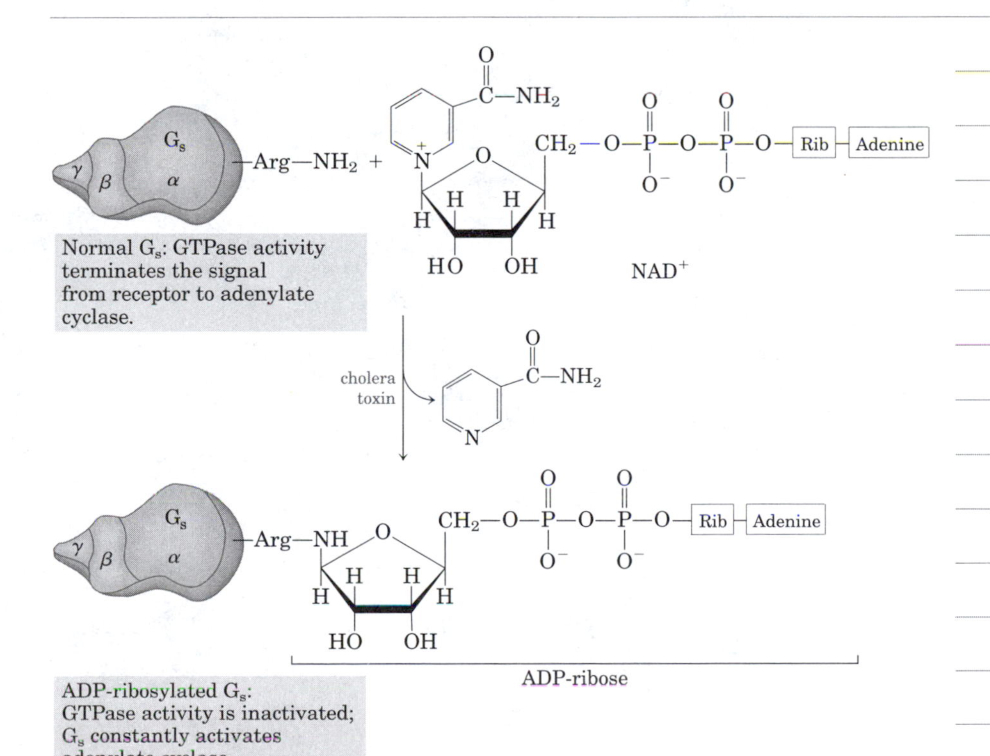

figure 13–28, page 467
Toxins produced by bacteria that cause cholera and whooping cough (pertussis).

① Hormone (H), carried to the target tissue on serum binding proteins, diffuses across the plasma membrane and binds to its specific receptor protein (Rec) in the nucleus.

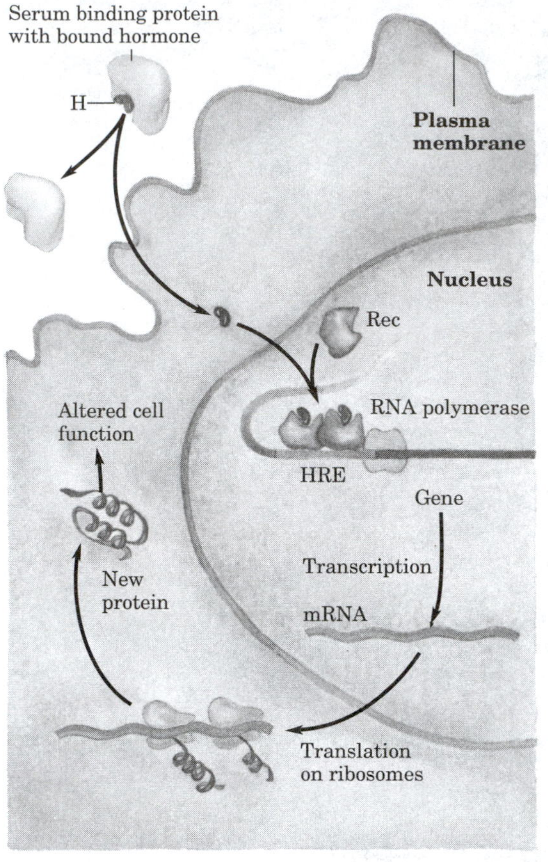

② Hormone binding changes the conformation of Rec; it forms homo- or heterodimers with other hormone-receptor complexes and binds to specific regulatory regions called hormone response elements (HREs) in the DNA adjacent to specific genes.

③ Binding regulates transcription of the adjacent gene(s), increasing or decreasing the rate of mRNA formation.

④ Altered levels of the hormone-regulated gene product produce the cellular response to the hormone.

figure 13–29, page 468
General mechanism by which steroid and thyroid hormones, retinoids, and vitamin D regulate gene expression.

CH_3, N, CH_3, O, CH_3

Tamoxifen

CH_3, N, CH_3, OH, $C\equiv C—CH_3$, O

RU486 (mifepristone)

page 469

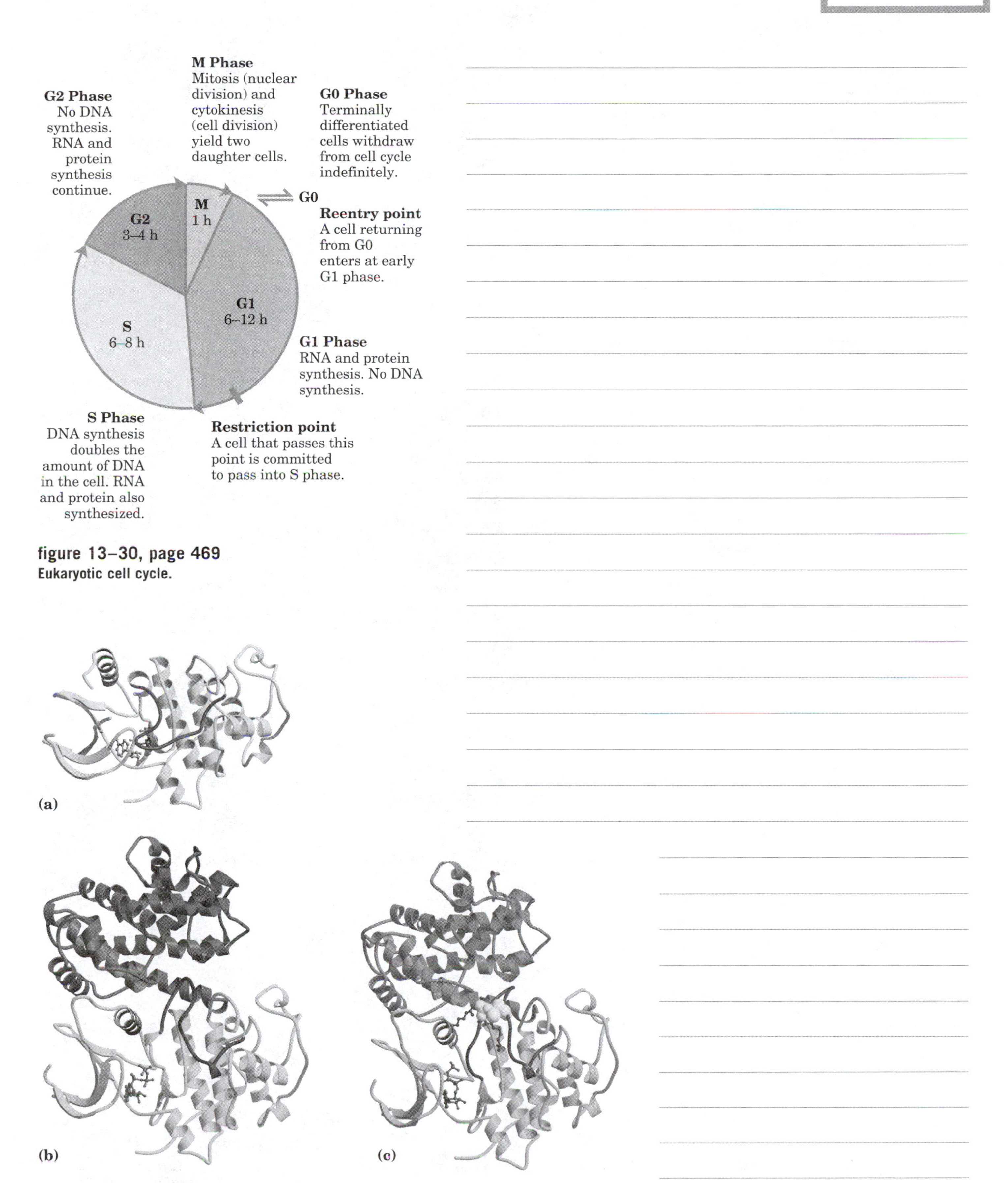

figure 13–30, page 469
Eukaryotic cell cycle.

figure 13–31, page 470
Activation of cyclin-dependent protein kinases (CDKs) by cyclin and phosphorylation.

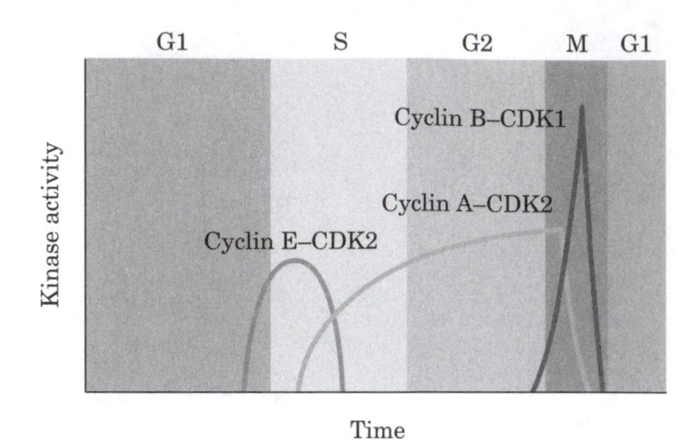

figure 13–32, page 471
Variations in the activities of specific CDKs during the cell cycle in animals.

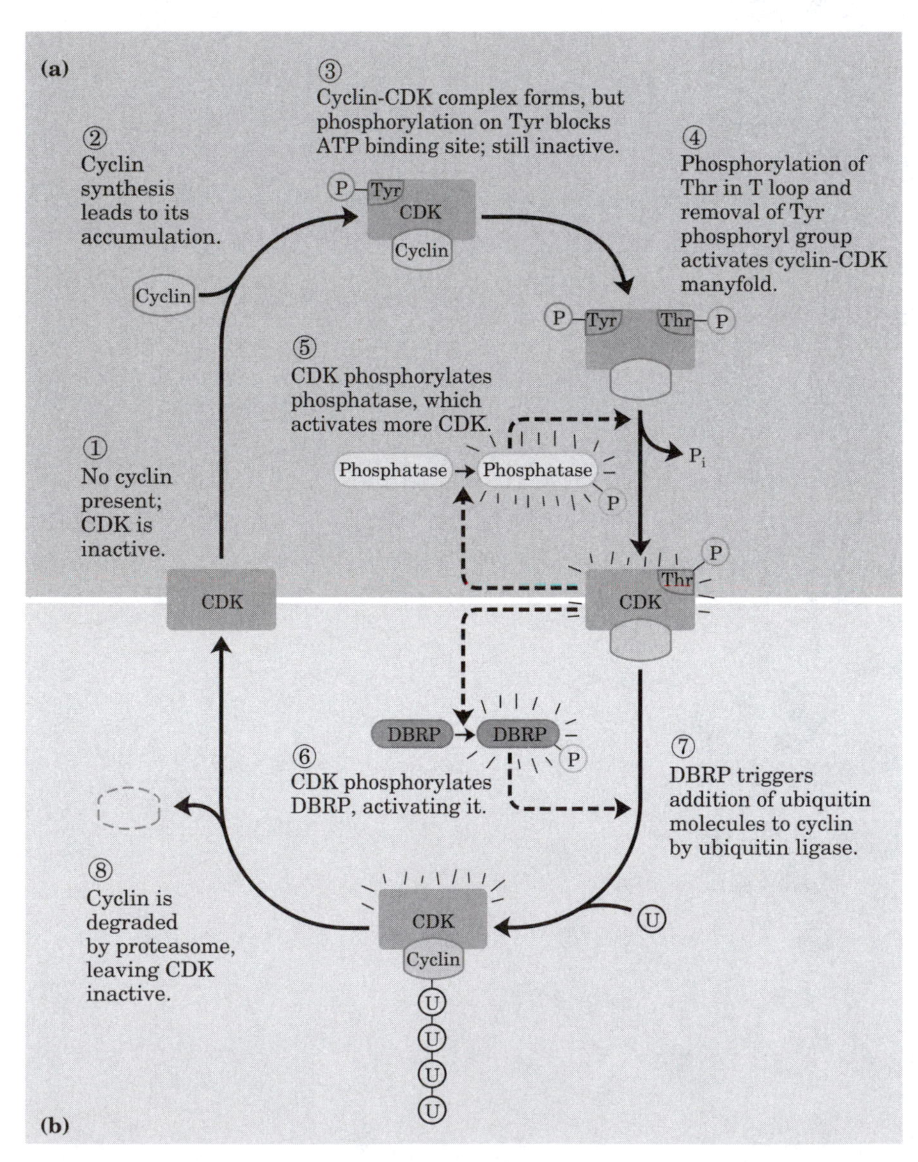

figure 13–33, page 471
Regulation of CDK by phosphorylation and proteolysis.

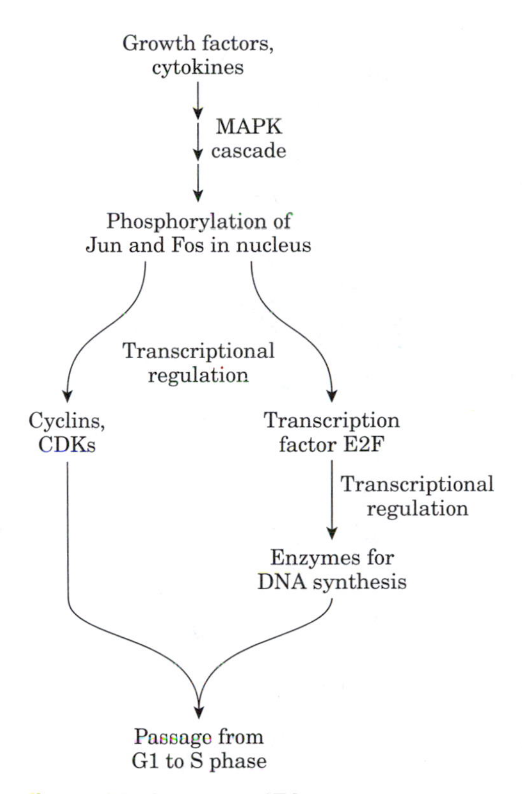

figure 13–34, page 472
Regulation of cell division by growth factors.

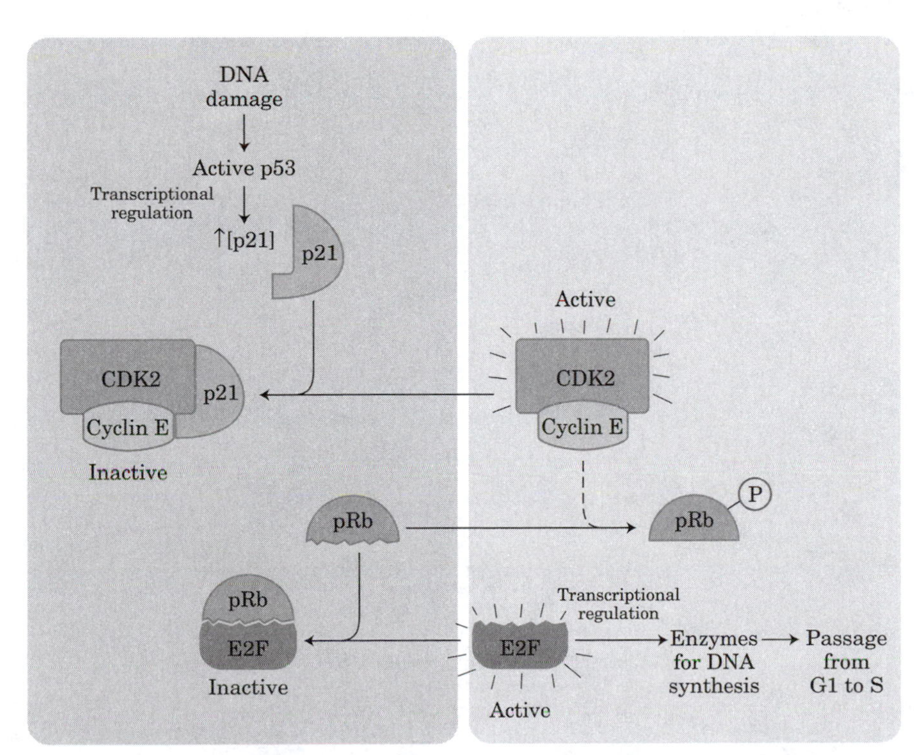

figure 13–35, page 473
Regulation of passage from G1 to S by phosphorylation of pRb.

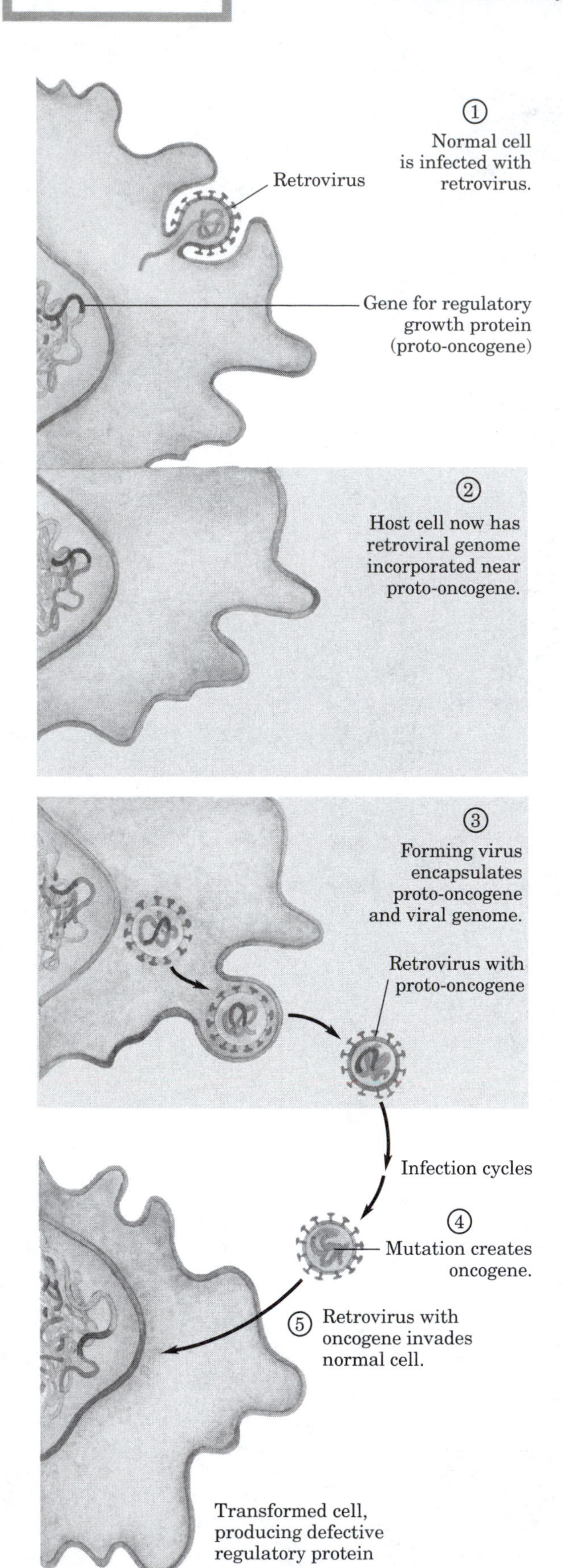

figure 13–36, page 474
Conversion of a regulatory gene into a viral oncogene.

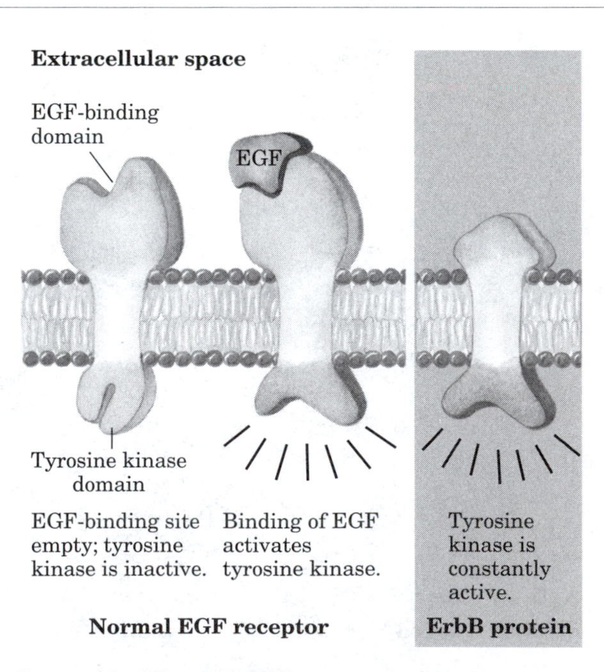

figure 13–37, page 475
Oncogene-encoded defective EGF receptor.

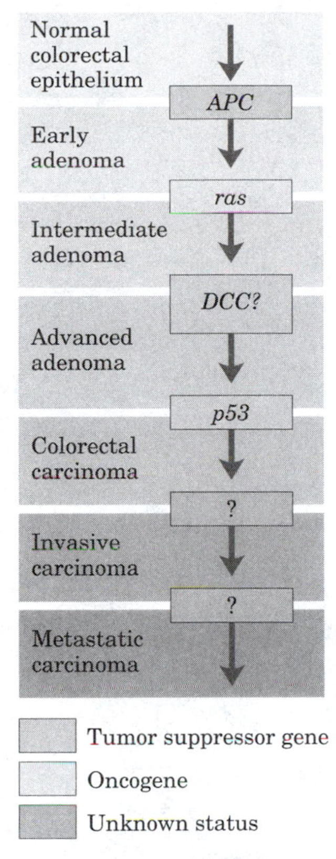

figure 13–38, page 476
From normal epithelial cell to colorectal cancer.

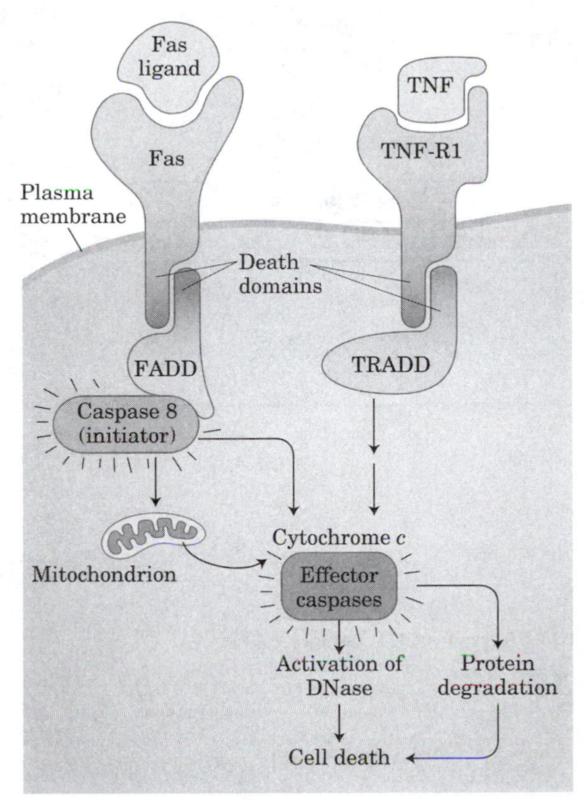

figure 13–39, page 477
Initial events of apoptosis.

PROBLEMS

5. Resting Membrane Potential, page 481.

Ion	Concentration (mM)	
	Intracellular	Extracellular
Na^+	440	50
K^+	20	400
Cl^-	560	21
Ca^{2+}	10	0.4

10. Effect of Dibutyryl cAMP versus cAMP on Intact Cells, page 482.

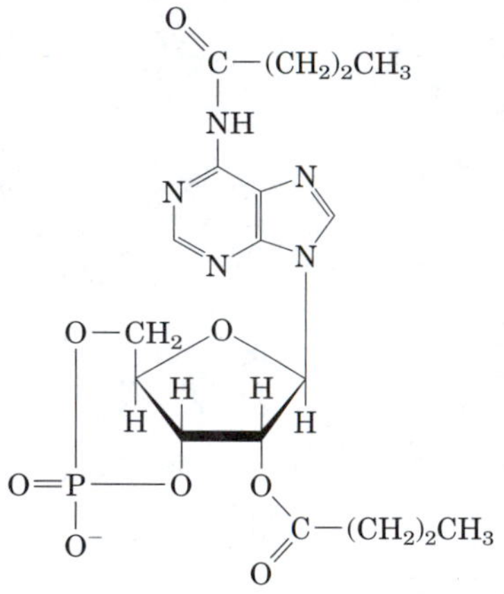

Dibutyryl cAMP
(N^6,$O^{2'}$-Dibutyryl adenosine 3′,5′-cyclic monophosphate)

11. Nonhydrolyzable GTP Analogs, page 482.

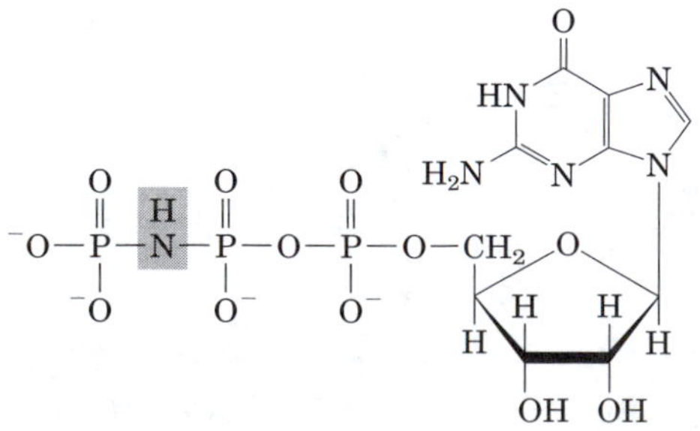

Gpp(NH)p
β,γ-Imidoguanosine 5′-triphosphate

13. EGTA Injection, page 483.

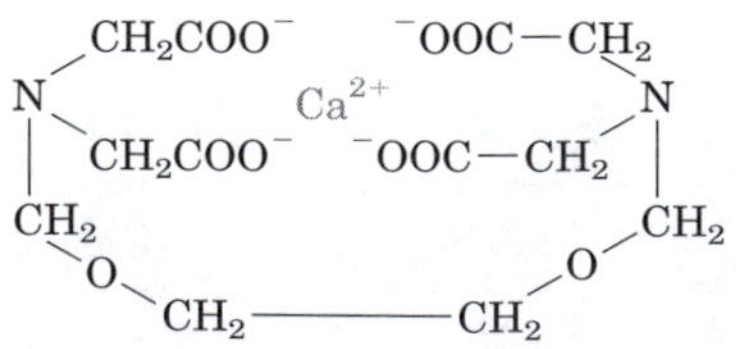

EGTA (with bound Ca^{2+})
(Ethylene glycol-bis(β-aminoethyl ether)-N,N,N′,N′-tetraacetic acid)

Bioenergetics and Metabolism

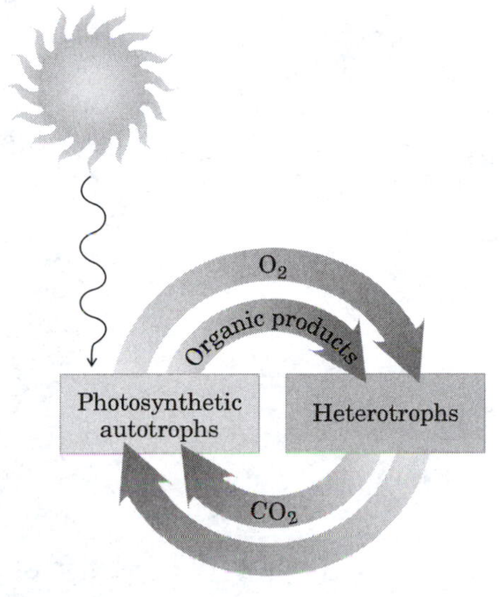

figure 1, page 486
Cycling of carbon dioxide and oxygen between the autotrophic (photosynthetic) and heterotrophic domains in the biosphere.

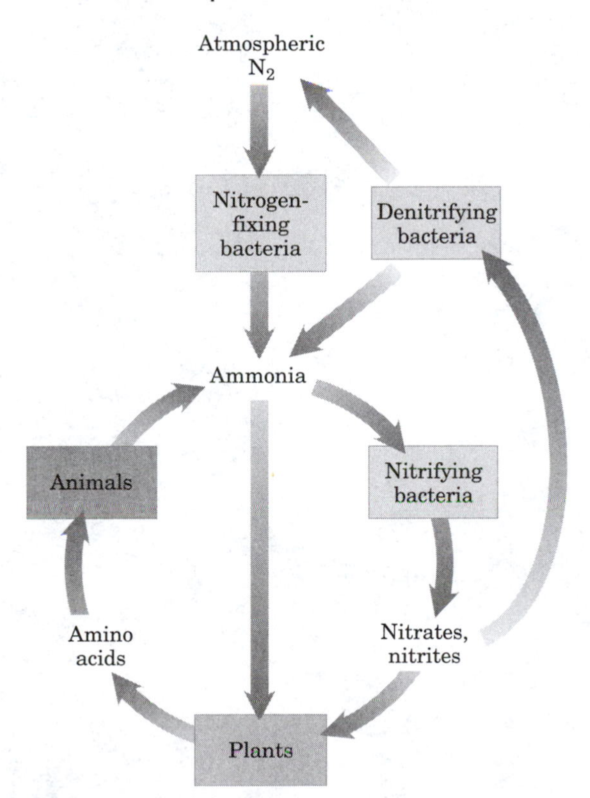

figure 2, page 486
Cycling of nitrogen in the biosphere.

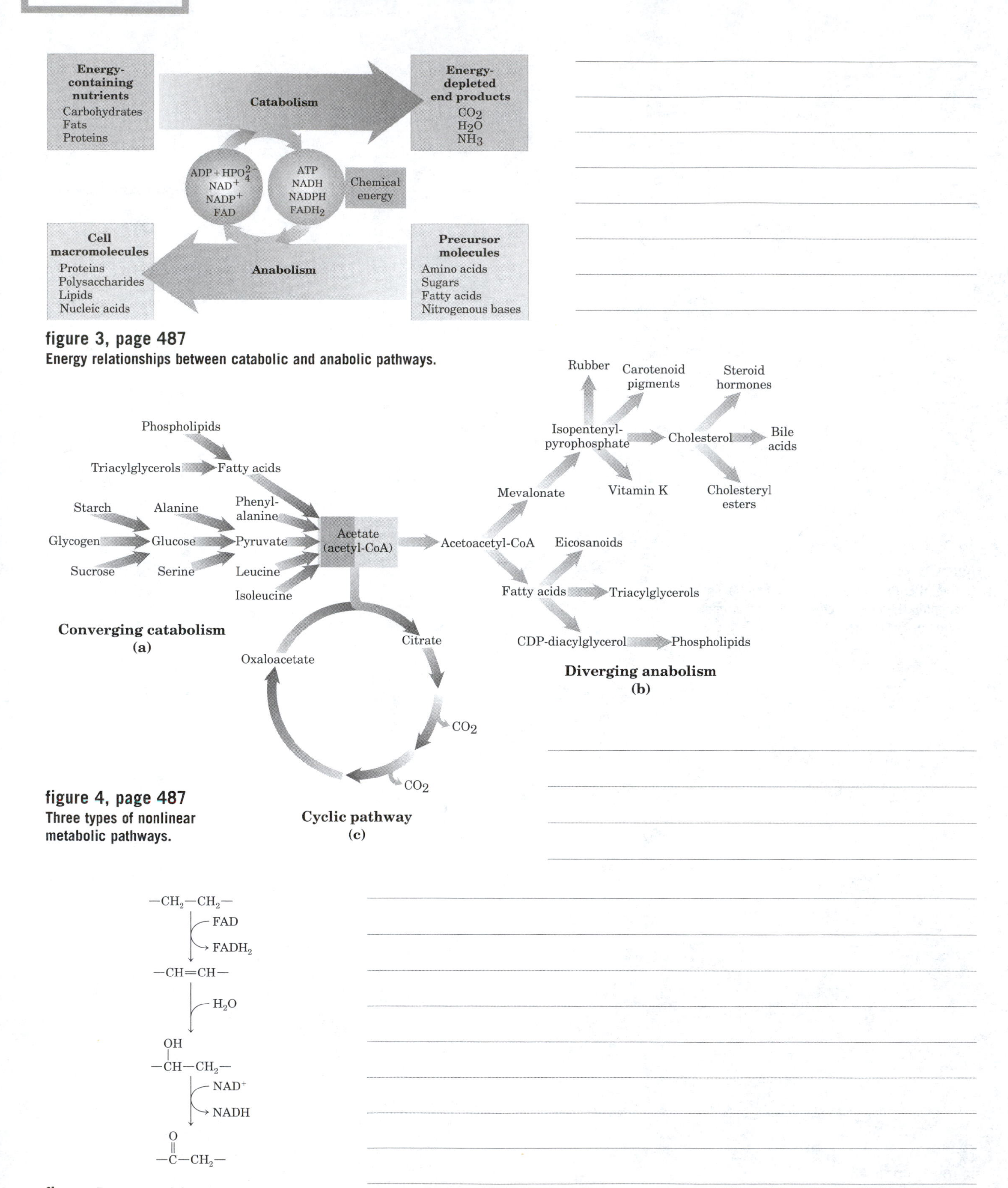

figure 3, page 487
Energy relationships between catabolic and anabolic pathways.

figure 4, page 487
Three types of nonlinear metabolic pathways.

figure 5, page 488
Common mechanism to oxidize an alkaline.

chapter

14

Bioenergetics and Metabolism

CD-ROM

When you study for exams, don't forget to review The Minicourses on the UNDERSTAND! BIOCHEMISTRY CD that came with your textbook.

Minicourses that apply to this Chapter include:

Molecules of Life
- Important Biomolecules
- An Interactive Gallery of Molecular Models

Bioenergetics
- Energetics
- Energetics of Metabolism
- Electron Transport — Oxidative Phosphorylation

table 14–1, page 491

Some Physical Constants and Units Used in Thermodynamics

Boltzmann constant, **k** = 1.381 × 10^{-23} J/K
Avogadro's number, N = 6.022 × 10^{23} mol^{-1}
Faraday constant, $\mathcal{F}$ = 96,480 J/V · mol
Gas constant, R = 8.315 J/mol · K
(= 1.987 cal/mol · K)

Units of ΔG and ΔH are J/mol (or cal/mol)
Units of ΔS are J/mol · K (or cal/mol · K)
1 cal = 4.184 J

Units of absolute temperature, T, are Kelvin, K
25 °C = 298 K
At 25 °C, RT = 2.479 kJ/mol
(= 0.592 kcal/mol)

$$K_{eq} = \frac{[C]^c[D]^d}{[A]^a[B]^b} \qquad \text{(Equation 14–2)}$$

Equilibrium constant, page 494

$$\Delta G'^{\circ} = -RT \ln K'_{eq}$$

Relationship between $\Delta G'^{\circ}$ and K'_{eq}, page 494

table 14–2, page 494

Relationship between the Equilibrium Constants and Standard Free-Energy Changes of Chemical Reactions

K'_{eq}	$\Delta G'^{\circ}$ (kJ/mol)	$\Delta G'^{\circ}$ (kcal/mol)*
10^3	−17.1	−4.1
10^2	−11.4	−2.7
10^1	−5.7	−1.4
1	0.0	0.0
10^{-1}	5.7	1.4
10^{-2}	11.4	2.7
10^{-3}	17.1	4.1
10^{-4}	22.8	5.5
10^{-5}	28.5	6.8
10^{-6}	34.2	8.2

*Although joules and kilojoules are the standard units of energy and are used throughout this text, biochemists sometimes express $\Delta G'^{\circ}$ values in kilocalories per mole. We have therefore included values in both kilojoules and kilocalories in this table and in Tables 14–4 and 14–6. To convert kilojoules to kilocalories, divide the number of kilojoules by 4.184.

table 14–3, page 495

Relationships among K'_{eq}, $\Delta G'^{\circ}$, and the Direction of Chemical Reactions under Standard Conditions

When K'_{eq} is	$\Delta G'^{\circ}$ is	Starting with 1 M components the reaction
>1.0	Negative	Proceeds forward
1.0	Zero	Is at equilibrium
<1.0	Positive	Proceeds in reverse

table 14–4, page 496

Standard Free-Energy Changes of Some Chemical Reactions at pH 7.0 and 25 °C (298 K)

Reaction type	$\Delta G'^{\circ}$ (kJ/mol)	$\Delta G'^{\circ}$ (kcal/mol)
Hydrolysis reactions		
Acid anhydrides		
Acetic anhydride + $H_2O \longrightarrow$ 2 acetate	−91.1	−21.8
ATP + $H_2O \longrightarrow$ ADP + P_i	−30.5	−7.3
ATP + $H_2O \longrightarrow$ AMP + PP_i	−45.6	−10.9
PP_i + $H_2O \longrightarrow 2P_i$	−19.2	−4.6
UDP-glucose + $H_2O \longrightarrow$ UMP + glucose 1-phosphate	−43.0	−10.3
Esters		
Ethyl acetate + $H_2O \longrightarrow$ ethanol + acetate	−19.6	−4.7
Glucose 6-phosphate + $H_2O \longrightarrow$ glucose + P_i	−13.8	−3.3
Amides and peptides		
Glutamine + $H_2O \longrightarrow$ glutamate + NH_4^+	−14.2	−3.4
Glycylglycine + $H_2O \longrightarrow$ 2 glycine	−9.2	−2.2
Glycosides		
Maltose + $H_2O \longrightarrow$ 2 glucose	−15.5	−3.7
Lactose + $H_2O \longrightarrow$ glucose + galactose	−15.9	−3.8
Rearrangements		
Glucose 1-phosphate $\longrightarrow$ glucose 6-phosphate	−7.3	−1.7
Fructose 6-phosphate $\longrightarrow$ glucose 6-phosphate	−1.7	−0.4
Elimination of water		
Malate $\longrightarrow$ fumarate + H_2O	3.1	0.8
Oxidations with molecular oxygen		
Glucose + $6O_2 \longrightarrow 6CO_2 + 6H_2O$	−2,840	−686
Palmitate + $23O_2 \longrightarrow 16CO_2 + 16H_2O$	−9,770	−2,338

$$\Delta G = \Delta G'^{\circ} + RT \ln \frac{[C][D]}{[A][B]} \qquad \text{(Equation 14–3)}$$

Relationship between ΔG and $\Delta G'^{\circ}$, page 497

$$\begin{array}{lll} (1) & A \longrightarrow B & \Delta G_1'^{\circ} \\ (2) & B \longrightarrow C & \Delta G_2'^{\circ} \\ \hline \textit{Sum:} & A \longrightarrow C & \Delta G_1'^{\circ} + \Delta G_2'^{\circ} \end{array}$$

Addition of $\Delta G'^{\circ}$, page 498

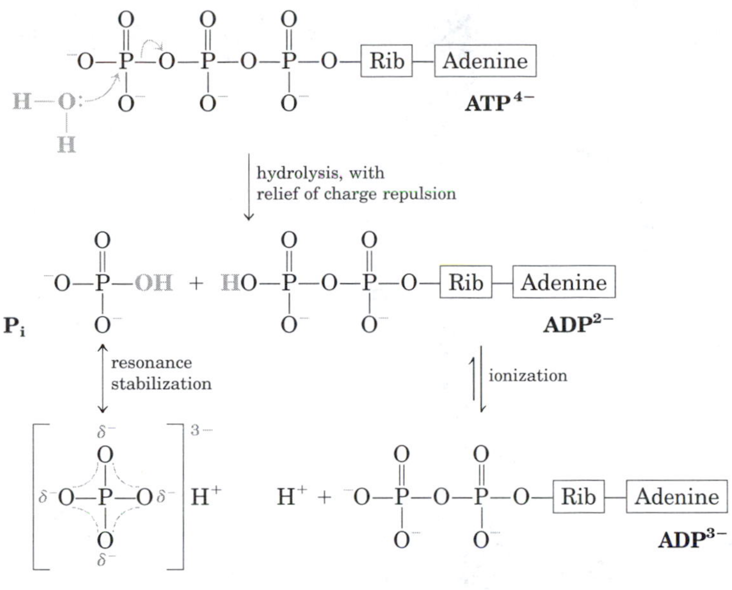

$$ATP^{4-} + H_2O \longrightarrow ADP^{3-} + P_i^{2-} + H^+$$
$$\Delta G'^\circ = -30.5 \text{ kJ/mol}$$

figure 14–1, page 500
Chemical basis for the large free-energy change associated with ATP hydrolysis.

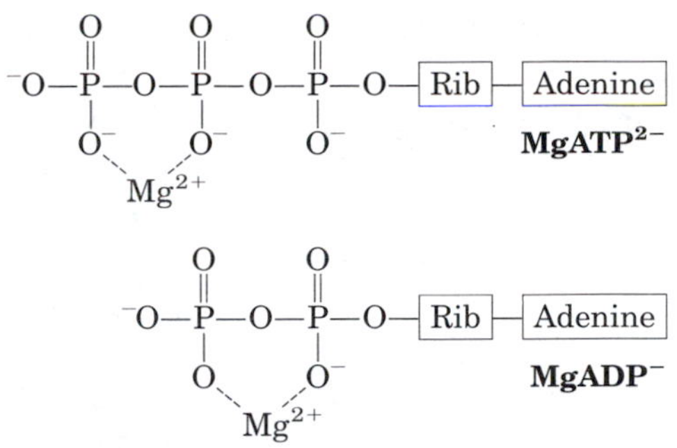

figure 14–2, page 500
Mg^{2+} and ATP.

table 14–5, page 501

Adenine Nucleotide, Inorganic Phosphate, and Phosphocreatine Concentrations in Some Cells*

	Concentration (mM)				
	ATP	ADP†	AMP	P_i	PCr
Rat hepatocyte	3.38	1.32	0.29	4.8	0
Rat myocyte	8.05	0.93	0.04	8.05	28
Rat neuron	2.59	0.73	0.06	2.72	4.7
Human erythrocyte	2.25	0.25	0.02	1.65	0
E. coli cell	7.90	1.04	0.82	7.9	0

*For erythrocytes the concentrations are those of the cytosol (human erythrocytes lack a nucleus and mitochondria). In the other types of cells the data are for the entire cell contents, although the cytosol and the mitochondria have very different concentrations of ADP. PCr is phosphocreatine, discussed on p. 502.

†This value reflects total concentration; the true value for free ADP may be much lower (see Box 14–2).

PEP — hydrolysis (H_2O, P_i) → Pyruvate (enol form) — tautomerization ⇌ Pyruvate (keto form)

$$PEP^{3-} + H_2O \longrightarrow pyruvate^{-} + P_i^{2-}$$
$$\Delta G'^{\circ} = -61.9 \text{ kJ/mol}$$

figure 14–3, page 502
Hydrolysis of phosphoenolpyruvate (PEP).

3-Phosphoglycerate

$$\text{1,3-Bisphosphoglycerate}^{4-} + H_2O \longrightarrow \text{3-phosphoglycerate}^{3-} + P_i^{2-} + H^+$$
$$\Delta G'^{\circ} = -49.3 \text{ kJ/mol}$$

figure 14–4, page 502
Hydrolysis of 1,3-bisphosphoglycerate.

Phosphocreatine — hydrolysis (H_2O, P_i) → Creatine — resonance stabilization

$$\text{Phosphocreatine}^{2-} + H_2O \longrightarrow \text{creatine} + P_i^{2-}$$
$$\Delta G'^{\circ} = -43.0 \text{ kJ/mol}$$

figure 14–5, page 502
Hydrolysis of phosphocreatine.

table 14–6, page 503

Standard Free Energies of Hydrolysis of Some Phosphorylated Compounds and Acetyl-CoA (a Thioester)

	$\Delta G'^{\circ}$ (kJ/mol)	$\Delta G'^{\circ}$ (kcal/mol)
Phosphoenolpyruvate	−61.9	−14.8
1,3-bisphosphoglycerate (→ 3-phosphoglycerate + P_i)	−49.3	−11.8
Phosphocreatine	−43.0	−10.3
ADP (→ AMP + P_i)	−32.8	−7.8
ATP (→ ADP + P_i)	−30.5	−7.3
ATP (→ AMP + PP_i)	−45.6	−10.9
AMP (→ adenosine + P_i)	−14.2	−3.4
PP_i (→ $2P_i$)	−19	−4.0
Glucose 1-phosphate	−20.9	−5.0
Fructose 6-phosphate	−15.9	−3.8
Glucose 6-phosphate	−13.8	−3.3
Glycerol 1-phosphate	−9.2	−2.2
Acetyl-CoA	−31.4	−7.5

Source: Data mostly from Jencks, W.P. (1976) in *Handbook of Biochemistry and Molecular Biology,* 3rd edn (Fasman, G.D., ed.), *Physical and Chemical Data,* Vol. I, pp. 296–304, CRC Press, Boca Raton, FL.

Acetyl-CoA

H_2O hydrolysis → CoASH

Acetic acid

ionization → H^+

Acetate

resonance stabilization

$$\text{Acetyl-CoA} + H_2O \longrightarrow \text{acetate}^- + \text{CoA} + H^+ \quad \Delta G'^{\circ} = -32.2 \text{ kJ/mol}$$

figure 14–6, page 503
Hydrolysis of acetyl-coenzyme A.

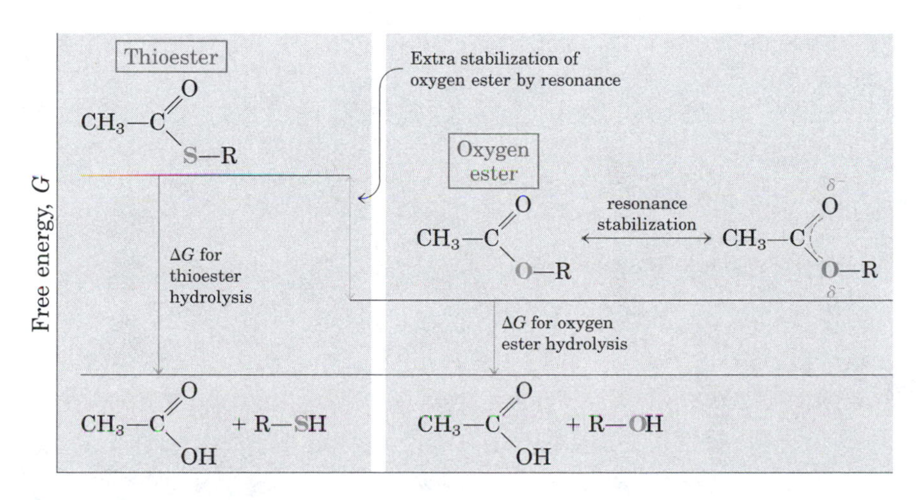

figure 14–7, page 503
Free energy of hydrolysis for thioesters and oxygen esters.

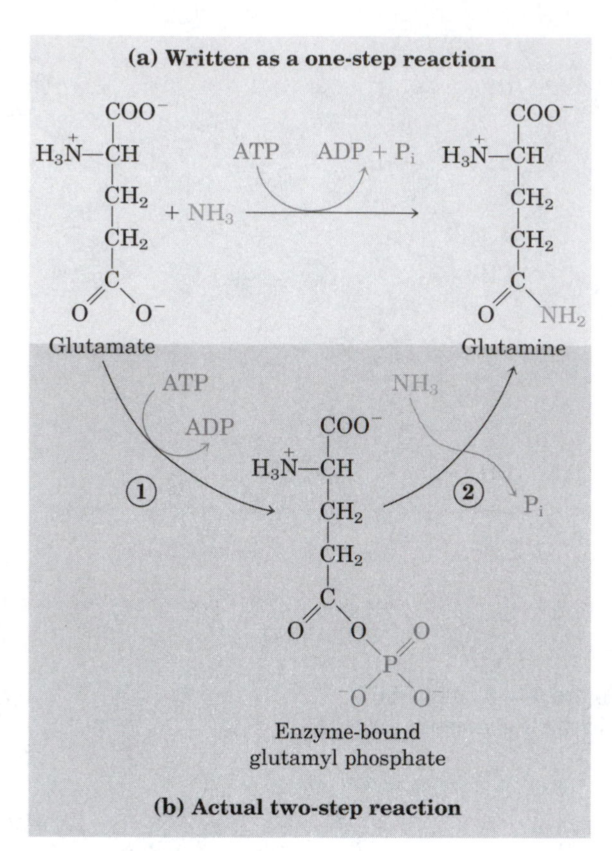

figure 14–8, page 504
ATP hydrolysis in two steps.

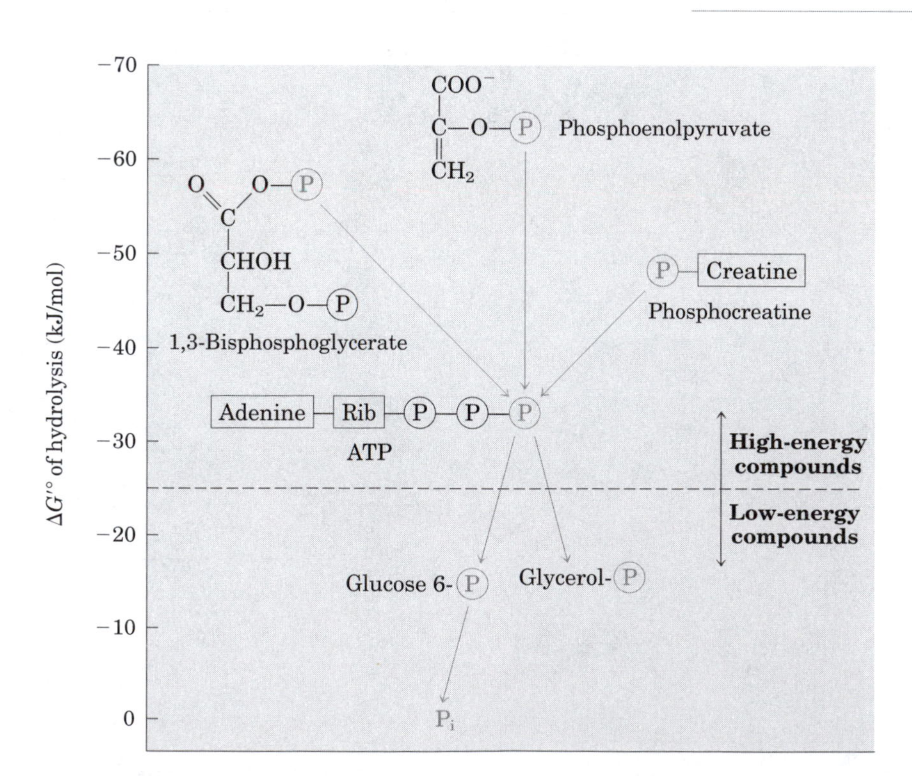

figure 14–9, page 505
Ranking of biological phosphate compounds by standard free energies of hydrolysis.

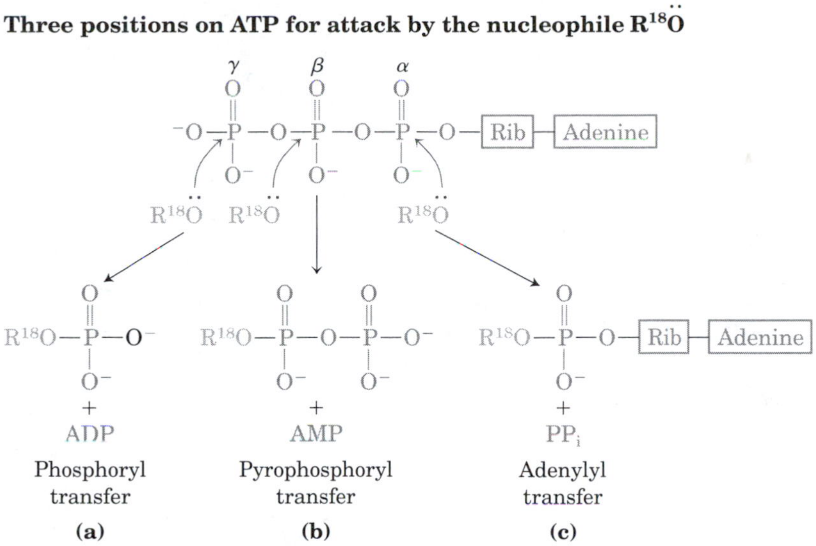

Three positions on ATP for attack by the nucleophile $R^{18}\ddot{O}$

γ β α

$^{-}O-P(=O)(O^{-})-O-P(=O)(O^{-})-O-P(=O)(O^{-})-O-$ Rib — Adenine

$R^{18}\ddot{O}$ $R^{18}\ddot{O}$ $R^{18}\ddot{O}$

$R^{18}O-P(=O)(O^{-})-O^{-}$ + ADP — Phosphoryl transfer **(a)**

$R^{18}O-P(=O)(O^{-})-O-P(=O)(O^{-})-O^{-}$ + AMP — Pyrophosphoryl transfer **(b)**

$R^{18}O-P(=O)(O^{-})-O-$ Rib — Adenine + PP_i — Adenylyl transfer **(c)**

figure 14–10, page 506
Nucleophilic displacement reactions of ATP.

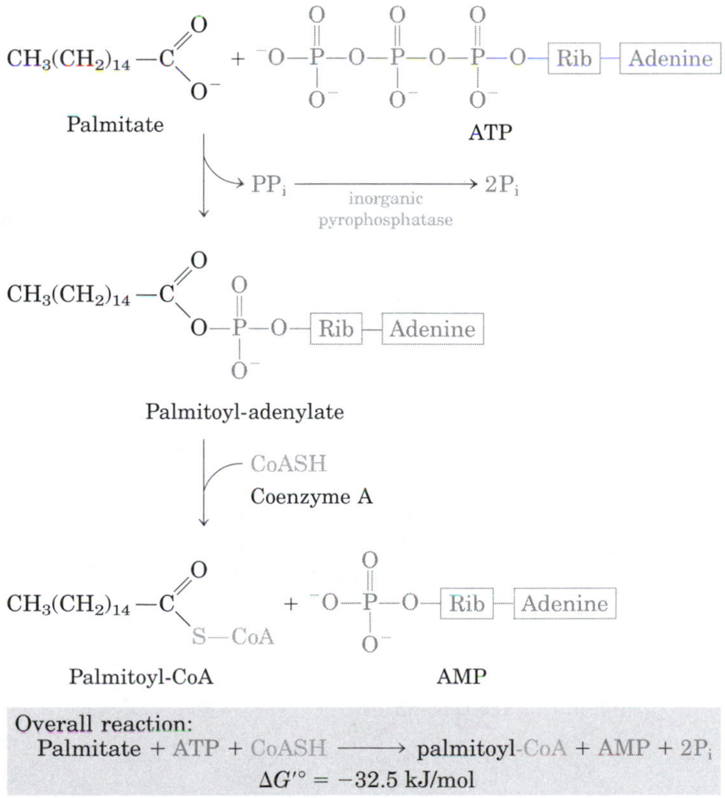

$CH_3(CH_2)_{14}-C(=O)O^{-}$ (Palmitate) + $^{-}O-P(=O)(O^{-})-O-P(=O)(O^{-})-O-P(=O)(O^{-})-O-$ Rib — Adenine (ATP)

↓ → PP_i ⟶ $2P_i$ (inorganic pyrophosphatase)

$CH_3(CH_2)_{14}-C(=O)-O-P(=O)(O^{-})-O-$ Rib — Adenine

Palmitoyl-adenylate

↓ CoASH (Coenzyme A)

$CH_3(CH_2)_{14}-C(=O)-S-CoA$ + $^{-}O-P(=O)(O^{-})-O-$ Rib — Adenine

Palmitoyl-CoA AMP

Overall reaction:
Palmitate + ATP + CoASH ⟶ palmitoyl-CoA + AMP + $2P_i$
$\Delta G'^{\circ} = -32.5$ kJ/mol

figure 14–11, page 507
Adenylylation reaction in activation of a fatty acid.

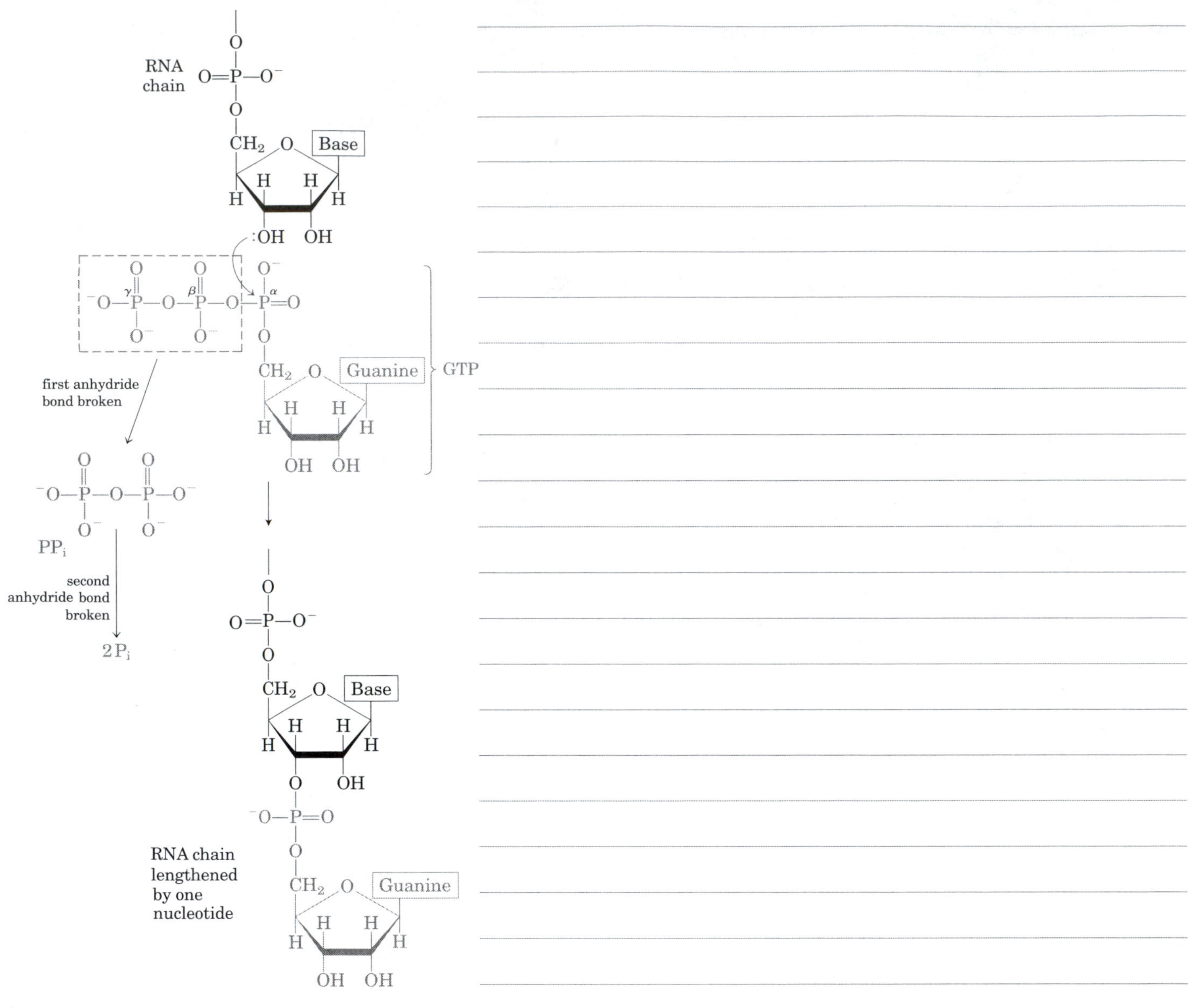

figure 14–12, page 508
Nucleoside triphosphates in RNA synthesis.

Firefly luciferin

Luciferyl adenylate

AMP

Luciferyl adenylate

O_2

luciferase

light

PP_i

ATP

CO_2 + AMP

Luciferin

Oxyluciferin

regenerating reactions

box 14–3, page 509
Important components in firefly bioluminescence, and the firefly bioluminescence cycle.

$$\text{ATP} + \text{NDP (or dNDP)} \xrightleftharpoons{\text{Mg}^{2+}} \text{ADP} + \text{NTP (or dNTP)} \qquad \Delta G'^{\circ} \approx 0$$

Nucleoside diphosphate kinase, page 510

$$2\text{ADP} \xrightleftharpoons{\text{Mg}^{2+}} \text{ATP} + \text{AMP} \qquad \Delta G'^{\circ} \approx 0$$

Adenylate kinase, page 510

$$\text{ATP} + \text{PCr} \xrightleftharpoons{\text{Mg}^{2+}} \text{ATP} + \text{Cr} \qquad \Delta G'^{\circ} = -12.5 \text{ kJ/mol}$$

Creatine kinase, page 510

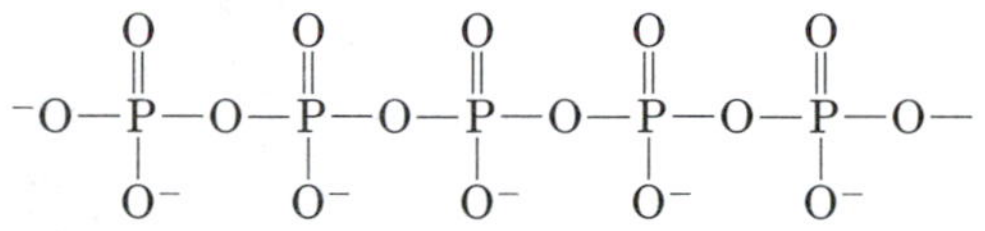

Inorganic polyphosphate (polyP), page 511

$$\text{ATP} + \text{polyP}_n \xrightleftharpoons{\text{Mg}^{2+}} \text{ADP} + \text{polyP}_{n+1} \qquad \Delta G'^{\circ} = -20 \text{ kJ/mol}$$

Polyphosphate kinase, page 511

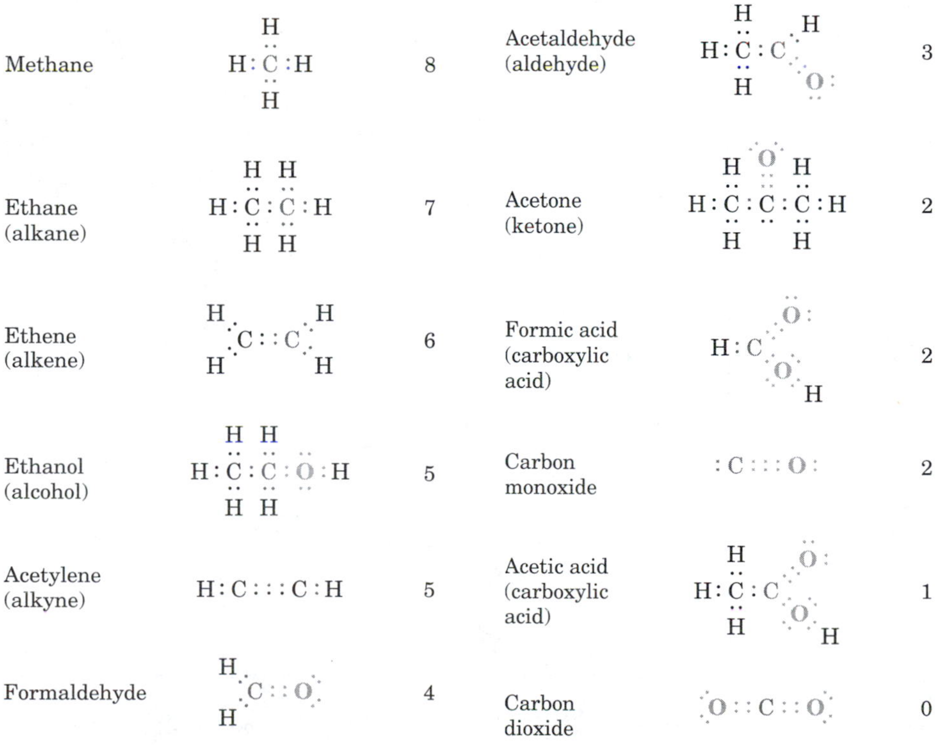
Methane
8
Ethane (alkane)
7
Ethene (alkene)
6
Ethanol (alcohol)
5
Acetylene (alkyne)
5
Formaldehyde
4
Acetaldehyde (aldehyde)
3
Acetone (ketone)
2
Formic acid (carboxylic acid)
2
Carbon monoxide
2
Acetic acid (carboxylic acid)
1
Carbon dioxide
0

figure 14–13, page 514
Oxidation states of carbon occurring in the biosphere.

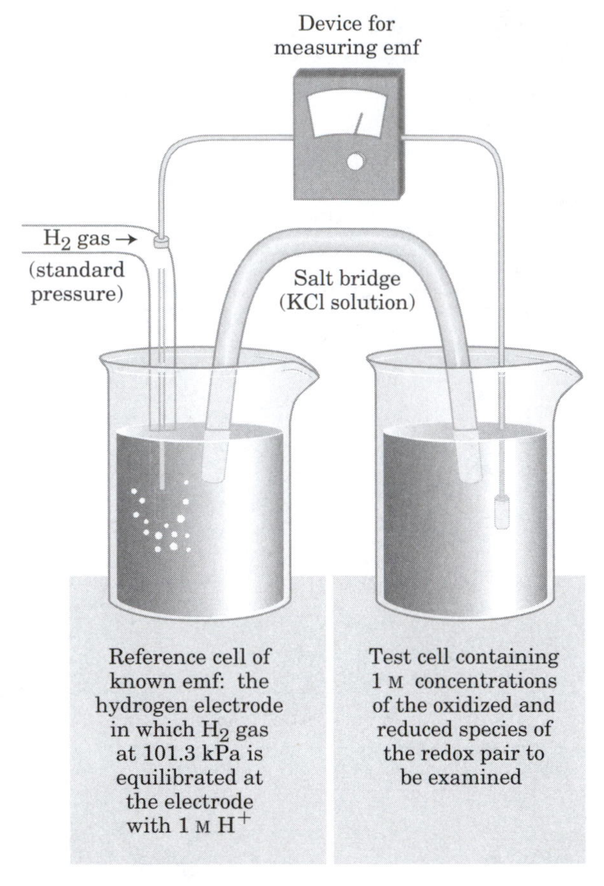

figure 14–14, page 515
Measurement of the standard reduction potential (E'°) of a redox pair.

$$E = E^\circ + \frac{RT}{n\mathcal{F}} \ln \frac{[\text{electron acceptor}]}{[\text{electron donor}]} \quad \text{(Equation 14–4)}$$

$$E = E^\circ + \frac{0.026\ \text{V}}{n} \ln \frac{[\text{electron acceptor}]}{[\text{electron donor}]} \quad \text{(Equation 14–5)}$$

Reduction potential, page 515

table 14–7, page 516

Standard Reduction Potentials of Some Biologically Important Half-Reactions, at 25 °C and pH 7

Half-reaction	E'° (V)
$\frac{1}{2}O_2 + 2H^+ + 2e^- \longrightarrow H_2O$	0.816
$Fe^{3+} + e^- \longrightarrow Fe^{2+}$	0.771
$NO_3^- + 2H^+ + 2e^- \longrightarrow NO_2^- + H_2O$	0.421
Cytochrome *f* (Fe^{3+}) + e^- ⟶ cytochrome *f* (Fe^{2+})	0.365
$Fe(CN)_6^{3-}$ (ferricyanide) + e^- ⟶ $Fe(CN)_6^{4-}$	0.36
Cytochrome a_3 (Fe^{3+}) + e^- ⟶ cytochrome a_3(Fe^{2+})	0.35
$O_2 + 2H^+ + 2e^- \longrightarrow H_2O_2$	0.295
Cytochrome *a* (Fe^{3+}) + e^- ⟶ cytochrome *a* (Fe^{2+})	0.29
Cytochrome *c* (Fe^{3+}) + e^- ⟶ cytochrome *c* (Fe^{2+})	0.254
Cytochrome c_1 (Fe^{3+}) + e^- ⟶ cytochrome c_1 (Fe^{2+})	0.22
Cytochrome *b* (Fe^{3+}) + e^- ⟶ cytochrome *b* (Fe^{2+})	0.077
Ubiquinone + $2H^+$ + $2e^-$ ⟶ ubiquinol + H_2	0.045
Fumarate^{2-} + $2H^+$ + $2e^-$ ⟶ succinate^{2-}	0.031
$2H^+ + 2e^- \longrightarrow H_2$ (at standard conditions, pH 0)	0.000
Crotonyl-CoA + $2H^+$ + $2e^-$ ⟶ butyryl-CoA	−0.015
Oxaloacetate^{2-} + $2H^+$ + $2e^-$ ⟶ malate^{2-}	−0.166
Pyruvate$^-$ + $2H^+$ + $2e^-$ ⟶ lactate$^-$	−0.185
Acetaldehyde + $2H^+$ + $2e^-$ ⟶ ethanol	−0.197
FAD + $2H^+$ + $2e^-$ ⟶ $FADH_2$	−0.219*
Glutathione + $2H^+$ + $2e^-$ ⟶ 2 reduced glutathione	−0.23
$S + 2H^+ + 2e^- \longrightarrow H_2S$	−0.243
Lipoic acid + $2H^+$ + $2e^-$ ⟶ dihydrolipoic acid	−0.29
NAD^+ + H^+ + $2e^-$ ⟶ NADH	−0.320
$NADP^+$ + H^+ + $2e^-$ ⟶ NADPH	−0.324
Acetoacetate + $2H^+$ + $2e^-$ ⟶ β-hydroxybutyrate	−0.346
α-Ketoglutarate + CO_2 + $2H^+$ + $2e^-$ ⟶ isocitrate	−0.38
$2H^+ + 2e^- \longrightarrow H_2$ (at pH 7)	−0.414
Ferredoxin (Fe^{3+}) + e^- ⟶ ferredoxin (Fe^{2+})	−0.432

Data mostly from Loach, P.A. (1976) In *Handbook of Biochemistry and Molecular Biology*, 3rd edn (Fasman, G.D., ed.), *Physical and Chemical Data*, Vol. I, pp. 122–130, CRC Press, Boca Raton, FL.

*This is the value for free FAD; FAD bound to a specific flavoprotein (for example succinate dehydrogenase) has a different E'°.

$\Delta G = -n\mathcal{F}\,\Delta E$, or $\Delta G'^{\circ} = -n\mathcal{F}\,\Delta E'^{\circ}$ (Figure 14–6)

Relationship between free-energy change and reduction potential, page 517

$NAD^+ + 2e^- + 2H^+ \longrightarrow NADH + H^+$
$NADP^+ + 2e^- + 2H^+ \longrightarrow NADPH + H^+$

NAD^+ / $NADP^+$, page 518

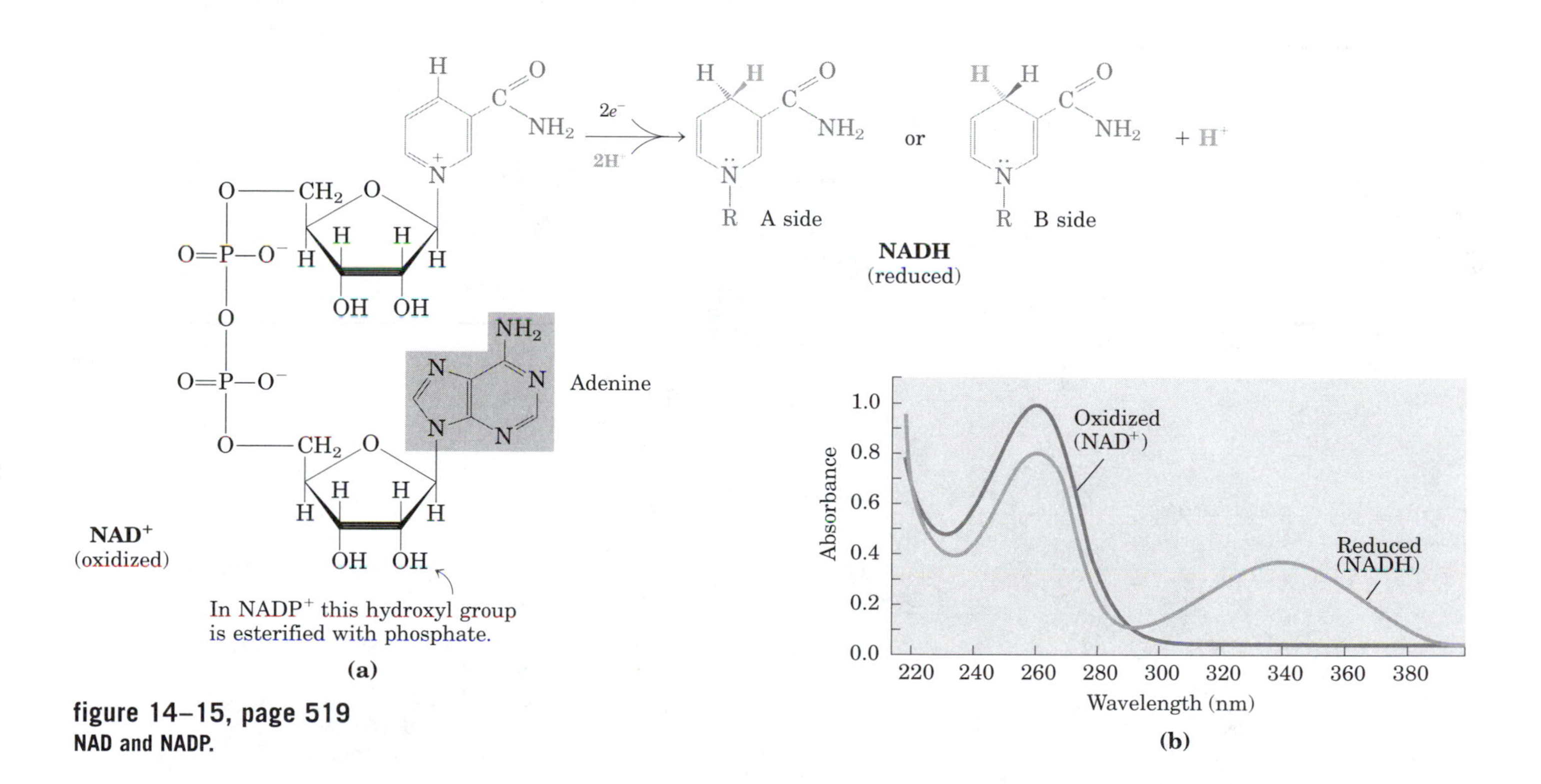

figure 14–15, page 519
NAD and NADP.

table 14–8, page 520

Stereospecificity of Dehydrogenases That Employ NAD^+ or $NADP^+$ as Coenzymes

Enzyme	Coenzyme	Stereochemical specificity for nicotinamide ring (A or B)
Isocitrate dehydrogenase	NAD^+	A
α-Ketoglutarate dehydrogenase	NAD^+	B
Glucose 6-phosphate dehydrogenase	$NADP^+$	B
Malate dehydrogenase	NAD^+	A
Glutamate dehydrogenase	NAD^+ or $NADP^+$	B
Glyceraldehyde 3-phosphate dehydrogenase	NAD^+	B
Lactate dehydrogenase	NAD^+	A
Alcohol dehydrogenase	NAD^+	A

table 14–9, page 520

Some Enzymes (Flavoproteins) That Employ Flavin Nucleotide Coenzymes

Enzyme	Flavin nucleotide
Fatty acyl–CoA dehydrogenase	FAD
Dihydrolipoyl dehydrogenase	FAD
Succinate dehydrogenase	FAD
Glycerol 3-phosphate dehydrogenase	FAD
Thioredoxin reductase	FAD
NADH dehydrogenase (Complex I)	FMN
Glycolate dehydrogenase	FMN

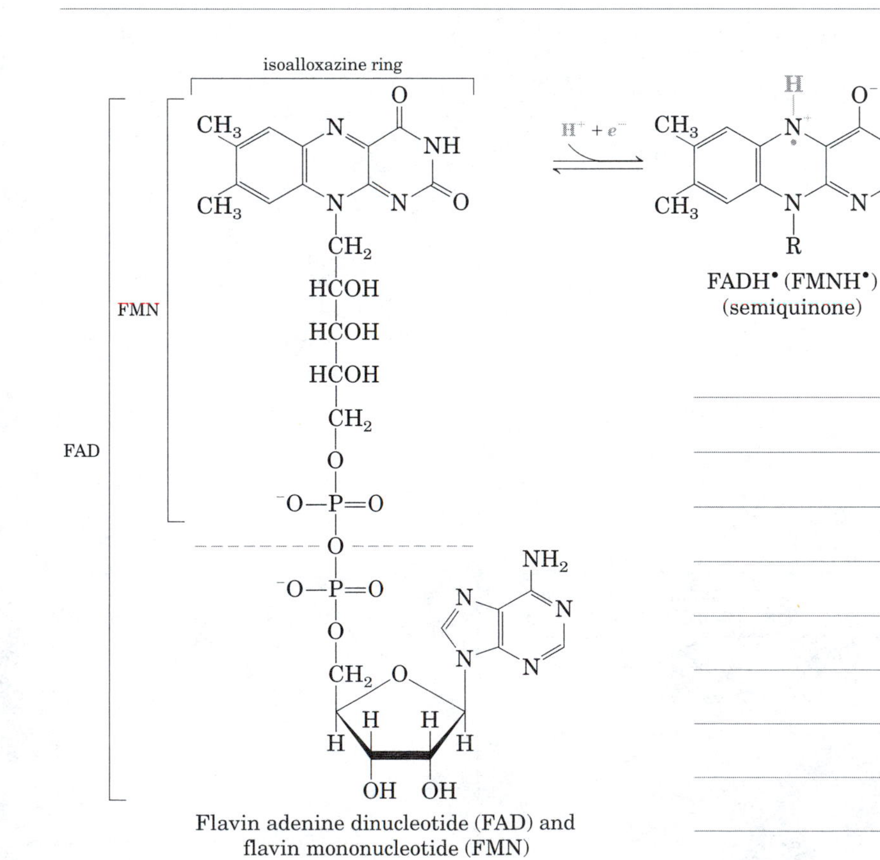

Flavin adenine dinucleotide (FAD) and flavin mononucleotide (FMN)

figure 14–16, page 521
Structures of oxidized and reduced FAD and FMN.

PROBLEMS

2. Calculation of $\Delta G'^\circ$ from Equilibrium Constants, page 524.

(a) Glutamate + oxaloacetate $\xrightleftharpoons{\text{aspartate aminotransferase}}$ aspartate + α-ketoglutarate $K'_{eq} = 6.8$

(b) Dihydroxyacetone phosphate $\xrightleftharpoons{\text{triose phosphate isomerase}}$ glyceraldehyde 3-phosphate $K'_{eq} = 0.0475$

(c) Fructose 6-phosphate + ATP $\xrightleftharpoons{\text{phosphofructokinase}}$ fructose 1,6-bisphosphate + ADP $K'_{eq} = 254$

3. Calculation of Equilibrium Constants from $\Delta G'^\circ$, page 524.

(a) Glucose 6-phosphate + H_2O $\xrightleftharpoons{\text{glucose 6-phosphatase}}$ glucose + P_i

(b) Lactose + H_2O $\xrightleftharpoons{\beta\text{-galactosidase}}$ glucose + galactose

(c) Malate $\xrightleftharpoons{\text{fumarase}}$ fumarate + H_2O

4. Experimental Determination of K'_{eq} and $\Delta G'^\circ$, page 524.

Glucose 1-phosphate $\rightleftharpoons$ glucose 6-phosphate
4.5×10^{-3} M $\quad$ 9.6×10^{-2} M

5. Experimental Determination of $\Delta G'^\circ$ for ATP Hydrolysis, page 524.

Glucose 6-phosphate + $H_2O \longrightarrow$ glucose + P_i
$K'_{eq} = 270$

ATP + glucose $\longrightarrow$ ADP + glucose 6-phosphate
$K'_{eq} = 890$

6. Difference between $\Delta G'^\circ$ and ΔG, page 524.

Fructose 6-phosphate $\rightleftharpoons$ glucose 6-phosphate
$K'_{eq} = 1.97$

9. Strategy for Overcoming an Unfavorable Reaction: ATP-Dependent Chemical Coupling, page 524.

Glucose + $P_i \longrightarrow$ glucose 6-phosphate + H_2O
$\Delta G'^\circ = 13.8$ kJ/mol

(1)	$\text{Glucose} + P_i \longrightarrow \text{glucose 6-phosphate} + H_2O$	$\Delta G'^{\circ} = 13.8\ \text{kJ/mol}$
(2)	$ATP + H_2O \longrightarrow ADP + P_i$	$\Delta G'^{\circ} = -30.5\ \text{kJ/mol}$
Sum:	$\text{Glucose} + ATP \longrightarrow$	$\text{glucose 6-phosphate} + ADP$

10. Calculations of $\Delta G'^{\circ}$ for ATP-Coupled Reactions, page 525.

(a) $\text{Phosphocreatine} + ADP \longrightarrow \text{creatine} + ATP$

(b) $ATP + \text{fructose} \longrightarrow ADP + \text{fructose 6-phosphate}$

11. Coupling ATP Cleavage to an Unfavorable Reaction, page 525.

$X + ATP + H_2O \longrightarrow Y + ADP + P_i$

12. Calculations of ΔG at Physiological Concentrations, page 525.

$\text{Phosphocreatine} + ADP \longrightarrow \text{creatine} + ATP$

13. Free Energy Required for ATP Synthesis under Physiological Conditions, page 525.

$$\frac{[ATP]}{[ADP][P_i]} = 5.33 \times 10^2\ \text{M}^{-1}$$

16. Cleavage of ATP to AMP and PP_i during Metabolism, page 526.

$\text{Acetate} + CoA + ATP \longrightarrow \text{acetyl-CoA} + AMP + PP_i$

18. Standard Reduction Potentials, page 526.

$\text{Oxidizing agent} + n\ \text{electrons} \longrightarrow \text{reducing agent}$

$\text{Pyruvate} + NADH + H^+ \rightleftharpoons \text{lactate} + NAD^+$

19. Energy Span of the Respiratory Chain, page 526.

$NADH + H^+ + \frac{1}{2}O_2$ 3 4 $H_2O + NAD^+$

22. Direction of Oxidation-Reduction Reactions, page 526.

(a) $\text{Malate} + NAD^+ \longrightarrow \text{oxaloacetate} + NADH + H^+$

(b) $\text{Acetoacetate} + NADH + H^+ \longrightarrow \beta\text{-hydroxybutyrate} + NAD^+$

(c) $\text{Pyruvate} + NADH + H^+ \longrightarrow \text{lactate} + NAD^+$

(d) $\text{Pyruvate} + \beta\text{-hydroxybutyrate} \longrightarrow \text{lactate} + \text{acetoacetate}$

(e) $\text{Malate} + \text{pyruvate} \longrightarrow \text{oxaloacetate} + \text{lactate}$

(f) $\text{Acetaldehyde} + \text{succinate} \longrightarrow \text{ethanol} + \text{fumarate}$

Glycolysis and the Catabolism of Hexoses

chapter 15

CD-ROM

When you study for exams, don't forget to review The Minicourses on the UNDERSTAND! BIOCHEMISTRY CD that came with your textbook.

Minicourses that apply to this Chapter include:

Molecules of Life
- An Interactive Gallery of Molecular Models
- An Interactive Gallery of Protein Structures

Bioenergetics
- Glycolysis and Fermentation
- Glycogen
- Pentose Phosphate Pathway

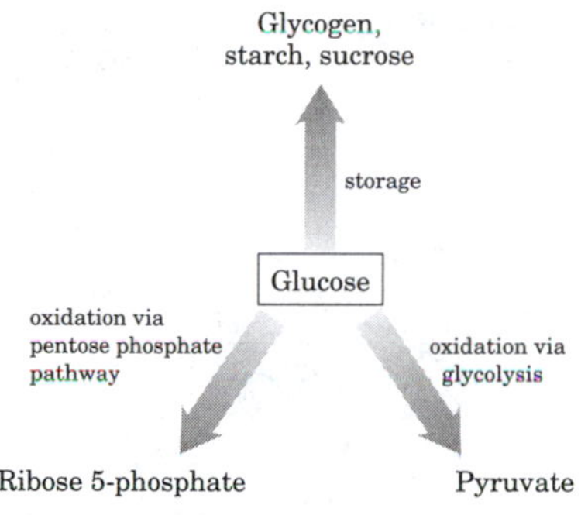

figure 15–1, page 527
Major pathways of glucose utilization in cells of higher plants and animals.

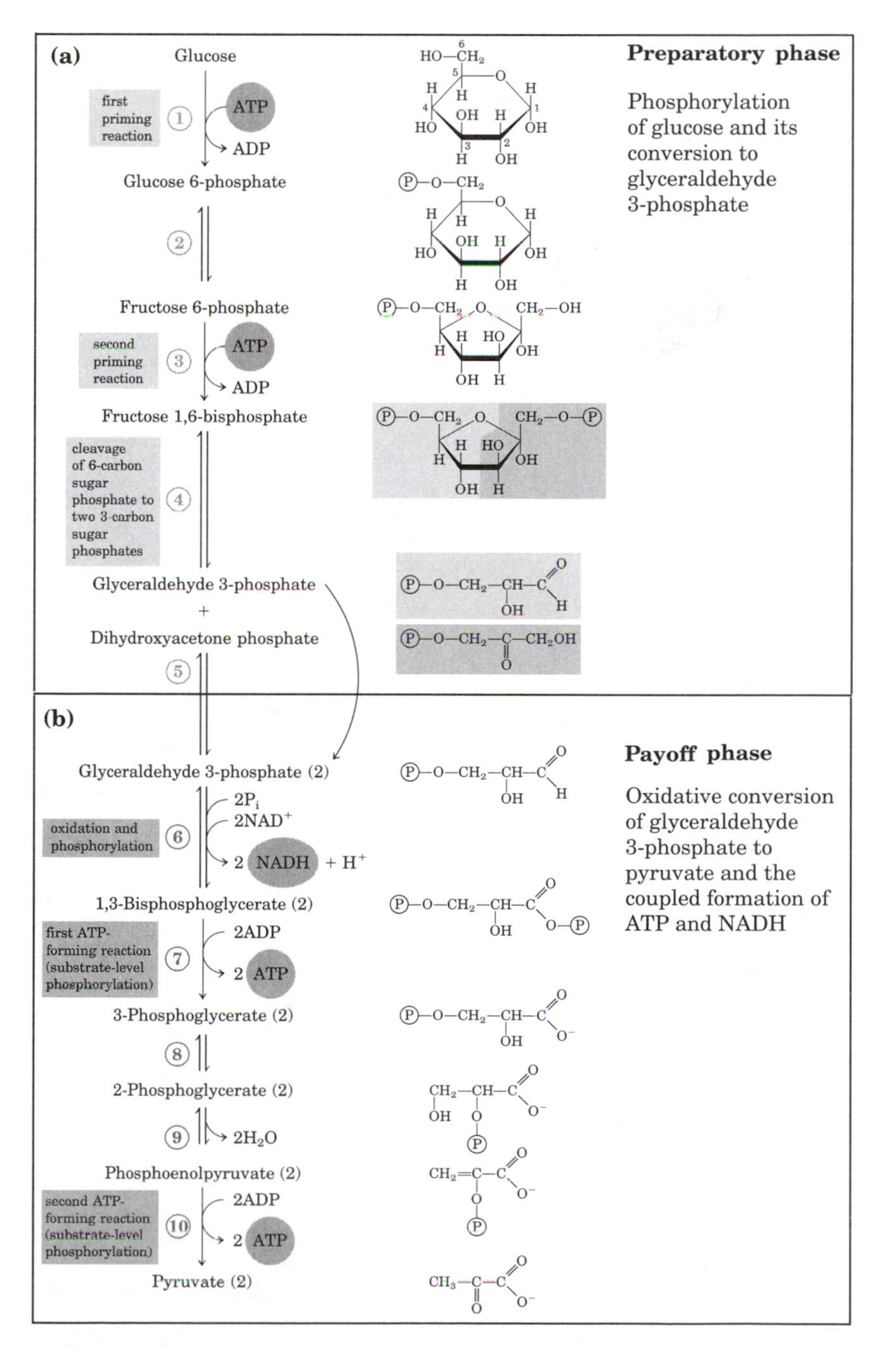

figure 15–2, page 529
The two phases of glycolysis.

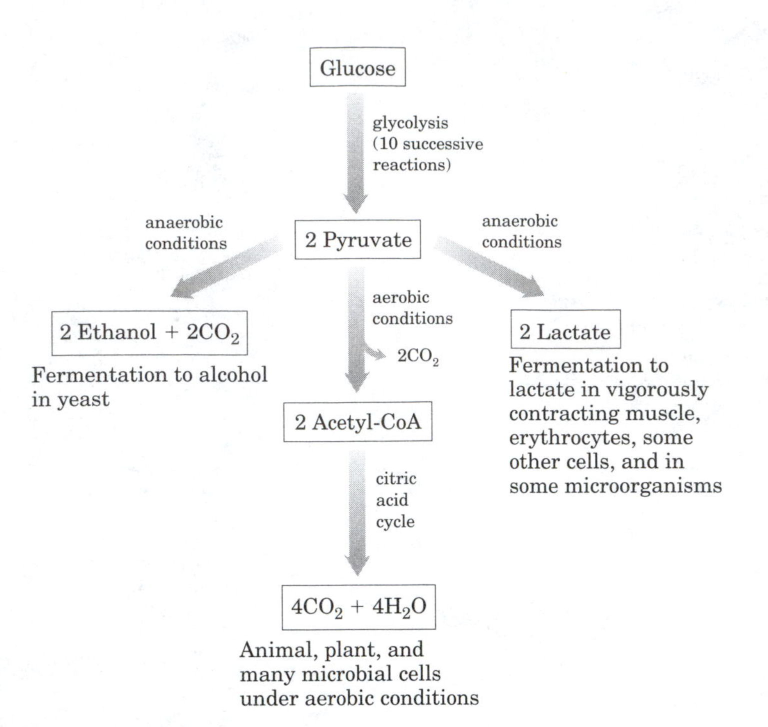

figure 15–3, page 530
Three possible catabolic fates of the pyruvate formed in glycolysis.

ATP ADP

Mg^{2+}

hexokinase

Glucose

Glucose 6-phosphate

$\Delta G'^{\circ} = -16.7$ kJ/mol

Hexokinase, page 532

Mg^{2+}

phosphohexose isomerase

Glucose 6-phosphate

Fructose 6-phosphate

$\Delta G'^{\circ} = 1.7$ kJ/mol

Phosphohexose isomerase, page 533

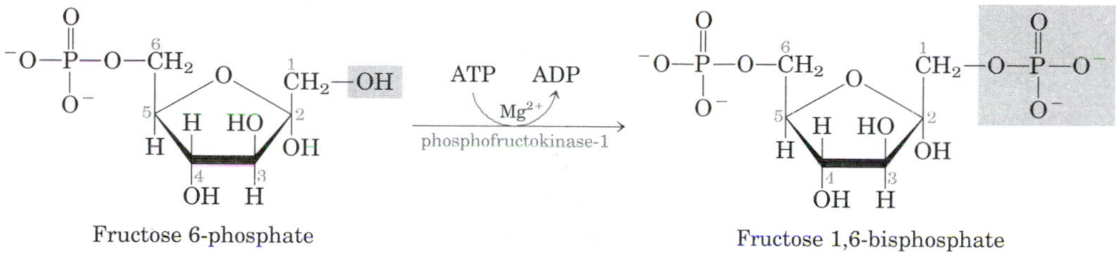

Fructose 6-phosphate → (ATP → ADP, Mg^{2+}, phosphofructokinase-1) → Fructose 1,6-bisphosphate

PFK-1, page 533

$\Delta G'^{\circ} = -14.2$ kJ/mol

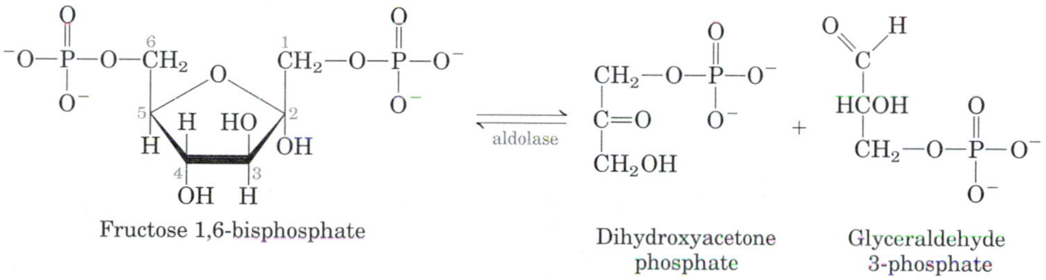

Fructose 1,6-bisphosphate ⇌ (aldolase) Dihydroxyacetone phosphate + Glyceraldehyde 3-phosphate

Aldolase, page 534

$\Delta G'^{\circ} = 23.8$ kJ/mol

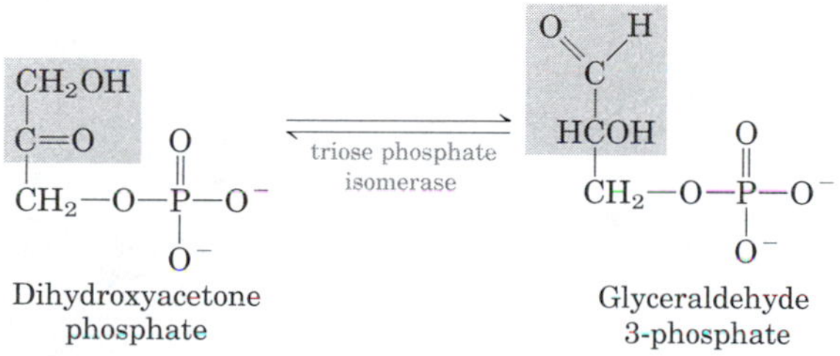

Dihydroxyacetone phosphate ⇌ (triose phosphate isomerase) Glyceraldehyde 3-phosphate

Triose phosphate isomerase, page 534

$\Delta G'^{\circ} = 7.5$ kJ/mol

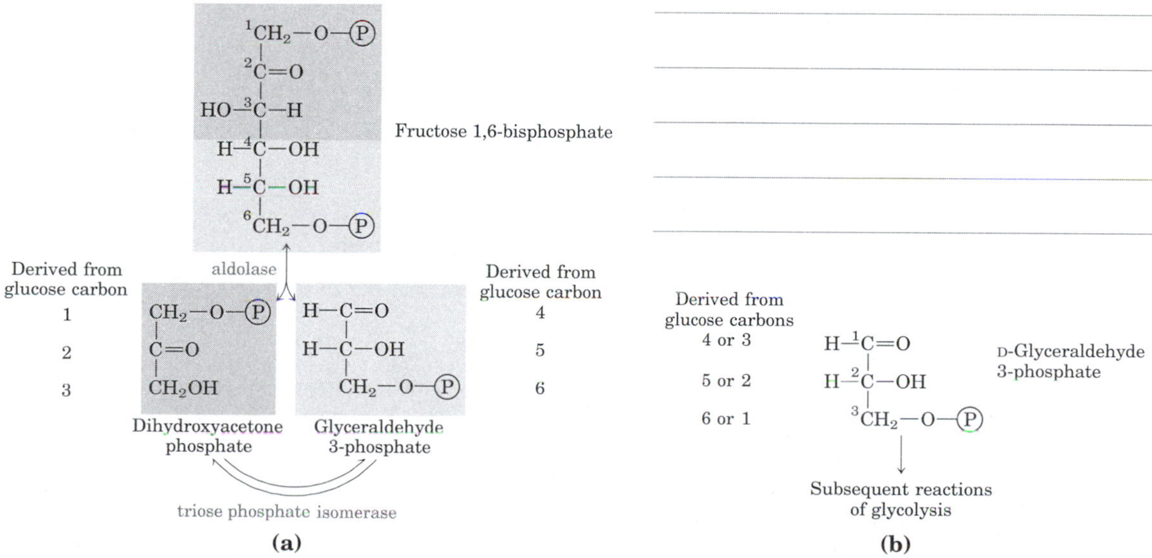

figure 15–4, page 535
Fate of the carbon atoms of glucose in the formation of glyceraldehyde 3-phosphate.

Glyceraldehyde 3-phosphate + Inorganic phosphate $\xrightarrow{NAD^+ \rightarrow NADH + H^+}$ 1,3-Bisphosphoglycerate

(glyceraldehyde 3-phosphate dehydrogenase)

$\Delta G'^{\circ} = 6.3$ kJ/mol

Glyceraldehyde 3-phosphate dehydrogenase, page 535

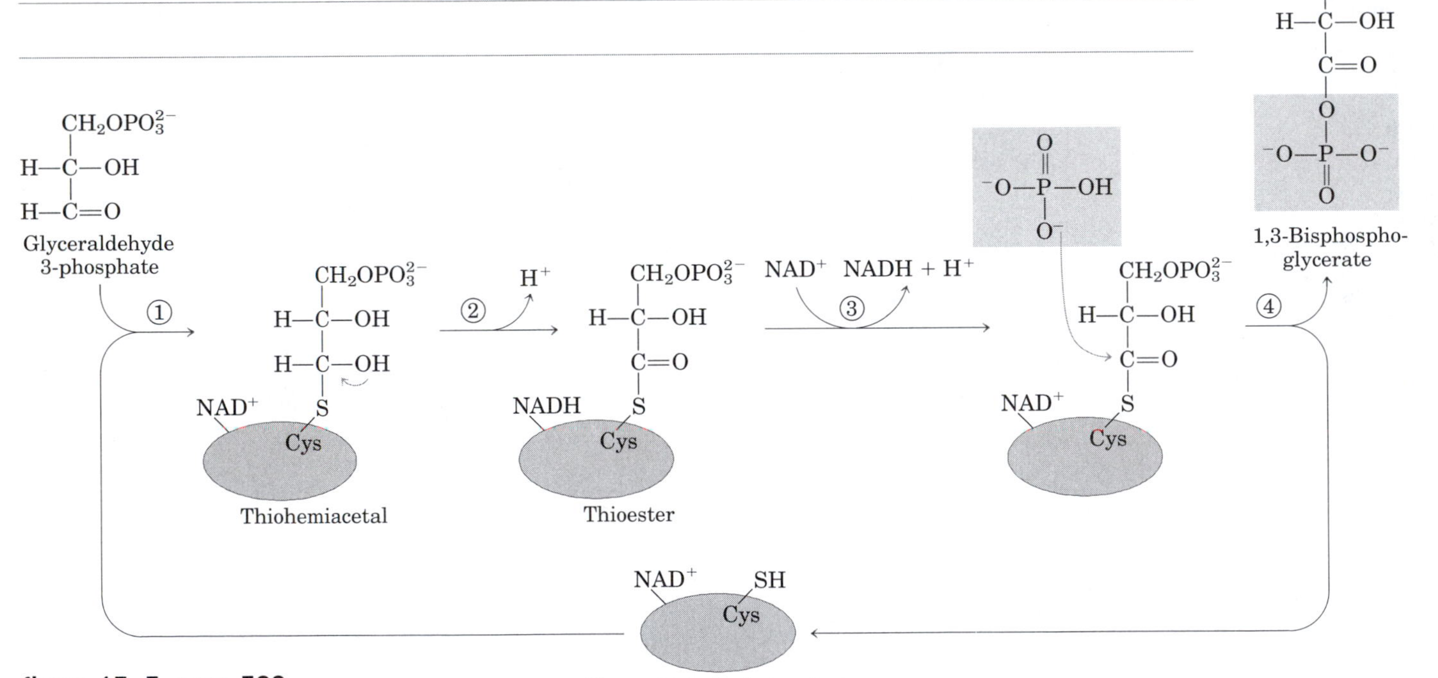

figure 15–5, page 536
Glyceraldehyde 3-phosphate dehydrogenase reaction: a more detailed representation.

Enzyme (NAD^+, Cys–SH) + ICH_2COO^- (Iodoacetate) $\xrightarrow{HI}$ Inactive enzyme (NAD^+, Cys–S–CH_2COO^-)

Action of iodoacetate, page 536

1,3-Bisphosphoglycerate + ADP $\xrightleftharpoons[\text{phosphoglycerate kinase}]{Mg^{2+}}$ 3-Phosphoglycerate + ATP

Phosphoglycerate kinase, page 537

$\Delta G'^{\circ} = -18.5$ kJ/mol

3-Phosphoglycerate $\xrightleftharpoons[\text{phosphoglycerate mutase}]{Mg^{2+}}$ 2-Phosphoglycerate

Phosphoglycerate mutase, page 538

$\Delta G'^{\circ} = 4.4$ kJ/mol

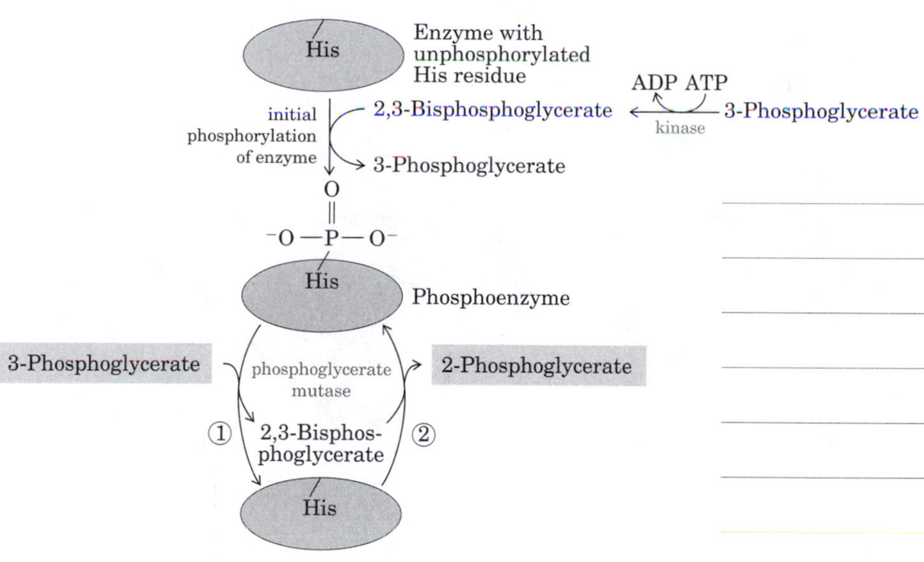

figure 15–6, page 538
Mechanism of the phosphoglycerate mutase reaction.

2-Phosphoglycerate $\xrightleftharpoons[\text{enolase}]{H_2O}$ Phosphoenolpyruvate

Enolase, page 539

$\Delta G'^{\circ} = 7.5$ kJ/mol

$$\text{Phosphoenolpyruvate} + \text{ADP} \xrightarrow[\text{pyruvate kinase}]{Mg^{2+},\ K^{+}} \text{Pyruvate} + \text{ATP}$$

Pyruvate kinase, page 539

$$\text{Pyruvate (enol form)} \underset{\text{tautomerization}}{\rightleftharpoons} \text{Pyruvate (keto form)}$$

Pyruvate tautomers, page 539

$$\text{Glucose} + 2\text{ATP} + 2\text{NAD}^+ + 4\text{ADP} + 2\text{P}_i \longrightarrow$$
$$2\ \text{pyruvate} + 2\text{ADP} + 2\text{NADH} + 2\text{H}^+ + 4\text{ATP} + 2\text{H}_2\text{O}$$

$$\text{Glucose} + 2\text{NAD}^+ + 2\text{ADP} + 2\text{P}_i \longrightarrow$$
$$2\ \text{pyruvate} + 2\text{NADH} + 2\text{H}^+ + 2\text{ATP} + 2\text{H}_2\text{O}$$

Glycolysis, page 540

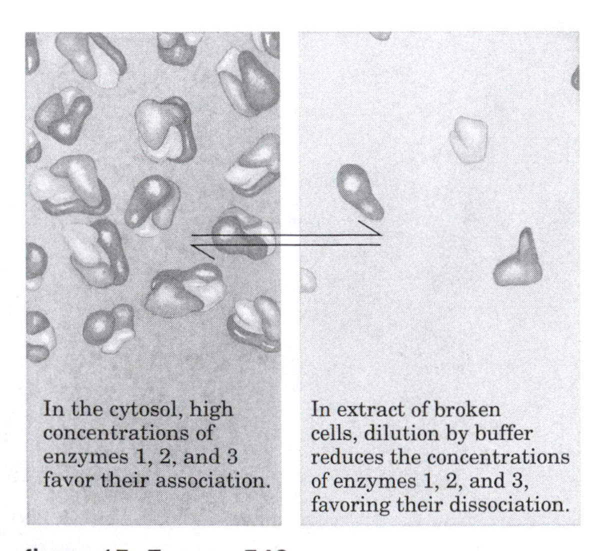

figure 15–7, page 540
Dilution of a solution containing a noncovalent protein complex favors dissociation of the complex into its constituents.

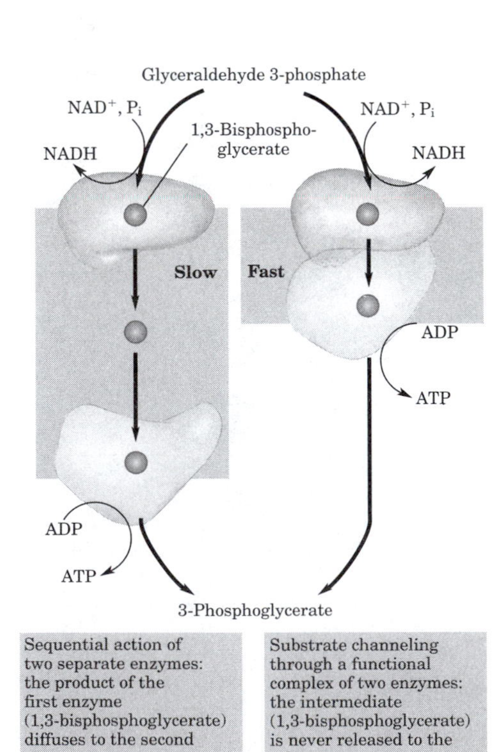

figure 15–8, page 541
Channeling of a substrate between two enzymes in the glycolytic pathway.

Pyruvate

lactate dehydrogenase

NADH + H^+

NAD^+

Lactate

$\Delta G'^{\circ} = -25.1$ kJ/mol

Glucose

2NAD$^+$

2NADH

2 Pyruvate → 2 Lactate

Lactate dehydrogenase, page 542

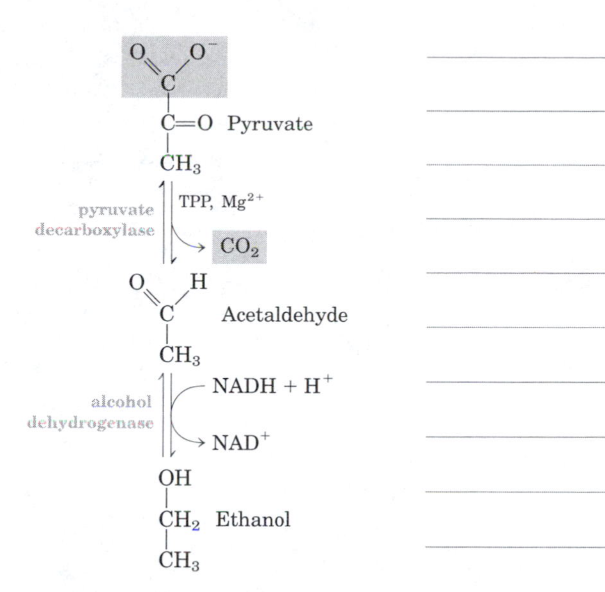

Formation of ethanol, page 544

thiazolium ring

Thiamine pyrophosphate (TPP)

(a)

active aldehyde

Hydroxyethyl thiamine pyrophosphate

(b)

Pyruvate

TPP carbanion

TPP

Acetaldehyde

Hydroxyethyl TPP

resonance stabilization

(c)

figure 15–9, page 545
Thiamine pyrophosphate (TPP) and its role in pyruvate decarboxylation.

table 15–1, page 546

Some TPP-Dependent Reactions

Enzyme	Pathway	Bond cleaved	Bond formed
Pyruvate decarboxylase	Alcohol fermentation	$R^1{-}C(=O){-}C(=O)O^-$	$R^1{-}C(=O)H$
Pyruvate dehydrogenase α-Ketoglutarate dehydrogenase	Synthesis of acetyl-CoA Citric acid cycle	$R^2{-}C(=O){-}C(=O)O^-$	$R^2{-}C(=O){-}S\text{-}CoA$
Transketolase	Carbon-fixation reactions of photosynthesis	$R^3{-}C(=O){-}C(OH)(H){-}R^4$	$R^3{-}C(=O){-}C(OH)(H){-}R^5$
Acetolactate synthetase	Valine, leucine biosynthesis	$R^6{-}C(=O){-}C(=O)O^-$	$R^6{-}C(=O){-}C(OH)(COO^-){-}$

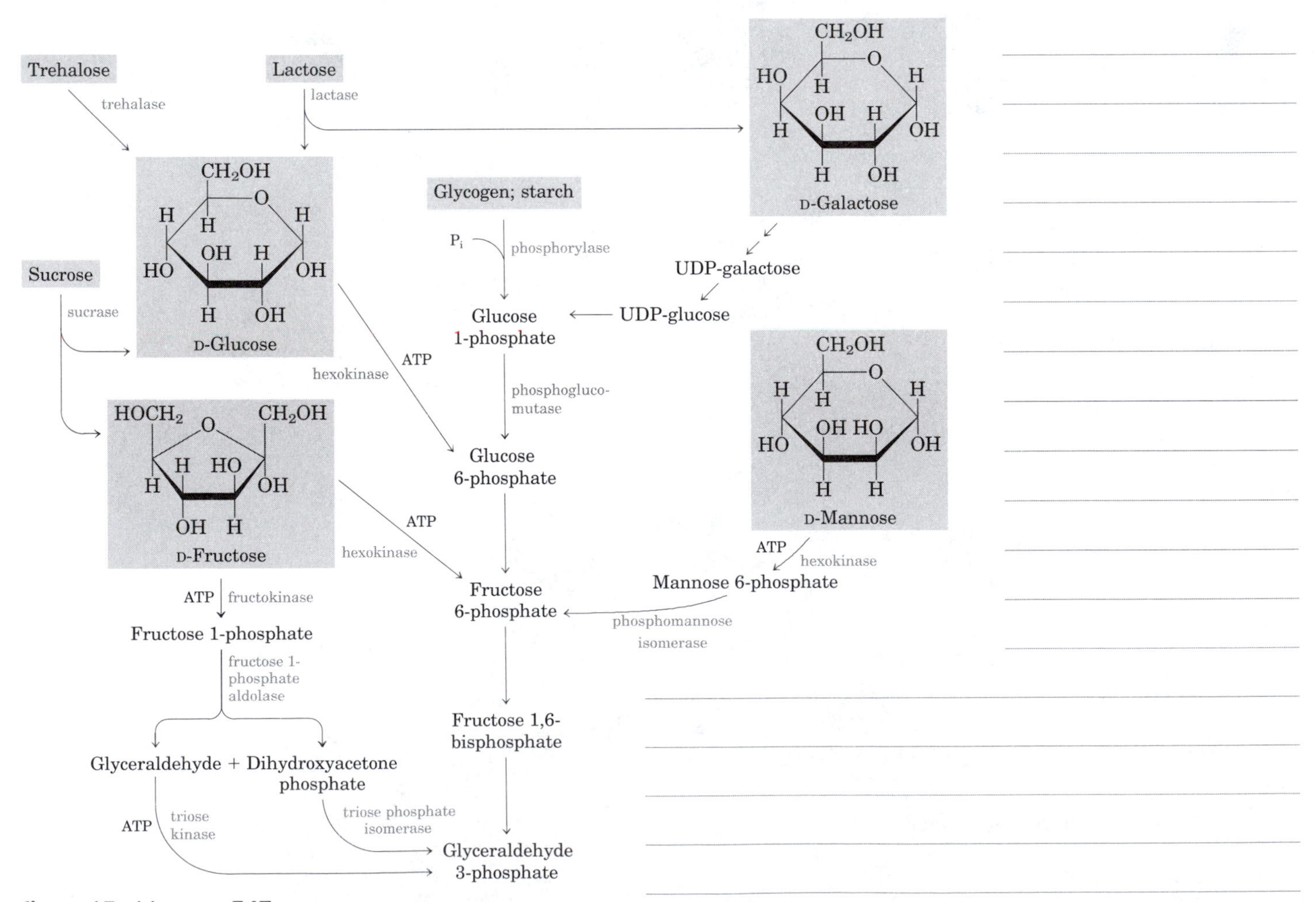

figure 15–11, page 547

Entry of glycogen, starch, disaccharides, and hexoses into the preparatory stage of glycolysis.

Pyridoxal phosphate, page 547

Glucose 1-phosphate

Glycogen shortened by one residue $(glucose)_{n-1}$

figure 15–12, page 548

Removal of a terminal glucose residue from the nonreducing end of a glycogen chain by glycogen phosphorylase.

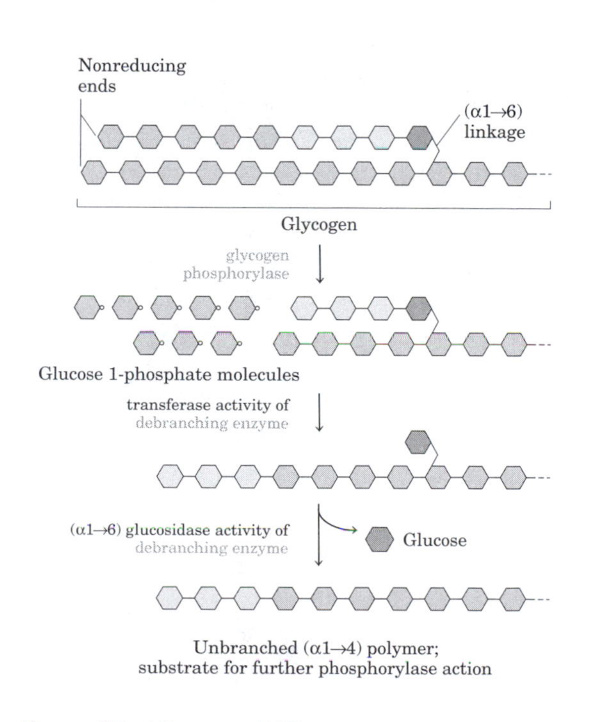

figure 15–13, page 548

Glycogen breakdown near an (α1→6) branch point.

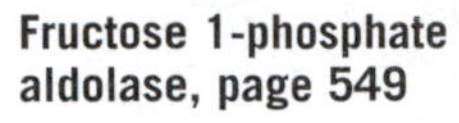

Fructose 1-phosphate aldolase, page 549

Fructose 1-phosphate $\xrightleftharpoons{\text{fructose 1-phosphate aldolase}}$ Dihydroxyacetone phosphate + Glyceraldehyde

Fructose 1-phosphate: $^{1}CH_2{-}O{-}P(=O)(O^-){-}O^-$, $^{2}C{=}O$, $^{3}HOCH$, $^{4}HCOH$, $^{5}HCOH$, $^{6}CH_2OH$

Dihydroxyacetone phosphate: $CH_2{-}O{-}P(=O)(O^-){-}O^-$, $C{=}O$, CH_2OH

Glyceraldehyde: $H{-}C{=}O$, $HCOH$, CH_2OH

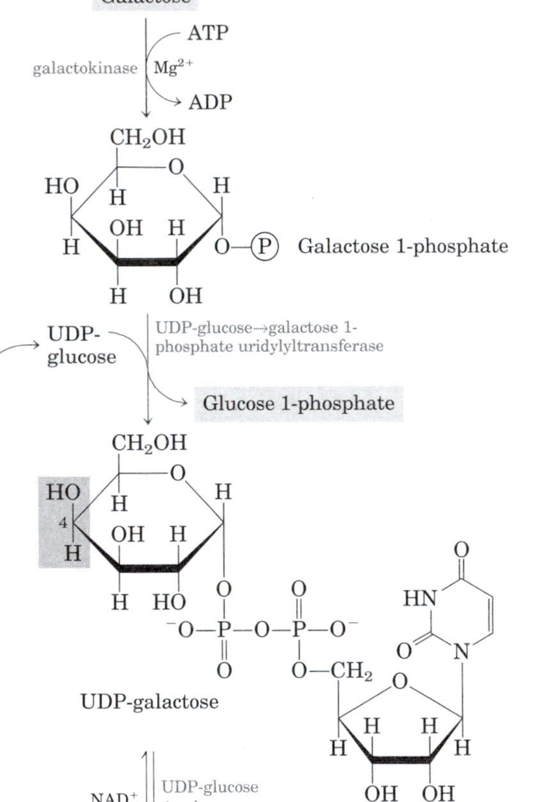

figure 15–14, page 549

Conversion of D-galactose 1-phosphate to D-glucose 1-phosphate.

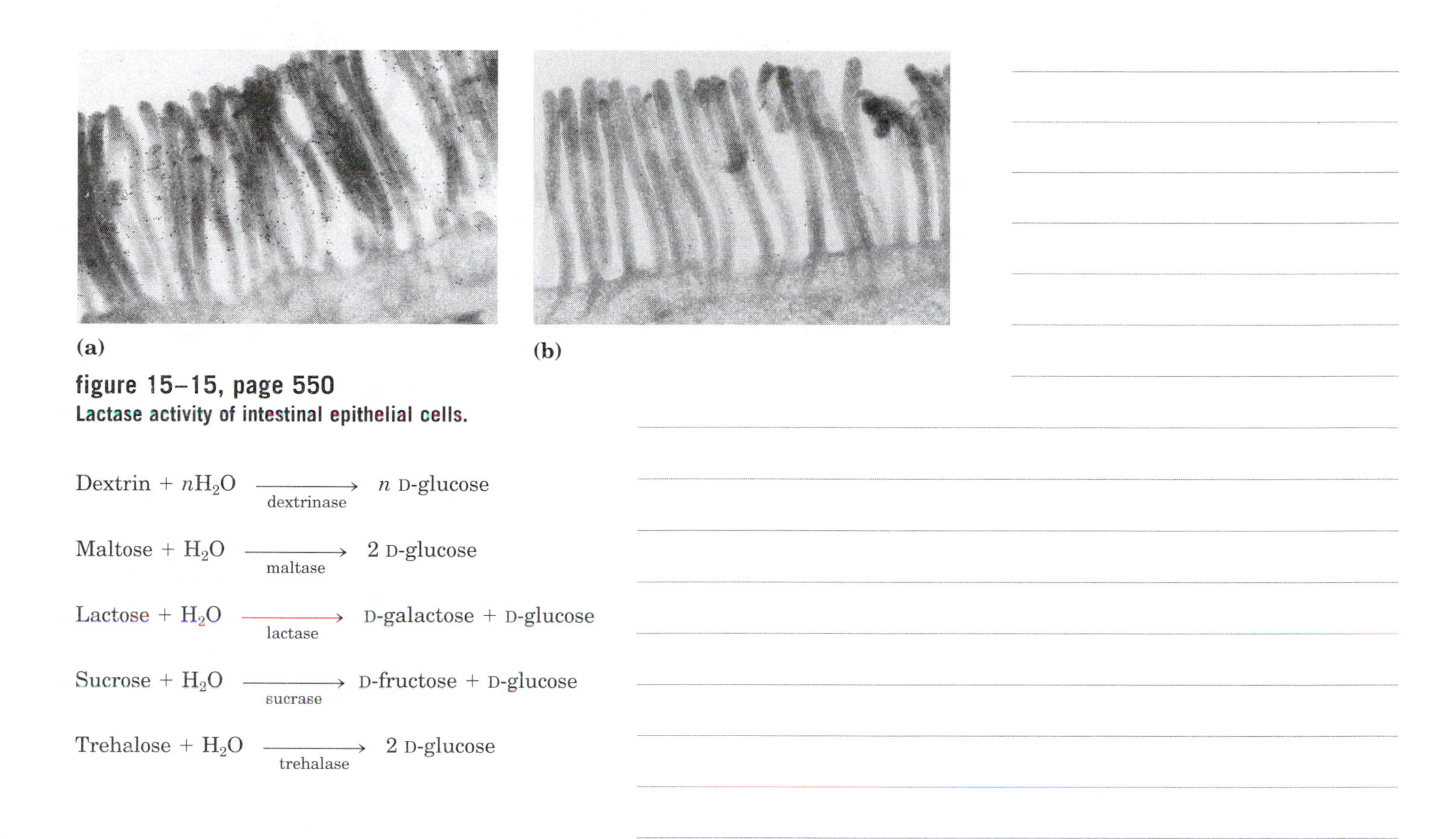

figure 15–15, page 550
Lactase activity of intestinal epithelial cells.

$$\text{Dextrin} + n\text{H}_2\text{O} \xrightarrow[\text{dextrinase}]{} n\ \text{D-glucose}$$

$$\text{Maltose} + \text{H}_2\text{O} \xrightarrow[\text{maltase}]{} 2\ \text{D-glucose}$$

$$\text{Lactose} + \text{H}_2\text{O} \xrightarrow[\text{lactase}]{} \text{D-galactose} + \text{D-glucose}$$

$$\text{Sucrose} + \text{H}_2\text{O} \xrightarrow[\text{sucrase}]{} \text{D-fructose} + \text{D-glucose}$$

$$\text{Trehalose} + \text{H}_2\text{O} \xrightarrow[\text{trehalase}]{} 2\ \text{D-glucose}$$

table 15–2, page 552

Cytosolic Concentrations of Enzymes and Intermediates of the Glycolytic Pathway in Skeletal Muscle

Enzyme	Concentration (μM)	Intermediate	Concentration (μM)
Aldolase	810	Glucose 6-phosphate	3,900
Triose phosphate isomerase	220	Fructose 6-phosphate	1,500
Glyceraldehyde 3-phosphate dehydrogenase	1,400	Fructose 1,6-bisphosphate	80
		Dihydroxyacetone phosphate	160
Phosphoglycerate kinase	130	Glyceraldehyde 3-phosphate	80
Phosphoglycerate mutase	240	1,3-Bisphosphoglycerate	50
Enolase	540	3-Phosphoglycerate	200
Pyruvate kinase	170	2-Phosphoglycerate	20
Lactate dehydrogenase	300	Phosphenolpyruvate	65
Phosphoglucomutase	32	Pyruvate	380
		Lactate	3,700
		ATP	8,000
		ADP	600
		P_i	8,000
		NAD^+	540
		NADH	50

Source: From Srivastava, D.K. & Bernhard, S.A. (1987) Biophysical chemistry of metabolic reaction sequences in concentrated solution and in the cell. *Annu. Rev. Biophys. Biophys. Chem.* **16,** 175–204.

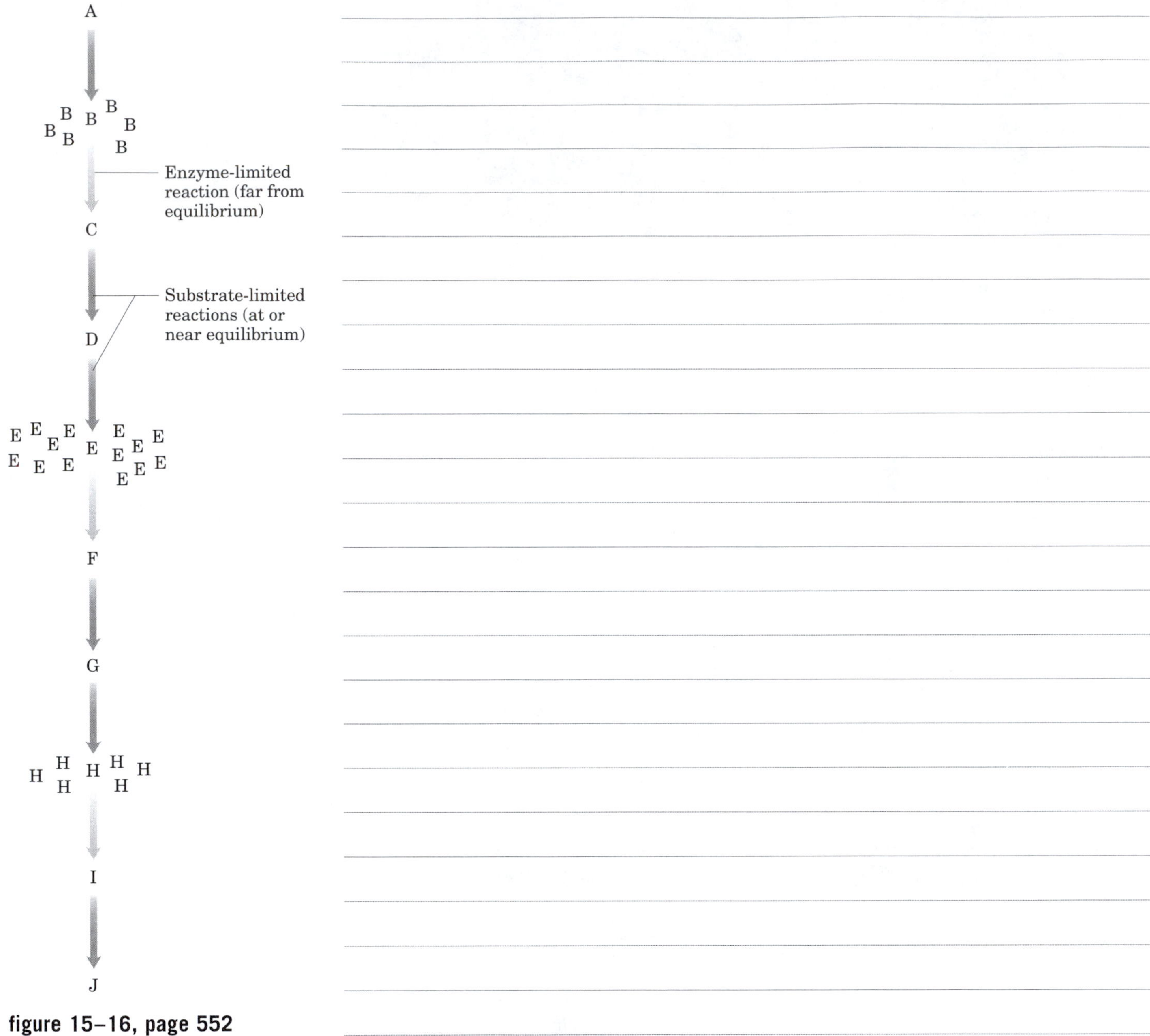

figure 15–16, page 552
Regulation of the flux through a multistep pathway.

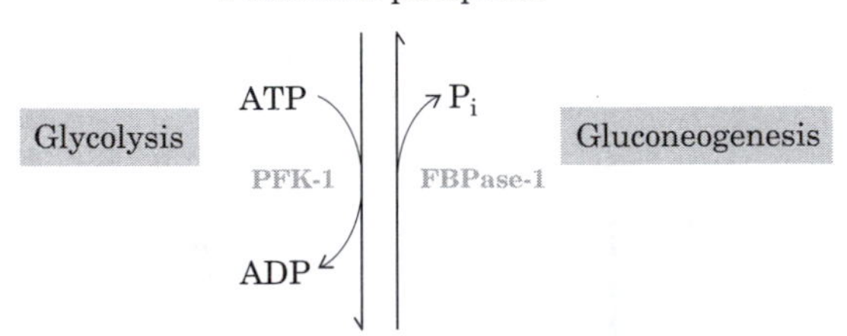

figure 15–17, page 553
The reaction of gluconeogenesis that bypasses the irreversible phosphofructokinase-1 reaction of glycolysis.

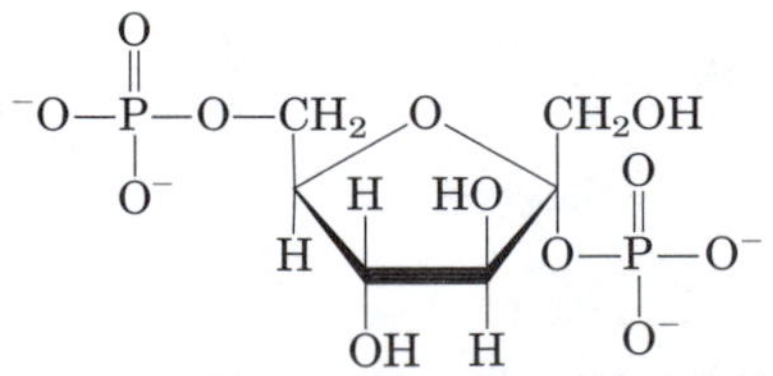

Fructose, 2,6-bisphosphate, page 554

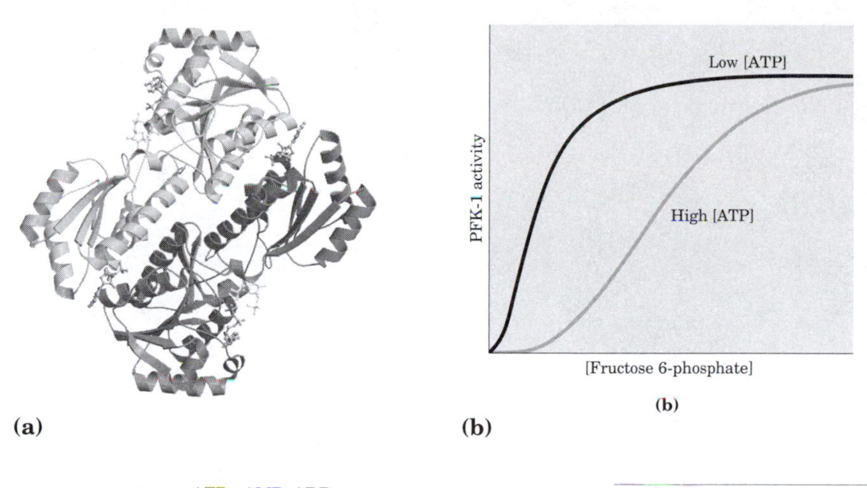

ATP AMP, ADP

Fructose 6-phosphate + ATP → Fructose 1,6-bisphosphate + ADP

citrate fructose 2,6-bisphosphate

(c)

figure 15–18, page 554
Phosphofructokinase-1 and its regulation.

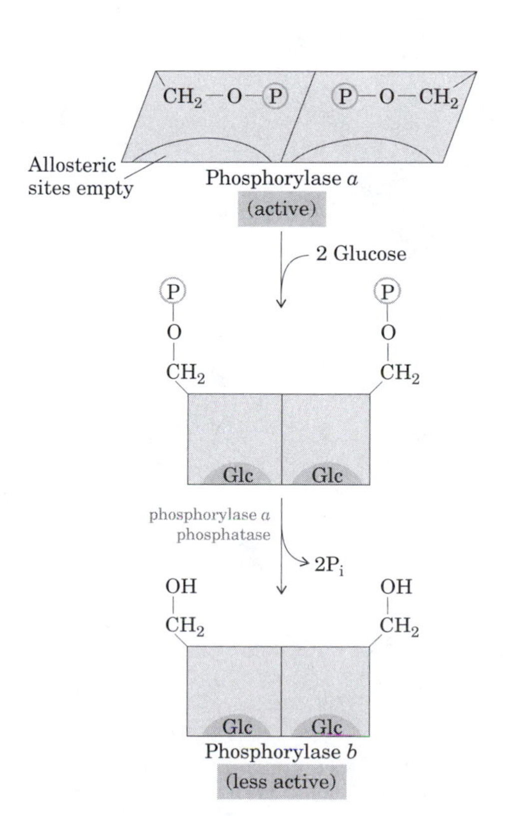

figure 15–19, page 557
Glycogen phosphorylase of liver as a glucose sensor.

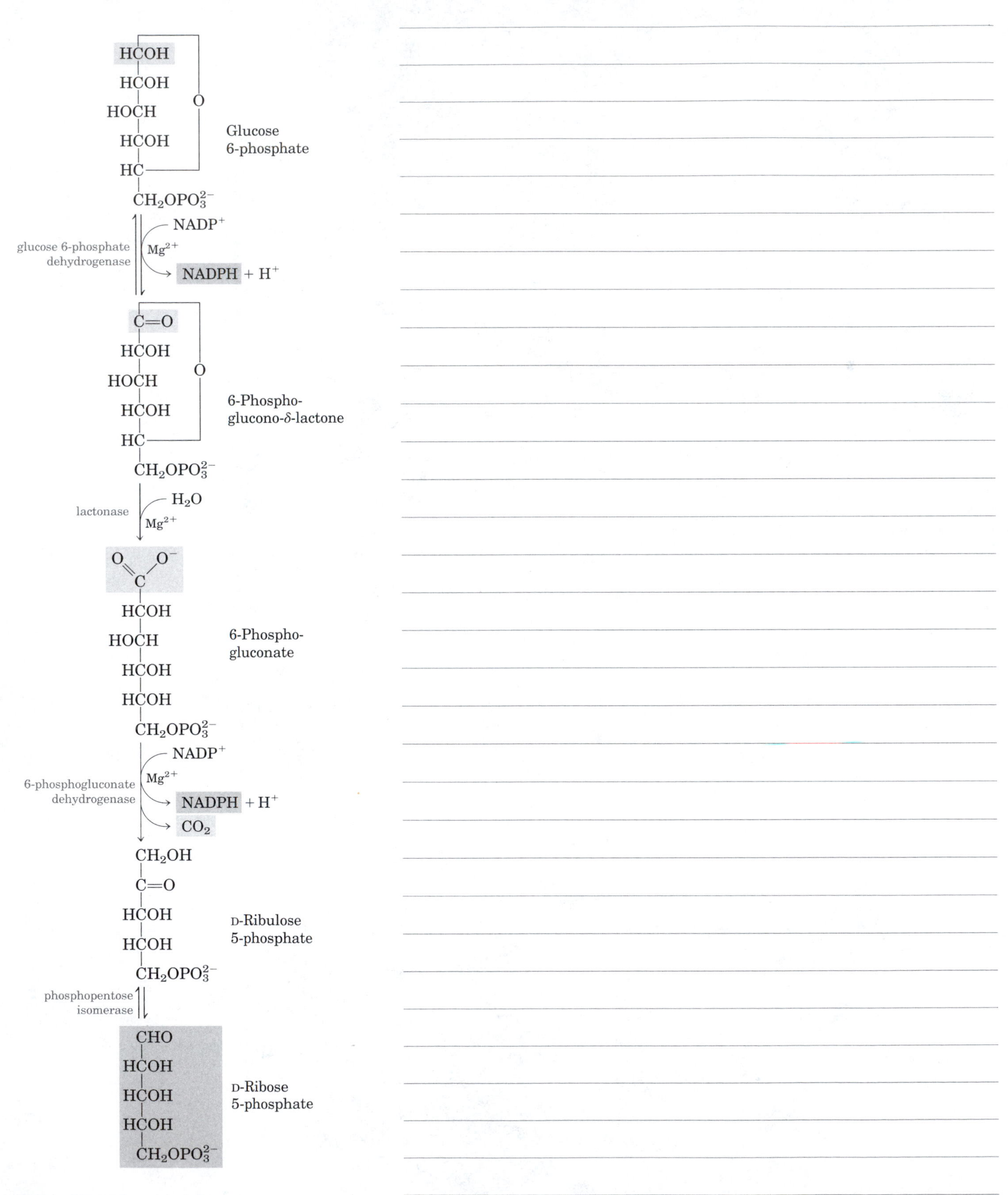

figure 15–20, page 558
Oxidative reactions of the pentose phosphate pathway.

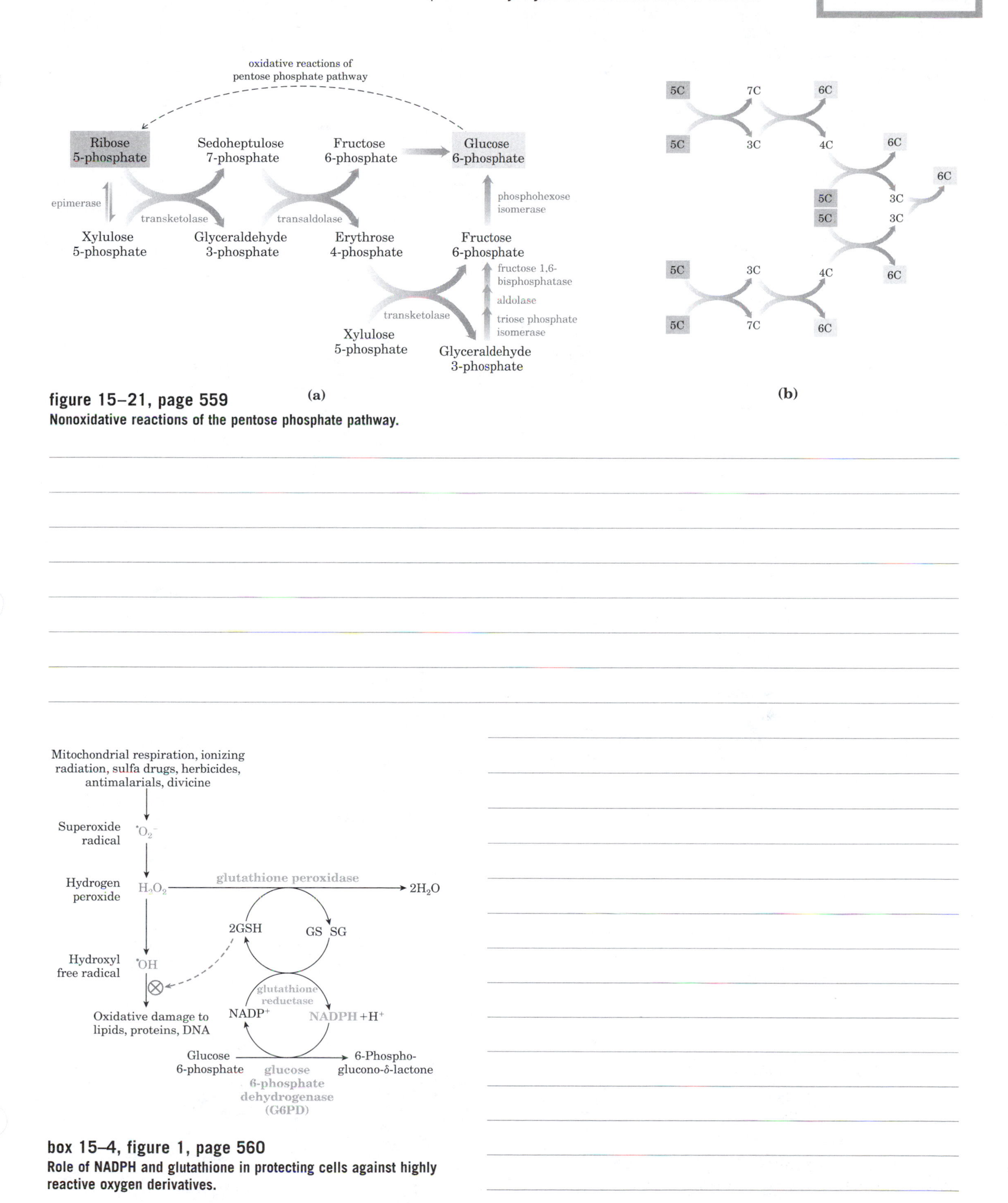

figure 15–21, page 559

Nonoxidative reactions of the pentose phosphate pathway.

box 15–4, figure 1, page 560

Role of NADPH and glutathione in protecting cells against highly reactive oxygen derivatives.

PROBLEMS

6. Glycolysis Shortcut, page 564.

$$\text{Glyceraldehyde 3-phosphate} + H_2 \xrightarrow{NAD^+ \;\rightarrow\; NADH + H^+} \text{3-phosphoglycerate}$$

11. Arsenate Poisoning, page 564.

$$R{-}\overset{\overset{O}{\|}}{C}{-}O{-}\underset{\underset{O^-}{|}}{\overset{\overset{O}{\|}}{As}}{-}O^- + H_2O \longrightarrow R{-}\overset{\overset{O}{\|}}{C}{-}O^- + HO{-}\underset{\underset{O^-}{|}}{\overset{\overset{O}{\|}}{As}}{-}O^- + H^+$$

15. Metabolism of Glycerol, page 565.

$$HOCH_2{-}\underset{\underset{H}{|}}{\overset{\overset{OH}{|}}{C}}{-}CH_2OH$$

Glycerol

16. Measurement of Intracellular Metabolite Concentrations, page 565.

Metabolite	Concentration (mM)*
Fructose 6-phosphate	0.087
Fructose 1,6-bisphosphate	0.022
ATP	11.4
ADP	1.32

18. Regulation of Phosphofructokinase-1, page 565.

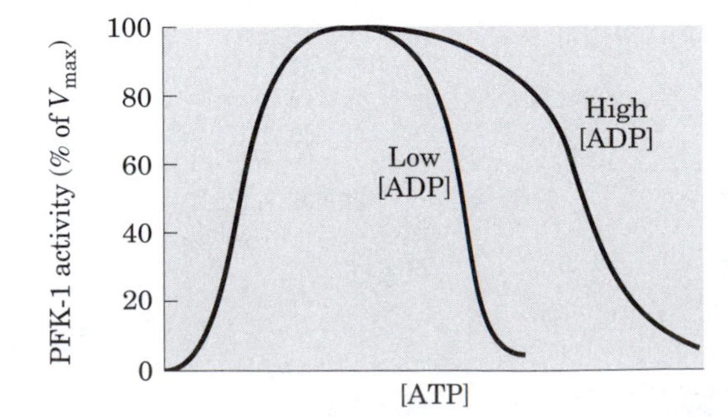

chapter

16

The Citric Acid Cycle

When you study for exams, don't forget to review The Minicourses on the UNDERSTAND! BIOCHEMISTRY CD that came with your textbook.

Minicourses that apply to this Chapter include:

Molecules of Life

An Interactive Gallery of Molecular Models

An Interactive Gallery of Protein Structures

Bioenergetics

Citric Acid Cycle

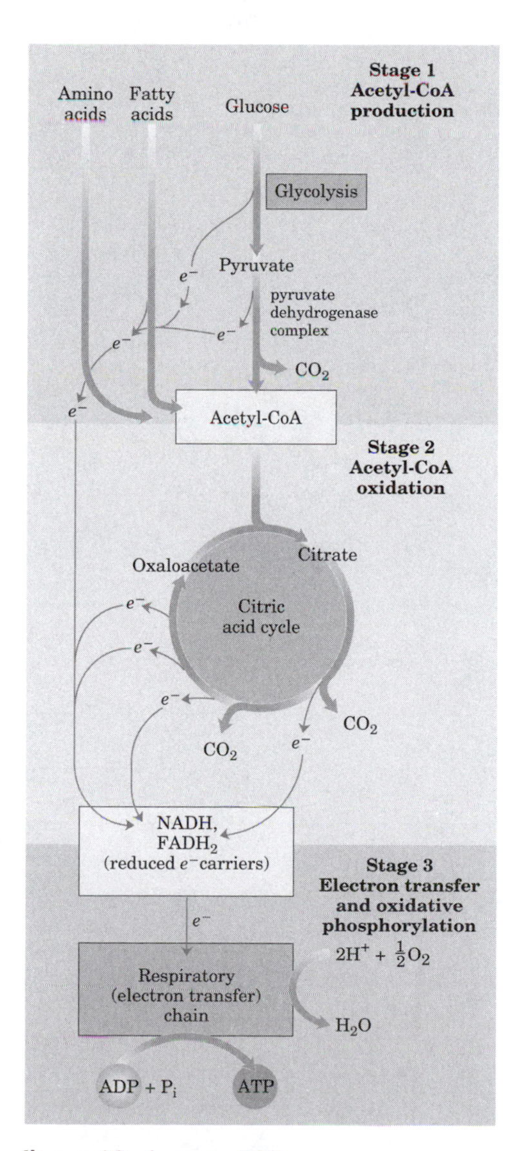

figure 16–1, page 567

Catabolism of proteins, fats, and carbohydrates in the three stages of cellular respiration.

figure 16–2, page 569

Overall reaction catalyzed by the pyruvate dehydrogenase complex.

Reactive thiol group

Coenzyme A

HS—CH$_2$—CH$_2$—N(H)—C(=O)—CH$_2$—CH$_2$—N(H)—C(=O)—C(H)(OH)—C(CH$_3$)(CH$_3$)—CH$_2$—O—P(O$^-$)(=O)—O—P(O$^-$)(=O)—O—CH$_2$ (5′)

β-Mercapto-ethylamine

Pantothenic acid

NH_2

Adenine

Ribose 3′-phosphate

O=P—O$^-$

3′-Phosphoadenosine diphosphate

$CH_3—C(=O)$—S-CoA

Acetyl-CoA

figure 16–3, page 569
Coenzyme A (CoA).

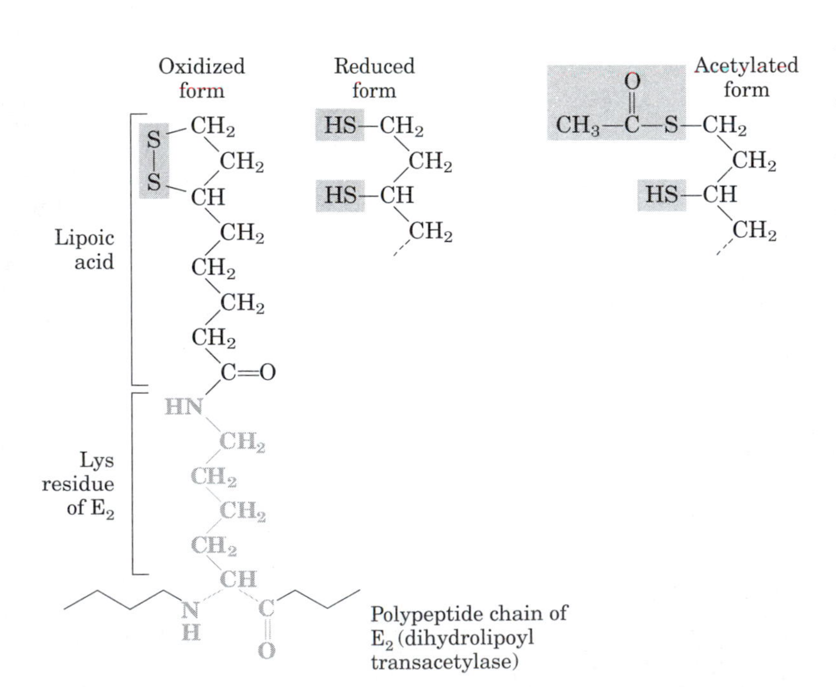

figure 16–4, page 570
Lipoic acid (lipoate) in amide linkage with the side chain of a Lys residue.

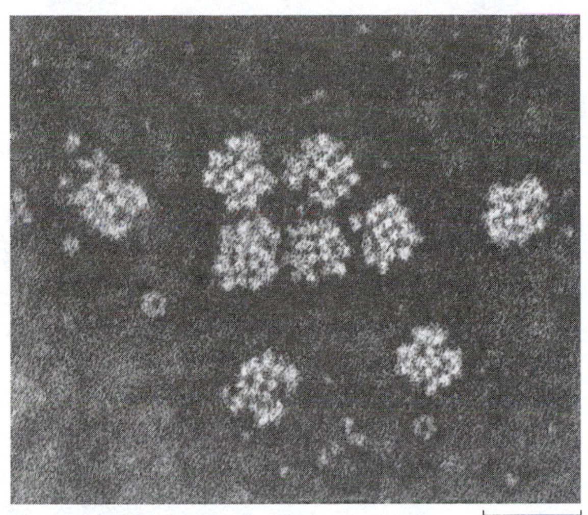

figure 16–5, page 570
Electron micrograph of pyruvate dehydrogenase complexes isolated from *E. coli*, showing the subunit structure.

E_1 pyruvate dehydrogenase
E_2 dehydrolipoyl transacetylase
E_3 dehydrolipoyl dehydrogenase

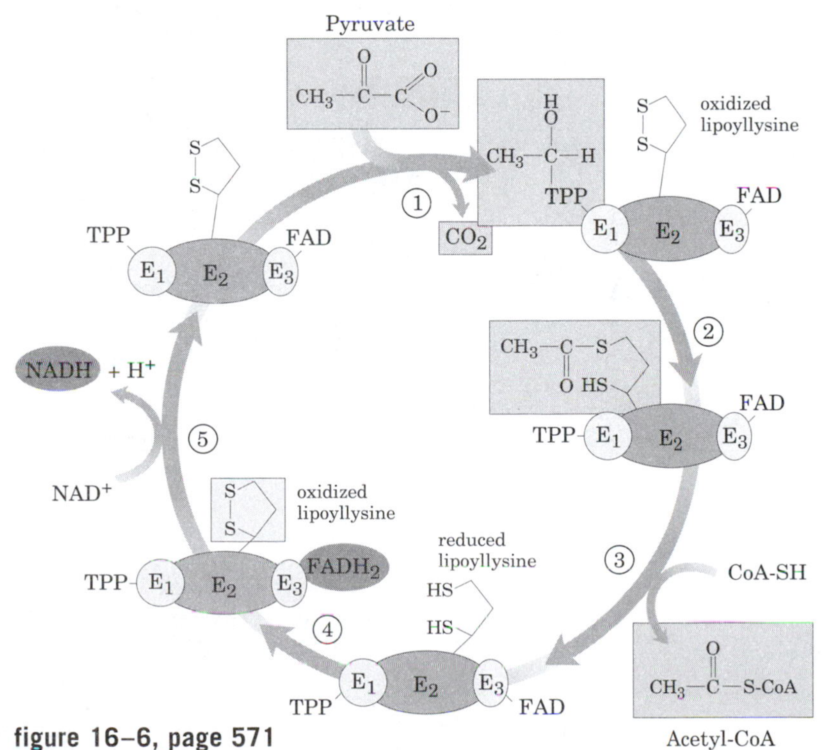

figure 16–6, page 571
Oxidative decarboxylation of pyruvate to acetyl-CoA by the pyruvate dehydrogenase complex.

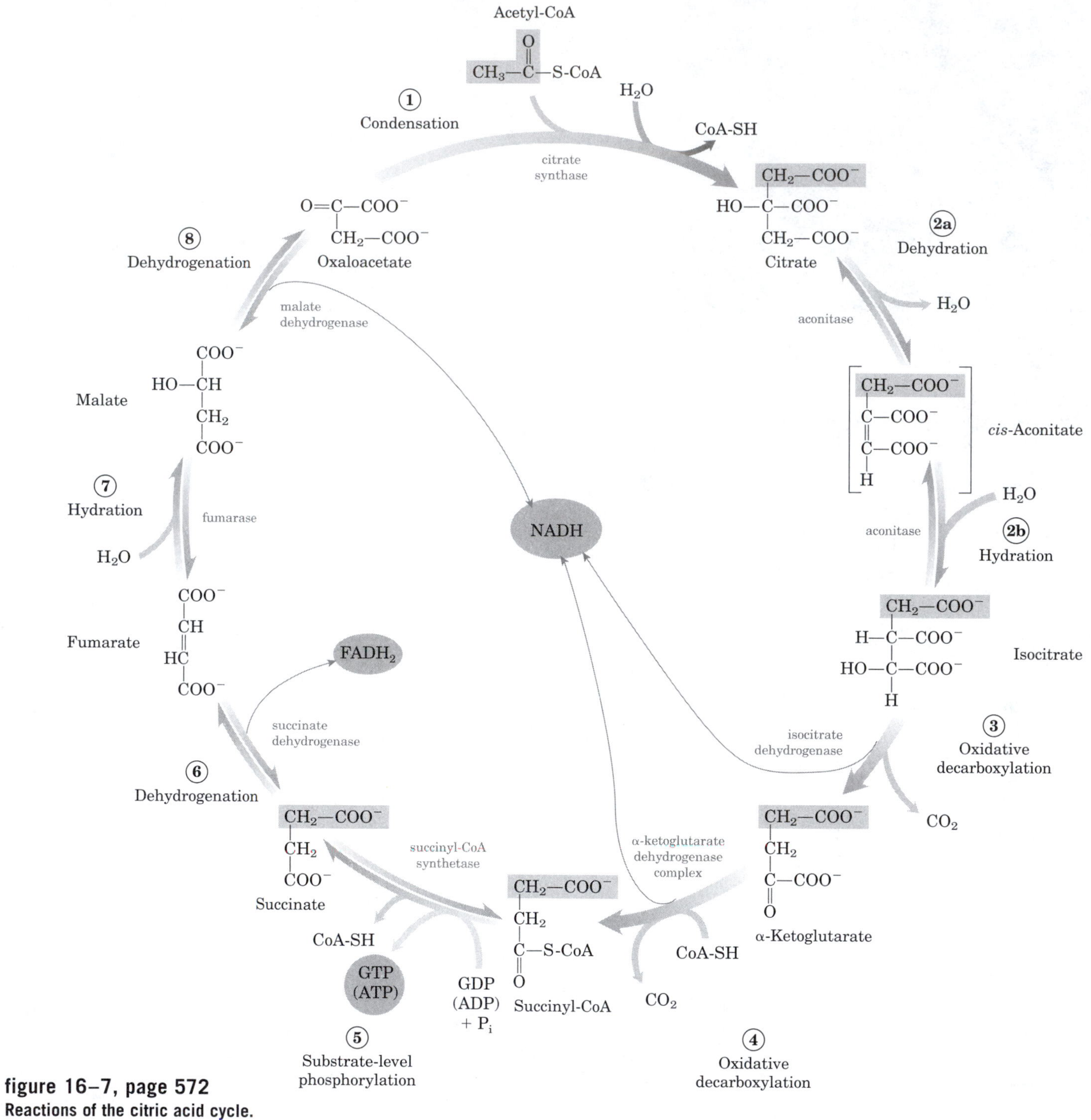

figure 16–7, page 572
Reactions of the citric acid cycle.

$$CH_3{-}C({=}O){-}S\text{-}CoA + O{=}C(COO^-){-}CH_2{-}COO^- \xrightarrow[\text{citrate synthase}]{H_2O \quad CoA\text{-}SH} HO{-}C(COO^-)(CH_2{-}COO^-){-}CH_2{-}C({=}O){-}O^-$$

Acetyl-CoA Oxaloacetate Citrate

Citrate synthase, page 573

$\Delta G'^{\circ} = -32.2$ kJ/mol

$$CH_2{-}C({=}O){-}S\text{-}CoA \quad HO{-}C{-}COO^- \quad CH_2{-}COO^-$$

Citroyl-CoA, page 573

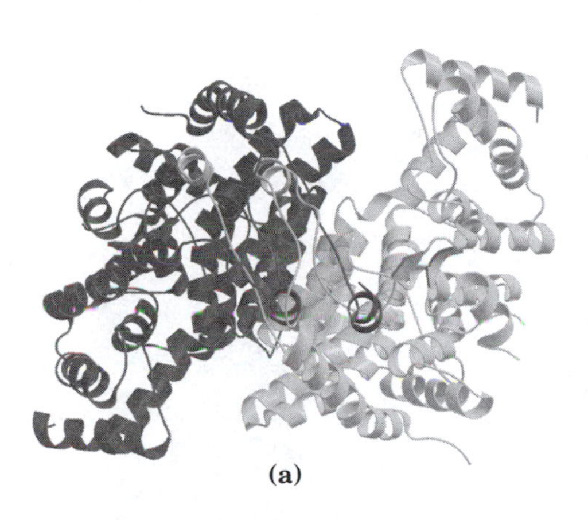

(a)

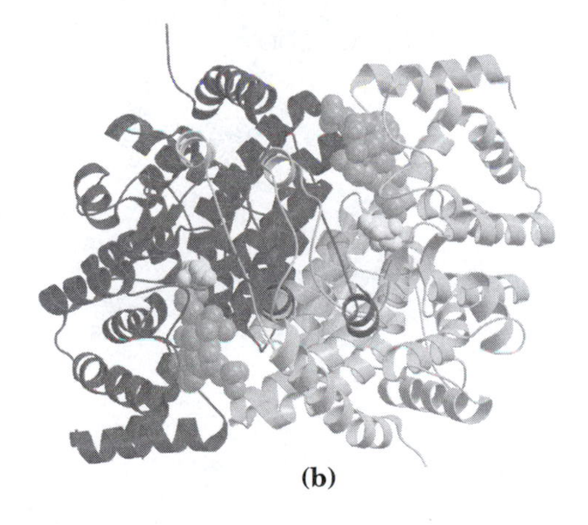

(b)

figure 16–8, page 574
Citrate synthase.

$$\text{Citrate} \underset{\text{aconitase}}{\overset{H_2O}{\rightleftharpoons}} [\textit{cis}\text{-Aconitate}] \underset{\text{aconitase}}{\overset{H_2O}{\rightleftharpoons}} \text{Isocitrate}$$

Citrate: $CH_2{-}COO^-$, $HO{-}C{-}COO^-$, $H{-}C{-}COO^-$, H

cis-Aconitate: $CH_2{-}COO^-$, $C{-}COO^-$ $\|$ $C{-}COO^-$, H

Isocitrate: $CH_2{-}COO^-$, $H{-}C{-}COO^-$, $HO{-}C{-}COO^-$, H

Aconitase, page 574

$\Delta G'^{\circ} = 13.3$ kJ/mol

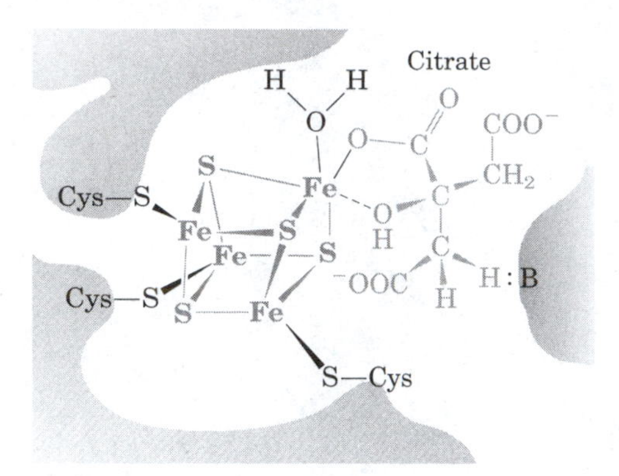

figure 16–9, page 574
Iron-sulfur center in aconitase.

$$\text{Isocitrate} \xrightleftharpoons[\text{isocitrate dehydrogenase}]{NAD(P)^+ \;\rightarrow\; NAD(P)H + H^+} \alpha\text{-Ketoglutarate} + CO_2$$

Isocitrate: $CH_2{-}COO^-$ / $H{-}C{-}COO^-$ / $HO{-}C{-}COO^-$ / H

α-Ketoglutarate: $CH_2{-}COO^-$ / CH_2 / $C(=O){-}COO^-$

Isocitrate dehydrogenase, page 574

$\Delta G'^{\circ} = -20.9$ kJ/mol

$$\alpha\text{-Ketoglutarate} \xrightarrow[\alpha\text{-ketoglutarate dehydrogenase complex}]{CoA\text{-}SH,\; NAD^+ \;\rightarrow\; NADH} \text{Succinyl-CoA} + CO_2$$

α-Ketoglutarate: $CH_2{-}COO^-$ / CH_2 / $C(=O){-}COO^-$

Succinyl-CoA: $CH_2{-}COO^-$ / CH_2 / $C(=O){-}S\text{-}CoA$

$\Delta G'^{\circ} = -33.5$ kJ/mol

α-Ketoglutarate dehydrogenase complex, page 575

$$\text{Succinyl-CoA} \xrightleftharpoons[\text{succinyl-CoA synthetase}]{GDP + P_i \;\rightarrow\; GTP,\; CoA\text{-}SH} \text{Succinate}$$

Succinyl-CoA: $CH_2{-}COO^-$ / CH_2 / $C(=O){-}S\text{-}CoA$

Succinate: COO^- / CH_2 / CH_2 / COO^-

Succinyl-CoA synthetase, page 575

$\Delta G'^{\circ} = -2.9$ kJ/mol

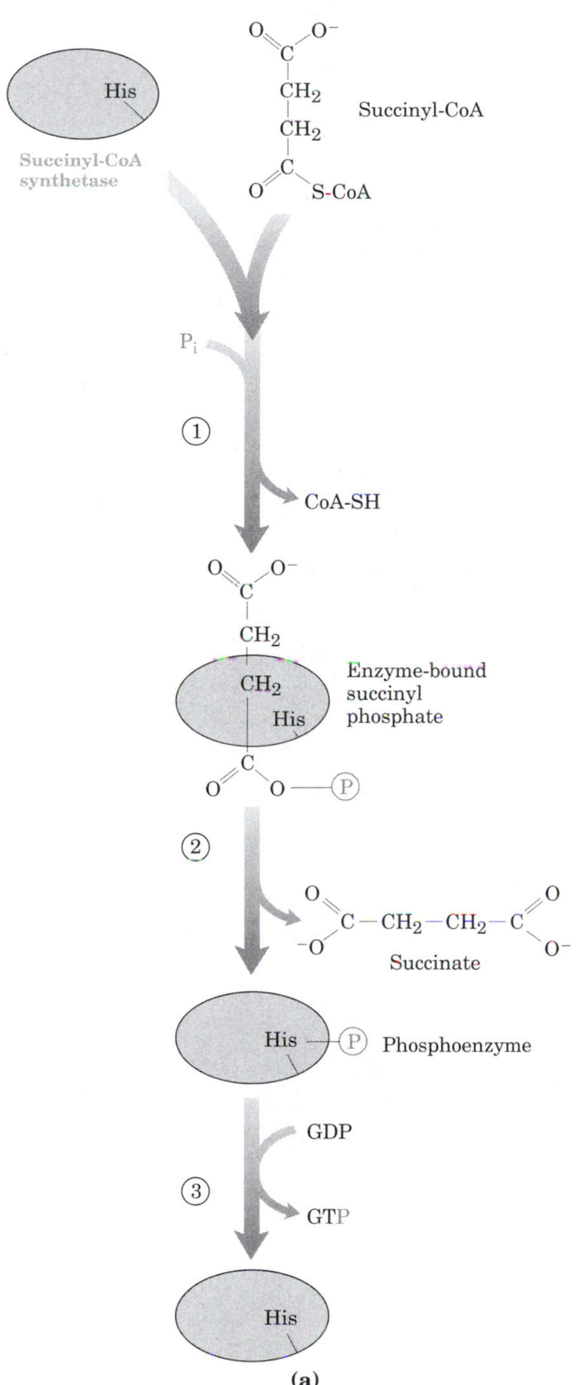

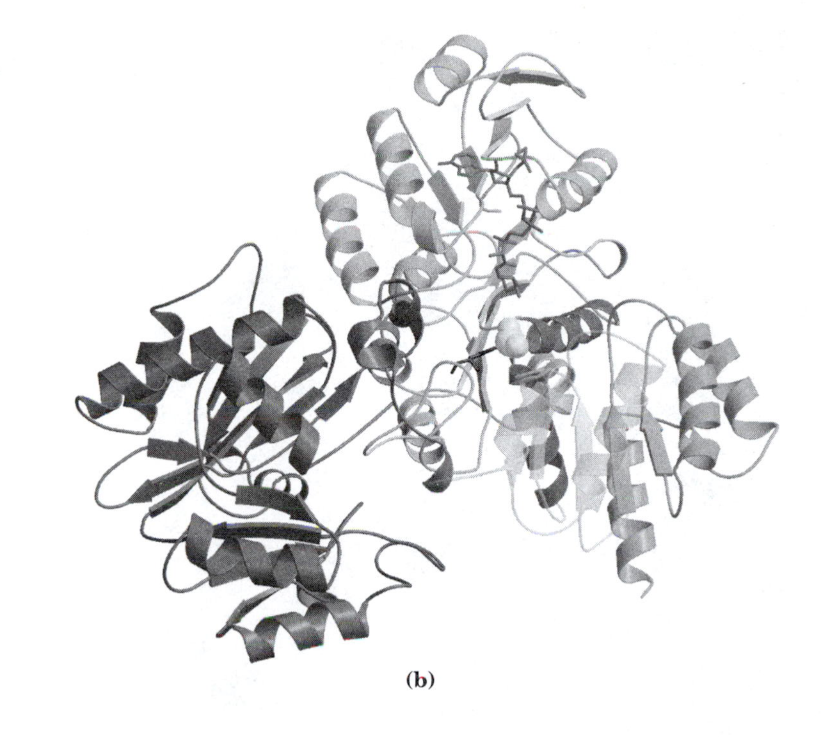

figure 16–10, page 577
The succinyl-CoA synthetase reaction.

$$\text{GTP} + \text{ADP} \xrightleftharpoons{\text{Mg}^{2+}} \text{GDP} + \text{ATP} \quad \Delta G'^{\circ} = 0 \text{ kJ/mol}$$

Nucleoside diphosphate kinase, page 578

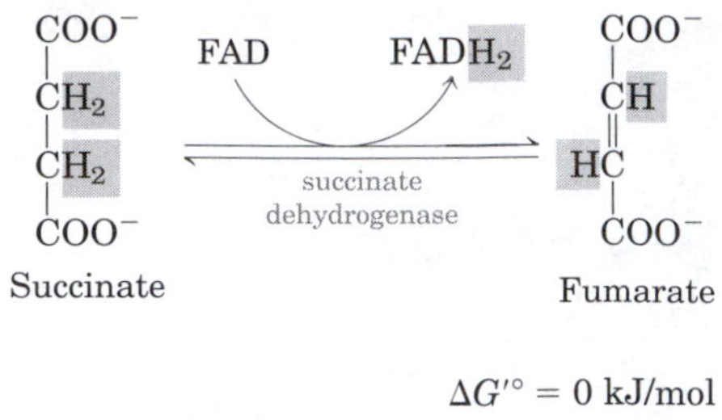

$$\begin{array}{c} COO^- \\ | \\ CH_2 \\ | \\ CH_2 \\ | \\ COO^- \end{array} \xrightleftharpoons[\text{succinate dehydrogenase}]{FAD \quad FADH_2} \begin{array}{c} COO^- \\ | \\ CH \\ \| \\ HC \\ | \\ COO^- \end{array}$$

Succinate → Fumarate

$\Delta G'^{\circ} = 0$ kJ/mol

Succinate dehydrogenase, page 578

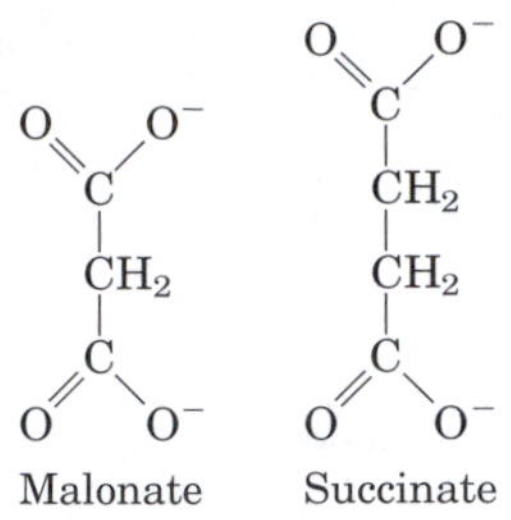

Malonate, Succinate

page 578

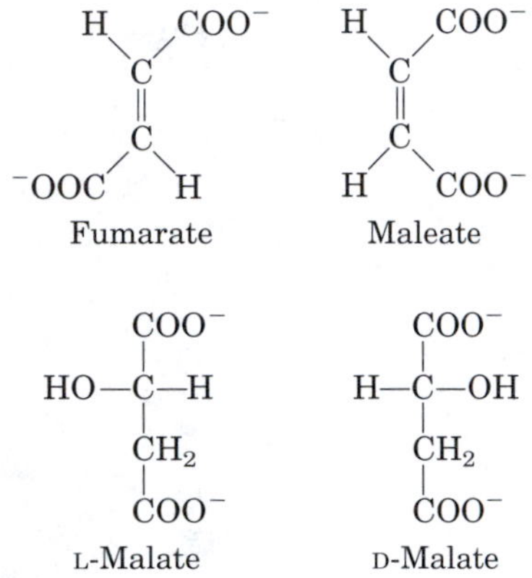

Fumarate, Maleate

L-Malate, D-Malate

page 578

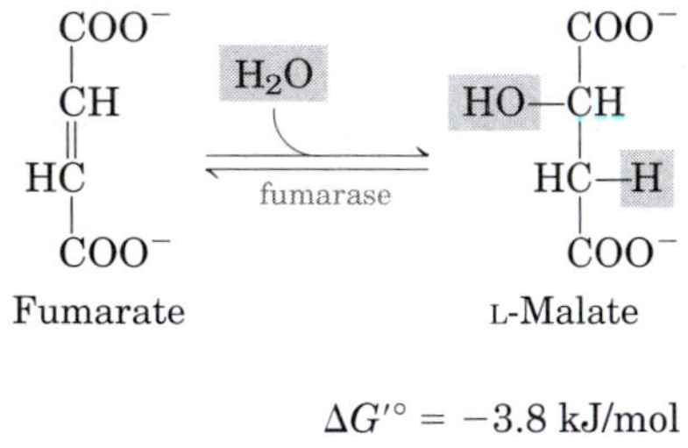

$$\begin{array}{c} COO^- \\ | \\ CH \\ \| \\ HC \\ | \\ COO^- \end{array} \xrightleftharpoons[\text{fumarase}]{H_2O} \begin{array}{c} COO^- \\ | \\ HO{-}CH \\ | \\ HC{-}H \\ | \\ COO^- \end{array}$$

Fumarate → L-Malate

$\Delta G'^{\circ} = -3.8$ kJ/mol

Fumarase, page 578

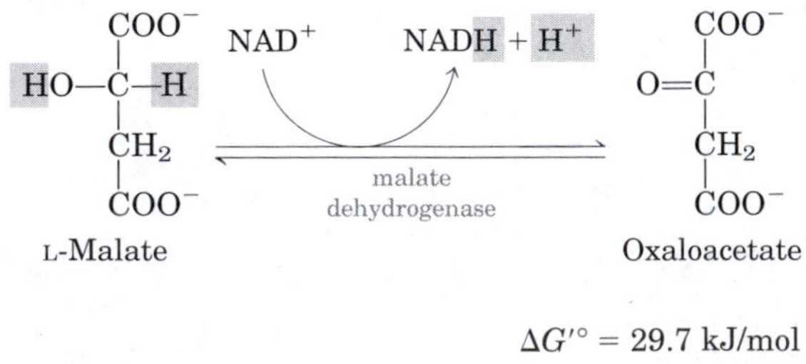

$$\begin{array}{c} COO^- \\ | \\ HO{-}C{-}H \\ | \\ CH_2 \\ | \\ COO^- \end{array} \xrightleftharpoons[\text{malate dehydrogenase}]{NAD^+ \quad NADH + H^+} \begin{array}{c} COO^- \\ | \\ O{=}C \\ | \\ CH_2 \\ | \\ COO^- \end{array}$$

L-Malate → Oxaloacetate

$\Delta G'^{\circ} = 29.7$ kJ/mol

Malate dehydrogenase, page 579

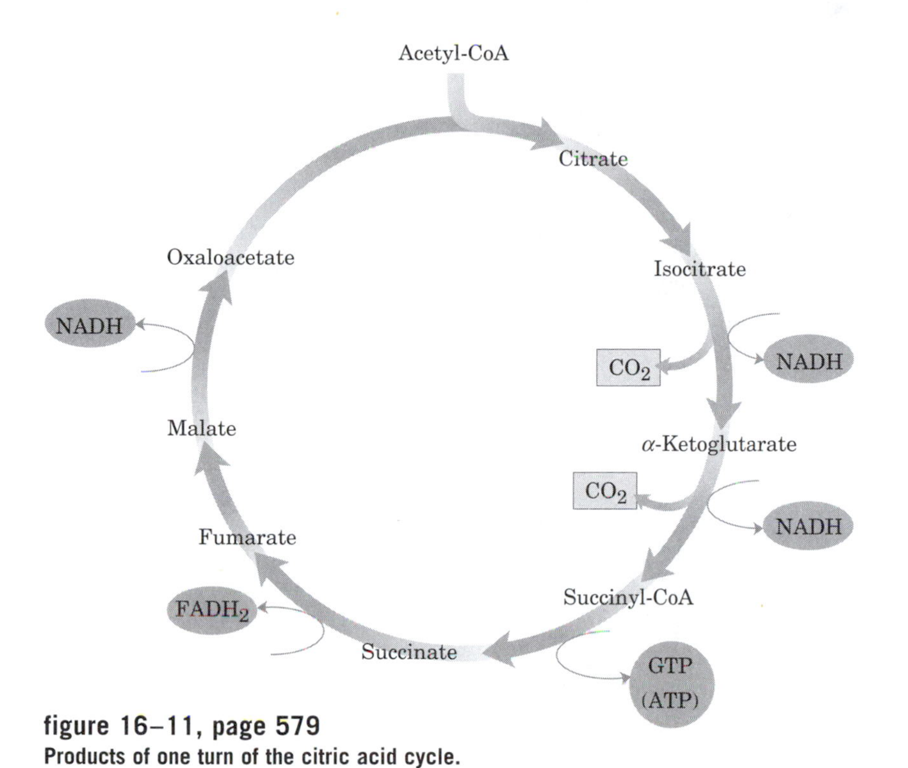

figure 16–11, page 579
Products of one turn of the citric acid cycle.

$CH_3-{}^{14}COO^-$
Labeled acetate
+
$O{=}C-COO^-$
CH_2-COO^-
Oxaloacetate

⟶

$CH_2-{}^{14}COO^-$
$HO-C-COO^-$
CH_2-COO^-
Labeled citrate

① ⟶ $CH_2-{}^{14}COO^-$ / $CH-COO^-$ / $HO-C-COO^-$ ⟶ ${}^{\gamma}CH_2-{}^{14}COO^-$ / ${}^{\beta}CH_2$ / $O{=}{}^{\alpha}C-COO^-$

Only this product was formed.

② ⟶ CH_2-COO^- / $CH-COO^-$ / $HO-CH-{}^{14}COO^-$ ⟶ ${}^{\gamma}CH_2-COO^-$ / ${}^{\beta}CH-COO^-$ / $O{=}{}^{\alpha}C-{}^{14}COO^-$

Isocitrate

This second form of labeled α-ketoglutarate was also expected, but was not formed.

box 16–2, figure 1, page 580
Incorporation of the isotopic carbon (^{14}C) of the labeled acetyl group into α-ketoglutarate by the citric acid cycle.

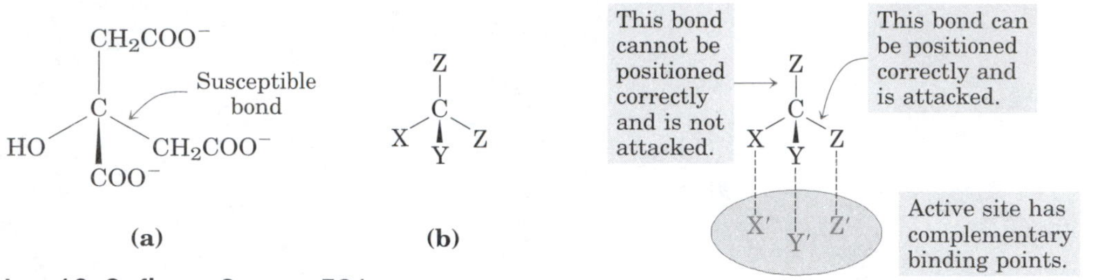

box 16–2, figure 2, page 581
The prochiral nature of citrate.

table 16–1, page 582

Stoichiometry of Coenzyme Reduction and ATP Formation in the Aerobic Oxidation of Glucose via Glycolysis, the Pyruvate Dehydrogenase Reaction, the Citric Acid Cycle, and Oxidative Phosphorylation

Reaction	Number of ATP or reduced coenzymes directly formed	Number of ATP ultimately formed*
Glucose ⟶ glucose 6-phosphate	−1 ATP	−1
Fructose 6-phosphate ⟶ fructose 1,6-bisphosphate	−1 ATP	−1
2 Glyceraldehyde 3-phosphate ⟶ 2 1,3-bisphosphoglycerate	2 NADH	3–5
2 1,3-Bisphosphoglycerate ⟶ 2 3-phosphoglycerate	2 ATP	2
2 Phosphoenolpyruvate ⟶ 2 pyruvate	2 ATP	2
2 Pyruvate ⟶ 2 acetyl-CoA	2 NADH	5
2 Isocitrate ⟶ 2 α-ketoglutarate	2 NADH	5
2 α-Ketoglutarate ⟶ 2 succinyl-CoA	2 NADH	5
2 Succinyl-CoA ⟶ 2 succinate	2 ATP (or 2 GTP)	2
2 Succinate ⟶ 2 fumarate	2 $FADH_2$	3
2 Malate ⟶ 2 oxaloacetate	2 NADH	5
Total		30–32

*This is calculated as 2.5 ATP per NADH and 1.5 ATP per $FADH_2$. A negative value indicates consumption.

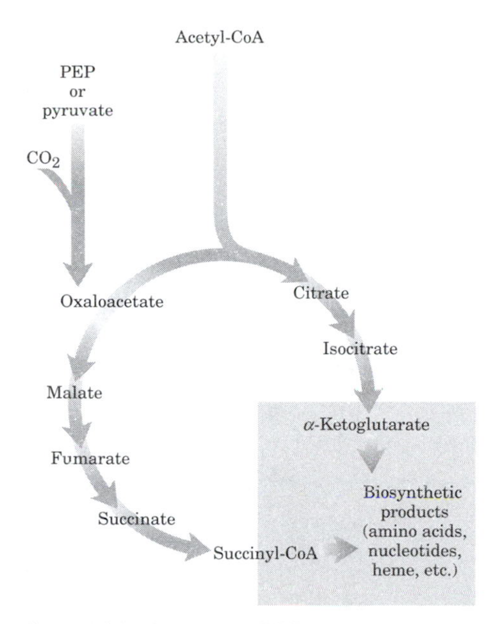

figure 16–12, page 582
Biosynthetic precursors produced by an incomplete citric acid cycle in anaerobic bacteria.

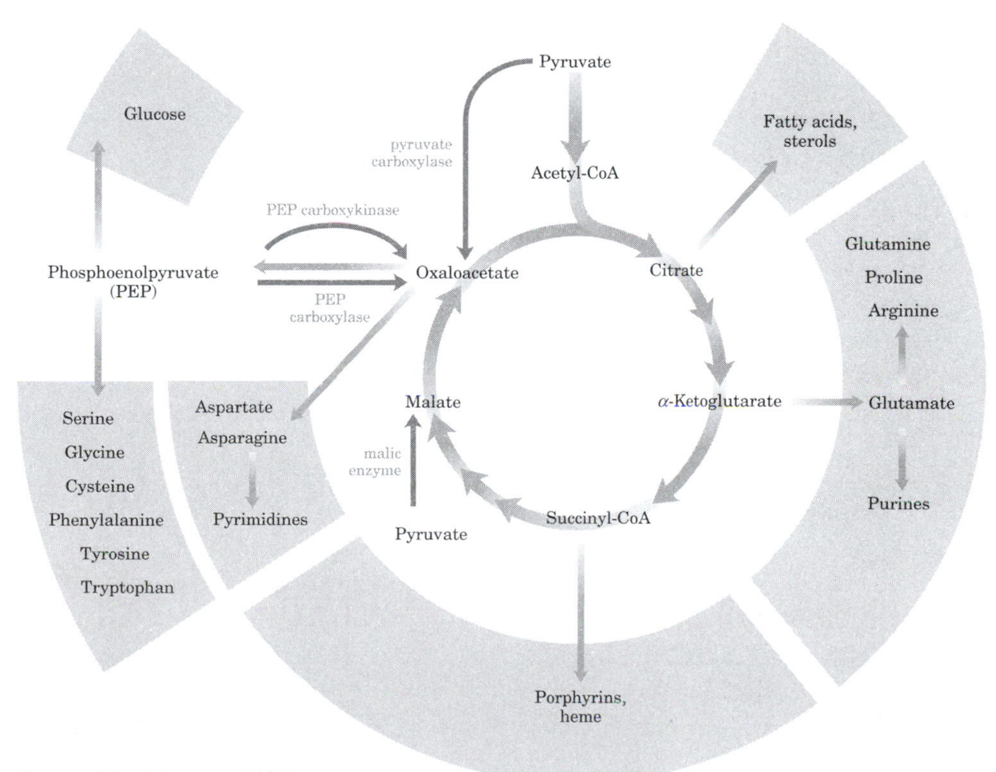

figure 16–13, page 584
Role of the citric acid cycle in anabolism.

table 16–2, page 585

Anaplerotic Reactions

Reaction	Tissue(s)/organism(s)
Pyruvate + HCO_3^- + ATP $\xrightleftharpoons{\text{pyruvate carboxylase}}$ oxaloacetate + ADP + P_i	Liver, kidney
Phosphoenolpyruvate + CO_2 + GDP $\xrightleftharpoons{\text{PEP carboxykinase}}$ oxaloacetate + GTP	Heart, skeletal muscle
Phosphoenolpyruvate + HCO_3^- $\xrightleftharpoons{\text{PEP carboxylase}}$ oxaloacetate + P_i	Higher plants, yeast, bacteria
Pyruvate + HCO_3^- + NAD(P)H $\xrightleftharpoons{\text{malic enzyme}}$ malate + $NAD(P)^+$	Widely distributed in eukaryotes and prokaryotes

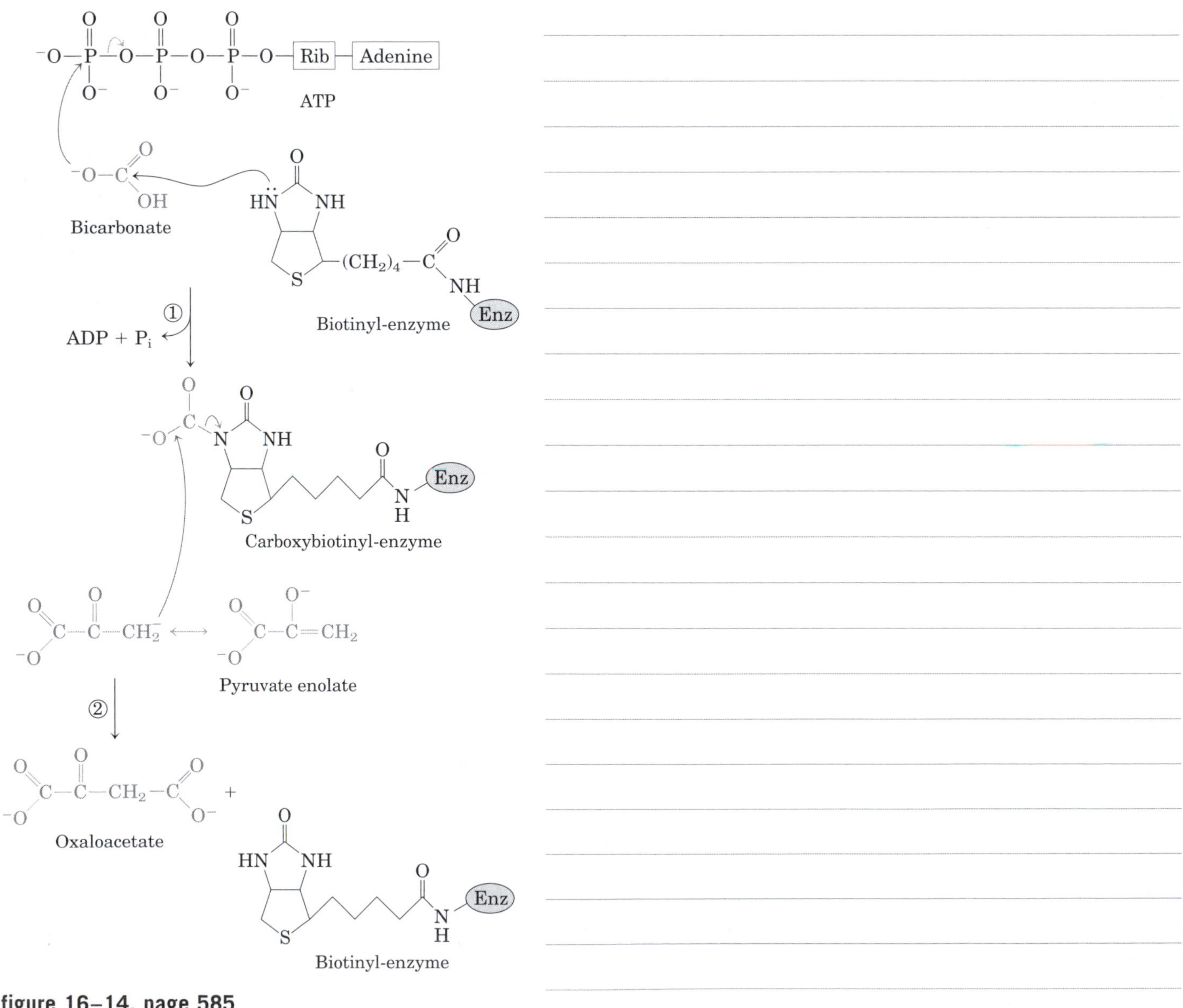

figure 16–14, page 585
Role of biotin in the pyruvate carboxylase reaction.

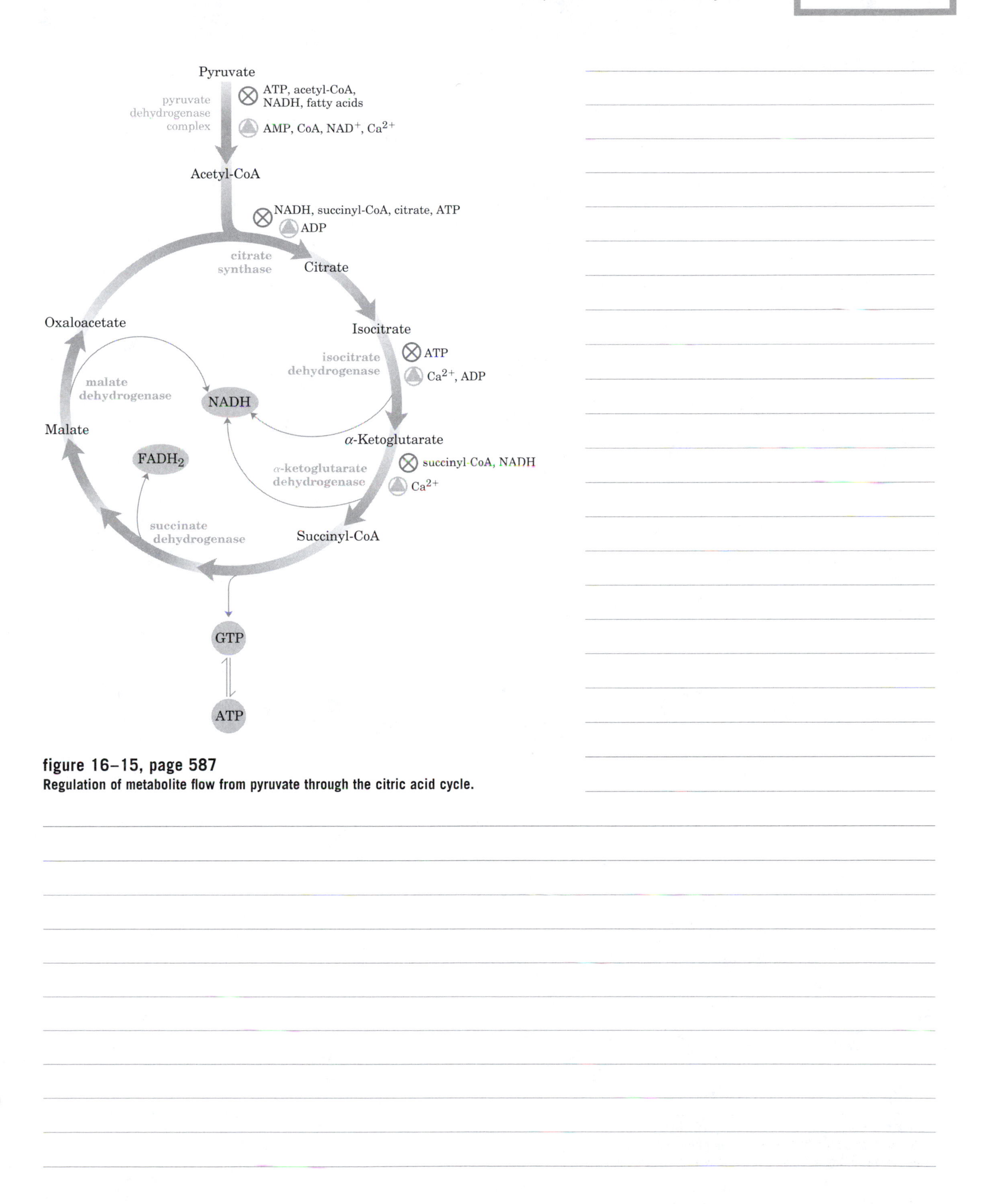

figure 16–15, page 587
Regulation of metabolite flow from pyruvate through the citric acid cycle.

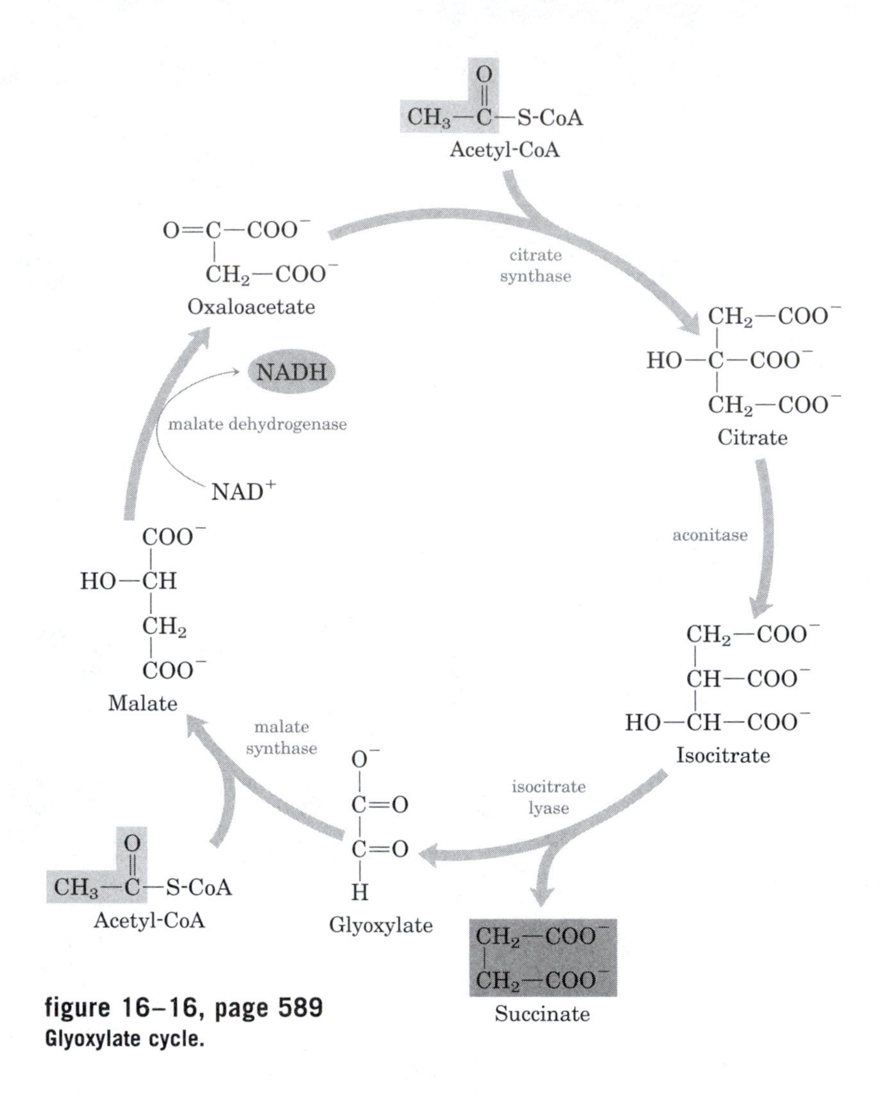

figure 16–16, page 589
Glyoxylate cycle.

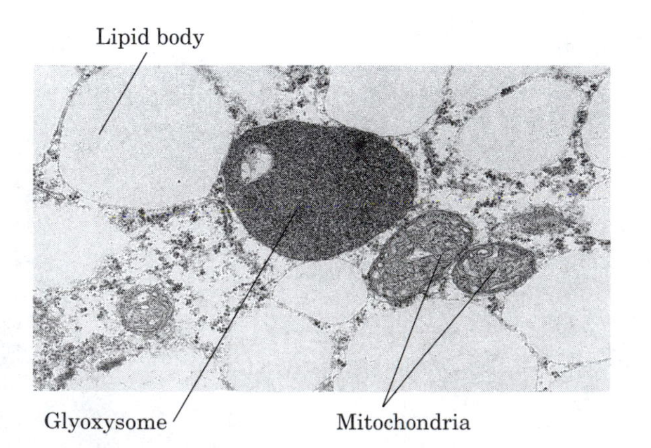

figure 16–17, page 590
Electron micrograph of a germinating cucumber seed, showing a glyoxysome, mitochondria, and surrounding lipid bodies.

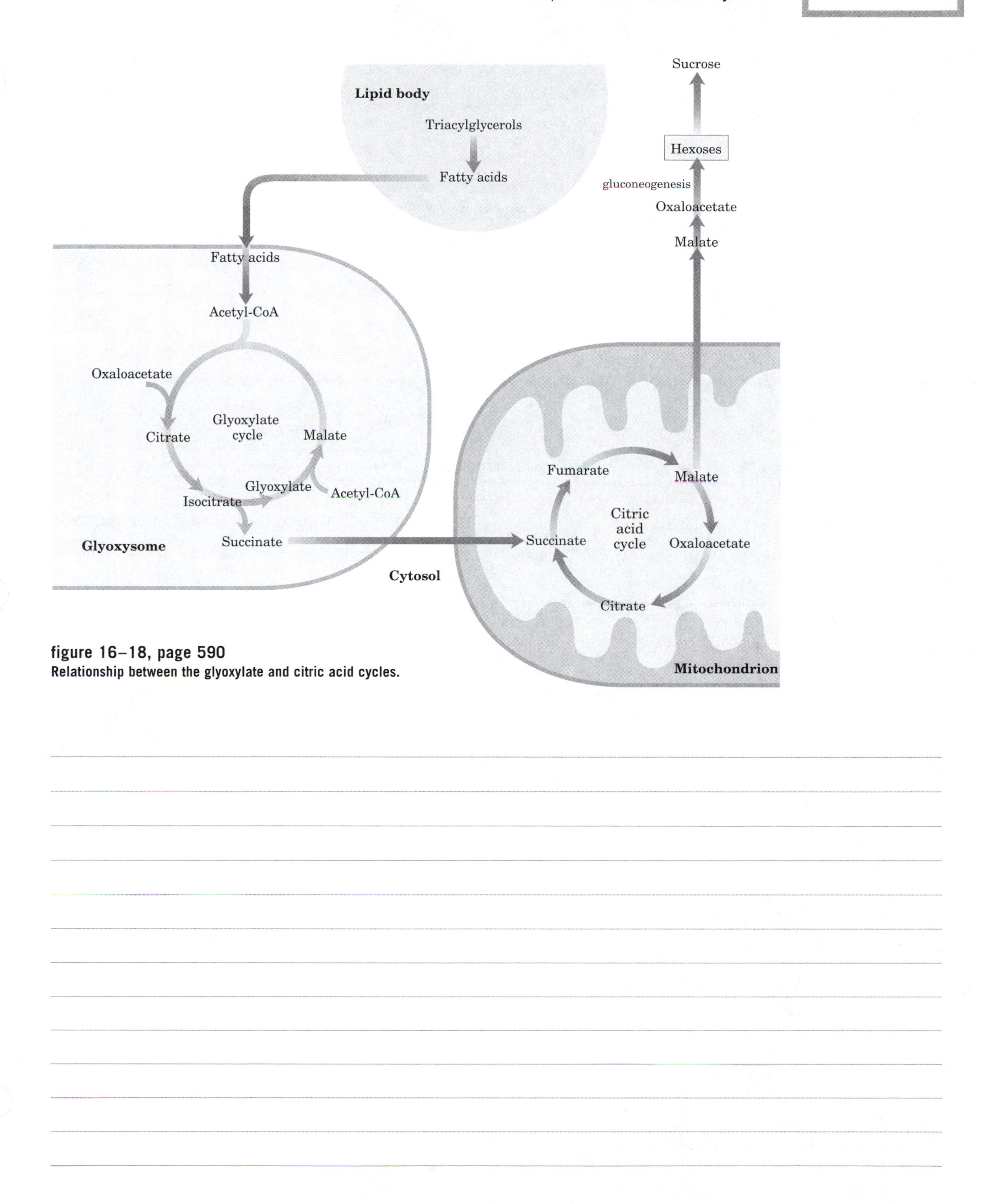

figure 16–18, page 590
Relationship between the glyoxylate and citric acid cycles.

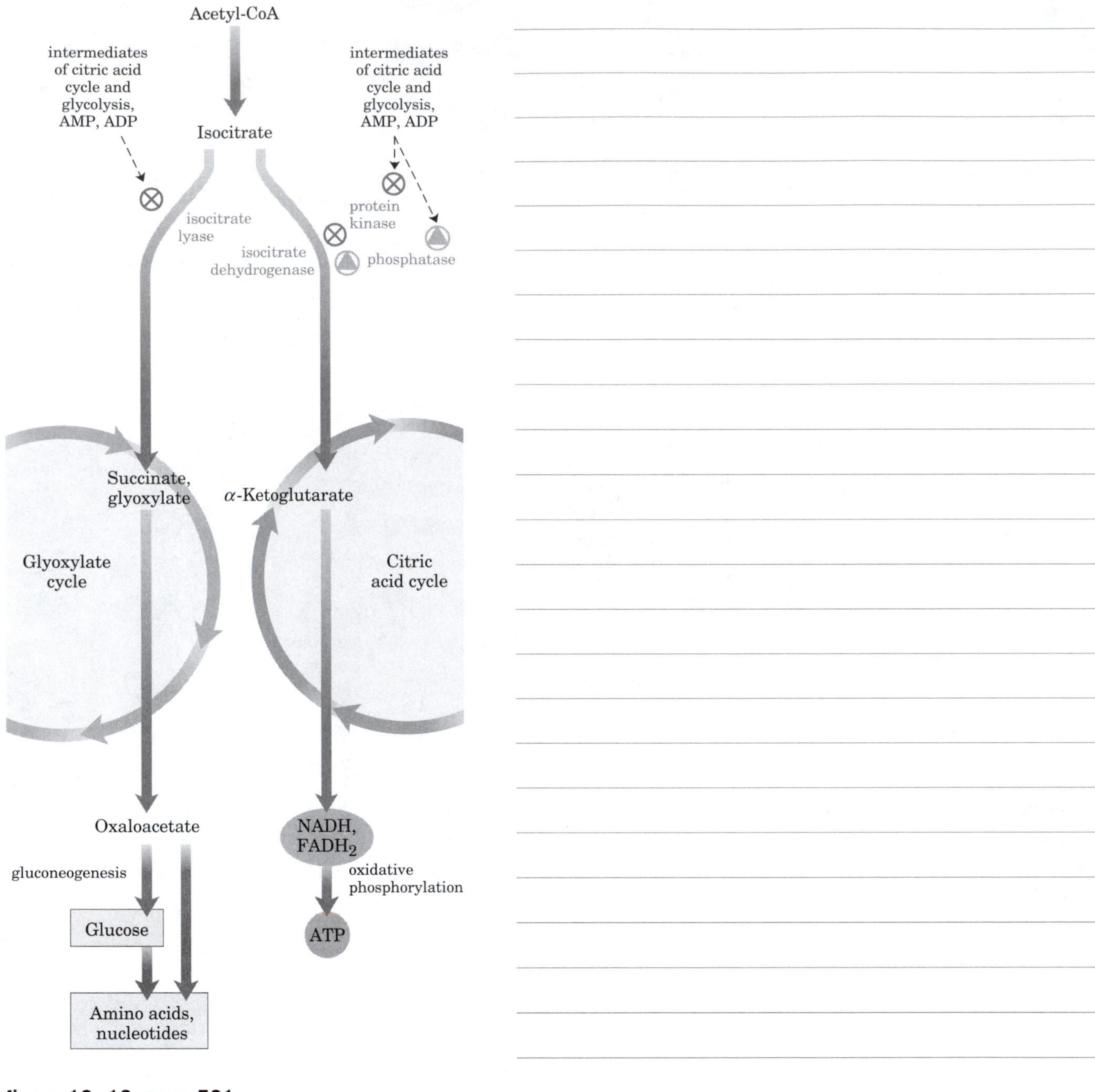

figure 16–19, page 591

Regulation of isocitrate dehydrogenase activity that determines partitioning of isocitrate between the glyoxylate and citric acid cycles.

PROBLEMS

2. Recognizing Oxidation and Reduction Reactions, page 594.

$$CH_3{-}\overset{\overset{\large O}{\|}}{C}{-}H + H{-}H \underset{\text{oxidation}}{\overset{\text{reduction}}{\rightleftharpoons}} \left[CH_3{-}\overset{\overset{\large O}{\|}}{C}(H) \cdots H{-}H \right] \underset{\text{oxidation}}{\overset{\text{reduction}}{\rightleftharpoons}} CH_3{-}\overset{O-H}{\underset{H}{C}}{-}H \qquad (1)$$

Acetaldehyde Ethanol

$$CH_3{-}C(=O)O^- + H^+ + H{-}H \underset{\text{oxidation}}{\overset{\text{reduction}}{\rightleftharpoons}} \left[CH_3{-}C(=O)O^- \;\; H{-}H \;\; H^+ \right] \underset{\text{oxidation}}{\overset{\text{reduction}}{\rightleftharpoons}} CH_3{-}\overset{\overset{\large O}{\|}}{C}{-}H + H{-}O{-}H \qquad (2)$$

Acetate Acetaldehyde

(a) $CH_3{-}OH \longrightarrow H{-}\overset{\overset{\large O}{\|}}{C}{-}H$

Methanol Formaldehyde

(b) $H{-}\overset{\overset{\large O}{\|}}{C}{-}H \longrightarrow H{-}C(=O)O^- + H^+$

Formaldehyde Formate

(c) $O{=}C{=}O \longrightarrow H{-}C(=O)O^- + H^+$

Carbon dioxide Formate

(d) $CH_2(OH){-}CH(OH){-}C(=O)O^- + H^+ \longrightarrow CH_2(OH){-}CH(OH){-}C(=O)H$

Glycerate Glyceraldehyde

(e) $CH_2(OH){-}CH(OH){-}CH_2(OH) \longrightarrow CH_2(OH){-}\overset{\overset{\large O}{\|}}{C}{-}CH_2(OH)$

Glycerol Dihydroxyacetone

(f) $C_6H_5{-}CH_3 \longrightarrow C_6H_5{-}C(=O)O^- + H^+$

Toluene Benzoate

(g) $^-O(O{=})C{-}CH_2{-}CH_2{-}C(=O)O^- \longrightarrow {}^-O{-}C(=O){-}CH{=}CH{-}C(=O)O^-$

Succinate Fumarate

(h) $CH_3{-}\overset{\overset{\large O}{\|}}{C}{-}C(=O)O^- \longrightarrow CH_3{-}C(=O)O^- + CO_2$

Pyruvate Acetate

4. Nicotinamide Coenzymes as Reversible Redox Carriers, page 595.

$$\underset{\text{Oxidized}}{\text{Substrate}} + \underset{\text{Reduced}}{\text{NADH}} + H^+ \rightleftharpoons \underset{\text{Reduced}}{\text{product}} + \underset{\text{Oxidized}}{\text{NAD}^+}$$

(a) $CH_3CH_2OH \longrightarrow CH_3{-}C(=O){-}H$

Ethanol → Acetaldehyde

(b) $^{2-}O_3PO{-}CH_2{-}CH(OH){-}C(=O){-}OPO_3^{2-} \longrightarrow {}^{2-}O_3PO{-}CH_2{-}CH(OH){-}C(=O){-}H + HPO_4^{2-}$

1,3-Bisphosphoglycerate → Glyceraldehyde 3-phosphate

(c) $CH_3{-}C(=O){-}C(=O){-}O^- \longrightarrow CH_3{-}C(=O^-){-}H + CO_2$

Pyruvate → Acetaldehyde

(d) $CH_3{-}C(=O){-}C(=O){-}O^- \longrightarrow CH_3{-}C(=O){-}O^- + CO_2$

Pyruvate → Acetate

(e) $^-OOC{-}CH_2{-}C(=O){-}COO^- \longrightarrow {}^-OOC{-}CH_2{-}CH(OH){-}COO^-$

Oxaloacetate → Malate

(f) $CH_3{-}C(=O){-}CH_2{-}C(=O){-}O^- + H^+ \longrightarrow CH_3{-}C(=O){-}CH_3 + CO_2$

Acetoacetate → Acetone

6. Formation of Oxaloacetate in a Mitochondrion, page 595.

L-Malate + $NAD^+ \longrightarrow$ oxaloacetate + NADH + H^+

$\Delta G'^{\circ} = 30$ kJ/mol

$$K'_{eq} = \frac{[\text{oxaloacetate}][\text{NADH}]}{[\text{L-malate}][\text{NAD}^+]}$$

13. Mode of Action of the Rodenticide Fluoroacetate, page 596.

$F{-}CH_2COO^- + \text{CoA-SH} + \text{ATP} \longrightarrow F{-}CH_2C(=O){-}\text{S-CoA} + \text{AMP} + PP_i$

17. Commercial Synthesis of Citric Acid, page 597.

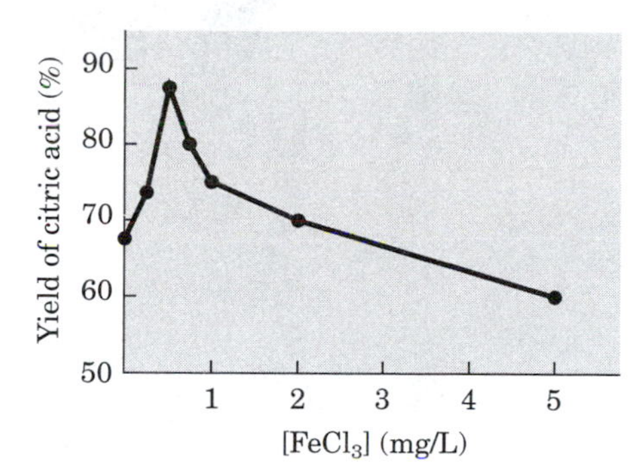

18. Regulation of Citrate Synthase, page 597.

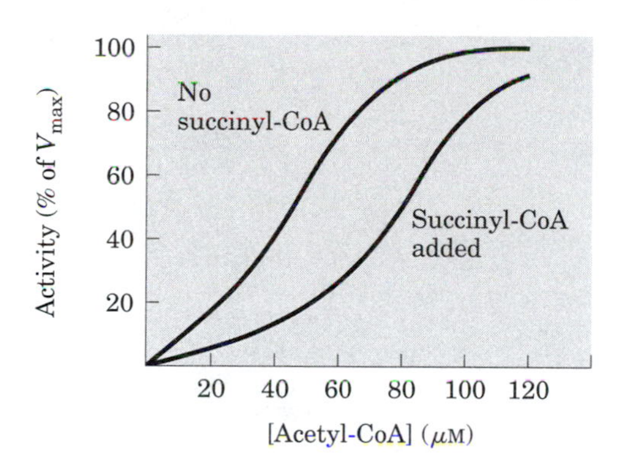

23. Thermodynamics of Citrate Synthase Reaction in Cells, page 597.

Oxaloacetate + acetyl-CoA + H_2O $\longrightarrow$ Citrate + CoA + H^+

chapter

17 Oxidation of Fatty Acids

CD-ROM

When you study for exams, don't forget to review The Minicourses on the UNDERSTAND! BIOCHEMISTRY CD that came with your textbook.

Minicourses that apply to this Chapter include:

Molecules of Life

An Interactive Gallery of Molecular Models

An Interactive Gallery of Protein Structures

Bioenergetics

Fatty Acid Breakdown

O
CH_3
C
N
SO_3^-
OH
H
CH_3
CH_3
HO
OH

Taurocholic acid

Taurocholic Acid, page 599

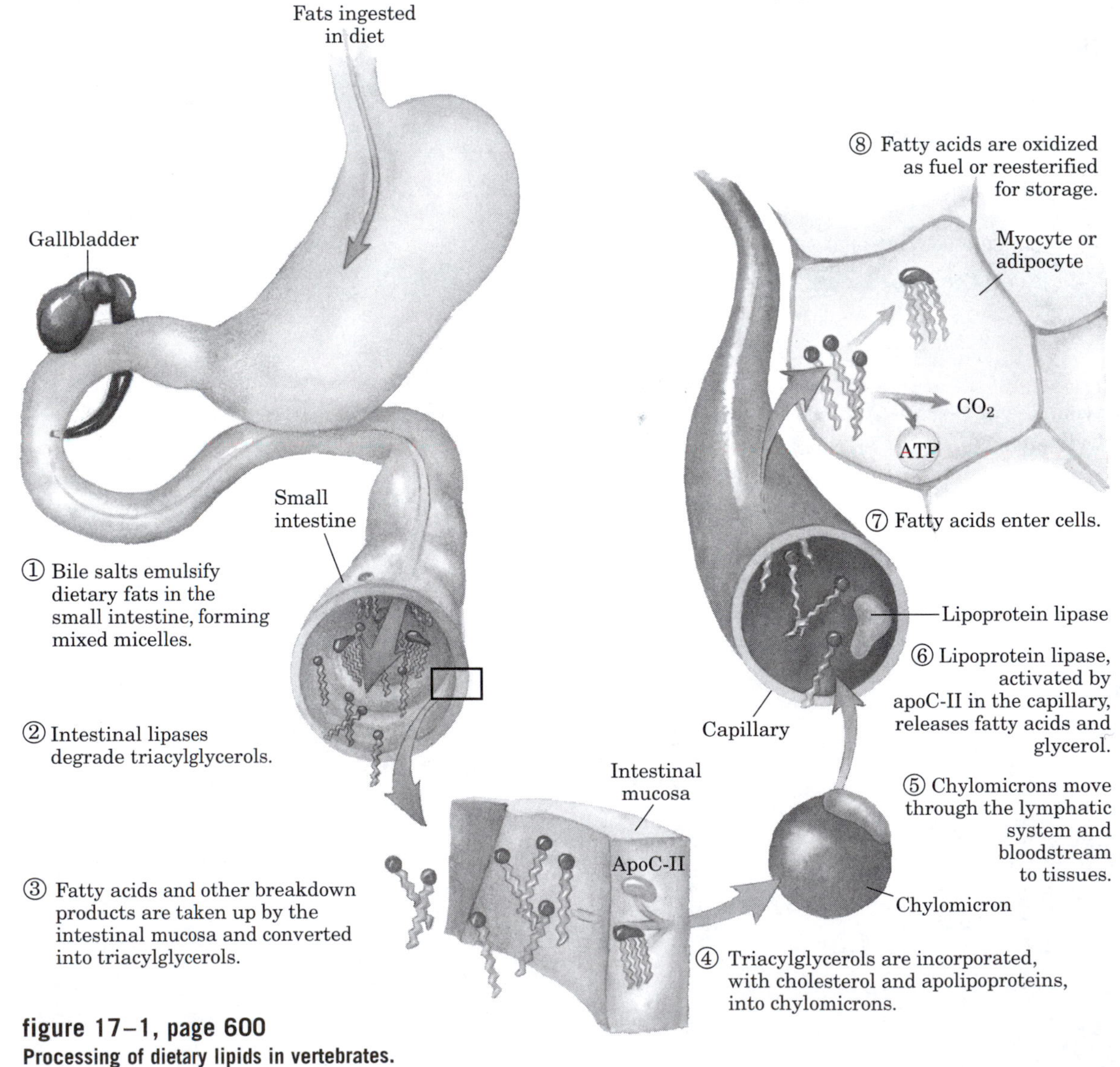

figure 17–1, page 600
Processing of dietary lipids in vertebrates.

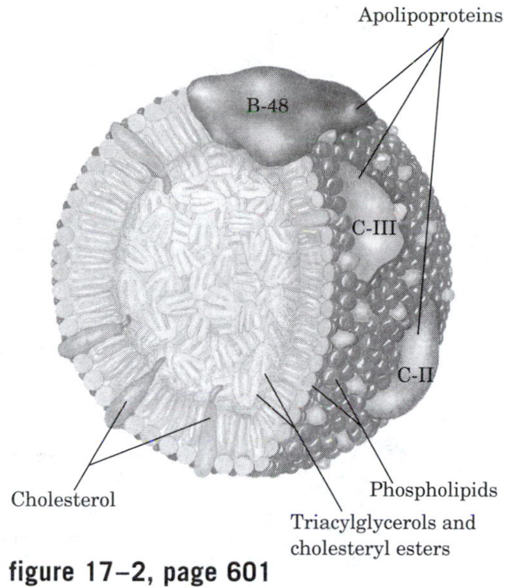

figure 17–2, page 601
Molecular structure of a chylomicron.

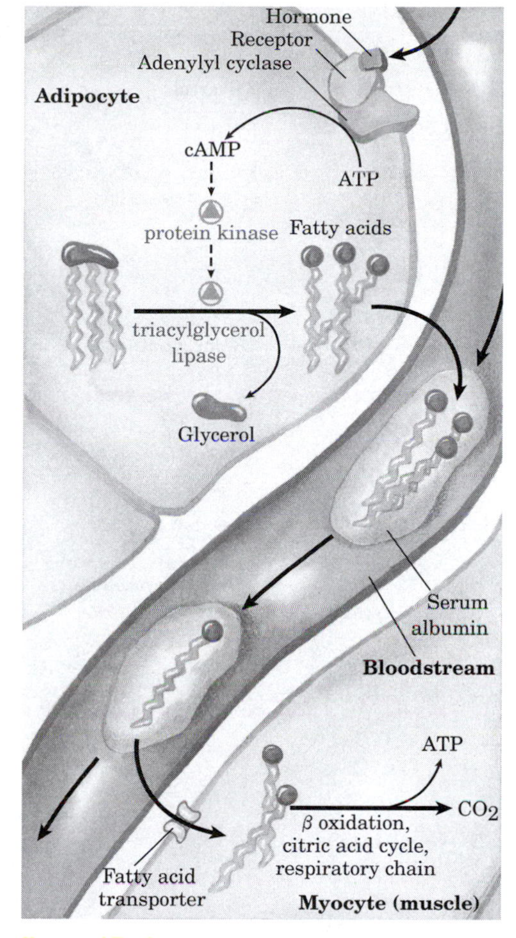

figure 17–3, page 601
Mobilization of triacylglycerols stored in adipose tissue.

Glycerol

glycerol kinase: ATP → ADP

L-Glycerol 3-phosphate

glycerol 3-phosphate dehydrogenase: NAD^+ → $NADH + H^+$

Dihydroxyacetone phosphate

triose phosphate isomerase

D-Glyceraldehyde 3-phosphate

↓

Glycolysis

figure 17–4, page 602
Entry of glycerol into the glycolytic pathway.

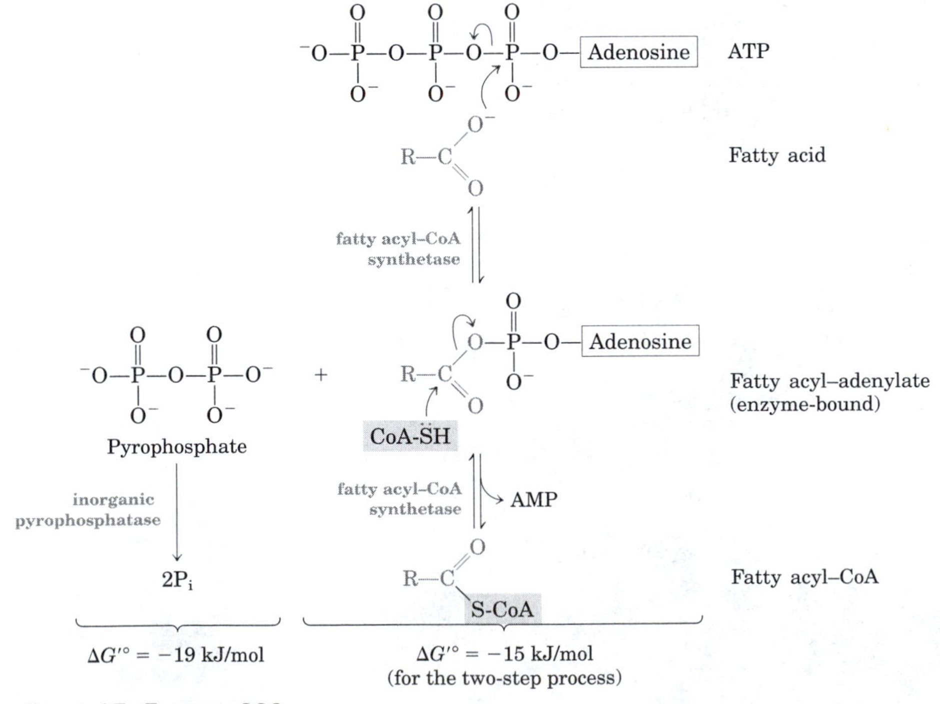

figure 17–5, page 602
Conversion of a fatty acid to a fatty acyl–CoA.

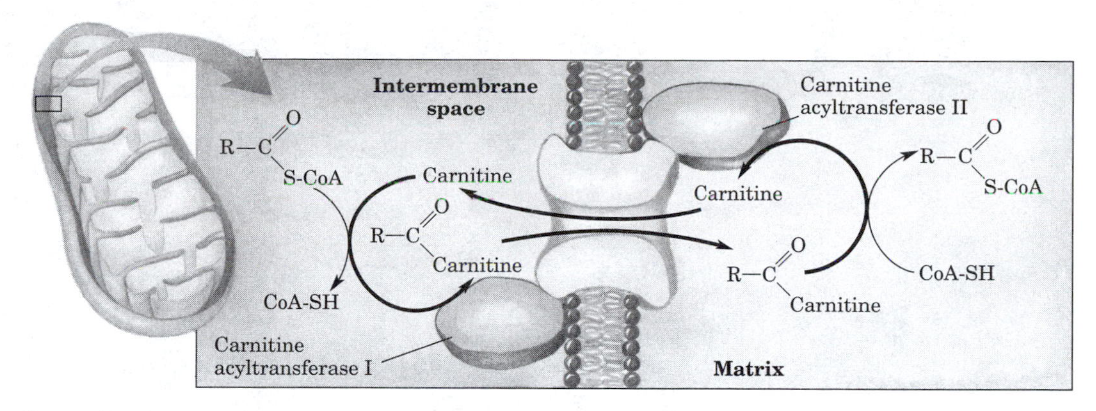

figure 17–6, page 603
Fatty acid entry into mitochondria via the acyl-carnitine/ carnitine transporter.

$$CH_3-\overset{\overset{\displaystyle CH_3}{|}}{\underset{\underset{\displaystyle CH_3}{|}}{N^+}}-CH_2-\underset{\underset{\displaystyle OH}{|}}{CH}-CH_2-COO^-$$

Carnitine, page 603

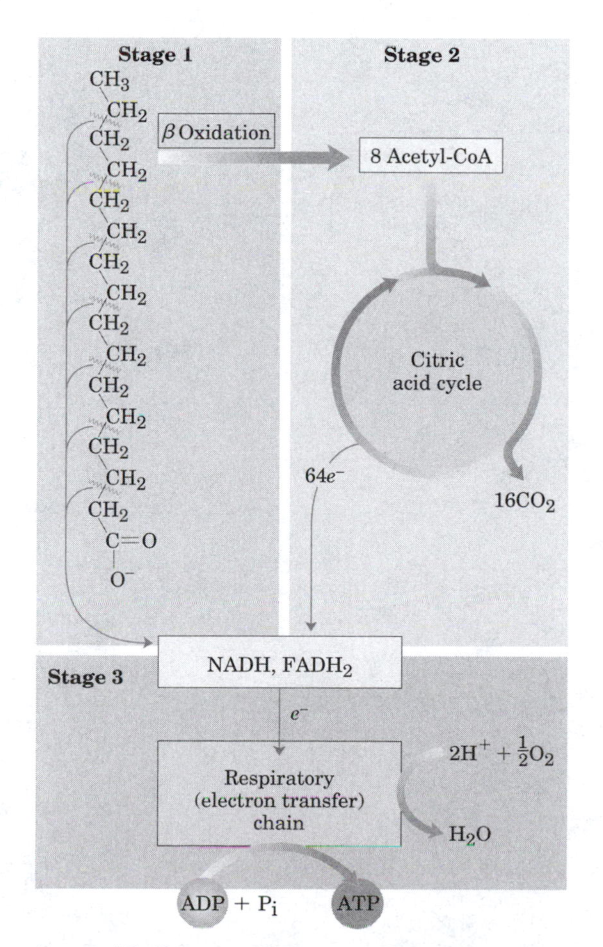

figure 17–7, page 604
Stages of fatty acid oxidation.

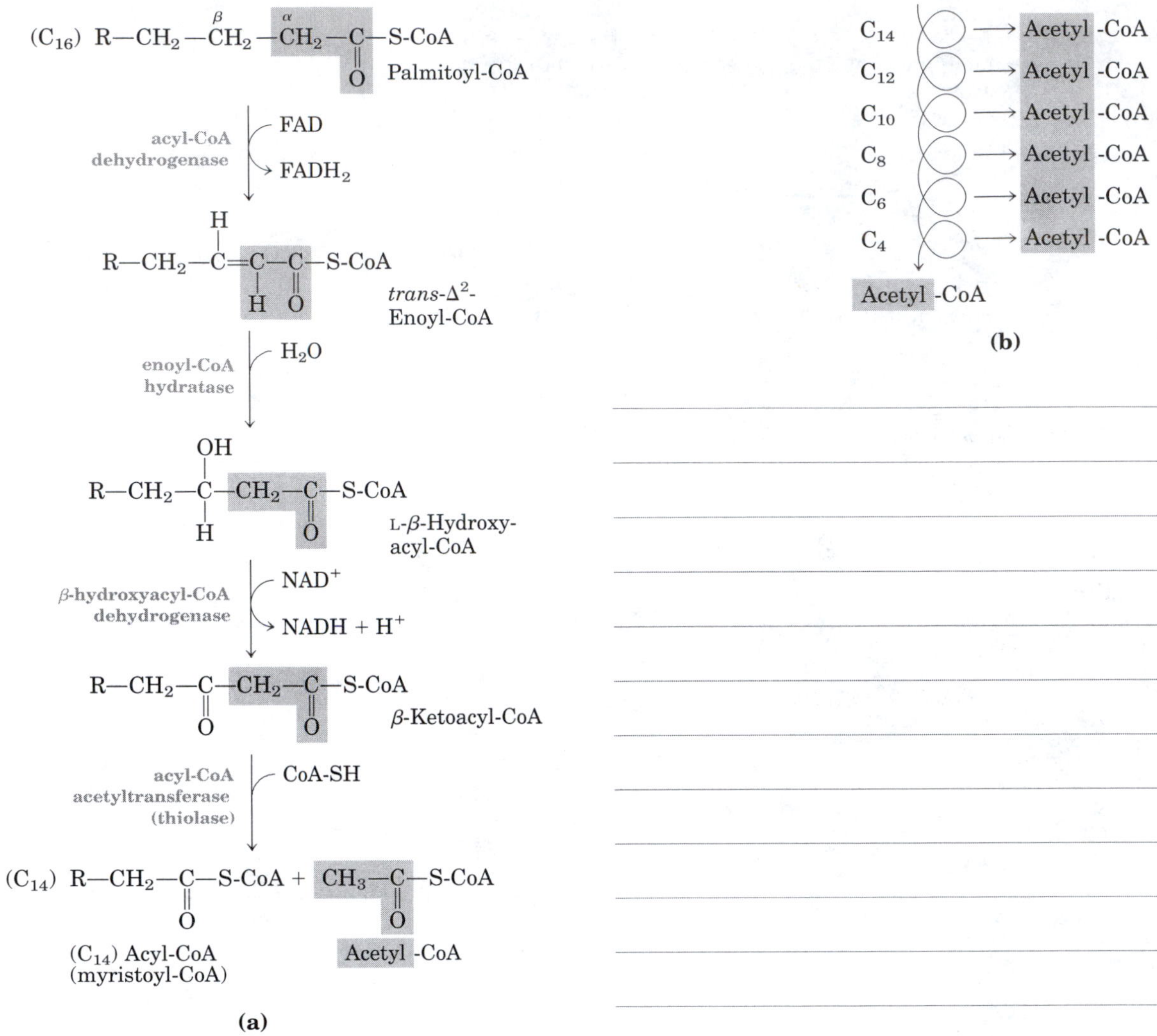

(a)

(C_{16}) R—CH_2—CH_2(β)—CH_2(α)—C(=O)—S-CoA Palmitoyl-CoA

acyl-CoA dehydrogenase: FAD → $FADH_2$

R—CH_2—C(H)=C(H)—C(=O)—S-CoA *trans*-Δ^2-Enoyl-CoA

enoyl-CoA hydratase: H_2O

R—CH_2—C(OH)(H)—CH_2—C(=O)—S-CoA L-β-Hydroxy-acyl-CoA

β-hydroxyacyl-CoA dehydrogenase: NAD^+ → NADH + H^+

R—CH_2—C(=O)—CH_2—C(=O)—S-CoA β-Ketoacyl-CoA

acyl-CoA acetyltransferase (thiolase): CoA-SH

(C_{14}) R—CH_2—C(=O)—S-CoA + CH_3—C(=O)—S-CoA

(C_{14}) Acyl-CoA (myristoyl-CoA) Acetyl -CoA

(b)

C_{14} → Acetyl -CoA
C_{12} → Acetyl -CoA
C_{10} → Acetyl -CoA
C_8 → Acetyl -CoA
C_6 → Acetyl -CoA
C_4 → Acetyl -CoA
Acetyl -CoA

figure 17–8, page 605
The β-oxidation pathway.

Palmitoyl-CoA + 7CoA + $7O_2$ + $28P_i$ + 28ADP ⟶
8 acetyl-CoA + 28ATP + $7H_2O$ (Equation 17–4)

8 Acetyl-CoA + $16O_2$ + $80P_i$ + 80ADP ⟶
8CoA + 80ATP + $16H_2O$ + $16CO_2$ (Equation 17–5)

Palmitoyl-CoA + $23O_2$ + $108P_i$ + 108ADP ⟶
CoA + 108ATP + $16CO_2$ + $23H_2O$ (Equation 17–6)

Oxidation of palmitoyl-CoA, pages 606–607

table 17–1, page 607

Yield of ATP during Oxidation of One Molecule of Palmitoyl-CoA to CO_2 and H_2O

Enzyme catalyzing the oxidation step	Number of NADH or $FADH_2$ formed	Number of ATP ultimately formed*
Acyl-CoA dehydrogenase	7 $FADH_2$	10.5
β-Hydroxyacyl-CoA dehydrogenase	7 NADH	17.5
Isocitrate dehydrogenase	8 NADH	20
α-Ketoglutarate dehydrogenase	8 NADH	20
Succinyl-CoA synthetase		8†
Succinate dehydrogenase	8 $FADH_2$	12
Malate dehydrogenase	8 NADH	20
Total		108

*These calculations assume that mitochondrial oxidative phosphorylation produces 1.5 ATP per $FADH_2$ oxidized and 2.5 ATP per NADH oxidized.

†GTP produced directly in this step yields ATP in the reaction catalyzed by nucleoside diphosphate kinase (p. 578).

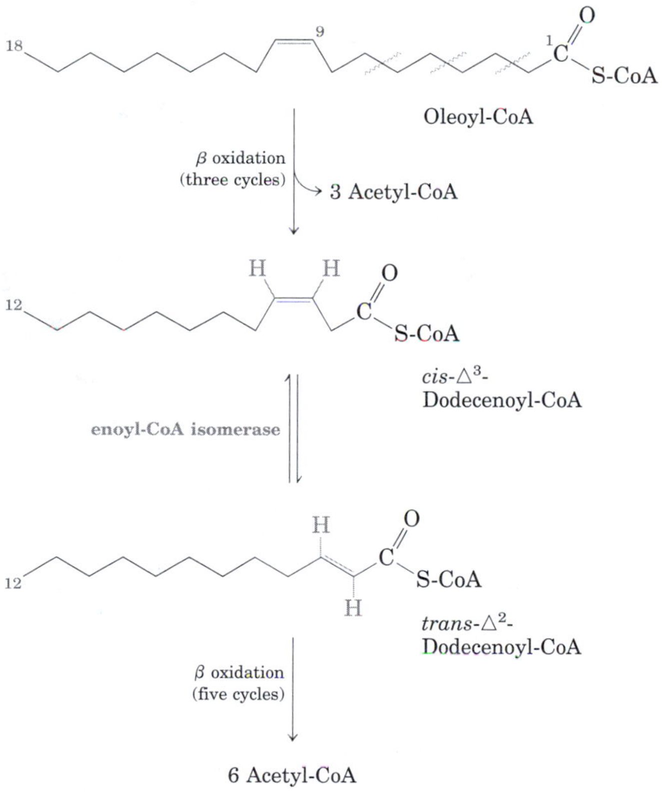

figure 17–9, page 608
Oxidation of a monounsaturated fatty acid.

$CH_3—CH_2—COO^-$

Propionate, page 608

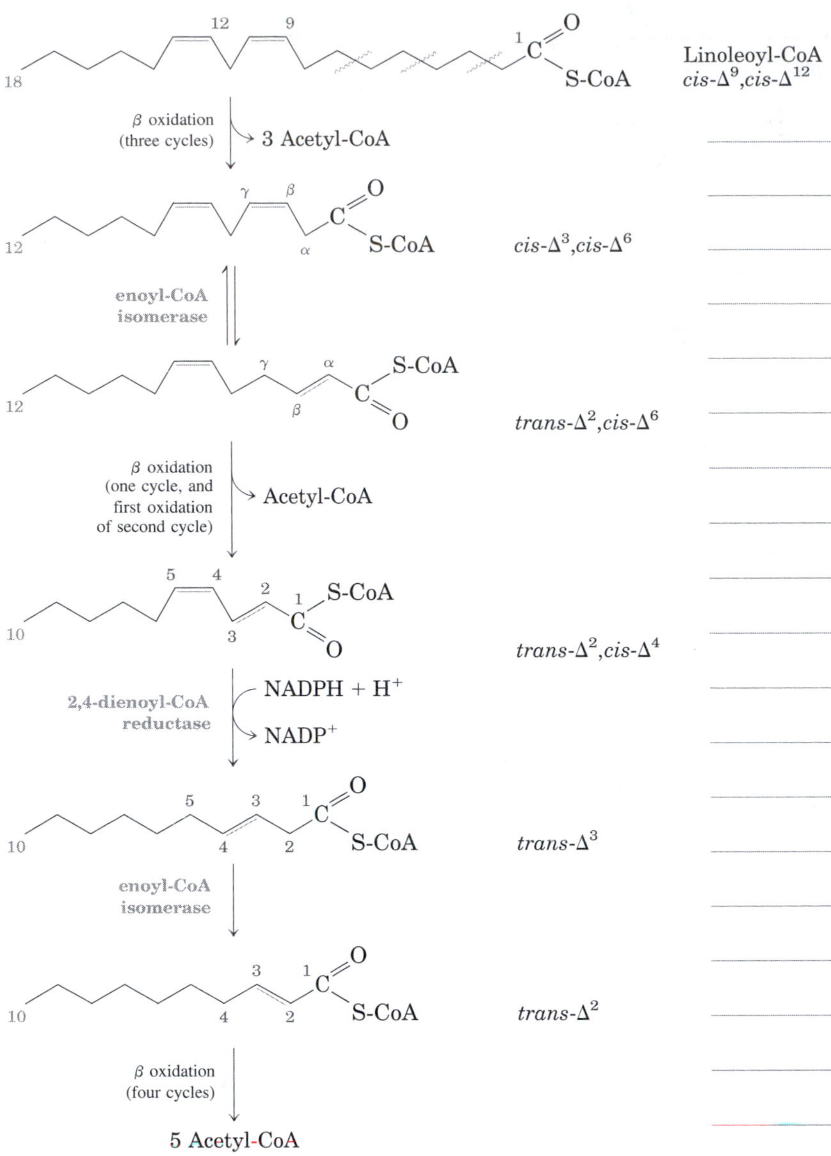

figure 17–10, page 609
Oxidation of a polyunsaturated fatty acid.

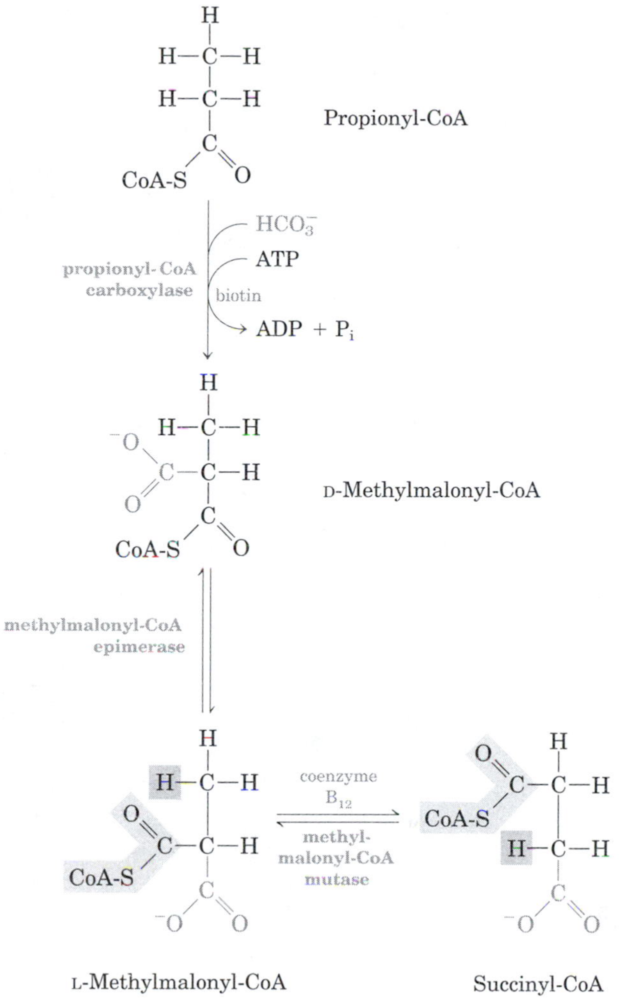

Propionyl-CoA

propionyl-CoA carboxylase

HCO_3^-

ATP

biotin

ADP + P_i

D-Methylmalonyl-CoA

methylmalonyl-CoA epimerase

coenzyme B_{12}

methylmalonyl-CoA mutase

L-Methylmalonyl-CoA

Succinyl-CoA

figure 17–11, page 609

Oxidation of propionyl-CoA produced by β oxidation of odd-number fatty acids.

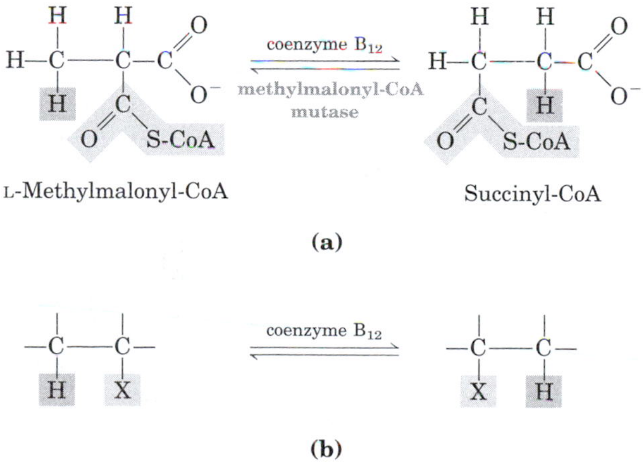

coenzyme B_{12}

methylmalonyl-CoA mutase

L-Methylmalonyl-CoA

Succinyl-CoA

(a)

coenzyme B_{12}

(b)

box 17–2, figure 1, page 610

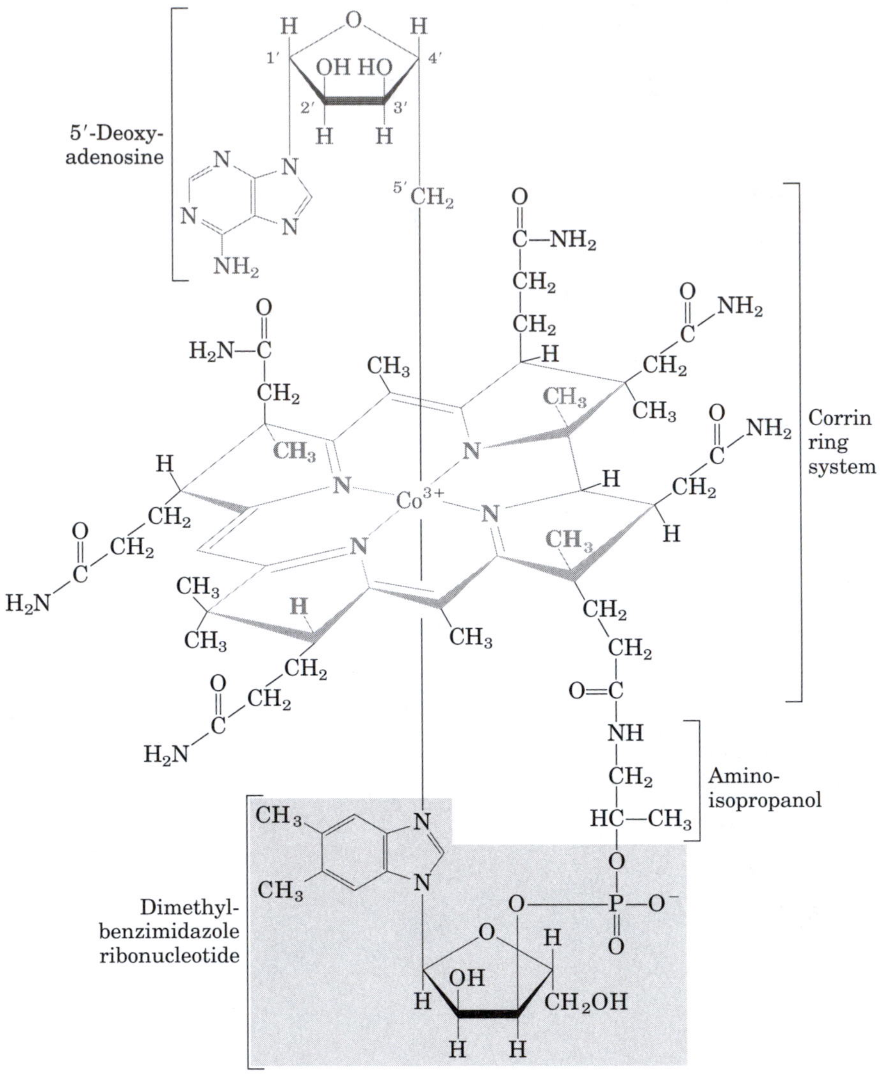

5′-Deoxy-adenosine

Corrin ring system

Amino-isopropanol

Dimethyl-benzimidazole ribonucleotide

box 17–2, figure 2, page 610

ATP

Cobalamin

Coenzyme B_{12}

box 17–2, figure 3, page 611

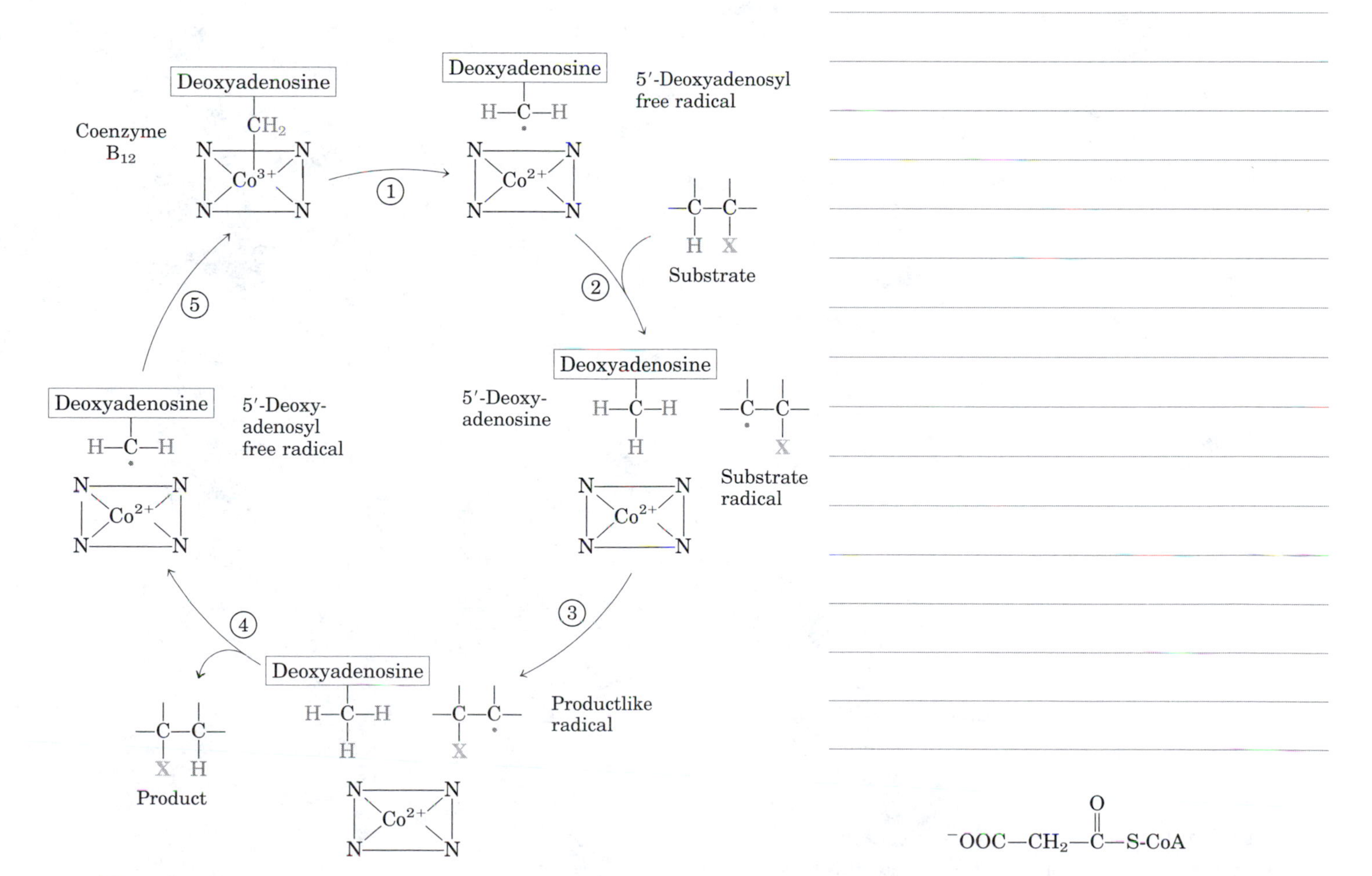

box 17–2, figure 4, page 611

Malonyl-CoA, page 612

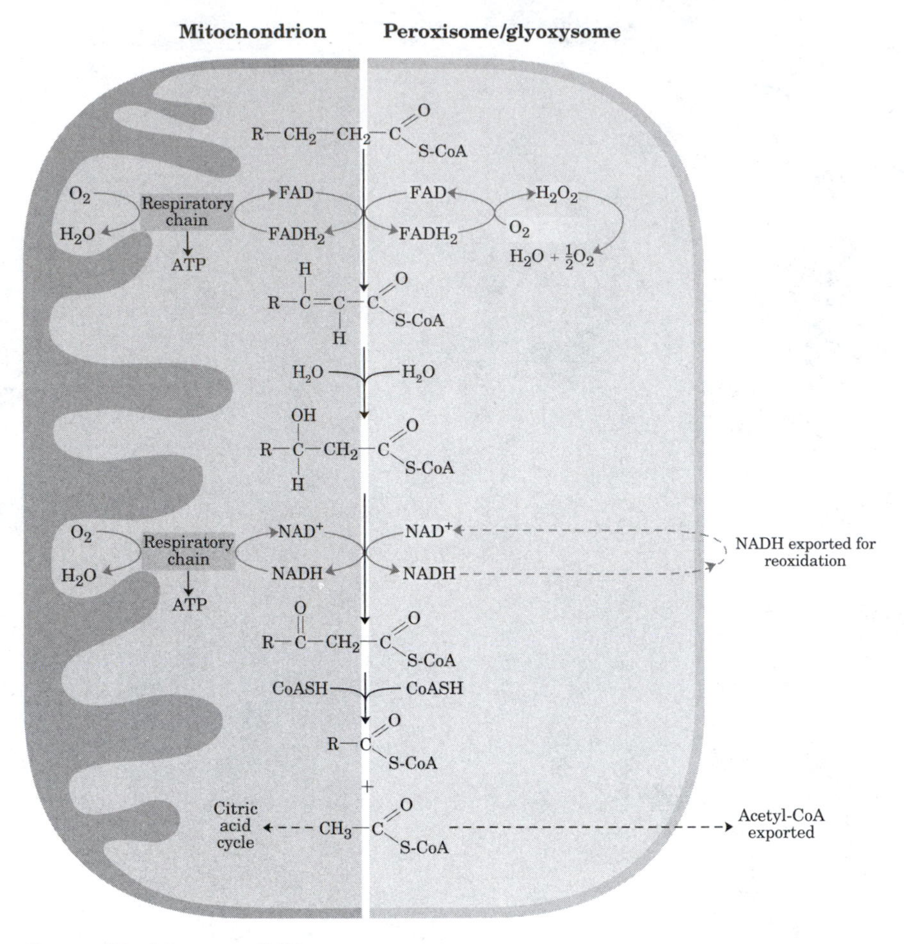

figure 17–12, page 613
Comparison of β oxidation in mitochondria and in peroxisomes and glyoxysomes.

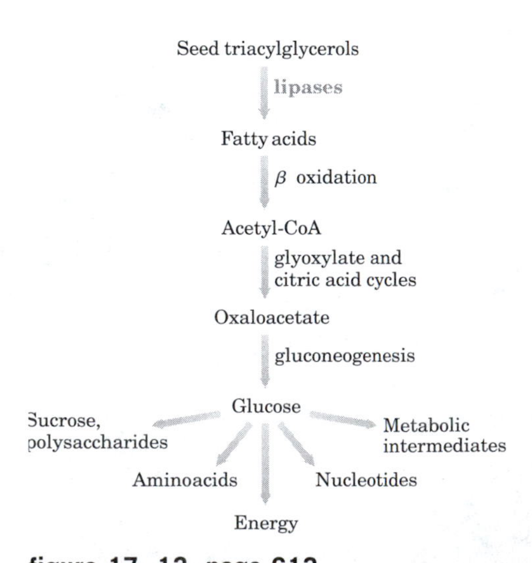

figure 17–13, page 613
Role of β oxidation in the conversion of stored triacylglycerols to glucose in germinating seeds.

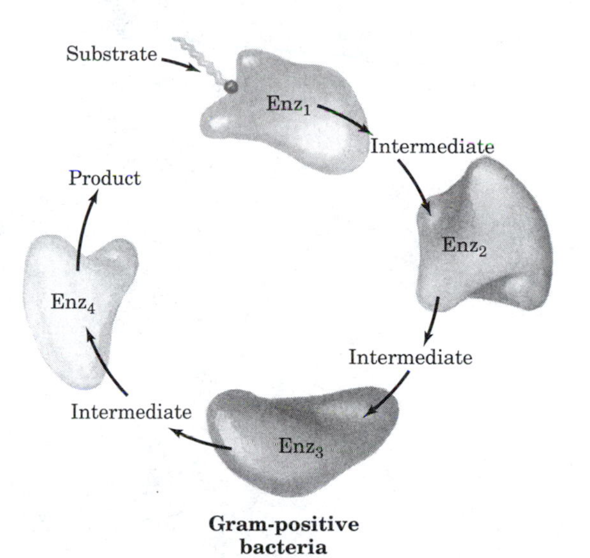

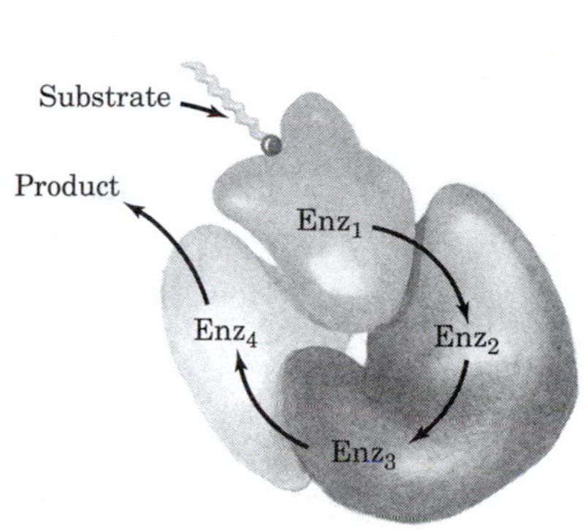

figure 17–14, page 614
Comparison of the enzymes of β oxidation in gram-positive and gram-negative bacteria.

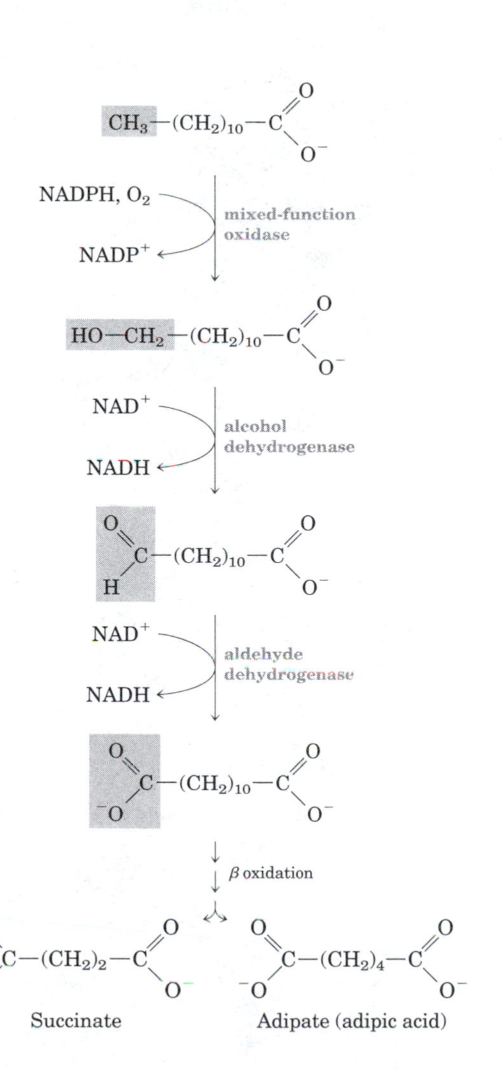

figure 17–15, page 615
Omega oxidation in the endoplasmic reticulum.

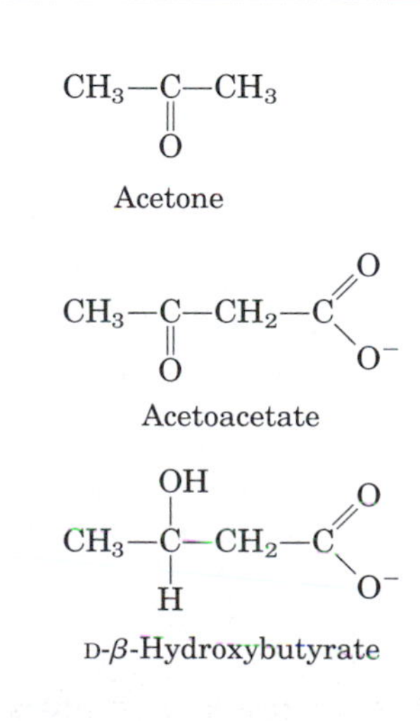

Ketone bodies, page 615

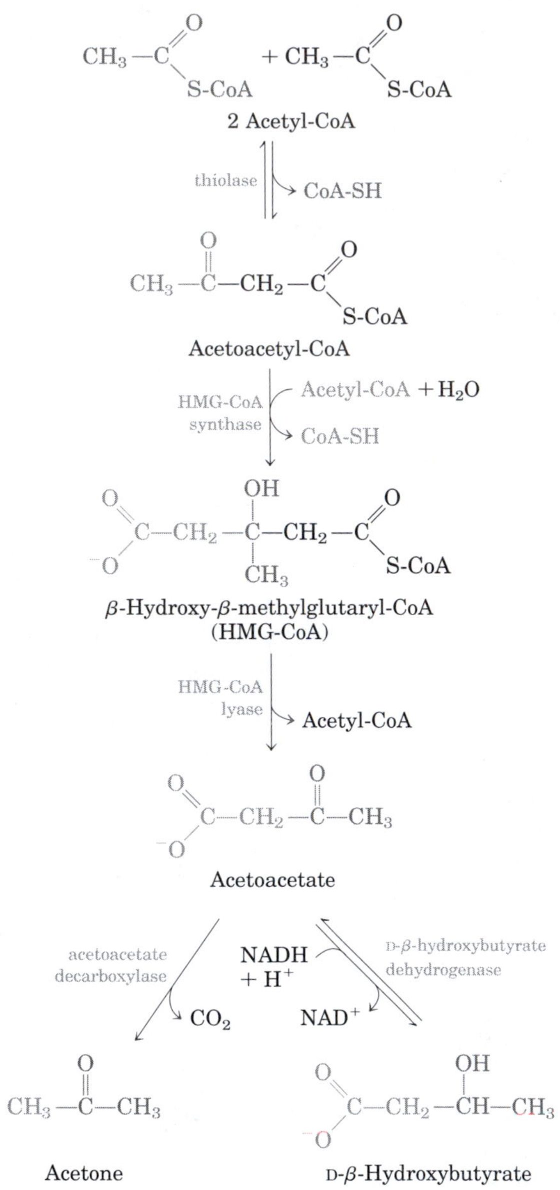

$CH_3-C(=O)-S\text{-}CoA + CH_3-C(=O)-S\text{-}CoA$

2 Acetyl-CoA

thiolase ⇅ → CoA-SH

$CH_3-C(=O)-CH_2-C(=O)-S\text{-}CoA$

Acetoacetyl-CoA

HMG-CoA synthase ↓ (Acetyl-CoA + H_2O → CoA-SH)

$^{-}O(O=)C-CH_2-C(OH)(CH_3)-CH_2-C(=O)-S\text{-}CoA$

β-Hydroxy-β-methylglutaryl-CoA (HMG-CoA)

HMG-CoA lyase ↓ → Acetyl-CoA

$^{-}O(O=)C-CH_2-C(=O)-CH_3$

Acetoacetate

acetoacetate decarboxylase ↙ → CO_2 | NADH + H^+ → NAD^+ ⇌ D-β-hydroxybutyrate dehydrogenase ↘

$CH_3-C(=O)-CH_3$

Acetone

$^{-}O(O=)C-CH_2-CH(OH)-CH_3$

D-β-Hydroxybutyrate

figure 17–16, page 616
Formation of ketone bodies from acetyl-CoA.

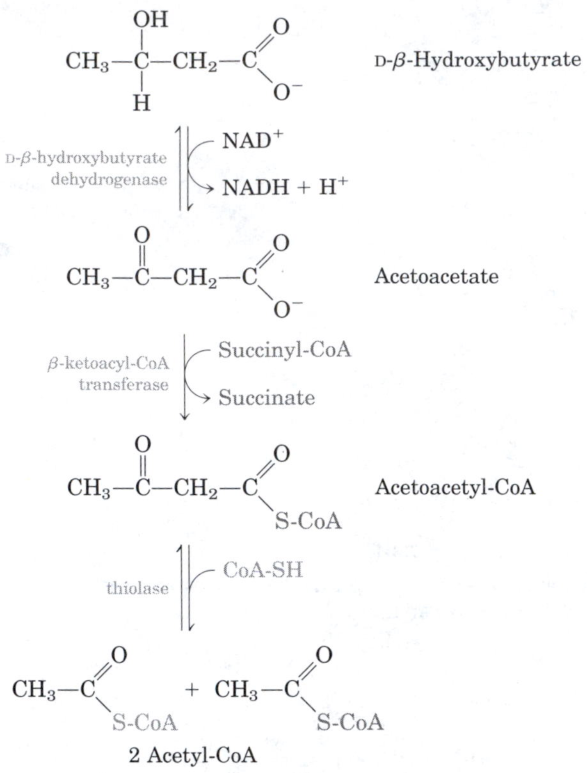

$CH_3-C(OH)(H)-CH_2-C(=O)-O^-$ D-β-Hydroxybutyrate

D-β-hydroxybutyrate dehydrogenase ⇅ (NAD^+ → NADH + H^+)

$CH_3-C(=O)-CH_2-C(=O)-O^-$ Acetoacetate

β-ketoacyl-CoA transferase ↓ (Succinyl-CoA → Succinate)

$CH_3-C(=O)-CH_2-C(=O)-S\text{-}CoA$ Acetoacetyl-CoA

thiolase ⇅ ← CoA-SH

$CH_3-C(=O)-S\text{-}CoA + CH_3-C(=O)-S\text{-}CoA$

2 Acetyl-CoA

figure 17–17, page 617
β-Hydroxybutyrate as a fuel.

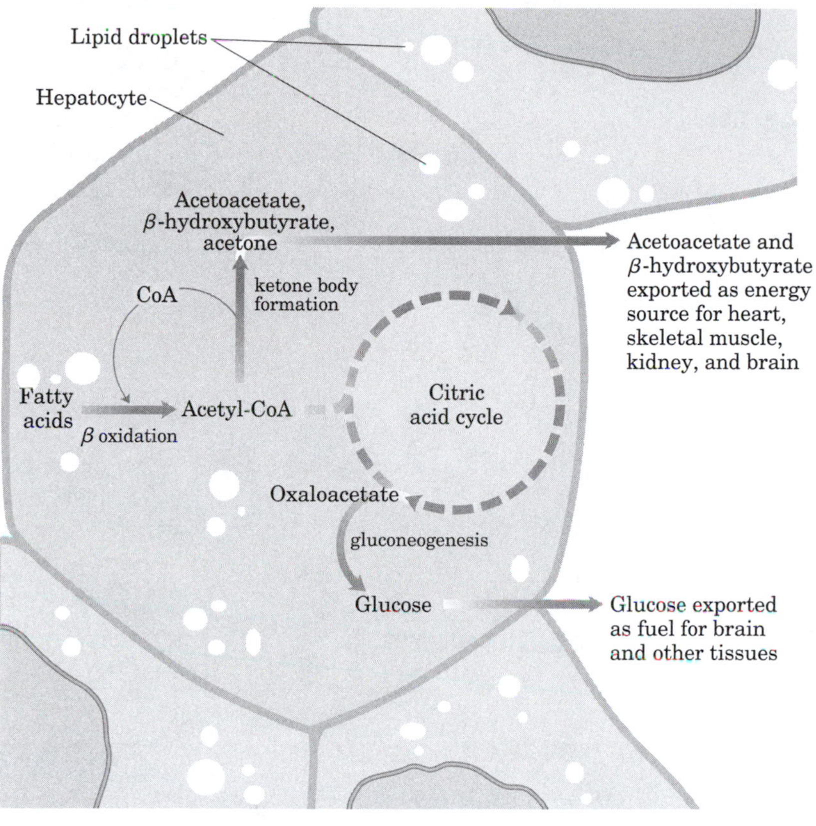

figure 17–18, page 617
Ketone body formation and export from the liver.

table 17–2, page 618

Ketone Body Accumulation in Diabetic Ketosis

	Urinary excretion (mg/24 h)	Blood concentration (mg/100 mL)
Normal	≤125	<3
Extreme ketosis (untreated diabetes)	5,000	90

PROBLEMS

4. Chemistry of the Acyl-CoA Synthetase Reaction, page 620.

$$R{-}COO^- + ATP + CoA \rightleftharpoons R{-}\overset{O}{\overset{\|}{C}}{-}CoA + AMP + PP_i$$

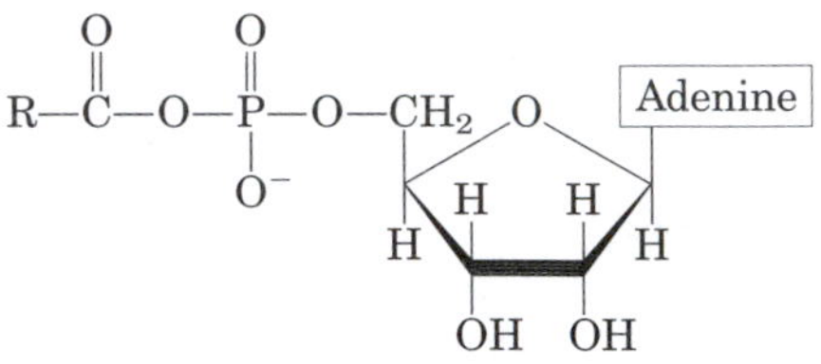

7. Comparative Biochemistry: Energy-Generating Pathways in Birds, page 620.

Enzyme	V_{max} (μmol substrate/min/g tissue)	
	Pigeon	Pheasant
Hexokinase	3.0	2.3
Glycogen phosphorylase	18.0	120.0
Phosphofructokinase-1	24.0	143.0
Citrate synthase	100.0	15.0
Triacylglycerol lipase	0.07	0.01

10. Petroleum as a Microbial Food Source, page 621.

$$CH_3(CH_2)_6CH_3 + NAD^+ + O_2 \rightleftharpoons CH_3(CH_2)_6COOH + NADH + H^+$$

11. Metabolism of a Straight-Chain Phenylated Fatty Acid, page 621.

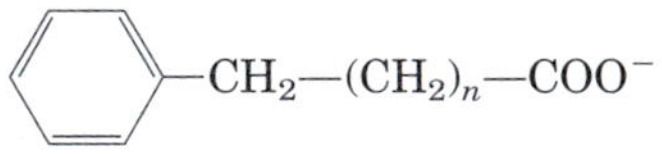

14. Metabolic Consequences of Ingesting ω-Fluorooleate, page 621.

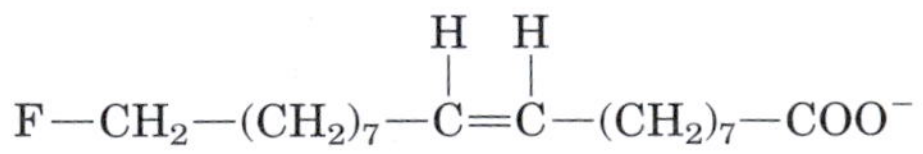

ω-Fluorooleate

18. Sources of H_2O Produced in β Oxidation, page 622.

$$\text{Palmitoyl-CoA} + 23O_2 + 108P_i + 108ADP \longrightarrow CoA + 16CO_2 + 108ATP + 23H_2O$$

$$ADP + P_i \longrightarrow ATP + H_2O$$

chapter 18

Amino Acid Oxidation and the Production of Urea

When you study for exams, don't forget to review The Minicourses on the UNDERSTAND! BIOCHEMISTRY CD that came with your textbook.

Minicourses that apply to this Chapter include:

Molecules of Life
- An Interactive Gallery of Molecular Models
- An Interactive Gallery of Protein Structures

Proteins in Action
- Catalysis and Regulation

Biosynthesis and Catabolism
- Nitrogen Metabolism
- Amino Acid Degradation

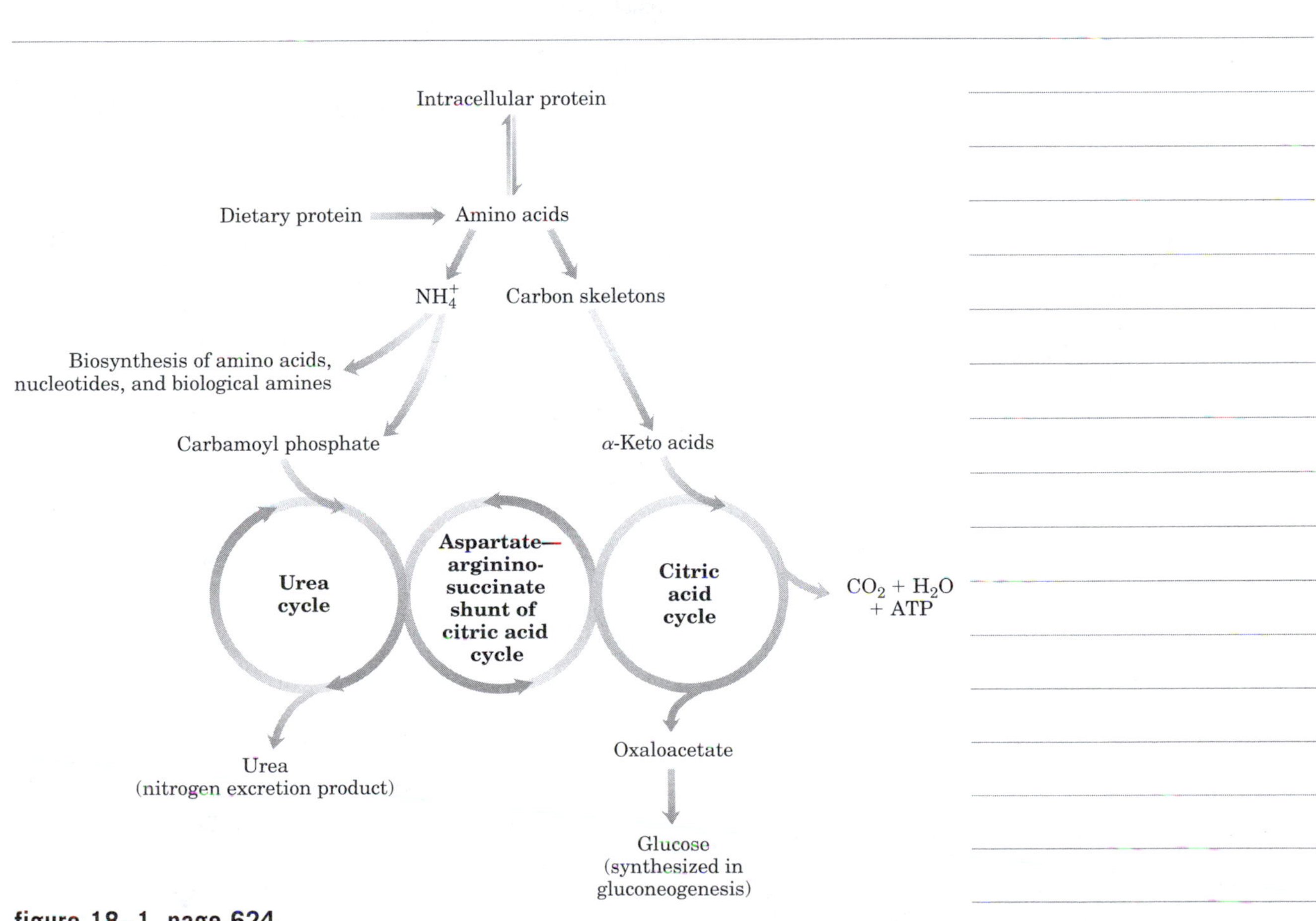

figure 18–1, page 624
Overview of the catabolism of amino acids in mammals.

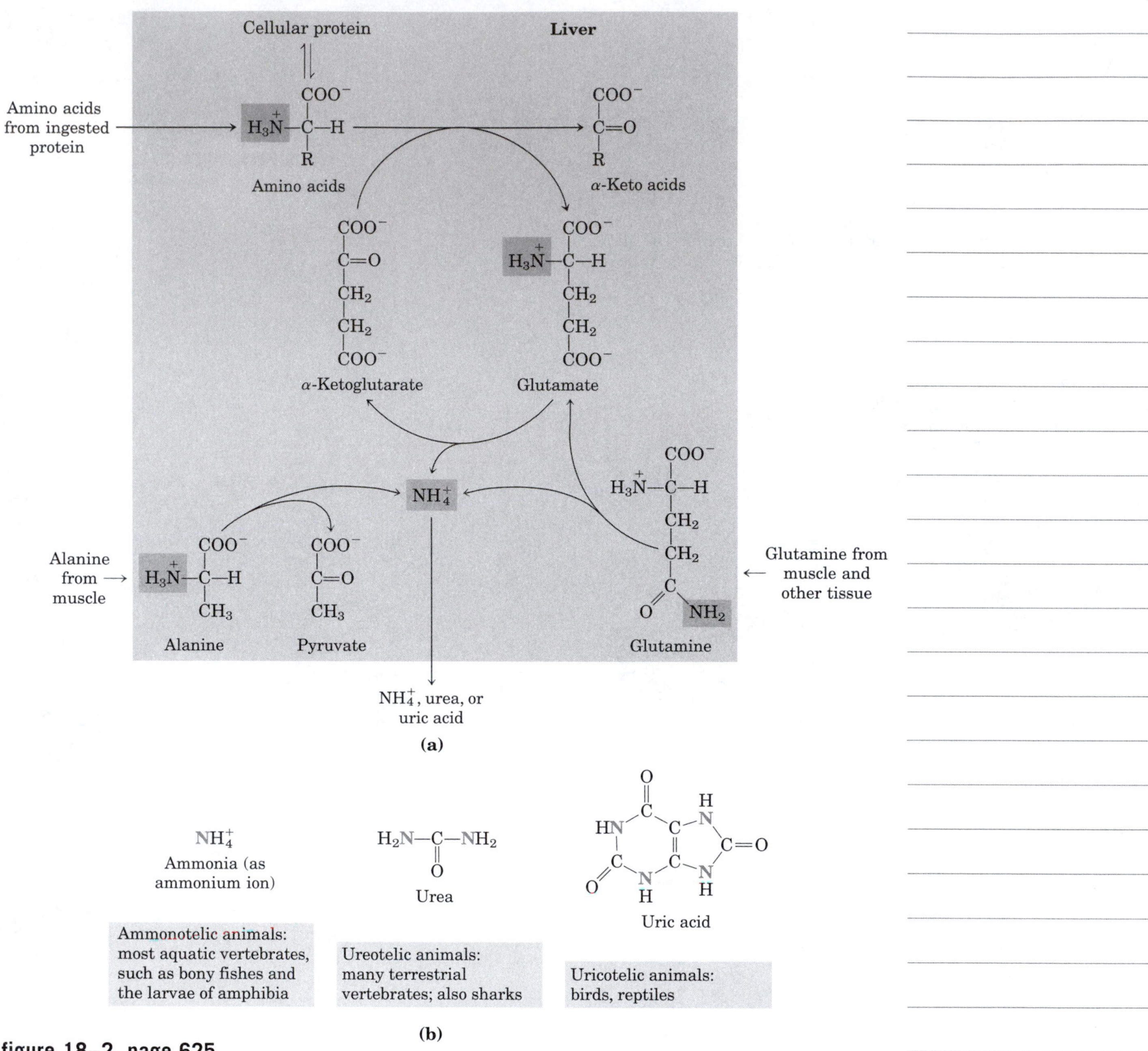

figure 18–2, page 625
Amino group catabolism.

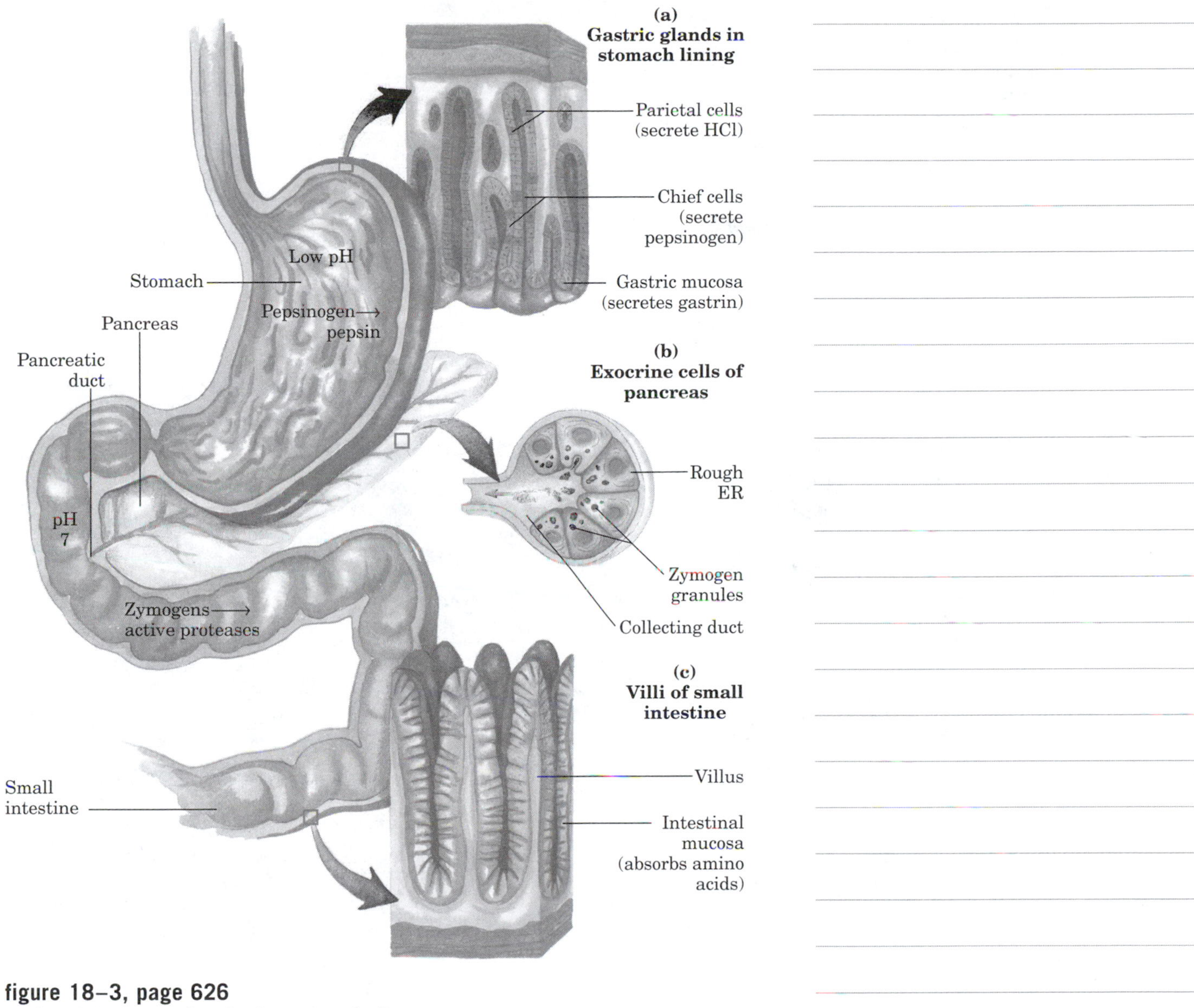

figure 18–3, page 626
Portion of the human digestive (gastrointestinal) tract.

α-Ketoglutarate + L-Amino acid $\xrightleftharpoons[\text{aminotransferase}]{\text{PLP}}$ L-Glutamate + α-Keto acid

figure 18–4, page 628
Enzyme-catalyzed transaminations.

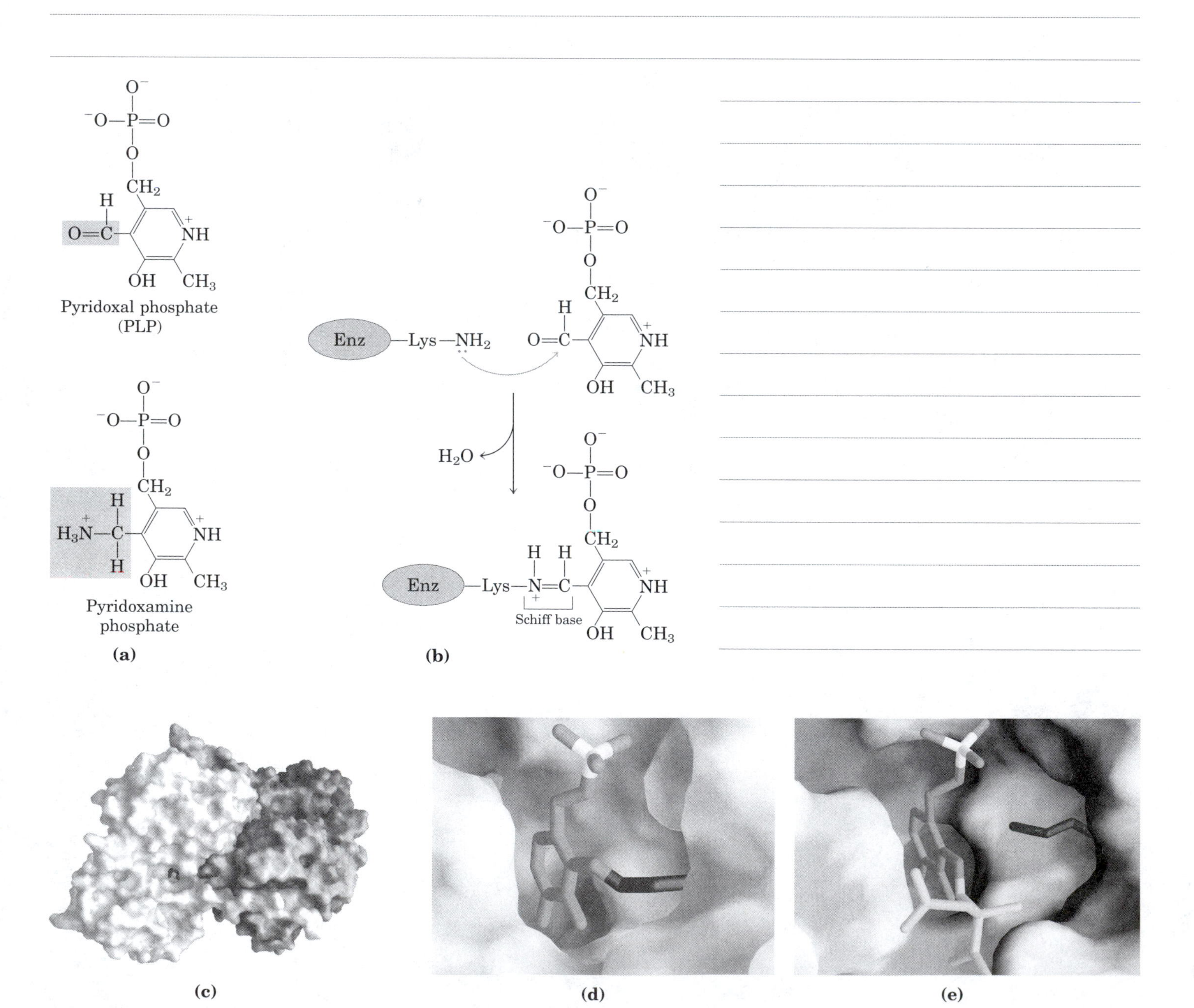

figure 18–5, page 629
Pyridoxal phosphate, the prosthetic group of aminotransferases.

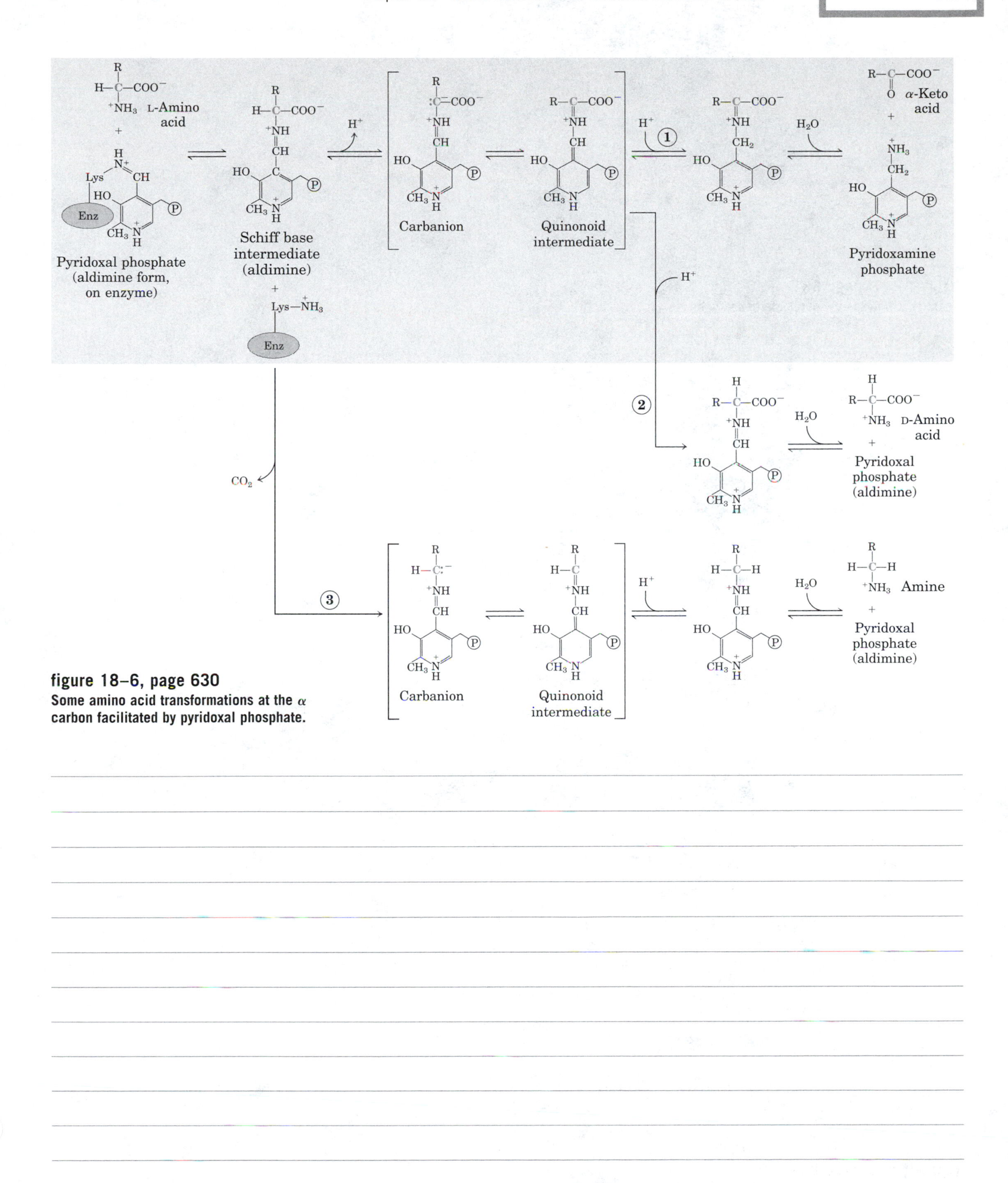

figure 18–6, page 630
Some amino acid transformations at the α carbon facilitated by pyridoxal phosphate.

$$\text{Glutamate } (\text{H}_3\overset{+}{\text{N}}\text{–CH(COO}^-\text{)–CH}_2\text{–CH}_2\text{–COO}^-) + H_2O \underset{\text{glutamate dehydrogenase}}{\overset{NAD(P)^+ \;\; NAD(P)H + H^+}{\rightleftharpoons}} \alpha\text{-Ketoglutarate } (\text{COO}^-\text{–C=O–CH}_2\text{–CH}_2\text{–COO}^-) + NH_4^+$$

figure 18–7, page 631
Reaction catalyzed by glutamate dehydrogenase.

$$^-OOC\text{–}CH_2\text{–}CH_2\text{–}CH(\overset{+}{N}H_3)\text{–}COO^- \quad \text{L-Glutamate}$$

$$^-O\text{–}P(=O)(O^-)\text{–}O\text{–}C(=O)\text{–}CH_2\text{–}CH_2\text{–}CH(\overset{+}{N}H_3)\text{–}COO^- \quad \gamma\text{-Glutamyl phosphate}$$

$$H_2N\text{–}C(=O)\text{–}CH_2\text{–}CH_2\text{–}CH(\overset{+}{N}H_3)\text{–}COO^- \quad \text{L-Glutamine}$$

$$^-O\text{–}C(=O)\text{–}CH_2\text{–}CH_2\text{–}CH(\overset{+}{N}H_3)\text{–}COO^- \quad \text{L-Glutamate}$$

Glutaminase, page 632

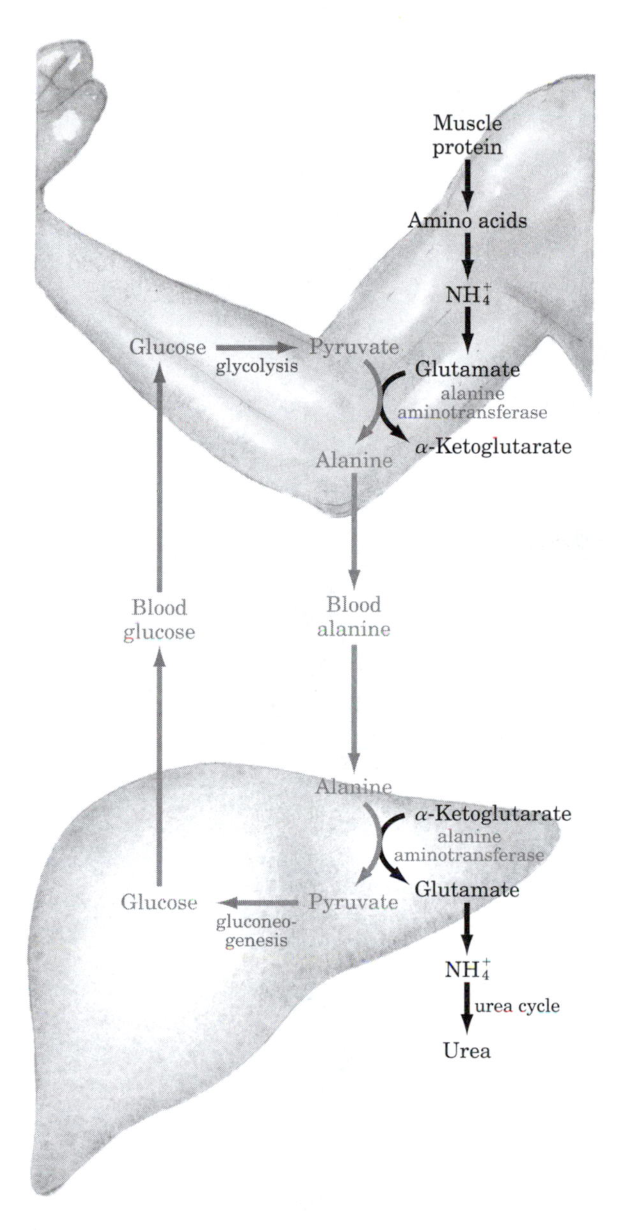

figure 18–8, page 633
Glucose-alanine cycle.

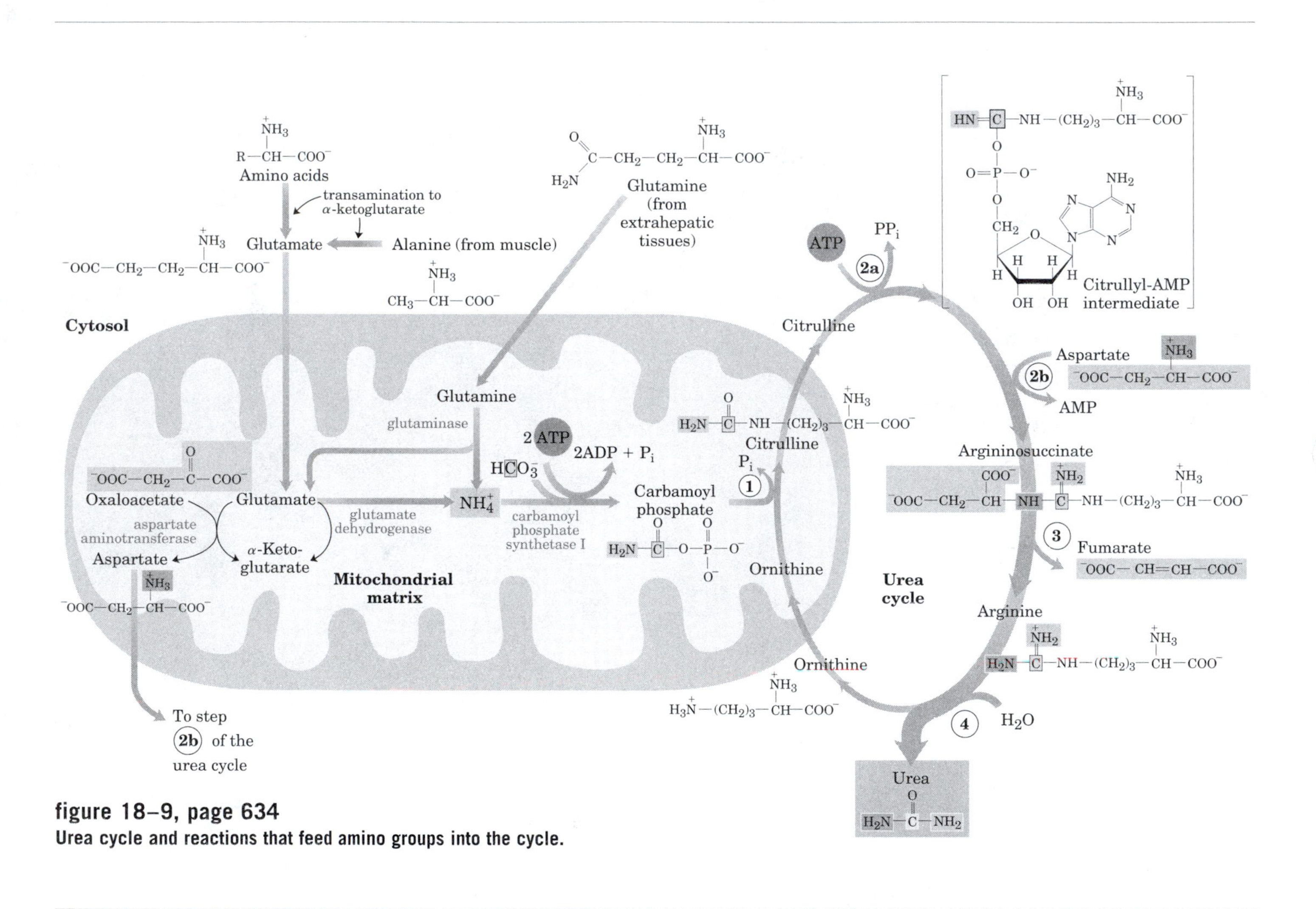

figure 18–9, page 634
Urea cycle and reactions that feed amino groups into the cycle.

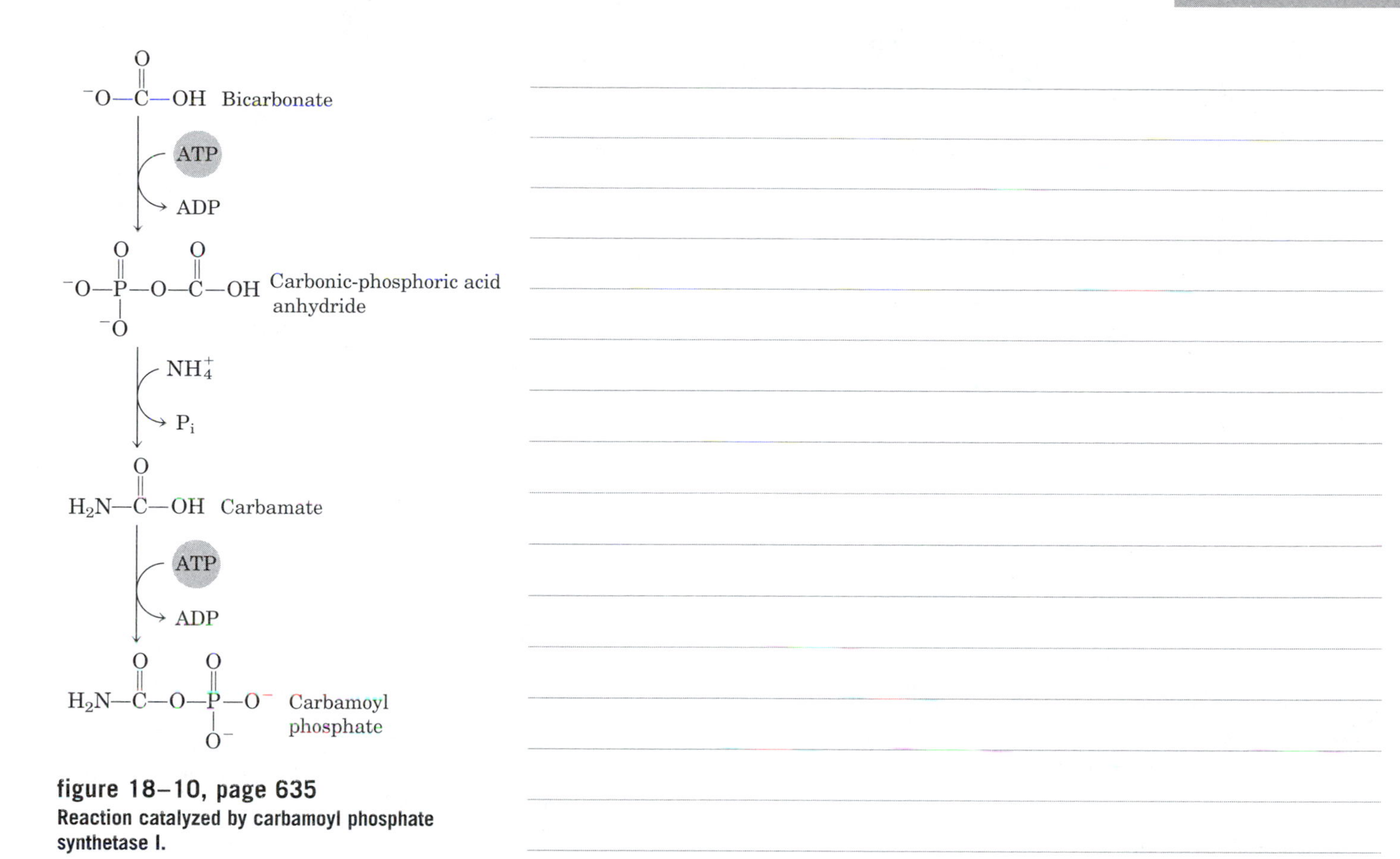

figure 18–10, page 635
Reaction catalyzed by carbamoyl phosphate synthetase I.

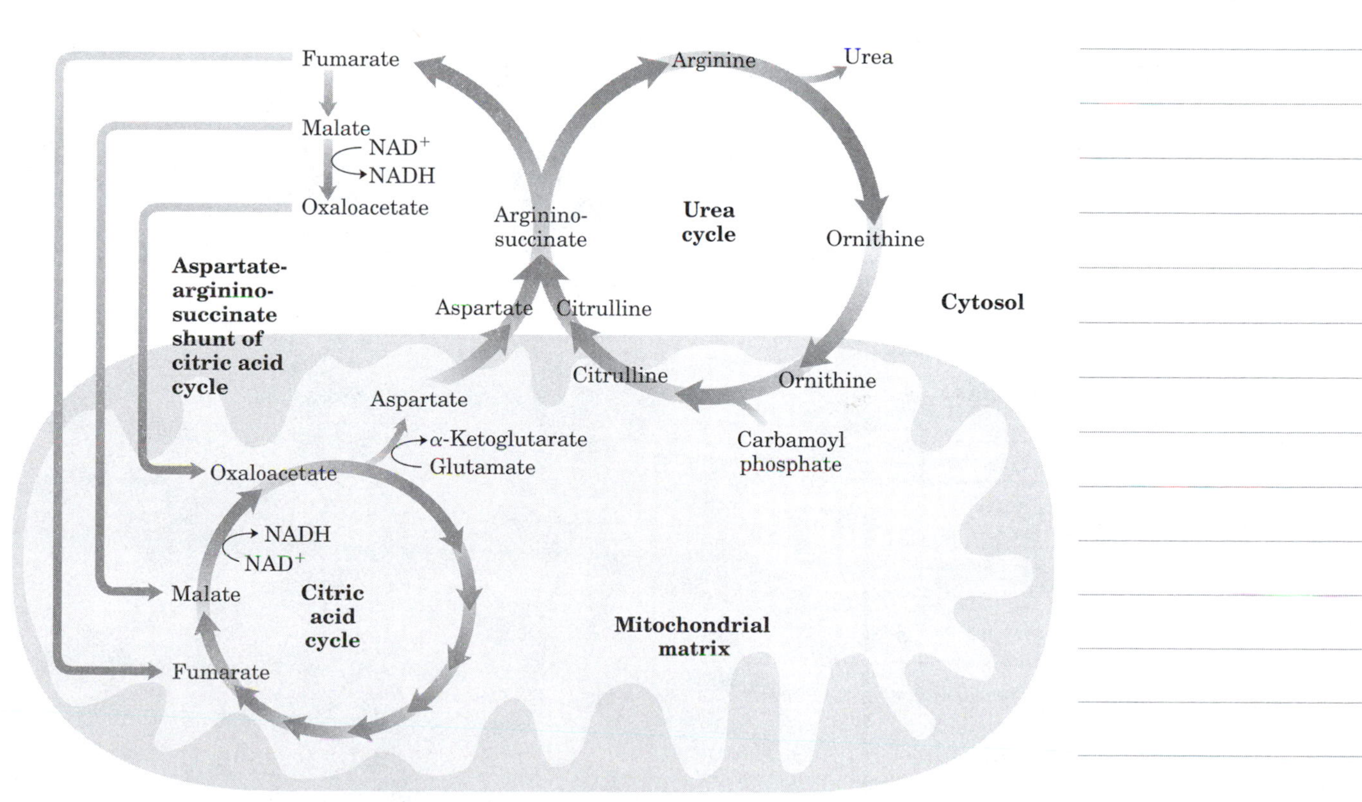

figure 18–11, page 636
Links between the urea cycle and citric acid cycle.

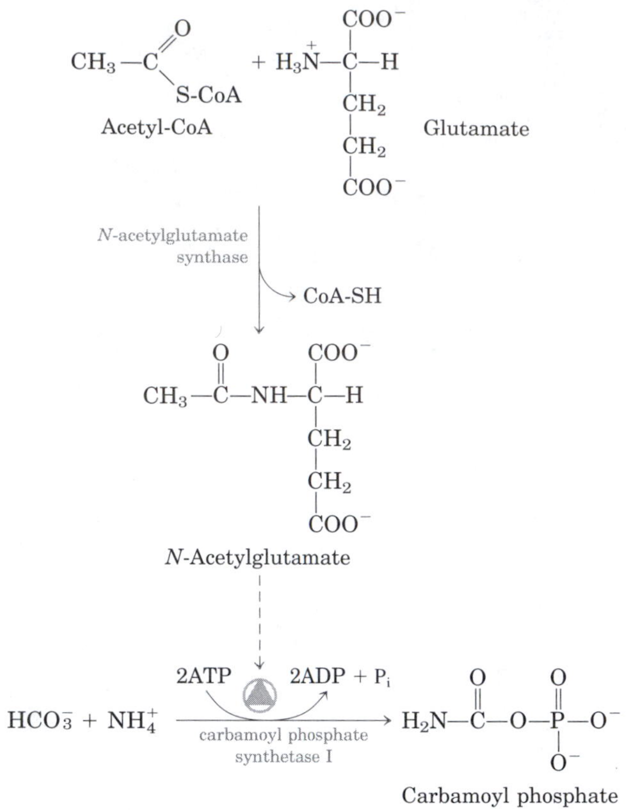

$$CH_3{-}C(=O){-}S\text{-}CoA \text{ (Acetyl-CoA)} + H_3\overset{+}{N}{-}CH(COO^-){-}CH_2{-}CH_2{-}COO^- \text{ (Glutamate)}$$

$$\xrightarrow[\;]{N\text{-acetylglutamate synthase}} CoA\text{-}SH$$

$$CH_3{-}C(=O){-}NH{-}CH(COO^-){-}CH_2{-}CH_2{-}COO^-$$

N-Acetylglutamate

$$HCO_3^- + NH_4^+ \xrightarrow[\text{carbamoyl phosphate synthetase I}]{2ATP \;\; 2ADP + P_i} H_2N{-}C(=O){-}O{-}P(=O)(O^-){-}O^-$$

Carbamoyl phosphate

figure 18–12, page 637
Synthesis of *N*-acetylglutamate and its activation of carbamoyl phosphate synthetase I.

table 18–1, page 637

Nonessential and Essential Amino Acids for Humans and the Albino Rat

Nonessential	Essential
Alanine	Arginine*
Asparagine	Histidine
Aspartate	Isoleucine
Cysteine	Leucine
Glutamate	Lysine
Glutamine	Methionine
Glycine	Phenylalanine
Proline	Threonine
Serine	Tryptophan
Tyrosine	Valine

*Essential in young, growing animals but not in adults.

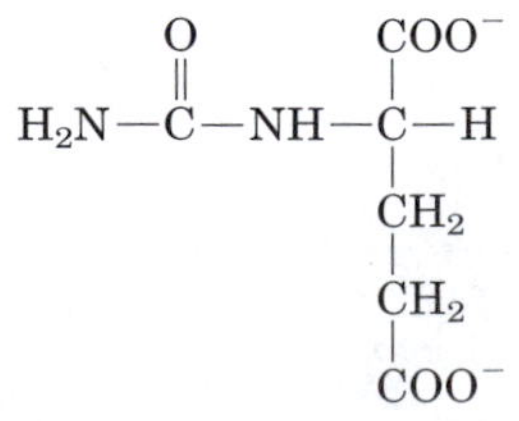

$$H_2N{-}C(=O){-}NH{-}CH(COO^-){-}CH_2{-}CH_2{-}COO^-$$

Carbamoyl glutamate, page 638

Benzoate + CoA-SH

↓ ATP → AMP + PP_i

Benzoyl-CoA (C_6H_5–C(=O)–S-CoA)

↓ Glycine → CoA-SH

Hippurate (benzoylglycine) (C_6H_5–C(=O)–NH–CH_2–COO^-)

Phenylacetate (C_6H_5–CH_2–COO^-) + CoA-SH

↓ ATP → AMP + PP_i

Phenylacetyl-CoA (C_6H_5–CH_2–C(=O)–S-CoA)

↓ Glutamine → CoA-SH

Phenylacetylglutamine (C_6H_5–CH_2–C(=O)–NH–CH(COO^-)–CH_2–CH_2–C(=O)–NH_2)

figure 18–13, page 638
Treatment for deficiencies in urea cycle enzymes.

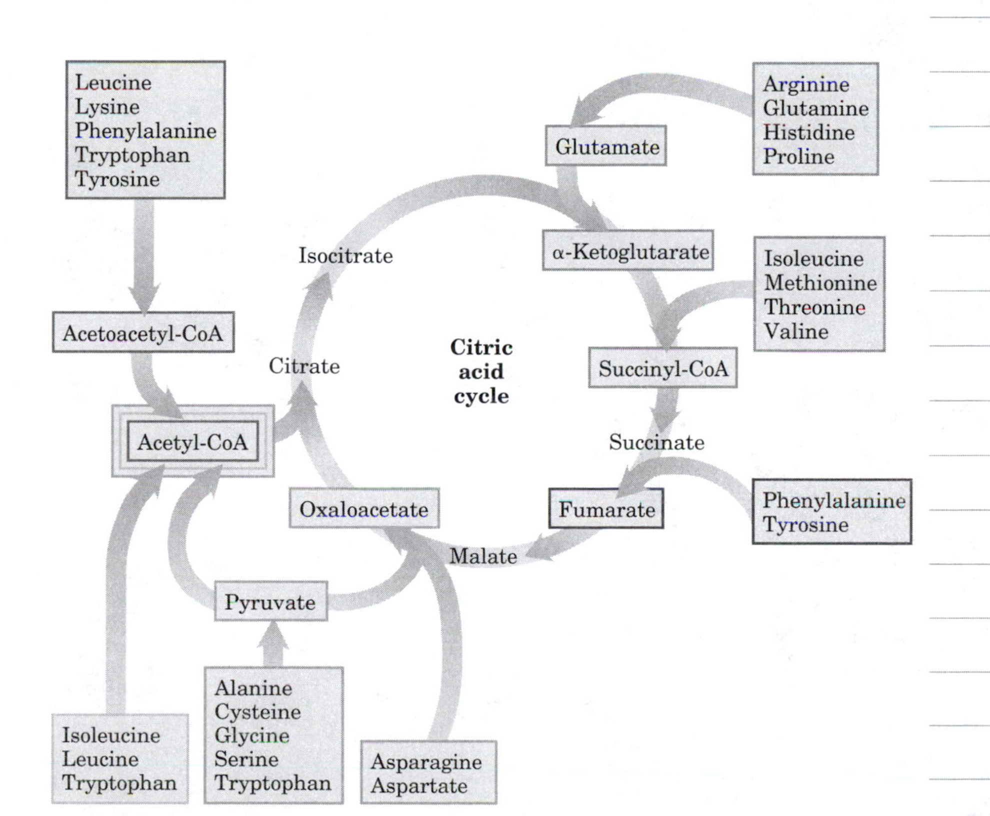

figure 18–14, page 639
Summary of the points of entry of the standard amino acids into the citric acid cycle.

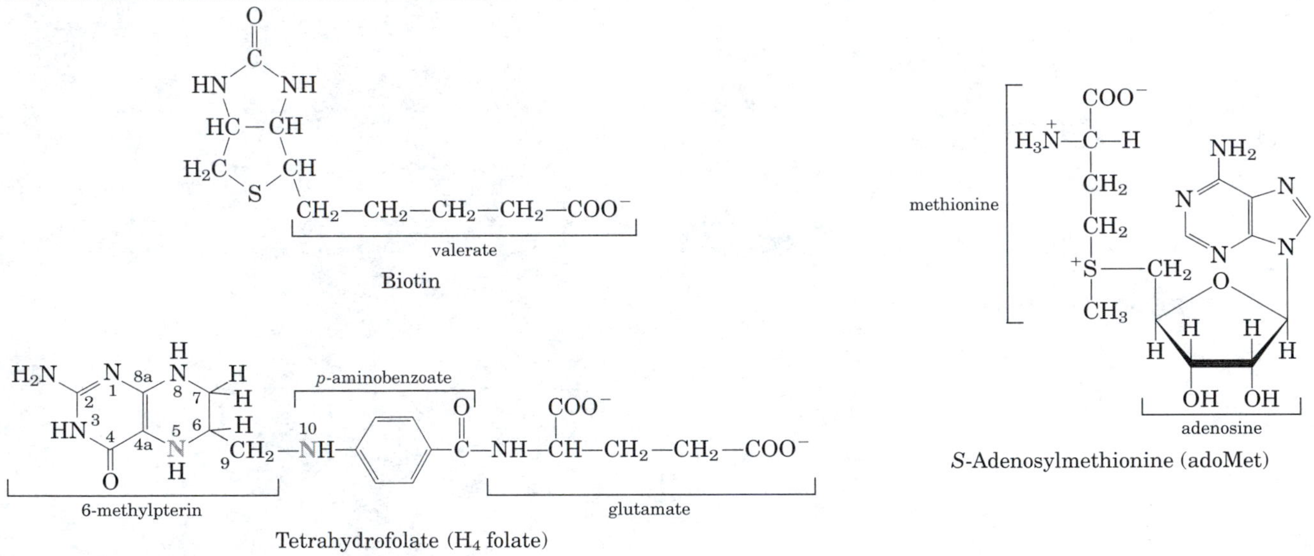

HN
NH
HC
CH
H$_2$C
S
CH
CH$_2$—CH$_2$—CH$_2$—CH$_2$—COO$^-$
valerate
Biotin
H$_2$N
8a
6-methylpterin
10
p-aminobenzoate
COO$^-$
C—NH—CH—CH$_2$—CH$_2$—COO$^-$
glutamate
Tetrahydrofolate (H$_4$ folate)
methionine
H$_3$N—C—H
NH$_2$
CH$_3$
OH OH
adenosine
S-Adenosylmethionine (adoMet)

figure 18–15, page 640
Some enzyme cofactors important in one-carbon transfer reactions.

H$_2$N
HN
R

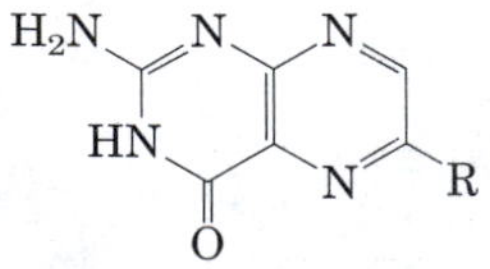

Pterin, page 640

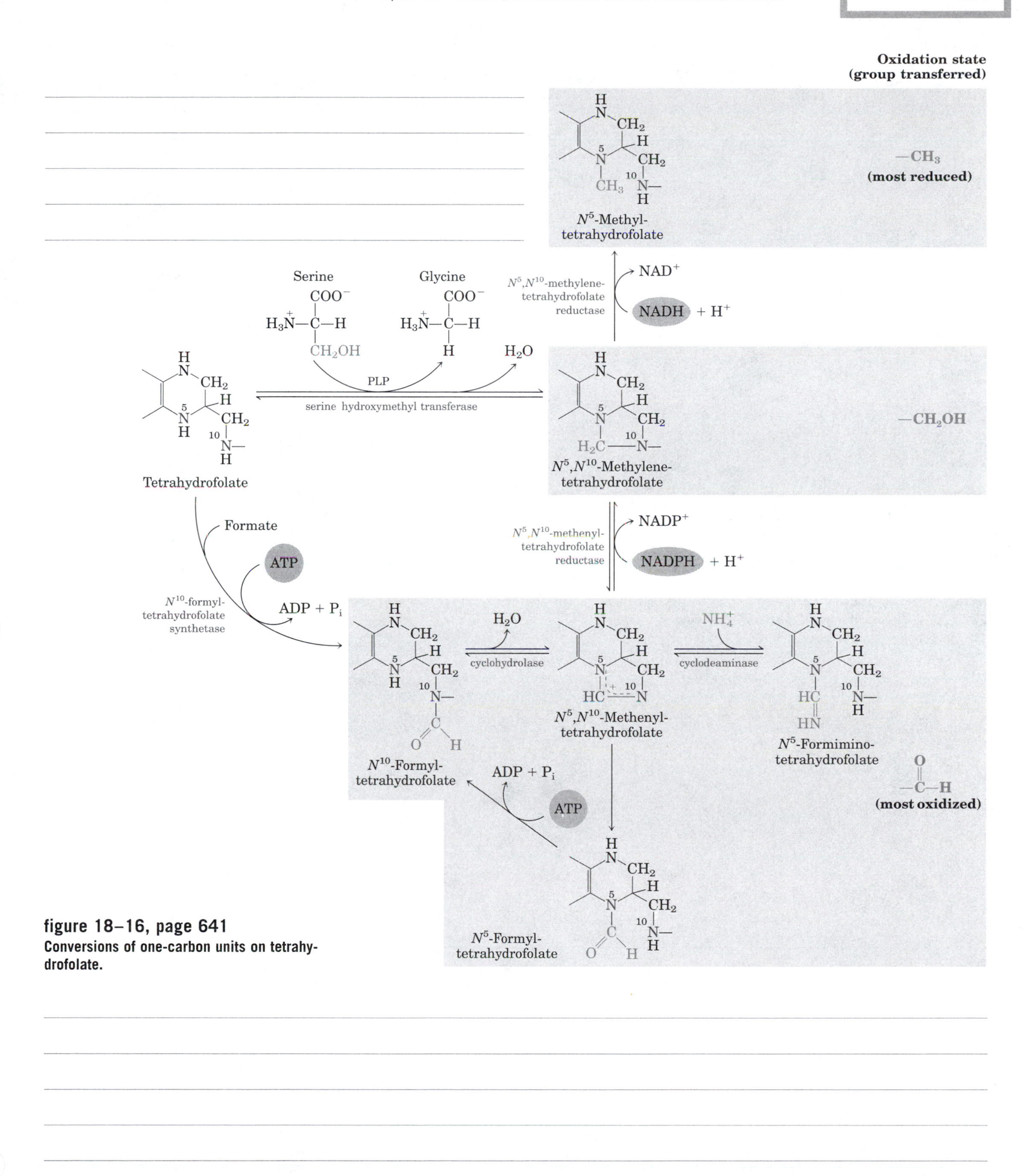

figure 18–16, page 641
Conversions of one-carbon units on tetrahydrofolate.

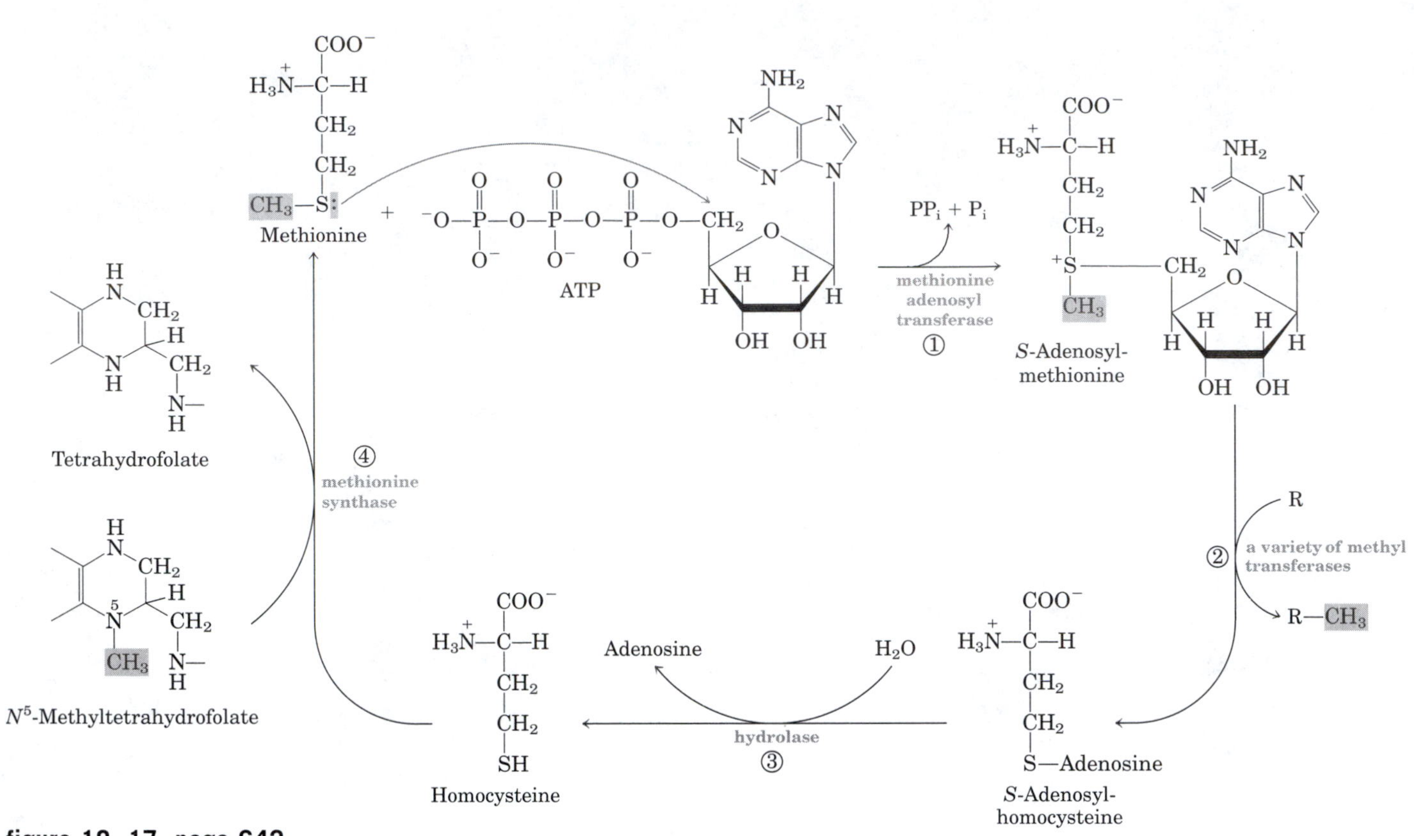

figure 18–17, page 642
Synthesis of methionine and *S*-adenosylmethionine in an activated-methyl cycle.

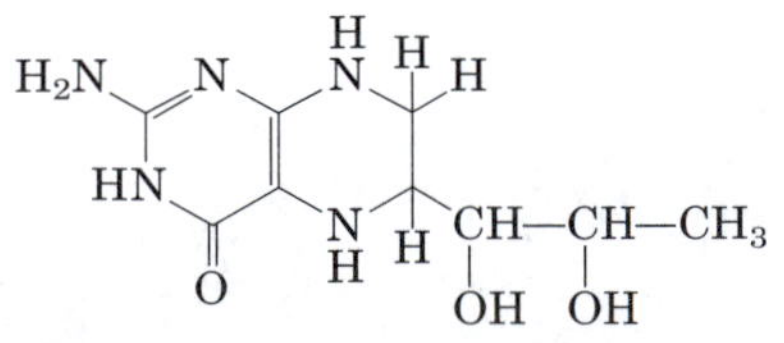

Tetrahydrobiopterin, page 643

figure 18–18, page 643
Catabolic pathways for alanine, glycine, serine, cysteine, tryptophan, and threonine.

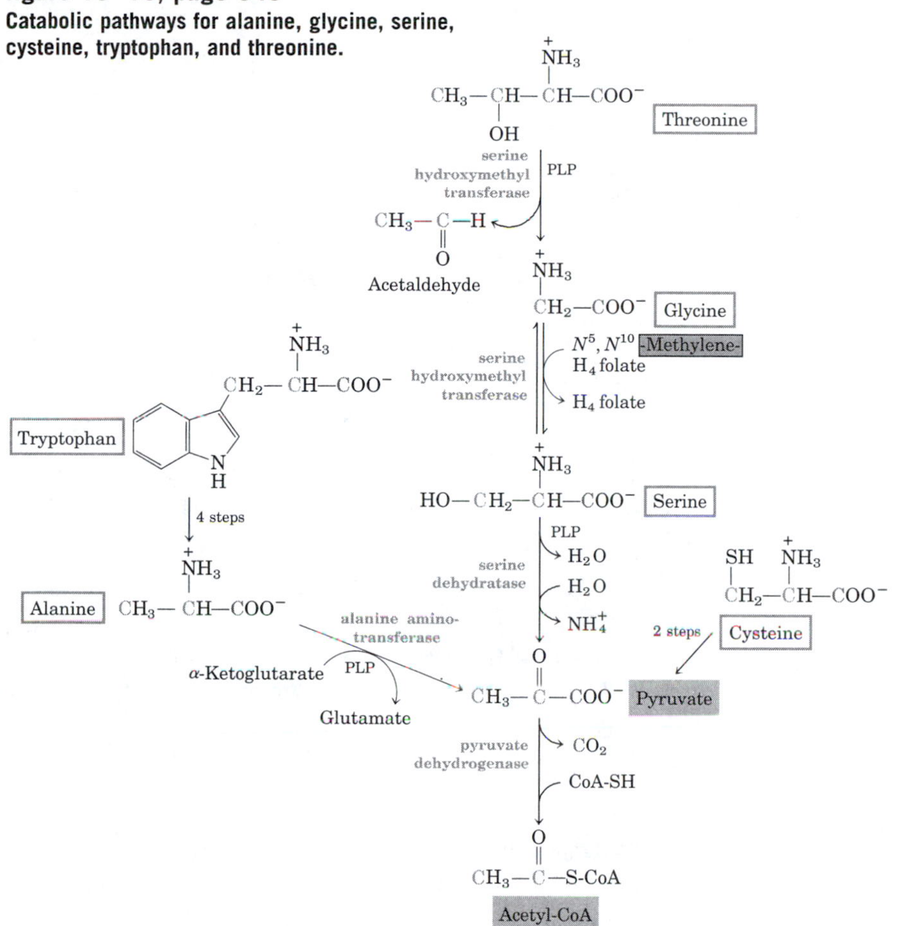

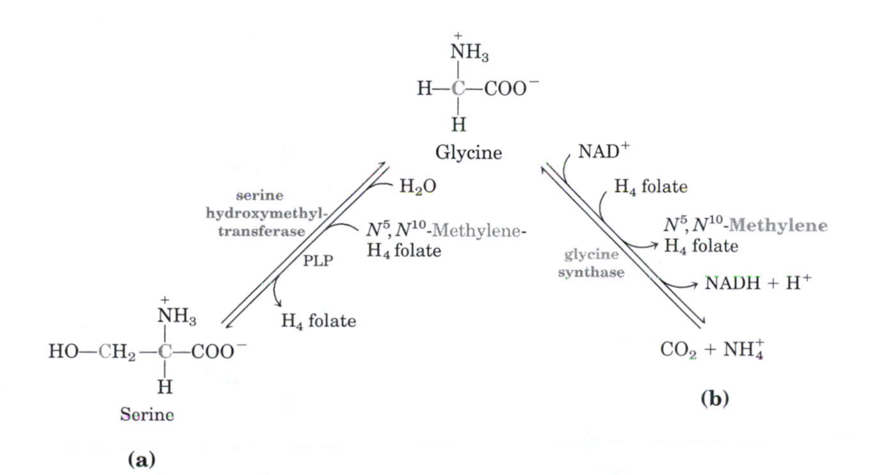

figure 18–19, page 644
Two metabolic fates of glycine.

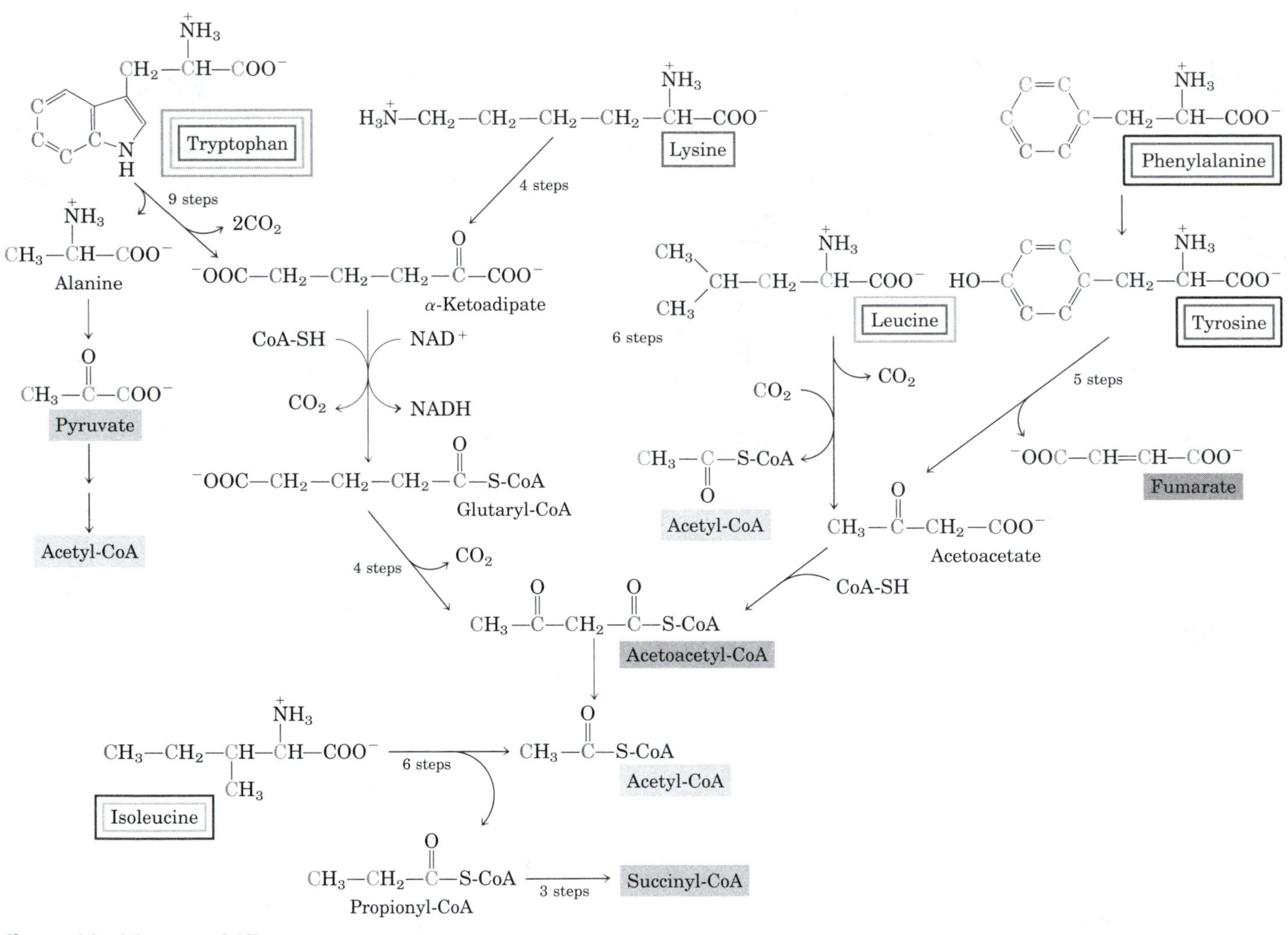

figure 18–20, page 645

Catabolic pathways for tryptophan, lysine, phenylalanine, tyrosine, leucine, and isoleucine.

Tryptophan

Nicotinate (niacin), a precursor of NAD and NADP

Indoleacetate, a plant growth factor

Serotonin, a neurotransmitter

figure 18–21, page 645
Tryptophan as precursor.

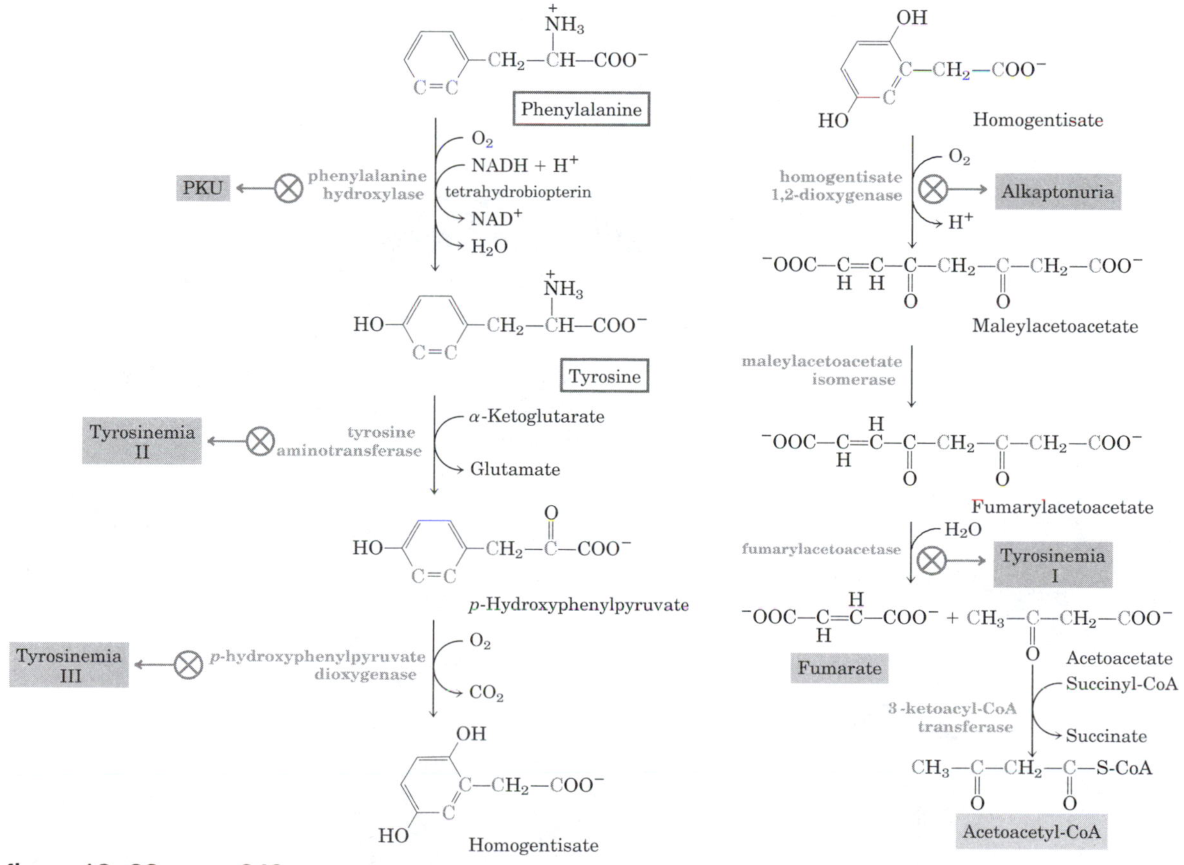

figure 18–22, page 646
Catabolic pathways for phenylalanine and tyrosine.

table 18–2, page 647

Some Human Genetic Disorders Affecting Amino Acid Catabolism

Medical condition	Approximate incidence (per 100,000 births)	Defective process	Defective enzyme	Symptoms and effects
Albinism	3	Melanin synthesis from tyrosine	Tyrosine 3-monooxygenase (tyrosinase)	Lack of pigmentation; white hair, pink skin
Alkaptonuria	0.4	Tyrosine degradation	Homogentisate 1,2-dioxygenase	Dark pigment in urine; late-developing arthritis
Argininemia	<0.5	Urea synthesis	Arginase	Mental retardation
Argininosuccinic acidemia	1.5	Urea synthesis	Argininosuccinate lyase	Vomiting, convulsions
Carbamoyl phosphate synthetase I deficiency	>0.5	Urea synthesis	Carbamoyl phosphate synthetase I	Lethargy, convulsions, early death
Homocystinuria	0.5	Methionine degradation	Cystathionine β-synthase	Faulty bone development, mental retardation
Maple syrup urine disease (branched-chain ketoaciduria)	0.4	Isoleucine, leucine, and valine degradation	Branched-chain α-keto acid dehydrogenase complex	Vomiting, convulsions, mental retardation, early death
Methylmalonic acidemia	<0.5	Conversion of propionyl-CoA to succinyl-CoA	Methylmalonyl-CoA mutase	Vomiting, convulsions, mental retardation, early death
Phenylketonuria	8	Conversion of phenylalanine to tyrosine	Phenylalanine hydroxylase	Neonatal vomiting; mental retardation

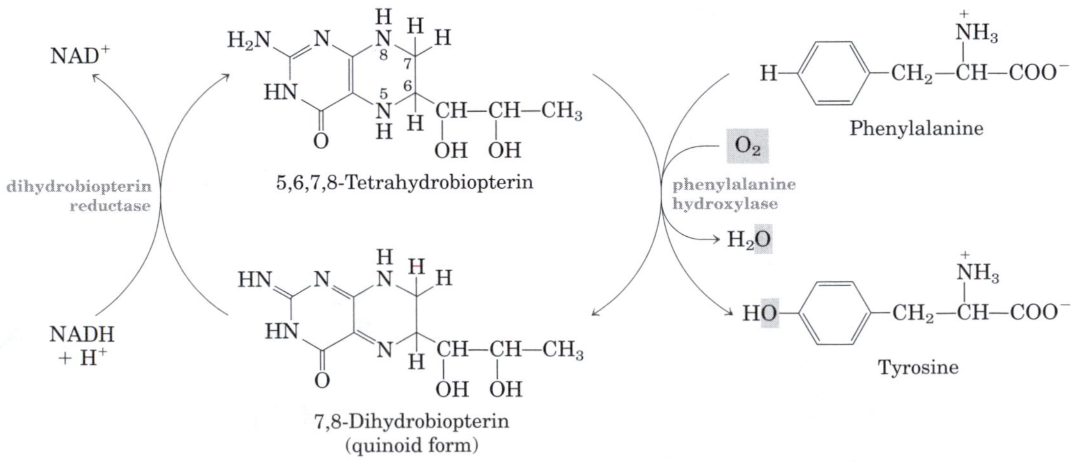

figure 18–23, page 647
Role of tetrahydrobiopterin in the phenylalanine hydroxylase reaction.

Phenylalanine
Pyruvate
aminotransferase
PLP
Alanine
Phenylpyruvate
CO_2
H_2O
Phenylacetate
Phenyllactate

figure 18–24, page 648
Alternative pathways for catabolism of phenylalanine in phenylketonuria.

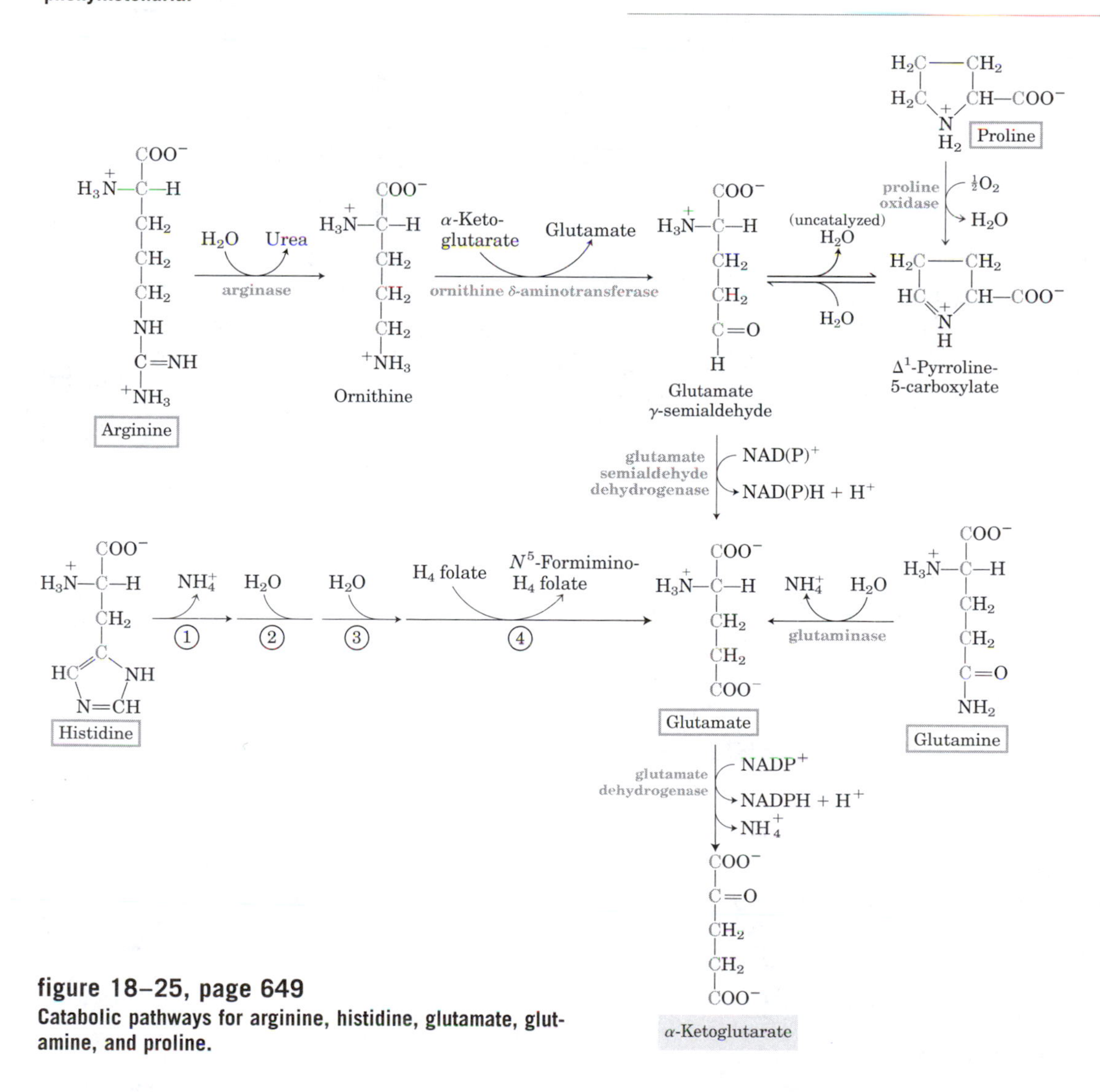

figure 18–25, page 649
Catabolic pathways for arginine, histidine, glutamate, glutamine, and proline.

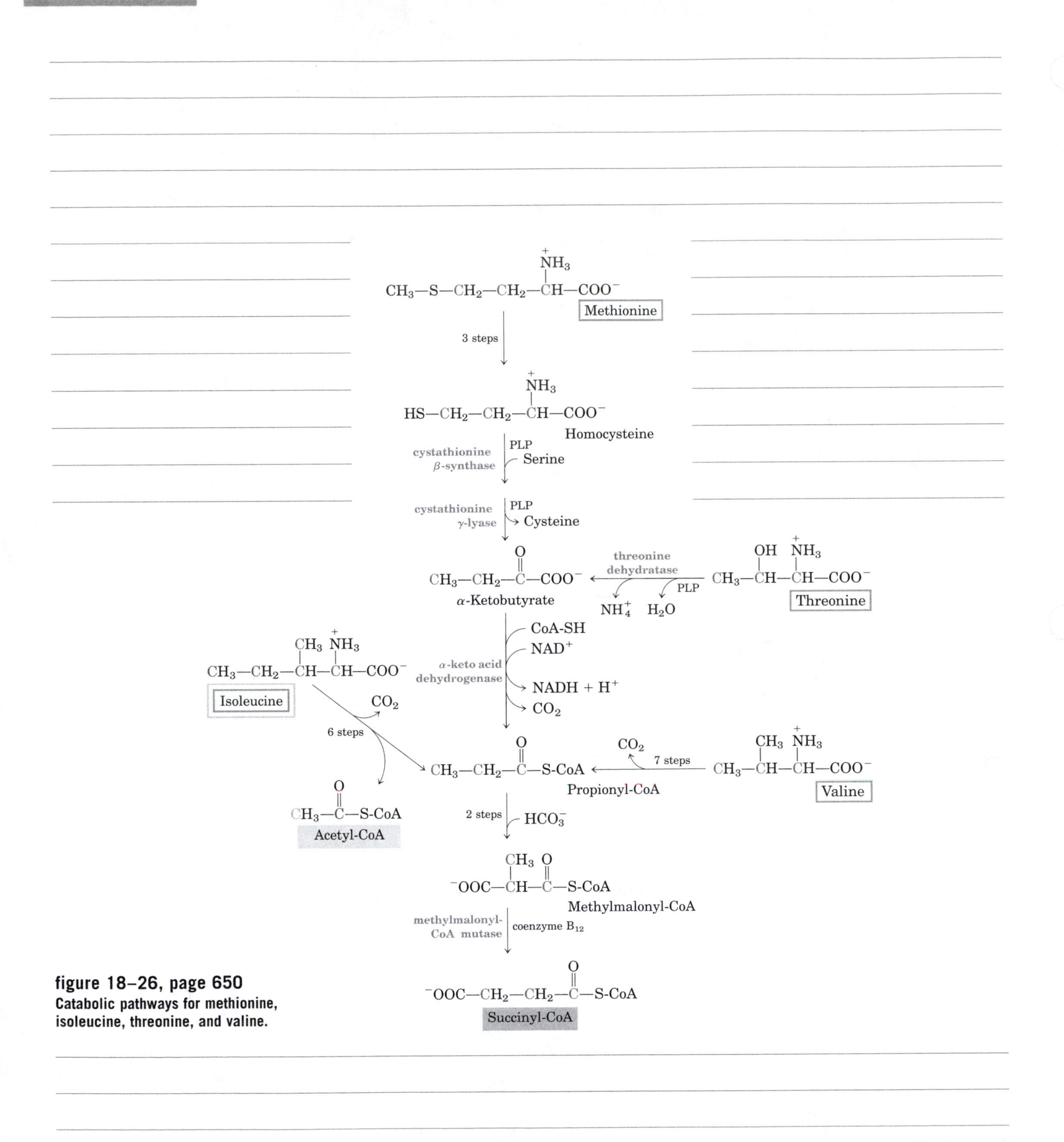

figure 18–26, page 650
Catabolic pathways for methionine, isoleucine, threonine, and valine.

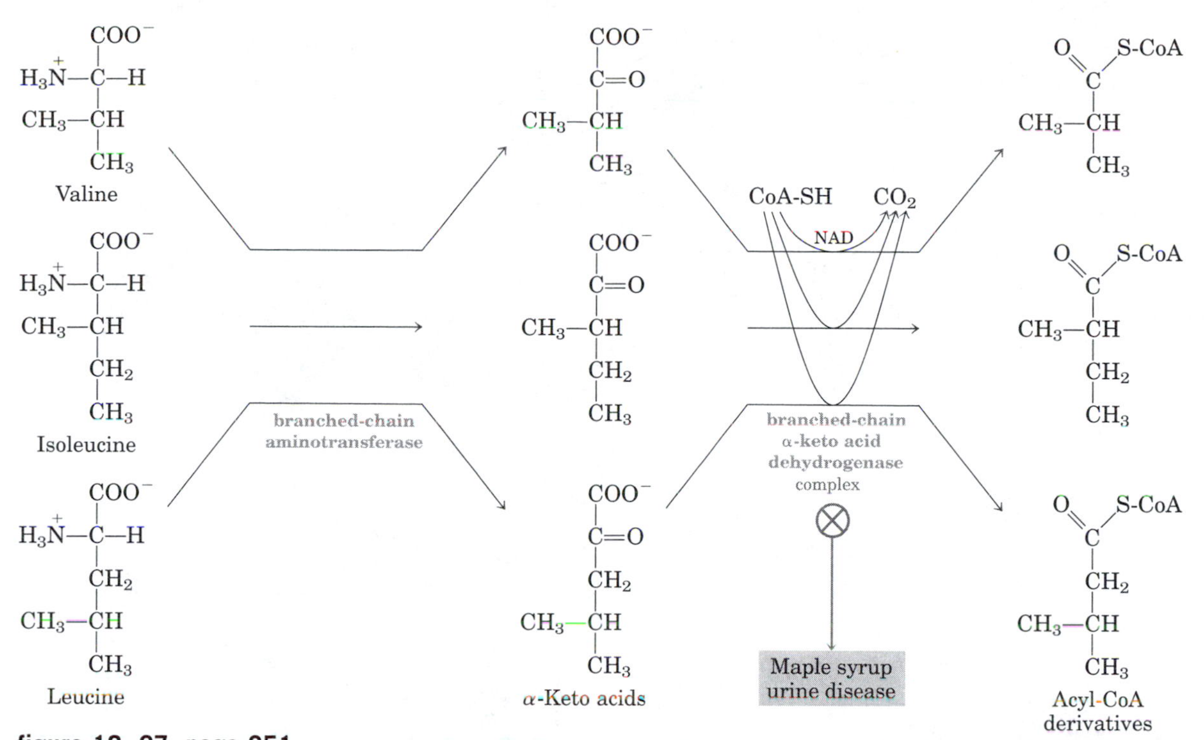

figure 18–27, page 651
Catabolic pathways for the three branched-chain amino acids: valine, isoleucine, and leucine.

figure 18–28, page 653
Catabolic pathway for asparagine and aspartate.

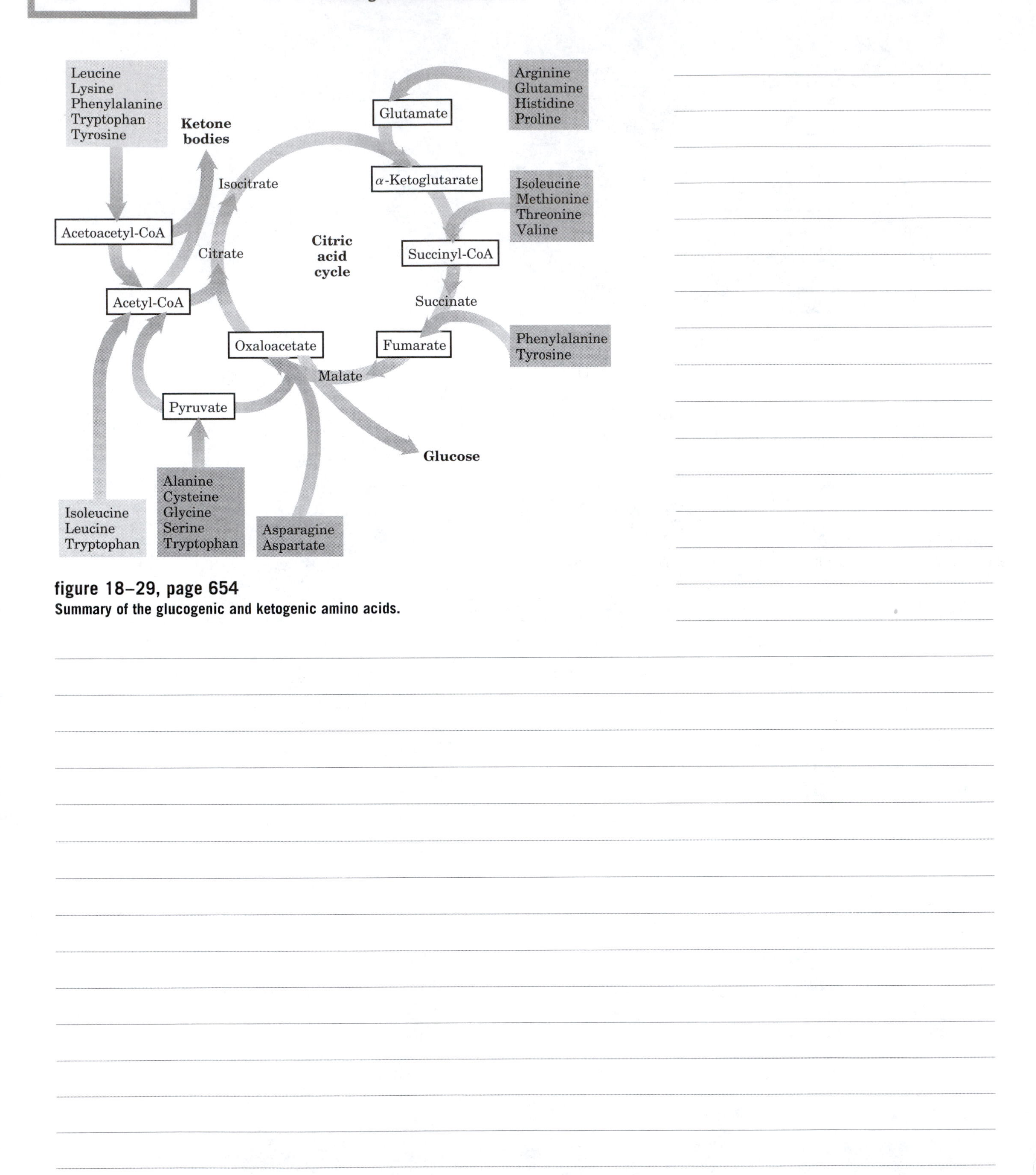

figure 18–29, page 654
Summary of the glucogenic and ketogenic amino acids.

PROBLEMS

4. A Genetic Defect in Amino Acid Metabolism: A Case History, page 656.

Substance	Concentration (mM)	
	Patient's urine	Normal urine
Phenylalanine	7.0	0.01
Phenylpyruvate	4.8	0
Phenyllactate	10.3	0

6. Lactate versus Alanine as Metabolic Fuel: The Cost of Nitrogen Removal, page 657.

COO^- / HO—C—H / H—C—H / H

Lactate

COO^- / $H_3\overset{+}{N}$—C—H / H—C—H / H

Alanine

7. Pathway of Carbon and Nitrogen in Glutamate Metabolism, page 657.

H—$^{15}\overset{+}{N}$(H)(H)—^{14}C(H)(COO^-)—CH_2—CH_2—COO^-

Glutamate

8. Chemical Strategy of Isoleucine Catabolism, page 657.

Isoleucine $\xrightarrow{\text{6 steps}}$ Propionyl-CoA + Acetyl-CoA

Isoleucine: $O=C(O^-)$—C(H)($H_3\overset{+}{N}$)—C(H)(CH_3)—CH_2—CH_3

Propionyl-CoA: $O=C(S\text{-}CoA)$—CH_2—CH_3

Acetyl-CoA: $O=C(S\text{-}CoA)$—CH_3

I: $O=C(S\text{-}CoA)$—C(CH_3)=C(H)—CH_3

II: $O=C(O^-)$—C=O—C(H)(CH_3)—CH_2—CH_3

III: $O=C(S\text{-}CoA)$—C(H)(CH_3)—C=O—CH_3

IV: $O=C(S\text{-}CoA)$—C(H)(CH_3)—CH_2—CH_3

V: $O=C(S\text{-}CoA)$—C(H)(CH_3)—C(H)(HO)—CH_3

12. Parallel Pathways for Amino Acid and Fatty Acid Degradation, page 658.

$CH_3-CH(CH_3)-CH_2-CH(\overset{+}{N}H_3)-COO^-$

Leucine

↓ (a)

$CH_3-CH(CH_3)-CH_2-C(=O)-COO^-$

α-Ketoisocaproate

↓ (b) CoA-SH → CO_2

$CH_3-CH(CH_3)-CH_2-C(=O)-S\text{-}CoA$

Isovaleryl-CoA

↓ (c)

$CH_3-C(H_3C)=C(H)-C(=O)-S\text{-}CoA$

β-Methylcrotonyl-CoA

↓ (d) HCO_3^-

$^-OOC-CH_2-C(H_3C)=C(H)-C(=O)-S\text{-}CoA$

β-Methylglutaconyl-CoA

↓ (e) H_2O

$^-OOC-CH_2-C(OH)(CH_3)-CH_2-C(=O)-S\text{-}CoA$

β-Hydroxy-β-methylglutaryl-CoA

↓ (f)

$^-OOC-CH_2-C(=O)-CH_3$ + $CH_3-C(=O)-S\text{-}CoA$

Acetoacetate + Acetyl-CoA

chapter

19

Oxidative Phosphorylation and Photophosphorylation

When you study for exams, don't forget to review The Minicourses on the UNDERSTAND! BIOCHEMISTRY CD that came with your textbook.

Minicourses that apply to this Chapter include:

Molecules of Life
- An Interactive Gallery of Protein Structures

Bioenergetics
- Energetics of Metabolism
- Electron Transport – Oxidative Phosphorylation
- Photophosphorylation

The Dividing Cell
- Cancer

ATP synthase (F_oF_1)

Cristae

Outer membrane

Freely permeable to small molecules and ions

Inner membrane

Impermeable to most small molecules and ions, including H^+

Contains:
- Respiratory electron carriers (Complexes I–IV)
- ADP-ATP translocases
- ATP synthase (F_oF_1)
- Other membrane transporters

Matrix

Contains:
- Pyruvate dehydrogenase complex
- Citric acid cycle enzymes
- Fatty acid β-oxidation enzymes
- Amino acid oxidation enzymes
- DNA, ribosomes
- Many other enzymes
- ATP, ADP, P_i, Mg^{2+}, Ca^{2+}, K^+
- Many soluble metabolic intermediates

Ribosomes

Porin channels

figure 19–1, page 660
Biochemical anatomy of a mitochondrion.

table 19–1, page 661

Some Important Reactions Catalyzed by NAD(P)H-Linked Dehydrogenases

Reaction*	Location†
NAD-linked	
α-Ketoglutarate + CoA + $NAD^+ \rightleftharpoons$ succinyl-CoA + CO_2 + NADH + H^+	M
L-Malate + $NAD^+ \rightleftharpoons$ oxaloacetate + NADH + H^+	M and C
Pyruvate + CoA + $NAD^+ \rightleftharpoons$ acetyl-CoA + CO_2 + NADH + H^+	M
Glyceraldehyde 3-phosphate + P_i + $NAD^+ \rightleftharpoons$ 1,3-bisphosphoglycerate + NADH + H^+	C
Lactate + $NAD^+ \rightleftharpoons$ pyruvate + NADH + H^+	C
β-Hydroxyacyl-CoA + $NAD^+ \rightleftharpoons$ β-ketoacyl-CoA + NADH + H^+	M
NADP-linked	
Glucose 6-phosphate + $NADP^+ \rightleftharpoons$ 6-phosphogluconate + NADPH + H^+	C
NAD- or NADP-linked	
L-Glutamate + H_2O + $NAD(P)^+ \rightleftharpoons$ α-ketoglutarate + NH_4^+ + NAD(P)H	M
Isocitrate + $NAD(P)^+ \rightleftharpoons$ α-ketoglutarate + CO_2 + NAD(P)H + H^+	M and C

*These reactions and their enzymes are discussed in Chapters 15 through 18.

†M designates mitochondria; C, cytosol.

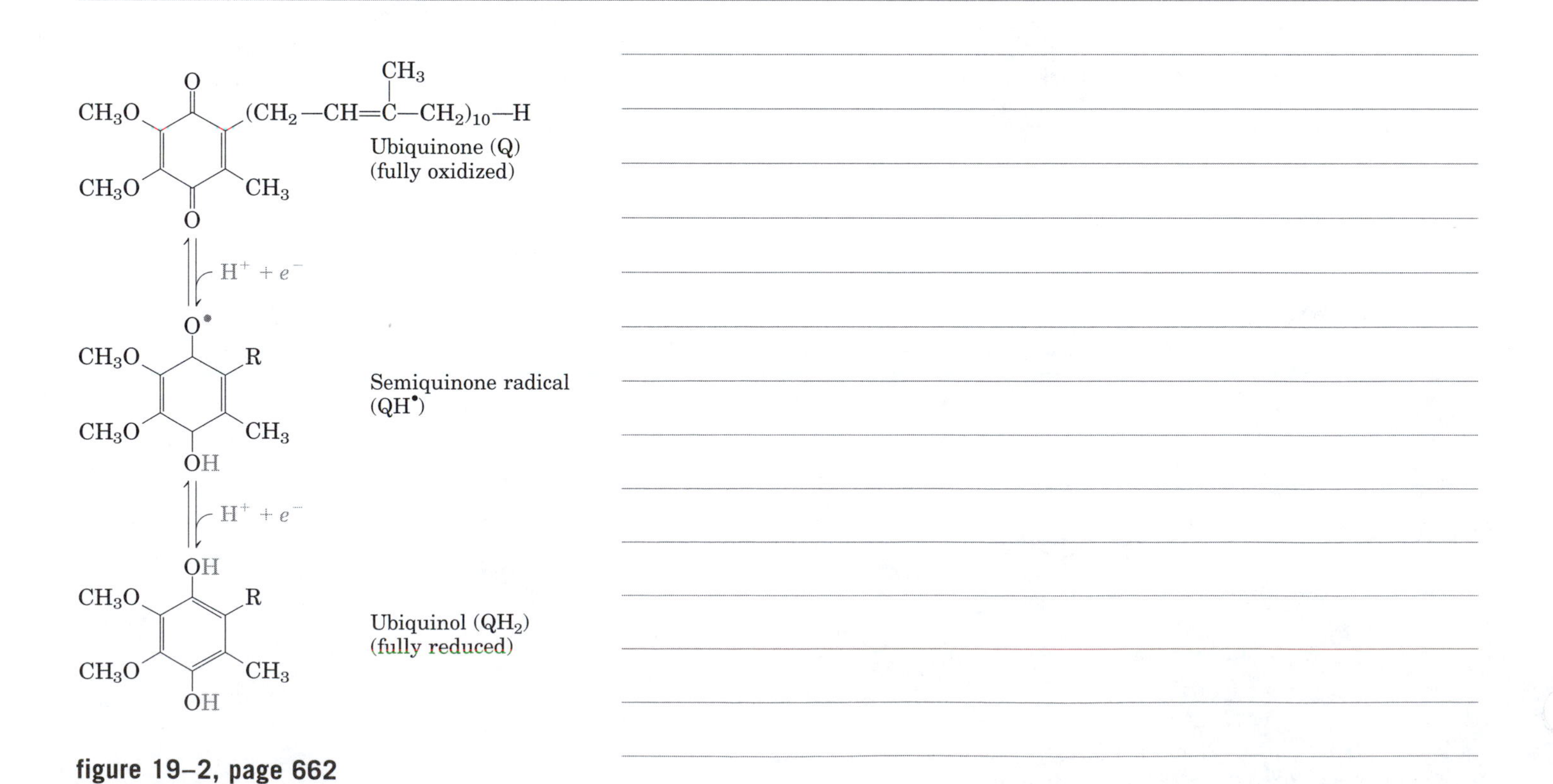

figure 19–2, page 662
Ubiquinone (Q, or coenzyme Q).

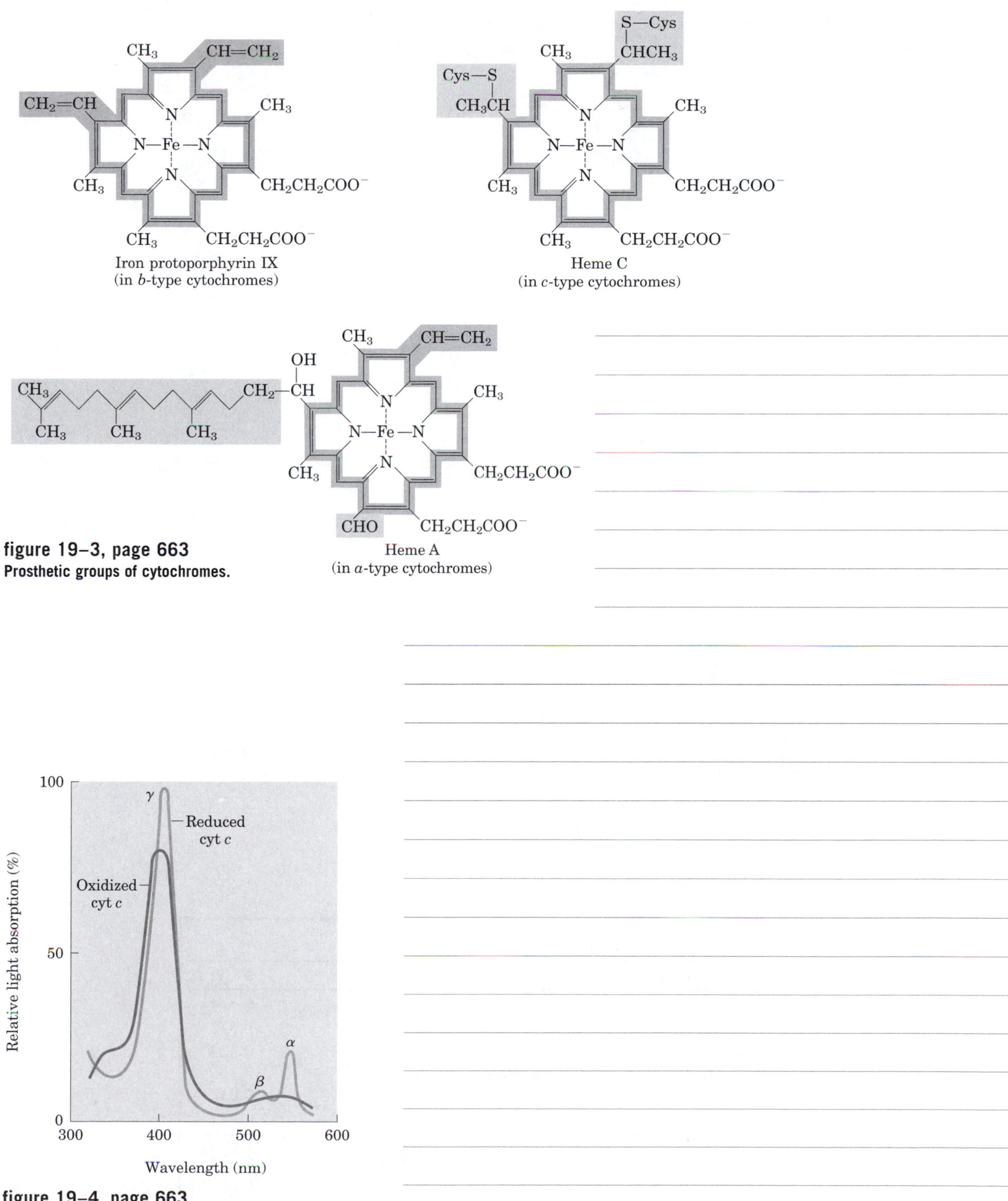

figure 19–3, page 663
Prosthetic groups of cytochromes.

figure 19–4, page 663
Absorption spectra of cytochrome *c* in its oxidized (dark gray) and reduced (light gray) forms. Also labeled are the characteristic α, β, and γ bands of the reduced form.

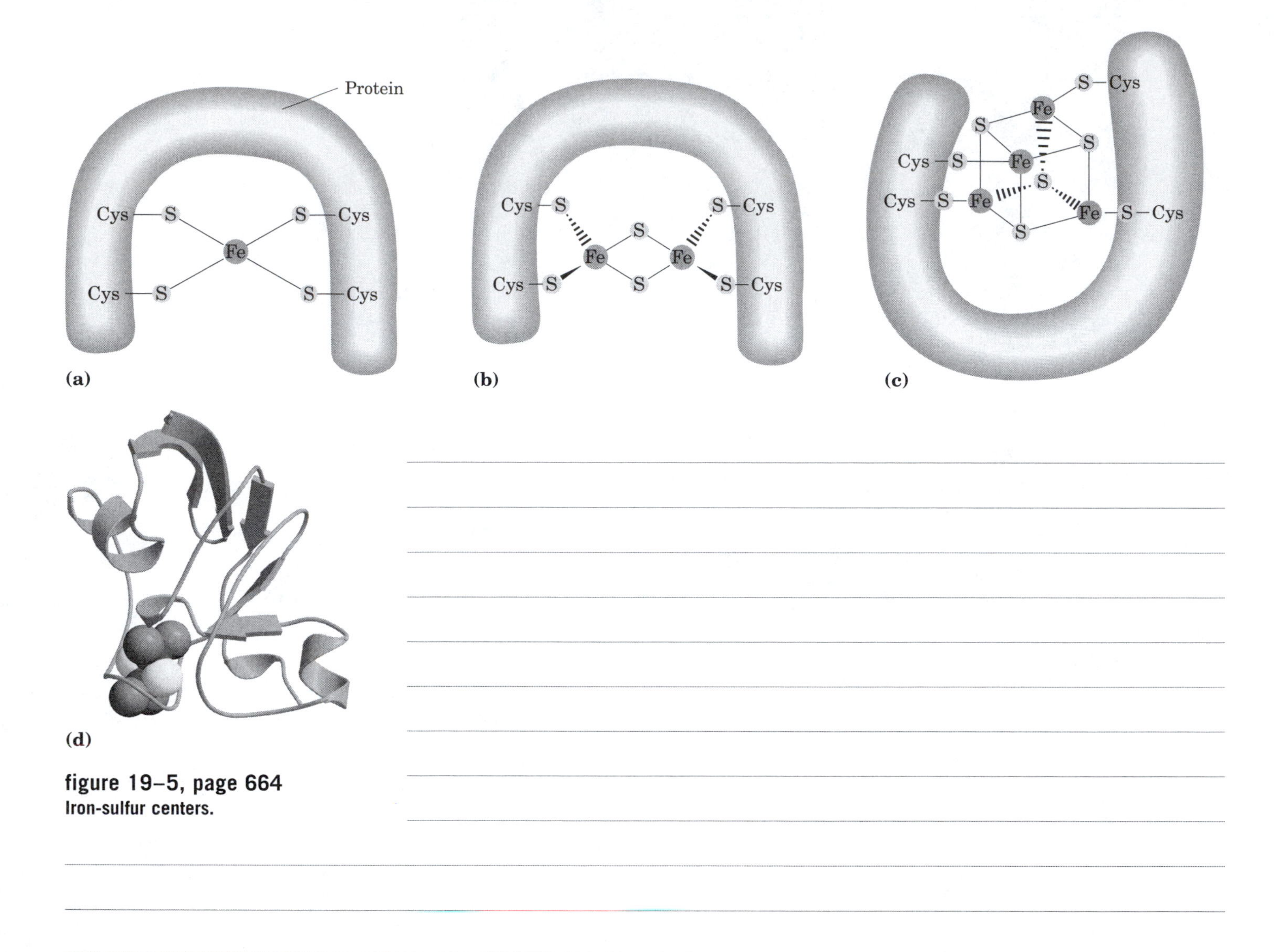

figure 19–5, page 664
Iron-sulfur centers.

table 19–2, page 665

Standard Reduction Potentials of Respiratory Chain and Related Electron Carriers

Redox reaction (half-reaction)	E'° (V)
$2H^+ + 2e^- \longrightarrow H_2$	−0.414
$NAD^+ + H^+ + 2e^- \longrightarrow$ NADH	−0.320
$NADP^+ + H^+ + 2e^- \longrightarrow$ NADPH	−0.324
NADH dehydrogenase (FMN) + $2H^+ + 2e^- \longrightarrow$ NADH dehydrogenase ($FMNH_2$)	−0.30
Ubiquinone + $2H^+ + 2e^- \longrightarrow$ ubiquinol	0.045
Cytochrome *b* (Fe^{3+}) + $e^- \longrightarrow$ cytochrome *b* (Fe^{2+})	0.077
Cytochrome c_1 (Fe^{3+}) + $e^- \longrightarrow$ cytochrome c_1 (Fe^{2+})	0.22
Cytochrome *c* (Fe^{3+}) + $e^- \longrightarrow$ cytochrome *c* (Fe^{2+})	0.254
Cytochrome *a* (Fe^{3+}) + $e^- \longrightarrow$ cytochrome *a* (Fe^{2+})	0.29
Cytochrome a_3 (Fe^{3+}) + $e^- \longrightarrow$ cytochrome a_3 (Fe^{2+})	0.55
$\frac{1}{2}O_2 + 2H^+ + 2e^- \longrightarrow H_2O$	0.816

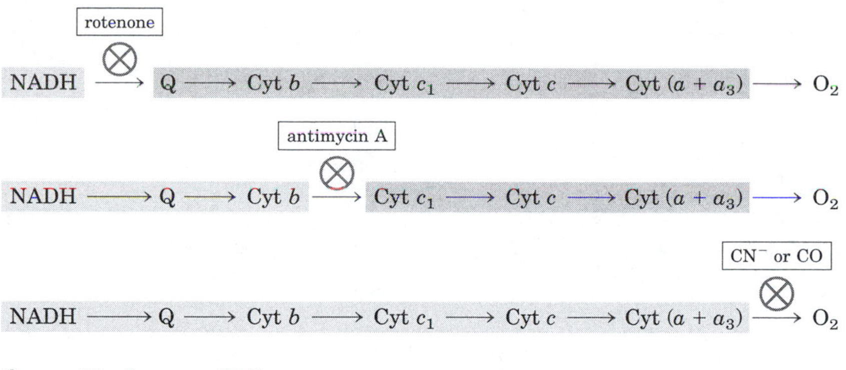

figure 19–6, page 665
Method for determining the sequence of electron carriers.

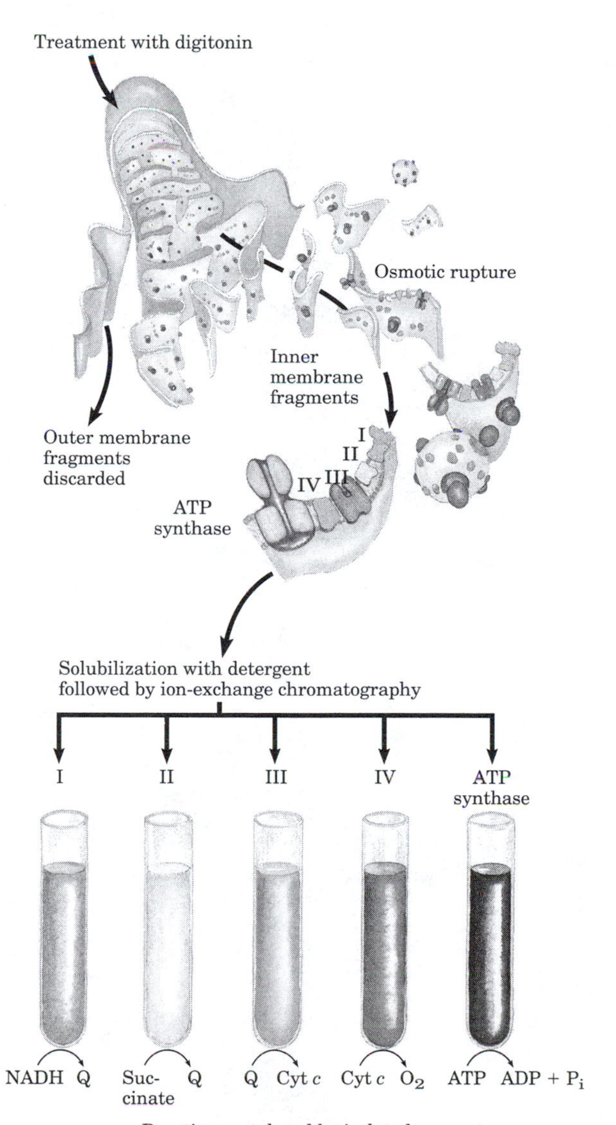

figure 19–7, page 666
Separation of functional complexes of the respiratory chain.

table 19–3, page 667

Protein Components of the Mitochondrial Electron-Transfer Chain

Enzyme complex	Mass (kDa)	Number of subunits*	Prosthetic group(s)
I NADH dehydrogenase	850	42 (14)	FMN, Fe-S
II Succinate dehydrogenase	140	5	FAD, Fe-S
III Ubiquinone: cytochrome *c* oxidoreductase	250	11	Hemes, Fe-S
Cytochrome *c*†	13	1	Heme
IV Cytochrome oxidase	160	13 (3–4)	Hemes; Cu_A, Cu_B

*Numbers of subunits in the bacterial equivalents in parentheses.

†Cytochrome *c* is not part of an enzyme complex; it moves between Complexes III and IV as a freely soluble protein.

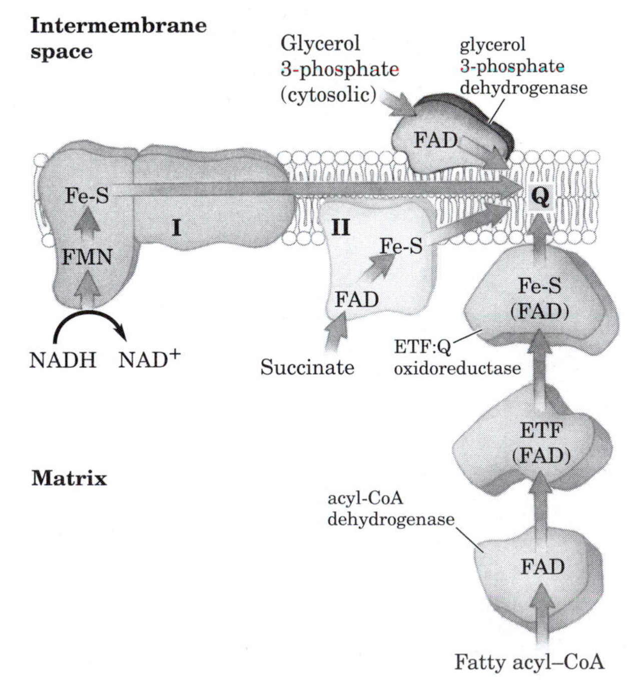

figure 19–8, page 667
Path of electrons from NADH, succinate, fatty acyl–CoA, and glycerol 3-phosphate to ubiquinone.

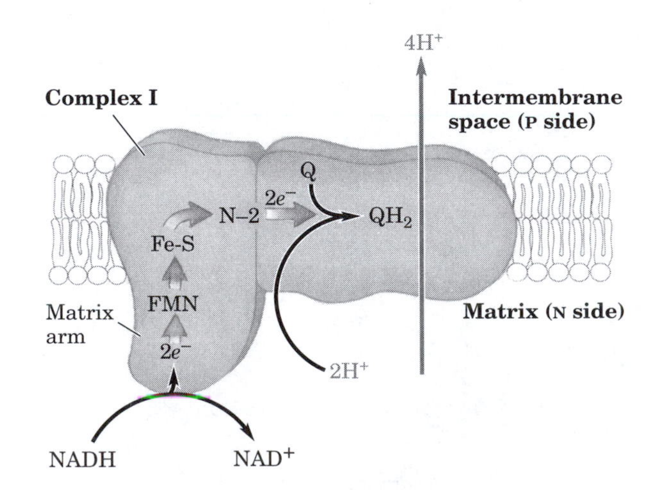

figure 19–9, page 667
NADH:ubiquinone oxidoreductase (Complex I).

table 19–4, page 668

Some Agents That Interfere with Oxidative Phosphorylation or Photophosphorylation

Type of interference	Compound*	Target/mode of action
Inhibition of electron transfer	Cyanide Carbon monoxide	Inhibit cytochrome oxidase
	Antimycin A	Blocks electron transfer from cytochrome *b* to cytochrome c_1
	Myxothiazol Rotenone Amytal Piericidin A	Prevent electron transfer from Fe-S center to ubiquinone
	DCMU	Competes with Q_B for binding site in PSII
Inhibition of ATP synthase	Aurovertin	Inhibits F_1
	Oligomycin Venturicidin	Inhibit F_o and CF_o
	DCCD	Blocks proton flow through F_o and CF_o
Uncoupling of phosphorylation from electron transfer	FCCP DNP	Hydrophobic proton carriers
	Valinomycin	K^+ ionophore
	Thermogenin	Forms proton-conducting pores in inner membrane of brown fat mitochondria
Inhibition of ATP-ADP exchange	Atractyloside	Inhibits adenine nucleotide translocase

*DCMU is 3-(3,4-dichlorophenyl)-1,1-dimethylurea; DCCD, dicyclohexylcarbodiimide; FCCP, cyanide-*p*-trifluoromethoxyphenylhydrazone; DNP, 2,4-dinitrophenol.

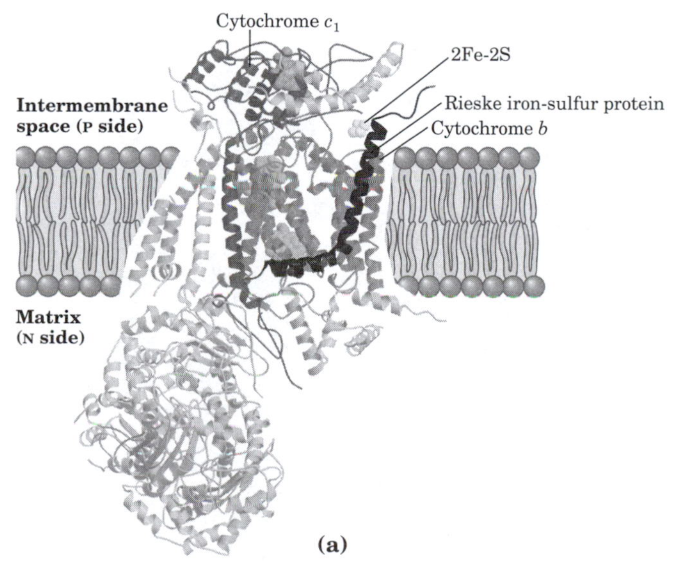

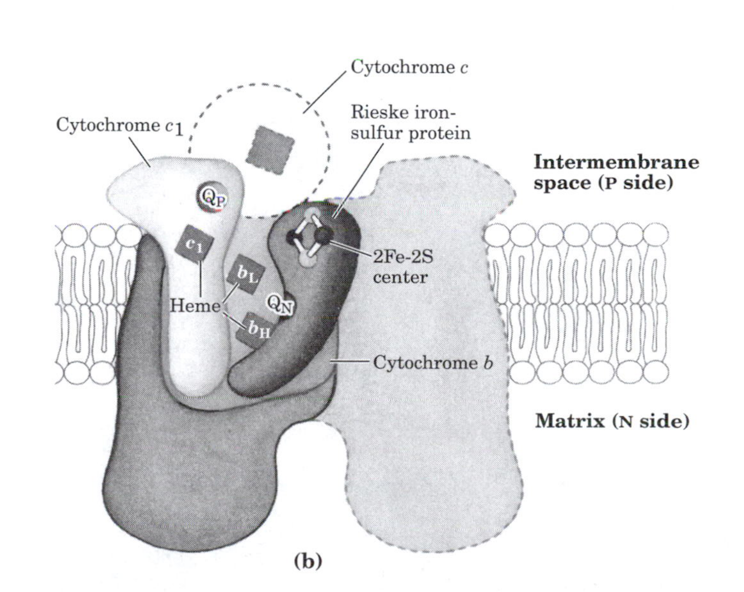

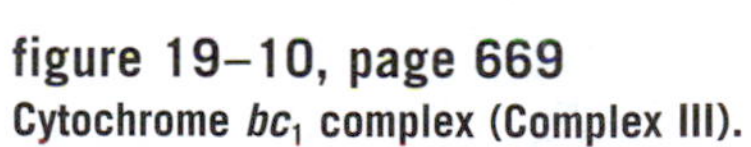
figure 19–10, page 669
Cytochrome bc_1 complex (Complex III).

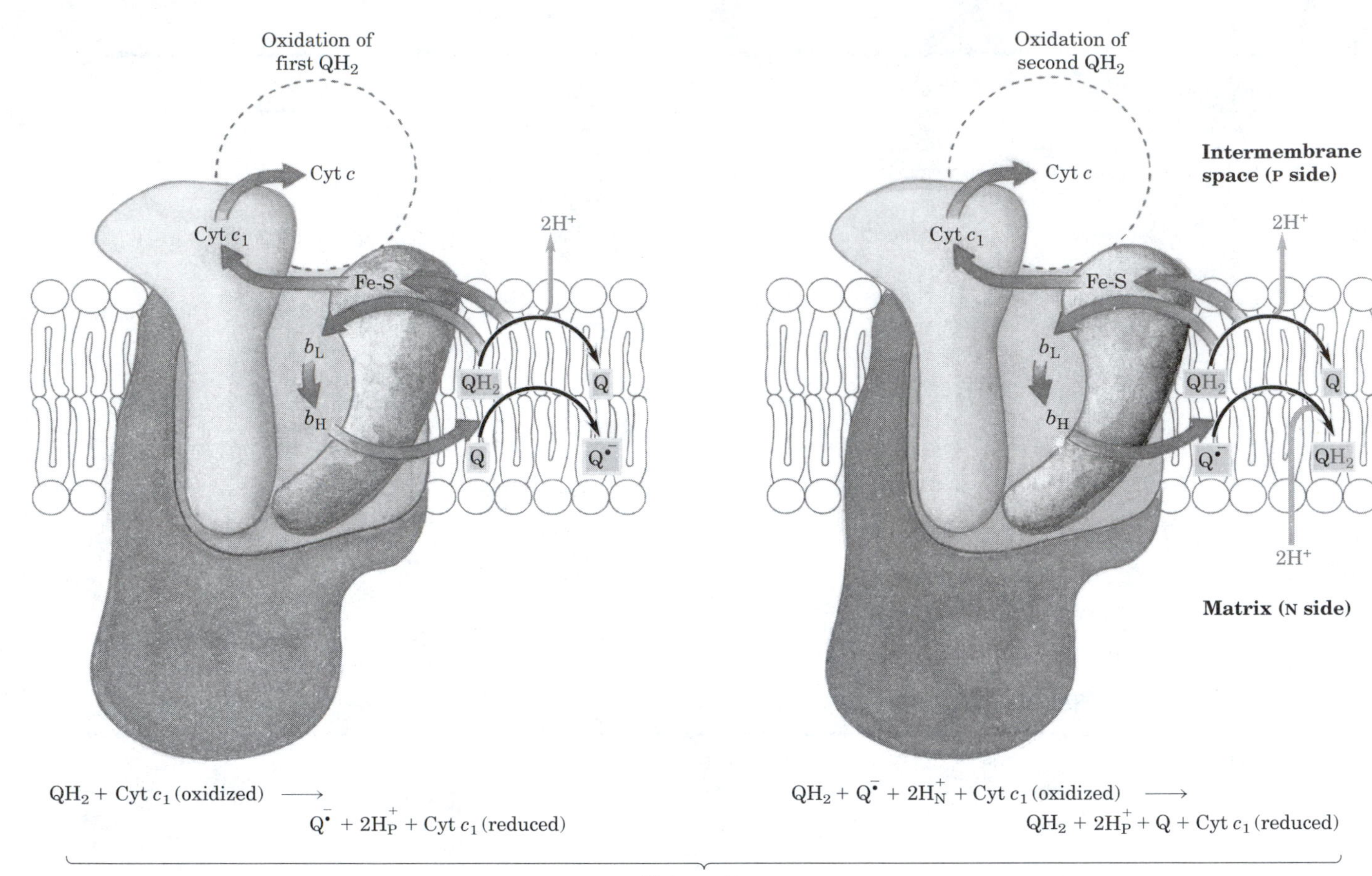

Net equation:

$$QH_2 + 2\ \text{Cyt}\ c_1\ (\text{oxidized}) + 2H_N^+ \longrightarrow Q + 2\ \text{Cyt}\ c_1\ (\text{reduced}) + 4H_P^+$$

figure 19–11, page 670
The Q cycle.

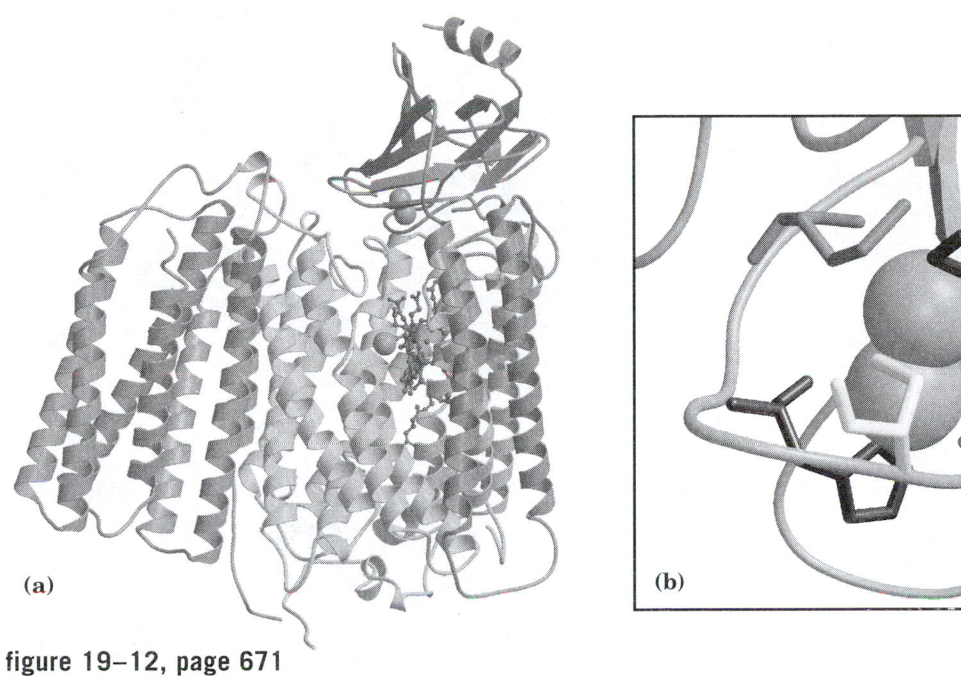

figure 19–12, page 671
Critical subunits of cytochrome oxidase (Complex IV).

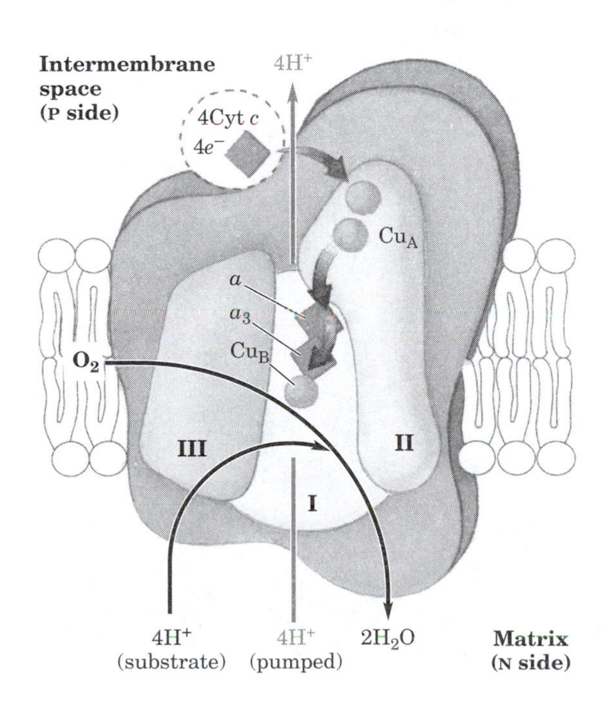

figure 19–13, page 671
Path of electrons through Complex IV.

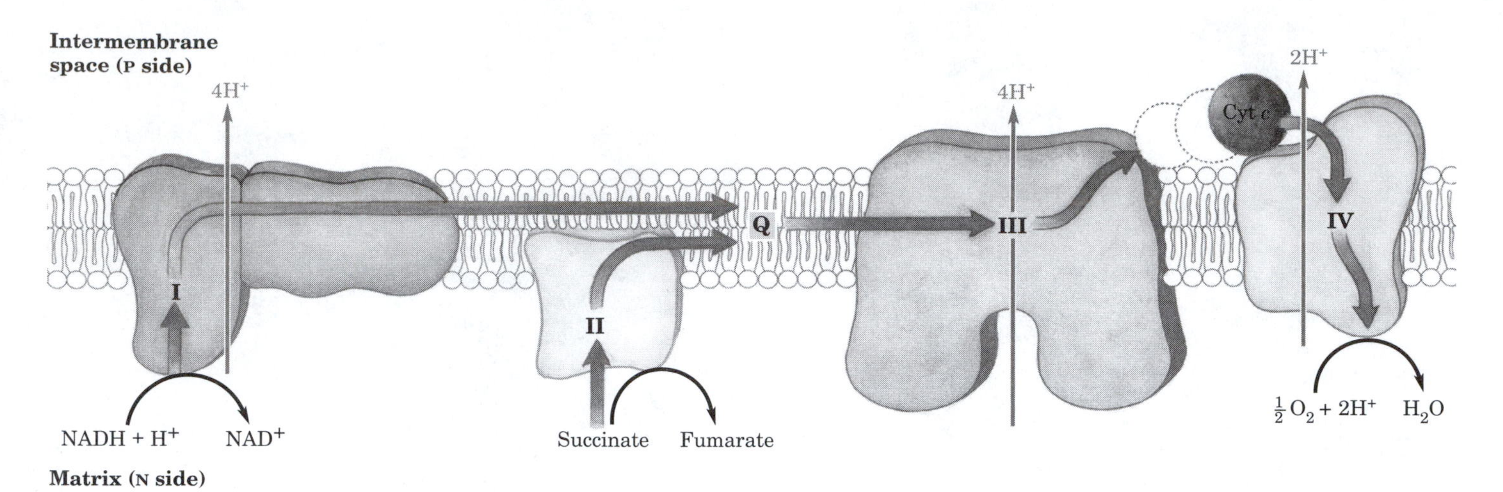

figure 19–14, page 672
Summary of the flow of electrons and protons through the four complexes of the respiratory chain.

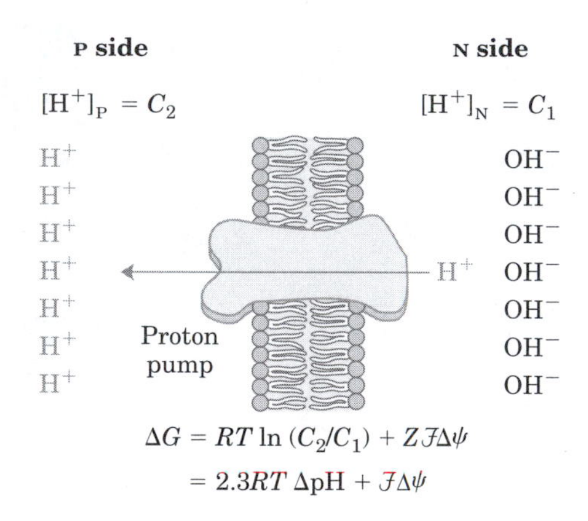

figure 19–15, page 673
Proton-motive force.

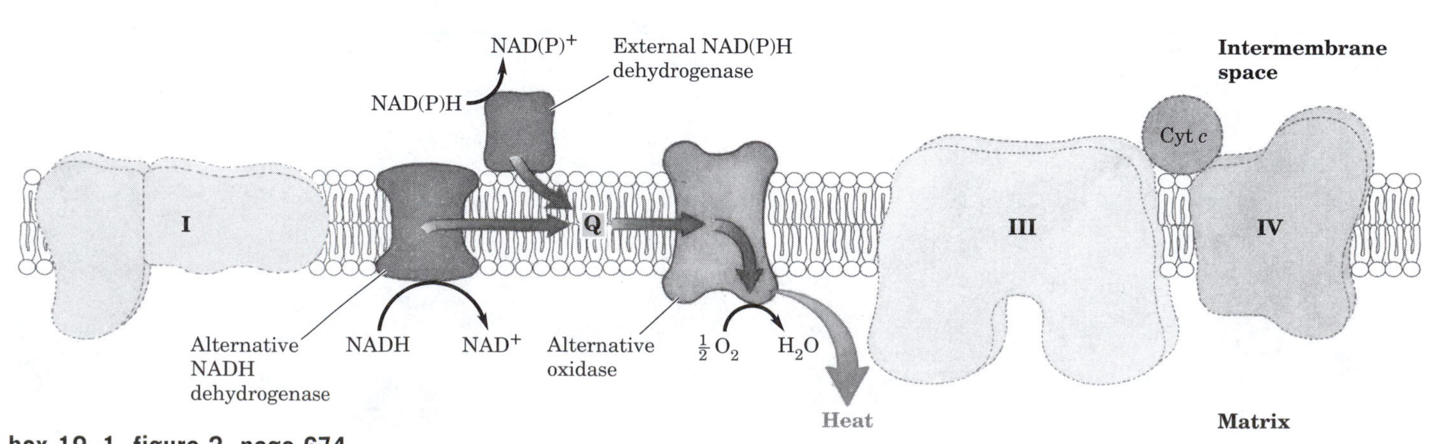

box 19–1, figure 2, page 674
Electron carriers of the inner membrane of plant mitochondria.

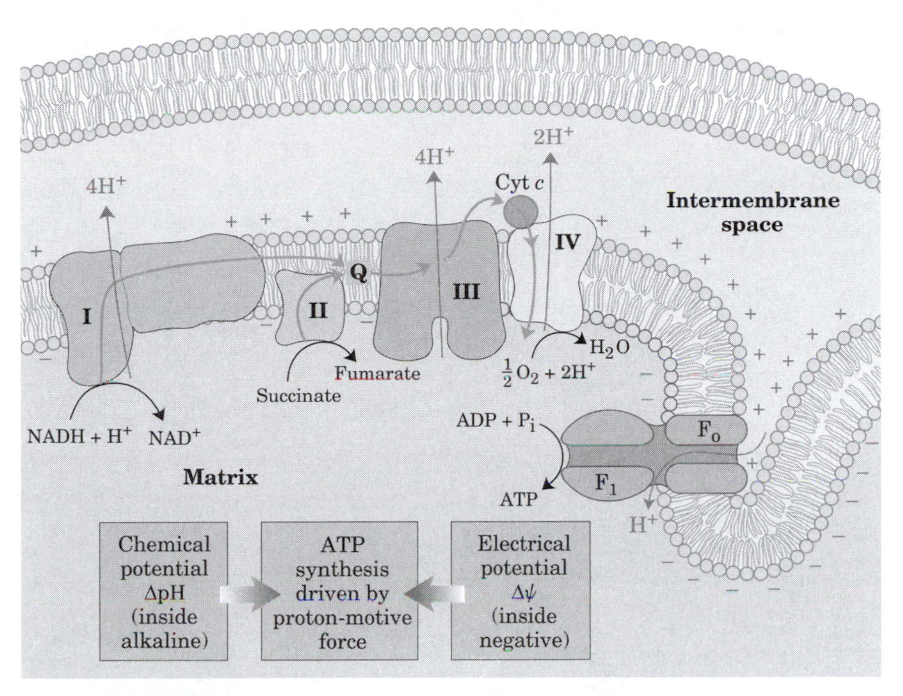

figure 19–16, page 675
Chemiosmotic model.

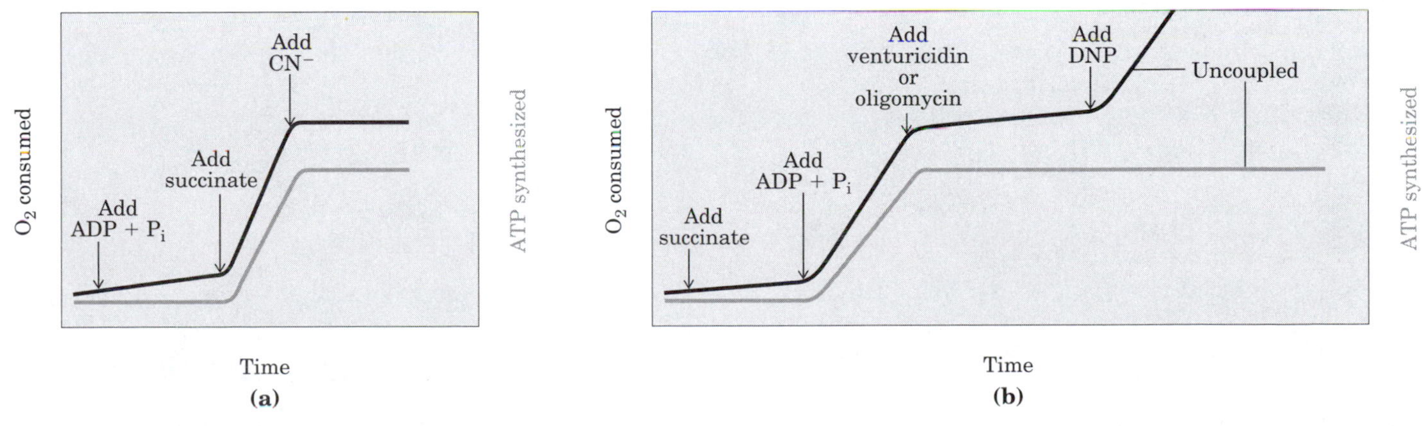

figure 19–17, page 676
Coupling of electron transfer and ATP synthesis in mitochondria.

2,4-Dinitrophenol
(DNP)

Carbonylcyanide-*p*-
trifluoromethoxyphenylhydrazone
(FCCP)

figure 19–18, page 677
Two chemical uncouplers of oxidative phosphorylation.

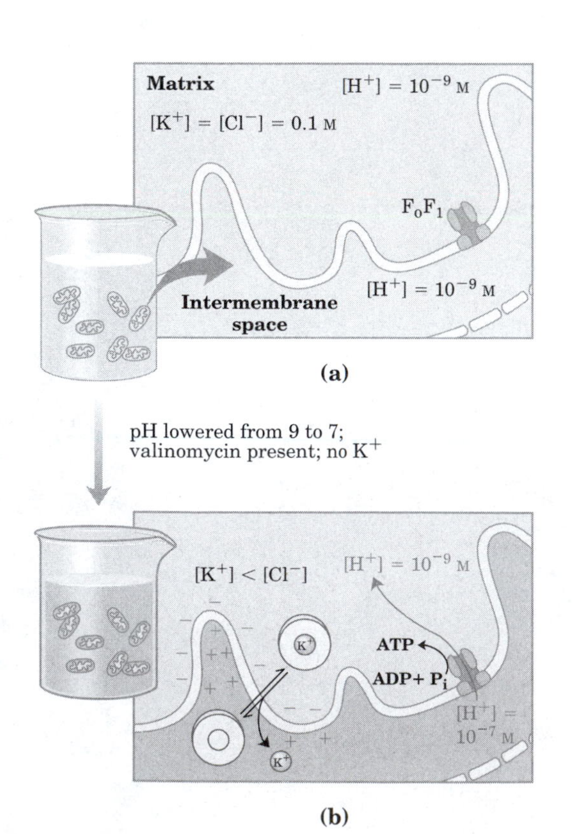

figure 19–19, page 677
Evidence for the role of a proton gradient in ATP synthesis.

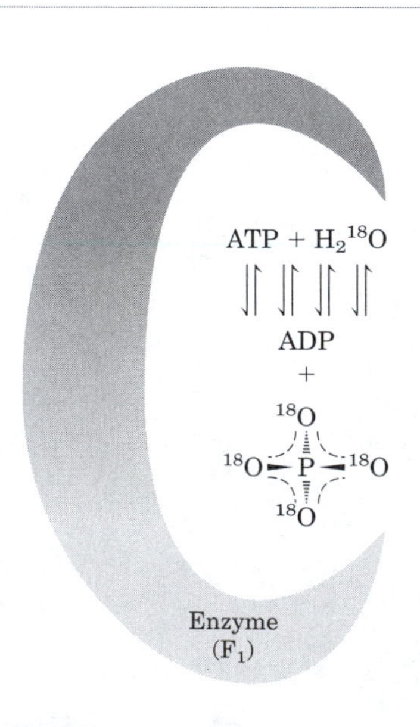

figure 19–20, page 678
Catalytic mechanism of F_1: ^{18}O exchange experiment.

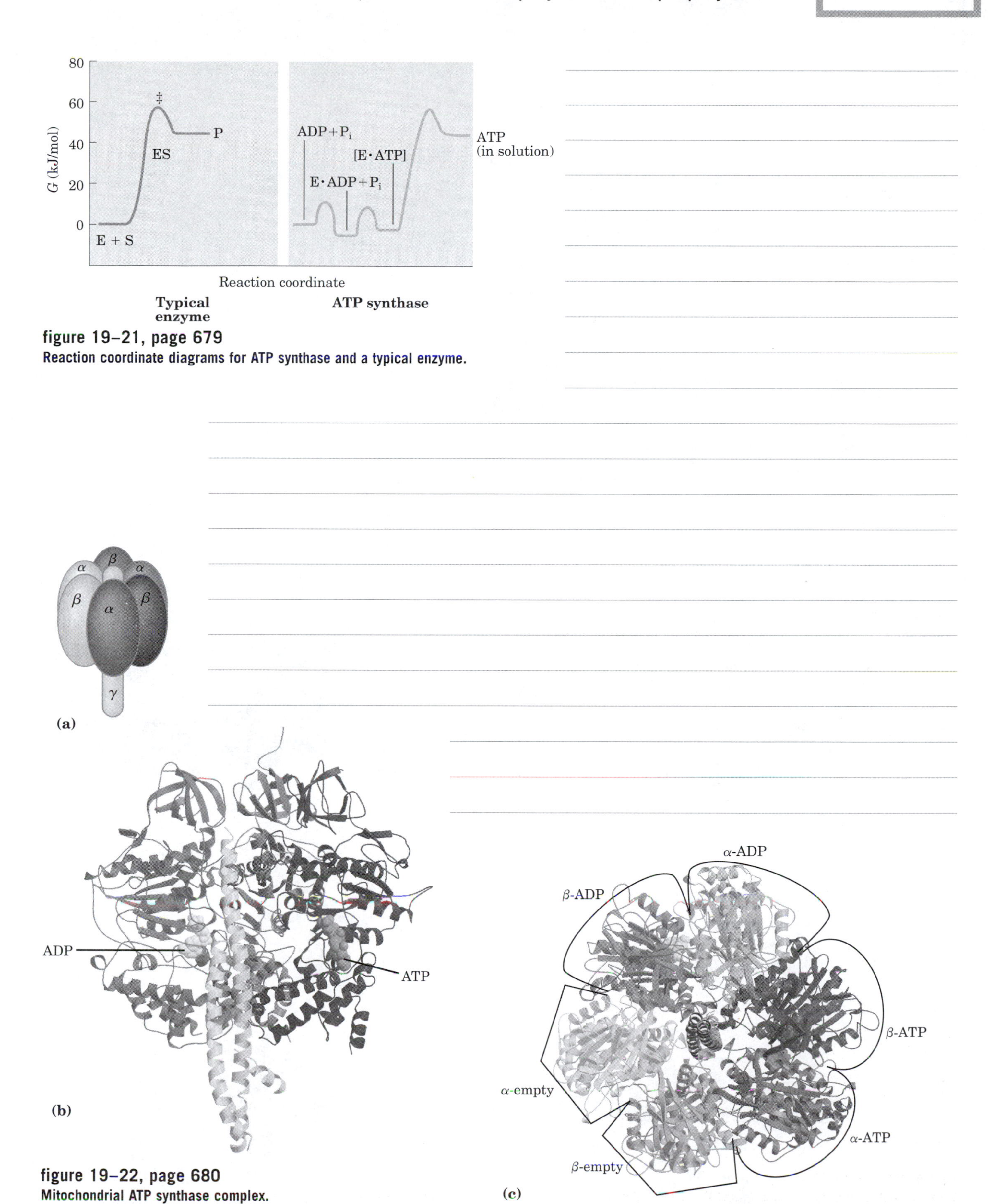

figure 19–21, page 679
Reaction coordinate diagrams for ATP synthase and a typical enzyme.

figure 19–22, page 680
Mitochondrial ATP synthase complex.

(β,γ-imidoadenosine 5′-triphosphate)

Nonhydrolyzable β-γ bond

App(NH)p, page 680

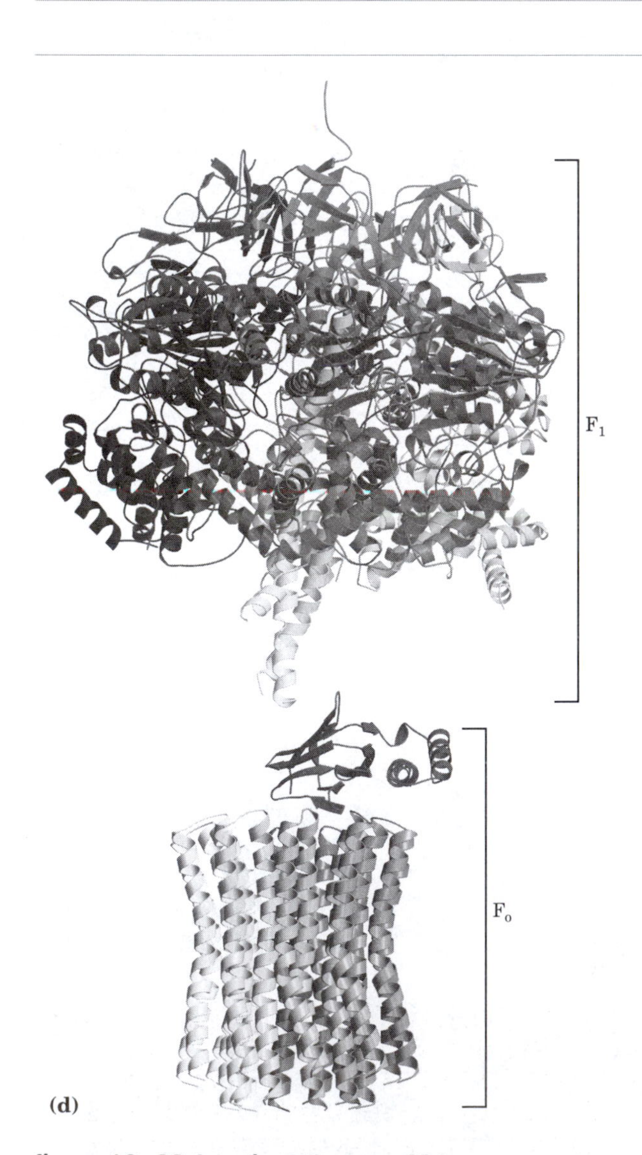

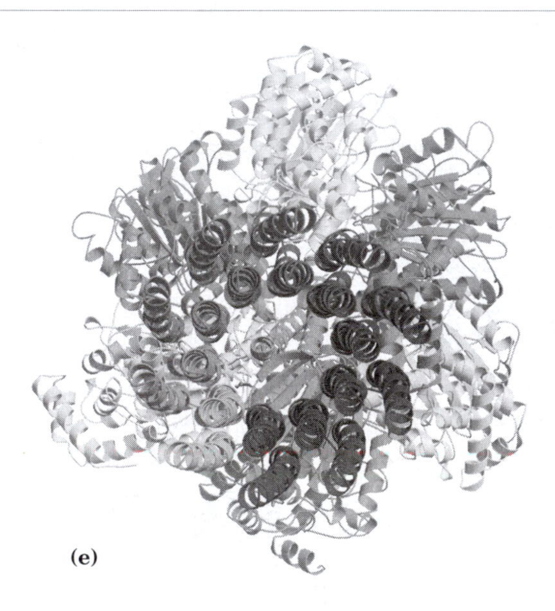

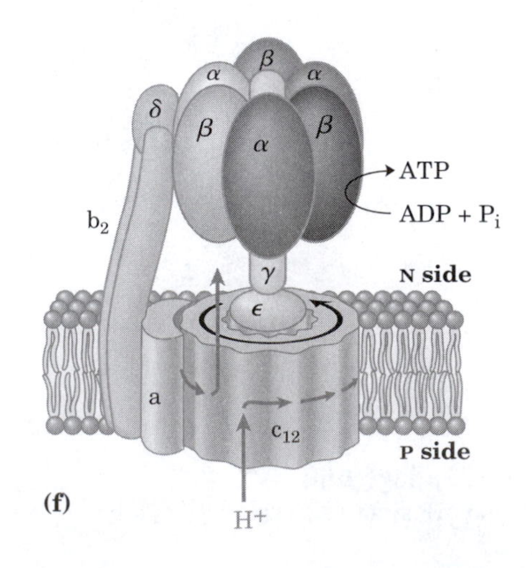

figure 19–22 (continued), page 681
Mitochondrial ATP synthase complex.

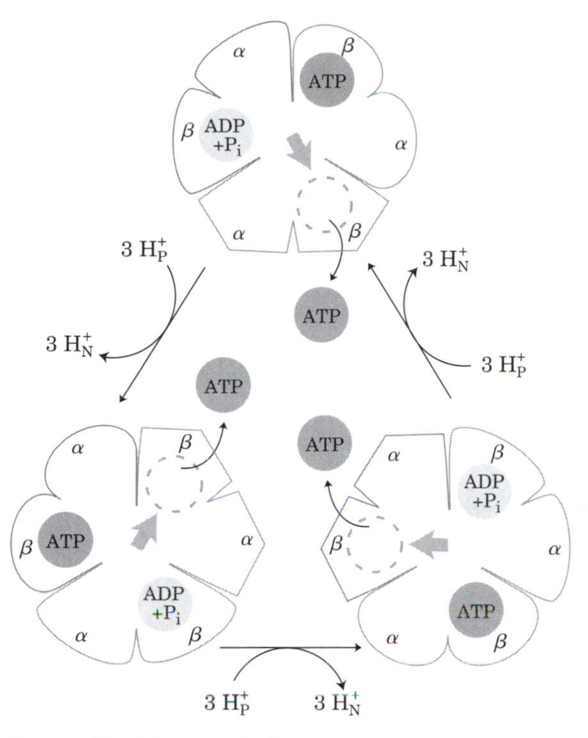

figure 19–23, page 682
Binding-change model for ATP synthase.

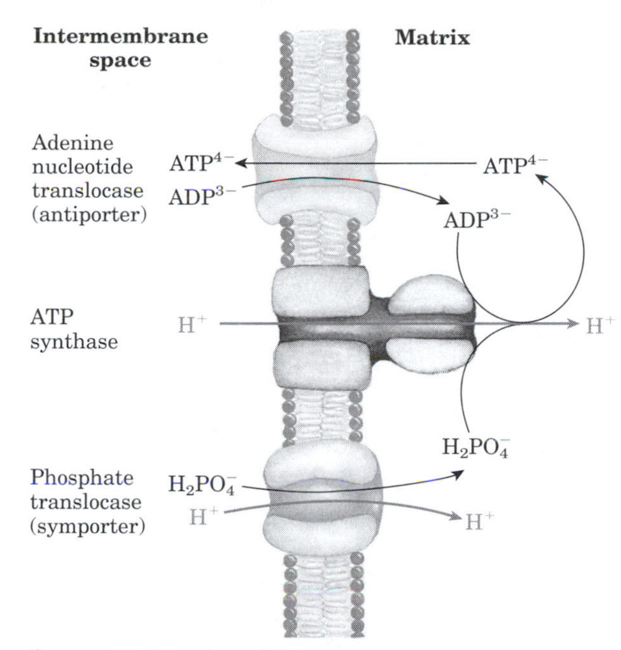

figure 19–25, page 684
Adenine nucleotide and phosphate translocases.

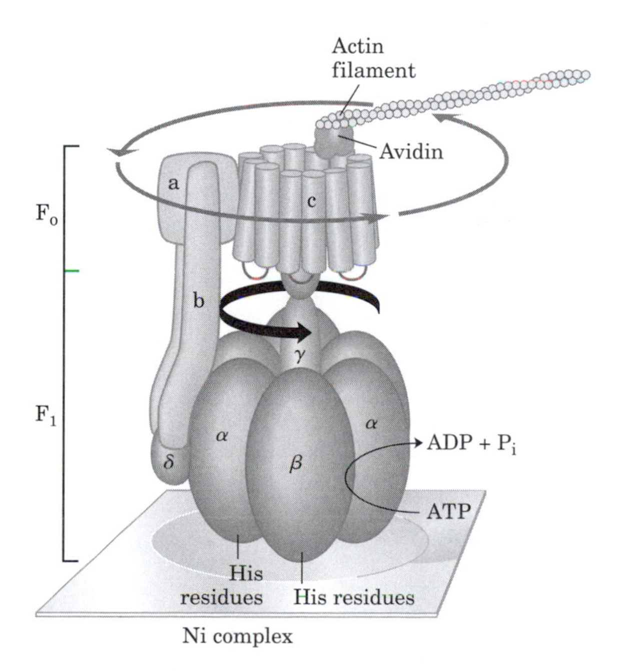

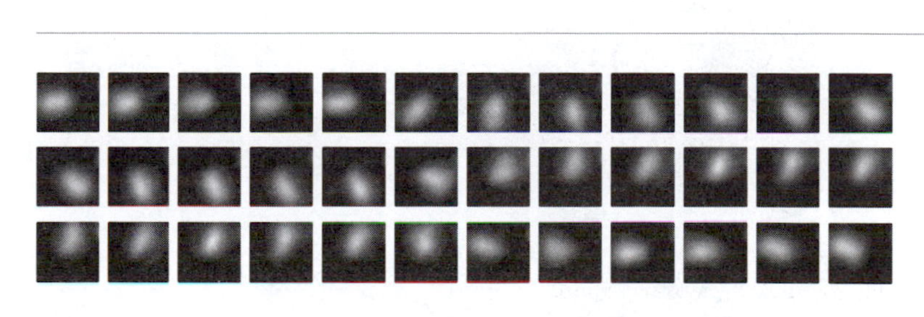

figure 19–24, page 683
Rotation of F$_o$ and γ experimentally demonstrated.

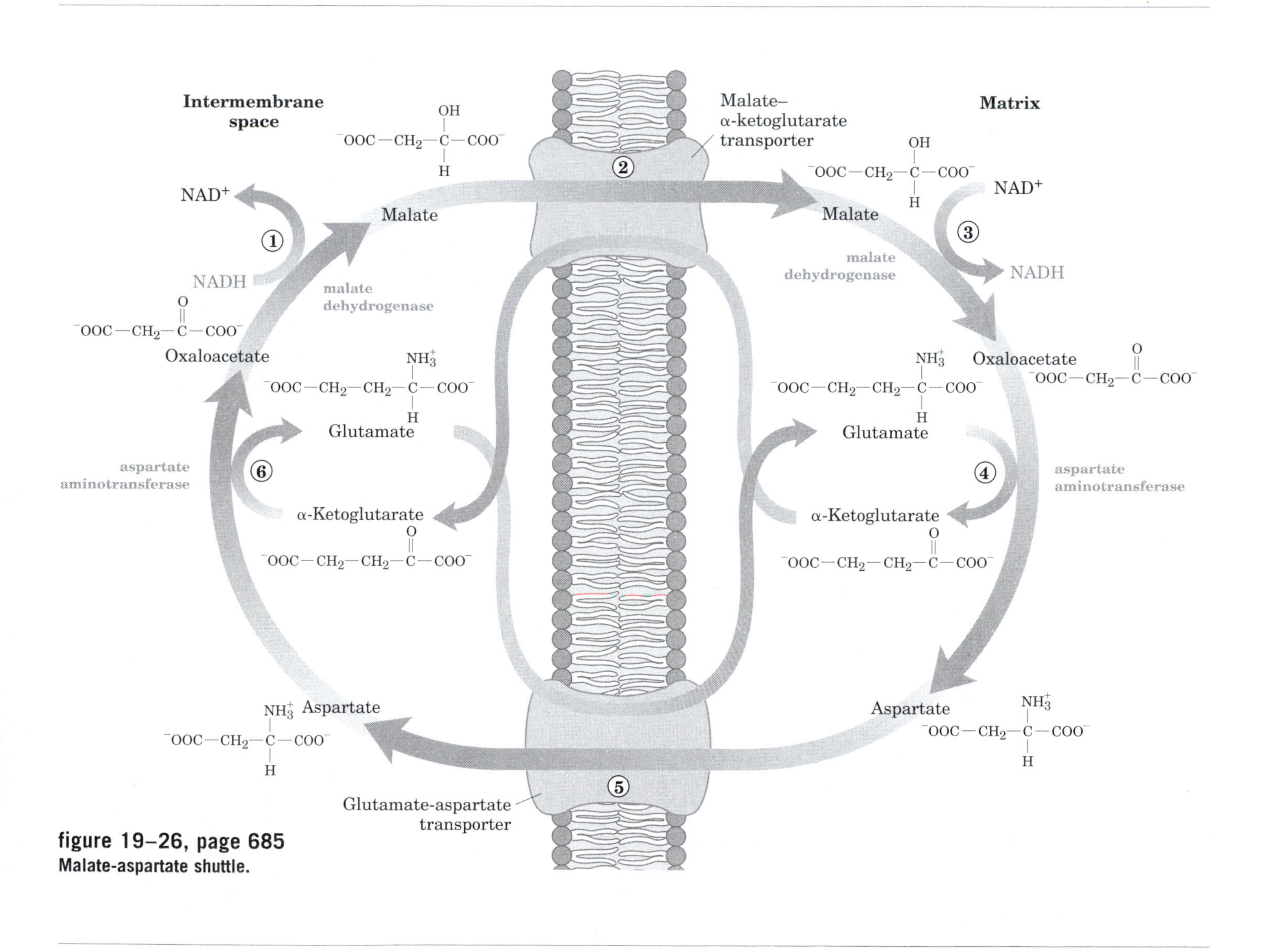

figure 19–26, page 685
Malate-aspartate shuttle.

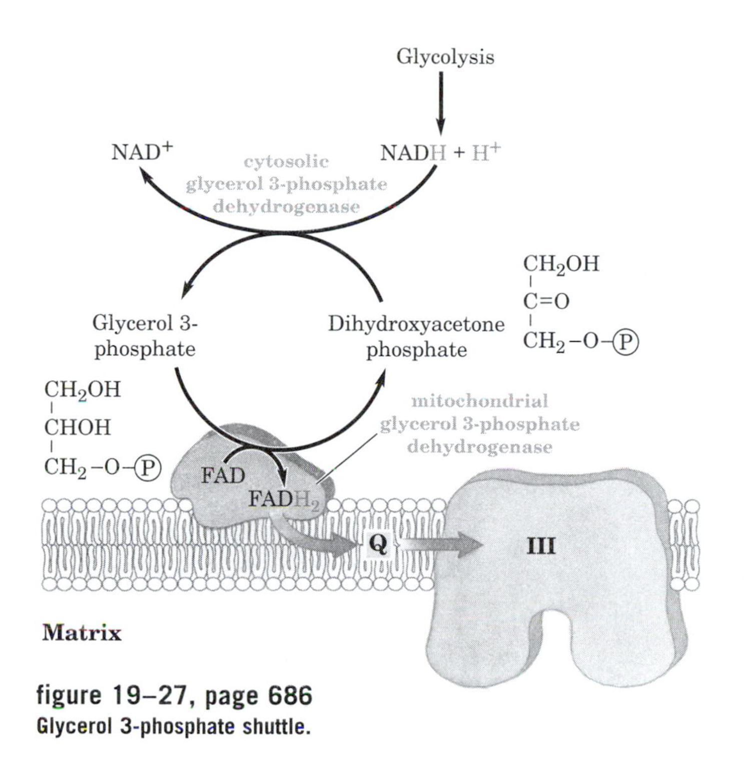

figure 19–27, page 686
Glycerol 3-phosphate shuttle.

table 19–5, page 686

ATP Yield from Complete Oxidation of Glucose

Process	Direct product	Final ATP
Glycolysis	2 NADH (cytosolic)	3 or 5*
	2 ATP	2
Pyruvate oxidation (two per glucose)	2 NADH (mitochondrial matrix)	5
Acetyl-CoA oxidation in citric acid cycle (two per glucose)	6 NADH (mitochondrial matrix)	15
	2 $FADH_2$	3
	2 ATP or 2 GTP	2
Total yield per glucose		30 or 32

*The number depends on which shuttle system transfers reducing equivalents into mitochondria.

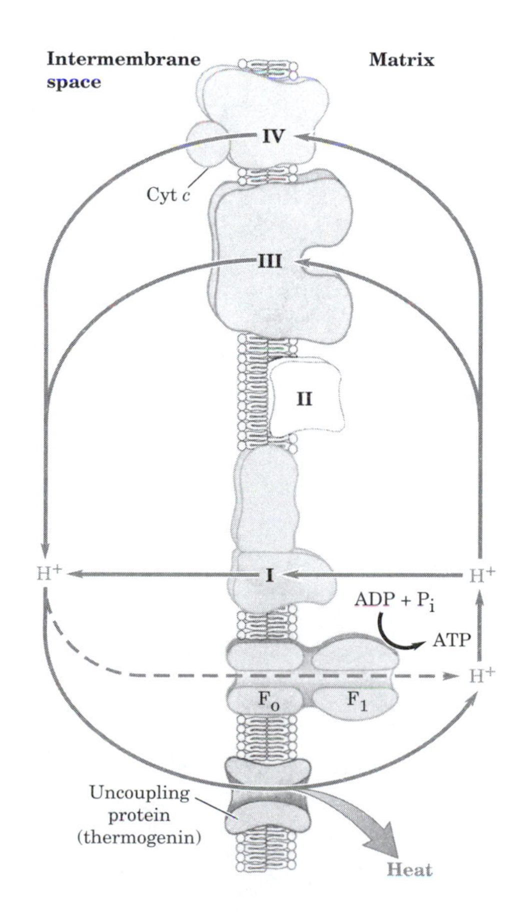

figure 19–28, page 687
Heat generation by uncoupled mitochondria.

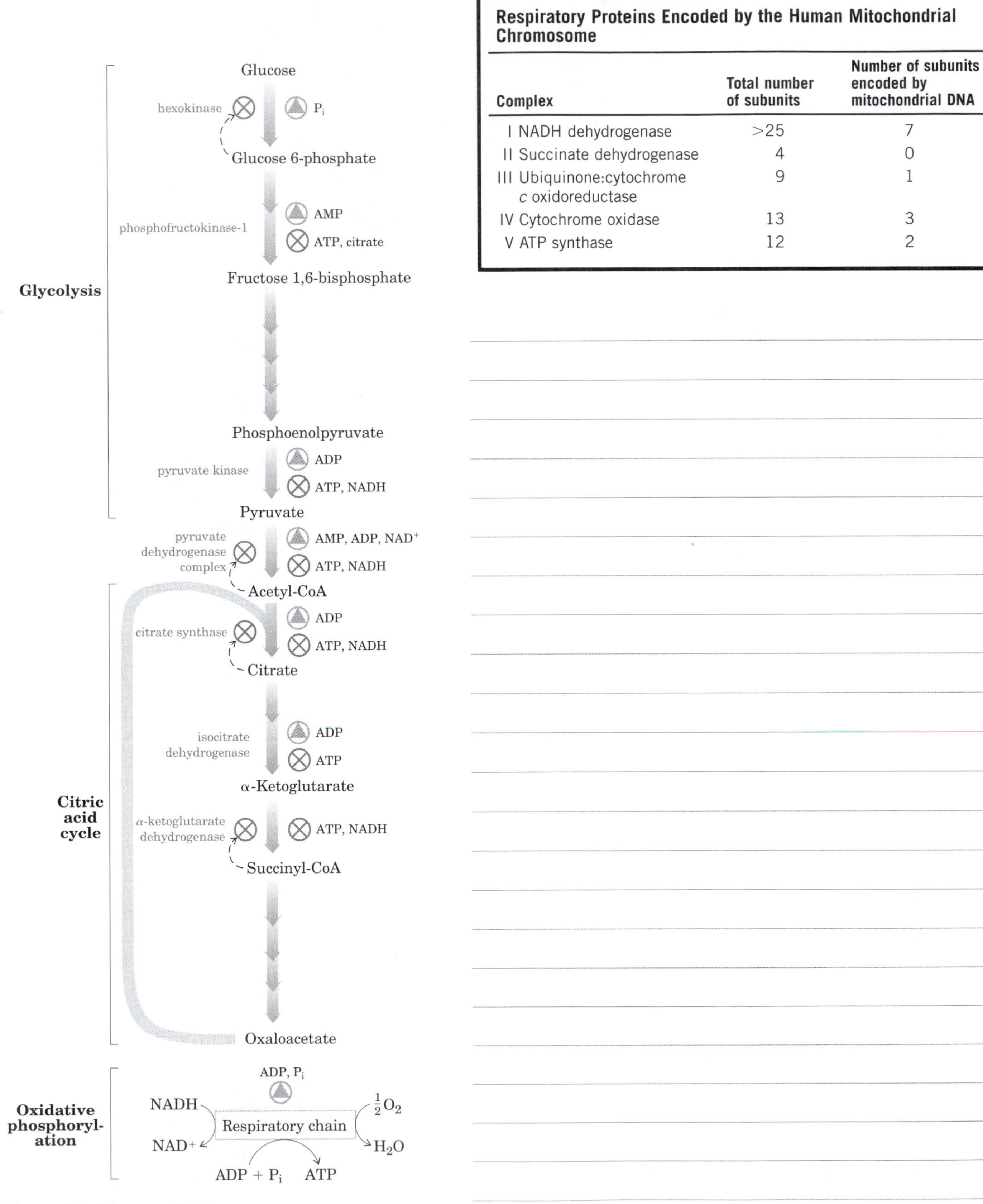

figure 19–29, page 688
Regulation of the ATP-producing pathways.

table 19–6, page 689

Respiratory Proteins Encoded by the Human Mitochondrial Chromosome

Complex	Total number of subunits	Number of subunits encoded by mitochondrial DNA
I NADH dehydrogenase	>25	7
II Succinate dehydrogenase	4	0
III Ubiquinone:cytochrome *c* oxidoreductase	9	1
IV Cytochrome oxidase	13	3
V ATP synthase	12	2

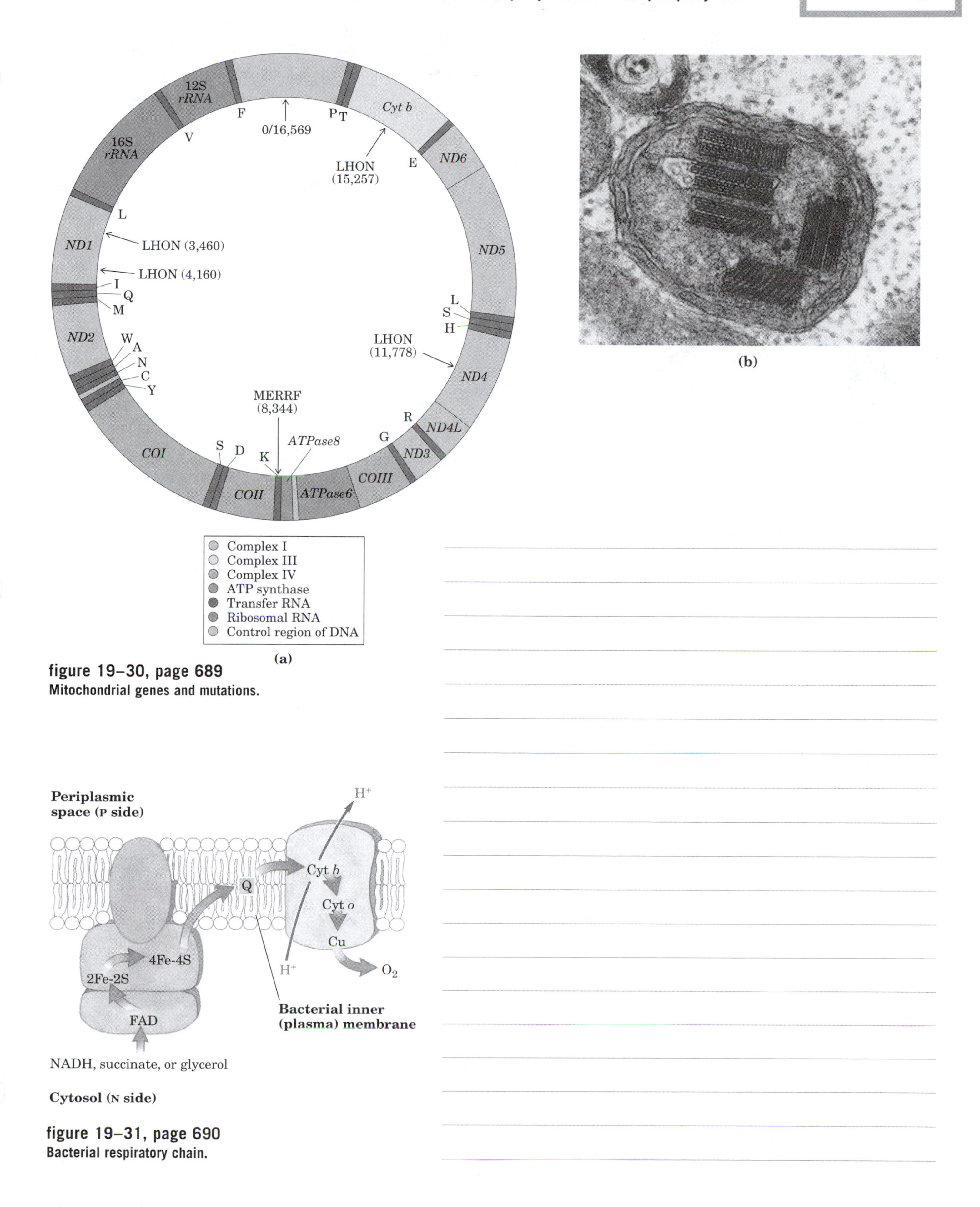

figure 19–30, page 689
Mitochondrial genes and mutations.

figure 19–31, page 690
Bacterial respiratory chain.

$(CH_2CH{=}\overset{CH_3}{C}{-}CH_2)_n{-}H$, CH_3, O, O

$n = 7–9$

Menaquinone, page 690

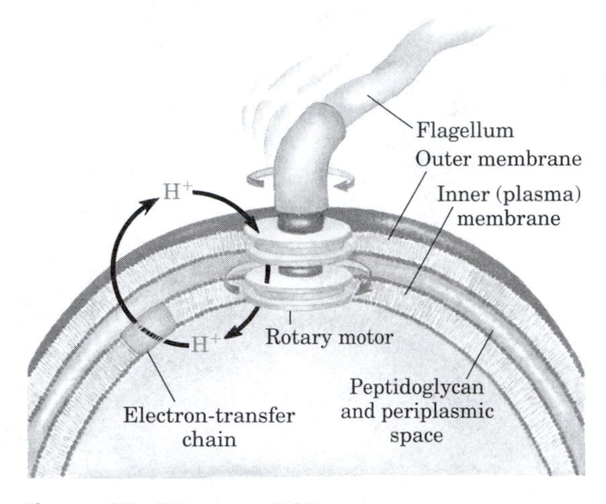

figure 19–32, page 690
Rotation of bacterial flagella by proton-motive force.

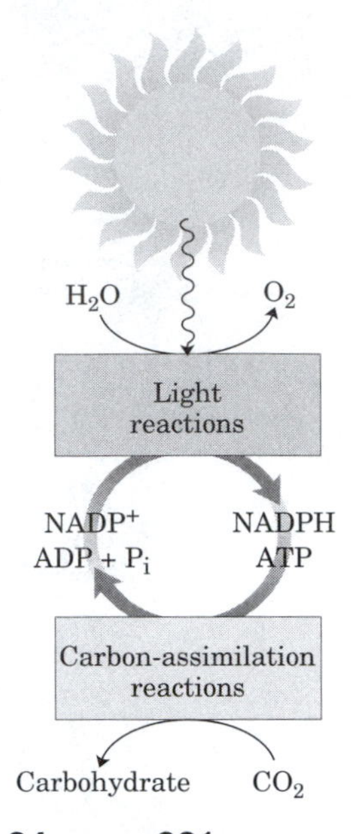

figure 19–34, page 691
The light reactions of photosynthesis generate energy-rich NADPH and ATP at the expense of solar energy.

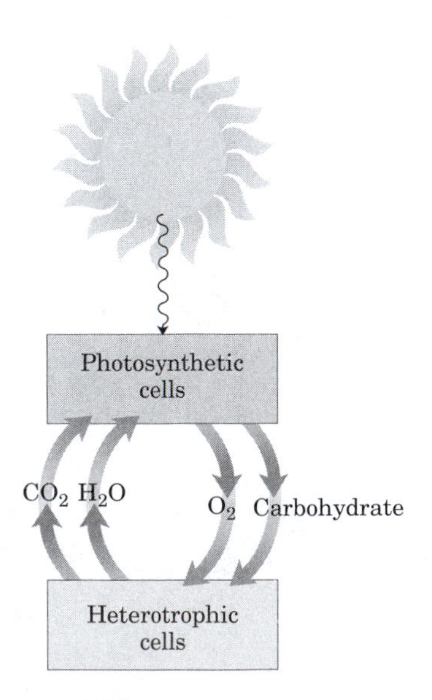

figure 19–33, page 691
Solar energy is the ultimate source of all biological energy.

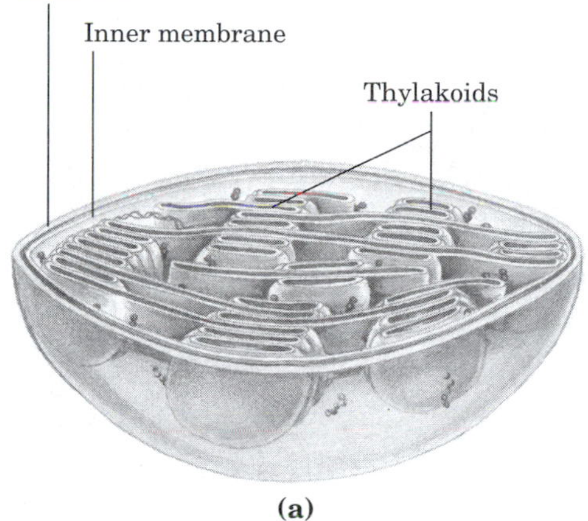

(a)

Grana (thylakoids)

Stroma

(b)

figure 19–35, page 692
Chloroplast.

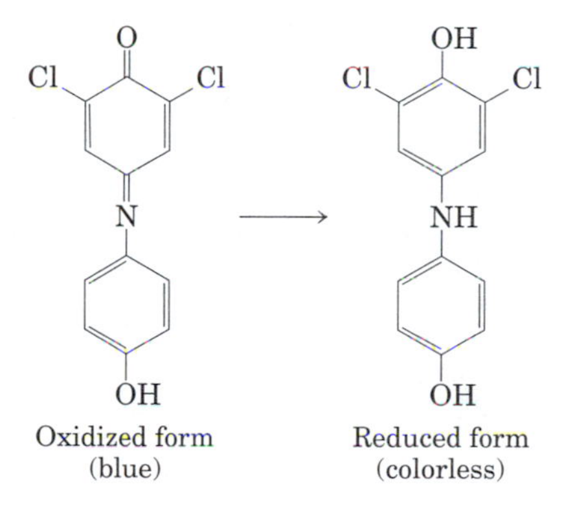

Dichlorophenolindophenol, page 692

Type of radiation: Gamma rays | X rays | UV | Infrared | Microwaves | Radio waves

Wavelength: <1 nm | 100 nm | <1 millimeter | 1 meter | Thousands of meters

Visible light

Violet Blue Cyan Green Yellow Orange Red

Wavelength (nm): 380 430 500 560 600 650 750

Energy (kJ/einstein): 300 240 200 170

figure 19–36, page 693
The spectrum of electromagnetic radiation, and the energy of photons in the visible range of the spectrum.

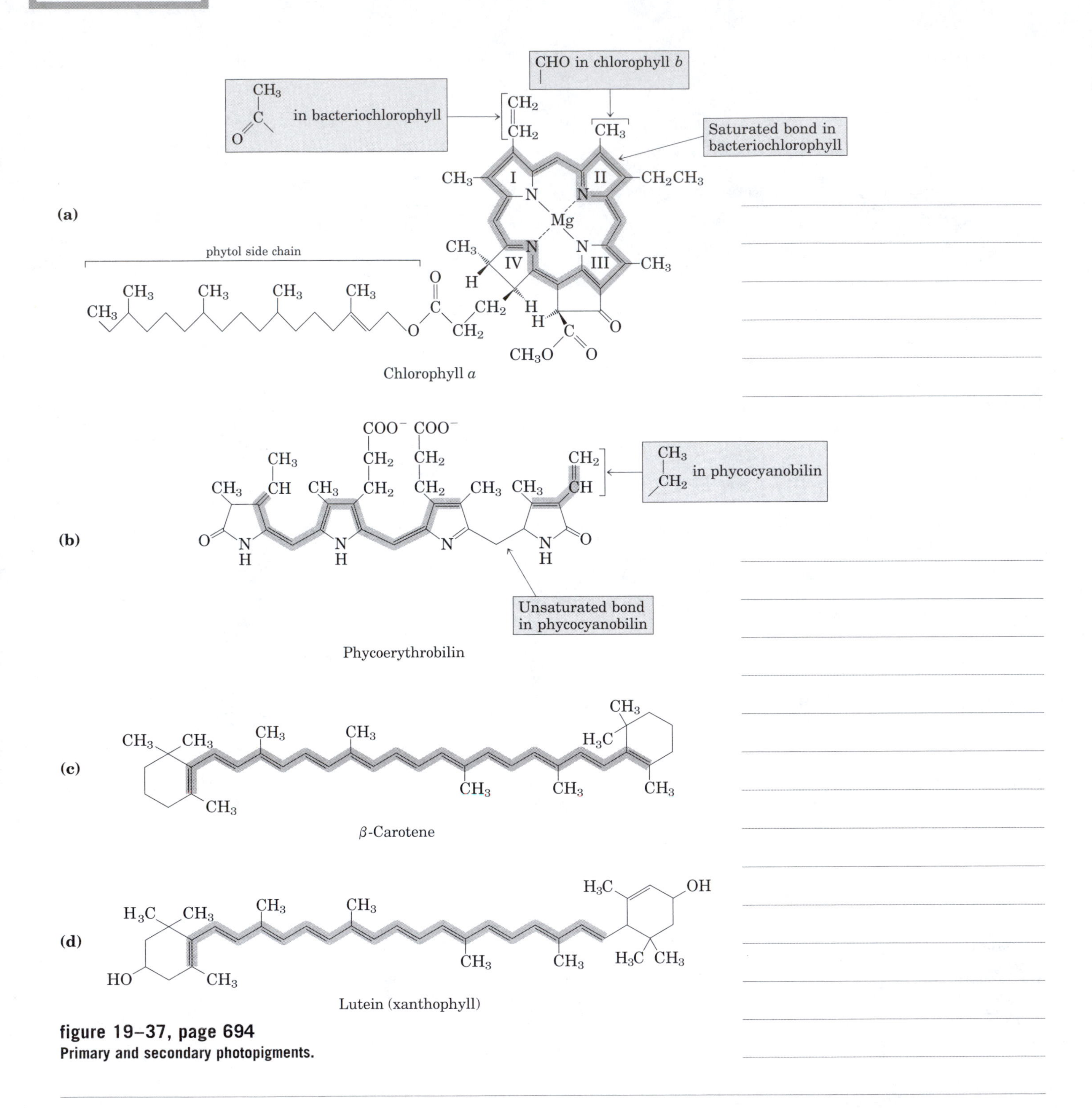

figure 19–37, page 694

Primary and secondary photopigments.

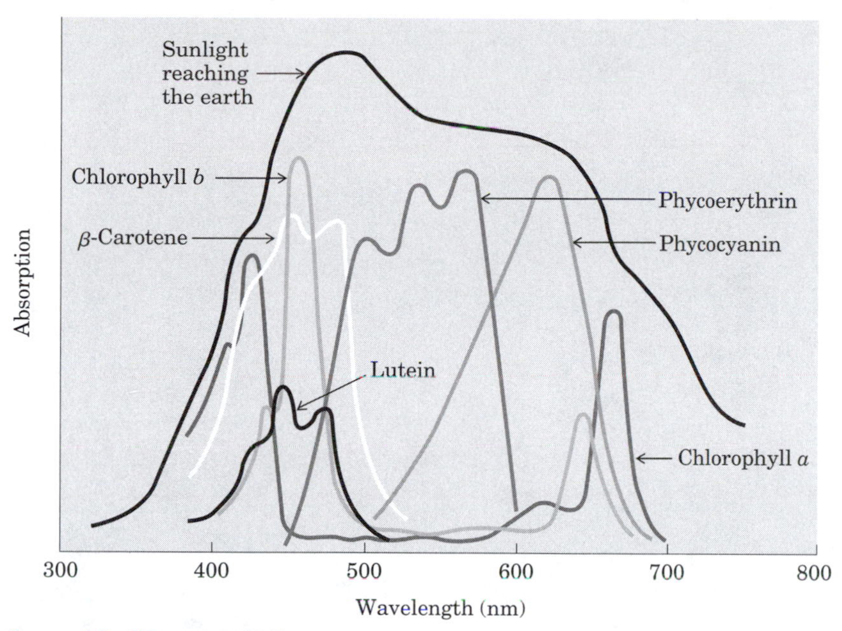

figure 19–38, page 695
Absorption of visible light by photopigments.

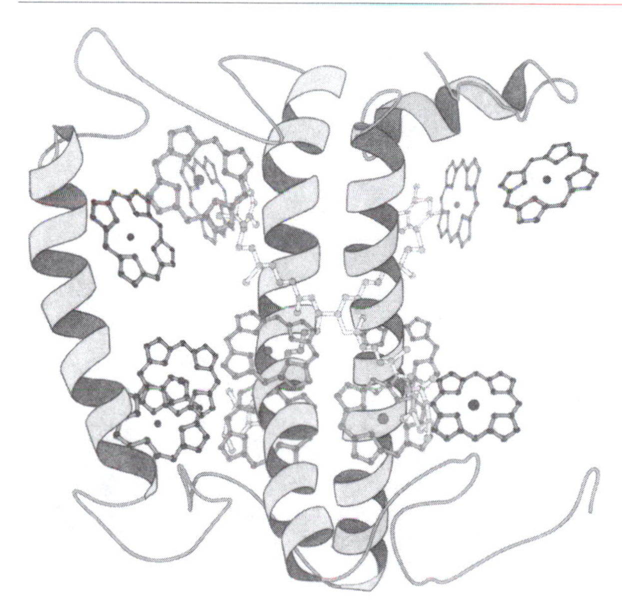

figure 19–39, page 695
A light-harvesting complex, LHCII.

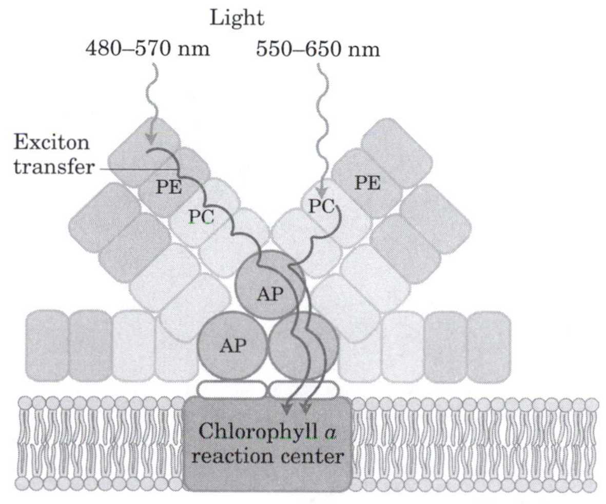

figure 19–40, page 696
A phycobilisome.

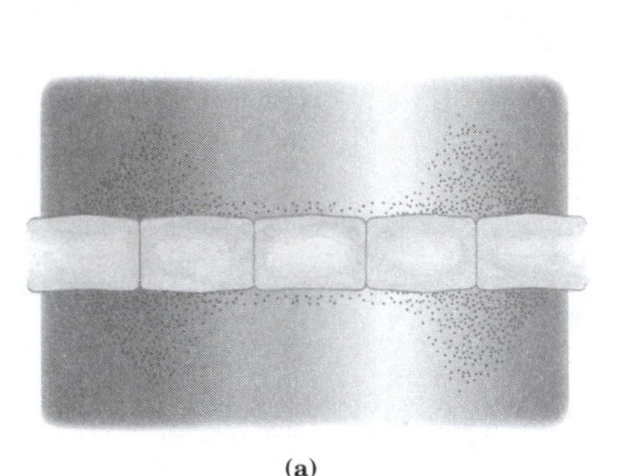

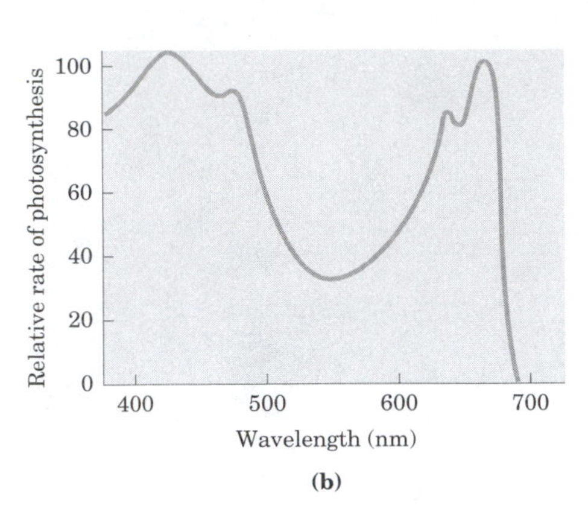

(a) (b)

figure 19–41, page 696
Two ways to determine the action spectrum for photosynthesis.

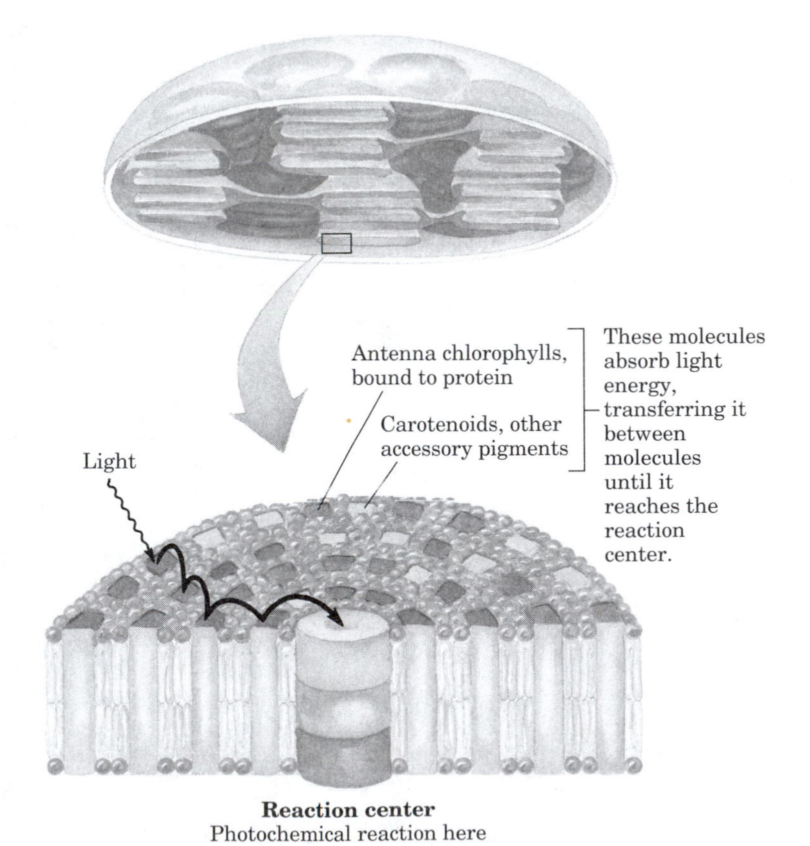

figure 19–42, page 697
Organization of photosystems in the thylakoid membrane.

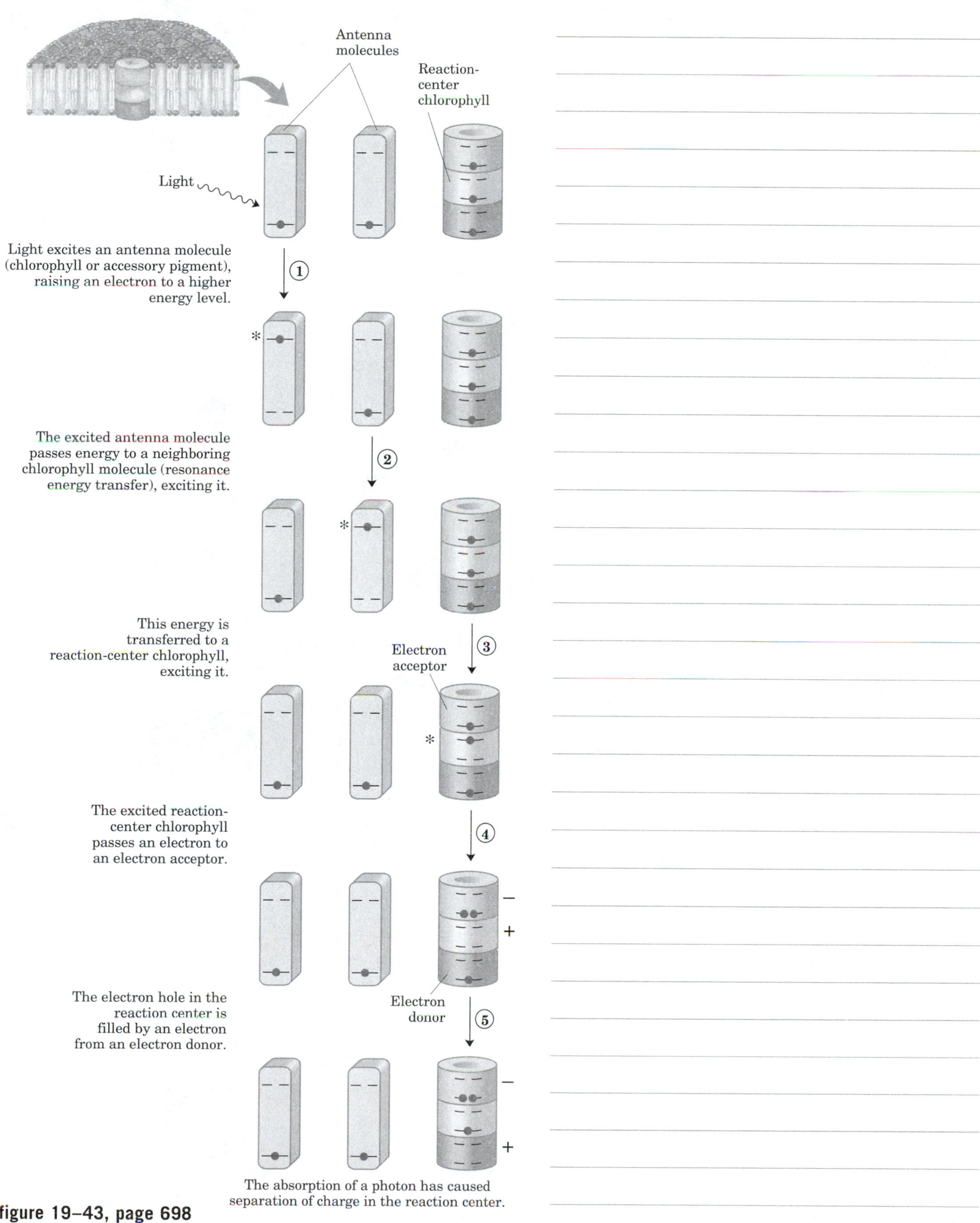

figure 19–43, page 698
Exciton transfer.

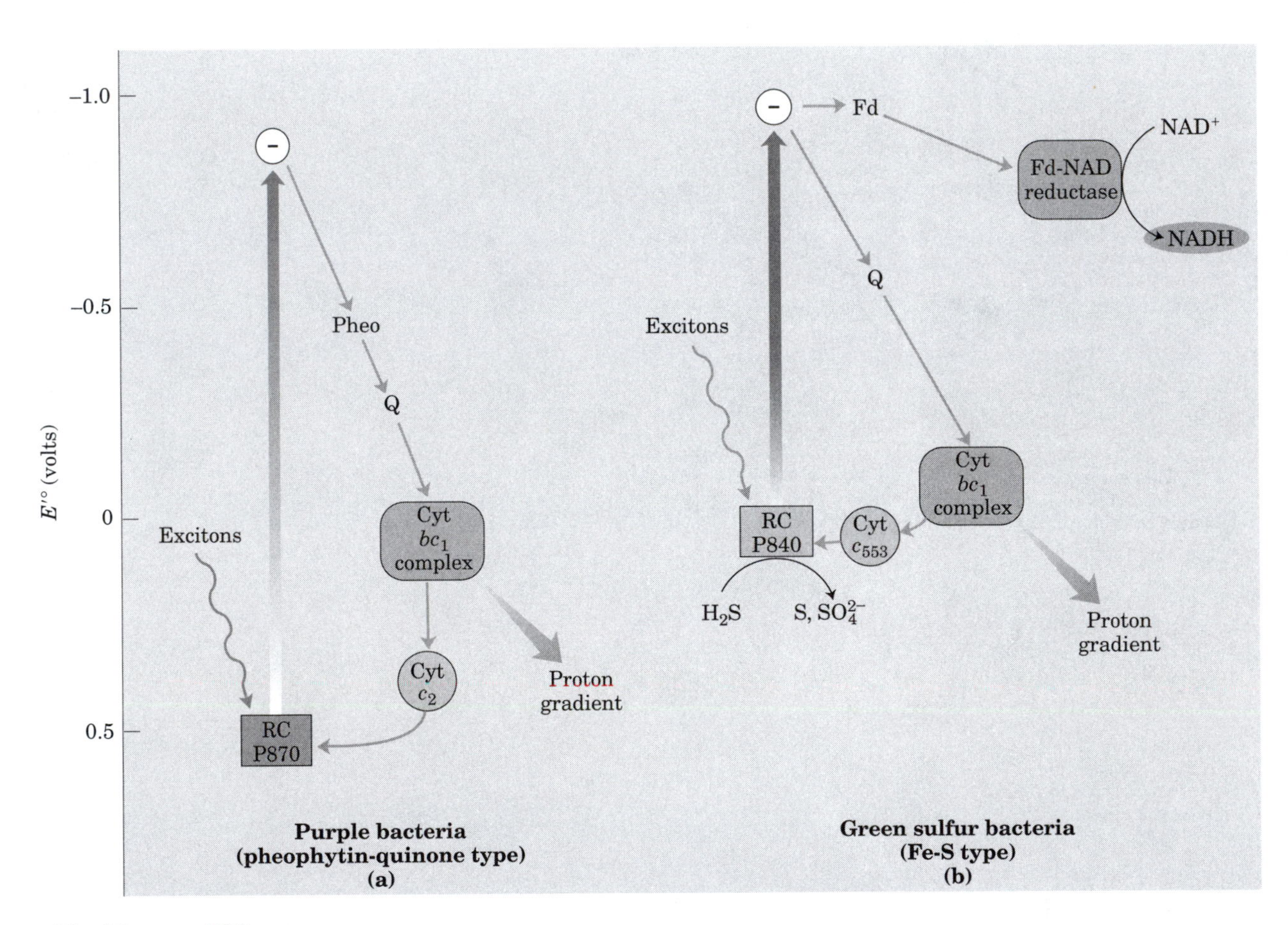

figure 19–44, page 700

Functional modules of photosynthetic machinery in purple bacteria and green sulfur bacteria.

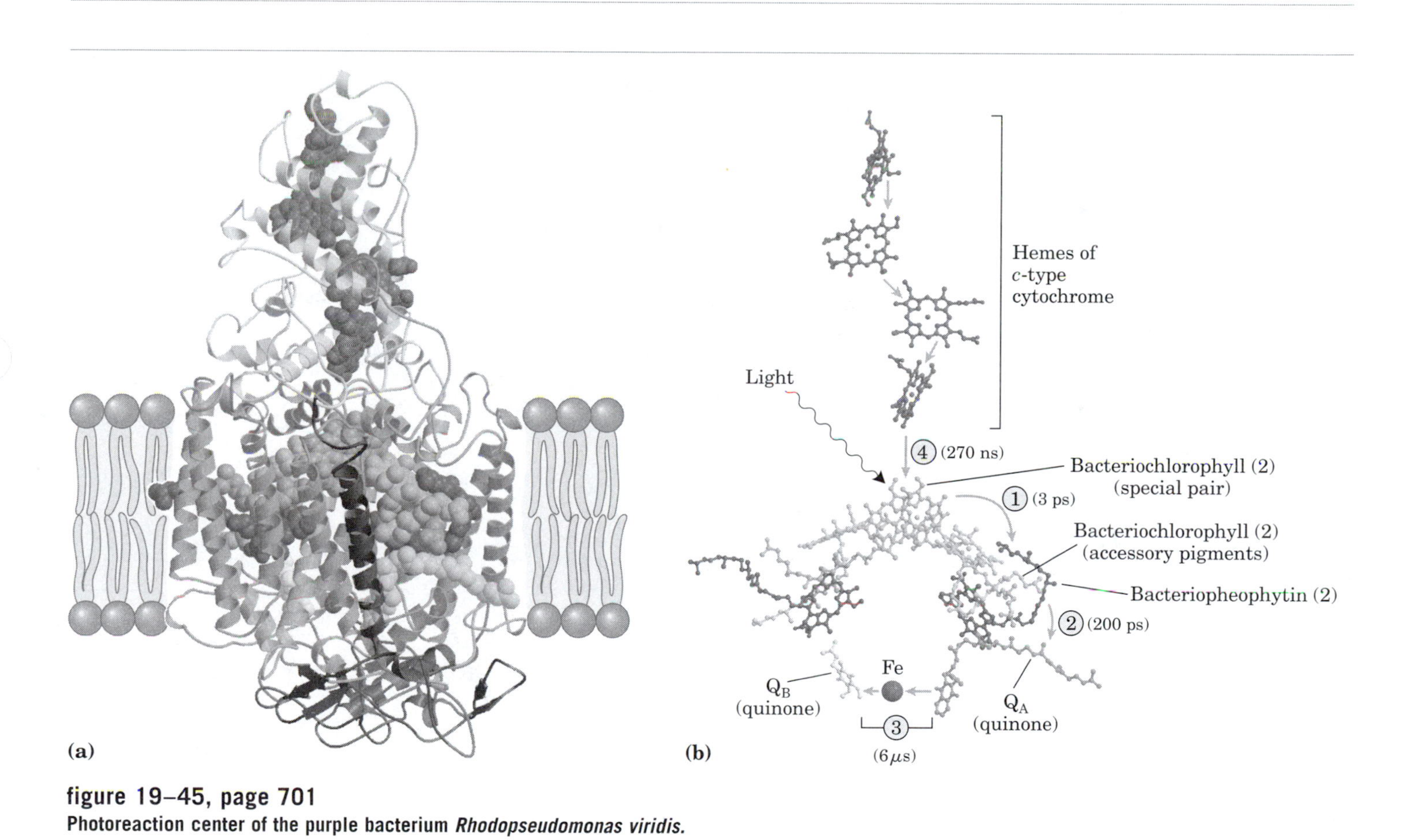

figure 19–45, page 701
Photoreaction center of the purple bacterium *Rhodopseudomonas viridis.*

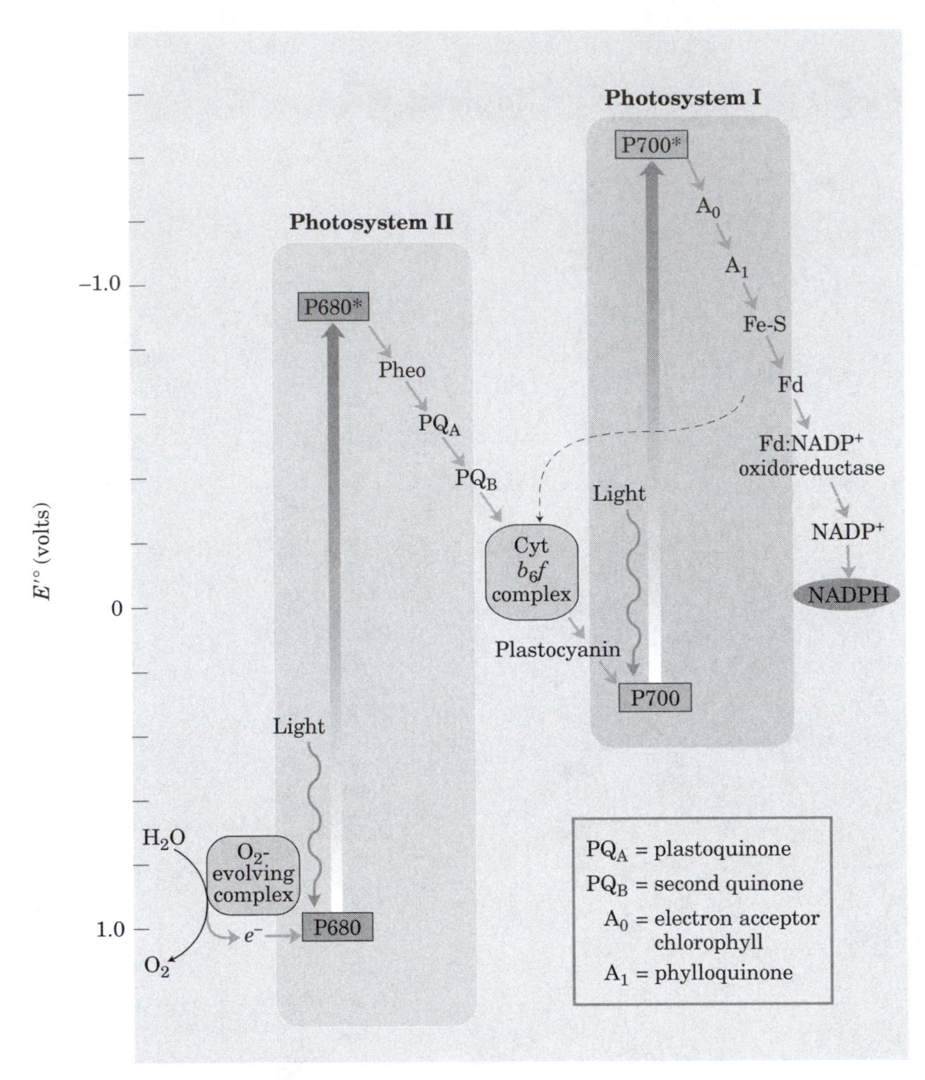

figure 19–46, page 703
Integration of photosystems I and II in chloroplasts.

$$\text{O}\quad \text{CH}_3 \quad \text{CH}_3$$
$$\text{CH}_2-\text{CH}=\text{C}-\text{CH}_2-(\text{CH}_2-\text{CH}_2-\text{CH}-\text{CH}_2)_3-\text{H}$$
$$\text{O}$$

Phylloquinone, page 704

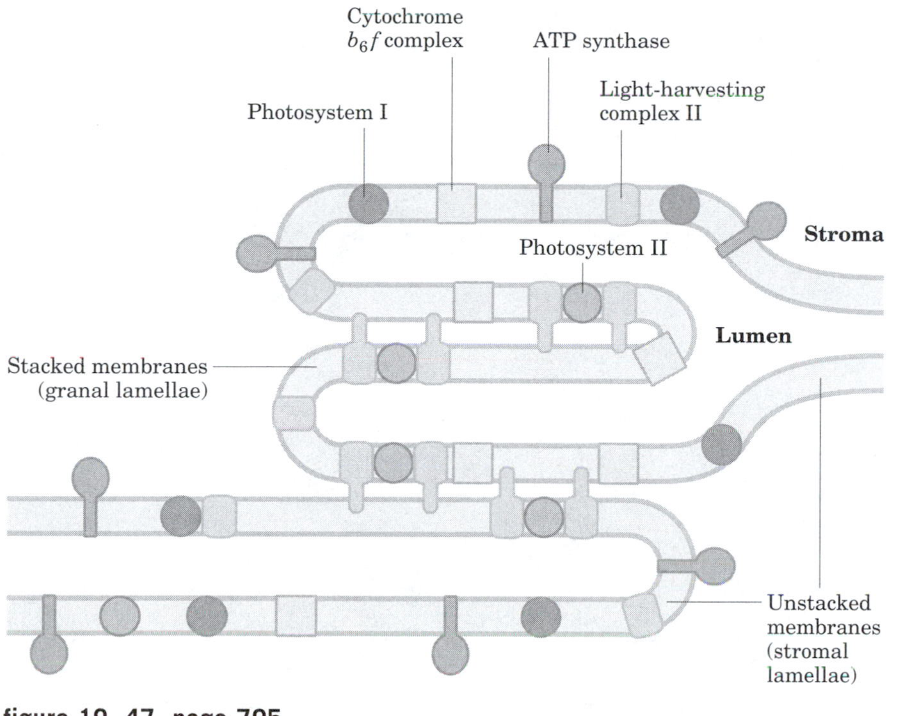

figure 19–47, page 705
Localization of PSI and PSII in thylakoid membranes.

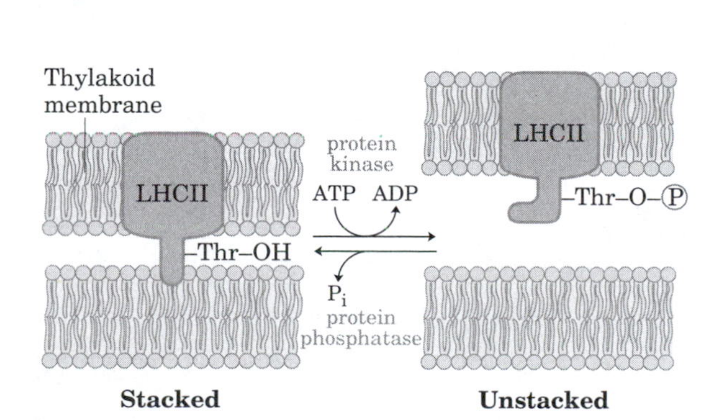

figure 19–48, page 705
Modulation of granal stacking equalizes electron flow in PSI and PSII.

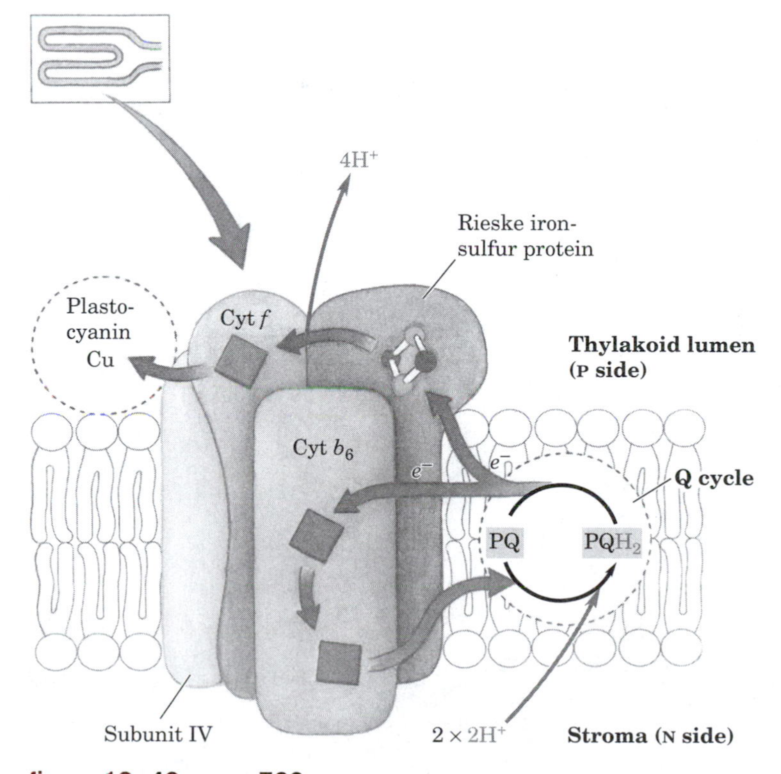

figure 19–49, page 706
Electron and proton flow through the cytochrome b_6f complex.

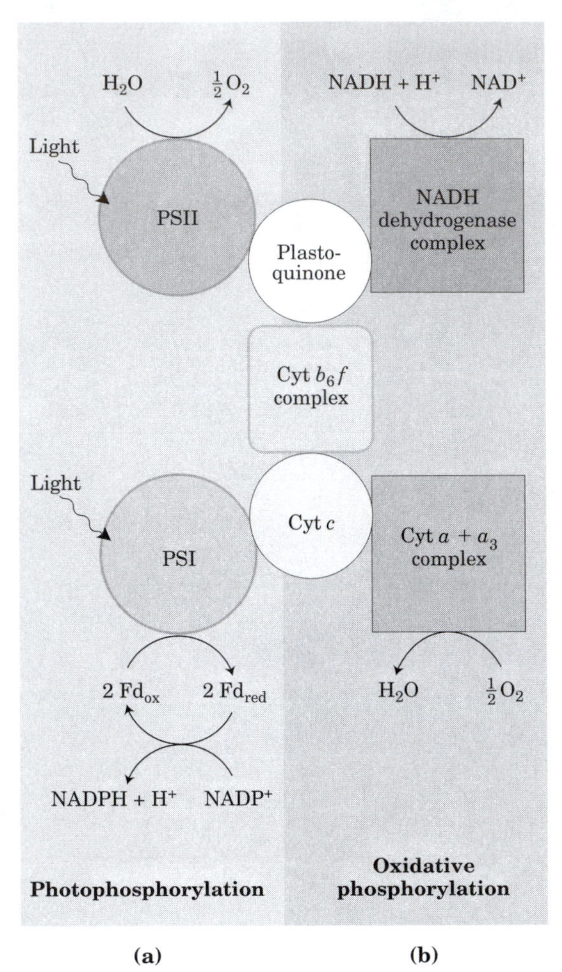

figure 19–50, page 707
Dual roles of cytochrome b_6f and cytochrome c in cyanobacteria.

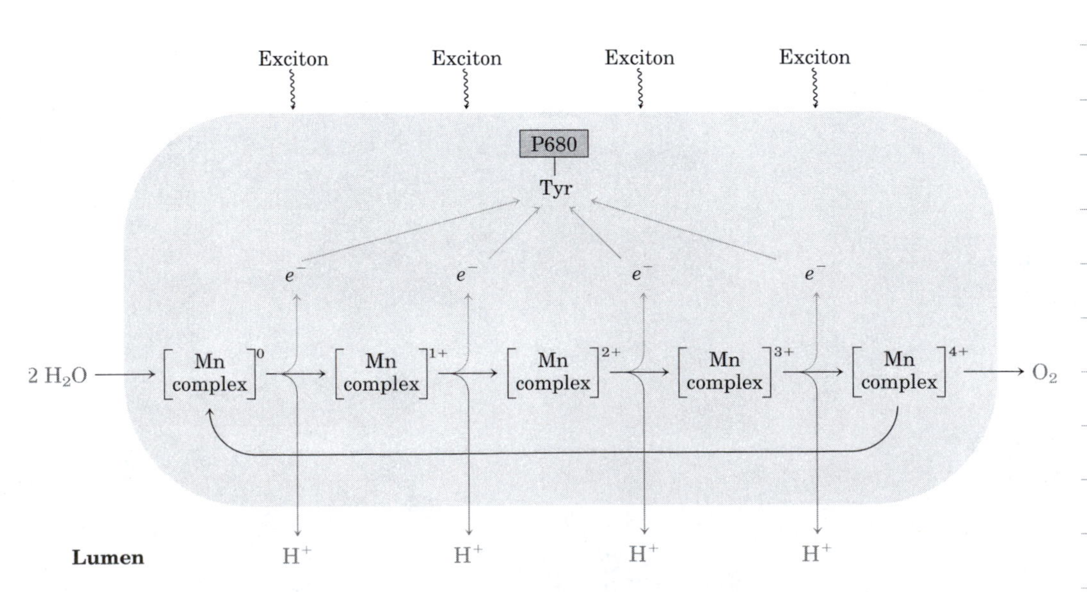

figure 19–51, page 707
Water-splitting activity of the oxygen-evolving complex.

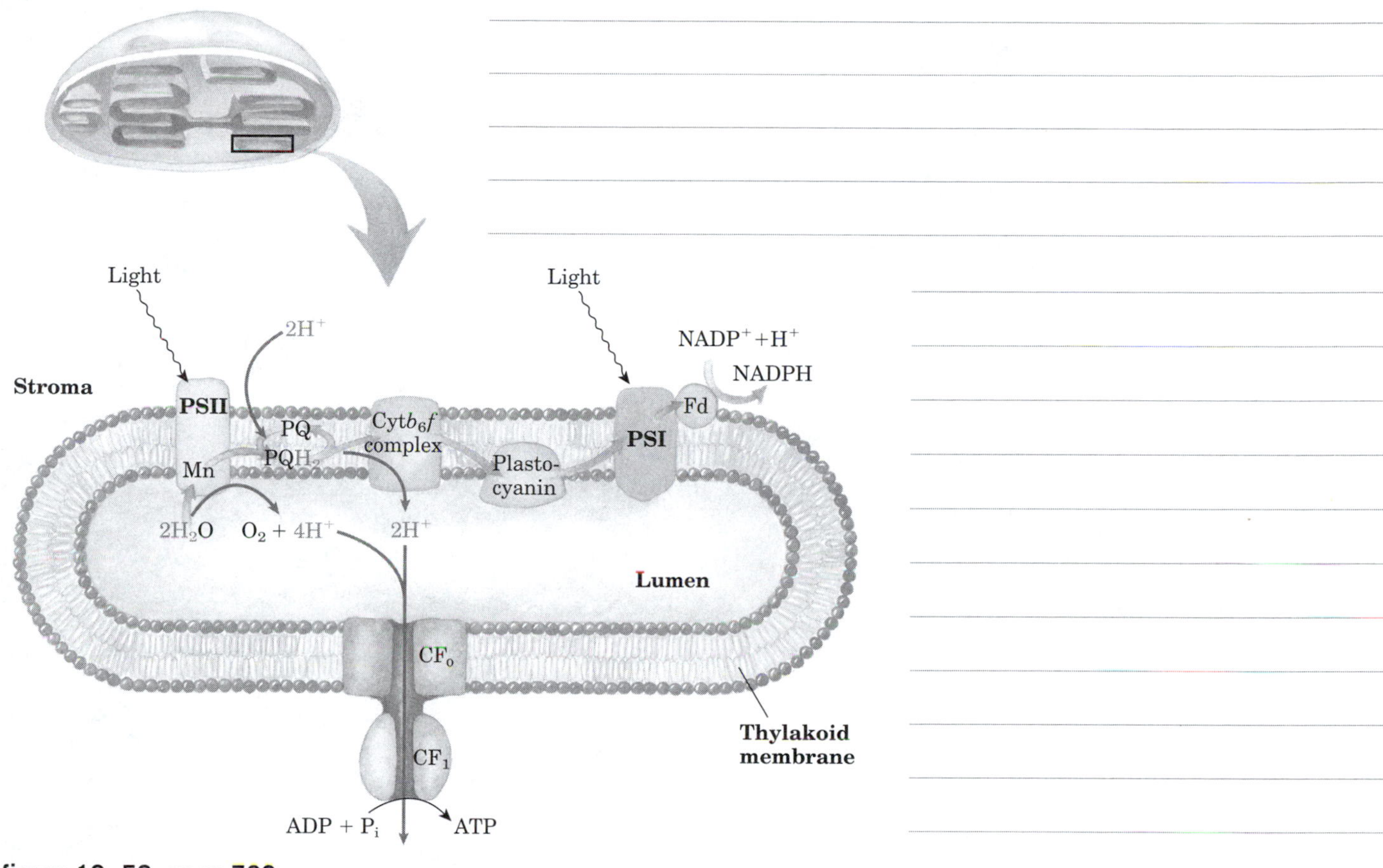

figure 19–52, page 709
Proton and electron circuits in thylakoids.

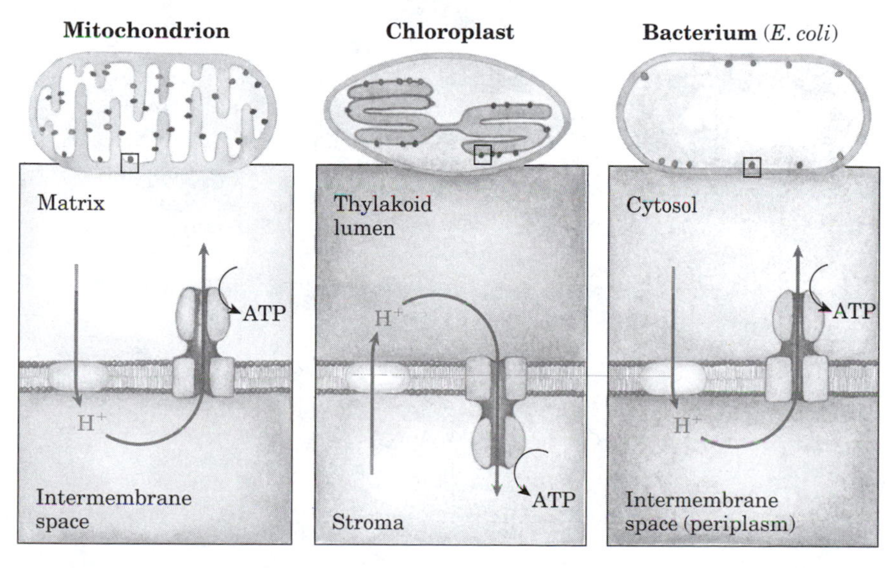

figure 19–53, page 711
Comparison of the topology of proton movement and ATP synthase orientation in the membranes of mitochondria, chloroplasts, and the bacterium *E. coli.*

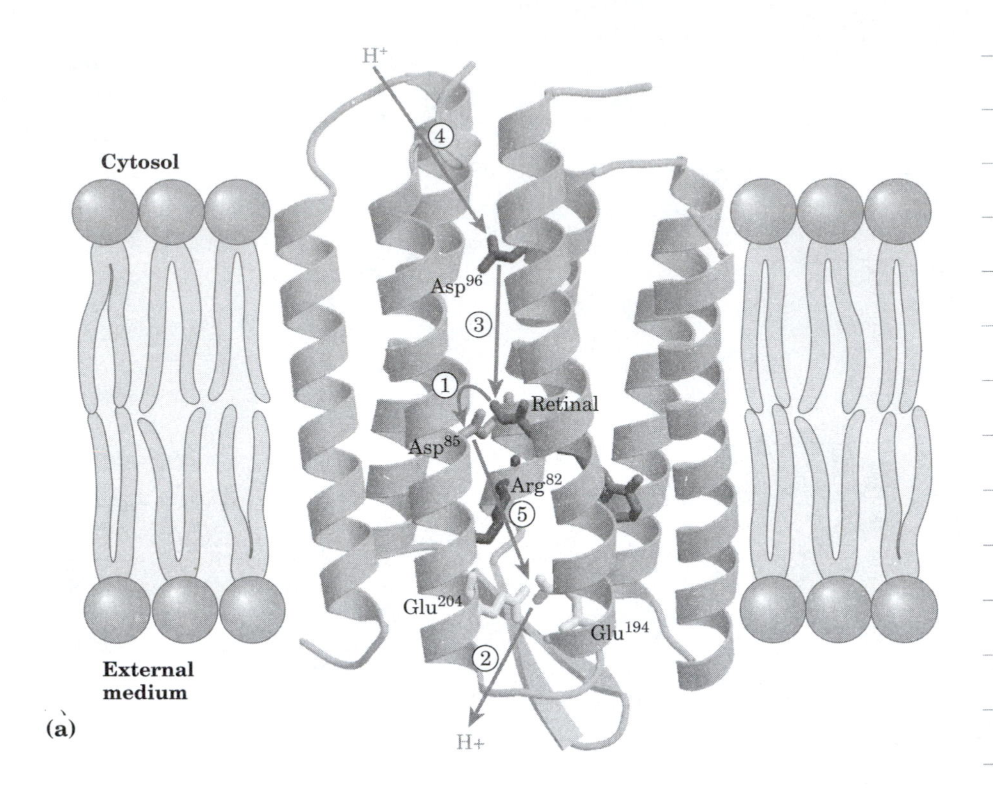

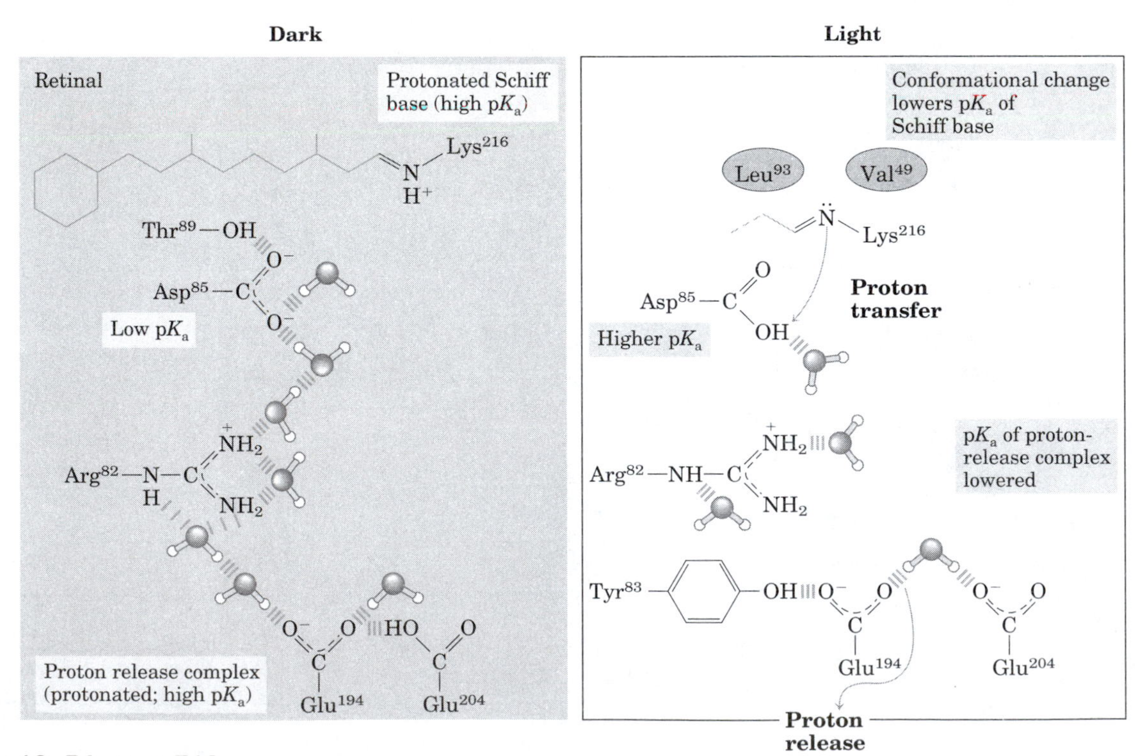

figure 19–54, page 713
Light-driven proton pumping by bacteriorhodopsin.

PROBLEMS

1. Oxidation-Reduction Reactions, page 718.

(1) $NADH + H^+ + E\text{-}FMN \rightarrow NAD^+ + E\text{-}FMNH_2$

(2) $E\text{-}FMNH_2 + 2Fe^{3+} \rightarrow E\text{-}FMN + 2Fe^{2+} + 2H^+$

(3) $2Fe^{2+} + 2H^+ + Q \rightarrow 2Fe^{3+} + QH_2$

Sum: $NADH + H^+ + Q \rightarrow NAD^+ + QH_2$

17. Transmembrane Movement of Reducing Equivalents, page 720.

[4-^{3}H]NADH

H H O ^{14}C NH_2 N R

[7-^{14}C]NADH

27. Equilibrium Constant for Water-Splitting Reactions, page 720.

$2H_2O + 2NADP^+ \longrightarrow 2NADPH + 2H^+ + O_2$

28. Energetics of Phototransduction, page 721.

$2H_2O + 2NADP^+ + 8\text{ photons} \longrightarrow 2NADPH + 2H^+ + O_2$

29. Electron Transfer to a Hill Reagent, page 721.

$2H_2O + 4Fe^{3+} \longrightarrow O_2 + 4H^+ + 4Fe^{2+}$

chapter

20 Carbohydrate Biosynthesis

CD-ROM

When you study for exams, don't forget to review The Minicourses on the UNDERSTAND! BIOCHEMISTRY CD that came with your textbook.

Minicourses that apply to this Chapter include:

Molecules of Life
- An Interactive Gallery of Molecular Models
- An Interactive Gallery of Protein Structures

Bioenergetics
- Glycogen
- Gluconeogenesis
- Pentose Phosphate Pathway
- Calvin Cycle
- C4 Photosynthesis

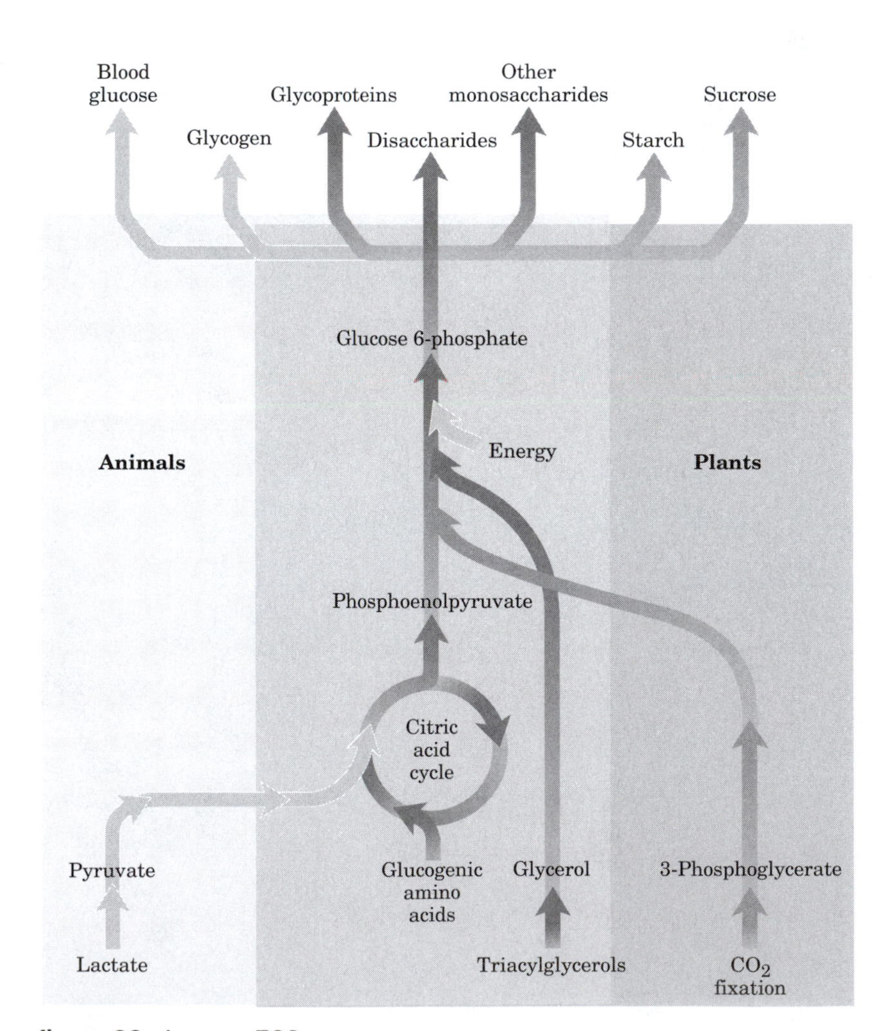

figure 20–1, page 723
Carbohydrate synthesis from simple precursors.

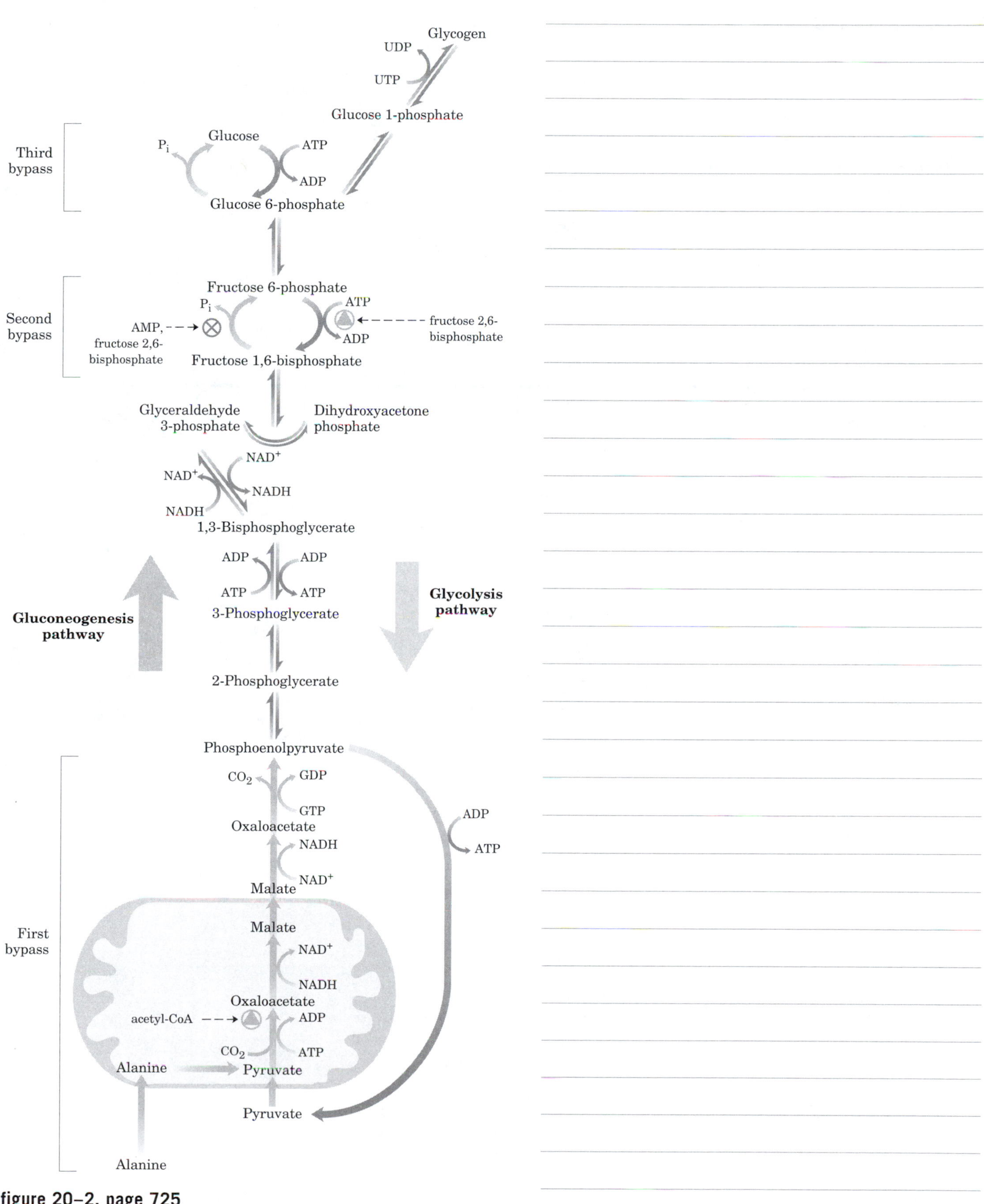

figure 20–2, page 725
Opposing pathways of glycolysis and gluconeogenesis in rat liver.

table 20–1, page 726

Free-Energy Changes of Glycolytic Reactions in Erythrocytes*

Glycolytic reaction step	$\Delta G'^{\circ}$ (kJ/mol)	ΔG (kJ/mol)
① Glucose + ATP ⟶ glucose 6-phosphate + ADP + H^+	**−16.7**	**−33.4**
② Glucose 6-phosphate ⇌ fructose 6-phosphate	1.7	−2.5
③ Fructose 6-phosphate + ATP ⟶ fructose 1,6-bisphosphate + ADP + H^+	**−14.2**	**−22.2**
④ Fructose 1,6-bisphosphate ⇌ dihydroxyacetone phosphate + glyceraldehyde 3-phosphate	23.8	−1.25
⑤ Dihydroxyacetone phosphate ⇌ glyceraldehyde 3-phosphate	7.5	2.5
⑥ Glyceraldehyde 3-phosphate + P_i + NAD^+ ⇌ 1,3-bisphosphoglycerate + NADH + H^+	6.3	−1.7
⑦ 1,3-Bisphosphoglycerate + ADP ⇌ 3-phosphoglycerate + ATP	−18.8	1.25
⑧ 3-Phosphoglycerate ⇌ 2-phosphoglycerate	4.4	0.8
⑨ 2-Phosphoglycerate ⇌ phosphoenolpyruvate + H_2O	7.5	−3.3
⑩ Phosphoenolpyruvate + ADP + H^+ ⟶ pyruvate + ATP	**−31.4**	**−16.7**

*$\Delta G'^{\circ}$ is the standard free-energy change, as defined in Chapter 14 (see p. 494). At pH 7.0, ΔG is the free-energy change calculated from the actual concentrations of glycolytic intermediates present under physiological conditions in erythrocytes. The glycolytic reactions bypassed in gluconeogenesis are boldfaced.

Pyruvate + HCO_3^- + ATP ⟶ oxaloacetate + ADP + P_i (Equation 20–1)

Oxaloacetate + NADH + H^+ ⇌ L-malate + NAD^+ (Equation 20–2)

Malate + NAD^+ ⟶ oxaloacetate + NADH + H^+ (Equation 20–3)

Oxaloacetate + GTP ⇌ phosphoenolpyruvate + CO_2 + GDP (Equation 20–4)

Pyruvate + ATP + GTP + HCO_3^- ⟶
phosphoenolpyruvate + ADP + GDP + P_i + CO_2 (Equation 20–5)
$\Delta G'^{\circ}$ = 0.9 kJ/mol

Pyruvate to phosphoenolpyruvate, pages 726–727

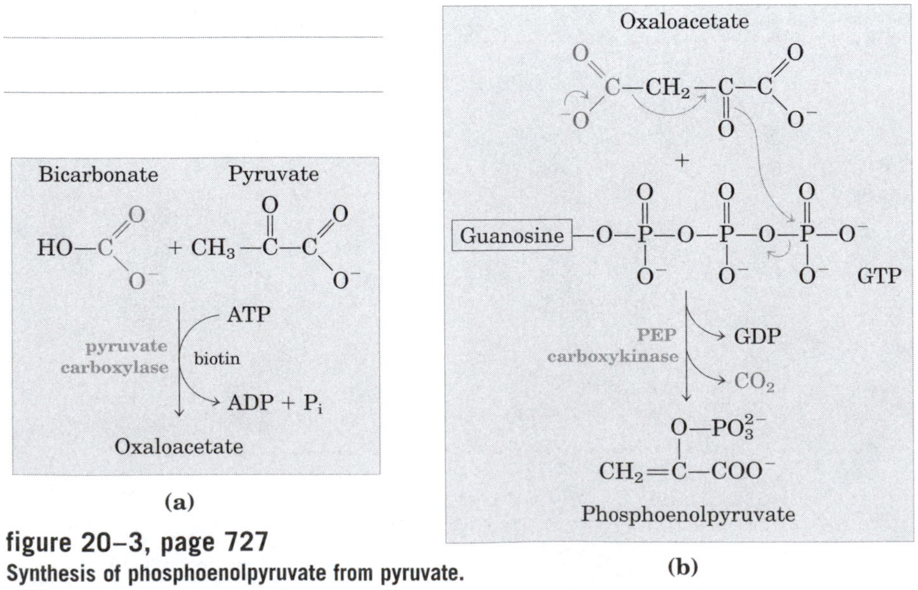

figure 20–3, page 727
Synthesis of phosphoenolpyruvate from pyruvate.

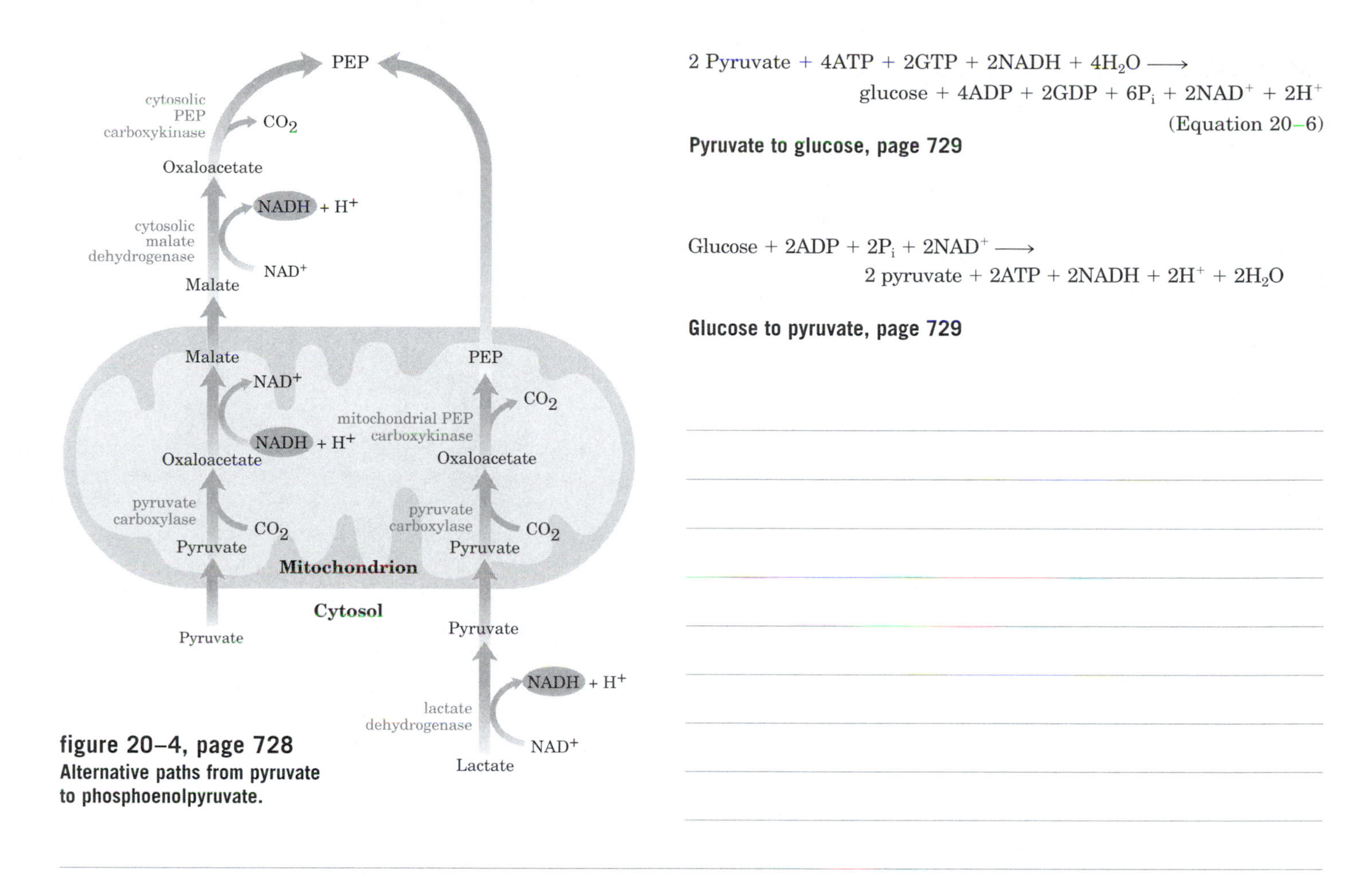

figure 20–4, page 728
Alternative paths from pyruvate to phosphoenolpyruvate.

$$2\text{ Pyruvate} + 4\text{ATP} + 2\text{GTP} + 2\text{NADH} + 4H_2O \longrightarrow \text{glucose} + 4\text{ADP} + 2\text{GDP} + 6P_i + 2\text{NAD}^+ + 2H^+ \quad \text{(Equation 20–6)}$$

Pyruvate to glucose, page 729

$$\text{Glucose} + 2\text{ADP} + 2P_i + 2\text{NAD}^+ \longrightarrow 2\text{ pyruvate} + 2\text{ATP} + 2\text{NADH} + 2H^+ + 2H_2O$$

Glucose to pyruvate, page 729

table 20–2, page 729

Sequential Reactions in Gluconeogenesis Starting from Pyruvate*

Reaction	
Pyruvate + HCO_3^- + ATP ⟶ oxaloacetate + ADP + P_i + H^+	×**2**
Oxaloacetate + GTP ⇌ phosphoenolpyruvate + CO_2 + GDP	×**2**
Phosphoenolpyruvate + H_2O ⇌ 2-phosphoglycerate	×2
2-Phosphoglycerate ⇌ 3-phosphoglycerate	×2
3-Phosphoglycerate + ATP ⇌ 1,3-bisphosphoglycerate + ADP + H^+	×2
1,3-Bisphosphoglycerate + NADH + H^+ ⇌ glyceraldehyde 3-phosphate + NAD^+ + P_i	×2
Glyceraldehyde 3-phosphate ⇌ dihydroxyacetone phosphate	
Glyceraldehyde 3-phosphate + dihydroxyacetone phosphate ⇌ fructose 1,6-bisphosphate	
Fructose 1,6-bisphosphate + H_2O ⟶ fructose 6-phosphate + P_i	
Fructose 6-phosphate ⇌ glucose 6-phosphate	
Glucose 6-phosphate + H_2O ⟶ glucose + P_i	
Sum: 2 Pyruvate + 4ATP + 2GTP + 2NADH + $4H_2O$ ⟶ glucose + 4ADP + 2GDP + $6P_i$ + $2NAD^+$ + $2H^+$	

*The bypass reactions are boldfaced; all other reactions are reversible steps of glycolysis. The figures at the right indicate that the reaction is to be counted twice, because two three-carbon precursors are required to make a molecule of glucose. Note that the reactions required to replace the cytosolic NADH consumed in the glyceraldehyde 3-phosphate dehydrogenase reaction (the conversion of lactate to pyruvate in the cytosol or the transport of reducing equivalents from mitochondria to the cytosol in the form of malate) are not considered in this summary.

table 20–3, page 730

Glucogenic Amino Acids, Grouped by Site of Entry*

Site of entry	Amino acids
Pyruvate	Alanine, Cysteine, Glycine, Serine, Tryptophan†
α-Ketoglutarate	Arginine, Glutamate, Glutamine, Histidine, Proline
Succinyl-CoA	Isoleucine†, Methionine, Threonine, Valine
Fumarate	Phenylalanine†, Tyrosine†
Oxaloacetate	Asparagine, Aspartate

*These amino acids are precursors of blood glucose or liver glycogen because they can be converted to pyruvate or citric acid cycle intermediates. Only leucine and lysine are unable to furnish carbon for net glucose synthesis.

†These amino acids are also ketogenic (see Fig. 18–19).

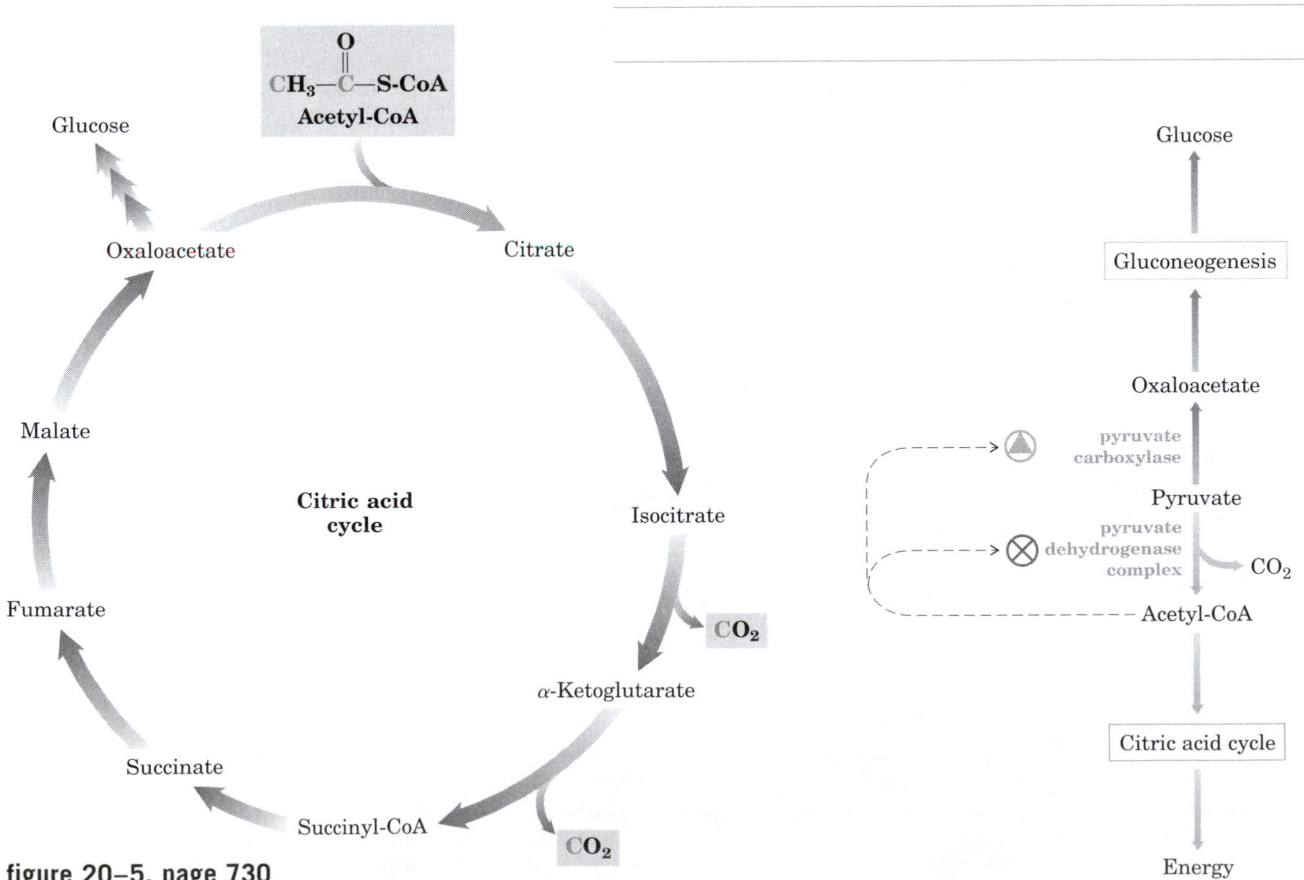

figure 20–5, page 730
Even-number fatty acids cannot be a source of carbon for the net synthesis of glucose in animals and microorganisms.

figure 20–6, page 731
Two alternative fates for pyruvate.

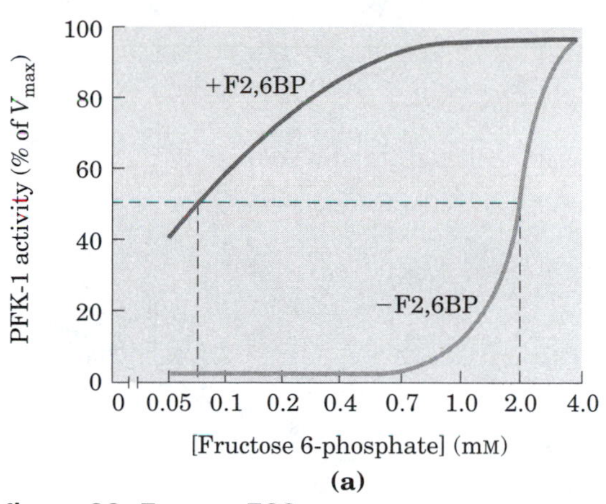

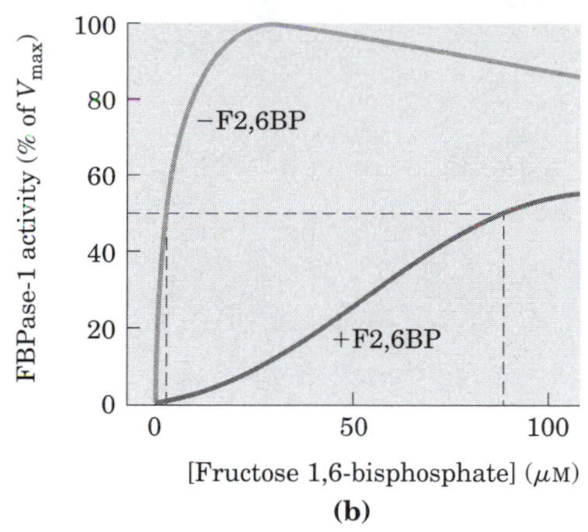

figure 20–7, page 732

Role of fructose 2,6-bisphosphate in regulation of glycolysis and gluconeogenesis.

Fructose 2,6-bisphosphate, page 732

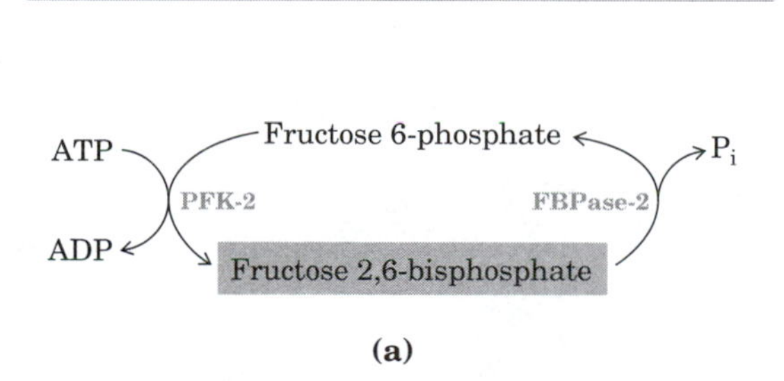

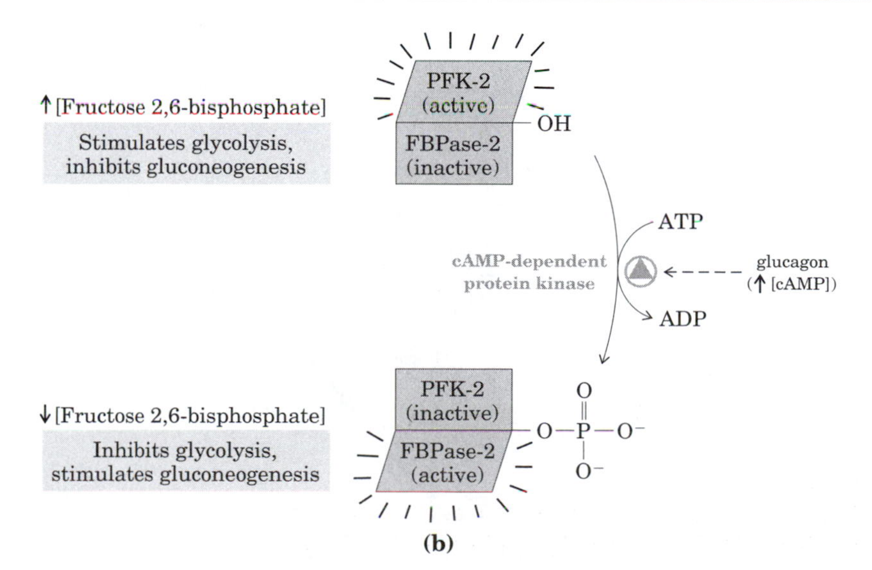

figure 20–8, page 733

Regulation of fructose 2,6-bisphosphate level.

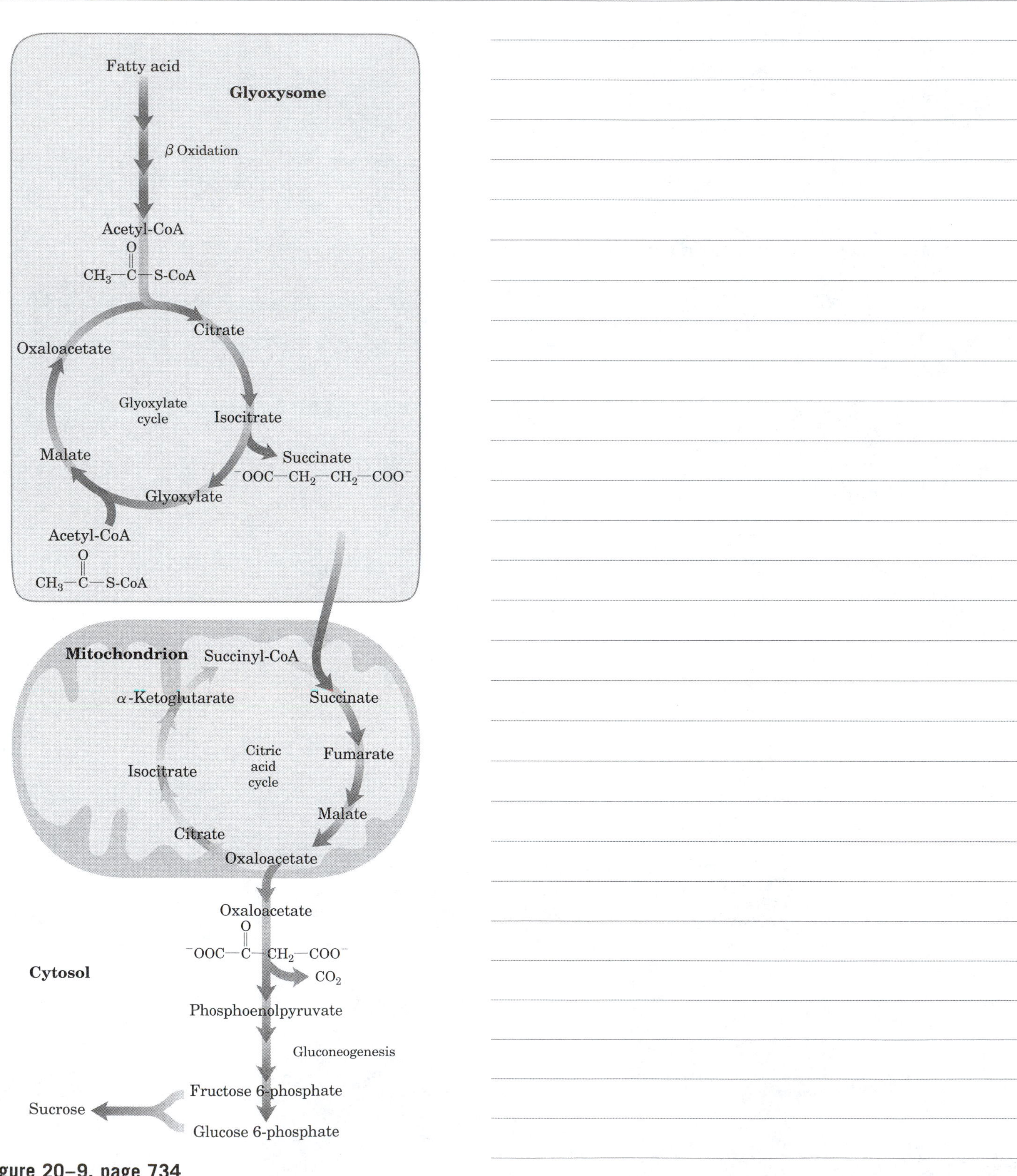

figure 20–9, page 734
Conversion of stored fatty acids to sucrose in germinating seeds.

Triacylglycerol

$CH_2-O-C(=O)-R$

$CH-O-C(=O)-R$

$CH_2-O-C(=O)-R$

lipase → Fatty acids —β oxidation→ $CH_3-C(=O)-S\text{-}CoA$ Acetyl-CoA

CH_2OH / $CHOH$ / CH_2OH Glycerol

glycerol kinase (ATP → ADP)

CH_2OH / $CHOH$ / CH_2O-Ⓟ Glycerol 3-phosphate

glycerol 3-phosphate dehydrogenase (NAD^+ → $NADH + H^+$)

CH_2OH / $C{=}O$ / CH_2O-Ⓟ Dihydroxyacetone phosphate

figure 20–10, page 734
Conversion of the glycerol moiety of triacylglycerols to sucrose in germinating seeds.

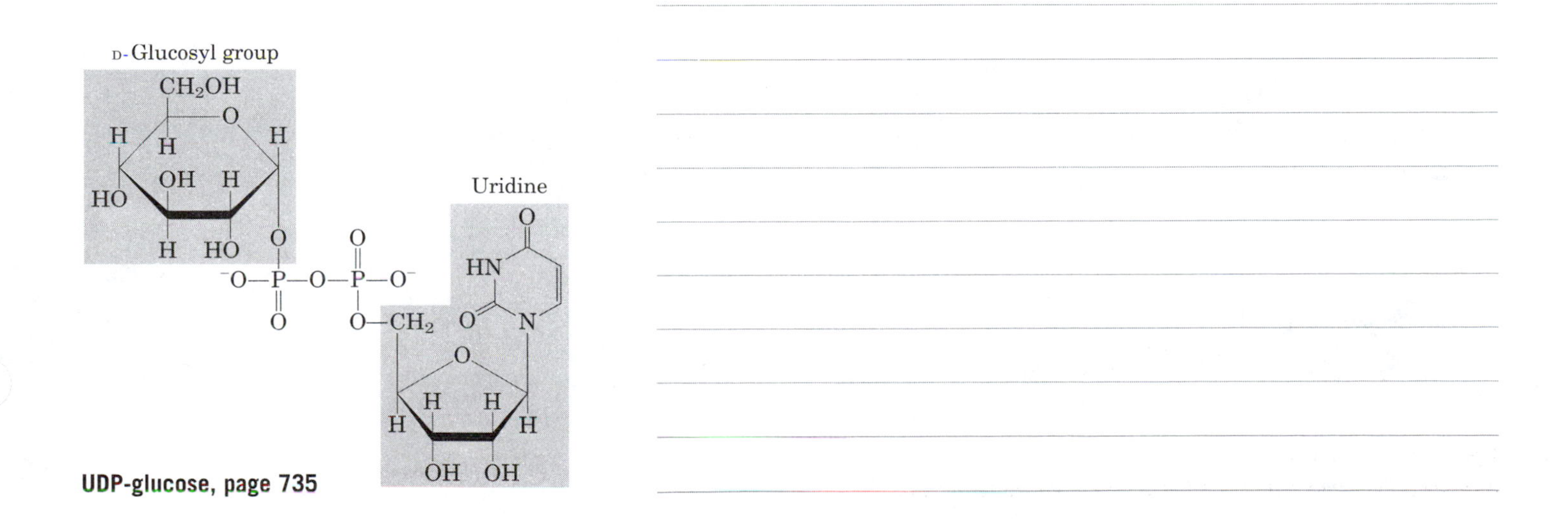

UDP-glucose, page 735

Sugar phosphate + NTP

NDP-sugar pyrophosphorylase

Pyrophosphate (PP_i) + Sugar nucleotide (NDP-sugar)

inorganic pyrophosphatase

2 Phosphate (P_i)

Net reaction: Sugar phosphate + NTP $\longrightarrow$ NDP-sugar + $2P_i$

figure 20–11, page 735
Formation of a sugar nucleotide.

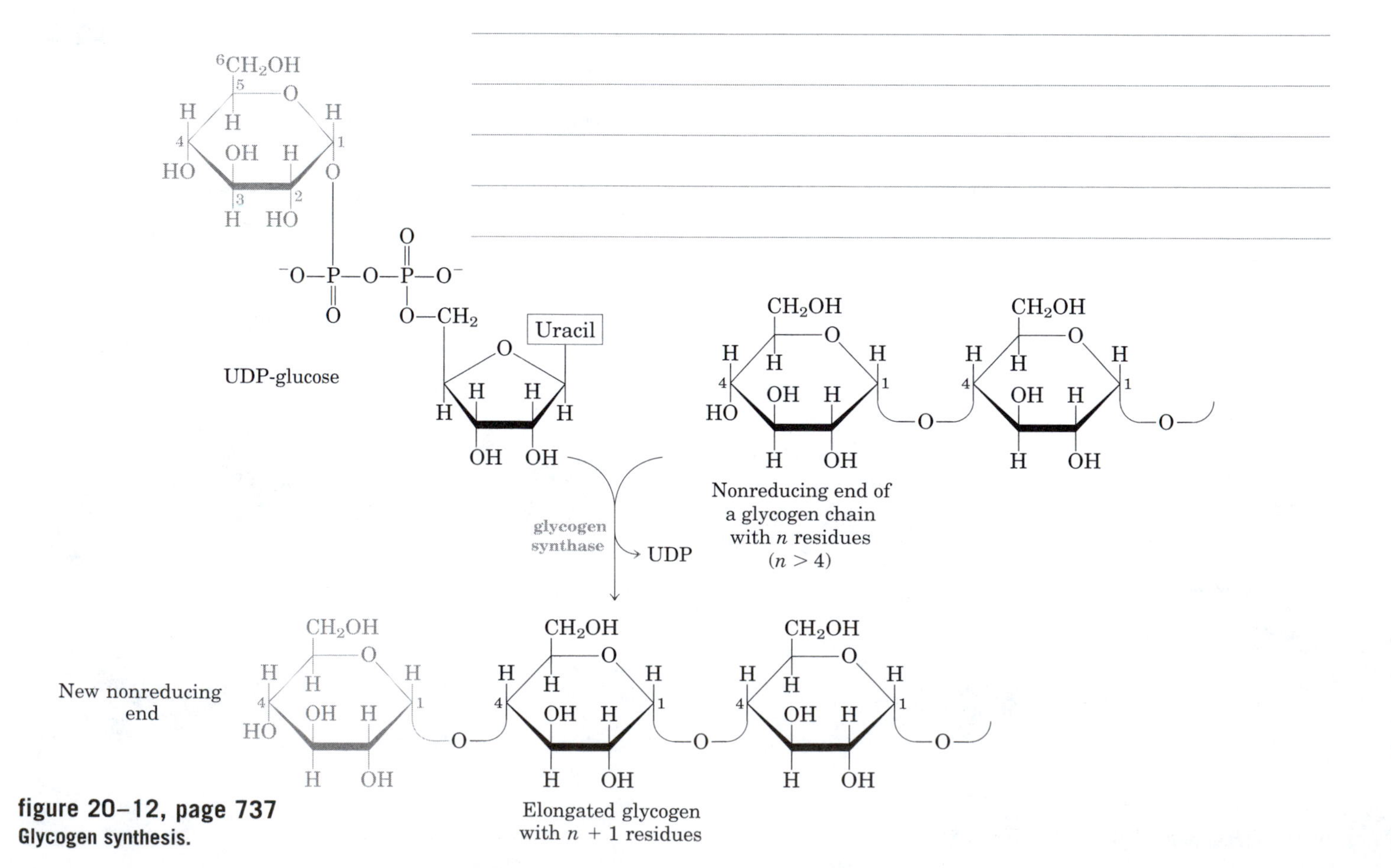

figure 20–12, page 737
Glycogen synthesis.

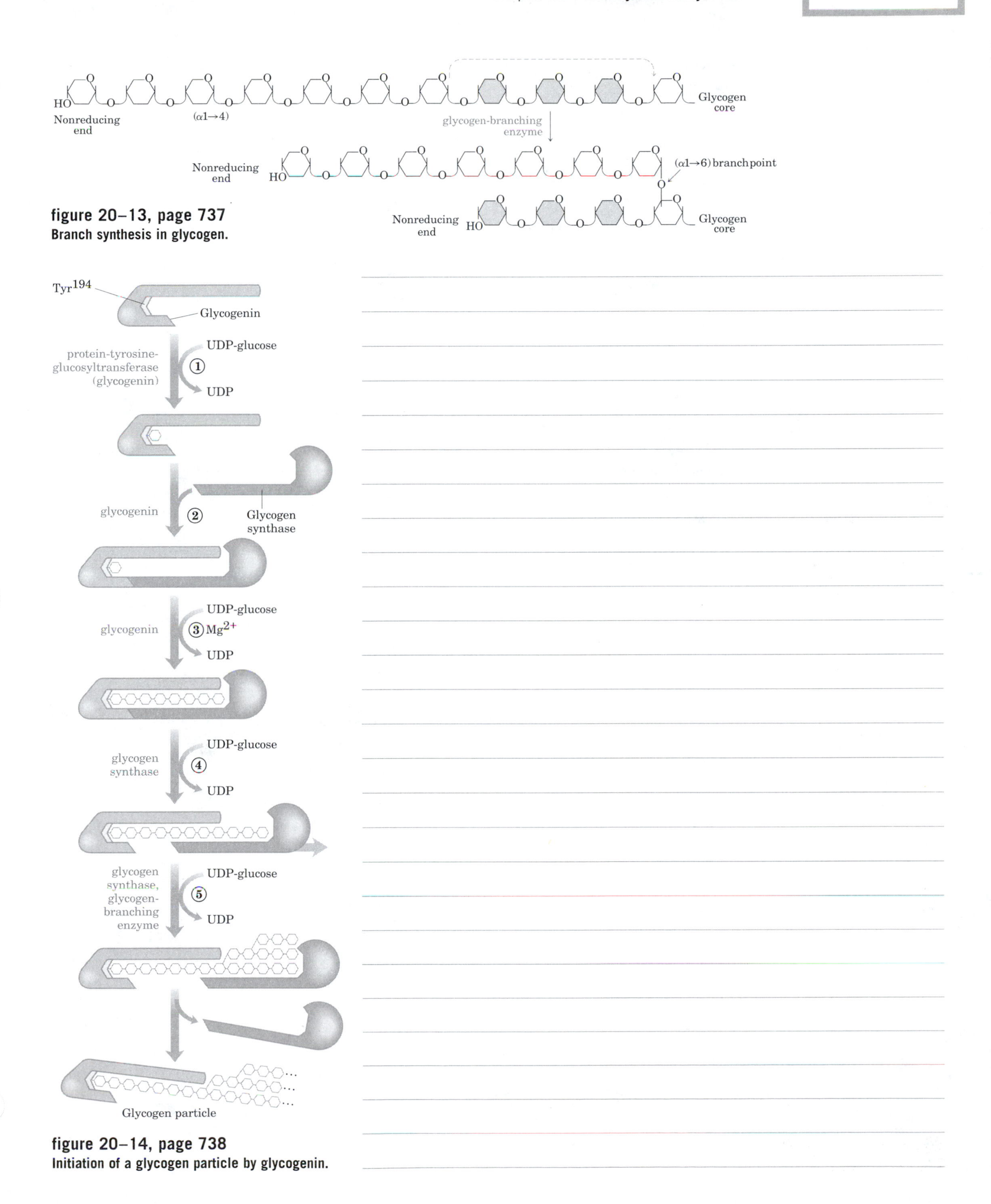

figure 20–13, page 737
Branch synthesis in glycogen.

figure 20–14, page 738
Initiation of a glycogen particle by glycogenin.

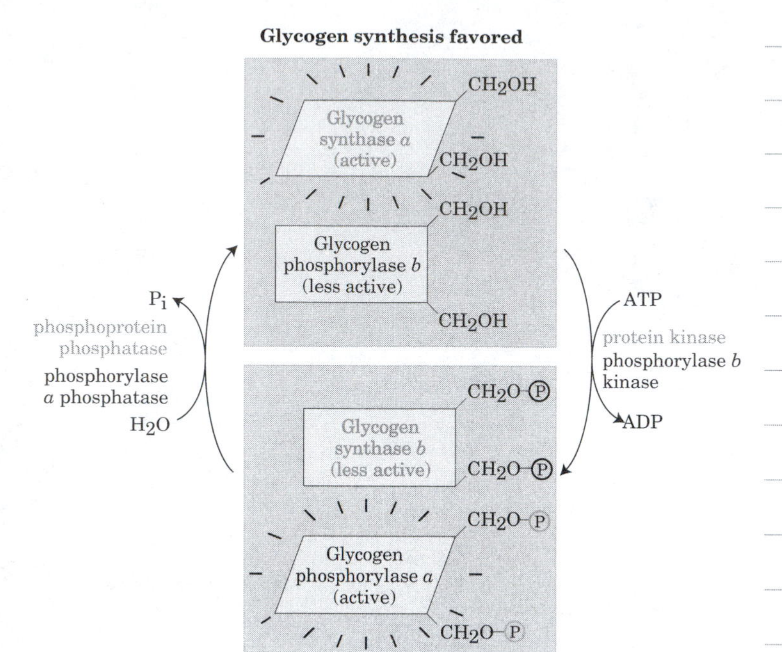

figure 20–15, page 739
Reciprocal regulation of glycogen synthase and glycogen phosphorylase.

Glucose 1-phosphate + ATP

ADP-glucose pyrophosphorylase

3-phosphoglycerate

PP_i

P_i

ADP-glucose

Adenine

Nonreducing end of a starch chain with n residues

starch synthase

ADP

New nonreducing end

Elongated starch with $n + 1$ residues

figure 20–16, page 740
Starch synthesis.

UDP-glucose + Fructose 6-phosphate

sucrose 6-phosphate synthase → UDP

Sucrose 6-phosphate

sucrose 6-phosphate phosphatase → P_i

Sucrose

figure 20–17, page 741
Sucrose synthesis.

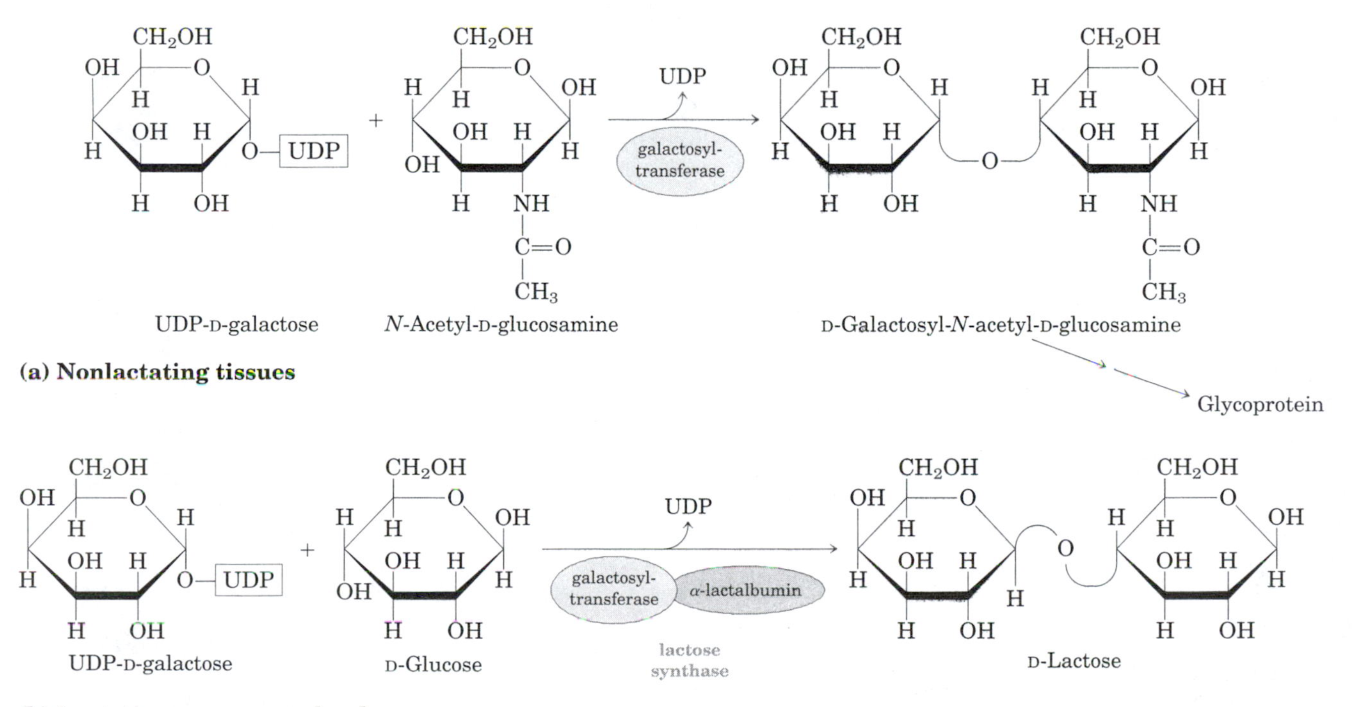

figure 20–18, page 742
Lactose synthesis.

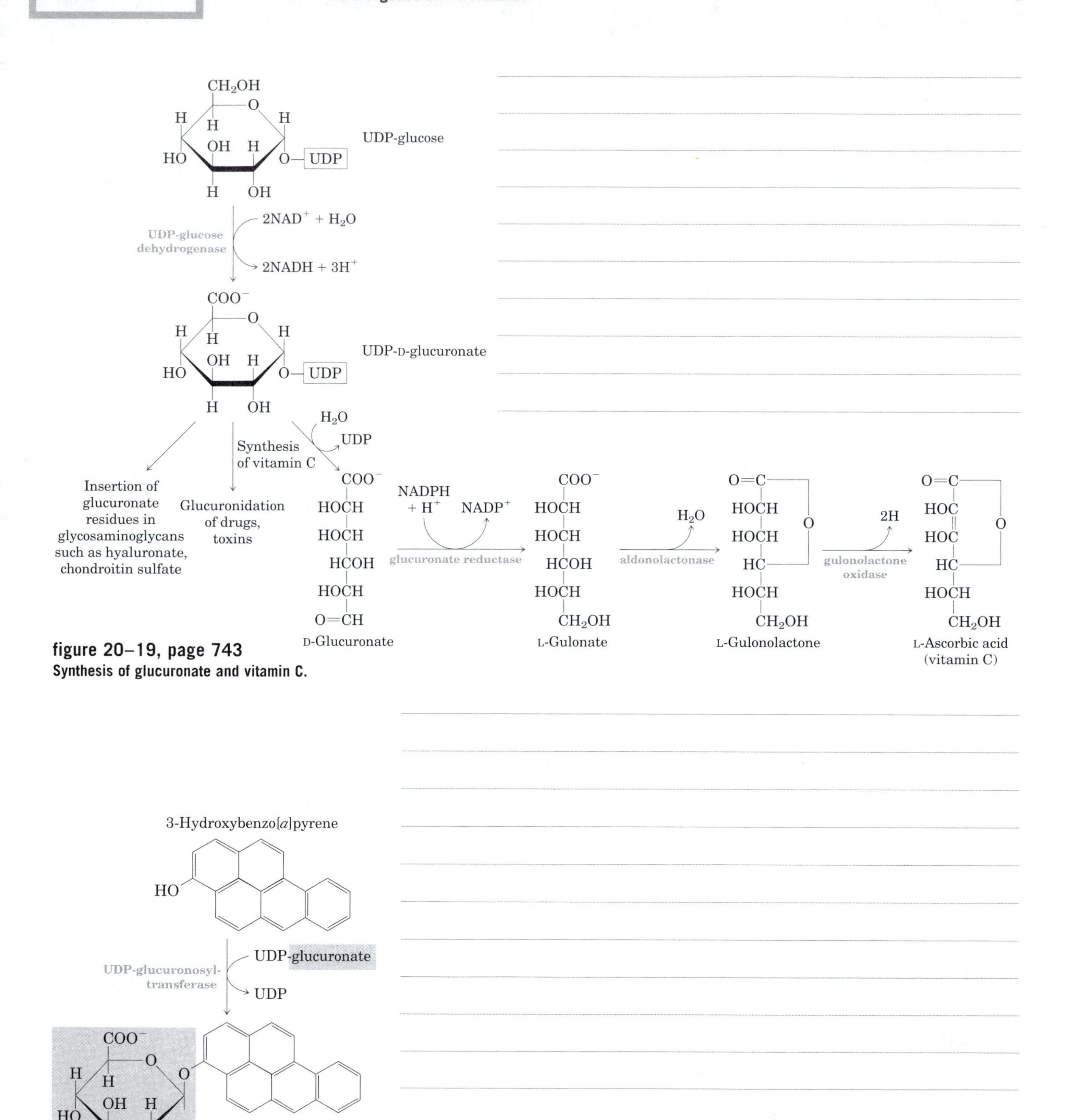

figure 20–19, page 743
Synthesis of glucuronate and vitamin C.

figure 20–20, page 744
Role of UDP-glucuronate in detoxification.

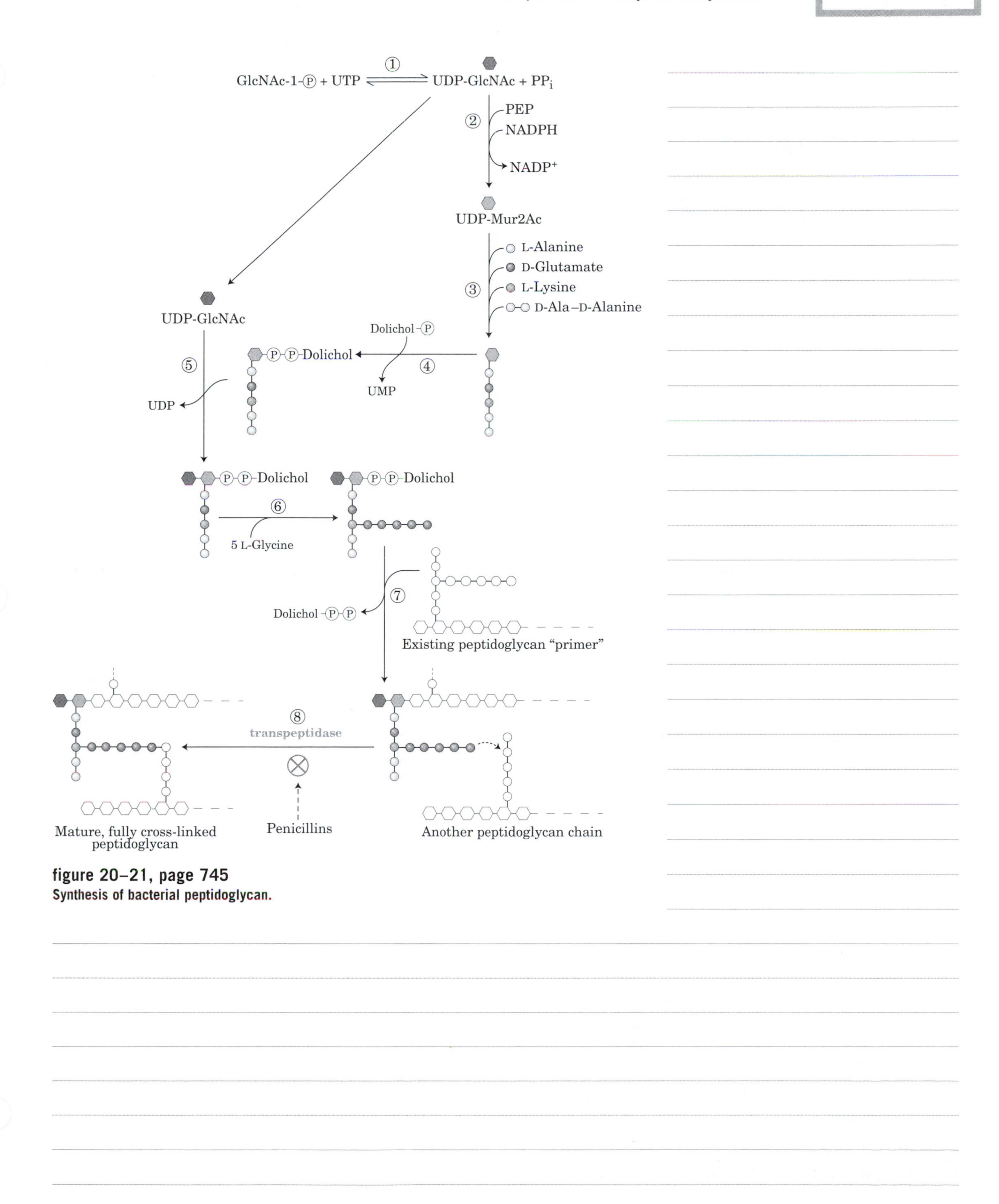

figure 20–21, page 745
Synthesis of bacterial peptidoglycan.

Side chain Thiazolidine ring

β-Lactam ring

General structure of penicillins

R =

$-CH_2-$ Penicillin G (Benzylpenicillin)

$-O-CH_2-$ Penicillin V

OCH_3 Methicillin OCH_3

box 20–1, figure 1, page 746

Enz—Ser—ÖH Penicillin

(a) (b)

Trans-peptidase—Ser—O

Stably derivatized, inactive transpeptidase

β-Lactamase—Ser—O

H_2O β-Lactamase

Inactive penicillin

box 20–1, figure 2, page 747
(a) Penicillin action. (b) Inactivation.

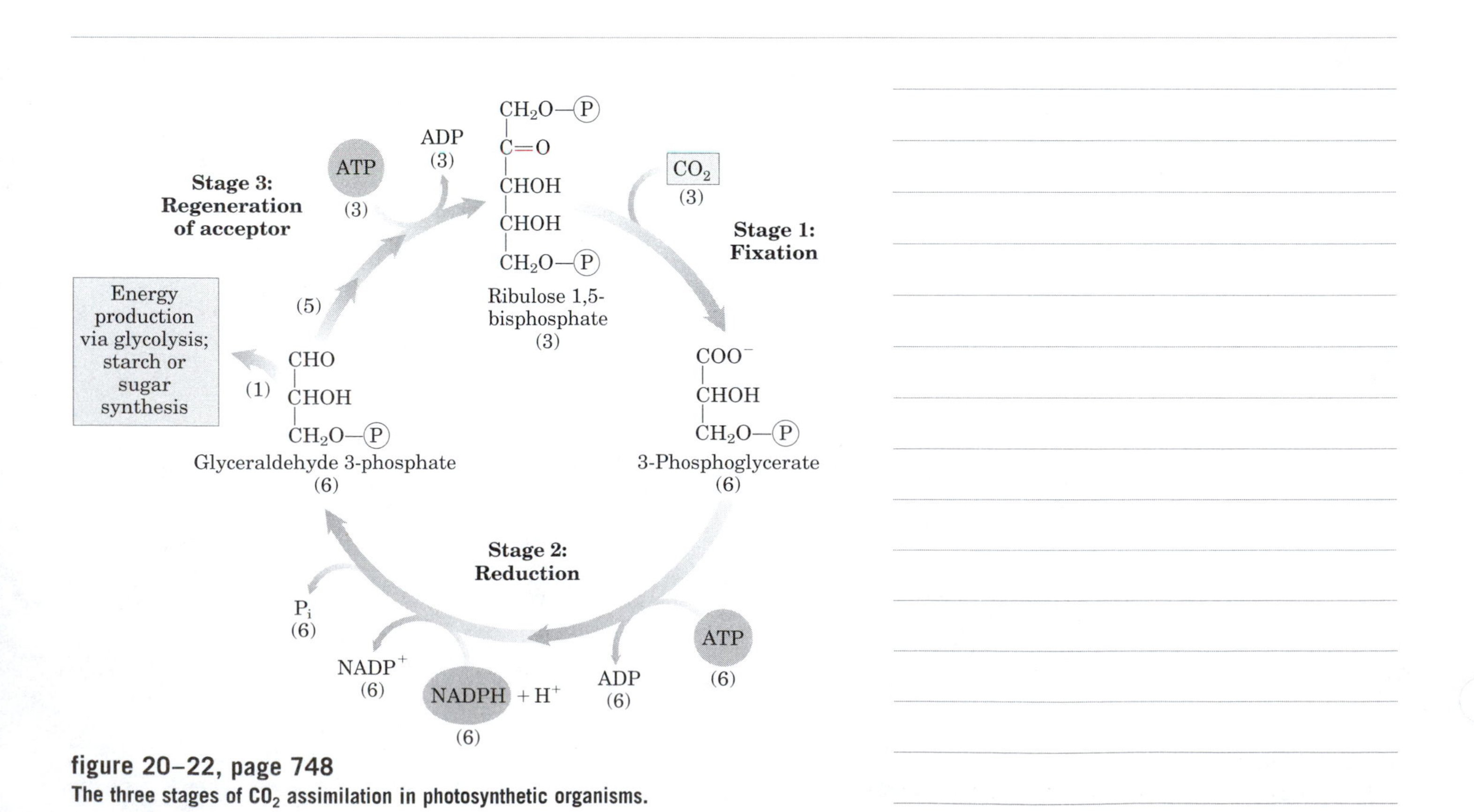

figure 20–22, page 748
The three stages of CO_2 assimilation in photosynthetic organisms.

Ribulose 1,5-bisphosphate

Enediol intermediate

2′-Carboxy-3-keto-D-arabinitol 1,5-bisphosphate (β-keto acid intermediate)

Hydrated intermediate

3-Phosphoglycerate

Carbanion + 3-Phosphoglycerate

Six-carbon reaction intermediates bound to ribulose 1,5-bisphosphate carboxylase (rubisco)

figure 20–23, page 749
First stage of CO_2 assimilation.

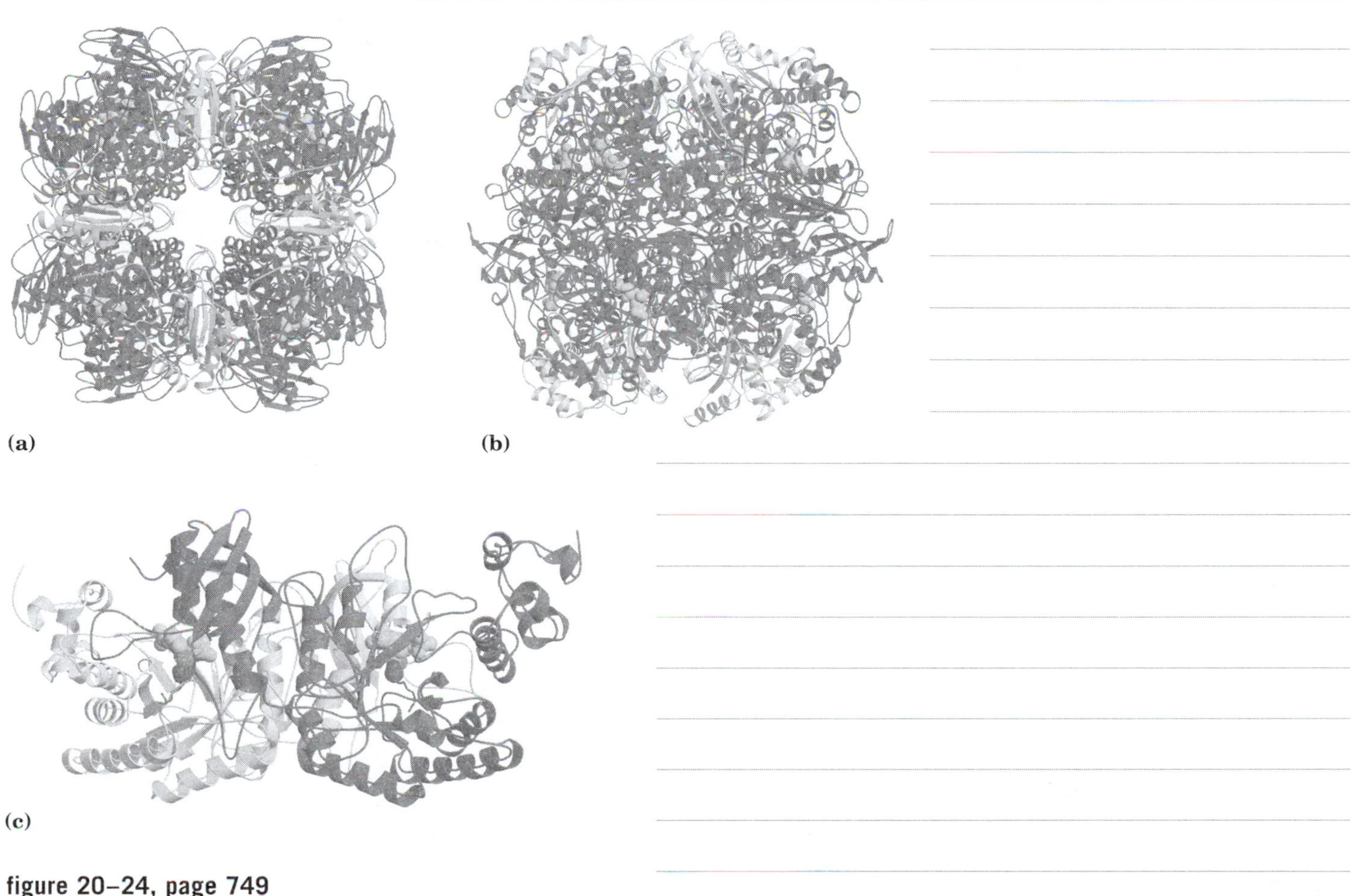

figure 20–24, page 749
Structure of ribulose 1,5-bisphosphate carboxylase (rubisco).

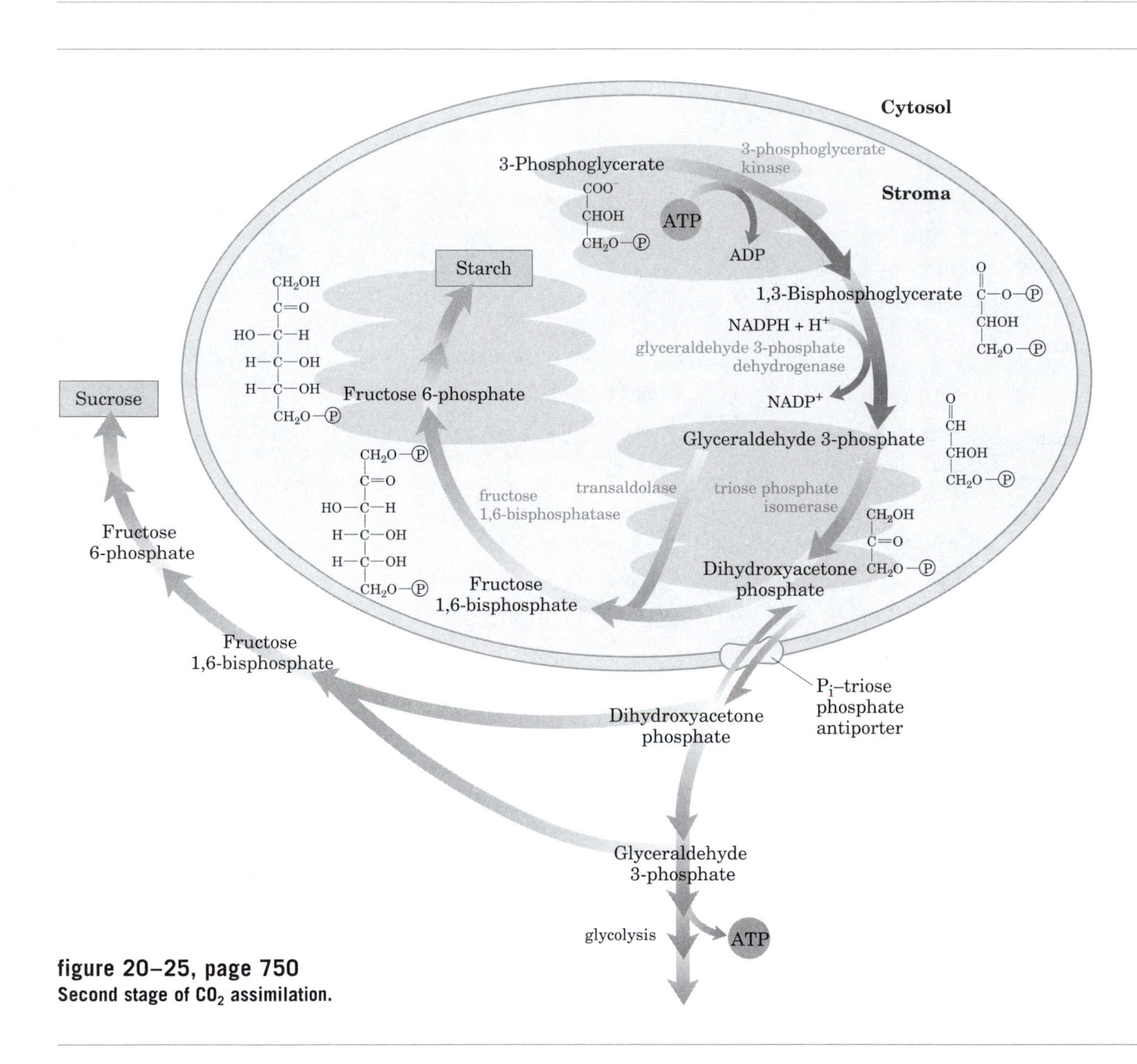

figure 20–25, page 750
Second stage of CO_2 assimilation.

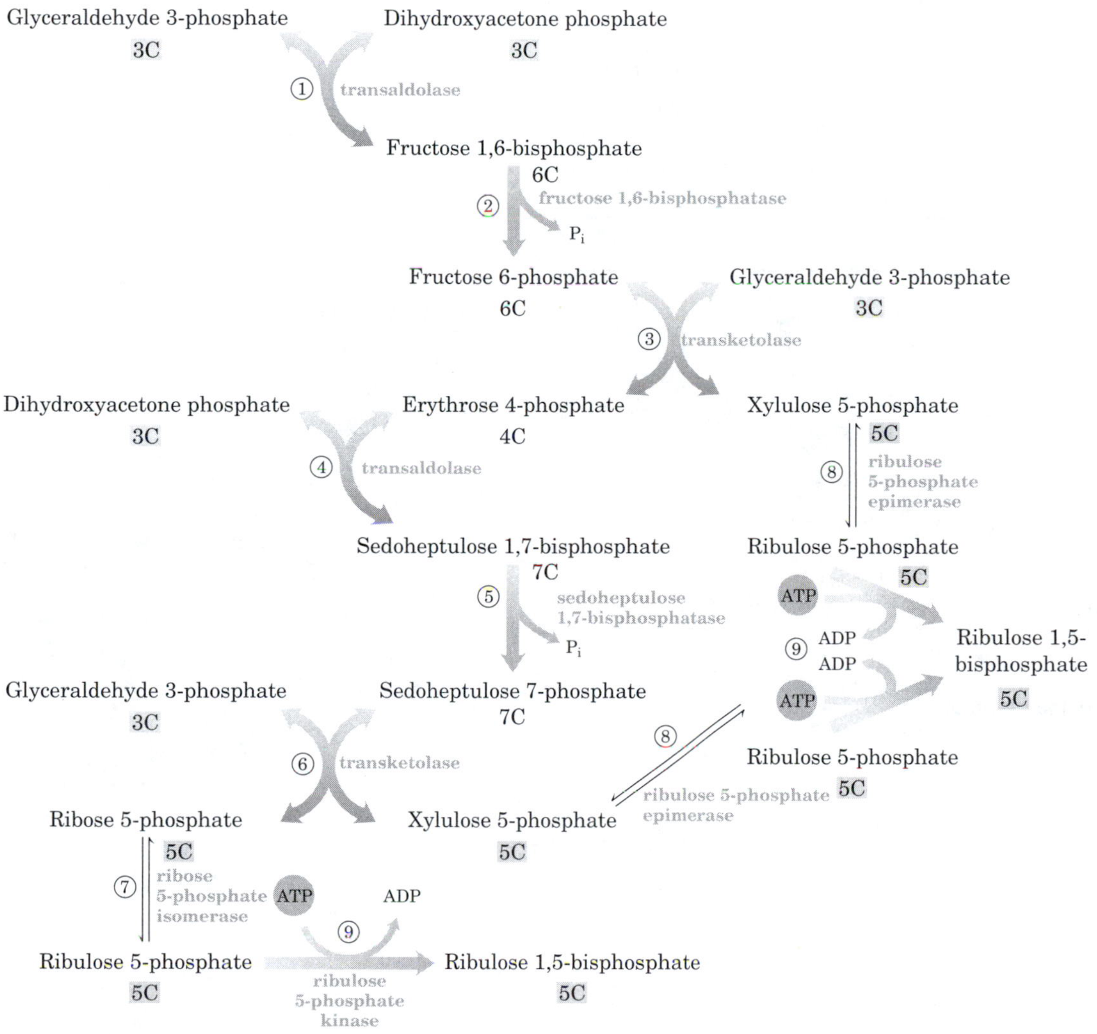

figure 20–26, page 751
Third stage of CO_2 assimilation.

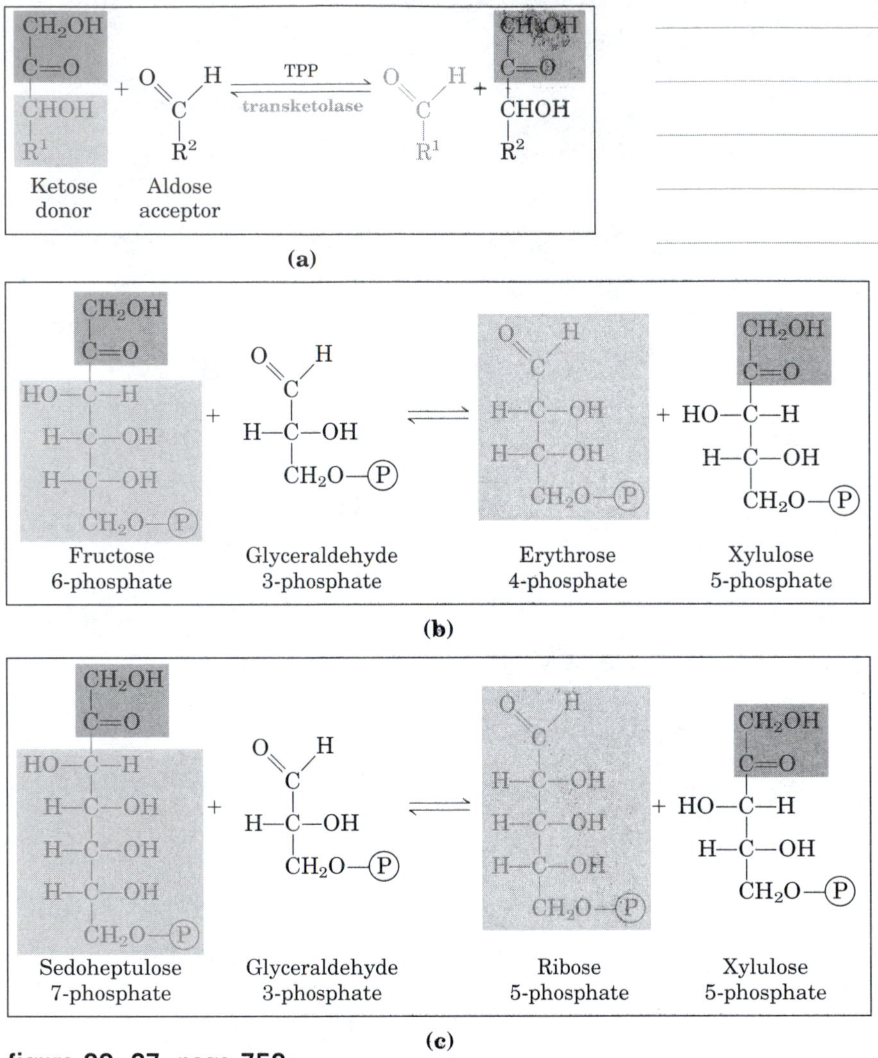

figure 20–27, page 752
Transketolase-catalyzed reactions of the Calvin cycle.

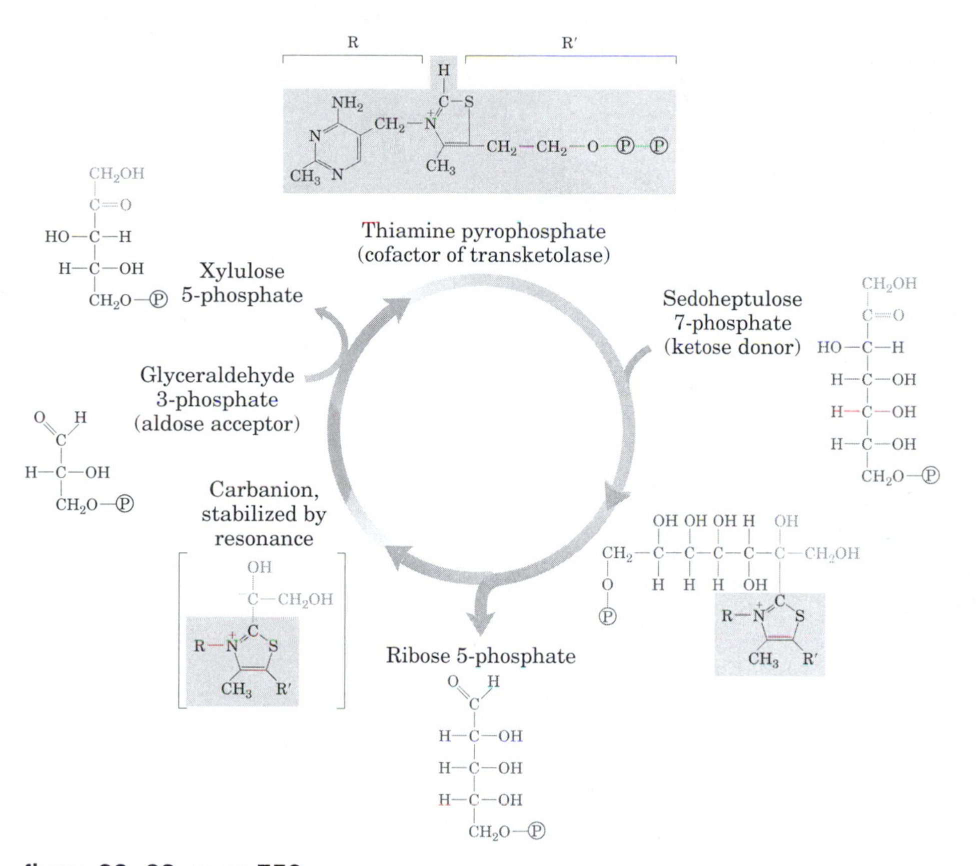

figure 20–28, page 753
Transketolase transfers a two-carbon group from sedoheptulose 7-phosphate to glyceraldehyde 3-phosphate, producing two pentose phosphates.

Ribose 5-phosphate

ribose 5-phosphate isomerase

Xylulose 5-phosphate

ribose 5-phosphate epimerase

Ribulose 5-phosphate

ATP ADP

ribulose 5-phosphate kinase

Ribulose 1,5-bisphosphate

figure 20–29, page 753
Regeneration of ribulose 1,5-bisphosphate.

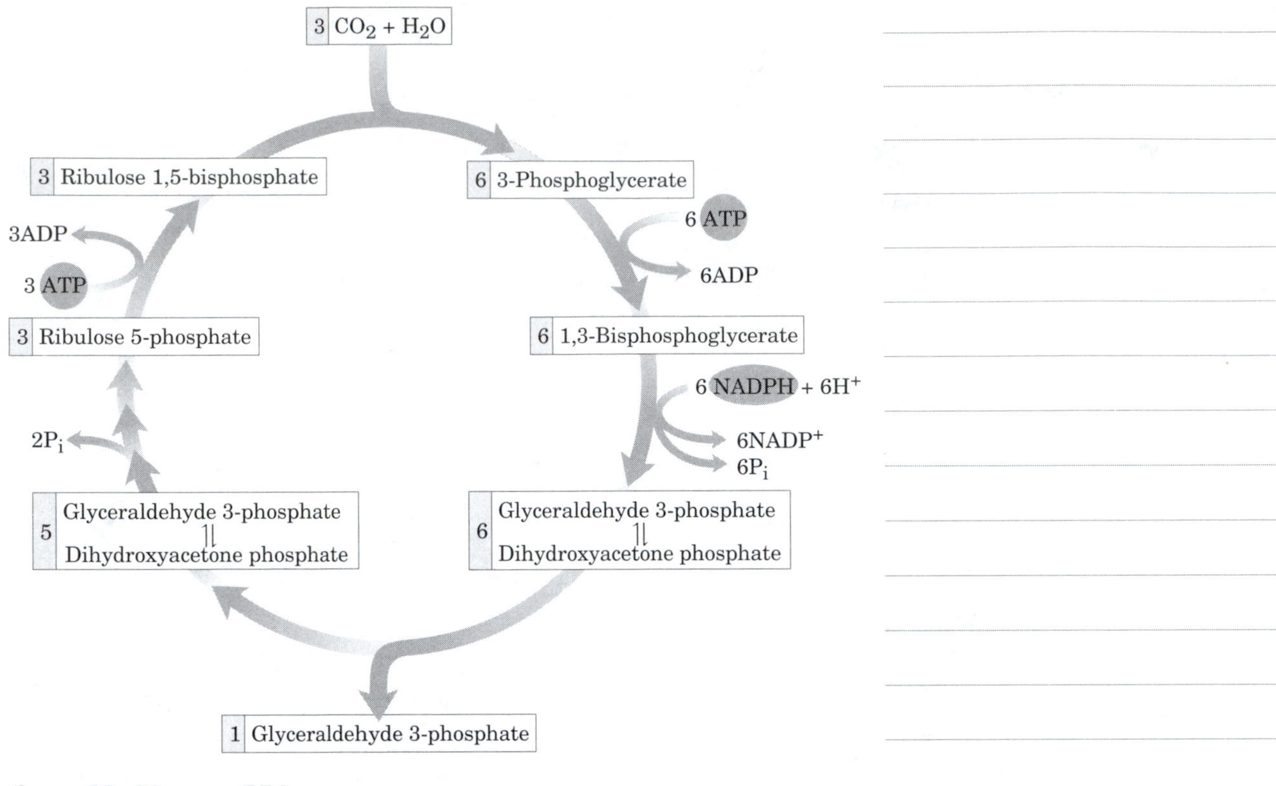

figure 20–30, page 754
Stoichiometry of CO_2 assimilation in the Calvin cycle.

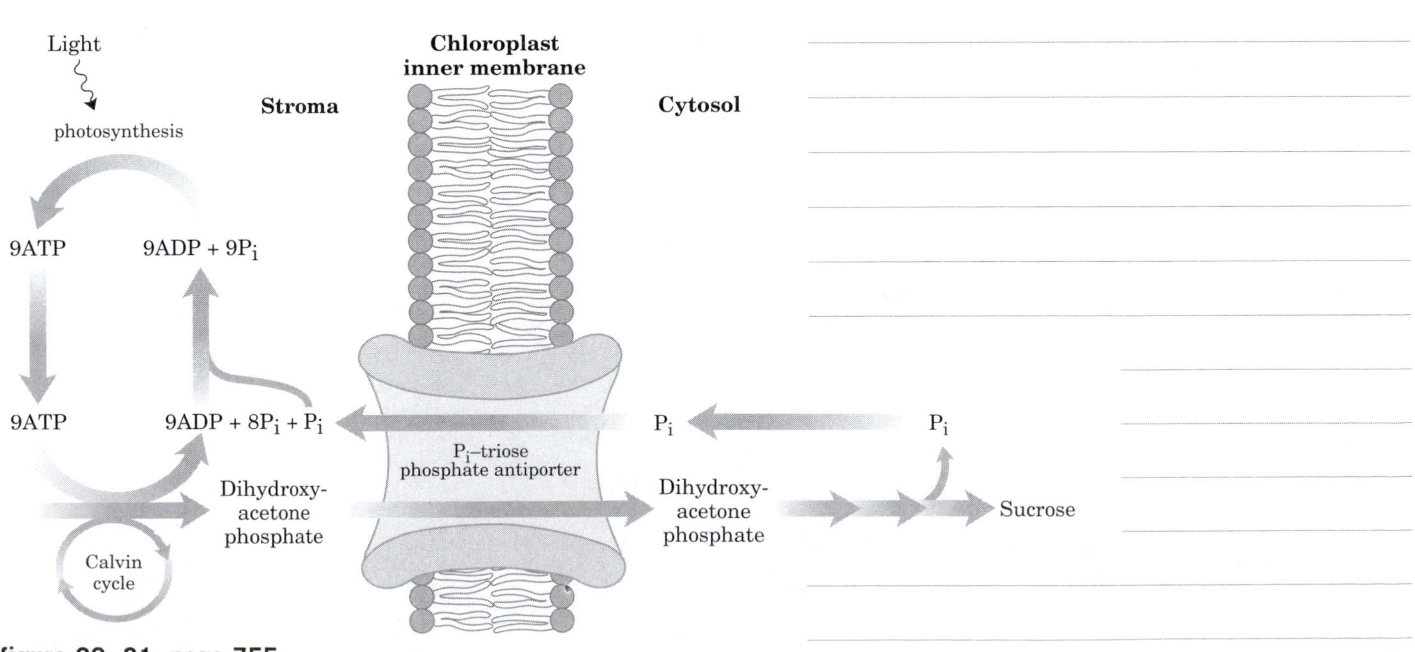

figure 20–31, page 755
The P_i–triose phosphate antiport system of the inner chloroplast membrane.

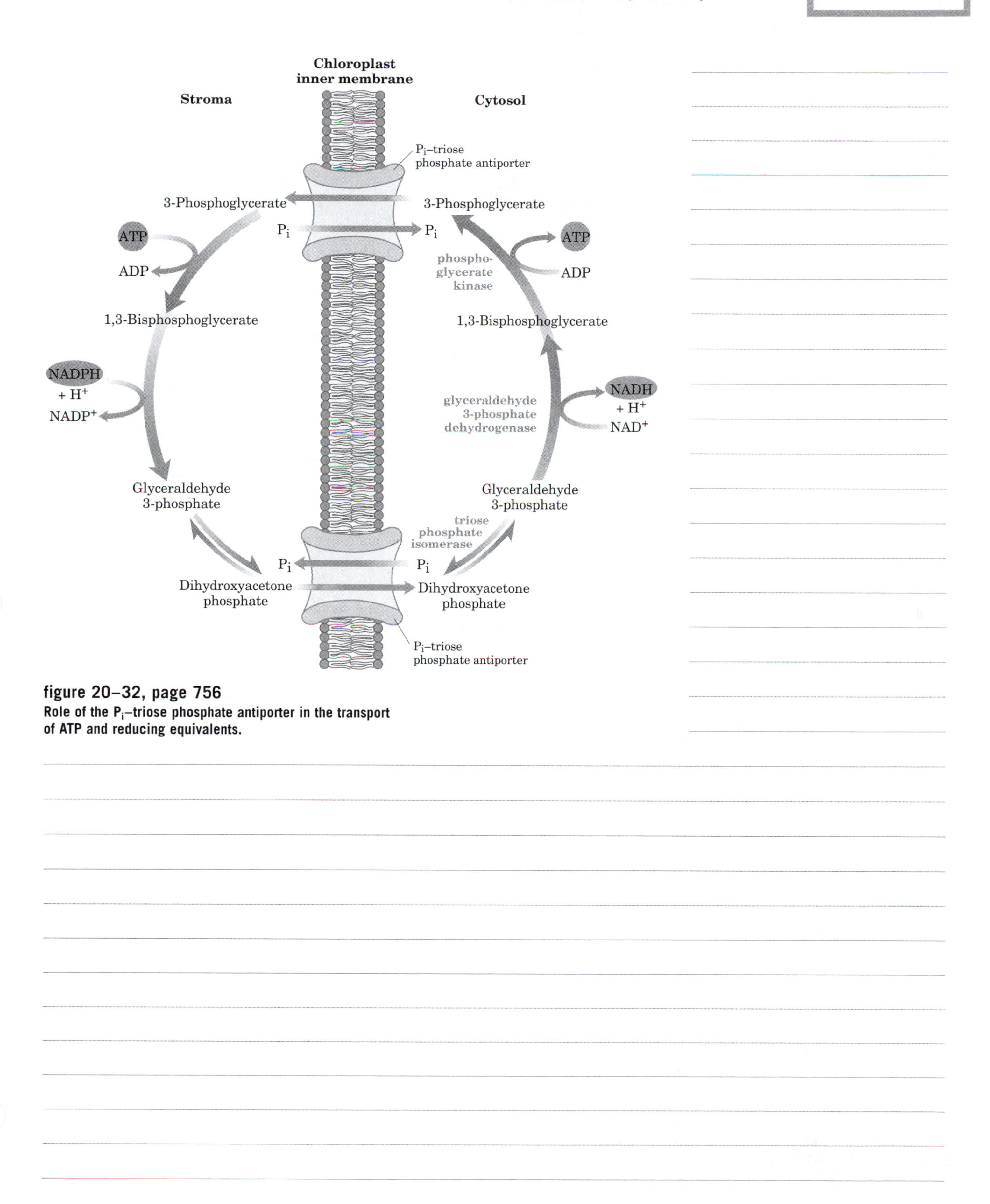

figure 20–32, page 756

Role of the P_i–triose phosphate antiporter in the transport of ATP and reducing equivalents.

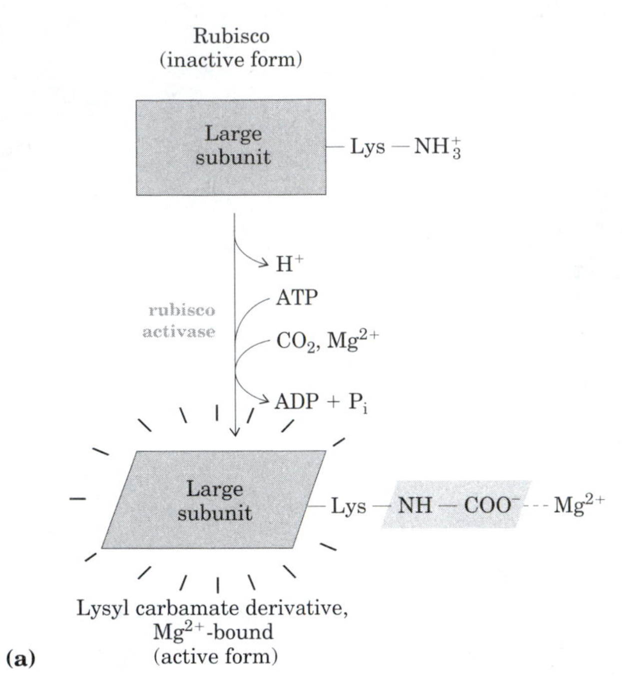

$CH_2O—PO_3^{2-}$, HO—C—COO^-, H—C—OH, H—C—OH, CH_2—OH
2-Carboxyarabinitol 1-phosphate

$CH_2O—PO_3^{2-}$, HO—C—COO^-, C=O, H—C—OH, $CH_2O—PO_3^{2-}$
β-Keto acid intermediate

(c)

figure 20–33, page 757
Regulation of rubisco.

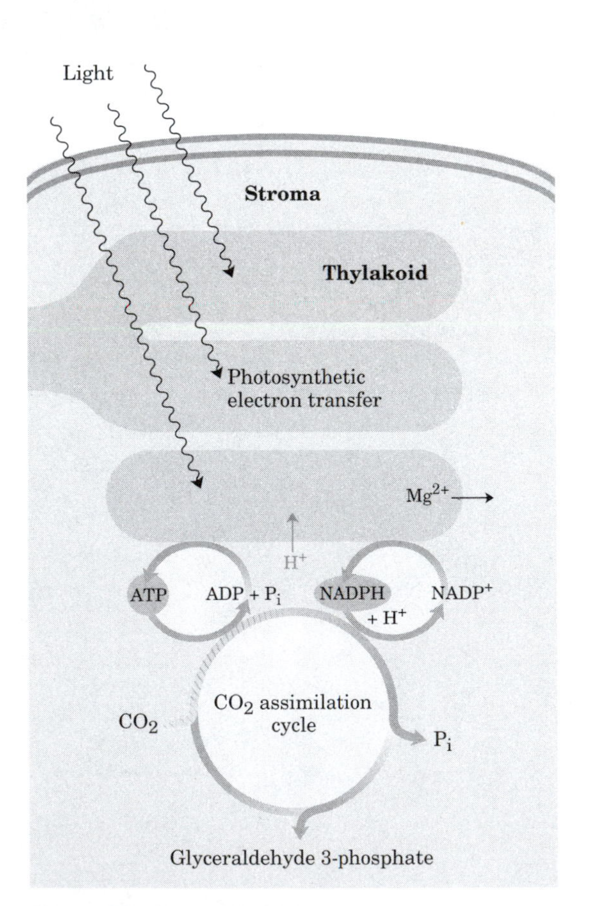

figure 20–34, page 758
ATP and NADPH produced by the light reactions are essential substrates for the reduction of CO_2.

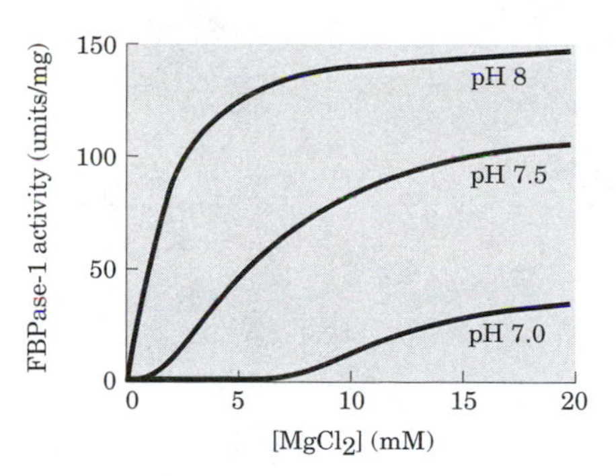

figure 20–35, page 758
Activation of chloroplast fructose 1,6-bisphosphatase.

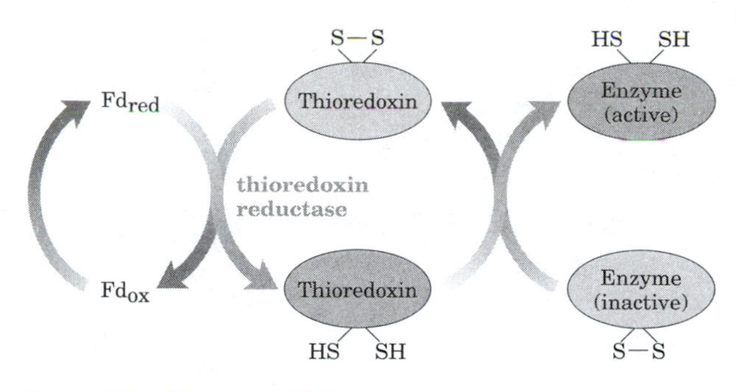

figure 20–36, page 758
Thioredoxin.

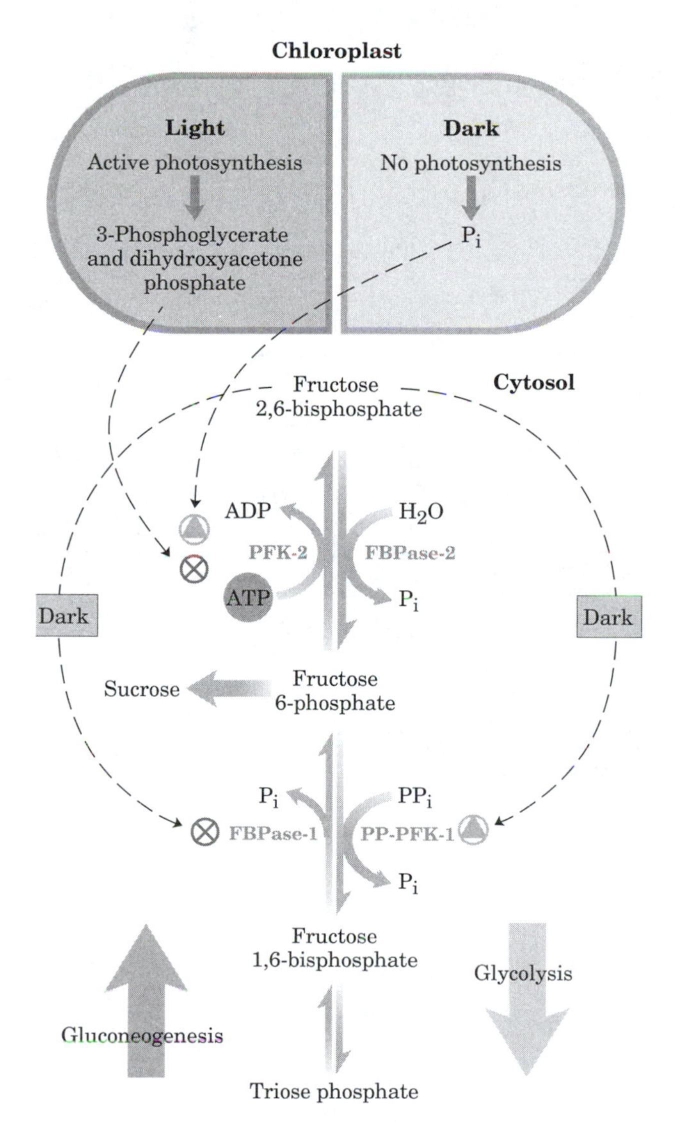

figure 20–37, page 759
Fructose 2,6-bisphosphate as regulator of sucrose synthesis.

$CH_2O{-}PO_3^{2-}$ / $C{=}O$ / $H{-}C{-}OH$ / $H{-}C{-}OH$ / $CH_2O{-}PO_3^{2-}$ — Ribulose 1,5-bisphosphate

$\rightleftharpoons$

$CH_2O{-}PO_3^{2-}$ / $C{-}OH$ / $\|$ / $C{-}OH$ / $H{-}C{-}OH$ / $CH_2O{-}PO_3^{2-}$ — Enediol form

$\downarrow O_2$

[$CH_2O{-}PO_3^{2-}$ / $H{-}O{-}O{-}C{-}OH$ / $C{=}O$ / $H{-}C{-}OH$ / $CH_2O{-}PO_3^{2-}$] — Enzyme-bound intermediate

$\downarrow OH^- \rightarrow H_2O$

$CH_2O{-}PO_3^{2-}$ / C (^-O, O) + ^-O, O / C / $H{-}C{-}OH$ / $CH_2O{-}PO_3^{2-}$

Phosphoglycolate 3-Phosphoglycerate

figure 20–38, page 760
Oxygenase activity of rubisco.

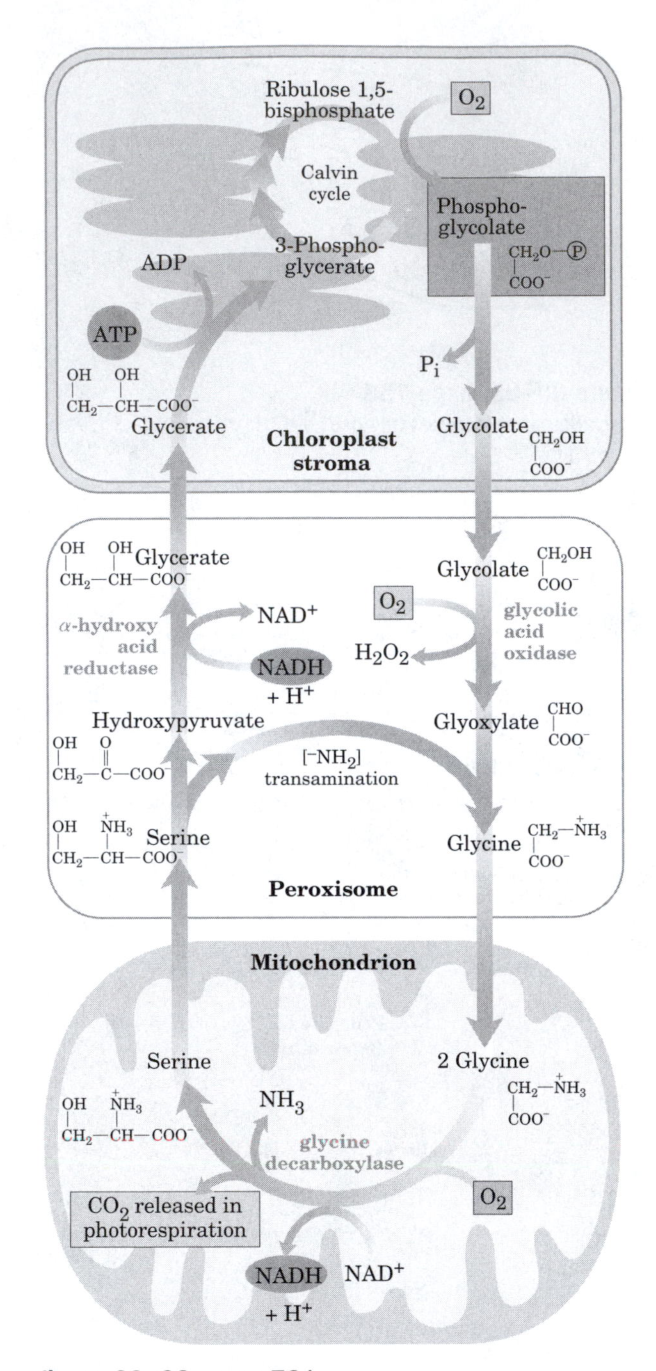

figure 20–39, page 761
Glycolate pathway.

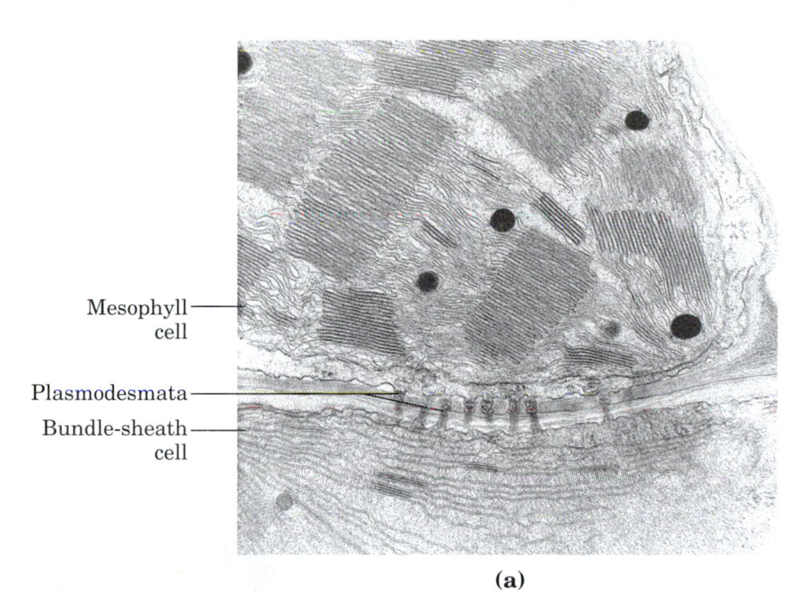

(a)

figure 20–40, page 762
Carbon assimilation in C_4 plants.

CO_2 (in air)
H_2O
H^+
HCO_3^-
Mesophyll cell
PEP carboxylase
PEP
Oxaloacetate
AMP + PP_i
pyruvate phosphate dikinase
ATP + P_i
malate dehydrogenase
NADPH + H^+
$NADP^+$
Pyruvate
Malate
Plasmodesmata
Plasma membranes
Pyruvate
Malate
malic enzyme
$NADP^+$
NADPH + H^+
CO_2
Ribulose 1,5-bisphosphate
3-Phosphoglycerate
Bundle-sheath cell
Triose phosphates

(b)

PROBLEMS

2. Pathway of Atoms in Gluconeogenesis, page 766.

(a) [^{14}C]Bicarbonate, HO—^{14}C(=O)O$^-$

(b) [1-^{14}C]Pyruvate, CH_3—C(=O)—$^{14}COO^-$

7. Glucogenic Substrates, page 767.

(a) $^-OOC_2$—CH_2—CH_2—COO^-
Succinate

(b) CH_2(OH)—C(OH)(H)—CH_2(OH)
Glycerol

(c) CH_3—C(=O)—S-CoA
Acetyl-CoA

(d) CH_3—C(=O)—COO^-
Pyruvate

(e) CH_3—CH_2—CH_2—COO^-
Butyrate

8. Blood Lactate Levels during Vigorous Exercise, page 767.

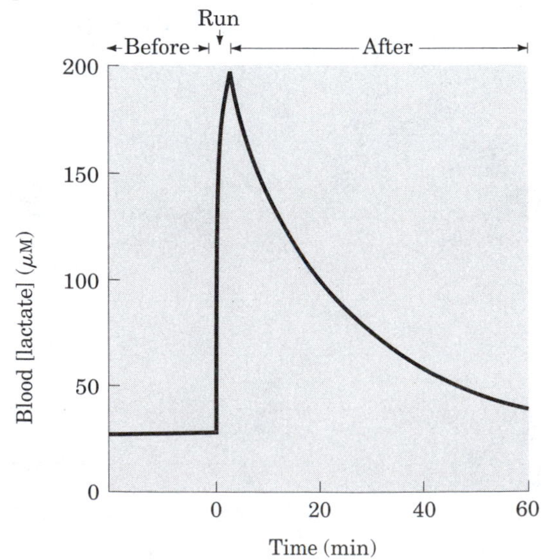

10. Ethanol Affects Blood Glucose Levels, page 767.

$$CH_3CH_2OH + NAD^+ \longrightarrow CH_3CHO + NADH + H^+$$

11. Effect of Phloridzin on Carbohydrate Metabolism, page 767.

Phloridzin

18. Identification of Key Intermediates in CO_2 Assimilation, page 768.

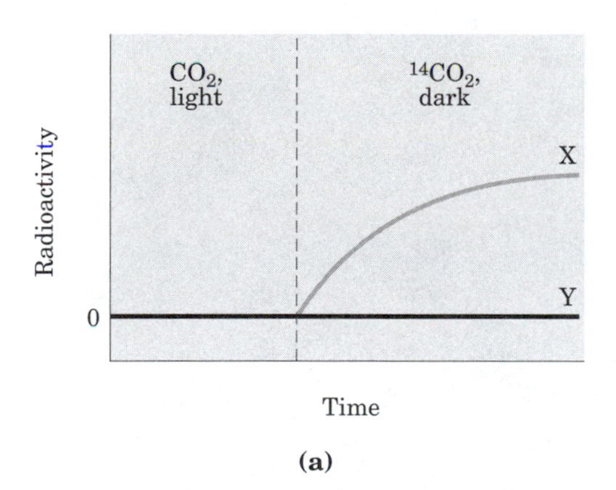

(a)

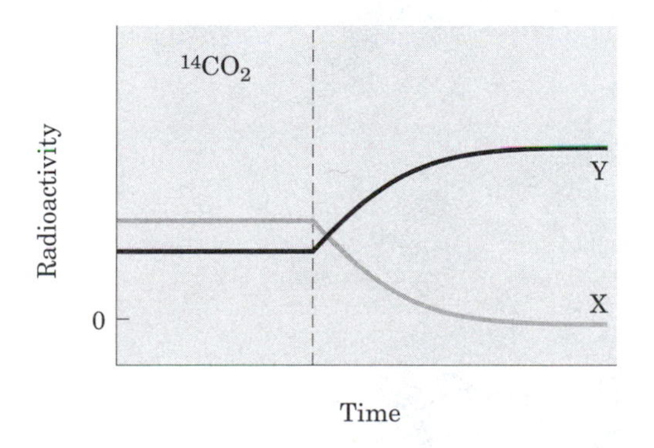

(b)

chapter

21 Lipid Biosynthesis

When you study for exams, don't forget to review The Minicourses on the UNDERSTAND! BIOCHEMISTRY CD that came with your textbook.

Minicourses that apply to this Chapter include:

Molecules of Life
An Interactive Gallery of Molecular Models
Bioenergetics
C4 Photosynthesis
Biosynthesis and Catabolism
Fatty Acid Biosynthesis
Steroid Biosynthesis

Malonyl-CoA, page 770

figure 21–1, page 771
The acetyl-CoA carboxylase reaction.

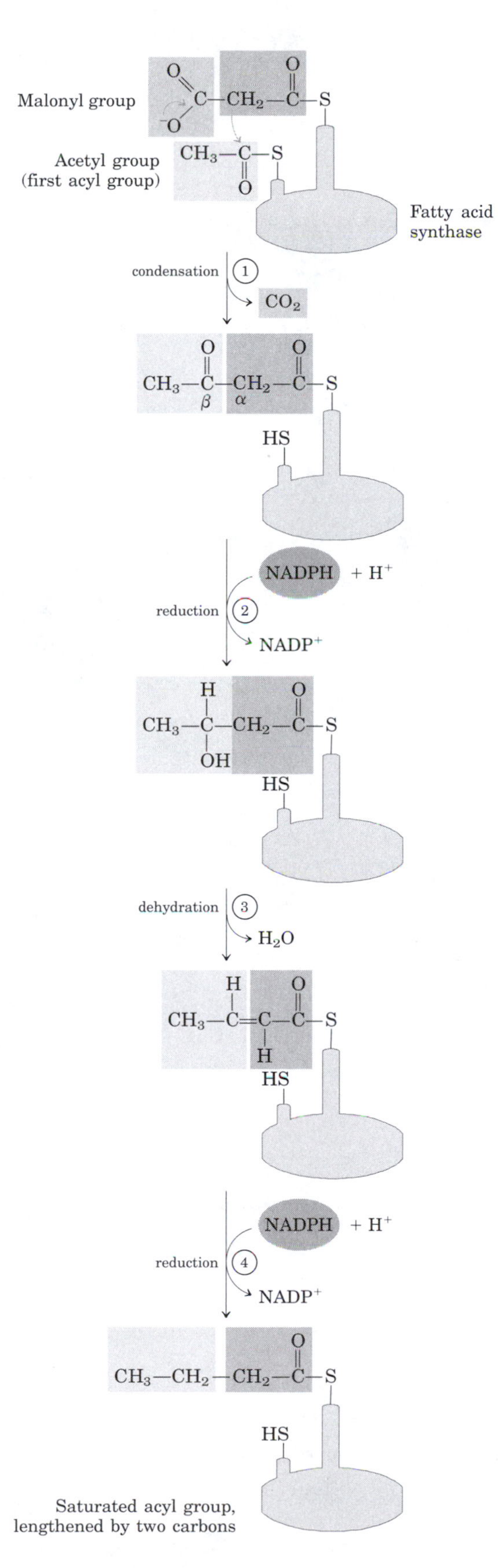

figure 21–2, page 772
A four-step sequence lengthens a growing fatty acyl chain by two carbons.

table 21–1, page 773

Proteins of the Fatty Acid Synthase Complex of *E. coli*

Protein	Role
Acyl carrier protein (ACP)	Carries acyl groups in thioester linkage
Acetyl-CoA–ACP transacetylase (AT)	Transfers acyl group from CoA to Cys residue of KS
β-Ketoacyl-ACP synthase (KS)	Condenses acyl and malonyl groups (there are at least three isozymes of KS)
Malonyl-CoA–ACP transferase (MT)	Transfers malonyl group from CoA to ACP
β-Ketoacyl-ACP reductase (KR)	Reduces β-keto group to β-hydroxy group
β-Hydroxyacyl-ACP dehydratase (HD)	Removes H_2O from β-hydroxyacyl-ACP, creating double bond
Enoyl-ACP reductase (ER)	Reduces double bond, forming saturated acyl-ACP

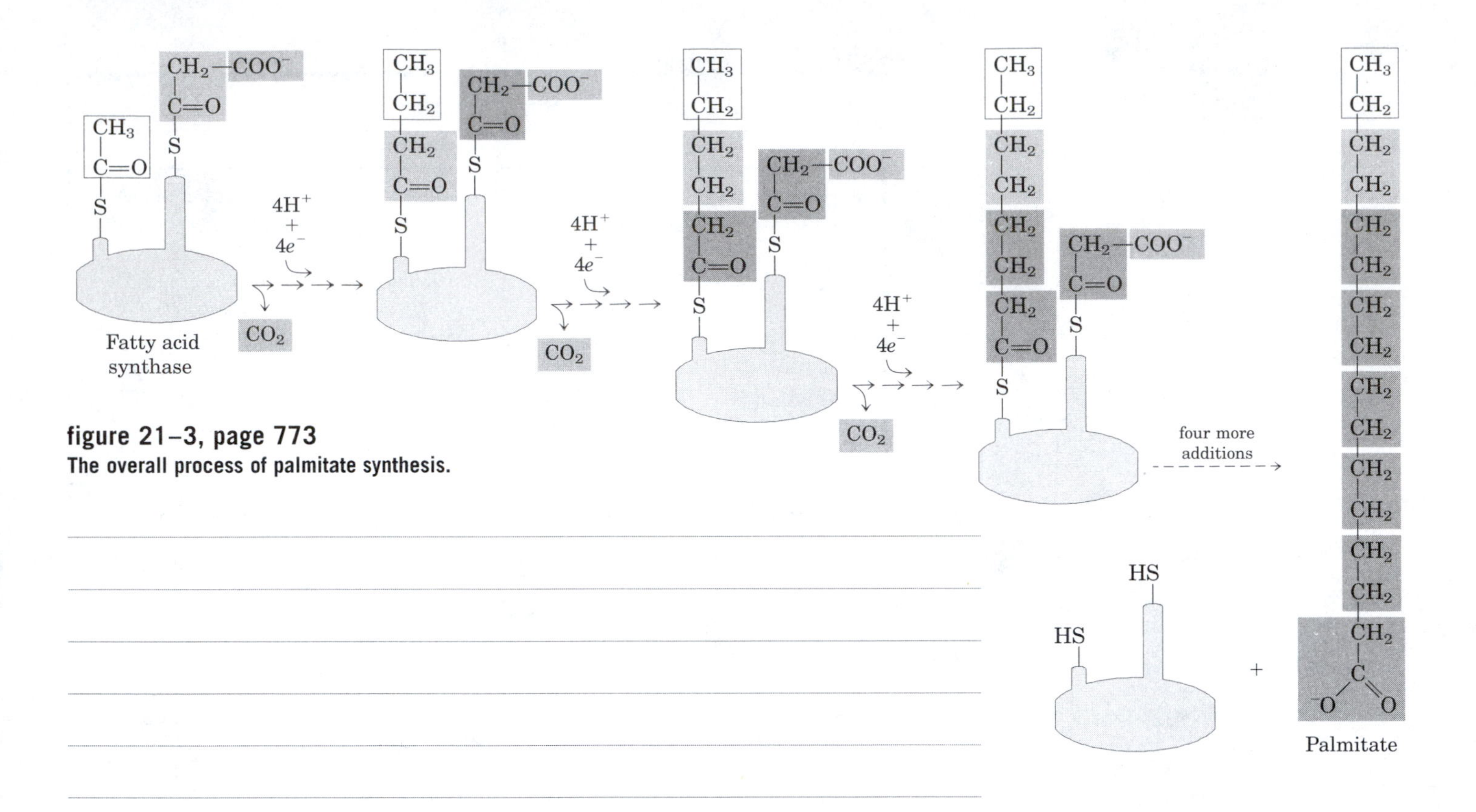

figure 21–3, page 773
The overall process of palmitate synthesis.

figure 21–4, page 773
Acyl carrier protein (ACP).

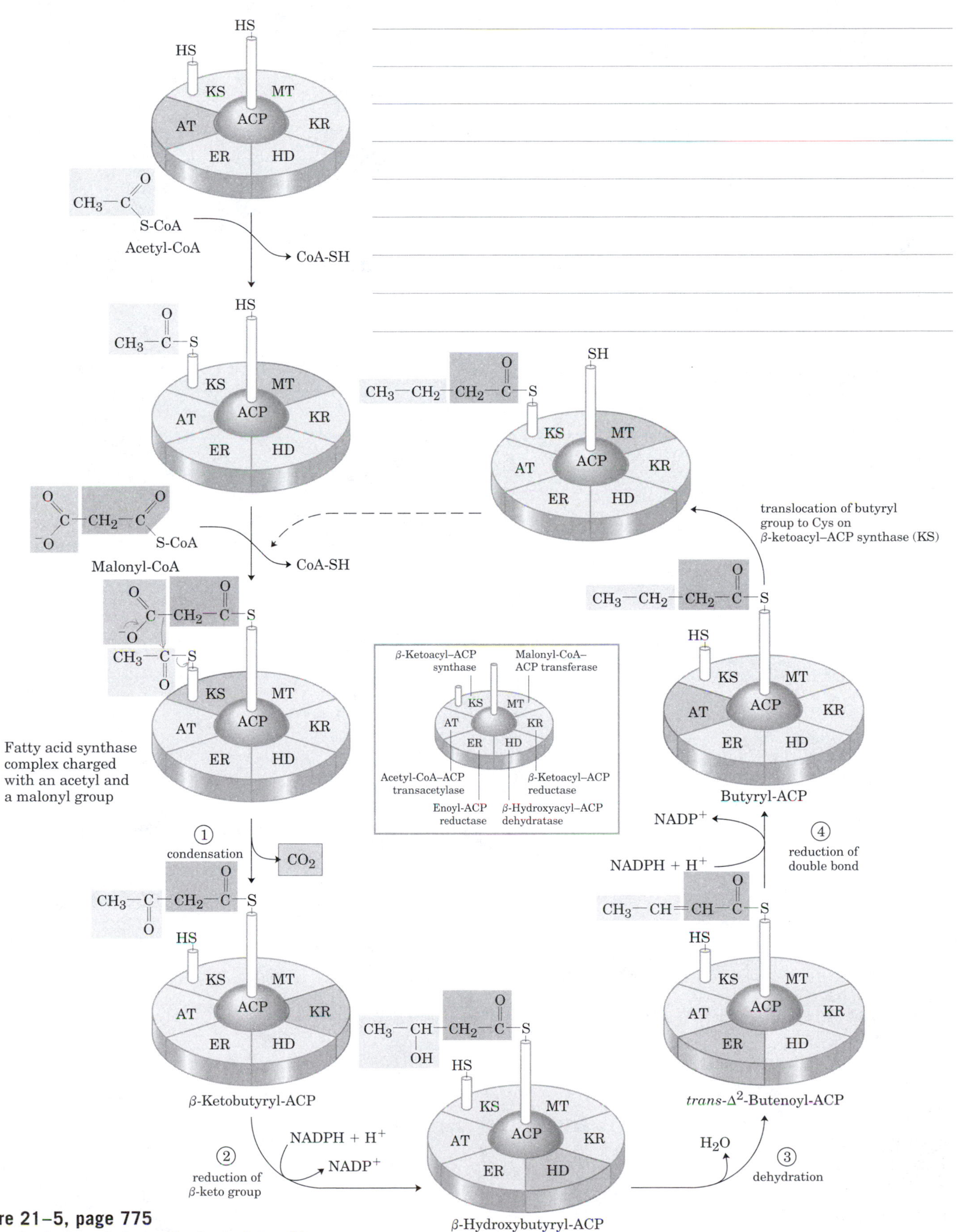

figure 21–5, page 775
Sequence of events during synthesis of a fatty acid.

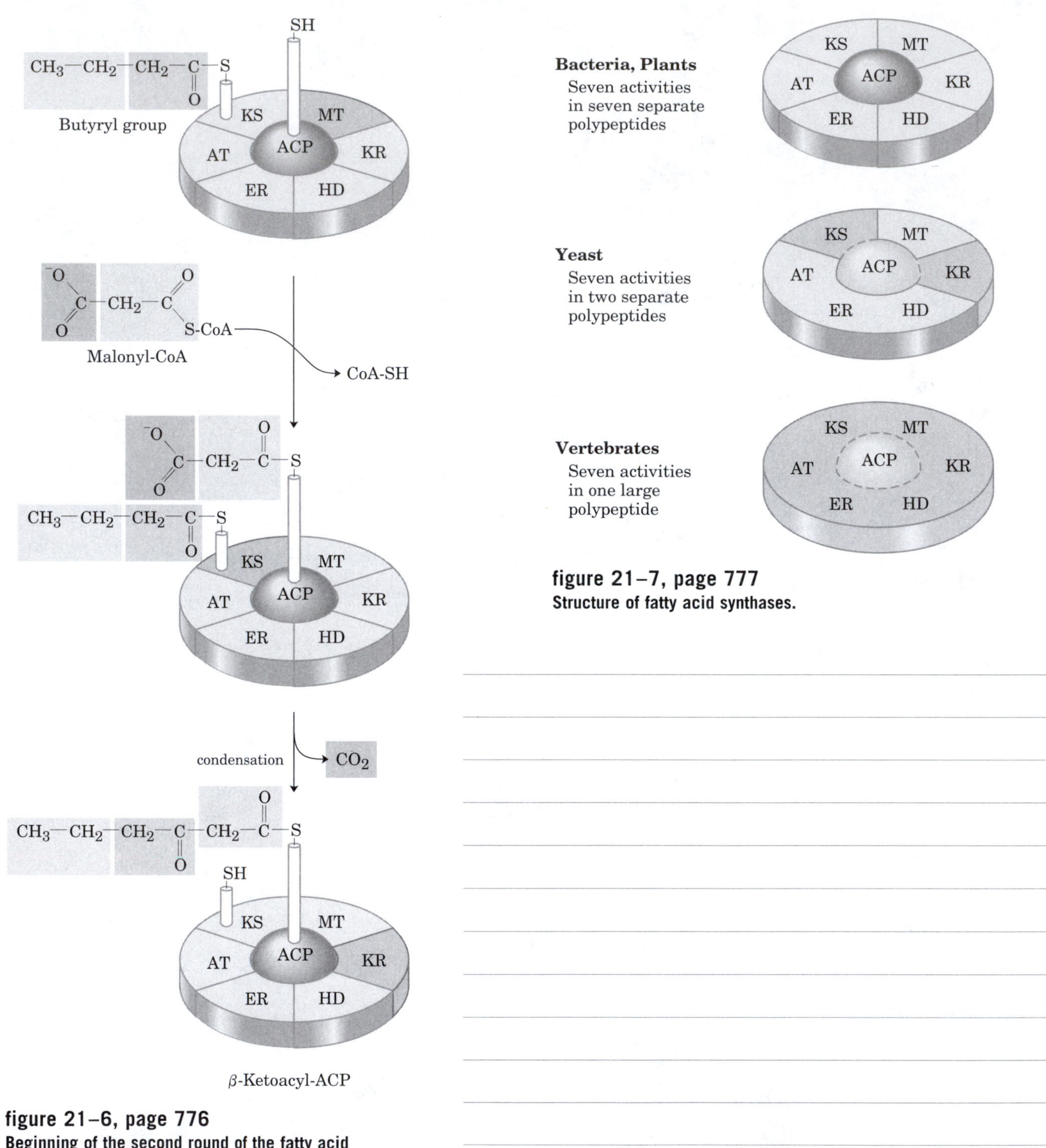

figure 21–6, page 776
Beginning of the second round of the fatty acid synthesis cycle.

figure 21–7, page 777
Structure of fatty acid synthases.

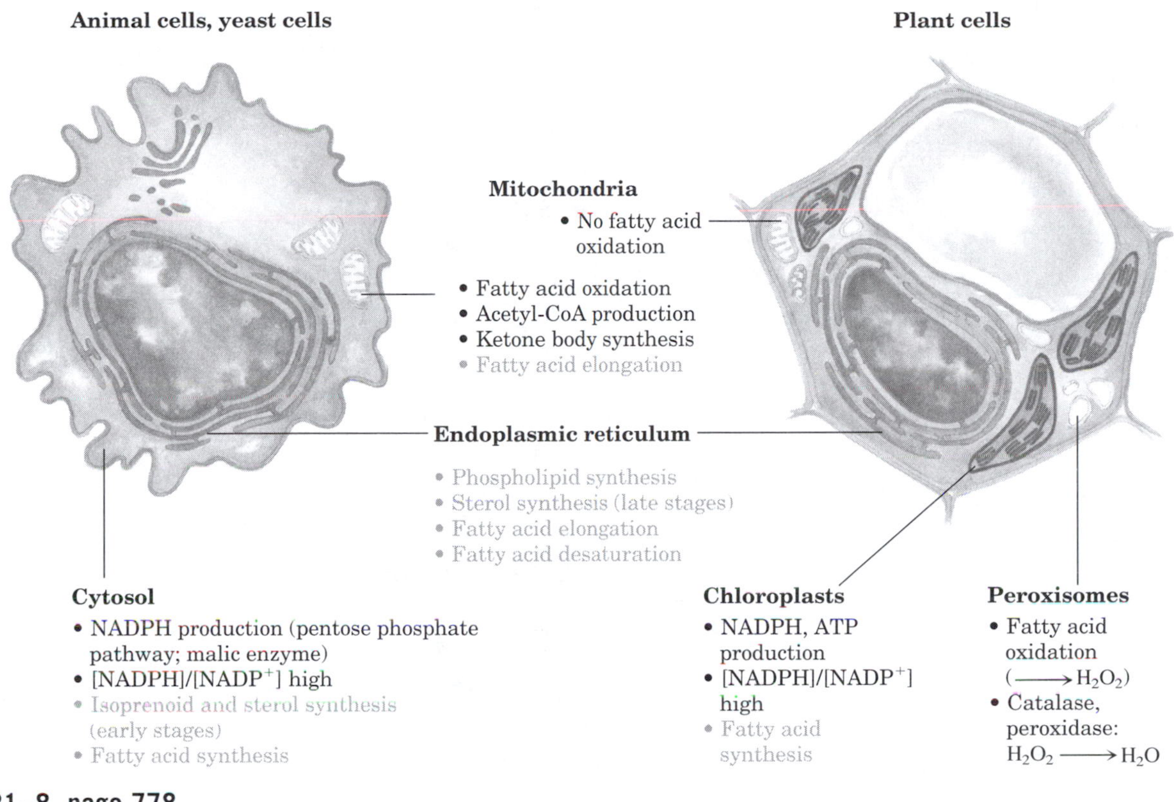

figure 21–8, page 778
Subcellular localization of lipid metabolism.

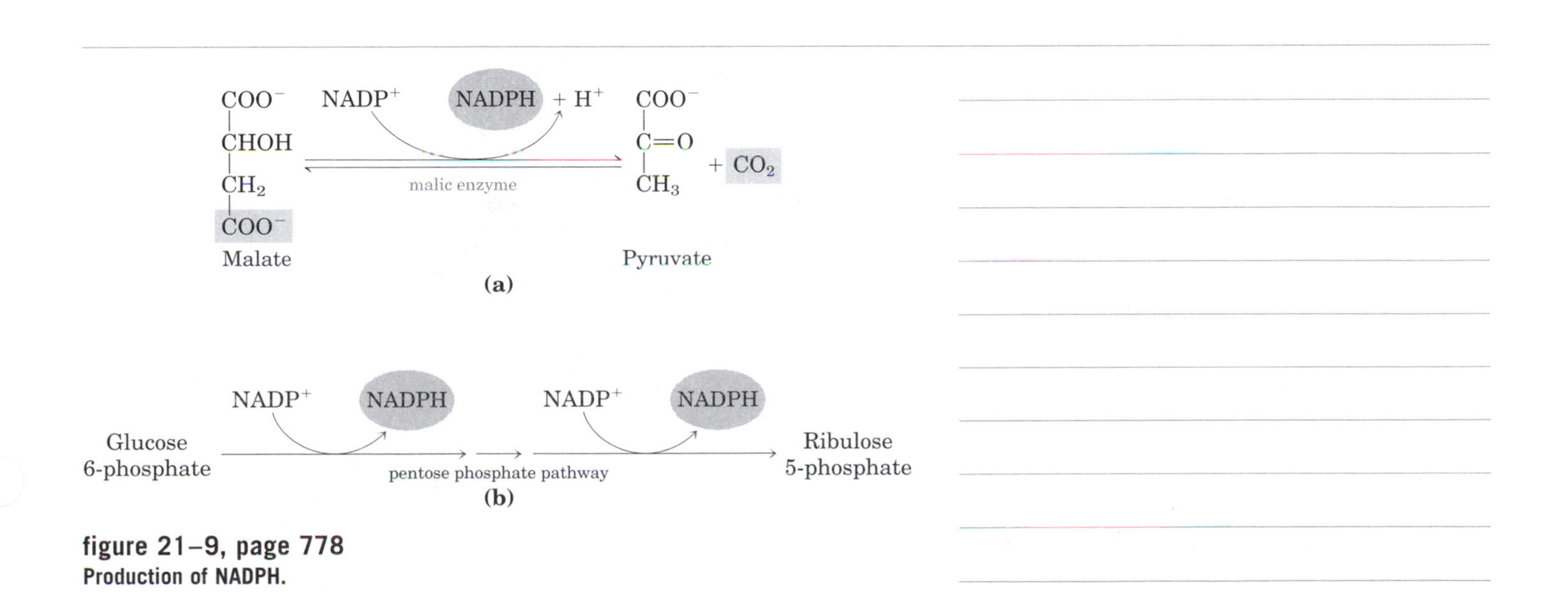

figure 21–9, page 778
Production of NADPH.

$$H_2O + NADP^+ \xrightarrow{\text{light}} \tfrac{1}{2}O_2 + NADPH + H^+$$

figure 21–10, page 779
Production of NADPH by photosynthesis.

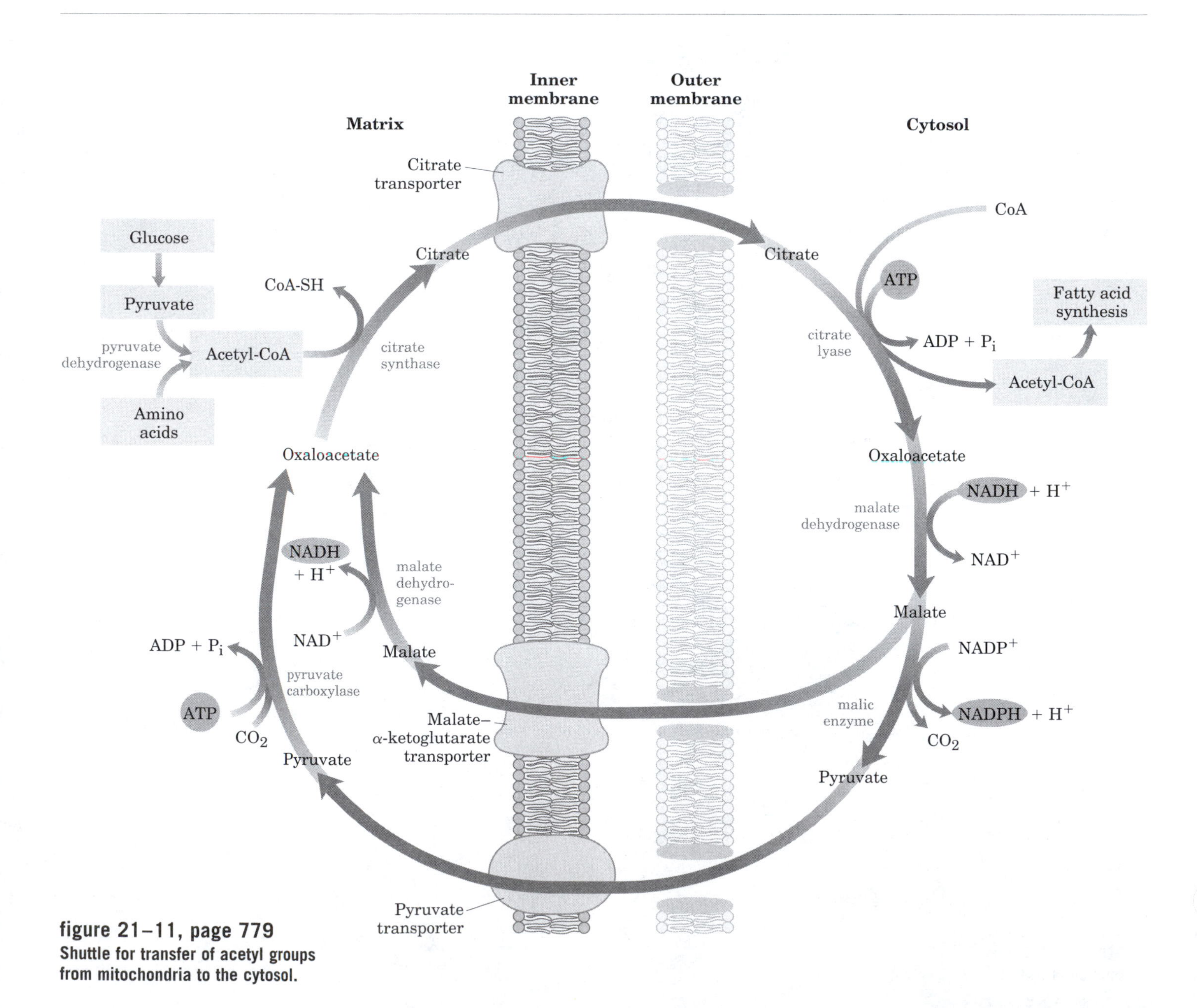

figure 21–11, page 779
Shuttle for transfer of acetyl groups from mitochondria to the cytosol.

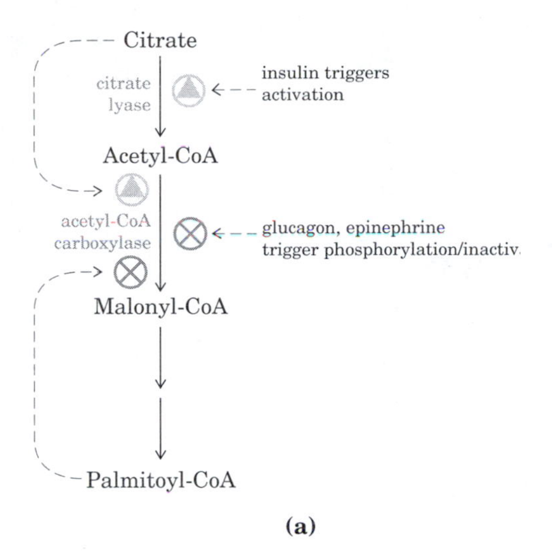

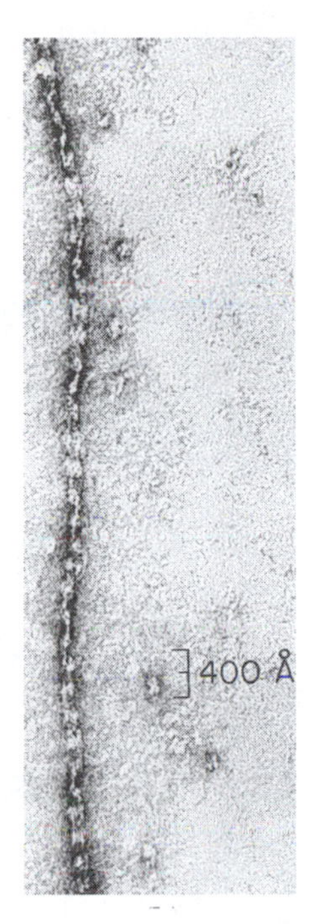

figure 21–12, page 780
Regulation of fatty acid synthesis.

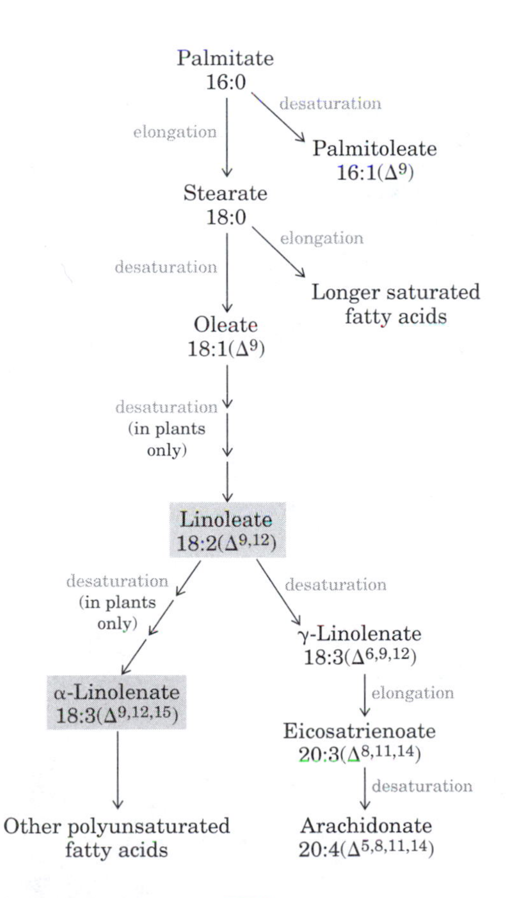

figure 21–13, page 781
Routes of synthesis of other fatty acids.

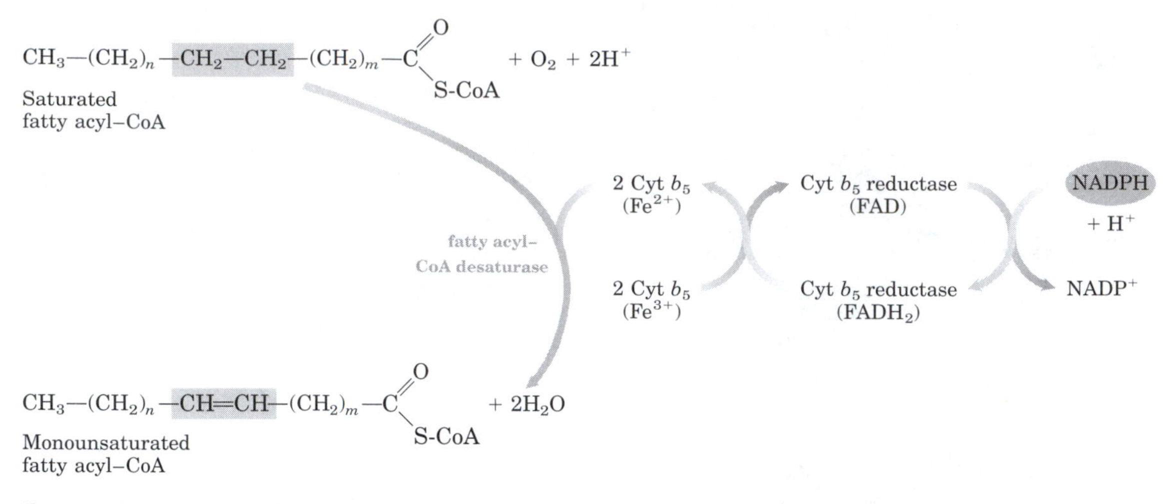

figure 21–14, page 781
Electron transfer in the desaturation of fatty acids in vertebrates.

Tryptophan

O_2

tryptophan 2,3-dioxygenase

N-Formylkynurenine

box 21–1, page 782
Tryptophan 2,3 Dioxygenase.

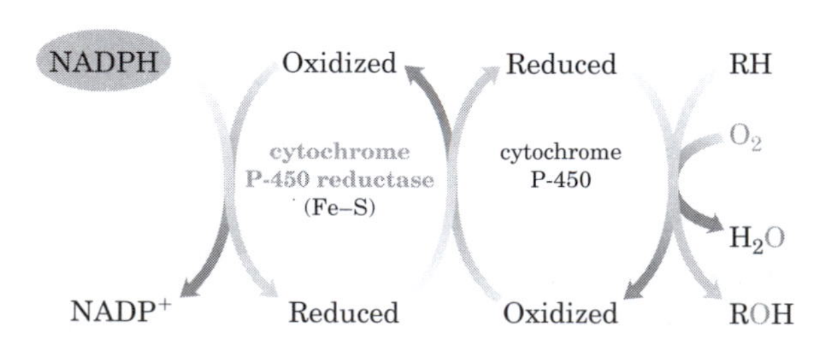

box 21–1, figure 1, page 783
Cytochrome P-450.

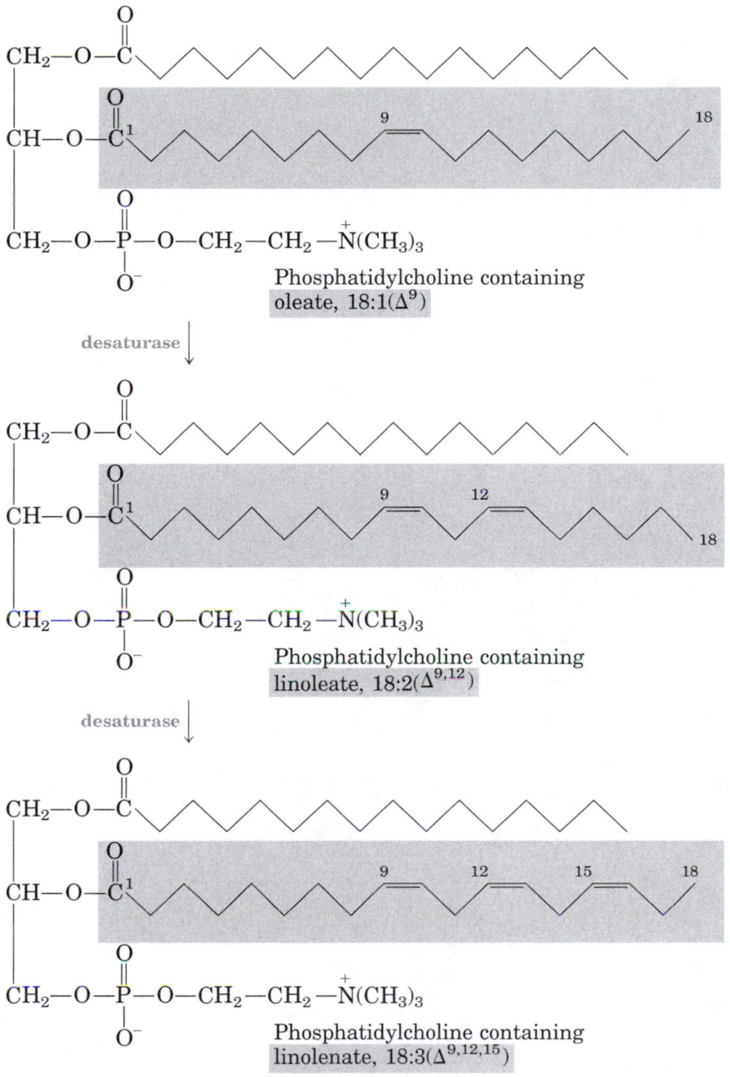

Phosphatidylcholine containing oleate, 18:1(Δ^9)

desaturase

Phosphatidylcholine containing linoleate, 18:2($\Delta^{9,12}$)

desaturase

Phosphatidylcholine containing linolenate, 18:3($\Delta^{9,12,15}$)

figure 21–15, page 784
Action of plant desaturases.

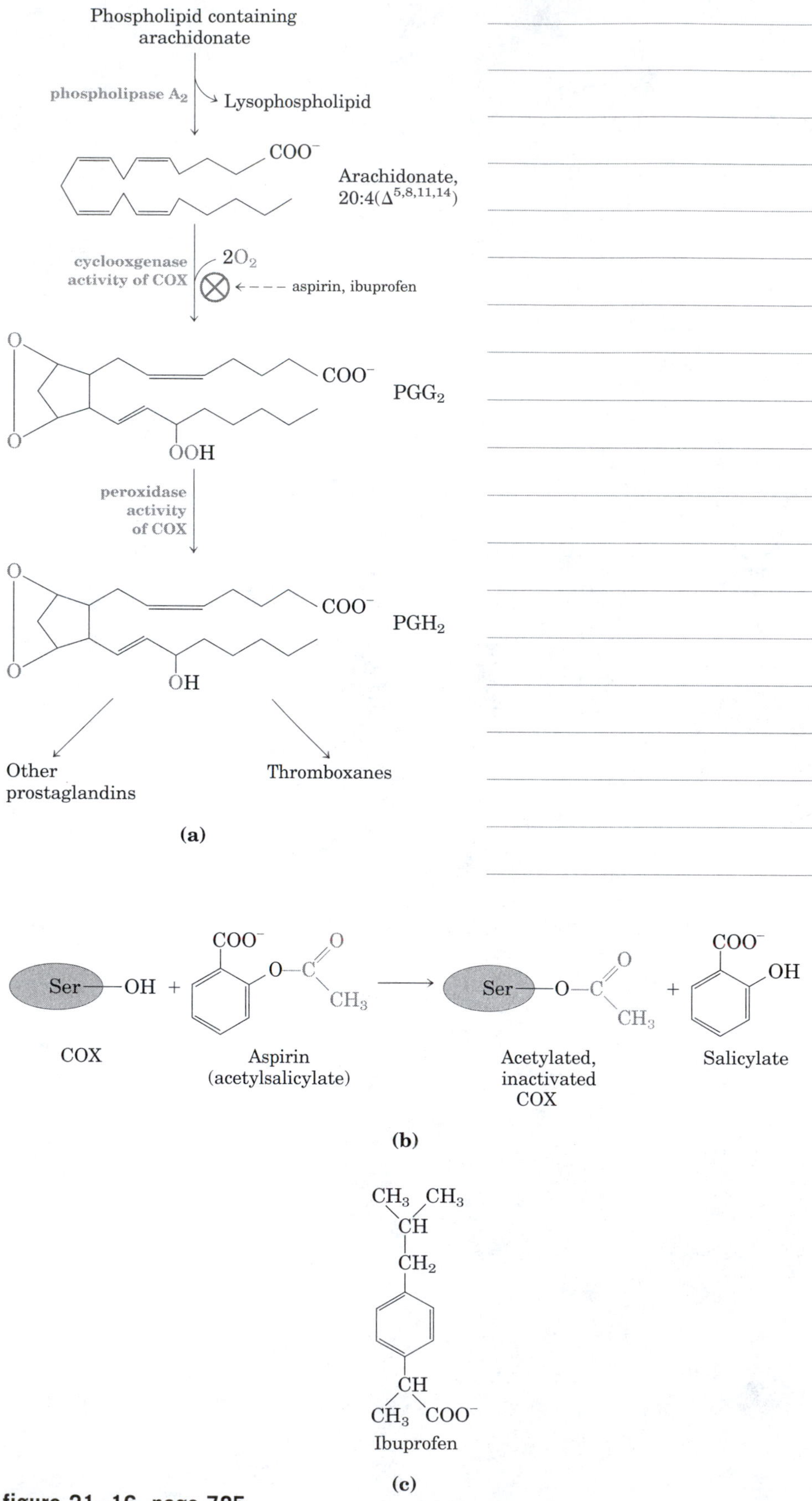

figure 21–16, page 785

The "cyclic" pathway from arachidonate to prostaglandins and thromboxanes.

Aspirin (acetylsalicylate)

Acetaminophen

Ibuprofen

Naproxen

box 21–2, figure 1, page 786

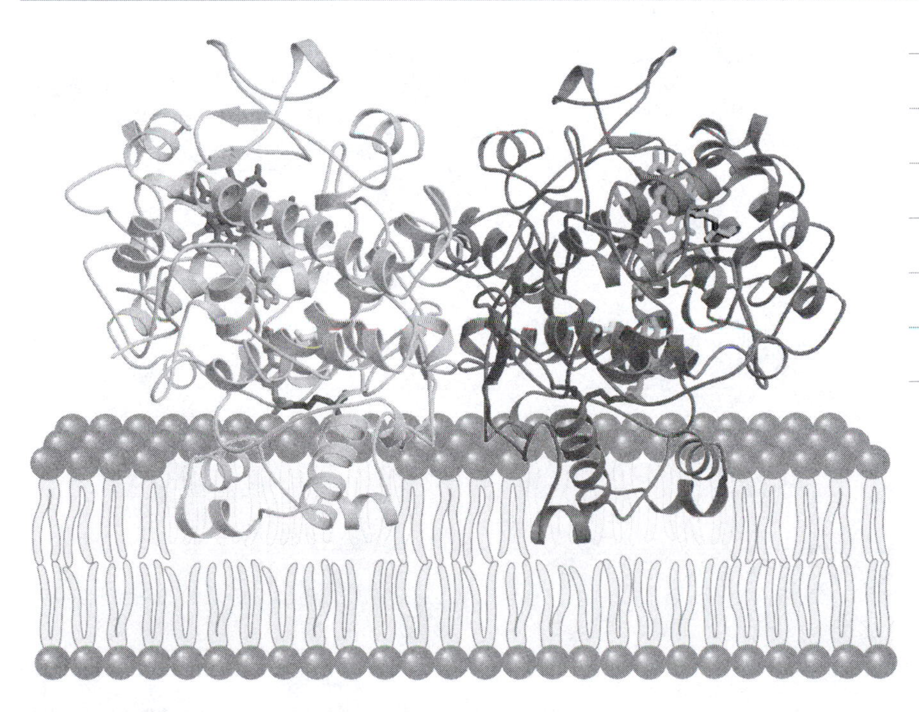

box 21–2, figure 2, page 787
Structure of COX-1, determined by x-ray crystallography.

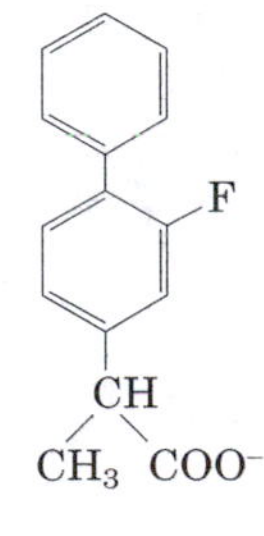

Flurbiprofen, page 787

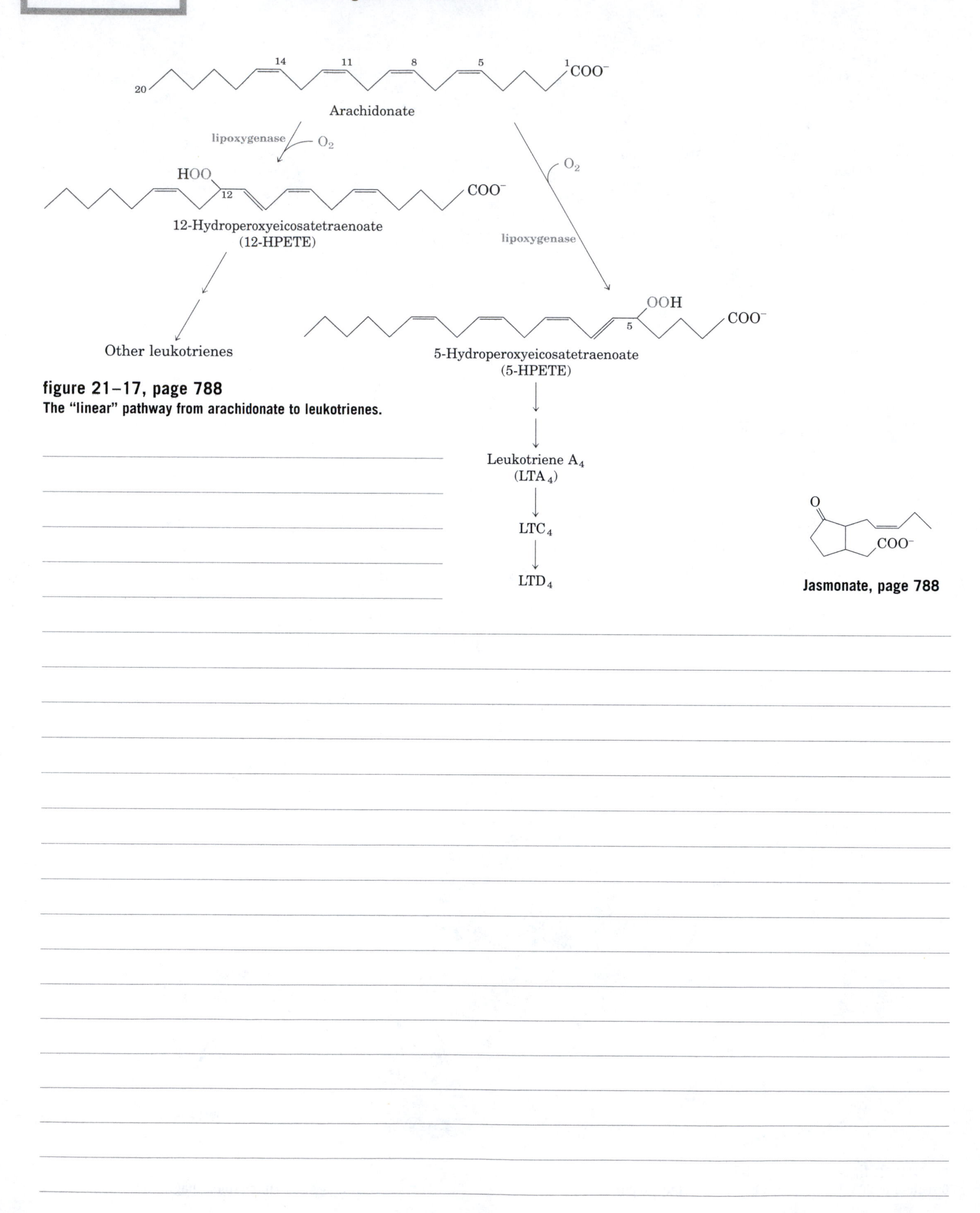

figure 21–17, page 788
The "linear" pathway from arachidonate to leukotrienes.

Jasmonate, page 788

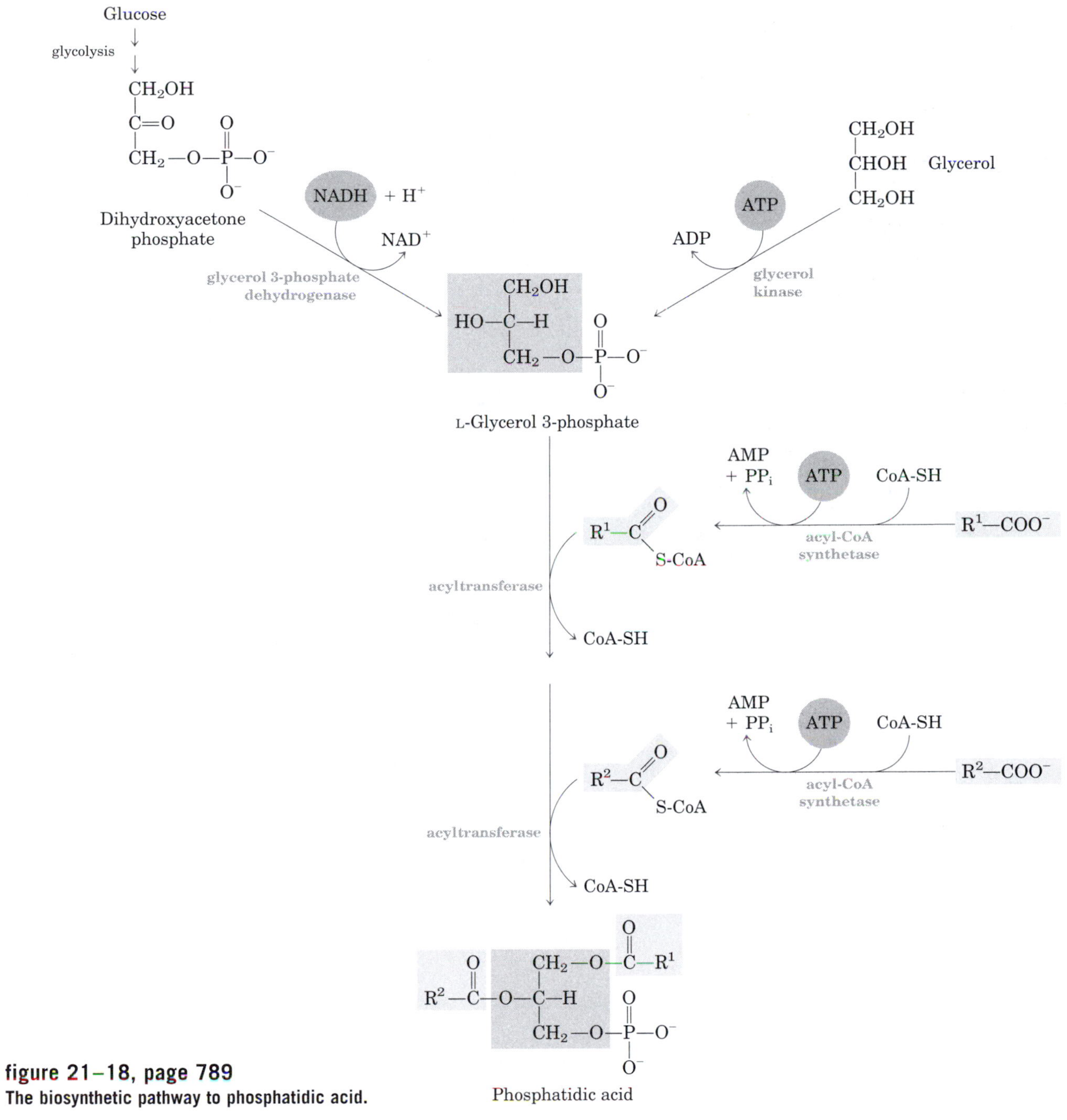

figure 21–18, page 789
The biosynthetic pathway to phosphatidic acid.

CH_2—O—C(=O)—R^1
CH—O—C(=O)—R^2 Phosphatidic acid
CH_2—O—P(=O)(—O^-)—O^-

phosphatidic acid phosphatase

attachment of head group (serine, choline, ethanolamine, etc.)

CH_2—O—C(=O)—R^1
CH—O—C(=O)—R^2
CH_2OH
1,2-Diacylglycerol

acyl transferase
R^3—C(=O)—S-CoA
CoA-SH

CH_2—O—C(=O)—R^1
CH—O—C(=O)—R^2
CH_2—O—P(=O)(—O^-)—O—Head group
Glycerophospholipid

CH_2—O—C(=O)—R^1
CH—O—C(=O)—R^2
CH_2—O—C(=O)—R^3
Triacylglycerol

figure 21–19, page 790
Phosphatidic acid in lipid biosynthesis.

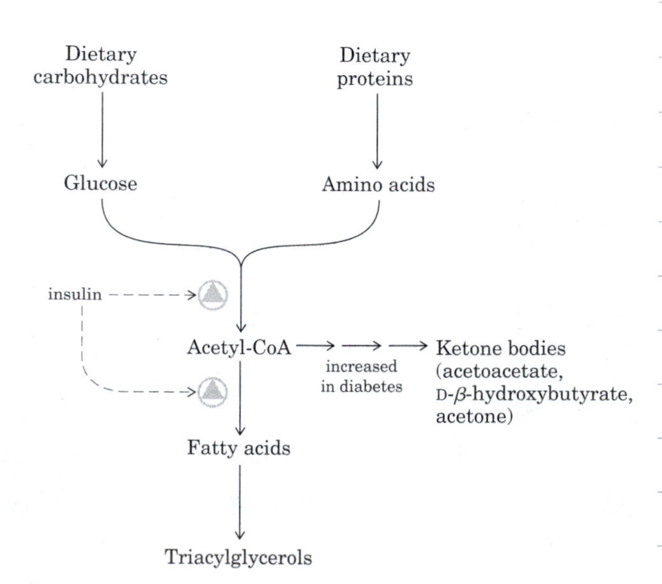

figure 21–20, page 790
Regulation of triacylglycerol synthesis by insulin.

O
CH_2—O—C—R^1
O
Diacylglycerol CH—O—C—R^2 O
CH_2—OH HO—P—OH HO— Head group
alcohol O⁻ alcohol
H_2O Phosphoric acid H_2O

O
CH_2—O—C—R^1
O
CH—O—C—R^2
O
CH_2—O—P—O— Head group
O⁻
phosphodiester
Glycerophospholipid

figure 21–21, page 791

The phospholipid head group is attached to a diacylglycerol by a phosphodiester bond, formed when phosphoric acid condenses with two alcohols, eliminating two molecules of H_2O.

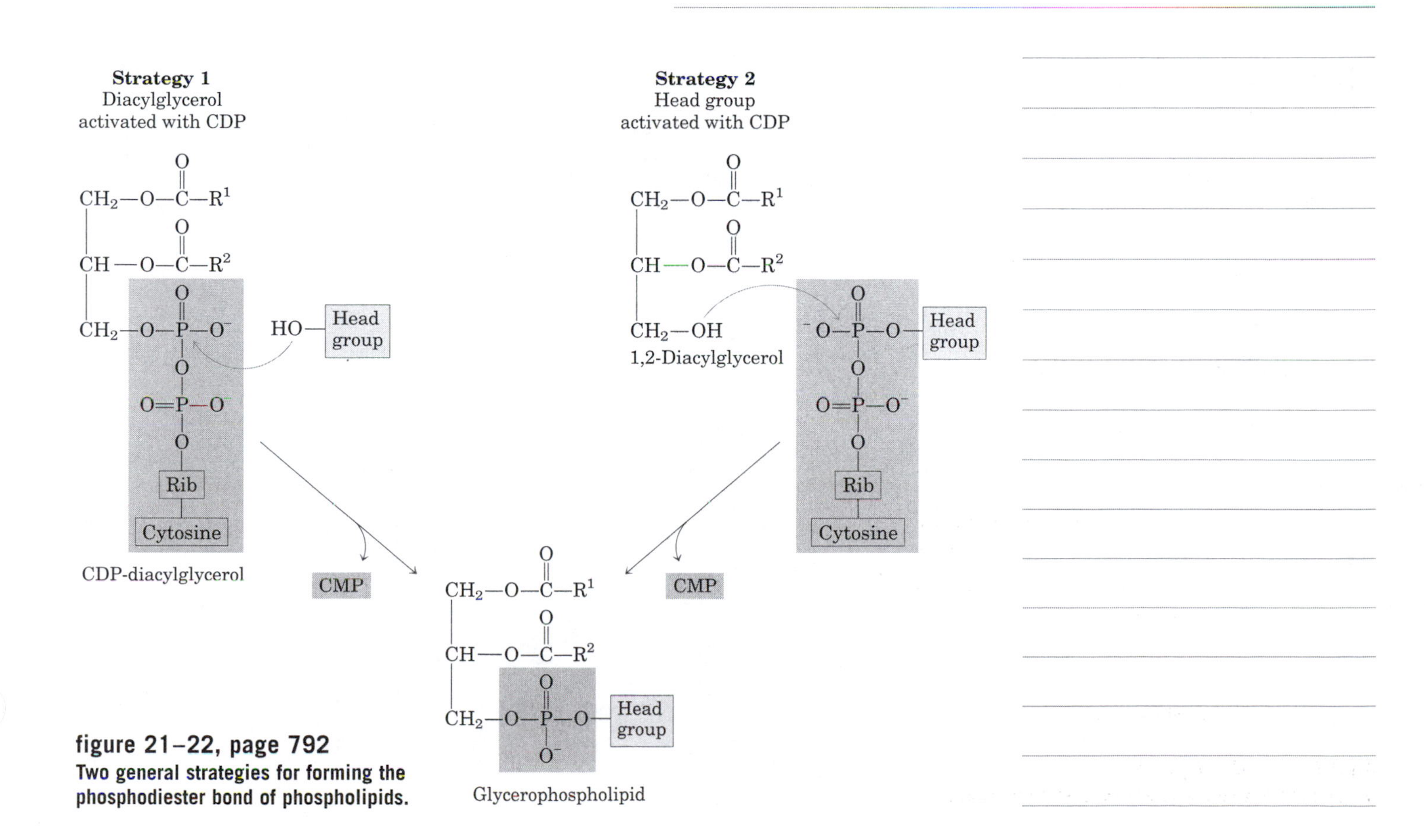

figure 21–22, page 792

Two general strategies for forming the phosphodiester bond of phospholipids.

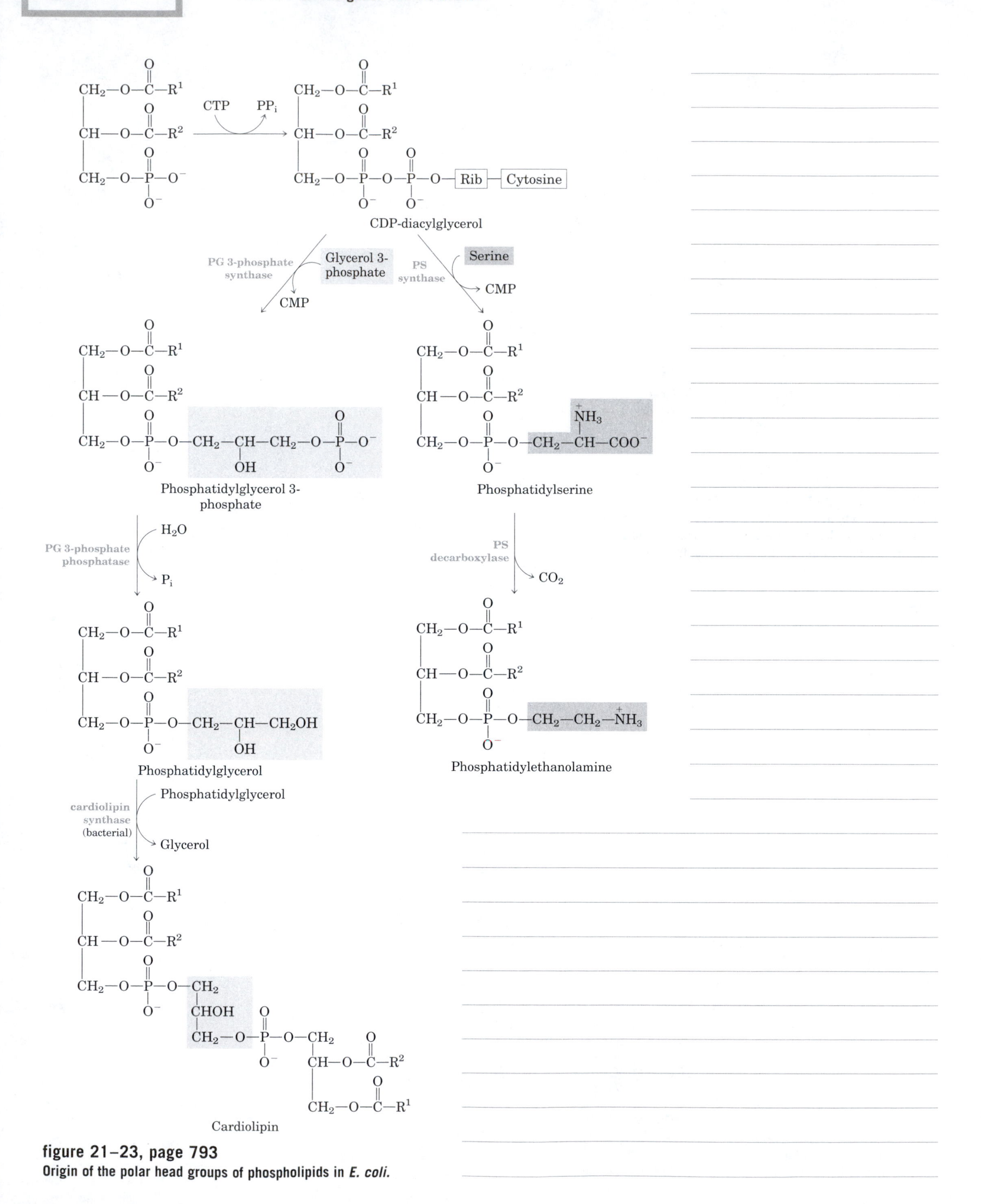

figure 21–23, page 793
Origin of the polar head groups of phospholipids in *E. coli*.

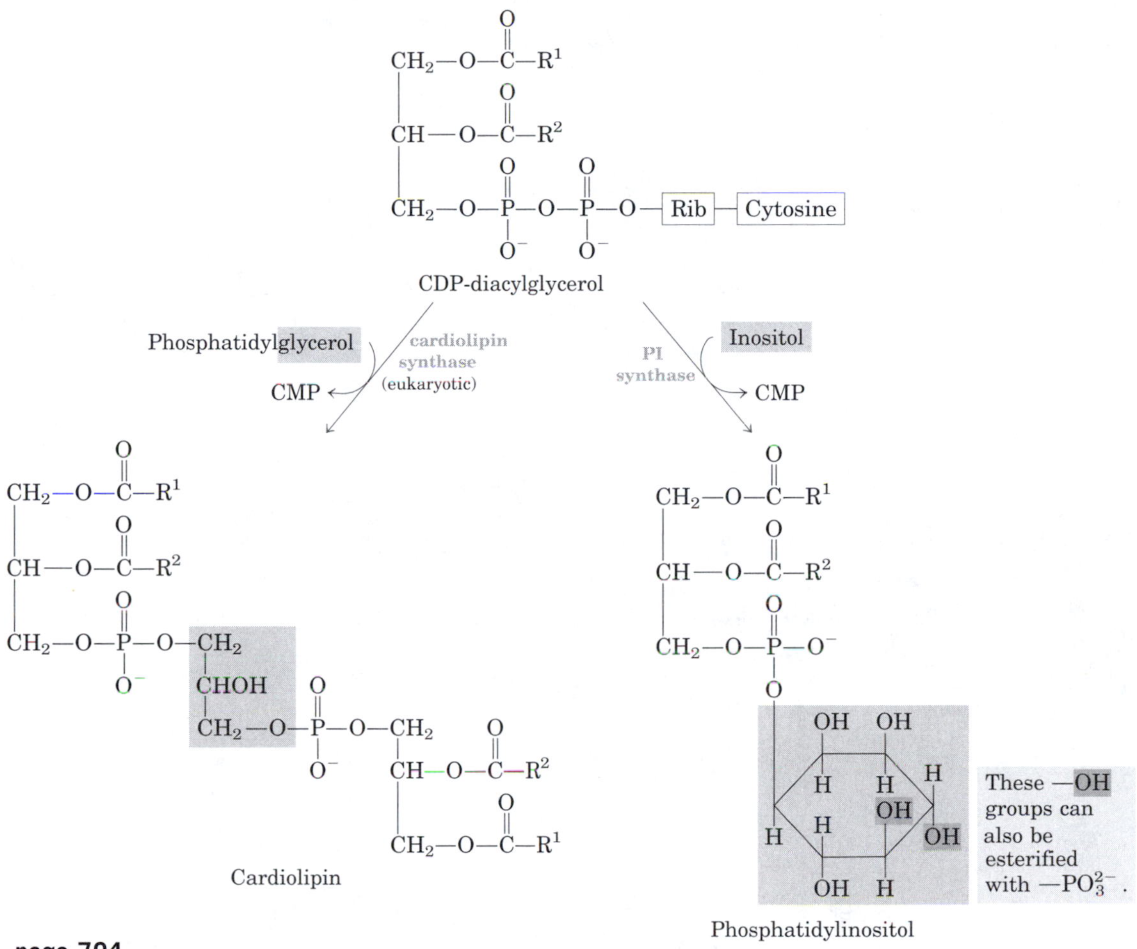

$CH_2—O—C(=O)—R^1$
$CH—O—C(=O)—R^2$
$CH_2—O—P(=O)(O^-)—O—P(=O)(O^-)—O—$ Rib — Cytosine

CDP-diacylglycerol

Phosphatidylglycerol

cardiolipin synthase (eukaryotic)

CMP

PI synthase

Inositol

CMP

$CH_2—O—C(=O)—R^1$
$CH—O—C(=O)—R^2$
$CH_2—O—P(=O)(O^-)—O—CH_2$
$CHOH$
$CH_2—O—P(=O)(O^-)—O—CH_2$
$CH—O—C(=O)—R^2$
$CH_2—O—C(=O)—R^1$

Cardiolipin

$CH_2—O—C(=O)—R^1$
$CH—O—C(=O)—R^2$
$CH_2—O—P(=O)(O^-)—O$

OH OH H H H OH H H OH OH H

These —OH groups can also be esterified with $—PO_3^{2-}$.

Phosphatidylinositol

figure 21–24, page 794

Synthesis of cardiolipin and phosphatidylinositol in eukaryotes.

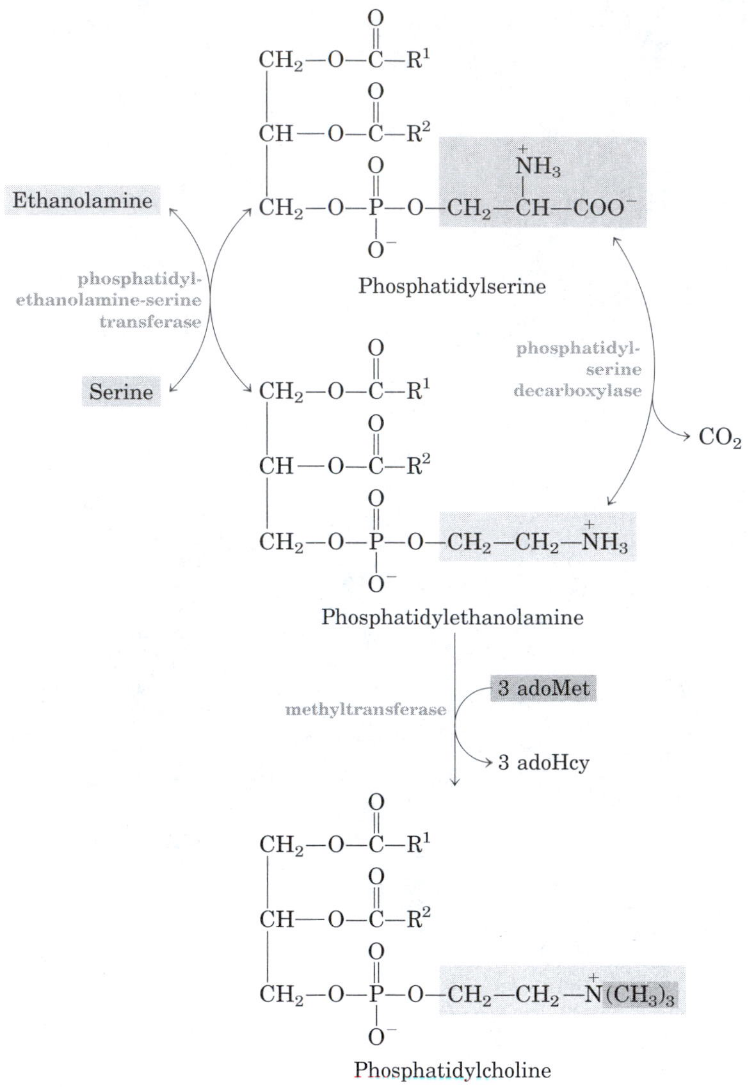

figure 21–25, page 795
The "salvage" pathway from phosphatidylserine to phophatidylethanolamine and phosphatidylcholine in yeast.

$HO-CH_2-CH_2-\overset{+}{N}(CH_3)_3$ Choline

choline kinase: ATP → ADP

$^{-}O-P(=O)(O^{-})-O-CH_2-CH_2-\overset{+}{N}(CH_3)_3$ Phosphocholine

CTP-choline cytidylyl transferase: CTP → PP_i

$^{-}O-P(=O)-O-CH_2-CH_2-\overset{+}{N}(CH_3)_3$
$|$
O
$|$
$O=P(O^{-})-O-$ Rib — Cytosine

CDP-choline

CDP-choline-diacylglycerol phosphocholine transferase: Diacylglycerol → CMP

$CH_2-O-C(=O)-R^1$
$CH-O-C(=O)-R^2$
$CH_2-O-P(=O)(O^{-})-O-CH_2-CH_2-\overset{+}{N}(CH_3)_3$

Phosphatidylcholine

figure 21–26, page 795
The pathway for phosphatidylcholine synthesis from choline in mammals.

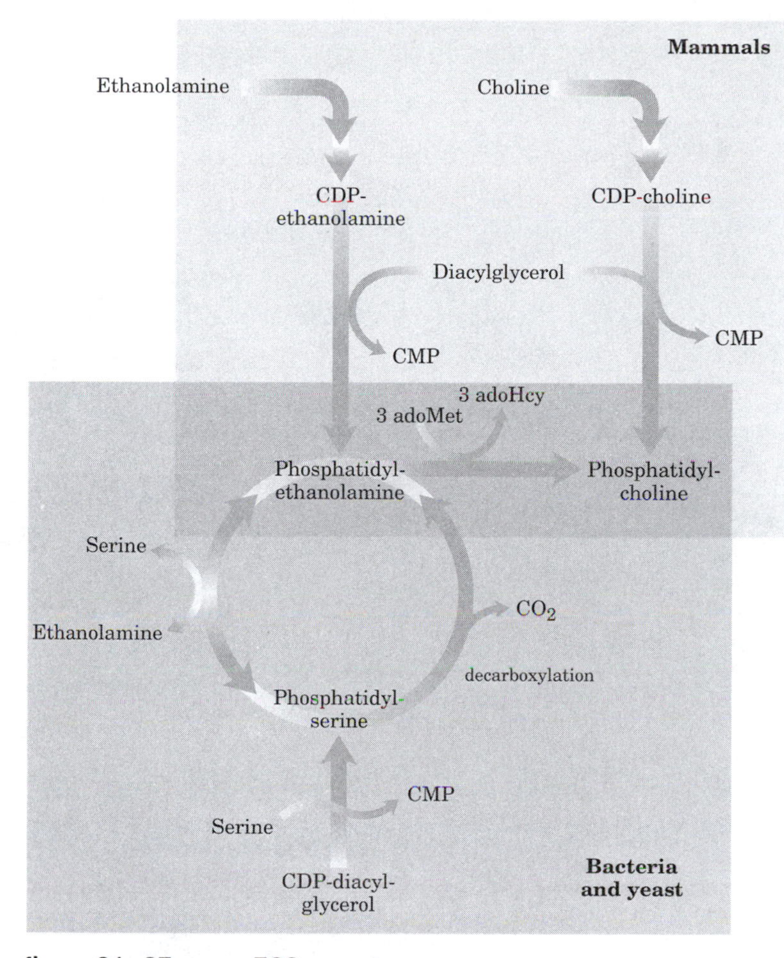

figure 21–27, page 796
Summary of the pathways to phosphatidylcholine and phosphatidylethanolamine.

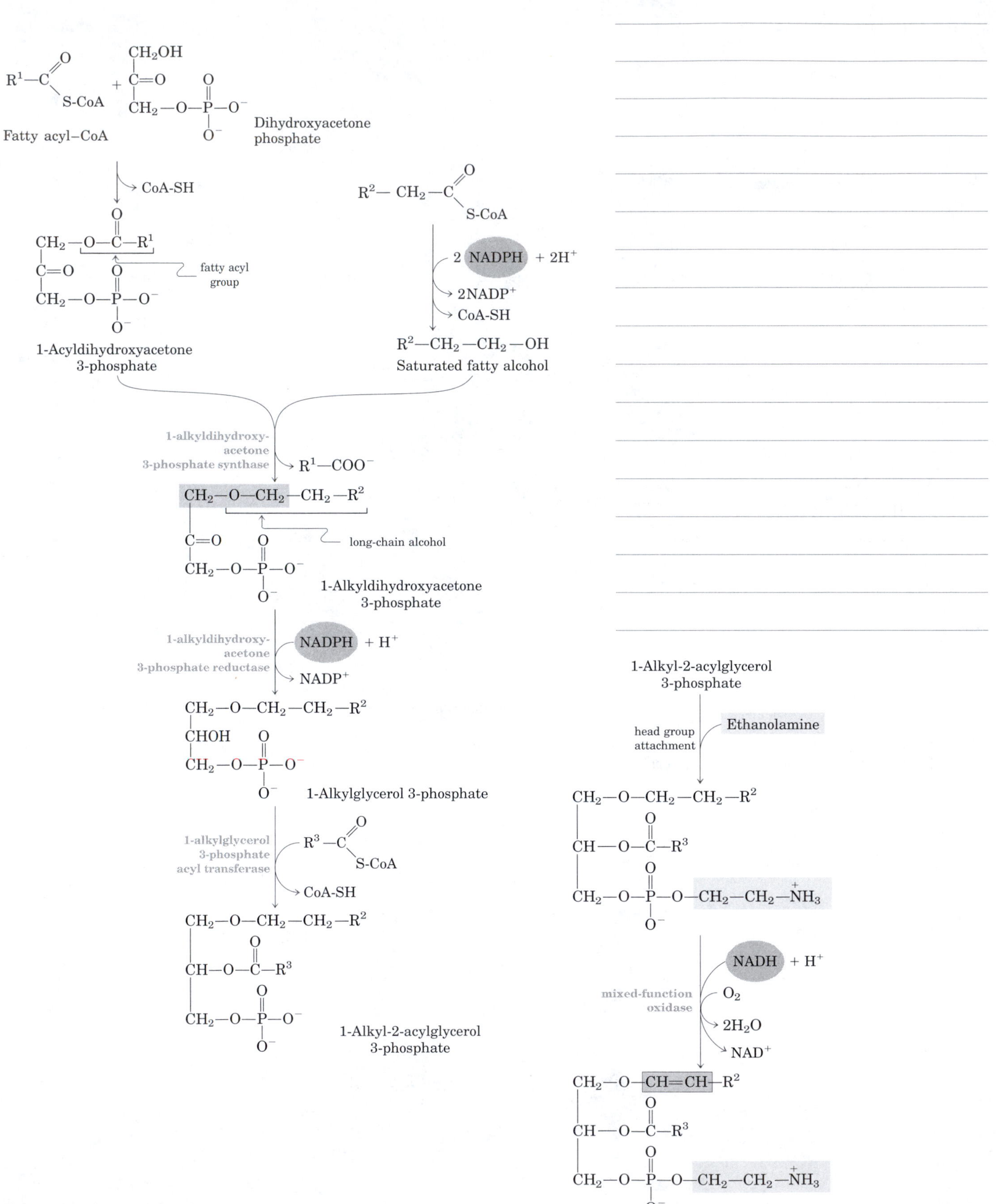

figure 21–28, page 797
Synthesis of ether lipids and plasmalogens.

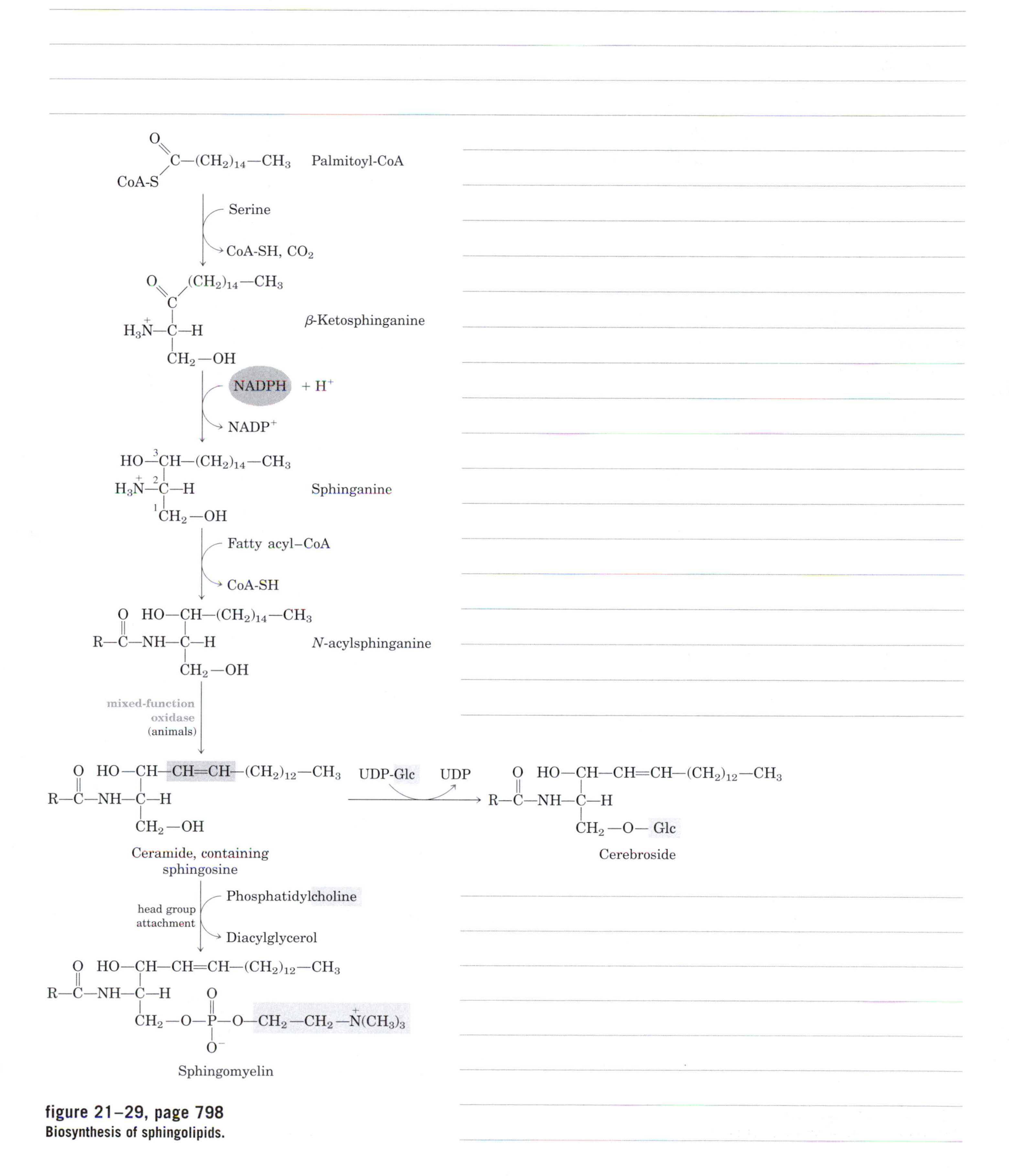

figure 21–29, page 798
Biosynthesis of sphingolipids.

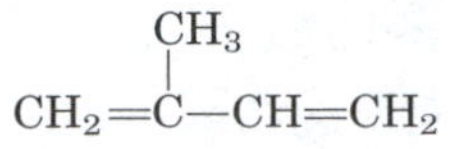

$CH_2{=}C(CH_3){-}CH{=}CH_2$

Isoprene, page 799

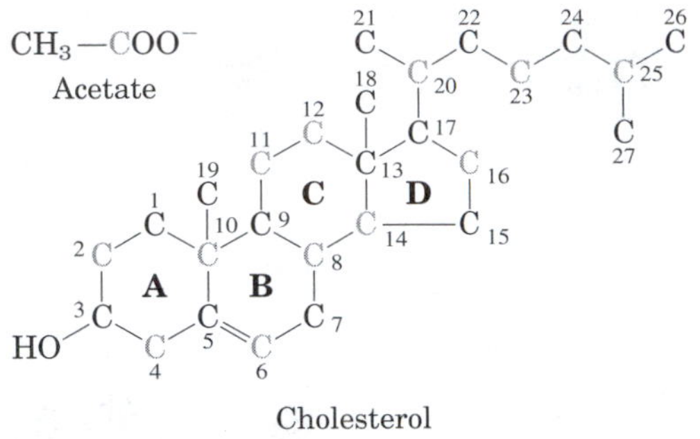

$CH_3{-}COO^-$
Acetate

Cholesterol

figure 21–30, page 799
The origin of the carbon atoms of cholesterol.

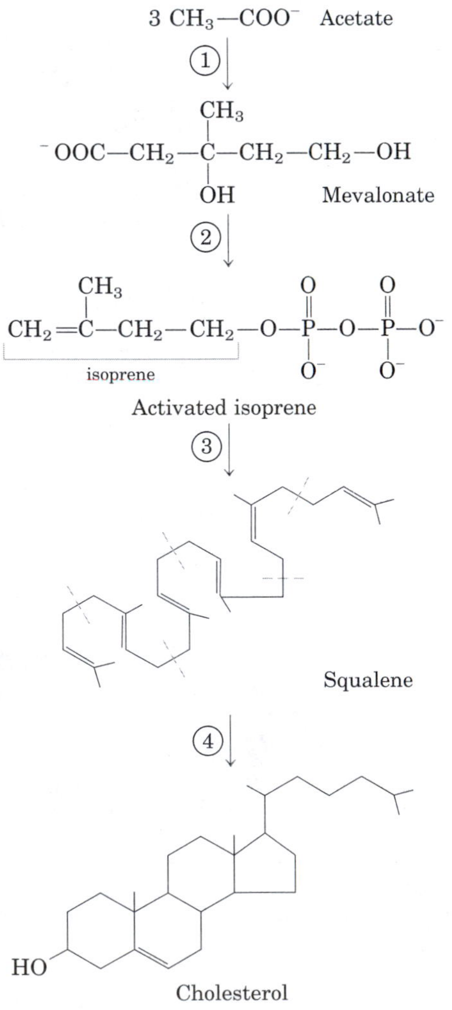

3 $CH_3{-}COO^-$ Acetate

(1)

$^-OOC{-}CH_2{-}C(CH_3)(OH){-}CH_2{-}CH_2{-}OH$ Mevalonate

(2)

$CH_2{=}C(CH_3){-}CH_2{-}CH_2{-}O{-}P(O)(O^-){-}O{-}P(O)(O^-){-}O^-$
isoprene
Activated isoprene

(3)

Squalene

(4)

Cholesterol

figure 21–31, page 800
A summary of cholesterol biosynthesis.

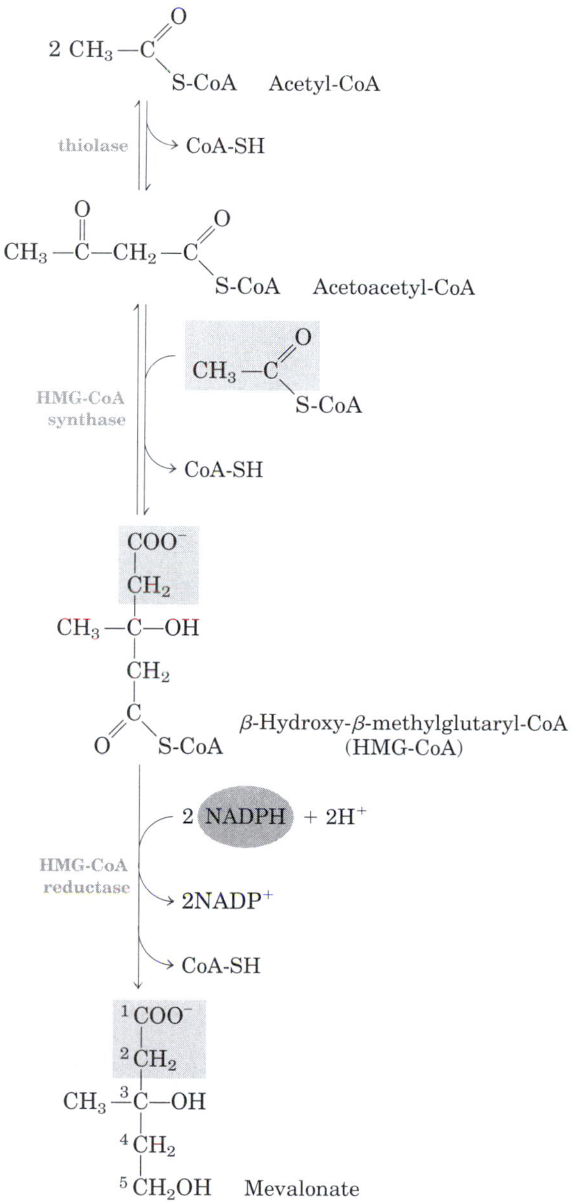

$$2\ CH_3-C(=O)-S\text{-}CoA \quad \text{Acetyl-CoA}$$

thiolase ⇌ $\rightarrow$ CoA-SH

$$CH_3-C(=O)-CH_2-C(=O)-S\text{-}CoA \quad \text{Acetoacetyl-CoA}$$

HMG-CoA synthase ⇌ $CH_3-C(=O)-S\text{-}CoA$ in, CoA-SH out

$$CH_3-C(OH)(CH_2COO^-)-CH_2-C(=O)-S\text{-}CoA$$

β-Hydroxy-β-methylglutaryl-CoA (HMG-CoA)

HMG-CoA reductase: 2 NADPH + $2H^+$ in, $2NADP^+$ out, CoA-SH out

$$\text{}^{1}COO^- - \text{}^{2}CH_2 - \text{}^{3}C(CH_3)(OH) - \text{}^{4}CH_2 - \text{}^{5}CH_2OH \quad \text{Mevalonate}$$

figure 21–32, page 800
Formation of mevalonate from acetyl-CoA.

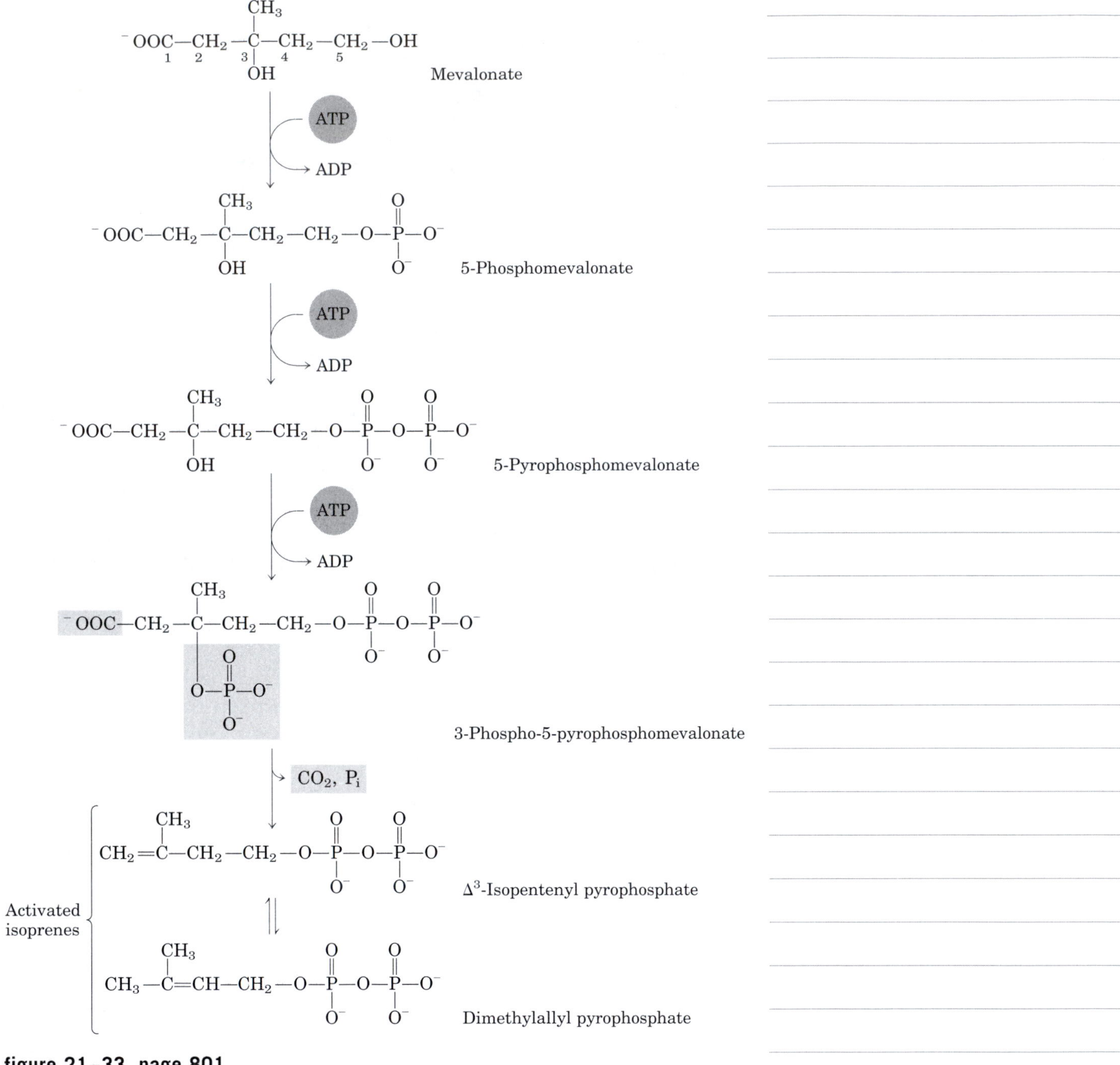

figure 21–33, page 801
Conversion of mevalonate into activated isoprene units.

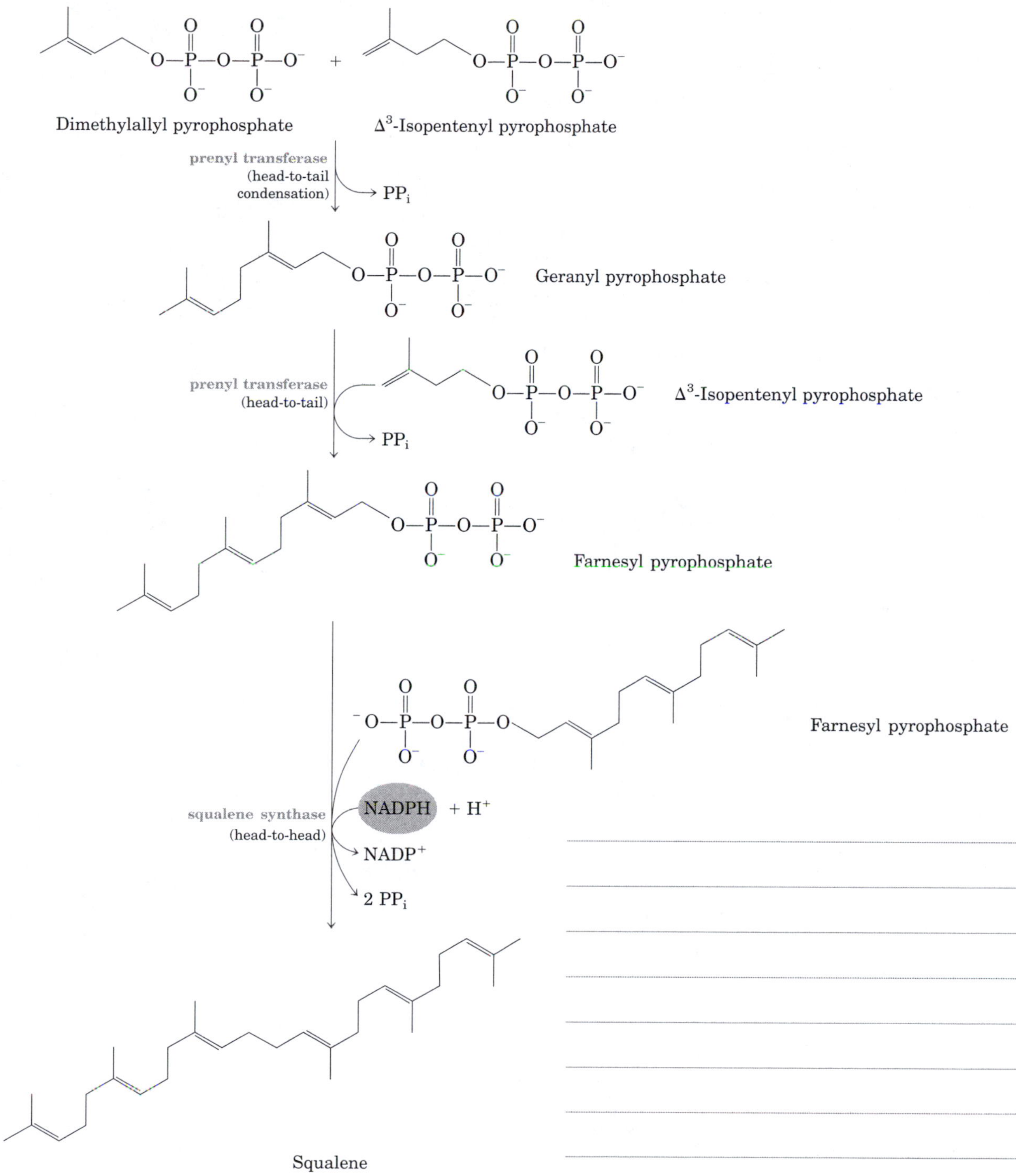

Dimethylallyl pyrophosphate + Δ^3-Isopentenyl pyrophosphate

prenyl transferase (head-to-tail condensation) → PP_i

Geranyl pyrophosphate

+ Δ^3-Isopentenyl pyrophosphate

prenyl transferase (head-to-tail) → PP_i

Farnesyl pyrophosphate

+ Farnesyl pyrophosphate

squalene synthase (head-to-head)

NADPH + H^+ → $NADP^+$

→ 2 PP_i

Squalene

figure 21–34, page 802
Formation of squalene.

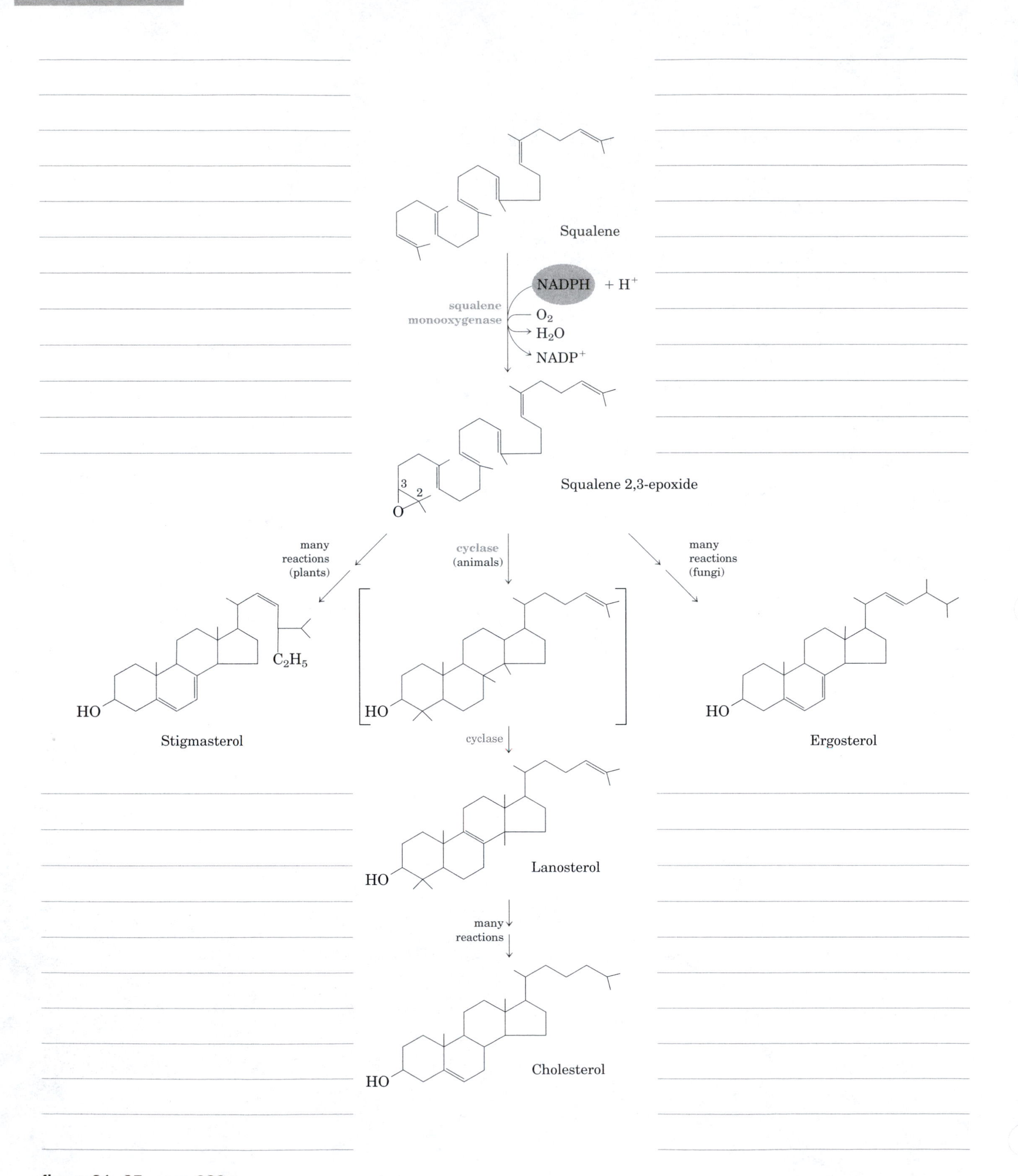

figure 21–35, page 803
Ring closure converts linear squalene into the condensed steroid nucleus.

HO

Cholesterol

acyl-CoA–cholesterol acyl transferase (ACAT)

Fatty acyl–CoA

CoA-SH

R—C(=O)—O

Cholesteryl ester

figure 21–36, page 804
Synthesis of cholesteryl esters.

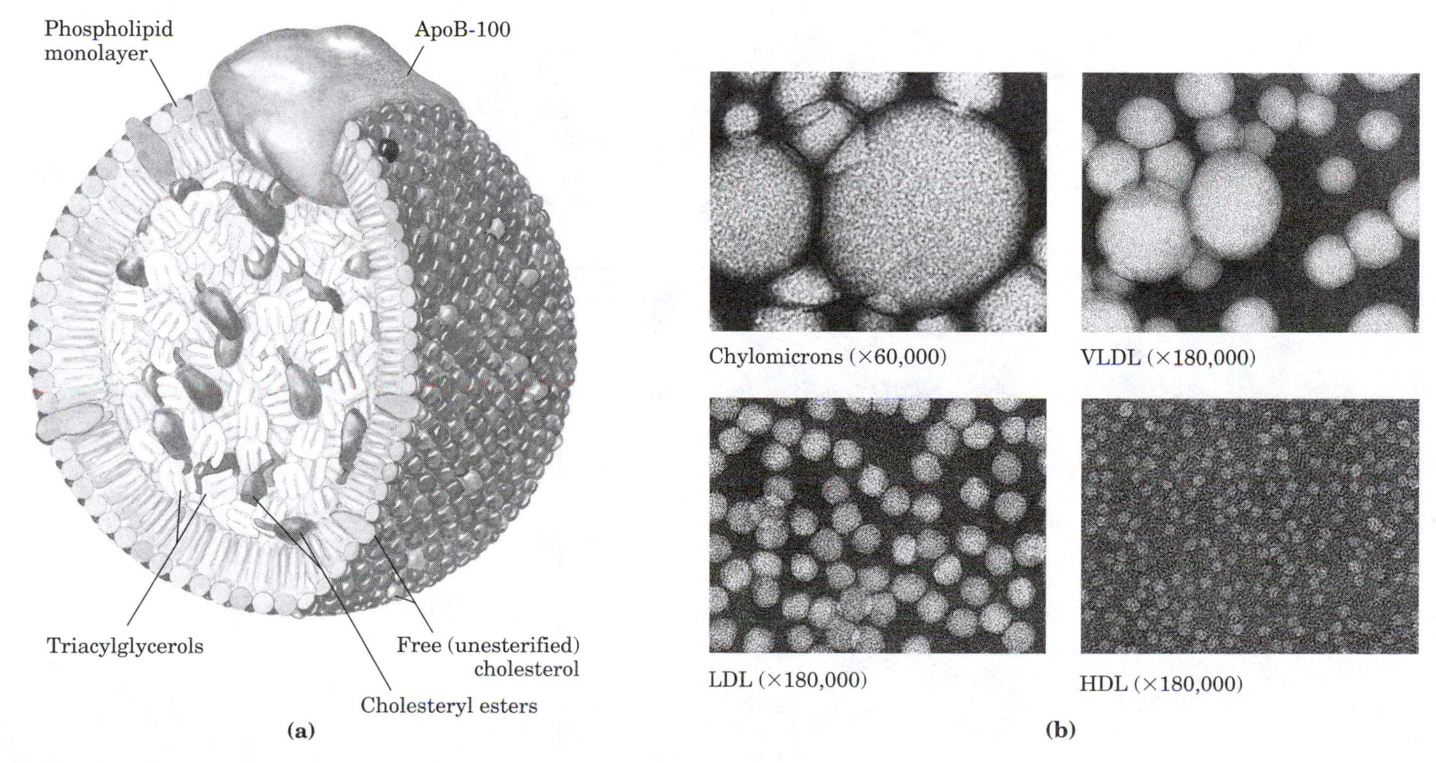

figure 21–37, page 805
(a) Structure of a low-density lipoprotein (LDL). (b) Four classes of lipoproteins visualized in the electron microscope after negative staining.

table 21–2, page 805

Major Classes of Human Plasma Lipoproteins: Some Properties

Lipoprotein	Density (g/mL)	Composition (wt %)				
		Protein	Phospholipids	Free cholesterol	Cholesteryl esters	Triacylglycerols
Chylomicrons	<1.006	2	9	1	3	85
VLDL	0.95–1.006	10	18	7	12	50
LDL	1.006–1.063	23	20	8	37	10
HDL	1.063–1.210	55	24	2	15	4

Source: Modified from Kritchevsky, D. (1986) Atherosclerosis and nutrition. *Nutr. Int.* **2,** 290–297.

table 21–3, page 806

Apolipoproteins of the Human Plasma Lipoproteins

Apolipoprotein	Molecular weight	Lipoprotein association	Function (if known)
ApoA-I	28,331	HDL	Activates LCAT; interacts with ABC transporter
ApoA-II	17,380	HDL	
ApoA-IV	44,000	Chylomicrons, HDL	
ApoB-48	240,000	Chylomicrons	
ApoB-100	513,000	VLDL, LDL	Binds to LDL receptor
ApoC-I	7,000	VLDL, HDL	
ApoC-II	8,837	Chylomicrons, VLDL, HDL	Activates lipoprotein lipase
ApoC-III	8,751	Chylomicrons, VLDL, HDL	Inhibits lipoprotein lipase
ApoD	32,500	HDL	
ApoE	34,145	Chylomicrons, VLDL, HDL	Triggers clearance of VLDL and chylomicron remnants

Source: Modified from Vance, D.E. & Vance, J.E. (eds) (1985) *Biochemistry of Lipids and Membranes.* The Benjamin/Cummings Publishing Company, Menlo Park, CA.

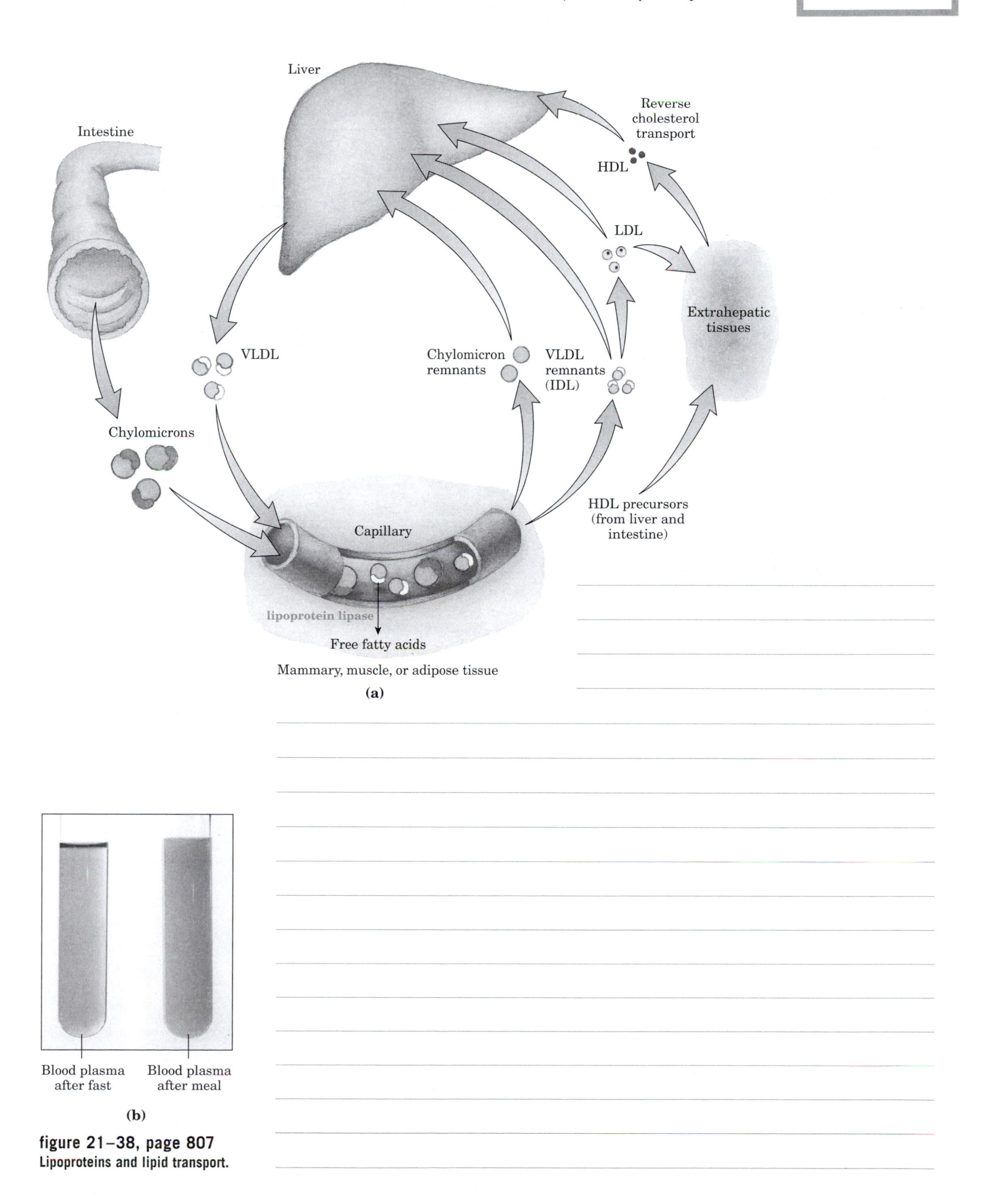

figure 21–38, page 807
Lipoproteins and lipid transport.

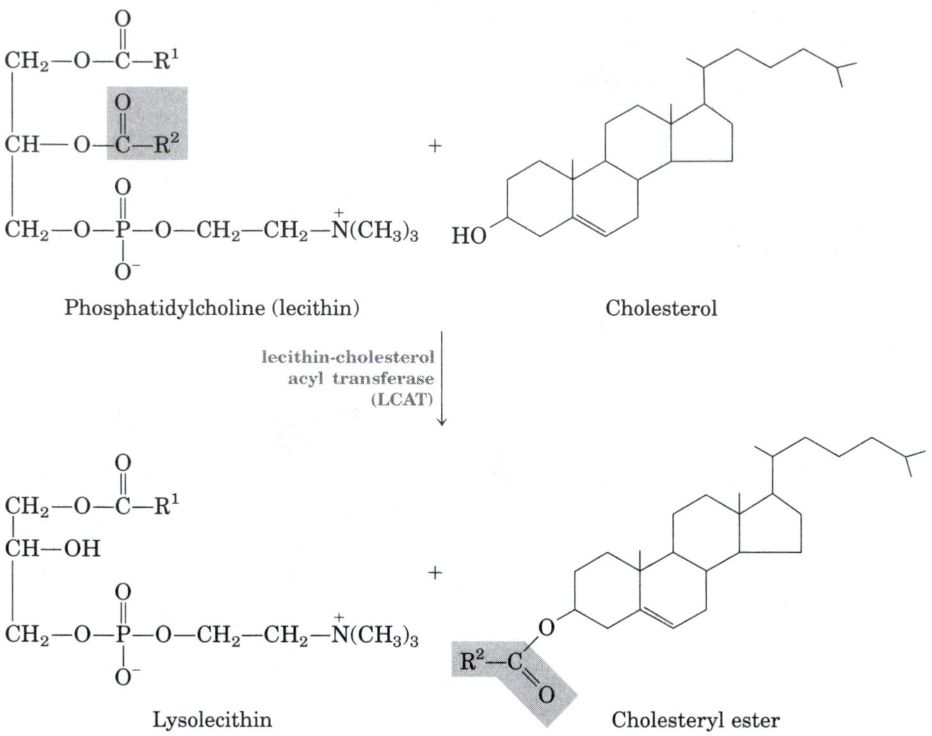

figure 21–39, page 809
Reaction catalyzed by lecithin-cholesterol acyl transferase (LCAT).

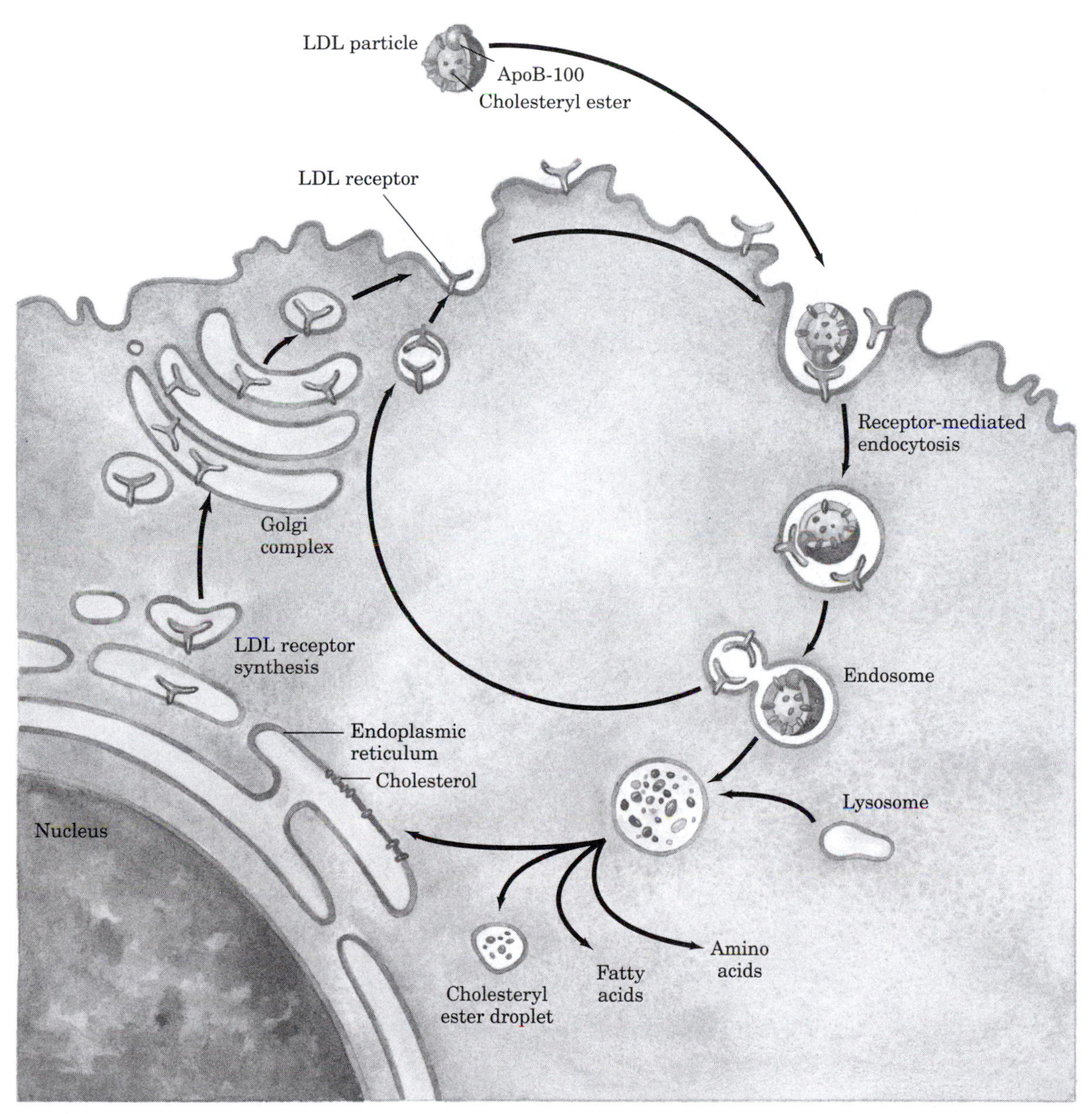

figure 21–40, page 810
Uptake of cholesterol by receptor-mediated endocytosis.

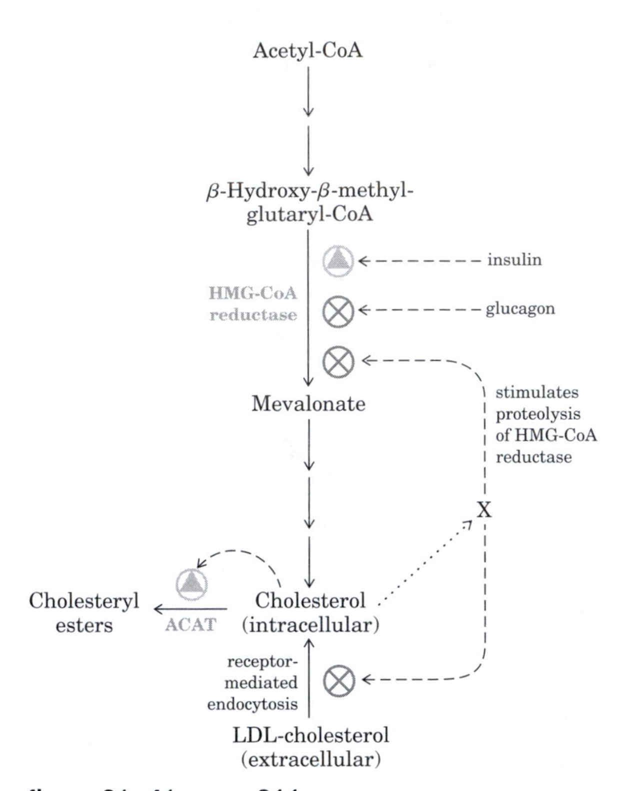

figure 21–41, page 811
Regulation of cholesterol biosynthesis balances synthesis with dietary uptake.

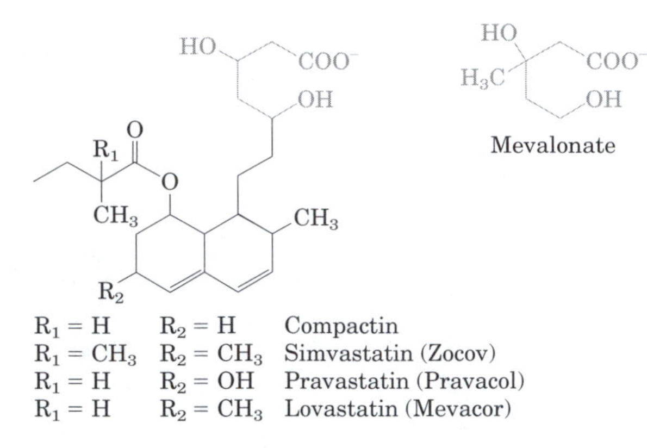

figure 21–42, page 811
Inhibitors of HMG-CoA synthase.

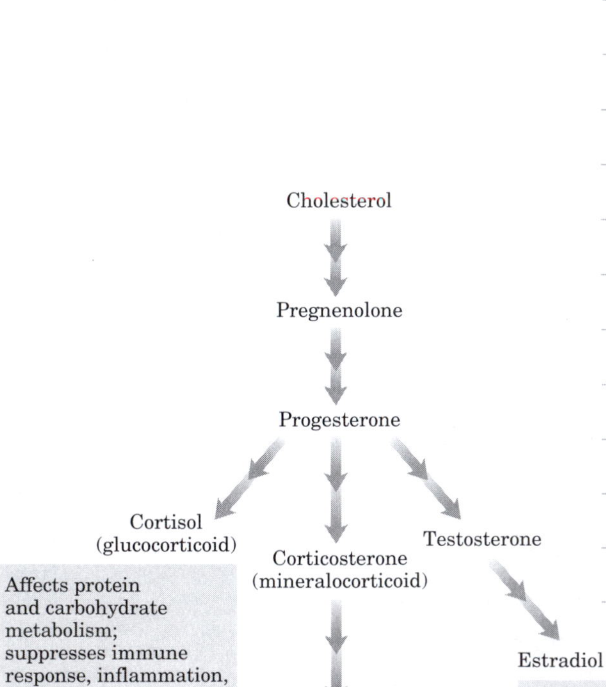

figure 21–43, page 812
Some steroid hormones derived from cholesterol.

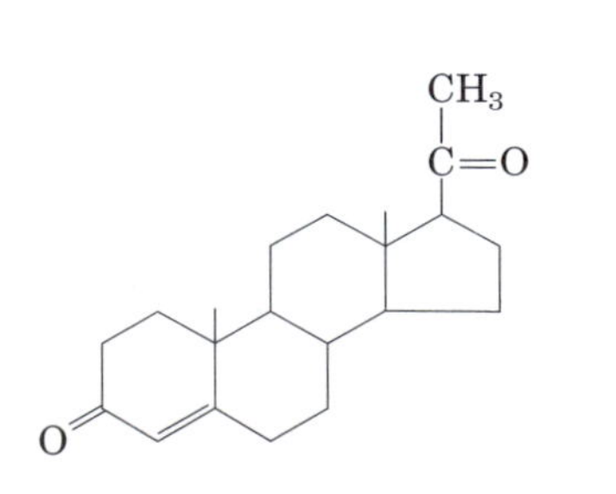

Progesterone, page 812

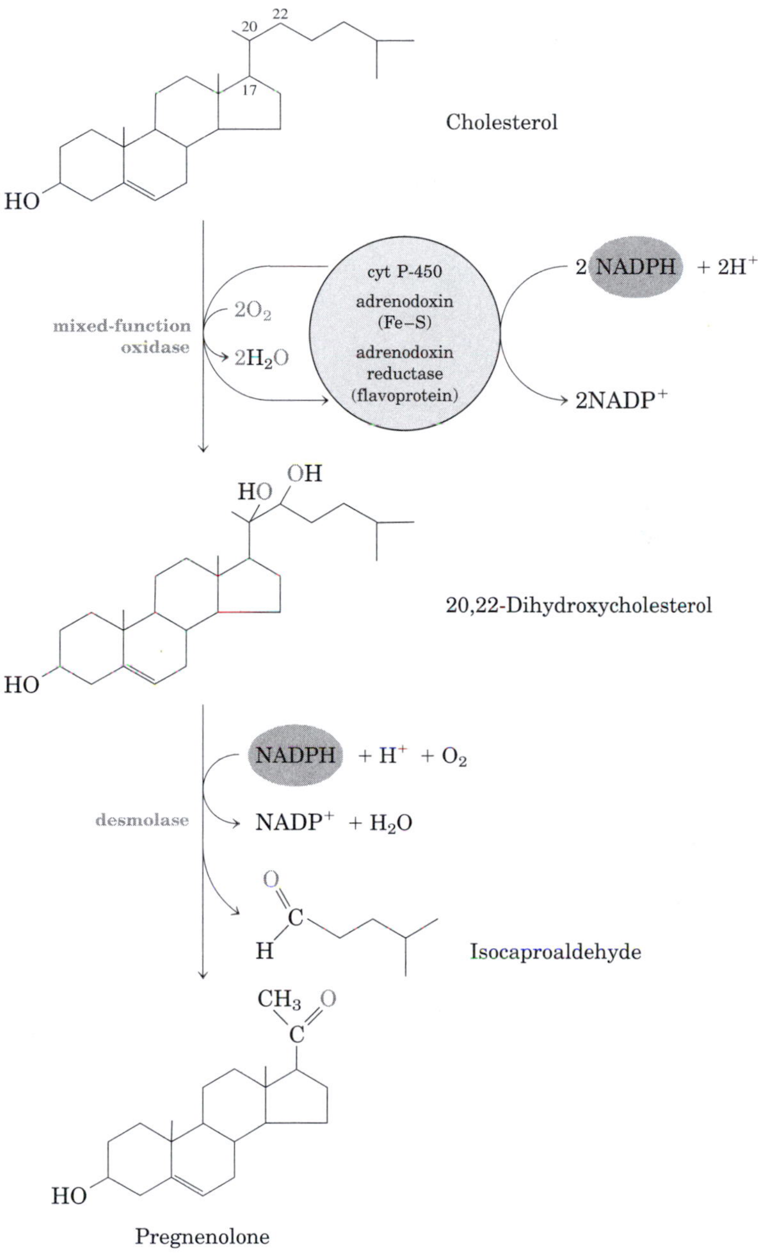

Cholesterol

mixed-function oxidase

$2O_2$ → $2H_2O$

cyt P-450, adrenodoxin (Fe–S), adrenodoxin reductase (flavoprotein)

2 NADPH + $2H^+$ → $2NADP^+$

20,22-Dihydroxycholesterol

desmolase

NADPH + H^+ + O_2 → $NADP^+$ + H_2O

Isocaproaldehyde

Pregnenolone

figure 21–44, page 813
Side chain cleavage in the synthesis of steroid hormones.

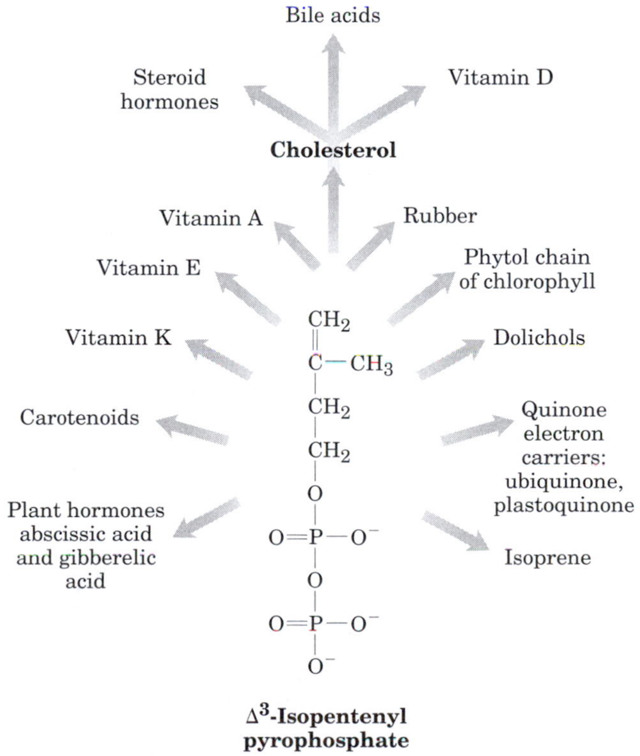

Bile acids

Steroid hormones

Vitamin D

Cholesterol

Vitamin A

Rubber

Vitamin E

Phytol chain of chlorophyll

Vitamin K

Dolichols

Carotenoids

Quinone electron carriers: ubiquinone, plastoquinone

Plant hormones abscissic acid and gibberelic acid

Isoprene

Δ^3-Isopentenyl pyrophosphate

figure 21–45, page 813
An overview of isoprenoid biosynthesis.

PROBLEMS

4. Pathway of Hydrogen in Fatty Acid Synthesis, page 817.

$$^{2}H_3C{-}C(=O){-}S\text{-}CoA$$

$$^{-}OOC{-}C(^{2}H)_2{-}C(=O){-}S\text{-}CoA$$

Biosynthesis of Amino Acids, Nucleotides, and Related Molecules

chapter 22

When you study for exams, don't forget to review The Minicourses on the UNDERSTAND! BIOCHEMISTRY CD that came with your textbook.

Minicourses that apply to this Chapter include:

Molecules of Life

An Interactive Gallery of Molecular Models

An Interactive Gallery of Protein Structures

Biosynthesis and Catabolism

Nitrogen Metabolism

Nuclear Biosynthesis and Breakdown

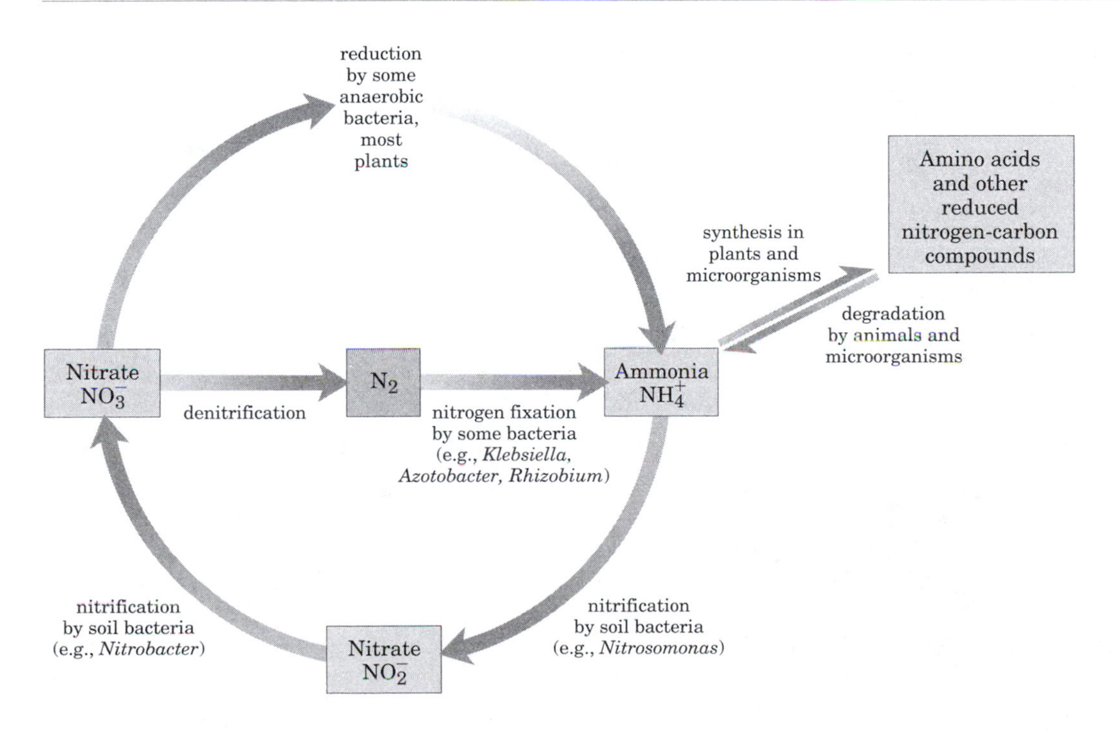

figure 22–1, page 819
The nitrogen cycle.

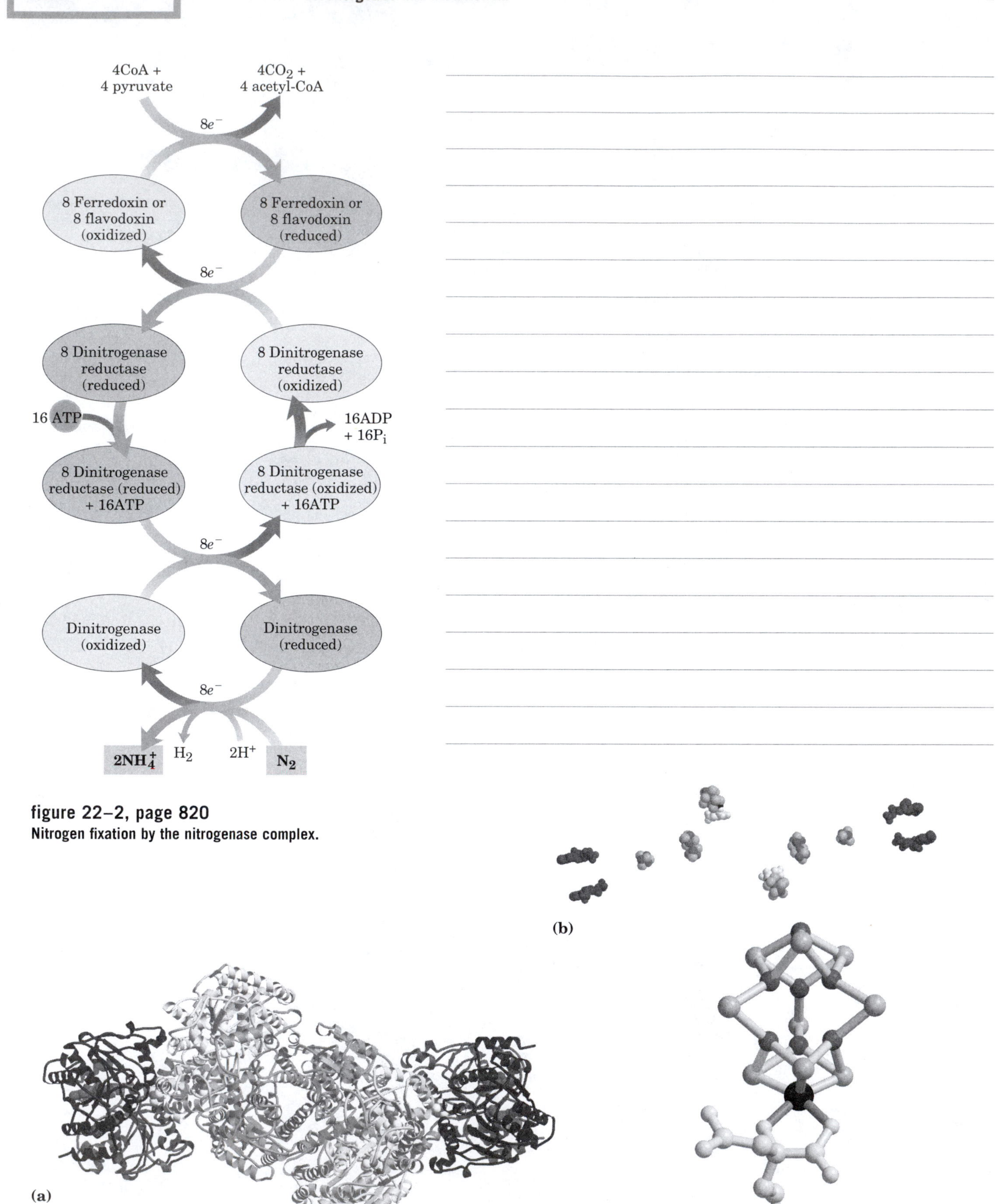

figure 22–2, page 820
Nitrogen fixation by the nitrogenase complex.

figure 22–3, page 821
Enzymes and cofactors of the nitrogenase complex.

(a)

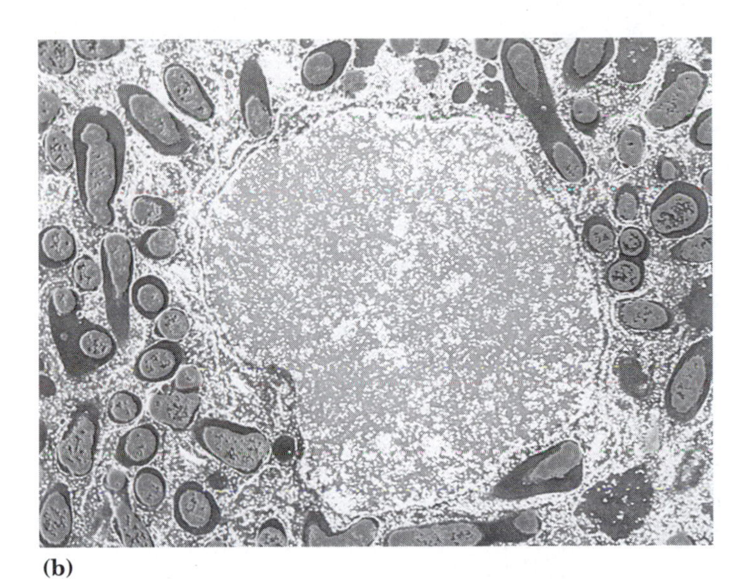

(b)

figure 22–4, page 822
Nitrogen-fixing nodules.

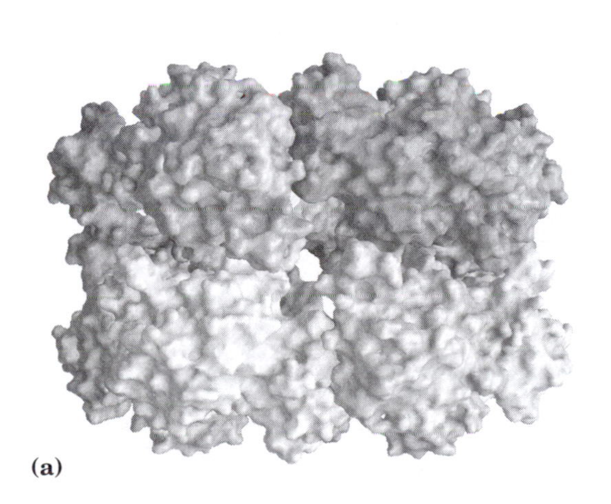

(a)

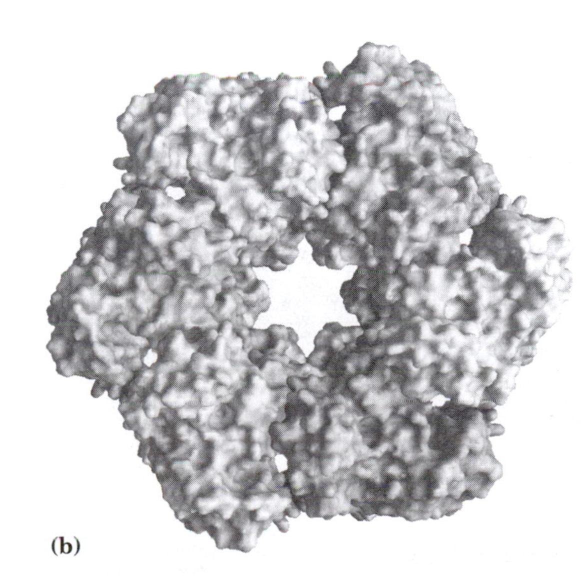

(b)

figure 22–5, page 824
Subunit structure of glutamine synthetase as determined by x-ray diffraction.

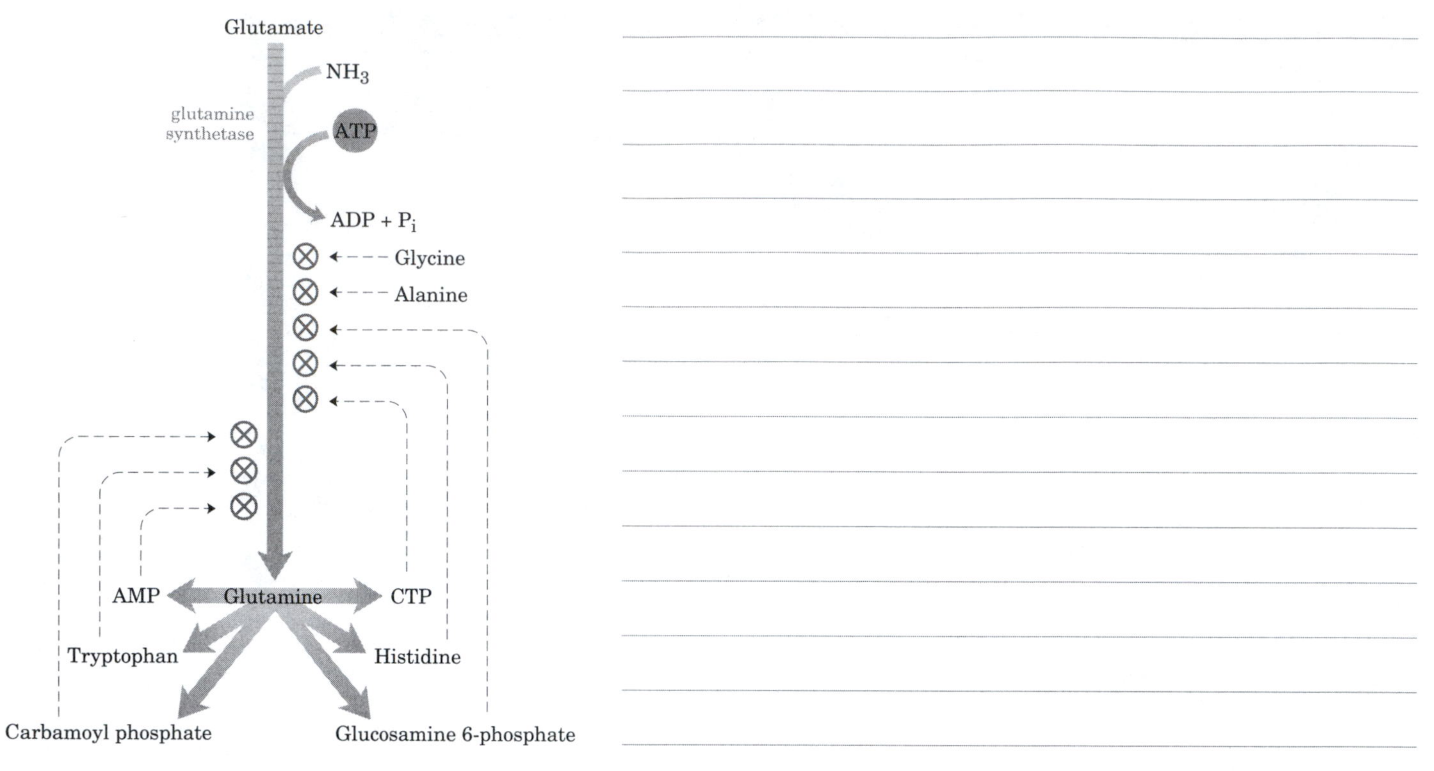

figure 22–6, page 824
Allosteric regulation of glutamine synthetase.

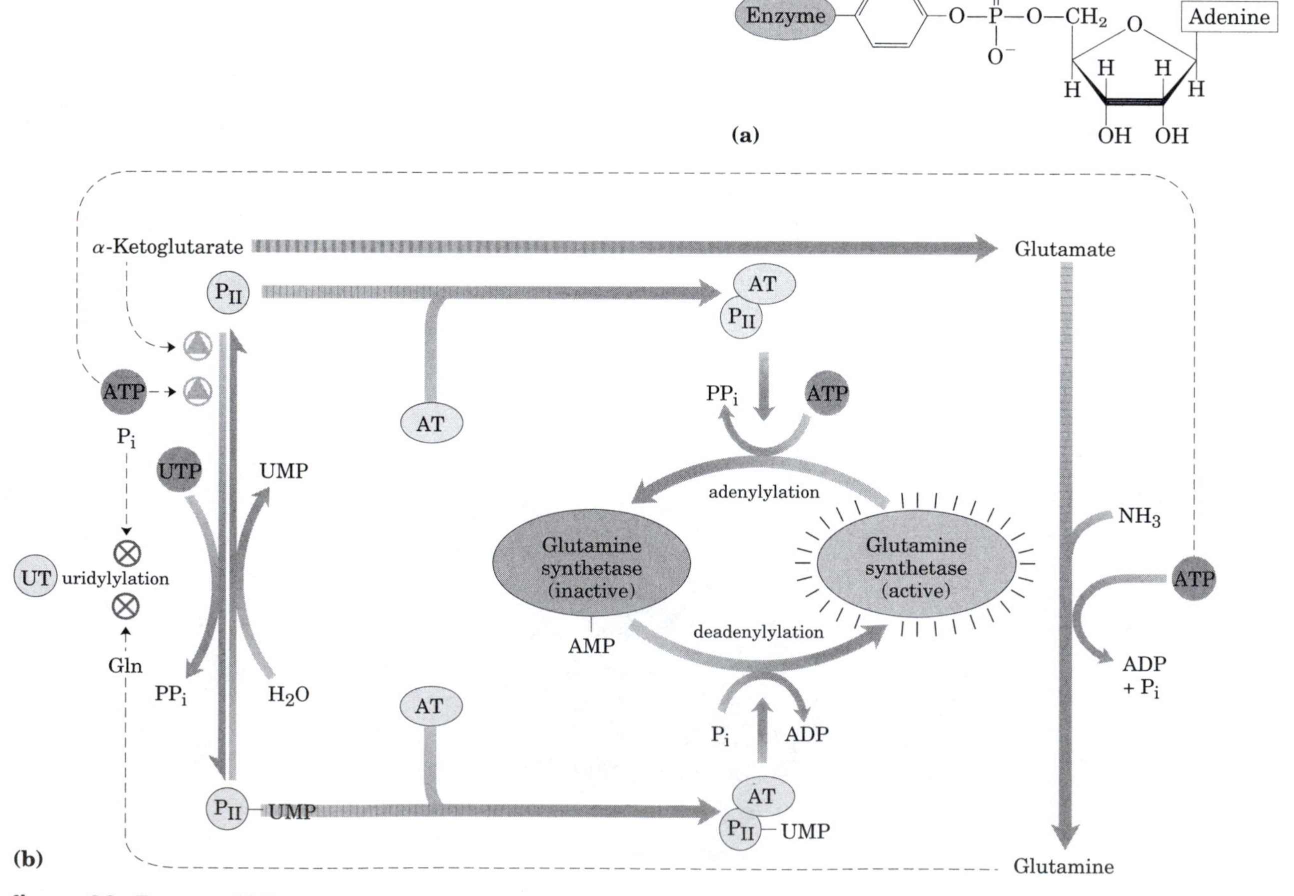

figure 22–7, page 825
Second level of regulation of glutamine synthetase: covalent modifications.

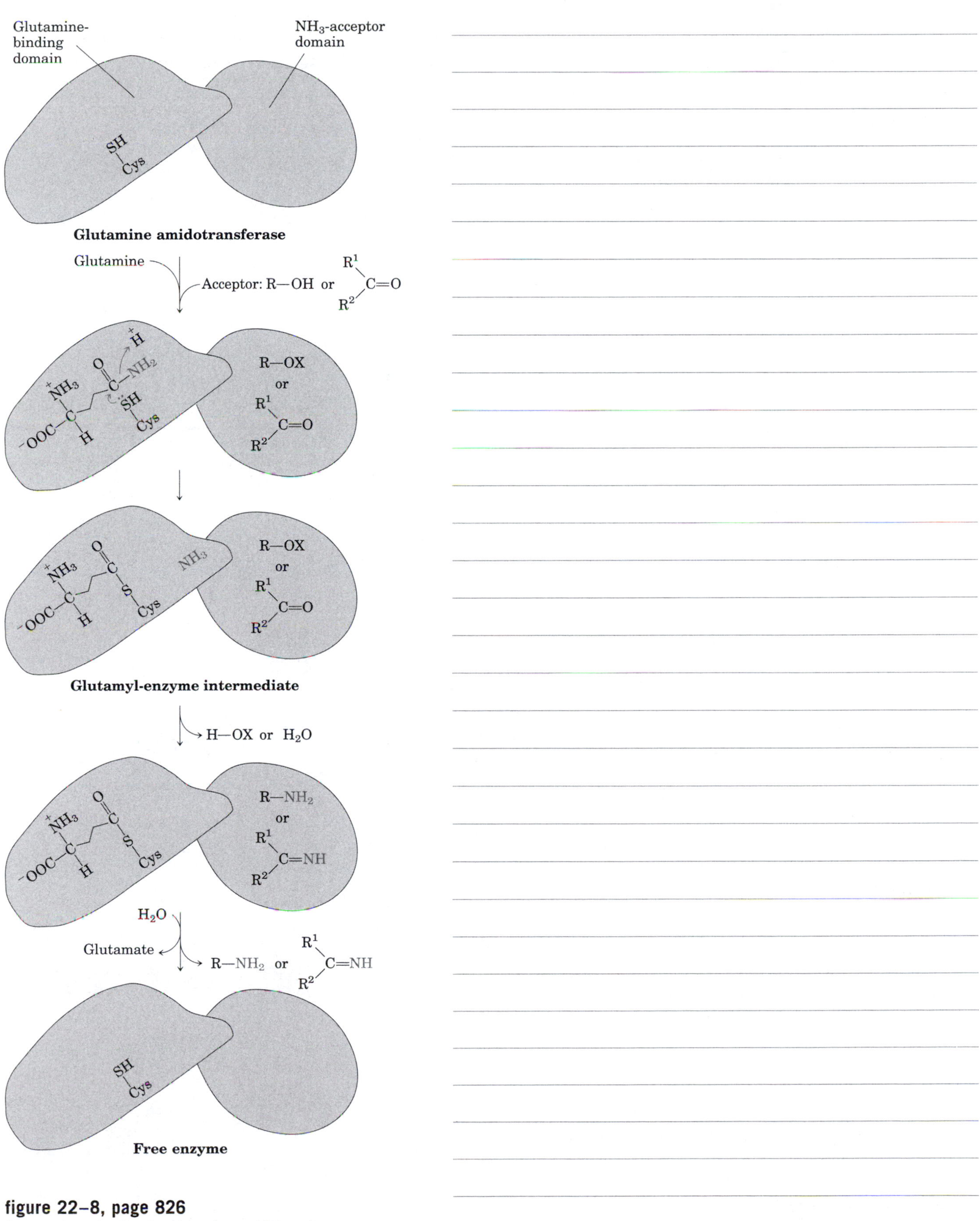

figure 22–8, page 826
Proposed mechanism for glutamine amidotransferases.

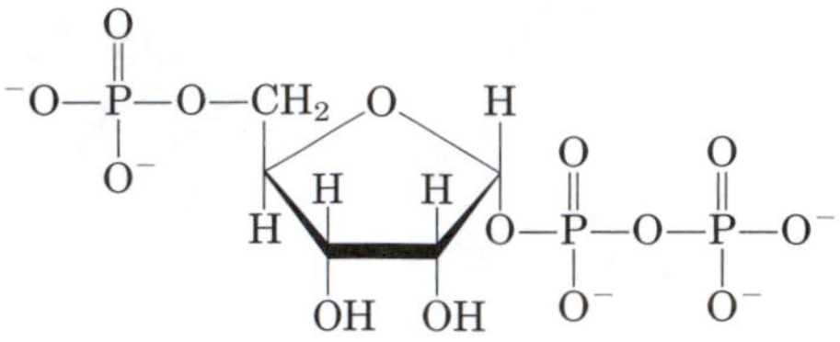

5-Phosphoribosyl-1-pyrophosphate (PRPP)

page 827

table 22–1, page 827

Amino Acid Biosynthetic Families, Grouped by Metabolic Precursor

α-Ketoglutarate	**Pyruvate**
Glutamate	Alanine
Glutamine	Valine†
Proline	Leucine†
Arginine*	
3-Phosphoglycerate	**Phosphoenolpyruvate and erythrose 4-phosphate**
Serine	Tryptophan†
Glycine	Phenylalanine†
Cysteine	Tyrosine‡
Oxaloacetate	**Ribose 5-phosphate**
Aspartate	Histidine†
Asparagine	
Methionine†	
Threonine†	
Lysine†	
Isoleucine†	

*Essential in young animals.
†Essential amino acids.
‡Derived from phenylalanine in mammals.

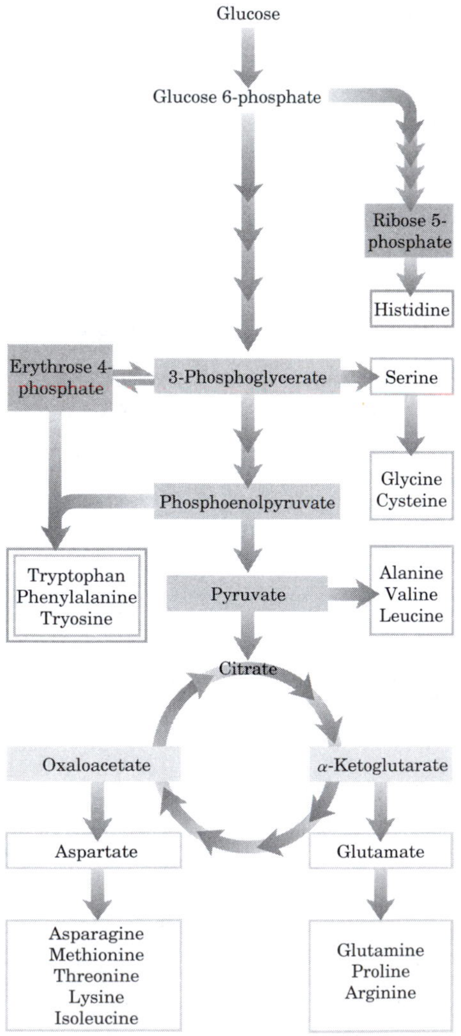

Glucose
Glucose 6-phosphate
Ribose 5-phosphate
Histidine
Erythrose 4-phosphate
3-Phosphoglycerate
Serine
Glycine
Cysteine
Phosphoenolpyruvate
Tryptophan
Phenylalanine
Tryosine
Pyruvate
Alanine
Valine
Leucine
Citrate
Oxaloacetate
α-Ketoglutarate
Aspartate
Glutamate
Asparagine
Methionine
Threonine
Lysine
Isoleucine
Glutamine
Proline
Arginine

figure 22–9, page 827
Overview of amino acid biosynthesis.

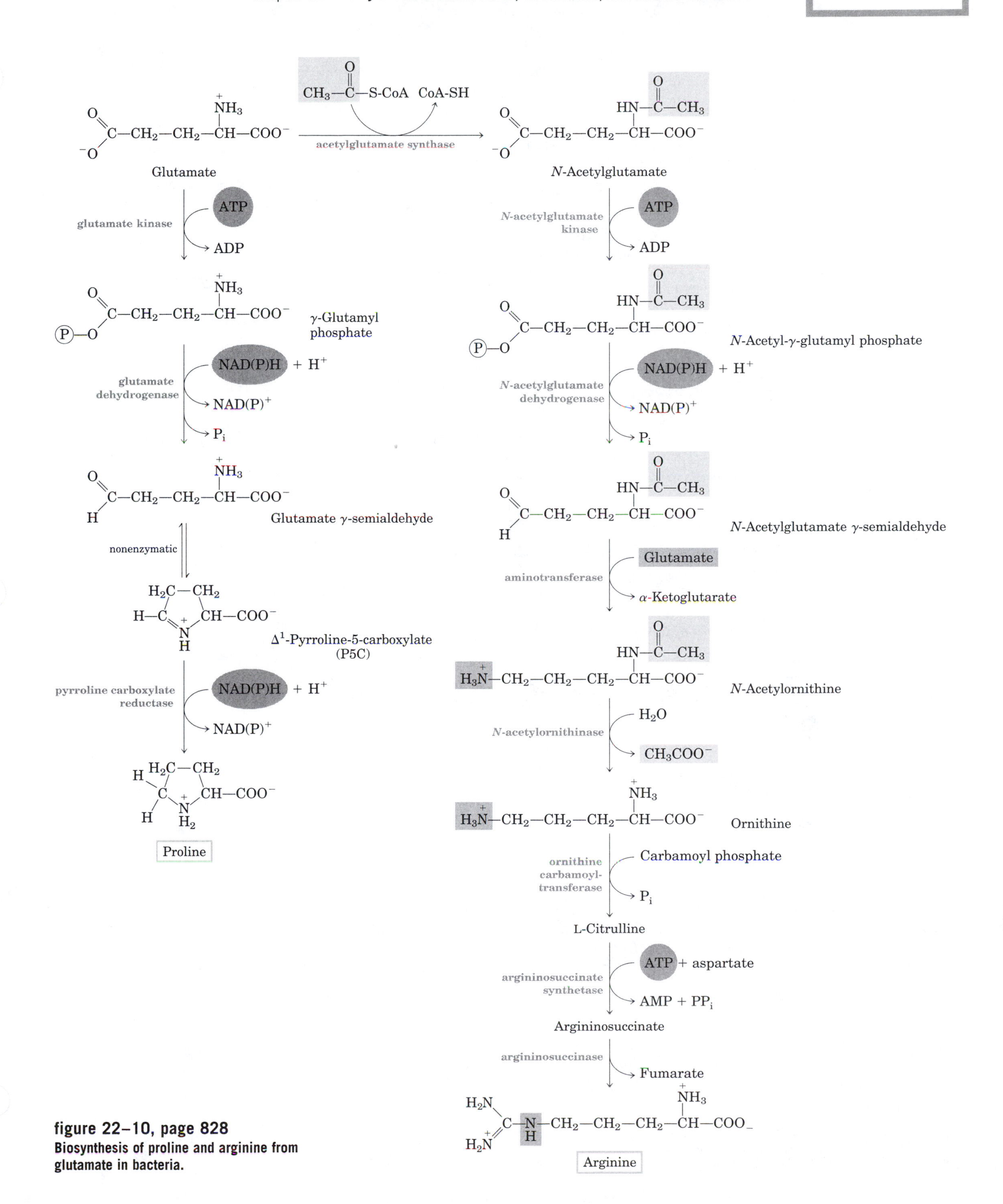

figure 22–10, page 828
Biosynthesis of proline and arginine from glutamate in bacteria.

Ornithine

α-Ketoglutarate → Glutamate

ornithine δ-aminotransferase

Glutamate γ-semialdehyde ⇌ Δ^1-Pyrroline-5-carboxylate (P5C)

figure 22–11, page 829

Ornithine δ-aminotransferase reaction: a step in the mammalian pathway to proline.

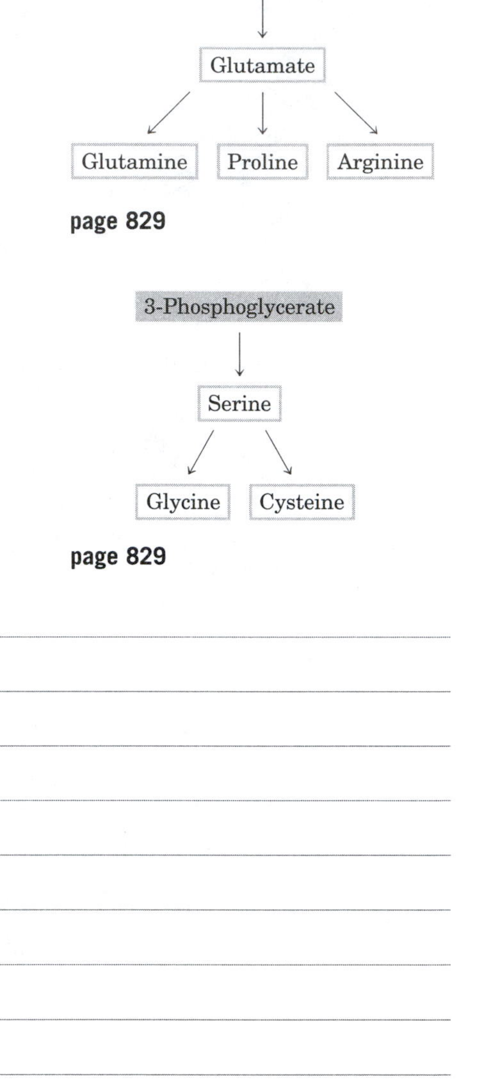

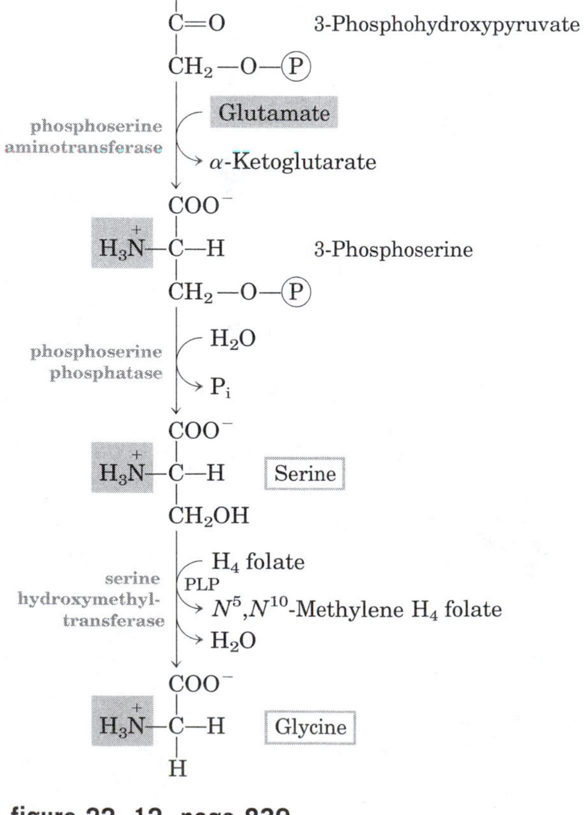

figure 22–12, page 830

Biosynthesis of serine from 3-phosphoglycerate and of glycine from serine in all organisms.

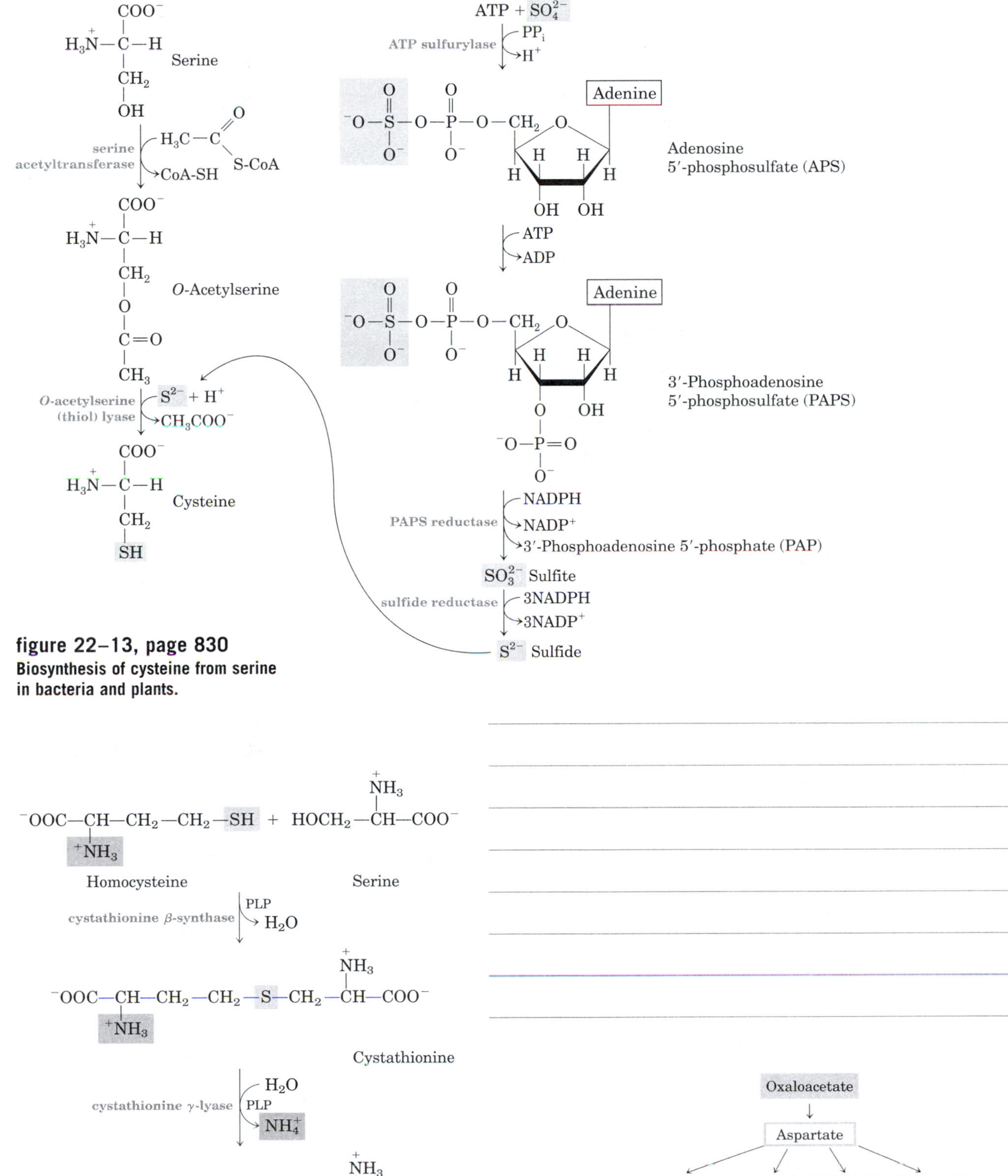

figure 22–13, page 830
Biosynthesis of cysteine from serine in bacteria and plants.

figure 22–14, page 831
Biosynthesis of cysteine from homocysteine and serine in mammals.

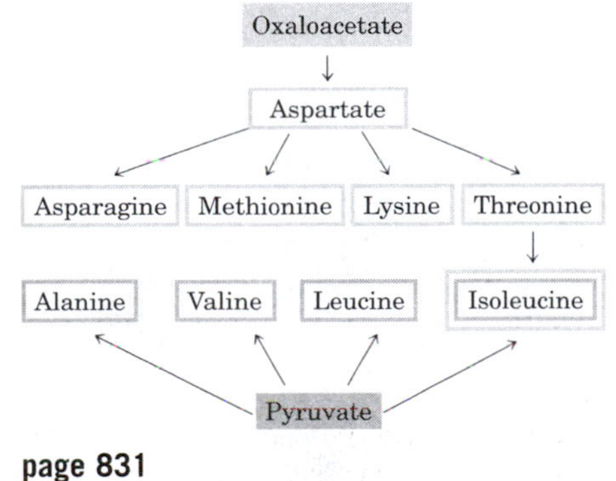

page 831

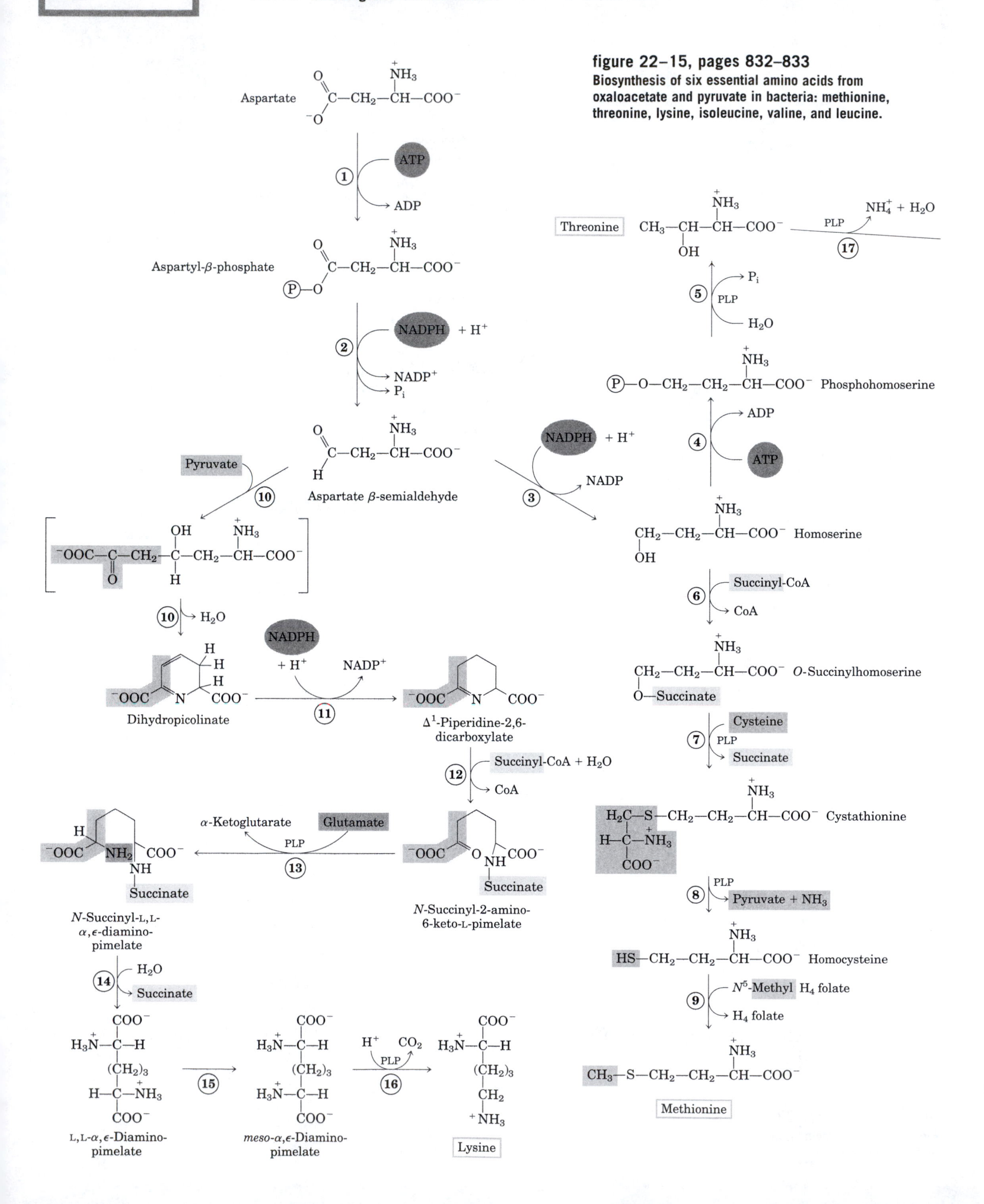

figure 22–15, pages 832–833
Biosynthesis of six essential amino acids from oxaloacetate and pyruvate in bacteria: methionine, threonine, lysine, isoleucine, valine, and leucine.

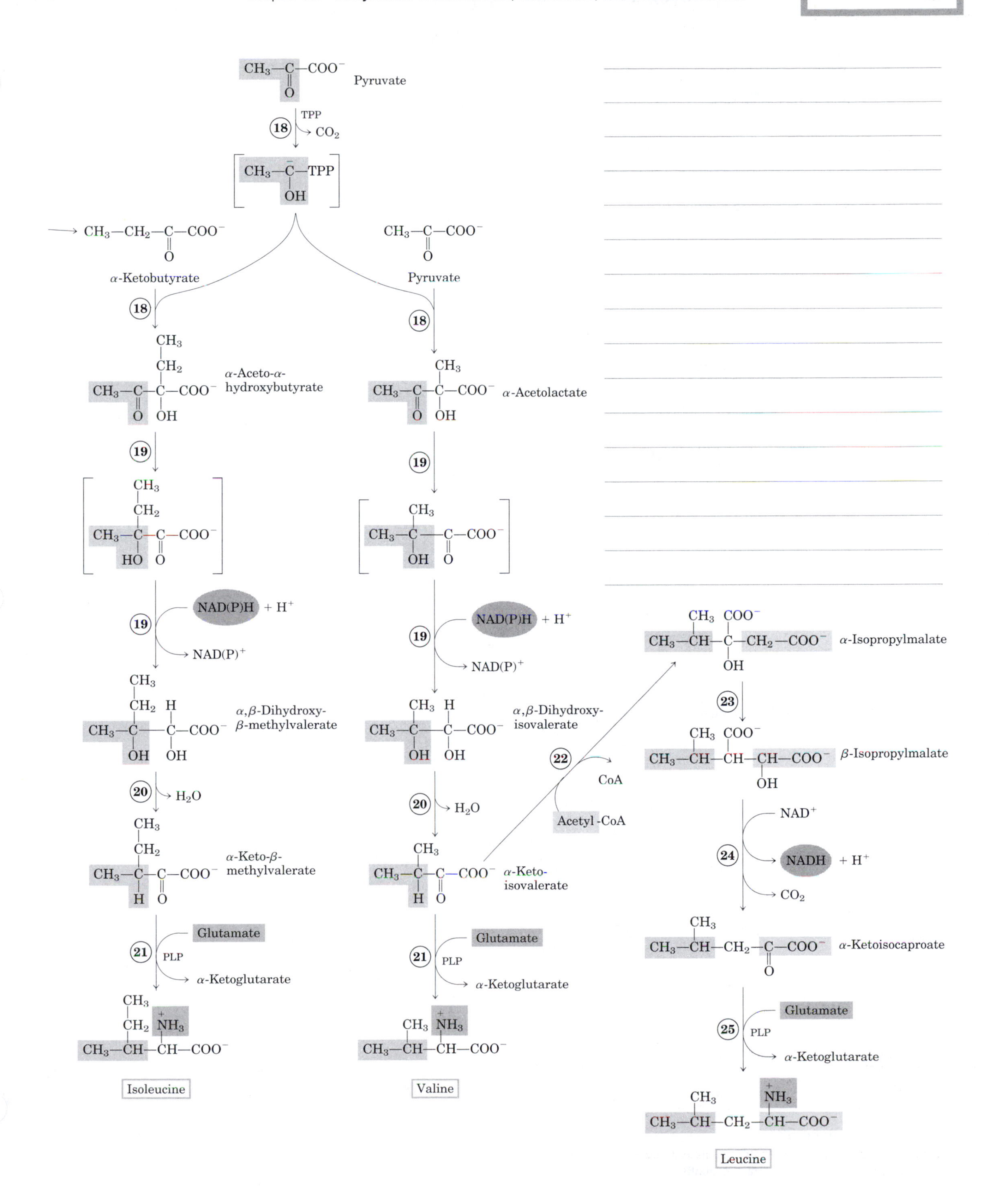
Pyruvate
TPP
18
CO2
CH3—C—TPP
OH
α-Ketobutyrate
Pyruvate
α-Aceto-α-hydroxybutyrate
α-Acetolactate
19
NAD(P)H + H+
NAD(P)+
α,β-Dihydroxy-β-methylvalerate
α,β-Dihydroxy-isovalerate
20
H2O
α-Keto-β-methylvalerate
α-Keto-isovalerate
22
CoA
Acetyl-CoA
α-Isopropylmalate
23
β-Isopropylmalate
24
NAD+
NADH + H+
CO2
α-Ketoisocaproate
21
25
Glutamate
PLP
α-Ketoglutarate
Isoleucine
Valine
Leucine

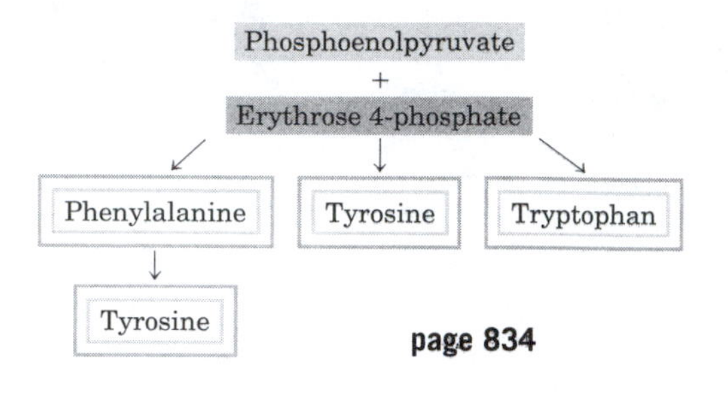

page 834

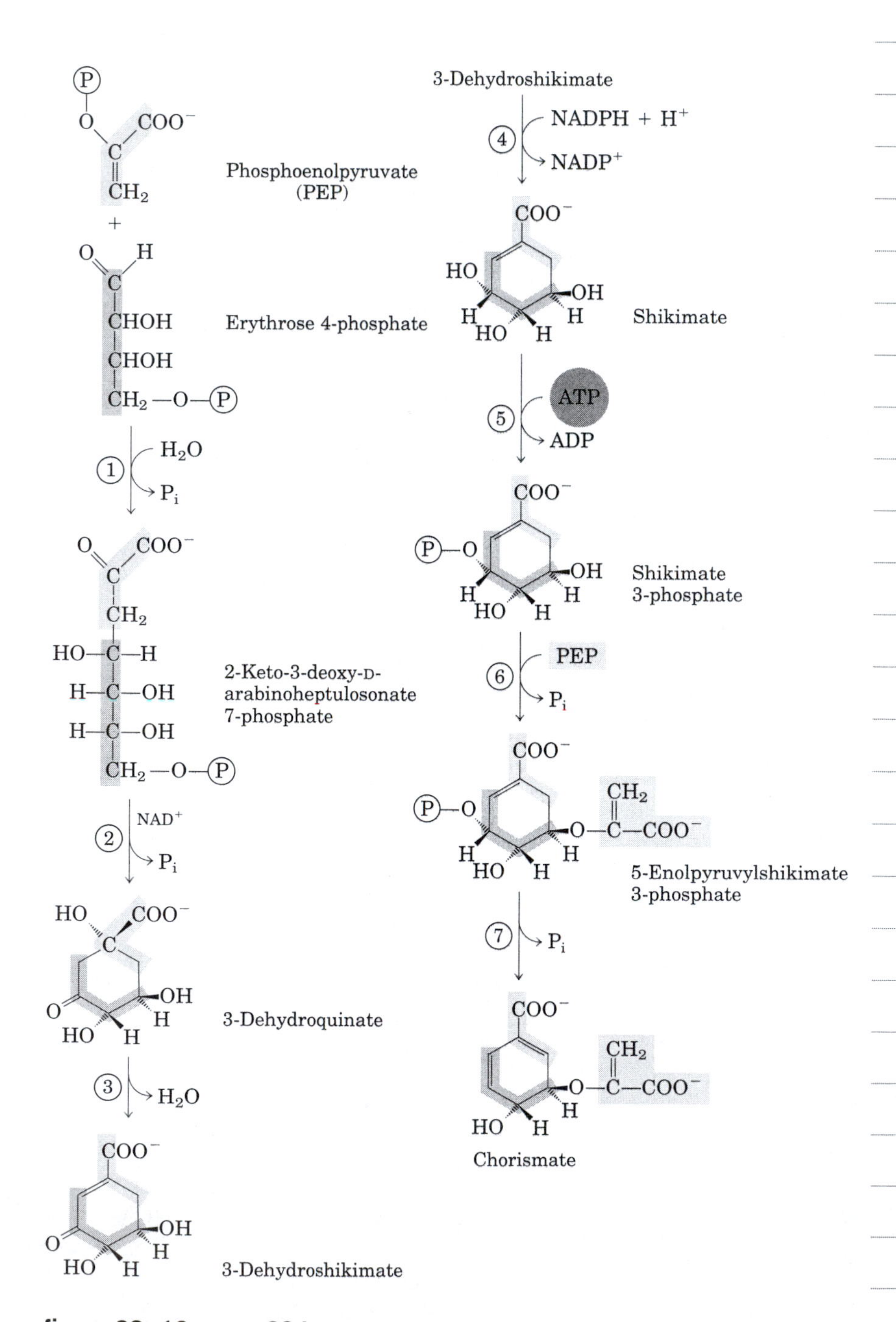

figure 22–16, page 834
Biosynthesis of chorismate, an intermediate in the synthesis of aromatic amino acids in bacteria and plants.

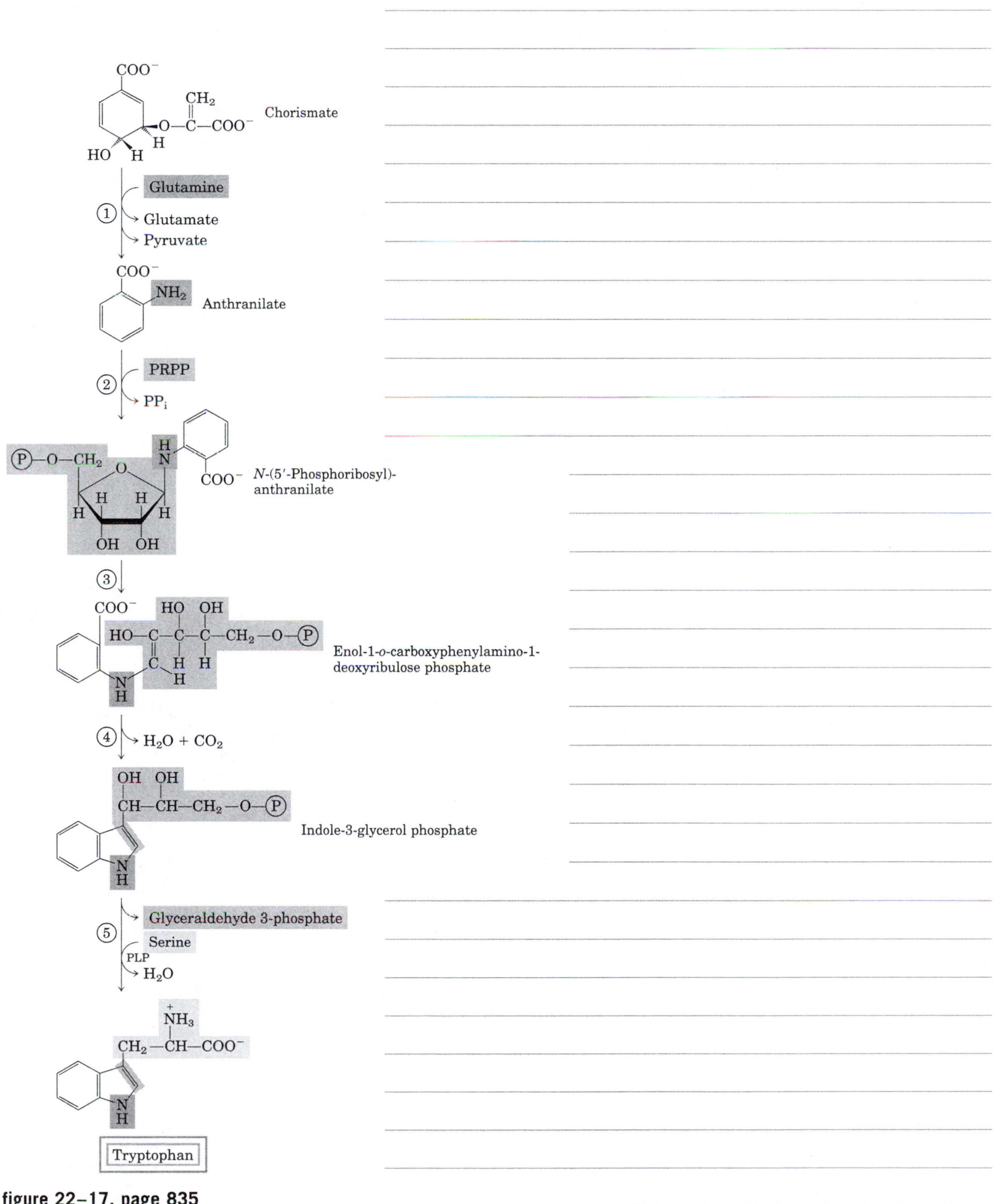

figure 22–17, page 835

Biosynthesis of tryptophan from chorismate in bacteria and plants.

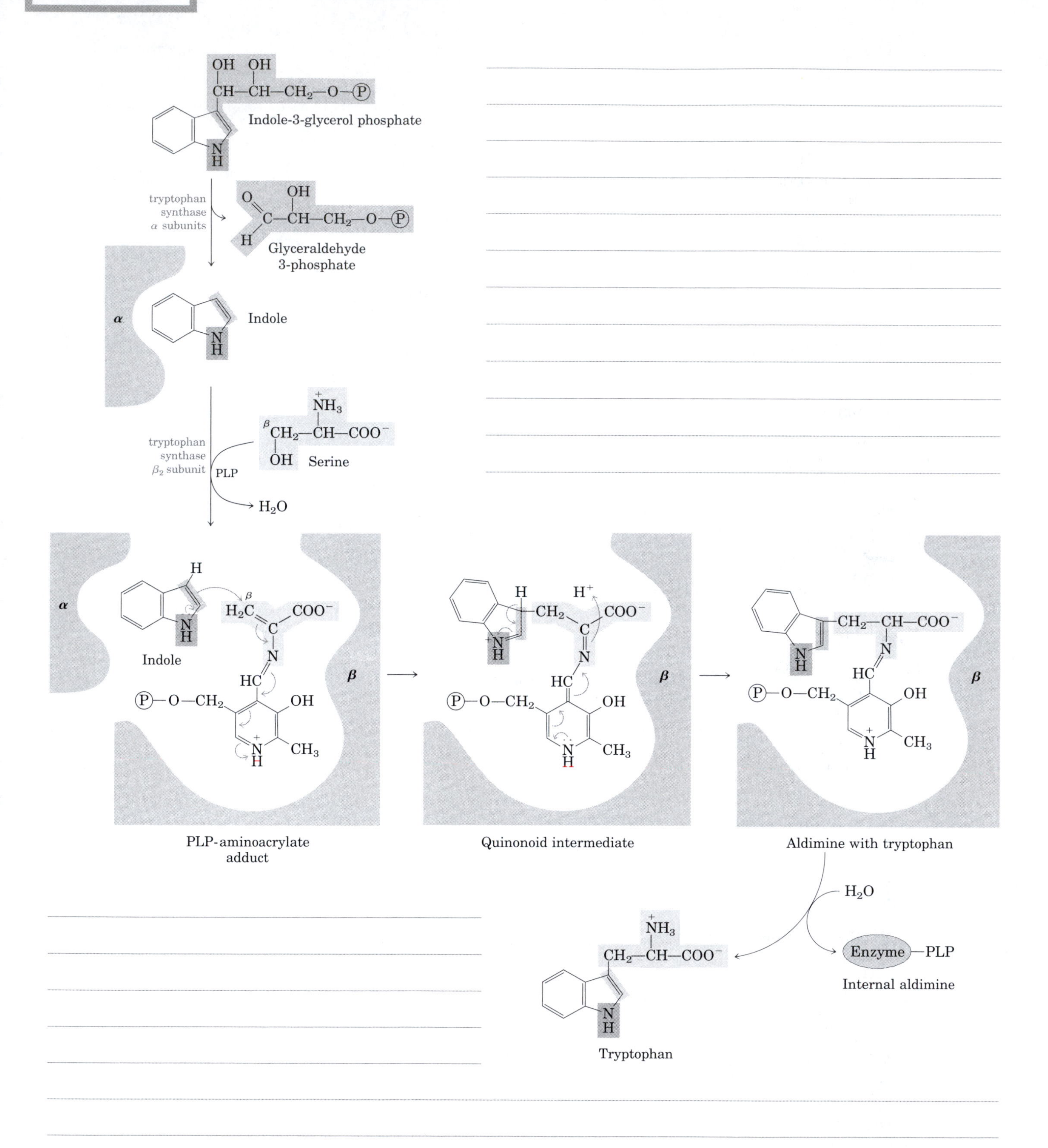

figure 22–18, page 836
Tryptophan synthase reaction.

$^{-}OOC—CH_2—NH—CH_2—PO_3^{2-}$

Glyphosate, page 837

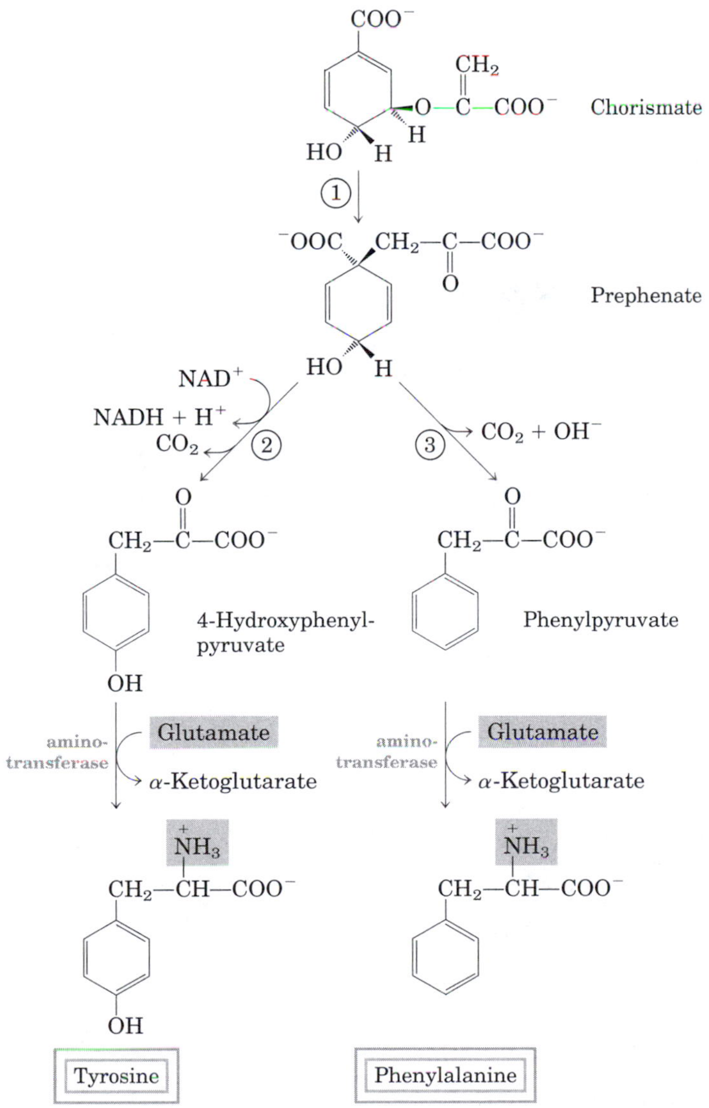

Chorismate

① ↓

Prephenate

NAD^+ → $NADH + H^+$, CO_2 ②

③ → $CO_2 + OH^-$

4-Hydroxyphenyl-pyruvate

Phenylpyruvate

aminotransferase: Glutamate → α-Ketoglutarate

aminotransferase: Glutamate → α-Ketoglutarate

Tyrosine

Phenylalanine

figure 22–19, page 837

Biosynthesis of phenylalanine and tyrosine from chorismate in bacteria and plants.

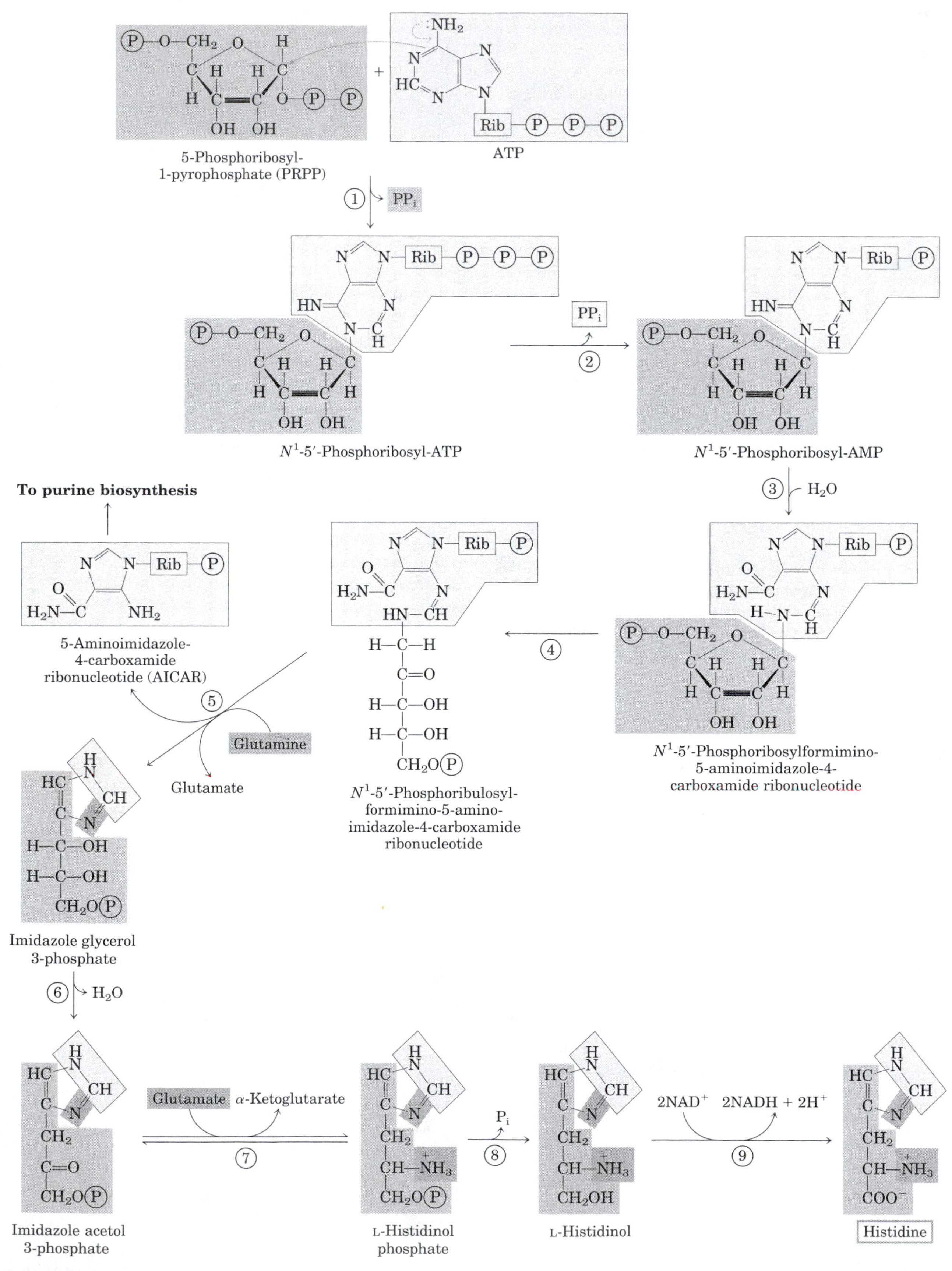

figure 22–20, page 838
Biosynthesis of histidine in bacteria and plants.

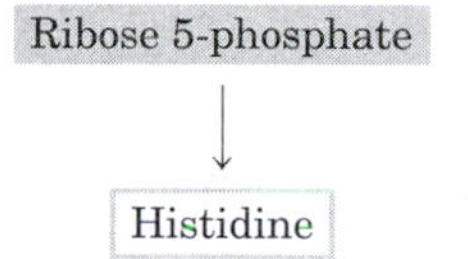

page 839

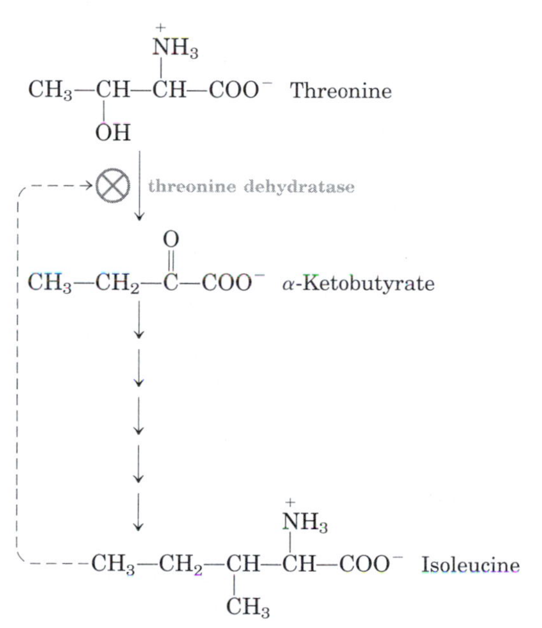

figure 22–21, page 839
Allosteric regulation of isoleucine biosynthesis.

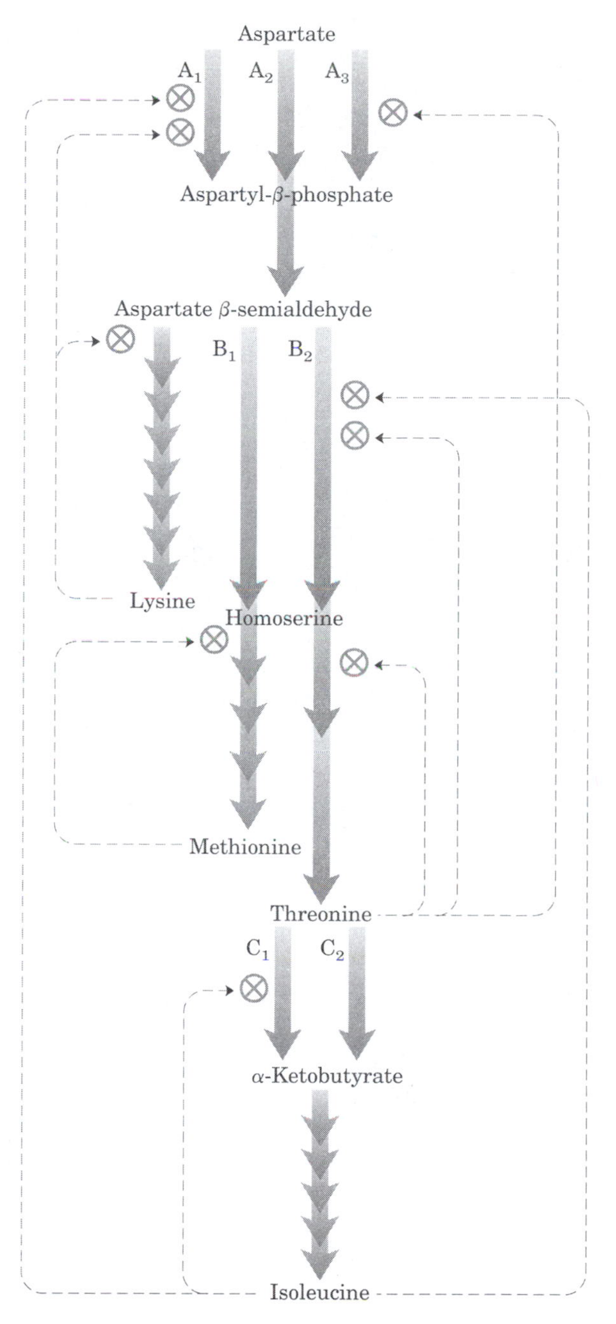

figure 22–22, page 840
Interlocking regulatory mechanisms in the biosynthesis of several amino acids derived from aspartate in *E. coli.*

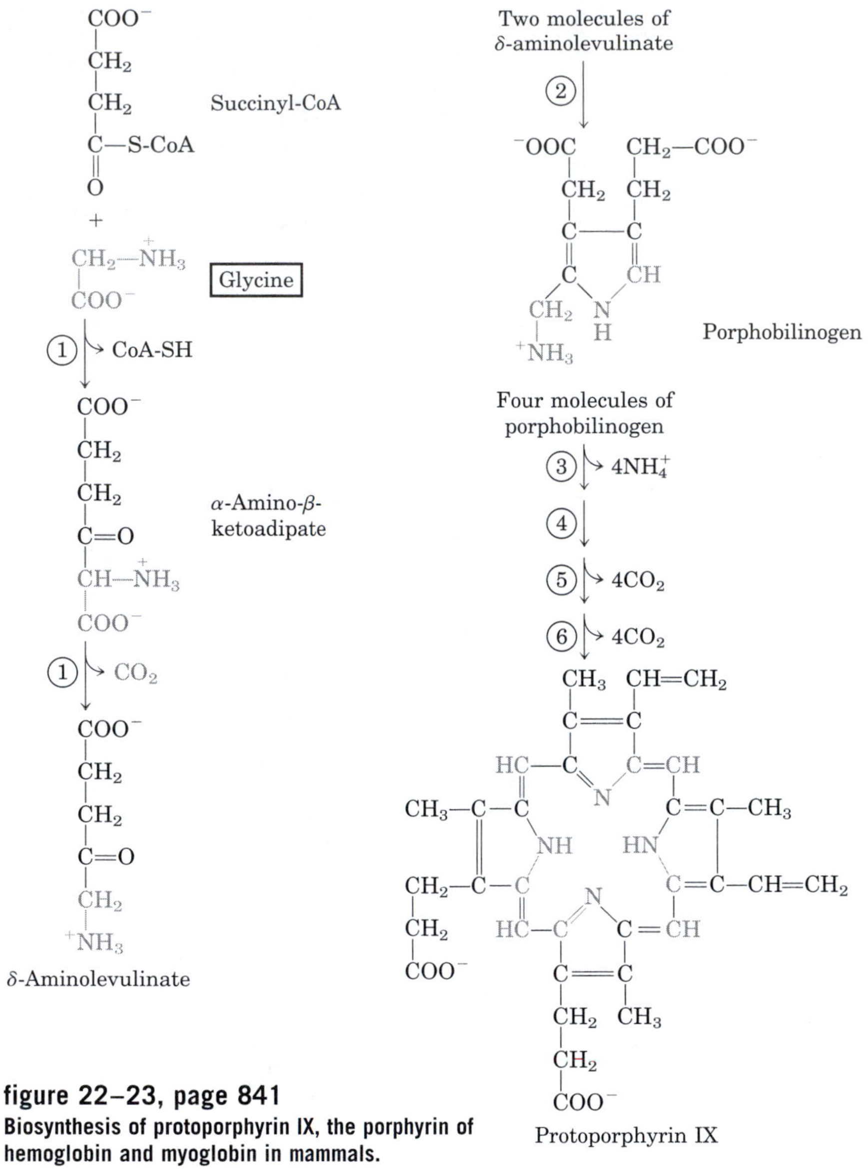

COO^- — CH_2 — CH_2 — $C(=O)$—S-CoA

Succinyl-CoA

+

CH_2—$\overset{+}{N}H_3$ — COO^-

Glycine

① → CoA-SH

COO^- — CH_2 — CH_2 — $C{=}O$ — CH—$\overset{+}{N}H_3$ — COO^-

α-Amino-β-ketoadipate

① → CO_2

COO^- — CH_2 — CH_2 — $C{=}O$ — CH_2 — $^+NH_3$

δ-Aminolevulinate

Two molecules of δ-aminolevulinate

②

Porphobilinogen

Four molecules of porphobilinogen

③ → $4NH_4^+$

④

⑤ → $4CO_2$

⑥ → $4CO_2$

Protoporphyrin IX

figure 22–23, page 841

Biosynthesis of protoporphyrin IX, the porphyrin of hemoglobin and myoglobin in mammals.

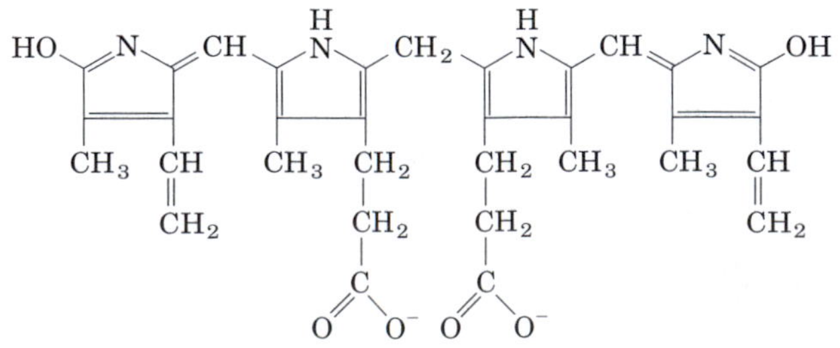

Bilirubin, page 842

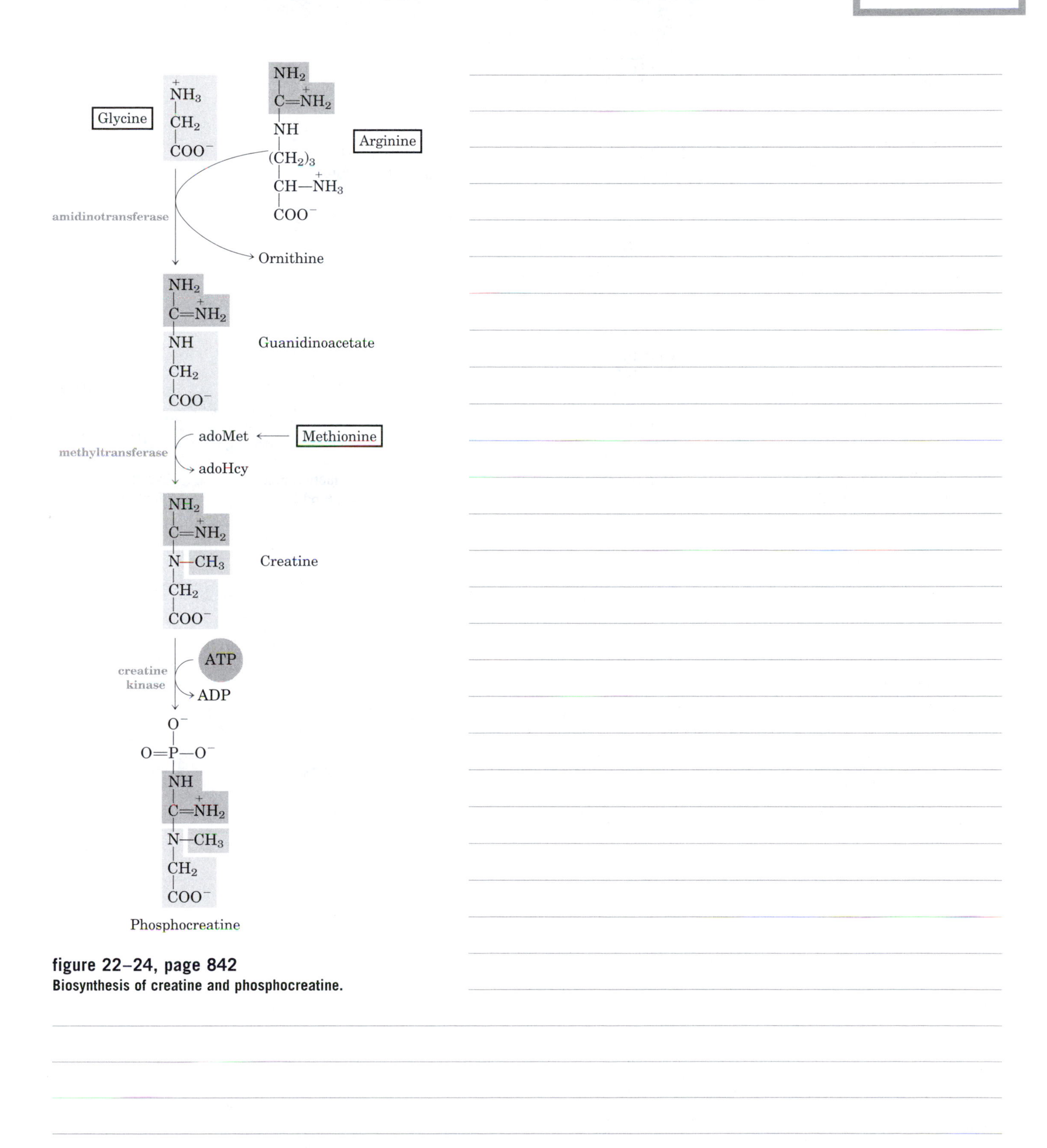

figure 22–24, page 842
Biosynthesis of creatine and phosphocreatine.

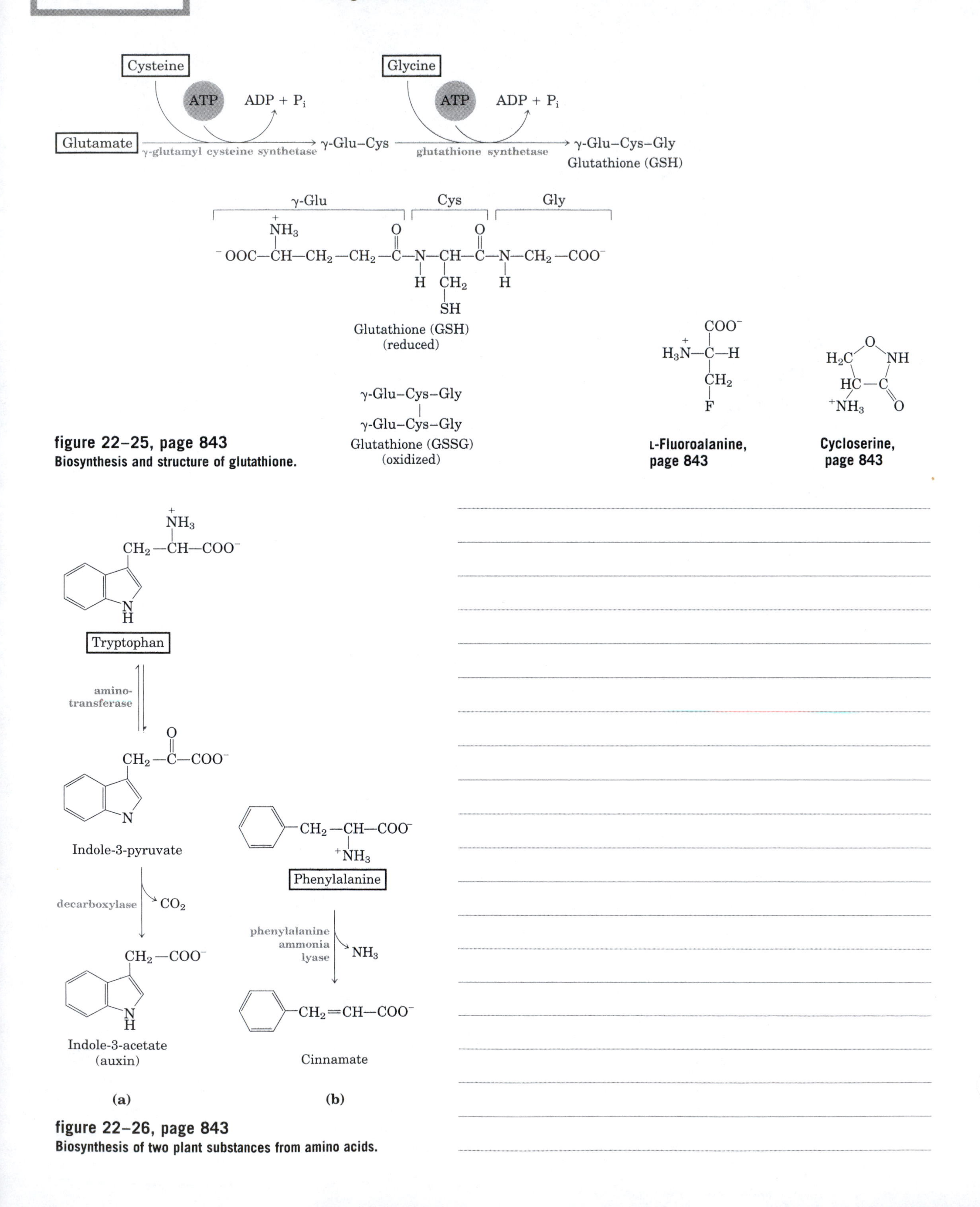

figure 22–25, page 843
Biosynthesis and structure of glutathione.

L-Fluoroalanine, page 843

Cycloserine, page 843

figure 22–26, page 843
Biosynthesis of two plant substances from amino acids.

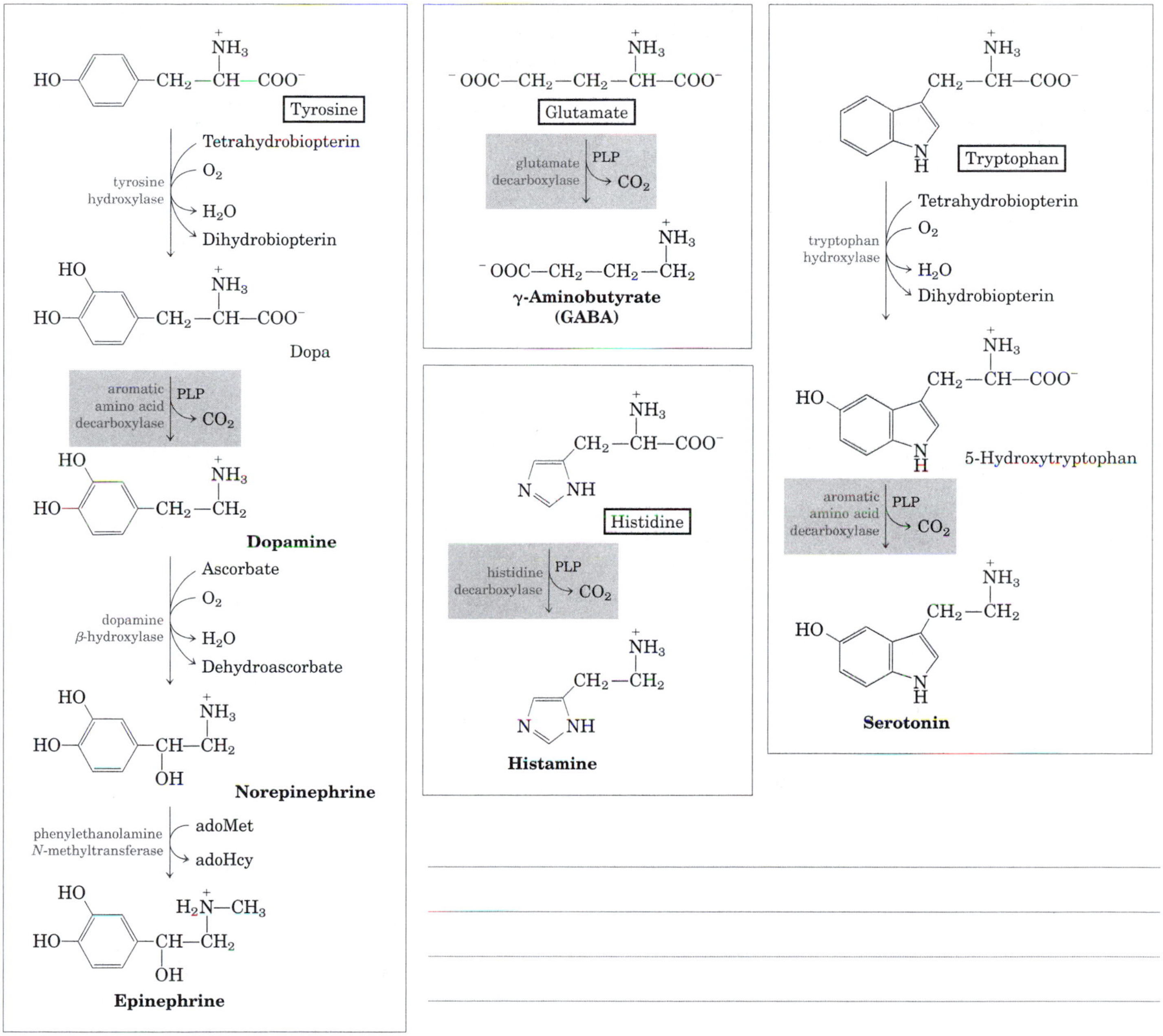

Tyrosine

tyrosine hydroxylase — Tetrahydrobiopterin, O_2 → H_2O, Dihydrobiopterin

Dopa

aromatic amino acid decarboxylase — PLP → CO_2

Dopamine

dopamine β-hydroxylase — Ascorbate, O_2 → H_2O, Dehydroascorbate

Norepinephrine

phenylethanolamine *N*-methyltransferase — adoMet → adoHcy

Epinephrine

Glutamate

glutamate decarboxylase — PLP → CO_2

γ-Aminobutyrate (GABA)

Histidine

histidine decarboxylase — PLP → CO_2

Histamine

Tryptophan

tryptophan hydroxylase — Tetrahydrobiopterin, O_2 → H_2O, Dihydrobiopterin

5-Hydroxytryptophan

aromatic amino acid decarboxylase — PLP → CO_2

Serotonin

figure 22–27, page 844
Biosynthesis of some neurotransmitters from amino acids.

CH_3 $CH_2—S—CH_2—CH_2—NH—C—NH—CH_3$
N NH N—C≡N

Cimetidine (Tagamet), page 845

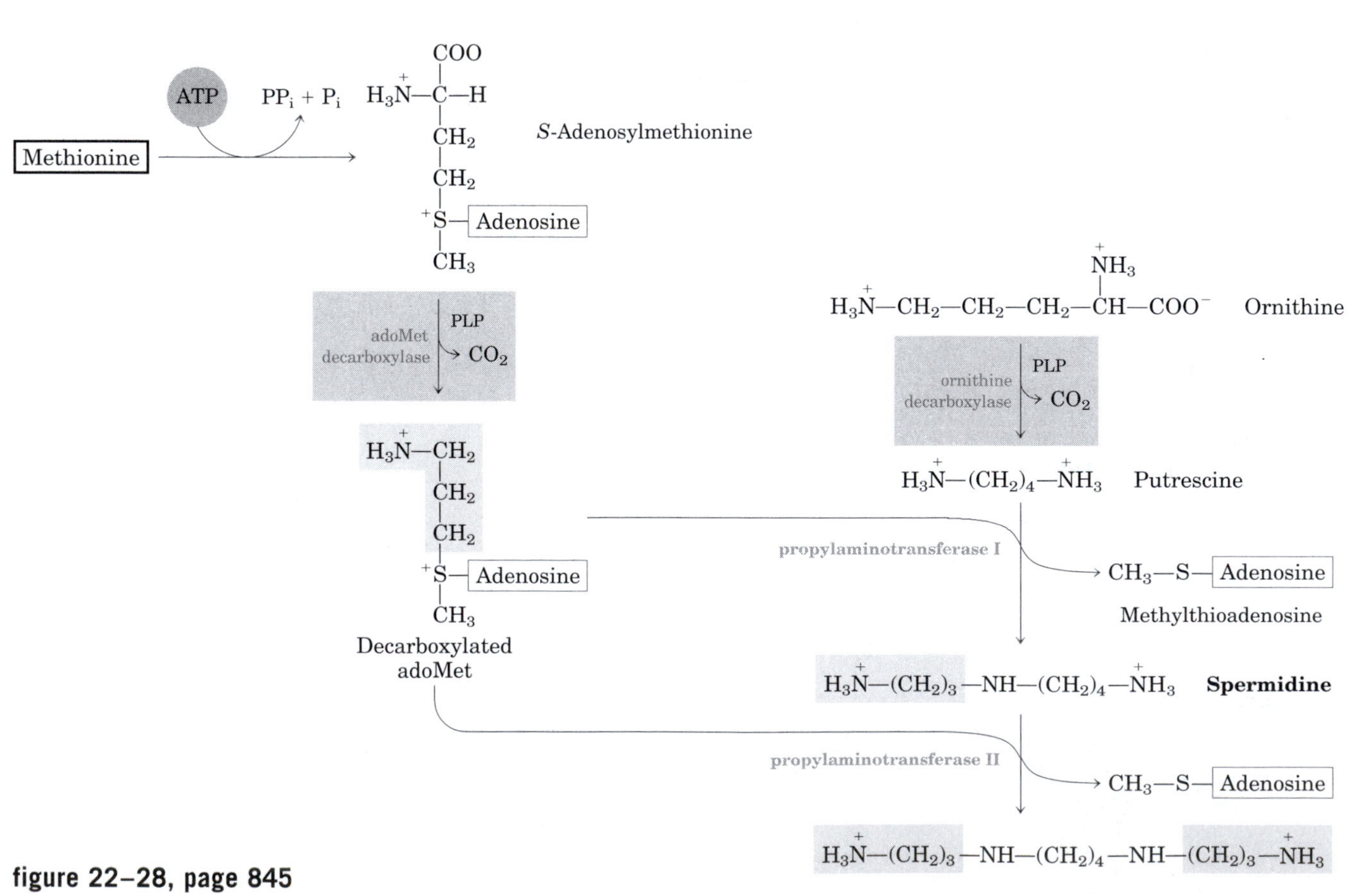

figure 22–28, page 845
Biosynthesis of spermidine and spermine.

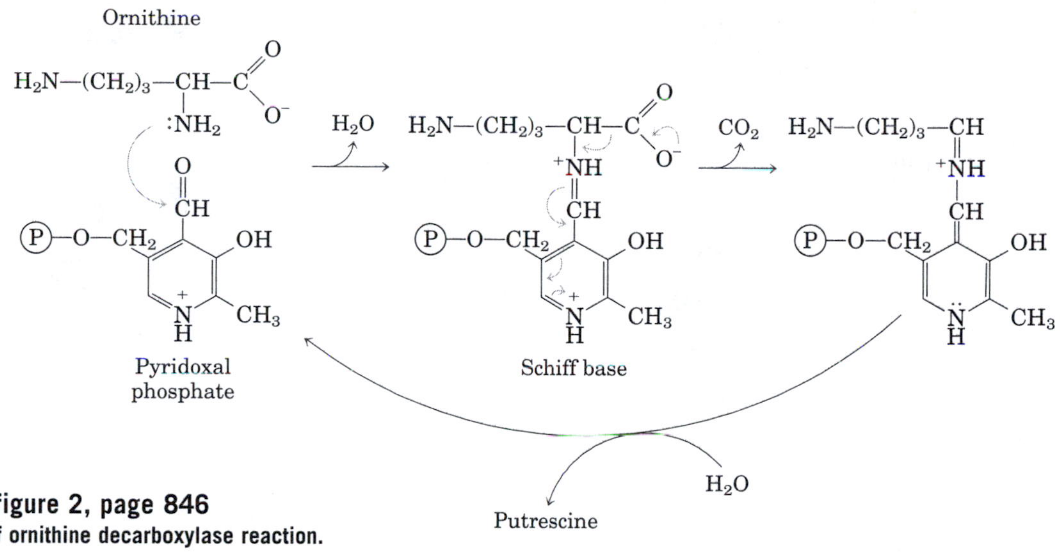

Ornithine

H_2O · CO_2 · H_2O

Pyridoxal phosphate

Schiff base

Putrescine

box 22–2, figure 2, page 846
Mechanism of ornithine decarboxylase reaction.

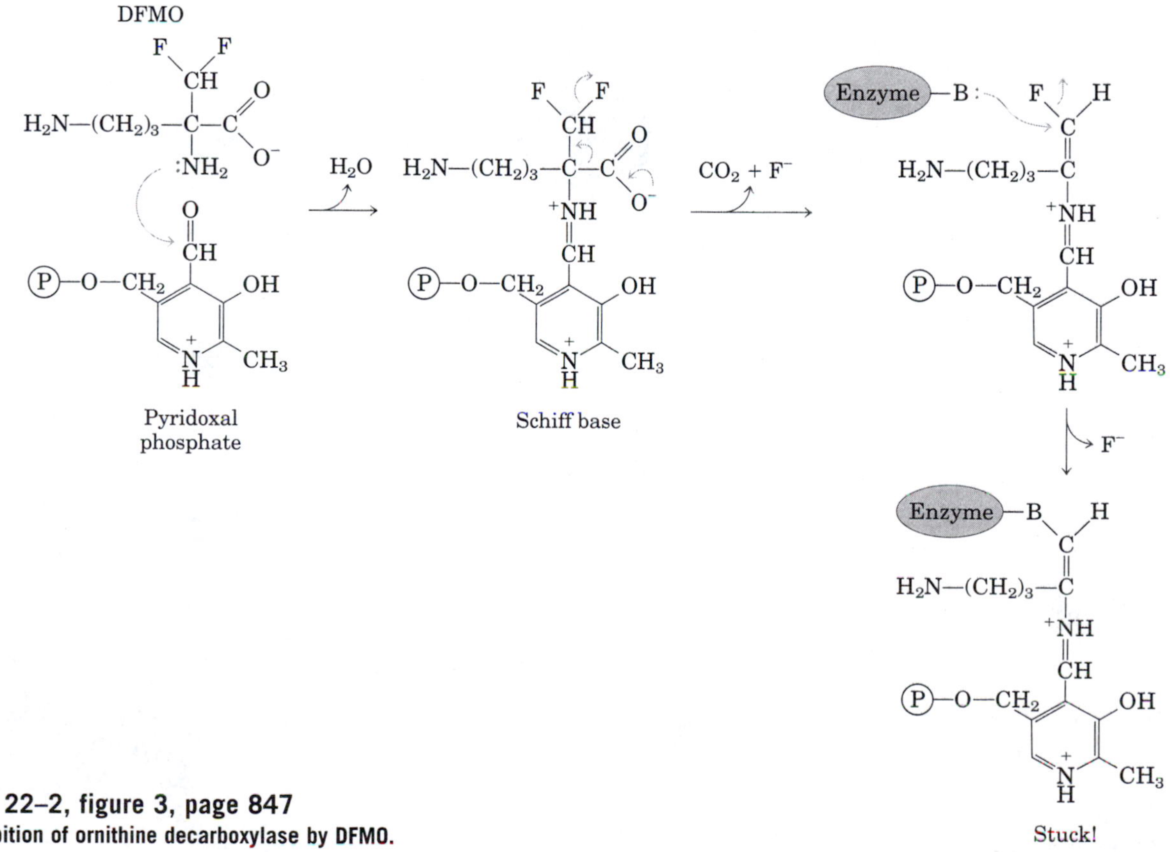

DFMO

H_2O · $CO_2 + F^-$ · F^-

Pyridoxal phosphate

Schiff base

Enzyme

Stuck!

box 22–2, figure 3, page 847
Inhibition of ornithine decarboxylase by DFMO.

$$\underset{\text{Arginine}}{H_3\overset{+}{N}-CH(COO^-)-CH_2-CH_2-CH_2-NH-C(=\overset{+}{N}H_2)-NH_2} \xrightarrow[\text{NADPH, } O_2 \;\to\; \text{NADP}^+,\ H_2O]{} \left[\underset{\text{Hydroxyarginine}}{H_3\overset{+}{N}-CH(COO^-)-CH_2-CH_2-CH_2-NH-C(=N-OH)-NH_2}\right] \xrightarrow[\frac{1}{2}\text{NADPH, } O_2 \;\to\; \frac{1}{2}\text{NADP}^+,\ H_2O]{} \underset{\text{Citrulline}}{H_3\overset{+}{N}-CH(COO^-)-CH_2-CH_2-CH_2-NH-C(=O)-NH_2} + \underset{\text{Nitric oxide}}{NO^{\bullet}}$$

figure 22–29, page 848
Biosynthesis of nitric oxide.

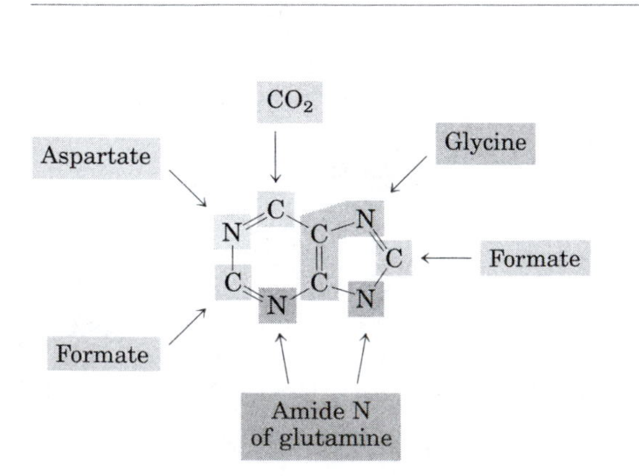

figure 22–30, page 849
Origin of the ring atoms of purines.

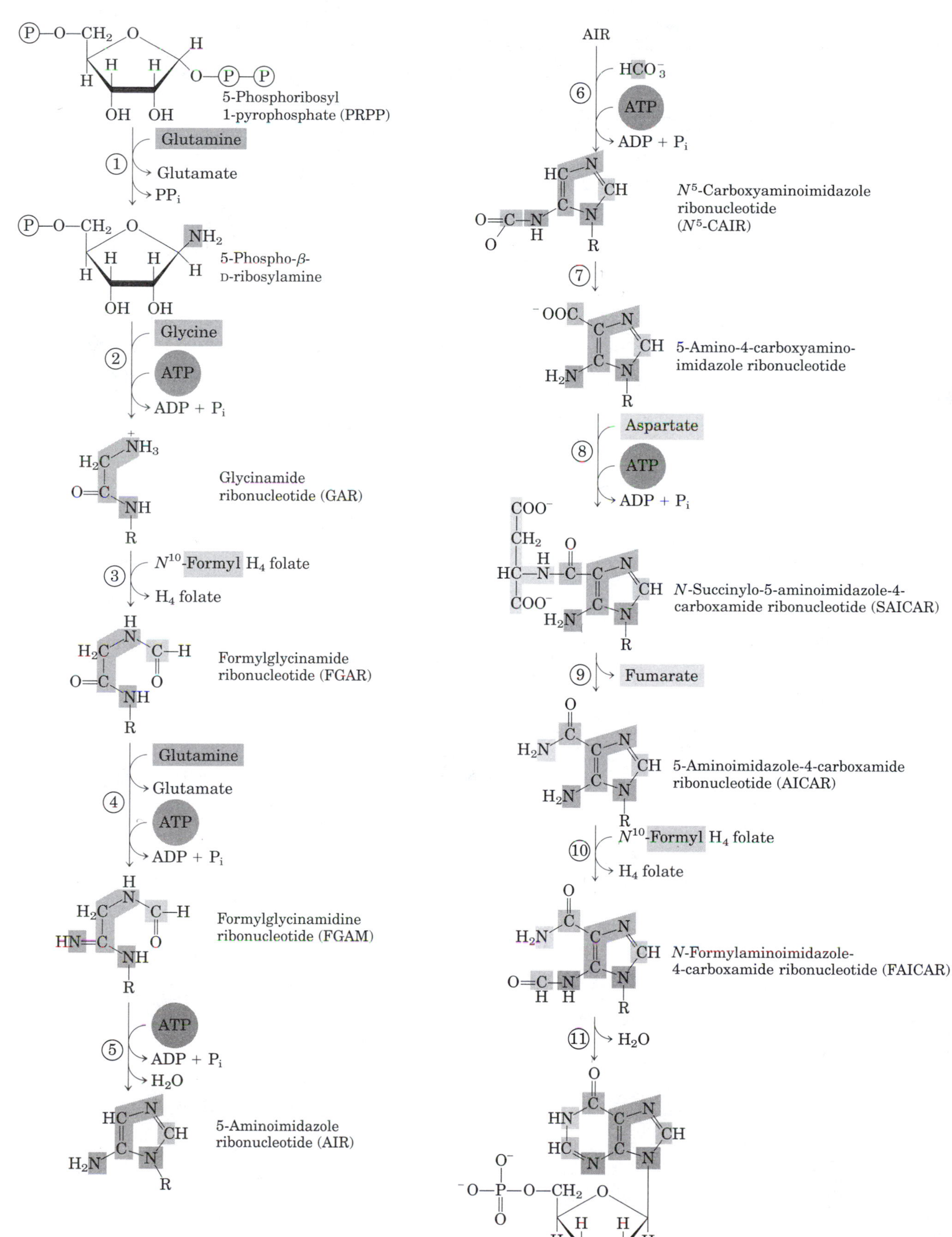

figure 22–31, page 851
De novo synthesis of purine nucleotides: construction of the purine ring of inosinate (IMP).

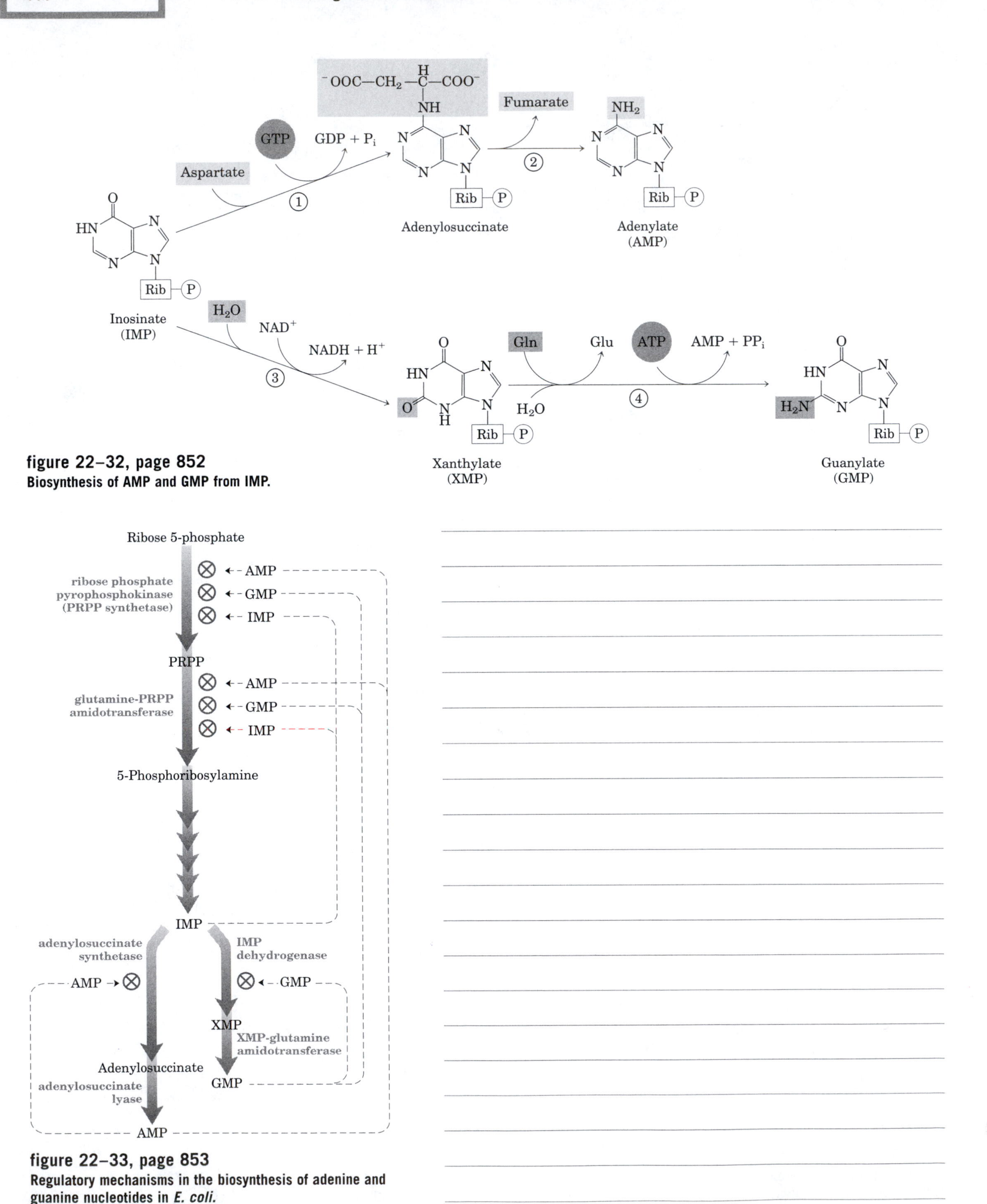

figure 22–32, page 852
Biosynthesis of AMP and GMP from IMP.

figure 22–33, page 853
Regulatory mechanisms in the biosynthesis of adenine and guanine nucleotides in *E. coli*.

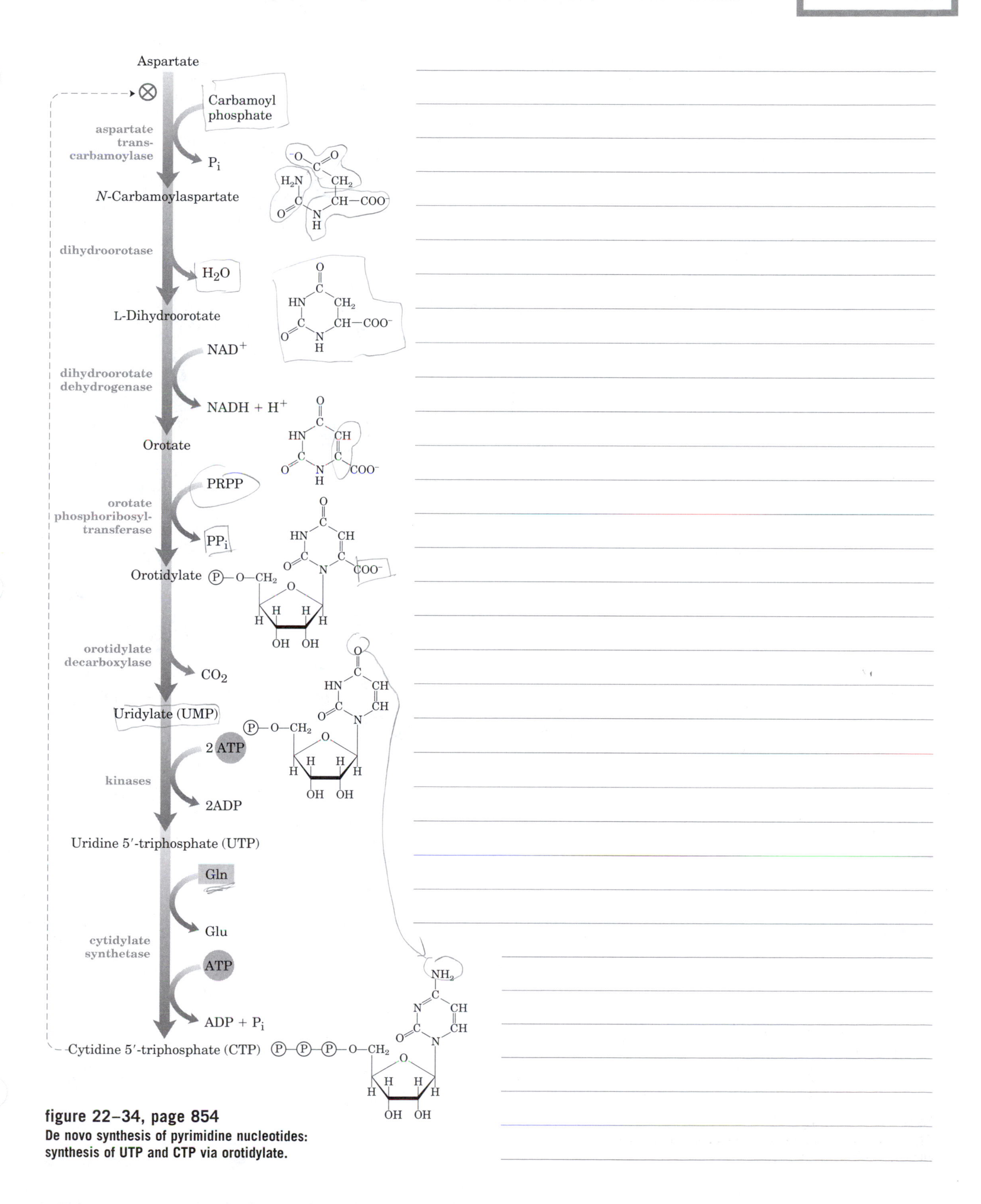

figure 22–34, page 854
De novo synthesis of pyrimidine nucleotides: synthesis of UTP and CTP via orotidylate.

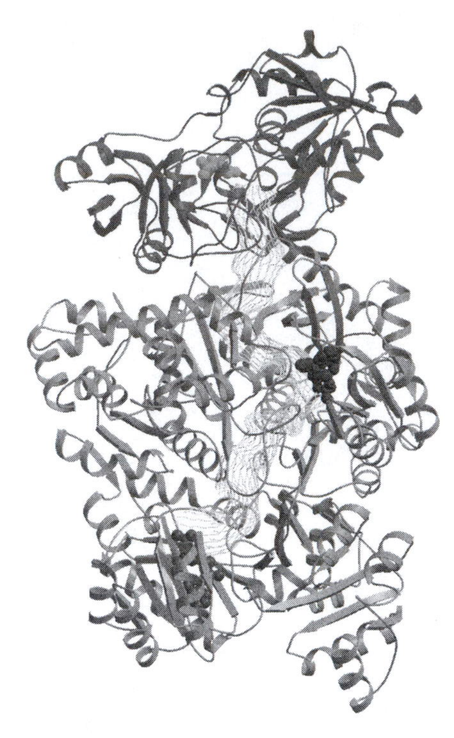

figure 22–35, page 855
Channeling of intermediates in bacterial carbamoyl phosphate synthetase.

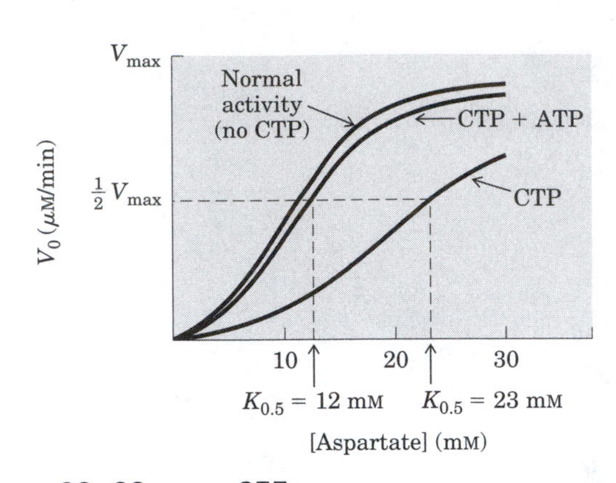

figure 22–36, page 855
Allosteric regulation of aspartate transcarbamoylase by CTP and ATP.

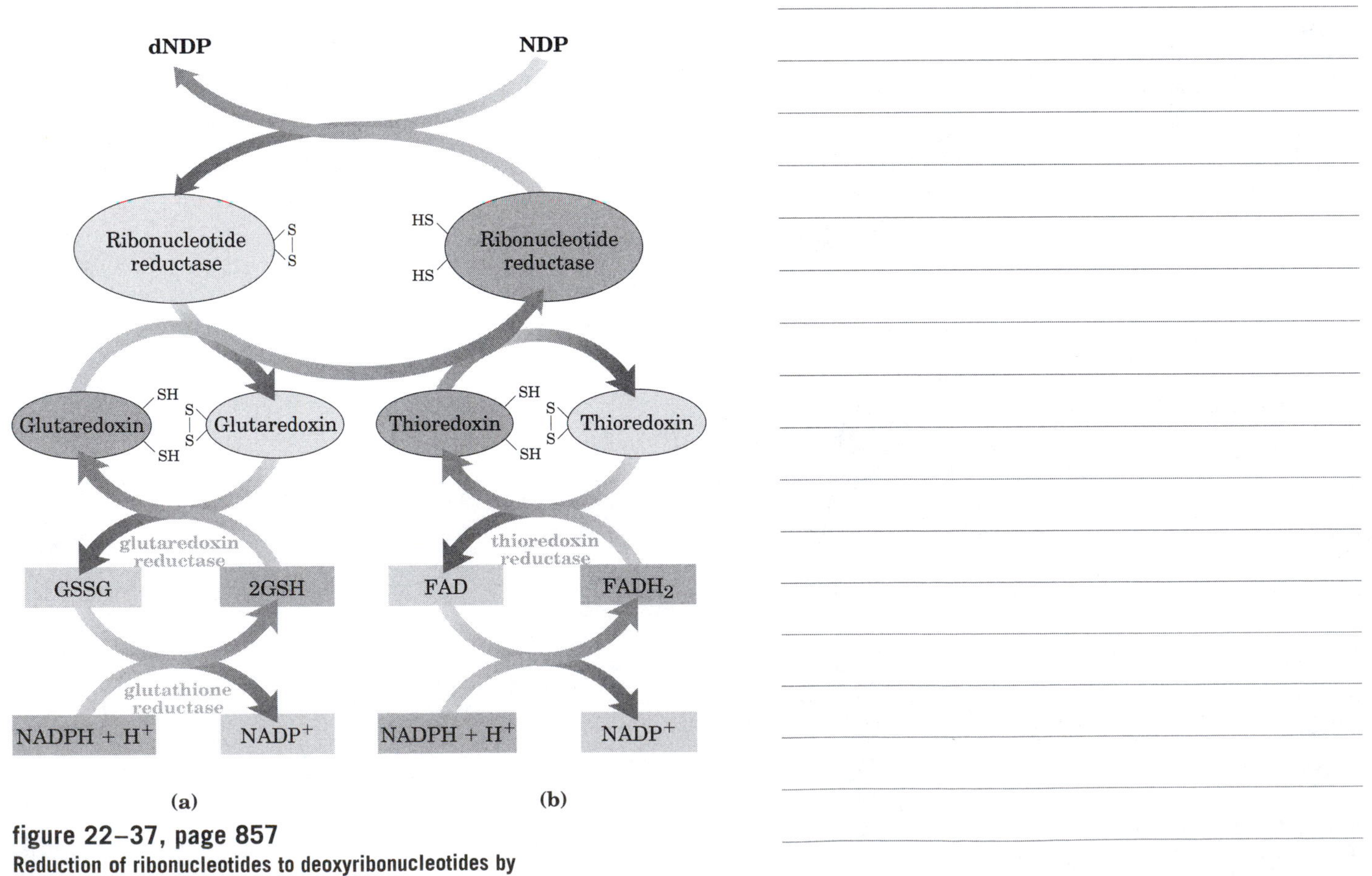

figure 22–37, page 857
Reduction of ribonucleotides to deoxyribonucleotides by ribonucleotide reductase.

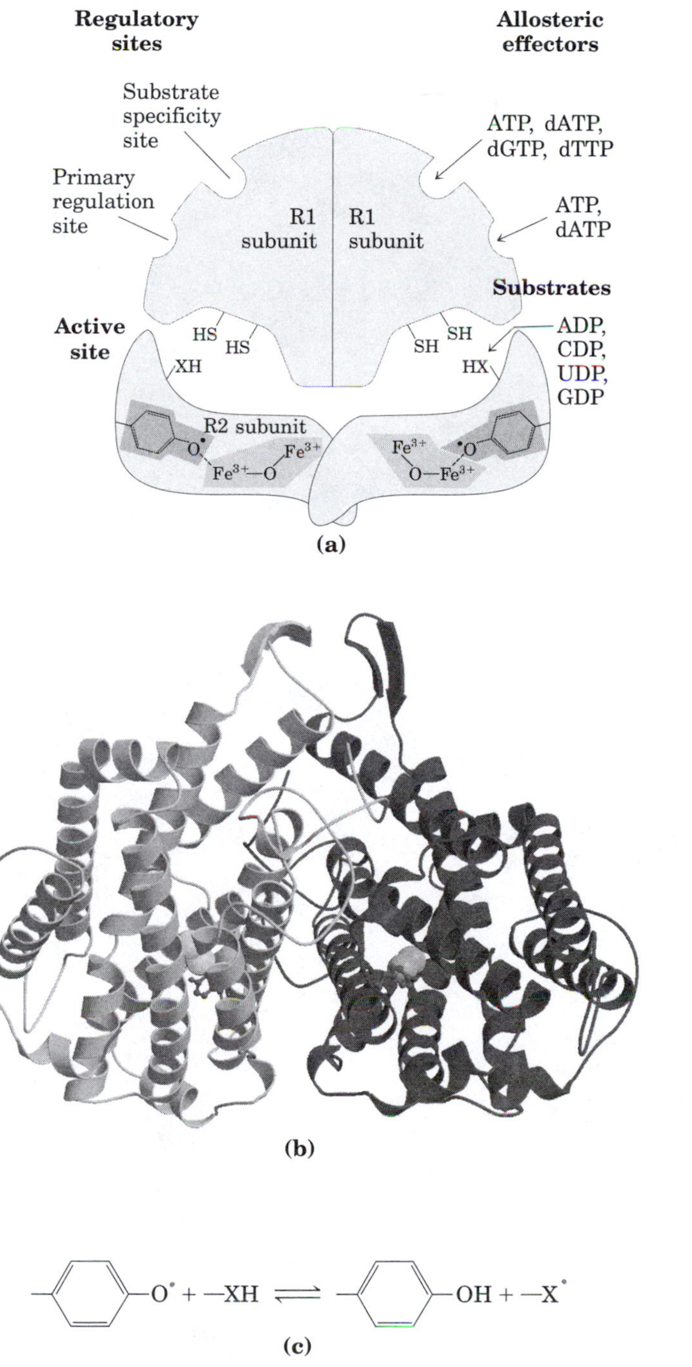

figure 22–38, page 857
Ribonucleotide reductase.

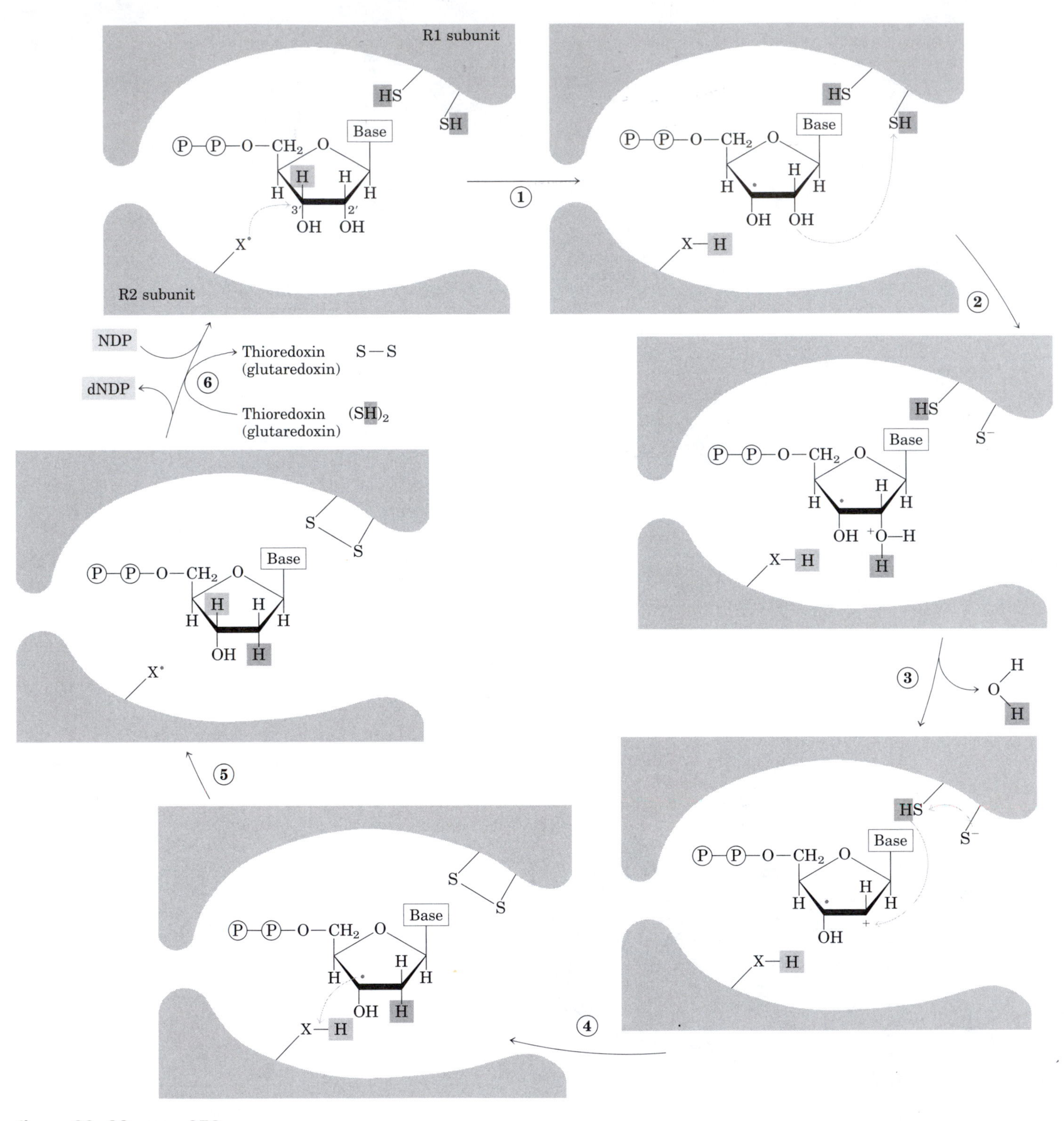

figure 22–39, page 858
Proposed mechanism for ribonucleotide reductase.

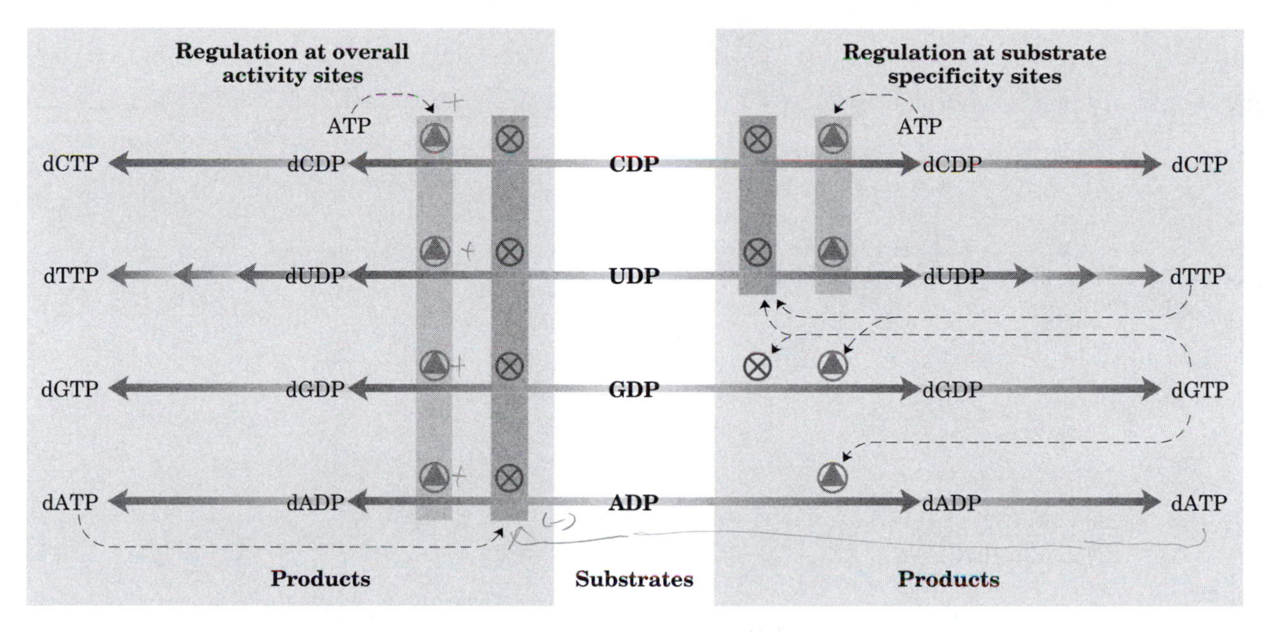

figure 22–40, page 859
Regulation of ribonucleotide reductase by deoxynucleoside triphosphates.

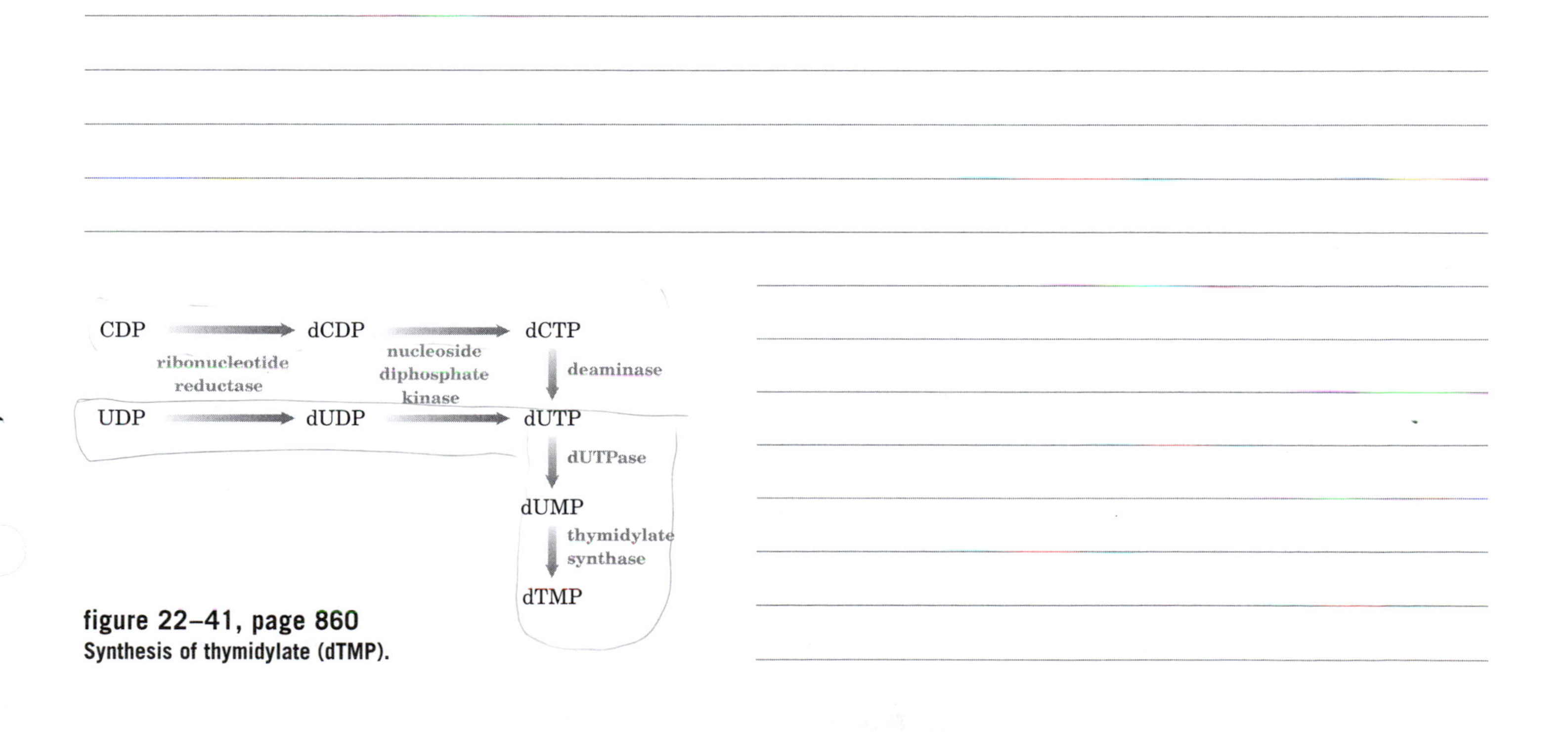

figure 22–41, page 860
Synthesis of thymidylate (dTMP).

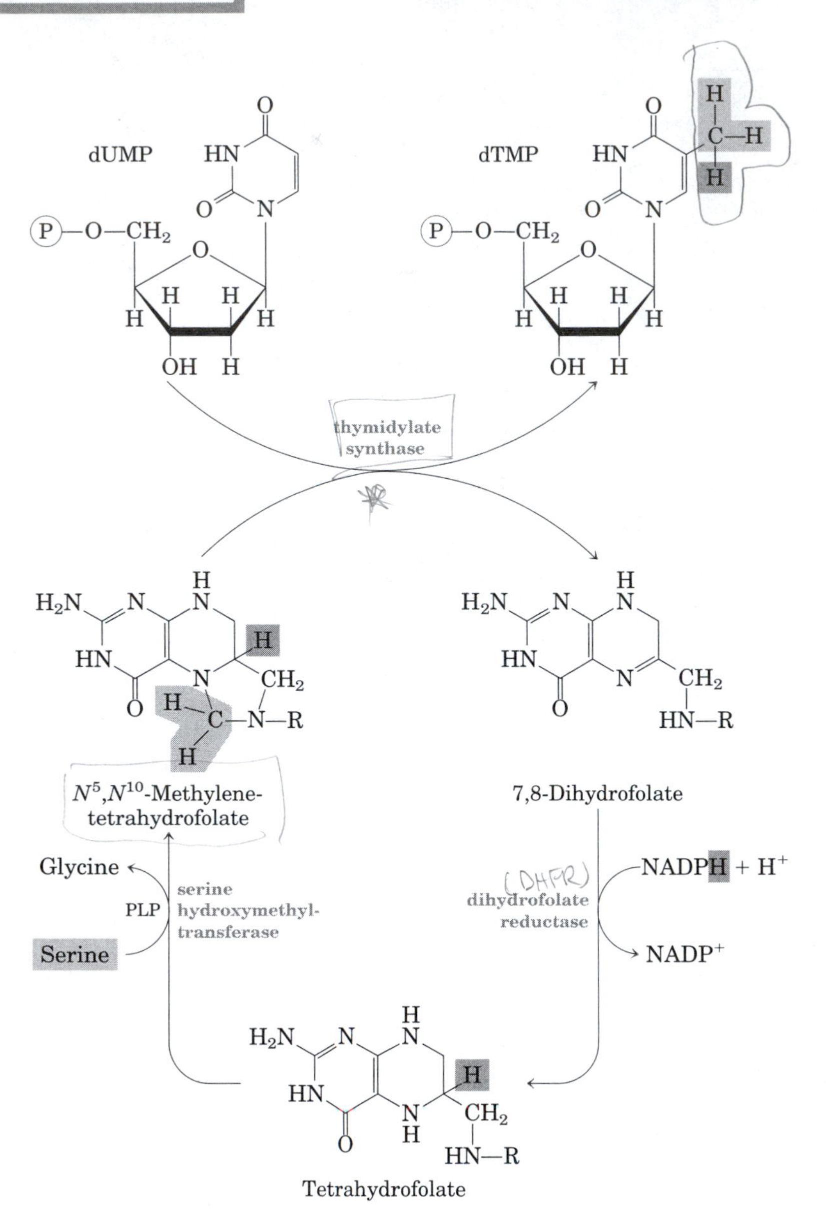

figure 22–42, page 860
Conversion of dUMP to dTMP by thymidylate synthase and dihydrofolate reductase.

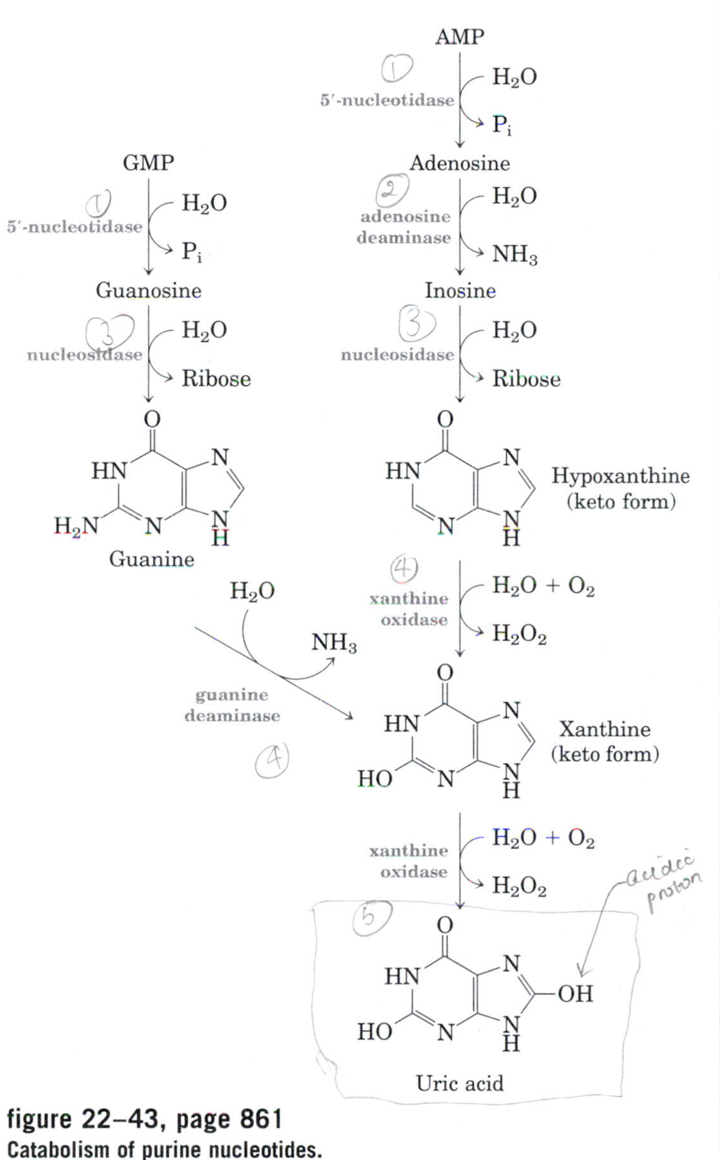

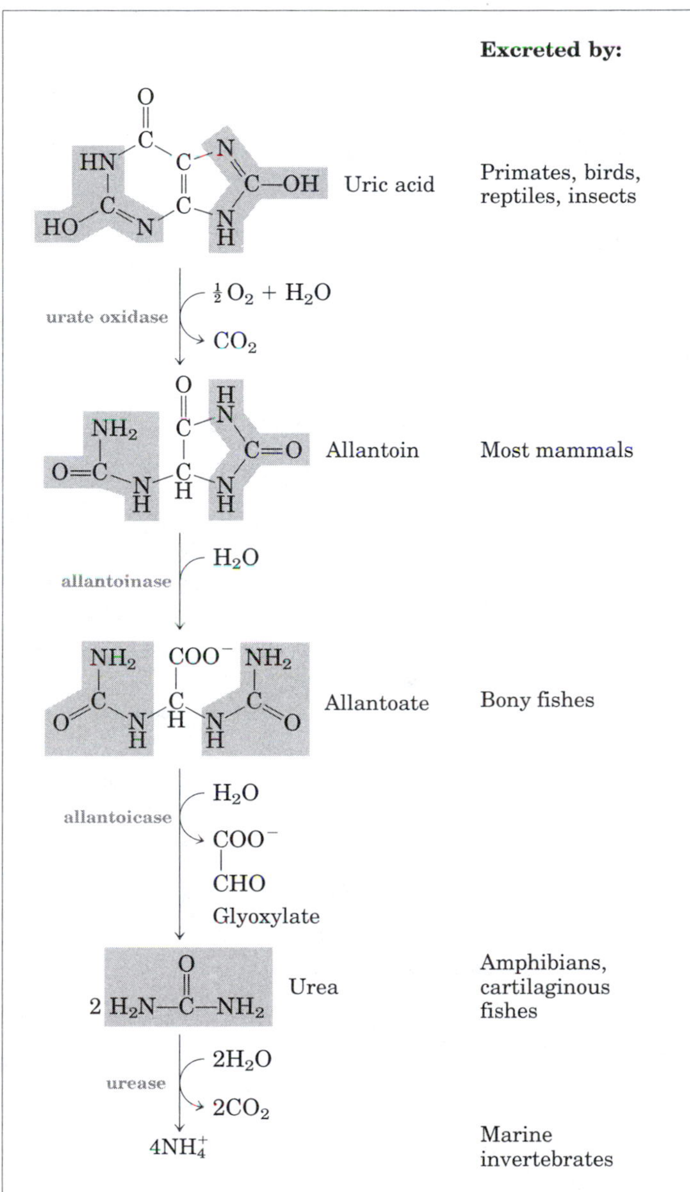

figure 22–43, page 861
Catabolism of purine nucleotides.

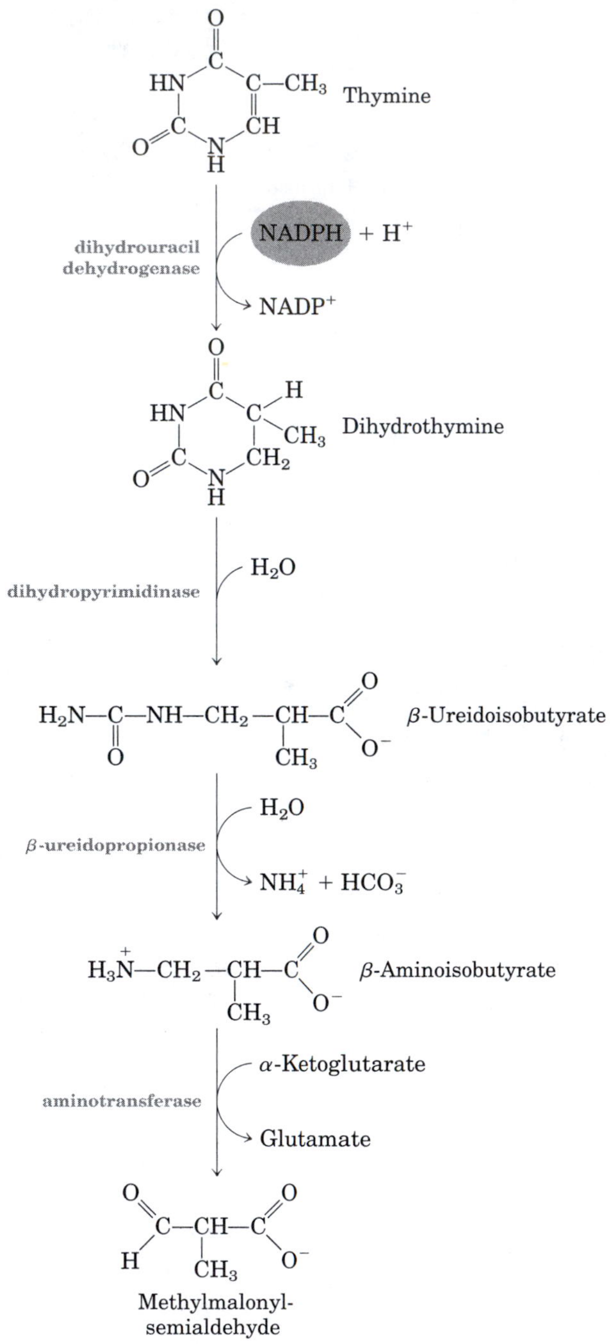

Thymine

dihydrouracil dehydrogenase

NADPH + H^+

$NADP^+$

Dihydrothymine

dihydropyrimidinase

H_2O

$H_2N{-}C(=O){-}NH{-}CH_2{-}CH(CH_3){-}COO^-$ β-Ureidoisobutyrate

β-ureidopropionase

H_2O

$NH_4^+ + HCO_3^-$

$H_3\overset{+}{N}{-}CH_2{-}CH(CH_3){-}COO^-$ β-Aminoisobutyrate

aminotransferase

α-Ketoglutarate

Glutamate

Methylmalonyl-semialdehyde

figure 22–44, page 862
Catabolism of a pyrimidine.

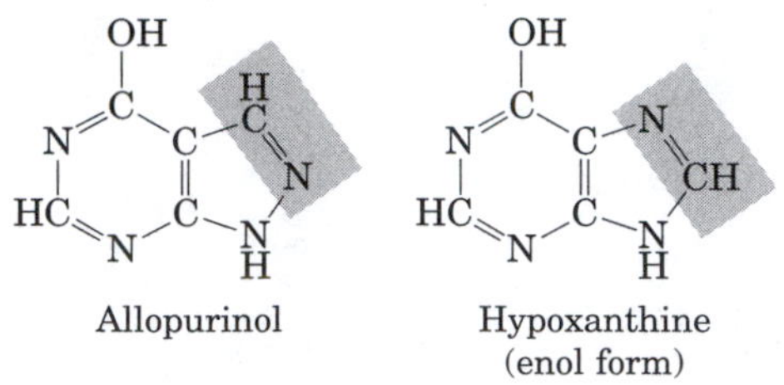

Allopurinol

Hypoxanthine (enol form)

figure 22–45, page 863
Allopurinol, an inhibitor of xanthine oxidase.

Azaserine

Acivicin

Glutamine

figure 22–46, page 863
Azaserine and acivicin, inhibitors of glutamine amidotransferases.

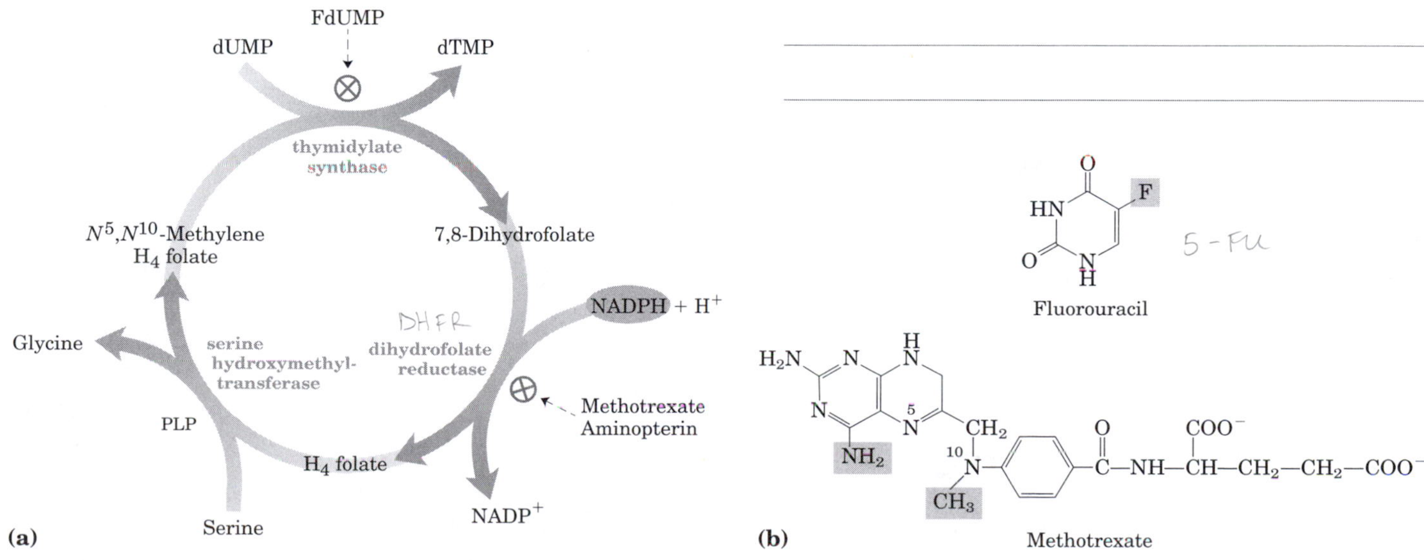

figure 22–47, page 864
Thymidylate synthesis and folate metabolism as targets of chemotherapy.

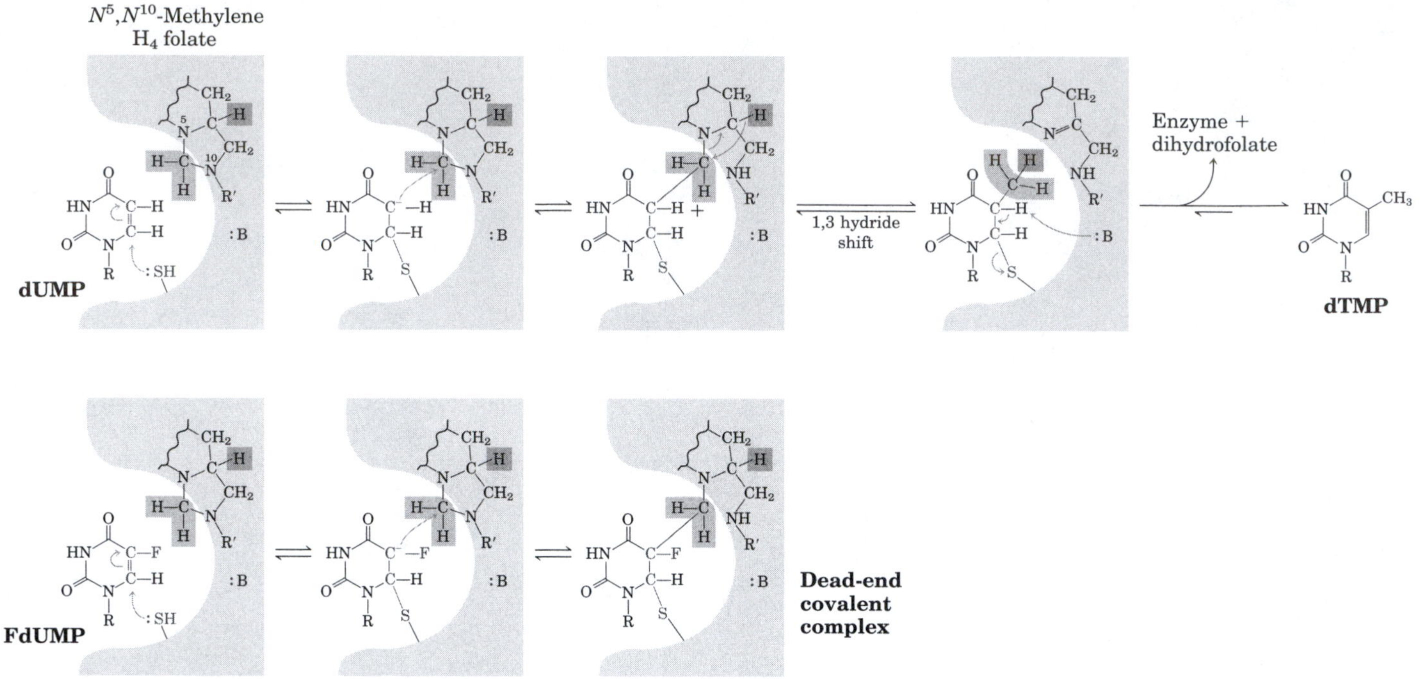

figure 22–48, page 865
Conversion of dUMP to dTMP and its inhibition by FdUMP.

PROBLEMS

11. Mode of Action of Sulfa Drugs, page 868.

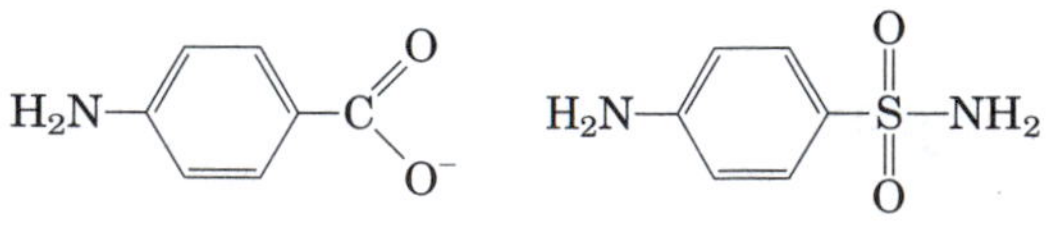

Integration and Hormonal Regulation of Mammalian Metabolism

chapter

23

CD-ROM

When you study for exams, don't forget to review The Minicourses on the UNDERSTAND! BIOCHEMISTRY CD that came with your textbook.

Minicourses that apply to this Chapter include:
Molecules of Life
An Interactive Gallery of Molecular Models
An Interactive Gallery of Protein Structures

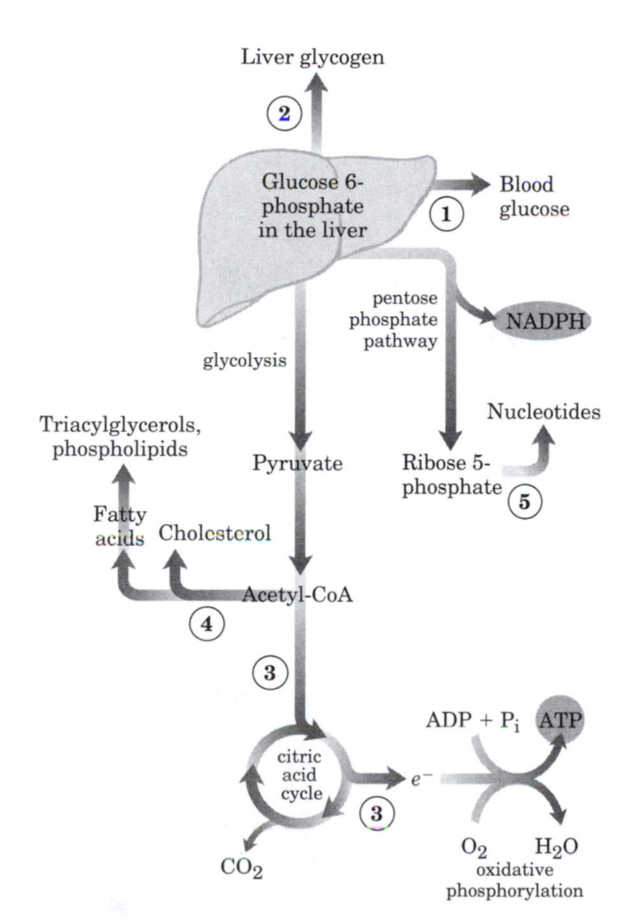

figure 23–1, page 870
Metabolic pathways for 6-phosphate in the liver.

table 23–1, page 871

Pathways of Carbohydrate, Amino Acid, and Fat Metabolism Shown in Earlier Figures

Pathway	Figure reference
Citric acid cycle: acetyl-CoA ⟶ $2CO_2$	16–7
Oxidative phosphorylation: ATP synthesis	19–16
Carbohydrate catabolism	
Glycogenolysis: glycogen ⟶ glucose 1-phosphate ⟶ blood glucose	15–11
Hexose entry into glycolysis: fructose, mannose, galactose ⟶ glucose 6-phosphate	15–11
Glycolysis: glucose ⟶ pyruvate	15–2
Pyruvate dehydrogenase reaction: pyruvate ⟶ acetyl-CoA	16–2
Lactic acid fermentation: glycogen ⟶ lactate + 2ATP	15–3
Pentose phosphate pathway: glucose 6-phosphate ⟶ pentose phosphates + NADPH	15–20
Carbohydrate anabolism	
Gluconeogenesis: citric acid cycle intermediates ⟶ glucose	20–2
Glucose-alanine cycle: glucose ⟶ pyruvate ⟶ alanine ⟶ glucose	18–8
Glycogen synthesis: glucose 6-phosphate ⟶ glucose 1-phosphate ⟶ glycogen	20–12
Amino acid and nucleotide metabolism	
Amino acid degradation: amino acids ⟶ acetyl-CoA, citric acid cycle intermediates	18–29
Amino acid synthesis	22–9
Urea cycle: NH_3 ⟶ urea	18–9
Glucose-alanine cycle: alanine ⟶ glucose	18–8
Nucleotide synthesis: amino acids ⟶ purines, pyrimidines	22–31; 22–34
Hormone and neurotransmitter synthesis	22–27
Fat catabolism	
β Oxidation of fatty acids: fatty acid ⟶ acetyl-CoA	17–8
Oxidation of ketone bodies: β-hydroxybutyrate ⟶ acetyl-CoA ⟶ CO_2	17–17
Fat anabolism	
Fatty acid synthesis: acetyl-CoA ⟶ fatty acids	21–5
Triacylglycerol synthesis: acetyl-CoA ⟶ fatty acids ⟶ triacylglycerol	21–18; 21–19
Ketone body formation: acetyl-CoA ⟶ acetoacetate, β-hydroxybutyrate	17–16
Cholesterol and cholesteryl ester synthesis: acetyl-CoA ⟶ cholesterol ⟶ cholesteryl esters	21–32 through 21–36
Phospholipid synthesis: fatty acids ⟶ phospholipids	21–27

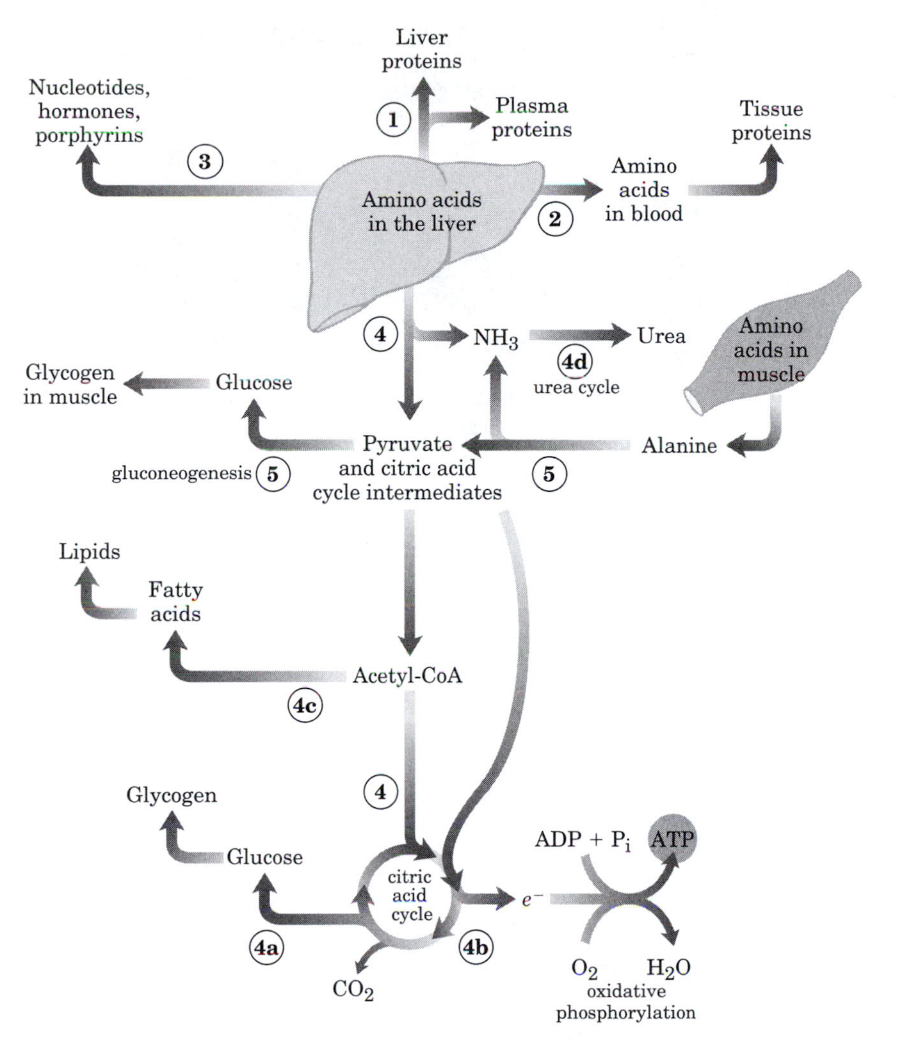

figure 23–2, page 872
Metabolism of amino acids in the liver.

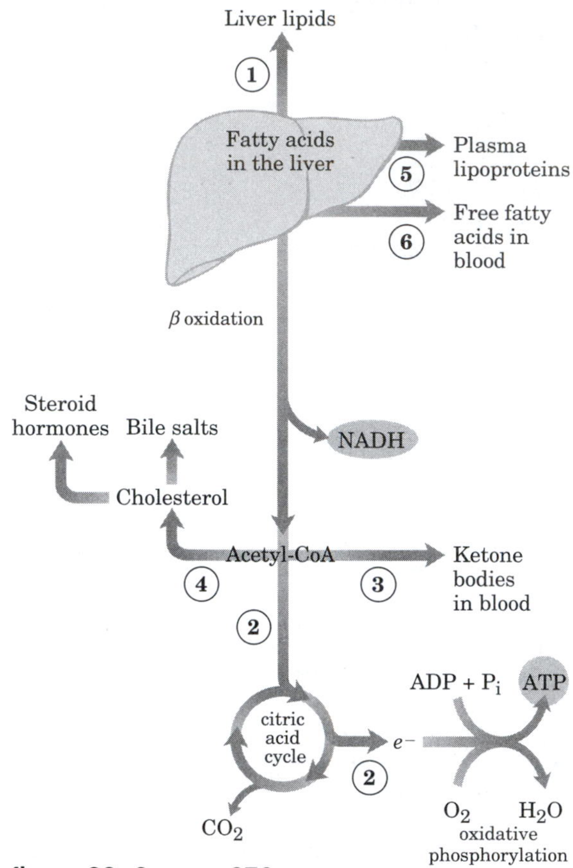

figure 23–3, page 873
Metabolism of fatty acids in the liver.

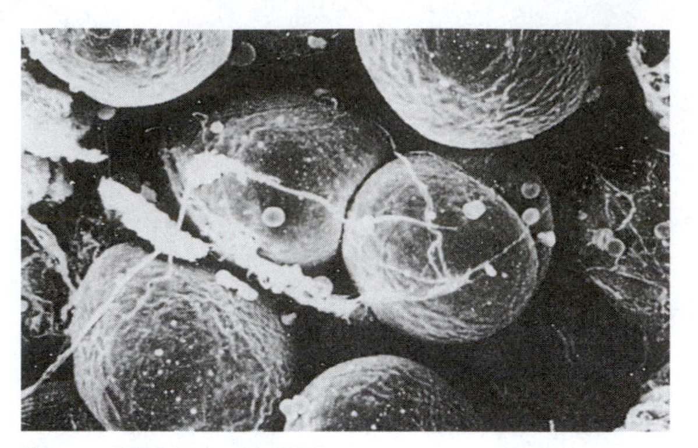

figure 23–4, page 873
Scanning electron micrograph of human adipocytes.

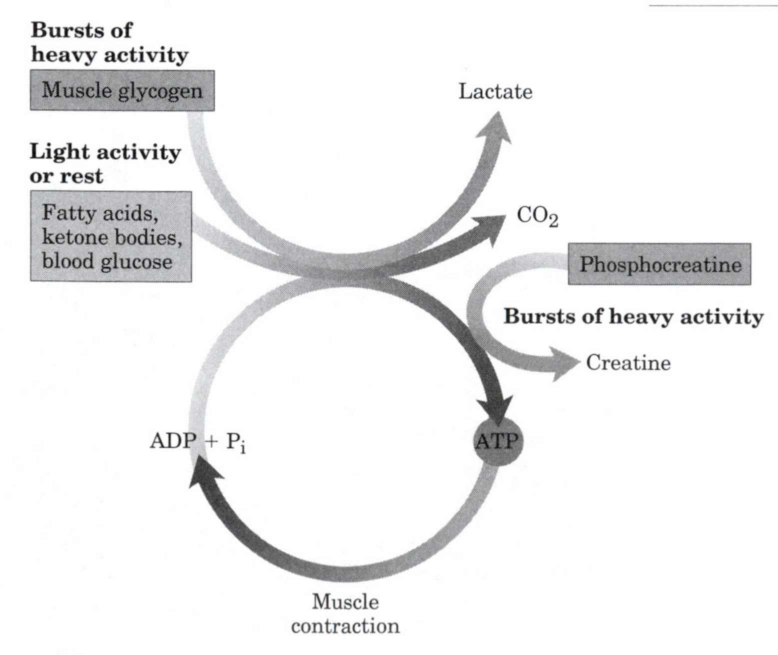

figure 23–5, page 874
Energy sources for muscle contraction.

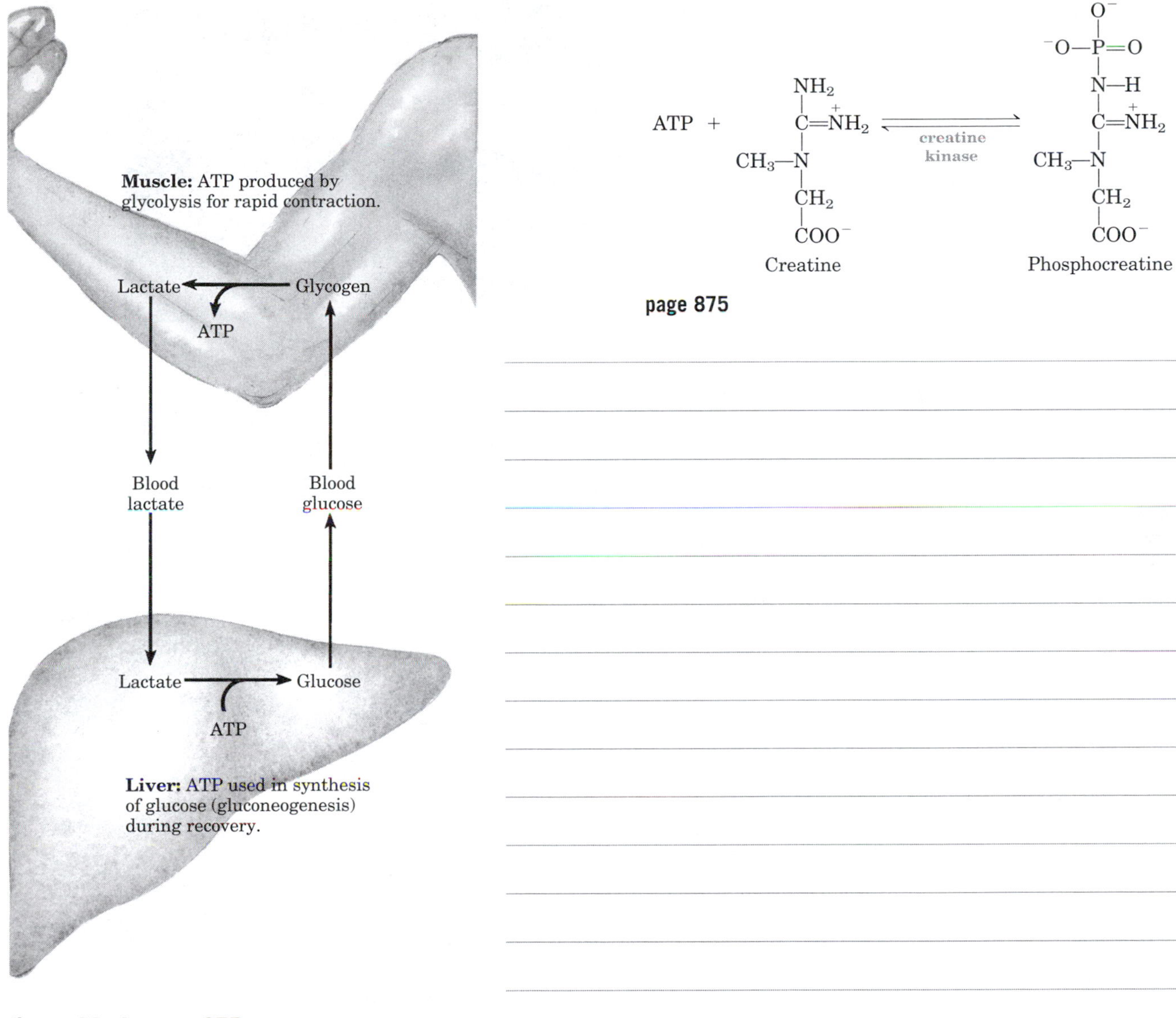

page 875

figure 23–6, page 875
Metabolic cooperation between skeletal muscle and the liver.

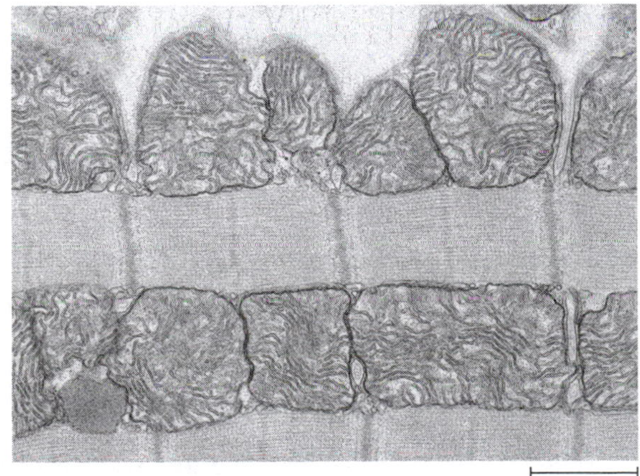

figure 23–7, page 876
Electron micrograph of heart muscle.

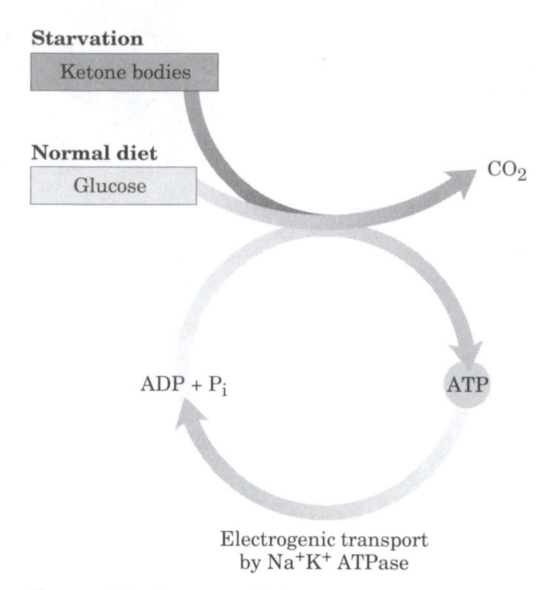

figure 23–8, page 876
Energy sources in the brain vary with nutritional state.

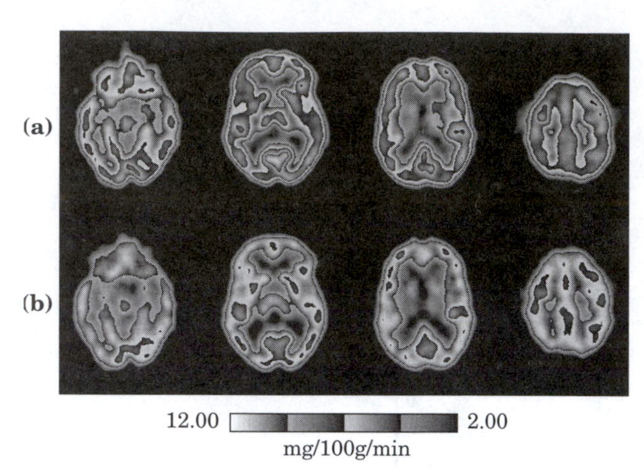

figure 23–9, page 876
Glucose metabolism in the brain.

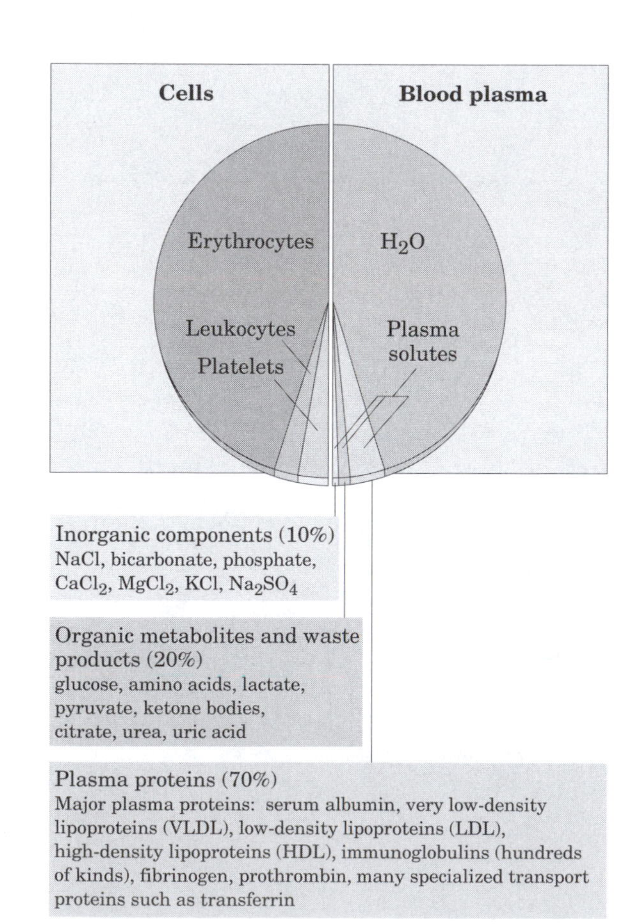

figure 23–10, page 877
The composition of blood.

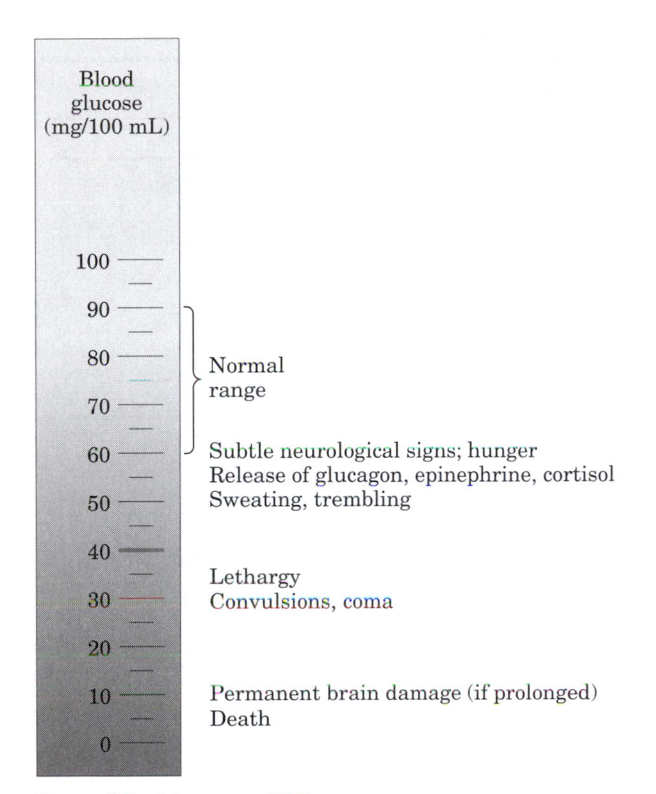

figure 23–11, page 878
Physiological effects of low blood glucose in humans.

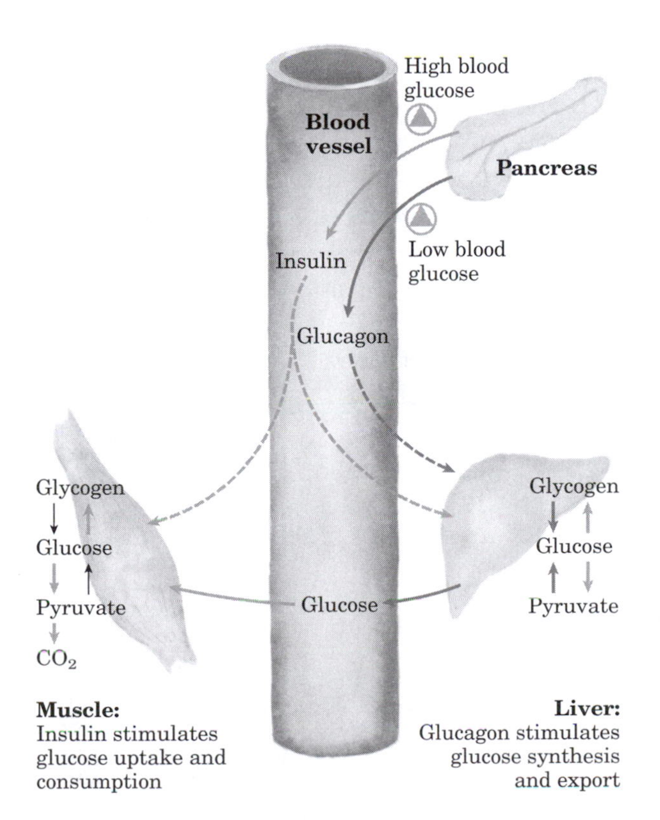

figure 23–12, page 879
Regulation of blood glucose by insulin and glucagon.

table 23–2, page 878

Physiological and Metabolic Effects of Epinephrine: Preparation for Action

Effect	Result
Physiological	
↑ Heart rate ↑ Blood pressure ↑ Dilation of respiratory passages	Increased delivery of O_2 to tissues (muscle)
Metabolic	
↑ Glycogen breakdown (muscle, liver) ↓ Glycogen synthesis (muscle, liver) ↑ Gluconeogenesis (liver)	Increased production of glucose for fuel
↑ Glycolysis (muscle)	Increased ATP production in muscle
↑ Fatty acid mobilization (adipose tissue)	Increased availability of fatty acids as fuel
↑ Glucagon secretion ↓ Insulin secretion	Reinforce metabolic effects of epinephrine

table 23–3, page 879

Effects of Glucagon on Blood Glucose: Production and Release of Glucose By the Liver

Metabolic effect	Effect on glucose metabolism	Target enzyme
↑ Glycogen breakdown (liver)	Glycogen ⟶ glucose	↑ Glycogen phosphorylase
↓ Glycogen synthesis (liver)	Less glucose stored as glycogen	↓ Glycogen synthase
↓ Glycolysis (liver)	Less glucose used as fuel in liver	↓ Phosphofructokinase-1
↑ Gluconeogenesis (liver)	Amino acids, Glycerol, Oxaloacetate ⟶ glucose	↑ Fructose 1,6-bisphosphatase ↓ Pyruvate kinase
↑ Fatty acid mobilization (adipose tissue)	Less glucose used as fuel by liver, muscle	↑ Triacylglycerol lipase

table 23–4, page 880

Available Metabolic Fuels in a Normal 70 kg Man and in an Obese 140 kg Man at the Beginning of a Fast

Type of fuel	Weight (kg)	Caloric equivalent [thousands of kcal (kJ)]	Estimated survival (months)*
Normal 70 kg man:			
Triacylglycerols (adipose tissue)	15	141 (589)	
Proteins (mainly muscle)	6	24 (100)	
Glycogen (muscle, liver)	0.225	0.90 (3.8)	
Circulating fuels (glucose, fatty acids, triacylglycerols, etc.)	0.023	0.10 (0.42)	
Total		166 (694)	3
Obese 140 kg man:			
Triacylglycerols (adipose tissue)	80	752 (3,140)	
Proteins (mainly muscle)	8	32 (134)	
Glycogen (muscle, liver)	0.23	0.92 (3.8)	
Circulating fuels	0.025	0.11 (0.46)	
Total		785 (3,280)	14

*Survival time is calculated on the assumption of a basal energy expenditure of 1,800 kcal/day.

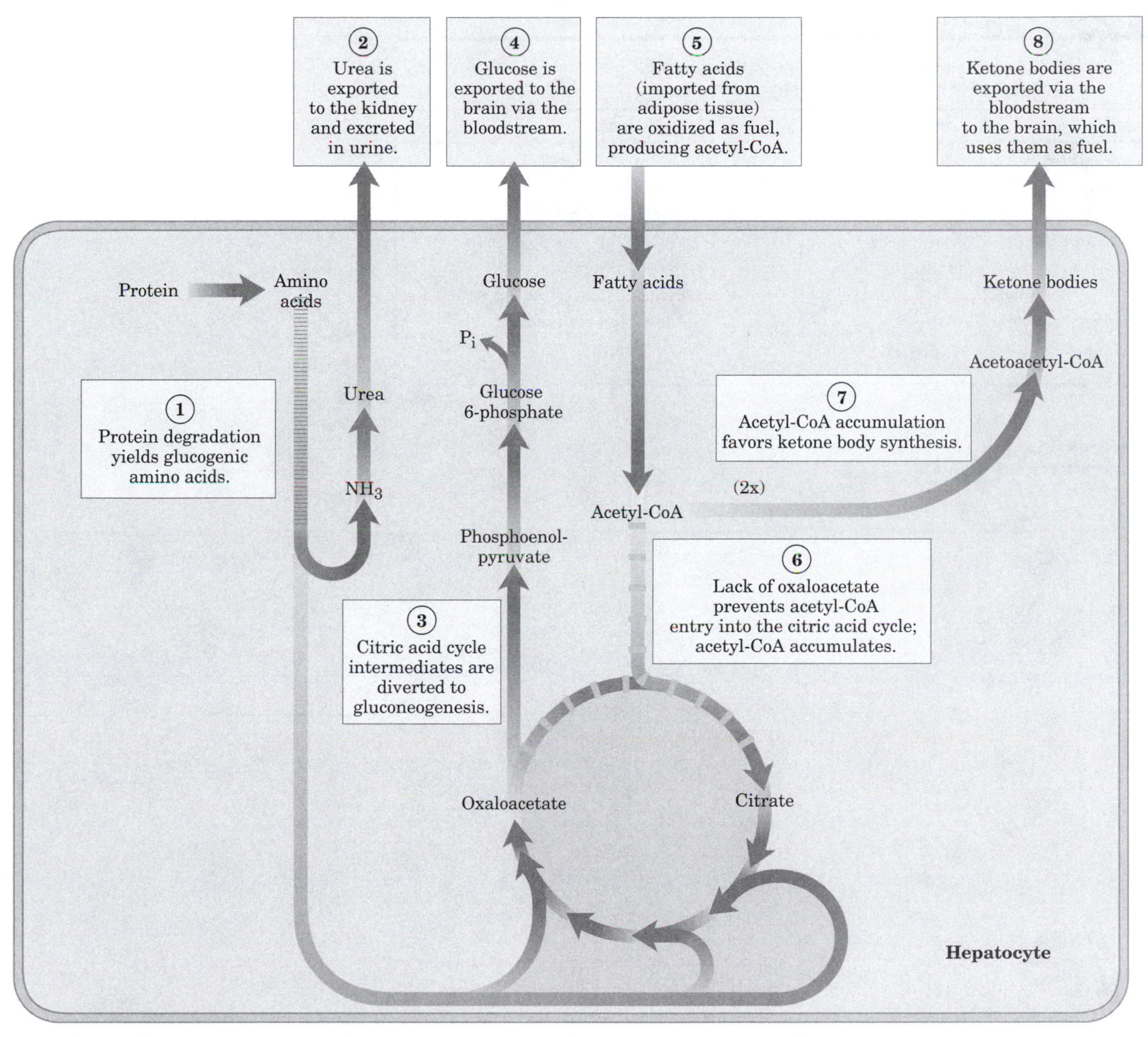

figure 23–13, page 881

Fuel metabolism in the liver during prolonged starvation.

table 23–5, page 882

Effect of Insulin on Blood Glucose: Uptake of Glucose by Cells and Storage as Triacylglycerols and Glycogen

Metabolic effect	Target enzyme
↑ Glucose uptake (muscle)	↑ Glucose transporter
↑ Glucose uptake (liver)	↑ Glucokinase
↑ Glycogen synthesis (liver, muscle)	↑ Glycogen synthase
↓ Glycogen breakdown (liver, muscle)	↓ Glycogen phosphorylase
↑ Glycolysis, acetyl-CoA production (liver, muscle)	↑ Phosphofructokinase-1
	↑ Pyruvate dehydrogenase complex
↑ Fatty acid synthesis (liver)	↑ Acetyl-CoA carboxylase
↑ Triacylglycerol synthesis (adipose tissue)	↑ Lipoprotein lipase

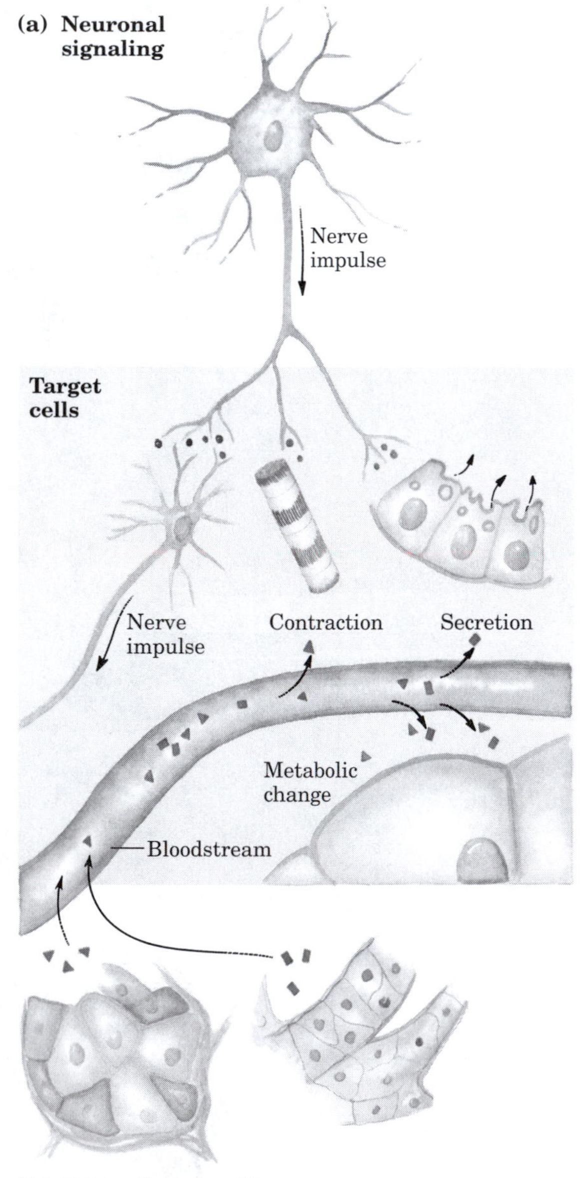

figure 23–14, page 884
Signaling by the neuroendocrine system.

Pyroglutamate Histidine Prolylamide

pyroGlu–His–Pro–NH_2

figure 23–15, page 886
The structure of thyrotropin-releasing hormone (TRH).

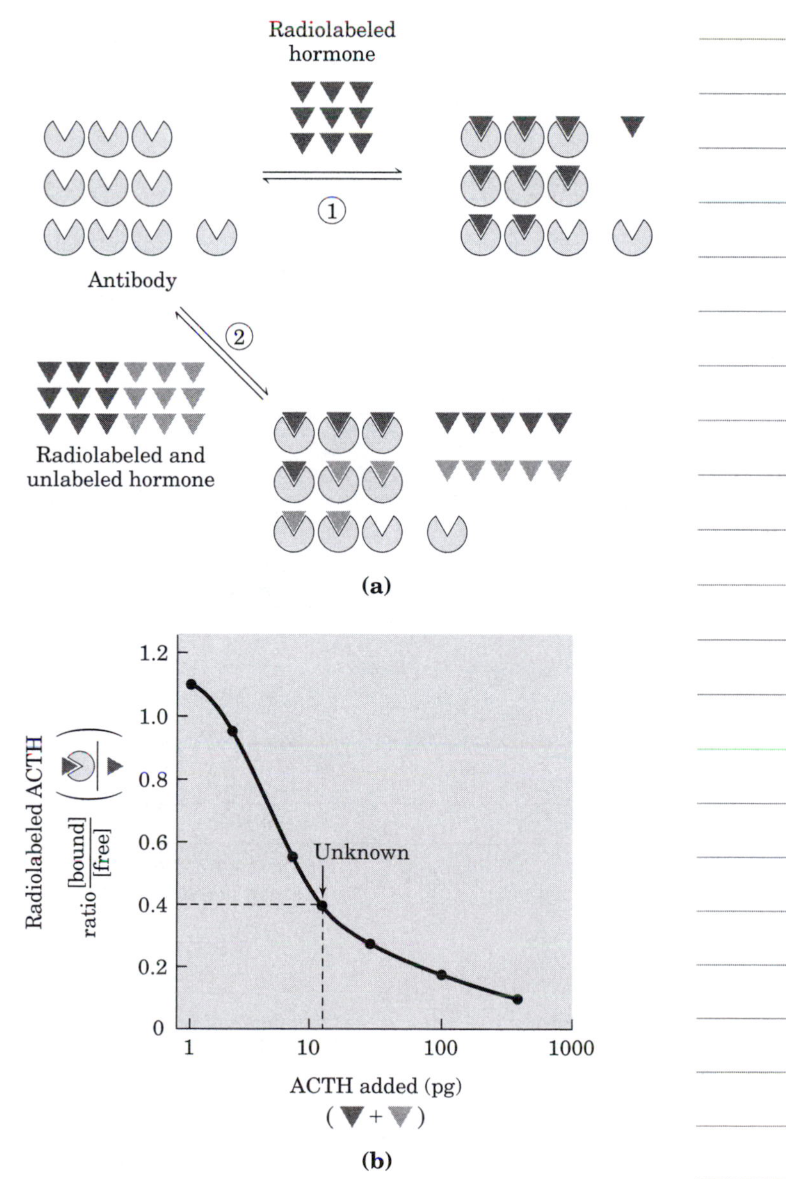

figure 23–16, page 887
The principle of the radioimmunoassay (RIA) and a real example.

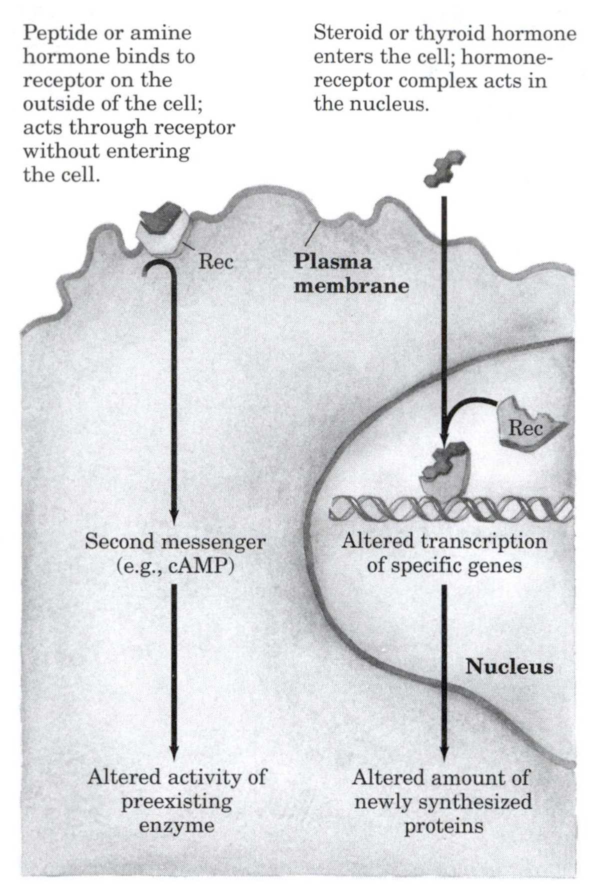

figure 23–17, page 888
Two general mechanisms of hormone action.

table 23–6, page 889

Classes of Hormones

Type	Example	Parent /origin	Synthetic path	Mode of action
Peptide	Leu-enkephalin	Tyr–Gly–Gly–Phe–Leu	Proteolytic processing of proenzyme	Plasma membrane receptors; second messengers
Catecholamine	Epinephrine	Tyrosine		
Eicosanoid	PGE_1	20:4 Fatty acid		
Steroid	Testosterone	Cholesterol		
Retinoid	Retinoic acid	Vitamin A		Nuclear receptors; transcriptional regulation
Thyroid	Triiodothyronine (T_3)	Tyr in thyroglobulin		
Vitamin D	1,25-dihydroxycholecalciferol	Cholesterol or vitamin D		
Nitric oxide	Nitric oxide	$NO^{\bullet}$	Arginine + O_2	Cytosolic receptor (guanylate cyclase) and second messenger (cGMP)

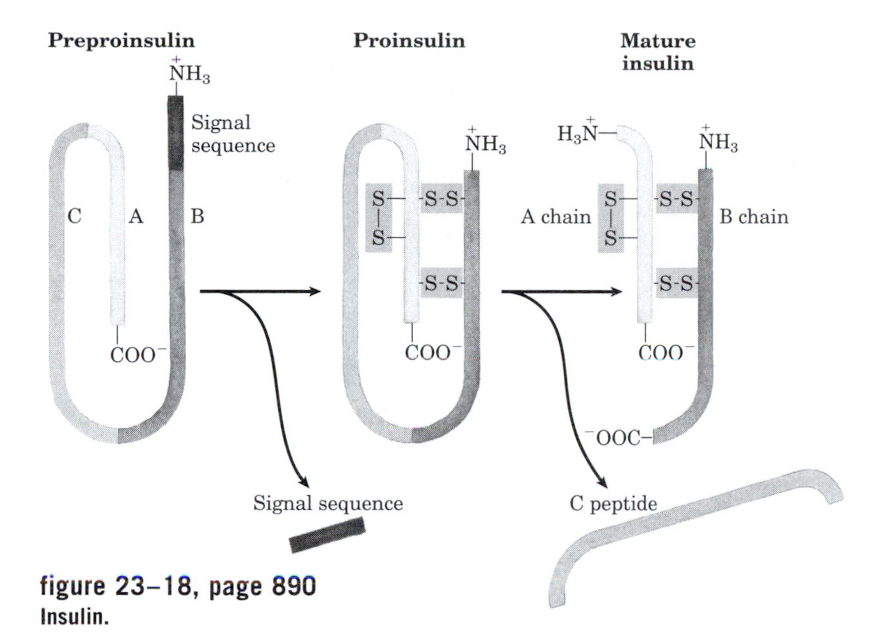

figure 23–18, page 890
Insulin.

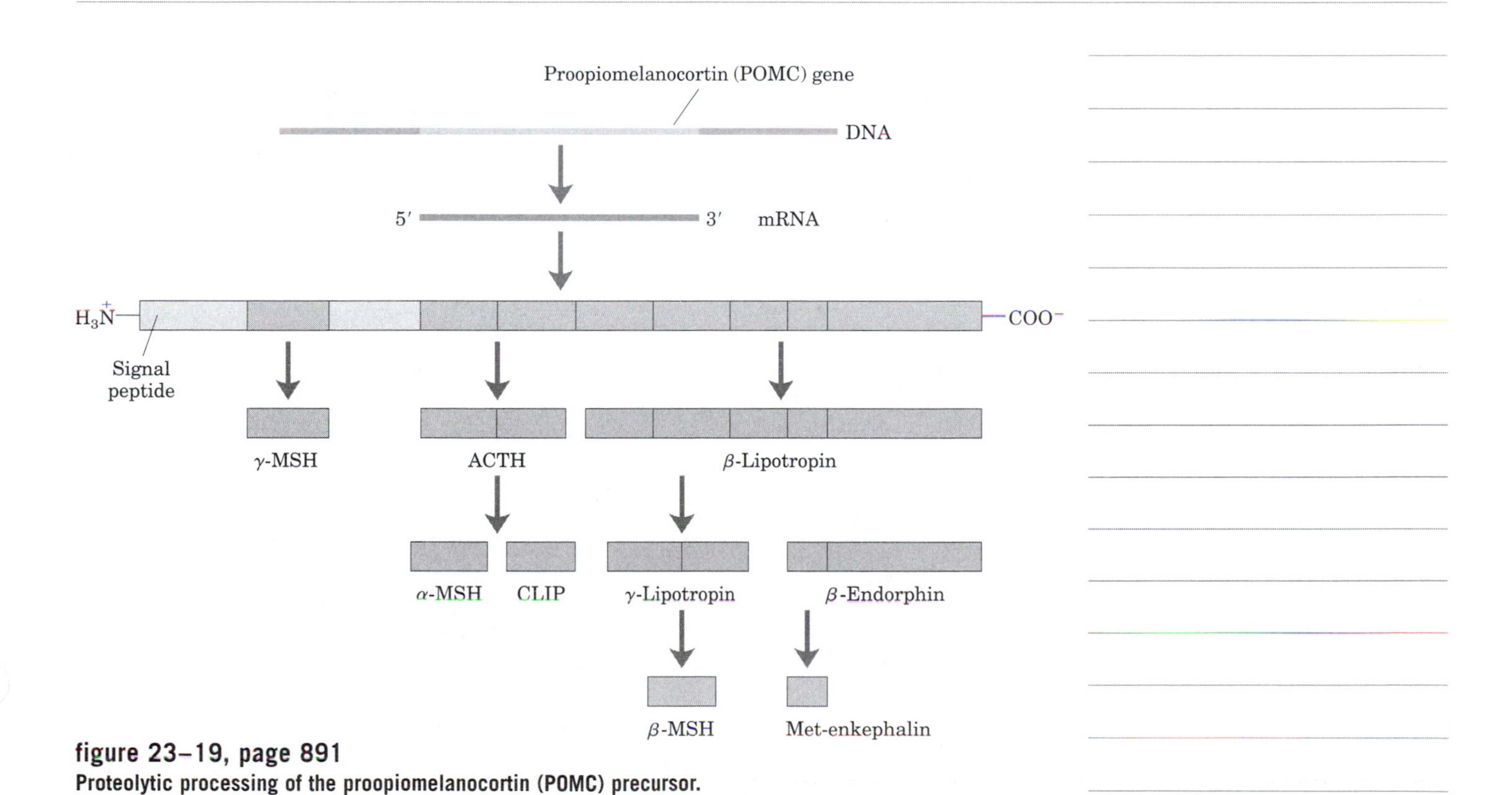

figure 23–19, page 891
Proteolytic processing of the proopiomelanocortin (POMC) precursor.

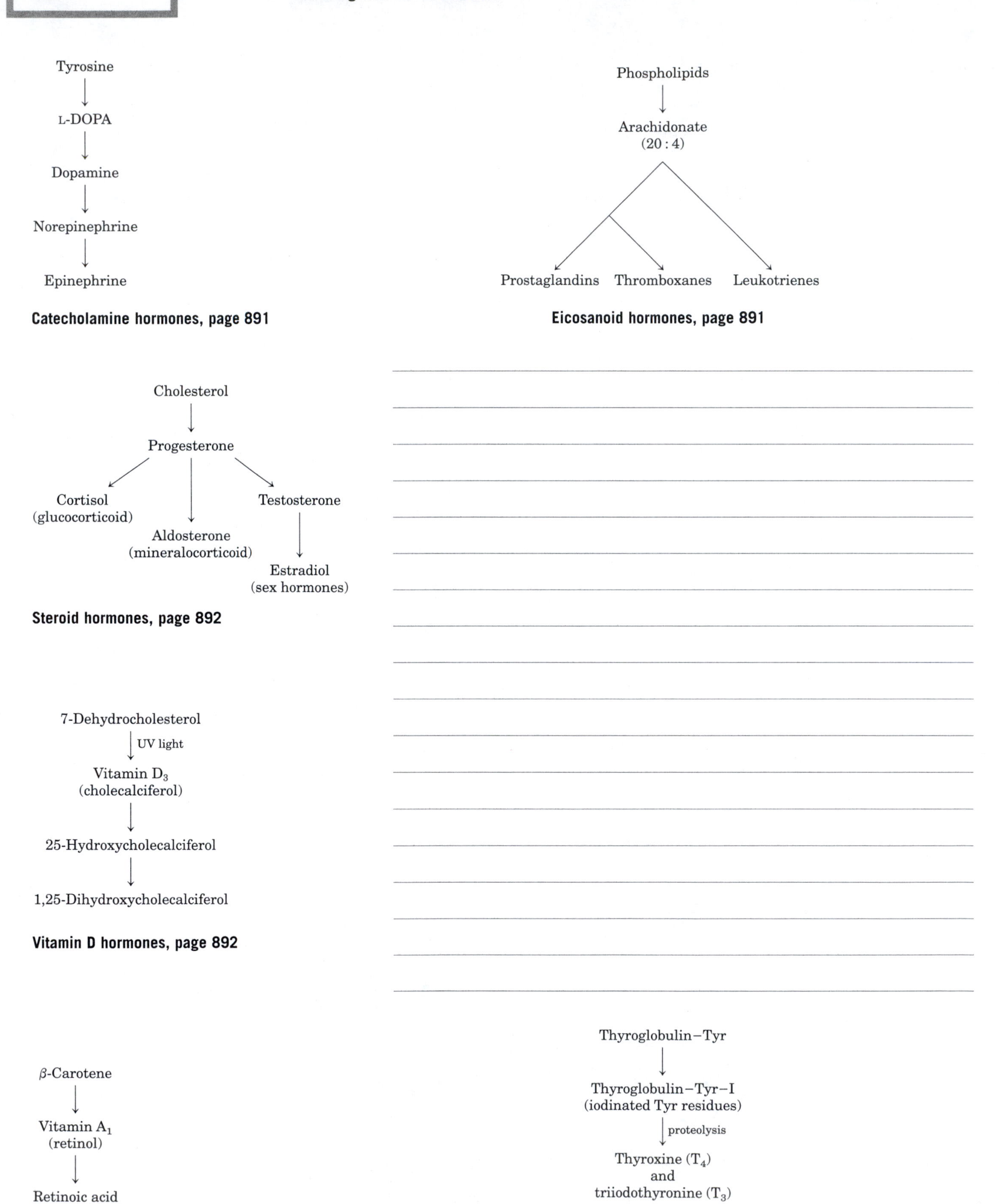

Catecholamine hormones, page 891

Eicosanoid hormones, page 891

Steroid hormones, page 892

Vitamin D hormones, page 892

Retinoid hormones, page 892

Thyroid hormones, page 892

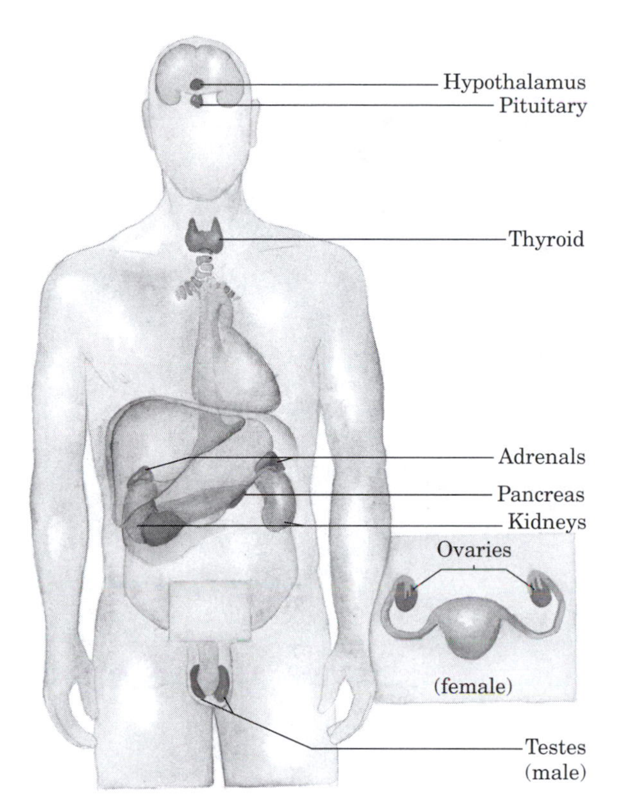

figure 23–20, page 893
The major endocrine glands.

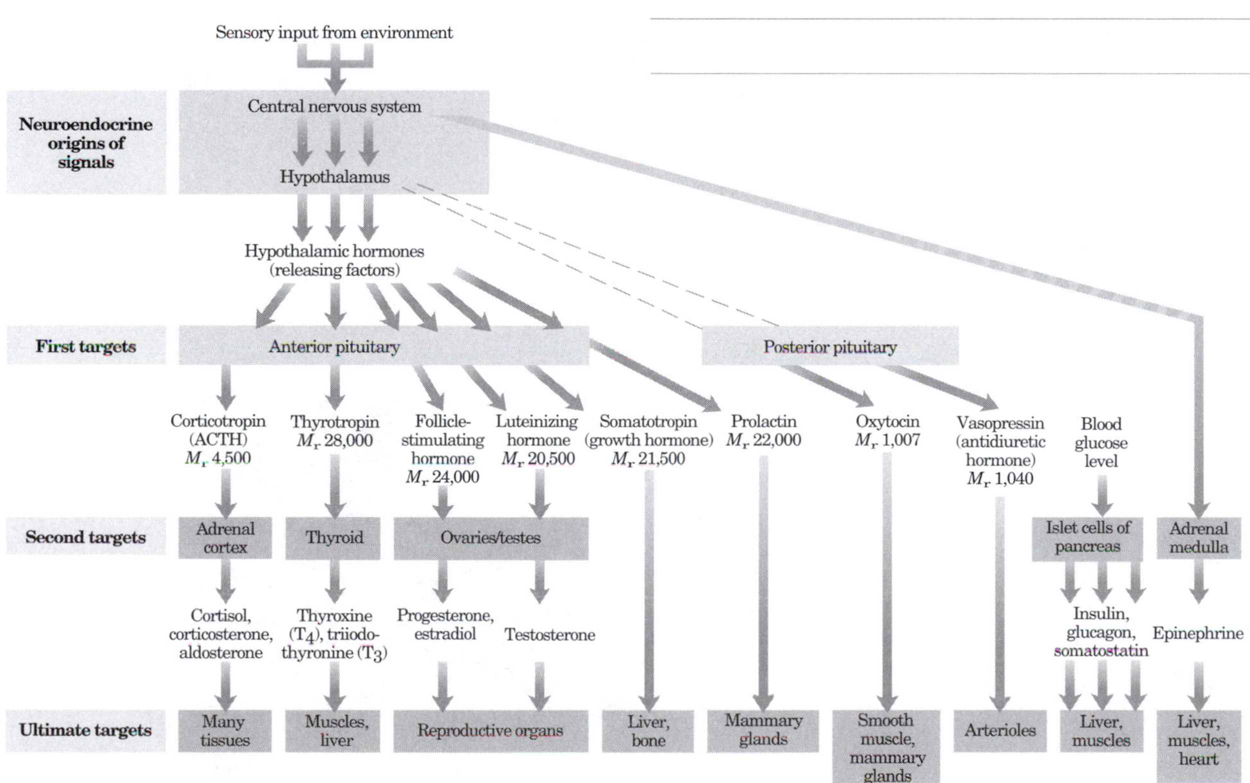

figure 23–21, page 893
The major endocrine systems and their target tissues.

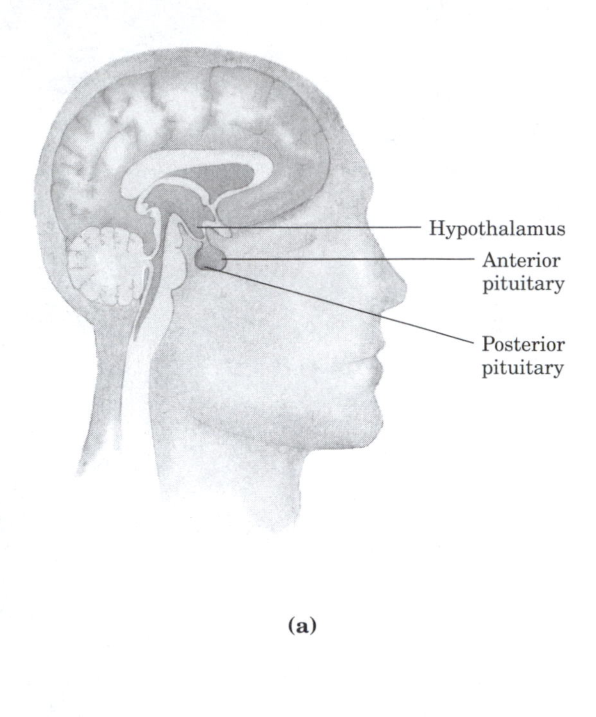

figure 23–22, page 894
Neuroendocrine origins of hormone signals.

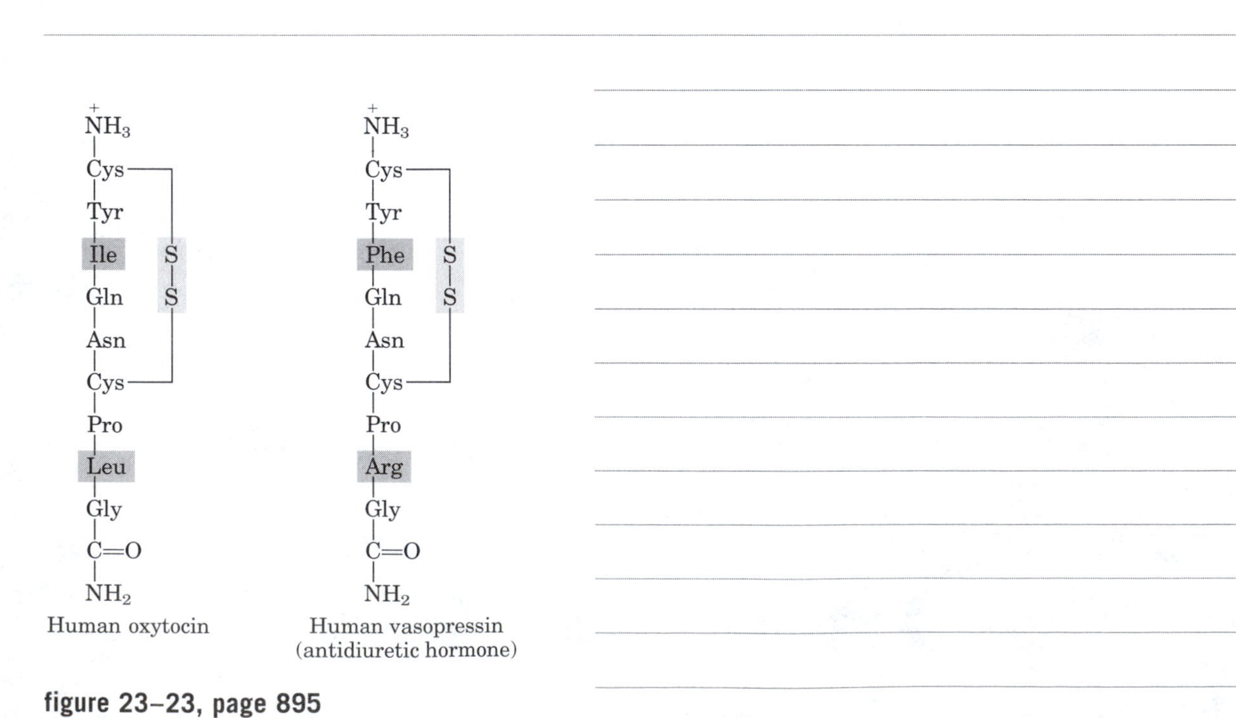

figure 23–23, page 895
Two hormones of the posterior pituitary gland.

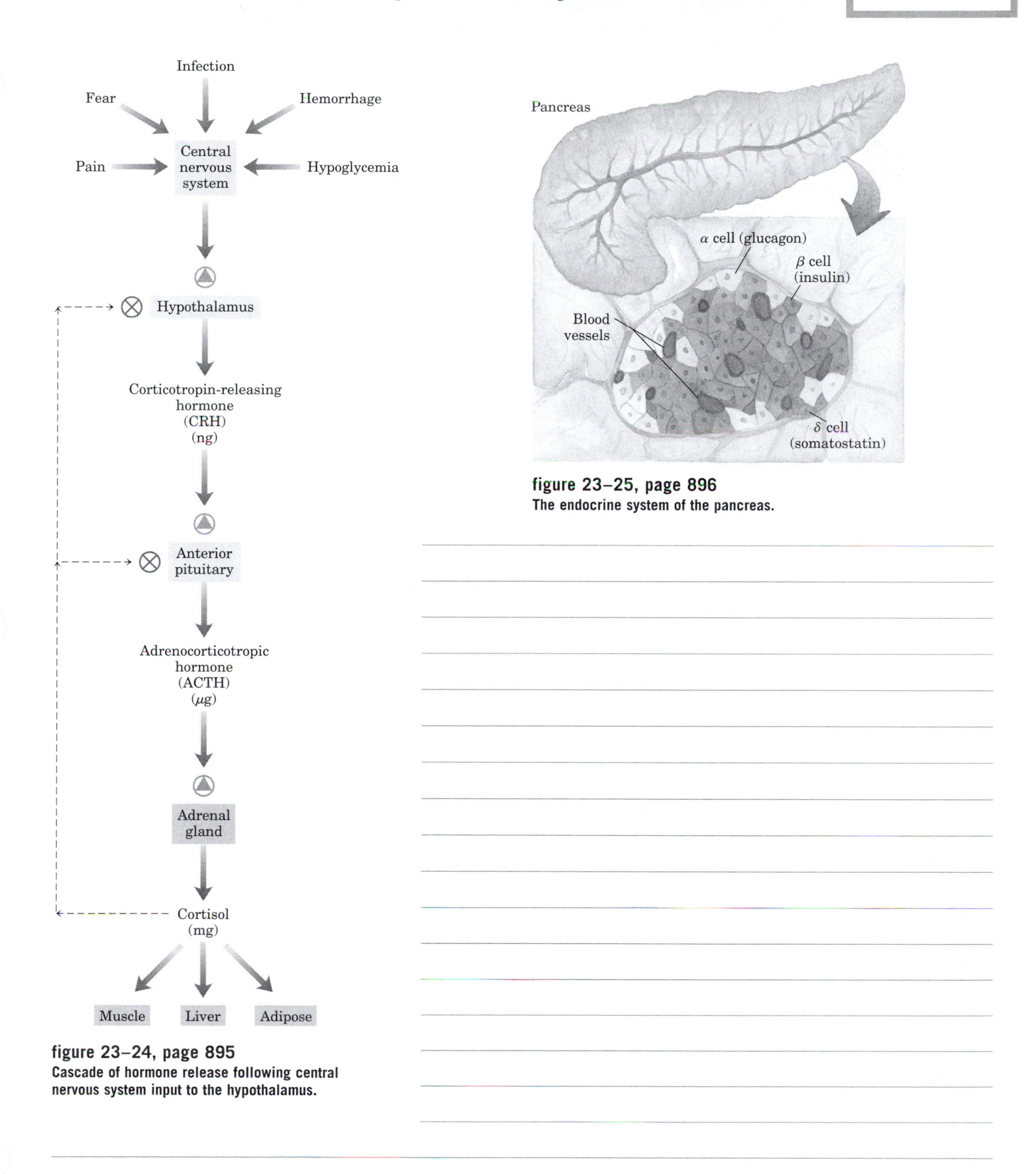

figure 23–24, page 895
Cascade of hormone release following central nervous system input to the hypothalamus.

figure 23–25, page 896
The endocrine system of the pancreas.

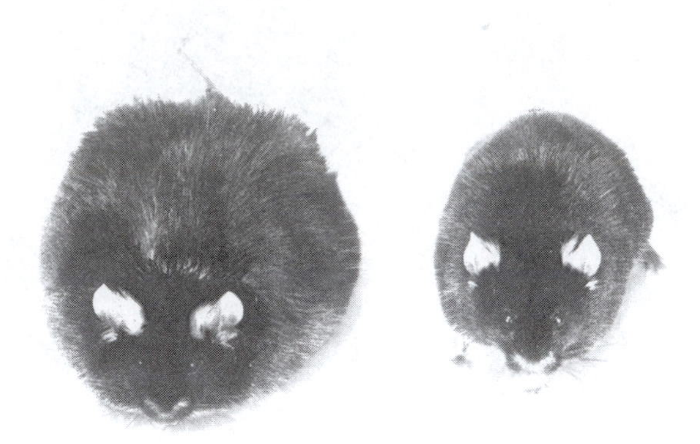

figure 23–26, page 896
A defect in leptin production leads to obesity.

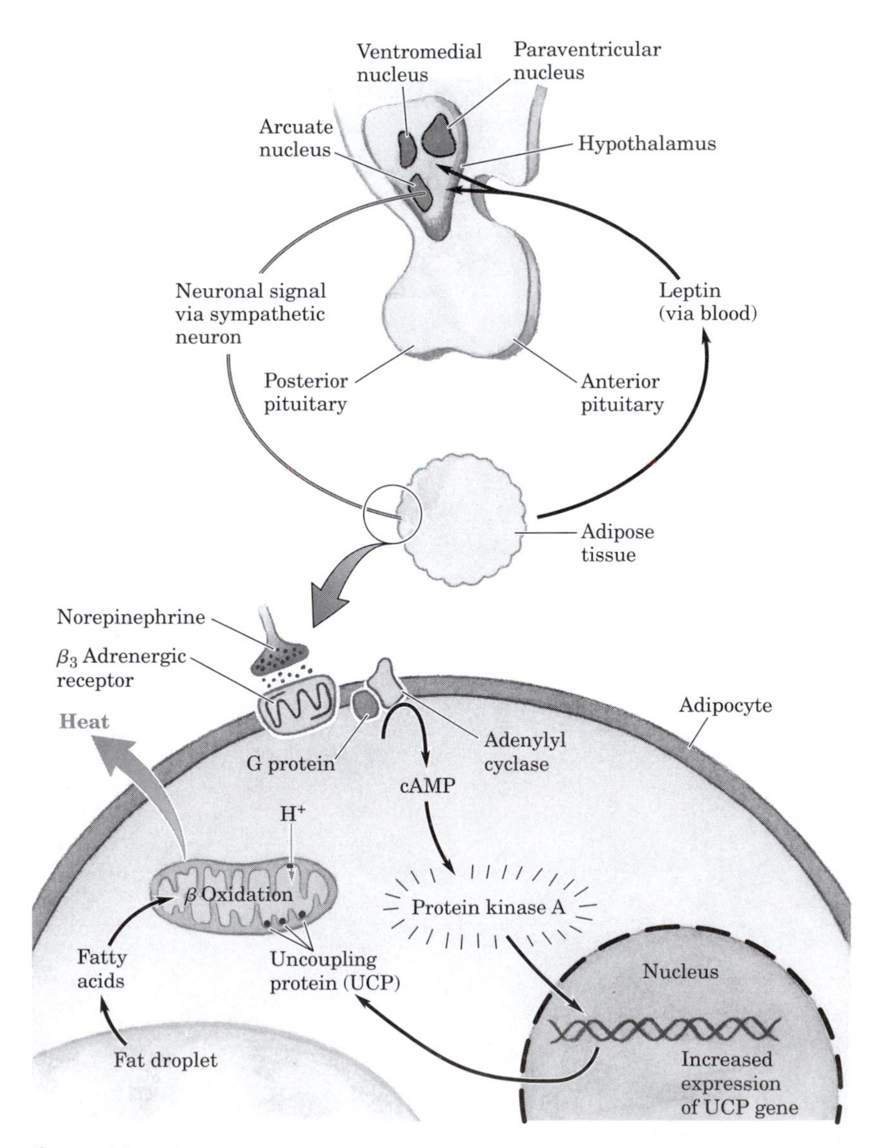

figure 23–27, page 897
The hypothalamus in regulation of food intake and energy expenditure.

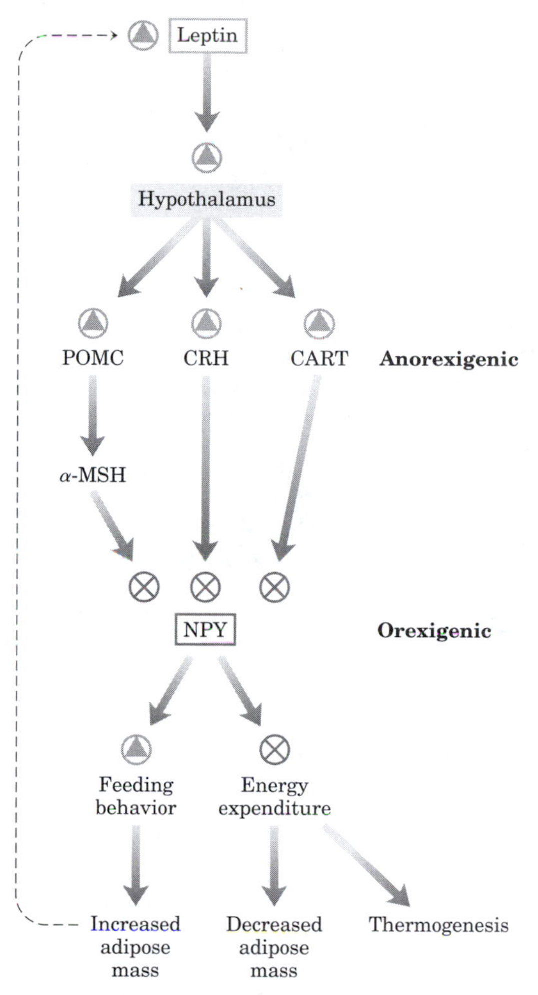

figure 23–28, page 898
The leptin cascade.

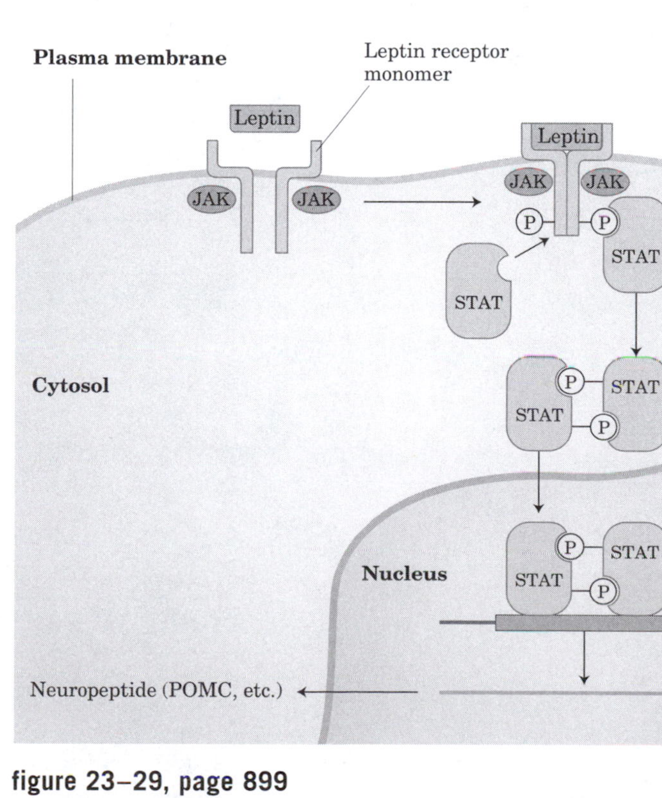

figure 23–29, page 899
The JAK-STAT mechanism by which the leptin signal is transduced in the hypothalamus.

part

IV Information Pathways

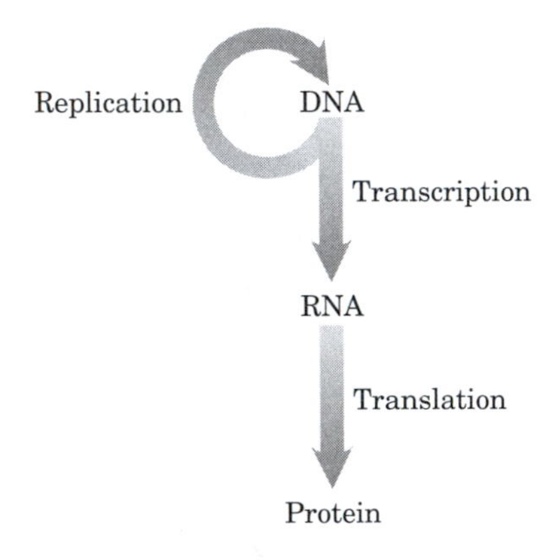

page 906

The central dogma of molecular biology, showing the general pathways of information flow via replication, transcription, and translation.

chapter

24

Genes and Chromosomes

When you study for exams, don't forget to review The Minicourses on the UNDERSTAND! BIOCHEMISTRY CD that came with your textbook.

Minicourses that apply to this Chapter include:

Background
  Fundamentals of Gene Action
Molecules of Life
  An Interactive Gallery of Protein Structures
Nucleic Acids and Their Expression
  DNA Topology

figure 24–1, page 907
Bacteriophage T2 protein coat surrounded by its single, linear molecule of DNA.

table 24–1, page 907

The Sizes of DNA and Viral Particles for Some Bacterial Viruses (Bacteriophages)

Virus	Number of base pairs in viral DNA*	Length of viral DNA (nm)	Long dimension of viral particle (nm)
ϕX174	5,386†	1,939†	25
T7	39,936	14,377	78
λ (lambda)	48,502	17,460	190
T4	168,889	60,800	210

* The complete base sequences of these bacteriophage genomes have been determined.
†Data are for the replicative form (double-stranded).

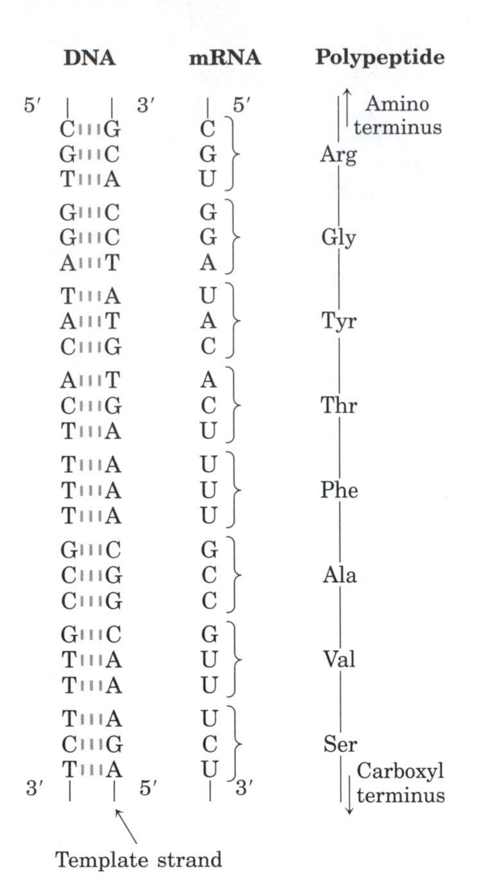

figure 24–2, page 908
Colinearity of the coding nucleotide sequences of DNA and mRNA and the amino acid sequence of a polypeptide chain.

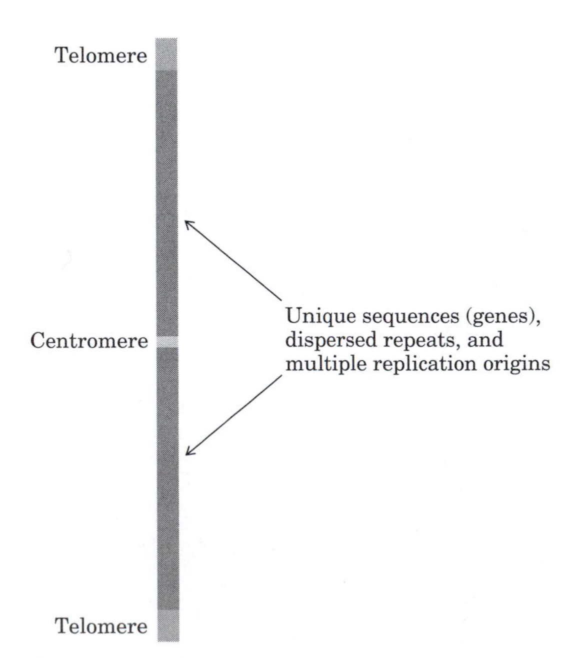

figure 24–3, page 909
Important structural elements of a yeast chromosome.

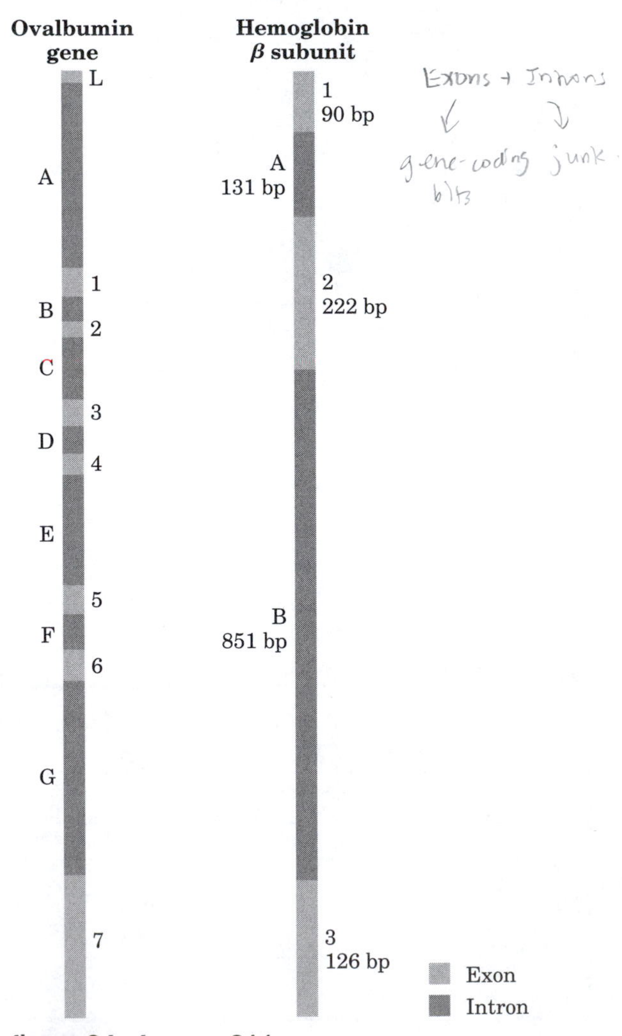

figure 24–4, page 911
Intervening sequences, or introns, in two eukaryotic genes.

figure 24–6, page 913
DNA from a lysed *E. coli* cell.

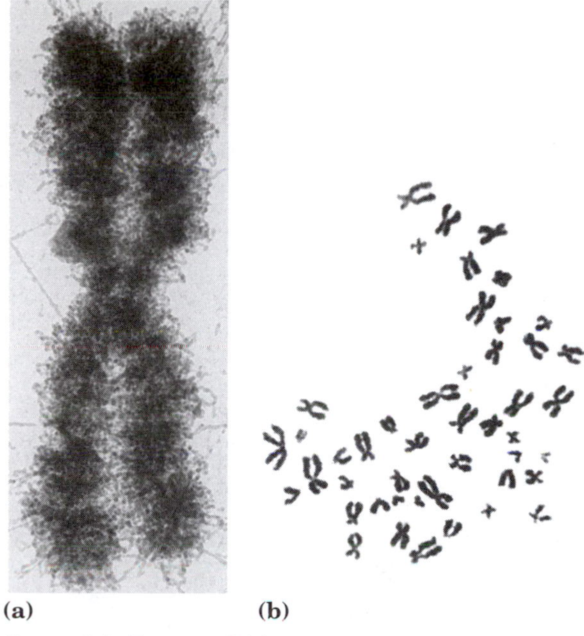

(a) (b)

figure 24–7, page 914
Eukaryotic chromosomes.

table 24–2, page 914

Normal Chromosome Number in Some Organisms*

Bacteria	1	Honeybee (female)	32
Fruit fly	8	Fox	34
Red clover	14	Cat	38
Garden pea	14	Mouse	40
Yeast	16[†]	Rat	42
Maize (corn)	20	Rabbit	44
Frog	26	Human	46
Hydra	30	Chicken	78

*The diploid chromosome number is given for all eukaryotes except yeast.

[†]This is the haploid chromosome number for the yeast *Saccharomyces cerevisiae*. Wild yeast strains generally have eight (octoploid) or more sets of these chromosomes.

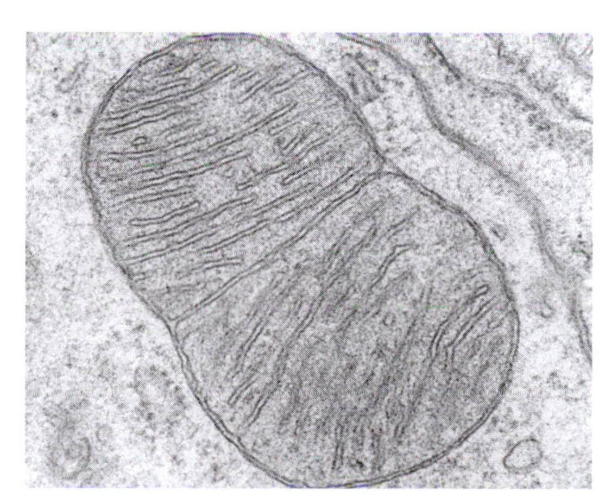

figure 24–8, page 915
A dividing mitochondrion.

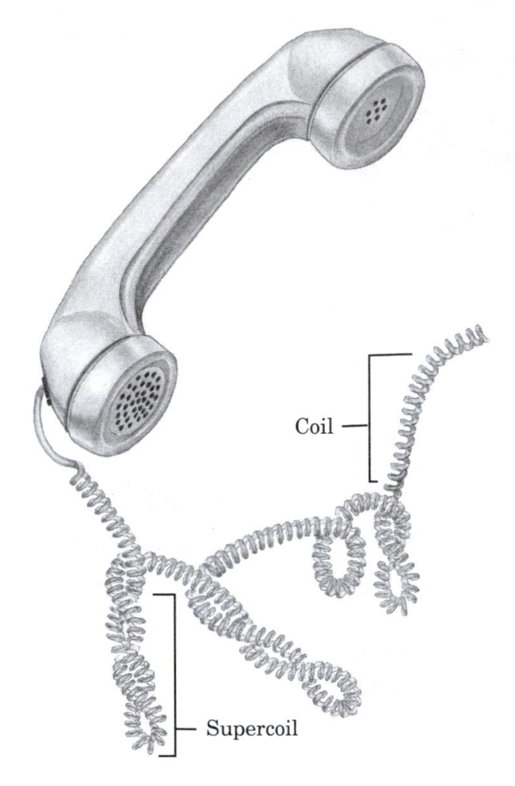

figure 24–9, page 916
Supercoils.

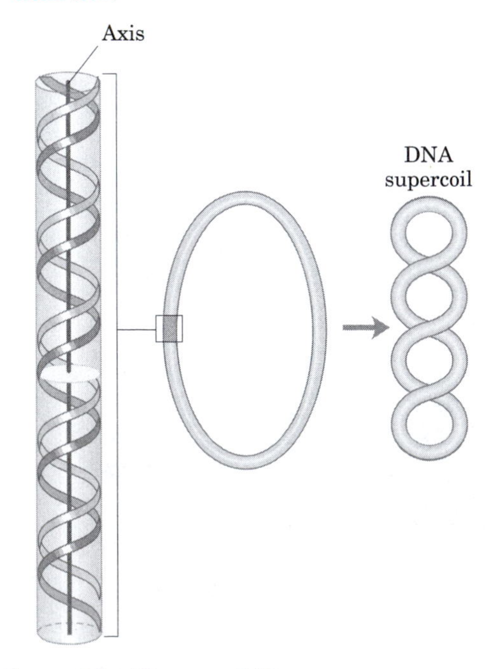

figure 24–10, page 916
Supercoiling of DNA.

figure 24–11, page 917
Supercoiling induced by separating the strands of a helical structure.

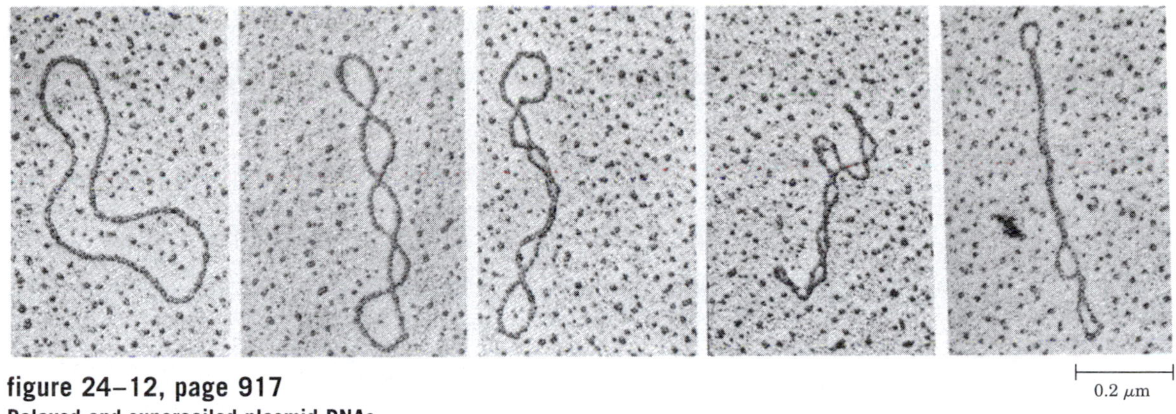

figure 24–12, page 917
Relaxed and supercoiled plasmid DNAs.

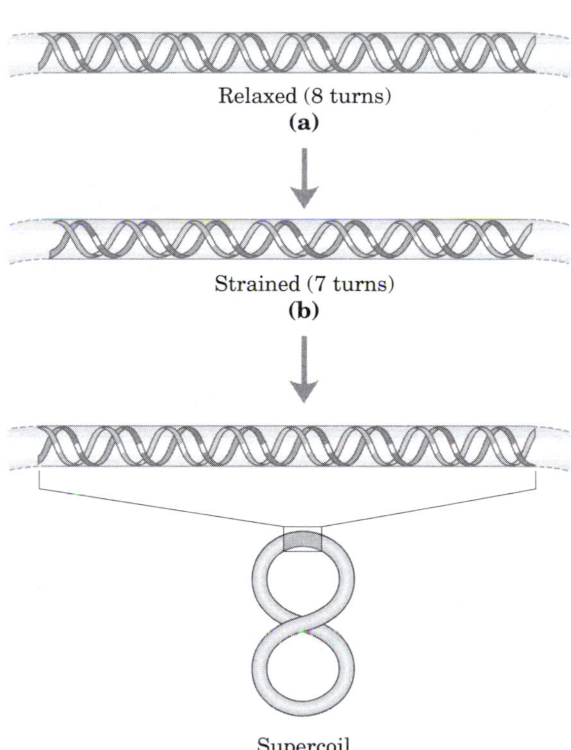

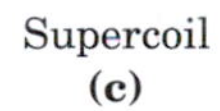

figure 24–13, page 918
Effects of DNA underwinding.

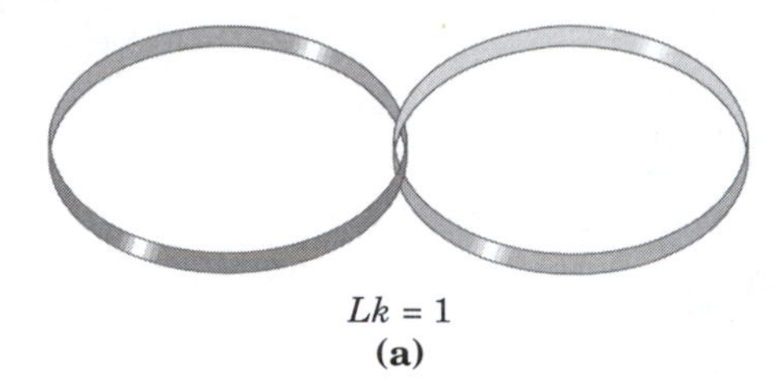

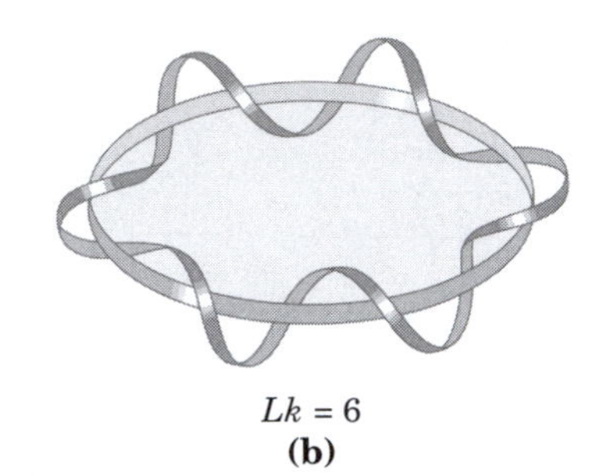

figure 24–14, page 919
Linking number, *Lk*.

$$\sigma = \frac{\Delta Lk}{Lk_0}$$

Superhelical density, page 919

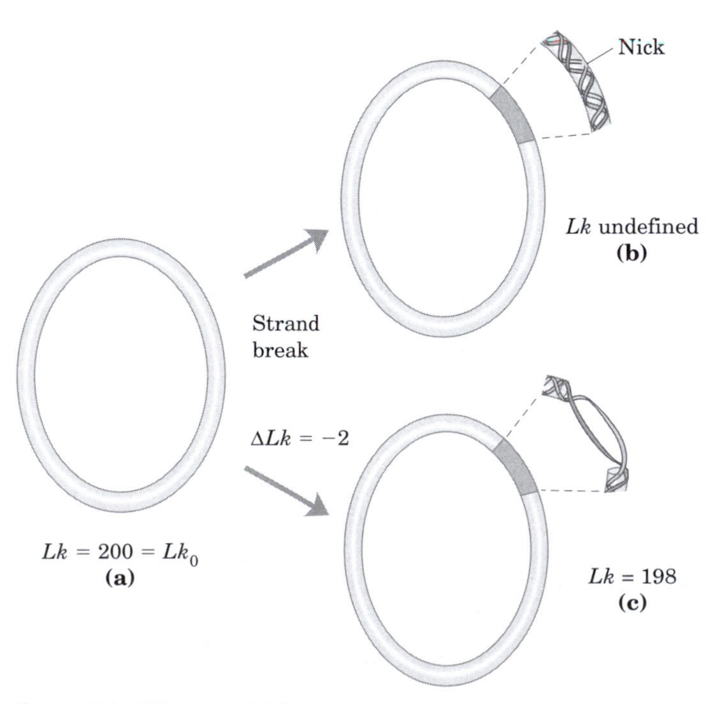

figure 24–15, page 919
Linking number applied to closed-circular DNA molecules.

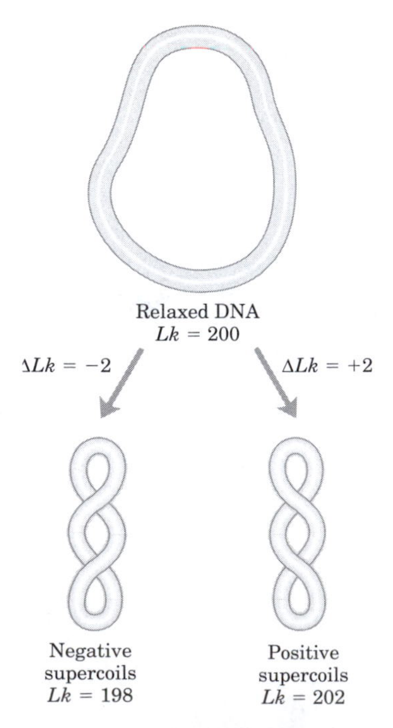

figure 24–16, page 920
Negative and positive supercoils.

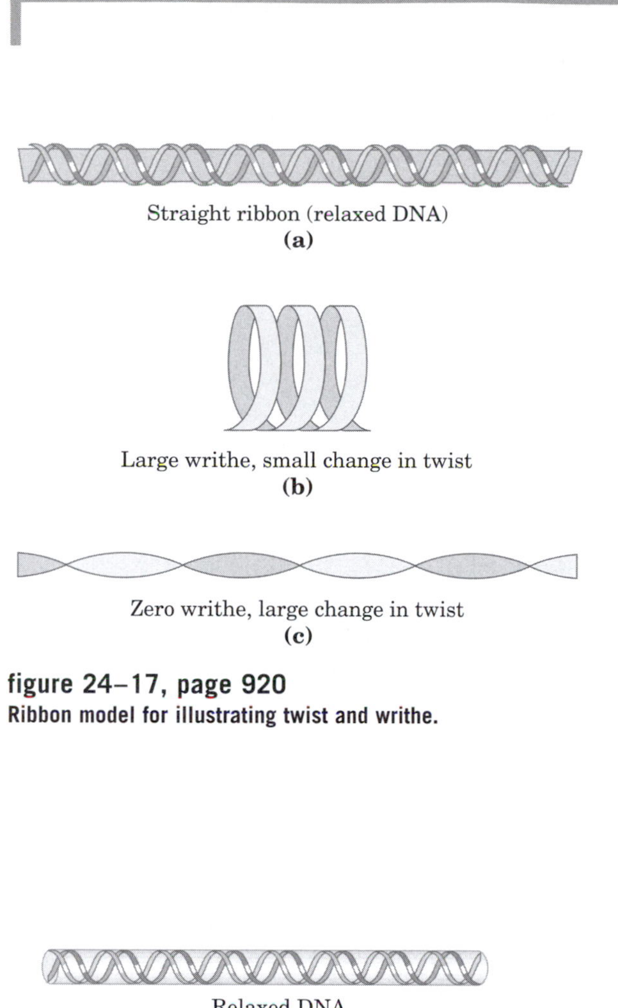

figure 24–17, page 920
Ribbon model for illustrating twist and writhe.

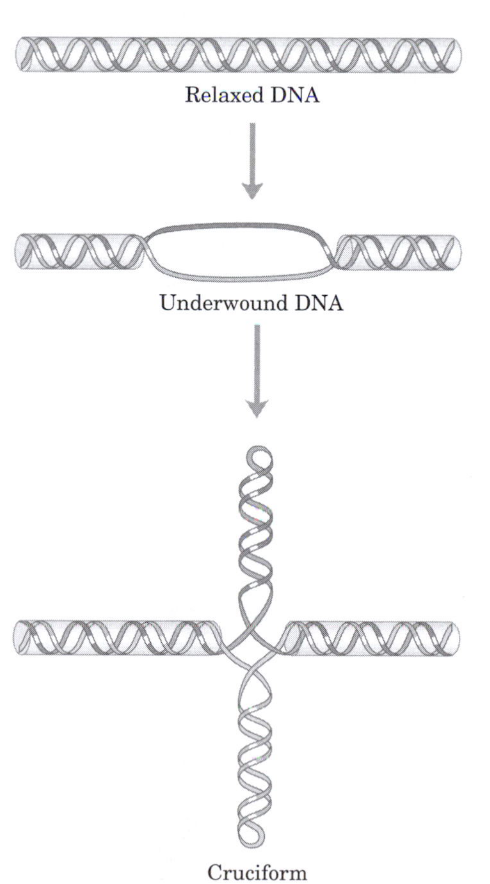

figure 24–18, page 921
Promotion of cruciform structures by DNA underwinding.

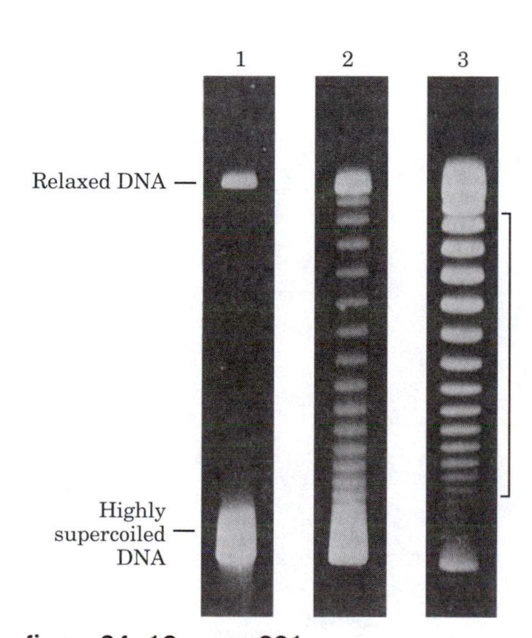

figure 24–19, page 921
Changes in linking number catalyzed by topoisomerases.

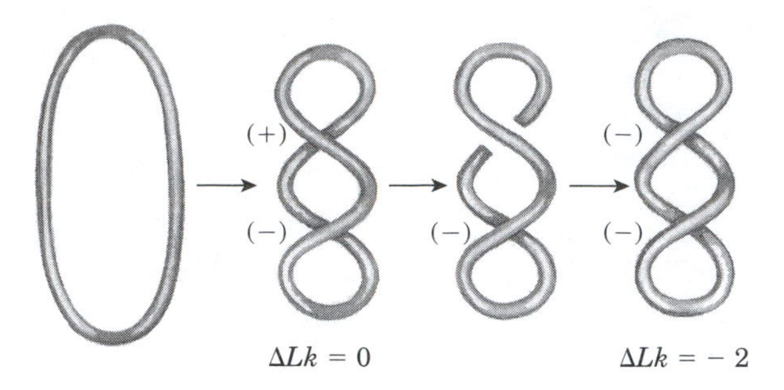

figure 24–20, page 922
Mechanism of DNA gyrase.

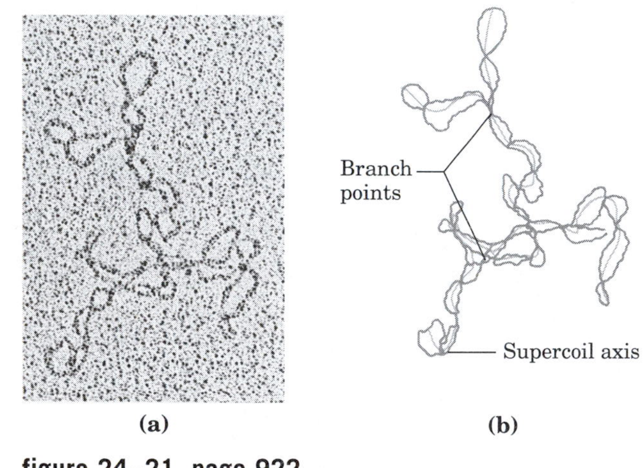

figure 24–21, page 922
Plectonemic supercoiling.

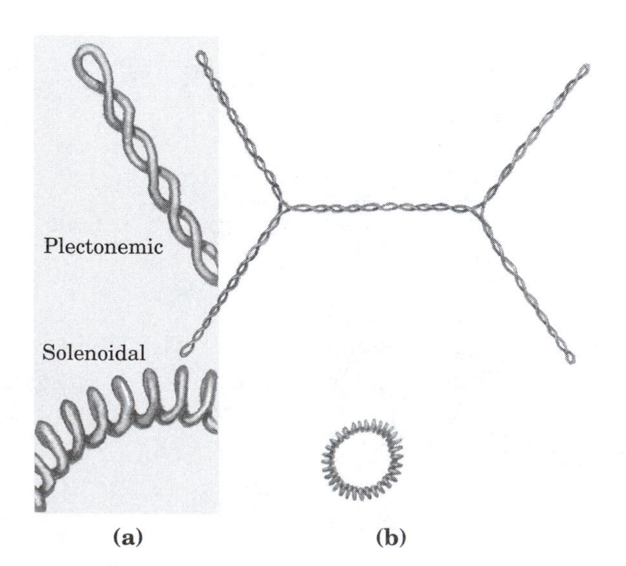

figure 24–22, page 923
Plectonemic and solenoidal supercoiling.

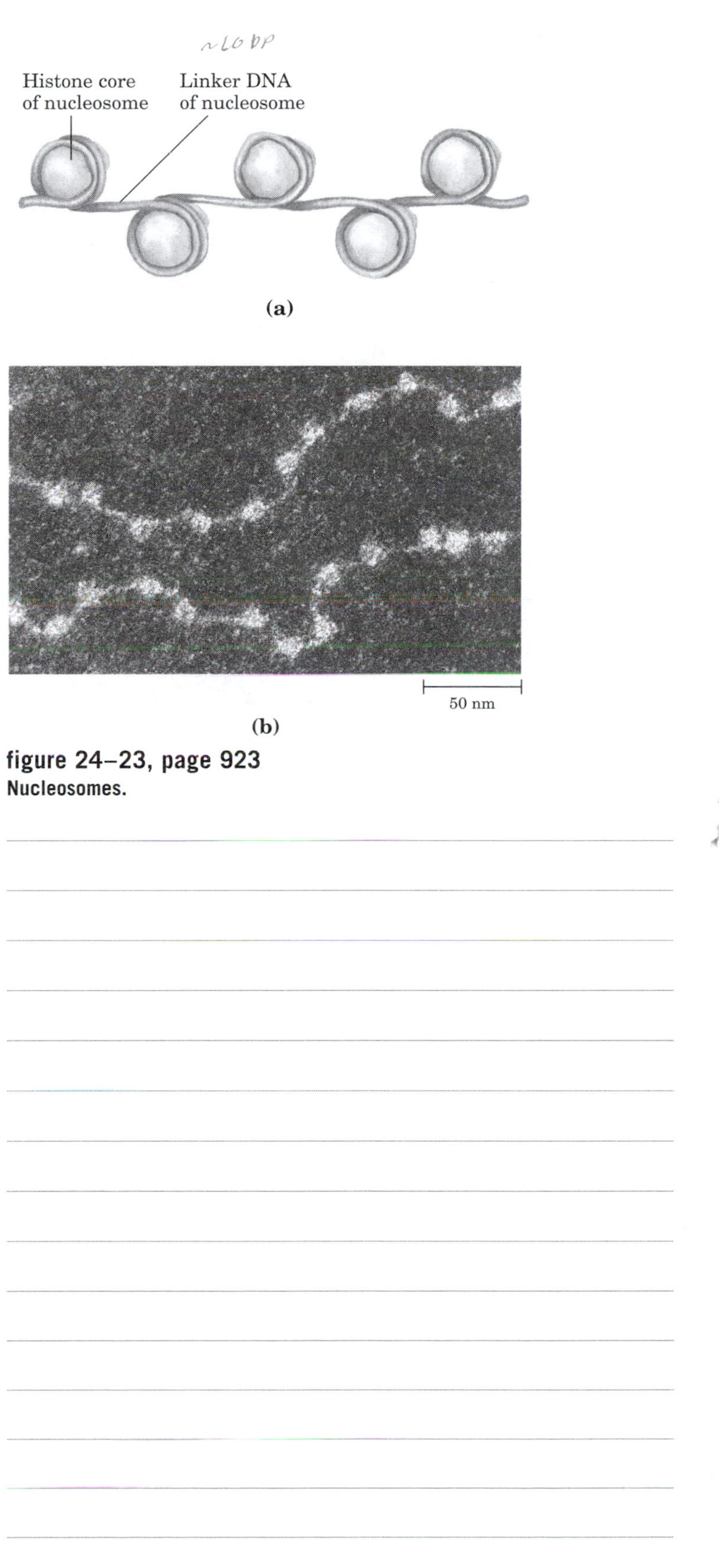

figure 24–23, page 923
Nucleosomes.

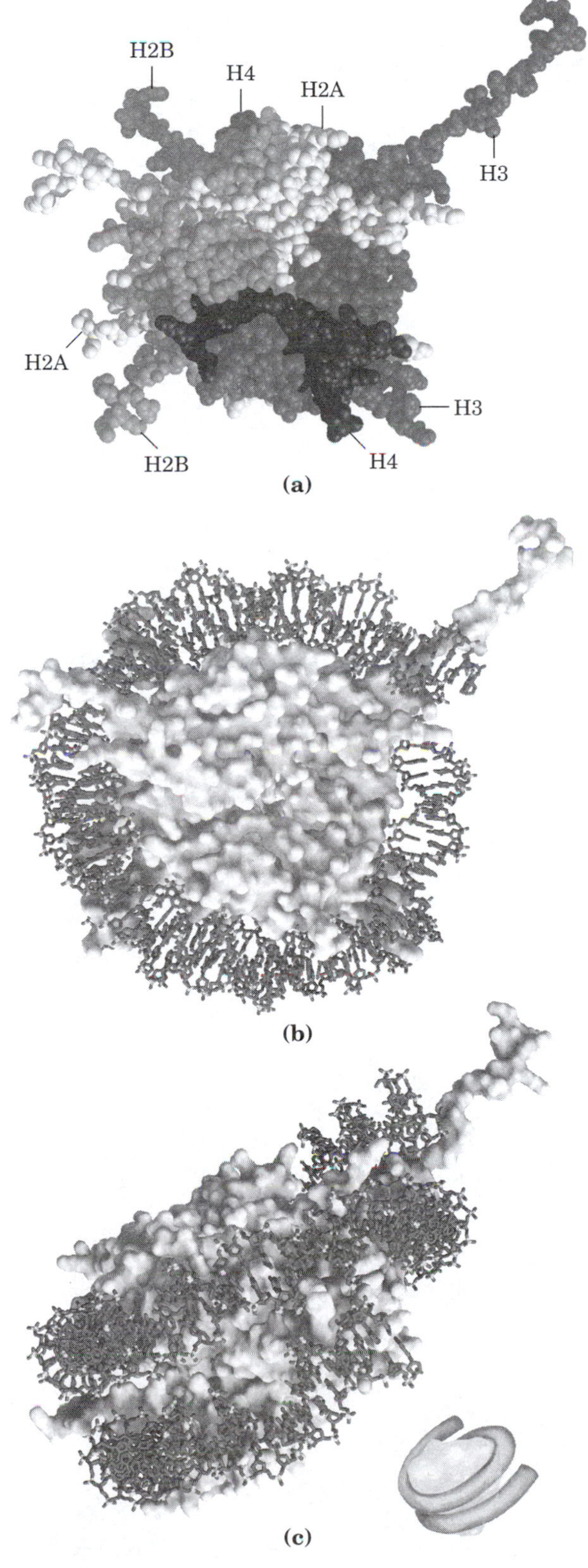

figure 24–24, page 924
DNA wrapped around a nucleosome core.

table 24–3, page 924

Types and Properties of Histones

Histone	Molecular weight	Number of amino acid residues	Content of basic amino acids (% of total)	
			Lys	Arg
H1*	21,130	223	29.5	1.3
H2A*	13,960	129	10.9	9.3
H2B*	13,774	125	16.0	6.4
H3	15,273	135	9.6	13.3
H4	11,236	102	10.8	13.7

*The sizes of these histones vary somewhat from species to species. The numbers given here are for bovine histones.

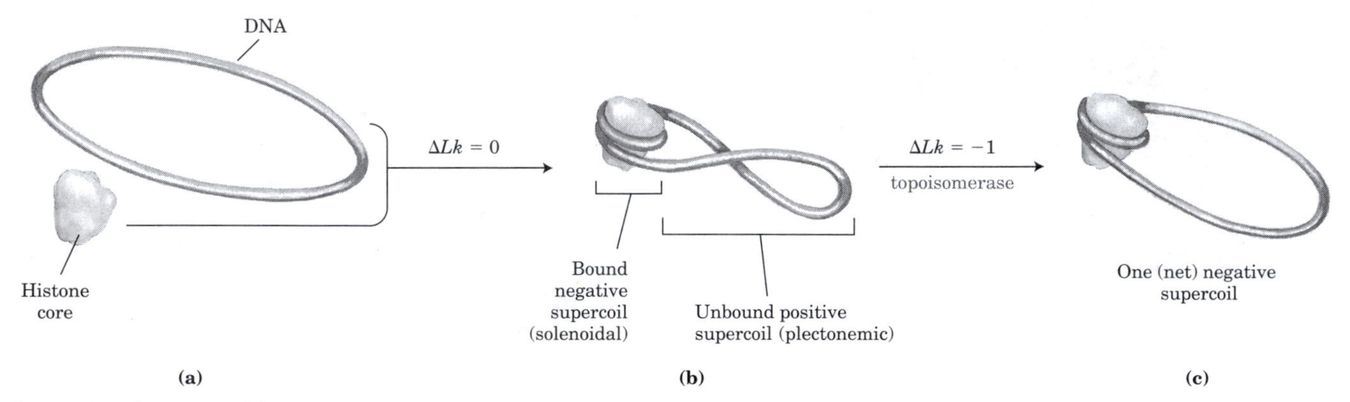

figure 24–25, page 925
Chromatin assembly.

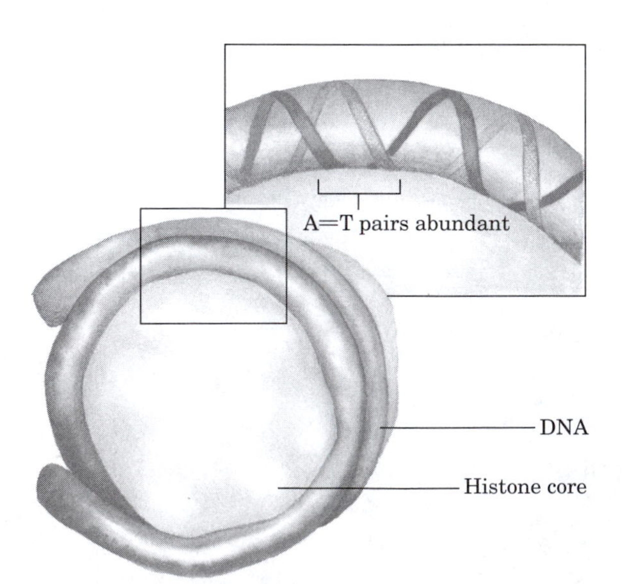

figure 24–26, page 925
Positioning of a nucleosome to make optimal use of A=T base pairs where the histone core is in contact with the minor groove of the DNA helix.

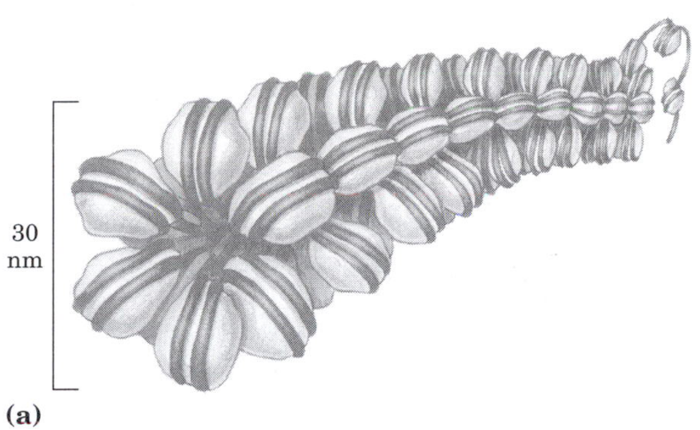

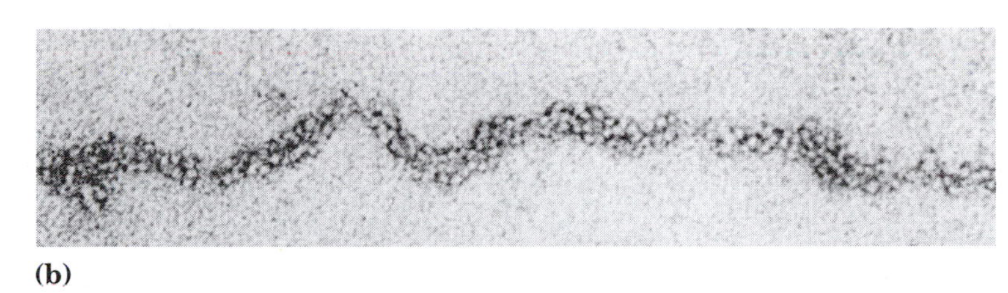

figure 24–27, page 926
The 30 nm fiber, a higher-order organization of nucleosomes.

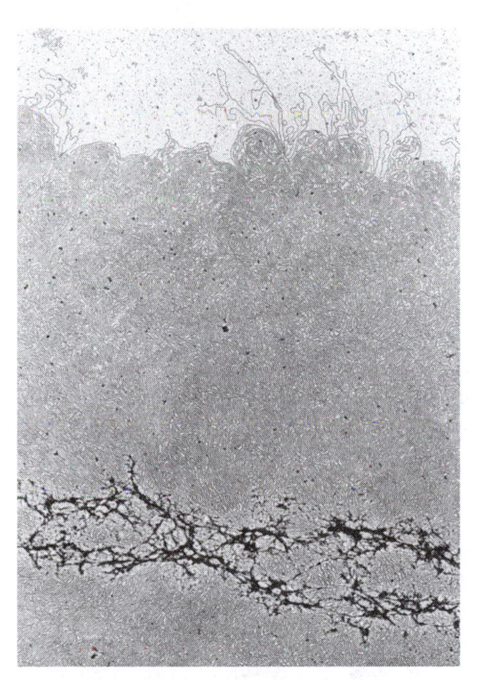

figure 24–28, page 926
A partially unraveled human chromosome, revealing numerous loops of DNA attached to a scaffoldlike structure.

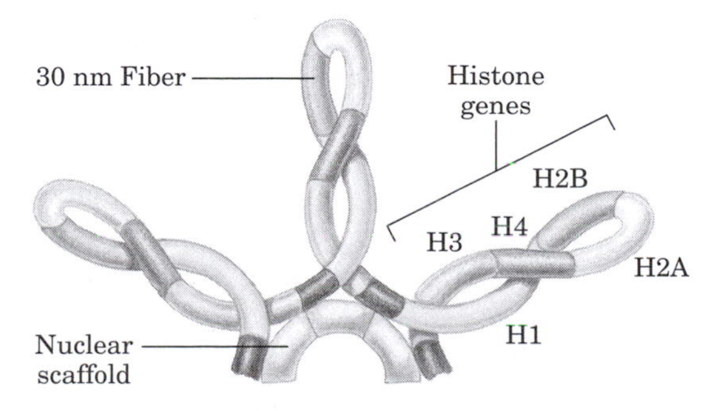

figure 24–29, page 926
Loops of chromosomal DNA attached to a nuclear scaffold.

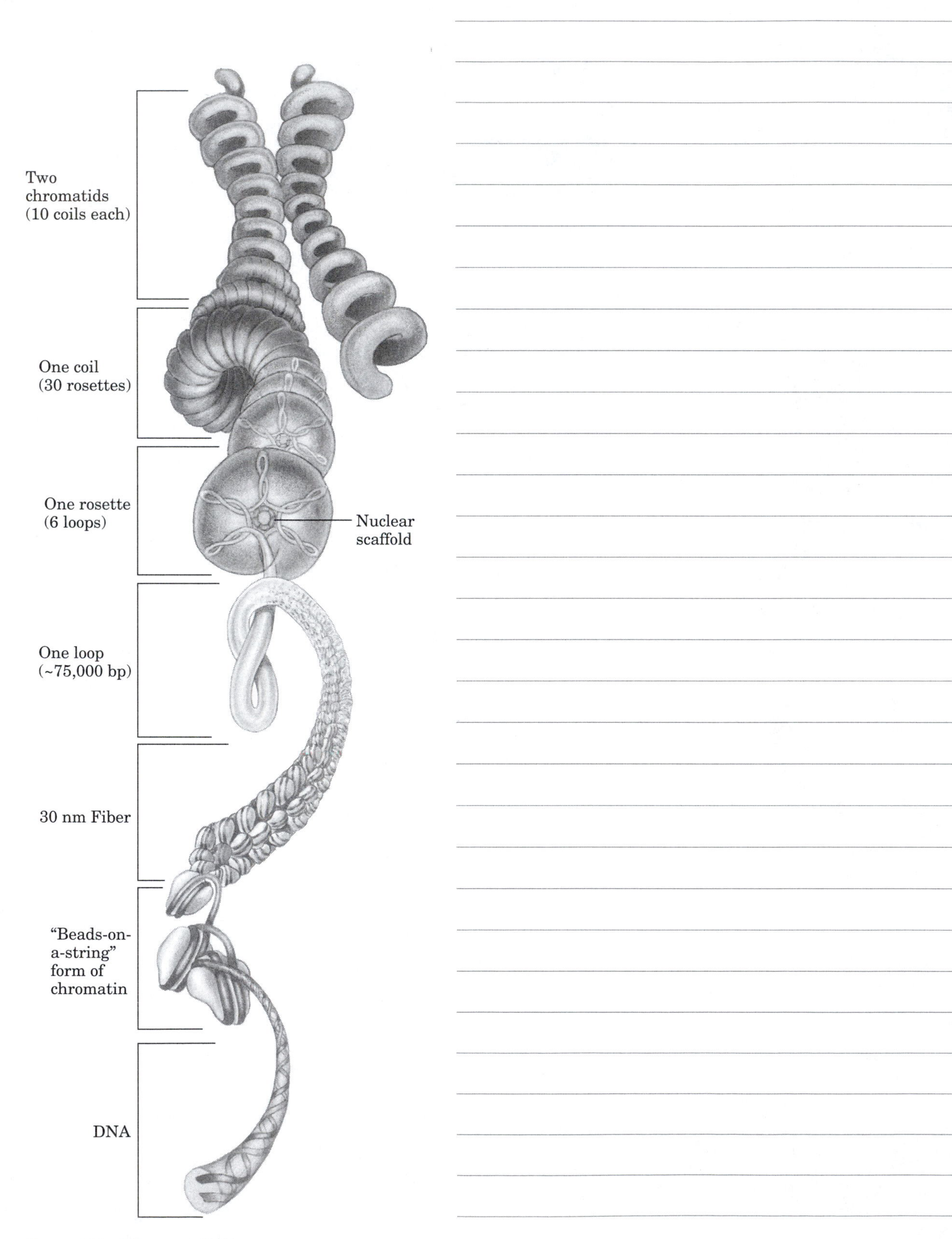

figure 24–30, page 927

Model for levels of organization that could provide DNA compaction in a eukaryotic chromosome.

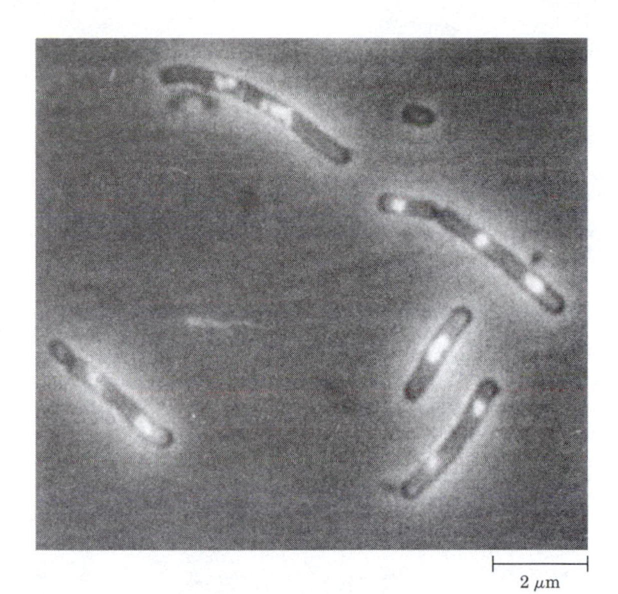

figure 24–31, page 927
E. coli cells showing nucleoids.

PROBLEMS

8. Chromatin, page 930.

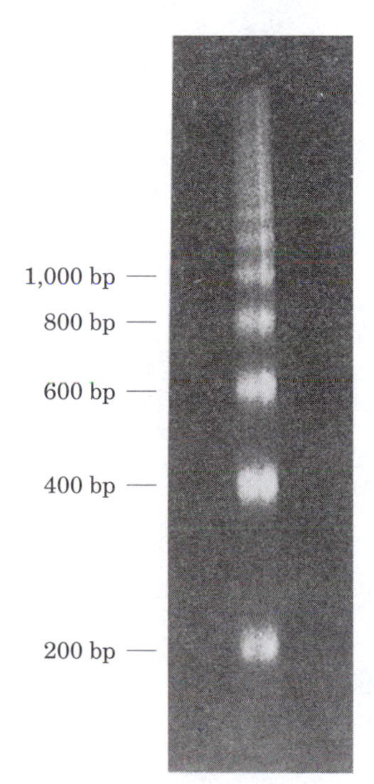

chapter

25 DNA Metabolism

When you study for exams, don't forget to review The Minicourses on the UNDERSTAND! BIOCHEMISTRY CD that came with your textbook.

Minicourses that apply to this Chapter include:

Molecules of Life
- An Interactive Gallery of Protein Structures

Proteins in Action
- Molecules of the Immune Systems

Nucleic Acids and Their Expression
- DNA Replication
- Mutation
- Mutagens
- DNA Repair
- Gene Regulation - Phage Lambda
- Recombination

The Dividing Cell
- Cancer

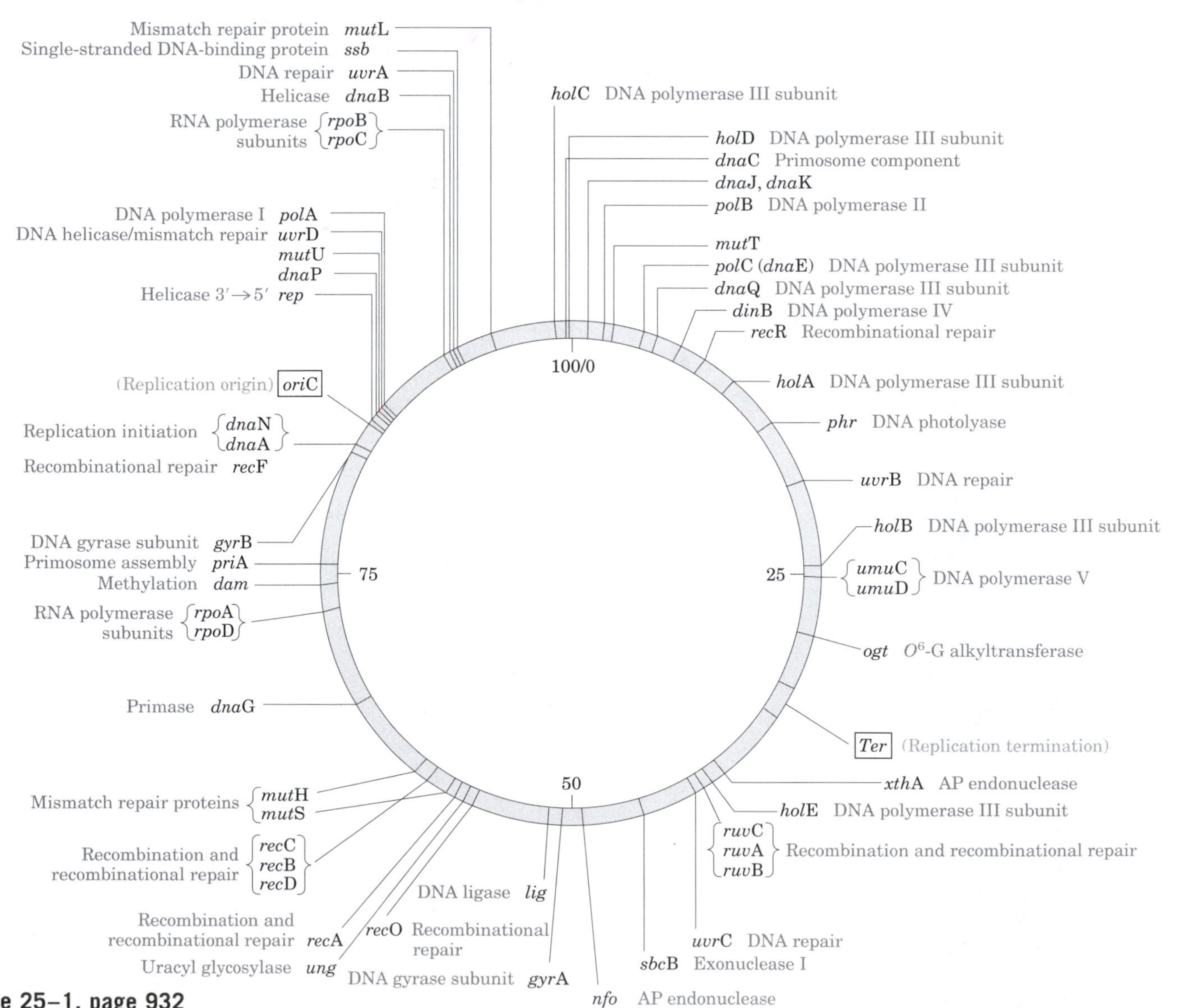

figure 25–1, page 932
Map of the *E. coli* chromosome.

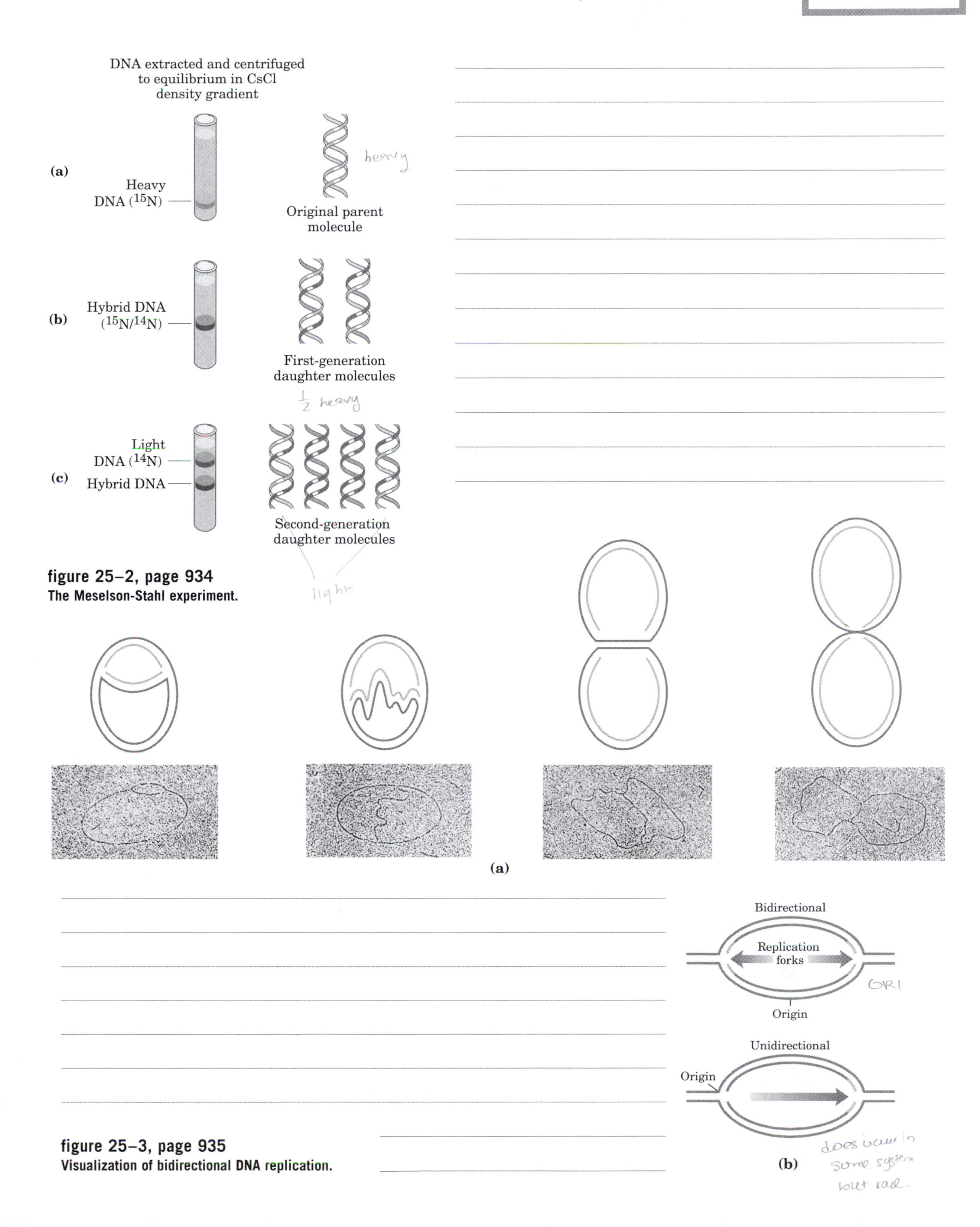

figure 25–2, page 934
The Meselson-Stahl experiment.

figure 25–3, page 935
Visualization of bidirectional DNA replication.

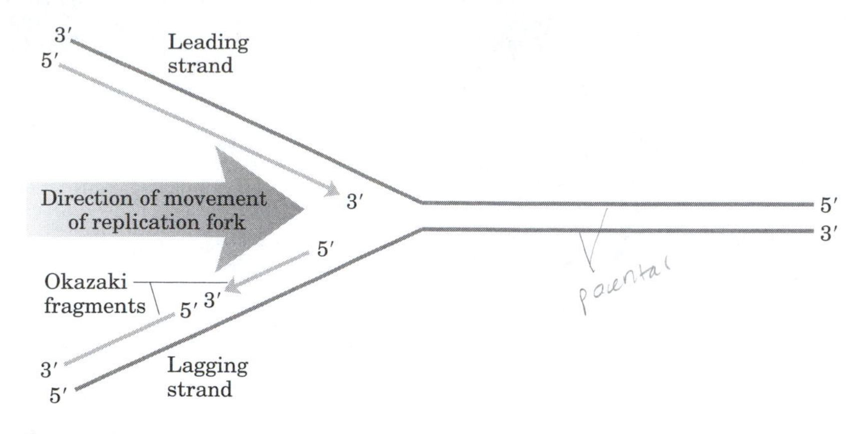

figure 25–4, page 936
Defining DNA strands at the replication fork.

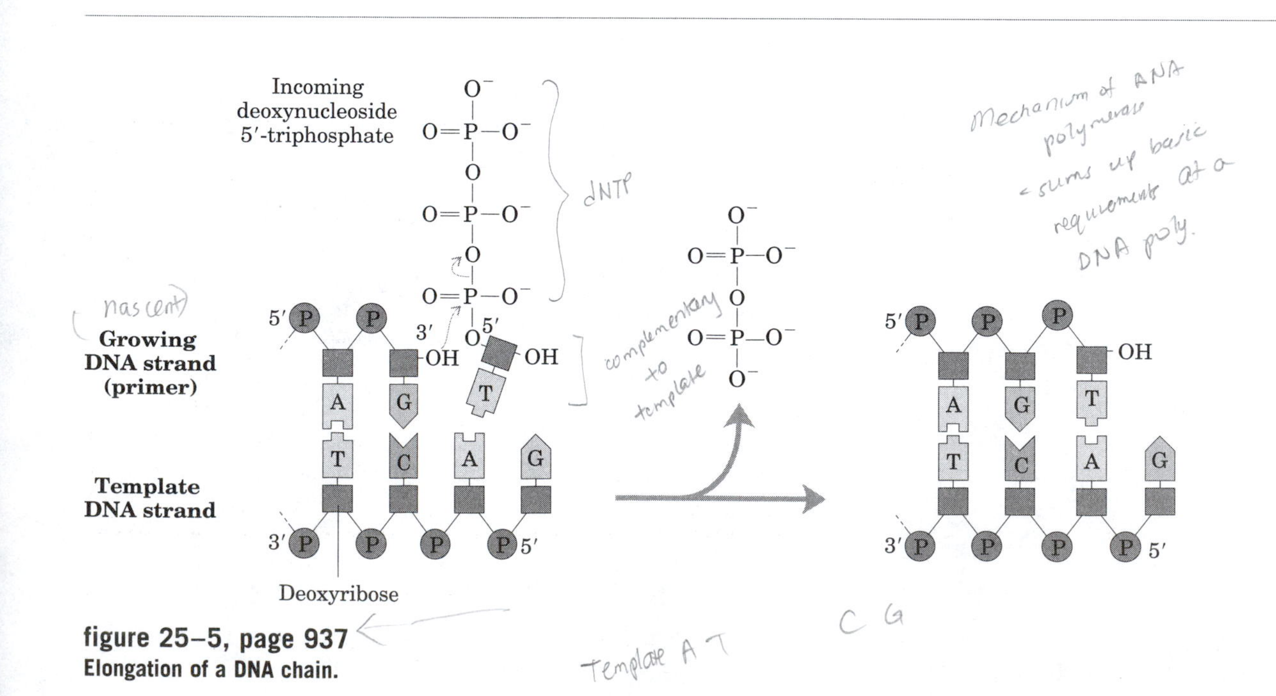

figure 25–5, page 937
Elongation of a DNA chain.

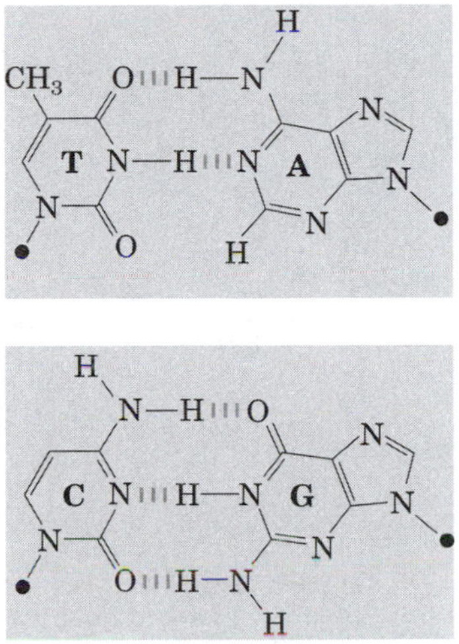

(a)

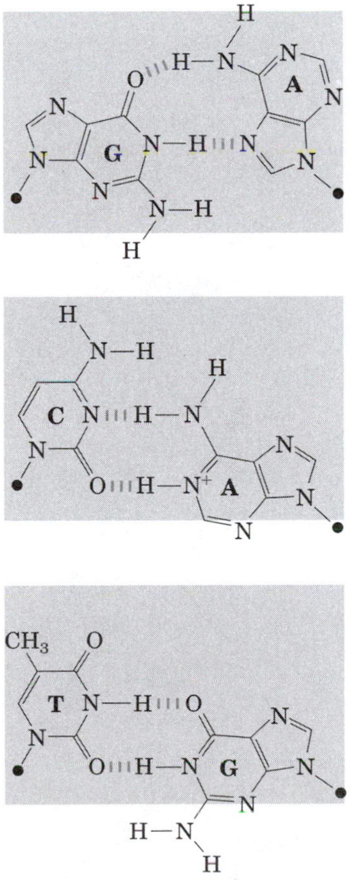

(b)

figure 25–6, page 938
Contribution of base pair geometry to the fidelity of DNA replication.

table 25–1, page 939

Comparison of DNA Polymerases of *E. coli*

	DNA polymerase		
	I	II	III
Structural gene*	*pol*A	*pol*B	*pol*C (*dna*E)
Subunits (number of different types)	1	≥4	≥10
M_r	103,000	88,000†	830,000
3′→5′ Exonuclease (proofreading)	Yes	Yes	Yes
5′→3′ Exonuclease	Yes	No	No
Polymerization rate (nucleotides/sec)	16–20	40	250–1,000
Processivity (nucleotides added before polymerase dissociates)	3–200	1,500	≥500,000

*For enzymes with more than one subunit, the gene listed here encodes the subunit with polymerization activity. Note that *dna*E is an earlier designation of the gene now referred to as *pol*C.

†Polymerization subunit only. DNA polymerase II shares several subunits with DNA polymerase III, including the β, γ, δ, δ', χ, and ψ subunits (see Table 25–2).

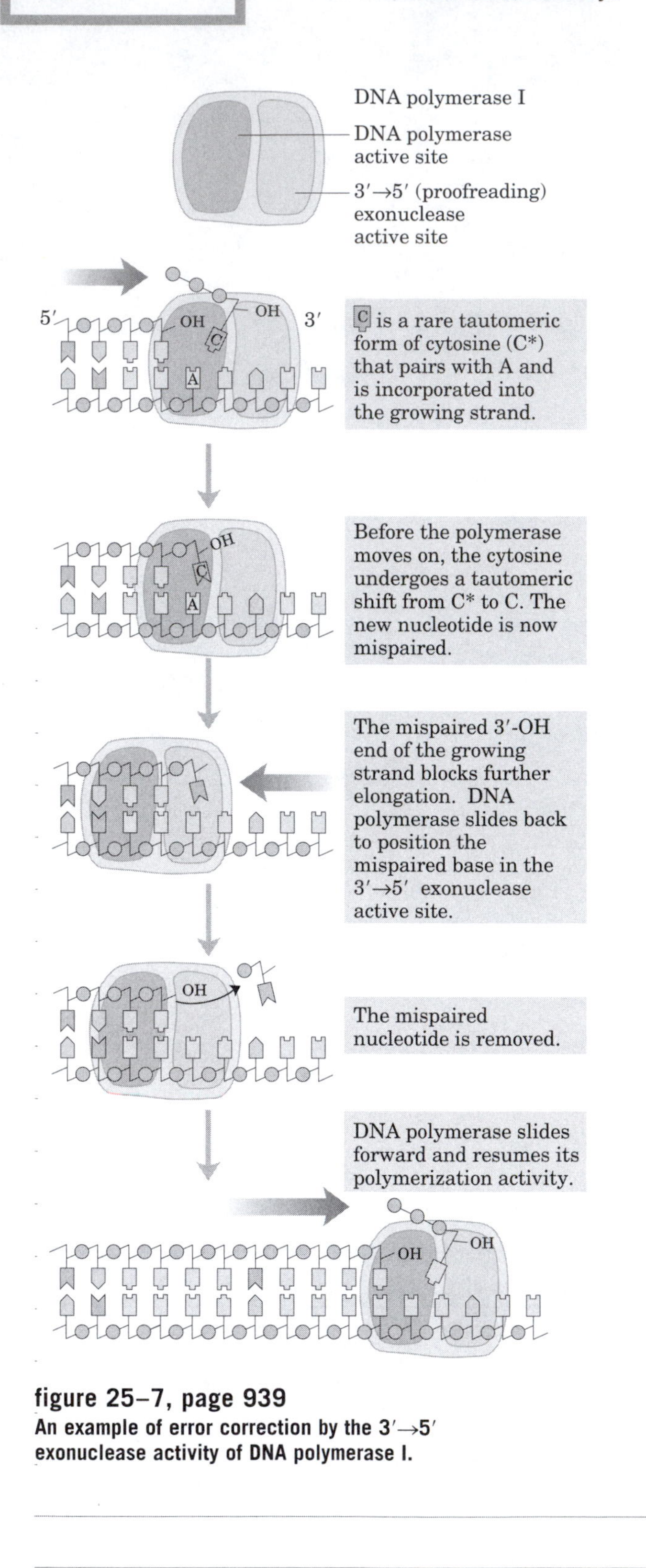

figure 25–7, page 939
An example of error correction by the 3′→5′ exonuclease activity of DNA polymerase I.

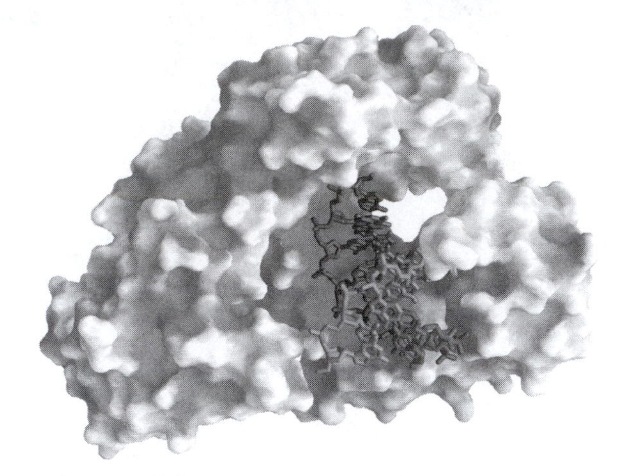

figure 25–8, page 940
Large (Klenow) fragment of DNA polymerase I.

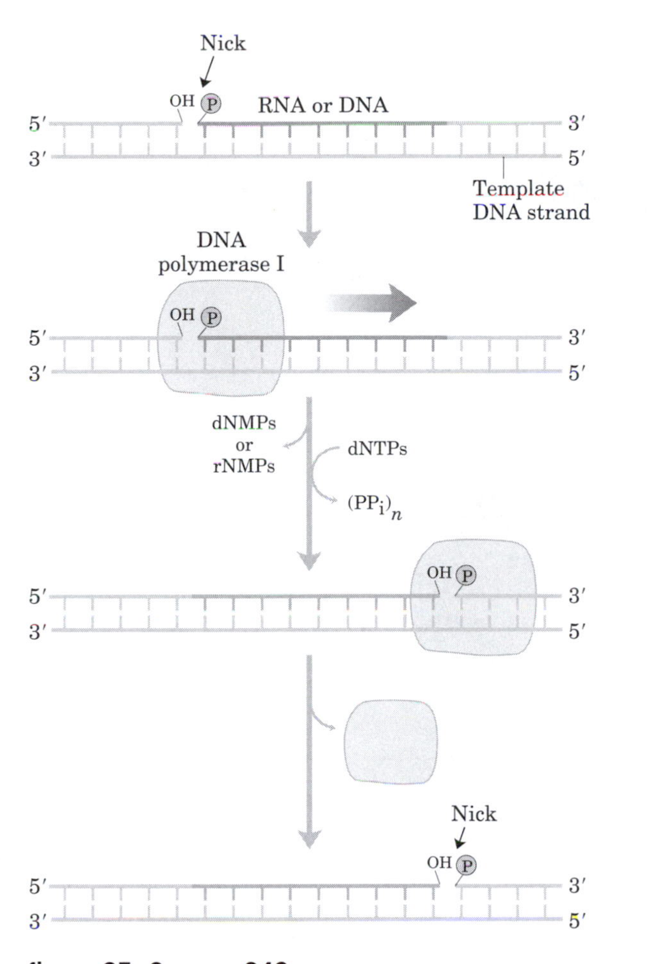

figure 25–9, page 940
Nick translation.

ignore

table 25–2, page 941

Subunits of DNA Polymerase III of *E. coli*

Subunit	Number of subunits per holoenzyme	M_r of subunit	Gene	Function of subunit	
α	2	132,000	*pol*C (*dna*E)	Polymerization activity	Core polymerase
ϵ	2	27,000	*dna*Q (*mut*D)	3′→5′ Proofreading exonuclease	
θ	2	10,000	*hol*E		
τ	2	71,000	*dna*X	Stable template binding; core enzyme dimerization	
γ	2	52,000	*dna*X*		
δ	1	35,000	*hol*A		
δ'	1	33,000	*hol*B		
χ	1	15,000	*hol*C		
ψ	1	12,000	*hol*D		
β	4	37,000	*dna*N	DNA clamp required for optimal processivity	

*The γ subunit is encoded by a portion of the gene for the τ subunit, such that the amino-terminal 80% of the τ subunit has the same amino acid sequence as the γ subunit. The γ subunit is generated by a translational frameshifting mechanism (see Box 28–1) that leads to premature translational termination.

figure 25–10, page 941
The two β subunits of *E. coli* polymerase III form a circular clamp that surrounds DNA.

figure 25–11, page 942
Arrangement of sequences in the *E. coli* replication origin, *ori*C.

table 25–3, page 943

Proteins Required to Initiate Replication at the *E. coli* Origin

Protein	M_r	Number of subunits	Function
DnaA protein	52,000	1	Recognizes origin sequence; opens duplex at specific sites in origin
DnaB protein (helicase)	300,000	6*	Unwinds DNA
DnaC protein	29,000	1	Required for DnaB binding at origin
HU	19,000	2	Histonelike protein; DNA bending protein; stimulates initiation
Primase (DnaG protein)	60,000	1	Synthesizes RNA primers
Single-stranded DNA-binding protein (SSB)	75,600	4*	Binds single-stranded DNA
RNA polymerase	454,000	5	Facilitates DnaA activity
DNA gyrase (DNA topoisomerase II)	400,000	4	Relieves torsional strain generated by DNA unwinding
Dam methylase	32,000	1	Methylates (5′)GATC sequences at *ori*C

*Subunits in these cases are identical.

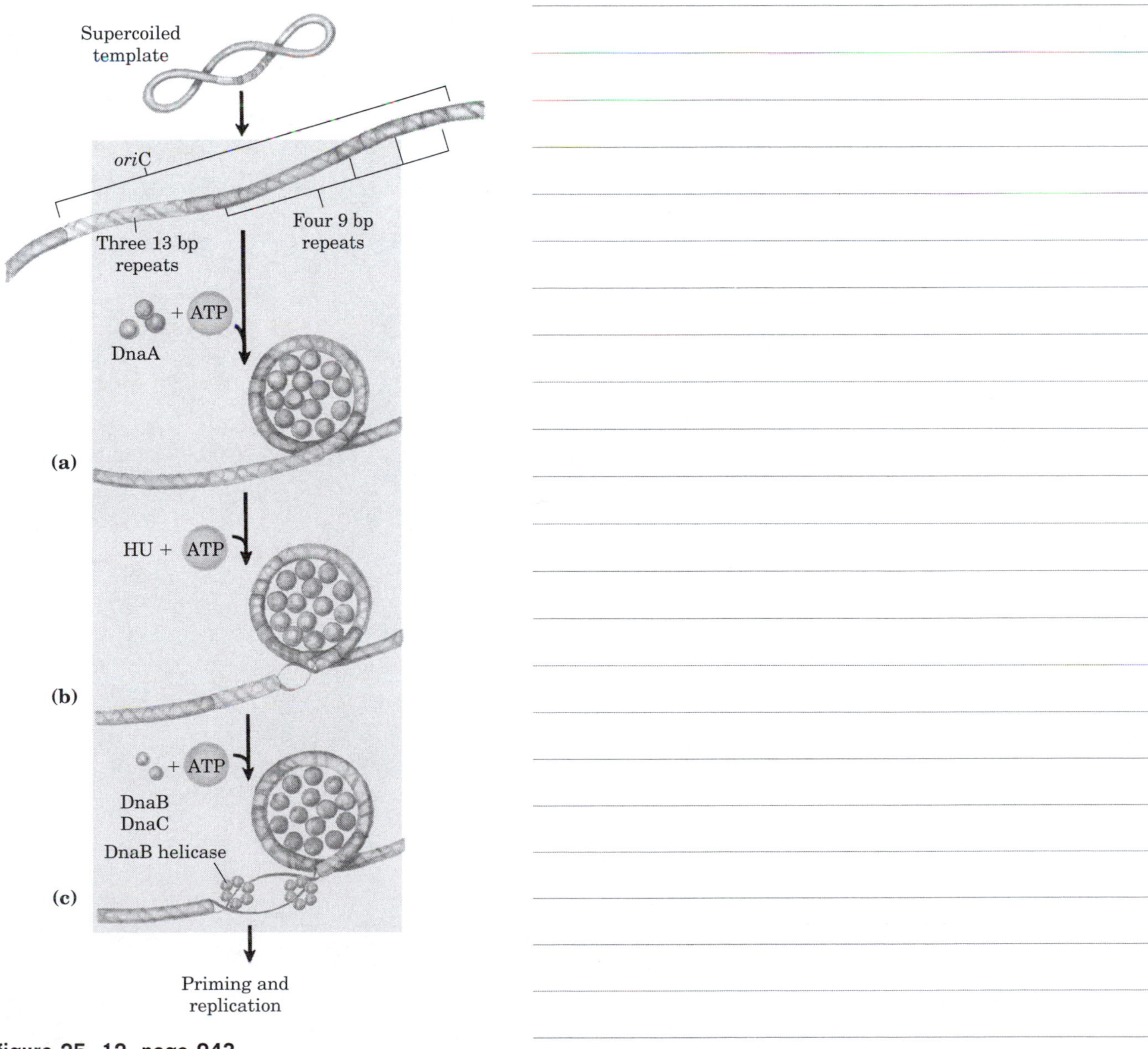

figure 25–12, page 943
Model for initiation of replication at the *E. coli* origin, *ori*C.

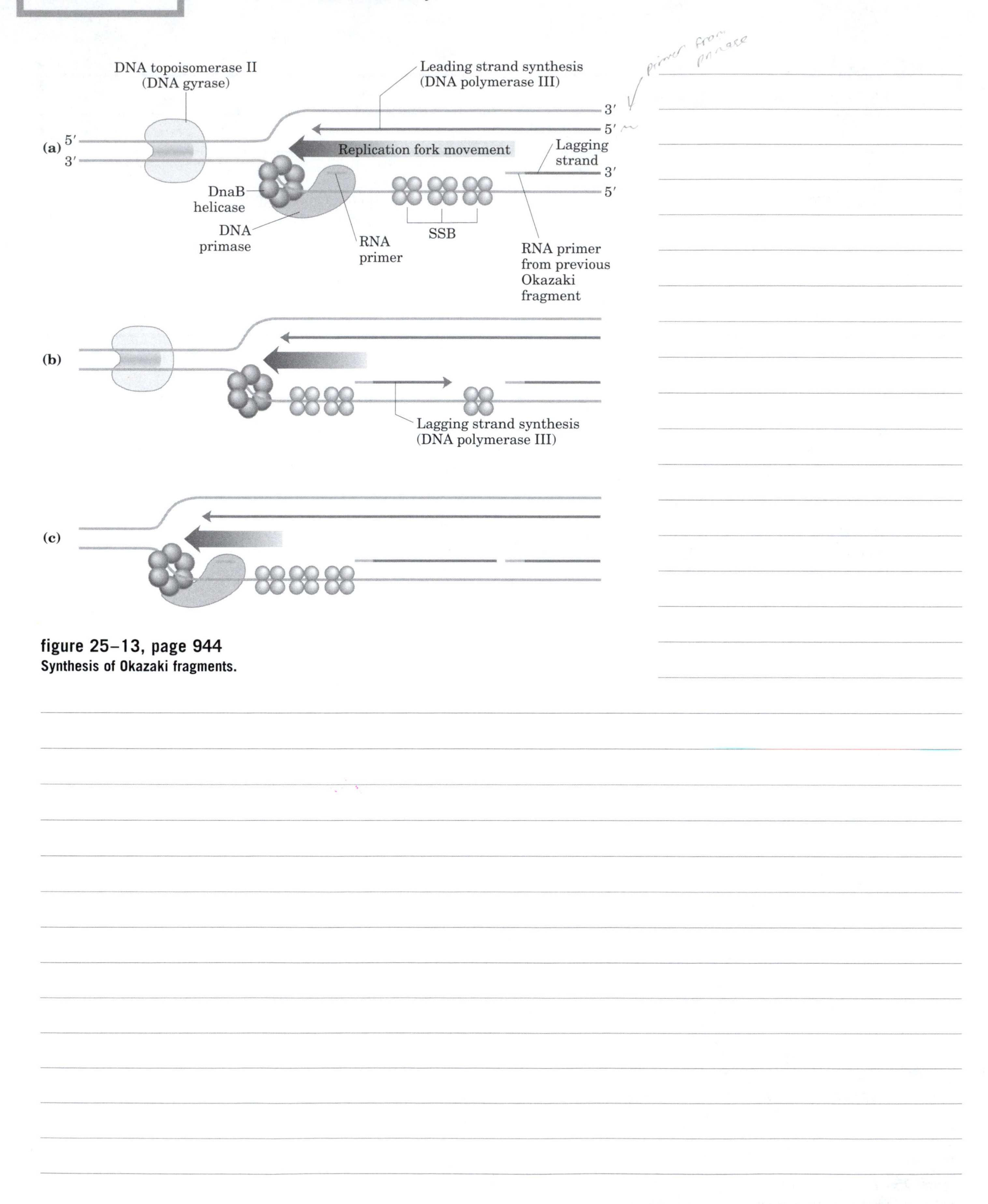

figure 25–13, page 944
Synthesis of Okazaki fragments.

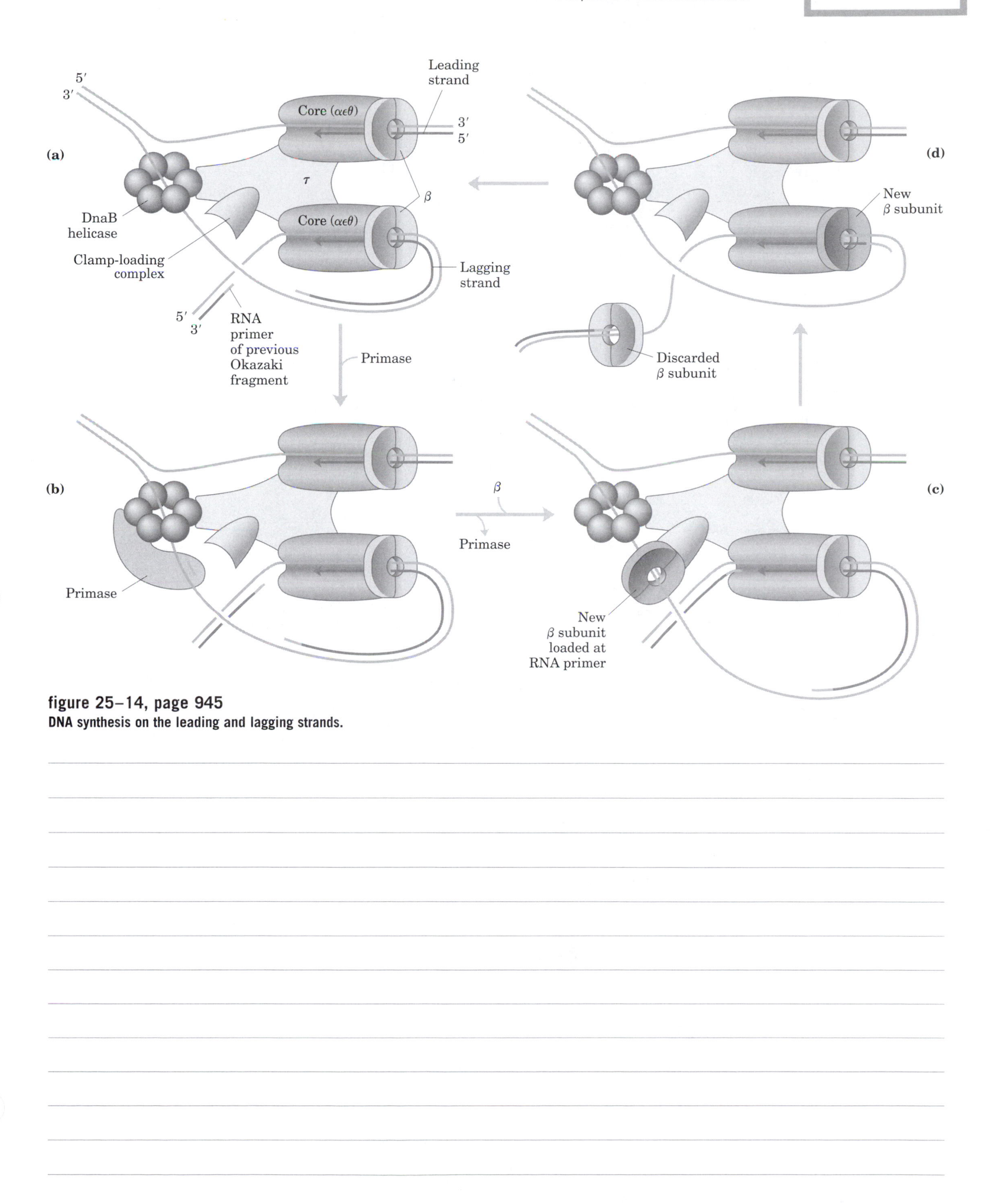

figure 25–14, page 945
DNA synthesis on the leading and lagging strands.

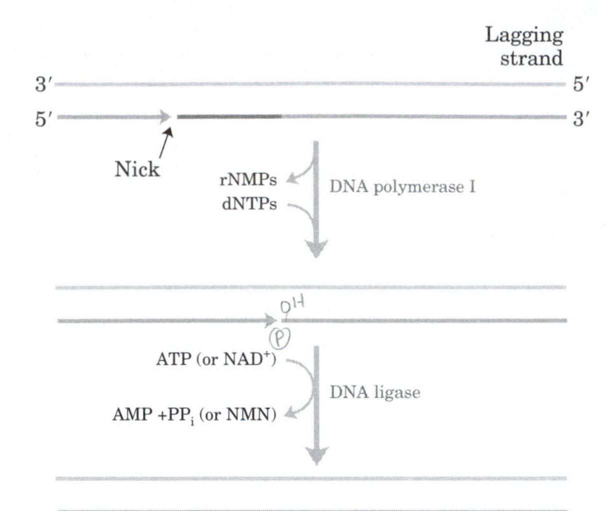

figure 25–15, page 946

RNA primers in the lagging strand are removed by the 5′→3′ exonuclease activity of DNA polymerase I and replaced with DNA by the same enzyme.

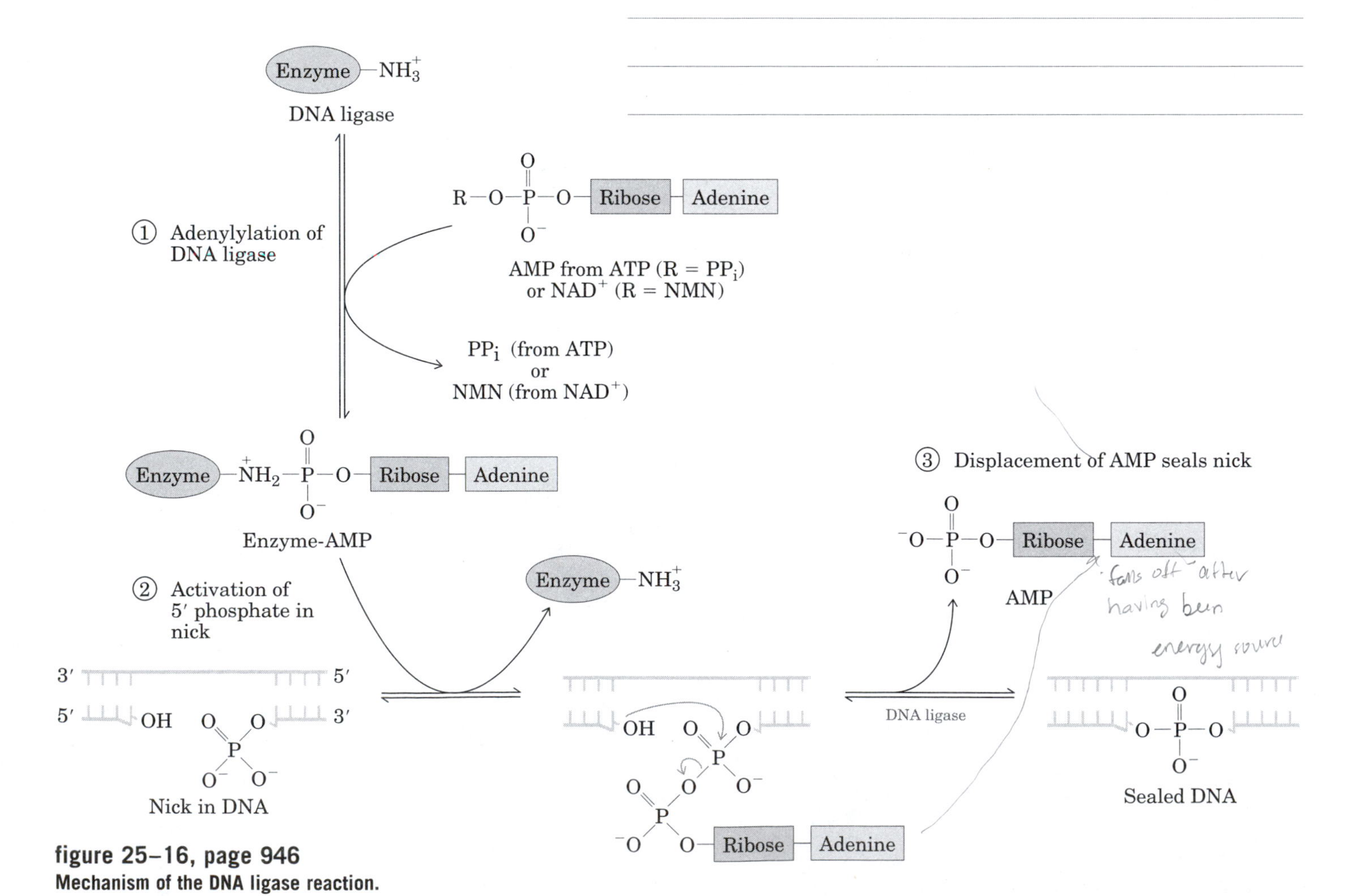

figure 25–16, page 946

Mechanism of the DNA ligase reaction.

table 25–4, page 947

Proteins at the *E. coli* Replication Fork

Protein	M_r	Number of subunits	Function
SSB	75,600	4	Binding to single-stranded DNA
DnaB protein (helicase)	300,000	6	DNA unwinding; primosome constituent
Primase (DnaG protein)	60,000	1	RNA primer synthesis; primosome constituent
DNA polymerase III	900,000	18–20	New strand elongation
DNA polymerase I	103,000	1	Filling of gaps, excision of primers
DNA ligase	74,000	1	Ligation
DNA gyrase (DNA topoisomerase II)	400,000	4	Supercoiling

Modified from Kornberg, A. (1982) *Supplement to DNA Replication,* Table S11–2, W.H. Freeman and Company, New York.

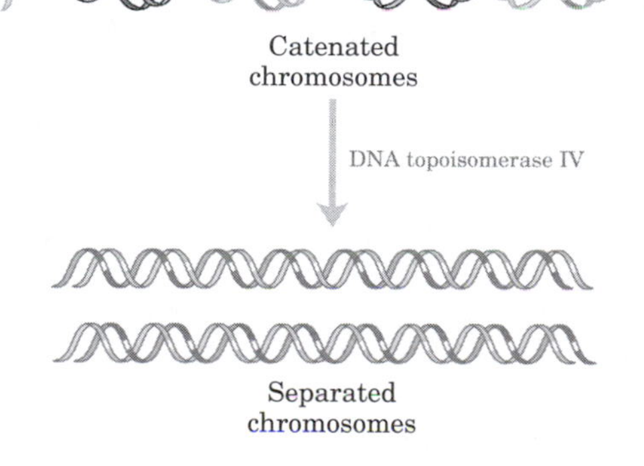

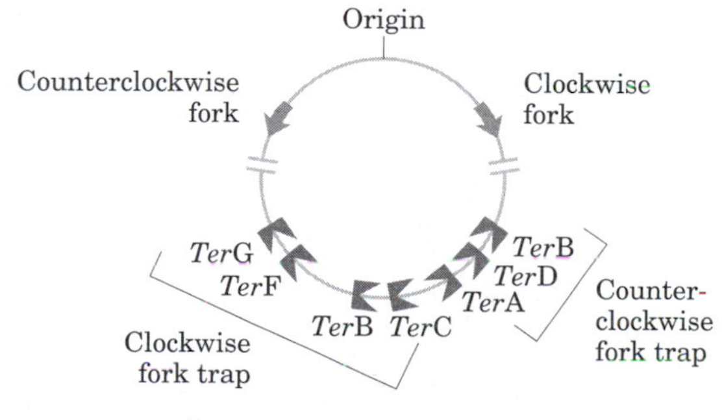

figure 25–17, page 947
Termination of chromosome replication in *E. coli*.

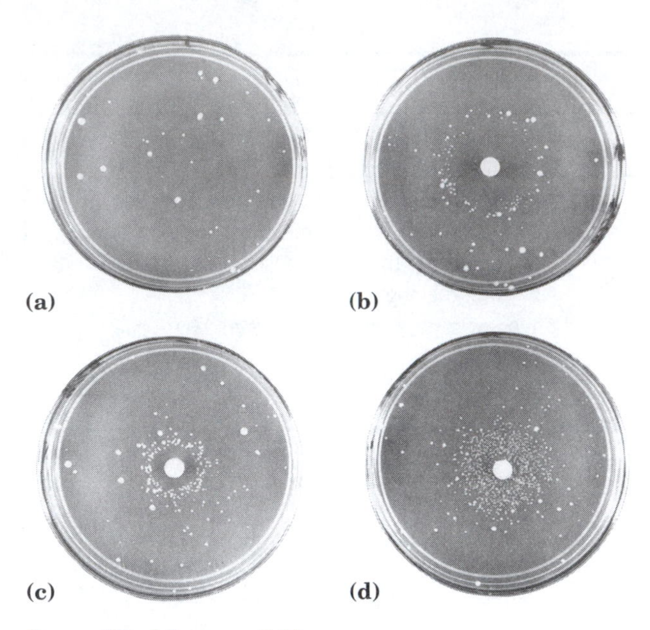

figure 25–18, page 949
Ames test for carcinogens, based on their mutagenicity.

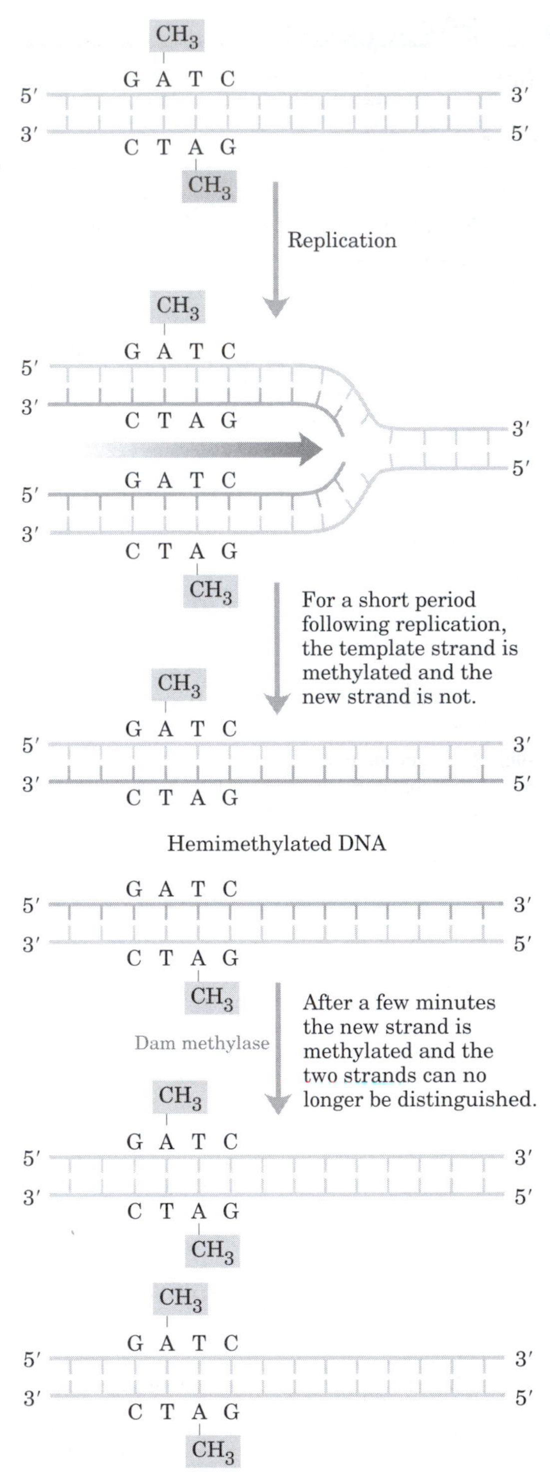

figure 25–19, page 950
Methylation and mismatch repair.

table 25–5, page 951

Types of DNA Repair Systems in *E. coli*

Enzymes/proteins	Type of damage
Mismatch repair	
Dam methylase	Mismatches
MutH, MutL, MutS proteins	
DNA helicase II	
SSB	
DNA polymerase III	
Exonuclease I	
Exonuclease VII	
RecJ nuclease	
Exonuclease X	
DNA ligase	
Base-excision repair	
DNA glycosylases	Abnormal bases (uracil, hypoxanthine, xanthine); alkylated bases; pyrimidine dimers in some other organisms
AP endonucleases	
DNA polymerase I	
DNA ligase	
Nucleotide-excision repair	
ABC excinuclease	DNA lesions that cause large structural changes (e.g., pyrimidine dimers)
DNA polymerase I	
DNA ligase	
Direct repair	
DNA photolyases	Pyrimidine dimers
O^6-Methylguanine-DNA methyltransferase	O^6-Methylguanine

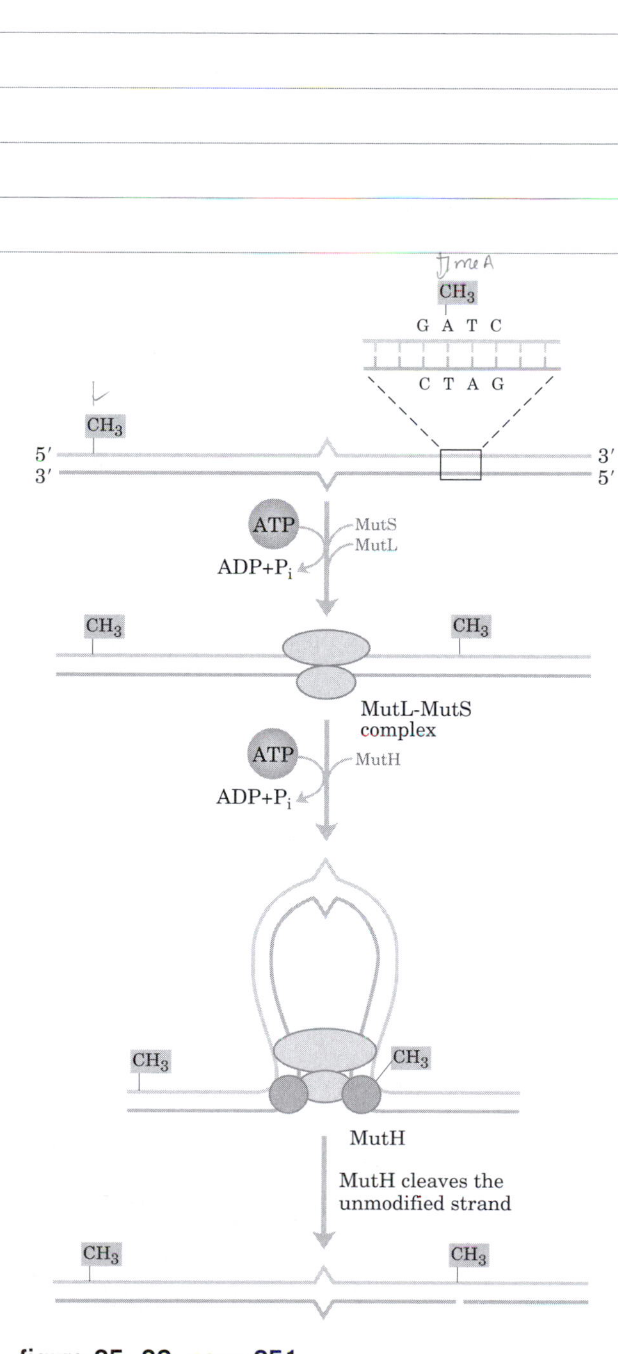

figure 25–20, page 951
Model for the early steps of methyl-directed mismatch repair.

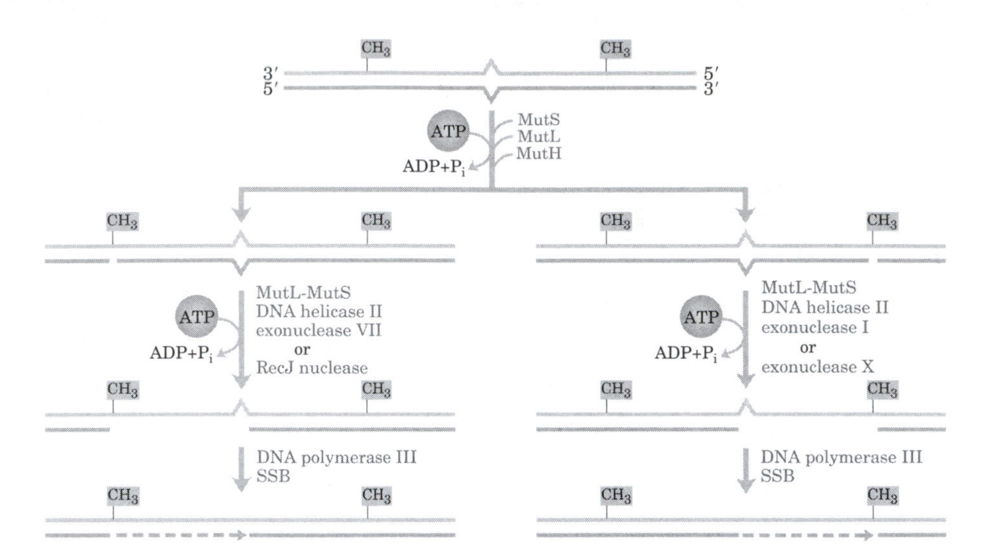

figure 25–21, page 952
Completing methyl-directed mismatch repair.

figure 25–22, page 954
DNA repair by the base-excision repair pathway.

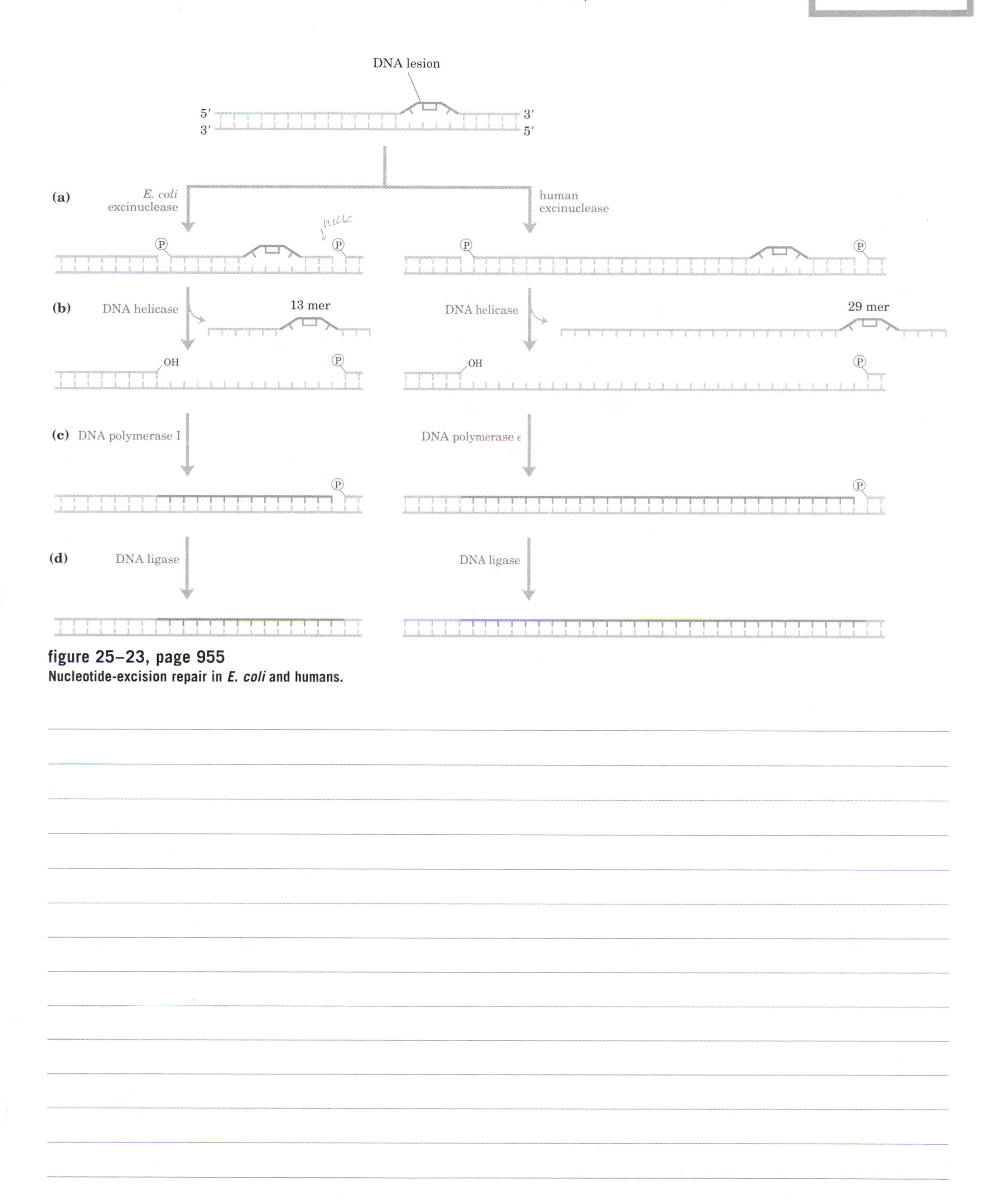

figure 25–23, page 955
Nucleotide-excision repair in *E. coli* and humans.

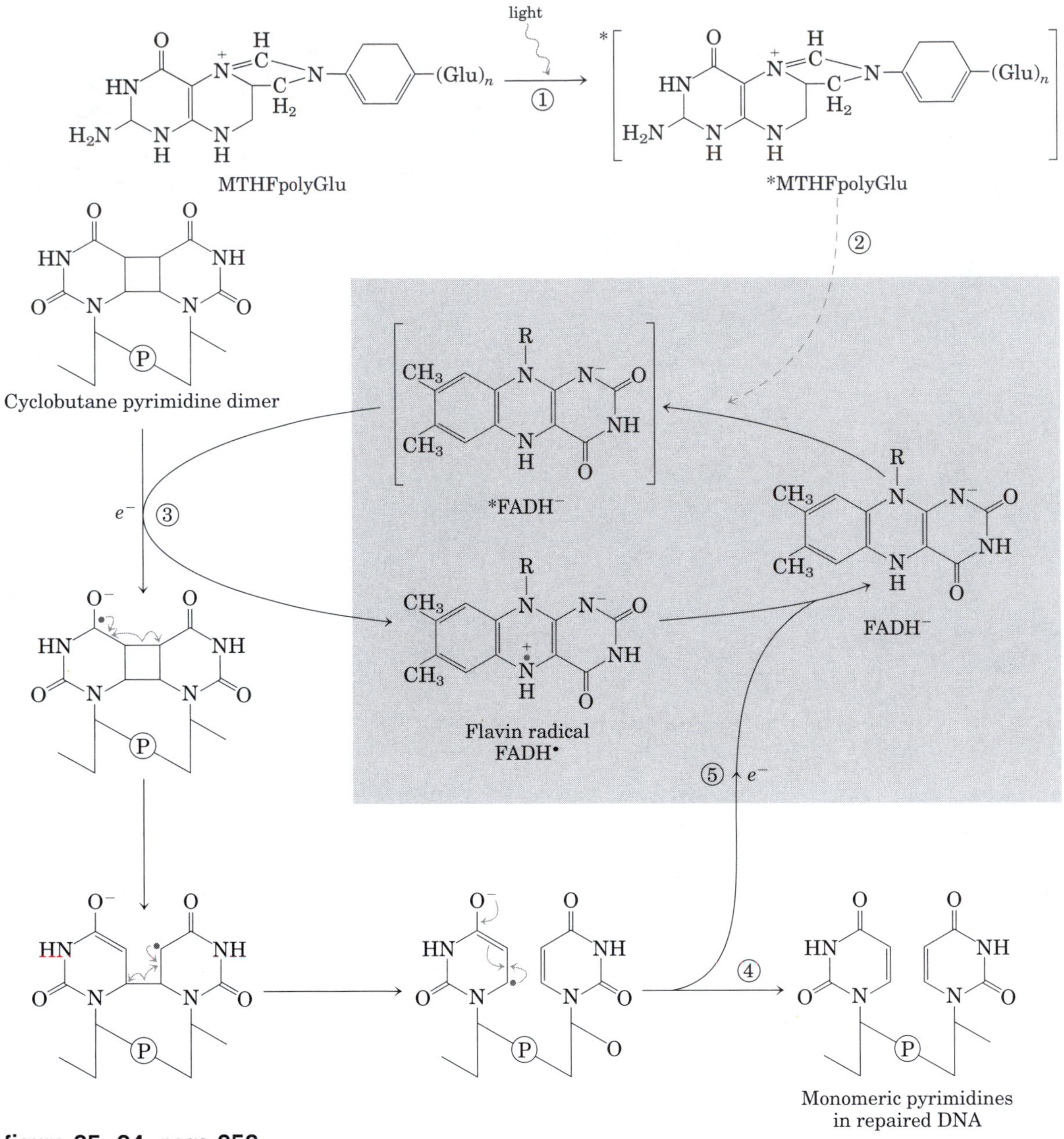

figure 25–24, page 956
Repair of pyrimidine dimers with photolyase.

Guanine

Cytosine

methylation and replication

O^6-Methylguanine

Thymine

(a)

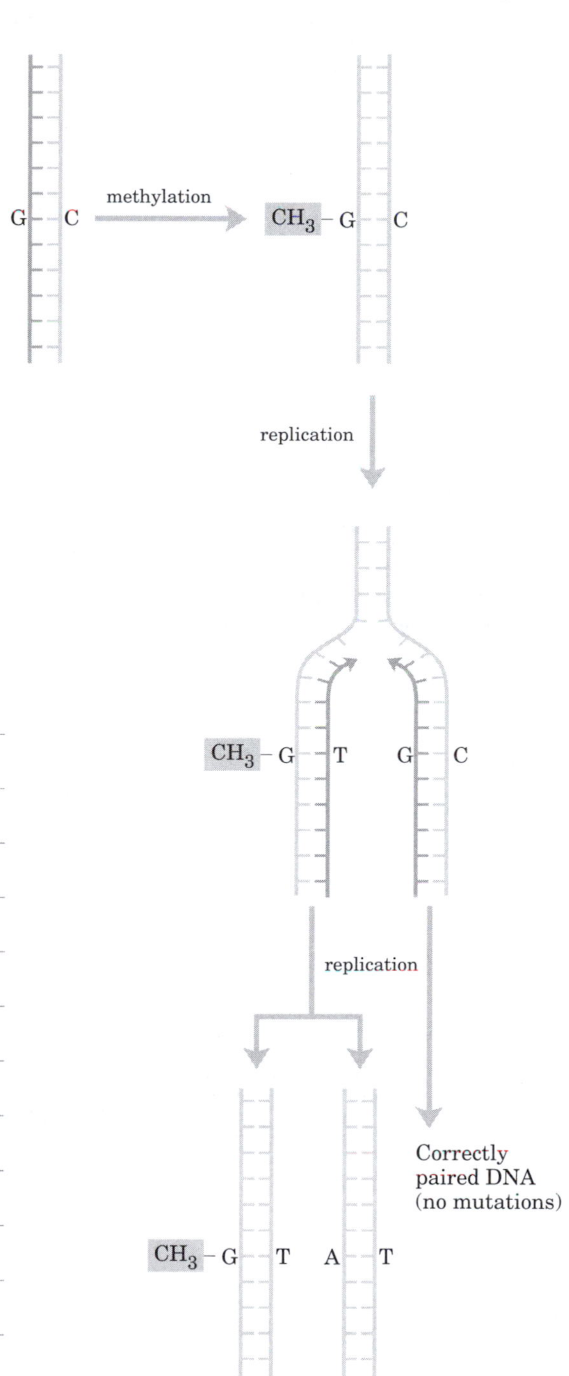

figure 25–25, page 957

Example of how DNA damage results in mutations.

Cys—SH

active

Cys—S—CH_3

inactive

methyltransferase

O^6-Methylguanine nucleotide

Guanine nucleotide

Methyltransferase, page 957

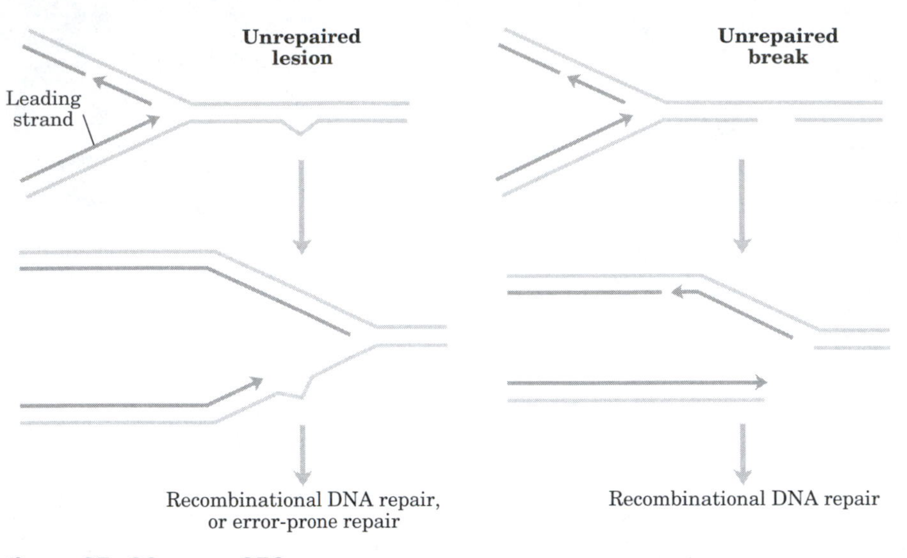

figure 25–26, page 958
DNA damage and its effect on DNA replication.

table 25–6, page 959

Genes Induced as Part of the SOS Response in *E. coli*

Gene name	Protein encoded and/or role in DNA repair
Genes of known function	
*pol*B (*din*A)	Encodes polymerization subunit of DNA polymerase II, required for replication restart in recombinational DNA repair
*uvr*A *uvr*B	Encode ABC excinuclease subunits UvrA and UvrB
*umu*C *umu*D	Encode DNA polymerase V
*sul*A	Encodes protein that inhibits cell division, possibly to allow time for DNA repair
*rec*A	Encodes RecA protein required for error-prone repair and recombinational repair
*din*B	Encodes DNA polymerase IV
Genes involved in DNA metabolism, but role in DNA repair unknown	
ssb	Encodes single-stranded DNA-binding protein (SSB)
*uvr*D	Encodes DNA helicase II (DNA-unwinding protein)
*him*A	Encodes subunit of integration host factor, involved in site-specific recombination, replication, transposition, regulation of gene expression
*rec*N	Required for recombinational repair
Genes of unknown function	
*din*D	
*din*F	

Note: Some of these genes and their functions are further discussed in Chapter 28.

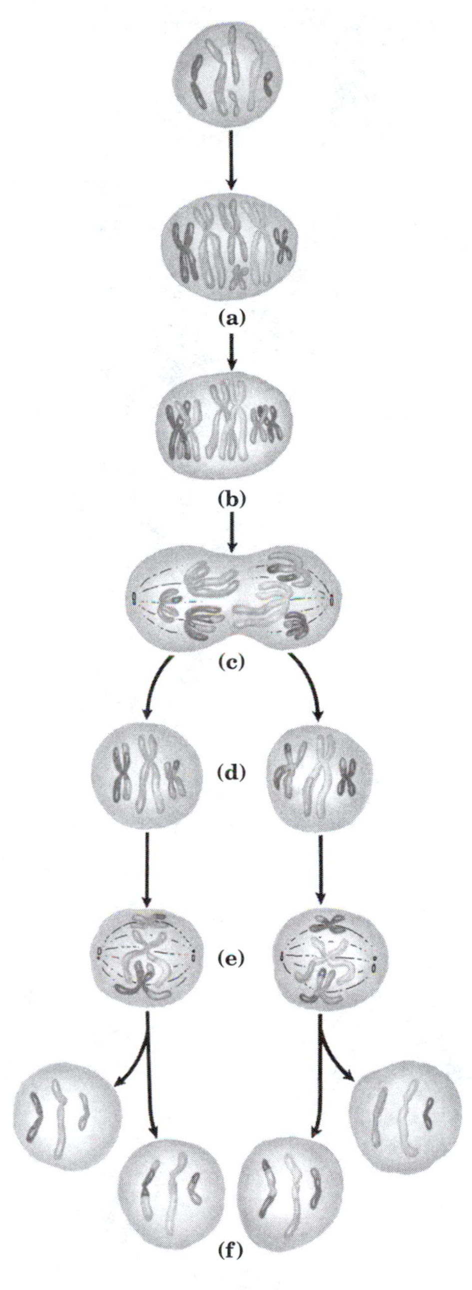

figure 25–27, page 960
Meiosis in eukaryotic germ-line cells.

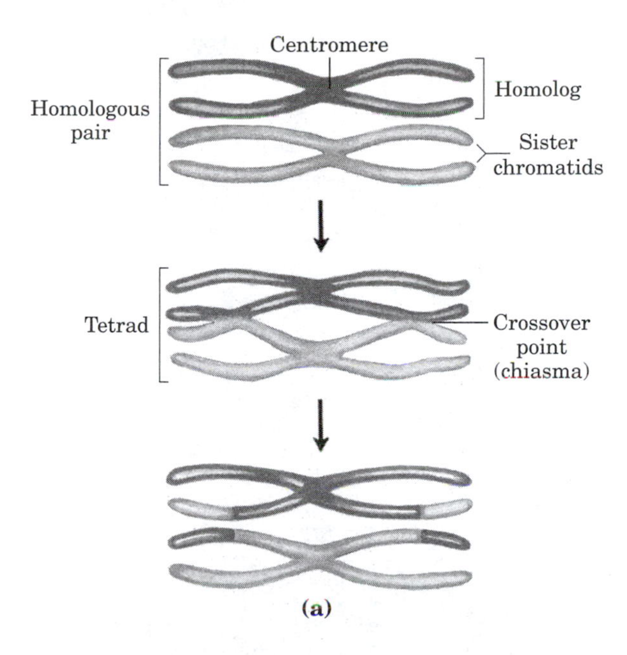

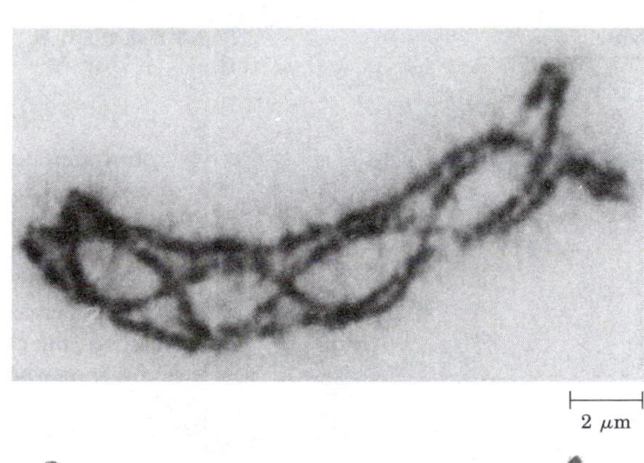

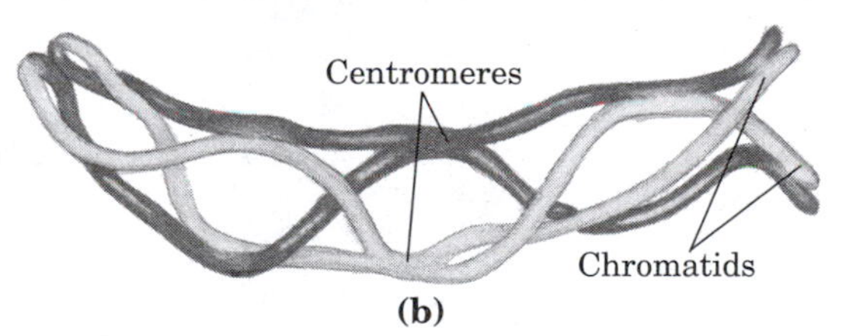

figure 25–28, page 961
Crossing over.

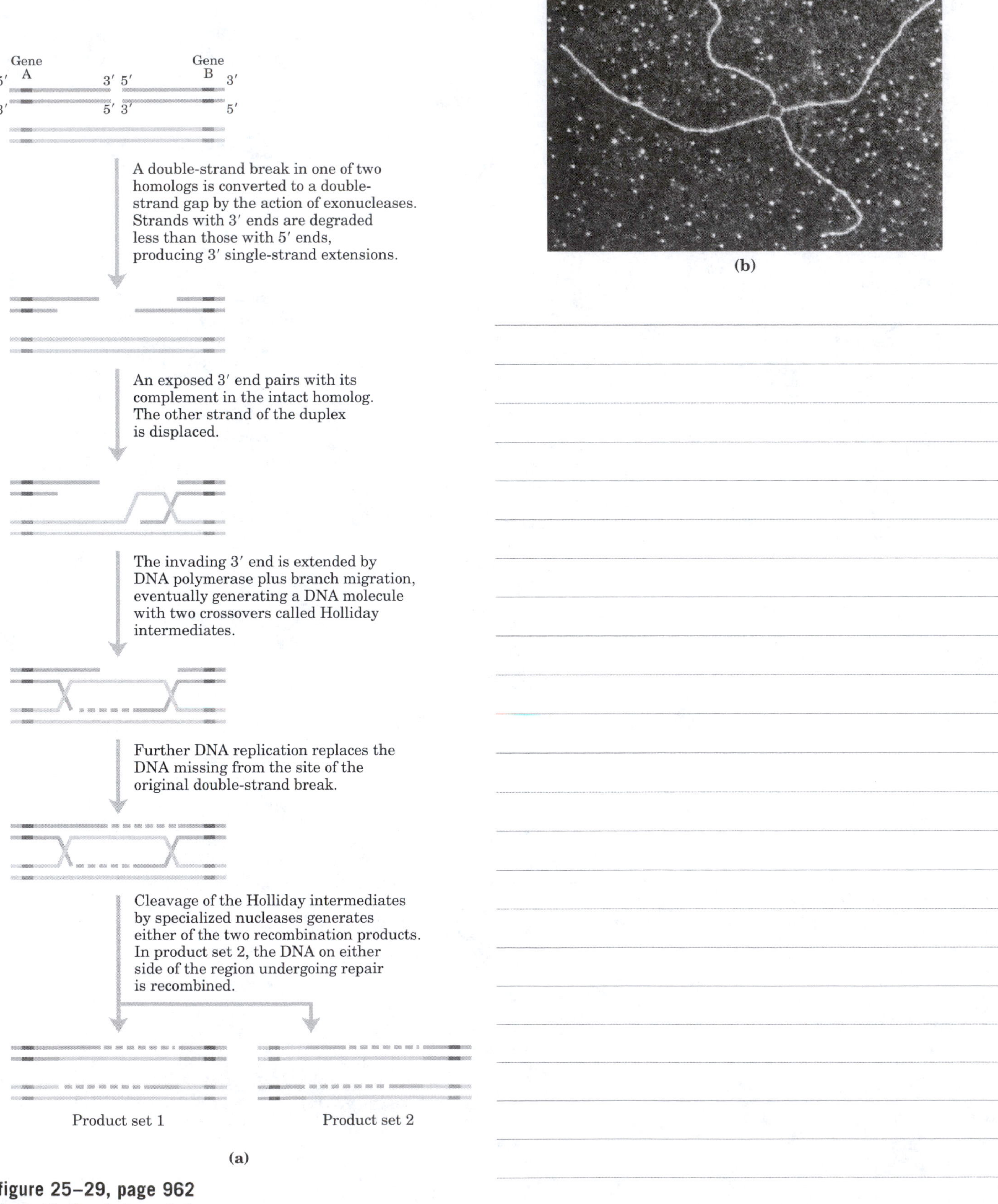

figure 25–29, page 962
Recombination during meiosis.

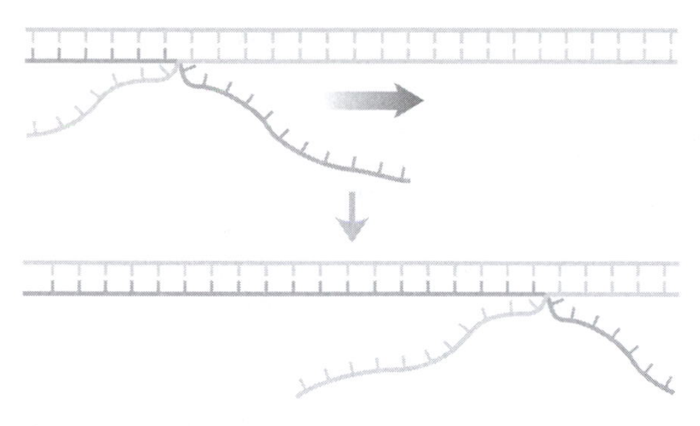

figure 25–30, page 963
Branch migration.

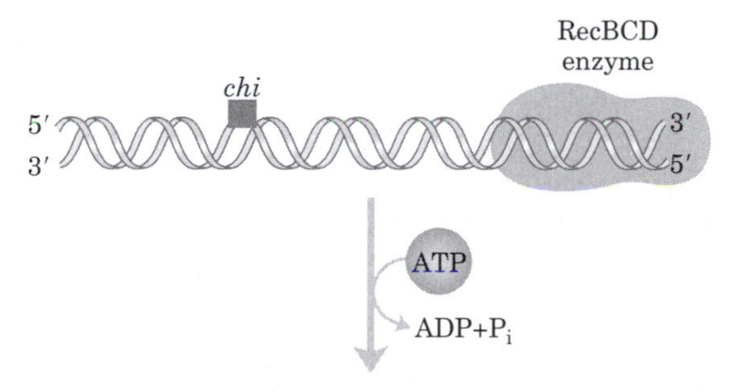

ATP

ADP+P_i

Helicase and nuclease activities of enzyme degrade the DNA.

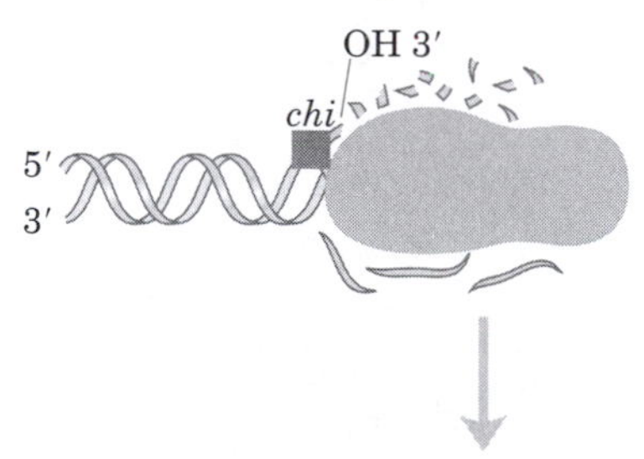

On reaching a *chi* sequence, nuclease activity on the strand with the 3′ end is suppressed. The other strand continues to be degraded, generating a 3′-terminal single-stranded end.

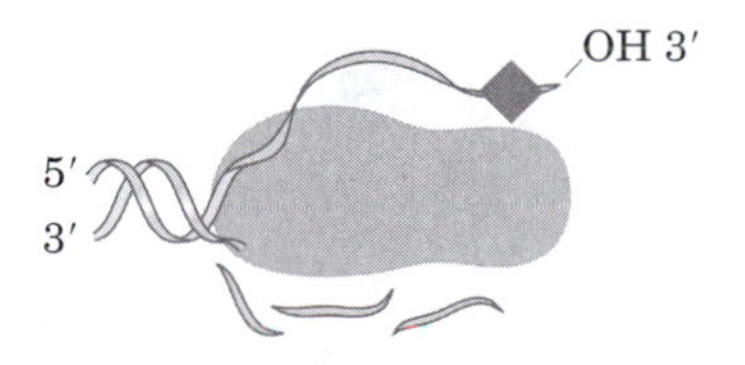

figure 25–31, page 964
Helicase and nuclease activities of the RecBCD enzyme.

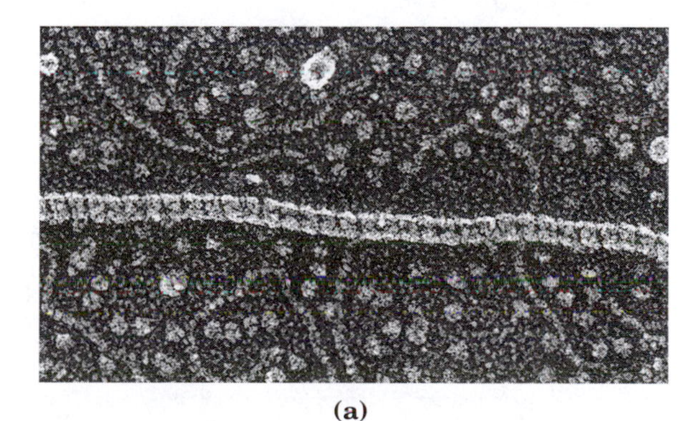

(a)

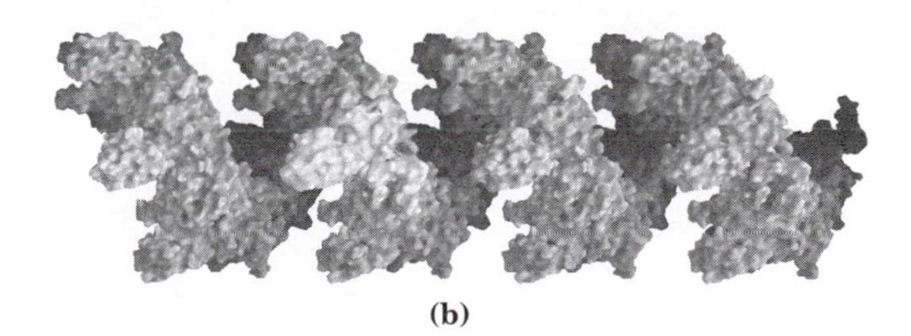

(b)

figure 25–32, page 964
RecA.

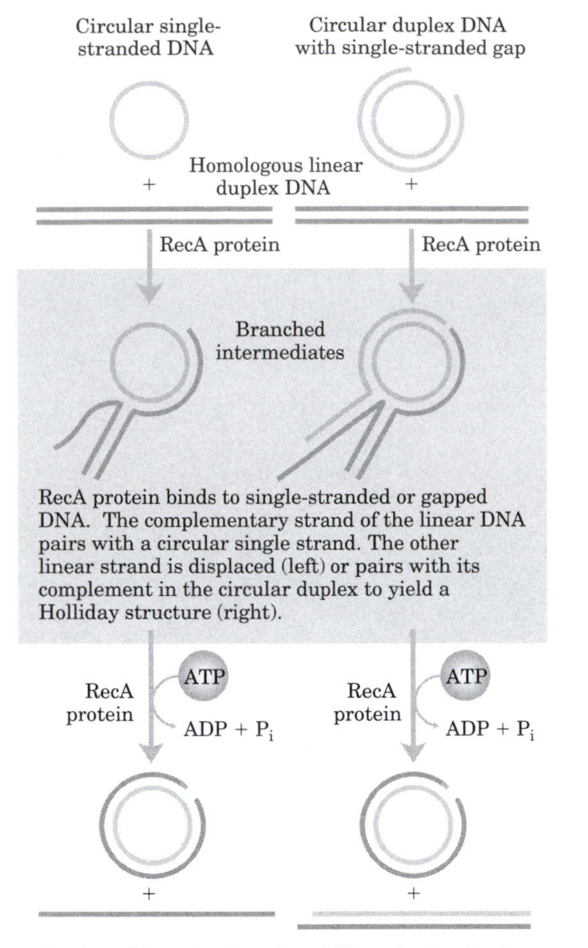

figure 25–33, page 965
DNA strand-exchange reactions promoted by RecA protein in vitro.

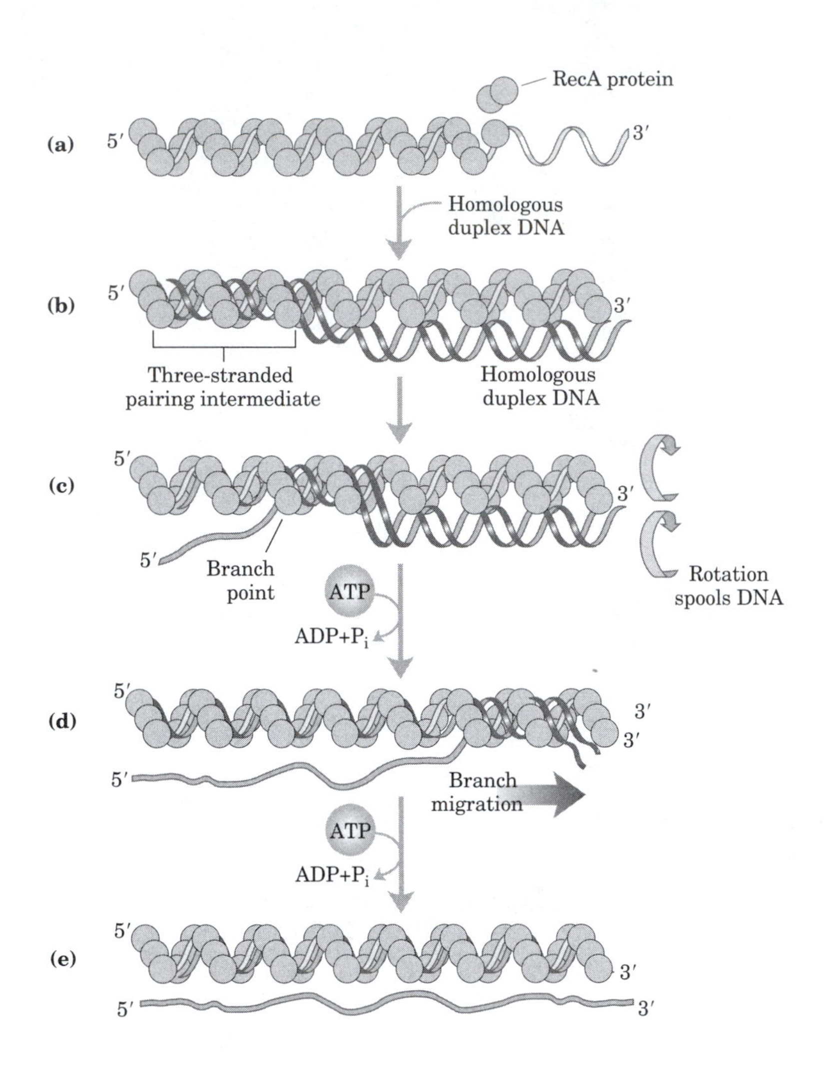

figure 25–34, page 965
Model for DNA strand exchange mediated by RecA protein.

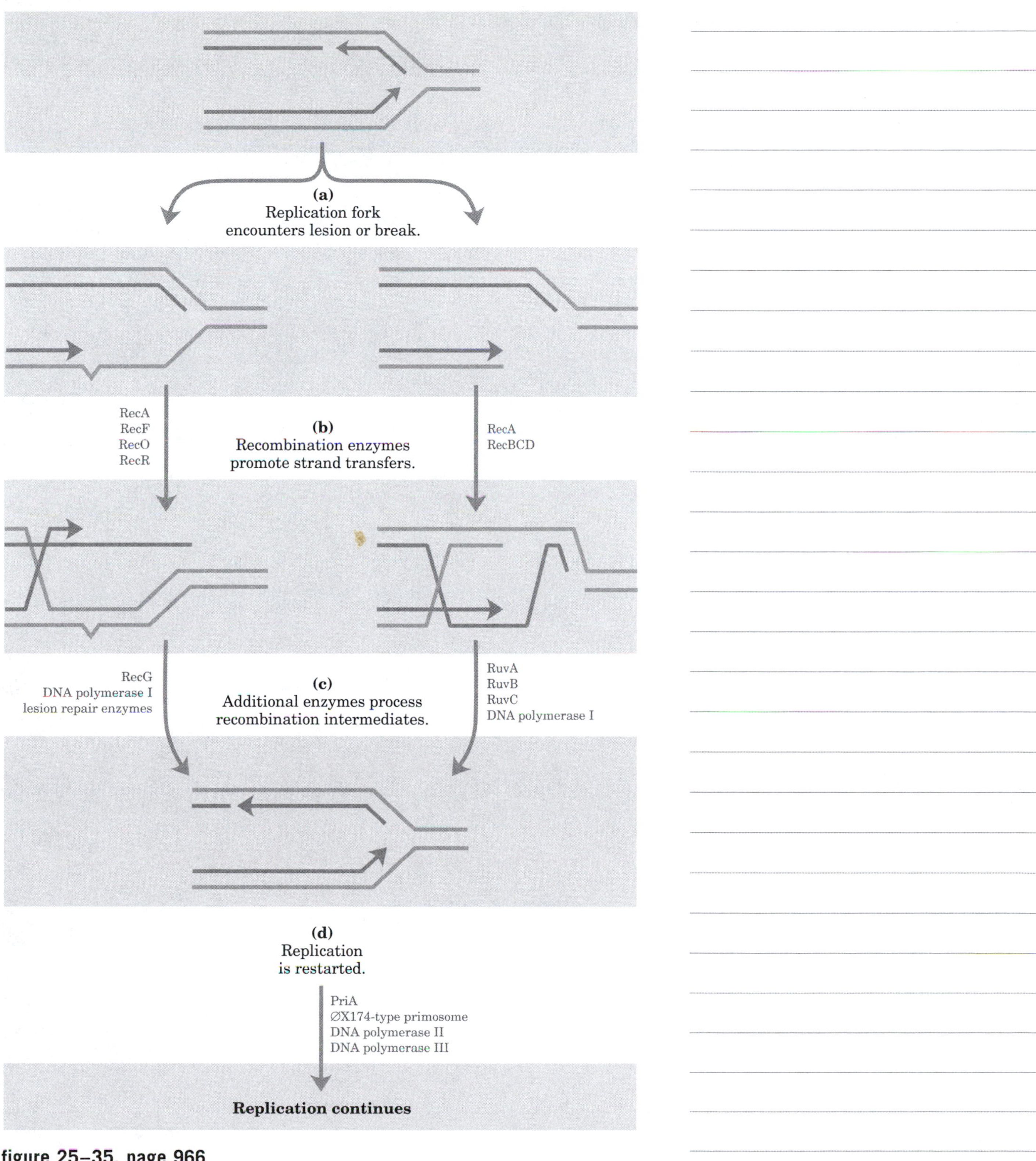

figure 25–35, page 966
Models for recombinational DNA repair of stalled replication forks.

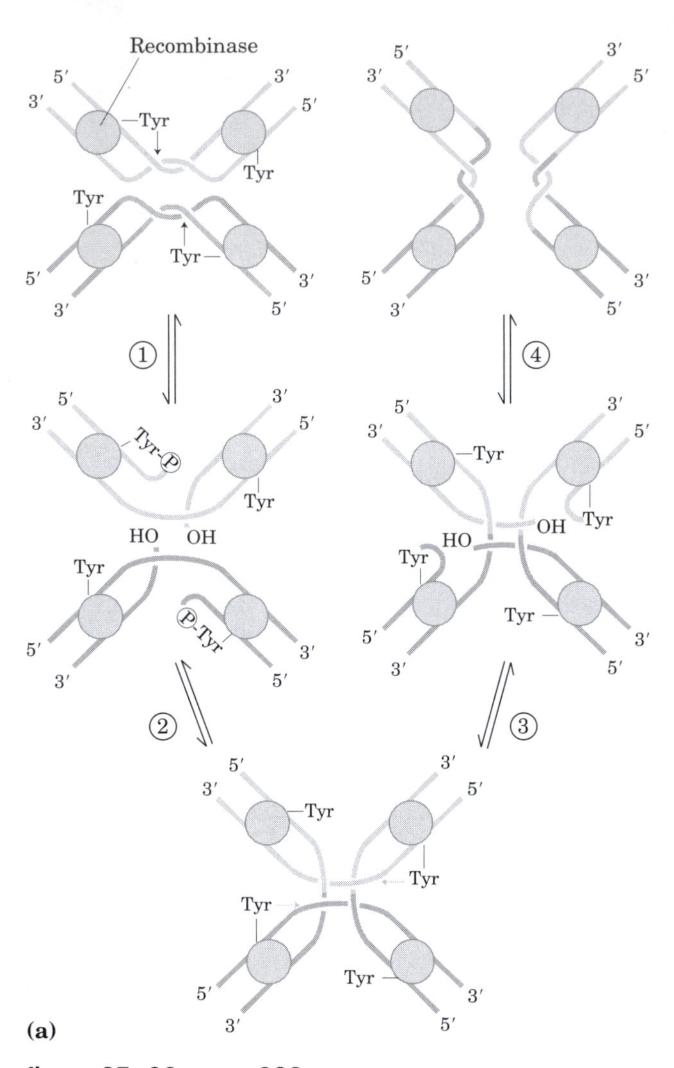

(b)

figure 25–36, page 968
A site-specific recombination reaction.

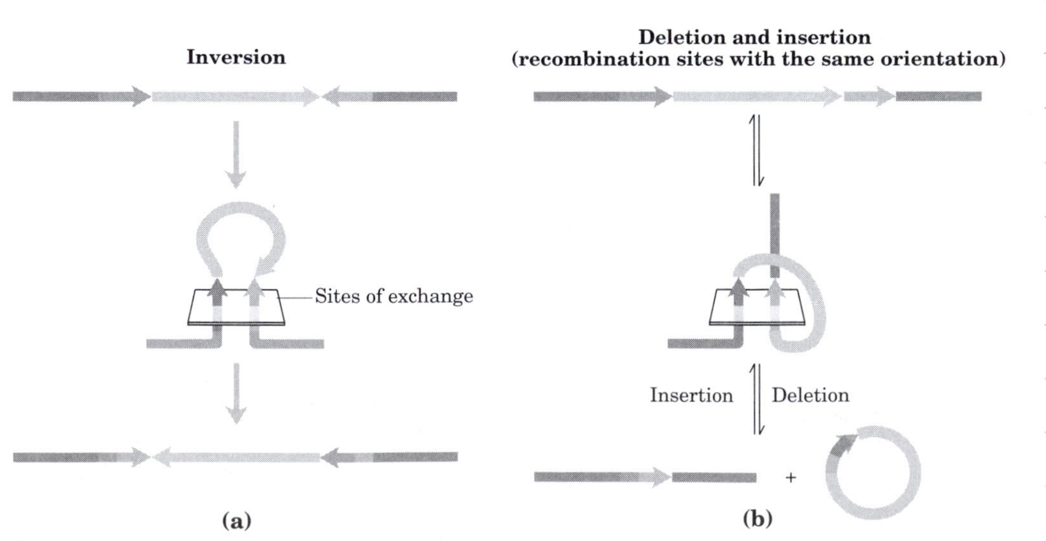

figure 25–37, page 969
Effects of site-specific recombination.

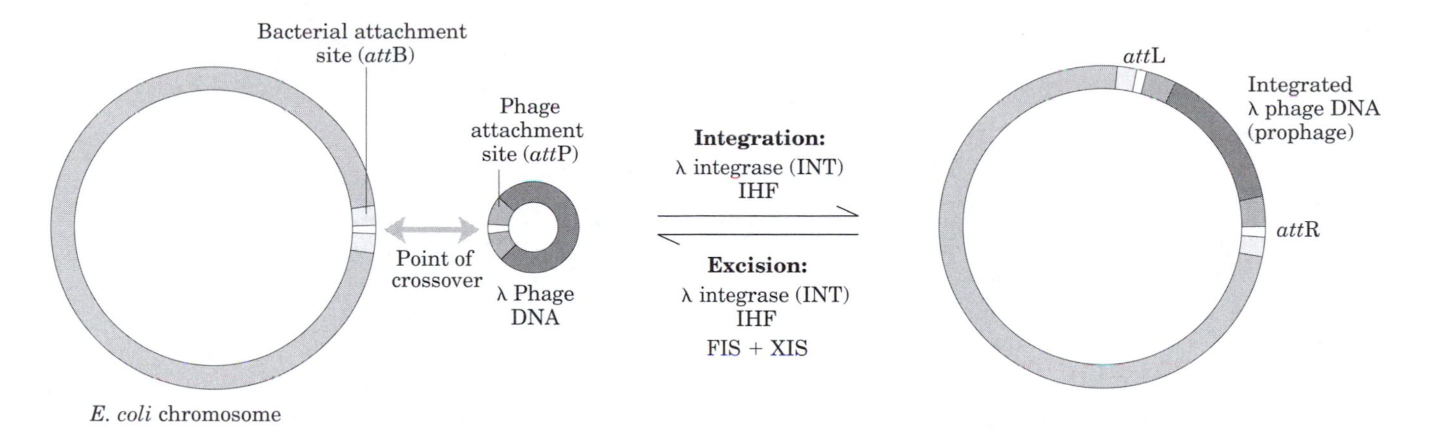

figure 25–38, page 969
Integration and excision of bacteriophage λ DNA at the chromosomal target site.

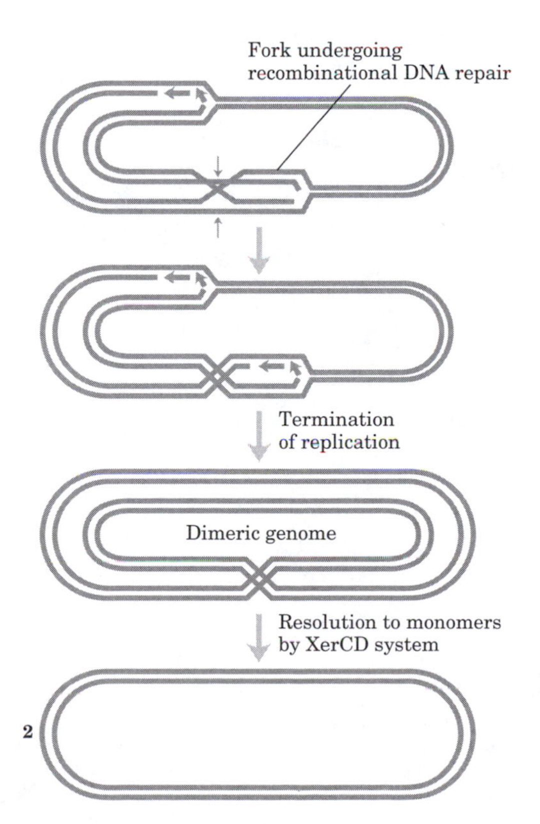

figure 25–39, page 970
DNA deletion to undo a deleterious effect of recombinational DNA repair.

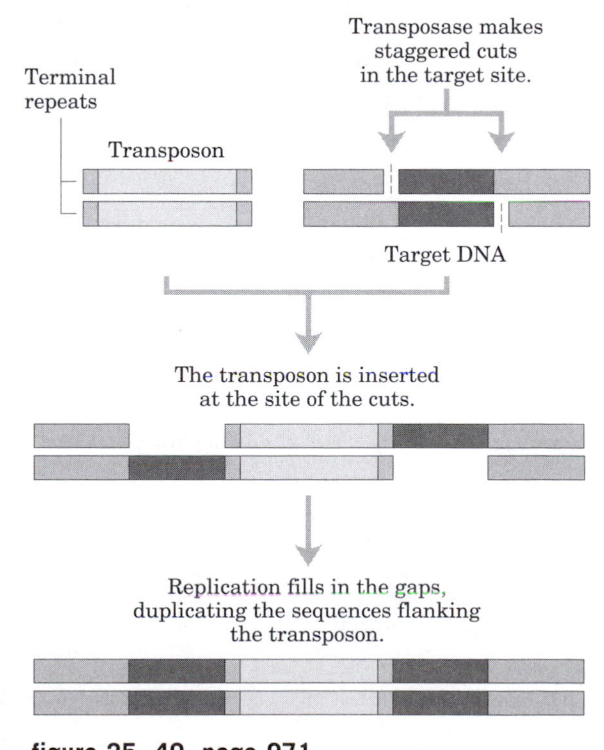

figure 25–40, page 971
Duplication of the DNA sequence at a target site when a transposon is inserted.

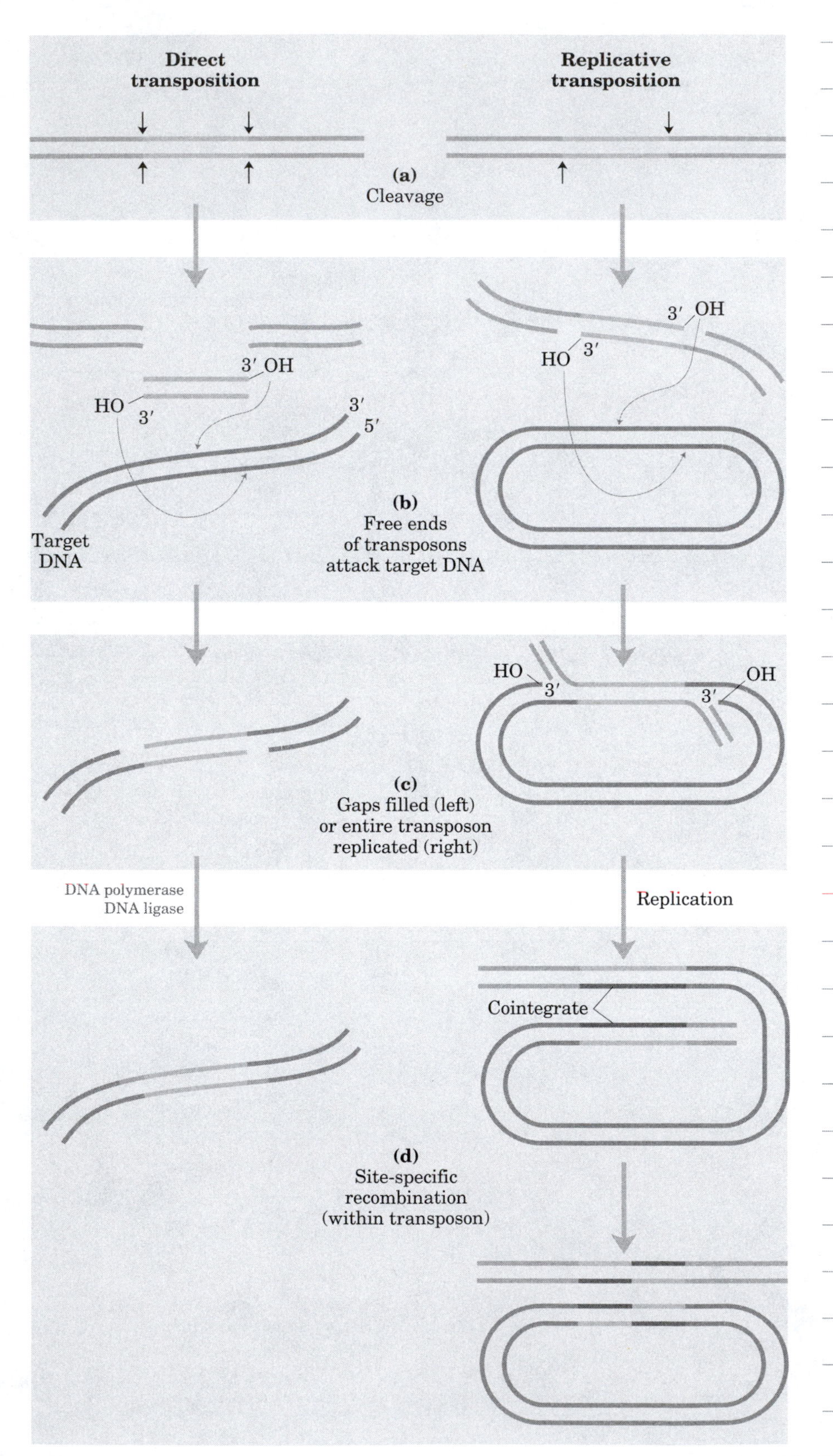

figure 25–41, page 972

Two general pathways for transposition: direct (simple) and replicative.

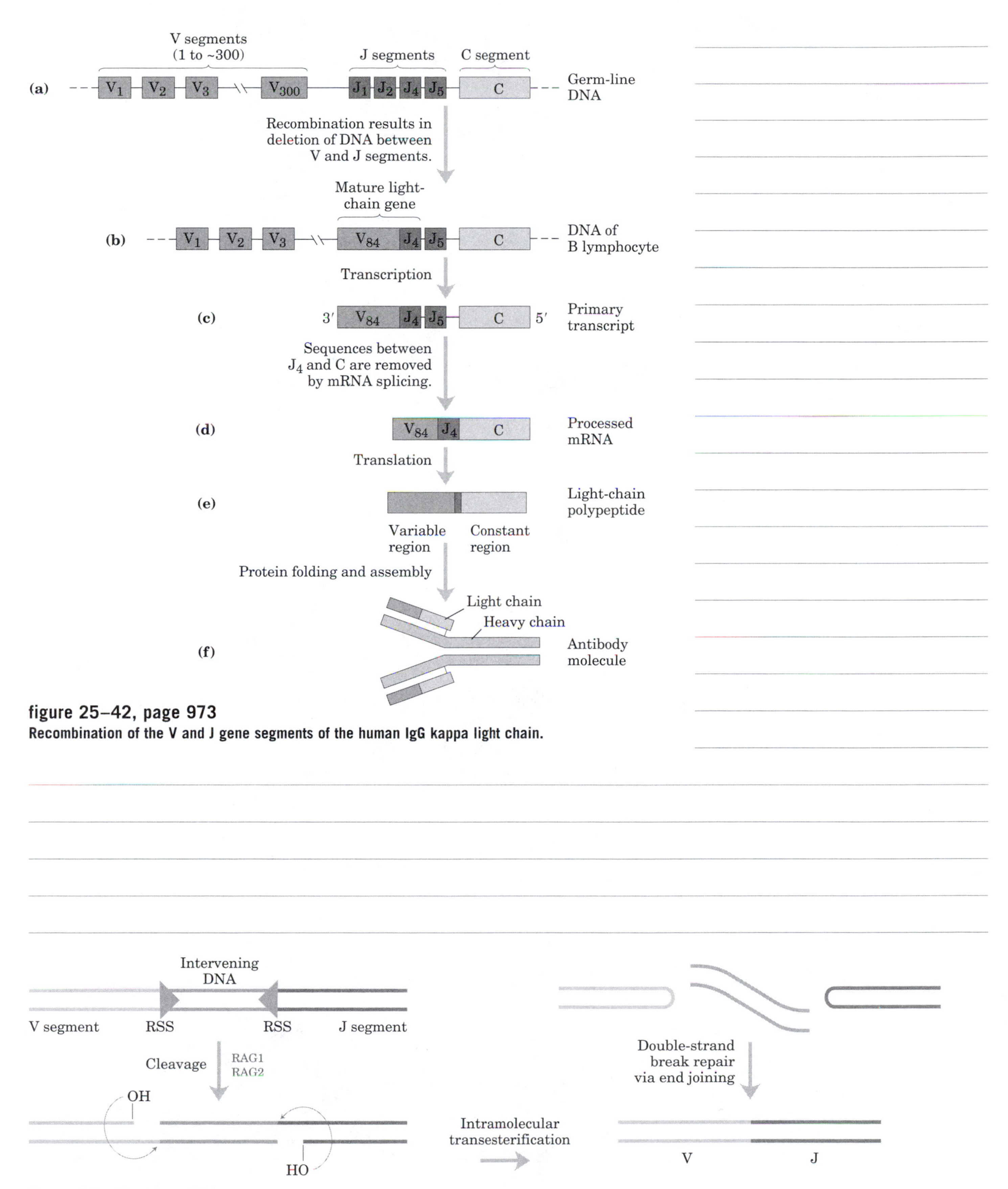

figure 25–42, page 973
Recombination of the V and J gene segments of the human IgG kappa light chain.

figure 25–43, page 974
Mechanism of immunoglobulin gene rearrangement.

chapter

26 RNA Metabolism

When you study for exams, don't forget to review The Minicourses on the UNDERSTAND! BIOCHEMISTRY CD that came with your textbook.

Minicourses that apply to this Chapter include:

Molecules of Life
- Fundamentals of Gene Action
- An Interactive Gallery of Protein Structures

Proteins in Action
- Catalysis and Regulation

Nucleic Acids and Their Expression
- DNA Binding Proteins
- Transcription
- Translation
- RNA Processing

The Dividing Cell
- Viruses

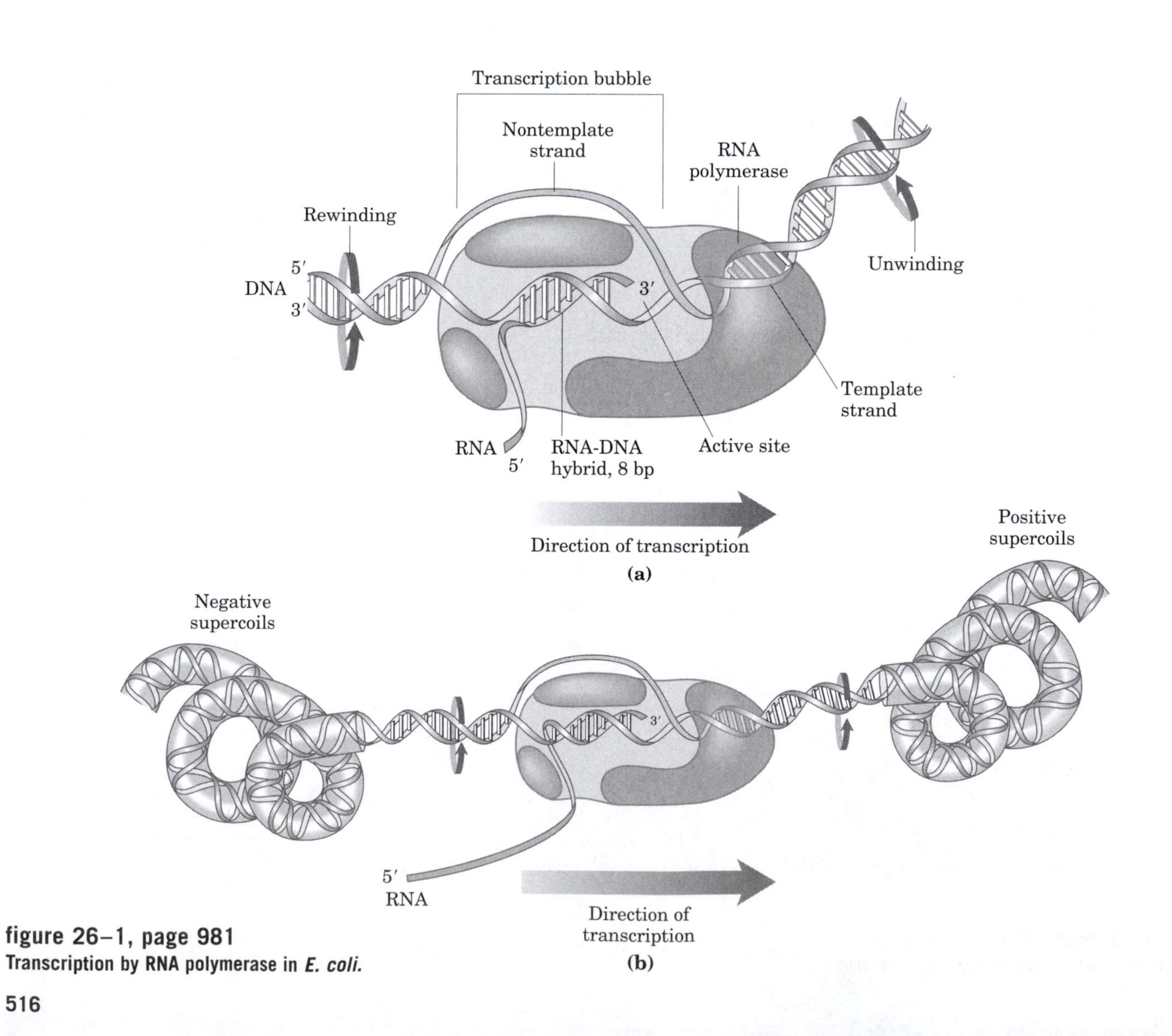

figure 26–1, page 981
Transcription by RNA polymerase in *E. coli.*

(5′) CGCTATAGCGTTT(3′)	DNA nontemplate (coding) strand
(3′) GCGATATCGCAAA(5′)	DNA template strand
(5′) CGCUAUAGCGUUU(3′)	RNA transcript

figure 26–2, page 982
Template and nontemplate (coding) DNA strands.

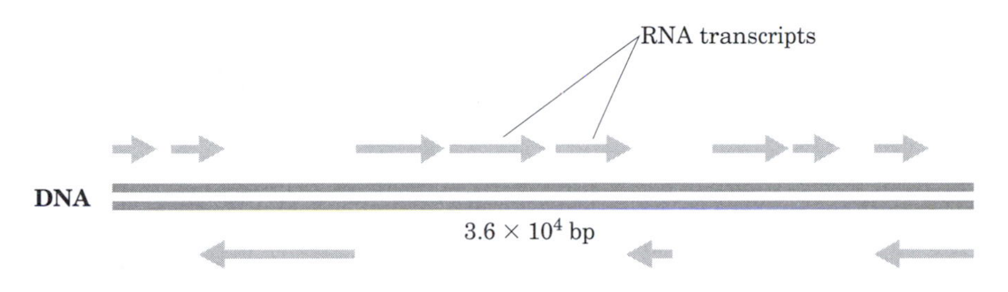

figure 26–3, page 982
The genetic information of the adenovirus genome (a conveniently simple example) is encoded by a double-stranded DNA molecule of 36,000 base pairs, both strands of which encode proteins.

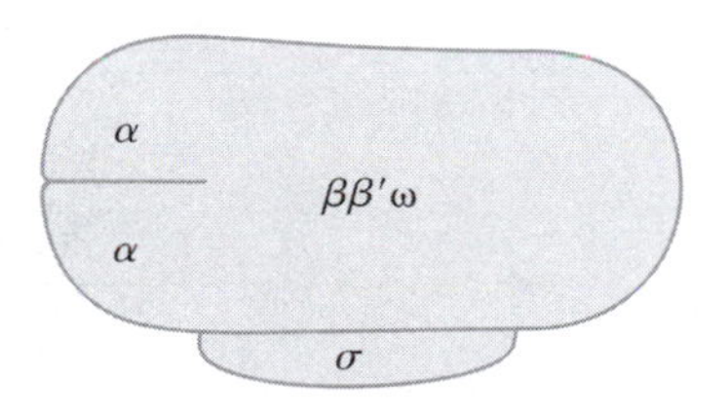

figure 26–4, page 982
Structure of *E. coli* RNA polymerase and its function in transcription.

	UP element		−35 Region	Spacer	−10 Region	Spacer	RNA start
							+1
Consensus sequence	NNAAA$^{AA}_{TT}$T$^{A}_{T}$TTTTNNAAAANNN	N	TTGACA	N_{17}	TATAAT	N_6	
*rrn*B P1	AGAAAATTATTTTAAATTTCCT	N	GTGTCA	N_{16}	TATAAT	N_8	A
trp			TTGACA	N_{17}	TTAACT	N_7	A
lac			TTTACA	N_{17}	TATGTT	N_6	A
*rec*A			TTGATA	N_{16}	TATAAT	N_7	A
*ara*BAD			CTGACG	N_{18}	TACTGT	N_6	A

figure 26–5, page 983
Typical *E. coli* promoters recognized by an RNA polymerase holoenzyme containing σ^{70}.

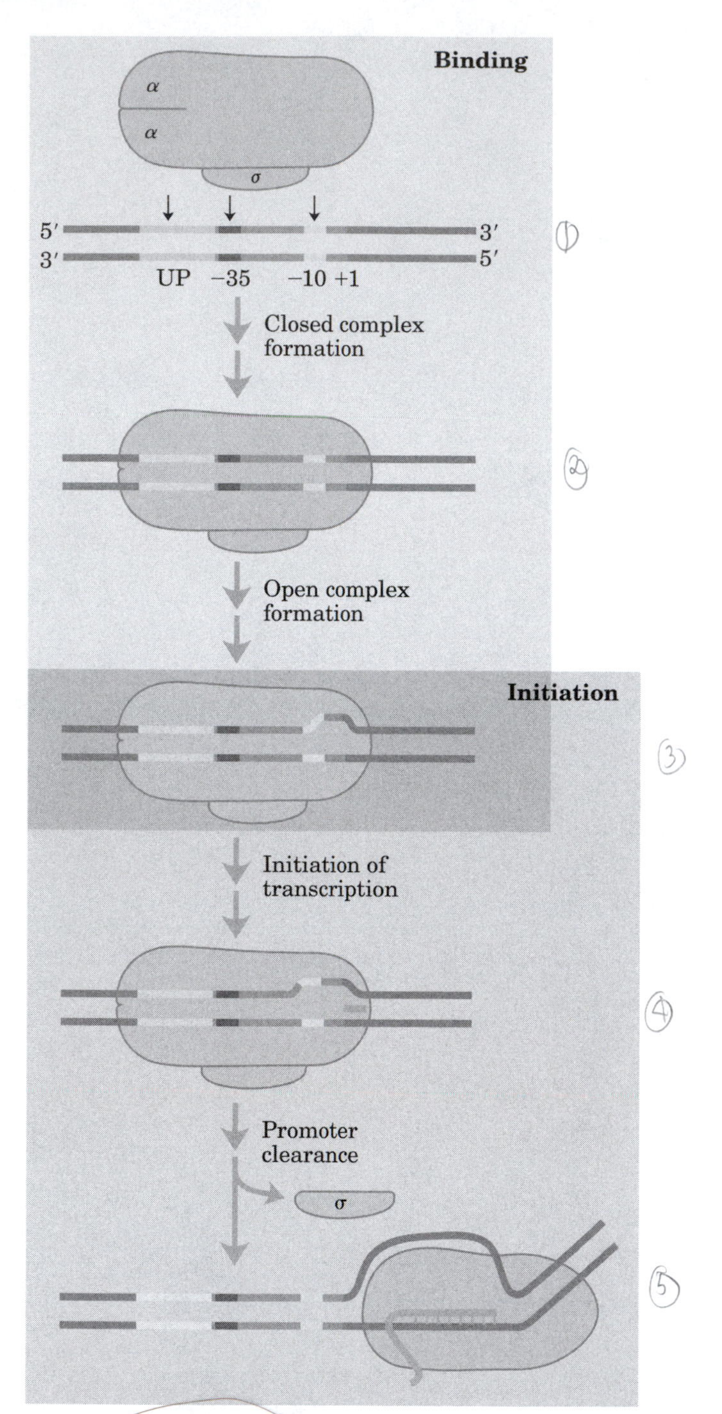

figure 26–6, page 984
Steps in the initiation of transcription by *E. coli* RNA polymerase.

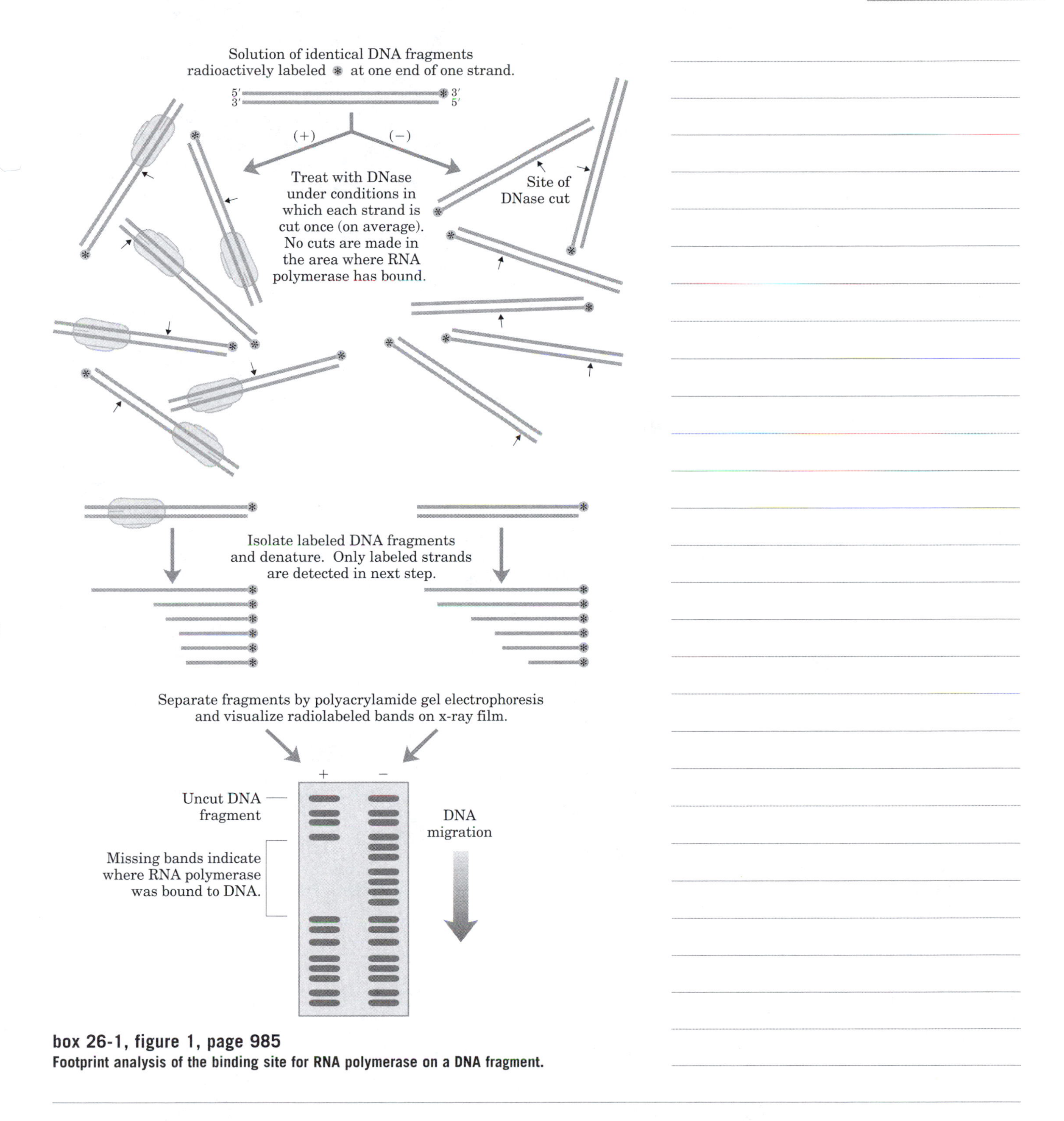

box 26-1, figure 1, page 985

Footprint analysis of the binding site for RNA polymerase on a DNA fragment.

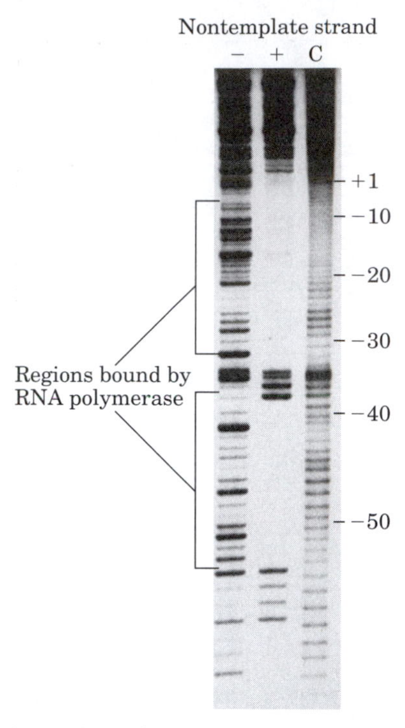

box 26-1, figure 2, page 985
Footprinting results of RNA polymerase binding to the *lac* promoter (see Fig. 26–5).

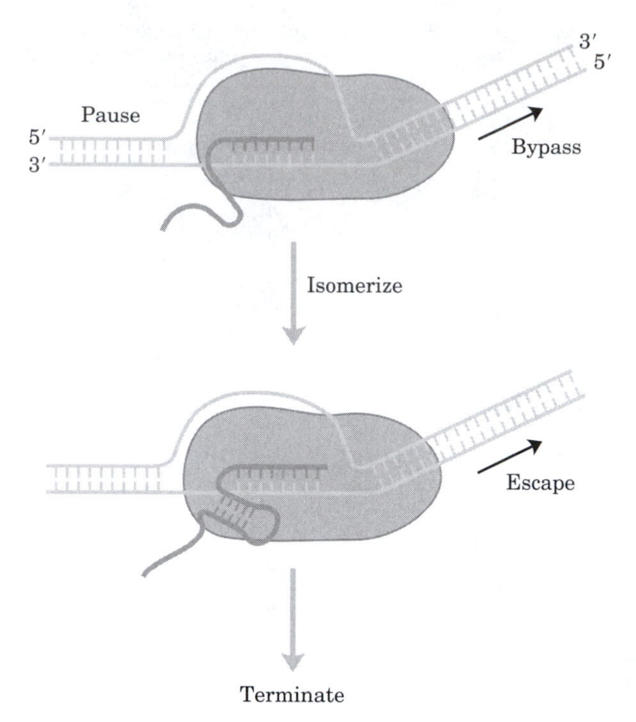

figure 26–7, page 986
A model for ρ-independent termination of transcription in *E. coli.*

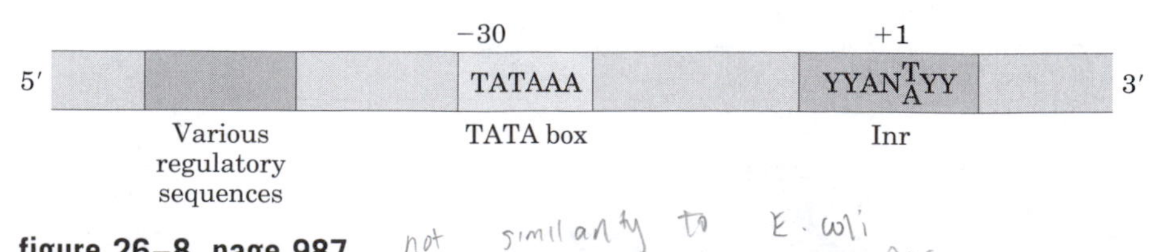

figure 26–8, page 987 not similarity to E. coli promoter
Common sequences in promoters recognized by eukaryotic RNA polymerase II.

* Know table

table 26–1, page 987

Proteins Required for Transcription at the RNA Polymerase II Promoters of Eukaryotes

Transcription factor	Number of subunits	Subunit M_r	Functions
Initiation			
RNA polymerase II	12	10,000–220,000	Catalyzes RNA synthesis
TBP (TATA-binding protein)	1	38,000	Specifically recognizes the TATA box
TFIIA	3	12,000, 19,000, 35,000	Stabilizes binding of TFIIB and TBP to the promoter
TFIIB	1	35,000	Binds to TBP; recruits RNA polymerase–TFIIF complex
TFIID	12	15,000–250,000	Interacts with positive and negative regulatory proteins
TFIIE	2	34,000, 57,000	Recruits TFIIH; ATPase and helicase activities
TFIIF	2	30,000, 74,000	Binds tightly to RNA polymerase II; binds to TFIIB and prevents binding of RNA polymerase to nonspecific DNA sequences
TFIIH	12	35,000–89,000	Unwinds DNA at promoter; phosphorylates RNA polymerase; recruits nucleotide-excision repair
Elongation*			
ELL†	1	80,000	
P-TEFb	2	43,000, 124,000	
SII (TFIIS)	1	38,000	
Elongin (SIII)	3	15,000, 18,000, 110,000	

*All elongation factors suppress the pausing or arrest of transcription by the RNA polymerase II–TFIIF complex.

†The name is derived from the term *e*leven-nineteen *l*ysine-rich *l*eukemia. The gene for the factor ELL is the site of chromosomal recombination events frequently associated with the cancerous condition known as acute myeloid leukemia.

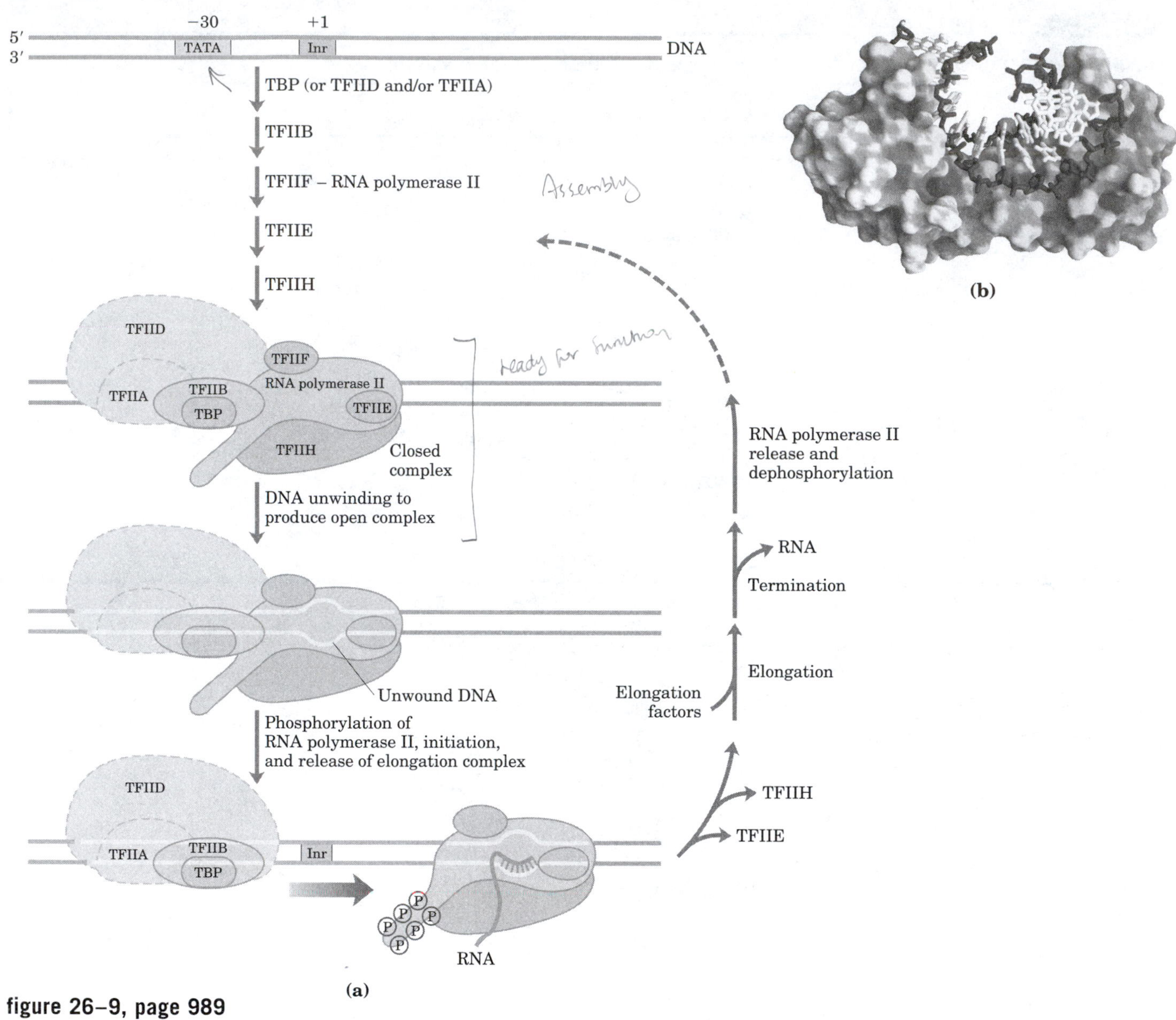

figure 26–9, page 989
Transcription at RNA polymerase II promoters.

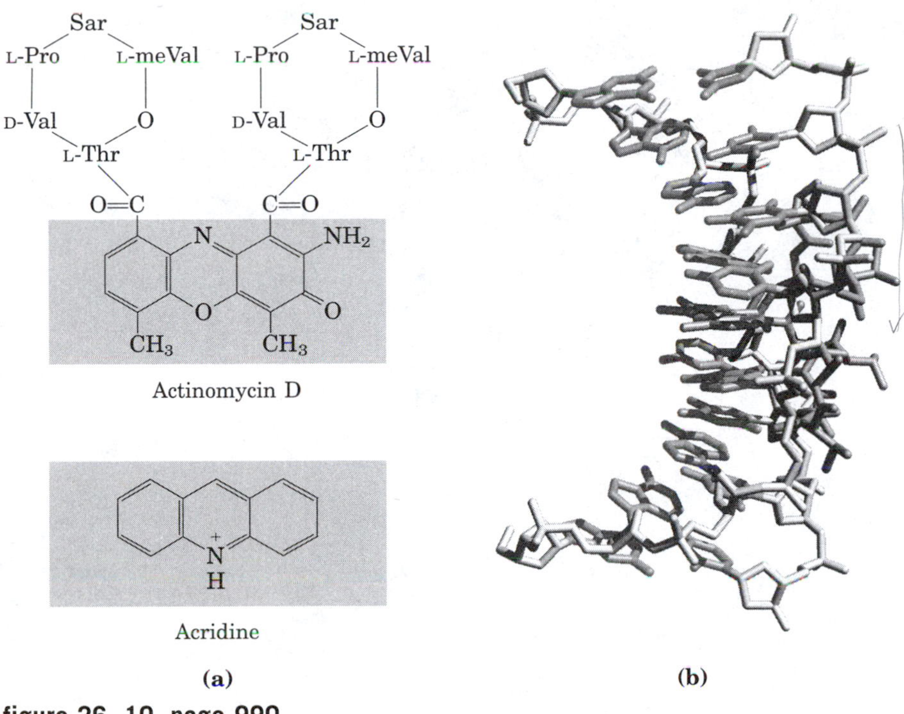

figure 26–10, page 990
Actinomycin D and acridine, inhibitors of DNA transcription.

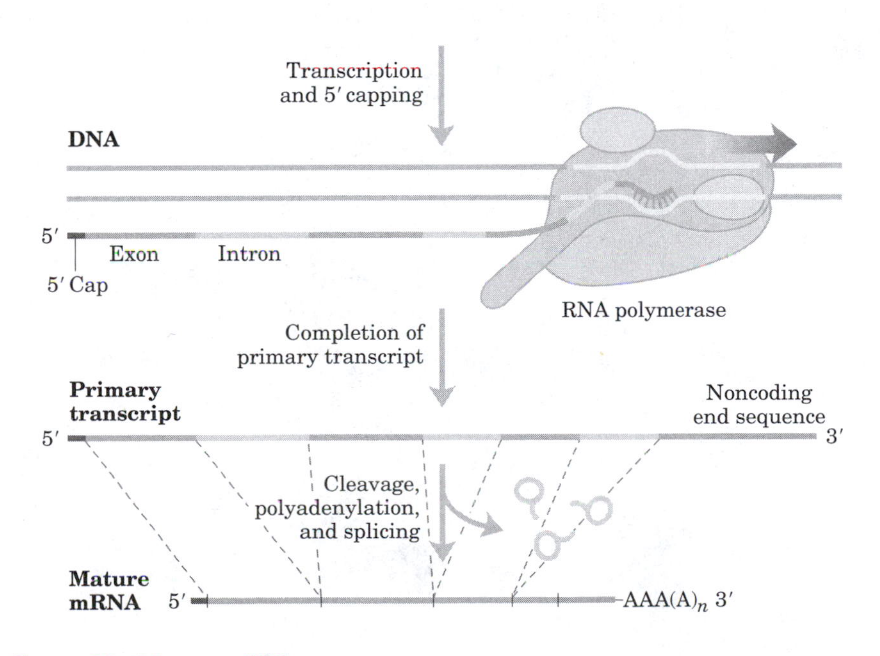

figure 26–11, page 991
Formation of the primary transcript and its processing during maturation of mRNA in a eukaryotic cell.

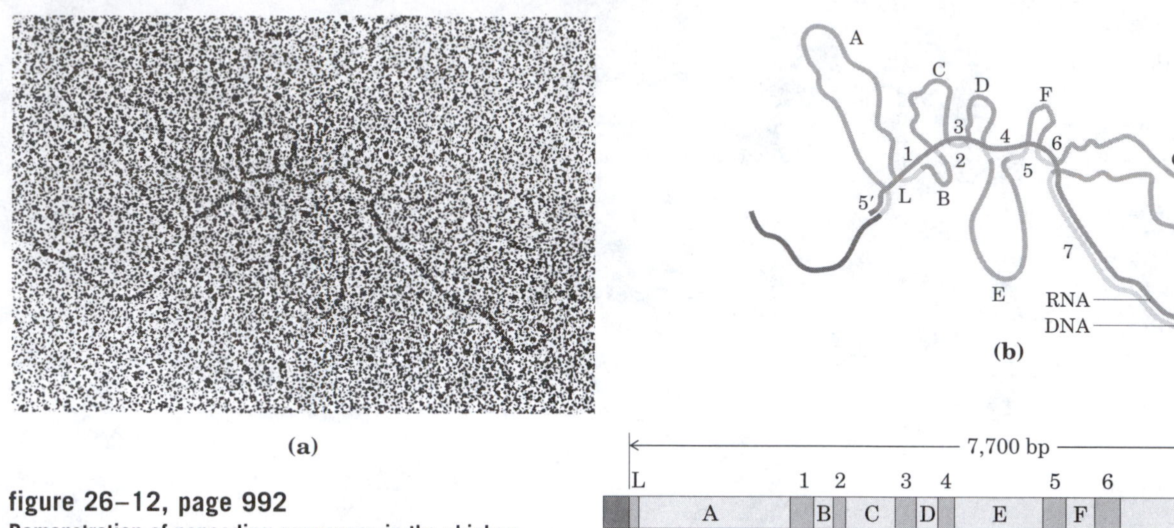

figure 26–12, page 992
Demonstration of noncoding sequences in the chicken ovalbumin gene by RNA-DNA hybridization.

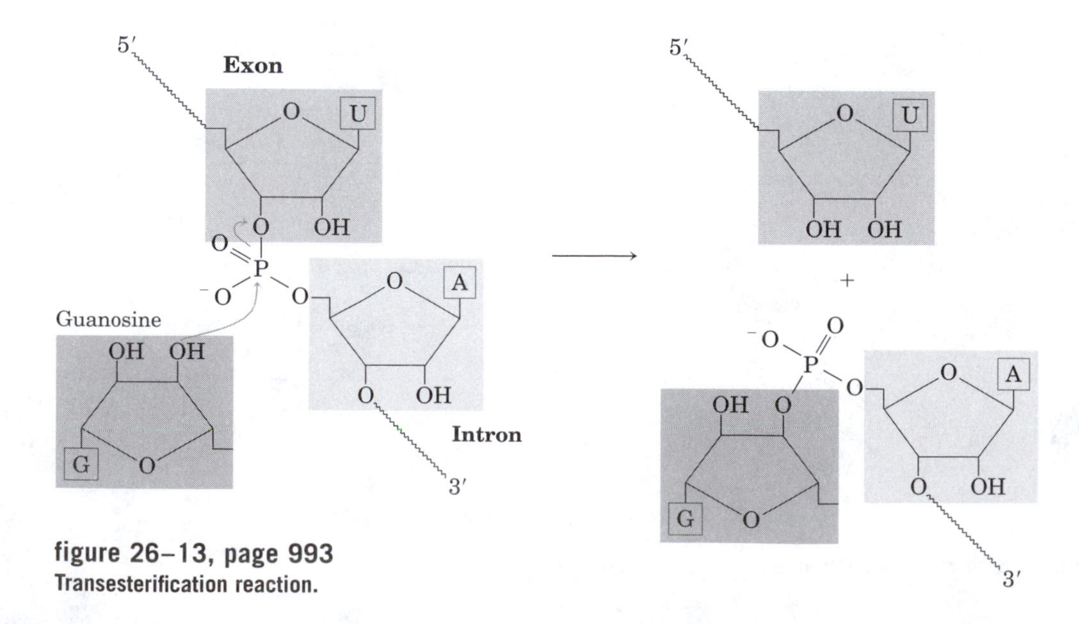

figure 26–13, page 993
Transesterification reaction.

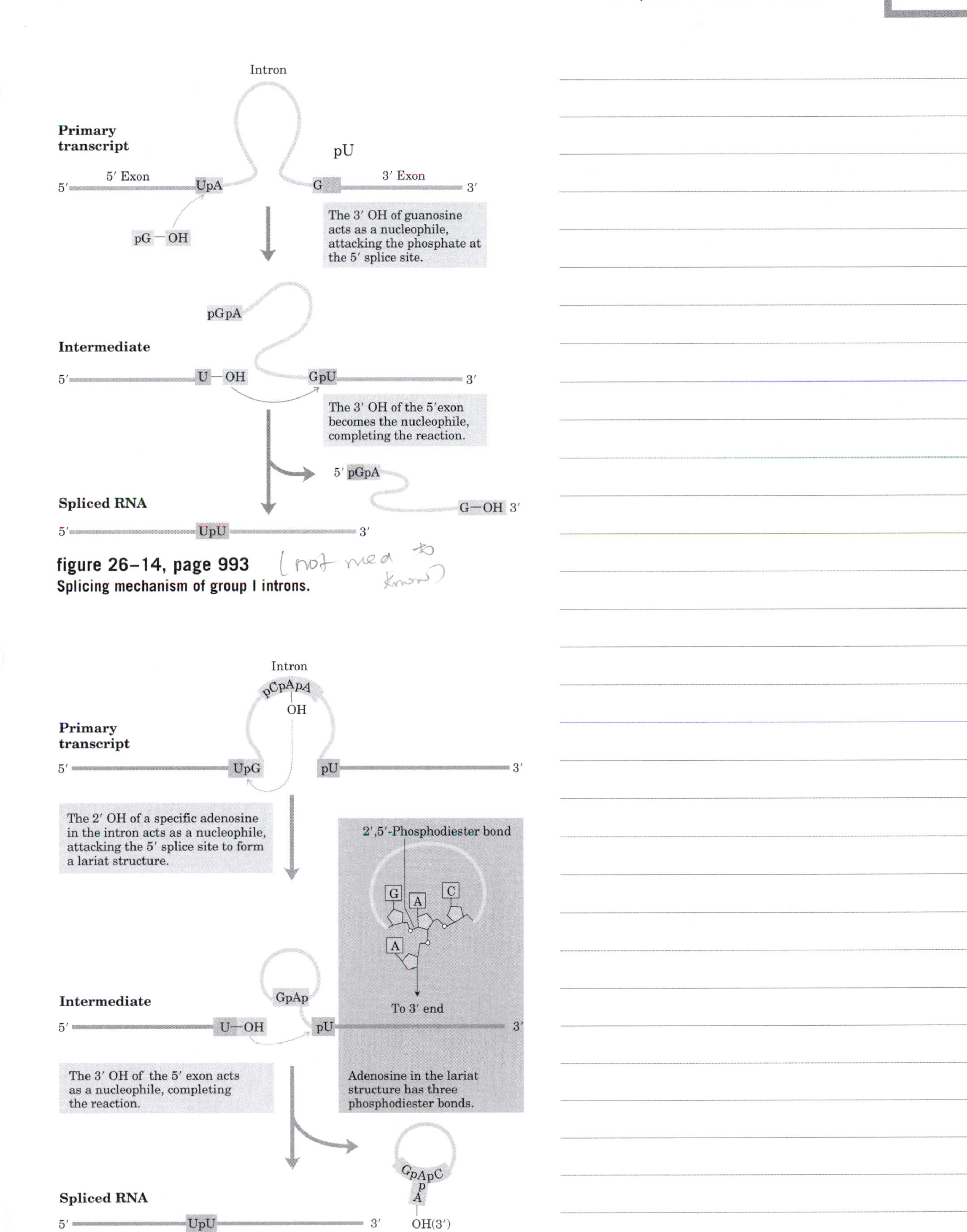

figure 26–14, page 993
Splicing mechanism of group I introns.

figure 26–15, page 994
Splicing mechanism of group II introns.

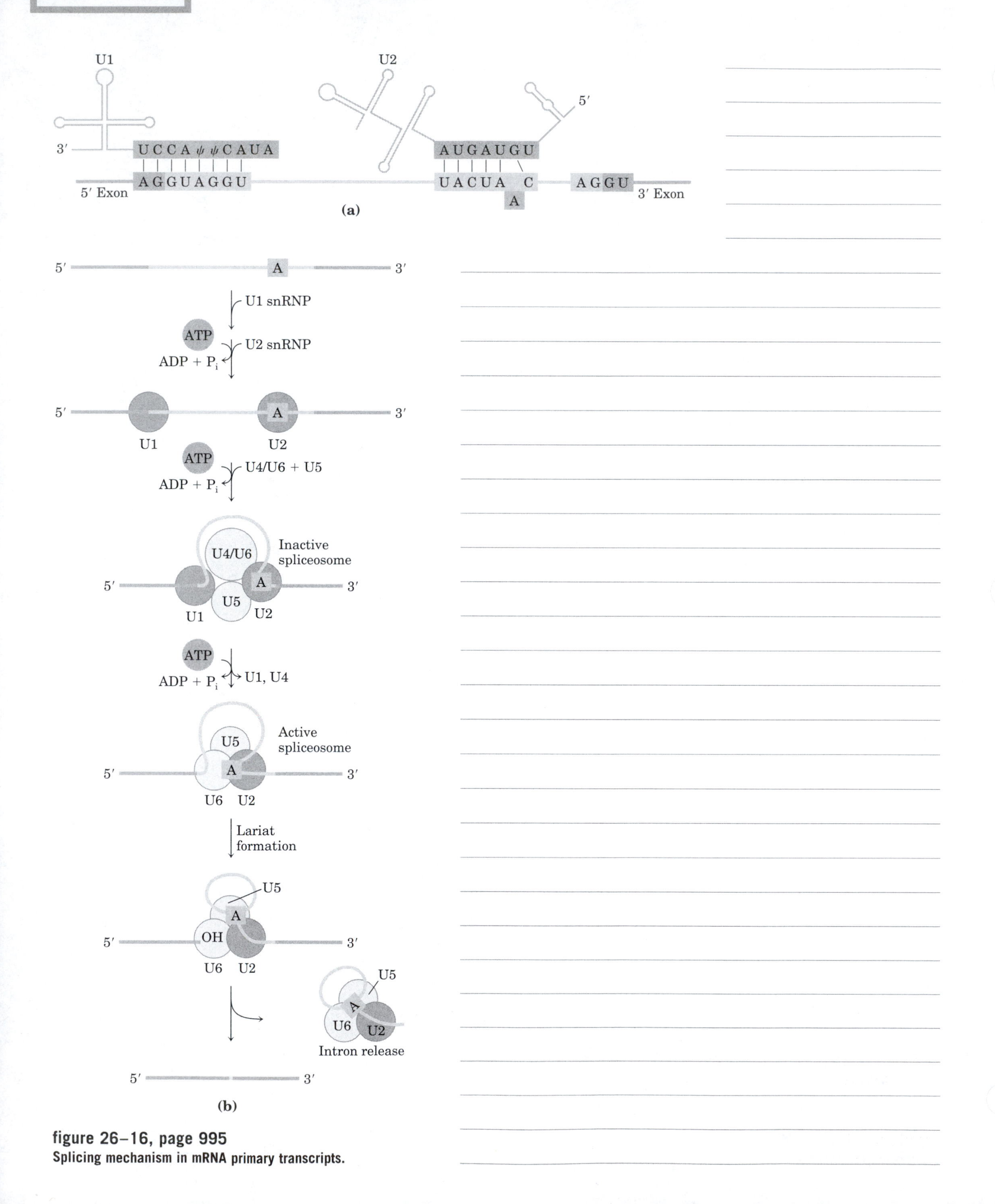

figure 26–16, page 995
Splicing mechanism in mRNA primary transcripts.

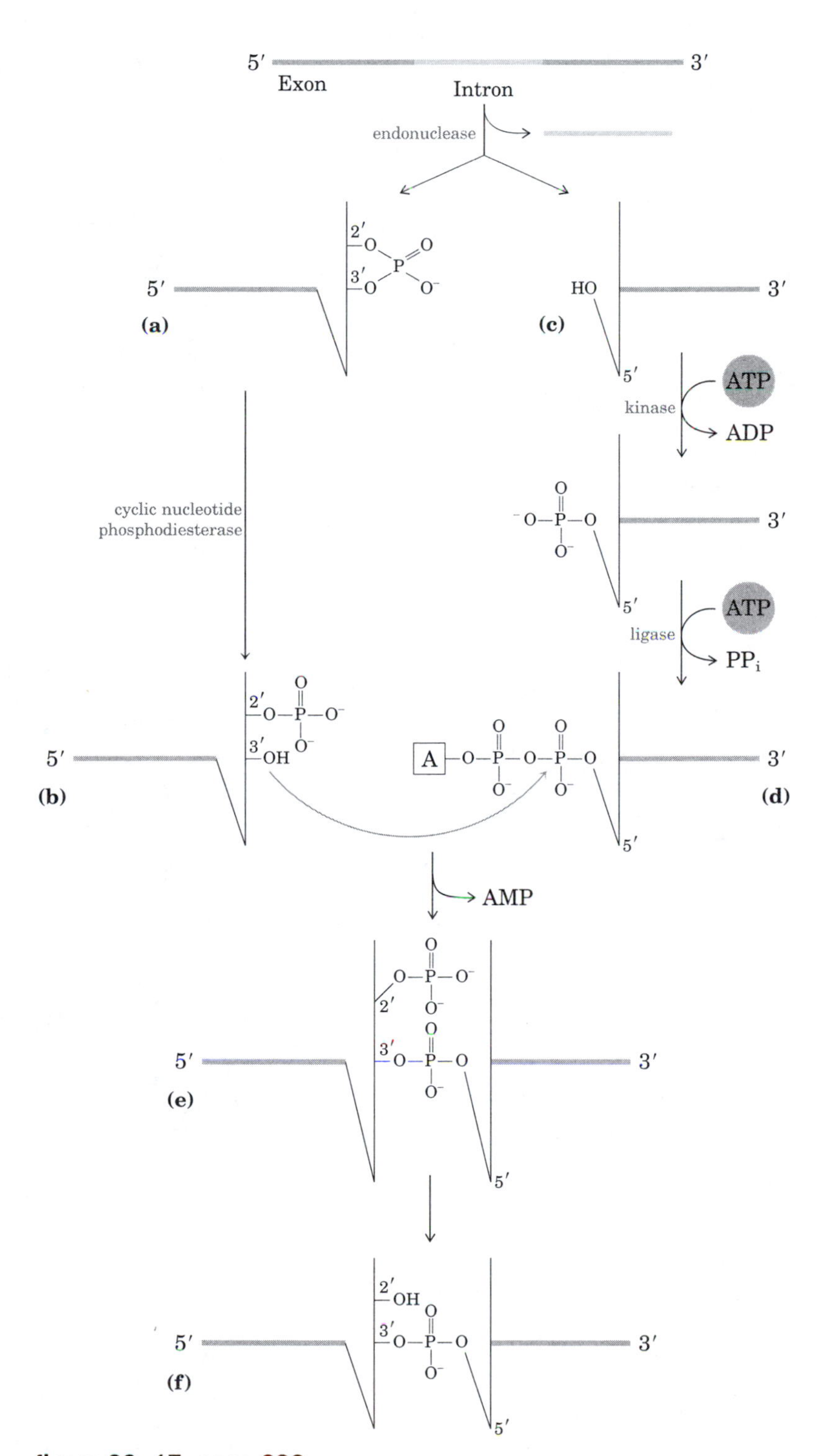

figure 26–17, page 996
Splicing of yeast tRNA.

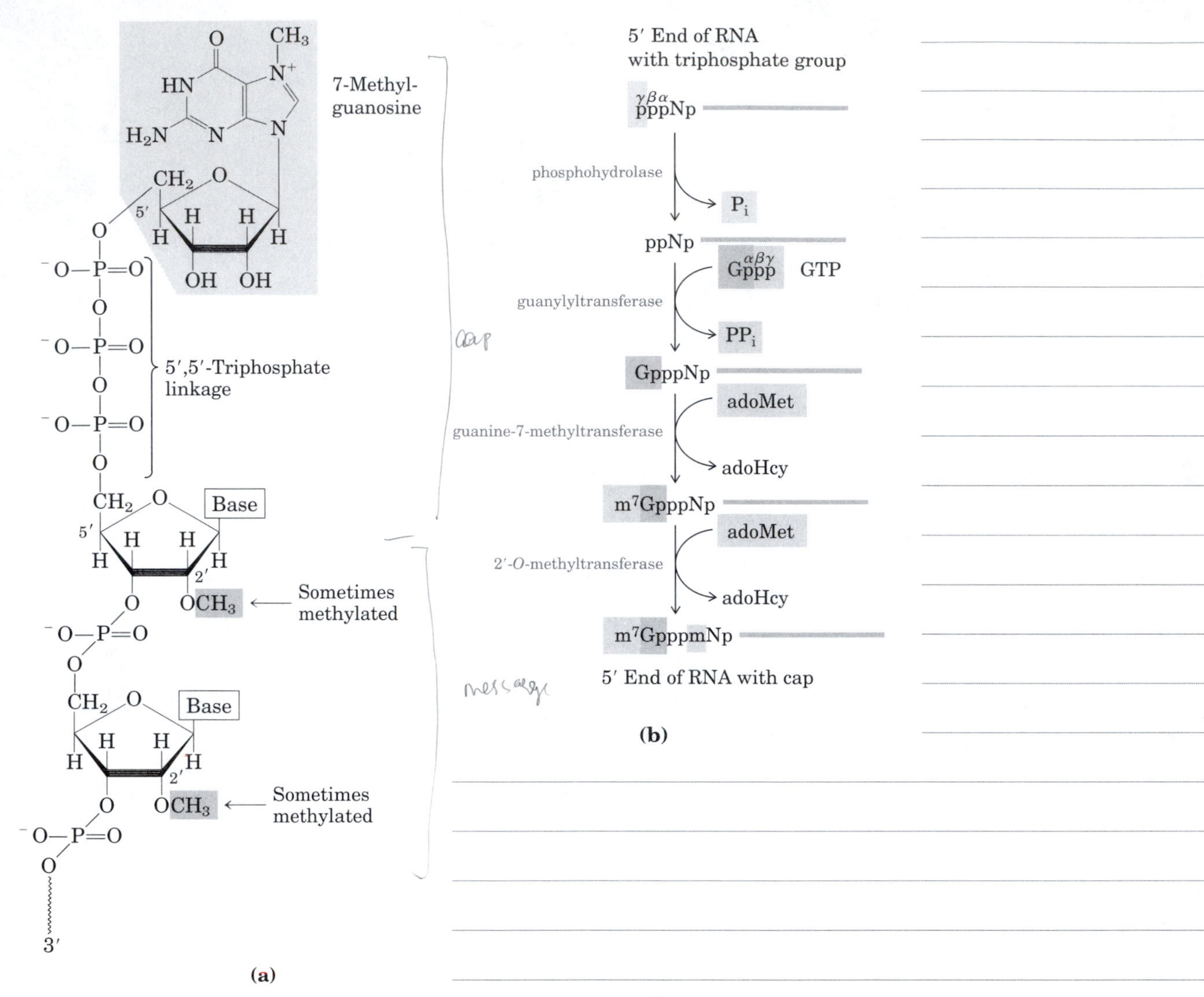

figure 26–18, page 997
The 5′ cap of mRNA.

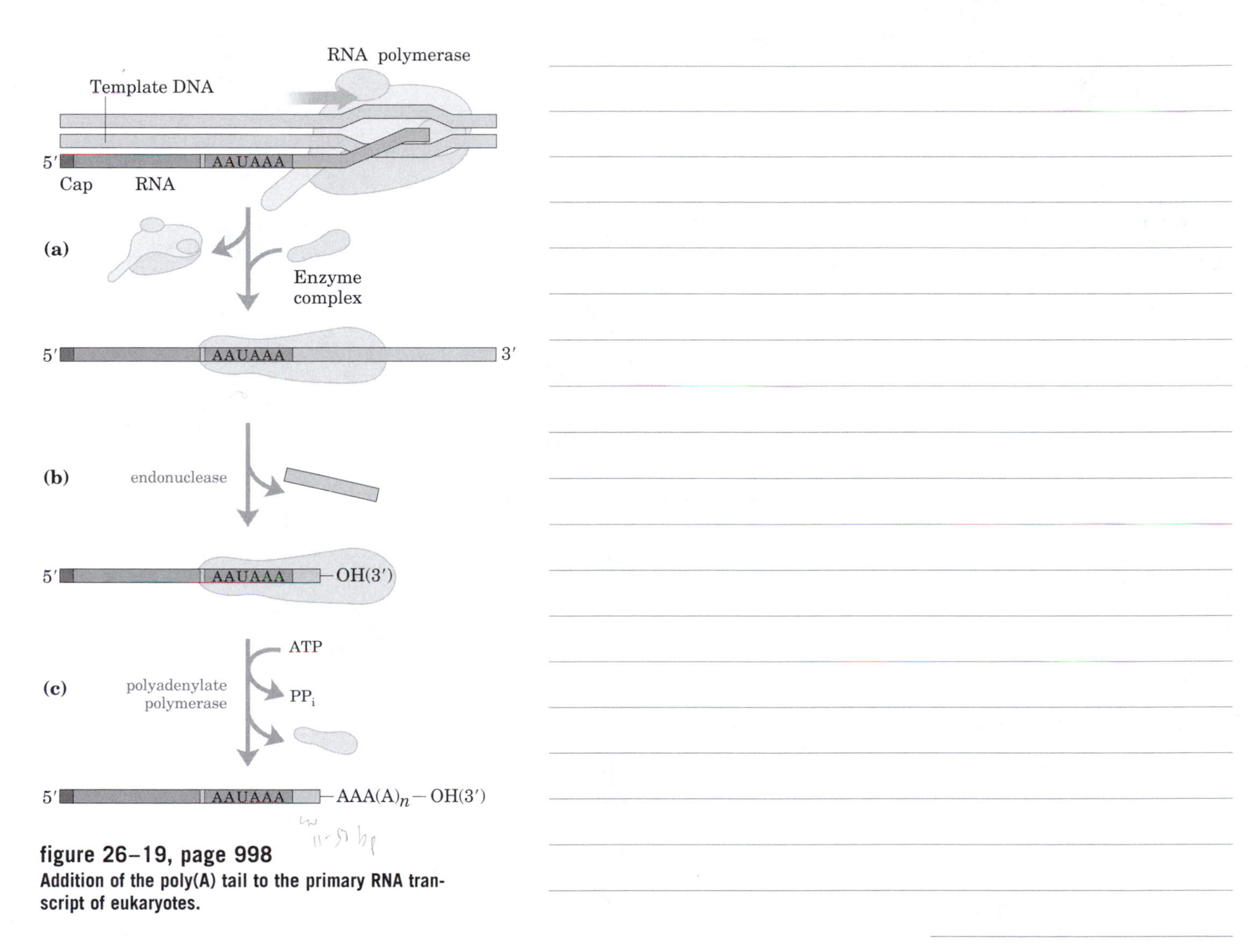

figure 26–19, page 998
Addition of the poly(A) tail to the primary RNA transcript of eukaryotes.

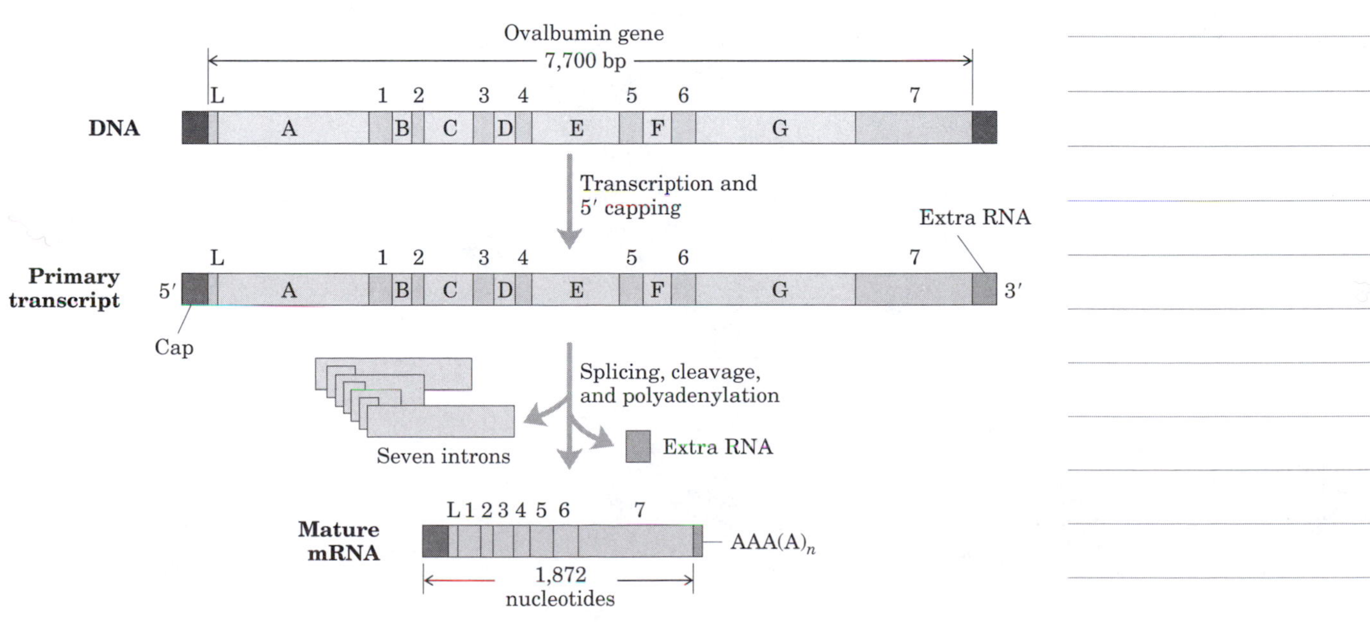

figure 26–20, page 998
Overview of the processing of a eukaryotic mRNA.

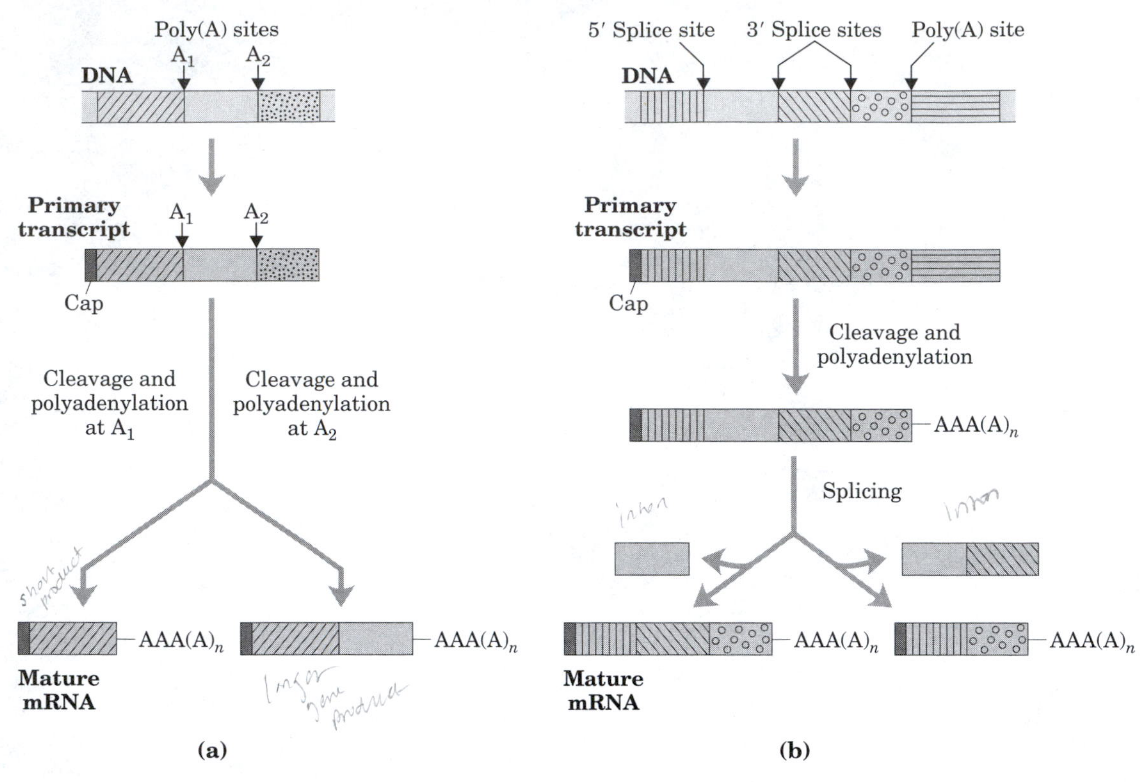

figure 26–21, page 999

Two mechanisms for the alternative processing of complex transcripts in eukaryotes.

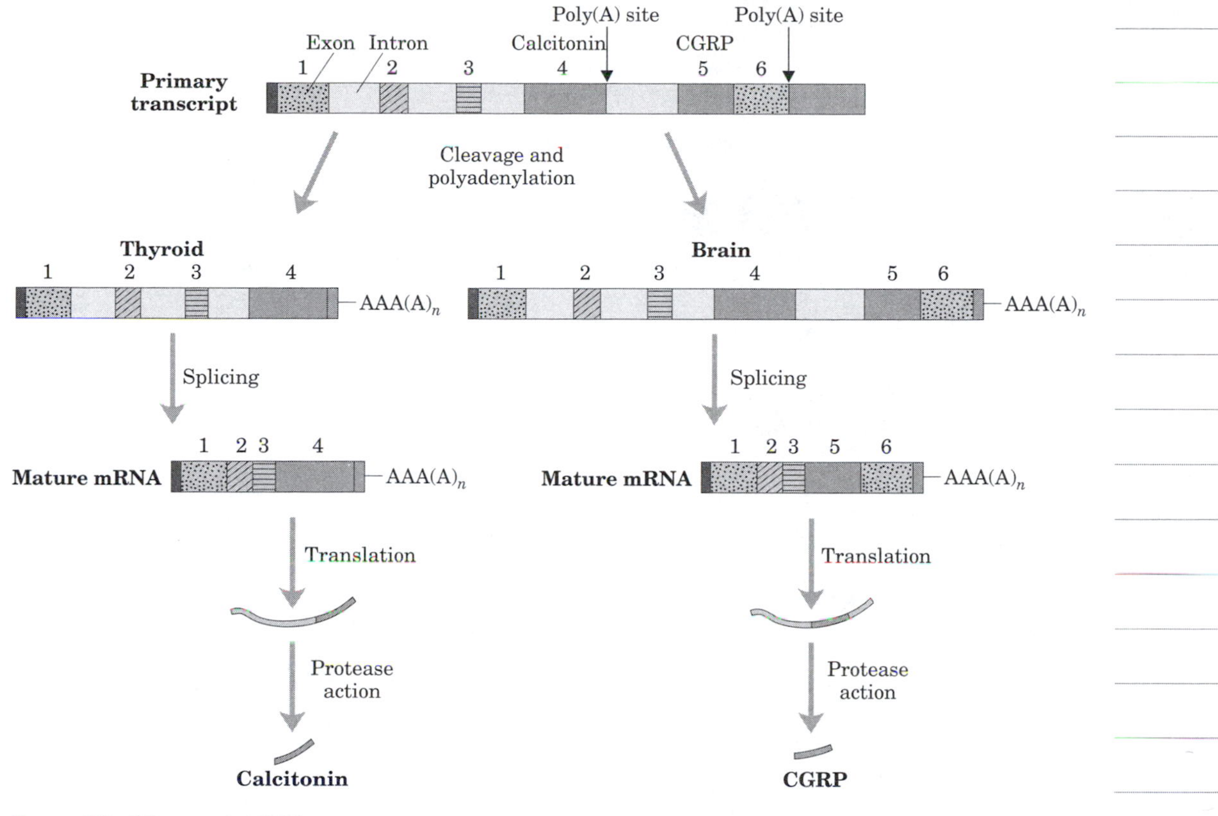

figure 26–22, page 1000

Alternative processing of the calcitonin gene transcript in rats.

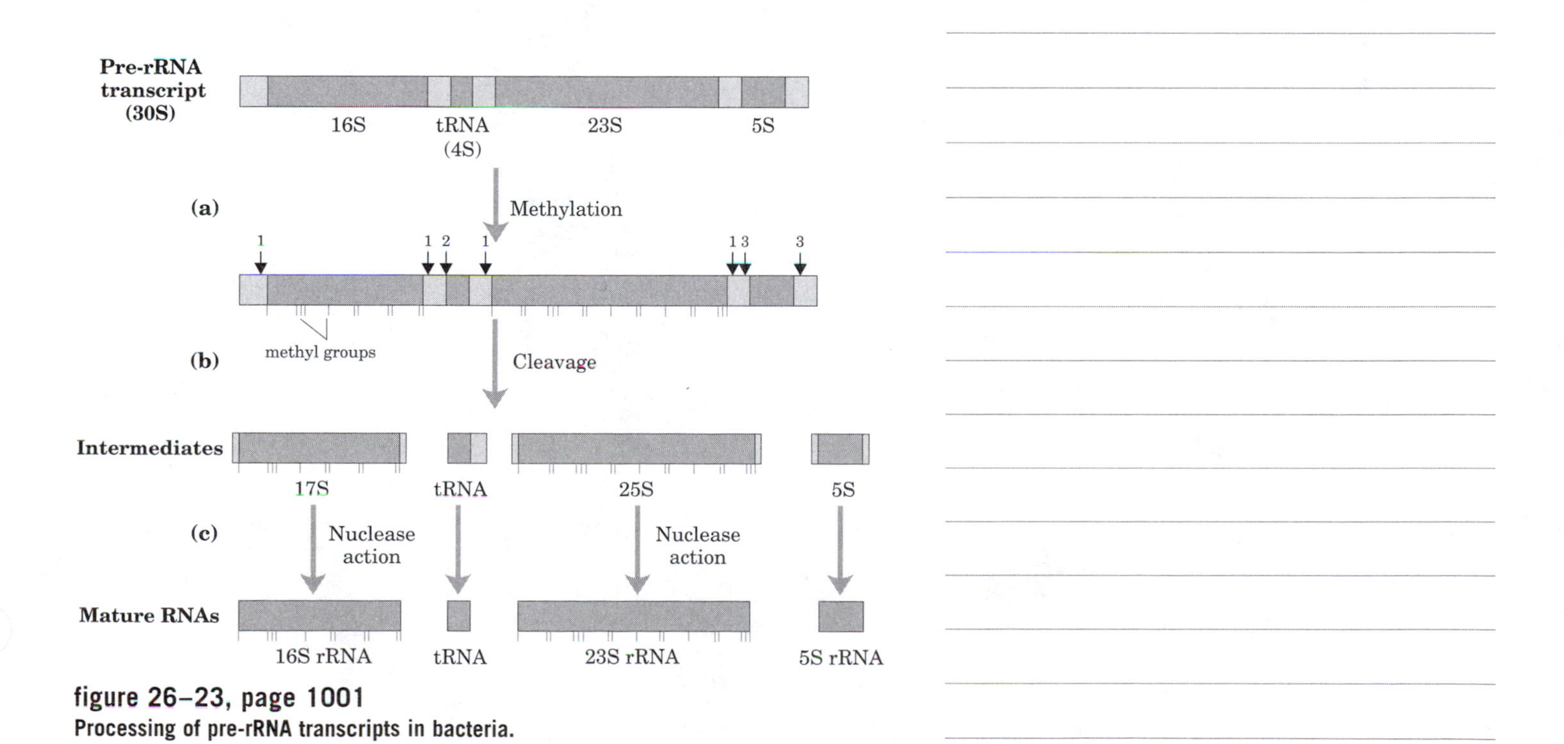

figure 26–23, page 1001

Processing of pre-rRNA transcripts in bacteria.

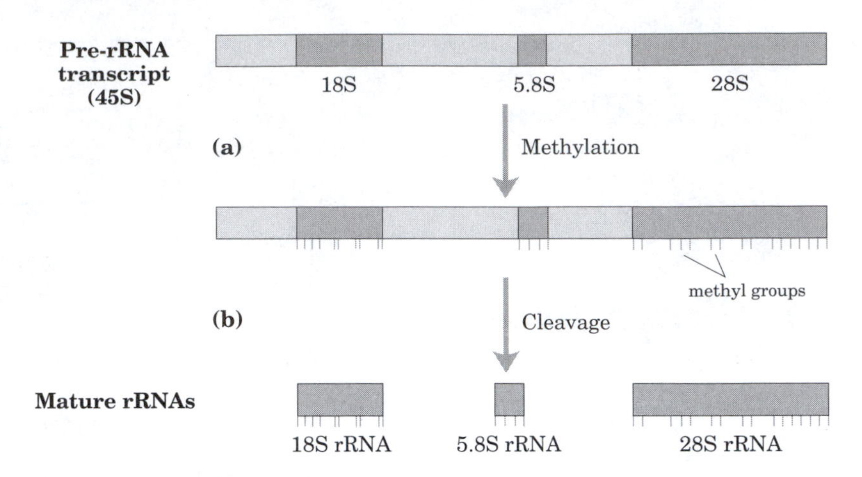

figure 26–24, page 1001
Processing of pre-rRNA transcripts in vertebrates.

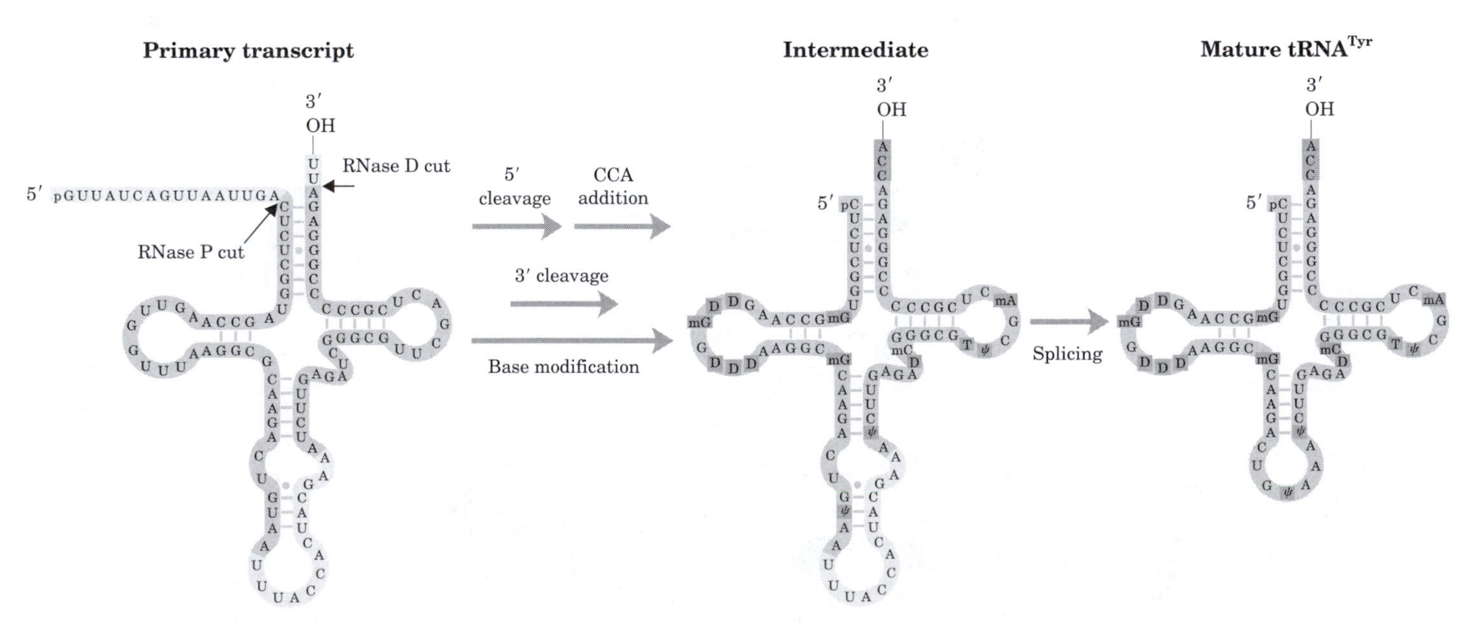

figure 26–25, page 1002
Processing of tRNAs in bacteria and eukaryotes.

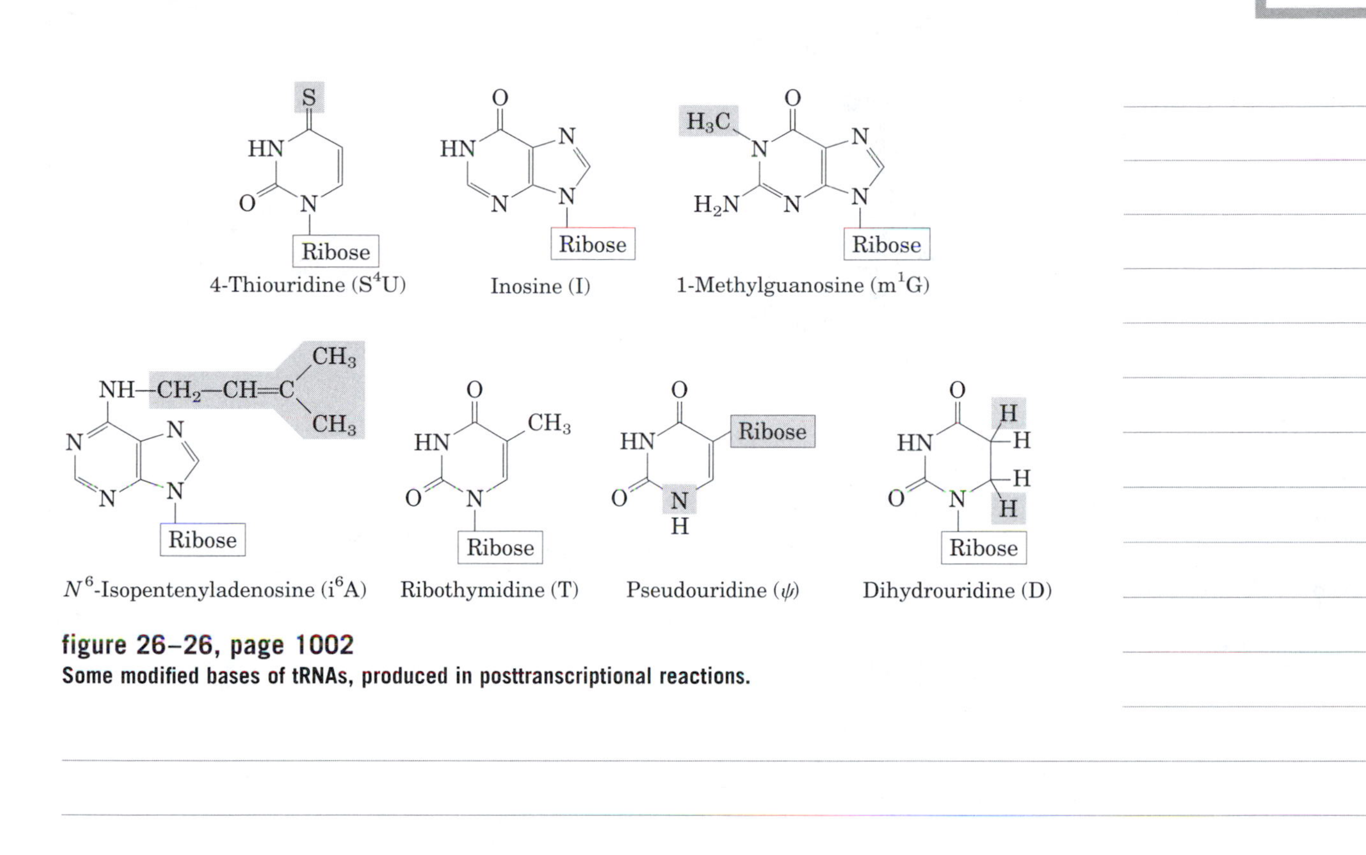

figure 26–26, page 1002
Some modified bases of tRNAs, produced in posttranscriptional reactions.

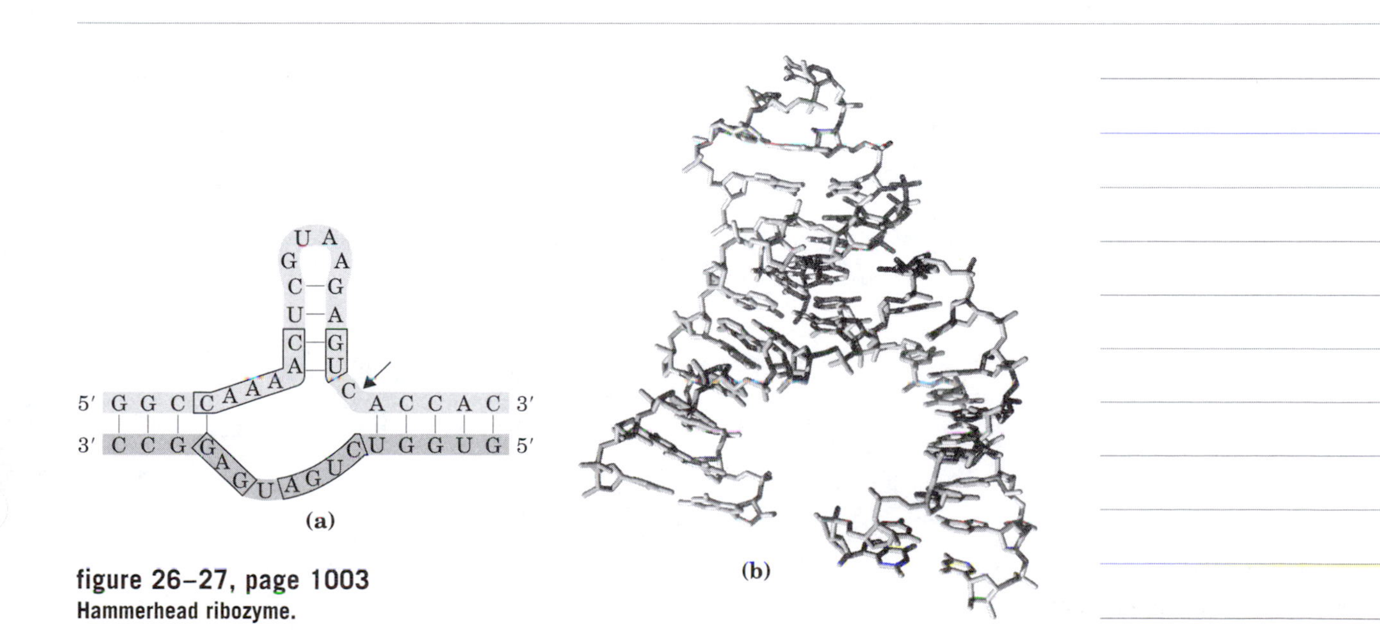

figure 26–27, page 1003
Hammerhead ribozyme.

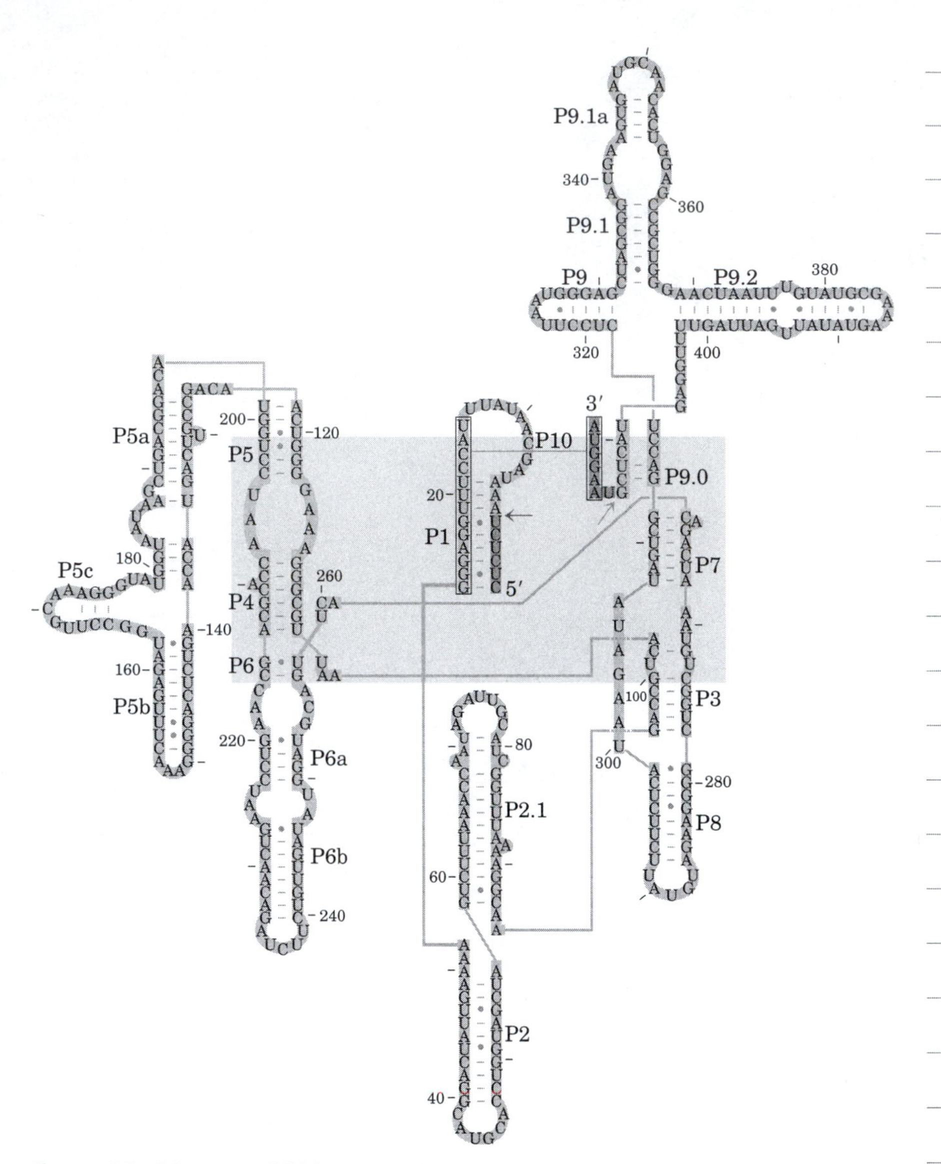

figure 26–28, page 1003
Secondary structure of the self-splicing rRNA intron from *Tetrahymena.*

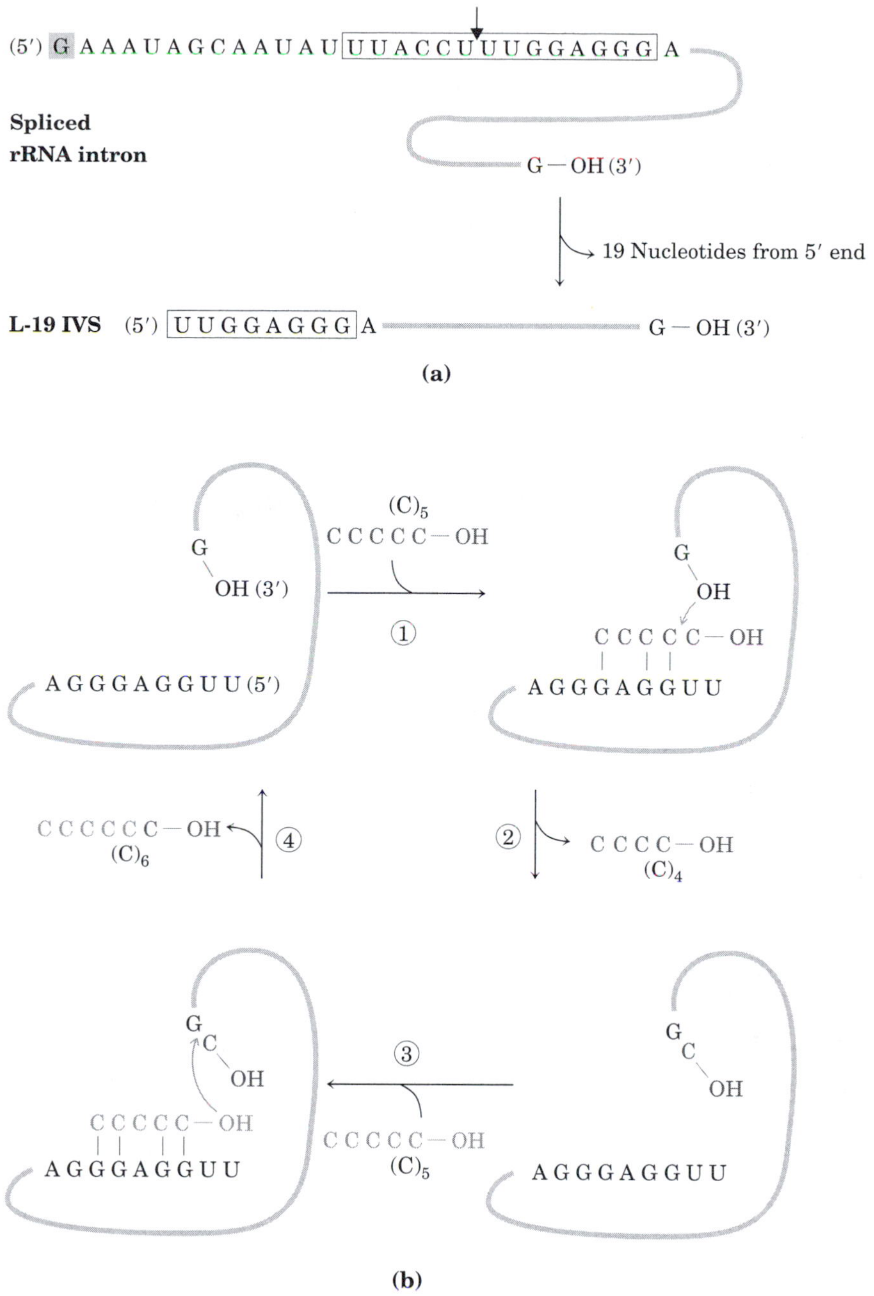

(5′) G A A A U A G C A A U A U U U A C C U U U G G A G G G A

Spliced rRNA intron

G — OH (3′)

19 Nucleotides from 5′ end

L-19 IVS (5′) U U G G A G G G A G — OH (3′)

(a)

(C)5 C C C C C — OH

G OH (3′)

A G G G A G G U U (5′)

①

G OH C C C C C — OH A G G G A G G U U

C C C C C C — OH (C)6 ④

② C C C C — OH (C)4

G C OH C C C C C — OH A G G G A G G U U

③ C C C C C — OH (C)5

G C OH A G G G A G G U U

(b)

figure 26–29, page 1005
In vitro catalytic activity of L-19 IVS.

$$(NMP)_n + NDP \rightleftharpoons \underset{\text{Lengthened polynucleotide}}{(NMP)_{n+1}} + P_i$$

Polynucleotide phosphorylase, page 1006

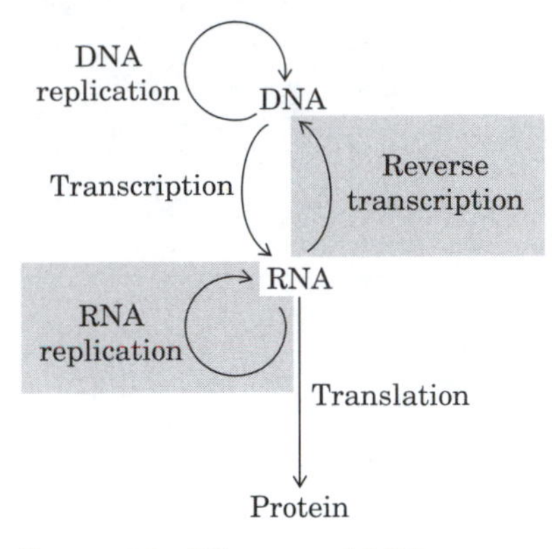

figure 26–30, page 1007
Extension of the central dogma to include RNA dependent synthesis of RNA and DNA.

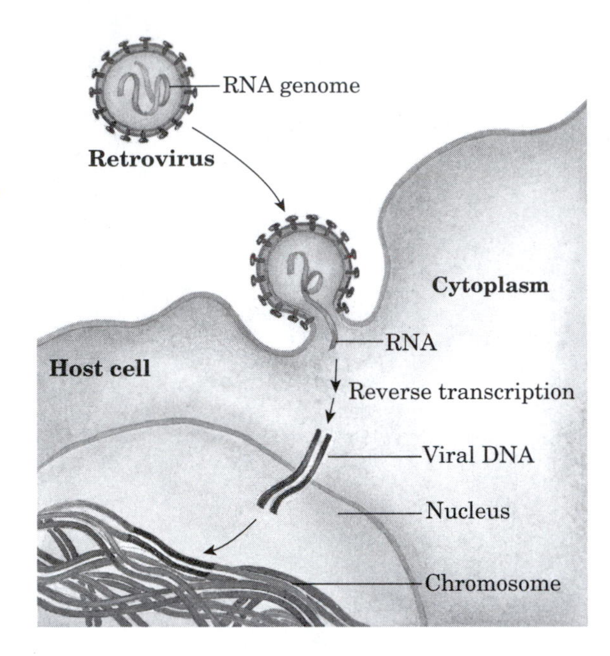

figure 26–31, page 1007
Retroviral infection of a mammalian cell and integration of the retrovirus into the host chromosome.

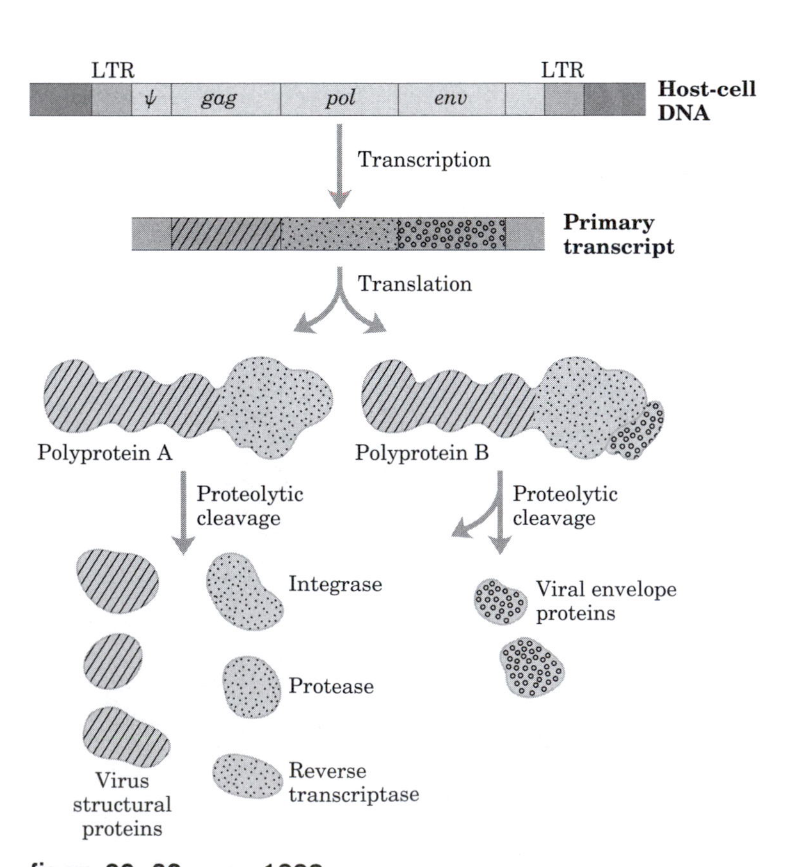

figure 26–32, page 1008
Structure and gene products of an integrated retrovirus genome.

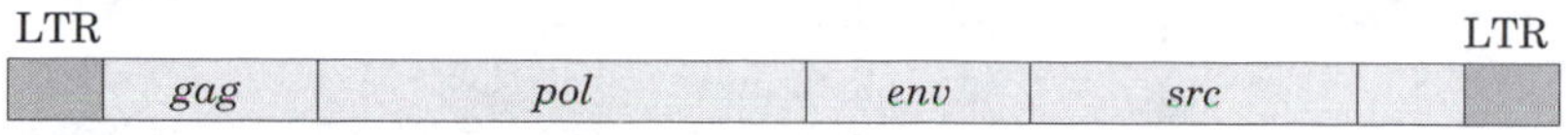

LTR
LTR
gag *pol* *env* *src*

figure 26–33, page 1009
Rous sarcoma virus genome.

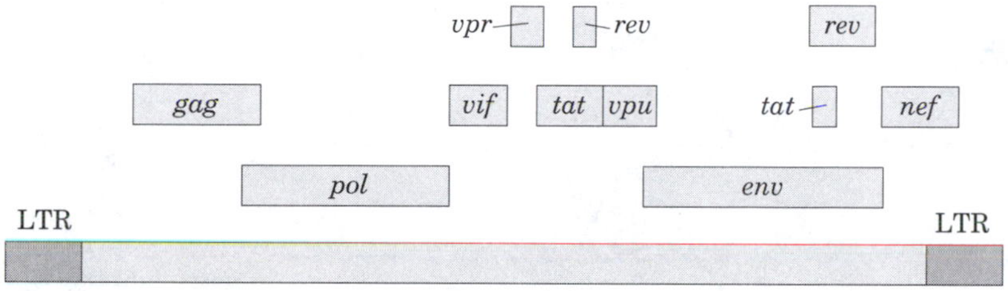

vpr *rev* *rev*
gag *vif* *tat* *vpu* *tat* *nef*
pol *env*
LTR
LTR

figure 26–34, page 1009
The genome of HIV, the virus that causes AIDS.

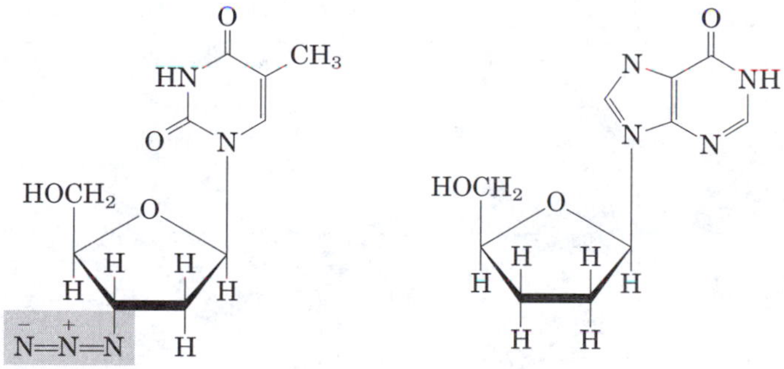

3′-Azido-2′,3′-dideoxythymidine (AZT)

2′,3′-Dideoxyinosine (DDI)

box 26-2, page 1010
AZT and DDDI.

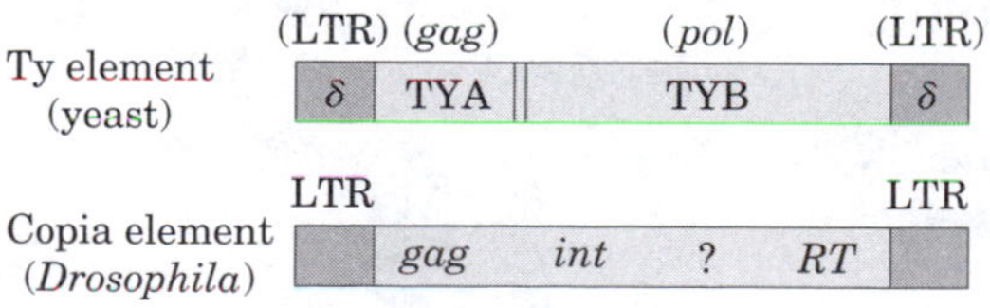

Ty element (yeast) (LTR) (*gag*) (*pol*) (LTR) δ TYA TYB δ

Copia element (*Drosophila*) LTR LTR *gag* *int* ? *RT*

figure 26–35, page 1010
Eukaryotic transposons.

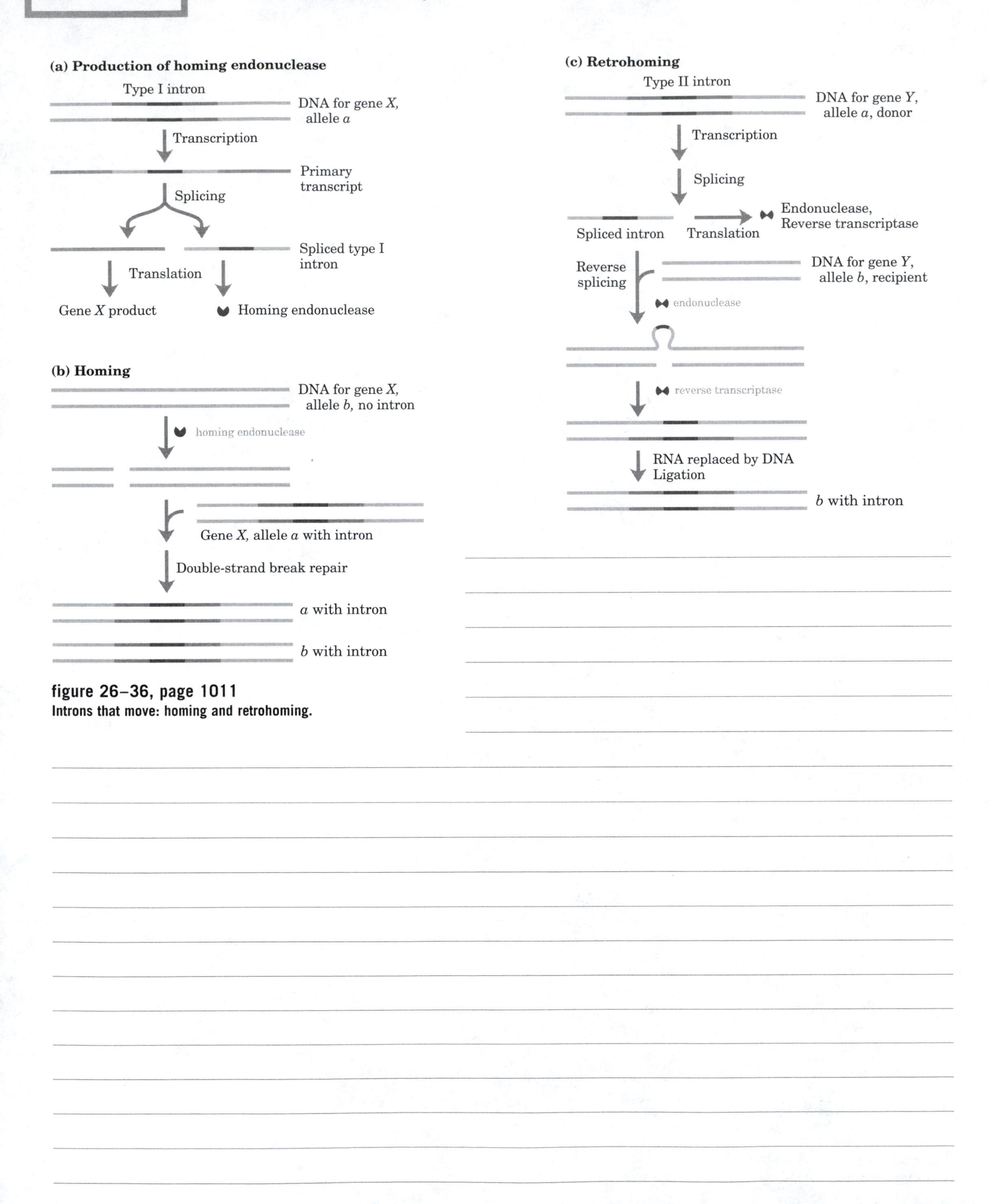

figure 26–36, page 1011
Introns that move: homing and retrohoming.

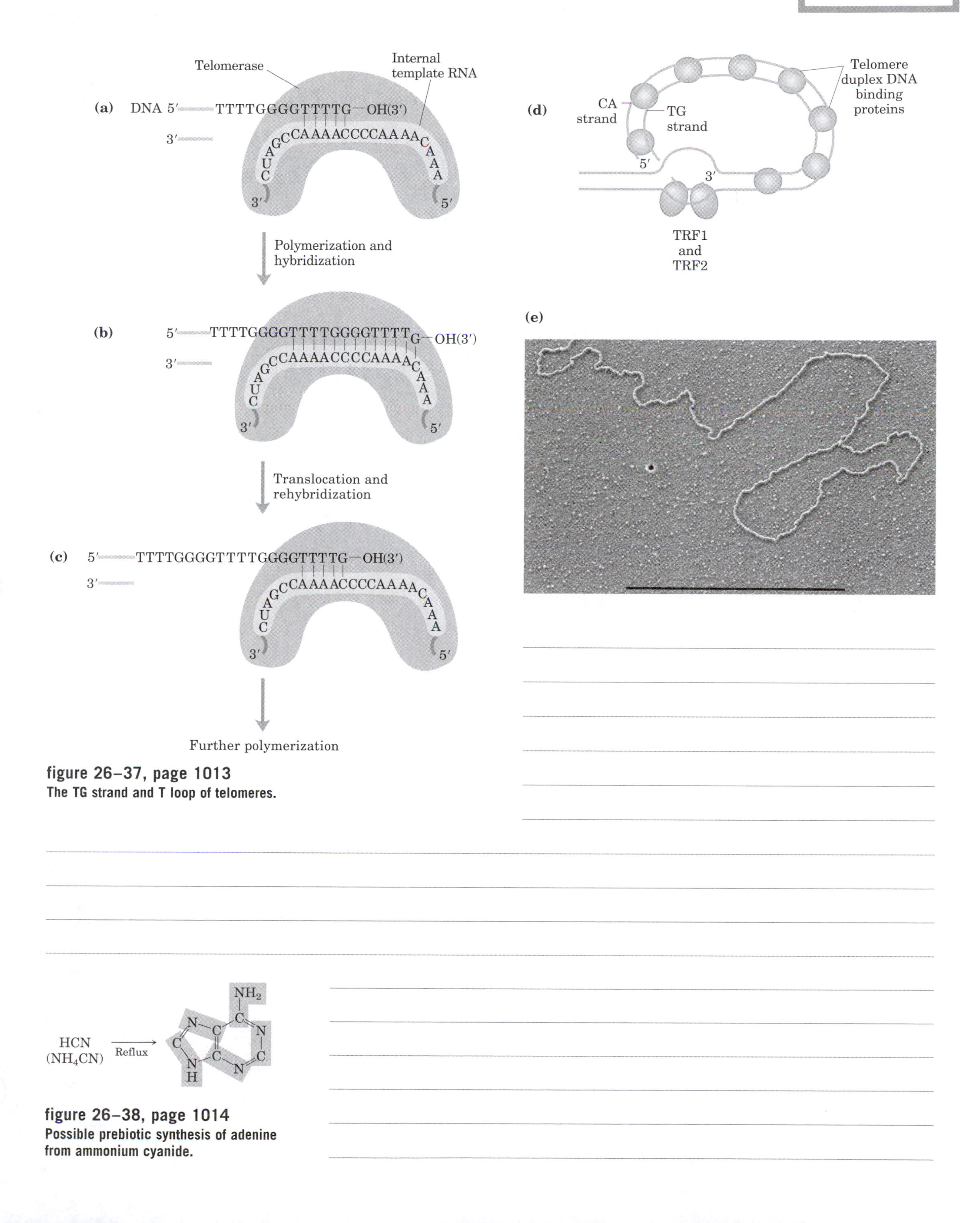

figure 26–37, page 1013
The TG strand and T loop of telomeres.

figure 26–38, page 1014
Possible prebiotic synthesis of adenine from ammonium cyanide.

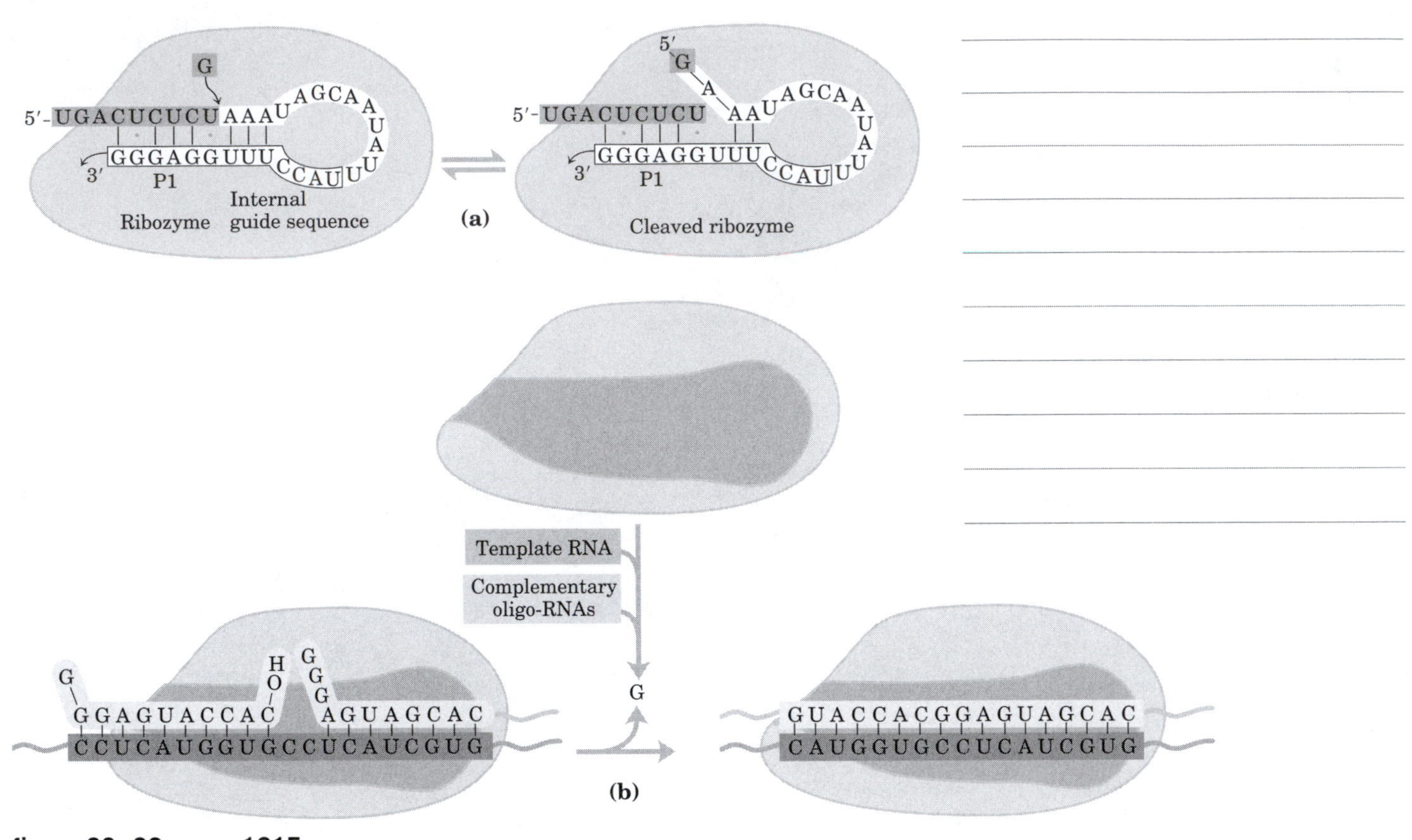

figure 26–39, page 1015
RNA-dependent synthesis of an RNA polymer from oligonucleotide precursors.

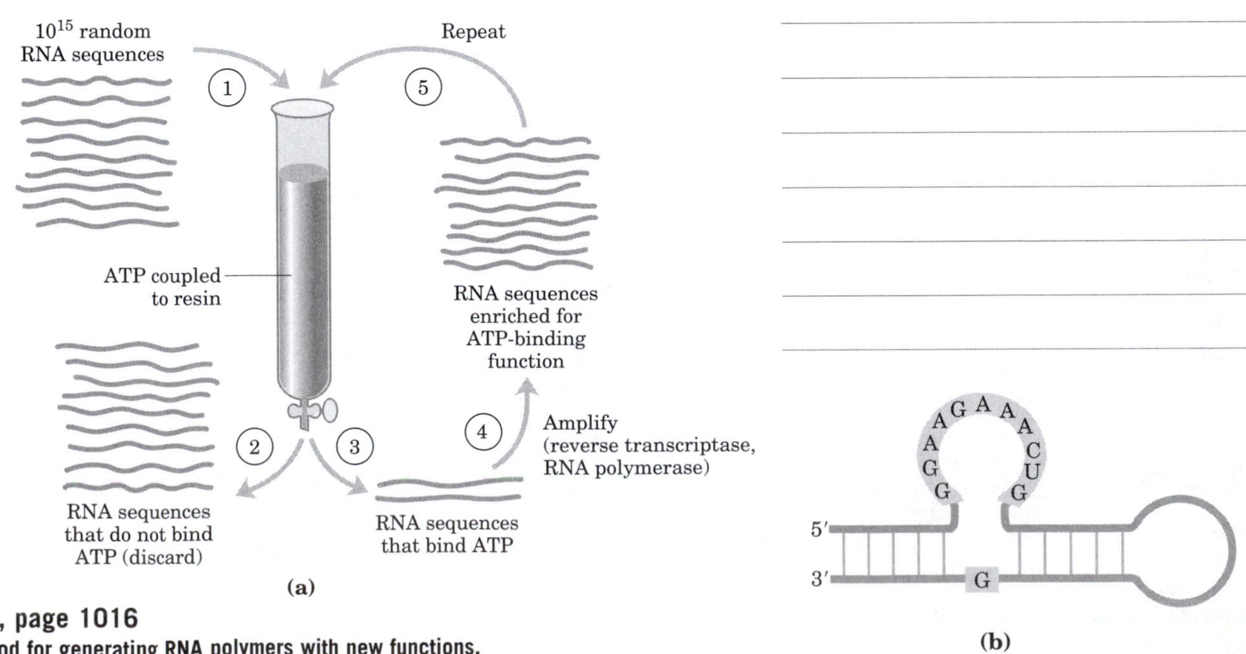

figure 26–40, page 1016
The SELEX method for generating RNA polymers with new functions.

chapter

27

Protein Metabolism

CD-ROM

When you study for exams, don't forget to review The Minicourses on the UNDERSTAND! BIOCHEMISTRY CD that came with your textbook.

Minicourses that apply to this Chapter include:

Background
- Fundamentals of Gene Action

Nucleic Acids and Their Expression
- Mutation
- Translation
- RNA Processing

Cellular Architecture and Traffic
- Protein Tarketing
- Protein Sorting in the ER and Golgi Appartus
- Nuclear Transport
- Vesicle Transport
- Sorting of Lysosomal Enzymes
- Protein Modifications

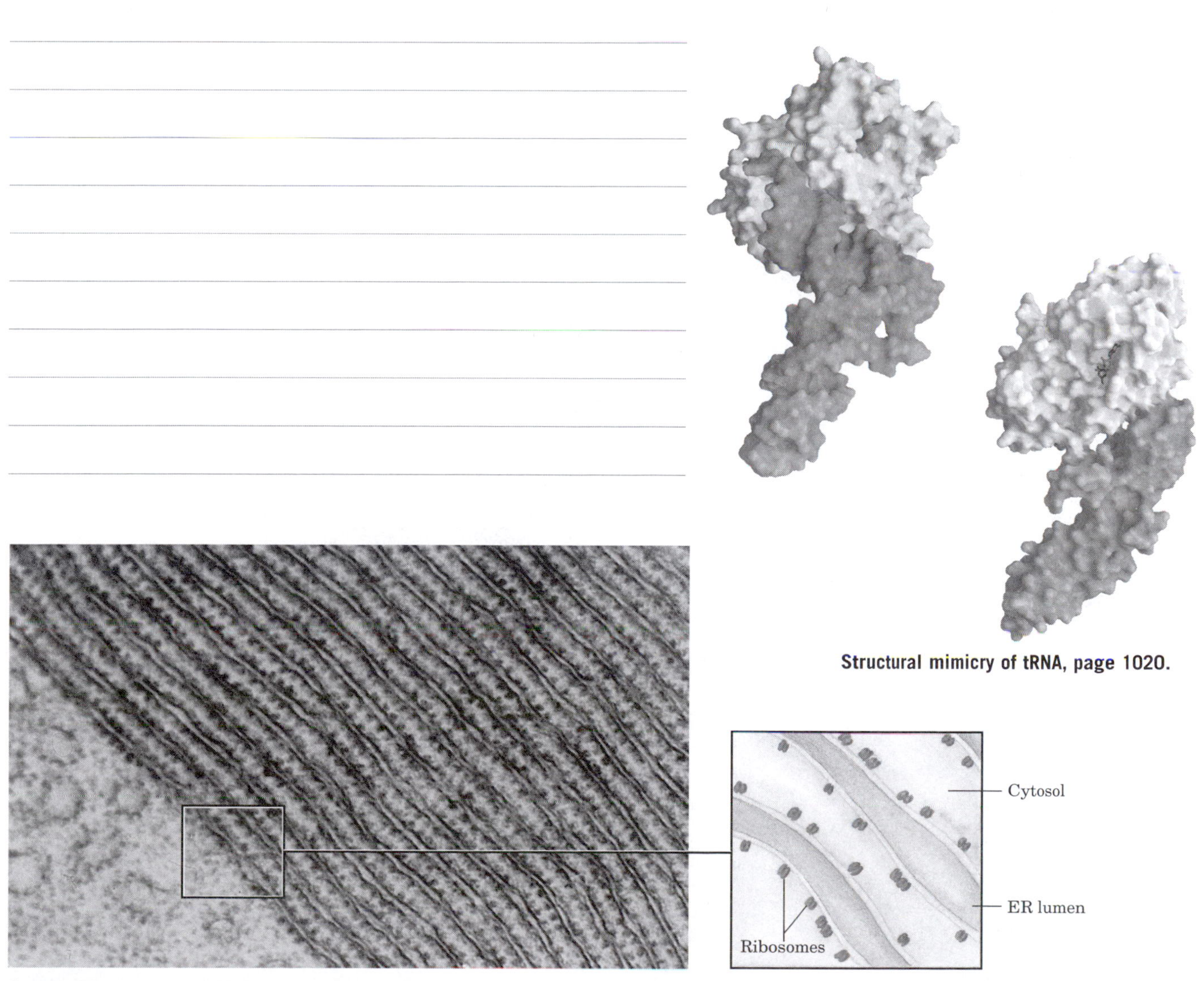

Structural mimicry of tRNA, page 1020.

figure 27–1, page 1021
Ribosomes and endoplasmic reticulum.

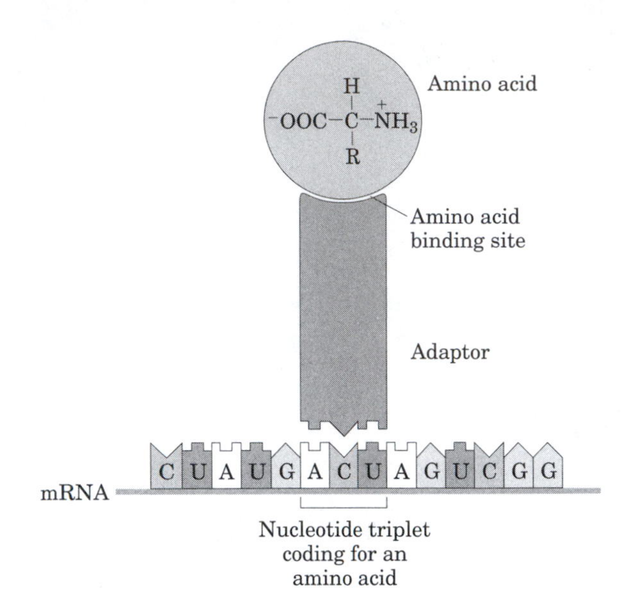

figure 27–2, page 1021
Crick's adaptor hypothesis.

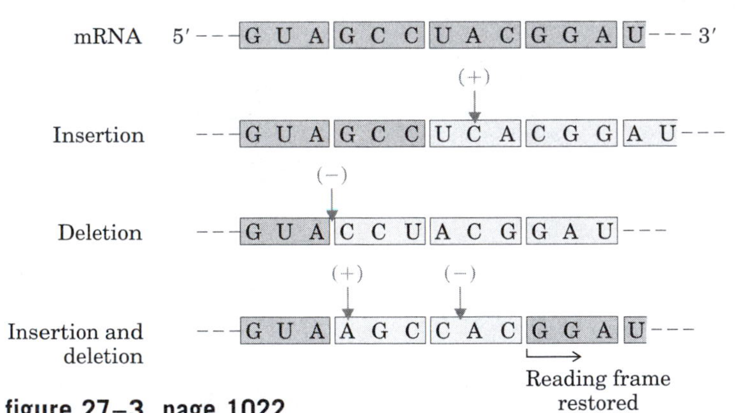

figure 27–3, page 1022
The triplet, nonoverlapping code.

Nonoverlapping code: A U A | C G A | G U C | - - - -
1 2 3

Overlapping code: A U A C G A G U C
1
2
3

figure 27–4, page 1022
Overlapping versus nonoverlapping genetic codes.

Reading frame 1	5′---UUC UCG GAC CUG GAG AUU CAC AGU---3′
Reading frame 2	---U UCU CGG ACC UGG AGA UUC ACA GU---
Reading frame 3	---UU CUC GGA CCU GGA GAU UCA CAG U---

figure 27–5, page 1022
In a triplet, nonoverlapping code, all mRNAs have three potential reading frames, shaded here in different colors. Note that the triplets, and hence the amino acids specified, are different in each reading frame.

table 27–1, page 1023

Incorporation of Amino Acids into Polypeptides in Response to Random Polymers of RNA*

Amino acid	Observed frequency of incorporation (Lys = 100)	Tentative assignment for nucleotide composition† of corresponding codon	Expected frequency of incorporation based on assignment (Lys = 100)
Asparagine	24	A_2C	20
Glutamine	24	A_2C	20
Histidine	6	AC_2	4
Lysine	100	AAA	100
Proline	7	AC_2, CCC	4.8
Threonine	26	A_2C, AC_2	24

*Presented here is a summary of data from one of the early experiments designed to elucidate the genetic code. A synthetic RNA containing only A and C residues in a 5:1 ratio directed polypeptide synthesis. Both the identity and the quantity of amino acids incorporated were determined. Based on the relative abundance of A and C residues in the synthetic RNA, and assigning the codon AAA (the most likely codon) a frequency of 100, there should be three different codons of composition A_2C, each at a relative frequency of 20; three codons of composition AC_2, each at a relative frequency of 4.0; and codon CCC at a relative frequency of 0.8. The CCC assignment was based on information derived from prior studies with poly(C). Where two tentative codon assignments are made, both are proposed to code for the same amino acid.

†Note that these designations of nucleotide composition contain no information on nucleotide sequence (except, of course, AAA and CCC).

table 27–2, page 1024

Trinucleotides Induce Specific Binding of Aminoacyl-tRNAs to Ribosomes

Trinucleotide	^{14}C-Labeled aminoacyl-tRNA bound to ribosome*		
	Phe-tRNA^{Phe}	Lys-tRNA^{Lys}	Pro-tRNA^{Pro}
UUU	4.6	0	0
AAA	0	7.7	0
CCC	0	0	3.1

Source: Modified from Nirenberg, M. & Leder, P. (1964) RNA code words and protein synthesis. *Science* **145,** 1399.

*Each number represents the factor by which the amount of bound ^{14}C increased when the indicated trinucleotide was present, relative to a control in which no trinucleotide was added.

Reading frame 1 5′---G U A|A G U|A A G|U A A|G U A|A G U|A A--- 3′

Reading frame 2 ---G|U A A|G U A|A G U|A A G|U A A|G U A|A---

Reading frame 3 ---G U|A A G|U A A|G U A|A G U|A A G|U A A|---

figure 27–6, page 1024

Effect of a termination codon in a repeating tetranucleotide.

Second letter of codon

First letter of codon (5′ end)	U	C	A	G
U	UUU Phe UUC Phe UUA Leu UUG Leu	UCU Ser UCC Ser UCA Ser UCG Ser	UAU Tyr UAC Tyr UAA Stop UAG Stop	UGU Cys UGC Cys UGA Stop UGG Trp
C	CUU Leu CUC Leu CUA Leu CUG Leu	CCU Pro CCC Pro CCA Pro CCG Pro	CAU His CAC His CAA Gln CAG Gln	CGU Arg CGC Arg CGA Arg CGG Arg
A	AUU Ile AUC Ile AUA Ile AUG Met	ACU Thr ACC Thr ACA Thr ACG Thr	AAU Asn AAC Asn AAA Lys AAG Lys	AGU Ser AGC Ser AGA Arg AGG Arg
G	GUU Val GUC Val GUA Val GUG Val	GCU Ala GCC Ala GCA Ala GCG Ala	GAU Asp GAC Asp GAA Glu GAG Glu	GGU Gly GGC Gly GGA Gly GGG Gly

figure 27–7, page 1025

"Dictionary" of amino acid code words as they occur in mRNAs.

table 27–3, page 1025

Polypeptides Produced in Response to Synthetic RNA Polymers with Repeating Sequences of Three and Four Bases

Polynucleotide	Polypeptide products
Trinucleotide repeats	
$(UUC)_n$	$(Phe)_n$, $(Ser)_n$, $(Leu)_n$
$(AAG)_n$	$(Lys)_n$, $(Arg)_n$, $(Glu)_n$
$(UUG)_n$	$(Leu)_n$, $(Cys)_n$, $(Val)_n$
$(CCA)_n$	$(Pro)_n$, $(His)_n$, $(Thr)_n$
$(GUA)_n$	$(Val)_n$, $(Ser)_n$, (chain terminator)*
$(UAC)_n$	$(Tyr)_n$, $(Thr)_n$, $(Leu)_n$
$(AUC)_n$	$(Ile)_n$, $(Ser)_n$, $(His)_n$
$(GAU)_n$	$(Asp)_n$, $(Met)_n$, (chain terminator)*
Tetranucleotide repeats	
$(UAUC)_n$	$(Tyr\text{–}Leu\text{–}Ser\text{–}Ile)_n$
$(UUAC)_n$	$(Leu\text{–}Leu\text{–}Thr\text{–}Tyr)_n$
$(GUAA)_n$	Di- and tripeptides*
$(AUAG)_n$	Di- and tripeptides*

*With these polynucleotides, the patterns of amino acid incorporation into polypeptides are affected by the presence of codons that are termination signals for protein biosynthesis. In the repeating trinucleotide sequences, one of the three reading frames includes only termination codons and thus only two homopolypeptides are observed (generated from the remaining two reading frames). In some of the repeating tetranucleotide sequences, every fourth codon is a termination codon in every reading frame, so only short peptides are produced. This is illustrated in Figure 27–6 for $(GUAA)_n$.

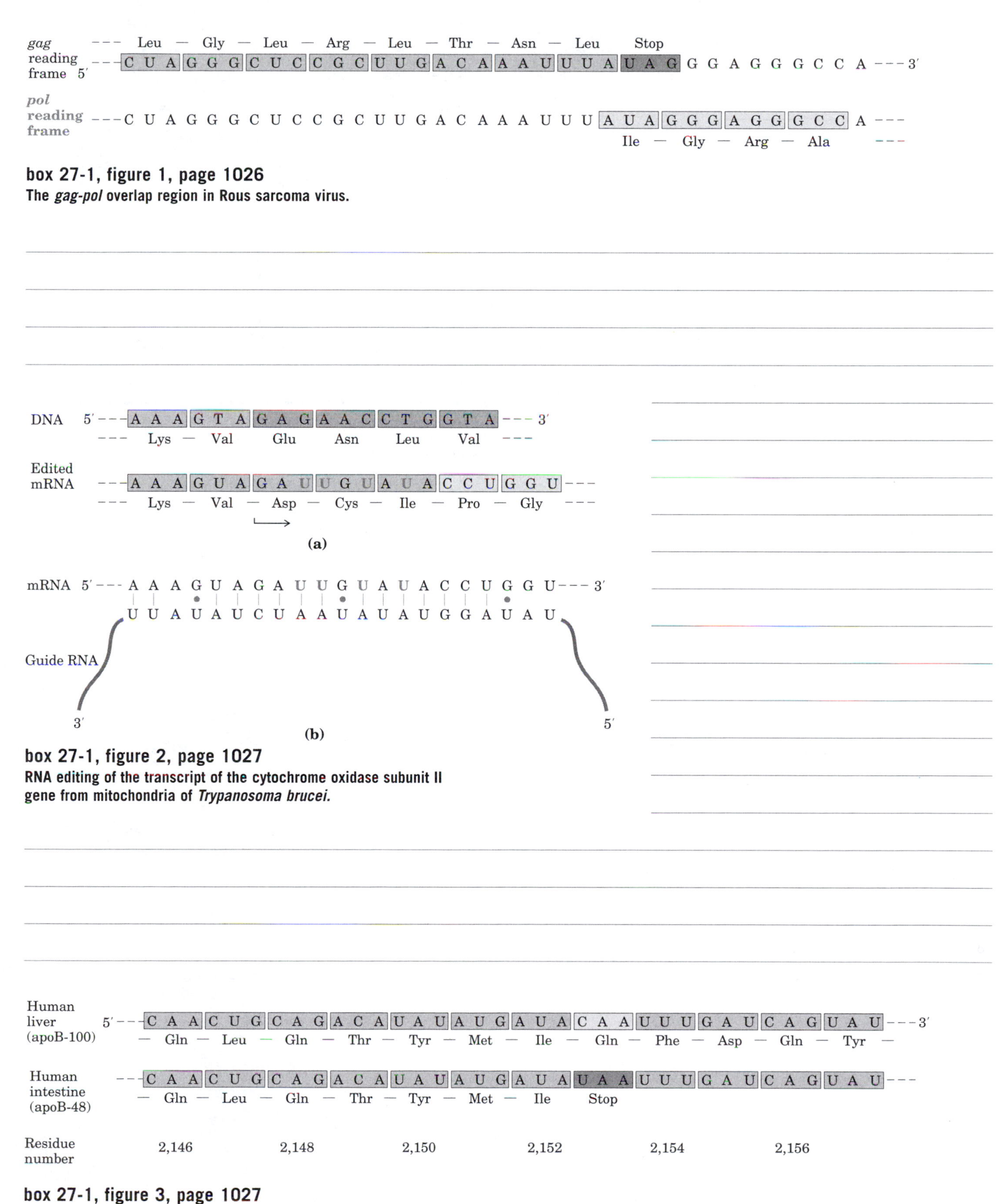

box 27-1, figure 1, page 1026
The *gag-pol* overlap region in Rous sarcoma virus.

box 27-1, figure 2, page 1027
RNA editing of the transcript of the cytochrome oxidase subunit II gene from mitochondria of *Trypanosoma brucei*.

box 27-1, figure 3, page 1027
RNA editing of the transcript of the gene for the apolipoprotein B-100 component of low-density lipoprotein.

table 27–4, page 1028

Degeneracy of the Genetic Code

Amino acid	Number of codons
Ala	4
Arg	6
Asn	2
Asp	2
Cys	2
Gln	2
Glu	2
Gly	4
His	2
Ile	3
Leu	6
Lys	2
Met	1
Phe	2
Pro	4
Ser	6
Thr	4
Trp	1
Tyr	2
Val	4

table 27–5, page 1029

How the Wobble Base of the Anticodon Determines the Number of Codons a tRNA Can Recognize*

1. One codon recognized:			
	Anticodon	(3′) X–Y–**C** (5′)	(3′) X–Y–**A** (5′)
	Codon	(5′) Y–X–**G** (3′)	(5′) Y–X–**U** (3′)
2. Two codons recognized:			
	Anticodon	(3′) X–Y–**U** (5′)	(3′) X–Y–**G** (5′)
	Codon	(5′) Y–X–**A/G** (3′)	(5′) Y–X–**C/U** (3′)
3. Three codons recognized:			
	Anticodon	(3′) X–Y–**I** (5′)	
	Codon	(5′) Y–X–**A/U/C** (3′)	

*X and Y denote complementary bases capable of strong Watson-Crick base pairing with each other. The bases in the wobble positions—the 3′ position of codons and 5′ position of anticodons—are shaded in red.

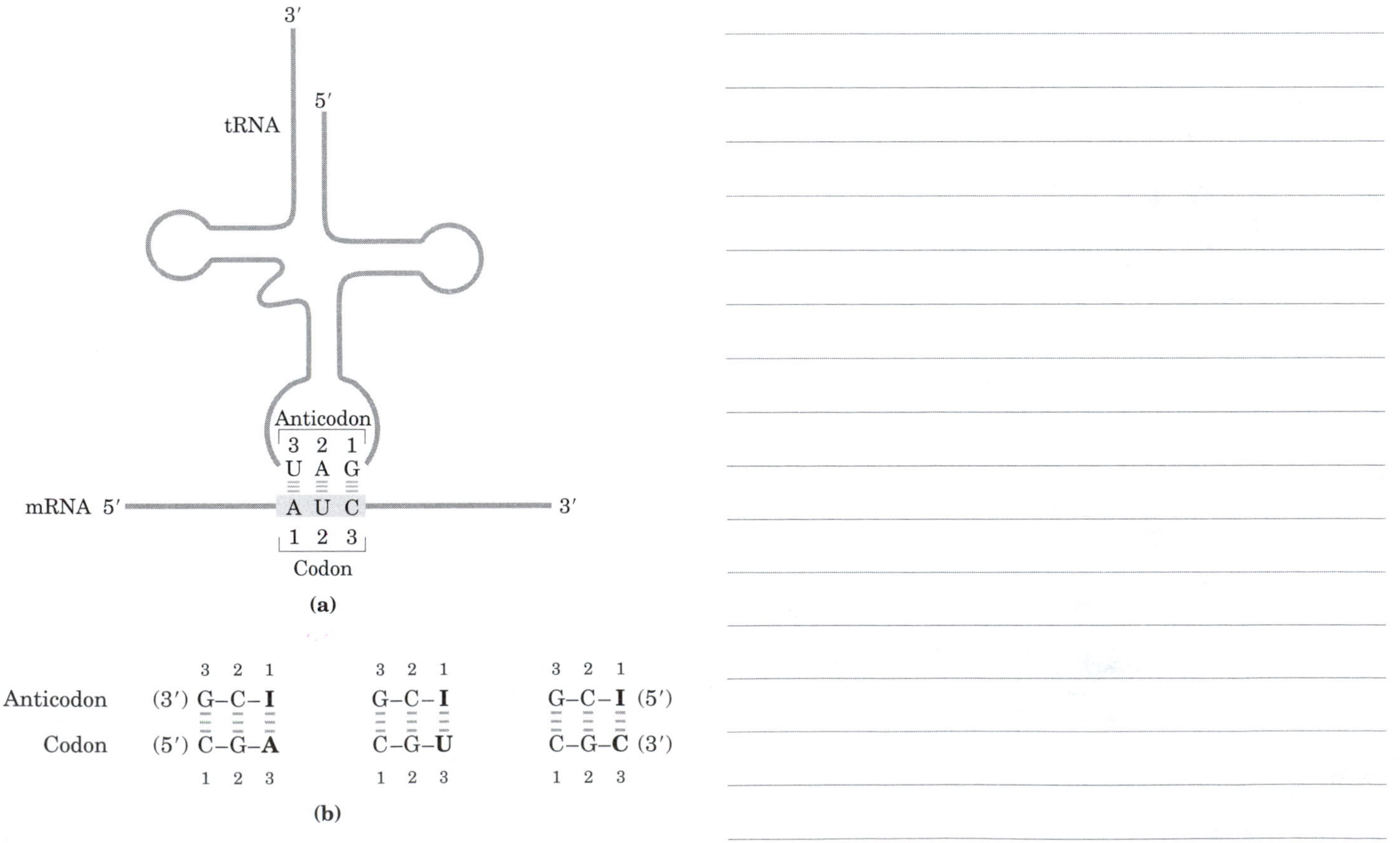

figure 27–8, page 1028
Pairing relationship of codon and anticodon.

box 27-2, table 1, page 1031

Known Variant Codon Assignments in Mitochondria

	Codons*				
	UGA	**AUA**	**AGA AGG**	**CUN**	**CGG**
Normal code assignment	**Stop**	**Ile**	**Arg**	**Leu**	**Arg**
Animals					
Vertebrates	Trp	Met	Stop	+	+
Drosophila	Trp	Met	Ser	+	+
Yeasts					
Saccharomyces cerevisiae	Trp	Met	+	Thr	+
Torulopsis glabrata	Trp	Met	+	Thr	?
Schizosaccharomyces pombe	Trp	+	+	+	+
Filamentous fungi	Trp	+	+	+	+
Trypanosomes	Trp	+	+	+	+
Higher plants	+	+	+	+	Trp
Chlamydomonas reinhardtii	?	+	+	+	?

*A question mark Indicates that the codon has not been observed in the indicated mitochondrial genome; N, any nucleotide; +, the codon has the same meaning as in the normal code.

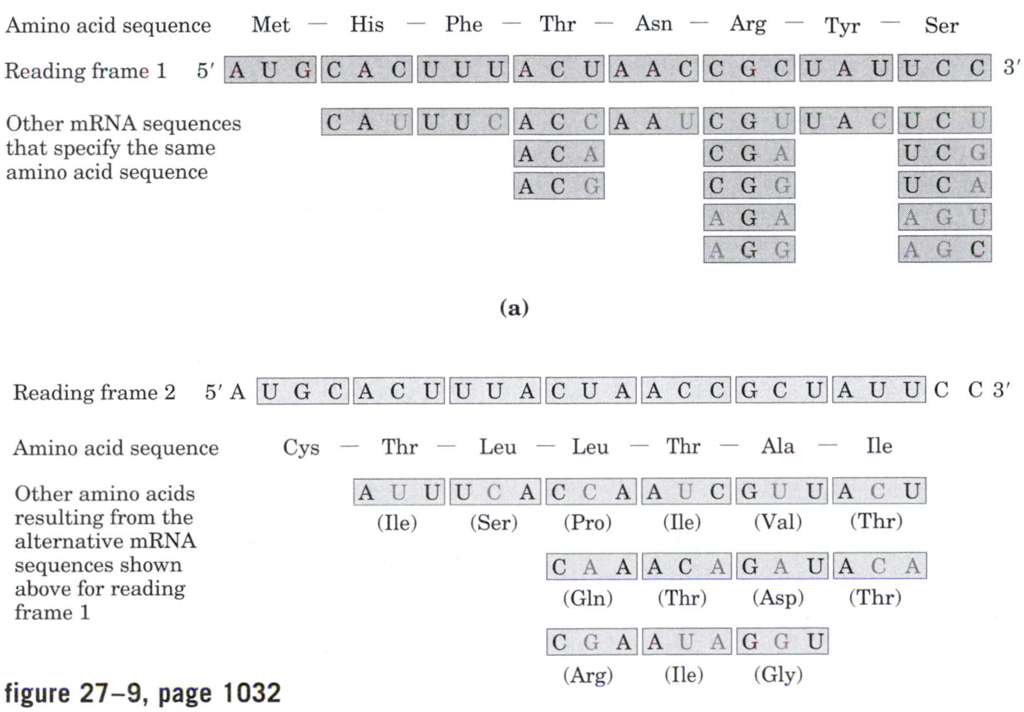

Amino acid sequence Met — His — Phe — Thr — Asn — Arg — Tyr — Ser

Reading frame 1 5′ AUG CAC UUU ACU AAC CGC UAU UCC 3′

Other mRNA sequences that specify the same amino acid sequence

His	Phe	Thr	Asn	Arg	Tyr	Ser
CAU	UUC	ACC	AAU	CGU	UAC	UCU
		ACA		CGA		UCG
		ACG		CGG		UCA
				AGA		AGU
				AGG		AGC

(a)

Reading frame 2 5′ A UGC ACU UUA CUA ACC GCU AUU C C 3′

Amino acid sequence Cys — Thr — Leu — Leu — Thr — Ala — Ile

Other amino acids resulting from the alternative mRNA sequences shown above for reading frame 1

AUU (Ile)	UCA (Ser)	CCA (Pro)	AUC (Ile)	GUU (Val)	ACU (Thr)
		CAA (Gln)	ACA (Thr)	GAU (Asp)	ACA (Thr)
		CGA (Arg)	AUA (Ile)	GGU (Gly)	

(b)

figure 27–9, page 1032
Reading frame and amino acid sequence.

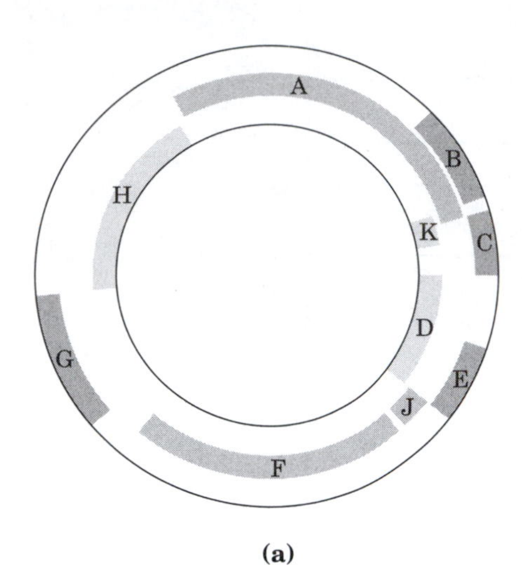

(a)

133
Gene C A U G A G A A A A U U C G A C C U A U --- 3′
Met – Arg – Lys – Phe – Asp – Leu ---

Gene A 5′--- G G C G G A A A A U G A
--- Gly – Gly – Lys – Stop

Gene K 5′--- G G C G G A A A A U G A G A A A A U U C G A C C U A U --- 3′
--- Ala – Glu – Asn – Glu – Lys – Ile – Arg – Pro – ---

(b)

figure 27–10, page 1033
Genes within genes.

table 27–6, page 1034

Components Required for the Five Major Stages of Protein Synthesis in *E. coli*

Stage	Essential components
1. Activation of amino acids	20 amino acids 20 aminoacyl-tRNA synthetases 20 or more tRNAs ATP Mg^{2+}
2. Initiation	mRNA *N*-Formylmethionyl-tRNA Initiation codon in mRNA (AUG) 30S ribosomal subunit 50S ribosomal subunit Initiation factors (IF-1, IF-2, IF-3) GTP Mg^{2+}
3. Elongation	Functional 70S ribosome (initiation complex) Aminoacyl-tRNAs specified by codons Elongation factors (EF-Tu, EF-Ts, EF-G) GTP Mg^{2+}
4. Termination and release	Termination codon in mRNA Polypeptide release factors (RF_1, RF_2, RF_3) ATP
5. Folding and posttranslational processing	Specific enzymes, cofactors, and other components for removal of initiating residues and signal sequences, additional proteolytic processing, modification of terminal residues, and attachment of phosphate, methyl, carboxyl, carbohydrate, or prosthetic groups

table 27–7, page 1035

RNA and Protein Components of the *E. coli* Ribosome

Subunit	Number of different proteins	Total number of proteins	Protein designations	Number and type of rRNAs
30S	21	21	S1–S21	1 (16S rRNA)
50S	33	36	L1–L36*	2 (5S and 23S rRNAs)

*The L1 to L36 protein designations do not correspond to 36 different proteins. The protein originally designated L7 is in fact a modified form of L12, and L8 is a complex of three other proteins. Also, L26 proved to be the same protein as S20 (and not part of the 50S subunit). This gives 33 different proteins in the large subunit. There are four copies of the L7/L12 protein, with the three extra copies bringing the total protein count to 36.

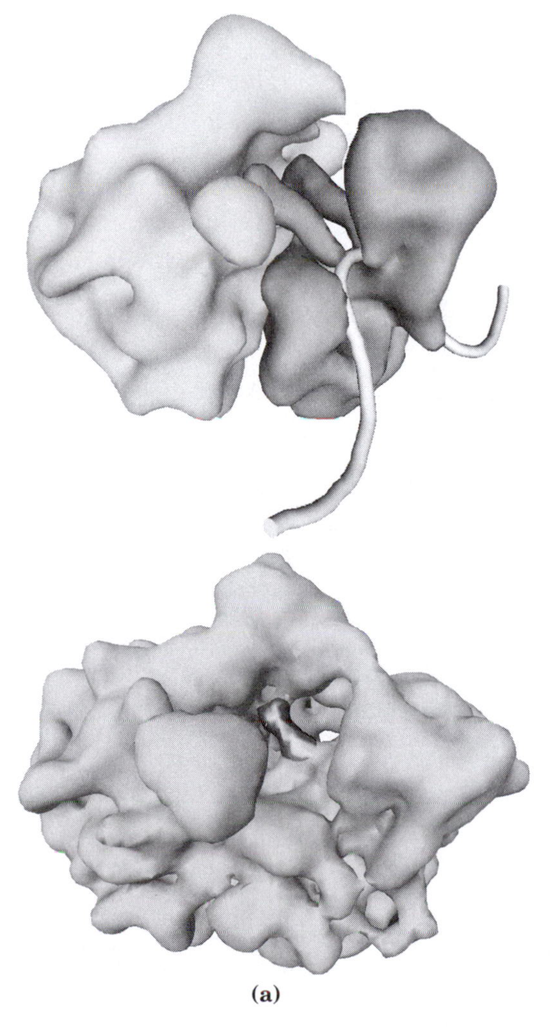

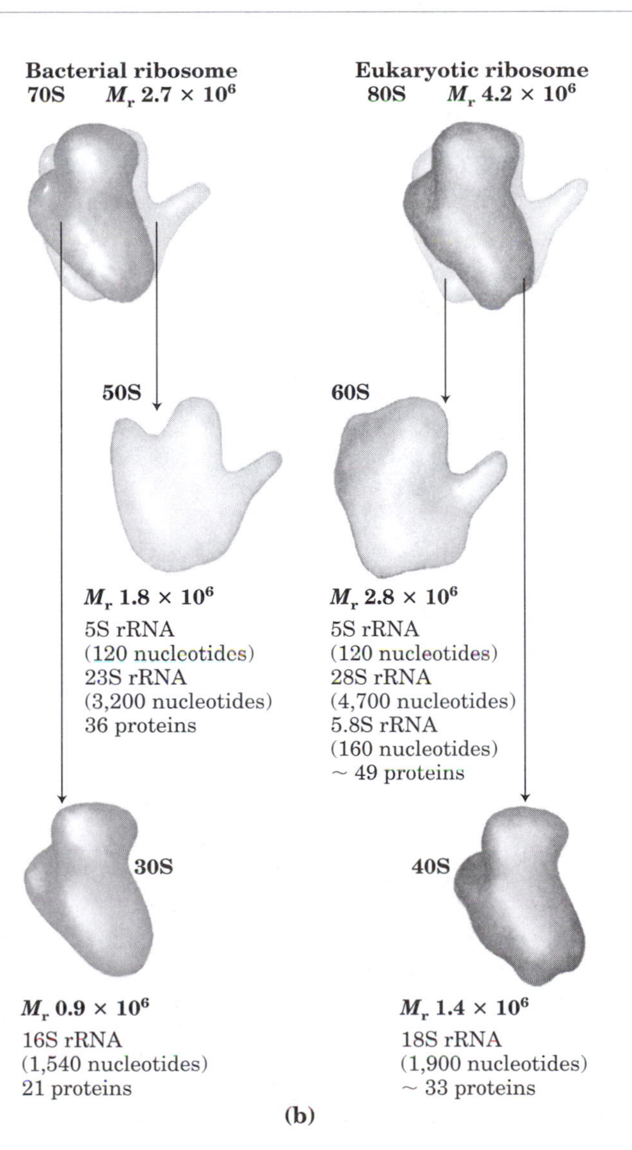

figure 27–11, page 1036
Ribosomes.

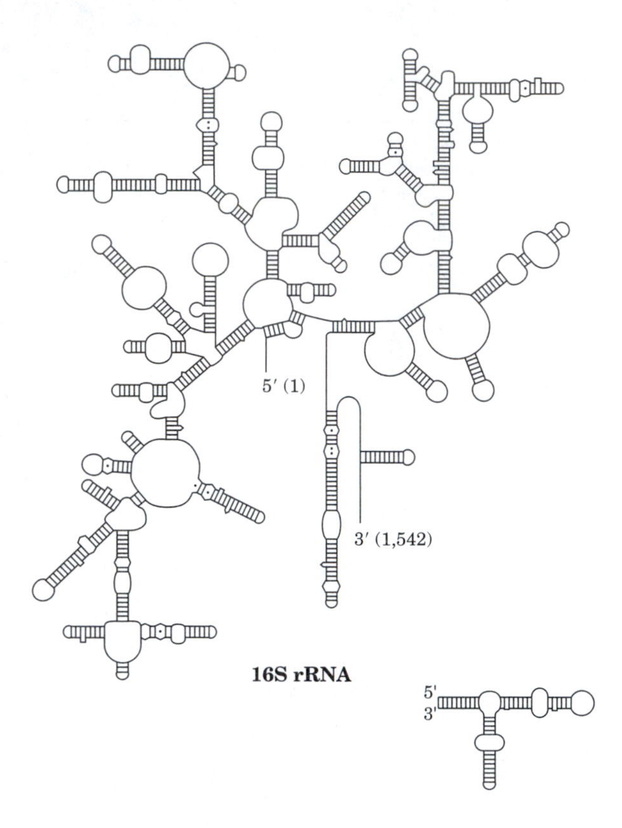

figure 27–12, page 1037
Bacterial rRNAs.

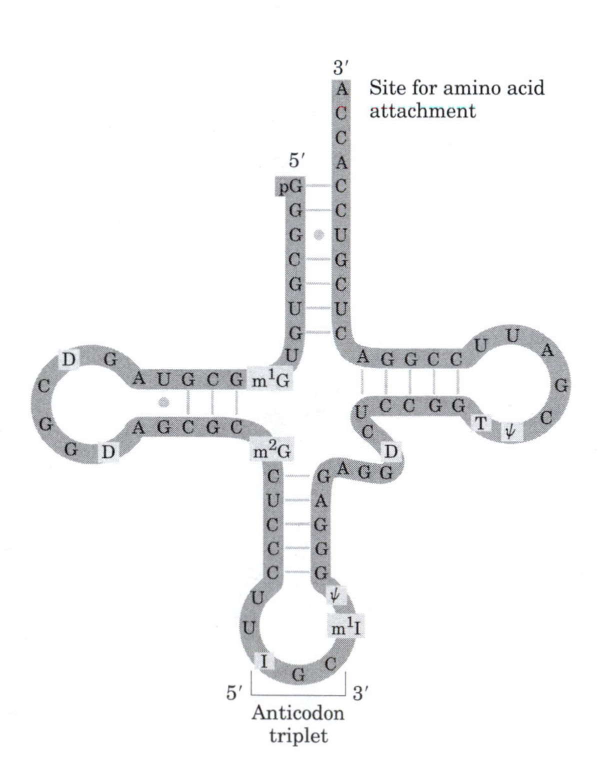

figure 27–13, page 1038
Nucleotide sequence of yeast $tRNA^{Ala}$.

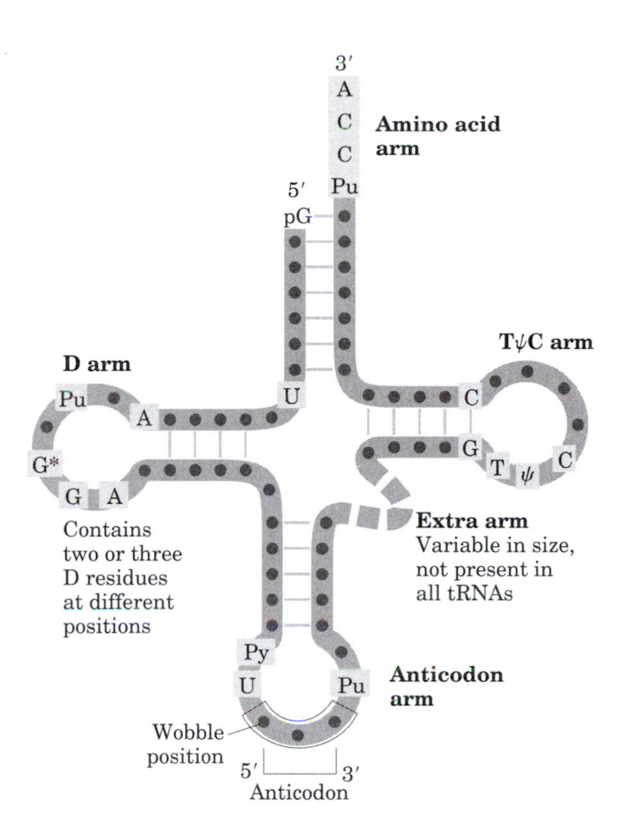

figure 27–14, page 1038
General cloverleaf secondary structure of all tRNAs.

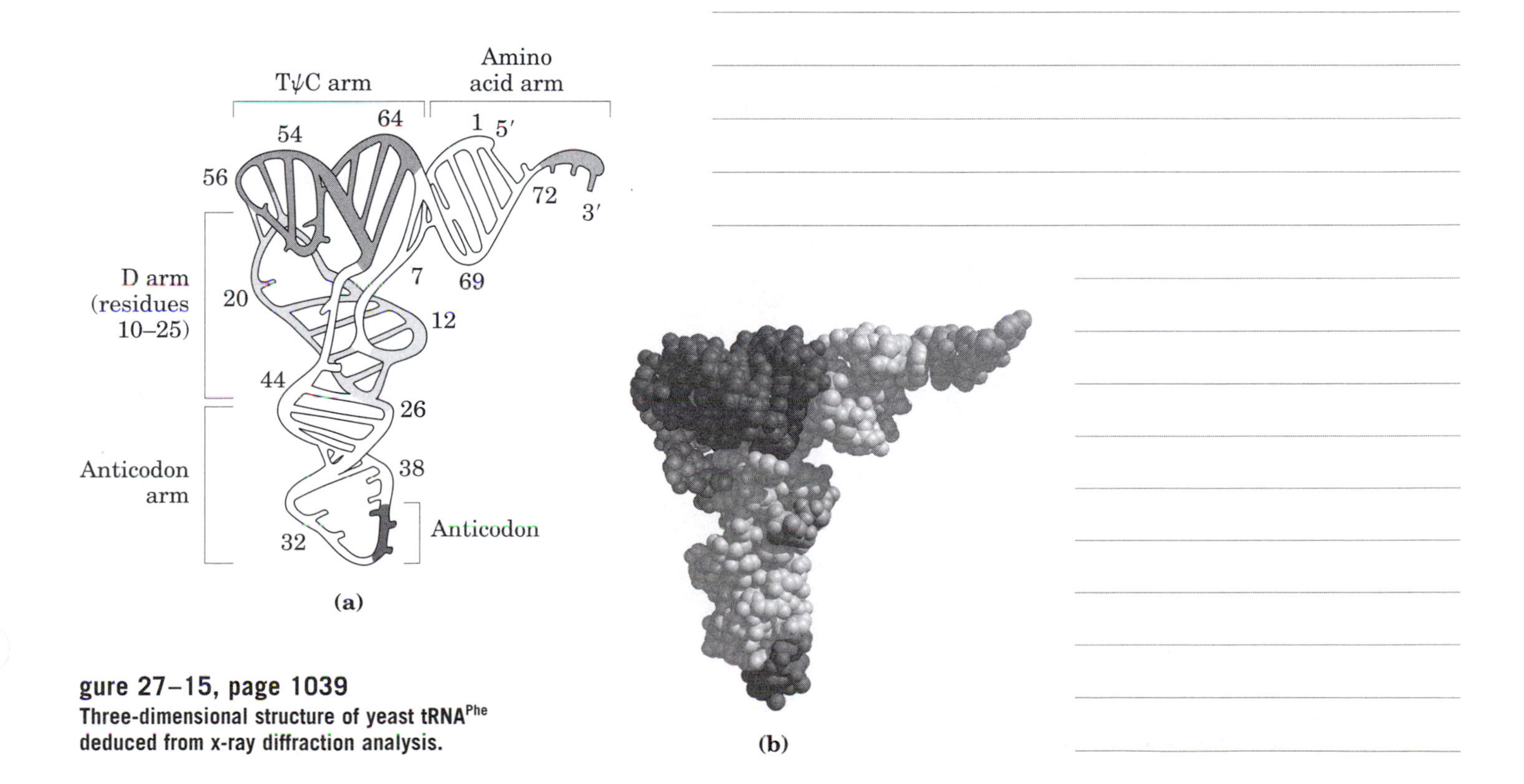

gure 27–15, page 1039
Three-dimensional structure of yeast tRNA^{Phe} deduced from x-ray diffraction analysis.

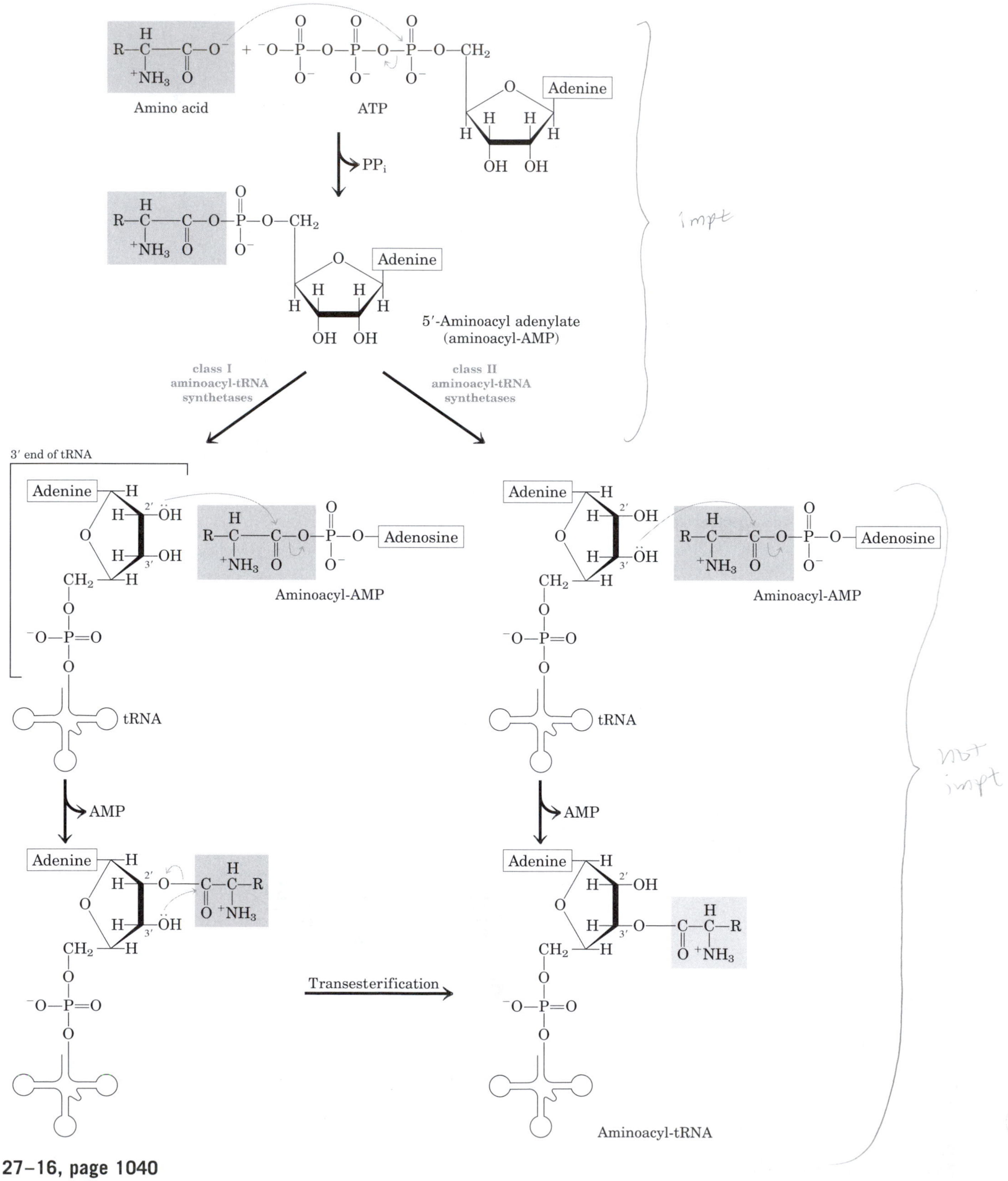

Amino acid

ATP

PP_i

5′-Aminoacyl adenylate
(aminoacyl-AMP)

class I
aminoacyl-tRNA
synthetases

class II
aminoacyl-tRNA
synthetases

3′ end of tRNA

Aminoacyl-AMP

tRNA

AMP

Transesterification

Aminoacyl-tRNA

figure 27–16, page 1040
Aminoacylation of tRNA by aminoacyl-tRNA synthetases.

table 27–8, page 1041

Two Classes of Aminoacyl-tRNA Synthetases*

Class I	Class II
Arg	Ala
Cys	Asn
Gln	Asp
Glu	Gly
Ile	His
Leu	Lys
Met	Phe
Trp	Pro
Tyr	Ser
Val	Thr

*Here, Arg represents arginyl-tRNA synthetase, and so forth. The classification applies to all organisms for which tRNA synthetases have been analyzed and is based on protein structural distinctions and on the mechanistic distinction outlined in Figure 27–16.

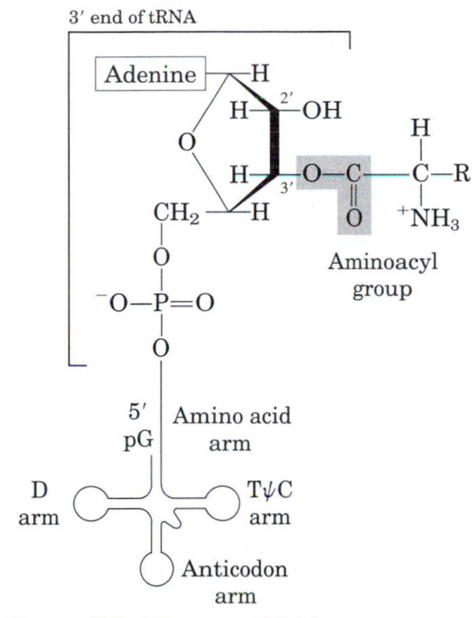

figure 27–17, page 1041
General structure of aminoacyl-tRNAs.

Valine: COO^- / $H_3\overset{+}{N}-C-H$ / $H-C-CH_3$ / CH_3

Isoleucine: COO^- / $H_3\overset{+}{N}-C-H$ / $H-C-CH_3$ / CH_2 / CH_3

page 1042

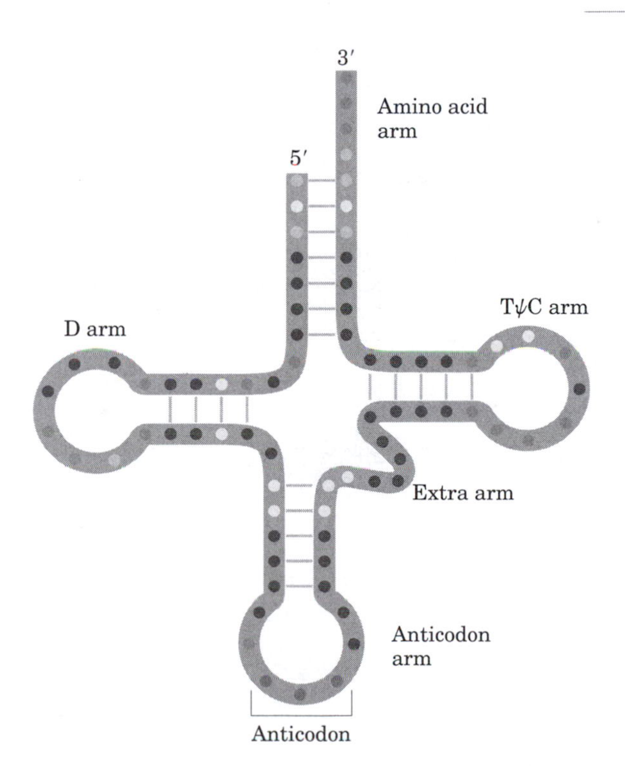

figure 27–18, page 1042
Known positions in tRNAs recognized by aminoacyl-tRNA synthetases.

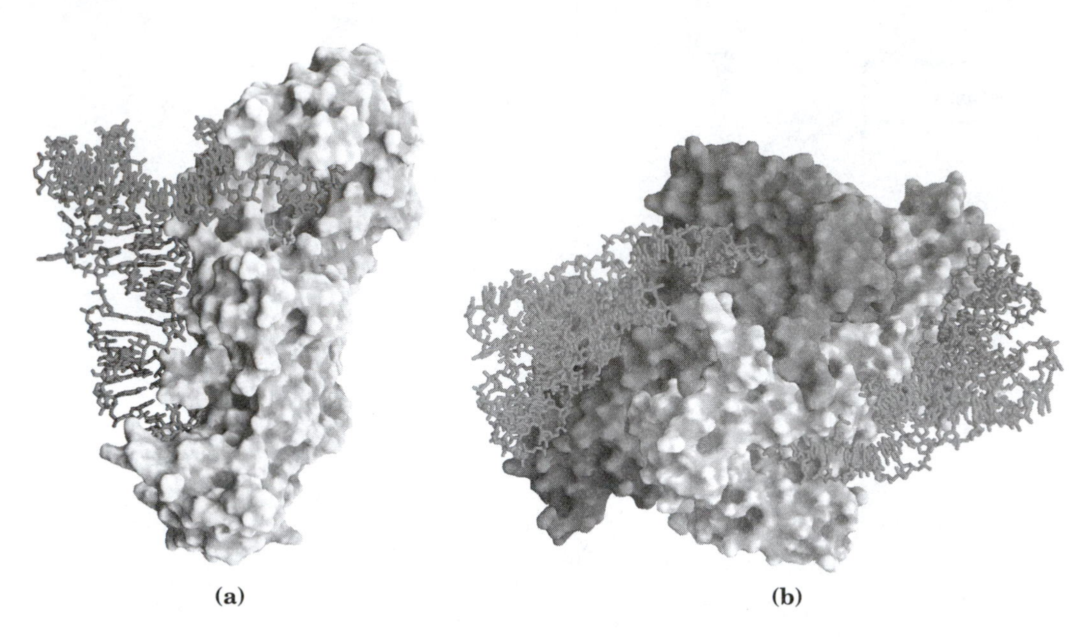

figure 27–19, page 1043
Aminoacyl-tRNA synthetases.

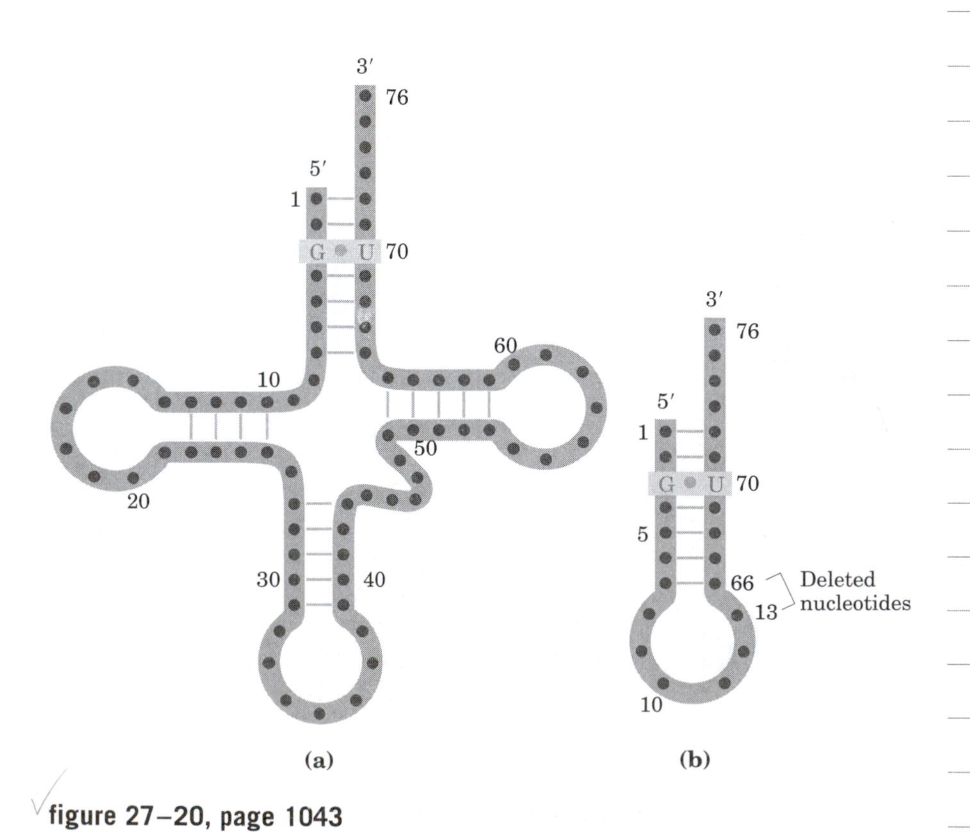

figure 27–20, page 1043

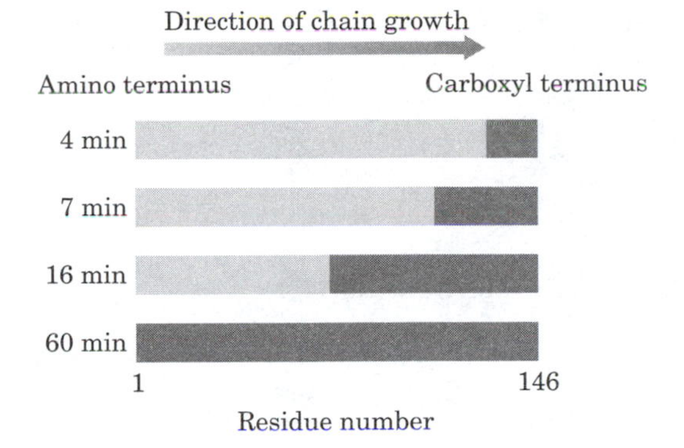

figure 27–21, page 1044
Proof that polypeptides grow by addition of new amino acid residues to the carboxyl end—the Dintzis experiment.

$$H{-}C({=}O){-}N(H){-}C(COO^-)(H){-}CH_2{-}CH_2{-}S{-}CH_3$$

***N*-Formylmethionine, page 1044**

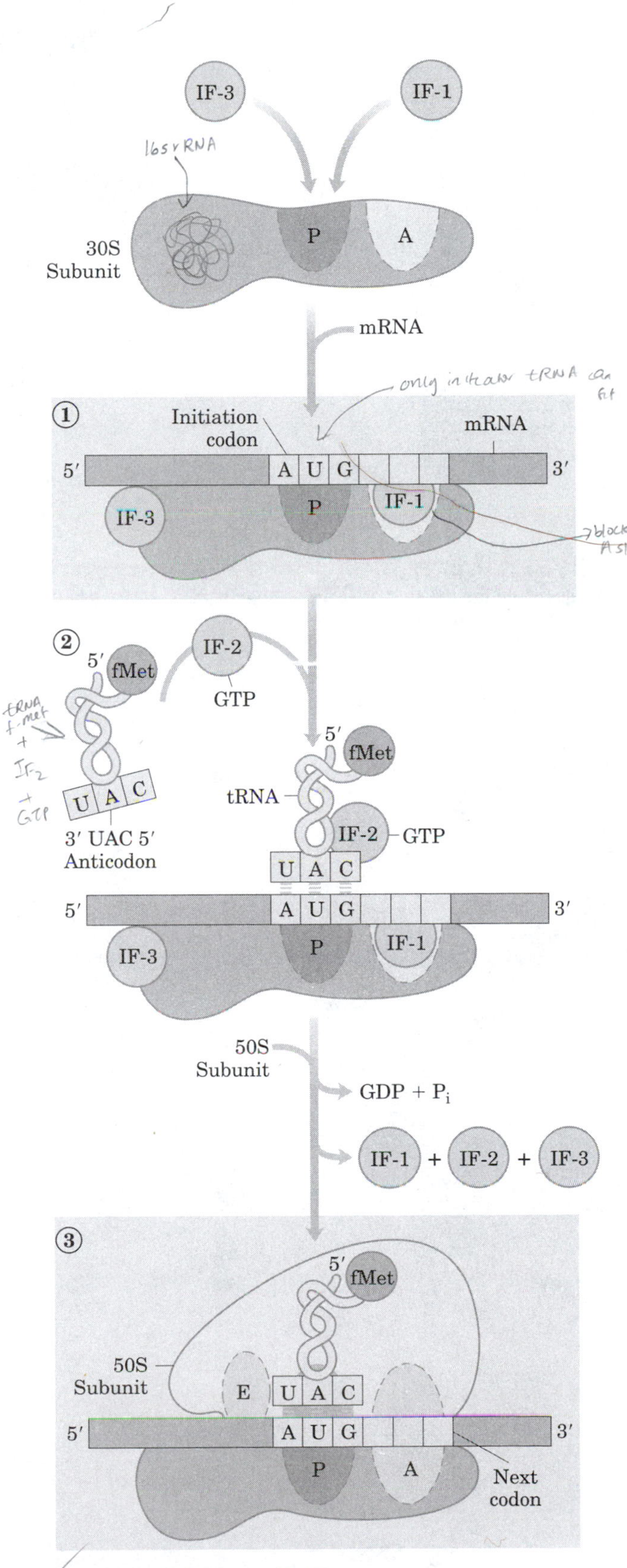

figure 27–22, page 1045
Formation of the initiation complex.

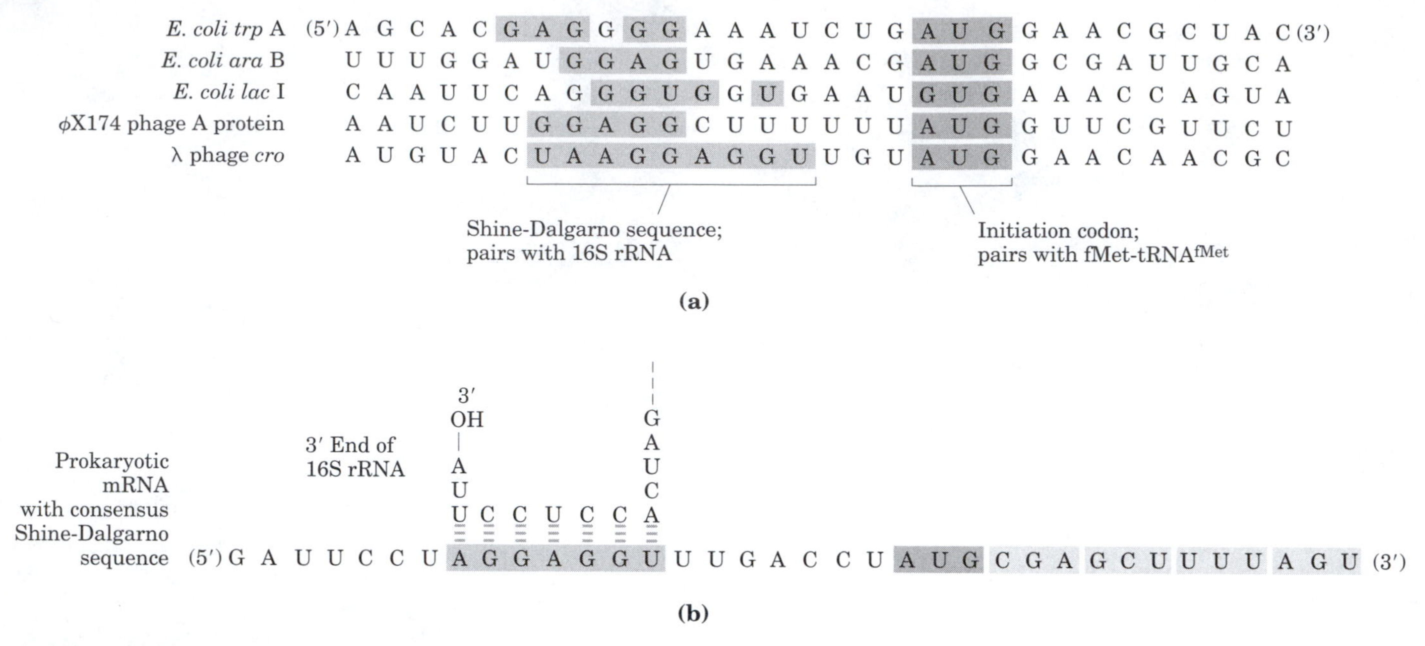

figure 27–23, page 1046
Sequences on the mRNA that serve as signals for initiation of protein synthesis in bacteria.

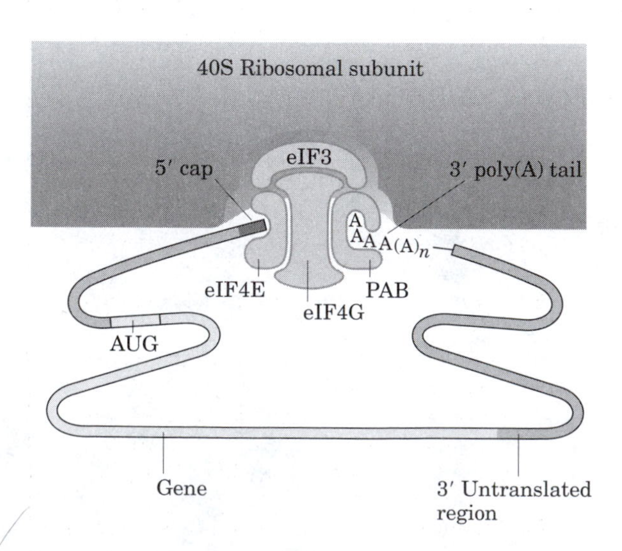

figure 27–24, page 1046
Protein complexes in the formation of a eukaryotic initiation complex.

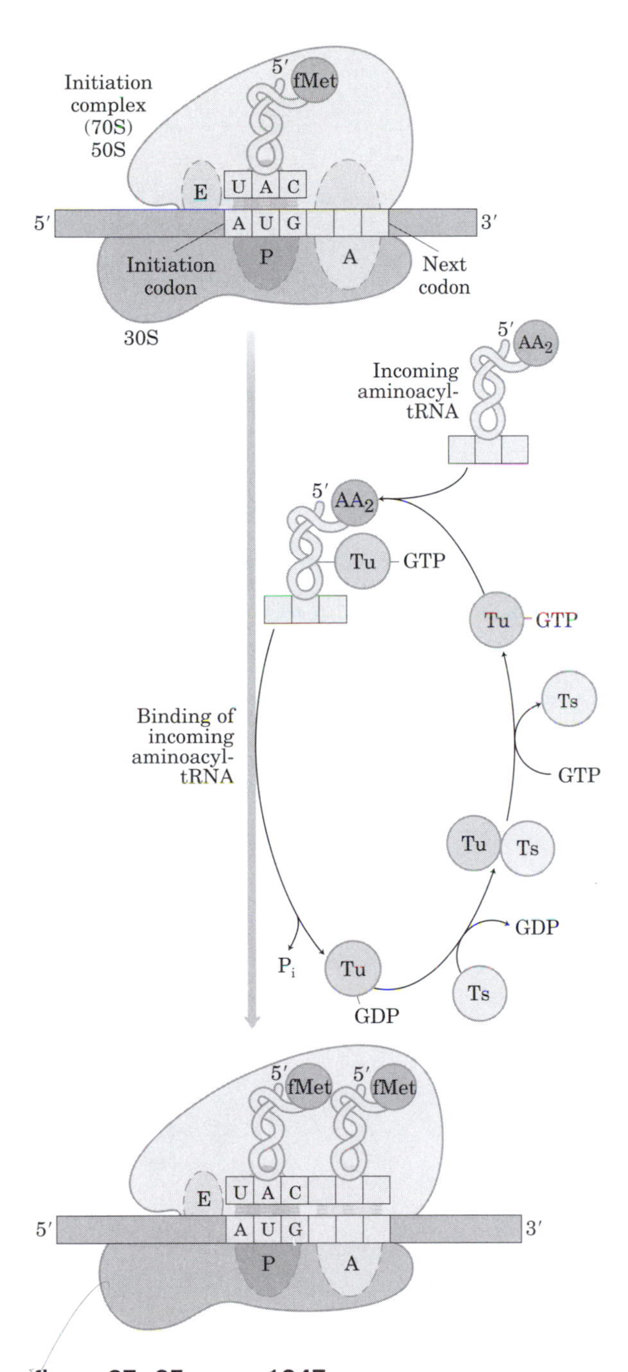

figure 27–25, page 1047
First step in elongation (bacteria): binding of the second aminoacyl-tRNA.

table 27–9, page 1047

Protein Factors Required for Initiation of Translation in Bacterial and Eukaryotic Cells

Bacterial

Factor	Function
IF-1	Prevents premature binding of tRNAs to A site
IF-2	Facilitates binding of fMet-tRNA^{fMet} to 30S ribosomal subunit
IF-3	Binds to 30S subunit; prevents premature association of 50S subunit; enhances specificity of P site for fMet-tRNA^{fMet}

Eukaryotic

Factor*	Function
eIF2	Facilitates binding of initiating Met-tRNA^{Met} to 40S ribosomal subunit
eIF2B, eIF3	First factors to bind 40S subunit; facilitate subsequent steps
eIF4A	RNA helicase activity removes secondary structure in the mRNA to permit binding to 40S subunit; part of the eIF4F complex
eIF4B	Binds to mRNA; facilitates scanning of mRNA to locate the first AUG
eIF4E	Binds to the 5′ cap of mRNA; part of the eIF4F complex
eIF4G	Binds to eIF4E and to poly(A) binding protein (PAB); part of the eIF4F complex
eIF5	Promotes dissociation of several other initiation factors from 40S subunit as a prelude to association of 60S subunit to form 80S initiation complex
eIF6	Facilitates dissociation of inactive 80S ribosome into 40S and 60S subunits

*The prefix "e" identifies these as eukaryotic factors.

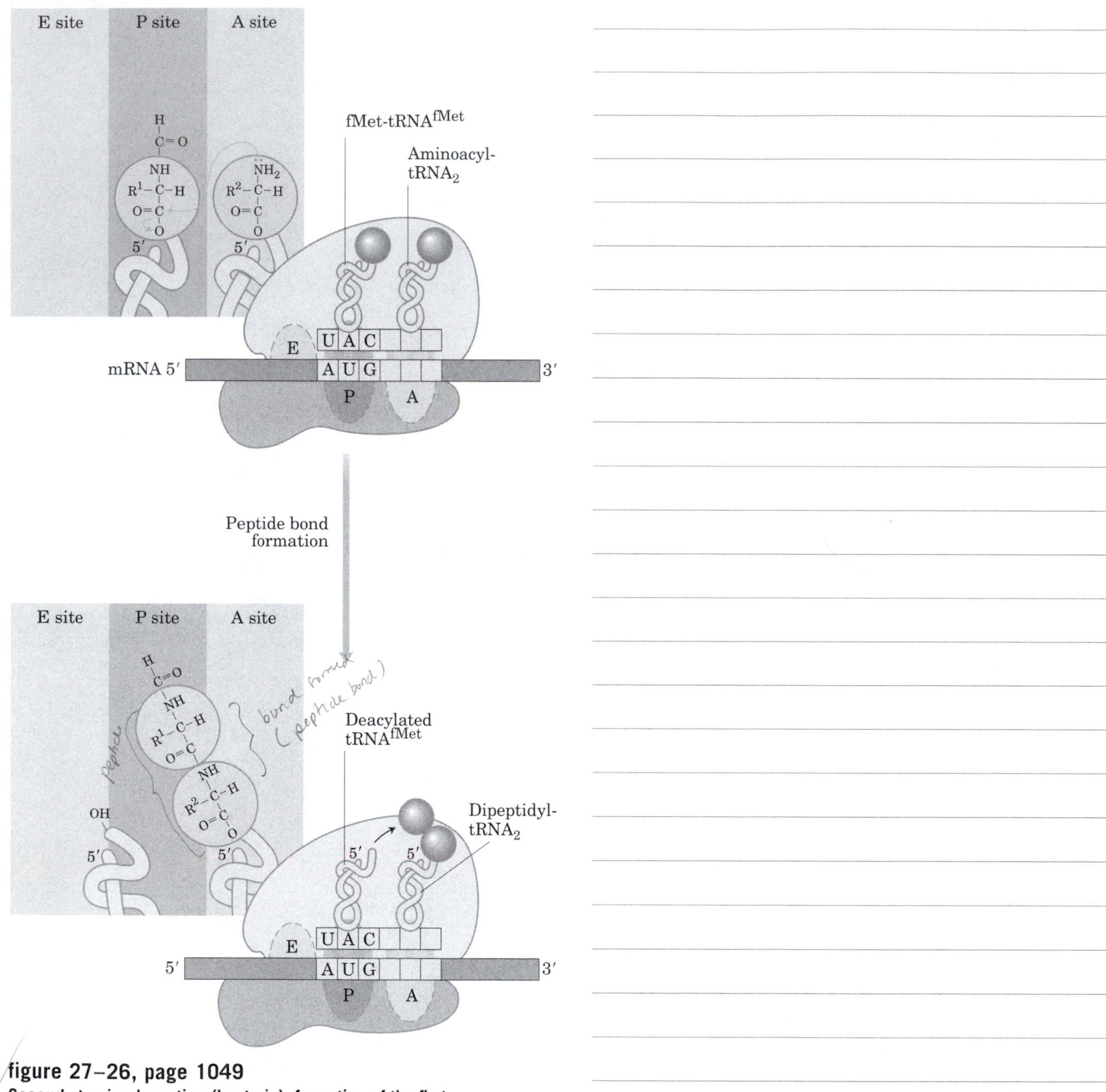

figure 27–26, page 1049
Second step in elongation (bacteria): formation of the first peptide bond.

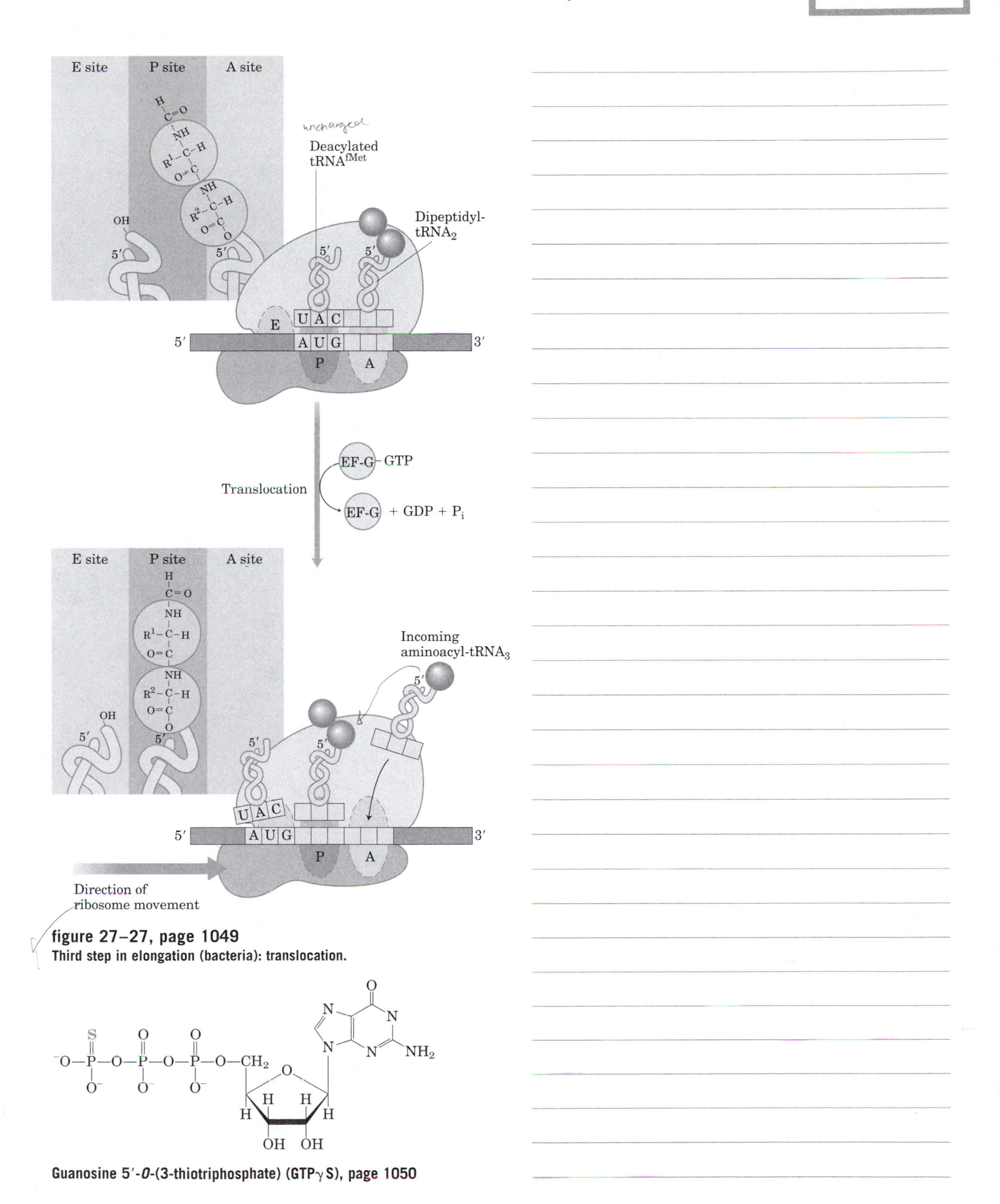

figure 27–27, page 1049
Third step in elongation (bacteria): translocation.

Guanosine 5′-*O*-(3-thiotriphosphate) (GTPγS), page 1050

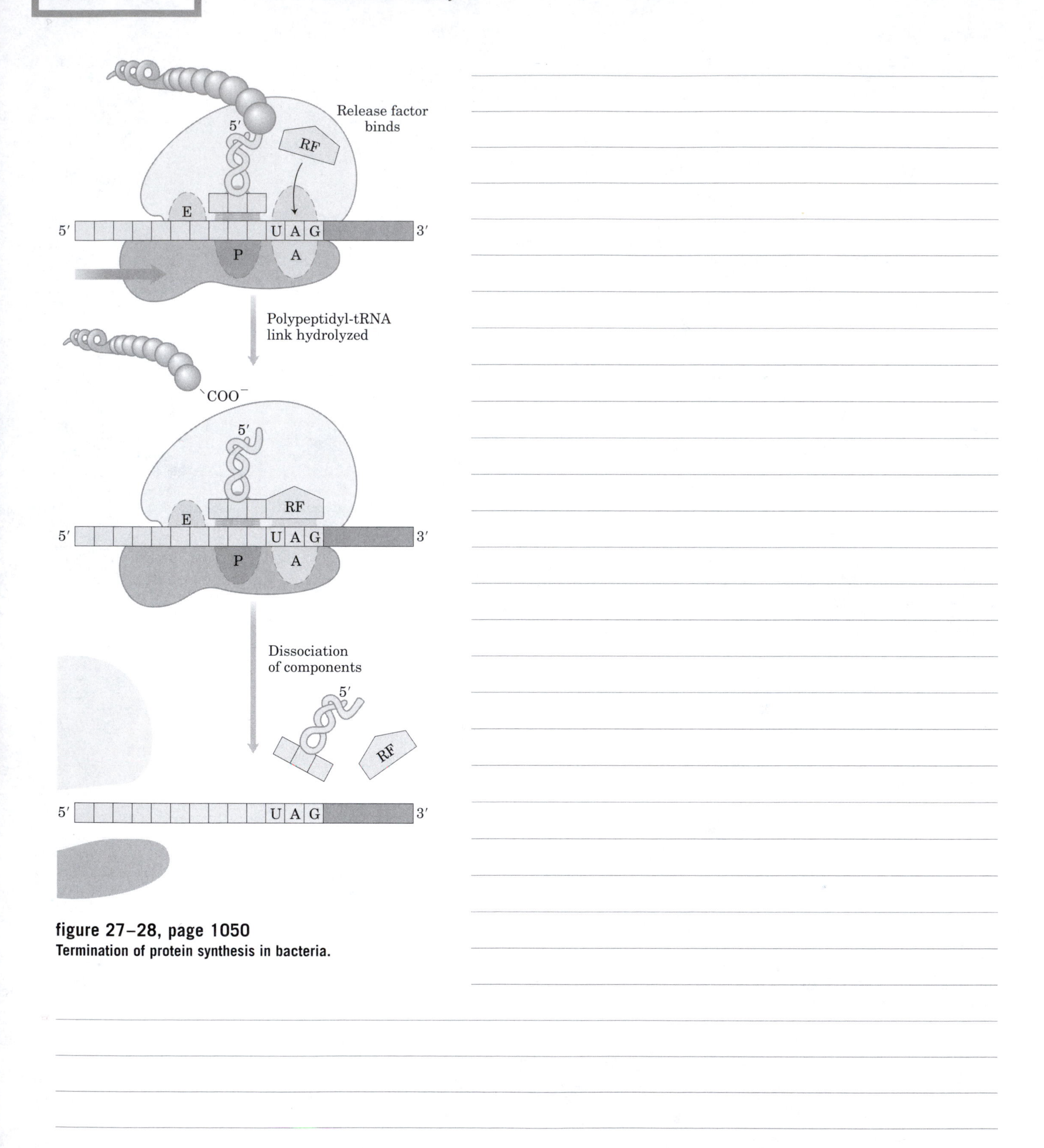

figure 27–28, page 1050
Termination of protein synthesis in bacteria.

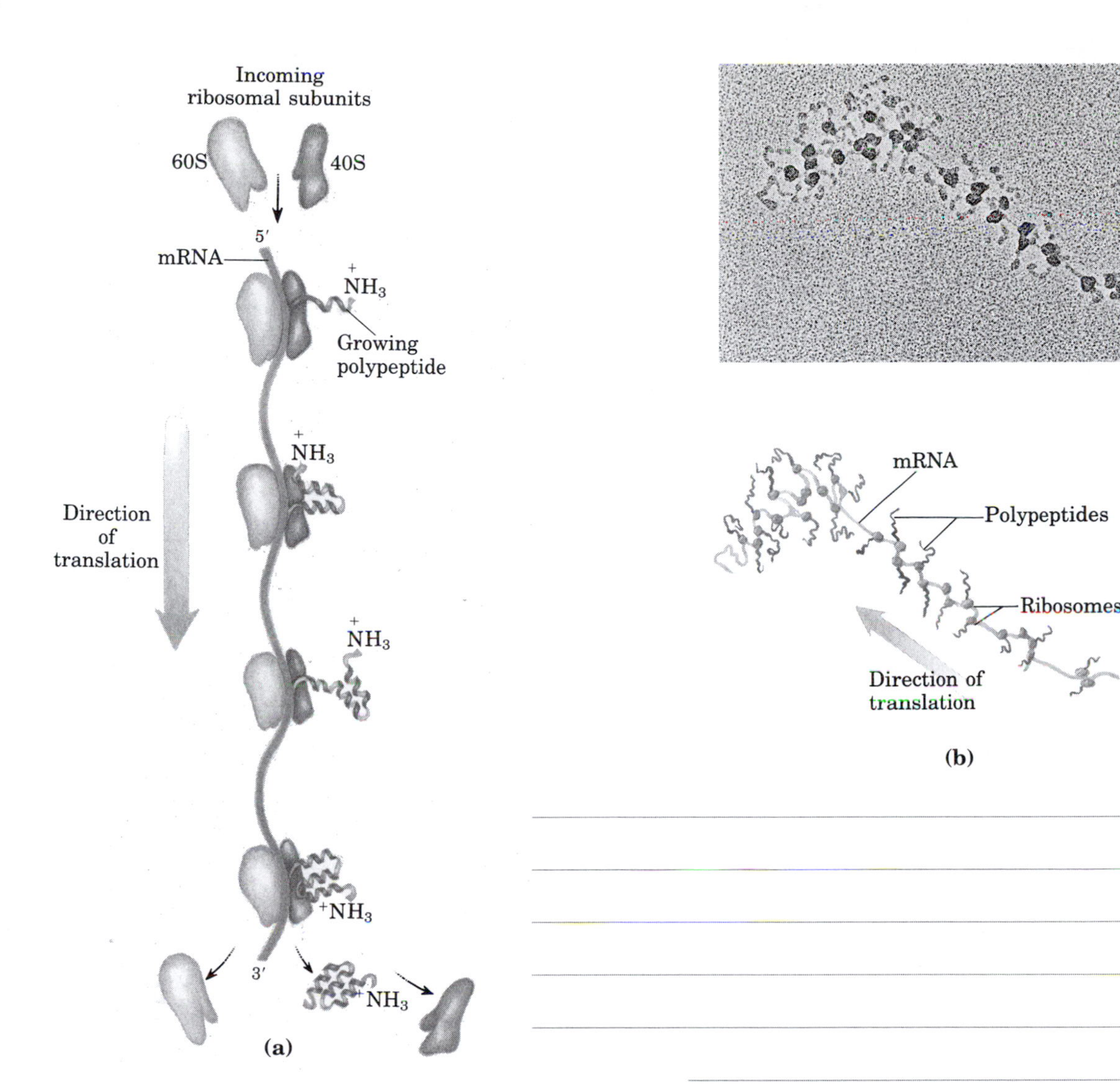

figure 27–29, page 1052
Polysome.

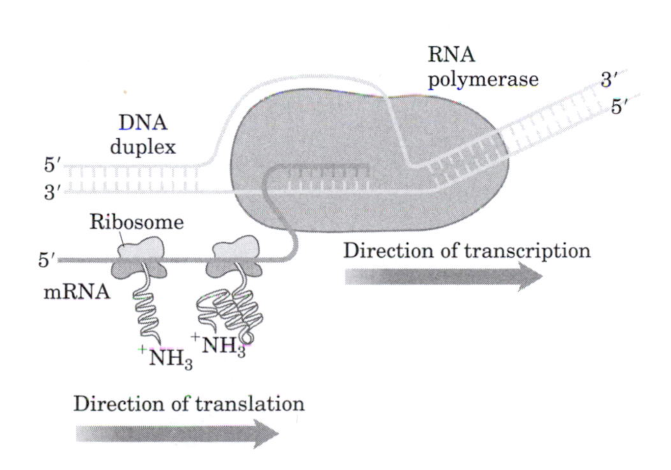

figure 27–30, page 1052
Coupling of transcription and translation in bacteria.

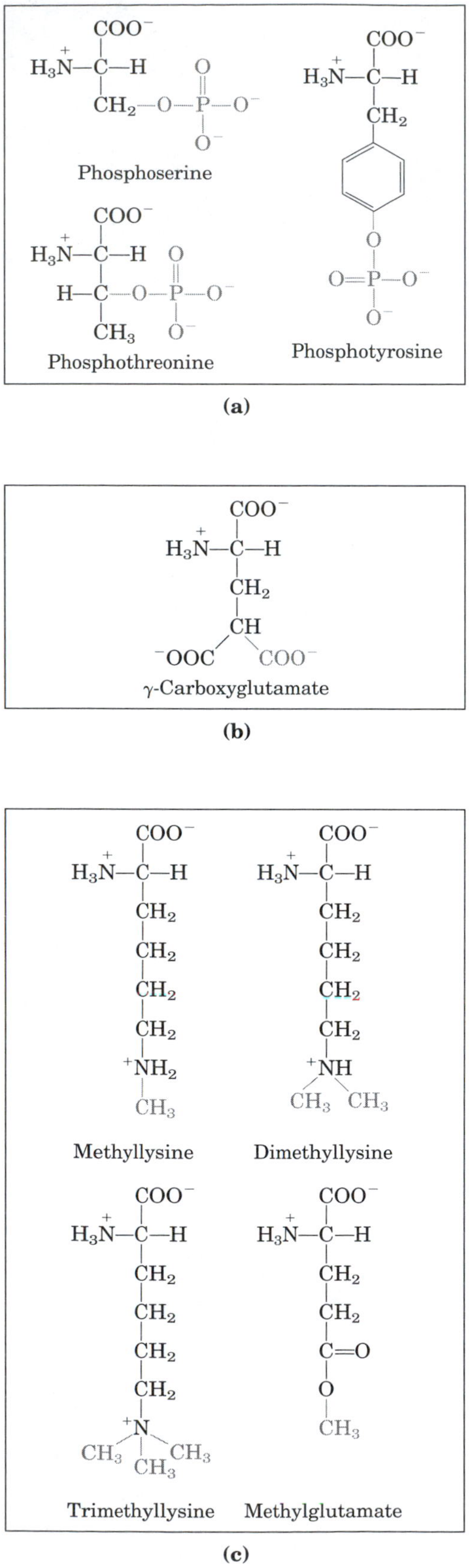

figure 27–31, page 1053
Some modified amino acid residues.

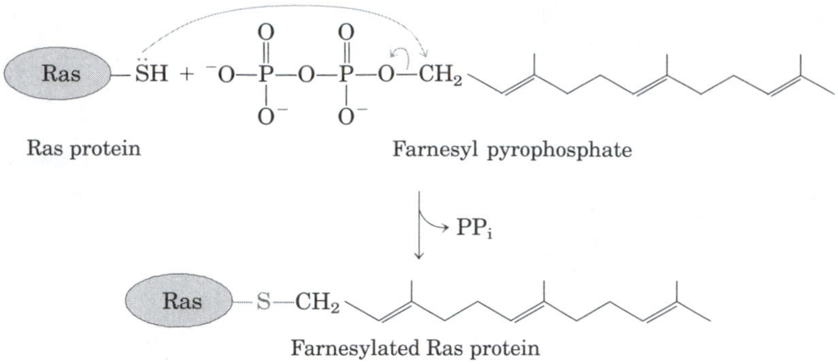

figure 27–32, page 1054
Farnesylation of a Cys residue on a protein.

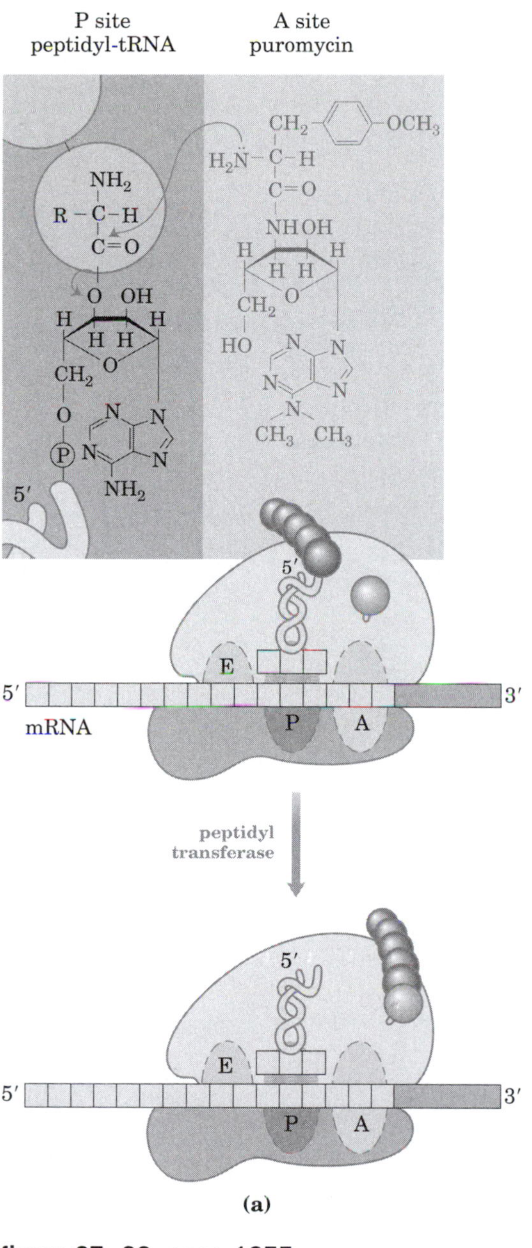

(b)

figure 27–33, page 1055
Disruption of peptide bond formation by puromycin.

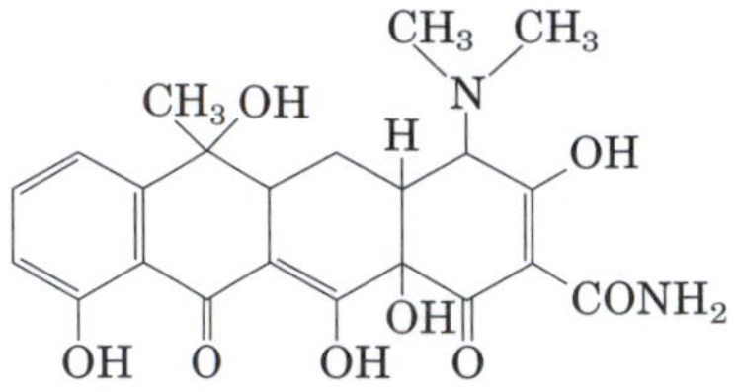

Tetracycline

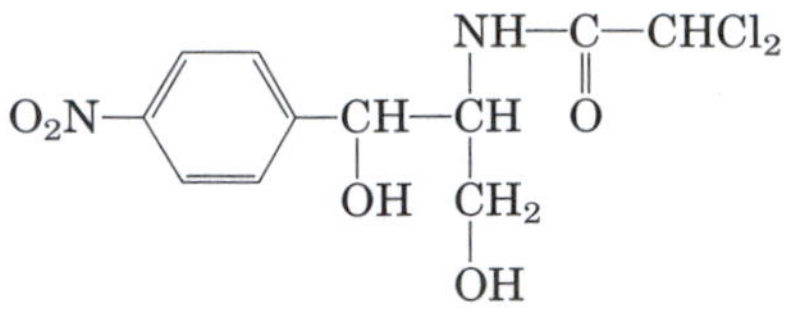

Chloramphenicol

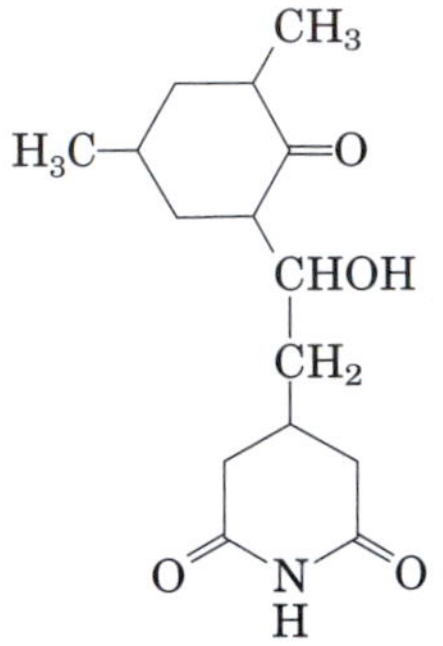

Cycloheximide

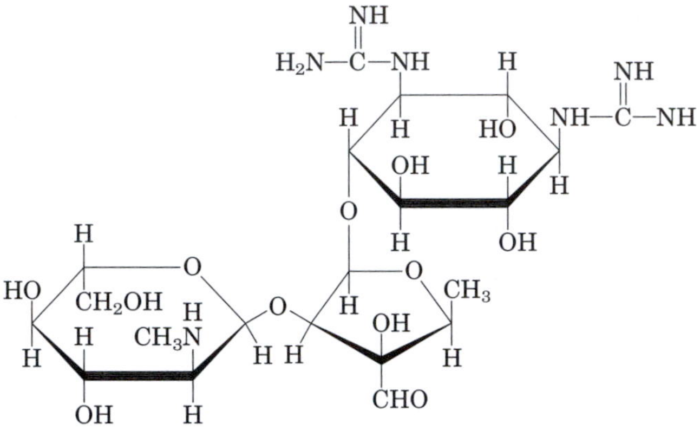

Streptomycin

Antibiotics, page 1056

cleavage site

Human influenza virus A: Met Lys Ala Lys Leu Leu Val Leu Leu Tyr Ala Phe Val Ala Gly ↓ Asp Gln --

Human preproinsulin: Met Ala Leu Trp Met Arg Leu Leu Pro Leu Leu Ala Leu Leu Ala Leu Trp Gly Pro Asp Pro Ala Ala Ala ↓ Phe Val --

Bovine growth hormone: Met Met Ala Ala Gly Pro Arg Thr Ser Leu Leu Leu Ala Phe Ala Leu Leu Cys Leu Pro Trp Thr Gln Val Val Gly ↓ Ala Phe --

Bee promellitin: Met Lys Phe Leu Val Asn Val Ala Leu Val Phe Met Val Val Tyr Ile Ser Tyr Ile Tyr Ala ↓ Ala Pro --

Drosophila glue protein: Met Lys Leu Leu Val Val Ala Val Ile Ala Cys Met Leu Ile Gly Phe Ala Asp Pro Ala Ser Gly ↓ Cys Lys --

figure 27–34, page 1057
Translocation into the ER directed by amino-terminal signal sequences of some eukaryotic proteins.

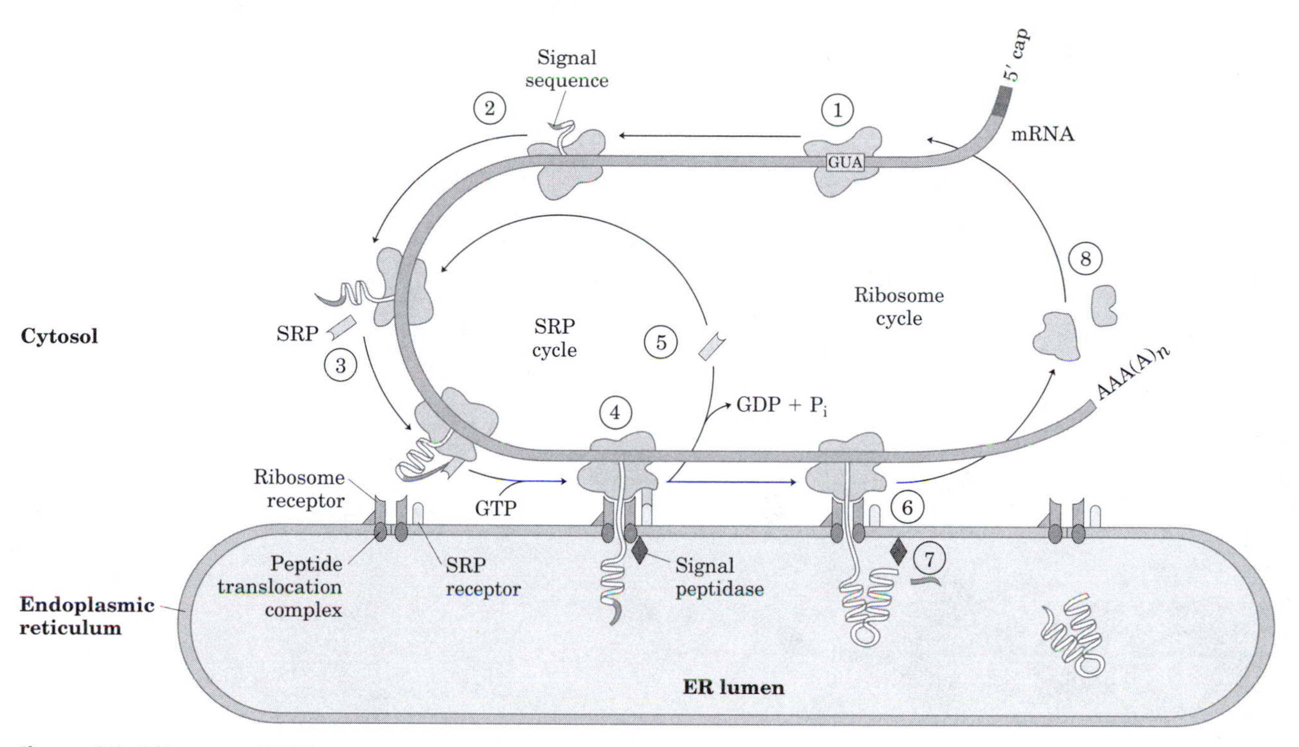

figure 27–35, page 1058
Directing eukaryotic proteins with the appropriate signals to the endoplasmic reticulum.

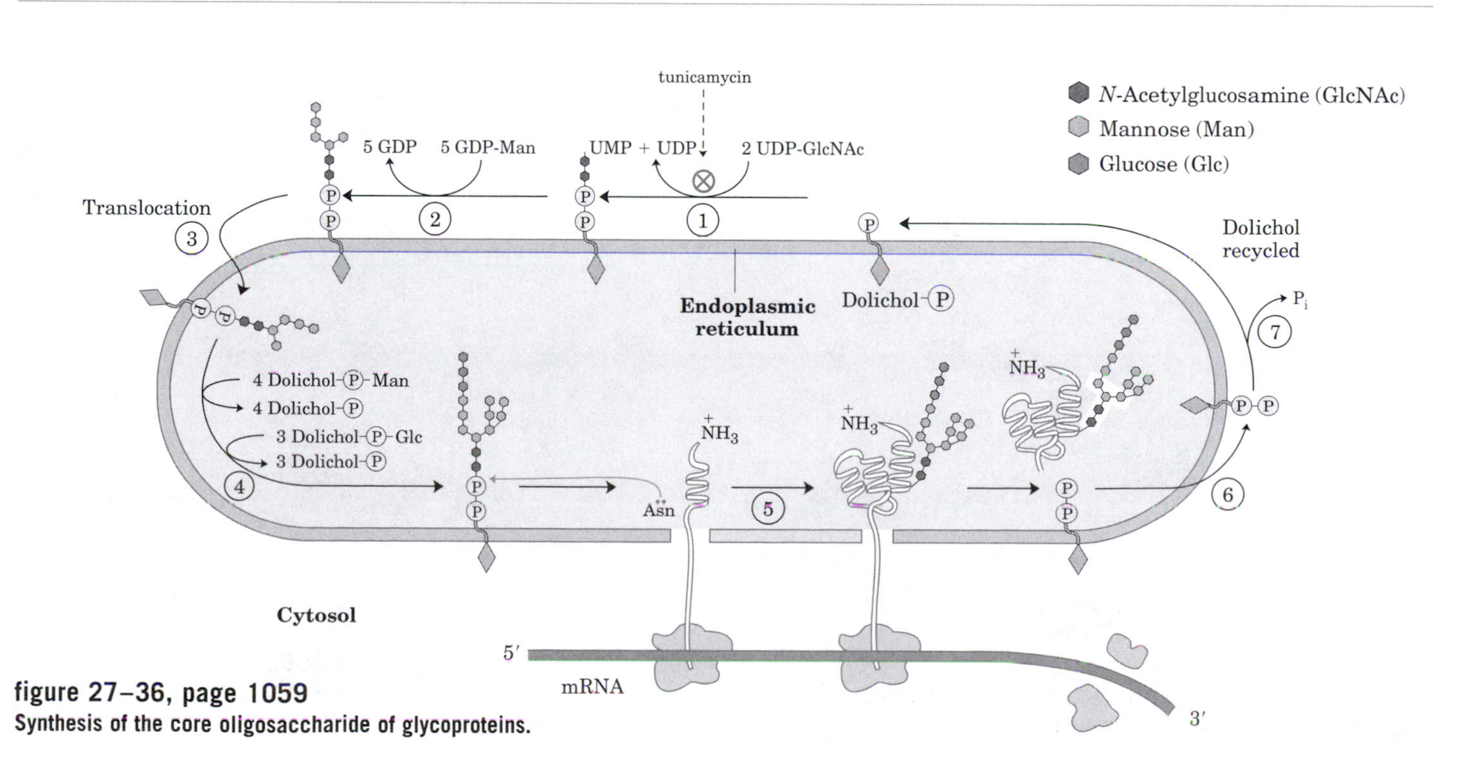

figure 27–36, page 1059
Synthesis of the core oligosaccharide of glycoproteins.

$$^{-}O-\overset{\overset{O}{\|}}{\underset{\underset{O^-}{|}}{P}}-O-CH_2-CH_2-\overset{\overset{CH_3}{|}}{\underset{\underset{H}{|}}{C}}-CH_2-\left(CH_2-CH=\overset{\overset{CH_3}{|}}{C}-CH_2\right)_n-CH_2-CH=\overset{\overset{CH_3}{|}}{C}-CH_3$$

Dolichol phosphate (*n* = 9–22), page 1059

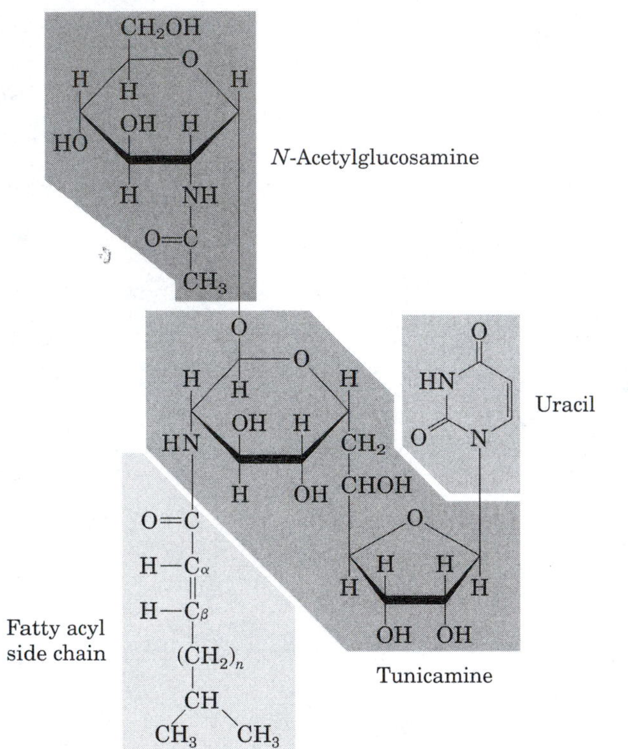

Tunicamycin, page 1060

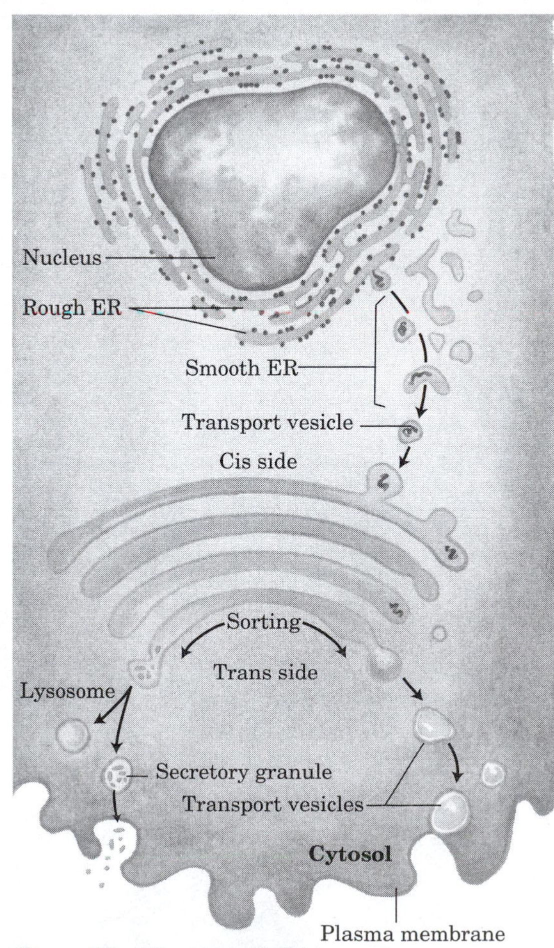

figure 27–37, page 1060
Pathway taken by proteins destined for lysosomes, the plasma membrane, or secretion.

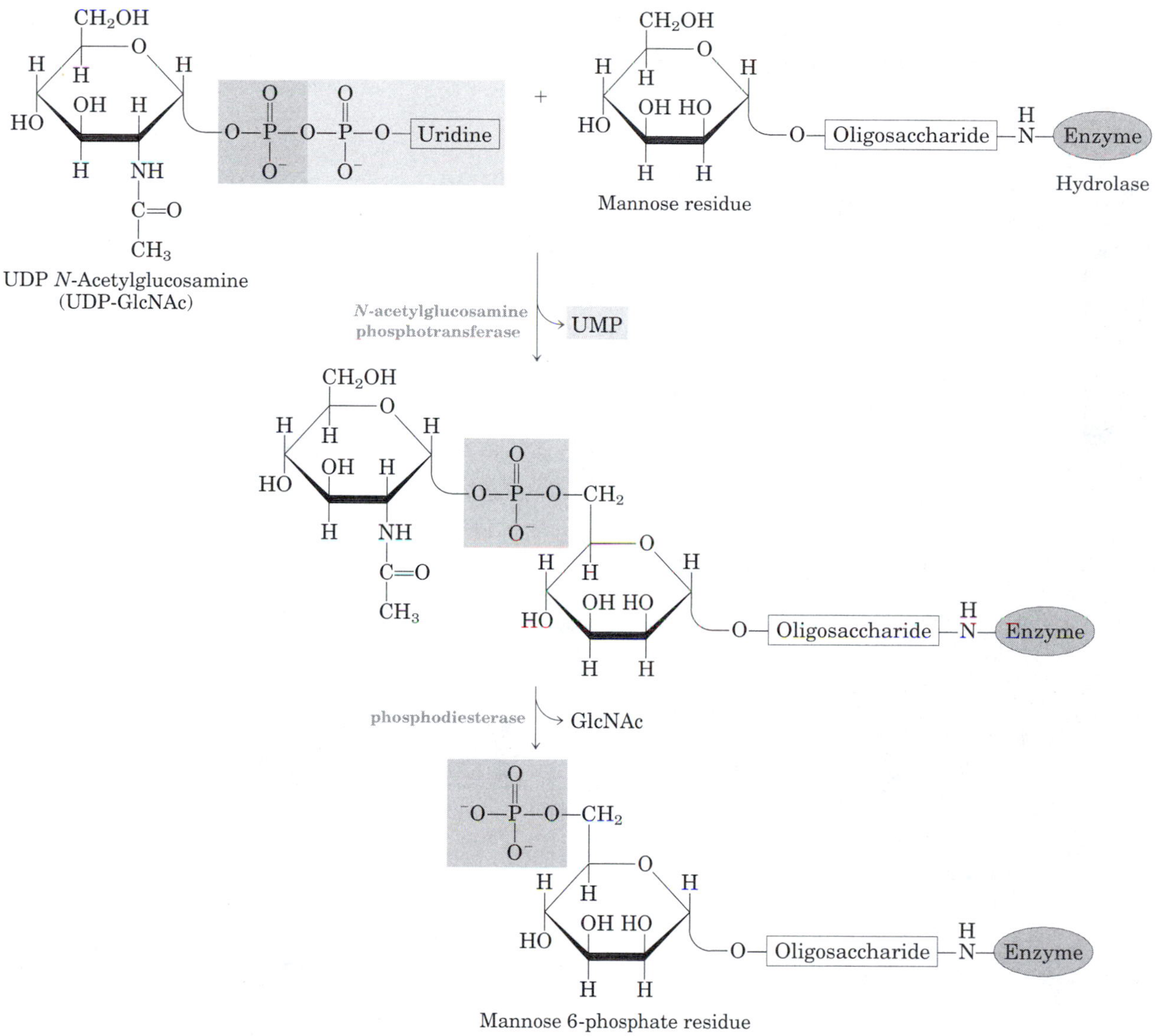

UDP *N*-Acetylglucosamine (UDP-GlcNAc)

Uridine

+

Mannose residue

Oligosaccharide — Enzyme

Hydrolase

N-acetylglucosamine phosphotransferase

UMP

phosphodiesterase

GlcNAc

Mannose 6-phosphate residue

figure 27–38, page 1061

Phosphorylation of mannose residues on lysosome-targeted enzymes.

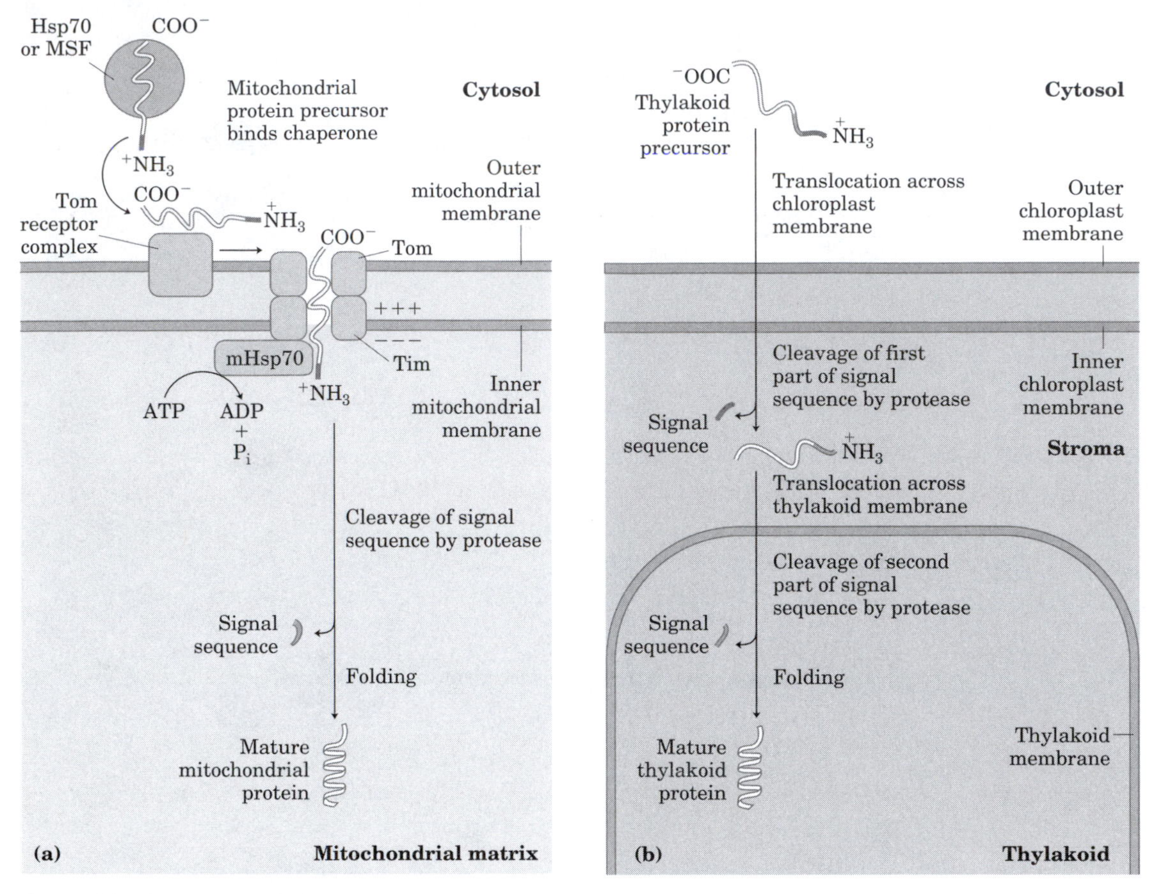

figure 27–39, page 1062

Protein targeting to mitochondria and chloroplasts.

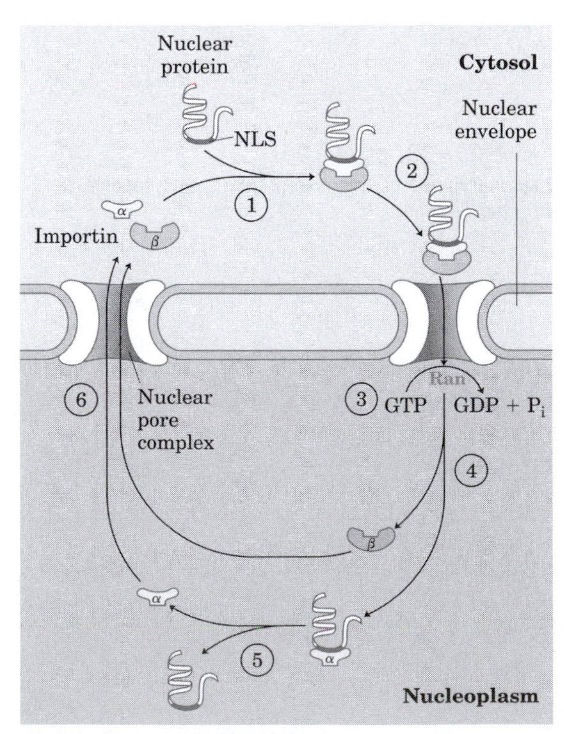

figure 27–40, page 1063

Targeting of nuclear proteins.

	Signal sequence (↓ = cleavage site)
Inner membrane proteins	
Phage fd, major coat protein	Met Lys Lys Ser Leu Val Leu Lys Ala Ser Val Ala Val Ala Thr Leu Val Pro Met Leu Ser Phe Ala ↓ Ala Glu -
Phage fd, minor coat protein	Met Lys Lys Leu Leu Phe Ala Ile Pro Leu Val Val Pro Phe Tyr Ser His Ser ↓ Ala Glu -
Periplasmic proteins	
Alkaline phosphatase	Met Lys Gln Ser Thr Ile Ala Leu Ala Leu Leu Pro Leu Leu Phe Thr Pro Val Thr Lys Ala ↓ Arg Thr -
Leucine-specific binding protein	Met Lys Ala Asn Ala Lys Thr Ile Ile Ala Gly Met Ile Ala Leu Ala Ile Ser His Thr Ala Met Ala ↓ Asp Asp -
β-Lactamase of pBR322	Met Ser Ile Gln His Phe Arg Val Ala Leu Ile Pro Phe Phe Ala Ala Phe Cys Leu Pro Val Phe Ala ↓ His Pro -
Outer membrane proteins	
Lipoprotein	Met Lys Ala Thr Lys Leu Val Leu Gly Ala Val Ile Leu Gly Ser Thr Leu Leu Ala Gly ↓ Cys Ser -
LamB	Leu Arg Lys Leu Pro Leu Ala Val Ala Val Ala Ala Gly Val Met Ser Ala Gln Ala Met Ala ↓ Val Asp -
OmpA	Met Met Ile Thr Met Lys Lys Thr Ala Ile Ala Ile Ala Val Ala Leu Ala Gly Phe Ala Thr Val Ala Gln Ala ↓ Ala Pro -

figure 27–41, page 1064
Signal sequences targeting different locations in bacteria.

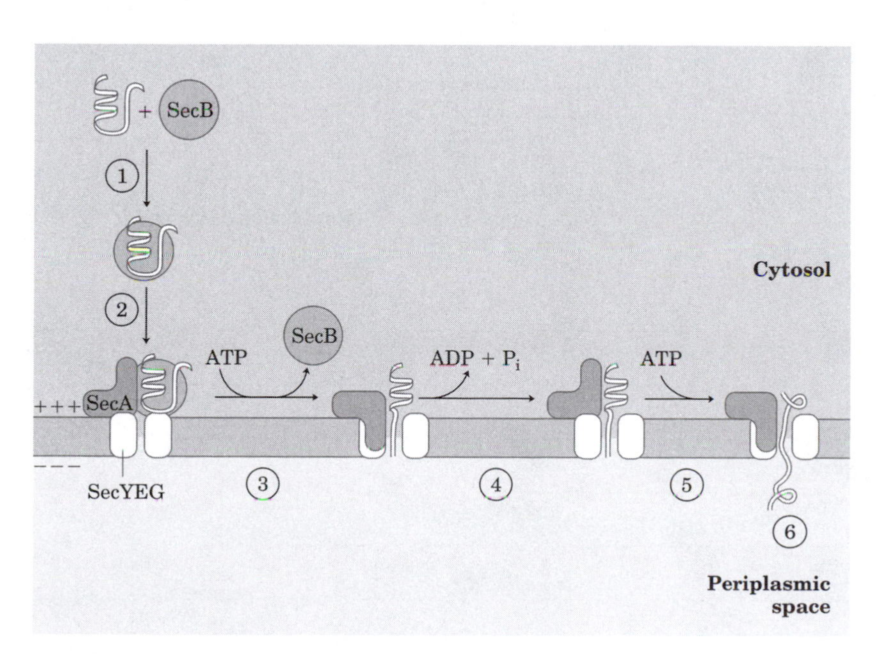

figure 27–42, page 1064
Model for protein export in bacteria.

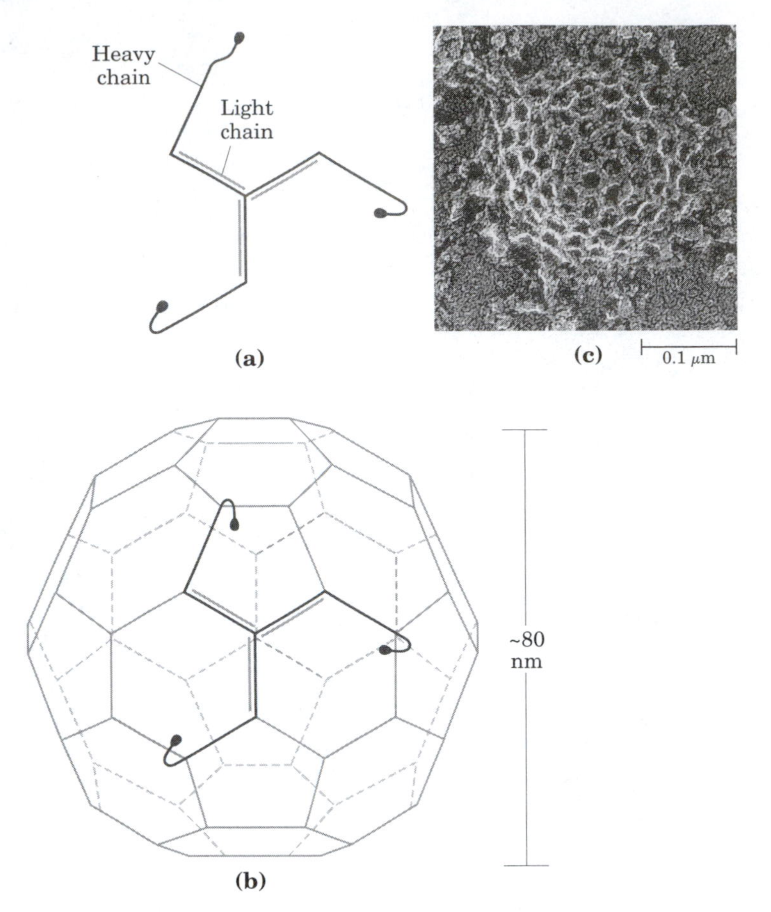

figure 27–43, page 1065
Clathrin.

Ubiquitin—$C(=O)O^-$

HS—E1 + ATP

→ AMP + PP_i

Ubiquitin—$C(=O)$—S—E1

HS—E2

→ HS—E1

Ubiquitin—$C(=O)$—S—E2

E3

H_2N—Lys—Target protein

→ HS—E2

Ubiquitin—$C(=O)$—NH—Lys—Target protein

Repeated cycles lead to attachment of additional ubiquitin

figure 27–44, page 1066
Three-step cascade pathway by which ubiquitin is attached to a protein.

table 27–10, page 1066

Relationship between Protein Half-Life and Amino-Terminal Amino Acid Residue

Amino-terminal residue	Half-life*
Stabilizing	
Met, Gly, Ala, Ser, Thr, Val	>20 h
Destabilizing	
Ile, Gln	~30 min
Tyr, Glu	~10 min
Pro	~7 min
Leu, Phe, Asp, Lys	~3 min
Arg	~2 min

Source: Modified from Bachmair, A., Finley, D., & Varshavsky, A. (1986) In vivo half-life of a protein is a function of its amino-terminal residue. *Science* **234,** 179–186.

*Half-lives were measured in yeast for a single protein modified so that in each experiment it had a different amino-terminal residue. (See Chapter 29 for a discussion of techniques used to engineer proteins with altered amino acid sequences.) Half-lives may vary for different proteins and in different organisms, but this general pattern appears to hold for all organisms: amino acids listed here as stabilizing when present at the amino terminus have a stabilizing effect on proteins in all cells.

PROBLEMS

15. Protein-Coding Capacity of a Viral DNA, page 1071.

Protein	Number of amino acid residues
A	455
B	120
C	86
D	152
E	91
F	427
G	175
H	328
J	38
K	56

chapter

28 Regulation of Gene Expression

When you study for exams, don't forget to review The Minicourses on the UNDERSTAND! BIOCHEMISTRY CD that came with your textbook.

Minicourses that apply to this Chapter include:

Molecules of Life
- An Interactive Gallery of Protein Structures

Nucleic Acids and Their Expression
- DNA Repair
- Transcription
- Gene Regulation – Prokaryotes
- Gene Regulation – Phage Lambda

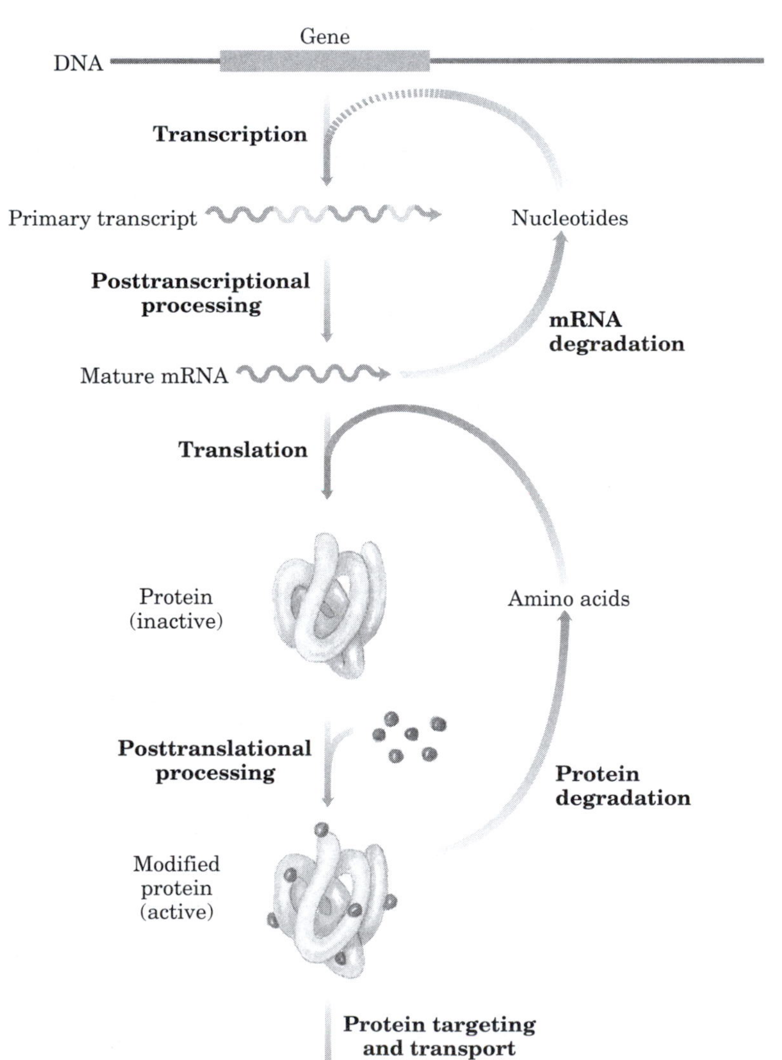

figure 28–1, page 1073
Seven processes that affect the steady-state concentration of a protein.

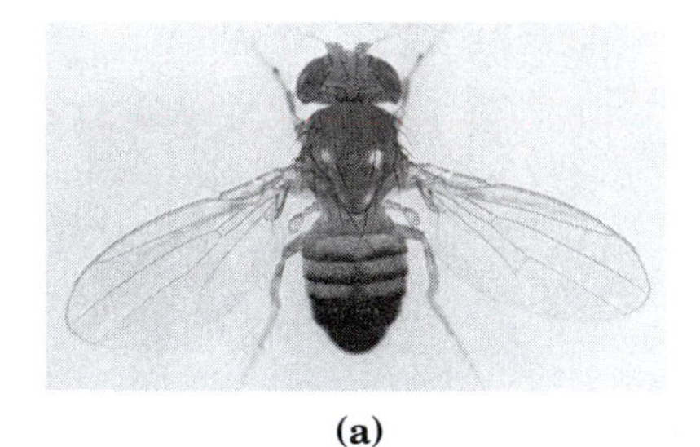

(a)

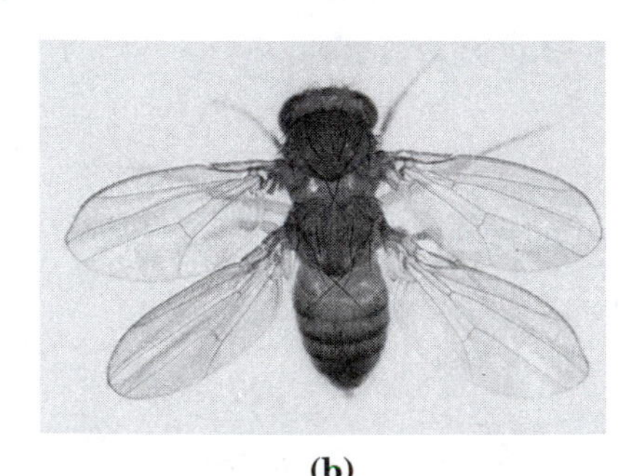

(b)

Defects in the regulation of gene expression can alter the developmental program of an organism. An example is this *bithorax* mutant of *Drosophila melanogaster* with an extra set of wings, page 1072.

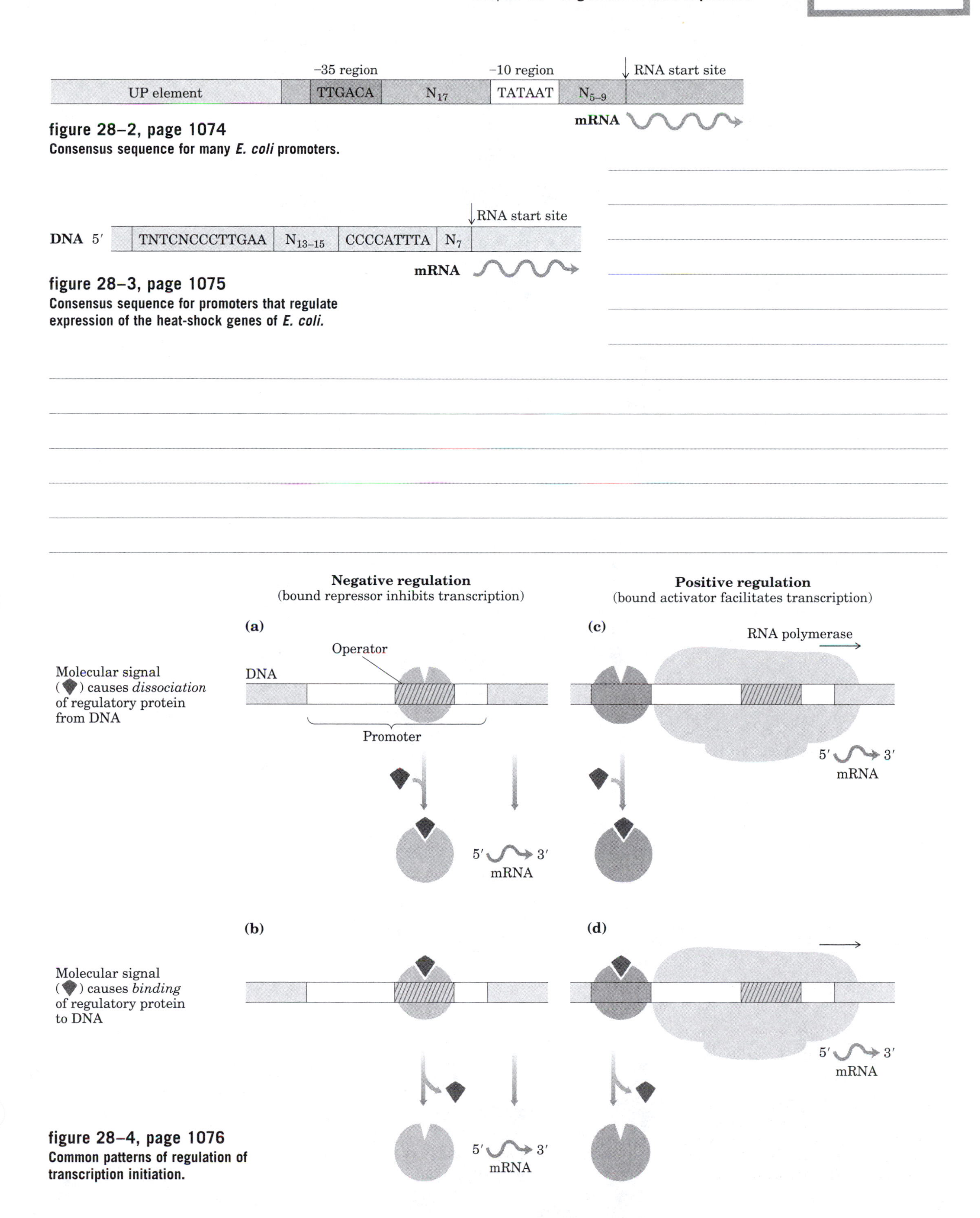

figure 28–2, page 1074
Consensus sequence for many *E. coli* promoters.

figure 28–3, page 1075
Consensus sequence for promoters that regulate expression of the heat-shock genes of *E. coli.*

figure 28–4, page 1076
Common patterns of regulation of transcription initiation.

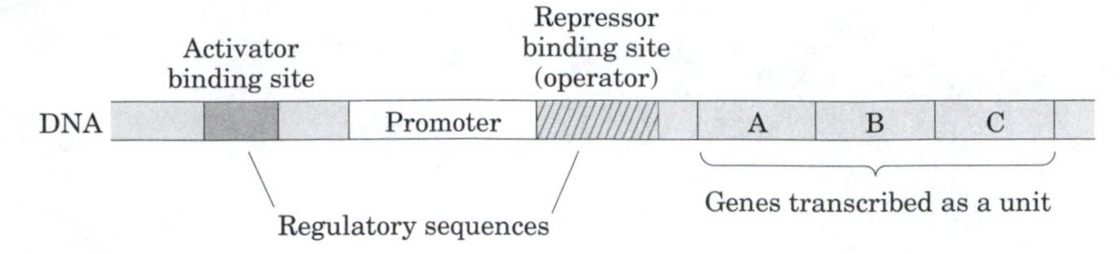

figure 28–5, page 1077
Representative prokaryotic operon.

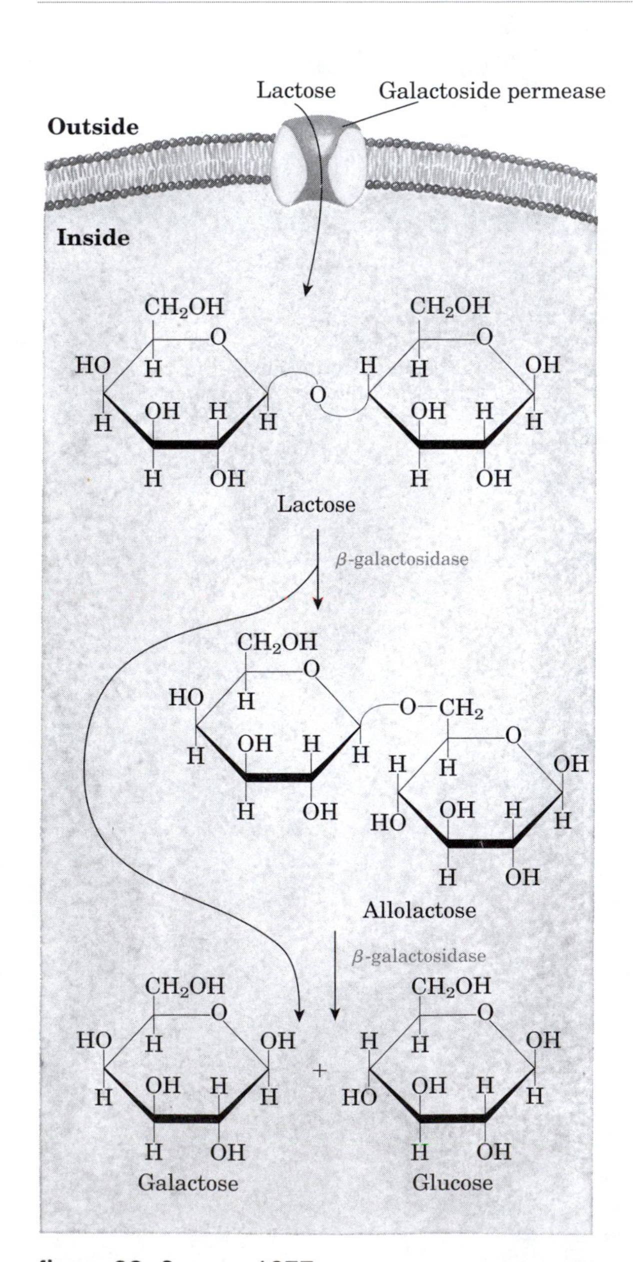

figure 28–6, page 1077
Lactose metabolism in *E. coli.*

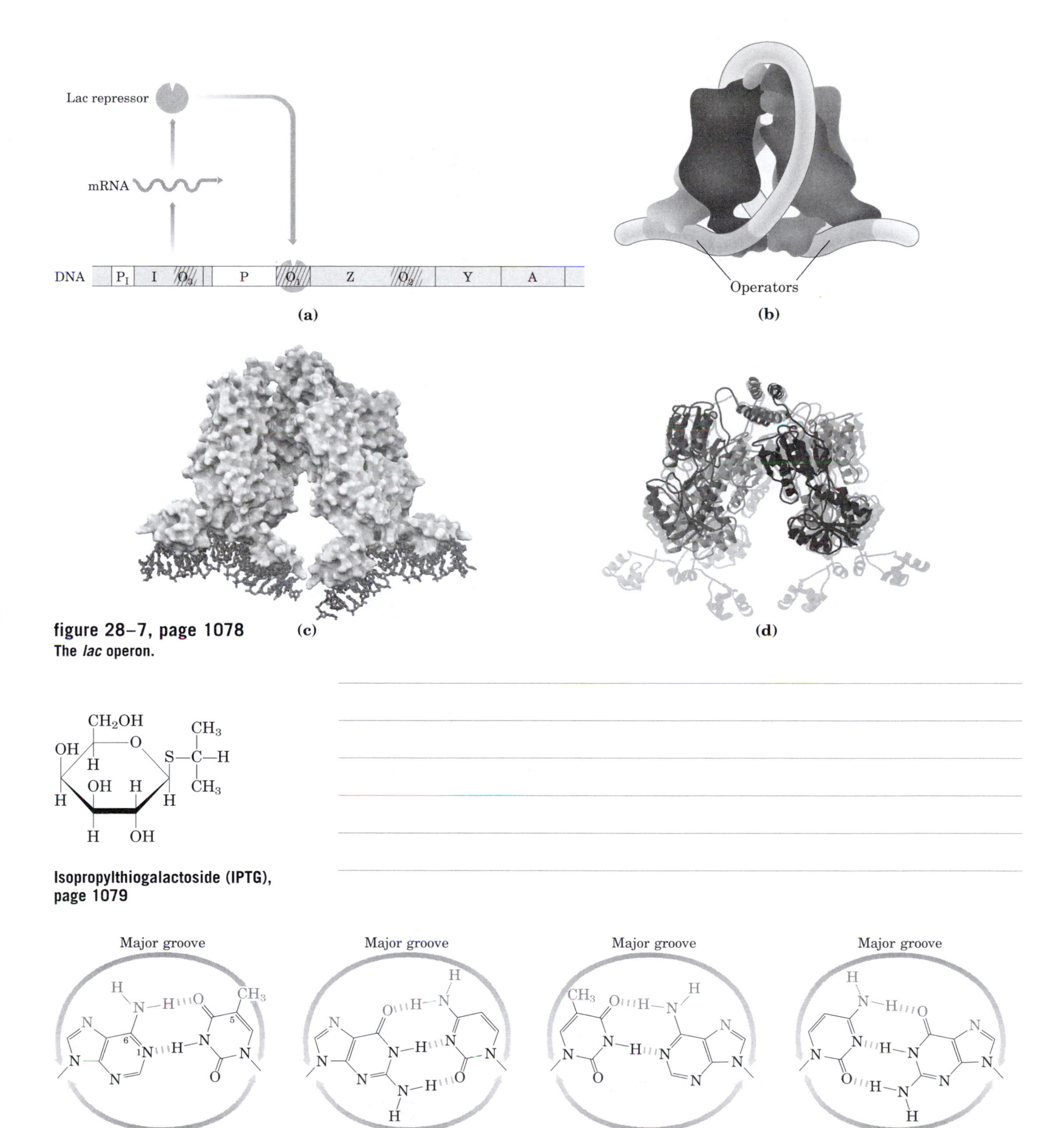

figure 28–7, page 1078
The *lac* operon.

Isopropylthiogalactoside (IPTG), page 1079

figure 28–8, page 1080
Groups in DNA available for protein binding.

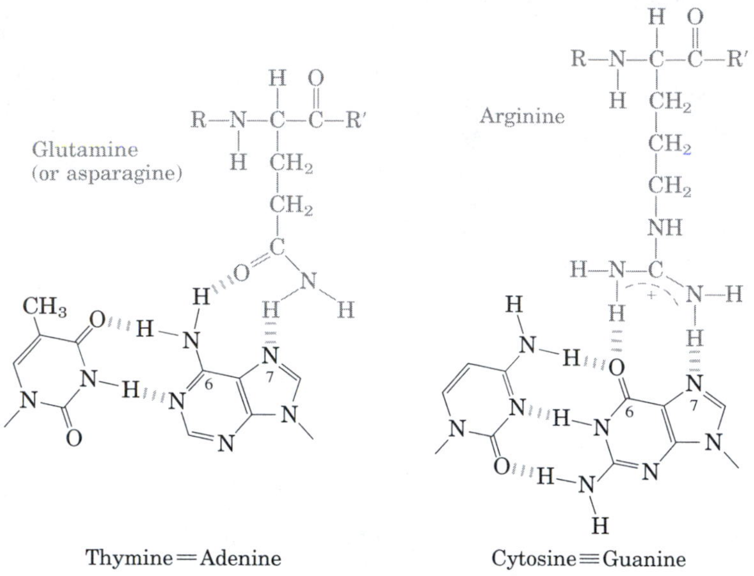

Glutamine (or asparagine)

Arginine

Thymine=Adenine

Cytosine≡Guanine

figure 28–9, page 1081
Two examples of specific amino acid–base pair interactions that have been observed in DNA-protein binding.

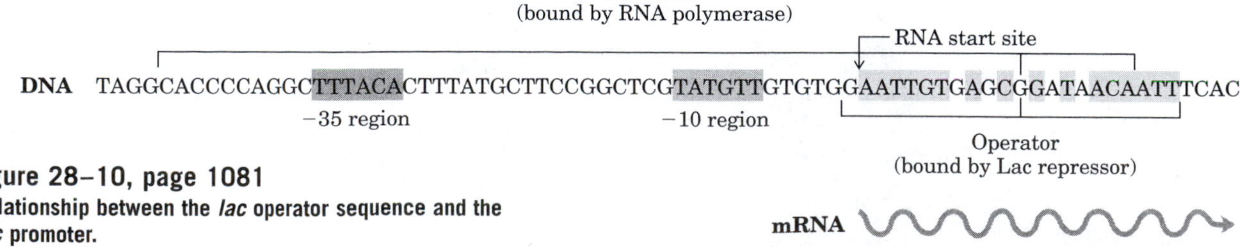

Promoter
(bound by RNA polymerase)

RNA start site

DNA TAGGCACCCCAGGCTTTACACTTTATGCTTCCGGCTCGTATGTTGTGTGGAATTGTGAGCGGATAACAATTTCAC

−35 region

−10 region

Operator
(bound by Lac repressor)

mRNA

figure 28–10, page 1081
Relationship between the *lac* operator sequence and the *lac* promoter.

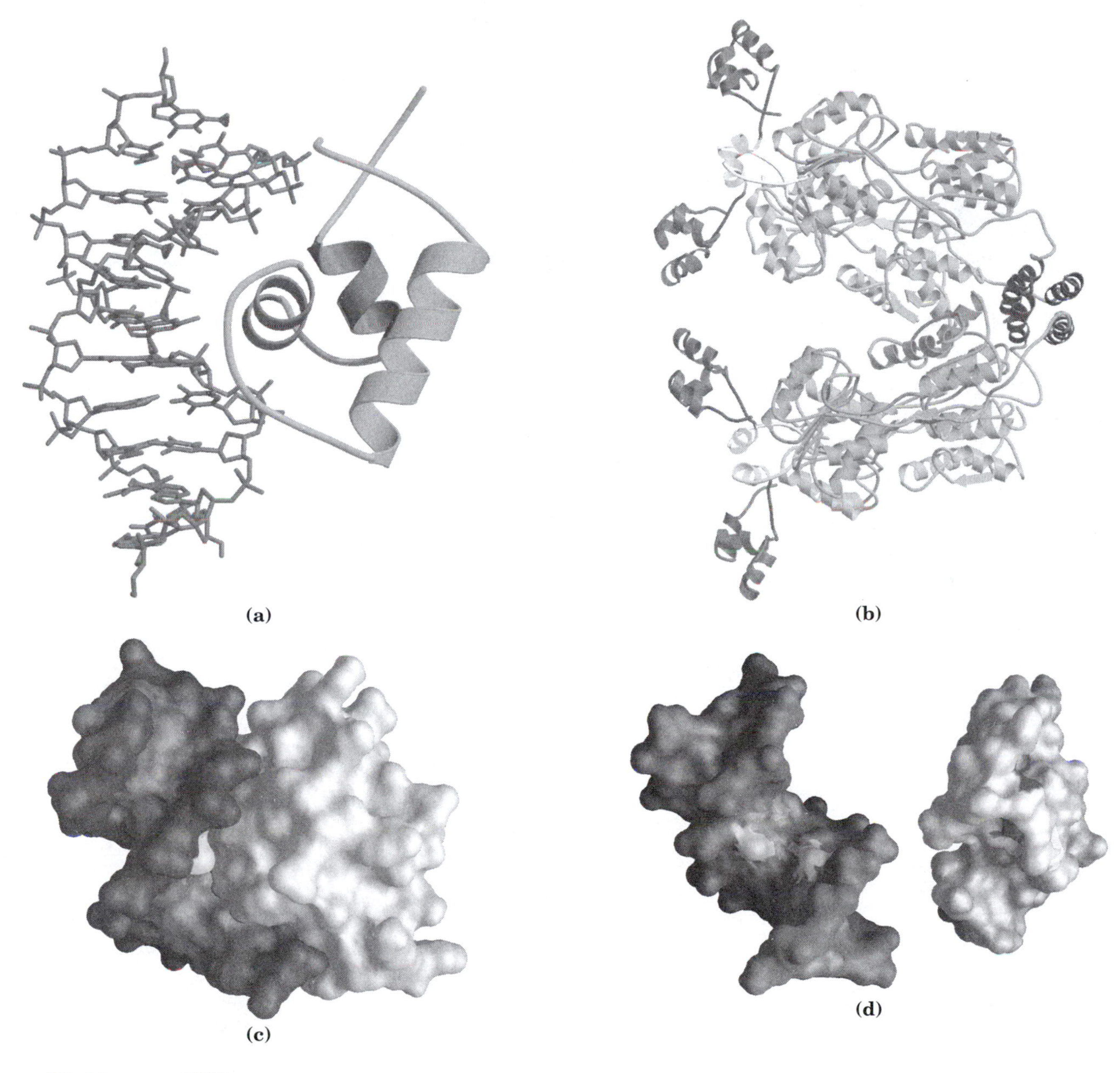

figure 28–11, page 1082
Helix-turn-helix.

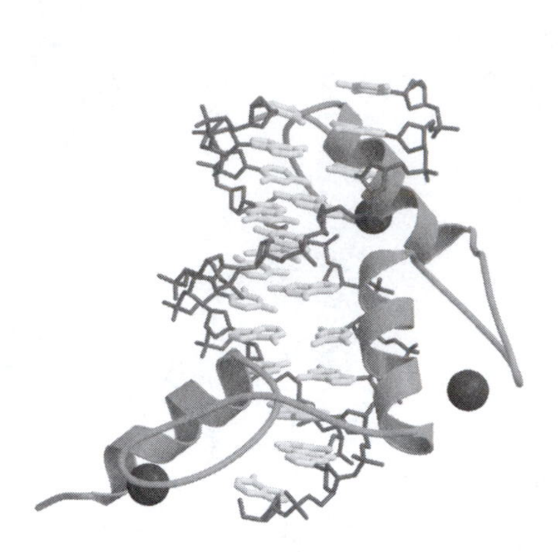

figure 28–12, page 1083
Zinc fingers.

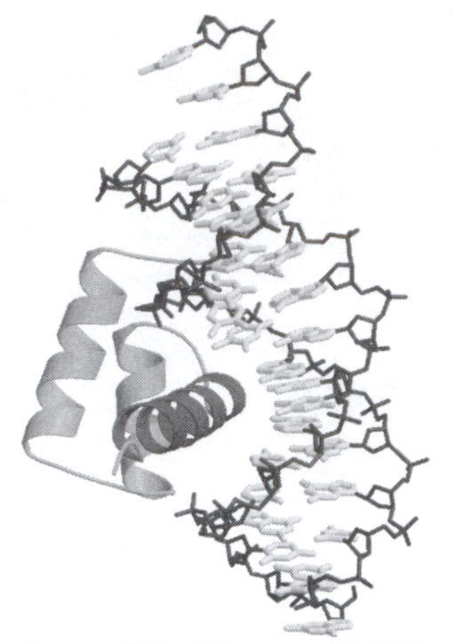

figure 28–13, page 1083
Homeodomain.

Source	Regulatory protein	Amino acid sequence: DNA-binding region	6-Amino acid connector	Leucine zipper
Mammal	C/EBP	DKNSNEYRVRRERNNIAVRKSRDKA	KQRNVE	TQQKVLELTSDNDRLRKRVEQLSRELDTLRG–
Mammal	Jun	SQERIKAERKRMRNRIAASKCRKRK	LERIAR	LEEKVKTLKAQNSELASTANMLTEQVAQLKQ–
Mammal	Fos	EERRRIRRIRRERNKMAAAKCRNRR	RELTDT	LQAETDQLEDKKSALQTEIANLLKEKEKLEF–
Yeast	GCN4	PESSDPAALKRARNTEAARRSRARK	LQRMKQ	LEDKVEELLSKNYHLENEVARLKKLVGER
Consensus molecule		RR/KK, R/K, N, R/K, R, RR/KK		L, L, L, L, L

Invariant Asn

(a)

(b)

figure 28–14, page 1084
Leucine zippers.

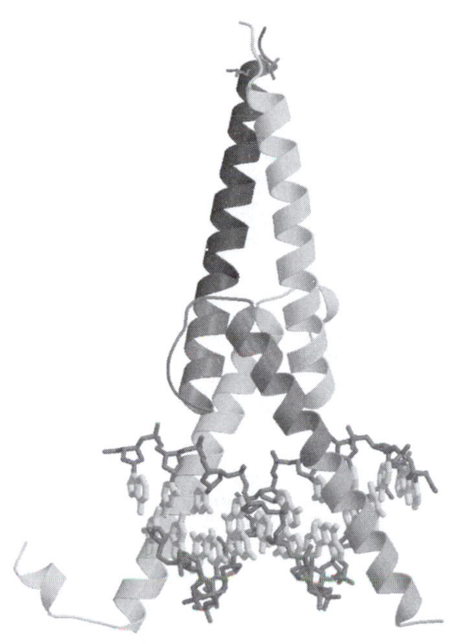

figure 28–15, page 1085
Helix-loop-helix.

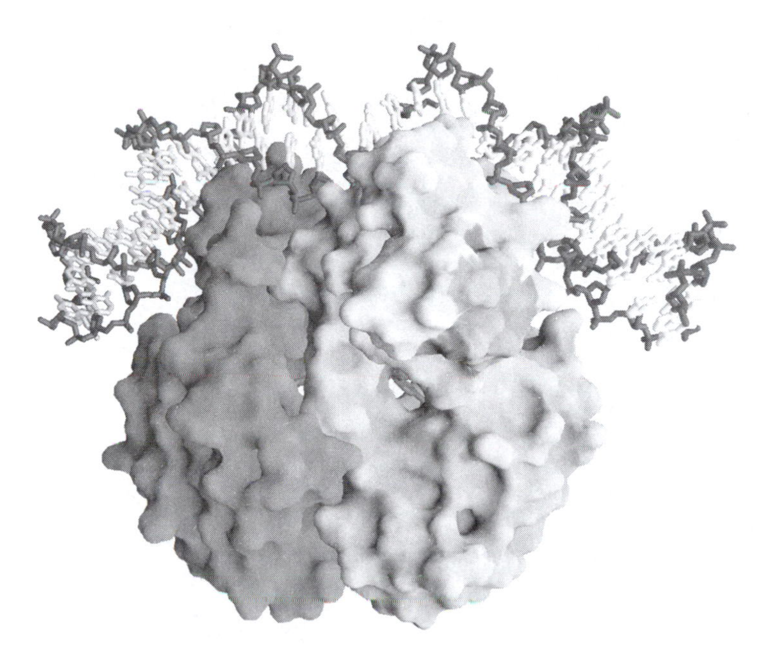

figure 28–16, page 1086
CRP homodimer.

CRP site

Bound by RNA polymerase 5′ → 3′ **mRNA**

DNA 5′–ATTAATGTGAGTTAGCTCACTCATTAGGCACCCCAGGCTTTACACTTTATGCTTCCGGCTCGTATGTTGTGTGGAATTGTGAGCGGATAACAATTTCACAC

−35 region

−10 region

Operator

(a)

lac promoter	TTTACA −35 region	TATGTT −10 region
Promoter consensus sequence	TTGACA	TATAAT

(b)

figure 28–17, page 1087
Activation of transcription of the *lac* operon by CRP.

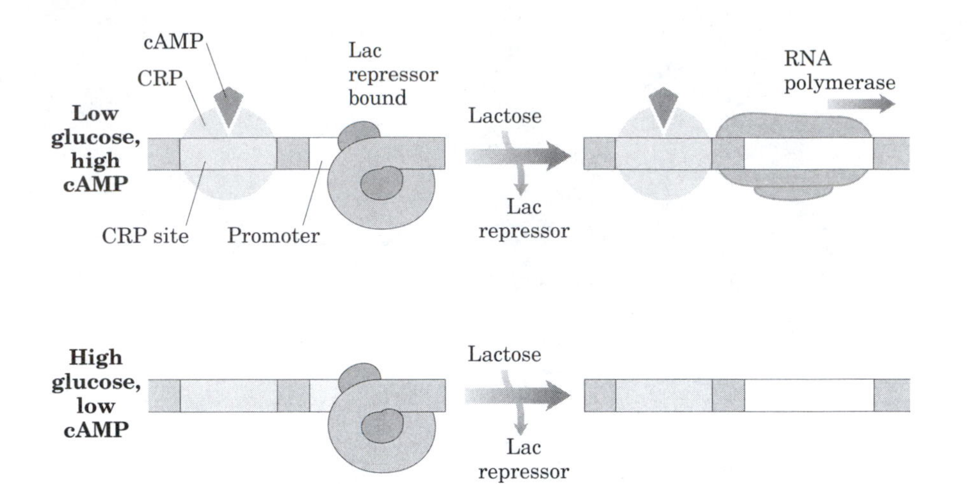

figure 28–18, page 1087
Combined effects of glucose and lactose on expression of the *lac* operon.

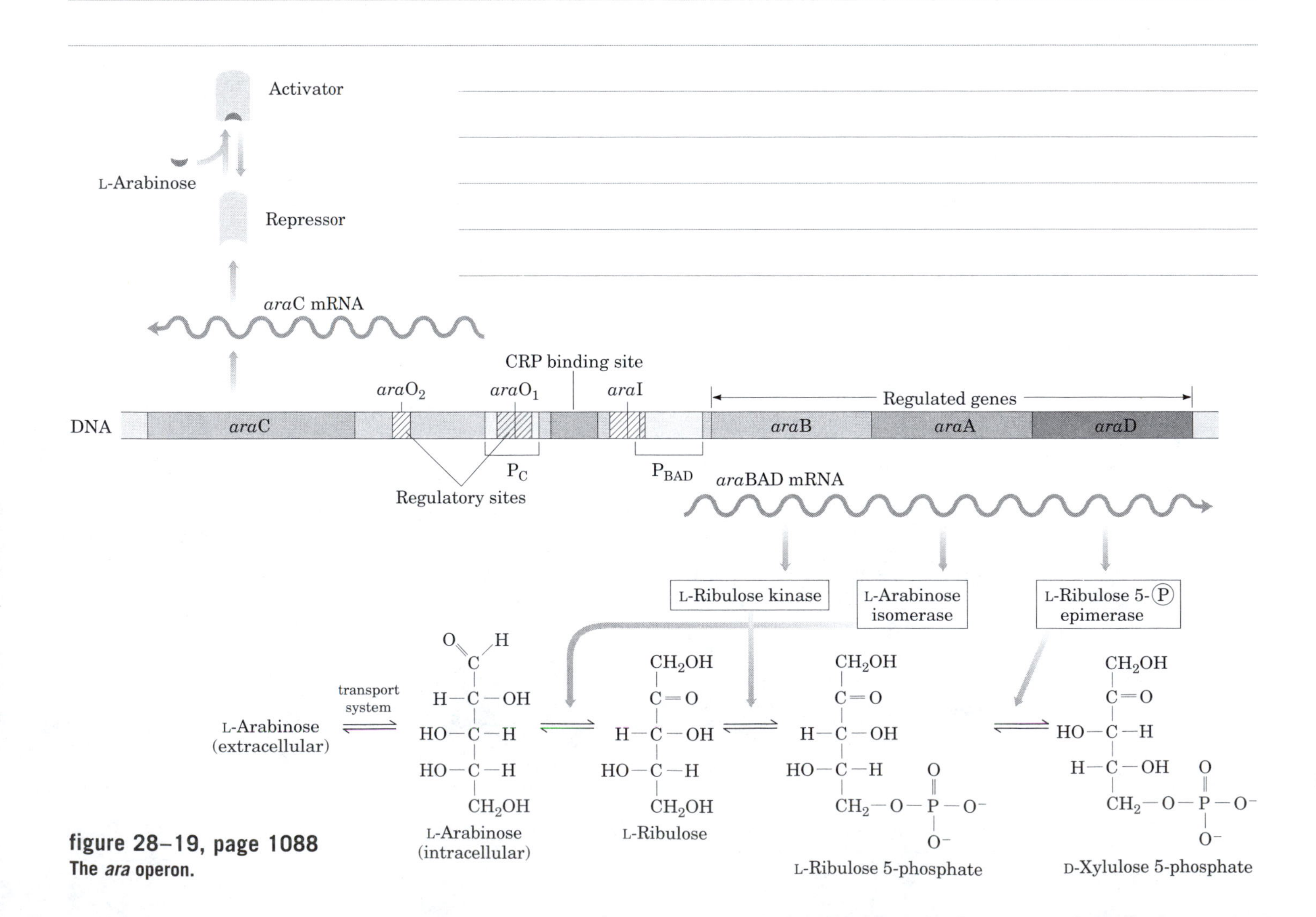

figure 28–19, page 1088
The *ara* operon.

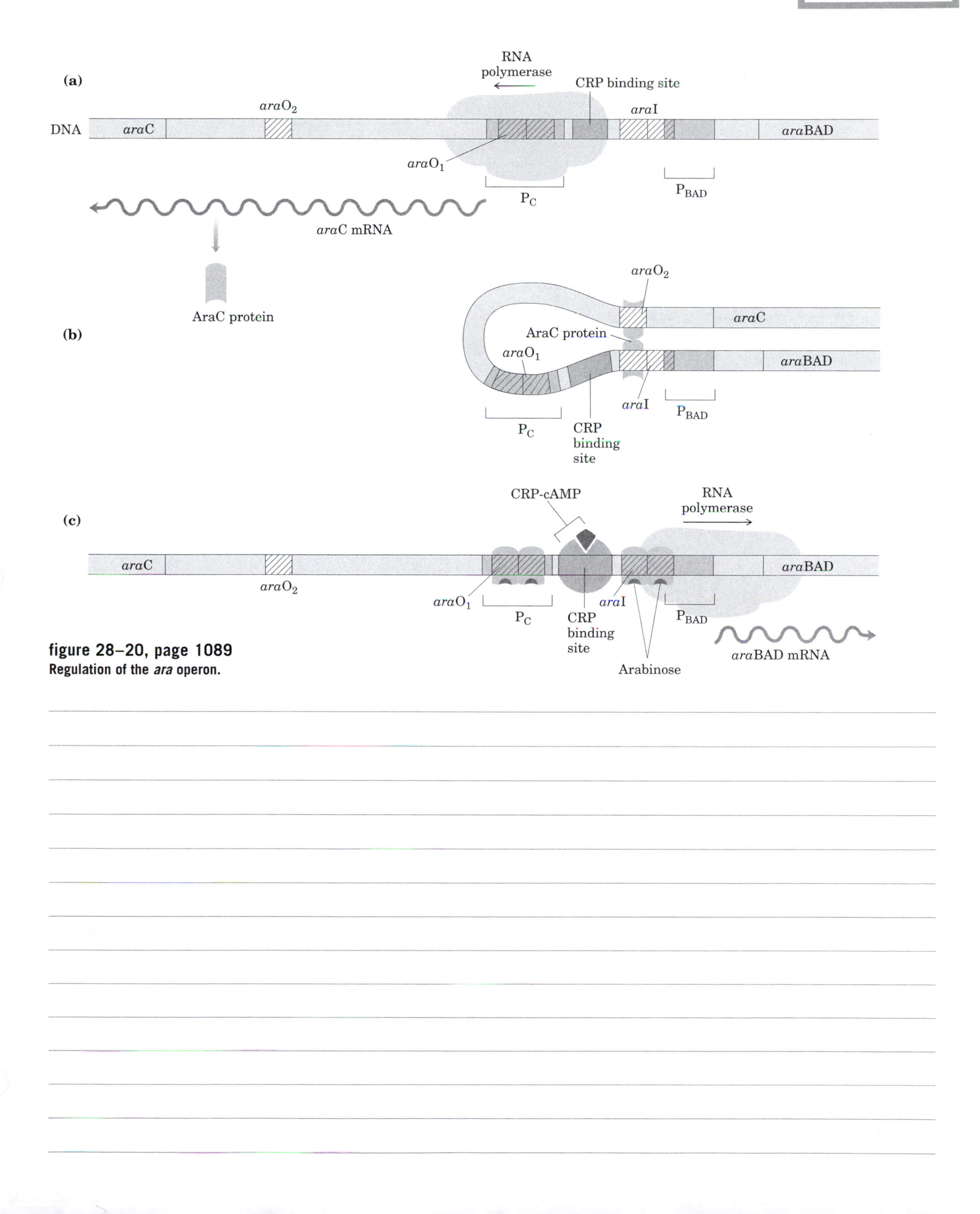

figure 28–20, page 1089
Regulation of the *ara* operon.

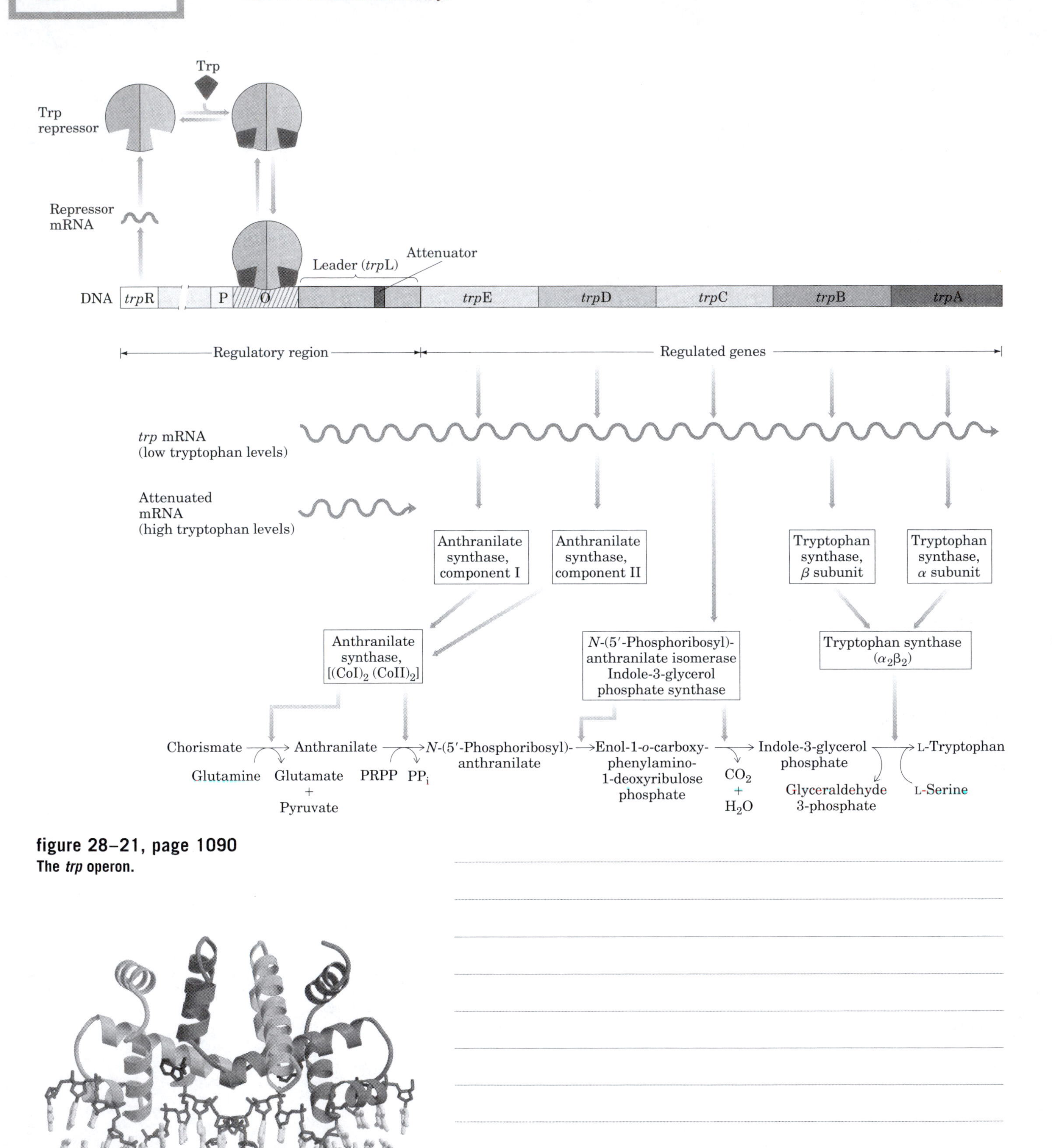

figure 28–21, page 1090
The *trp* operon.

figure 28–22, page 1091
Trp repressor.

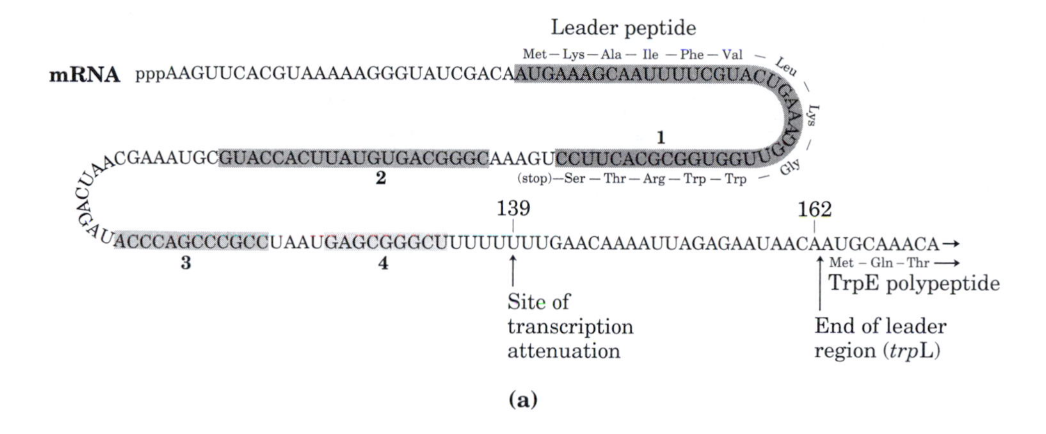

(a)

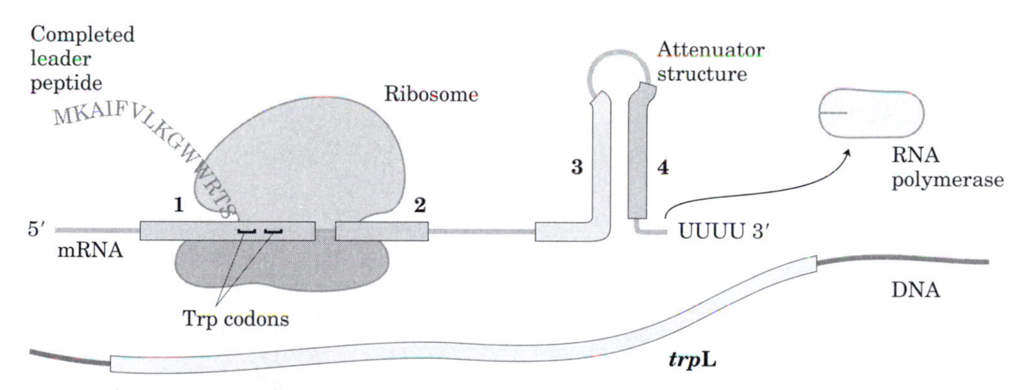

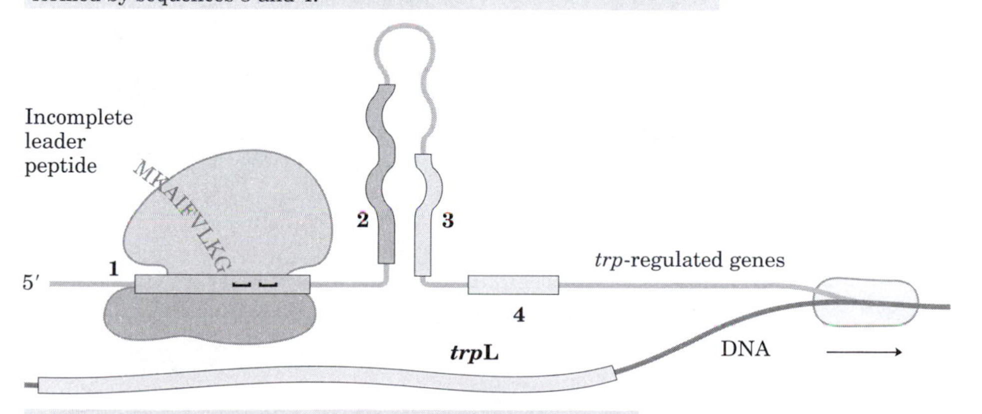

(b)

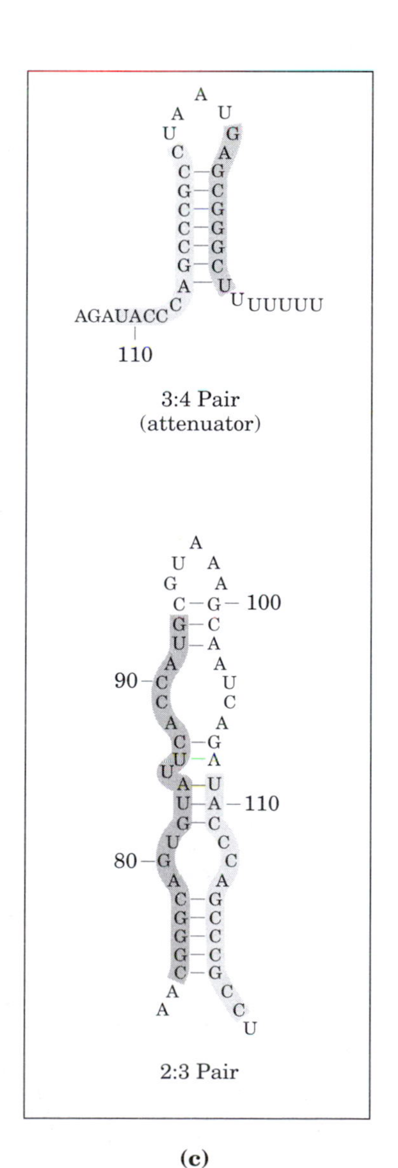

(c)

figure 28–23, pages 1092–1093
Transcriptional attenuation in the *trp* operon.

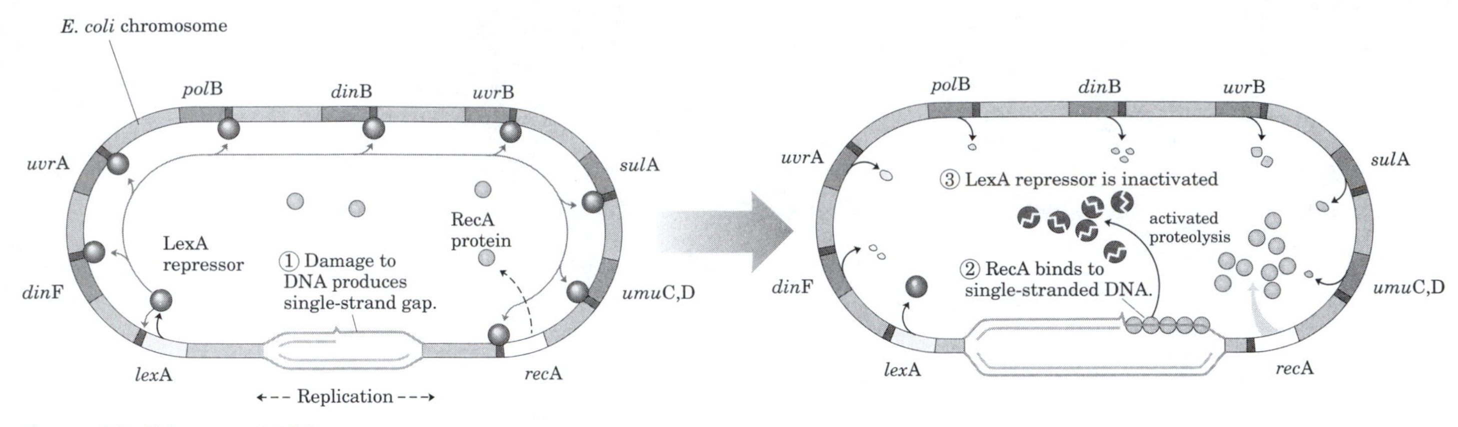

figure 28–24, page 1094
SOS response in *E. coli*.

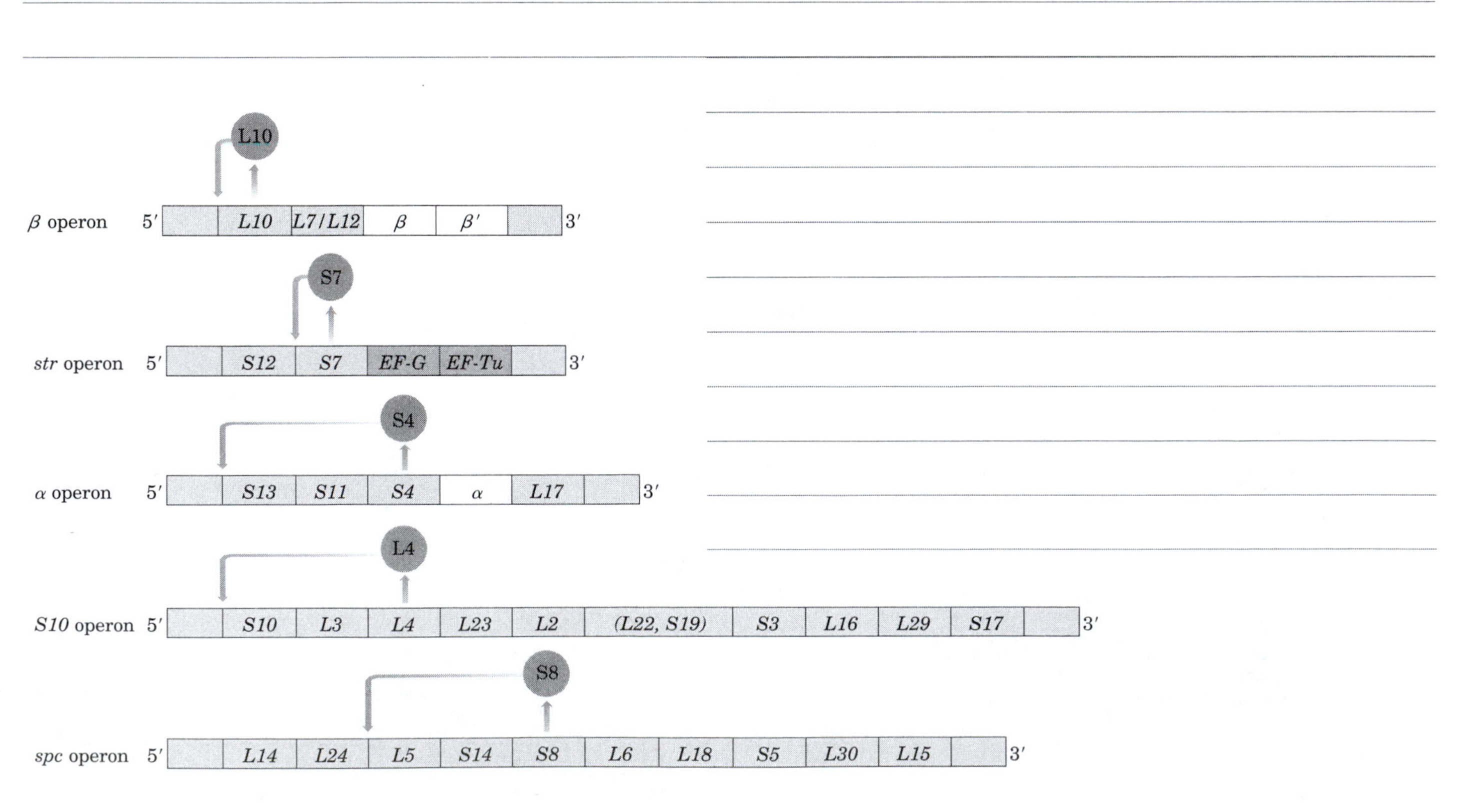

figure 28–25, page 1096
Some mRNA transcripts of ribosomal protein operons.

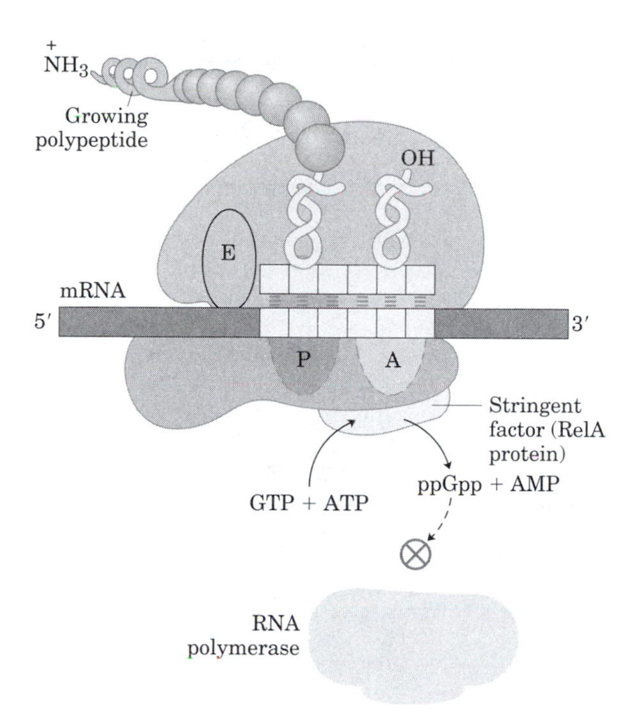

figure 28–26, page 1097
Stringent response in *E. coli*.

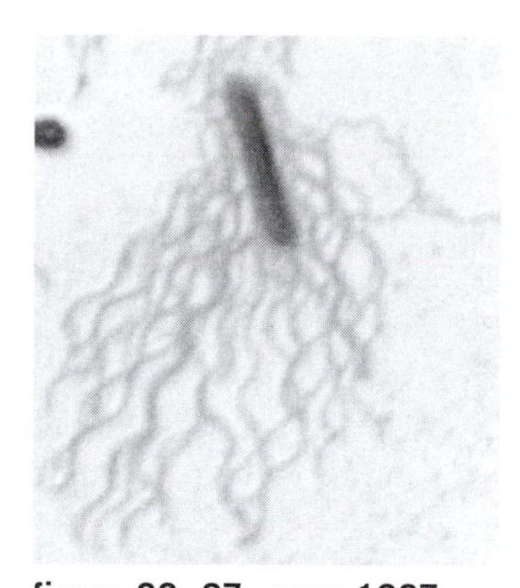

figure 28–27, page 1097
***Salmonella typhimurium*, with flagella evident.**

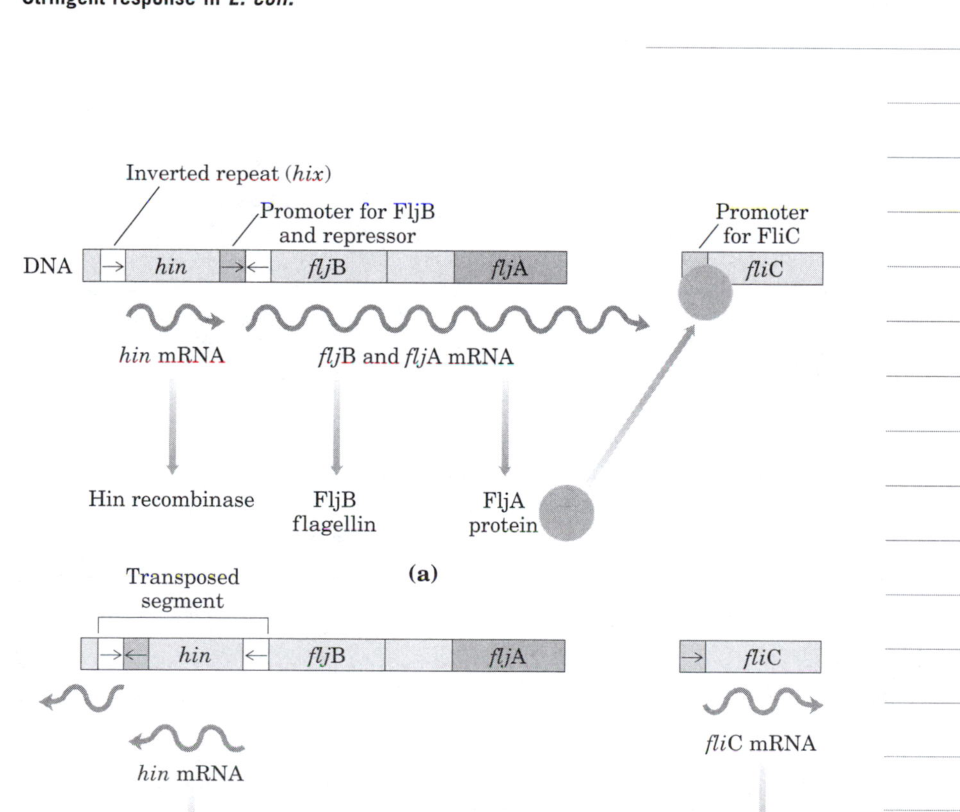

figure 28–28, page 1098
Regulation of flagellin genes in *Salmonella:* phase variation.

table 28–1, page 1099

Examples of Gene Regulation by Recombination

System	Recombinase/ recombination site	Type of recombination	Function
Phase variation (*Salmonella*)	Hin/*hix*	Site-specific	Alternative expression of two flagellin genes allows evasion of host immune response.
Host range (bacteriophage μ)	Gin/*gix*	Site-specific	Alternative expression of two sets of tail fiber genes affects host range.
Mating type switch (yeast)	HO endonuclease, RAD52 protein, other proteins/*MAT*	Nonreciprocal gene conversion*	Alternative expression of two mating types of yeast, a and α, creates cells of different mating types that can mate and undergo meiosis.
Antigenic variation (trypanosomes)†	Varies	Nonreciprocal gene conversion*	Successive expression of different genes encoding the variable surface glycoproteins (VSGs) allows evasion of host immune response.

*Nonreciprocal gene conversion is a class of recombination events not discussed in Chapter 25. Genetic information is moved from one part of the genome (where it is silent) to another (where it is expressed) in a reaction similar to replicative transposition (see Fig. 25–41).

†Trypanosomes cause African sleeping sickness and other diseases (see Box 22–2). The outer surface of a trypanosome is made up of multiple copies of a single VSG, the major surface antigen. A cell can change surface antigens to more than 100 different forms, precluding an effective defense by the host immune system. Trypanosome infections are chronic and, if untreated, result in death.

table 28–2, page 1101

Some Enzyme Complexes Catalyzing Chromatin Structural Changes during Transcription

Enzyme complex*	Oligomeric structure	Source	Activities
GCN5-ADA2-ADA3	3 polypeptides	Yeast	GCN5 has type A HAT activity
SAGA/PCAF	>20 polypeptides	Eukaryotes	Includes GCN5-ADA2-ADA3
SWI/SNF	>11 polypeptides; M_r 2×10^6	Eukaryotes	ATP-dependent nucleosome remodeling
NURF	4 polypeptides; M_r 500,000	*Drosophila*	ATP-dependent nucleosome remodeling
CAFI	>2 polypeptides	Humans; *Drosophila*	Responsible for binding histones H3 and H4 to DNA
NAP1	1 polypeptide; M_r 125,000	Widely distributed in eukaryotes	Responsible for binding histones H2A and H2B to DNA

*The abbreviations used to identify eukaryotic genes and proteins are often more confusing or obscure than those used for bacteria. The complex of GCN5 (*g*eneral *c*ontrol *n*onderepressible) and ADA (*a*lteration/*d*eficiency *a*ctivation) proteins was discovered during investigation of the regulation of the genes of nitrogen metabolism in yeast. These proteins can be part of the larger SAGA complex. SAGA (*S*PF, *A*DA2,3, *G*CN5, *a*cetyltransferase) is a yeast complex; its equivalent in humans is PCAF (*p*300/*C*BP-*a*ssociated *f*actor). SWI was discovered as a protein required for expression of certain genes involved in mating type *swi*tching in yeast, and SNF (sucrose *n*on*f*ermenting) as a factor for expression of the yeast gene for sucrase. Subsequent studies revealed multiple SWI and SNF proteins that acted in a complex. The SWI/SNF complex has a role in the expression of a wide range of genes and has been found in many eukaryotes, including humans. NURF is *nu*clear *r*emodeling *f*actor; CAF1, *c*hromatin *a*ssembly *f*actor; and NAP1, *n*ucleosome *a*ssembly *p*rotein.

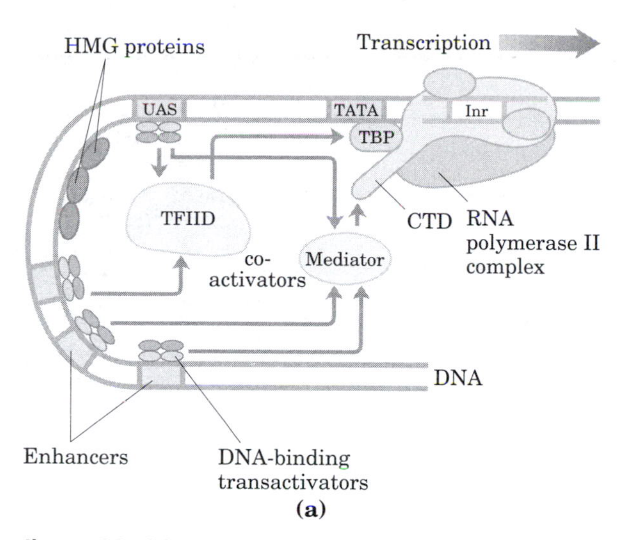

figure 28–29, page 1103
Eukaryotic promoters and regulatory proteins.

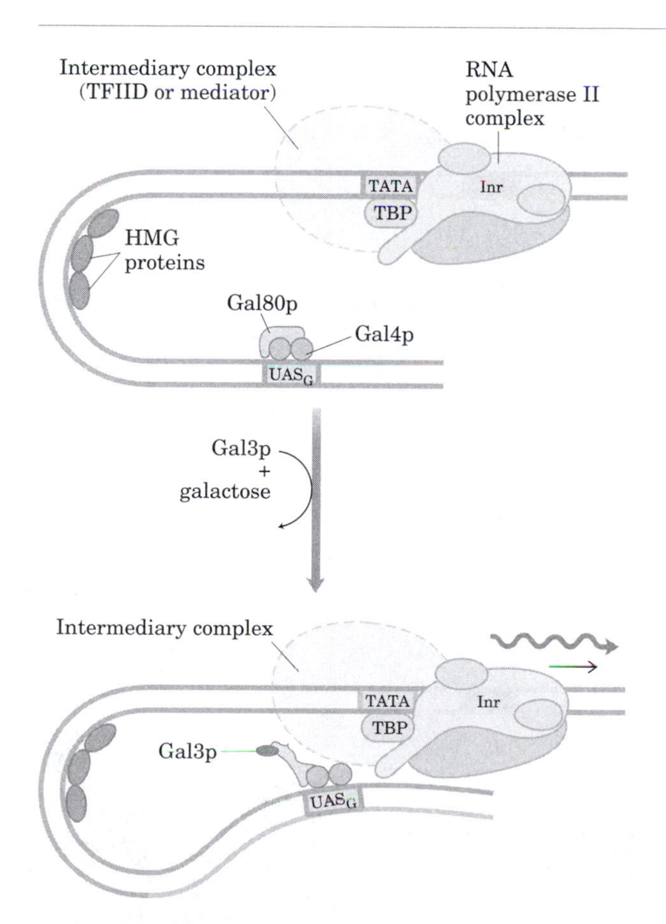

figure 28–30, page 1104
Regulation of transcription at genes of galactose metabolism in yeast.

table 28–3, page 1105

Genes of Galactose Metabolism in Yeast

	Protein function	Chromosomal location	Protein size (number of residues)	Relative protein expression in different carbon sources: Glucose	Glycerol	Galactose
Regulated genes						
GAL1	Galactokinase	II	528	–	–	+++
GAL2	Galactose permease	XII	574	–	–	+++
PGM2	Phosphoglucomutase	XIII	569	+	+	++
GAL7	Galactose 1-phosphate uridylyltransferase	II	365	–	–	+++
GAL10	UDP-glucose 4-epimerase	II	699	–	–	+++
MEL1	α-Galactosidase	II	453	–	+	++
Regulatory genes						
GAL3	Inducer	IV	520	–	+	++
GAL4	Transcriptional activator	XVI	881	+/–	+	+
GAL80	Transcriptional inhibitor	XIII	435	+	+	++

Adapted from Reece, R. & Platt, A. (1997) Signaling activation and repression of RNA polymerase II transcription in yeast. *Bioessays* **19**, 1001–1010.

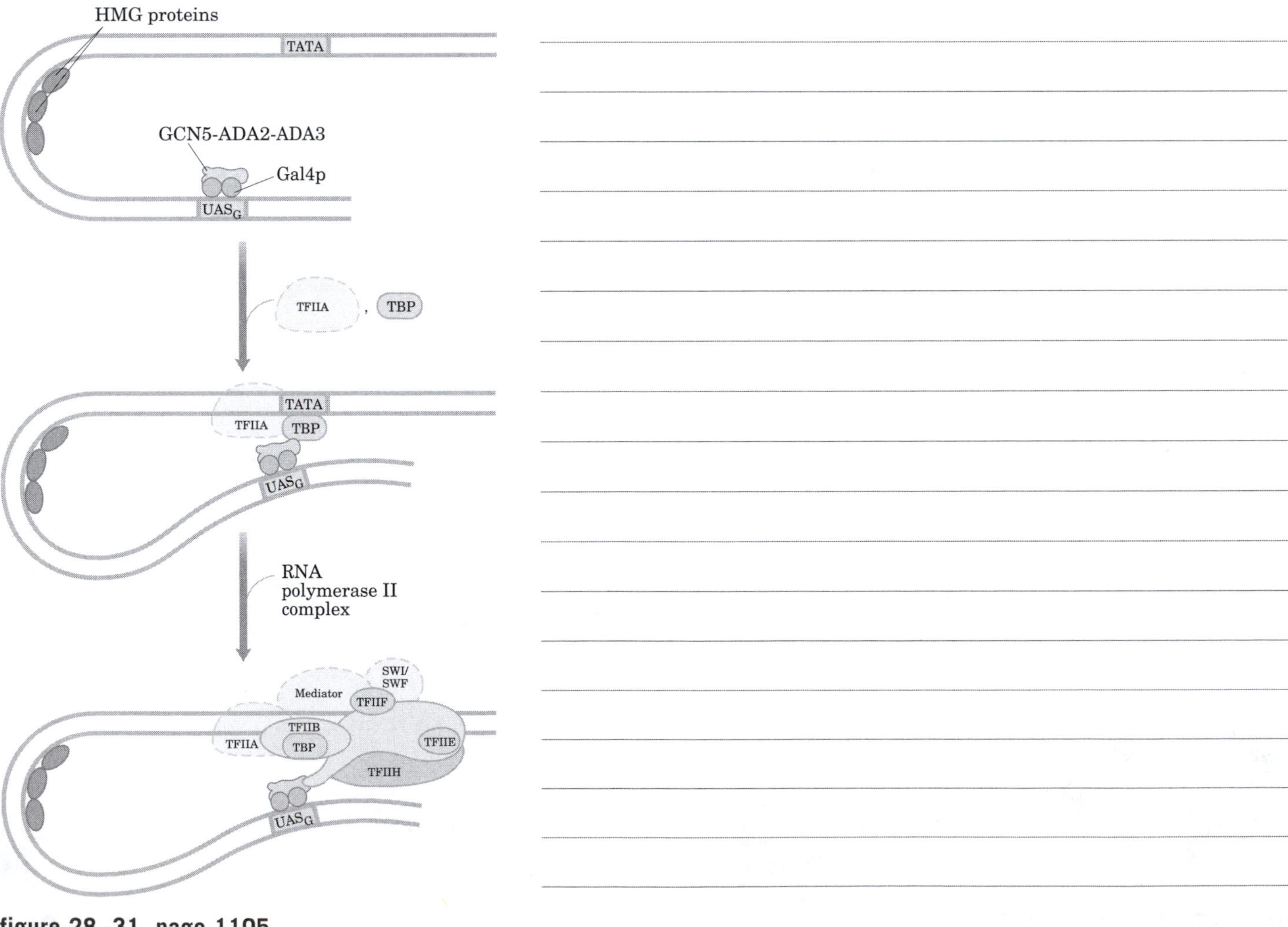

figure 28–31, page 1105
Protein complexes involved in transcription activation of a group of related eukaryotic genes.

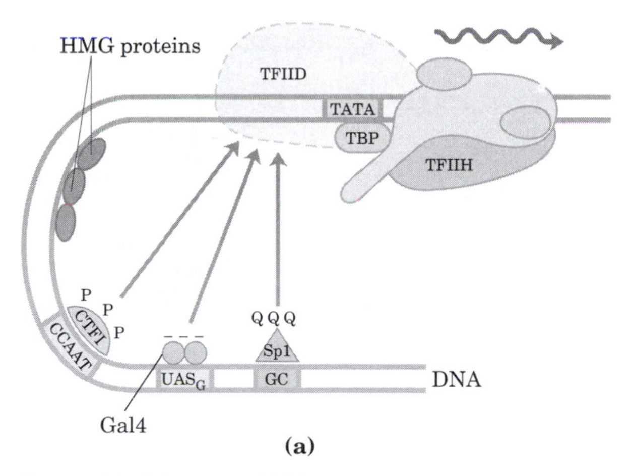

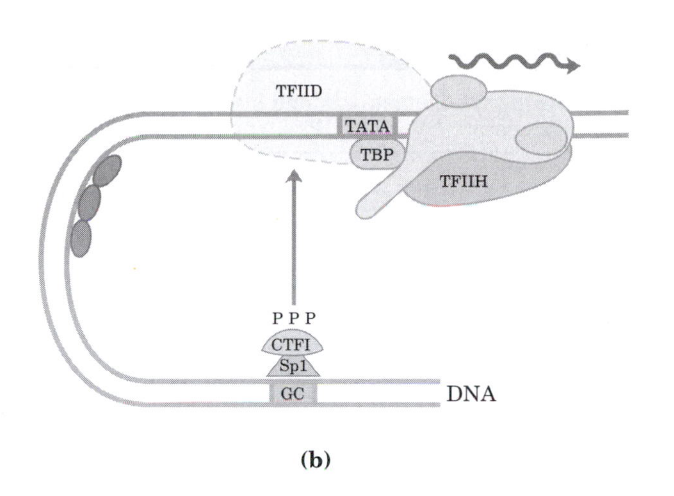

figure 28–32, page 1106
DNA-binding transactivators.

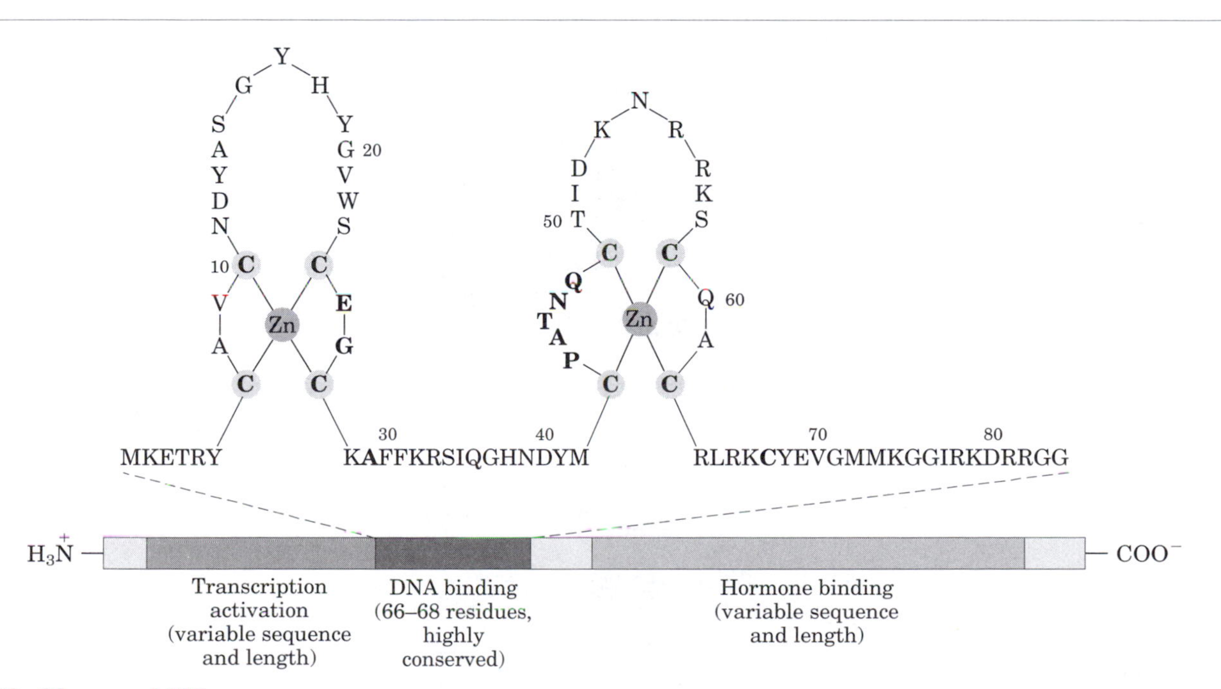

figure 28–33, page 1107
Typical steroid hormone receptors.

table 28–4, page 1107

Hormone Response Elements Bound by Steroid-Type Hormone Receptors

Receptor	Consensus sequence bound*
Androgen	$GG^{A}/_{T}ACAN_2TGTTCT$
Glucocorticoid	$GGTACAN_3TGTTCT$
Retinoic acid (some)	$AGGTCAN_5AGGTCA$
Vitamin D	$AGGTCAN_3AGGTCA$
Thyroid hormone	$AGGTCAN_3AGGTCA$
RX†	AGGTCANAGGTCANAGGTCANAGGTCA

*N represents any nucleotide.
†Forms dimer with retinoic acid receptor or vitamin D receptor.

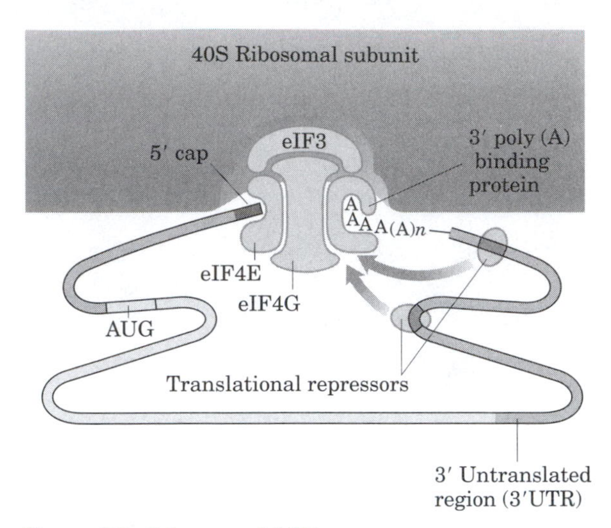

figure 28–34, page 1109
Translational regulation of eukaryotic mRNA.

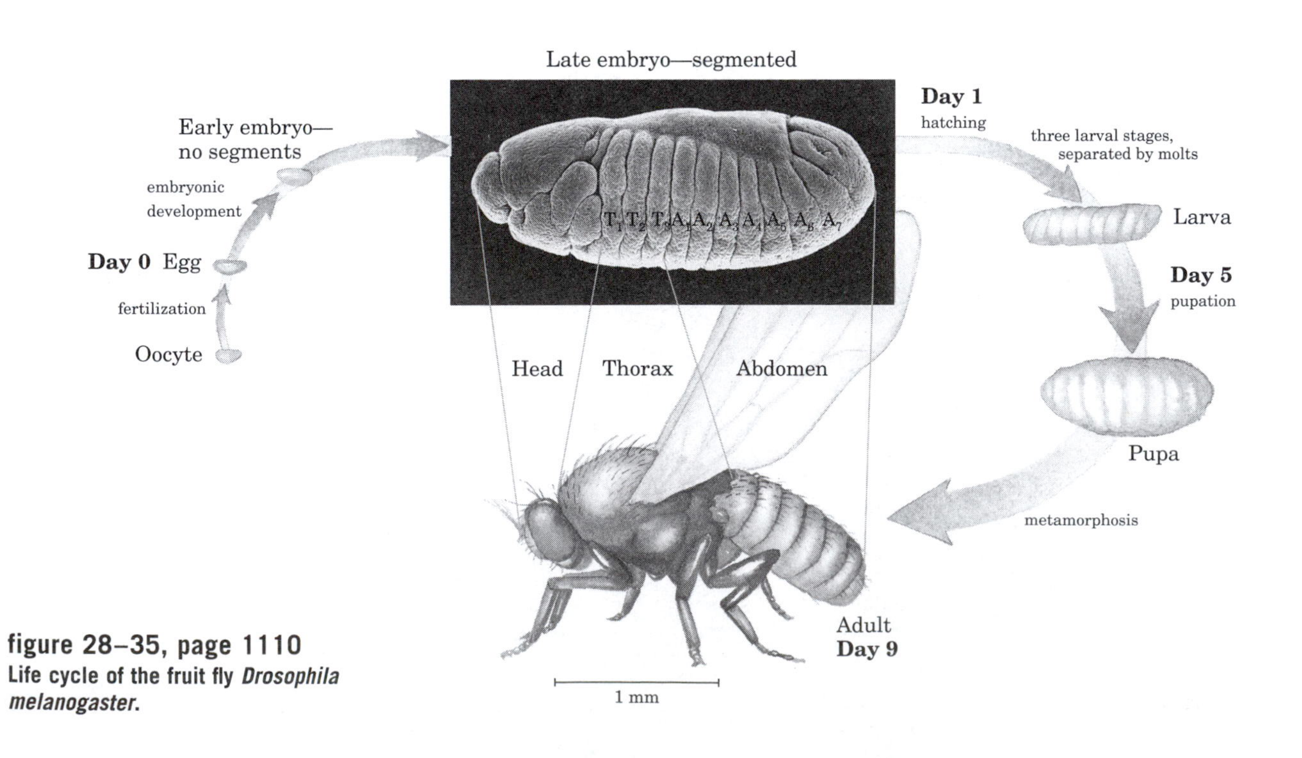

figure 28–35, page 1110
Life cycle of the fruit fly *Drosophila melanogaster*.

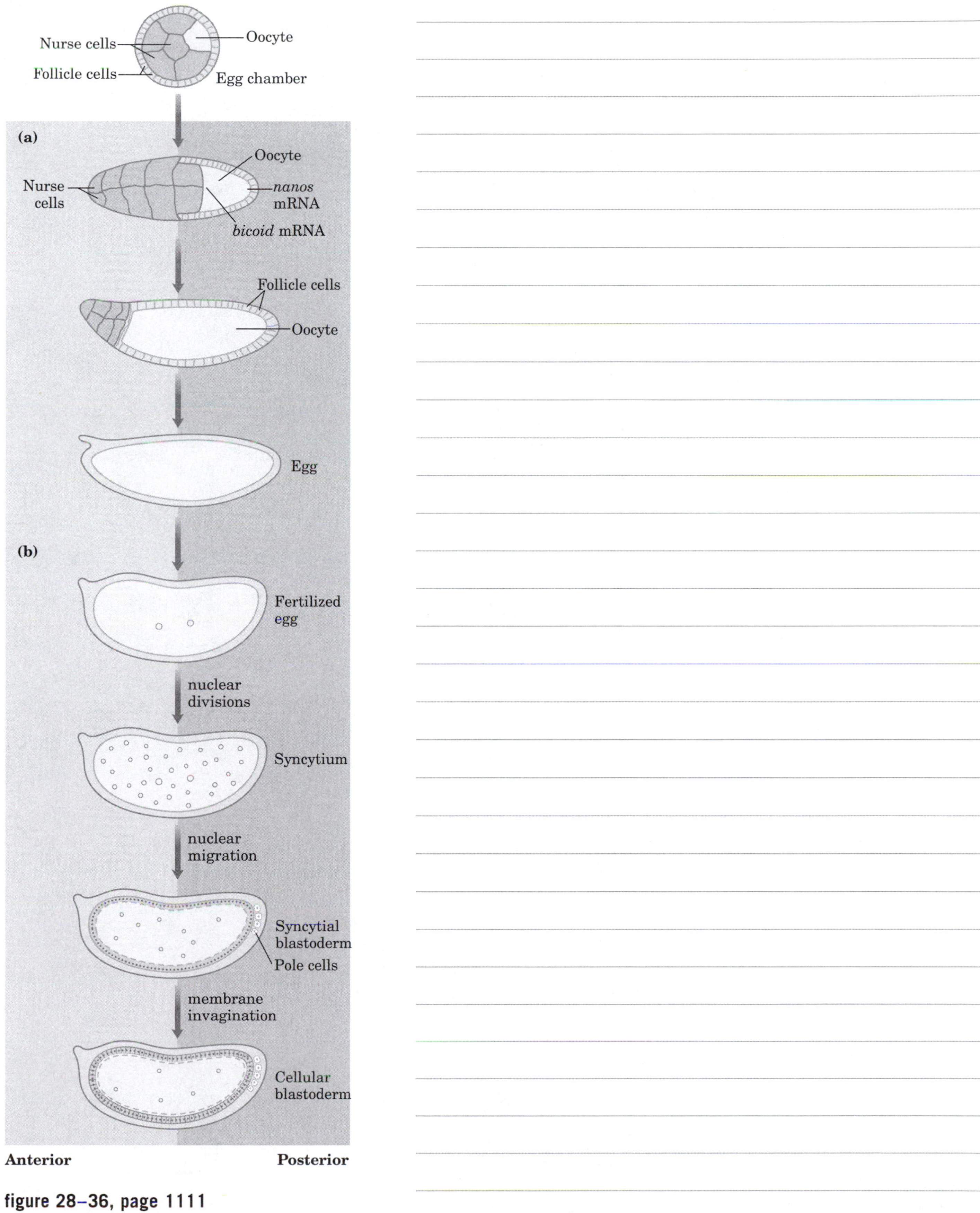

figure 28–36, page 1111
Early development in *Drosophila*.

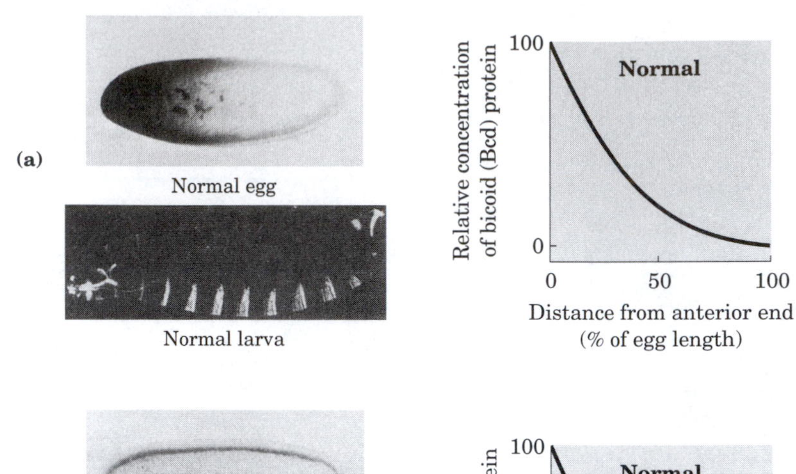

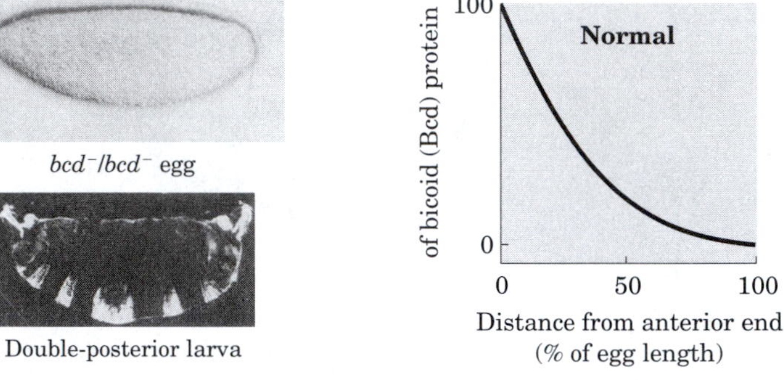

figure 28–37, page 1112
Distribution of a maternal gene product in a *Drosophila* egg.

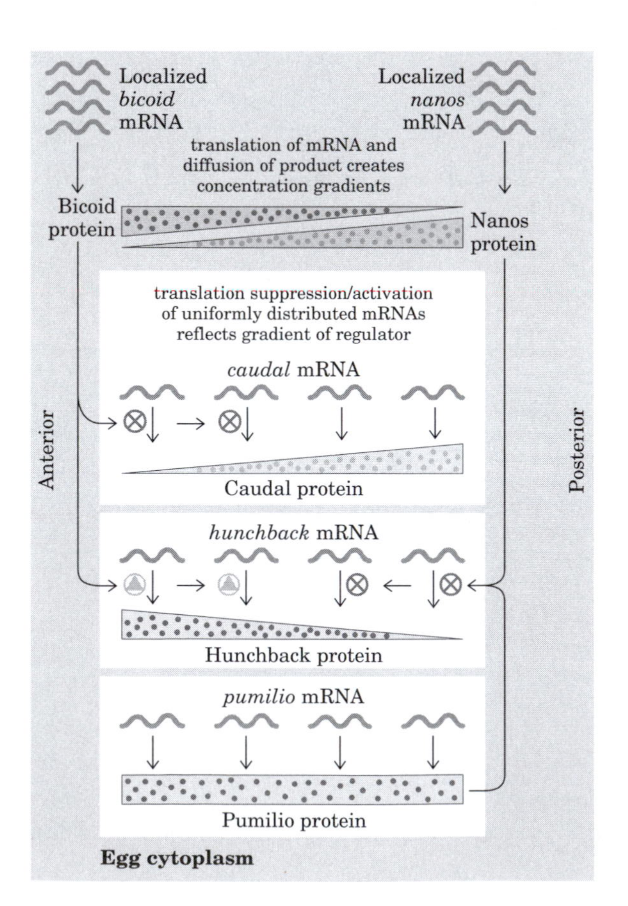

figure 28–38, page 1113
Regulatory circuits of the anterior-posterior axis in a *Drosophila* egg.

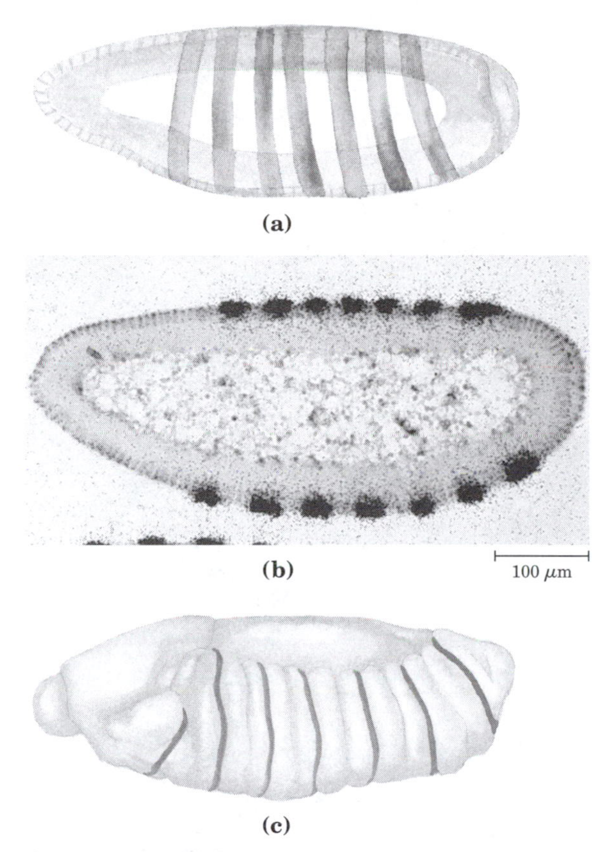

figure 28–39, page 1114
Distribution of the *fushi tarazu (ftz)* gene product in early *Drosophila* embryos.

(a)

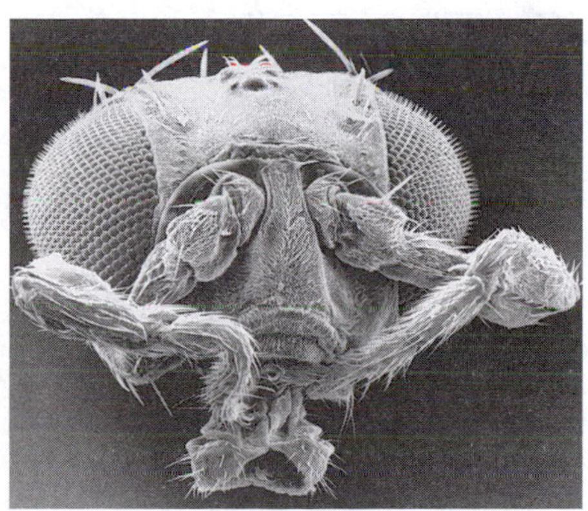

(b)

figure 28–40, page 1114
Effects of mutations in homeotic genes in *Drosophila*.

chapter

29 Recombinant DNA Technology

CD-ROM

When you study for exams, don't forget to review The Minicourses on the UNDERSTAND! BIOCHEMISTRY CD that came with your textbook.

Minicourses that apply to this Chapter include:
Some Important Techniques
DNA Cloning and Sequencing
Nucleic Acid Analysis
Protein Separation and Anaylsis

table 29–1, page 1120

Some Enzymes Used in Recombinant DNA Technology

Enzyme(s)	Function
Type II restriction endonucleases	Cleave DNAs at specific base sequences
DNA ligase	Joins two DNA molecules or fragments
DNA polymerase I (*E. coli*)	Fills gaps in duplexes by stepwise addition of nucleotides to 3′ ends
Reverse transcriptase	Makes a DNA copy of an RNA molecule
Polynucleotide kinase	Adds a phosphate to the 5′-OH end of a polynucleotide to label it or permit ligation
Terminal transferase	Adds homopolymer tails to the 3′-OH ends of a linear duplex
Exonuclease III	Removes nucleotide residues from the 3′ ends of a DNA strand
Bacteriophage λ exonuclease	Removes nucleotides from the 5′ ends of a duplex to expose single-stranded 3′ ends
Alkaline phosphatase	Removes terminal phosphates from either the 5′ or 3′ end (or both)

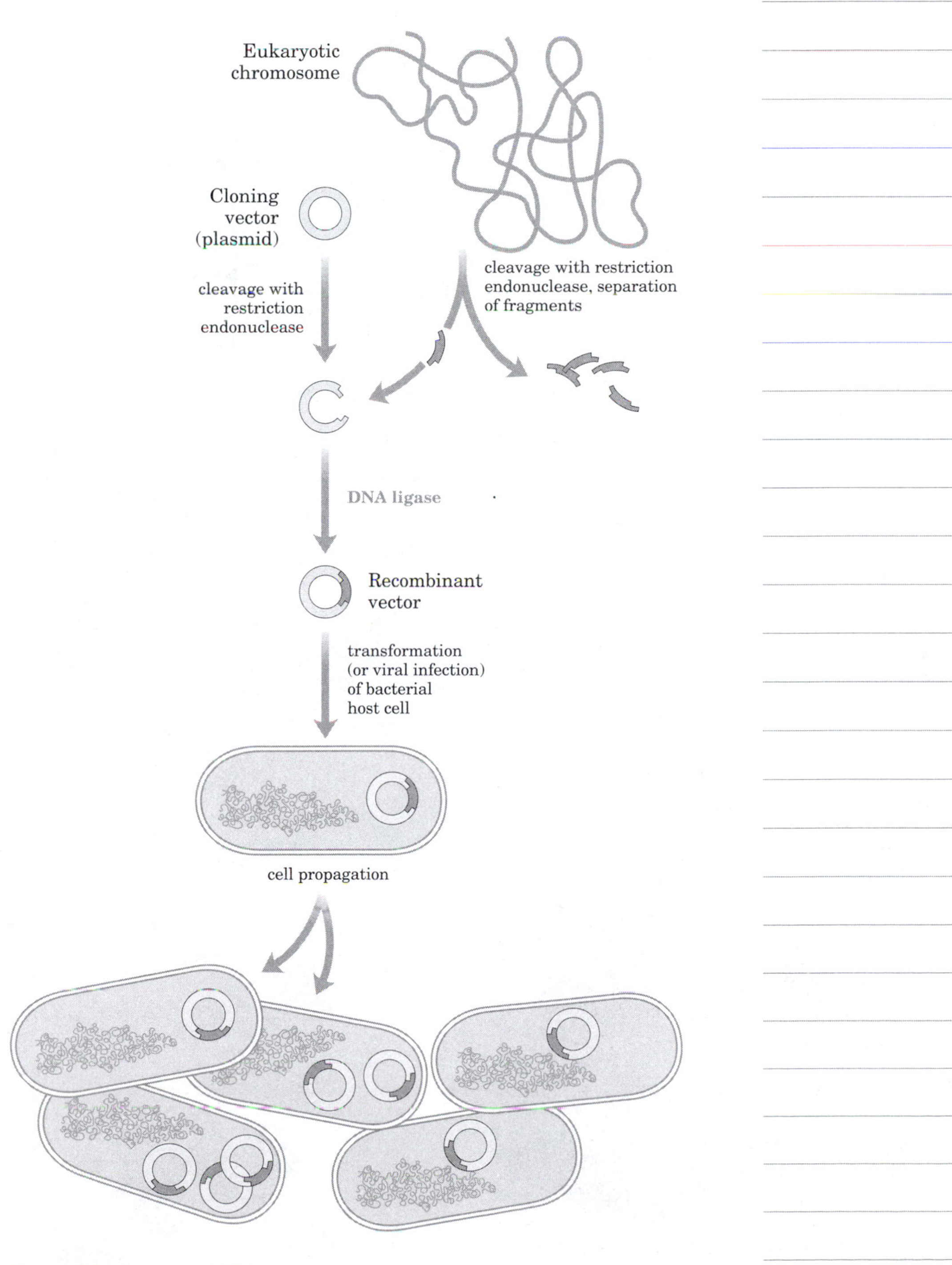

figure 29–1, page 1121
Schematic illustration of DNA cloning.

table 29–2, page 1122

Recognition Sequences for Some Type II Restriction Endonucleases

Enzyme	Recognition sequence	Enzyme	Recognition sequence
*Bam*HI	↓ * (5′) G G A T C C (3′) C C T A G G * ↑	*Hind*III	(5′) A A G C T T (3′) T T C G A A ↑
*Cla*I	↓ * (5′) A T C G A T (3′) T A G C T A * ↑	*Not*I	↓ (5′) G C G G C C G C (3′) C G C C G G C G ↑
*Eco*RI	↓ * (5′) G A A T T C (3′) C T T A A G * ↑	*Pst*I	*↓ (5′) C T G C A G (3′) G A C G T C ↑*
*Eco*RV	↓ (5′) G A T A T C (3′) C T A T A G ↑	*Pvu*II	↓ (5′) C A G C T G (3′) G T C G A C ↑
*Hae*III	↓* (5′) G G C C (3′) C C G G *↑ ↓	*Tth*111I	↓ (5′) G A C N N N G T C (3′) C T G N N N C A G ↑

Arrows indicate the phosphodiester bonds cleaved by each restriction endonuclease. Asterisks indicate bases that are methylated by the corresponding methylase (where known). N denotes any base. Note that the name of each enzyme consists of a three-letter abbreviation of the bacterial species from which it is derived (e.g., *Bam* for *B*acillus *am*yloliquefaciens, *Eco* for *E*scherichia *co*li). The Roman numerals included in the enzyme names (e.g., *Bam*HI) distinguish different restriction endonucleases isolated from the same bacterial species rather than the type of restriction enzyme.

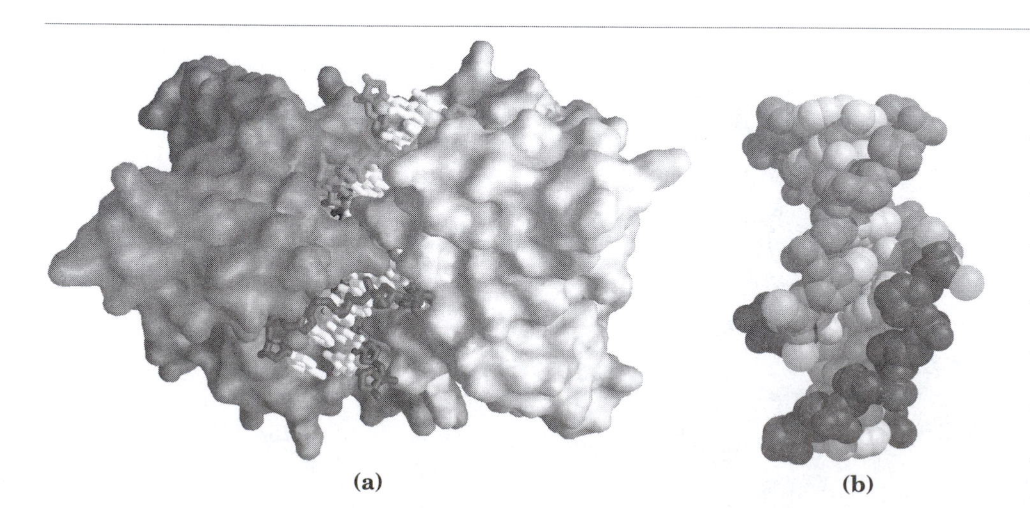

figure 29–2, page 1122
Interaction of *Eco*RV restriction endonuclease with its target sequence.

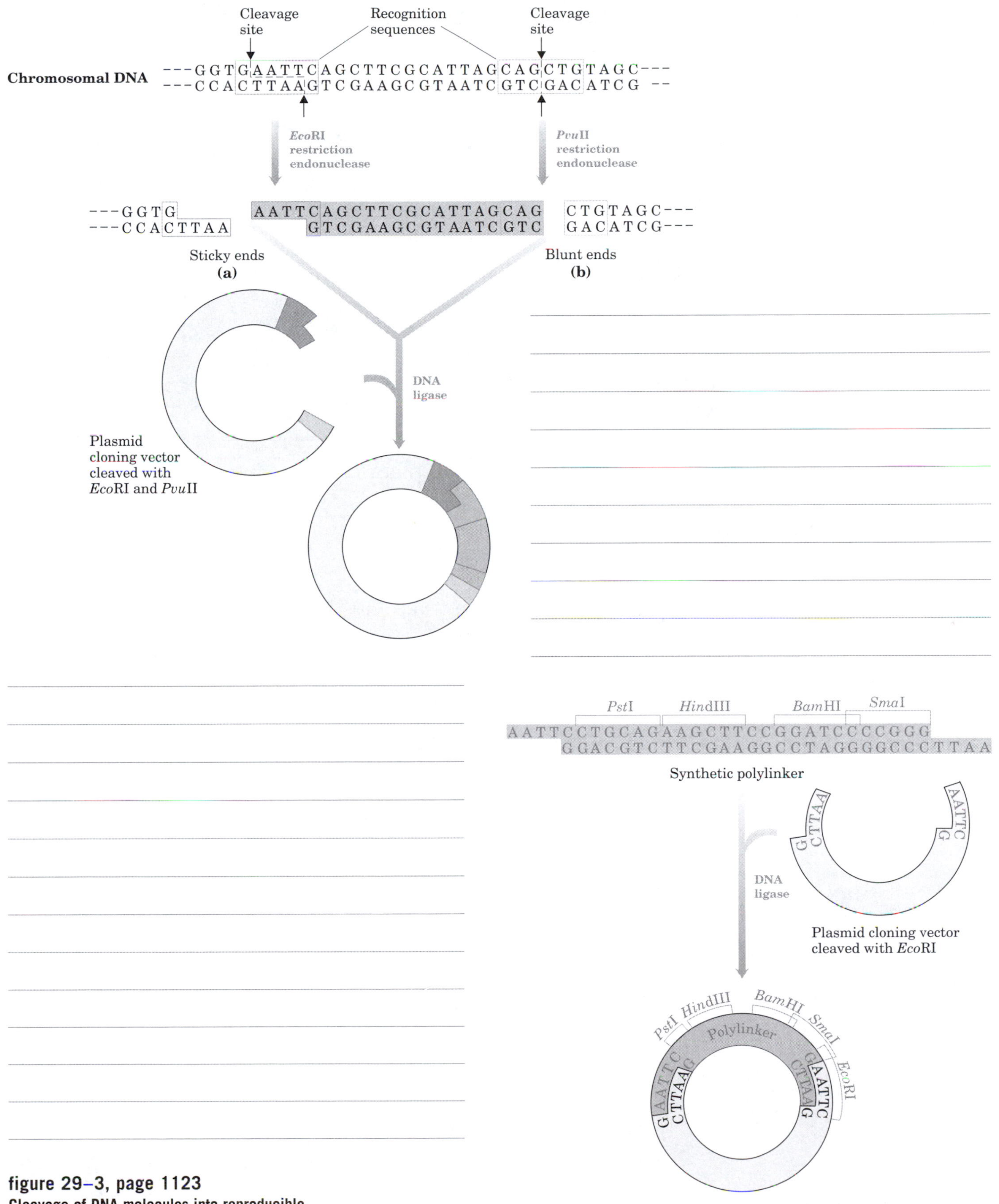

figure 29–3, page 1123
Cleavage of DNA molecules into reproducible fragments by restriction endonucleases.

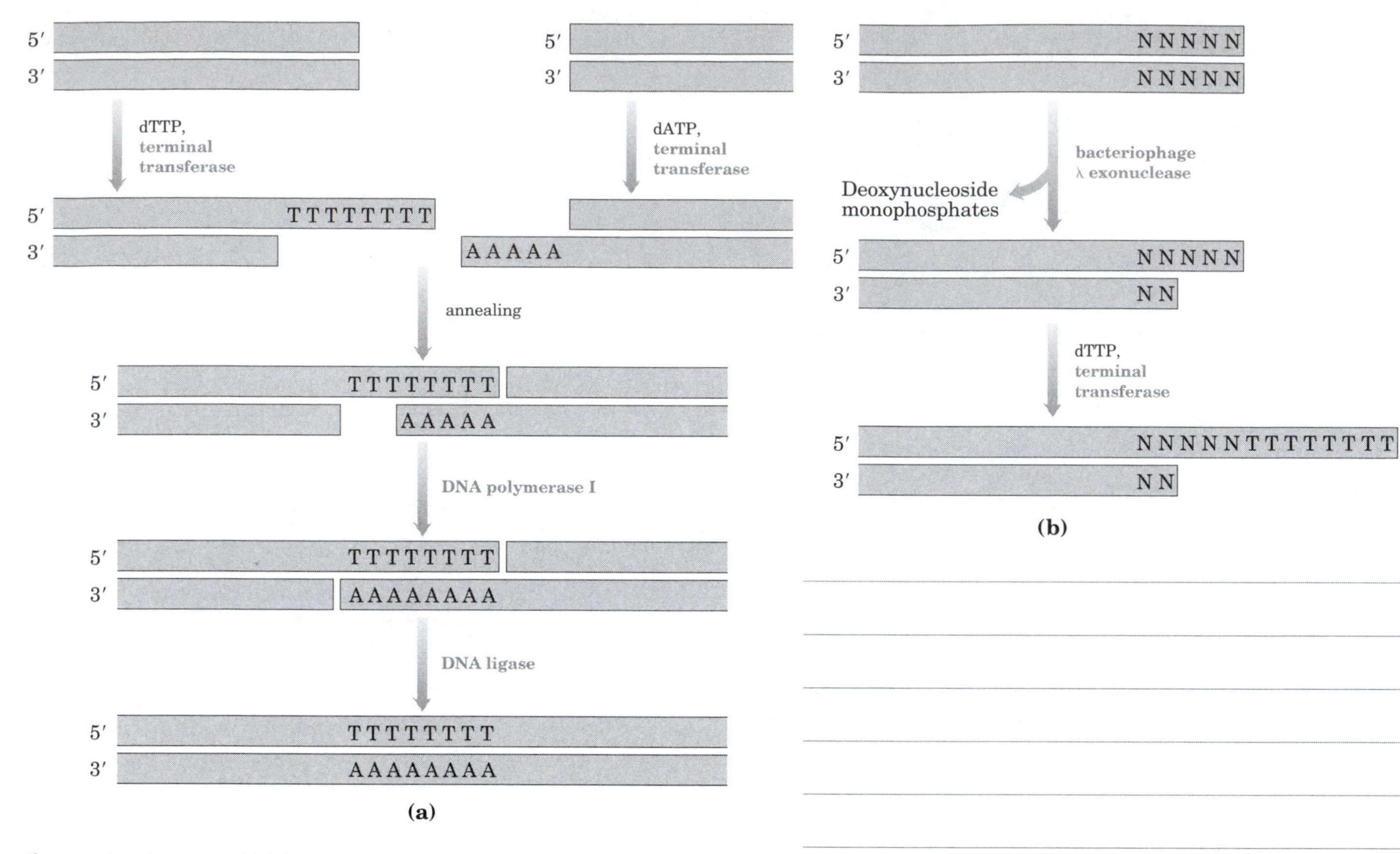

figure 29–4, page 1125
Terminal transferase can be used to generate sticky ends for joining two DNA fragments.

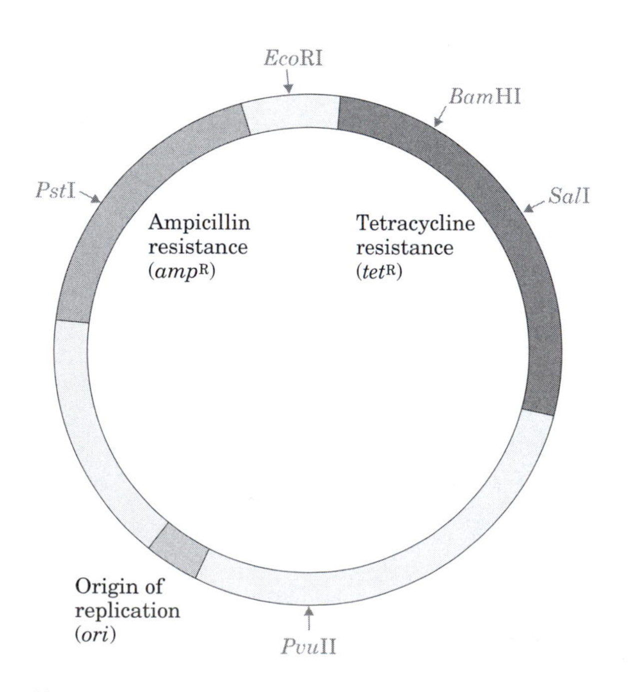

figure 29–5, page 1126
The constructed *E. coli* plasmid pBR322.

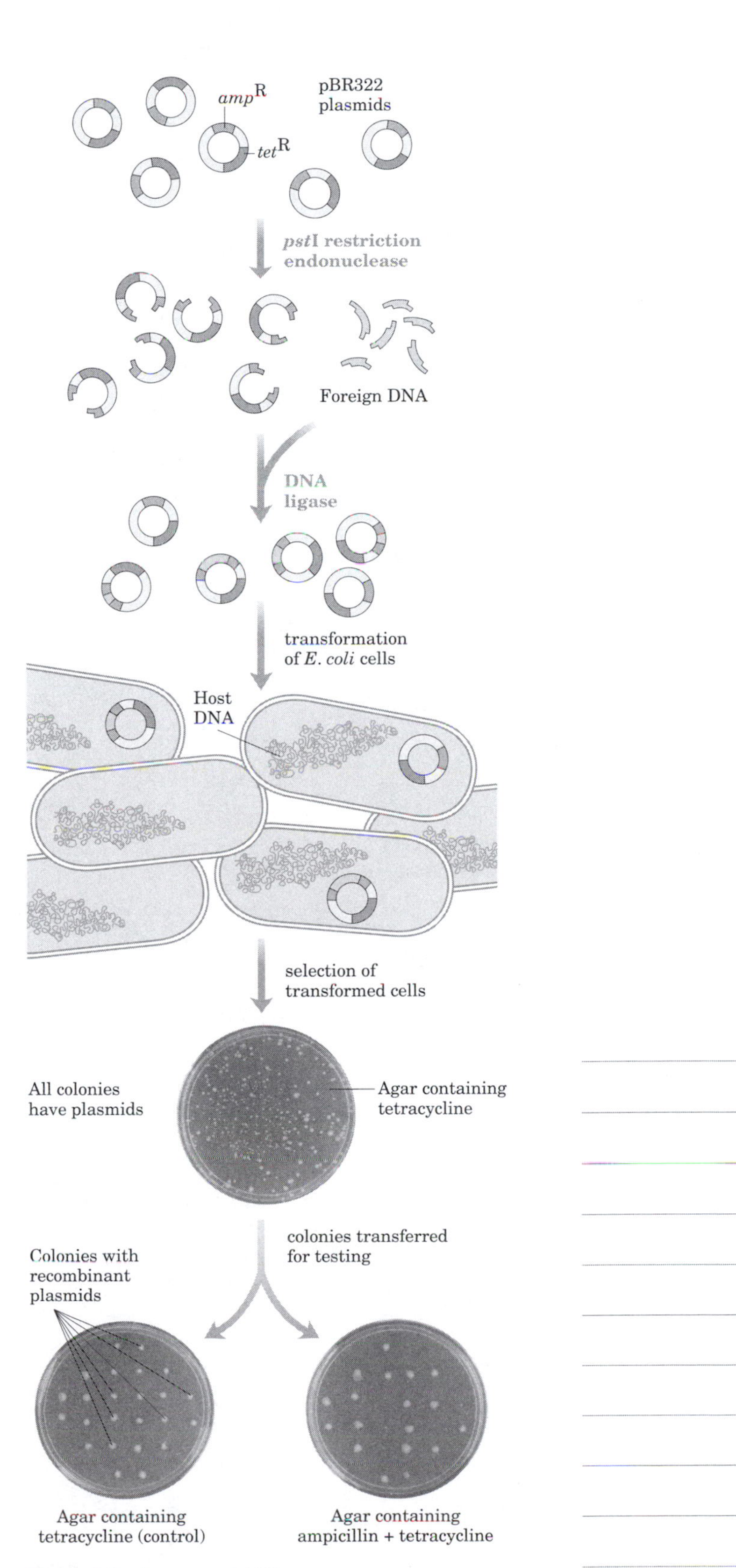

figure 29–6, page 1126
Cloning foreign DNA in *E. coli* with pBR322.

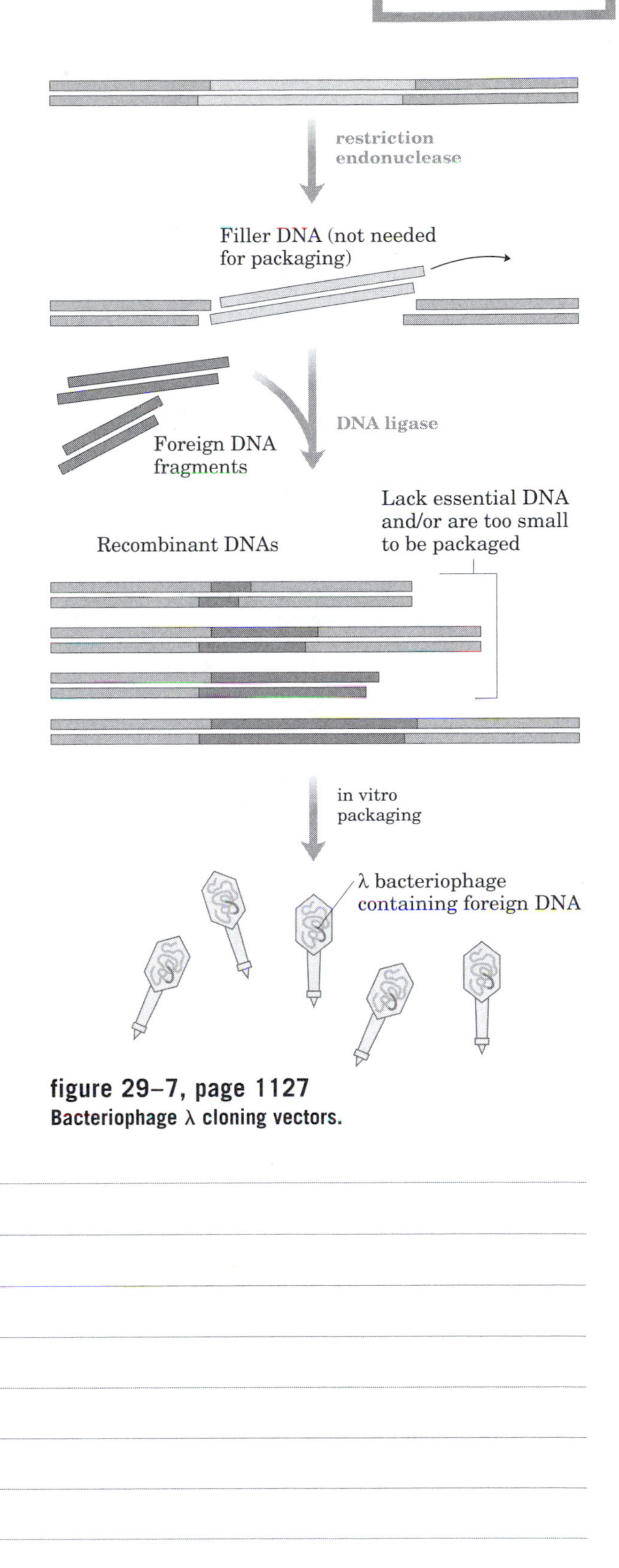

figure 29–7, page 1127
Bacteriophage λ cloning vectors.

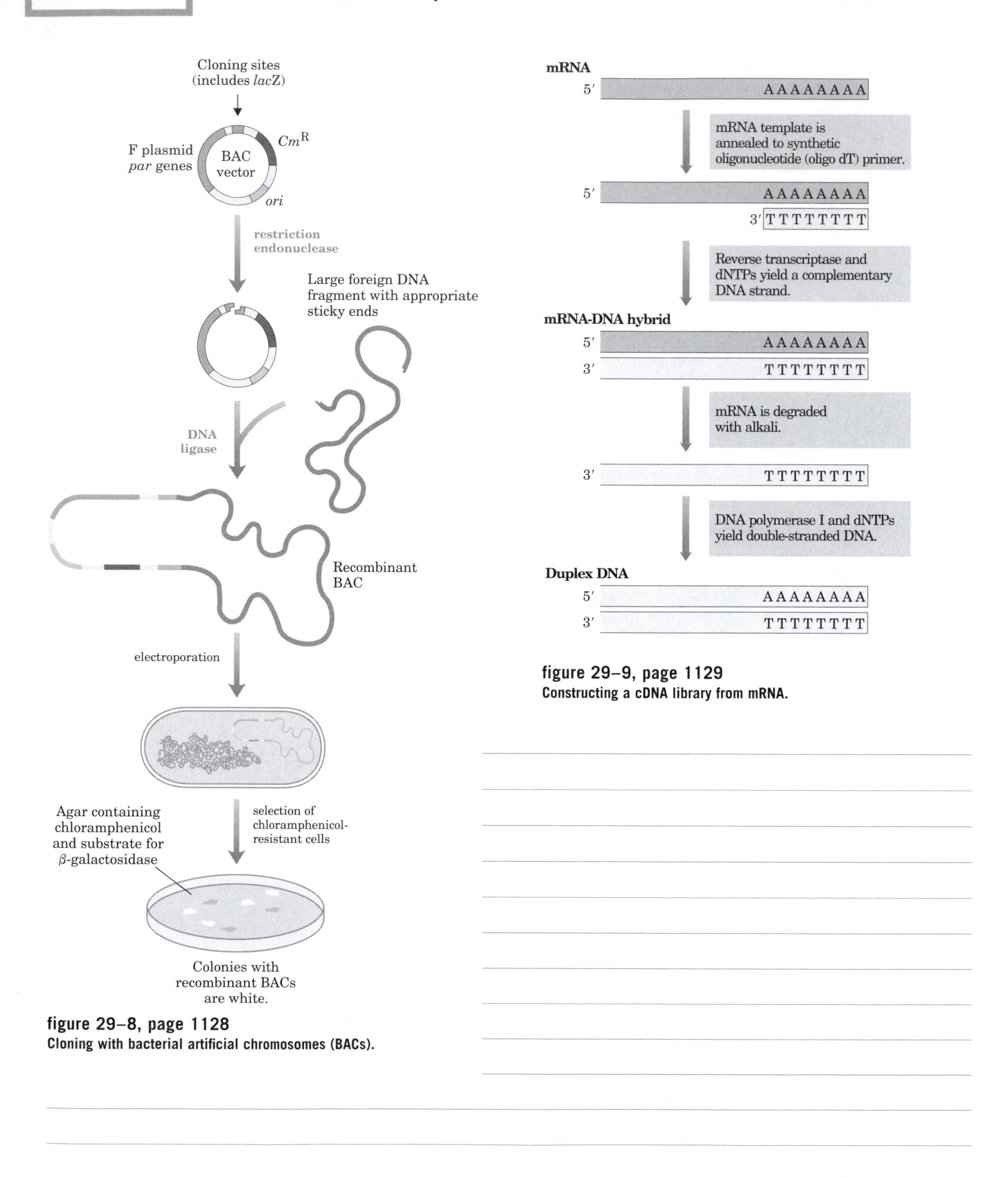

figure 29–8, page 1128
Cloning with bacterial artificial chromosomes (BACs).

figure 29–9, page 1129
Constructing a cDNA library from mRNA.

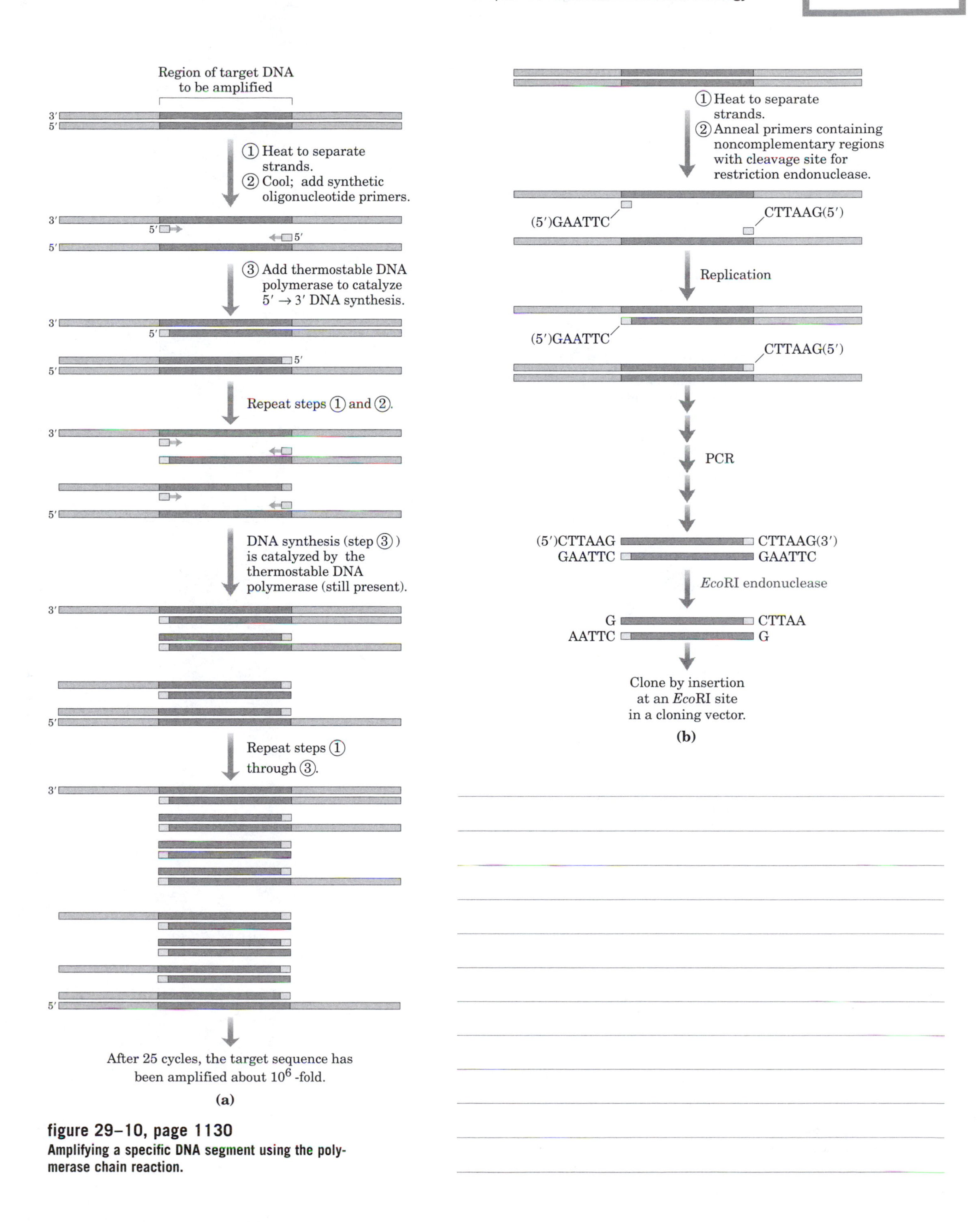

figure 29–10, page 1130
Amplifying a specific DNA segment using the polymerase chain reaction.

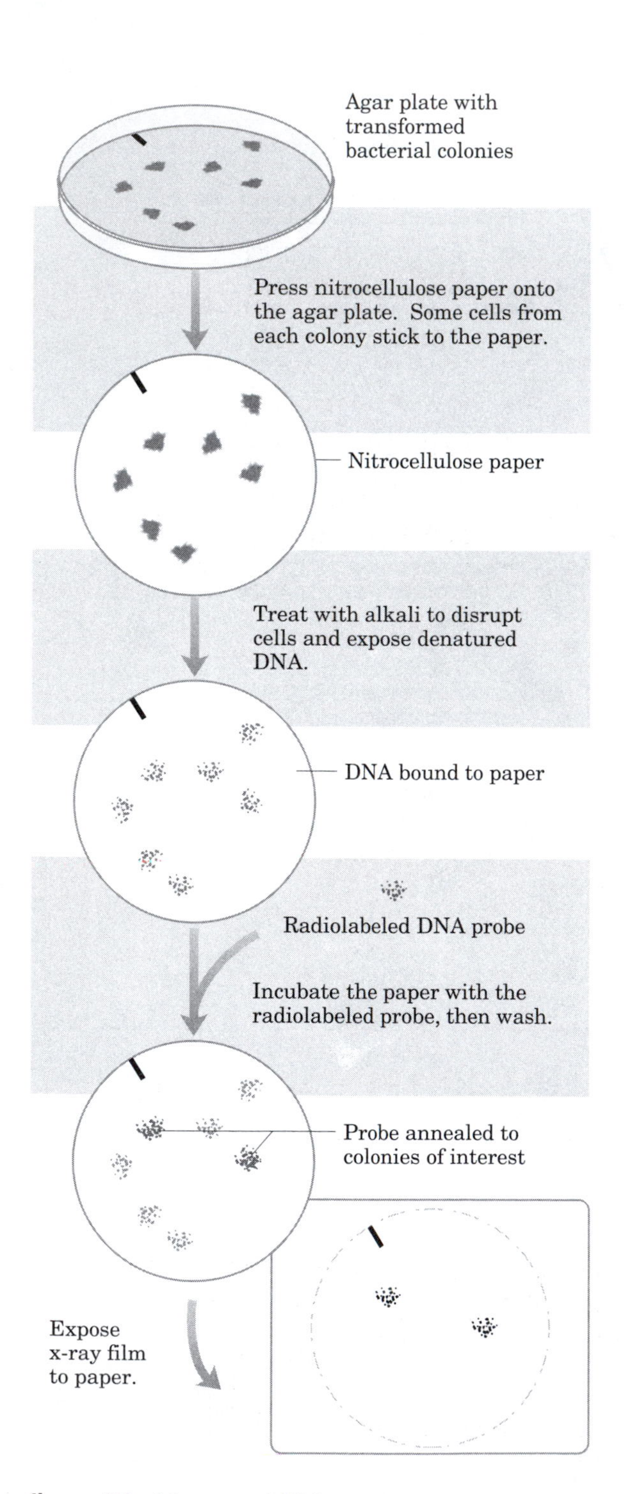

figure 29–11, page 1131
Identifying a clone with a particular DNA segment using hybridization.

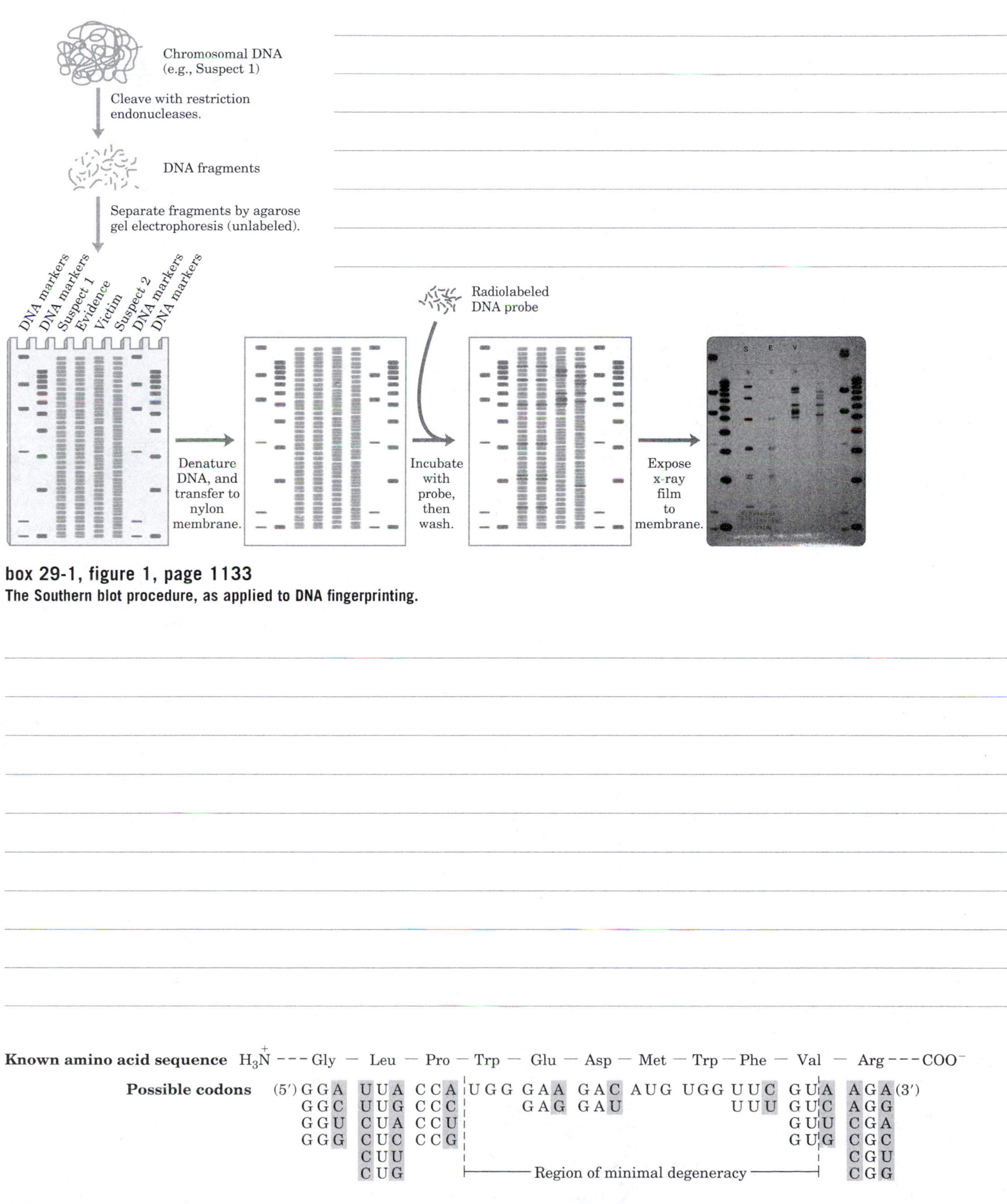

box 29-1, figure 1, page 1133
The Southern blot procedure, as applied to DNA fingerprinting.

figure 29–12, page 1133
Designing a probe to detect the gene for a protein of known amino acid sequence.

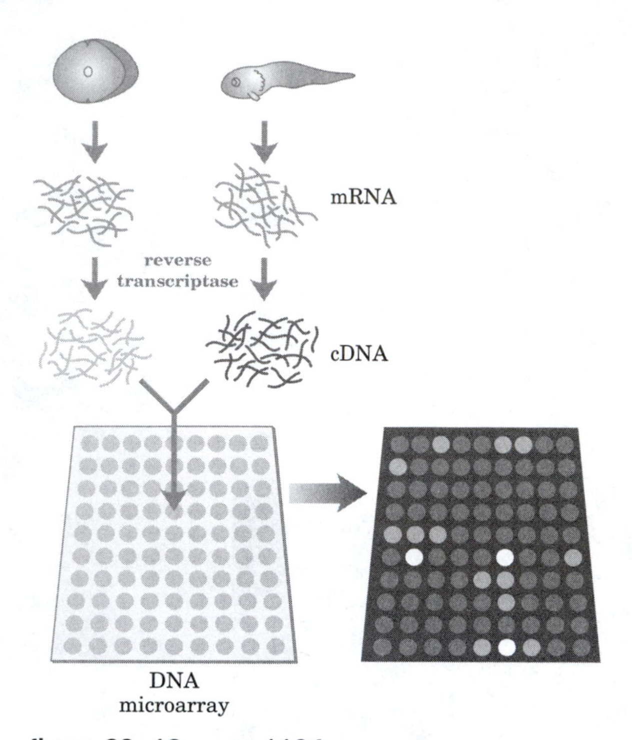

figure 29–13, page 1134
Constructing a DNA microarray.

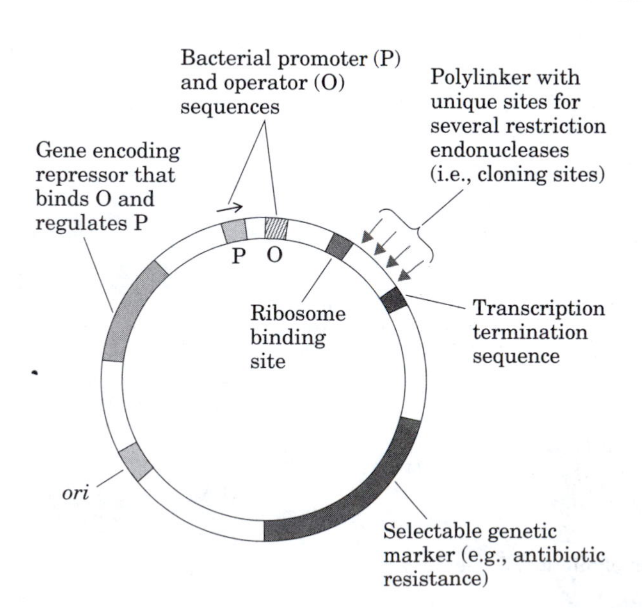

figure 29–14, page 1135
Types of DNA sequences found in a typical *E. coli* expression vector.

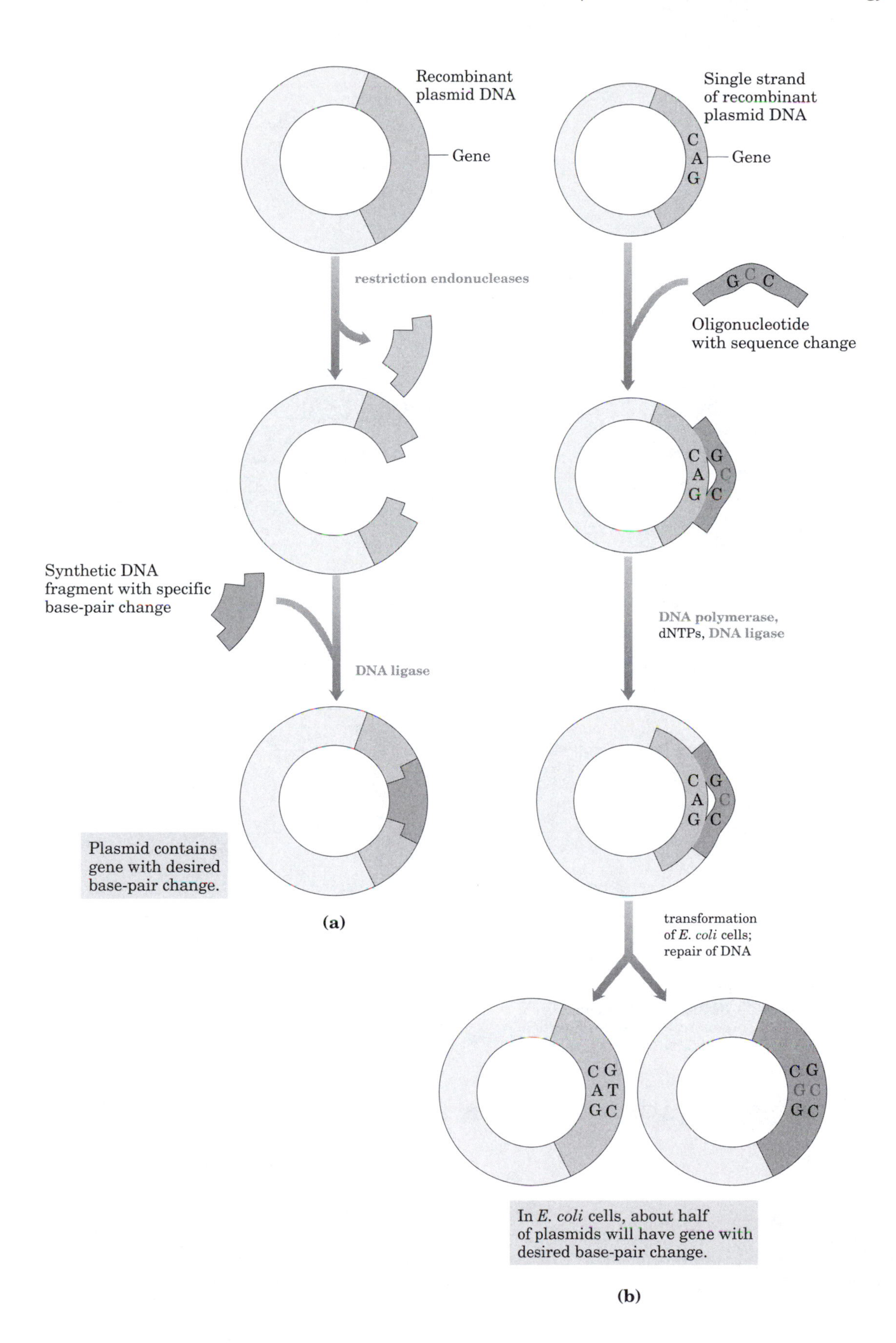

figure 29–15, page 1137
Two approaches to site-directed mutagenesis.

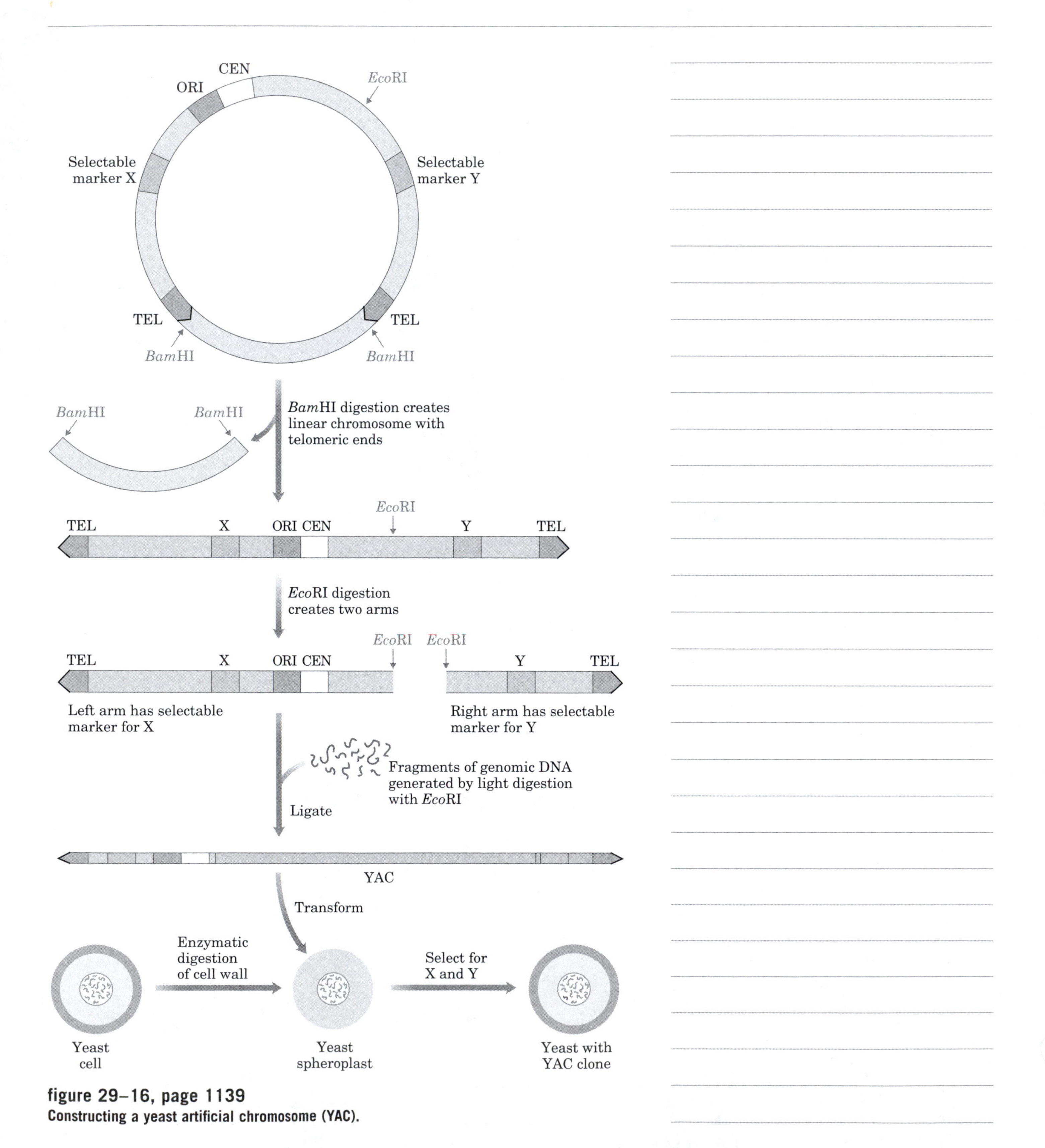

figure 29–16, page 1139
Constructing a yeast artificial chromosome (YAC).

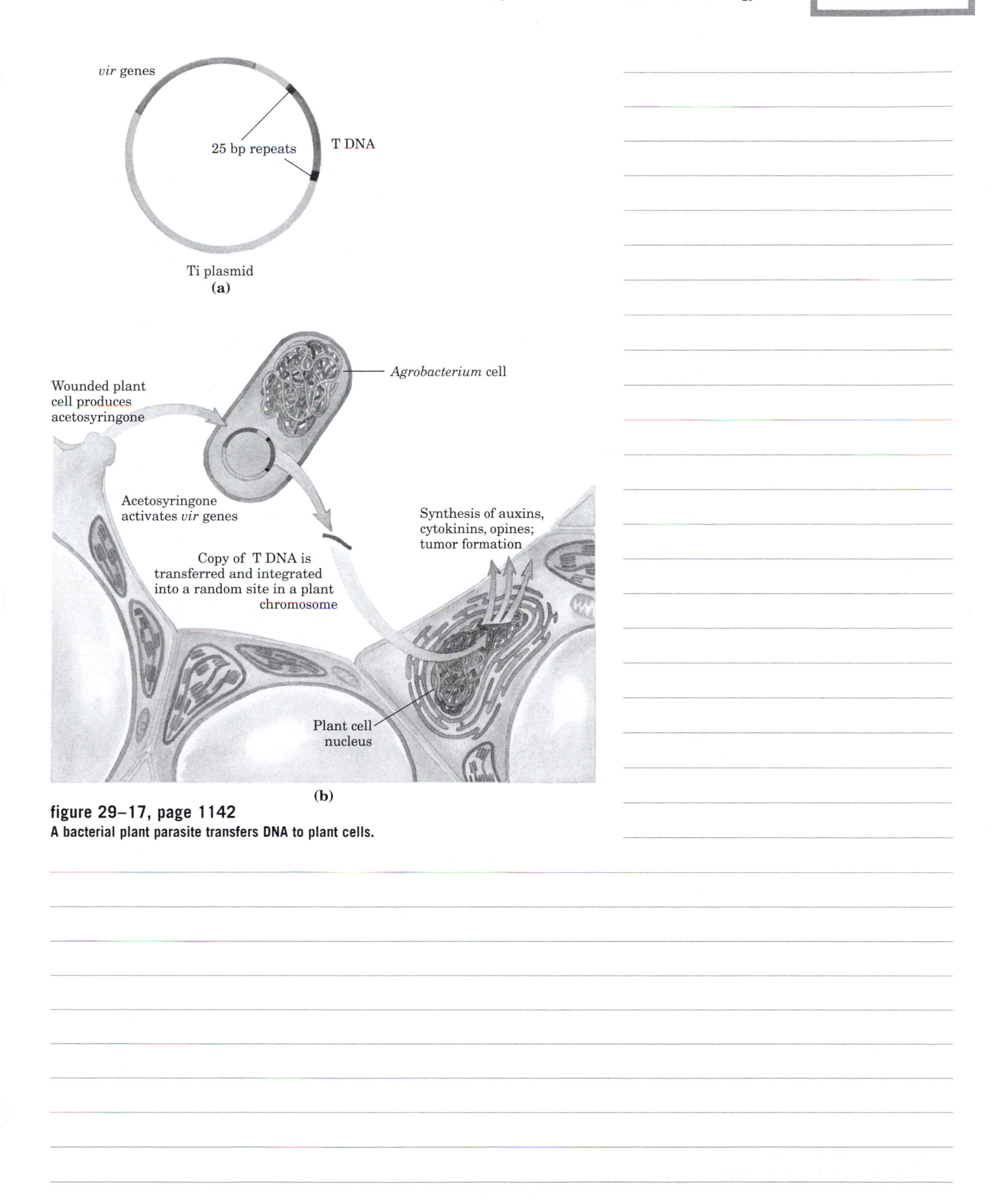

figure 29–17, page 1142
A bacterial plant parasite transfers DNA to plant cells.

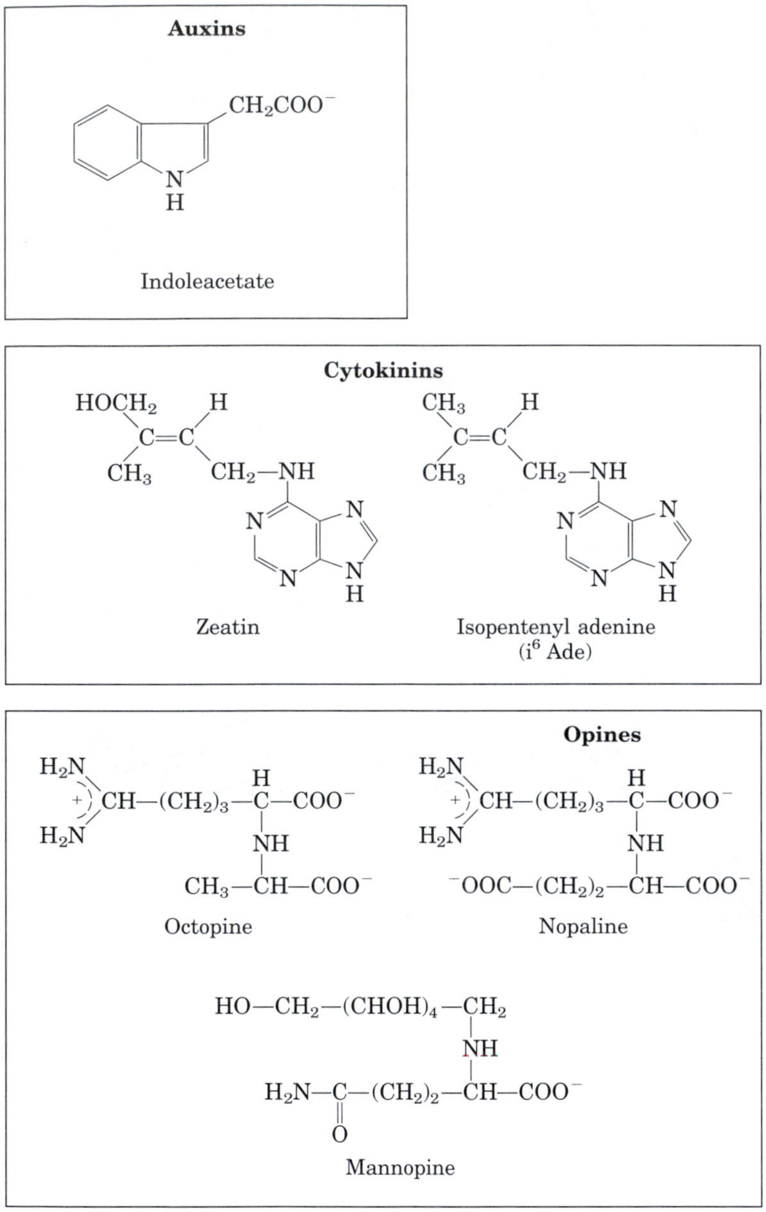

Auxins

CH_2COO^-

N
H

Indoleacetate

Cytokinins

$HOCH_2$ H
C=C
CH_3 CH_2—NH
N N
N N
H

Zeatin

CH_3 H
C=C
CH_3 CH_2—NH
N N
N N
H

Isopentenyl adenine
(i^6 Ade)

Opines

H_2N
+ CH—$(CH_2)_3$—C(H)—COO^-
H_2N
NH
CH_3—CH—COO^-

Octopine

H_2N
+ CH—$(CH_2)_3$—C(H)—COO^-
H_2N
NH
^-OOC—$(CH_2)_2$—CH—COO^-

Nopaline

HO—CH_2—$(CHOH)_4$—CH_2
NH
H_2N—C(=O)—$(CH_2)_2$—CH—COO^-

Mannopine

figure 29–18, page 1143
Metabolites produced in *Agrobacterium*-infected plant cells.

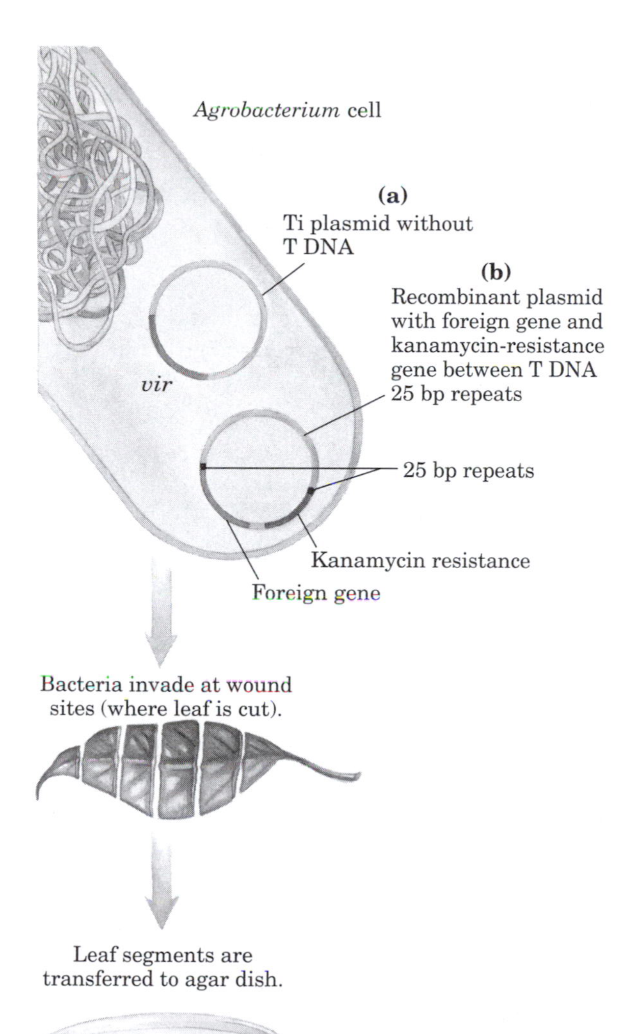

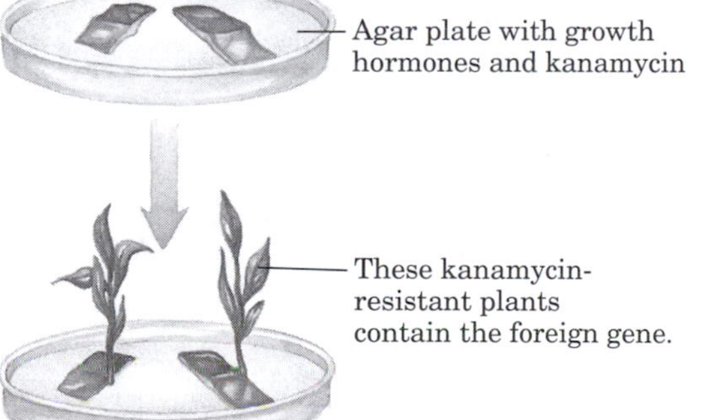

figure 29–19, page 1143
A two-plasmid strategy to create a recombinant plant.

figure 29–2, page 1144
A tobacco plant in which the gene for firefly luciferase is expressed.

figure 29–21, page 1144
Tomato plants engineered to be resistant to some insect larvae.

(a)

(b)

figure 29–22, page 1145
Glyphosate-resistant soybean plants growing in a field in Wisconsin.

$$^{-}O-\overset{\overset{O^{-}}{\parallel}}{\underset{\underset{O^{-}}{|}}{P}}-CH_2-NH-CH_2-COO^{-}$$

Glyphosate, page 1145

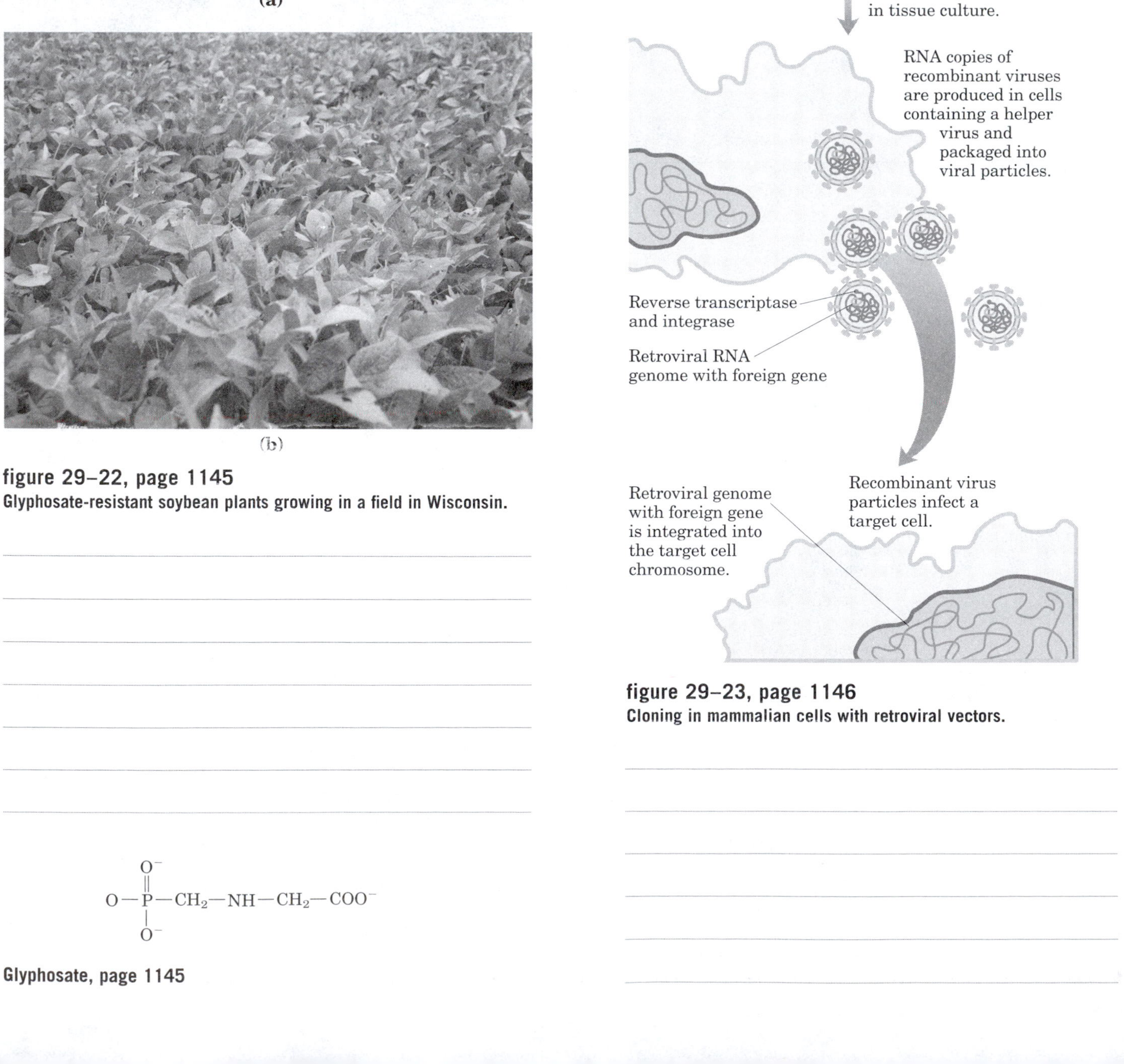

figure 29–23, page 1146
Cloning in mammalian cells with retroviral vectors.

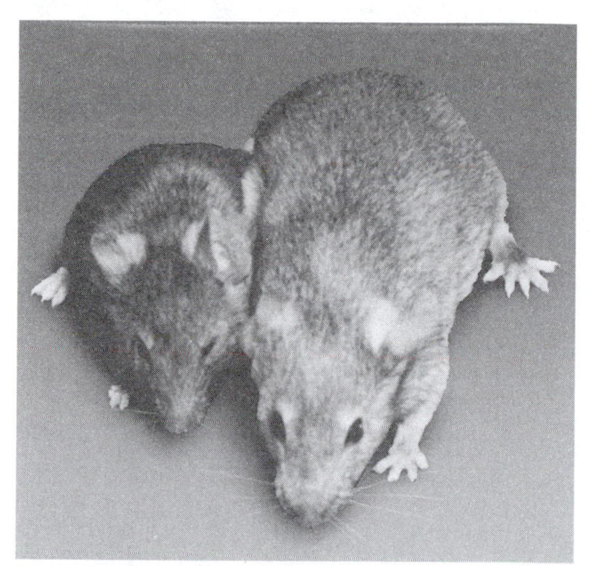

figure 29–24, page 1147
Cloning in mice.

table 29–3, page 1147

Some Recombinant DNA Products in Medicine

Product category	Examples/uses
Anticoagulants	Tissue plasminogen activator (TPA) activates plasmin, an enzyme involved in dissolving clots; effective in treating heart attack victims.
Blood factors	Factor VIII promotes clotting and is deficient in hemophiliacs; use of factor VIII produced by recombinant DNA technology eliminates infection risks associated with blood transfusions.
Colony stimulating factors	Immune system growth factors that stimulate leukocyte production; used to treat immune deficiencies and to fight infections.
Erythropoietin	Stimulates erythrocyte production; used to treat anemia in patients with kidney disease.
Growth factors	Stimulate differentiation and growth of various cell types; used to promote wound healing.
Human growth hormone	Used to treat dwarfism.
Human insulin	Used to treat diabetes.
Interferons	Interfere with viral reproduction; used to treat some cancers.
Interleukins	Activate and stimulate different classes of leukocytes; possible uses in treating wounds, HIV infection, cancer, and immune deficiencies.
Monoclonal antibodies	Extraordinary binding specificity is used in: diagnostic tests; targeted transport (of drugs, toxins, or radioactive compounds to tumors as a cancer therapy); many other applications.
Superoxide dismutase	Prevents tissue damage from reactive oxygen species when tissues briefly deprived of O_2 during surgery suddenly have blood flow restored.
Vaccines	Proteins derived from viral coats are as effective in "priming" an immune system as the killed virus more traditionally used for vaccines, but are safer; first developed was the vaccine for hepatitis B.

PROBLEMS

3. DNA Cloning, page 1151.

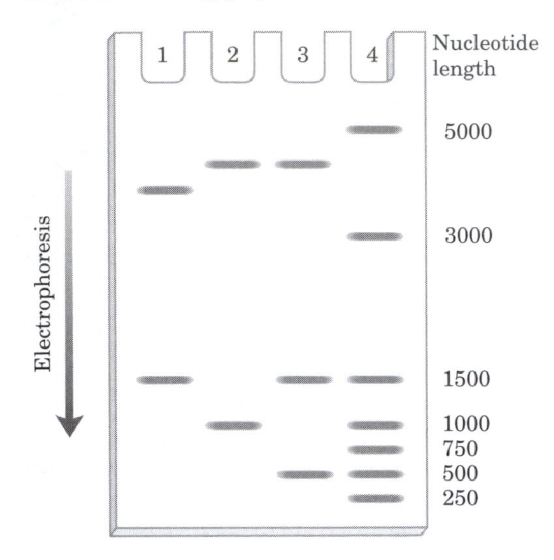

4. Expressing a Cloned Gene, page 1151.

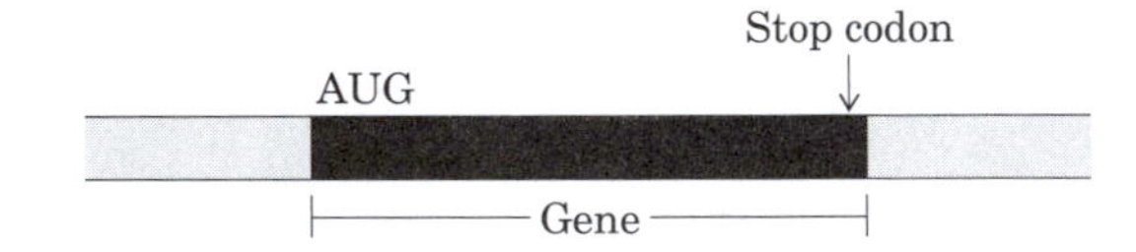

5. Identifying the Gene for a Protein with a Known Amino Acid Sequence, page 1151.

$H_3\overset{+}{N}$–Ala–Pro–Met–Thr–Trp–Tyr–Cys–
Met–Asp–Trp–Ile–Ala–Gly–Gly–Pro–
Trp–Phe–Arg–Lys–Asn–Thr–Lys–

8. Designing a Diagnostic Test for a Genetic Disease, pages 1151–1152.

307 ATGGCGACCCTGGAAAAGCTGATGAAGGCCTTCGAGTCCCTCAAGTCCTTC
1 **M A T L E K L M K A F E S L K S F**

358 CAGCAGTTCCAGCAGCAGCAGCAGCAGCAGCAGCAGCAGCAGCAGCAGCAG
18 **Q Q F Q Q Q Q Q Q Q Q Q Q Q Q Q Q**

409 CAGCAGCAGCAGCAGCAGCAGCAACAGCCGCCACCGCCGCCGCCGCCGCCG
35 **Q Q Q Q Q Q Q Q Q P P P P P P P P**

460 CCGCCTCCTCAGCTTCCTCAGCCGCCGCCG
52 **P P P Q L P Q P P P**

Source: The Huntington's Disease Collaborative Research Group (1993) A novel gene containing a trinucleotide repeat that is expanded and unstable on Huntington's disease chromosomes. *Cell* **72,** 971–983.

9. Using PCR to Detect Circular DNA Molecules, page 1152.

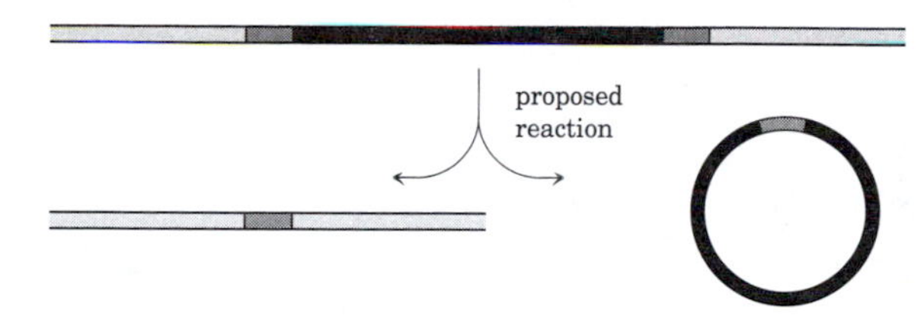

10. DNA Fingerprinting and RFLP Analysis, page 1152.

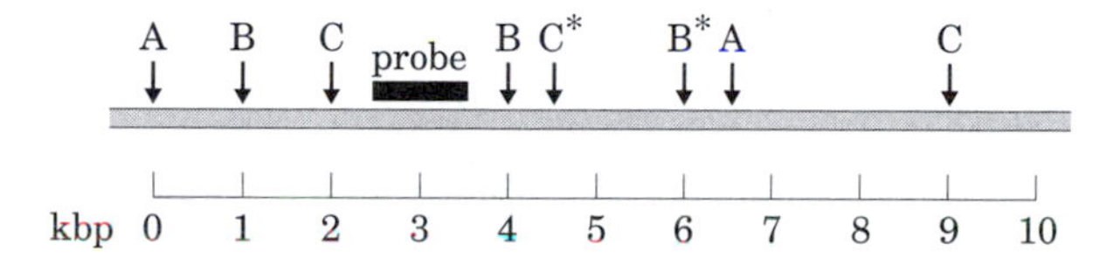

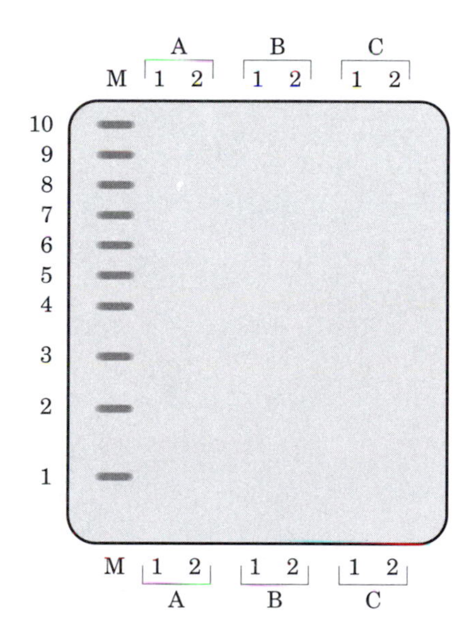

11. RFLP Analysis for Paternity Testing, page 1152.

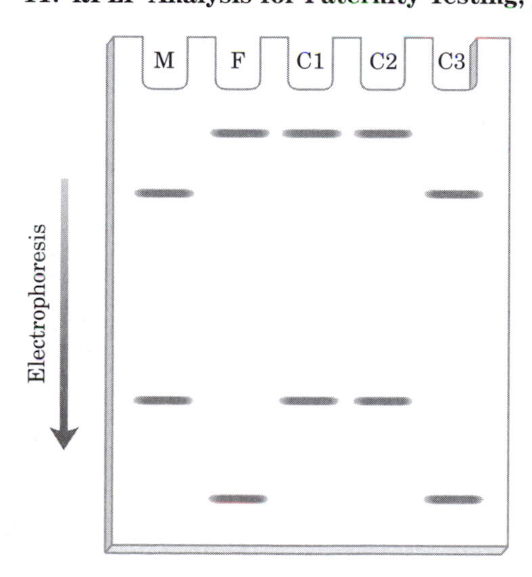